McGraw-Hill

Making a Difference in Teaching and Learning

Delivering tools to assist instructors
Making learning easier for students

Learn how McGraw-Hill technology can make a difference for you and your students.

MathZone

A course management site for McGraw-Hill math titles, MathZone™ combines book-specific practice and tutorial content with automatic, online assessment. You can utilize MathZone's algorithmic capabilities to generate multiple versions of assignments and quizzes, edit the problems and exercises we've provided, or create your own. MathZone's automatic gradebook function makes tracking student progress a snap. Learn more about MathZone by visiting www.mathzone.com.

ALEKS® individualizes assessment and learning by recognizing that students learn math in different ways, at different speeds. It is an artificial intelligence system that helps students take "ownership" of their learning process by enabling them to experience success, rather than feeling lost. It's like a tutor who teaches what each student is most ready to learn.

To request your FREE 24-hour trial of ALEKS, visit our website at www.highed.aleks.com/guest.html.

NetTutor™ is an online tutoring system where your students can get live, one-on-one help. It requires no special software or downloads on the part of the student. All students need to do is connect to a website. NetTutor™ distinguishes itself from other online tutoring systems in its richness of special mathematical symbols and graphs, which allow the students to use the exact notation presented in their books.

A refreshing new math series

Ignacio Bello, from Hillsborough Community College, brings a breath of fresh air to math textbooks. Students will appreciate his down-to-earth explanations, patient skillbuilding, and interesting and realistic applications, as well as Bello's recognition of different learning styles. Contact your McGraw-Hill sales representative or visit www.mhhe.com to request your complimentary copy.

BASIC COLLEGE MATHEMATICS with MATHZONE™, Second Edition

by Ignacio Bello, ©2006
Softcover
ISBN 0-07-299098-8

INTRODUCTORY ALGEBRA with MATHZONE™, Second Edition

by Ignacio Bello, ©2006
Softcover
ISBN 0-07-299099-6

INTERMEDIATE ALGEBRA with with MATHZONE™, Second edition

by Ignacio Bello and Fran Hopf, ©2006
Softcover
ISBN 0-07-299100-3

The McGraw·Hill Companies

ISBN 0-07-319219-8, Bello

Live Tutoring

Students use NetTutor™ to access an online tutoring service to get help when they need it. This is especially great for commuter schools and distance education schools.

NetTutor™ is a revolutionary new online learning environment for the live distribution of mathematical content. NetTutor™ includes web-based graphical chat, threaded and platform independent, allowing students to use their own computers to access learning materials in a nonlinear fashion or in real-time with live corresponding tutors.

Some features of NetTutor™ include

- ***Whiteboard.*** The whiteboard enables students and tutors to "speak" math. They are able to use mathematical symbols as well as circle, highlight, and pull out important steps. ***Real-time discussion on the whiteboard*** is the key to this service by allowing tutors to explain in mathematical symbols as well as to "converse" in real time with the student to answer questions.
- ***Students register online*** and there are no passwords for instructors or McGraw-Hill to maintain.
- ***Free!*** Students simply need to tell the tutor which McGraw-Hill text they are using, the page number, and the exercise.
- Tutors will work on ***only odd-numbered problems.*** The even-numbered problems remain available for faculty for assigned homework.
- ***Seventy-two hours weekly*** of live tutors online. Tutors are placed at peak hours, and usage is monitored frequently to see if hours need to be changed.
- ***Students may access archives*** of previous sessions saved by other students to see if there is a similar problem/exercise.
- If students are not able to enter a tutorial session (not enough time or the tutor is not available), they can ***post a question and it will be answered within 24 hours.***

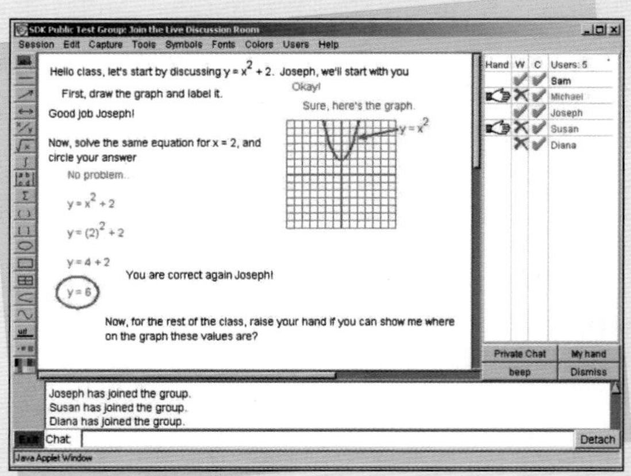

To learn more about NetTutor™, or any other McGraw-Hill product, please contact your sales representative. To locate your sales representative's contact information, visit www.mhhe.com and click on "find my sales rep" in the help center.

The McGraw·Hill Companies

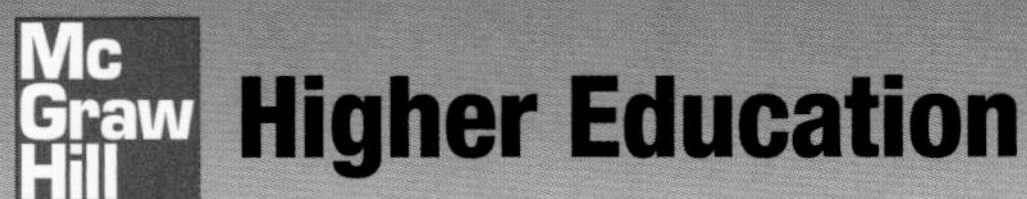

ALEKS® Makes the Grade!

Would your students appreciate their own personal math tutor available 24 hours per day, 7 days a week? A tutor who teaches what they're most ready to learn? And a tutor who analyzes their answers to problems and responds with specific advice when they make a mistake?

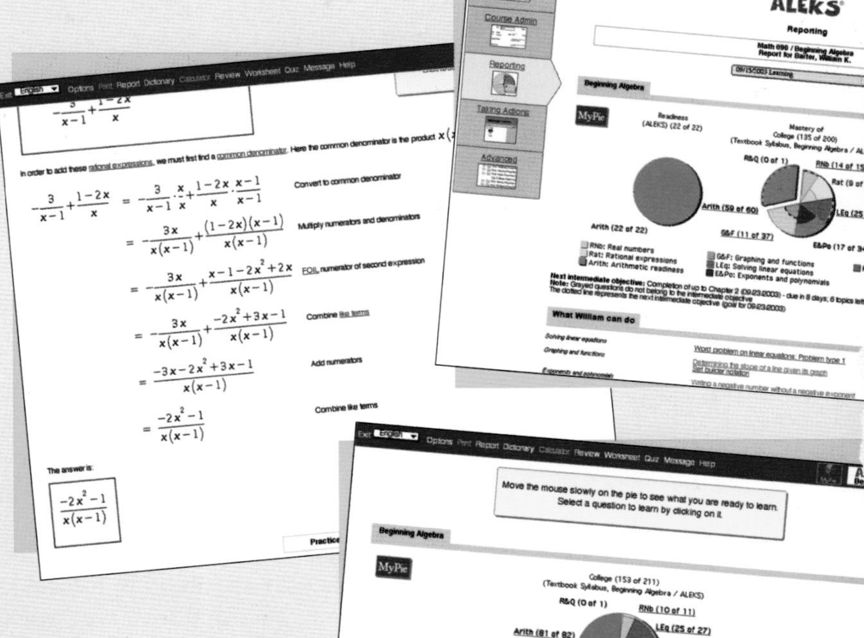

ALEKS individualizes assessment and learning, is customizable for each course topic, and provides a course management system telling you exactly what your students know and don't know.

Be Our Guest

To request your FREE 24-hour trial of ALEKS, visit our website at www.highed.aleks.com/guest.html. Click on "login" under the "as a student" heading.

To learn more about ALEKS, visit the website at www.highed.aleks.com or contact your local McGraw-Hill representative. To locate your sales representative's contact information, visit www.mhhe.com and click on "find my sales rep" in the help center.

Here's what professors are saying about ALEKS . . .

"The system appears to offer many advantages over anything else currently offered."

"Good product. It's practical and logical—based on a model that makes sense."

"This is the first time I have gotten excited about using a computer-based approach in our classes. ALEKS eliminates the equipment and cost barriers we face with other systems."

Here's what students are sayi[ng] about ALEKS . . .

"This program helped me dramatically with math from getting D's on tests to receiving an A+ on m[y] I actually scored a 200/200 on a final."

"My math abilities really have progressed. AL[EKS] makes math a much easier course."

Models of Classroom Integration

- Supervised Math Lab
- Math Lab in Structured Course
- Small-Group Instruction
- Self-Paced Learning
- Distance Learning

In some ways, planning a course in which ALEKS is to be used is simpler than planning other kinds of courses. The instructor may assume complete freedom in planning lectures, lessons, and assignments, while ALEKS ensures that students can progress toward mastery, regardless of their level of preparation.

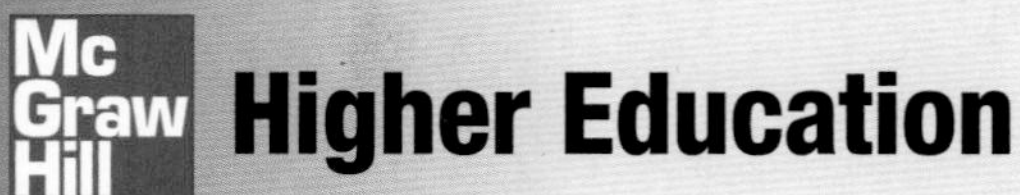

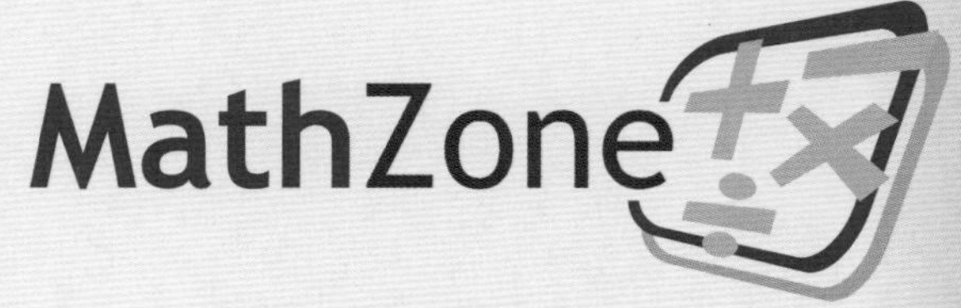

MathZone™ Free Easy Has It All

MathZone's™ powerful feature set includes book-specific, assignable algorithmic content; ADA-compliant videos; and e-Professor animated solutions. MathZone provides students virtually unlimited practice through algorithmic quizzing and testing, and free live tutoring via NetTutor's™ whiteboard technology, and instructors can track their students' progress in the online gradebook.

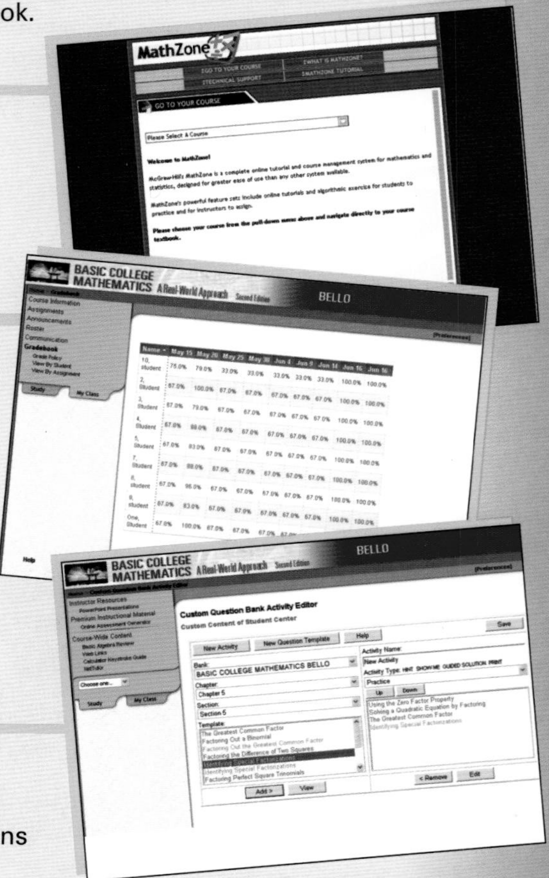

Free to you and your students

MathZone is provided for free with the purchase of a new McGraw-Hill math textbook. Each text is packaged with a registration code for the website at www.mathzone.com and comes with a MathZone CD-ROM.

Easy to use

All functions for students and instructors are located within one login. The online course management system offers simple functionality. The gradebook provides detailed feedback on student responses and the problem-solving process that students performed to select those responses. In addition to tracking student performance on problems, the gradebook also tracks students' usage of nonassessed elements, such as video lecture or e-Professor. The gradebook can be easily exported to Excel, allowing the instructor the flexibility to use MathZone in a variety of ways.

Has it all

MathZone is tailored to a McGraw-Hill textbook, so every assignment, question, e-Professor tutorial, and video-lecture piece is derived directly from text-specific materials. Instructors can modify questions and assignments and create their own from scratch.

With just a few clicks of the mouse, you can share your MathZone course with an unlimited number of colleagues. Instructors can share assignments, algorithmically generated questions, or static questions with simplicity and speed.

Instructor benefits

- Offers complete course management.
- Track students' progress with unprecedented depth.
- Includes 100% algorithmic homework, quizzing, and testing.
- Edit, change, and create algorithmic questions.
- Assign all course resources, including homework, video lectures and e-Professor.
- Live, book-specific tutoring saves instructors time.
- Provides unmatched ease-of-use in one system.

Student benefits

- Includes video-lectures that are ADA-compliant.
- Provides e-Professor-guided solutions.
- Provides prompted, step-by-step problem solving.
- Includes online algorithmic practice and assessment.
- Students have access to all resources, whether assigned or not.
- The end result is a better understanding and an improved grade.

Introductory Algebra

Introductory Algebra

A Real-World Approach

Second Edition

Ignacio Bello
Hillsborough Community College
Tampa, Florida

Boston Burr Ridge, IL Dubuque, IA Madison, WI New York San Francisco St. Louis
Bangkok Bogotá Caracas Kuala Lumpur Lisbon London Madrid Mexico City
Milan Montreal New Delhi Santiago Seoul Singapore Sydney Taipei Toronto

Higher Education

INTRODUCTORY ALGEBRA: A REAL-WORLD APPROACH, SECOND EDITION

Published by McGraw-Hill, a business unit of The McGraw-Hill Companies, Inc., 1221 Avenue of the Americas, New York, NY 10020.

 This book is printed on recycled, acid-free paper containing 10% postconsumer waste.

1 2 3 4 5 6 7 8 9 0 QPD/QPD 0 9 8 7 6 5
1 2 3 4 5 6 7 8 9 0 QPD/QPD 0 9 8 7 6 5

ISBN 0–07–283105–7
ISBN 0–07–298477–5 (Annotated Instructor's Edition)

Publisher, Mathematics and Statistics: *William K. Barter*
Publisher, Developmental Mathematics: *Elizabeth J. Haefele*
Director of Development: *David Dietz*
Senior Developmental Editor: *Randy Welch*
Marketing Manager: *Steven R. Stembridge*
Senior Project Manager: *Vicki Krug*
Senior Production Supervisor: *Sherry L. Kane*
Senior Media Project Manager: *Sandra M. Schnee*
Lead Media Technology Producer: *Jeff Huettman*
Senior Designer: *David W. Hash*
Cover/Interior Designer: *Rokusek Design*
(USE) Cover Image: ©*Paul Steel/CORBIS, Fireworks over Sydney Harbor Bridge, Australia*
Senior Photo Research Coordinator: *Lori Hancock*
Photo Research: *LouAnn K. Wilson*
Supplement Producer: *Brenda A. Ernzen*
Compositor: *Interactive Composition Corporation*
Typeface: *10/12 New Times Roman*
Printer: *Quebecor World Dubuque, IA*

The credits section for this book begins on page C-1 and is considered an extension of the copyright page.

www.mhhe.com

About the Author

Ignacio Bello attended the University of South Florida, where he earned a B.A. and M.A. in Mathematics. He began teaching at USF in 1967, and in 1971 he became a member of the faculty and Coordinator of the Math and Sciences Department at Hillsborough Community College (HCC). Professor Bello instituted the USF/HCC remedial program, which started with 17 students taking Intermediate Algebra and grew to more than 800 students with courses covering Developmental English, Reading, and Mathematics.

Aside from the present series of books (*Basic College Mathematics, Introductory Algebra,* and *Intermediate Algebra*), Professor Bello is the author of more than 40 textbooks, including *Topics in Contemporary Mathematics, College Algebra, Algebra and Trigonometry,* and *Business Mathematics.* Many of these textbooks have been translated into Spanish. With Professor Fran Hopf, Bello started the Algebra Hotline, the only live, college-level television help program in Florida.

Professor Bello is featured in three television programs on the award-winning Education Channel. He has helped create and develop the USF Mathematics Department website (http://mathcenter.usf.edu), which serves as support for the Finite Math, College Algebra, Intermediate Algebra, Introductory Algebra, and CLAST classes at USF. You can see Professor Bello's presentations and streaming videos at that website, as well as at http://www.ibello.com. Professor Bello is a member of the MAA and AMATYC. He has given many presentations regarding the teaching of mathematics at the local, state, and national levels.

Contents

R Prealgebra Review

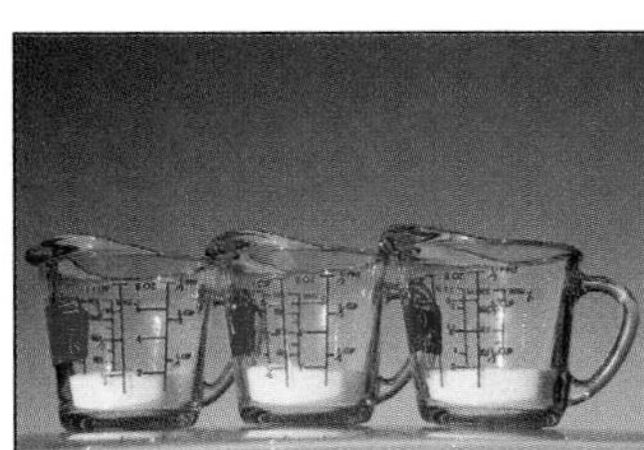

1 Real Numbers and Their Properties

2 Equations, Problem Solving, and Inequalities

3 Graphs of Linear Equations

4 Exponents and Polynomials

5 Factoring

6 Rational Expressions

7 Graphs, Slopes, Inequalities, and Applications

8 Solving Systems of Linear Equations and Inequalities

9 Roots and Radicals

10 Quadratic Equations

Preface

FROM THE AUTHOR

The Inspiration for My Teaching

I was born in Havana, Cuba, and encountered the same challenges of mathematics that many other students do: I failed freshman math. However, perseverance was one of my traits: I made 100% on the final exam the second time around.

I might still be in Cuba had not a police officer kindly informed my family that the members of a club to which I belonged were in jeopardy. An ability to figure out the obvious was another one of my traits: I left for the United States.

I came to the United States and, yes, I did know some English. After working in various jobs (roofer, sheetrock installer, dock worker), I went back to school, finished high school in one year, and received a college academic scholarship. I enrolled in Calculus and made a C. Never one to be discouraged, I persevered, became a math major, and learned to excel in the courses that had previously frustrated me. While a graduate student at the University of South Florida (USF), I taught at a technical school, an experience that contributed to my resolve to teach math and to make it come alive for my students the way brilliant instructors such as Jack Britton, Donald Rose, and Frank Cleaver had done for me.

A Lively Approach to Reach Today's Students

Teaching math at the University of South Florida was a great new career for me, but I was disappointed by the materials I had to use. A rather imposing, mathematically correct but boring book was in vogue. Students hated it, professors hated it, and administrators hated it. I took the challenge to write a better book, a book that was not only mathematically correct, but **student-oriented** with **interesting applications—**many suggested by the students themselves—and even, dare we say, entertaining! That book's approach and philosophy proved an instant success and was a precursor to my current series.

Students fondly called my class "The Bello Comedy Hour," but they worked hard, and they performed well. Because my students always ranked among the highest on the common final exam at USF, I knew I had found a way to motivate them through **commonsense language** and humorous, **realistic math applications.** I also wanted to show students they could overcome the same obstacles I had in math and become successful, too.

If math has just never been a subject that some of your students have felt comfortable with, then they're not alone! I wrote this book with the **math-anxious** student in mind, so they'll find my tone is jovial, my explanations are patient, and instead of making math seem mysterious, I make it down-to-earth and easily digestible. For example, after I've explained the different methods for simplifying fractions, I speak directly to readers: "Which method should you use to simplify fractions? The way you understand!" Once students realize that math is within their grasp and not a foreign language, they'll be surprised at how much more confident they feel.

A Real-World Approach: Applications, Student Motivation, and Problem Solving

What is a "real-world approach"? I have found that most textbooks put forth "real-world" applications that mean nothing to the "real world" of my students. How many of my students would really need to calculate the speed of a bullet (unless they are in its path) or care to know when two trains traveling in different directions would pass by each other (disaster will certainly occur if they are on the same track)? For my students, both traditional and nontraditional, the real world consists of questions such as "How do I find the best cell phone plan?" and "How will I pay my tuition and fees if they increase by $x\%$?" That is why I introduce mathematical concepts through everyday applications with **real data** and give homework using similar, well-grounded situations (see the "Getting Started" application that introduces every section's topic and the word problems in every exercise section).

Putting math in a real-world context has helped me to overcome one of the problems we all face as math educators: **student motivation.** Seeing math in the real world makes students perk up in a math class in a way I have never seen before, and realism has proven to be the best motivator I've ever used. In addition, the real-world approach has enabled me to enhance students' **problem-solving skills** since they are far more likely to tackle a real-world problem that matters to them than to attempt a problem that seems contrived.

Diverse Students and Multiple Learning Styles

We live in a pluralistic society, so how does one write one textbook for everyone? The answer is to build a flexible set of teaching tools that instructors and students can adapt to their own situations. Are any of your students members of a **cultural minority?** So am I! Did they learn **English as a second language?** So did I! You'll find my book speaks directly to them in a way that no other book ever has, and fuzzy explanations in other books will be clear and comprehensible in mine.

Do all your students have the same **learning style?** Of course not. That's why I wrote a book that will help students learn mathematics no matter what their personal learning style is. **Visual learners** will benefit from the text's clean page layout, careful use of color highlighting, "Web It" features and the video lectures on the text's website. **Auditory learners** will profit from the audio "e-Professor" lectures on the text's website, and both **auditory** and **social learners** will be aided by the "Collaborative Learning" projects. **Applied** and **pragmatic learners** will find a bonanza of features geared to help them: pretests, practice problems alongside every example, and mastery tests, to name just a few. **Spatial learners** will find the "Chapter Summary" is designed especially for them, while **creative learners** will find the "Research Questions" to be a natural fit. Finally, **conceptual learners** will feel at home with features such as "The Human Side of Algebra" and the "Write On" exercises. Every student who is accustomed to opening a math book and feeling as if they've run into a brick wall will find in my books that a number of doors are standing open and inviting them inside.

Listening to Student and Instructor Concerns

McGraw-Hill has given me a wonderful resource for making my textbook more responsive to the immediate concerns of students and faculty. In addition to sending my manuscript out for review by instructors at many different colleges, several times a year McGraw-Hill holds symposia and focus groups with math instructors where the emphasis is *not* on selling products but instead on **listening** to the needs of faculty and their students. These encounters have provided me with a wealth of ideas on how to improve my chapter organization, make the page layout of my books more readable, and fine-tune exercises in every chapter. As a result, students and faculty will feel comfortable using my book because it incorporates their specific suggestions and anticipates their needs.

IMPROVEMENTS IN THE SECOND EDITION

Based on the valuable feedback of numerous reviewers and users over the years, the following improvements were made to the Second Edition of *Introductory Algebra*.

Organizational Changes

- A new section, Operations with Fractions, has been added to the review chapter.
- A new section, Introduction to Algebra, has been added to Chapter 1.
- Graphing linear equations has been moved from Chapter 6 to Chapter 3.

- Factoring coverage has been expanded. The previous Factoring Trinomials section is now two sections, the first with $a = 1$ and the second with $a \neq 1$. There is also a new section on the Applications of Quadratics in the factoring chapter.
- Chapter 7 includes two new sections: Applications of Equations of Lines, and Direct and Inverse Variation.
- The section on Functions has been moved from the end of Chapter 6 to the end of Chapter 10.

Pedagogical Changes

- Many examples, applications, and real-data problems have been added or updated to keep the book's content current.
- The book is now produced as a paperback workbook to encourage students to write in their books as they do their homework.
- *Practice problems* with answers at the bottom of the page now appear adjacent to each example to give students immediate reinforcement of their own skills after they have read through the step-by-step solutions of the example.
- *Web Its* have been added to encourage students to visit math sites and discover the many informative and creative websites that are dedicated to stimulating better education in math.
- *Pretests* with answer grids immediately following them have been added to the beginning of each chapter to serve as a diagnostic tool.
- *Calculate Its* have been updated with recent information and keystrokes relevant to currently popular calculators.
- *Applications* have been titled where appropriate to help orient students to the kind of word problem they are about to solve.
- The RSTUV approach to problem solving has been expanded and used throughout this edition in response to positive comments from users of the previous edition.
- *Collaborative Learning* exercises have been added to encourage students to work in teams to solve fun and thought-provoking projects.
- A *Cumulative Test* has been added to the end of each chapter to continually reinforce material students have previously learned.

ACKNOWLEDGMENTS

I would like to thank the following people at McGraw-Hill (in order of appearance):

David Dietz, my sponsoring editor, who provided the necessary incentives and encouragement for creating this series with the cooperation of Bill Barter; Christien Shangraw, my first developmental editor, who worked many hours getting reviewers and gathering responses into concise and usable reports; Randy Welch, who continued and expanded the Christien tradition into a well-honed editing engine with many features, including humor, organization, and very hard work; Liz Haefele, my new editor and publisher, who was encouraging and always on the lookout for new markets; Lori Hancock and her many helpers (LouAnn, Emily, David), who always get the picture; Dr. Tom Porter, of Photos at Your Place, who improved on the pictures I provided; Vicki Krug, one of the most exacting persons at McGraw-Hill, who will always give you the time of day, and then solve the problem; Hal Whipple, for his help in preparing the answers manuscript; Cindy Trimble, for the accuracy of the text; Jeff Huettman, one of the 100 best producers in the United States, who learned Spanish in anticipation of this project; Marie Bova, for her detective work in tracking down permission rights; Steve Stembridge and Barbara Owca, for their help and enthusiasm in marketing the Bello series; and to Professor Nancy Mills, for her expert advice on how my books address multiple learning styles. Finally, thanks to our attack secretary, Beverly DeVine, who managed to send all materials back to the publisher on time. To all of them, my many thanks.

I would also like to extend my gratitude to the following reviewers of the Bello series for their many helpful suggestions and insights that helped me to write better textbooks:

Tony Akhlaghi, *Bellevue Community College*
Theresa Allen, *University of Idaho*
John Anderson, *San Jacinto College–South Campus*
Keith A. Austin, *Devry University–Arlington*
Sohrab Bakhtyari, *St. Petersburg College–Clearwater*
Fatemah Bicksler, *Delgado Community College*
Ann Brackebusch, *Olympic College*
Gail G. Burkett, *Palm Beach Community College*
Linda Burton, *Miami Dade Community College*
Judy Carlson, *Indiana University–Purdue University Indianapolis*
Randall Crist, *Creighton University*
Mark Czerniak, *Moraine Valley Community College*
Parsla Dineen, *University of Nebraska–Omaha*
Sue Duff, *Guilford Technical Community College*
Lynda Fish, *St. Louis Community College–Forest Park*
Donna Foster, *Piedmont Technical College*
Jeanne H. Gagliano, *Delgado Community College*
Debbie Garrison, *Valencia Community College*
Donald K. Gooden, *Northern Virginia Community College–Woodbridge*
Ken Harrelson, *Oklahoma City Community College*
Joseph Lloyd Harris, *Gulf Coast Community College*
Tony Hartman, *Texarkana College*
Susan Hitchcock, *Palm Beach Community College*
Patricia Carey Horacek, *Pensacola Junior College*
Peter Intarapanich, *Southern Connecticut State University*
Judy Ann Jones, *Madison Area Technical College*
Linda Kass, *Bergen Community College*
Joe Kemble, *Lamar University*
Joanne Kendall, *Blinn College–Brenham*
Bernadette Kocyba, *J S Reynolds Community College*
Marie Agnes Langston, *Palm Beach Community College*
Kathryn Lavelle, *Westchester Community College*
Angela Lawrenz, *Blinn College–Bryan*
Richard Leedy, *Polk Community College*
Judith L. Maggiore, *Holyoke Community College*
Timothy Magnavita, *Bucks Community College*
Tsun-Zee Mai, *University of Alabama*
Harold Mardones, *Community College of Denver*
Lois Martin, *Massasoit Community College*
Gary McCracken, *Shelton State Community College*
Tania McNutt, *Community College of Aurora*
Barbara Miller, *Lexington Community College*
Danielle Morgan, *San Jacinto College–South Campus*
Joanne Peeples, *El Paso Community College*
Faith Peters, *Miami Dade College–Wolfson*
Jane Pinnow, *University of Wisconsin–Parkside*
Janice F. Rech, *University of Nebraska–Omaha*
Libbie Reeves, *Mitchell Community College*
Karen Roothaan, *Harold Washington College*
Don Rose, *College of the Sequoias*
Pascal Roubides, *Miami Dade College–Wolfson*
Juan Saavedra, *Albuquerque Technical Vocational Institute*
Judith Salmon, *Fitchburg State College*
Mansour Samimi, *Winston-Salem State University*
Susan Santolucito, *Delgado Community College*
Ellen Sawyer, *College of DuPage*

A COMMITMENT TO ACCURACY

You have a right to expect an accurate textbook, and McGraw-Hill invests considerable time and effort to make sure that we deliver one. Listed below are the many steps we take to make sure this happens.

OUR ACCURACY VERIFICATION PROCESS

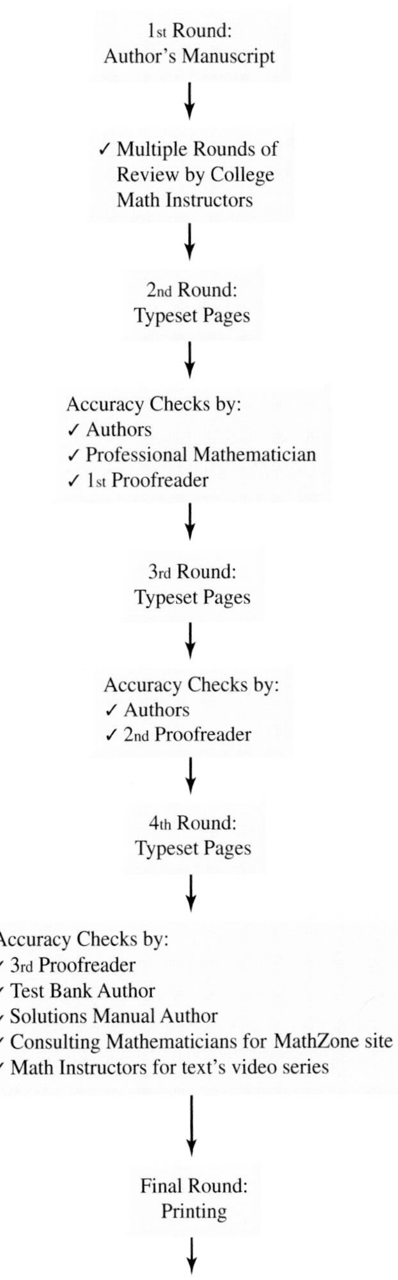

First Round

Step 1: Numerous **college math instructors** review the manuscript and report on any errors that they may find, and the authors make these corrections in their final manuscript.

Second Round

Step 2: Once the manuscript has been typeset, the **authors** check their manuscript against the first page proofs to ensure that all illustrations, graphs, examples, exercises, solutions, and answers have been correctly laid out on the pages, and that all notation is correctly used.

Step 3: An outside, **professional mathematician** works through every example and exercise in the page proofs to verify the accuracy of the answers.

Step 4: A **proofreader** adds a triple layer of accuracy assurance in the first pages by hunting for errors, then a second, corrected round of page proofs is produced.

Third Round

Step 5: The **author team** reviews the second round of page proofs for two reasons: 1) to make certain that any previous corrections were properly made, and 2) to look for any errors they might have missed on the first round.

Step 6: A **second proofreader** is added to the project to examine the new round of page proofs to double check the author team's work and to lend a fresh, critical eye to the book before the third round of paging.

Fourth Round

Step 7: A **third proofreader** inspects the third round of page proofs to verify that all previous corrections have been properly made and that there are no new or remaining errors.

Step 8: Meanwhile, in partnership with **independent mathematicians,** the text accuracy is verified from a variety of fresh perspectives:

- The **test bank author** checks for consistency and accuracy as they prepare the computerized test item file.
- The **solutions manual author** works every single exercise and verifies their answers, reporting any errors to the publisher.
- A **consulting group of mathematicians,** who write material for the text's MathZone site, notifies the publisher of any errors they encounter in the page proofs.
- A video production company employing **expert math instructors** for the text's videos will alert the publisher of any errors they might find in the page proofs.

Final Round

Step 9: The **project manager,** who has overseen the book from the beginning, performs a **fourth proofread** of the textbook during the printing process, providing a final accuracy review.

$\Rightarrow$ What results is a mathematics textbook that is as accurate and error-free as is humanly possible, and our authors and publishing staff are confident that our many layers of quality assurance have produced textbooks that are the leaders of the industry for their integrity and correctness.

Features and Supplements

• Motivation for a Diverse Student Audience

A number of features exist in every chapter to motivate students' interest in the topic and thereby increase their performance in the course:

The Human Side of Algebra

To personalize the subject of algebra, the origins of algebraic notation, concepts, and methods are introduced through the lives of real people solving ordinary problems.

> **The Human Side of Algebra**
>
> The digits 1–9 originated with the Hindus and were passed on to us by the Arabs. Mohammed ibn Musa al-Khowarizmi (ca. A.D. 780–850) wrote two books, one on algebra (*Hisak al-jabr w'almuqabala,* "the science of equations"), from which the name *algebra* was derived, and one dealing with the Hindu numeration system. The oldest dated European manuscript containing the Hindu-Arabic numerals is the *Codex Vigilanus* written in Spain in A.D. 976, which used the nine symbols
>
> 1 2 3 4 5 6 7 8 9
>
> Zero, on the other hand, has a history of its own. According to scholars, "At the time of the birth of Christ, the idea of zero as a symbol or number had never occurred to anyone." So, who invented zero? An unknown Hindu who wrote a symbol of his own, a dot he called *sunya,* to indicate a column with no beads on his counting board. The Hindu notation reached Europe thanks to the Arabs, who called it *sifr*. About A.D. 150, the Alexandrian astronomer Ptolemy began using *o* (omicron), the first letter in the Greek word for *nothing,* in the manner of our zero.

> **GETTING STARTED**
>
> **Don't Forget the Tip!**
>
> Jasmine is a server at CDB restaurant. Aside from her tips, she gets \$2.88/hour. In 1 hour, she earns \$2.88; in 2 hr, she earns \$5.76; in 3 hr, she earns \$8.64, and so on. We can form the set of ordered pairs (1, 2.88), (2, 5.76), (3, 8.64) using the number of hours she works as the first coordinate and the amount she earns as the second coordinate. Note that the ratio of second coordinates to first coordinates is the same number:
>
> $\frac{2.88}{1} = 2.88, \quad \frac{5.76}{2} = 2.88, \quad \frac{8.64}{3} = 2.88,$

Getting Started

Each topic is introduced in a setting familiar to students' daily lives, making the subject personally relevant and more easily understood.

Web It

Appearing in margins where relevant, these boxes refer students to the abundance of resources available on the Web that can show them fun, alternative explanations and demonstrations of important topics.

> **Web It**
>
> To review how to translate and evaluate expressions, go to link 1-1-1 at the Bello Website at mhhe.com/bello.
>
> You can also try link 1-1-2.

Write On

Writing exercises give students the opportunity to express mathematical concepts and procedures in their own words, thereby internalizing what they have learned.

WRITE ON

71. In the expression "$\frac{1}{2}$ of x," what operation does the word *of* signify?

72. Most people believe that the word *and* always means addition.

a. In the expression "the sum of x and y," does "and" signify the operation of addition? Explain.

b. In the expression "the product of 2 and three more than a number," does "and" signify the operation of addition? Explain.

73. Explain the difference between "x divided by y" and "x divided into y."

74. Explain the difference between "a less than b" and "a less b."

Collaborative Learning

Concluding the chapter are exercises for collaborative learning that promote teamwork by students on interesting and enjoyable exploration projects.

COLLABORATIVE LEARNING

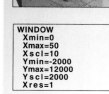

This exercise concerns an emergency ejection from a fighter jet and how it could relate to the present discussion of completing the square.

The height H (in feet relative to the ground) of a pilot after t seconds is given by

$$H = -16t^2 + 608t + 4482$$

At about 10,000 feet a malfunction occurs. Eject! Eject!

We want to answer three questions:

1. What is the maximum height (relative to the ground) reached by the pilot after being ejected?
2. How many feet above the jet was the pilot ejected?
3. When did the pilot reach the ground with the aid of his or her parachute?

```
WINDOW
 Xmin=0
 Xmax=50
 Xscl=10
 Ymin=-2000
 Ymax=12000
 Yscl=2000
 Xres=1
```

Form three teams to investigate the incident.

Team 1 is in charge of graphing the path of the pilot and the point of ejection using a graphing calculator. A possible viewing window to graph the path is shown. What expression Y_1 does the team have to graph? What is the equation of the line Y_2 shown in the diagram and representing the point at which the pilot ejected?

Team 2 is in charge of answering the first two questions.

Start with: $H = -16(\quad)^2 + C$

Here is a hint: $H = -16(t^2 - 38t) + 4482$

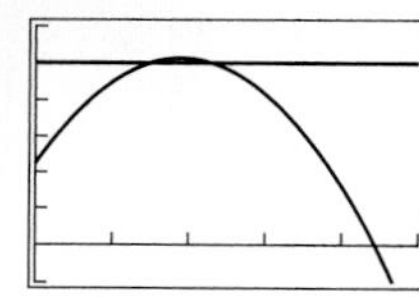

Now, complete the square inside the parentheses to put the equation in the standard form for a parabola. What is the maximum for this parabola? This answers question 1. Since we know that the pilot ejected at 10,000 feet, how many feet above the jet was the pilot ejected? This answers question 2.

Team 3 has to answer the third question. When the pilot reaches the ground, what is H? The point (t, H) is a solution (zero) of the equation whose graph is shown and it occurs when $H = 0$. To find this point press 2nd TRACE 2. The value of t is the time it took the pilot to land back on earth!

Research Questions

Research questions provide students with additional opportunities to explore interesting areas of math where they may find that the questions can lead to surprising answers.

1. In the *Human Side of Algebra* at the beginning of this chapter, we mentioned the Hindu numeration system. The Egyptians and Babylonians also developed numeration systems. Write a report about each of these numeration systems, detailing the symbols used for the digits 1–9, the base used, and the manner in which fractions were written.
2. Write a report on the life and works of Mohammed al-Khowarizmi, with special emphasis on the books he wrote.
3. We have now studied the four fundamental operations. But do you know where the symbols used to indicate these operations originated?

• Abundant Practice and Problem Solving

Bello offers students numerous opportunities and different paths for developing their problem-solving skills.

Pretest

An optional pretest that begins each chapter is especially helpful for students taking the course as a review who may remember some concepts but not others. The answer grid that follows immediately afterward gives students the page number, section, and example to study in case they missed a question.

Pretest for Chapter 1

(*Answers on page 43*)

1. Write in symbols:

a. The sum of m and n

b. $7m$ plus $3n$ minus 4

c. 4 times m

d. $\frac{1}{9}$ of m

2. Write in symbols:

a. The quotient of m and 3

b. The quotient of 3 and n

c. The sum of m and n, divided by the difference of m and n

3. For $m = 6$ and $n = 3$, evaluate:

a. $m + n$

b. $m - n$

c. $2m - 3n$

d. $\frac{2m + n}{n}$

Answers to Pretest

ANSWER			IF YOU MISSED	REVIEW		
			QUESTION	SECTION	EXAMPLES	PAGE
1. a. $m + n$	**b.** $7m + 3n - 4$		1	1.1	1, 2	46
c. $4m$	**d.** $\frac{1}{9}m$					
2. a. $\frac{m}{3}$	**b.** $\frac{3}{n}$	**c.** $\frac{m + n}{m - n}$	2	1.1	3	47
3. a. 9	**b.** 3		3	1.1	4	47
c. 3	**d.** 5					

Paired Examples/Problems

Examples are provided adjacent to similar problems intended for students to obtain immediate reinforcement of the skill they have just observed. These are especially effective for students who learn by doing and who benefit from frequent practice of important methods. Answers to the problems appear at the bottom of the page.

EXAMPLE 4 **Evaluating algebraic expressions**

Evaluate the given expressions by substituting 10 for x and 5 for y.

a. $x + y$ **b.** $x - y$

c. $4y$ **d.** $\frac{x}{y}$

e. $3x - 2y$

SOLUTION

a. Substitute 10 for x and 5 for y in $x + y$.

We obtain: $x + y = 10 + 5 = 15$.

The number 15 is called the **value** of $x + y$.

b. $x - y = 10 - 5 = 5$ **c.** $4y = 4(5) = 20$

d. $\frac{x}{y} = \frac{10}{5} = 2$ **e.** $3x - 2y = 3(10) - 2(5) = 30 - 10 = 20$

PROBLEM 4

Evaluate the expressions by substituting 22 for a and 3 for b.

a. $a + b$ **b.** $2a - b$

c. $5b$ **d.** $\frac{2a}{b}$

e. $2a - 3b$

Answers

3. a. $\frac{a}{b}$ **b.** $\frac{b}{a}$ **c.** $\frac{x - y}{z}$ **d.** $\frac{x - y}{x + y}$ **4. a.** 25 **b.** 41 **c.** 15 **d.** $\frac{44}{3}$ **e.** 35

PROCEDURE

RSTUV Method for Solving Word Problems

1. **R**ead the problem carefully and decide what is asked for (the unknown).
2. **S**elect a variable to represent this unknown.
3. **T**hink of a plan to help you write an equation.
4. **U**se algebra to solve the resulting equation.
5. **V**erify the answer.

RSTUV Method

The easy-to-remember **"RSTUV" method** gives students a reliable and helpful tool in demystifying word problems so that they can more readily translate them into equations they can recognize and solve.

- **R**ead the problem carefully and decide what is being asked for (the unknown).
- **S**elect a variable to represent this unknown.
- **T**hink of a plan to help you write an equation.
- **U**se algebra to solve the resulting equation.
- **V**erify the answer.

Exercises

A wealth of exercises for each section is organized according to the learning objectives for that section, giving students a reference to study if they need extra help.

Exercises 1.1

A In Problems 1–40, write each expression in symbols.

1. The sum of a and c
2. The sum of u and v
3. The sum of $3x$ and y
4. The sum of 8 and x
5. $9x$ plus $17y$
6. $5a$ plus $2b$

Boost *your* GRADE at mathzone.com!

- Practice Problems
- Self-Tests
- Videos
- NetTutor
- e-Professors

Applications

Students will enjoy the exceptionally creative applications in most sections that bring math alive and demonstrate that it can even be performed with a sense of humor.

APPLICATIONS

41. *Gasoline octane rating* Have you noticed the octane rating of gasoline at the gas pump? This octane rating is given by

$$\frac{R+M}{2}$$

where R is a number measuring the performance of gasoline using the Research Method and M is a number measuring the performance of gasoline using the Motor Method. If a certain gasoline has $R = 92$ and $M = 82$, what is its octane rating?

42. *Gasoline octane rating* If a gasoline has $R = 97$ and $M = 89$, what is its octane rating?

43. *Exercise pulse rate* If A is your age, the minimum pulse rate you should maintain during aerobic activities is $0.72(220 - A)$. What is the minimum pulse rate you should maintain if you are

    a. 20 years old? **b.** 45 years old?

44. *Exercise pulse rate* If A is your age, the maximum pulse rate you should maintain during aerobic activities is $0.88(220 - A)$. What is the maximum pulse rate you should maintain if you are

    a. 20 years old? **b.** 45 years old?

USING YOUR KNOWLEDGE

Children's Dosages

What is the corresponding dose (amount) of medication for children when the adult dosage is known? There are several formulas that tell us.

51. *Friend's rule* (for children under 2 years):

    (Age in months · Adult dose) ÷ 150 = Child's dose

    Suppose a child is 10 months old and the adult dose of aspirin is a 75-milligram tablet. What is the child's dose? [*Hint:* Simplify (10 · 75) ÷ 150.]

52. *Clarke's rule* (for children over 2 years):

    (Weight of child · Adult dose) ÷ 150 = Child's dose

    If a 7-year-old child weighs 75 pounds and the adult dose is 4 tablets a day, what is the child's dose? [*Hint:* Simplify (75 · 4) ÷ 150.]

Using Your Knowledge

Optional, extended applications give students an opportunity to practice what they've learned in a multistep problem requiring reasoning skills in addition to algebraic operations.

• Study Aids to Make Math Accessible

Since some students confront math anxiety as soon as they sign up for the course, the Bello system provides numerous study aids to make their learning easier.

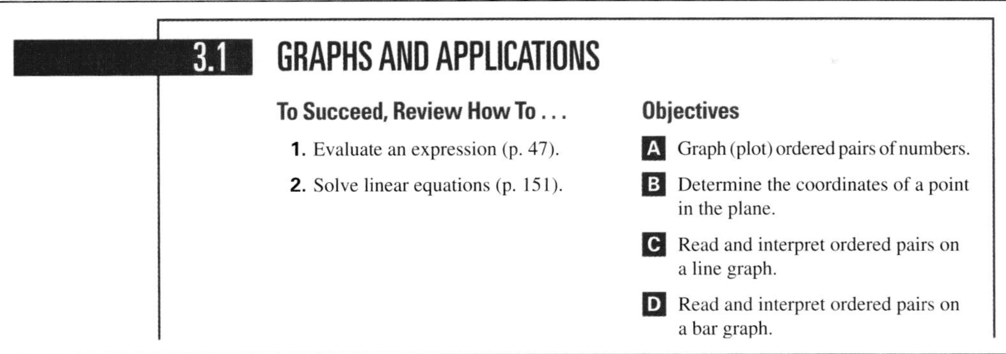

3.1 GRAPHS AND APPLICATIONS

To Succeed, Review How To . . .

1. Evaluate an expression (p. 47).
2. Solve linear equations (p. 151).

Objectives

A Graph (plot) ordered pairs of numbers.

B Determine the coordinates of a point in the plane.

C Read and interpret ordered pairs on a line graph.

D Read and interpret ordered pairs on a bar graph.

Reviews

Every section begins with "To succeed, review how to . . . ," which directs students to specific pages to study key topics they need to understand to successfully begin that section.

Objectives

The objectives for each section not only identify the specific tasks students should be able to perform, but they organize the section itself with letters corresponding to each section heading, making it easy to follow.

Calculate It

Adding Integers

To enter $-3 + (-2)$ using a scientific calculator, enter

3 [+/−] [+] 2 [+/−] [ENTER]

If your calculator has a set of parentheses, then you can enter parentheses around the -3 and -2.

Calculate It

Appearing in margins where relevant, these boxes give students optional advice on how to use calculators to reinforce their understanding of algebra and check their work.

Skill Checkers

These brief exercises help students keep their math skills well honed in preparation for the next section.

SKILL CHECKER

Try the Skill Checker Exercises so you'll be ready for the next section.

Find:

45. $11 \cdot \frac{1}{11}$

46. $17 \cdot \frac{1}{17}$

47. $-2.3 + 2.3$

48. $-1.7 + 1.7$

49. $-2.5 \cdot \frac{1}{-2.5}$

50. $-3.7 \cdot \frac{1}{-3.7}$

MASTERY TEST

If you know how to do these problems, you have learned your lesson!

Find the value of:

57. $63 \div 9 - (2 + 5)$

58. $-64 \div 8 - (6 - 2)$

59. $16 + 2^3 + 3 - 9$

60. $3 \cdot 4 - 18$

Mastery Tests

Brief tests in every section give students a quick checkup to make sure they're ready to go on to the next topic.

Summary

SECTION	ITEM	MEANING	EXAMPLE
1.1A	Arithmetic expressions	Expressions containing numbers and operation signs	$3 + 4 - 8, 9 \cdot 4 \div 6$ are arithmetic expressions.
1.1A	Algebraic expressions	Expressions containing numbers, operation signs, and variables	$3x - 4y + 3z, 2x \div y + 9z$, and $7x - 9y \div 3z$ are algebraic expressions.
1.1A	Factor	The items to be multiplied	In the expression $4xy$, the factors are 4, x, and y.
1.1B	Evaluate	To substitute a value for one or more of the variables in an expression	Evaluating $3x - 4y + 3z$ when $x = 1, y = 2$, and $z = 3$ yields $3 \cdot 1 - 4 \cdot 2 + 3 \cdot 3$ or 4.

Summary

An easy-to-read grid summarizes the essential chapter information by section, providing an item, its meaning, and an example to help students connect concepts with their concrete occurrences.

Review Exercises

Chapter review exercises are coded by section number and give students extra reinforcement and practice to boost their confidence.

Review Exercises

(If you need help with these exercises, look in the section indicated in brackets.)

1. [1.1A] Write in symbols:
   a. The sum of a and b
   b. a minus b
   c. $7a$ plus $2b$ minus 8

2. [1.1A] Write using juxtaposition:
   a. 3 times m
   b. $-m$ times n times r
   c. $\frac{1}{7}$ of m
   d. The product of 8 and m

Practice Test

(*Answers on pages 116–117*)

1. Write in symbols:
   a. The sum of g and h
   b. g minus h
   c. $6g$ plus $3h$ minus 8

2. Write using juxtaposition:
   a. 3 times g
   b. $-g$ times h times r
   c. $\frac{1}{5}$ of g
   d. The product of 9 and g

Answers to Practice Test

ANSWER	IF YOU MISSED	REVIEW		
	QUESTION	SECTION	EXAMPLES	PAGE
1. a. $g + h$ b. $g - h$ c. $6g + 3h - 8$	1	1.1	1	46
2. a. $3g$ b. $-ghr$ c. $\frac{1}{5}g$ d. $9g$	2	1.1	2	46

Practice Test with Answers

The chapter Practice Test offers students a nonthreatening environment to review the material and determine whether they are ready to take a test given by their instructor. The answers to the Practice Test give students immediate feedback on their performance, and the answer grid gives them specific guidance on which section, example, and pages to review for any answers they may have missed.

Cumulative Test

The Cumulative Test covers material from the present chapter and any prior chapters. It can be used for extra homework or for student review to improve retention of important skills and concepts.

Cumulative Review Chapters 1–2

1. Find the additive inverse (opposite) of -7.
2. Find: $\left|-9\frac{9}{10}\right|$
3. Find: $-\frac{2}{7} + \left(-\frac{2}{9}\right)$
4. Find: $-0.7 - (-8.9)$
5. Find: $(-2.4)(3.6)$
6. Find: $-(2^4)$

• Supplements for Instructors

Annotated Instructor's Edition

This version of the student text contains **answers** to all odd- and even-numbered exercises in addition to helpful **teaching tips.** The answers are printed on the same page as the exercises themselves so that there is no need to consult a separate appendix or answer key.

Instructor's Testing and Resource CD

This cross-platform CD-ROM provides a wealth of resources for the instructor. Supplements featured on this CD-ROM include a **computerized test bank** utilizing Brownstone Diploma® **algorithm-based** testing software to quickly create customized exams. This user-friendly program allows instructors to search for questions by topic, format, or difficulty level; edit existing questions or add new ones; and scramble questions and answer keys for multiple versions of the same test.

Instructor's Solutions Manual

This supplement contains detailed solutions to all exercises in the text. The methods used to solve the problems in the manual are the same as those used to solve the examples in the textbook.

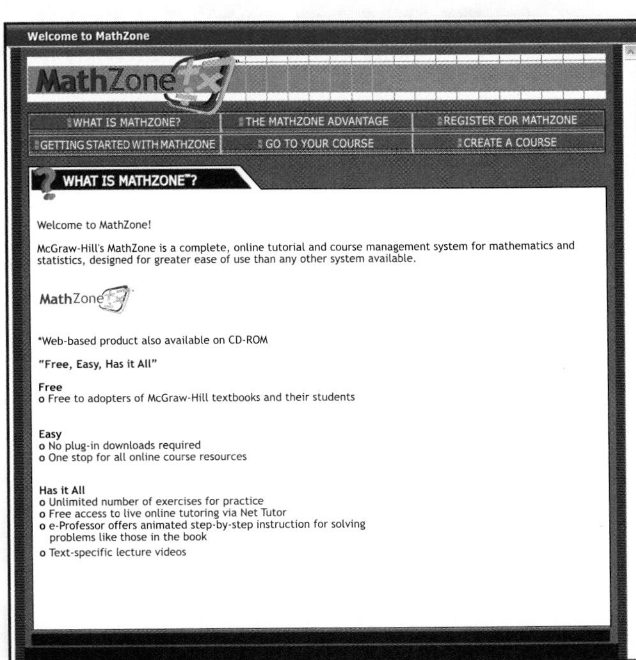

McGraw-Hill's **MathZone** is a complete, online tutorial and course management system for mathematics and statistics, designed for greater ease of use than any other system available. **Free** upon adoption of a McGraw-Hill title, the system allows instructors to create and share courses and assignments with colleagues and adjuncts with only a few

*Web-based product also available on CD-ROM

clicks of the mouse. All assignments, questions, e-Professors, online tutoring, and video lectures are directly tied to text-specific materials in Bello, *Introductory Algebra,* 2nd Edition. MathZone courses are customized to your textbook, but you can **edit** questions and algorithms, **import** your own content, and **create** announcements and due dates for assignments. MathZone has **automatic grading** and reporting of easy-to-assign algorithmically generated homework, quizzing, and testing. All student activity within MathZone is automatically recorded and available to you through a **fully integrated grade book** that can be downloaded to Excel.

ALEKS® v2.0

ALEKS® (**A**ssessment and **LE**arning in **K**nowledge **S**paces) is an artificial intelligence-based system for individualized math learning, available over the Web. ALEKS delivers precise, qualitative diagnostic assessments of students' math knowledge, guides them in the selection of appropriate new study material, and records their progress toward mastery of curricular goals in a robust classroom management system. See page xxix for more details regarding ALEKS.

PageOut

PageOut is McGraw-Hill's unique, intuitive tool enabling instructors to create a full-featured, professional quality course website *without* being a technical expert. With PageOut you can post your syllabus online, assign content from the Bello MathZone site, add links to important off-site resources, and maintain student results in the online grade book. PageOut is free for every McGraw-Hill Higher Education user and, if you're short on time, we even have a team ready to help you create your site. Contact your McGraw-Hill representative for further information.

• Supplements for Students

Student's Solutions Manual

This supplement contains complete worked-out solutions to all odd-numbered exercises and all odd- and even-numbered problems in the Review Exercises and Cumulative Reviews in the textbook. The methods used to solve the problems in the manual are the same as those used to solve the examples in the textbook. This tool can be an invaluable aid to students who want to check their work and improve their grades by comparing their own solutions to those found in the manual and finding specific areas where they can do better.

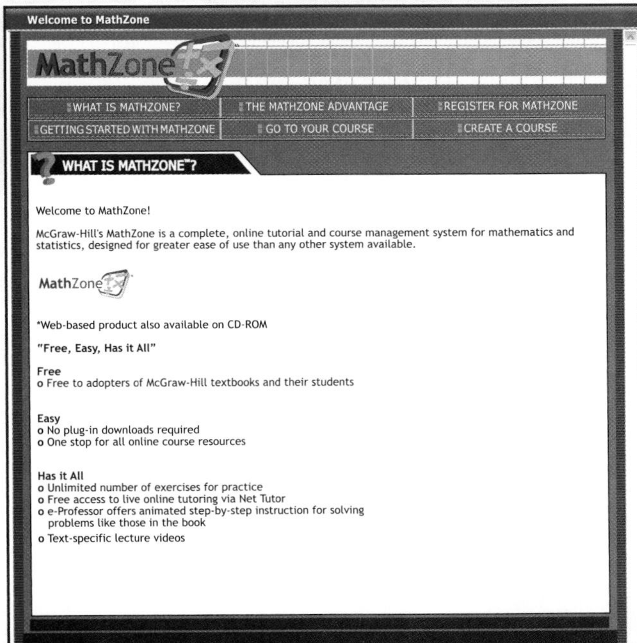

Welcome to MathZone
MathZone
WHAT IS MATHZONE?
THE MATHZONE ADVANTAGE
REGISTER FOR MATHZONE
GETTING STARTED WITH MATHZONE
GO TO YOUR COURSE
CREATE A COURSE
WHAT IS MATHZONE™?
Welcome to MathZone!
McGraw-Hill's MathZone is a complete, online tutorial and course management system for mathematics and statistics, designed for greater ease of use than any other system available.
MathZone
*Web-based product also available on CD-ROM
"Free, Easy, Has it All"
Free
o Free to adopters of McGraw-Hill textbooks and their students
Easy
o No plug-in downloads required
o One stop for all online course resources
Has it All
o Unlimited number of exercises for practice
o Free access to live online tutoring via Net Tutor
o e-Professor offers animated step-by-step instruction for solving problems like those in the book
o Text-specific lecture videos

McGraw-Hill's MathZone is a powerful new online tutorial for homework, quizzing, testing, and interactive applications. MathZone offers:

- **Practice exercises** based on the text and generated in an unlimited number for as much practice as needed to master any topic you study.
- **Videos** of classroom instructors giving lectures and showing you how to solve exercises from the text.
- **e-Professors** to take you through animated, step-by-step instructions (delivered via on-screen text and synchronized audio) for solving problems in the book, allowing you to digest each step at your own pace.
- **NetTutor,** which offers live, personalized tutoring via the Internet.
- Every assignment, question, e-Professor, and video lecture is derived directly from Bello, *Introductory Algebra,* 2nd Edition.

---

*Web-based product also available on CD-ROM

NetTutor

Also available separately from MathZone, NetTutor is a revolutionary system that enables students to interact with a live tutor over the Web by using NetTutor's Web-based, graphical chat capabilities. Students can also submit questions and receive answers, browse previously answered questions, and view previous live chat sessions. NetTutor can be accessed on the text's MathZone site through the Student Edition.

ALEKS® v2.0

ALEKS® (**A**ssessment and **LE**arning in **K**nowledge **S**paces) is an artificial intelligence-based system for individualized math learning, available over the Web. ALEKS delivers precise, qualitative diagnostic assessments of students' math knowledge, guides them in the selection of appropriate new study material, and records their progress toward mastery of curricular goals in a robust classroom management system. See page xxix for more details regarding ALEKS.

Bello Video Series

The video series is available on DVD and VHS tape and features an instructor introducing topics and working through selected exercises from the text, explaining how to complete them step-by-step. The DVDs are **closed-captioned** for the hearing-impaired and also **subtitled in Spanish.**

Math for the Anxious: Building Basic Skills, by Rosanne Proga

Math for the Anxious: Building Basic Skills is written to provide a practical approach to the problem of math anxiety. By combining strategies for success with a pain-free introduction to basic math content, students will overcome their anxiety and find greater success in their math courses.

ALEKS is an artificial intelligence-based system for individualized math learning, available for Higher Education from McGraw-Hill over the World Wide Web.

ALEKS delivers precise assessments of math knowledge, guides the student in the selection of appropriate new study material, and records student progress toward mastery of goals.

ALEKS interacts with a student much as a skilled human tutor would, moving between explanation and practice as needed, correcting and analyzing errors, defining terms and changing topics on request. By accurately assessing a student's knowledge, ALEKS can focus clearly on what the student is ready to learn next, helping to master the course content more quickly and easily.

ALEKS is:

- **A comprehensive course management system.** It tells the instructor exactly what students know and don't know.
- **Artificial intelligence.** It totally individualizes assessment and learning.
- **Customizable.** ALEKS can be set to cover the material in your course.
- **Web-based.** It uses a standard browser for easy Internet access.
- **Inexpensive.** There are no setup fees or site license fees.

ALEKS 2.0 adds the following new features:

- **Automatic Textbook Integration**
- **New Instructor Module**
- **Instructor-Created Quizzes**
- **New Message Center**

ALEKS maintains the features that have made it so popular including:

- **Web-Based Delivery** No complicated network or lab setup
- **Immediate Feedback** for students in learning mode
- **Integrated Tracking of Student Progress and Activity**
- **Individualized Instruction** which gives students problems they are ***Ready to Learn***

For more information please contact your McGraw-Hill Sales Representative or visit ALEKS at http://www.highedmath.aleks.com.

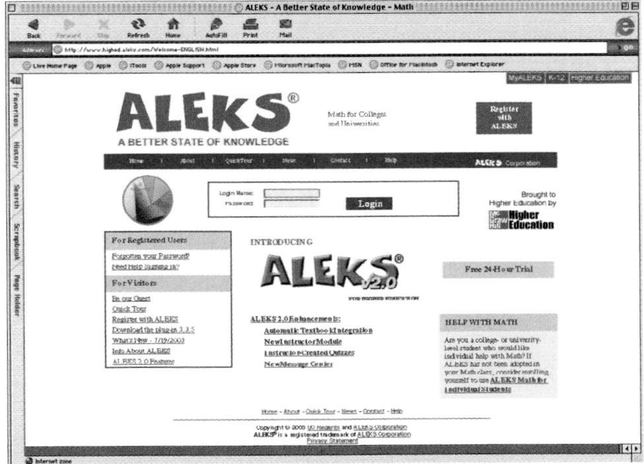

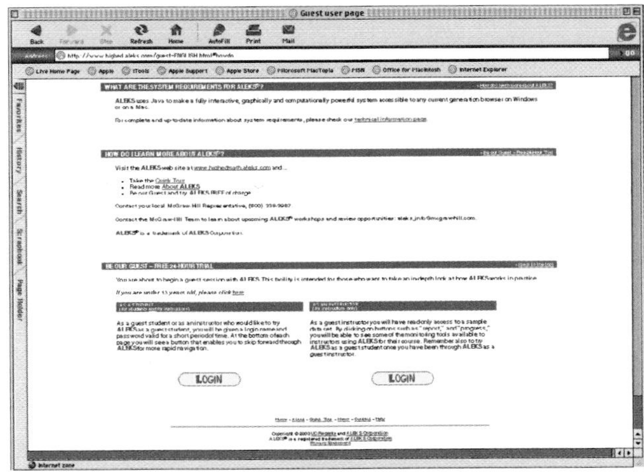

Applications Index

Business and Economics

Consumer Issues

Distance, Rate, and Speed

Education

Engineering

Entertainment

Finance and Investment

Food

Geometry

History

Law Enforcement

Medicine and Health

Science, Nature, and Environment

Social Sciences

Sports

Prealgebra Review

Pretest for Chapter R

(Answers on page 3)

1. Write -8 as a fraction with a denominator of 1.

2. Find a fraction equivalent to $\frac{7}{5}$ with a denominator of 20.

3. Find a fraction equivalent to $\frac{15}{21}$ with a denominator of 7.

4. Reduce $\frac{45}{36}$ to lowest terms.

5. Find: $5\frac{1}{4} \cdot \frac{32}{21}$

6. Find: $3\frac{1}{2} \div \frac{7}{10}$

7. Find: $2\frac{1}{10} + \frac{1}{12}$

8. Find: $2\frac{1}{6} - 1\frac{9}{10}$

9. Write 41.017 in expanded form.

10. Write 0.035 as a reduced fraction.

11. Write 2.12 as a reduced fraction.

12. Add: $9.91 + 7.698$

13. Subtract: $479.9 - 183.21$

14. Multiply: 2.05×13.61

15. Divide: $\frac{4.2}{0.070}$

16. **a.** Write $\frac{1}{6}$ as a decimal.

b. Round the answer to two decimal places.

17. Write 8.6% as a decimal.

18. Write 0.31 as a percent.

19. Write 48% as a fraction.

20. Write $\frac{1}{6}$ as a percent.

Answers to Pretest

ANSWER	IF YOU MISSED	REVIEW		
	QUESTION	SECTION	EXAMPLES	PAGE
1. $\frac{-8}{1}$	1	R.1	1	5
2. $\frac{28}{20}$	2	R.1	2	6
3. $\frac{5}{7}$	3	R.1	3	6
4. $\frac{5}{4}$	4	R.1	4	7
5. 8	5	R.2	1, 2, 3	12–13
6. 5	6	R.2	4, 5	14–15
7. $\frac{131}{60}$	7	R.2	6	17
8. $\frac{4}{15}$	8	R.2	7, 8, 9	19–21
9. $40 + 1 + \frac{1}{100} + \frac{7}{1000}$	9	R.3	1	25
10. $\frac{7}{200}$	10	R.3	2	25
11. $\frac{53}{25}$	11	R.3	3	26
12. 17.608	12	R.3	4	26
13. 296.69	13	R.3	5, 6	27
14. 27.9005	14	R.3	7	28
15. 60	15	R.3	8	29
16. a. $0.1\overline{6}$ **b.** 0.17	16	R.3	9, 15	30, 34–35
17. 0.086	17	R.3	10	31
18. 31%	18	R.3	11	31
19. $\frac{12}{25}$	19	R.3	12	32
20. $16\frac{2}{3}\%$	20	R.3	13, 14	33–34

R.1 FRACTIONS

To Succeed, Review How To . . .

1. Add, subtract, multiply, and divide natural numbers.

Objectives

A Write an integer as a fraction.

B Find a fraction equivalent to a given one, but with a specified denominator.

C Reduce fractions to lowest terms.

GETTING STARTED

Algebra and Arithmetic

The symbols on the sundial and the Roman clock have something in common: they both use numerals to name the numbers from 1 to 12. Algebra and arithmetic also have something in common: they use the same numbers and the same rules.

Teaching Tip

Some books call the natural numbers the *counting* numbers.

In arithmetic you learned about the counting numbers. The numbers used for counting are the **natural numbers:**

1, 2, 3, 4, 5, and so on

These numbers are also used in algebra. We use the **whole numbers**

0, 1, 2, 3, 4, and so on

as well. Later on, you probably learned about the **integers.** The integers include the **positive integers,**

$+1, +2, +3, +4, +5$, and so on   Read "positive one, positive two," and so on.

Teaching Tip

Point out that the integers consist of the *positive* integers, the *negative* integers, and 0, which is neither positive nor negative.

the **negative integers,**

$-1, -2, -3, -4, -5$, and so on   Read "negative one, negative two," and so on.

and the number **0**, which is neither positive nor negative. Thus the integers are

$$\ldots, -2, -1, 0, 1, 2, \ldots$$

where the dots (. . .) indicate that the enumeration continues without end. Note that $+1 = 1$, $+2 = 2$, $+3 = 3$, and so on. Thus, the positive integers are the natural numbers.

Writing Integers as Fractions

All the numbers discussed can be written as **common fractions** of the form:

$$\text{Fraction bar} \rightarrow \frac{a \leftarrow \text{Numerator}}{b \leftarrow \text{Denominator}}$$

When a and b are integers and b is not 0, this ratio is called a **rational number.** For example, $\frac{1}{3}$, $\frac{-5}{2}$, and $\frac{0}{7}$ are rational numbers. In fact, all natural numbers, all whole numbers, and all integers are also rational numbers.

When the numerator a of a fraction is smaller than the denominator b, the fraction $\frac{a}{b}$ is a **proper fraction.** Otherwise, the fraction is **improper.** Improper fractions are often written as **mixed numbers.** Thus $\frac{9}{5}$ may be written as $1\frac{4}{5}$, and $\frac{13}{4}$ may be written as $3\frac{1}{4}$.

Of course, any integer can be written as a fraction by writing it with a denominator of 1. For example,

$$4 = \frac{4}{1}, \quad 8 = \frac{8}{1}, \quad 0 = \frac{0}{1}, \quad \text{and} \quad -3 = \frac{-3}{1}$$

EXAMPLE 1 **Writing integers as fractions**

Write as fractions with a denominator of 1:

a. 10 **b.** -15

SOLUTION

a. $10 = \frac{10}{1}$ **b.** $-15 = \frac{-15}{1}$

PROBLEM 1

Write as a fraction with a denominator of 1:

a. 18 **b.** -24

The rational numbers we have discussed are part of a larger set of numbers, the set of *real numbers.* The **real numbers** include the *rational numbers* and the *irrational numbers.* The **irrational numbers** are numbers that cannot be written as the ratio of two integers. For example, $\sqrt{2}$, π, $-\sqrt[3]{10}$, and $\frac{\sqrt{3}}{2}$ are irrational numbers. Thus each real number is either rational or irrational. We shall say more about the irrational numbers in Chapter 1.

B Equivalent Fractions

In Example 1(a) we wrote 10 as $\frac{10}{1}$. Can you find other ways of writing 10 as a fraction? Here are some:

$$10 = \frac{10}{1} = \frac{10 \times 2}{1 \times 2} = \frac{20}{2}$$

$$10 = \frac{10}{1} = \frac{10 \times 3}{1 \times 3} = \frac{30}{3}$$

and

$$10 = \frac{10}{1} = \frac{10 \times 4}{1 \times 4} = \frac{40}{4}$$

Note that $\frac{2}{2} = 1$, $\frac{3}{3} = 1$, and $\frac{4}{4} = 1$. As you can see, the fraction $\frac{10}{1}$ is **equivalent to** (has the same value as) many other fractions. We can always obtain other fractions equivalent to any given fraction by *multiplying* the numerator and denominator of the original fraction

Answers

1. a. $\frac{18}{1}$ **b.** $\frac{-24}{1}$

by the *same* nonzero number, a process called **building up** the fraction. This is the same as multiplying the fraction by 1, where 1 is written as $\frac{2}{2}, \frac{3}{3}, \frac{4}{4}$, and so on. For example,

$$\frac{3}{5} = \frac{3 \times 2}{5 \times 2} = \frac{6}{10}$$

$$\frac{3}{5} = \frac{3 \times 3}{5 \times 3} = \frac{9}{15}$$

and

$$\frac{3}{5} = \frac{3 \times 4}{5 \times 4} = \frac{12}{20}$$

Teaching Tip

In Example 2 we are "building up" a fraction, a process we need when adding fractions.

EXAMPLE 2 **Finding equivalent fractions**

Find a fraction equivalent to $\frac{3}{5}$ with a denominator of 20.

SOLUTION We must solve the problem

$$\frac{3}{5} = \frac{?}{20}$$

Note that the denominator, 5, was multiplied by 4 to get 20. So, we must also multiply the numerator, 3, by 4.

$$\frac{3}{5} = \frac{?}{20}$$ If you multiply the denominator by 4,

Multiply by 4.

Multiply by 4.

$$\frac{3}{5} = \frac{12}{20}$$ you have to multiply the numerator by 4.

Thus the equivalent fraction is $\frac{12}{20}$.

PROBLEM 2

Find a fraction equivalent to $\frac{4}{7}$ with a denominator of 21.

Teaching Tip

In Example 3 we are "reducing" a fraction.

Here is a slightly different problem. Can we find a fraction equivalent to $\frac{15}{20}$ with a denominator of 4? We do this in the next example.

EXAMPLE 3 **Finding equivalent fractions**

Find a fraction equivalent to $\frac{15}{20}$ with a denominator of 4.

SOLUTION We proceed as before.

$$\frac{15}{20} = \frac{?}{4}$$ 20 was divided by 5 to get 4.

Divide by 5.

Divide by 5.

$$\frac{15}{20} = \frac{3}{4}$$ 15 was divided by 5 to get 3.

PROBLEM 3

Find a fraction equivalent to $\frac{24}{30}$ with a denominator of 5.

Answers

2. $\frac{12}{21}$ **3.** $\frac{4}{5}$

We can summarize our work with equivalent fractions in the following procedure.

PROCEDURE

Obtaining Equivalent Fractions

To obtain an **equivalent fraction,** *multiply* or *divide* both numerator and denominator of the fraction by the *same* nonzero number.

C Reducing Fractions to Lowest Terms

The preceding rule can be used to *reduce* (simplify) fractions to lowest terms. A fraction is *reduced* to lowest terms (simplified) when there is no number (except 1) that will divide the numerator and the denominator exactly. The procedure is as follows.

PROCEDURE

Reducing Fractions to Lowest Terms

To **reduce** a fraction to *lowest* terms, divide the numerator and denominator by the *largest* natural number that will divide them exactly.

"Divide them exactly" means that both remainders are zero.

To reduce $\frac{12}{30}$ to lowest terms, we divide the numerator and denominator by 6, the *largest* natural number that divides 12 and 30 exactly. [6 is sometimes called the *greatest common divisor* (GCD) of 12 and 30.] Thus,

$$\frac{12}{30} = \frac{12 \div 6}{30 \div 6} = \frac{2}{5}$$

This reduction is sometimes shown like this:

$$\frac{\overset{2}{\cancel{12}}}{\underset{5}{\cancel{30}}} = \frac{2}{5}$$

EXAMPLE 4 Reducing fractions

Reduce to lowest terms:

a. $\frac{15}{20}$ **b.** $\frac{30}{45}$ **c.** $\frac{60}{48}$

SOLUTION

a. The *largest* natural number exactly dividing 15 and 20 is 5. Thus,

$$\frac{15}{20} = \frac{15 \div 5}{20 \div 5} = \frac{3}{4}$$

b. The *largest* natural number exactly dividing 30 and 45 is 15. Hence,

$$\frac{30}{45} = \frac{30 \div 15}{45 \div 15} = \frac{2}{3}$$

c. The *largest* natural number dividing 60 and 48 is 12. Therefore,

$$\frac{60}{48} = \frac{60 \div 12}{48 \div 12} = \frac{5}{4}$$

PROBLEM 4

Reduce to lowest terms:

a. $\frac{30}{50}$ **b.** $\frac{45}{60}$ **c.** $\frac{84}{72}$

Answers

4. a. $\frac{3}{5}$ **b.** $\frac{3}{4}$ **c.** $\frac{7}{6}$

Teaching Tip

Point out to students that if they write $\frac{30}{45} = \frac{6}{9}$, the fraction is *not* reduced to lowest terms.

What if you are unable to see at once that the *largest* natural number dividing the numerator and denominator in, say, $\frac{30}{45}$, is 15? No problem; it just takes a little longer. Suppose you notice that 30 and 45 are both divisible by 5. You then write

$$\frac{30}{45} = \frac{30 \div 5}{45 \div 5} = \frac{6}{9}$$

Now you can see that 6 and 9 are both divisible by 3. Thus,

$$\frac{6}{9} = \frac{6 \div 3}{9 \div 3} = \frac{2}{3}$$

which is the same answer we got in Example 4(b). The whole procedure can be written as

$$\frac{\overset{\overset{2}{\cancel{6}}}{\cancel{30}}}{\underset{\underset{3}{\cancel{9}}}{\cancel{45}}} = \frac{2}{3}$$

We can also reduce $\frac{15}{20}$ using *prime factorization* by writing

$$\frac{15}{20} = \frac{3 \cdot 5}{2 \cdot 2 \cdot 5} = \frac{3}{4}$$

(The dot, $\cdot$, indicates multiplication.) Note that 15 is written as the *product* of 3 and 5, and 20 is written as the product $2 \cdot 2 \cdot 5$. A **product** is the answer to a multiplication problem. When 15 is written as the product $3 \cdot 5$, 3 and 5 are the **factors** of 15. As a matter of fact, 3 and 5 are **prime numbers,** so $3 \cdot 5$ is the prime factorization of 15; similarly, $2 \cdot 2 \cdot 5$ is the prime factorization of 20, because 2 and 5 are primes. In general,

Web It

Do you want to find the prime numbers up to 500? You do not have to do it manually. Go to the Bello Website at mhhe.com/bello and access link R-1-1.

Simply indicate that you want all the primes up to 500 by entering 500 then pressing "Calculate Primes."

For a lot more information about primes, go to link R-1-2.

> A natural number is prime if it has *exactly two different factors,* itself and 1.

The first few primes are

$$2, 3, 5, 7, 11, 13, 17, 19, 23, 29, 31, \quad \text{and so on.}$$

A natural number that is *not* prime is called a **composite number.** Thus, the numbers missing in the previous list, 4, 6, 8, 9, 10, 12, 14, 15, 16, and so on, are **composite.** Note that 1 is considered neither prime nor composite.

When using prime factorization, we can keep better track of the factors by using a **factor tree.** For example, to reduce $\frac{30}{45}$ using prime factorization, we make two trees to factor 30 and 45 as shown:

Divide 30 by the smallest prime, 2. $\quad 30 = 2 \cdot 15$

Now, divide 15 by 3 to find the factors of 15. $\quad 30 = 2 \cdot 3 \cdot 5$

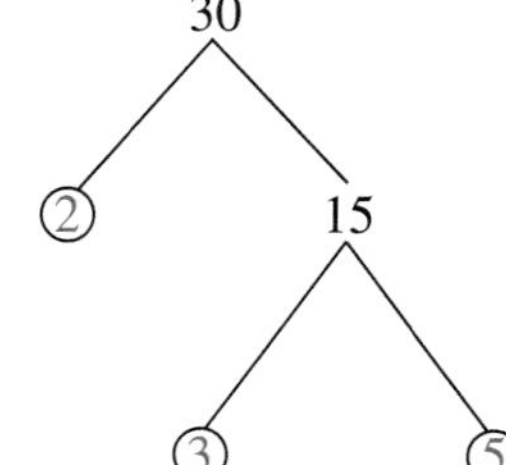

Web It

There are different ways to factor a number into a product of primes. You can see some of these methods by going to the Bello Website at mhhe.com/bello and accessing link R-1-3.

Similarly,

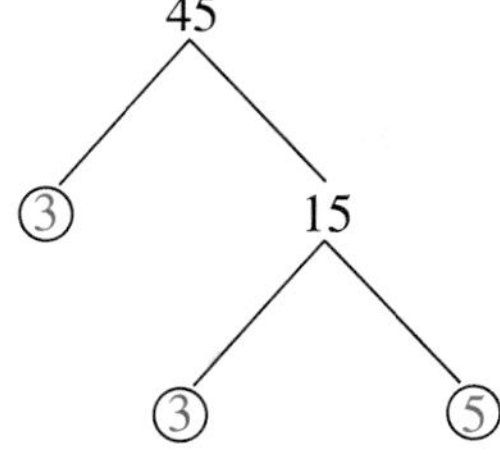

Divide 45 by 3. $45 = 3 \cdot 15$

Now, divide 15 by 3 to find the factors of 15. $45 = 3 \cdot 3 \cdot 5$

Thus,

$$\frac{30}{45} = \frac{2 \cdot 3 \cdot 5}{3 \cdot 3 \cdot 5} = \frac{2}{3} \quad \text{as before.}$$

Caution: When reducing a fraction to lowest terms, it is easier to look for the *largest common factor* in the numerator and denominator, but if no largest factor is obvious, *any common factor* can be used and the simplification can be done in stages. For example,

$$\frac{200}{250} = \frac{20 \cdot 10}{25 \cdot 10} = \frac{4 \cdot 5 \cdot 10}{5 \cdot 5 \cdot 10} = \frac{4}{5}$$

Which way should you simplify fractions? The way you understand! Make sure you follow the instructor's preferences regarding the procedure used.

Exercises R.1

A In Problems 1–8, write the given number as a fraction with a denominator of 1.

1. 28 $\frac{28}{1}$ **2.** 93 $\frac{93}{1}$ **3.** -42 $\frac{-42}{1}$

4. -86 $\frac{-86}{1}$ **5.** 0 $\frac{0}{1}$ **6.** 1 $\frac{1}{1}$

7. -1 $\frac{-1}{1}$ **8.** -17 $\frac{-17}{1}$

B In Problems 9–30, find the missing number.

9. $\frac{1}{8} = \frac{?}{24}$ 3 **10.** $\frac{7}{1} = \frac{?}{2}$ 14 **11.** $\frac{7}{1} = \frac{?}{6}$ 42 **12.** $\frac{5}{6} = \frac{?}{48}$ 40

13. $\frac{5}{3} = \frac{?}{15}$ 25 **14.** $\frac{9}{8} = \frac{?}{32}$ 36 **15.** $\frac{7}{11} = \frac{?}{33}$ 21 **16.** $\frac{11}{7} = \frac{?}{35}$ 55

17. $\frac{1}{8} = \frac{4}{?}$ 32 **18.** $\frac{3}{5} = \frac{27}{?}$ 45 **19.** $\frac{5}{6} = \frac{5}{?}$ 6 **20.** $\frac{9}{10} = \frac{9}{?}$ 10

21. $\frac{8}{7} = \frac{16}{?}$ 14 **22.** $\frac{9}{5} = \frac{36}{?}$ 20 **23.** $\frac{6}{5} = \frac{36}{?}$ 30 **24.** $\frac{5}{3} = \frac{45}{?}$ 27

25. $\frac{21}{56} = \frac{?}{8}$ 3 **26.** $\frac{12}{18} = \frac{?}{3}$ 2 **27.** $\frac{36}{180} = \frac{?}{5}$ 1 **28.** $\frac{8}{24} = \frac{4}{?}$ 12

29. $\frac{18}{12} = \frac{3}{?}$ 2 **30.** $\frac{56}{49} = \frac{8}{?}$ 7

C In Problems 31–40, reduce the fraction to lowest terms by writing the numerator and denominator as products of primes.

31. $\frac{15}{12}$ $\frac{5}{4}$ **32.** $\frac{30}{28}$ $\frac{15}{14}$ **33.** $\frac{13}{52}$ $\frac{1}{4}$ **34.** $\frac{27}{54}$ $\frac{1}{2}$

35. $\frac{56}{24}$ $\frac{7}{3}$ **36.** $\frac{56}{21}$ $\frac{8}{3}$ **37.** $\frac{22}{33}$ $\frac{2}{3}$ **38.** $\frac{26}{39}$ $\frac{2}{3}$

39. $\frac{100}{25}$ $\frac{4}{1} = 4$ **40.** $\frac{21}{3}$ $\frac{7}{1} = 7$

APPLICATIONS

41. *AOL ad* If you take advantage of the AOL offer and get 1000 hours free for 45 days:

a. How many hours will you get per day? (Assume you use the same number of hours each day and do not simplify your answer.) $\frac{1000}{45}$

b. Write the answer to part **a** in lowest terms. $\frac{200}{9}$

c. Write the answer to part **a** as a mixed number. $22\frac{2}{9}$

d. If you used AOL 24 hours a day, for how many days would the 1000 free hours last? $\frac{1000}{24} = \frac{125}{3} = 41\frac{2}{3}$

42. *Census data* ***Are you "poor"?*** In 1959, the U.S. Census reported that about 40 million of the 180 million people living in the United States were "poor." In the year 2000, about 30 million out of 275 million were "poor" (income less than $8794).

a. What reduced fraction of the people was poor in the year 1959? $\frac{2}{9}$

b. What reduced fraction of the people was poor in the year 2000? $\frac{6}{55}$

43. *High school completion rate* In a recent year, about 41 out of 100 persons in the United States had completed 4 years of high school or more. What fraction of the people is that? $\frac{41}{100}$

44. *High school completion rates* In a recent year, about 84 out of 100 Caucasians, 76 out of 100 African-Americans, and 56 out of 100 Hispanics had completed 4 years of high school or more. What reduced fractions of the Caucasians, African-Americans, and Hispanics had completed 4 years of high school or more? $\frac{21}{25}, \frac{19}{25}, \frac{14}{25}$

45. *Pizza consumption* The pizza shown here consists of six pieces. Alejandro ate $\frac{1}{3}$ of the pizza.

a. Write $\frac{1}{3}$ with a denominator of 6. $\frac{2}{6}$

b. How many pieces did Alejandro eat? 2 pieces

c. If Cindy ate two pieces of pizza, what fraction of the pizza did she eat? $\frac{2}{6} = \frac{1}{3}$

d. Who ate more pizza, Alejandro or Cindy? They ate the same amount.

Problems 46–50 refer to the photo of the fuel gauge. What fraction of the tank is full if the needle:

46. Points to the line midway between E and $\frac{1}{2}$ full? $\frac{1}{4}$

47. Points to the first line to the right of empty (E)? $\frac{1}{8}$

48. Is midway between empty and $\frac{1}{8}$ of a tank? $\frac{1}{16}$

49. Is one line past the $\frac{1}{2}$ mark? $\frac{5}{8}$

50. Is one line before the $\frac{1}{2}$ mark? $\frac{3}{8}$

USING YOUR KNOWLEDGE

During the 1953–1954 basketball season, the NBA (National Basketball Association) had a problem with the game: it was boring. Danny Biasone, the owner of the Syracuse Nationals, thought that limiting the time a team could have the ball should encourage more shots. Danny figured out that in a fast-paced game, each team should take 60 shots during the 48 minutes the game lasted (4 quarters of 12 minutes each). He then looked at the fraction $\frac{\text{Seconds}}{\text{Shots}}$.

51. a. How many seconds does the game last? 2880 seconds

b. How many total shots are to be taken in the game? 120 shots

c. What is the reduced fraction $\frac{\text{Seconds}}{\text{Shots}}$? 24

Now you know where the $\frac{\text{Seconds}}{\text{Shots}}$ clock came from!

You have learned how to work with fractions. Now you can use your knowledge to interpret fractions from diagrams. As you see from the diagram (circle graph), "CNN Headline News" devotes the first quarter of every hour $\left(\frac{15}{60} = \frac{1}{4}\right)$ to National and International News. What total fraction of the hour is devoted to:

52. Dollars & Sense? $\frac{1}{6}$

53. Sports? $\frac{2}{15}$

54. Local News or People & Places? $\frac{1}{5}$

55. Which feature uses the most time, and what fraction of an hour does it use? Nat'l & Int'l News, $\frac{1}{2}$

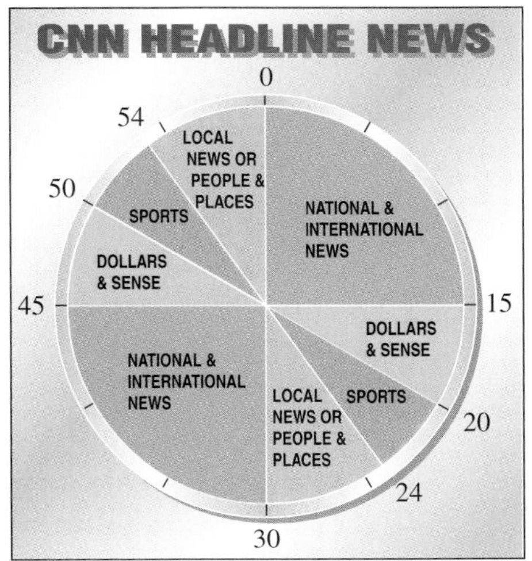

Source: Data from CNN.

As you can see from the illustration, "Bay News 9" devotes $8 + 6 = 14$ minutes of the hour (60 minutes) to News. This represents $\frac{14}{60} = \frac{7}{30}$ of the hour. What fraction of the hour is devoted to:

56. Traffic? $\frac{1}{10}$

57. Weather? $\frac{7}{30}$

58. News & Beyond the Bay? $\frac{13}{30}$

59. Which features use the most and least time, and what fraction of the hour does each one use? Most: News & Beyond the Bay, $\frac{13}{30}$; Least: Traffic, $\frac{1}{10}$

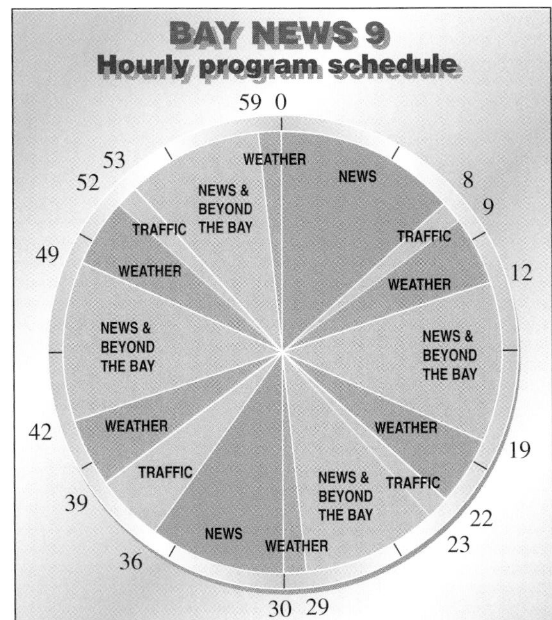

Source: Data from Bay News 9.

WRITE ON

60. Write the procedure you use to reduce a fraction to lowest terms. Answers may vary.

61. What are the advantages and disadvantages of writing the numerator and denominator of a fraction as a product of primes before reducing the fraction? Answers may vary.

62. Write a procedure to determine whether two fractions are equivalent. Reduce both to lowest terms, then compare.

R.2 OPERATIONS WITH FRACTIONS

To Succeed, Review How To . . .

1. Reduce fractions (pp. 7–9).
2. Write a number as a product of primes (pp. 8–9).

Objectives

A Add, subtract, multiply, and divide fractions.

GETTING STARTED

A Sweet Problem

Each of the cups contains $\frac{1}{4}$ cup of sugar. How much sugar do the cups contain altogether?

To find the answer, we can multiply 3 by $\frac{1}{4}$; that is, we can find $3 \cdot \frac{1}{4}$.

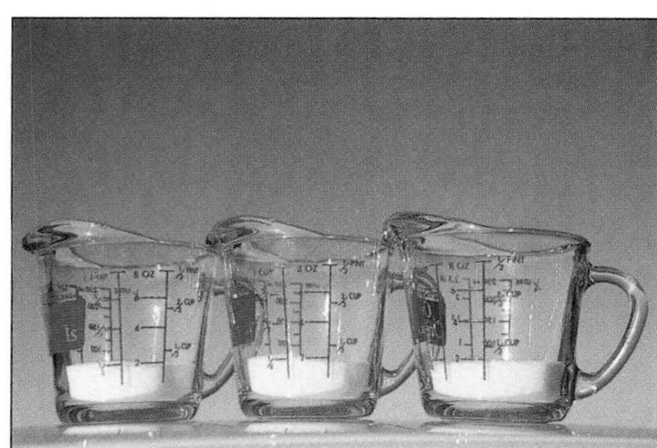

A Adding, Subtracting, Multiplying, and Dividing Fractions

How do we perform the multiplication $3 \cdot \frac{1}{4}$ required to determine the total amount of sugar in the three cups? Note that

$$3 \cdot \frac{1}{4} = \frac{3}{1} \cdot \frac{1}{4} = \frac{3 \cdot 1}{1 \cdot 4} = \frac{3}{4}$$

Similarly,

$$\frac{4}{9} \cdot \frac{2}{5} = \frac{4 \cdot 2}{9 \cdot 5} = \frac{8}{45}$$

Here is the general rule for multiplying fractions.

Calculate It

Multiplying Fractions

To multiply 3 by $\frac{1}{4}$ using a calculator with an x/y key, enter

3 × 1 x/y 4 ENTER 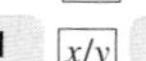 

and you get $\frac{3}{4}$. If your calculator has a / key, the strokes will be

3 × 1 / 4 ENTER    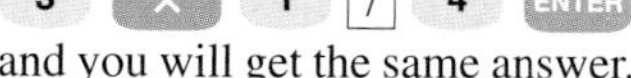

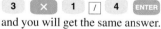

and you will get the same answer.

Teaching Tip

The rule applies to fractions, *not* mixed numbers.

RULE

Multiplying Fractions

$$\frac{a}{b} \cdot \frac{c}{d} = \frac{a \cdot c}{b \cdot d}$$

EXAMPLE 1 Multiplying fractions

Multiply: $\frac{9}{5} \cdot \frac{3}{4}$

SOLUTION We use our rule for multiplying fractions:

$$\frac{9}{5} \cdot \frac{3}{4} = \frac{9 \cdot 3}{5 \cdot 4} = \frac{27}{20}$$

PROBLEM 1

Multiply: $\frac{9}{7} \cdot \frac{3}{5}$

Answer

1. $\frac{27}{35}$

Teaching Tip

You can only divide out factors common to both the numerator *and* denominator.

Yes	No
$\frac{1}{3}\cdot\frac{\overset{2}{\cancel{4}}}{\underset{3}{\cancel{6}}}$	$\frac{\overset{1}{\cancel{2}}}{7}\cdot\frac{\overset{2}{\cancel{4}}}{5}$

When multiplying fractions, we can save time if we divide out common factors *before* we multiply. For example,

$$\frac{2}{5}\cdot\frac{5}{7}=\frac{2\cdot 5}{5\cdot 7}=\frac{10}{35}=\frac{2}{7}$$

We can save time by writing

$$\frac{2}{\underset{1}{\cancel{5}}}\cdot\frac{\overset{1}{\cancel{5}}}{7}=\frac{2\cdot 1}{1\cdot 7}=\frac{2}{7}$$

This can be done because $\frac{5}{5}=1$.

CAUTION

Only factors that are common to *both* numerator *and* denominator can be divided out.

EXAMPLE 2 Multiplying fractions with common factors

Multiply:

a. $\frac{3}{7}\cdot\frac{7}{8}$ **b.** $\frac{5}{8}\cdot\frac{4}{15}$

SOLUTION

a. $\frac{3}{\underset{1}{\cancel{7}}}\cdot\frac{\overset{1}{\cancel{7}}}{8}=\frac{3\cdot 1}{1\cdot 8}=\frac{3}{8}$

b. $\frac{\overset{1}{\cancel{5}}}{\underset{2}{\cancel{8}}}\cdot\frac{\overset{1}{\cancel{4}}}{\underset{3}{\cancel{15}}}=\frac{1\cdot 1}{2\cdot 3}=\frac{1}{6}$

PROBLEM 2

Multiply:

a. $\frac{4}{9}\cdot\frac{9}{11}$ **b.** $\frac{5}{14}\cdot\frac{7}{20}$

If we wish to multiply a fraction by a **mixed number,** such as $3\frac{1}{4}$, we must convert the mixed number to a fraction first. The number $3\frac{1}{4}$ (read "3 and $\frac{1}{4}$") means $3+\frac{1}{4}=\frac{12}{4}+\frac{1}{4}=\frac{13}{4}$. (This addition will be clearer to you after studying the addition of fractions.) For now, we can shorten the procedure by using the following diagram.

$$3\frac{1}{4}=\frac{13}{4}$$

+ ×

Same denominator

Work clockwise. *First* multiply the denominator 4 by the whole number part 3; add the numerator. This is the new numerator. Use the same denominator.

EXAMPLE 3 Multiplying fractions and mixed numbers

Multiply: $5\frac{1}{3}\cdot\frac{9}{16}$

SOLUTION We first convert the mixed number to a fraction:

$$5\frac{1}{3}=\frac{3\cdot 5+1}{3}=\frac{16}{3}$$

Thus,

$$5\frac{1}{3}\cdot\frac{9}{16}=\frac{\overset{1}{\cancel{16}}}{\underset{1}{\cancel{3}}}\cdot\frac{\overset{3}{\cancel{9}}}{\underset{1}{\cancel{16}}}=\frac{1\cdot 3}{1\cdot 1}=3$$

PROBLEM 3

Multiply: $2\frac{3}{4}\cdot\frac{8}{11}$

Answers

2. a. $\frac{4}{11}$ **b.** $\frac{1}{8}$ **3.** 2

If we wish to divide one number by another, we can indicate the division by a fraction. Thus, to divide 2 by 5 we can write

Multiply.

$$2 \div 5 = \frac{2}{5} = 2 \cdot \frac{1}{5}$$

The divisor 5 is inverted.

Note that to divide 2 by 5 we multiplied 2 by $\frac{1}{5}$, where the fraction $\frac{1}{5}$ was obtained by *inverting* $\frac{5}{1}$ to obtain $\frac{1}{5}$. (In mathematics, $\frac{5}{1}$ and $\frac{1}{5}$ are called **reciprocals.**) *Note:* Only the 5 (the divisor) is replaced by its reciprocal, $\frac{1}{5}$.

Now let's try the problem $5 \div \frac{5}{7}$. If we do it like the preceding example, we write

Multiply.

$$5 \div \frac{5}{7} = 5 \cdot \frac{7}{5} = \frac{\overset{1}{\cancel{5}}}{1} \cdot \frac{7}{\underset{1}{\cancel{5}}} = 7$$

Invert.

In general, to divide $\frac{a}{b}$ by $\frac{c}{d}$, we multiply $\frac{a}{b}$ by the *reciprocal* of $\frac{c}{d}$—that is, $\frac{d}{c}$. Here is the rule.

Teaching Tip

Point out that $\frac{a}{b}$ remained the same.

RULE

Dividing Fractions

$$\frac{a}{b} \div \frac{c}{d} = \frac{a}{b} \cdot \frac{d}{c}$$

EXAMPLE 4 Dividing fractions

Divide:

a. $\frac{3}{5} \div \frac{2}{7}$ **b.** $\frac{4}{9} \div 5$

SOLUTION

a. $\frac{3}{5} \div \frac{2}{7} = \frac{3}{5} \cdot \frac{7}{2} = \frac{21}{10}$ **b.** $\frac{4}{9} \div 5 = \frac{4}{9} \cdot \frac{1}{5} = \frac{4}{45}$

PROBLEM 4

Divide:

a. $\frac{3}{4} \div \frac{5}{7}$ **b.** $\frac{3}{7} \div 5$

As in the case of multiplication, if the problem involves mixed numbers, we change them to fractions first, like this:

Change.

$$2\frac{1}{4} \div \frac{3}{5} = \frac{9}{4} \div \frac{3}{5} = \frac{\overset{3}{\cancel{9}}}{4} \cdot \frac{5}{\underset{1}{\cancel{3}}} = \frac{15}{4}$$

Invert.

Answers

4. a. $\frac{21}{20}$ **b.** $\frac{3}{35}$

EXAMPLE 5 Dividing fractions and mixed numbers

Divide:

a. $3\frac{1}{4} \div \frac{7}{8}$ **b.** $\frac{11}{12} \div 7\frac{1}{3}$

SOLUTION

a. $3\frac{1}{4} \div \frac{7}{8} = \frac{13}{4} \div \frac{7}{8} = \frac{13}{\cancel{4}_1} \cdot \frac{\cancel{8}^2}{7} = \frac{26}{7}$

b. $\frac{11}{12} \div 7\frac{1}{3} = \frac{11}{12} \div \frac{22}{3} = \frac{\cancel{11}^1}{\cancel{12}_4} \cdot \frac{\cancel{3}^1}{\cancel{22}_2} = \frac{1}{8}$

PROBLEM 5

Divide:

a. $2\frac{1}{5} \div \frac{7}{10}$ **b.** $\frac{11}{15} \div 7\frac{1}{3}$

Now we are ready to *add* fractions.

The photo shows that 1 quarter plus 2 quarters equals 3 quarters. In symbols,

$$\frac{1}{4} + \frac{2}{4} = \frac{1+2}{4} = \frac{3}{4}$$

In general, to add fractions with the *same* denominator, we add the numerators and keep the denominator.

RULE

Adding Fractions with the Same Denominator

$$\frac{a}{b} + \frac{c}{b} = \frac{a+c}{b}$$

Thus,

$$\frac{1}{5} + \frac{2}{5} = \frac{1+2}{5} = \frac{3}{5}$$

and

$$\frac{3}{8} + \frac{1}{8} = \frac{3+1}{8} = \frac{4}{8} = \frac{1}{2}$$

NOTE

In the last addition we reduced $\frac{4}{8}$ to $\frac{1}{2}$ by dividing the numerator and denominator by 4. When a result involves fractions, we always reduce to lowest terms.

Now suppose you wish to add $\frac{5}{12}$ and $\frac{1}{18}$. Since these two fractions do not have the same denominators, the rule does not work. To add them, you must learn how to write $\frac{5}{12}$ and $\frac{1}{18}$ as equivalent fractions with the same denominator. To keep things simple, you should also try to make this denominator as small as possible; that is, you should first find the **least common denominator (LCD)** of the fractions.

Answers

5. a. $\frac{22}{7}$ **b.** $\frac{1}{10}$

The LCD of two fractions is the **least** (*smallest*) **common multiple (LCM)** of their denominators. Thus, to find the LCD of $\frac{5}{12}$ and $\frac{1}{18}$, we find the LCM of 12 and 18. There are several ways of doing this. One way is to select the larger number (18) and find its multiples. The first multiple of 18 is $2 \cdot 18 = 36$, and 12 divides into 36. Thus, 36 is the LCD of 12 and 18.

Unfortunately, this method is not practical in algebra. A more convenient method consists of writing the denominators 12 and 18 in completely *factored* form. We can start by writing 12 as $2 \cdot 6$. In turn, $6 = 2 \cdot 3$; thus

$$12 = 2 \cdot 2 \cdot 3$$

Similarly,

$$18 = 2 \cdot 9, \quad \text{or} \quad 2 \cdot 3 \cdot 3$$

Now we can see that the smallest number that is a multiple of 12 and 18 must have at least two 2's (there are two 2's in 12) and two 3's (there are two 3's in 18). Thus, the LCD of $\frac{5}{12}$ and $\frac{1}{18}$ is

$$\underbrace{2 \cdot 2}_{\text{Two 2's}} \cdot \underbrace{3 \cdot 3}_{\text{Two 3's}} = 36$$

Fortunately, there is an even shorter way of finding the LCM of 12 and 18. Note that what we want to do is find the *common* factors in 12 and 18, so we can divide 12 and 18 by the smallest divisor common to both numbers (2) and then the next (3), and so on. If we multiply these divisors by the final quotient, the result is the LCM. Here is the shortened version.

Divide by 2. 2) 12 18
Divide by 3. 3) 6 9
2 — 3 ⟶ Multiply $2 \cdot 3 \cdot 2 \cdot 3$.

The LCM is $2 \cdot 3 \cdot 2 \cdot 3 = 36$.

In general, we use the following procedure to find the LCD of two fractions.

PROCEDURE

Finding the LCD of Two Fractions

1. Write the denominators in a horizontal row and divide each number by a *divisor* common to both numbers.
2. Continue the process until the resulting quotients have no common divisor (except 1).
3. The product of the *divisors* and the *final quotients* is the LCD.

Teaching Tip

Show students how to write equivalent fractions in algebra by leaving the denominator in factored form and multiplying by the "missing factor."

Example:

$$\frac{5 \cdot \textcircled{3}}{2 \cdot 2 \cdot 2 \cdot \textcircled{3}} = \frac{5 \cdot ?}{2 \cdot 2 \cdot 2 \cdot \textcircled{3}}$$

$$\frac{1}{2 \cdot 3 \cdot 3 \cdot \textcircled{2}} = \frac{1 \cdot ?}{2 \cdot 3 \cdot \textcircled{2} \cdot 3}$$

Now that we know that the LCD of $\frac{5}{12}$ and $\frac{1}{18}$ is 36, we can add the two fractions by writing each as an equivalent fraction with a denominator of 36. We do this by multiplying numerators and denominators of $\frac{5}{12}$ by 3 and of $\frac{1}{18}$ by 2. Thus, we get

$$\frac{5}{12} = \frac{5 \cdot 3}{12 \cdot 3} = \frac{15}{36} \longrightarrow \frac{15}{36}$$

$$\frac{1}{18} = \frac{1 \cdot 2}{18 \cdot 2} = \frac{2}{36} \longrightarrow + \frac{2}{36}$$

$$\frac{17}{36}$$

We can also write the results as

$$\frac{5}{12} + \frac{1}{18} = \frac{15}{36} + \frac{2}{36} = \frac{15 + 2}{36} = \frac{17}{36}$$

EXAMPLE 6 Adding fractions with different denominators

Add: $\frac{1}{20} + 1\frac{1}{18}$

SOLUTION We first find the LCM of 20 and 18.

Divide by 2. $2\,)\,\underline{20 \quad 18}$
$\quad 10 — 9 \longrightarrow$ Multiply $2 \cdot 10 \cdot 9$.

Since 10 and 9 have no common divisor, the LCM is

$$2 \cdot 10 \cdot 9 = 180$$

(You could also find the LCM by writing $20 = 2 \cdot 2 \cdot 5$ and $18 = 2 \cdot 3 \cdot 3$; the LCM is $2 \cdot 2 \cdot 3 \cdot 3 \cdot 5 = 180$.)

Now, write the fractions with denominators of 180 and add.

$$\frac{1}{20} = \frac{1 \cdot 9}{20 \cdot 9} = \frac{9}{180} \longrightarrow \frac{9}{180}$$

$$1\frac{1}{18} = \frac{19}{18} = \frac{19 \cdot 10}{18 \cdot 10} = \frac{190}{180} \longrightarrow + \frac{190}{180}$$

$$\frac{199}{180}$$

If you prefer, you can write the procedure like this:

$$\frac{1}{20} + 1\frac{1}{18} = \frac{9}{180} + \frac{190}{180} = \frac{199}{180}, \quad \text{or} \quad 1\frac{19}{180}$$

PROBLEM 6

Add: $\frac{3}{20} + 1\frac{1}{16}$

Can we use the same procedure to add three or more fractions? Almost; but we need to know how to find the LCD for three or more fractions. The procedure is very similar to that used for finding the LCM of two numbers. If we write 15, 21, and 28 in factored form,

$$15 = 3 \cdot 5$$

$$21 = 3 \cdot 7$$

$$28 = 2 \cdot 2 \cdot 7$$

we see that the LCM must contain at least $2 \cdot 2$, 3, 5, and 7. (Note that we select each factor the *greatest* number of times it appears in any factorization.) Thus, the LCM of 15, 21, and 28 is $2 \cdot 2 \cdot 3 \cdot 5 \cdot 7 = 420$.

We can also use the following shortened procedure.

Answer

6. $1\frac{17}{80}$

PROCEDURE

Finding the LCD of Three or More Fractions

1. Write the denominators in a horizontal row and divide the numbers by a divisor common to two or more of the numbers. If any of the other numbers is not divisible by this divisor, *circle* the number and carry it to the next line.

   Divide by 3. $3\,)\,\underline{15 \quad 21 \quad (28)}$

2. Repeat step 1 with the quotients and carrydowns until *no two numbers* have a common divisor (except 1).

   Divide by 7. $7\,)\,\underline{(5) \quad 7 \quad 28}$
   $\quad 5 \quad 1 \quad 4$

   Multiply $3 \cdot 7 \cdot 5 \cdot 1 \cdot 4$.

3. The LCD is the *product* of the divisors from the preceding steps and the numbers in the final row.

   The LCD is $3 \cdot 7 \cdot 5 \cdot 1 \cdot 4 = 420$.

Thus to add

$$\frac{1}{16} + \frac{3}{10} + \frac{1}{28}$$

we first find the LCD of 16, 10, and 28. We can use either method.

Method 1. Find the LCD by writing

$$16 = 2 \cdot 2 \cdot 2 \cdot 2$$
$$10 = 2 \cdot 5$$
$$28 = 2 \cdot 2 \cdot 7$$

and selecting each factor the *greatest* number of times it appears in any factorization. Since 2 appears four times and 5 and 7 appear once, the LCD is

$$\underbrace{2 \cdot 2 \cdot 2 \cdot 2}_{\text{four times}} \cdot \underbrace{5 \cdot 7}_{\text{once}} = 560$$

Method 2. Use the three-step procedure.

Step 1. Divide by 2. $2\overline{)\,16 \quad 10 \quad 28}$

Step 2. Divide by 2. $2\overline{)\,8 \quad \textcircled{5} \quad 14}$
$\quad 4 \quad 5 \quad 7$

Multiply $2 \cdot 2 \cdot 4 \cdot 5 \cdot 7$.

Step 3. The LCD is

$$2 \cdot 2 \cdot 4 \cdot 5 \cdot 7 = 560.$$

We then write $\frac{1}{16}$, $\frac{3}{10}$, and $\frac{1}{28}$ with denominators of 560.

$$\frac{1}{16} = \frac{1 \cdot 35}{16 \cdot 35} = \frac{35}{560} \longrightarrow \quad \frac{35}{560}$$
$$\frac{3}{10} = \frac{3 \cdot 56}{10 \cdot 56} = \frac{168}{560} \longrightarrow \quad + \frac{168}{560}$$
$$\frac{1}{28} = \frac{1 \cdot 20}{28 \cdot 20} = \frac{20}{560} \longrightarrow \quad + \frac{20}{560}$$
$$\frac{223}{560}$$

Or, if you prefer,

$$\frac{1}{16} + \frac{3}{10} + \frac{1}{28} = \frac{35}{560} + \frac{168}{560} + \frac{20}{560} = \frac{223}{560}$$

Teaching Tip

Stress Method 1!

NOTE

Method 1 for finding the LCD is preferred because it is more easily generalized to algebraic fractions.

Now that you know how to add fractions, *subtraction* is no problem. All the rules we have mentioned still apply! For example, we subtract fractions with the same denominator as shown.

$$\frac{5}{8} - \frac{2}{8} = \frac{5-2}{8} = \frac{3}{8}$$

$$\frac{7}{9} - \frac{1}{9} = \frac{7-1}{9} = \frac{6}{9} = \frac{2}{3}$$

The next example shows how to subtract fractions involving *different* denominators.

EXAMPLE 7 **Subtracting fractions with different denominators**

Subtract: $\frac{7}{12} - \frac{1}{18}$

SOLUTION We first find the LCD of the fractions.

Method 1. Write

$$12 = 2 \cdot 2 \cdot 3$$

$$18 = 2 \cdot 3 \cdot 3$$

The LCD is $2 \cdot 2 \cdot 3 \cdot 3 = 36$.

Method 2.

Step 1. $2\,\underline{)\,12 \quad 18}$

Step 2. $3\,\underline{)\;6 \quad 9}$
$\quad\;\; 2 \quad 3$

Step 3. The LCD is $2 \cdot 3 \cdot 2 \cdot 3 = 36$.

We then write each fraction with 36 as the denominator.

$$\frac{7}{12} = \frac{7 \cdot 3}{12 \cdot 3} = \frac{21}{36} \quad \text{and} \quad \frac{1}{18} = \frac{1 \cdot 2}{18 \cdot 2} = \frac{2}{36}$$

Thus,

$$\frac{7}{12} - \frac{1}{18} = \frac{21}{36} - \frac{2}{36} = \frac{21 - 2}{36} = \frac{19}{36}$$

PROBLEM 7

Subtract: $\frac{7}{18} - \frac{1}{12}$

The rules for adding fractions also apply to subtraction. If mixed numbers are involved, we can use horizontal or vertical subtraction, as illustrated next.

EXAMPLE 8 **Subtracting mixed numbers**

Subtract: $3\frac{1}{6} - 2\frac{5}{8}$

SOLUTION *Horizontal subtraction.* We first find the LCM of 6 and 8, which is 24. Then we convert the mixed numbers to improper fractions.

$$3\frac{1}{6} = \frac{19}{6} \quad \text{and} \quad 2\frac{5}{8} = \frac{21}{8}$$

Thus,

$$\begin{aligned} 3\frac{1}{6} - 2\frac{5}{8} &= \frac{19}{6} - \frac{21}{8} \\ &= \frac{19 \cdot 4}{6 \cdot 4} - \frac{21 \cdot 3}{8 \cdot 3} \\ &= \frac{76}{24} - \frac{63}{24} \\ &= \frac{13}{24} \end{aligned}$$

Vertical subtraction. Some students prefer to subtract mixed numbers by setting the problem in a column, as shown:

$$\begin{array}{r} 3\frac{1}{6} \\ -2\frac{5}{8} \\ \hline \end{array}$$

PROBLEM 8

Subtract: $4\frac{1}{8} - 2\frac{5}{6}$

Answers

7. $\frac{11}{36}$ **8.** $1\frac{7}{24}$

The fractional part is subtracted first, followed by the whole number part. Unfortunately, $\frac{5}{8}$ cannot be subtracted from $\frac{1}{6}$ at this time, so we rename $1 = \frac{6}{6}$ and rewrite the problem as

$$\begin{array}{r} 2\frac{7}{6} \\ -2\frac{5}{8} \\ \hline \end{array} \qquad 3\frac{1}{6} = 2 + \frac{6}{6} + \frac{1}{6} = 2\frac{7}{6}$$

Since the LCM is 24, we rewrite $\frac{7}{6}$ and $\frac{5}{8}$ with 24 as the denominator.

$$\begin{array}{r} 2\frac{7 \cdot 4}{6 \cdot 4} = 2\frac{28}{24} \\ -2\frac{5 \cdot 3}{8 \cdot 3} = 2\frac{15}{24} \\ \hline \frac{13}{24} \end{array}$$

The complete procedure can be written as follows:

$$\begin{array}{ccccc} 3\frac{1}{6} & & 2\frac{7}{6} & & 2\frac{28}{24} \\ & \longrightarrow & & \longrightarrow & \\ -2\frac{5}{8} & & -2\frac{5}{8} & & -2\frac{15}{24} \\ \hline & & & & \frac{13}{24} \end{array}$$

Of course, the answer is the same as before.

Now we can do a problem involving both addition and subtraction of fractions.

EXAMPLE 9 Adding and subtracting fractions

Perform the indicated operations: $1\frac{1}{10} + \frac{5}{21} - \frac{3}{28}$

SOLUTION We first write $1\frac{1}{10}$ as $\frac{11}{10}$ and then find the LCM of 10, 21, and 28. (Use either method.)

Method 1.

$10 = 2 \cdot 5$

$21 = 3 \cdot 7$

$28 = 2 \cdot 2 \cdot 7$

The LCD is $2 \cdot 2 \cdot 3 \cdot 5 \cdot 7 = 420$.

Method 2.

Step 1. 2) 10 (21) 28

Step 2. 7) (5) 21 14
5 3 2

Step 3. The LCD is
$2 \cdot 7 \cdot 5 \cdot 3 \cdot 2 = 420$.

PROBLEM 9

Perform the indicated operations:
$1\frac{1}{8} + \frac{7}{21} - \frac{3}{28}$

Answer

9. $1\frac{59}{168}$

Now

$$1\frac{1}{10} = \frac{11}{10} = \frac{11 \cdot 42}{10 \cdot 42} = \frac{462}{420} \longrightarrow \quad \frac{462}{420}$$

$$\frac{5}{21} = \frac{5 \cdot 20}{21 \cdot 20} = \frac{100}{420} \longrightarrow \quad +\frac{100}{420}$$

$$\frac{3}{28} = \frac{3 \cdot 15}{28 \cdot 15} = \frac{45}{420} \longrightarrow \quad -\frac{45}{420}$$

$$\frac{517}{420}$$

Horizontally, we have

$$1\frac{1}{10} + \frac{5}{21} - \frac{3}{28} = \frac{462}{420} + \frac{100}{420} - \frac{45}{420}$$

$$= \frac{462 + 100 - 45}{420}$$

$$= \frac{517}{420}$$

EXAMPLE 10 **Application: Operations with fractions**

How many total cups of water and oat bran should be mixed to prepare 3 servings? (Refer to the instructions in the illustration.)

SOLUTION We need $3\frac{3}{4}$ cups of water and $1\frac{1}{2}$ cups of oat bran, a total of $3\frac{3}{4} + 1\frac{1}{2}$ cups.

Write as improper fractions. $\frac{15}{4} + \frac{3}{2}$

The LCD is 4. Rewrite as $\frac{15}{4} + \frac{6}{4}$

Add the fractions. $\frac{21}{4}$

Rewrite as a mixed number. $5\frac{1}{4}$

Note that this time it is more appropriate to write the answer as a *mixed number*. In general, it is quite proper to write the result of an operation involving fractions as an *improper fraction*.

PROBLEM 10

How many total cups of water and oats should be mixed to prepare 1 serving?

Hot Cereal

MICROWAVE AND STOVE TOP AMOUNTS

SERVINGS	1	2	3
WATER	1-1/4 cups	2-1/2 cups	3-3/4 cups
OAT BRAN	1/2 cup	1 cup	1-1/2 cups
SALT (optional)	dash	dash	1/8 tsp

Web It

To review and practice operations with fractions, try link R-2-1 at the Bello Website, mhhe.com/bello.

For an interactive website where you can enter your problem and receive the explanation, go to link R-2-2.

Finally, for all the fraction rules you will ever need, try link R-2-3.

Answer

10. $1\frac{3}{4}$

Exercises R.2

A In Problems 1–50, perform the indicated operations and reduce your answers to lowest terms.

1. $\frac{2}{3}\cdot\frac{7}{3}$ $\frac{14}{9}$

2. $\frac{3}{4}\cdot\frac{7}{8}$ $\frac{21}{32}$

3. $\frac{6}{5}\cdot\frac{7}{6}$ $\frac{7}{5}$

4. $\frac{2}{5}\cdot\frac{5}{3}$ $\frac{2}{3}$

5. $\frac{7}{3}\cdot\frac{6}{7}$ $\frac{2}{1}$ or 2

6. $\frac{5}{6}\cdot\frac{3}{5}$ $\frac{1}{2}$

7. $7\cdot\frac{8}{7}$ $\frac{8}{1}$ or 8

8. $10\cdot 1\frac{1}{5}$ $\frac{12}{1}$ or 12

9. $2\frac{3}{5}\cdot 2\frac{1}{7}$ $\frac{39}{7}$

10. $2\frac{1}{3}\cdot 4\frac{1}{2}$ $\frac{21}{2}$

11. $7\div\frac{3}{5}$ $\frac{35}{3}$

12. $5\div\frac{2}{3}$ $\frac{15}{2}$

13. $\frac{3}{5}\div\frac{9}{10}$ $\frac{2}{3}$

14. $\frac{2}{3}\div\frac{6}{7}$ $\frac{7}{9}$

15. $\frac{9}{10}\div\frac{3}{5}$ $\frac{3}{2}$

16. $\frac{3}{4}\div\frac{3}{4}$ $\frac{1}{1}$ or 1

17. $1\frac{1}{5}\div\frac{3}{8}$ $\frac{16}{5}$

18. $3\frac{3}{4}\div 3$ $\frac{5}{4}$

19. $2\frac{1}{2}\div 6\frac{1}{4}$ $\frac{2}{5}$

20. $3\frac{1}{8}\div 1\frac{1}{3}$ $\frac{75}{32}$

21. $\frac{1}{5}+\frac{2}{5}$ $\frac{3}{5}$

22. $\frac{1}{3}+\frac{1}{3}$ $\frac{2}{3}$

23. $\frac{3}{8}+\frac{5}{8}$ $\frac{1}{1}$ or 1

24. $\frac{2}{9}+\frac{4}{9}$ $\frac{2}{3}$

25. $\frac{7}{8}+\frac{3}{4}$ $\frac{13}{8}$

26. $\frac{1}{2}+\frac{1}{6}$ $\frac{2}{3}$

27. $\frac{5}{6}+\frac{3}{10}$ $\frac{17}{15}$

28. $3+\frac{2}{5}$ $\frac{17}{5}$

29. $2\frac{1}{3}+1\frac{1}{2}$ $\frac{23}{6}$

30. $1\frac{3}{4}+2\frac{1}{6}$ $\frac{47}{12}$

31. $\frac{1}{5}+2+\frac{9}{10}$ $\frac{31}{10}$

32. $\frac{1}{6}+\frac{1}{8}+\frac{1}{3}$ $\frac{5}{8}$

33. $3\frac{1}{2}+1\frac{1}{7}+2\frac{1}{4}$ $\frac{193}{28}$

34. $1\frac{1}{3}+2\frac{1}{4}+1\frac{1}{5}$ $\frac{287}{60}$

35. $\frac{5}{8}-\frac{2}{8}$ $\frac{3}{8}$

36. $\frac{3}{7}-\frac{1}{7}$ $\frac{2}{7}$

37. $\frac{1}{3}-\frac{1}{6}$ $\frac{1}{6}$

38. $\frac{5}{12}-\frac{1}{4}$ $\frac{1}{6}$

39. $\frac{7}{10}-\frac{3}{20}$ $\frac{11}{20}$

40. $\frac{5}{20}-\frac{7}{40}$ $\frac{3}{40}$

41. $\frac{8}{15}-\frac{2}{25}$ $\frac{34}{75}$

42. $\frac{7}{8}-\frac{5}{12}$ $\frac{11}{24}$

43. $2\frac{1}{5}-1\frac{3}{4}$ $\frac{9}{20}$

44. $4\frac{1}{2}-2\frac{1}{3}$ $\frac{13}{6}$

45. $3-1\frac{3}{4}$ $\frac{5}{4}$

46. $2-1\frac{1}{3}$ $\frac{2}{3}$

47. $\frac{5}{6}+\frac{1}{9}-\frac{1}{3}$ $\frac{11}{18}$

48. $\frac{3}{4}+\frac{1}{12}-\frac{1}{6}$ $\frac{2}{3}$

49. $1\frac{1}{3}+2\frac{1}{3}-1\frac{1}{5}$ $\frac{37}{15}$

50. $3\frac{1}{2}+1\frac{1}{7}-2\frac{1}{4}$ $\frac{67}{28}$

APPLICATIONS

51. *Weights on Earth and the moon* The weight of an object on the moon is $\frac{1}{6}$ of its weight on Earth. How much did the Lunar Rover, weighing 450 pounds on Earth, weigh on the moon? 75 lb

52. *Do you want to meet a millionaire?* Your best bet is to go to Idaho. In this state, $\frac{1}{38}$ of the population happens to be millionaires! If there are about 912,000 persons in Idaho, about how many millionaires are there in the state? 24,000

53. *Union membership* The Actors Equity has 28,000 members, but only $\frac{1}{5}$ of the membership is working at any given moment. How many working actors is that? 5600

54. *Battery voltage* The voltage of a standard C battery is $1\frac{1}{2}$ volts (V). What is the total voltage of 6 such batteries? 9 V

55. *Excavation* An earthmover removed $66\frac{1}{2}$ cubic yards of sand in 4 hours. How many cubic yards did it remove per hour? $16\frac{5}{8}$

56. *Bond prices* A bond is selling for $\$3\frac{1}{2}$. How many can be bought with $98? 28

57. *Household incomes* In a recent survey, it was found that $\frac{9}{50}$ of all American households make more than $25,000 annually and that $\frac{3}{25}$ make between $20,000 and $25,000. What total fraction of American households make more than $20,000? $\frac{3}{10}$

58. *Immigration* Between 1971 and 1977, $\frac{1}{5}$ of the immigrants coming to America were from Europe and $\frac{9}{20}$ came from this hemisphere. What total fraction of the immigrants were in these two groups? $\frac{13}{20}$

59. *Employment* U.S. workers work an average of $46\frac{3}{5}$ hours per week, while Canadians work $38\frac{9}{10}$. How many more hours per week do U.S. workers work? $7\frac{7}{10}$

60. *Bone composition* Human bones are $\frac{1}{4}$ water, $\frac{3}{10}$ living tissue, and the rest minerals. Thus, the fraction of the bone that is minerals is $1 - \frac{1}{4} - \frac{3}{10}$. Find this fraction. $\frac{9}{20}$

USING YOUR KNOWLEDGE

Hot Dogs and Buns

In this section we learned that the LCM of two numbers is the *smallest* multiple of the two numbers. Can you ever apply this idea to anything besides adding and subtracting fractions? You bet! Hot dogs and buns. Have you noticed that hot dogs come in packages of 10 but buns come in packages of 8 (or 12)?

61. What is the smallest number of packages of hot dogs (10 to a package) and buns (8 to a package) you must buy so that you have as many hot dogs as you have buns? (*Hint:* Think of multiples.) 4 pkg of hot dogs, 5 pkg of buns

62. If buns are sold in packages of 12, what is the smallest number of packages of hot dogs (10 to a package) and buns you must buy so that you have as many hot dogs as you have buns? 6 pkg of hot dogs, 5 pkg of buns

WRITE ON

63. Write the procedure you use to find the LCM of two numbers. Answers may vary.

64. Write the procedure you use to convert a mixed number to an improper fraction. Answers may vary.

65. Write the procedure you use to

a. Multiply two fractions. Answers may vary.

b. Divide two fractions. Answers may vary.

66. Write the procedure you use to

a. Add two fractions with the same denominator. Answers may vary.

b. Add two fractions with different denominators. Answers may vary.

c. Subtract two fractions with the same denominator. Answers may vary.

d. Subtract two fractions with different denominators. Answers may vary.

SKILL CHECKER

The *Skill Checker* exercises review skills previously studied to help you maintain the concepts you have mastered. They will appear from now on in the exercise sets.

67. Divide 128 by 16. 8

68. Divide 2100 by 35. 60

69. Reduce: $\frac{20}{100}$ $\frac{1}{5}$

70. Reduce: $\frac{75}{100}$ $\frac{3}{4}$

R.3 DECIMALS AND PERCENTS

To Succeed, Review How To . . .

1. Divide one whole number by another one.
2. Reduce fractions (pp. 7–9).

Objectives

A Write a decimal in expanded form.

B Write a decimal as a fraction.

C Add and subtract decimals.

D Multiply and divide decimals.

E Write a fraction as a decimal.

F Convert decimals to percents and percents to decimals.

G Convert fractions to percents and percents to fractions.

H Round numbers to a specified number of decimal places.

GETTING STARTED

Decimals at the Gas Pump

The price of gas is $1.939 (read "one point nine three nine"). This number is called a **decimal.** The word *decimal* means that we count by *tens*—that is, we use *base ten.* The dot in 1.939 is called the *decimal point.* The decimal 1.939 consists of two parts: the whole-number part (the number to the *left* of the decimal point) and the decimal part, .939.

1.939

Whole number: 1

Decimal: .939

Writing Decimals in Expanded Form

The number 11.664 can be written in a diagram, as shown. With the help of this diagram, we can write the number in **expanded form,** like this:

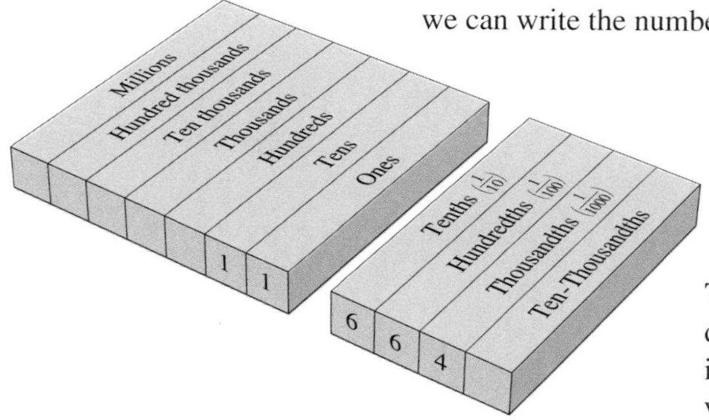

11.664

$$10 + 1 + \frac{6}{10} + \frac{6}{100} + \frac{4}{1000}$$

The number 11.664 is read by reading the number to the left of the decimal (*eleven*), using the word *and* for the decimal point, and reading the number to the right of the decimal point followed by the place value of the *last* digit: *eleven and six hundred sixty-four thousandths.*

EXAMPLE 1 **Writing decimals in expanded form**

Write in expanded form: 35.216

SOLUTION

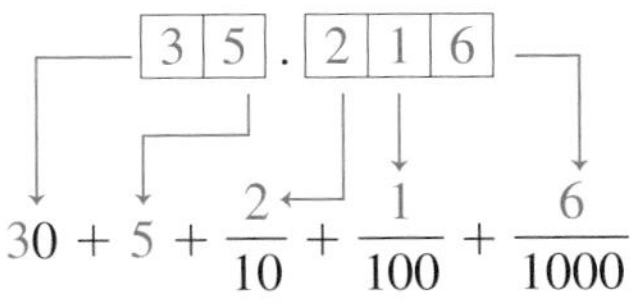

$$30 + 5 + \frac{2}{10} + \frac{1}{100} + \frac{6}{1000}$$

PROBLEM 1

Write in expanded form: 46.325

B Writing Decimals as Fractions

Decimals can be converted to fractions. For example,

$$0.3 = \frac{3}{10}$$

$$0.18 = \frac{18}{100} = \frac{9}{50}$$

and

$$0.150 = \frac{150}{1000} = \frac{3}{20}$$

Here is the procedure for changing a decimal to a fraction.

PROCEDURE

Changing Decimals to Fractions

1. Write the nonzero digits of the number, omitting the decimal point, as the *numerator* of the fraction.
2. The *denominator* is a 1 *followed by as many zeros as there are decimal digits in the decimal.*
3. Reduce, if possible.

EXAMPLE 2 **Changing decimals to fractions**

Write as a reduced fraction:

a. 0.035 **b.** 0.0275

SOLUTION

a. $0.\underbrace{035}_{\text{3 digits}} = \frac{35}{\underbrace{1000}_{\text{3 zeros}}} = \frac{7}{200}$

b. $0.\underbrace{0275}_{\text{4 digits}} = \frac{275}{\underbrace{10{,}000}_{\text{4 zeros}}} = \frac{11}{400}$

PROBLEM 2

Write as a reduced fraction:

a. 0.045 **b.** 0.0375

If the decimal has a whole-number part, we write all the digits of the number, omitting the decimal point, as the numerator of the fraction. For example, to write 4.23 as a fraction, we write

$$4.\underbrace{23}_{\text{2 digits}} = \frac{423}{\underbrace{100}_{\text{2 zeros}}}$$

Answers

1. $40 + 6 + \frac{3}{10} + \frac{2}{100} + \frac{5}{1000}$

2. a. $\frac{9}{200}$ **b.** $\frac{3}{80}$

EXAMPLE 3 **Changing decimals with whole-number parts**

Write as a reduced fraction:

a. 3.11 **b.** 5.15

SOLUTION

a. $3.11 = \frac{311}{100}$ **b.** $5.15 = \frac{515}{100} = \frac{103}{20}$

PROBLEM 3

Write as a reduced fraction:

a. 3.13 **b.** 5.25

C Adding and Subtracting Decimals

Calculate It

Decimal Points

If you have a calculator, you do not have to worry about placing the decimal point. Simply enter

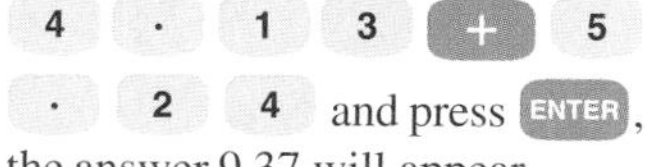

and press ENTER, the answer 9.37 will appear.

Teaching Tip

Make sure students align the decimal point.

Adding decimals is similar to adding whole numbers, as long as we first line up (align) the decimal points by writing them in the *same* column and then make sure that digits of the same place value are in the same column. We then add the columns from right to left, as usual. For example, to add 4.13 + 5.24, we write

The result is 9.37.

EXAMPLE 4 **Adding decimals**

Add: 5 + 18.15

SOLUTION Note that 5 = 5. and attach 00 to the 5. so that both addends have the same number of decimal digits.

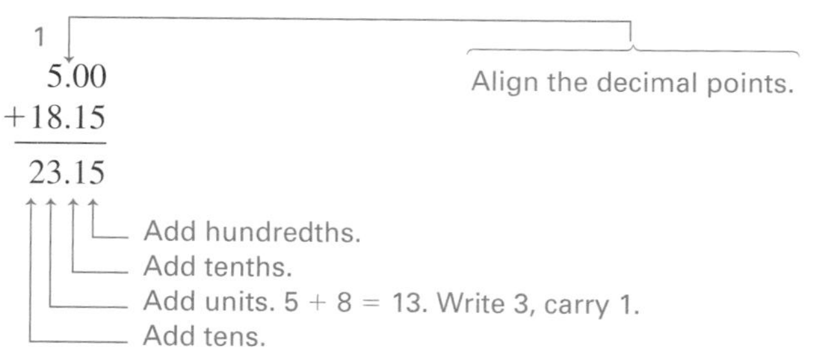

The result is 23.15.

PROBLEM 4

Add: 6 + 17.25

Answers

3. a. $\frac{313}{100}$ **b.** $\frac{21}{4}$ **4.** 23.25

Subtraction of decimals is also like subtraction of whole numbers as long as you remember to align the decimal points and attach any needed zeros so that both numbers have

Calculate It

Subtracting Decimals

Subtraction with a calculator is quite easy. Just enter the numbers 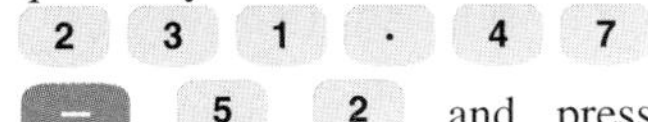 5 2 and press ENTER. You do not have to worry about adding two zeros to the 52—the calculator does it!

the same number of decimal digits. For example, if you earned \$231.47 and you made a \$52 payment, you have \$231.47 − \$52. To find how much you have left, we write

```
   1 12 11
 $ 2 3 1 . 4 7    Align the decimal points.
 −   5 2 . 0 0    Attach two zeros ($52 is the same as 52 dollars and no cents).
   1 7 9 . 4 7
         ↑ The decimal point is in the same column.
```

Thus you have \$179.47 left. You can check this answer by adding 52 and 179.47 to obtain 231.47.

EXAMPLE 5 Subtracting decimals

Subtract: 383.43 − 17.5

SOLUTION

```
   7 12  14
 3 8 3 . 4 3    Align the decimal points.
 −  1 7 . 5 0   Attach one zero.
 3 6 5 . 9 3    Subtract.
       ↑ The decimal point is in the same column.
```

PROBLEM 5

Subtract: 273.32 − 14.5

EXAMPLE 6 Subtracting decimals

Subtract: 347.8 − 182.231

SOLUTION

```
               9
  2 14     7 10 10
  3 4 7 . 8 0 0    Align the decimal points.
− 1 8 2 . 2 3 1    Attach two zeros.
  1 6 5 . 5 6 9    Subtract.
        ↑ The decimal point is in the same column.
```

PROBLEM 6

Subtract: 458.6 − 193.341

D Multiplying and Dividing Decimals

When multiplying decimals, the number of decimal digits in the product is the *sum* of the numbers of decimal digits in the factors. For example, since 0.3 has *one* decimal digit and 0.0007 has *four,* the product

$$0.3 \times 0.0007 = \frac{3}{10} \times \frac{7}{10{,}000} = \frac{21}{100{,}000} = \underbrace{0.00021}_{1+4=5 \text{ digits}}$$

has 1 + 4 = 5 decimal digits.

Answers

5. 258.82 **6.** 265.259

Here is the procedure used to multiply decimals.

PROCEDURE

Multiplying Decimals

1. *Multiply* the two decimal numbers *as if they were whole numbers.*
2. The *number of decimal digits* in the product is the *sum* of the numbers of decimal digits in the factors.

EXAMPLE 7 Multiplying Decimals

Multiply:

a. 5.102×21.03 **b.** 5.213×0.0012

SOLUTION

a.

$$\begin{array}{r} 5.102 \\ \times 21.03 \\ \hline 15306 \\ 51020 \\ 10204 \\ \hline 107.29506 \end{array}$$

5.102: 3 decimal digits
21.03: 2 decimal digits
107.29506: 3 + 2 = 5 decimal digits

b.

$$\begin{array}{r} 5.213 \\ \times 0.0012 \\ \hline 10426 \\ 5213 \\ \hline 0.0062556 \end{array}$$

5.213: 3 decimal digits
0.0012: 4 decimal digits
0.0062556: 3 + 4 = 7 decimal digits

Attach two zeros to obtain 7 decimal digits.

PROBLEM 7

Multiply:

a. 6.203×31.03 **b.** 3.123×0.0015

Now suppose you want to divide 52 by 6.5. You can write

$$\frac{52}{6.5}$$ ⟵ This is called the dividend. (52) ⟵ This is called the divisor. (6.5)

If we multiply the numerator and denominator of this fraction by 10, we obtain

$$\frac{52}{6.5} = \frac{52 \times 10}{6.5 \times 10} = \frac{520}{65} = 8$$

Thus $\frac{52}{6.5} = 8$, as can be easily checked, since $52 = 6.5 \times 8$. This problem can be shortened by using the following steps.

Step 1. Write the problem in the usual long division form. — $6.5\overline{)52}$ (Divisor ↑ ↑ Dividend)

Step 2. Move the decimal point in the divisor, 6.5, to the right until a whole number is obtained. (This is the same as multiplying 6.5 by 10.) — $65.\overline{)52}$ (↑ Multiply by 10.)

Step 3. Move the decimal point in the dividend the *same* number of places as in step 2. (This is the same as multiplying the dividend 52 by 10.) Attach zeros if necessary. — $65.\overline{)520}$ (↑ ↑ Multiply by 10.)

Step 4. Place the decimal point in the answer directly above the new decimal point in the dividend. — $65\overline{)520.}$

Step 5. Divide exactly as you would divide whole numbers. The result is 8, as before.

$$\begin{array}{r} 8. \\ 65\overline{)520.} \\ 520 \\ \hline 0 \end{array}$$

Teaching Tip

Remind students that a *product* is the answer to a multiplication of *factors*.

Answers

7. a. 192.47909 **b.** 0.0046845

Calculate It

Dividing Decimals

To divide decimals using a calculator, recall that the bar in $\frac{1.28}{1.6}$ means to divide. Accordingly, we enter 1.28 ÷ 1.6 ENTER. Do not worry about the decimal point; the answer comes out as 0.8 as before.

Here is another example. Divide

$$\frac{1.28}{1.6}$$

Step 1. Write the problem in the usual long division form. $1.6\overline{)1.28}$

Step 2. Move the decimal point in the divisor to the right until a whole number is obtained. $16.\overline{)1.28}$

Step 3. Move the decimal point in the dividend the *same* number of places as in step 2. $16.\overline{)12.8}$

Step 4. Place the decimal point in the answer directly above the new decimal point in the dividend. $16\overline{)12.8}$

Step 5. Divide exactly as you would divide whole numbers.

```
   0.8
16)12.8
   12.8
      0
```

Thus

$$\frac{1.28}{1.6} = 0.8$$

EXAMPLE 8 **Dividing decimals**

Divide: $\frac{2.1}{0.035}$

SOLUTION We move the decimal point in the divisor (and also in the dividend) three places to the right. To do this, we need to attach two zeros to 2.1.

$$035.\overline{)2100.}$$

Multiply by 1000.

We next place the decimal point in the answer directly above the one in the dividend and proceed in the usual manner.

```
    60.
35)2100.
   210
    00
```

The answer is 60; that is,

$$\frac{2.1}{0.035} = 60$$

CHECK $0.035 \times 60 = 2.100$

PROBLEM 8

Divide: $\frac{1.8}{0.045}$

Web It

If you want to practice some more and review the operations with decimals, go to link R-3-1 at the Bello Website at mhhe.com/bello. There you will find additional examples and practice problems.

Writing Fractions as Decimals

Since a fraction indicates a division, we can write a fraction as a decimal by dividing. For example, $\frac{3}{4}$ means $3 \div 4$, so

```
  0.75
4)3.00     Note that 3 = 3.00.
  2 8
    20
    20
     0
```

Answer

8. 40

Here the division **terminates** (has a 0 remainder). Thus, $\frac{3}{4} = 0.75$, and 0.75 is called a **terminating decimal.**

Since $\frac{2}{3}$ means $2 \div 3$,

$$\begin{array}{r} 0.666\ldots \\ 3\overline{)2.000} \\ \underline{1\,8} \\ 20 \\ \underline{18} \\ 20 \\ \underline{18} \\ 2 \end{array}$$

The ellipsis (. . .) means that the 6 repeats.

The division continues because the remainder 2 repeats.

Hence $\frac{2}{3} = 0.666\ldots$. Such decimals (called **repeating decimals**) may be written by placing a bar over the repeating digits; that is, $\frac{2}{3} = 0.\overline{6}$.

Web It

If you want additional work changing fractions to decimals, go to link R-3-2 at the Bello Website at mhhe.com/bello.

Calculate It Changing Fractions to Decimals

To change a fraction to a decimal using a calculator, just perform the indicated operations. Thus, to convert $\frac{2}{3}$ to a decimal, enter 2 ÷ 3 ENTER and you get the answer 0.666666666. Note that you get nine 6's. How do you know the 6 repeats indefinitely? At this point, you do not. We shall discuss this in the Collaborative Learning section.

EXAMPLE 9 Changing fractions to decimals

Write as a decimal:

a. $\frac{5}{8}$ **b.** $\frac{2}{11}$

SOLUTION

a. $\frac{5}{8}$ means $5 \div 8$. Thus,

$$\begin{array}{r} 0.625 \\ 8\overline{)5.000} \\ \underline{4\,8} \\ 20 \\ \underline{16} \\ 40 \\ \underline{40} \\ 0 \end{array}$$

The remainder is 0. The answer is a *terminating decimal.*

$$\frac{5}{8} = 0.625$$

b. $\frac{2}{11}$ means $2 \div 11$. Thus,

$$\begin{array}{r} 0.1818\ldots \\ 11\overline{)2.0000} \\ \underline{1\,1} \\ 90 \\ \underline{88} \\ 20 \\ \underline{11} \\ 90 \\ \underline{88} \\ 20 \end{array}$$

The remainders 90 and 20 alternately repeat. The answer is a *repeating decimal.*

$$\frac{2}{11} = 0.\overline{18}$$

The bar is written over the digits that repeat.

PROBLEM 9

Write as a decimal:

a. $\frac{5}{6}$ **b.** $\frac{1}{4}$

Teaching Tip

Make sure students are aware that the bar over 18 means that the 18 repeats.

Answers

9. a. $0.8\overline{3}$ **b.** 0.25

Converting Decimals and Percents

An important application of decimals is the study of **percents.** The word *percent* means "per one hundred" and is written using the symbol %. Thus,

$$29\% = \frac{29}{100}$$

or "twenty-nine hundredths," that is, 0.29.

Similarly,

$$43\% = \frac{43}{100} = 0.43, \quad 87\% = \frac{87}{100} = 0.87, \quad \text{and} \quad 4.5\% = \frac{4.5}{100} = \frac{45}{1000} = 0.045$$

Note that in each case, the number is divided by 100, which is equivalent to moving the decimal point two places to the *left.*

Here is the general procedure.

Teaching Tip

Show examples involving multiplication and division by multiples of 10.

×	÷
1.42 × 10 = 14.2	1.42 ÷ 10 = 0.142
1.42 × 100 = 142	1.42 ÷ 100 = 0.0142
1.42 × 1000 = 1420	1.42 ÷ 1000 = 0.00142

PROCEDURE

Converting Percents to Decimals

Move the decimal point in the number *two* places to the *left* and omit the % symbol.

EXAMPLE 10 Converting percents to decimals

Write as a decimal:

a. 98% **b.** 34.7% **c.** 7.2%

SOLUTION

a. 98% = 98. = 0.98 Recall that a decimal point follows every whole number.

b. 34.7% = 34.7 = 0.347

c. 7.2% = 07.2 = 0.072 Note that a 0 was inserted so we could move the decimal point two places to the left.

PROBLEM 10

Write as a decimal:

a. 86% **b.** 48.9% **c.** 8.3%

As you can see from Example 10, 0.98 = 98% and 0.347 = 34.7%. Thus, we can reverse the previous procedure to convert decimals to percents. Here is the way we do it.

PROCEDURE

Converting Decimals to Percents

Move the decimal point *two* places to the *right* and attach the % symbol.

EXAMPLE 11 Converting decimals to percents

Write as a percent:

a. 0.53 **b.** 3.19 **c.** 64.7

SOLUTION

a. 0.53 = 0.53% = 53%

b. 3.19 = 3.19% = 319%

c. 64.7 = 64.70% = 6470% Note that a 0 was inserted so we could move the decimal point two places to the right.

PROBLEM 11

Write as a percent:

a. 0.49 **b.** 4.17 **c.** 89.2

Answers

10. a. 0.86 **b.** 0.489 **c.** 0.083 **11. a.** 49% **b.** 417% **c.** 8920%

G Converting Fractions and Percents

Since % means *per hundred,*

$$5\% = \frac{5}{100}$$

$$7\% = \frac{7}{100}$$

$$23\% = \frac{23}{100}$$

$$4.7\% = \frac{4.7}{100}$$

$$134\% = \frac{134}{100}$$

Thus we can convert a number written as a percent to a fraction by using the following procedure.

PROCEDURE

Converting Percents to Fractions

Write the *number* over 100, omit the % sign, and *reduce* the fraction, if possible.

EXAMPLE 12 **Converting percents to fractions**

Write as a fraction:

a. 49% **b.** 75%

SOLUTION

a. $49\% = \frac{49}{100}$ **b.** $75\% = \frac{75}{100} = \frac{3}{4}$

PROBLEM 12

Write as a fraction:

a. 37% **b.** 25%

How do we write a fraction as a percent? If the fraction has a denominator that is a factor of 100, it's easy. To write $\frac{1}{5}$ as a percent, we first multiply numerator and denominator by a number that will make the denominator 100. Thus,

$$\frac{1}{5} = \frac{1 \times 20}{5 \times 20} = \frac{20}{100} = 20\%$$

Similarly,

$$\frac{3}{4} = \frac{3 \times 25}{4 \times 25} = \frac{75}{100} = 75\%$$

Note that in both cases the denominator of the fraction was a factor of 100.

Answers

12. a. $\frac{37}{100}$ **b.** $\frac{1}{4}$

EXAMPLE 13 **Converting fractions to percents**

Write as a percent: $\frac{4}{5}$

SOLUTION We multiply by $\frac{20}{20}$ to get 100 in the denominator.

$$\frac{4}{5} = \frac{4 \times 20}{5 \times 20} = \frac{80}{100} = 80\%$$

PROBLEM 13

Write as a percent: $\frac{3}{5}$

In the preceding examples the denominator of the given fraction was a *factor* of 100. Suppose we wish to change $\frac{1}{6}$ to a percent. The problem here is that 6 is *not* a factor of 100. But don't panic! We can write $\frac{1}{6}$ as a percent by dividing the numerator by the denominator. Thus we divide 1 by 6, continuing the division until we have two decimal digits.

```
   0.16
6)1.00
  6
   40
   36
    4   Remainder
```

The answer is 0.16 with remainder 4; that is,

$$\frac{1}{6} = 0.16\frac{4}{6} = 0.16\frac{2}{3} = \frac{16\frac{2}{3}}{100} = 16\frac{2}{3}\%$$

Similarly, we can write $\frac{2}{3}$ as a percent by dividing 2 by 3, obtaining

```
   0.66
3)2.00
  1 8
    20
    18
     2   Remainder
```

Thus

$$\frac{2}{3} = 0.66\frac{2}{3} = \frac{66\frac{2}{3}}{100} = 66\frac{2}{3}\%$$

Here is the procedure we use.

PROCEDURE

Converting Fractions to Percents

Divide the *numerator* by the *denominator* (carry the division to two decimal places), convert the resulting decimal to a percent by *moving the decimal point two places to the right,* and attach the % symbol. (Don't forget to include the remainder, if there is one.)

Web It

To review the percent topics we have studied, go to the Bello Website at mhhe.com/bello and try link R-3-3.

Answer

13. 60%

EXAMPLE 14 **Converting fractions to percents**

Write as a percent: $\frac{5}{8}$

SOLUTION Dividing 5 by 8, we have

$$\begin{array}{r} 0.62 \\ 8\overline{)5.00} \\ \underline{4\,8} \\ 20 \\ \underline{16} \\ 4 \end{array}$$

Note that the remainder is $\frac{4}{8} = \frac{1}{2}$.

Thus

$$\frac{5}{8} = 0.62\frac{1}{2} = 0.62\frac{1}{2} = 62\frac{1}{2}\%$$

PROBLEM 14

Write as a percent: $\frac{3}{8}$

Rounding Numbers

We can shorten some of the decimals we have considered in this section if we use **rounding.** For example, the numbers 107.29506 and 0.062556 can be rounded to three decimal places—that is, to the nearest *thousandth.* Here is the rule we use.

RULE

To Round Off Numbers

Step 1. **Underline** the digit in the place to which you are rounding.

Step 2. If the first digit to the right of the underlined digit is *5 or more,* add 1 to the underlined digit. Otherwise, do not change the underlined digit.

Step 3. **Drop** all the numbers to the right of the underlined digit (attach 0's to fill in the place values if necessary).

EXAMPLE 15 **Rounding off decimals**

Round to three decimal places:

a. 107.29506 **b.** 0.062556

SOLUTION

a. Step 1. Underline the 5. $107.29\underline{5}06$

Step 2. The digit to the right of 5 is 0, so we do not change the 5. $107.29\underline{5}06$

Step 3. Drop all numbers to the right of 5. $107.29\underline{5}$

Thus, when 107.29506 is rounded to three decimal places, that is, to the nearest thousandth, the answer is 107.295.

PROBLEM 15

Round to two decimal places:

a. $0.\overline{6}$ **b.** $0.\overline{18}$

Answers

14. $37\frac{1}{2}\%$ **15. a.** 0.67 **b.** 0.18

b. Step 1. Underline the 2. 0.06$\underline{2}$556

Step 2. The digit to the right of 2 is 5, so we add 1 to the underlined digit. 0.06$\overset{3}{\underline{2}}$556

Step 3. Drop all numbers to the right of 3. 0.06$\underline{3}$

Thus, when 0.062556 is rounded to three decimal places, that is, to the nearest thousandth, the answer is 0.063.

Exercises R.3

A In Problems 1–10, write the given decimal in expanded form.

1. 4.7 $4 + \frac{7}{10}$

2. 3.9 $3 + \frac{9}{10}$

3. 5.62 $5 + \frac{6}{10} + \frac{2}{100}$

4. 9.28 $9 + \frac{2}{10} + \frac{8}{100}$

5. 16.123 $10 + 6 + \frac{1}{10} + \frac{2}{100} + \frac{3}{1000}$

6. 18.845 $10 + 8 + \frac{8}{10} + \frac{4}{100} + \frac{5}{1000}$

7. 49.012 $40 + 9 + \frac{1}{100} + \frac{2}{1000}$

8. 93.038 $90 + 3 + \frac{3}{100} + \frac{8}{1000}$

9. 57.104 $50 + 7 + \frac{1}{10} + \frac{4}{1000}$

10. 85.305 $80 + 5 + \frac{3}{10} + \frac{5}{1000}$

B In Problems 11–20, write the given decimal as a reduced fraction.

11. 0.9 $\frac{9}{10}$

12. 0.7 $\frac{7}{10}$

13. 0.06 $\frac{3}{50}$

14. 0.08 $\frac{2}{25}$

15. 0.12 $\frac{3}{25}$

16. 0.18 $\frac{9}{50}$

17. 0.054 $\frac{27}{500}$

18. 0.062 $\frac{31}{500}$

19. 2.13 $\frac{213}{100}$

20. 3.41 $\frac{341}{100}$

C In Problems 21–40, add or subtract as indicated.

21. \$648.01 + \$341.06 \$989.07

22. \$237.49 + \$458.72 \$696.21

23. 72.03 + 847.124 919.154

24. 13.12 + 108.138 121.258

25. 104 + 78.103 182.103

26. 184 + 69.572 253.572

27. 0.35 + 3.6 + 0.127 4.077

28. 5.2 + 0.358 + 21.005 26.563

29. 27.2 − 0.35 26.85

30. 4.6 − 0.09 4.51

31. \$19 − \$16.62 \$2.38

32. \$99 − \$0.61 \$98.39

33. 9.43 − 6.406 3.024

34. 9.08 − 3.465 5.615

35. 8.2 − 1.356 6.844

36. 6.3 − 4.901 1.399

37. 6.09 + 3.0046 9.0946

38. 2.01 + 1.3045 3.3145

39. 4.07 + 8.0035 12.0735

40. 3.09 + 5.4895 8.5795

D In Problems 41–60, multiply or divide as indicated.

41. 9.2×0.613 5.6396
42. 0.514×7.4 3.8036
43. 8.7×11 95.7
44. 78.1×108 8434.8
45. 7.03×0.0035 0.024605
46. 8.23×0.025 0.20575
47. 3.0012×4.3 12.90516
48. 6.1×2.013 12.2793
49. 0.0031×0.82 0.002542
50. 0.51×0.0045 0.002295
51. $15\overline{)9}$ 0.6
52. $48\overline{)6}$ 0.125
53. $5\overline{)32}$ 6.4
54. $8\overline{)36}$ 4.5
55. $8.5 \div 0.005$ 1700
56. $4.8 \div 0.003$ 1600
57. $4 \div 0.05$ 80
58. $18 \div 0.006$ 3000
59. $2.76 \div 60$ 0.046
60. $31.8 \div 30$ 1.06

E In Problems 61–76, write the given fraction as a decimal.

61. $\frac{1}{5}$ 0.2
62. $\frac{1}{2}$ 0.5
63. $\frac{7}{8}$ 0.875
64. $\frac{1}{8}$ 0.125
65. $\frac{3}{16}$ 0.1875
66. $\frac{5}{16}$ 0.3125
67. $\frac{2}{9}$ $0.\overline{2}$
68. $\frac{4}{9}$ $0.\overline{4}$
69. $\frac{6}{11}$ $0.\overline{54}$
70. $\frac{7}{11}$ $0.\overline{63}$
71. $\frac{3}{11}$ $0.\overline{27}$
72. $\frac{5}{11}$ $0.\overline{45}$
73. $\frac{1}{6}$ $0.1\overline{6}$
74. $\frac{5}{6}$ $0.8\overline{3}$
75. $\frac{10}{9}$ $1.\overline{1}$
76. $\frac{11}{9}$ $1.\overline{2}$

F In Problems 77–86, write the given percent as a decimal.

77. 33% 0.33
78. 52% 0.52
79. 5% 0.05
80. 9% 0.09
81. 300% 3
82. 500% 5
83. 11.8% 0.118
84. 89.1% 0.891
85. 0.5% 0.005
86. 0.7% 0.007

In Problems 87–96, write the given decimal as a percent.

87. 0.05 5%
88. 0.07 7%
89. 0.39 39%
90. 0.74 74%
91. 0.416 41.6%
92. 0.829 82.9%
93. 0.003 0.3%
94. 0.008 0.8%
95. 1.00 100%
96. 2.1 210%

G In Problems 97–106, write the given percent as a reduced fraction.

97. 30% $\frac{3}{10}$
98. 40% $\frac{2}{5}$
99. 6% $\frac{3}{50}$
100. 2% $\frac{1}{50}$
101. 7% $\frac{7}{100}$
102. 19% $\frac{19}{100}$
103. $4\frac{1}{2}\%$ $\frac{9}{200}$
104. $2\frac{1}{4}\%$ $\frac{9}{400}$
105. $1\frac{1}{3}\%$ $\frac{1}{75}$
106. $5\frac{2}{3}\%$ $\frac{17}{300}$

In Problems 107–116, write the given fraction as a percent.

107. $\frac{3}{5}$ 60% **108.** $\frac{4}{25}$ 16% **109.** $\frac{1}{2}$ 50% **110.** $\frac{3}{50}$ 6%

111. $\frac{5}{6}$ $83\frac{1}{3}\%$ **112.** $\frac{1}{3}$ $33\frac{1}{3}\%$ **113.** $\frac{4}{8}$ 50% **114.** $\frac{7}{8}$ $87\frac{1}{2}\%$

115. $\frac{4}{3}$ $133\frac{1}{3}\%$ **116.** $\frac{7}{6}$ $116\frac{2}{3}\%$

H In Problems 117–126, round each number to the given number of decimal places.

117. 27.6263; 1 27.6 **118.** 99.6828; 1 99.7 **119.** 26.746706; 4 26.7467 **120.** 54.037755; 4 54.0378

121. 35.24986; 3 35.250 **122.** 69.24851; 3 69.249 **123.** 52.378; 2 52.38 **124.** 6.724; 2 6.72

125. 74.846008; 5 74.84601 **126.** 39.948712; 5 39.94871

SKILL CHECKER

Try the Skill Checker Exercises so you'll be ready for the next section.

Reduce:

127. $\frac{36}{100}$ $\frac{9}{25}$ **128.** $\frac{64}{1000}$ $\frac{8}{125}$ **129.** $\frac{480}{1000}$ $\frac{12}{25}$

130. Write $\frac{42.1}{100}$ as a fraction with a denominator of 1000. $\frac{421}{1000}$

USING YOUR KNOWLEDGE

The Pursuit of Happiness

Psychology Today conducted a survey about happiness. Here are some conclusions from that report.

131. Seven out of ten people said they had been happy over the last six months. What percent of the people is that? 70%

132. Of the people surveyed, 70% expected to be happier in the future than now. What fraction of the people is that? $\frac{7}{10}$

133. The survey also showed that 0.40 of the people felt lonely.

a. What percent of the people is that? 40%

b. What fraction of the people is that? $\frac{2}{5}$

134. Only 4% of the men were ready to cry. Write this percent as a decimal. 0.04

135. Of the people surveyed, 49% were single. Write this percent as

a. A fraction. $\frac{49}{100}$

b. A decimal. 0.49

Do you wonder how they came up with some of these percents? They used their knowledge. You do the same and fill in the spaces in the accompanying table, which refers to the marital status of the 52,000 people surveyed. For example, the first line shows that 25,480 persons out of 52,000 were single. This is

$$\frac{25{,}480}{52{,}000} = 49\%$$

Fill in the percents in the last column.

	Marital Status	Number	Percent
	Single	25,480	49%
136.	Married (first time)	15,600	30%
137.	Remarried	2600	5%
138.	Divorced, separated	5720	11%
139.	Widowed	520	1%
140.	Cohabiting	2080	4%

COLLABORATIVE LEARNING

Form several groups of four or five students each. The first group will convert the fractions $\frac{1}{2}, \frac{1}{3}, \frac{1}{4}, \frac{1}{5}$, and $\frac{1}{6}$ to decimals. The second group will convert the fractions $\frac{1}{7}, \frac{1}{8}, \frac{1}{9}, \frac{1}{10}$, and $\frac{1}{11}$ to decimals. The next group will convert $\frac{1}{12}, \frac{1}{13}, \frac{1}{14}, \frac{1}{15}$, and $\frac{1}{16}$ to decimals. Ask each group to show the fractions that have terminating decimal representations. You should have $\frac{1}{2}, \frac{1}{4}$, and so on. Now, look at the fractions whose decimal representations are not terminating. You should have $\frac{1}{3}, \frac{1}{6}$, and so on. Can you tell which fractions will have terminating decimal representations? (*Hint:* Look at the denominators of the terminating fractions!)

Practice Test R

(Answers on page 39)

1. Write -18 as a fraction with a denominator of 1.

2. Find a fraction equivalent to $\frac{3}{7}$ with a denominator of 21.

3. Find a fraction equivalent to $\frac{9}{15}$ with a denominator of 5.

4. Reduce $\frac{27}{54}$ to lowest terms.

5. Find: $5\frac{1}{4} \cdot \frac{32}{21}$.

6. Find: $2\frac{1}{4} \div \frac{3}{8}$.

7. Find: $1\frac{1}{10} + \frac{5}{12}$.

8. Find: $4\frac{1}{6} - 1\frac{9}{10}$.

9. Write 68.428 in expanded form.

10. Write 0.045 as a reduced fraction.

11. Write 3.12 as a reduced fraction.

12. Add: $847.18 + 29.365$.

13. Subtract: $447.58 - 27.6$.

14. Multiply: 4.315×0.0013.

15. Divide: $\frac{4.2}{0.035}$.

16. **a.** Write $\frac{8}{11}$ as a decimal.
b. Round the answer to two decimal places.

17. Write 84.8% as a decimal.

18. Write 0.69 as a percent.

19. Write 52% as a fraction.

20. Write $\frac{5}{9}$ as a percent.

Answers to Practice Test

ANSWER	IF YOU MISSED	REVIEW		
	QUESTION	SECTION	EXAMPLES	PAGE
1. $\frac{-18}{1}$	1	R.1	1	5
2. $\frac{9}{21}$	2	R.1	2	6
3. $\frac{3}{5}$	3	R.1	3	6
4. $\frac{1}{2}$	4	R.1	4	7
5. 8	5	R.2	1, 2, 3	12–13
6. 6	6	R.2	4, 5	14–15
7. $\frac{91}{60}$	7	R.2	6	17
8. $\frac{34}{15}$	8	R.2	7, 8, 9	19–21
9. $60 + 8 + \frac{4}{10} + \frac{2}{100} + \frac{8}{1000}$	9	R.3	1	25
10. $\frac{9}{200}$	10	R.3	2	25
11. $\frac{78}{25}$	11	R.3	3	26
12. 876.545	12	R.3	4	26
13. 419.98	13	R.3	5, 6	27
14. 0.0056095	14	R.3	7	28
15. 120	15	R.3	8	29
16. a. $0.\overline{72}$ **b.** 0.73	16	R.3	9, 15	30, 34–35
17. 0.848	17	R.3	10	31
18. 69%	18	R.3	11	31
19. $\frac{13}{25}$	19	R.3	12	32
20. $55\frac{5}{9}\%$	20	R.3	13, 14	33–34

1 Real Numbers and Their Properties

The Human Side of Algebra

The digits 1–9 originated with the Hindus and were passed on to us by the Arabs. Mohammed ibn Musa al-Khowarizmi (ca. A.D. 780–850) wrote two books, one on algebra (*Hisak al-jabr w'almuqabala,* "the science of equations"), from which the name *algebra* was derived, and one dealing with the Hindu numeration system. The oldest dated European manuscript containing the Hindu-Arabic numerals is the *Codex Vigilanus* written in Spain in A.D. 976, which used the nine symbols

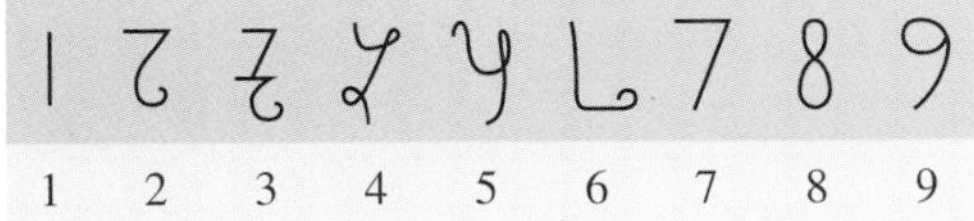

Zero, on the other hand, has a history of its own. According to scholars, "At the time of the birth of Christ, the idea of zero as a symbol or number had never occurred to anyone." So, who invented zero? An unknown Hindu who wrote a symbol of his own, a dot he called *sunya,* to indicate a column with no beads on his counting board. The Hindu notation reached Europe thanks to the Arabs, who called it *sifr*. About A.D. 150, the Alexandrian astronomer Ptolemy began using *o* (omicron), the first letter in the Greek word for *nothing,* in the manner of our zero.

Pretest for Chapter 1

(Answers on page 43)

1. Write in symbols:

a. The sum of m and n

b. $7m$ plus $3n$ minus 4

c. 4 times m

d. $\frac{1}{9}$ of m

2. Write in symbols:

a. The quotient of m and 3

b. The quotient of 3 and n

c. The sum of m and n, divided by the difference of m and n

3. For $m = 6$ and $n = 3$, evaluate:

a. $m + n$

b. $m - n$

c. $2m - 3n$

d. $\frac{2m + n}{n}$

4. Find the additive inverse (opposite) of:

a. -8

b. $\frac{2}{5}$

c. $0.666\ldots$

5. Find:

a. $\left|-\frac{1}{3}\right|$

b. $|12|$

c. $-|0.82|$

6. Consider the set $\{-9, \frac{5}{3}, \sqrt{2}, 0, 8.4, 0.333\ldots, -3\frac{1}{2}, 8, 0.123\ldots\}$. List the numbers in the set that are:

a. Natural numbers

b. Whole numbers

c. Integers

d. Rational numbers

e. Irrational numbers

f. Real numbers

7. Find:

a. $8 + (-7)$

b. $(-0.9) + (-0.7)$

c. $-9.3 - (-3.3)$

d. $\frac{1}{6} - \frac{7}{4}$

e. $10 - (-15) + 13 - 15 - 7$

8. Find:

a. $-6 \cdot 3$ **b.** $-\frac{4}{7}\left(-\frac{3}{5}\right)$

c. $(-8)^2$ **d.** -7^2

e. $-14 \div (-2)$ **f.** $-\frac{3}{5} \div \left(-\frac{7}{9}\right)$

9. Find the value of:

a. $56 \div 8 - (4 - 9)$

b. $36 \div 2^2 + 5$

c. $30 \div 5 + \{3 \cdot 4 - [2 + (8 - 10)]\}$

d. $-8^2 + \frac{4(3 - 9)}{2} + 12 \div (-4)$

10. The handicap H of a bowler with average A is $H = 0.80(200 - A)$. What is the handicap of a bowler whose average is 160?

11. Name the property illustrated in the statement.

$$6 \cdot (8 \cdot 7) = 6 \cdot (7 \cdot 8)$$

12. Simplify:

a. $9 - 4x - 12$

b. $-7 + 6x + 10 - 13x$

13. Fill in the blank so that the result is a true statement.

a. $8.2 + \_\_ = 0$ **b.** $\_\_ \cdot \frac{3}{2} = 1$

c. $\_\_ + \frac{8}{7} = \frac{8}{7}$ **d.** $-8(x + \_\_) = -8x + (-16)$

14. Use the distributive property to multiply.

a. $-3(a - 4)$ **b.** $-(x - 9)$

15. Combine like terms.

a. $-8x + (-9x)$ **b.** $-9ab^2 - (-3ab^2)$

16. Remove parentheses and combine like terms.

a. $(6a - 6) - (8a + 2)$

b. $8x - 3(x - 2) - (x + 2)$

17. Write in symbols and indicate the number of terms to the right of the equals sign:

The temperature F (in degrees Fahrenheit) can be found by finding the sum of 37 and one-quarter the number of chirps C a cricket makes in 1 min.

Answers to Pretest

ANSWER	IF YOU MISSED	REVIEW		
	QUESTION	SECTION	EXAMPLES	PAGE
1. a. $m + n$ **b.** $7m + 3n - 4$ **c.** $4m$ **d.** $\frac{1}{9}m$	1	1.1	1, 2	46
2. a. $\frac{m}{3}$ **b.** $\frac{3}{n}$ **c.** $\frac{m+n}{m-n}$	2	1.1	3	47
3. a. 9 **b.** 3 **c.** 3 **d.** 5	3	1.1	4	47
4. a. 8 **b.** $-\frac{2}{5}$ **c.** $-0.666\ldots$	4	1.2	1, 2	52–53
5. a. $\frac{1}{3}$ **b.** 12 **c.** -0.82	5	1.2	3, 4	54
6. a. 8 **b.** 0, 8 **c.** $-9, 0, 8$ **d.** $-9, \frac{5}{3}, 0, 8.4, 0.333\ldots, -3\frac{1}{2}, 8$ **e.** $\sqrt{2}, 0.123\ldots$ **f.** All	6	1.2	5	56
7. a. 1 **b.** -1.6 **c.** -6 **d.** $-\frac{19}{12}$ **e.** 16	7	1.3	1–8	62–66
8. a. -18 **b.** $\frac{12}{35}$ **c.** 64 **d.** -49 **e.** 7 **f.** $\frac{27}{35}$	8	1.4	1, 2, 3, 4, 5, 6	71–75
9. a. 12 **b.** 14 **c.** 18 **d.** -79	9	1.5	1, 2, 3, 4, 5	81–84
10. 32	10	1.5	6	84
11. Commutative property of multiplication	11	1.6	1, 2	89–90
12. a. $-4x - 3$ **b.** $3 - 7x$	12	1.6	3	90
13. a. -8.2 **b.** $\frac{2}{3}$ **c.** 0 **d.** 2	13	1.6	4, 5, 7, 8	91–94
14. a. $-3a + 12$ **b.** $-x + 9$	14	1.6	7, 8	93–94
15. a. $-17x$ **b.** $-6ab^2$	15	1.7	1, 2	100–101
16. a. $-2a - 8$ **b.** $4x + 4$	16	1.7	3, 4, 5	101–102
17. $F = 37 + \frac{1}{4}C$; two terms	17	1.7	6, 8	103–105

1.1 INTRODUCTION TO ALGEBRA

To Succeed, Review How To . . .

1. Add, subtract, multiply, and divide numbers (see Chapter R).

Objectives

A Translate words into algebraic expressions.

B Evaluate algebraic expressions.

GETTING STARTED

When getting Z's, keep your mouth shut!

The poster uses the language of algebra to tell you how to be successful. The letters X, Y, and Z are used as *placeholders, unknowns,* or *variables.* The letters t, u, v, w, x, y, and z are frequently used as unknowns, and the letter x is used most often (x, y, and z are used in algebra because they are seldom used in ordinary words).

Source: C. Brown.

R-I-S-E to Success in Math

Why are some students more successful in math than others? Oftentimes it is because they know how to manage their time and have a plan for action. Use models similar to the tables shown here to make a weekly schedule of your time (classes, study, work, personal, etc.)

Weekly Time Schedule

Time	S	M	T	W	R	F	S
8:00							
9:00							
10:00							
11:00							
12:00							
1:00							
2:00							
3:00							
4:00							
5:00							
6:00							
7:00							
8:00							
9:00							
10:00							
11:00							

Semester Calendar

Wk	M	T	W	R	F
1					
2					
3					
4					
5					
6					
7					
8					
9					
10					
11					
12					
13					
14					
15					
16					

and a semester calendar indicating major course events such as tests, papers, etc. Then, try to do as many of the suggestions on the "**R-I-S-E**" list as possible.

R—Read and/or view the material before and after each class. This includes the textbook, the videos that come with the book, and any special material given to you by your instructor.

I—Interact and/or practice using the tutorial software that comes with the book and Web exercises suggested in the sections, or seeking tutoring from your school.

S—Study and/or discuss your homework and class notes with a study partner/group, with your instructor, or on a discussion board if available.

E—Evaluate your progress by checking the odd homework questions with the answer key in the back of the book, using the Mastery Test questions in each section of the book as a self-test, and using the Chapter Reviews and Chapter Practice Tests as practice before taking the actual test.

Translating into Algebraic Expressions

In arithmetic, we express ideas using **arithmetic expressions.** How do we express our ideas in algebra? We use **algebraic expressions,** which contain the variables x, y, z, and so on, of course. Here is a comparison.

Arithmetic	Algebra
$9 + 51$	$9 + x$
$4.3 - 2$	$4.3 - y$
7×8.4	$7 \times y$ or $7y$
$\frac{3}{4}$	$\frac{a}{b}$

In algebra, it is better to write $7y$ instead of $7 \times y$ because the multiplication sign $\times$ can be easily confused with the letter x. Besides, look at the confusion that would result if we wrote x multiplied by x as $x \times x$! (You probably know that $x \times x$ is written as x^2.)

There are many words that indicate an addition or a subtraction. We list some of these here for your convenience.

Add (+)	Subtract (−)
Plus, sum, increase, more than	**Minus, difference, decrease, less than**
$a + b$ (read "a plus b") means:	$a - b$ (read "a minus b") means:
1. The sum of a and b	**1.** The difference of a and b
2. a increased by b	**2.** a decreased by b
3. b more than a	**3.** b less than a
4. b added to a	**4.** b subtracted from a

With this in mind, try the next example.

EXAMPLE 1 **Translations involving addition and subtraction**

Write in symbols.

a. The sum of x and y **b.** x minus y **c.** $7x$ plus $2a$ minus 3

SOLUTION

a. $x + y$ **b.** $x - y$ **c.** $7x + 2a - 3$

PROBLEM 1

Write in symbols.

a. The sum of p and q

b. q minus p

c. $3q$ plus $5y$ minus 2

How do we write multiplication problems in algebra? We use the raised dot (·) or parentheses, ( ). Here are some ways of writing the product of a and b.

A raised dot:	$a \cdot b$
Parentheses:	$a(b)$, $(a)b$, or $(a)(b)$
Juxtaposition (writing a and b next to each other):	ab

In each of these cases, a and b (the things to be multiplied) are called **factors.** Of course, the last notation must not be used when multiplying specific numbers because then "5 times 8" would be written as "58," which looks like fifty-eight. Here are some words that indicate a multiplication.

Multiply (× or ·)
times, of, product

We will use them in Example 2.

EXAMPLE 2 **Writing products using juxtaposition**

Write the indicated products using juxtaposition.

a. 8 times x **b.** $-x$ times y times z

c. 4 times x times x **d.** $\frac{1}{5}$ of x

e. The product of 3 and x

SOLUTION

a. 8 times x is written as $8x$. **b.** $-x$ times y times z is written as $-xyz$.

c. 4 times x times x is written as $4xx$. **d.** $\frac{1}{5}x$

e. $3x$

PROBLEM 2

Write using juxtaposition.

a. -3 times x

b. a times b times c

c. 5 times a times a

d. $\frac{1}{2}$ of a

e. The product of 5 and a

What about division? In arithmetic we use the division (÷) sign to indicate division. In algebra we usually use fractions to indicate division. Thus, in arithmetic we write $15 \div 3$ (or $3\overline{)15}$) to indicate the quotient of 15 and 3. However, in algebra usually we write

$$\frac{15}{3}$$

Similarly, "the quotient of x and y" is written as

$$\frac{x}{y}$$

Answers

1. a. $p + q$ **b.** $q - p$ **c.** $3q + 5y - 2$
2. a. $-3x$ **b.** abc **c.** $5aa$ **d.** $\frac{1}{2}a$ **e.** $5a$

[We avoid writing $\frac{x}{y}$ as x/y because more complicated expressions such as $\frac{x}{y+z}$ then need to be written as $x/(y + z)$.]

Here are some words that indicate a division.

Divide (÷ or fraction bar —)
Divided by, quotient

We will use them in Example 3.

EXAMPLE 3 **Translations involving division**

Write in symbols.

a. The quotient of x and 7

b. The quotient of 7 and x

c. The quotient of $(x + y)$ and z

d. The sum of a and b, divided by the difference of a and b

SOLUTION

a. $\frac{x}{7}$ **b.** $\frac{7}{x}$ **c.** $\frac{x+y}{z}$ **d.** $\frac{a+b}{a-b}$

PROBLEM 3

Write in symbols.

a. The quotient of a and b

b. The quotient of b and a

c. The quotient of $(x - y)$ and z

d. The difference of x and y, divided by the sum of x and y

B Evaluating Algebraic Expressions

As we have seen, algebraic expressions contain variables, operation signs, and numbers. If we substitute a value for one or more of the variables, we say that we are **evaluating the expression.** We shall see how this works in Example 4.

EXAMPLE 4 **Evaluating algebraic expressions**

Evaluate the given expressions by substituting 10 for x and 5 for y.

a. $x + y$ **b.** $x - y$

c. $4y$ **d.** $\frac{x}{y}$

e. $3x - 2y$

SOLUTION

a. Substitute 10 for x and 5 for y in $x + y$.

We obtain: $x + y = 10 + 5 = 15$.

The number 15 is called the **value** of $x + y$.

b. $x - y = 10 - 5 = 5$ **c.** $4y = 4(5) = 20$

d. $\frac{x}{y} = \frac{10}{5} = 2$

e. $3x - 2y = 3(10) - 2(5)$
$= 30 - 10$
$= 20$

PROBLEM 4

Evaluate the expressions by substituting 22 for a and 3 for b.

a. $a + b$ **b.** $2a - b$

c. $5b$ **d.** $\frac{2a}{b}$

e. $2a - 3b$

Answers

3. a. $\frac{a}{b}$ **b.** $\frac{b}{a}$ **c.** $\frac{x-y}{z}$ **d.** $\frac{x-y}{x+y}$ **4. a.** 25 **b.** 41 **c.** 15 **d.** $\frac{44}{3}$ **e.** 35

Web It

To review how to translate and evaluate expressions, go to link 1-1-1 at the Bello Website at mhhe.com/bello.

You can also try link 1-1-2.

The terminology of this section and evaluating expressions are extremely important concepts in everyday life. Examine the federal income tax form shown in the figure. Can you find the words *subtract* and *multiply?* Suppose we use the variables A, S, and E to represent the adjusted gross income, the standard deduction, and the total *number* of exemptions, respectively. What expression will represent the taxable income, and how can we evaluate it? See Example 5!

Form 1040A (2003) Page **2**

Tax, credits, and payments	**22**	Enter the amount from line 21 (adjusted gross income).	22	A
	23a	Check if: ☐ **You** were born before January 2, 1939, ☐ Blind; ☐ **Spouse** was born before January 2, 1939, ☐ Blind. **Total boxes checked ▶** 23a ☐		
Standard Deduction for—	**b**	If you are married filing separately and your spouse itemizes deductions, see page 32 and check here ▶ 23b ☐		
	24	Enter your **standard deduction** (see left margin).	24	S
• People who checked any box on line 23a or 23b **or** who can be	**25**	Subtract line 24 from line 22. If line 24 is more than line 22, enter -0-.	25	
	26	Multiply $3,050 by the total number of exemptions claimed on line 6d.	26	E
	27	Subtract line 26 from line 25. If line 26 is more than line 25, enter -0-. This is your **taxable income.** ▶	27	

Source: Internal Revenue Service.

EXAMPLE 5 Application: Evaluating expressions

Look at part of Form 1040A, where A represents the adjusted gross income; S, the standard deduction; and E, the total *number* of exemptions.

a. What algebraic expression will go on line 25?

b. What algebraic expression will go on line 26?

c. What algebraic expression will go on line 27?

d. Suppose your adjusted gross income is $25,000, the standard deduction is $4750, and you have 6 exemptions. What will be the taxable income entered on line 27?

SOLUTION

a. Line 25 directs you to subtract line 24 (S) from line 22 (A). Thus, the expression $A - S$ will go on line 25.

b. Line 26 says to multiply $3050 by the total *number* of exemptions E. Thus, the entry on line 26 is $3050E$.

c. Line 27 says to subtract line 26, which is $3050E$, from line 25, which is $A - S$. The entry will be $A - S - 3050E$.

d. The taxable income is $A - S - 3050E$, where $A = \$25{,}000$, $S = \$4750$, and $E = 6$. Evaluating this expression, we obtain

$$\begin{aligned} A - S - 3050E &= 25{,}000 - 4750 - 3050(6) \\ &= 25{,}000 - 4750 - 18{,}300 \\ &= \$1950 \end{aligned}$$

PROBLEM 5

Suppose A, the adjusted gross income, was $20,000.

a. What algebraic expression would go on line 27?

b. Assuming that you still have 6 exemptions, what arithmetic expression would go on line 27? (Do not simplify.)

c. According to the instructions, what number should you enter on line 27?

Answers

5. a. $A - S - 3050E$
b. $20{,}000 - 4750 - 18{,}300$
c. 0

Exercises 1.1

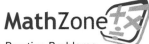

A In Problems 1–40, write each expression in symbols.

1. The sum of a and c $a + c$

2. The sum of u and v $u + v$

3. The sum of $3x$ and y $3x + y$

4. The sum of 8 and x $8 + x$

5. $9x$ plus $17y$ $9x + 17y$

6. $5a$ plus $2b$ $5a + 2b$

7. The difference of $3a$ and $2b$ $3a - 2b$

8. The difference of $6x$ and $3y$ $6x - 3y$

9. $-2x$ less 5 $-2x - 5$

10. $-7y$ less $3x$ $-7y - 3x$

11. 7 times a $7a$

12. -9 times y $-9y$

13. $\frac{1}{7}$ of a $\frac{1}{7}a$

14. $\frac{1}{9}$ of y $\frac{1}{9}y$

15. The product of b and d bd

16. The product of 4 and c $4c$

17. xy multiplied by z xyz

18. $-a$ multiplied by bc $-abc$

19. $-b$ times $(c + d)$ $-b(c + d)$

20. $(p + q)$ multiplied by r $(p + q)r$

21. $(a - b)$ times x $(a - b)x$

22. $(a + d)$ times $(x - y)$ $(a + d)(x - y)$

23. The product of $(x - 3y)$ and $(x + 7y)$ $(x - 3y)(x + 7y)$

24. The product of $(a - 2b)$ and $(2a - 3b)$ $(a - 2b)(2a - 3b)$

25. $(c - 4d)$ times $(x + y)$ $(c - 4d)(x + y)$

26. x divided by $2y$ $\frac{x}{2y}$

27. y divided by $3x$ $\frac{y}{3x}$

28. The quotient of $2a$ and b $\frac{2a}{b}$

29. The quotient $2b$ and a $\frac{2b}{a}$

30. The quotient of $2b$ and ac $\frac{2b}{ac}$

31. The quotient of a and the sum of x and y $\frac{a}{x + y}$

32. The quotient of $(a + b)$ and c $\frac{a + b}{c}$

33. The quotient of the difference of a and b, and c $\frac{a - b}{c}$

34. The sum of a and b, divided by the difference of x and y $\frac{a + b}{x - y}$

35. The quotient when x is divided into y $\frac{y}{x}$

36. The quotient when y is divided into x $\frac{x}{y}$

37. The quotient when the sum of p and q is divided into the difference of p and q $\frac{p - q}{p + q}$

38. The quotient when the difference of $3x$ and y is divided into the sum of x and $3y$ $\frac{x + 3y}{3x - y}$

39. The quotient obtained when the sum of x and $2y$ is divided by the difference of x and $2y$ $\frac{x + 2y}{x - 2y}$

40. The quotient obtained when the difference of x and $3y$ is divided by the sum of x and $3y$ $\frac{x - 3y}{x + 3y}$

B In Problems 41–60, evaluate the expression for the given values.

41. The sum of a and c for $a = 7$ and $c = 9$ 16

42. The sum of u and v for $u = 15$ and $v = 23$ 38

43. $9x$ plus $17y$ for $x = 3$ and $y = 2$ 61

44. $5a$ plus $2b$ for $a = 5$ and $b = 2$ 29

45. The difference of $3a$ and $2b$ for $a = 5$ and $b = 3$ 9

46. The difference of $6x$ and $3y$ for $x = 3$ and $y = 6$ 0

47. $2x$ less 5 for $x = 4$ 3

48. $7y$ less $3x$ for $x = 3$ and $y = 7$ 40

49. 7 times ab for $a = 2$ and $b = 4$ 56

50. 9 times yz for $y = 2$ and $z = 4$ 72

51. The product of b and d for $b = 3$ and $d = 2$ 6

52. The product of 4 and c for $c = 5$ 20

53. xy multiplied by z for $x = 10$, $y = 5$, and $z = 1$ 50

54. a multiplied by bc for $a = 5$, $b = 7$, and $c = 3$ 105

55. The quotient of a and the sum of x and y for $a = 3$, $x = 1$, and $y = 2$ 1

56. The quotient of a plus b divided by c for $a = 10$, $b = 2$, and $c = 3$ 4

57. The quotient when x is divided into y for $x = 2$ and $y = 8$ 4

58. The quotient when $2y$ is divided into x for $x = 3$ and $y = 5$ $\frac{3}{10}$

59. The quotient when x is divided by y for $x = 2$ and $y = 8$ $\frac{1}{4}$

60. The quotient when $2y$ is divided by x for $x = 3$ and $y = 5$ $\frac{10}{3}$

USING YOUR KNOWLEDGE

The words we have studied are used in many different fields. Perhaps you have seen some of the following material in your classes! Use the knowledge gained in this section to write it in symbols. The word in italics indicates the field from which the material is taken.

61. *Electricity* The voltage V across any part of a circuit is the product of the current I and the resistance R. $V = IR$

62. *Economics* The total profit TP equals the total revenue TR minus the total cost TC. $TP = TR - TC$

63. *Chemistry* The total pressure P in a container filled with gases A, B, and C is equal to the sum of the partial pressures P_A, P_B, and P_C. $P = P_A + P_B + P_C$

64. *Psychology* The intelligence quotient (IQ) for a child is obtained by multiplying his or her mental age M by 100 and dividing the result by his or her chronological age C. $\text{IQ} = \frac{100M}{C}$

65. *Physics* The distance D traveled by an object moving at a constant rate R is the product of R and the time T. $D = RT$

66. *Astronomy* The square of the period P of a planet's orbit equals the product of a constant C and the cube of the planet's distance R from the sun. $P^2 = CR^3$

67. *Physics* The energy E of an object equals the product of its mass m and the square of the speed of light c. $E = mc^2$

68. *Engineering* The depth h of a gear tooth is found by dividing the difference between the major diameter D and the minor diameter d by 2. $h = \frac{D - d}{2}$

69. *Geometry* The square of the hypotenuse c of a right triangle equals the sum of the squares of the sides a and b. $c^2 = a^2 + b^2$

70. *Auto mechanics* The horsepower (hp) of an engine is obtained by multiplying 0.4 by the square of the diameter D of each cylinder and by the number N of cylinders. $\text{hp} = 0.4D^2N$

WRITE ON

71. In the expression "$\frac{1}{2}$ of x," what operation does the word *of* signify? Multiplication

72. Most people believe that the word *and* always means addition.
   a. In the expression "the sum of x and y," does "and" signify the operation of addition? Explain. Yes; Answers may vary.
   b. In the expression "the product of 2 and three more than a number," does "and" signify the operation of addition? Explain. No; Answers may vary.

73. Explain the difference between "x divided by y" and "x divided into y." Answers may vary.

74. Explain the difference between "a less than b" and "a less b." "a less than b" is $b - a$, while "a less b" is $a - b$.

MASTERY TEST

If you know how to do these problems, you have learned your lesson!

In Problems 75–79, write in symbols.

75. The product of 3 and xy $3xy$

76. The difference of $2x$ and y $2x - y$

77. The quotient of $3x$ and $2y$ $\frac{3x}{2y}$

78. The sum of $7x$ and $4y$ $7x + 4y$

79. The difference of b and c divided by the sum of b and c $\frac{b - c}{b + c}$

80. Evaluate the expression $2x + y - z$ for $x = 3$, $y = 4$, and $z = 5$. 5

81. Evaluate the expression $\frac{p - q}{3}$ for $p = 9$ and $q = 3$. 2

82. Evaluate the expression $\frac{2x - 3y}{x + y}$ for $x = 10$ and $y = 5$. $\frac{1}{3}$

1.2 THE REAL NUMBERS

To Succeed, Review How To...

1. Recognize a rational number (see Chapter R).

Objectives

A Find the additive inverse (opposite) of a number.

B Find the absolute value of a number.

C Classify numbers as natural, whole, integer, rational, or irrational.

D Solve applications using real numbers.

GETTING STARTED Temperatures and Integers

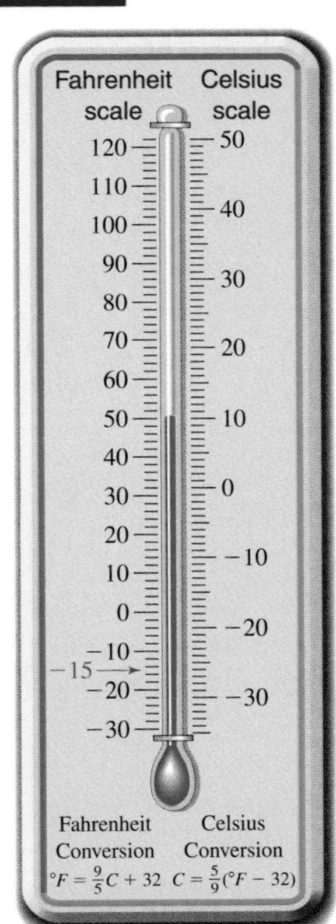

To study algebra we need to know about numbers. In this section we examine *sets* of numbers that are related to each other and learn how to classify them. To visualize these sets more easily and to study some of their properties, we represent (graph) them on a number line. For example, the thermometer shown uses *integers* (not fractions) to measure temperature; the integers are . . . , $-3, -2, -1, 0, 1, 2, 3, \ldots$. You will find that you use integers every day. For example, when you earn \$20, you *have* 20 dollars; we write this as +20 dollars. When you spend \$5, you *no longer* have it; we write this as −5 dollars. The number 20 is a *positive integer,* and the number −5 (read "negative 5") is a *negative integer.* Here are some other quantities that can be represented by positive and negative integers.

A loss of \$25	−25	A \$25 gain	25
10 ft below sea level	−10	10 ft above sea level	10
15° below zero	−15	15° above zero	15

These quantities are examples of *real numbers.* There are other types of real numbers; we shall learn about them later in this section.

A Finding Additive Inverses (Opposites)

The temperature 15 degrees below zero is highlighted on the Fahrenheit scale of the thermometer in the *Getting Started.* If we take the scale on this thermometer and turn it sideways so that the positive numbers are on the right, the resulting scale is called a **number line** (see Figure 1). Clearly, on a number line the positive integers are to the right of 0, the negative integers are to the left of 0, and 0 itself is called the **origin.**

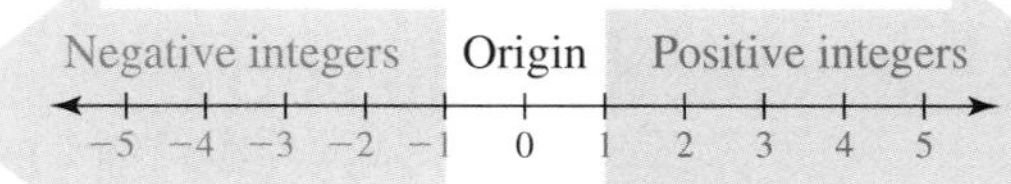

Figure 1

Note that the number line in Figure 1 is drawn a little over 5 units long on each side; the arrows at each end indicate that the line could be drawn to any desired length. Moreover, for every *positive* integer, there is a corresponding *negative* integer. Thus for the positive integer 4, we have the negative integer -4. Since 4 and -4 are the same distance from the origin but in opposite directions, 4 and -4 are called **opposites.** Moreover, since $4 + (-4) = 0$, we call 4 and -4 **additive inverses.** Similarly, the additive inverse (opposite) of -3 is 3, and the additive inverse (opposite) of 2 is -2. Note that $-3 + 3 = 0$ and $2 + (-2) = 0$. In general, we have

$$a + (-a) = (-a) + a = 0 \quad \text{for any integer } a$$

Figure 2 shows the relation between the negative and the positive integers.

Teaching Tip

Students should note three uses for the minus sign:

1. Subtraction; $7 - 3$
2. Negative number; -4
3. Additive inverse; $-(-2)$

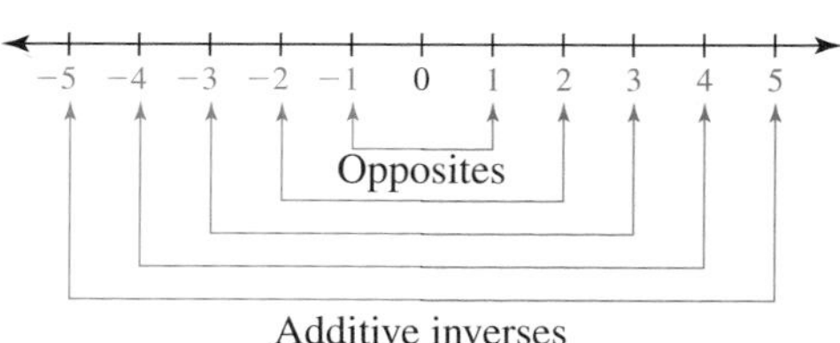

Figure 2

ADDITIVE INVERSE The **additive inverse** (opposite) of any number a is $-a$.

You can read $-a$ as "the opposite of a" or "the additive inverse of a." Note that a and $-a$ are additive inverses of each other. Thus 10 and -10 are additive inverses of each other, and -7 and 7 are additive inverses of each other.

EXAMPLE 1 Finding additive inverses of integers

Find the additive inverse (opposite) of:

a. 5 **b.** -4 **c.** 0

SOLUTION

a. The additive inverse of 5 is -5 (see Figure 3).

b. The additive inverse of -4 is $-(-4) = 4$ (see Figure 3).

c. The additive inverse of 0 is 0.

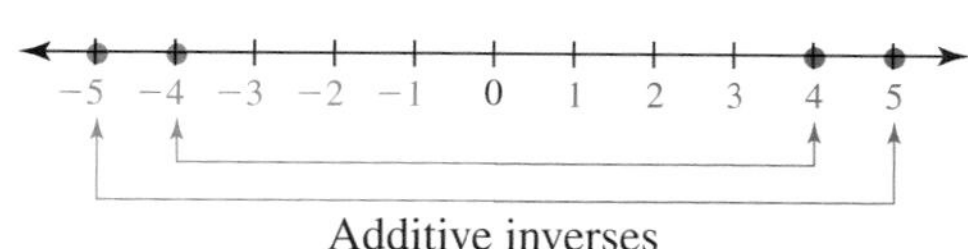

Figure 3

PROBLEM 1

Find the additive inverse of:

a. -8 **b.** 9 **c.** -3

Answers

1. a. 8 **b.** -9 **c.** 3

Note that $-(-4) = 4$ and $-(-8) = 8$. In general, we have

$$-(-a) = a \quad \text{for any number } a$$

As with the integers, every rational number—that is, every fraction written as the ratio of two integers and every decimal—has an *additive inverse* (*opposite*). Here are some rational numbers and their additive inverses.

Rational Number	Additive Inverse (Opposite)
$\frac{9}{2}$	$-\frac{9}{2}$
$-\frac{3}{4}$	$-\left(-\frac{3}{4}\right) = \frac{3}{4}$
2.9	-2.9
-1.8	$-(-1.8) = 1.8$

EXAMPLE 2 **Finding additive inverses of fractions and decimals**

Find the additive inverse (opposite) of:

a. $\frac{5}{2}$ **b.** -4.8 **c.** $-3\frac{1}{3}$ **d.** 1.2

SOLUTION

a. $-\frac{5}{2}$

b. $-(-4.8) = 4.8$

c. $-\left(-3\frac{1}{3}\right) = 3\frac{1}{3}$

d. -1.2

These rational numbers and their inverses are graphed on the number line in Figure 4. Note that to locate $\frac{5}{2}$, it is easier to first write $\frac{5}{2}$ as the mixed number $2\frac{1}{2}$.

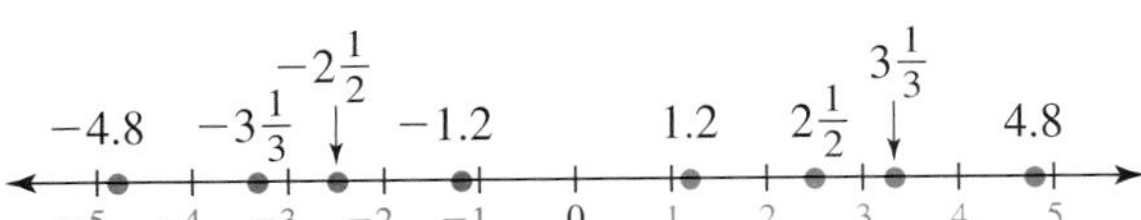

Figure 4

PROBLEM 2

Find the additive inverse of:

a. $\frac{3}{11}$ **b.** -7.4

c. $-9\frac{8}{13}$ **d.** 3.4

B Finding the Absolute Value of a Number

Let's look at the number line again. What is the distance between 3 and 0? The answer is 3 units. What about the distance between -3 and 0? The answer is *still* 3 units.

The distance between any number n and 0 is called the *absolute value* of the number and is denoted by $|n|$. Thus $|-3| = 3$ and $|3| = 3$. (See Figure 5.)

3 units 3 units

-4 -3 -2 -1 0 1 2 3 4

Figure 5

ABSOLUTE VALUE

The **absolute value** of a number n is its distance from 0 and is denoted by $|n|$.

Answers

2. a. $-\frac{3}{11}$ **b.** 7.4 **c.** $9\frac{8}{13}$ **d.** -3.4

You can think of the absolute value of a number as the number of units it represents disregarding its sign. For example, suppose Pedro and Tamika leave class together. Pedro walks 2 miles east while Tamika walks 2 miles west. Who walked farther? They walked the same distance! They are both 2 miles from the starting point.

> **NOTE**
>
> Since the absolute value of a number represents a distance and a distance is **never** negative, the absolute value $|a|$ of a nonzero number a is *always* positive. Because of this, $-|a|$ is *always* negative. Thus $-|5| = -5$ and $-|-3| = -3$.

EXAMPLE 3 Finding absolute values of integers

Find:

a. $|-8|$ **b.** $|7|$ **c.** $|0|$ **d.** $-|-3|$

SOLUTION

a. $|-8| = 8$ −8 is 8 units from 0.

b. $|7| = 7$ 7 is 7 units from 0.

c. $|0| = 0$ 0 is 0 units from 0.

d. $-|-3| = -3$ −3 is 3 units from 0.

PROBLEM 3

Find:

a. $|-5|$ **b.** $|10|$

c. $|2|$ **d.** $-|-5|$

Every fraction and decimal also has an absolute value, which is its distance from zero. Thus $|-\frac{1}{2}| = \frac{1}{2}$, $|3.8| = 3.8$, and $|-1\frac{1}{7}| = 1\frac{1}{7}$, as shown in Figure 6.

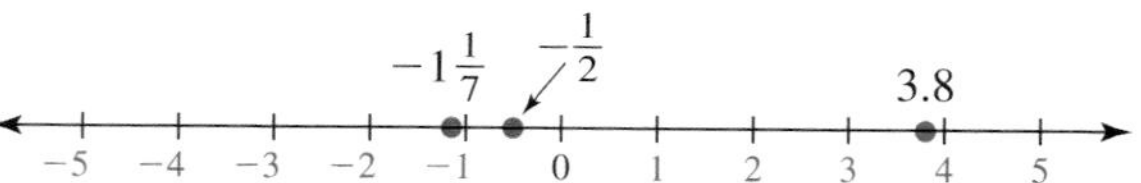

Figure 6

EXAMPLE 4 Finding absolute values of rational numbers

Find:

a. $\left|-\frac{3}{7}\right|$ **b.** $|2.1|$ **c.** $\left|-2\frac{1}{2}\right|$

d. $|-4.1|$ **e.** $-\left|\frac{1}{4}\right|$

SOLUTION

a. $\left|-\frac{3}{7}\right| = \frac{3}{7}$ **b.** $|2.1| = 2.1$ **c.** $\left|-2\frac{1}{2}\right| = 2\frac{1}{2}$

d. $|-4.1| = 4.1$ **e.** $-\left|\frac{1}{4}\right| = -\frac{1}{4}$

PROBLEM 4

Find:

a. $\left|-\frac{5}{7}\right|$ **b.** $|3.4|$ **c.** $\left|-3\frac{1}{8}\right|$

d. $|-3.8|$ **e.** $-\left|-\frac{1}{5}\right|$

Answers

3. a. 5 **b.** 10 **c.** 2 **d.** −5
4. a. $\frac{5}{7}$ **b.** 3.4 **c.** $3\frac{1}{8}$ **d.** 3.8 **e.** $-\frac{1}{5}$

C Classifying Numbers

The real-number line is a picture (graph) used to represent the *set* of real numbers. A **set** is a collection of objects called the **members** or **elements** of the set. If the elements can be listed, a pair of braces { } encloses the list with individual elements separated by commas. Here are some sets of numbers contained in the real-number line:

The set of **natural** numbers $\{1, 2, 3, \ldots\}$

The set of **whole** numbers $\{0, 1, 2, 3, \ldots\}$

The set of **integers** $\{\ldots, -2, -1, 0, 1, 2, 3, \ldots\}$

The three dots (an ellipsis) at the end or beginning of a list of elements indicates that the list continues indefinitely in the same manner.

As you can see, every natural number is a whole number and every whole number is an integer. In turn, every integer is a *rational number,* a number that can be written in the form $\frac{a}{b}$, where a and b are integers and b is not zero. Thus, an integer n can always be written as the rational number $\frac{n}{1} = n$. (The word *rational* comes from the word *ratio,* which indicates a quotient.) Since the fraction $\frac{a}{b}$ can be written as a decimal by dividing the numerator a by the denominator b, to obtain either a terminating (as in $\frac{3}{4} = 0.75$) or a repeating (as in $\frac{1}{3} = 0.\overline{3}$) decimal, all terminating or repeating decimals are also rational numbers. The set of rational numbers is described next in words, since it's impossible to make a list containing all the rational numbers.

RATIONAL NUMBERS

The set of **rational numbers** consists of all numbers that can be written as quotients $\frac{a}{b}$, where a and b are integers and b is not 0.

The set of *irrational numbers* is the set of all real numbers that are not rational.

IRRATIONAL NUMBERS

The set of **irrational numbers** consists of all real numbers that *cannot* be written as the quotient of two integers.

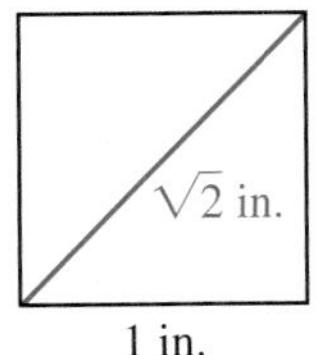

Figure 7

For example, if you draw a square 1 inch on a side, the length of its diagonal is $\sqrt{2}$ (read "the square root of 2"), as shown in Figure 7. This number cannot be written as a quotient of integers. It is irrational. Since a rational number $\frac{a}{b}$ can be written as a fraction that terminates or repeats, the decimal form of an irrational number never terminates and never repeats. Here are some irrational numbers:

$$\sqrt{2}, \quad -\sqrt{50}, \quad 0.123\ldots, \quad -5.1223334444\ldots, \quad 8.101001000\ldots, \quad \text{and} \quad \pi$$

All the numbers we have mentioned are *real numbers,* and their relationship is shown in Figure 8. Of course, you may also represent (graph) these numbers on the number line. Note that a number may belong to more than one category. For example, -15 is an integer, a rational number, and a real number.

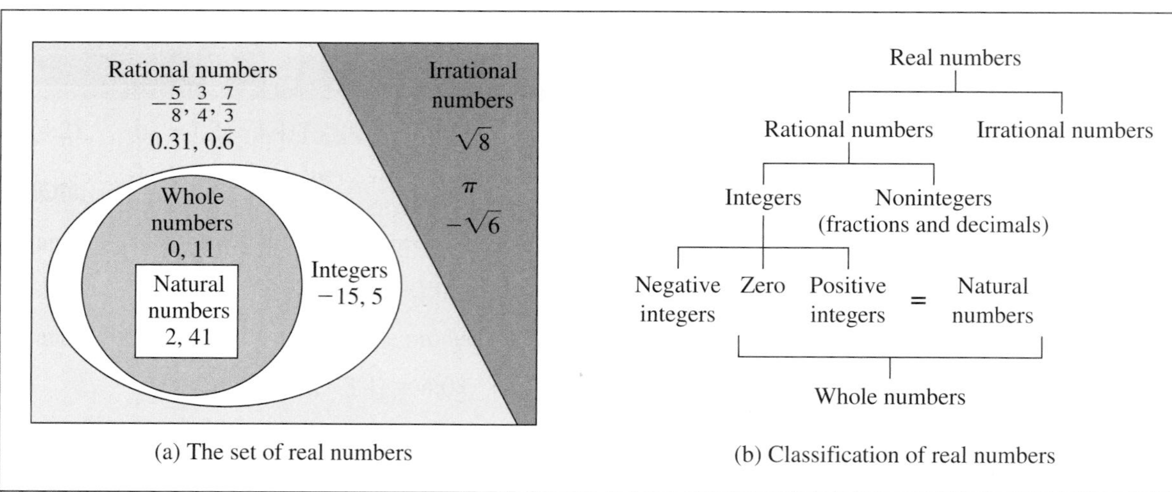

(a) The set of real numbers (b) Classification of real numbers

Figure 8

EXAMPLE 5 **Classifying numbers**

Classify as whole, integer, rational, irrational, or real:

a. -3 **b.** 0 **c.** $\sqrt{5}$

d. 0.3 **e.** 0.101001000. . . **f.** 0.101001000

SOLUTION

a. -3 is an integer, a rational number, and a real number.

b. 0 is a whole number, an integer, a rational number, and a real number.

c. $\sqrt{5}$ is an irrational number and a real number.

d. 0.3 is a terminating decimal, so it is a rational number and a real number.

e. 0.101001000. . . never terminates and never repeats, so it is an irrational number and a real number.

f. 0.101001000 terminates, so it is a rational number and a real number.

PROBLEM 5

Classify as whole, integer, rational, irrational, or real:

a. -9 **b.** 200

c. $\sqrt{7}$ **d.** 0.9

e. 0.010010001. . . **f.** 0.010010001

D Solving Applications

EXAMPLE 6 **Using real numbers**

Use real numbers to write the quantities in the following applications:

a. The price of a fund *went down* $\$\frac{3}{4}$.

b. A record low of 128.6 degrees Fahrenheit *below* zero was recorded at Vostok, Antarctica, on July 21, 1983.

c. The oldest mathematical puzzle is contained in the Rhind Papyrus and dates back to 1650 B.C.

SOLUTION

a. $-\frac{3}{4}$ **b.** -128.6°F **c.** -1650

PROBLEM 6

Use real numbers to write the quantities in the applications:

a. The price of a fund *went up* $\$\frac{1}{2}$.

b. A record low of 80 degrees Fahrenheit *below* zero was recorded at Prospect Creek, Alaska, on Jan. 23, 1971.

c. The Moscow Papyrus was copied by an unknown scribe around 1850 B.C.

EXAMPLE 7 Using + or − to indicate height or depth

The following table lists the altitude (the distance above or below sea level or zero altitude) of various locations around the world.

Location	Altitude (feet)
Mount Everest	+29,029
Mount McKinley	+20,320
Sea level	0
Caspian Sea	−92
Marianas Trench (deepest descent)	−35,813

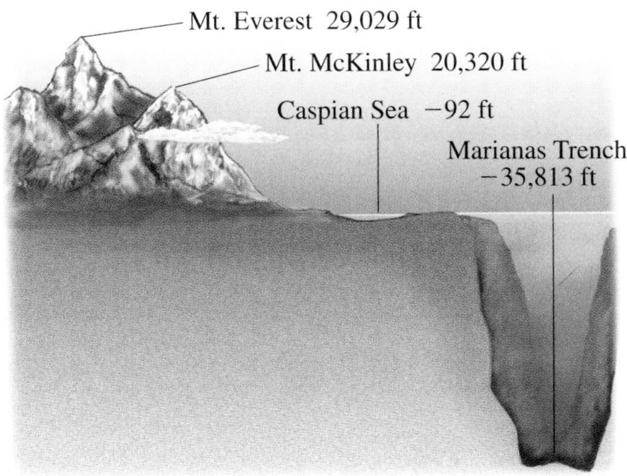

Use this information to answer the following questions:

a. How far above sea level is Mount Everest?

b. How far below sea level is the Caspian Sea?

c. How far below sea level was the deepest descent made?

SOLUTION

a. 29,029 feet above sea level

b. 92 feet below sea level

c. 35,813 feet below sea level

PROBLEM 7

a. Refer to the table. How far above sea level is Mount McKinley?

b. The Dead Sea, Israel-Jordan, is located at altitude −1349 feet. How far below sea level is that?

c. Some scientists claim that the altitude of the Marianas Trench is really −36,198 feet. How far below sea level is that?

Web It

To learn more about absolute values, go to link 1-2-1 at the Bello Website at mhhe.com/bello.

To practice with absolute values try link 1-2-2.

Calculate It Additive Inverse and Absolute Value

The additive inverse and absolute value of a number are so important that graphing calculators have special keys to handle them. To find the additive inverse, press (−). Don't confuse the additive inverse key with the minus sign key. (Operation signs usually have color keys; the (−) key is gray.)

To find absolute values with a TI-83 Plus calculator, you have to do some math, so press MATH. Next, you have to deal with a special type of number, absolute value, so press ▸ to highlight the NUM menu at the top of the screen. Next press 1, which tells the calculator you want an absolute value; finally, enter the number whose absolute value you want, and close the parentheses. The display window shows how to calculate the additive inverse of −5, the absolute value of 7, and the absolute value of −4. Check your results with your calculator when you work the exercise sets.

```
--5
                5
abs (7)
                7
abs (-4)
                4
```

Answers

5. a. Integer, rational, real **b.** Whole number, integer, rational, real **c.** Irrational, real **d.** Rational, real **e.** Irrational, real **f.** Rational, real
6. a. $+\frac{1}{2}$ **b.** −80°F **c.** −1850 **7. a.** 20,320 feet above sea level **b.** 1349 feet below sea level **c.** 36,198 feet below sea level

Exercises 1.2

A In Problems 1–18, find the additive inverse (opposite) of the given number.

1. 4 -4

2. 11 -11

3. -49 49

4. -56 56

5. $\frac{7}{3}$ $-\frac{7}{3}$

6. $-\frac{8}{9}$ $\frac{8}{9}$

7. -6.4 6.4

8. -2.3 2.3

9. $3\frac{1}{7}$ $-3\frac{1}{7}$

10. $-4\frac{1}{8}$ $4\frac{1}{8}$

11. 0.34 -0.34

12. 0.85 -0.85

13. $-0.\overline{5}$ $0.\overline{5}$

14. $-3.\overline{7}$ $3.\overline{7}$

15. $\sqrt{7}$ $-\sqrt{7}$

16. $-\sqrt{17}$ $\sqrt{17}$

17. π $-\pi$

18. $\frac{\pi}{2}$ $-\frac{\pi}{2}$

B In Problems 19–38, find:

19. $|-2|$ 2

20. $|-6|$ 6

21. $|48|$ 48

22. $|78|$ 78

23. $|-(-3)|$ 3

24. $|-(-17)|$ 17

25. $\left|-\frac{4}{5}\right|$ $\frac{4}{5}$

26. $\left|-\frac{9}{2}\right|$ $\frac{9}{2}$

27. $|-3.4|$ 3.4

28. $|-2.1|$ 2.1

29. $\left|-1\frac{1}{2}\right|$ $1\frac{1}{2}$

30. $\left|-3\frac{1}{4}\right|$ $3\frac{1}{4}$

31. $-\left|-\frac{3}{4}\right|$ $-\frac{3}{4}$

32. $-\left|-\frac{1}{5}\right|$ $-\frac{1}{5}$

33. $-|-0.\overline{5}|$ $-0.\overline{5}$

34. $-|-3.\overline{7}|$ $-3.\overline{7}$

35. $-|-\sqrt{3}|$ $-\sqrt{3}$

36. $-|-\sqrt{6}|$ $-\sqrt{6}$

37. $-|-\pi|$ $-\pi$

38. $-\left|-\frac{\pi}{2}\right|$ $-\frac{\pi}{2}$

C In Problems 39–54, classify the given numbers. (See Figure 8; some numbers belong in more than one category.)

39. 17 Natural, whole, integer, rational, real

40. -8 Integer, rational, real

41. $-\frac{4}{5}$ Rational, real

42. $-\frac{7}{8}$ Rational, real

43. 0 Whole, integer, rational, real

44. 0.37 Rational, real

45. 3.76 Rational, real

46. $3.\overline{8}$ Rational, real

47. $17.\overline{28}$ Rational, real

48. $\sqrt{10}$ Irrational, real

49. $-\sqrt{3}$ Irrational, real

50. $0.777\ldots$ Rational, real

51. $-0.888\ldots$ Rational, real

52. $0.202002000\ldots$ Irrational, real

53. 0.202002000 Rational, real

54. $\frac{\pi}{2}$ Irrational, real

In Problems 55–62, consider the set $\{-5, \frac{1}{5}, 0, 8, \sqrt{11}, 0.\overline{1}, 2.505005000\ldots, 3.666\ldots\}$. List the numbers in the set that are

55. Natural numbers 8

56. Whole numbers 0, 8

57. Positive integers 8

58. Negative integers -5

59. Nonnegative integers 0, 8

60. Irrational $\sqrt{11}$, 2.505005000. . .

61. Rational numbers $-5, \frac{1}{5}, 0, 8, 0.\overline{1}, 3.666\ldots$

62. Real numbers All are real.

In Problems 63–74, determine whether the statement is true or false. If false, give an example that shows it is false.

63. The opposite of any positive number is negative. True

64. The opposite of any negative number is positive. True

65. The absolute value of any real number is positive. False; $|0| = 0$

66. The negative of the absolute value of a number is equal to the absolute value of its negative. False; $-|4| \neq |-4|$

67. The absolute value of a number is equal to the absolute value of its opposite. True

68. Every integer is a rational number. True

69. Every rational number is an integer. False; $\frac{3}{5}$ is not an integer.

70. Every terminating decimal is rational. True

71. Every nonterminating decimal is rational. False; 0.12345. . . is not rational.

72. Every nonterminating and nonrepeating decimal is irrational. True

73. A decimal that never repeats and never terminates is a real number. True

74. The decimal representation of a real number never terminates and never repeats. False; 0.12 is real and terminates.

APPLICATIONS

In Problems 75–80, use real numbers to write the given quantities.

75. A 20-yard *gain* in a football play $+20$ yards

76. A 10-yard *loss* in a football play -10 yards

77. The Dead Sea is 1312 feet *below* sea level -1312 feet from sea level

78. Mount Everest reaches a height of 29,029 feet *above* sea level $+29{,}029$ feet from sea level

79. On January 22, 1943, the temperature in Spearfish, South Dakota, rose from 4°F *below* zero to 45°F *above* zero in a period of 2 minutes. -4°F to $+45$°F

80. Every hour, there are 460 *births* and 250 *deaths* in the United States. $+460, -250$

SKILL CHECKER

Try the Skill Checker Exercises so you'll be ready for the next section.

Find:

81. $8.6 - 3.4$ 5.2

82. $2.3 + 4.1$ 6.4

83. $\frac{5}{8} - \frac{2}{5}$ $\frac{9}{40}$

84. $\frac{5}{6} + \frac{7}{4}$ $\frac{31}{12}$

85. $3.1(4.2)$ 13.02

86. $1.2(3.4)$ 4.08

87. $\frac{3}{4}\left(\frac{5}{2}\right)$ $\frac{15}{8}$

88. $\frac{5}{6}\left(\frac{3}{7}\right)$ $\frac{5}{14}$

USING YOUR KNOWLEDGE

Let's Talk about the Weather

According to *USA Today Weather Almanac,* the coldest city in the United States (based on its average annual temperature in degrees Fahrenheit) is International Falls, Minnesota (see Figure 9). Note that the lowest ever temperature there was −46°F.

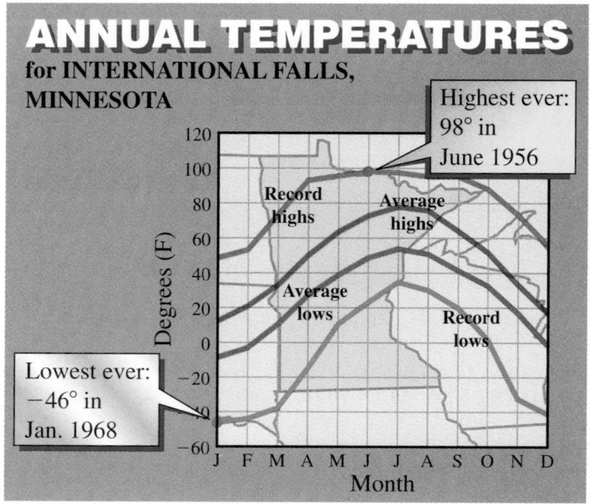

Figure 9
Annual temperatures for International Falls, Minnesota (°F)
Source: Data from USA Today Weather Almanac.

89. What was the next lowest ever temperature (February) in International Falls? −43°F

90. What was the highest record low? 35°F

91. What was the record low in December? −40°F

92. What was the lowest average low? −10°F

WRITE ON

93. What do we mean by
 a. the additive inverse of a number? Answers may vary.
 b. the absolute value of a number? Answers may vary.

94. Explain why every integer is a rational number. Every integer a can be written as $\frac{a}{1}$.

95. The rational numbers have been defined as the set of numbers that can be written in the form $\frac{a}{b}$, where a and b are integers and b is not 0. Define the rational numbers in terms of their decimal representation. All numbers with either terminating or repeating decimal representations.

96. Define the set of irrational numbers in terms of their decimal representation. All numbers whose decimal representation neither terminates nor repeats.

97. Write a paragraph explaining the set of real numbers as it relates to the natural numbers, the integers, the rational numbers, and the irrational numbers. Answers may vary.

MASTERY TEST

If you know how to do these problems, you have learned your lesson!

Find the additive inverse of:

98. $0.\overline{7}$ $-0.\overline{7}$

99. $-\sqrt{19}$ $\sqrt{19}$

100. $\frac{\pi}{3}$ $-\frac{\pi}{3}$

101. $-8\frac{1}{4}$ $8\frac{1}{4}$

Find:

102. $|\sqrt{23}|$ $\sqrt{23}$

103. $|-0.\overline{4}|$ $0.\overline{4}$

104. $-\left|\frac{3}{5}\right|$ $-\frac{3}{5}$

105. $-\left|-\frac{1}{2}\right|$ $-\frac{1}{2}$

Classify as a whole, integer, rational, irrational, or real number. (*Hint:* More than one category may apply.)

106. $\sqrt{21}$ Irrational, real

107. −2 Integer, rational, real

108. 0.010010001. . . Irrational, real

109. $4\frac{1}{3}$ Rational, real

110. 0.333. . . Rational, real

111. 0 Whole, integer, rational, real

112. 39 Natural, whole, integer, rational, real

1.3 ADDING AND SUBTRACTING REAL NUMBERS

To Succeed, Review How To . . .

1. Add and subtract fractions (pp. 15–21).
2. Add and subtract decimals (pp. 26–27).

Objectives

A Add two real numbers.

B Subtract one real number from another.

C Add and subtract several real numbers.

D Solve an application.

GETTING STARTED

Signed Numbers and Population Changes

Now that we know what real numbers are, we will use the real-number line to visualize the process used to add and subtract them. For example, what is the U.S. population change per hour? To find out, examine the graph and add the births (+456), the deaths (−273), and the new immigrants (+114). The result is

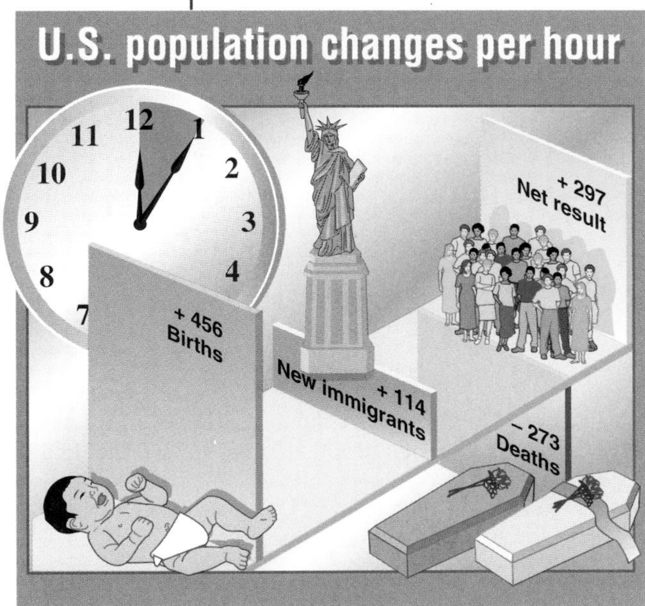

Source: Population Reference Bureau (2000, 2001, and 2002 data).

$$456 + (-273) + 114 = 570 + (-273)$$ Add 456 and 114.

$$= +297$$ Subtract 273 from 570.

This answer means that the U.S. population is increasing by 297 persons every hour! Note that

1. The addition of 456 and −273 is written as $456 + (-273)$, instead of the confusing $456 + -273$.
2. To add 570 and −273, we *subtracted* 273 from 570 because subtracting 273 from 570 is the *same* as adding 570 and −273. We will use this idea to define subtraction.

If you want to know what the population change is per day or per minute, you must learn how to multiply and divide real numbers, as we will do in Section 1.4.

A Adding Real Numbers

The number line we studied in Section 1.2 can help us add real numbers. Here's how we do it.

PROCEDURE

Adding on the Number Line

To add $a + b$ on the number line,

1. Start at zero and move to a (to the *right* if a is *positive,* to the *left* if a is *negative*).
2. A. If b is *positive,* move *right* b units.
   B. If b is *negative,* move *left* $|b|$ units.
   C. If b is zero, stay at a.

For example, the sum $2 + 4$, or $(+2) + (+4)$, is found by starting at zero, moving 2 units to the right, followed by 4 more units to the right. Thus $2 + 4 = 6$, as shown in Figure 10.

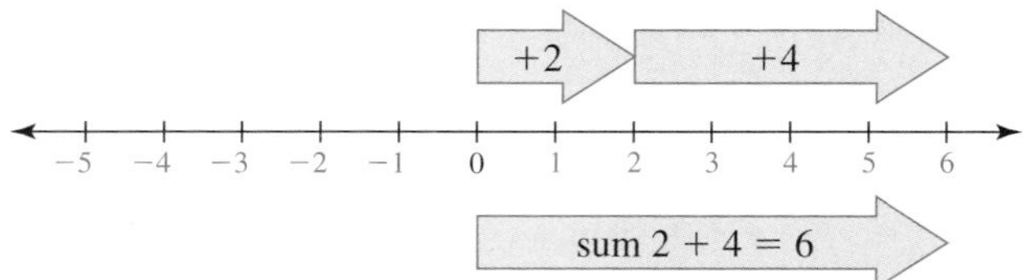

Figure 10

A number along with the sign (+ or −) indicating a direction on the number line is called a **signed number.**

EXAMPLE 1 Adding integers with different signs

Find: $5 + (-3)$

SOLUTION Start at zero. Move 5 units to the right and then 3 units to the left. The result is 2. Thus $5 + (-3) = 2$, as shown in Figure 11.

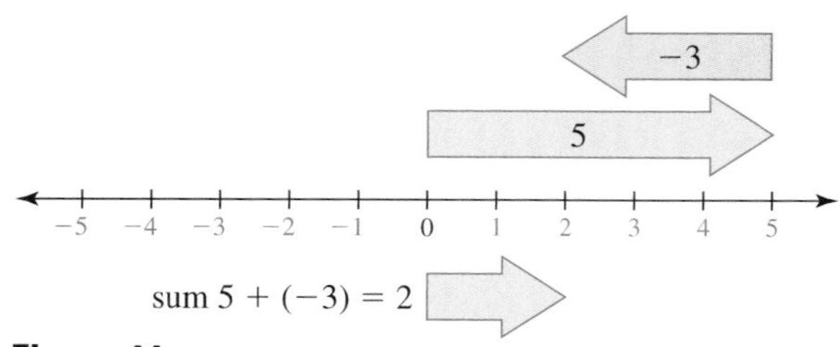

Figure 11

PROBLEM 1

Find: $4 + (-2)$

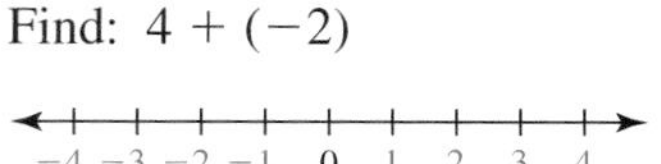

Calculate It

Adding Integers

To enter $-3 + (-2)$ using a scientific calculator, enter

3 [+/−] [+] 2 [+/−] [ENTER]

If your calculator has a set of parentheses, then you can enter parentheses around the −3 and −2.

This same procedure can be used to add two negative numbers. However, we have to be careful when writing such problems. For example, to add −3 and −2, we should write

$$-3 + (-2)$$

Why the parentheses? Because writing

$$-3 + -2$$

is confusing. *Never* write two signs together without parentheses.

EXAMPLE 2 Adding integers with the same sign

Find: $-3 + (-2)$

SOLUTION Start at zero. Move 3 units left and then 2 more units left. The result is 5 units left of zero; that is, $-3 + (-2) = -5$, as shown in Figure 12.

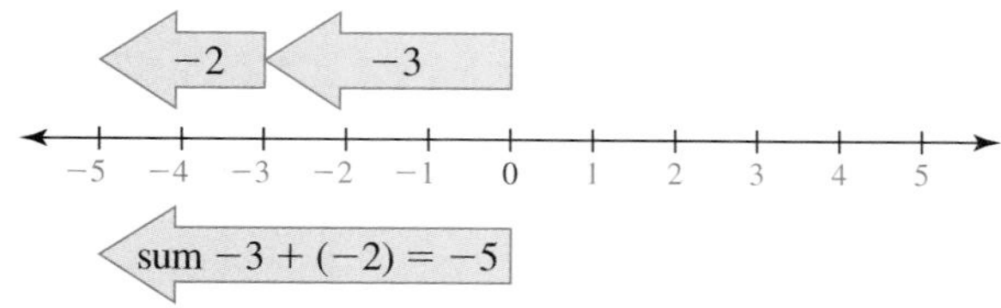

Figure 12

PROBLEM 2

Find: $-2 + (-1)$

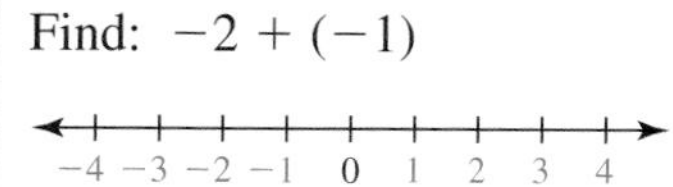

Answers

1. 2 **2.** −3

As you can see from Example 2, if we add numbers with the *same* sign (both + or both −), the result is a number with the same sign. Thus

$$2 + 4 = 6 \quad \text{and} \quad -3 + (-2) = -5$$

If we add numbers with *different* signs, the answer carries the sign of the number with the larger absolute value. Hence, in Example 1,

$5 + (-3) = 2$ The answer is positive because 5 has a larger absolute value than −3. (See Figure 11.)

but note that

$-5 + 3 = -2$ The answer is negative because −5 has a larger absolute value than 3.

The following rules summarize this discussion.

RULES

Adding Signed Numbers

1. With the **same (like)** sign: *Add* their absolute values and give the sum the *common* sign.
2. With **different (unlike)** signs: *Subtract* their absolute values and give the difference the sign of the number with the *larger* absolute value.

For example, to add $8 + 5$ or $-8 + (-5)$, we note that the 8 and 5, and −8 and −5 have the same signs. So we add their absolute values and give the sum the common sign. Thus

$$8 + 5 = 13 \quad \text{and} \quad -8 + (-5) = -13$$

To add $-8 + 5$, we first notice that the numbers have *different* signs. So we subtract their absolute values and give the difference the sign of the number with the larger absolute value. Thus

$$-8 + 5 = -(8 - 5) = -3$$

Use the sign of the number with the larger absolute value. Subtract the smaller number from the larger one.

Similarly,

$$8 + (-5) = +(8 - 5) = 3$$

Here we have used the sign of the number with the larger absolute value, 8, which is understood to be +.

EXAMPLE 3 Adding integers with different signs

Find:

a. $-14 + 6$ **b.** $14 + (-6)$

SOLUTION

a. $-14 + 6 = -(14 - 6) = -8$ **b.** $14 + (-6) = +(14 - 6) = 8$

PROBLEM 3

Find:

a. $-10 + 4$ **b.** $10 + (-4)$

Answers

3. a. -6 **b.** 6

The addition of rational numbers uses the same rules for signs, as we illustrate in Examples 4 and 5.

EXAMPLE 4 **Adding decimals with different signs**

Find:

a. $-8.6 + 3.4$ **b.** $6.7 + (-9.8)$

c. $-2.3 + (-4.1)$

SOLUTION

a. $-8.6 + 3.4 = -(8.6 - 3.4) = -5.2$

b. $6.7 + (-9.8) = -(9.8 - 6.7) = -3.1$

c. $-2.3 + (-4.1) = -(2.3 + 4.1) = -6.4$

PROBLEM 4

Find:

a. $-7.8 + 2.5$ **b.** $5.4 + (-7.8)$

c. $-3.4 + (-5.1)$

EXAMPLE 5 **Adding fractions with different signs**

Find:

a. $-\frac{3}{7} + \frac{5}{7}$ **b.** $\frac{2}{5} + \left(-\frac{5}{8}\right)$

SOLUTION

a. Note that $\left|\frac{5}{7}\right|$ is larger than $\left|-\frac{3}{7}\right|$; hence

$$-\frac{3}{7} + \frac{5}{7} = +\left(\frac{5}{7} - \frac{3}{7}\right) = \frac{2}{7}$$

b. As usual, we must first find the LCM of 5 and 8, which is 40. We then write

$$\frac{2}{5} = \frac{16}{40} \quad \text{and} \quad -\frac{5}{8} = -\frac{25}{40}$$

Thus

$$\frac{2}{5} + \left(-\frac{5}{8}\right) = \frac{16}{40} + \left(-\frac{25}{40}\right) = -\left(\frac{25}{40} - \frac{16}{40}\right) = -\frac{9}{40}$$

PROBLEM 5

Find:

a. $-\frac{4}{9} + \frac{5}{9}$ **b.** $\frac{2}{5} + \left(-\frac{3}{4}\right)$

B Subtracting Real Numbers

We are now ready to subtract signed numbers. Suppose you use *positive* integers to indicate money *earned* and *negative* integers to indicate money *spent* (expenditures). If you earn \$10 and then spend \$12, you owe \$2. Thus

$$\underbrace{10}_{\text{Earn \$10.}} \underbrace{- 12}_{\text{Spend \$12.}} = 10 + (-12) = \underbrace{-2}_{\text{Owe \$2.}}$$

Also,

$$-5 - 10 = -5 + (-10) = -15$$

To take away (subtract) earned money is the same as adding an expenditure.

because if you spend \$5 and then spend \$10 more, you now owe \$15. What about $-10 - (-3)$? We claim that

$$-10 - (-3) = -7$$

Answers

4. a. -5.3 **b.** -2.4 **c.** -8.5

5. a. $\frac{1}{9}$ **b.** $-\frac{7}{20}$

because if you spend \$10 and then subtract (take away) a \$3 expenditure (represented by −3), you save \$3; that is,

$$-10 - (-3) = -10 + 3 = -7$$

When you take away (subtract) a \$3 expenditure, you save (add) \$3.

In general, we have the following definition.

Teaching Tip

Remind students that "inverse" in this definition refers to additive inverse or opposite in sign.

SUBTRACTION

$$a - b = a + (-b)$$

To **subtract** a number b, add its inverse $(-b)$.

Thus

To subtract 8, add its inverse.

$$5 - 8 = 5 + (-8) = -3$$
$$7 - 3 = 7 + (-3) = 4$$
$$-4 - 2 = -4 + (-2) = -6$$
$$-6 - (-4) = -6 + 4 = -2$$

EXAMPLE 6 Subtracting integers

Find:

a. $17 - 6$ **b.** $-21 - 4$

c. $-11 - (-5)$ **d.** $-4 - (-6)$

SOLUTION

a. $17 - 6 = 17 + (-6) = 11$ **b.** $-21 - 4 = -21 + (-4) = -25$

c. $-11 - (-5) = -11 + 5 = -6$ **d.** $-4 - (-6) = -4 + 6 = 2$

PROBLEM 6

Find:

a. $15 - 8$ **b.** $-23 - 5$

c. $-12 - (-4)$ **d.** $-5 - (-7)$

EXAMPLE 7 Subtracting decimals or fractions

Find:

a. $-4.2 - (-3.1)$ **b.** $-2.5 - (-7.8)$

c. $\frac{2}{9} - \left(-\frac{4}{9}\right)$ **d.** $-\frac{5}{6} - \frac{7}{4}$

SOLUTION

a. $-4.2 - (-3.1) = -4.2 + 3.1 = -1.1$ Note that $-(-3.1) = 3.1$.

b. $-2.5 - (-7.8) = -2.5 + 7.8 = 5.3$ Note that $-(-7.8) = 7.8$.

c. $\frac{2}{9} - \left(-\frac{4}{9}\right) = \frac{2}{9} + \frac{4}{9} = \frac{6}{9} = \frac{2}{3}$ Note that $-(-\frac{4}{9}) = \frac{4}{9}$.

d. The LCD is 12. Now,

$$-\frac{5}{6} = -\frac{10}{12} \quad \text{and} \quad \frac{7}{4} = \frac{21}{12}$$

Thus

$$-\frac{5}{6} - \frac{7}{4} = -\frac{5}{6} + \left(-\frac{7}{4}\right) = -\frac{10}{12} + \left(-\frac{21}{12}\right) = -\frac{31}{12}$$

PROBLEM 7

Find:

a. $3.2 - (-2.1)$ **b.** $-3.4 - (-6.9)$

c. $\frac{2}{7} - \left(-\frac{3}{7}\right)$ **d.** $-\frac{5}{8} - \frac{9}{4}$

Answers

6. a. 7 **b.** −28 **c.** −8 **d.** 2 **7. a.** 5.3 **b.** 3.5 **c.** $\frac{5}{7}$ **d.** $-\frac{23}{8}$

C Adding and Subtracting Several Real Numbers

Suppose you wish to find $18 - (-10) + 12 - 10 - 17$. Using the fact that $a - b = a + (-b)$, we write

$$18 - (-10) + 12 - 10 - 17 = 18 + 10 + 12 + (-10) + (-17)$$

$10 + (-10) = 0$

$$= 18 + 12 + (-17)$$
$$= 30 + (-17)$$
$$= 13$$

Teaching Tip

When adding more than two terms, students should look for additive inverses like 10 and −10, which will always equal 0.

EXAMPLE 8 Adding and subtracting numbers

Find: $12 - (-13) + 10 - 25 - 13$

SOLUTION First, rewrite as an addition.

$$12 - (-13) + 10 - 25 - 13 = 12 + 13 + 10 + (-25) + (-13)$$

$13 + (-13) = 0$

$$= 12 + 10 + (-25)$$
$$= 22 + (-25)$$
$$= -3$$

PROBLEM 8

Find: $14 - (-15) + 10 - 23 - 15$

Teaching Tip

Draw a thermometer scale to help students see why "subtraction" turns into addition.

44°, 0°, −56°; 44, 56; Total fall = 44° + 56°

D Solving an Application

EXAMPLE 9 Finding temperature differences

The greatest temperature variation in a 24-hour period occurred in Browning, Montana, January 23 to 24, 1916. The temperature fell from 44°F to −56°F. How many degrees did the temperature fall?

SOLUTION We have to find the difference between 44 and −56; that is, we have to find $44 - (-56)$:

$$44 - (-56) = 44 + 56$$
$$= 100$$

Thus the temperature fell 100°F.

PROBLEM 9

The greatest temperature variation in a 12-hour period occurred in Fairfield, Montana. The temperature fell from 63°F at noon to −21°F at midnight. How many degrees did the temperature fall? (When did this happen? December 24, 1924. They certainly did have a cool Christmas!)

Web It

If you want to review operations with integers, go to link 1-3-1 at the Bello Website (mhhe.com/bello) and practice your skills.

For a lesson involving integers, go to link 1-3-2.

To practice adding integers, go to link 1-3-3.

Finally, to review the subtraction of integers, go to link 1-3-4.

Answers

8. 1 **9.** 84°F

Exercises 1.3

A In Problems 1–40, do the indicated operations (verify your answer using a number line).

1. $3 + 3$ 6

2. $2 + 1$ 3

3. $-5 + 1$ -4

4. $-4 + 3$ -1

5. $6 + (-5)$ 1

6. $5 + (-1)$ 4

7. $-2 + (-5)$ -7

8. $-3 + (-3)$ -6

9. $3 + (-3)$ 0

10. $-4 + 4$ 0

11. $-18 + 21$ 3

12. $-3 + 5$ 2

13. $19 + (-6)$ 13

14. $8 + (-1)$ 7

15. $-9 + 11$ 2

16. $-8 + 13$ 5

17. $-18 + 9$ -9

18. $-17 + 4$ -13

19. $-17 + (+5)$ -12

20. $-4 + (+8)$ 4

21. $-3.8 + 6.9$ 3.1

22. $-4.5 + 7.8$ 3.3

23. $-7.8 + (3.1)$ -4.7

24. $-6.7 + (2.5)$ -4.2

25. $3.2 + (-8.6)$ -5.4

26. $4.1 + (-7.9)$ -3.8

27. $-3.4 + (-5.2)$ -8.6

28. $-7.1 + (-2.6)$ -9.7

29. $-\frac{2}{7} + \frac{5}{7}$ $\frac{3}{7}$

30. $-\frac{5}{11} + \frac{7}{11}$ $\frac{2}{11}$

31. $-\frac{3}{4} + \frac{1}{4}$ $-\frac{1}{2}$

32. $-\frac{5}{6} + \frac{1}{6}$ $-\frac{2}{3}$

33. $\frac{3}{4} + \left(-\frac{5}{6}\right)$ $-\frac{1}{12}$

34. $\frac{5}{6} + \left(-\frac{7}{8}\right)$ $-\frac{1}{24}$

35. $-\frac{1}{6} + \frac{3}{4}$ $\frac{7}{12}$

36. $-\frac{1}{8} + \frac{7}{6}$ $\frac{25}{24}$

37. $-\frac{1}{3} + \left(-\frac{2}{7}\right)$ $-\frac{13}{21}$

38. $-\frac{4}{7} + \left(-\frac{3}{8}\right)$ $-\frac{53}{56}$

39. $-\frac{5}{6} + \left(-\frac{8}{9}\right)$ $-\frac{31}{18}$

40. $-\frac{4}{5} + \left(-\frac{7}{8}\right)$ $-\frac{67}{40}$

B In Problems 41–60, do the indicated operations.

41. $-5 - 11$ -16

42. $-4 - 7$ -11

43. $-4 - 16$ -20

44. $-9 - 11$ -20

45. $7 - 13$ -6

46. $8 - 12$ -4

47. $9 - (-7)$ 16

48. $8 - (-4)$ 12

49. $0 - 4$ -4

50. $0 - (-4)$ 4

51. $-3.8 - (-1.2)$ -2.6

52. $-6.7 - (-4.3)$ -2.4

53. $-3.5 - (-8.7)$ 5.2

54. $-6.5 - (-9.9)$ 3.4

55. $4.5 - 8.2$ -3.7

56. $3.7 - 7.9$ -4.2

57. $\frac{3}{7} - \left(-\frac{1}{7}\right)$ $\frac{4}{7}$

58. $\frac{5}{6} - \left(-\frac{1}{6}\right)$ 1

59. $-\frac{5}{4} - \frac{7}{6}$ $-\frac{29}{12}$

60. $-\frac{2}{3} - \frac{3}{4}$ $-\frac{17}{12}$

C In Problems 61–66, do the indicated operations.

61. $8 - (-10) + 5 - 20 - 10$ -7

62. $15 - (-9) + 8 - 2 - 9$ 21

63. $-15 + 12 - 8 - (-15) + 5$ 9

64. $-12 + 14 - 7 - (-12) + 3$ 10

65. $-10 + 9 - 14 - 3 - (-14)$ -4

66. $-7 + 2 - 6 - 8 - (-6)$ -13

APPLICATIONS

67. *Earth temperatures* The temperature in the center core of the earth reaches +5000°C. In the thermosphere (a region in the upper atmosphere), the temperature is +1500°C. Find the difference in temperature between the center of the earth and the thermosphere. 3500°C

68. *Extreme temperatures* The record high temperature in Calgary, Alberta, is +99°F. The record low temperature is −46°F. Find the difference between these extremes. 145°F

69. *Stock price fluctuations* The price of a certain stock at the beginning of the week was $47. Here are the changes in price during the week: +1, +2, −1, −2, −1. What is the price of the stock at the end of the week? $46

70. *Stock price fluctuations* The price of a stock on Friday was $37. On the following Monday, the price went up $2; on Tuesday, it went down $3; and on Wednesday, it went down another $1. What was the price of the stock then? $35

71. *Temperature variations* Here are the temperature changes (in degrees Celsius) by the hour in a certain city:

1 P.M.	+2
2 P.M.	+1
3 P.M.	−1
4 P.M.	−3

If the temperature was initially 15°C, what was it at 4 P.M.? 14°C

SKILL CHECKER

Try the Skill Checker Exercises so you'll be ready for the next section.

Find:

72. $5\frac{1}{4} \cdot 3\frac{1}{8}$ $\frac{525}{32}$ or $16\frac{13}{32}$

73. $6\frac{1}{6} \cdot 5\frac{7}{10}$ $\frac{703}{20}$ or $35\frac{3}{20}$

74. $6\frac{1}{4} \cdot \frac{20}{21}$ $\frac{125}{21}$ or $5\frac{20}{21}$

75. $\frac{18}{11} \div 5\frac{1}{2}$ $\frac{36}{121}$

76. $2\frac{1}{4} \div 1\frac{3}{8}$ $\frac{18}{11}$ or $1\frac{7}{11}$

USING YOUR KNOWLEDGE

A Little History

The following chart contains some important historical dates.

Important Historical Dates

323 B.C.	Alexander the Great died
216 B.C.	Hannibal defeated the Romans
A.D. 476	Fall of the Roman Empire
A.D. 1492	Columbus landed in America
A.D. 1776	The Declaration of Independence signed
A.D. 1939	World War II started
A.D. 1988	Reagan–Gorbachev summit

We can use negative integers to represent years B.C. For example, the year Alexander the Great died can be written as −323, whereas the fall of the Roman Empire occurred in +476 (or simply 476). To find the number of years that elapsed between the fall of the Roman Empire and their defeat by Hannibal, we write

$$476 - (-216) = 476 + 216 = 692$$

↑ Fall of the Roman Empire (A.D. 476)
↑ Hannibal defeats the Romans (216 B.C.)
↑ Years elapsed

Use these ideas to find the number of years elapsed between the following:

77. The fall of the Roman Empire and the death of Alexander the Great 799 years

78. Columbus's landing in America and Hannibal's defeat of the Romans 1708 years

79. The discovery of America and the signing of the Declaration of Independence 284 years

80. The year of the Reagan–Gorbachev summit and the signing of the Declaration of Independence 212 years

81. The start of World War II and the death of Alexander the Great 2262 years

WRITE ON

82. Explain what the term *signed numbers* means and give examples. Answers may vary.

83. State the rule you use to add signed numbers. Explain why the sum is sometimes positive and sometimes negative. Answers may vary.

84. State the rule you use to subtract signed numbers. How do you know whether the answer is going to be positive? How do you know whether the answer is going to be negative? Answers may vary.

85. The definition of subtraction is as follows: To subtract a number, add its inverse. Use a number line to explain why this works. Answers may vary.

MASTERY TEST

If you know how to do these problems, you have learned your lesson!

Find:

86. $7 - 13$ -6

87. $3.8 - 6.9$ -3.1

88. $\frac{1}{5} - \frac{3}{4}$ $-\frac{11}{20}$

89. $-3.5 - 4.2$ -7.7

90. $-\frac{2}{3} - \frac{4}{5}$ $-\frac{22}{15}$

91. $-5 - 15$ -20

92. $-6 - (-4)$ -2

93. $-3.4 - (-4.6)$ 1.2

94. $-\frac{3}{4} - \left(-\frac{1}{5}\right)$ $-\frac{11}{20}$

95. $3.9 + (-4.2)$ -0.3

96. $\frac{3}{4} + \left(-\frac{1}{6}\right)$ $\frac{7}{12}$

97. $-3.2 + (-2.5)$ -5.7

98. $7 - (-11) + 13 - 11 - 15$ 5

99. $\frac{3}{4} - \left(-\frac{2}{5}\right) + \frac{4}{5} - \frac{2}{5} + \frac{5}{4}$ $\frac{14}{5}$

100. The temperature in Verkhoyansk, Siberia, ranges from 98°F to −94°F. What is the difference between these temperatures? 192°F

1.4 MULTIPLYING AND DIVIDING REAL NUMBERS

To Succeed, Review How To . . .

1. Multiply and divide whole numbers, decimals, and fractions (pp. 12–15, 27–29).
2. Find the reciprocal of a number (p. 14).

Objectives

A Multiply two real numbers.

B Evaluate expressions involving exponents.

C Divide one real number by another.

D Solve an application.

GETTING STARTED

A Stock Market Loss

In Section 1.3, we learned how to add and subtract real numbers. Now we will learn how to multiply and divide them. Pay particular attention to the notation used when multiplying a number by itself and also to the different applications of the multiplication and division of real numbers. For example, suppose you own 4 shares of Eastman Kodak (East Kodak), and the closing price today is *down* \$3 (written as -3). Your loss then is

$$4 \cdot (-3) \quad \text{or} \quad 4(-3)$$

How do we multiply positive and negative integers? As you may recall, the result of a multiplication is a *product,* and the numbers being multiplied (4 and -3) are called *factors.* Now you can think of multiplication as *repeated addition.* Thus

$$4 \cdot (-3) = \underbrace{(-3) + (-3) + (-3) + (-3)}_{\text{four } (-3)\text{'s}} = -12$$

Also note that

$$(-3) \cdot 4 = -12$$

So your stock has gone down \$12. As you can see, the product of a *negative* integer and a *positive* integer is negative. What about the product of two negative integers, say $-4 \cdot (-3)$? Look for the pattern in the following table:

The number in this column decreases by 1. ↓ The number in this column increases by 3. ↓

$$\begin{aligned} 4 \cdot (-3) &= -12 \\ 3 \cdot (-3) &= -9 \\ 2 \cdot (-3) &= -6 \\ 1 \cdot (-3) &= -3 \\ 0 \cdot (-3) &= 0 \\ -1 \cdot (-3) &= 3 \\ -2 \cdot (-3) &= 6 \\ -3 \cdot (-3) &= 9 \\ -4 \cdot (-3) &= 12 \end{aligned}$$

You can think of $-4 \cdot (-3)$ as subtracting -3 four times; that is,

$$-(-3) - (-3) - (-3) - (-3) = 3 + 3 + 3 + 3 = 12$$

So, when we multiply two integers with *different* (*unlike*) signs, the product is *negative.* If we multiply two integers with the *same* (*like*) signs, the product is *positive.* This idea can be generalized to include the product of any two real numbers, as you will see later.

Multiplying Real Numbers

Here are the rules that we used in the *Getting Started.*

RULES

Multiplying Signed Numbers

1. When two numbers with the *same (like)* sign are multiplied, the product is *positive* (+).
2. When two numbers with *different (unlike)* signs are multiplied, the product is *negative* (−).

Here are some examples.

$9 \cdot 4 = 36$, $-9 \cdot (-4) = 36$ — 9 and 4 have the same sign (+); −9 and −4 have the same sign (−); thus the product is positive.

$-9 \cdot 6 = -54$, $9 \cdot (-6) = -54$ — −9 and 6 have different signs; 9 and −6 have different signs; thus the product is negative.

EXAMPLE 1 **Finding products of integers**

Find:

a. $7 \cdot 8$ **b.** $-8 \cdot 6$

c. $4 \cdot (-3)$ **d.** $-7 \cdot (-9)$

SOLUTION

a. $7 \cdot 8 = 56$

b. $-8 \cdot 6 = -48$

Different signs Negative product

c. $4 \cdot (-3) = -12$

Different signs Negative product

d. $-7 \cdot (-9) = 63$

Same sign Positive product

PROBLEM 1

Find:

a. $9 \cdot 6$ **b.** $-7 \cdot 8$

c. $5 \cdot (-4)$ **d.** $-5 \cdot (-6)$

Answers

1. a. 54 **b.** −56 **c.** −20 **d.** 30

The multiplication of rational numbers also uses the same rules for signs, as we illustrate in Example 2. Remember that $-3.1(4.2)$ means $-3.1 \cdot 4.2$. Parentheses are one of the ways we indicate multiplication.

EXAMPLE 2 **Finding products of decimals and fractions**

Find:

a. $-3.1(4.2)$ **b.** $-1.2(-3.4)$ **c.** $-\frac{3}{4}\left(-\frac{5}{2}\right)$ **d.** $\frac{5}{6}\left(-\frac{4}{7}\right)$

SOLUTION

a. -3.1 and 4.2 have different signs. The product is *negative.* Thus

$$-3.1(4.2) = -13.02$$

b. -1.2 and -3.4 have the same sign. The product is *positive.* Thus

$$-1.2(-3.4) = 4.08$$

c. $-\frac{3}{4}$ and $-\frac{5}{2}$ have the same sign. The product is *positive.* Thus

$$-\frac{3}{4}\left(-\frac{5}{2}\right) = \frac{15}{8}$$

d. $\frac{5}{6}$ and $-\frac{4}{7}$ have different signs. The product is *negative.* Thus

$$\frac{5}{6}\left(-\frac{4}{7}\right) = -\frac{20}{42} = -\frac{10}{21}$$

PROBLEM 2

Find:

a. $-4.1 \cdot (3.2)$ **b.** $-1.3(-4.2)$

c. $-\frac{3}{5}\left(-\frac{7}{10}\right)$ **d.** $\frac{5}{9}\left(-\frac{6}{7}\right)$

B Evaluating Expressions Involving Exponents

Sometimes a number is used several times as a factor. For example, we may wish to find the products

$$3 \cdot 3 \quad \text{or} \quad 4 \cdot 4 \cdot 4 \quad \text{or} \quad 5 \cdot 5 \cdot 5 \cdot 5$$

In the expression $3 \cdot 3$, the 3 is used as a factor twice. In such cases it's easier to use **exponents** to indicate how many times the number is used as a factor. We then write

3^2 (read "3 squared") instead of $3 \cdot 3$

4^3 (read "4 cubed") instead of $4 \cdot 4 \cdot 4$

5^4 (read "5 to the fourth") instead of $5 \cdot 5 \cdot 5 \cdot 5$

The expression 3^2 uses the exponent 2 to indicate how many times the **base** 3 is used as a factor. Similarly, in the expression 5^4, the 5 is the base and the 4 is the exponent. Now,

$$3^2 = \underbrace{3 \cdot 3} = 9$$ 3 is used as a factor 2 times.

$$4^3 = \underbrace{4 \cdot 4 \cdot 4} = 64$$ 4 is used as a factor 3 times.

and

$$\left(\frac{1}{5}\right)^4 = \underbrace{\frac{1}{5} \cdot \frac{1}{5} \cdot \frac{1}{5} \cdot \frac{1}{5}} = \frac{1}{625}$$ $\frac{1}{5}$ is used as a factor 4 times.

What about $(-2)^2$? Using the definition of exponents, we have

$$(-2)^2 = (-2) \cdot (-2) = 4$$ -2 and -2 have the same sign; thus their product is positive.

Moreover, -2^2 means $-(2 \cdot 2)$. To emphasize that the multiplication is to be done *first,* we put parentheses around the $2 \cdot 2$. Thus the placing of the parentheses in the expression $(-2)^2$ is very important. Clearly, since $(-2)^2 = 4$ and $-2^2 = -(2 \cdot 2) = -4$,

$$(-2)^2 \neq -2^2$$ "$\neq$" means "is not equal to."

Answers

2. a. -13.12 **b.** 5.46 **c.** $\frac{21}{50}$ **d.** $-\frac{10}{21}$

EXAMPLE 3 **Evaluating expressions involving exponents**

Find:

a. $(-4)^2$ **b.** -4^2 **c.** $\left(-\frac{1}{3}\right)^2$ **d.** $-\left(\frac{1}{3}\right)^2$

SOLUTION

a. $(-4)^2 = (-4)(-4) = 16$ Note that the base is -4.

b. $-4^2 = -(4 \cdot 4) = -16$ Here the base is 4.

c. $\left(-\frac{1}{3}\right)^2 = \left(-\frac{1}{3}\right)\left(-\frac{1}{3}\right) = \frac{1}{9}$ The base is $-\frac{1}{3}$.

d. $-\left(\frac{1}{3}\right)^2 = -\left(\frac{1}{3}\right)\left(\frac{1}{3}\right) = -\frac{1}{9}$ The base is $\frac{1}{3}$.

PROBLEM 3

Find:

a. $(-7)^2$ **b.** -7^2

c. $\left(-\frac{1}{4}\right)^2$ **d.** $-\left(\frac{1}{4}\right)^2$

Teaching Tip

Teach students to *read* $(-4)^2$ as "negative four squared" and -4^2 as "the opposite (inverse) of four squared."

EXAMPLE 4 **Evaluating expressions involving exponents**

Find:

a. $(-2)^3$ **b.** -2^3

SOLUTION

a. $(-2)^3 = \underbrace{(-2)\cdot(-2)}\cdot(-2) = 4 \cdot(-2) = -8$

b. $-2^3 = -\underbrace{(2\cdot 2\cdot 2)} = -(8) = -8$

PROBLEM 4

Find:

a. $(-4)^3$ **b.** -4^3

Note that

$(-4)^2 = 16$ Negative number raised to an even power; positive result.

but

$(-2)^3 = -8$ Negative number raised to an odd power; negative result.

C Dividing Real Numbers

What about the rules for division? As you recall, a division problem can always be checked by multiplication. Thus the division

$$3\overline{)18} = 6 \quad (-18,\ 0) \quad \text{or} \quad \frac{18}{3} = 6$$

is correct because $18 = 3 \cdot 6$. In general, we have the following definition for division.

Answers

3. a. 49 **b.** -49 **c.** $\frac{1}{16}$ **d.** $-\frac{1}{16}$

4. a. -64 **b.** -64

DIVISION

If b is not zero,

$$\frac{a}{b} = c \quad \text{means} \quad a = b \cdot c$$

Note that the operation of division is defined using multiplication. Because of this, the same rules of sign that apply to the multiplication of real numbers also apply to the division of real numbers. Note that the division of two numbers is called the **quotient.**

Teaching Tip

Use the following table to summarize sign rules.

	Same Signs	Different Signs
Add	Add Keep sign	Subtract. Use the sign of the number with the larger absolute value*
Multiply or divide	+	−

*Rewrite subtraction as addition.

RULES

Dividing with Signed Numbers

1. When dividing two numbers with the *same (like)* sign, give the quotient a *positive* (+) sign.
2. When dividing two numbers with *different (unlike)* signs, give the quotient a *negative* (−) sign.

Here are some examples:

$\frac{24}{6} = 4$ 24 and 6 have the same sign; the quotient is positive.

$\frac{-18}{-9} = 2$ −18 and −9 have the same sign; the quotient is positive.

$\frac{-32}{4} = -8$ −32 and 4 have different signs; the quotient is negative.

$\frac{35}{-7} = -5$ 35 and −7 have different signs; the quotient is negative.

Teaching Tip

To remember about division with zero use the following:

$\frac{O}{K} = 0$ "ok"

$\frac{N}{O}$ (undefined) "no"

EXAMPLE 5 **Finding quotients of integers**

Find:

a. $48 \div 6$ **b.** $\frac{54}{-9}$ **c.** $\frac{-63}{-7}$ **d.** $-28 \div 4$ **e.** $5 \div 0$

SOLUTION

a. $48 \div 6 = 8$ 48 and 6 have the same sign; the quotient is positive.

b. $\frac{54}{-9} = -6$ 54 and −9 have different signs; the quotient is negative.

c. $\frac{-63}{-7} = 9$ −63 and −7 have the same sign; the quotient is positive.

d. $-28 \div 4 = -7$ −28 and 4 have different signs; the quotient is negative.

e. $5 \div 0$ is not defined. Note that if we let $5 \div 0$ equal any number, such as a, we have

$\frac{5}{0} = a$ This means $5 = a \cdot 0 = 0$ or $5 = 0$.

which, of course, is impossible. Thus $\frac{5}{0}$ is *not defined.*

PROBLEM 5

Find:

a. $56 \div 8$ **b.** $\frac{36}{-4}$ **c.** $\frac{-49}{-7}$

d. $-24 \div 6$ **e.** $-7 \div 0$

Answers

5. a. 7 **b.** −9 **c.** 7
d. −4 **e.** Not defined

If the division involves real numbers written as fractions, we use the following procedure.

PROCEDURE

Dividing Fractions

To divide $\frac{a}{b}$ by $\frac{c}{d}$, multiply $\frac{a}{b}$ by the reciprocal of $\frac{c}{d}$, that is,

$$\frac{a}{b} \div \frac{c}{d} = \frac{a}{b} \cdot \frac{d}{c} = \frac{ad}{bc}$$

Of course, the rules of signs still apply!

EXAMPLE 6 **Finding quotients of fractions**

Find:

a. $\frac{2}{5} \div \left(-\frac{3}{4}\right)$ **b.** $-\frac{5}{6} \div \left(-\frac{7}{2}\right)$ **c.** $-\frac{3}{7} \div \frac{6}{7}$

SOLUTION

a. $\frac{2}{5} \div \left(-\frac{3}{4}\right) = \frac{2}{5} \cdot \left(-\frac{4}{3}\right) = -\frac{8}{15}$

Different signs — Negative quotient

b. $-\frac{5}{6} \div \left(-\frac{7}{2}\right) = -\frac{5}{6} \cdot \left(-\frac{2}{7}\right) = \frac{10}{42} = \frac{5}{21}$

Same sign — Positive quotient

c. $-\frac{3}{7} \div \frac{6}{7} = -\frac{3}{7} \cdot \frac{7}{6} = -\frac{21}{42} = -\frac{1}{2}$

Different signs — Negative quotient

PROBLEM 6

Find:

a. $\frac{4}{5} \div \left(-\frac{3}{4}\right)$ **b.** $-\frac{5}{6} \div \left(-\frac{7}{4}\right)$

c. $-\frac{4}{7} \div \frac{8}{7}$

D Solving an Application

When you are driving and push down or let up on the gas or brake pedal, your car changes speed. This change in speed over a period of time is called *acceleration* and is given by

$$a = \frac{f - s}{t}$$

Final speed (f), Starting speed (s), Acceleration (a), Time period (t)

Answers

6. a. $-\frac{16}{15}$ **b.** $\frac{10}{21}$ **c.** $-\frac{1}{2}$

We use this idea in Example 7.

Web It

To review the multiplication rules and do some interactive problems, go to link 1-4-1 on the Bello Website at mhhe.com/bello.

For a more detailed lesson and practice problems on the rules for multiplying integers, go to link 1-4-2.

Finally, to review division and practice some more division problems involving integers, visit link 1-4-3.

EXAMPLE 7 **Acceleration and deceleration**

You are driving at 55 miles per hour (mi/hr), and for the next 10 seconds:

a. you *increase* your speed to 65 miles per hour. What is your acceleration?

b. you *decrease* your speed to 40 miles per hour. What is your acceleration?

SOLUTION Your starting speed is $s = 55$ miles per hour and the time period is $t = 10$ seconds (sec).

a. Your final speed is $f = 65$ miles per hour, so

$$a = \frac{(65 - 55)\text{ mi/hr}}{10\text{ sec}} = \frac{10\text{ mi/hr}}{10\text{ sec}} = 1\,\frac{\text{mi/hr}}{\text{sec}}$$

Thus your acceleration is 1 mile per hour each second.

b. Here the final speed is 40, so

$$a = \frac{(40 - 55)\text{ mi/hr}}{10\text{ sec}} = \frac{-15\text{ mi/hr}}{10\text{ sec}} = -1\frac{1}{2}\,\frac{\text{mi/hr}}{\text{sec}}$$

Thus your acceleration is $-1\frac{1}{2}$ miles per hour every second. When acceleration is *negative,* it is called *deceleration.* Deceleration can be thought of as negative acceleration, so your deceleration is $1\frac{1}{2}$ miles per hour each second.

PROBLEM 7

a. Find your acceleration if you increase your speed from 55 miles per hour to 70 miles an hour for the next 5 seconds.

b. Find your acceleration if you decrease your speed from 50 miles per hour to 40 miles per hour for the next 5 seconds.

Answers

7. a. $3\,\frac{\text{mi/hr}}{\text{sec}}$ **b.** $-2\,\frac{\text{mi/hr}}{\text{sec}}$

Exercises 1.4

A In Problems 1–20, perform the indicated operation.

1. $4 \cdot 9$ 36 **2.** $16 \cdot 2$ 32 **3.** $-10 \cdot 4$ -40

4. $-7 \cdot 8$ -56 **5.** $-9 \cdot 9$ -81 **6.** $-2 \cdot 5$ -10

7. $-6 \cdot (-3)(-2)$ -36 **8.** $-4 \cdot (-5)(-3)$ -60 **9.** $-9 \cdot (-2)(-3)$ -54 **10.** $-7 \cdot (-10)(-2)$ -140

11. $-2.2(3.3)$ -7.26 **12.** $-1.4(3.1)$ -4.34 **13.** $-1.3(-2.2)$ 2.86 **14.** $-1.5(-1.1)$ 1.65

15. $\frac{5}{6}\left(-\frac{5}{7}\right)$ $-\frac{25}{42}$

16. $\frac{3}{8}\left(-\frac{5}{7}\right)$ $-\frac{15}{56}$

17. $-\frac{3}{5}\left(-\frac{5}{12}\right)$ $\frac{1}{4}$

18. $-\frac{4}{7}\left(-\frac{21}{8}\right)$ $\frac{3}{2}$

19. $-\frac{6}{7}\left(\frac{35}{8}\right)$ $-\frac{15}{4}$

20. $-\frac{7}{5}\left(\frac{15}{28}\right)$ $-\frac{3}{4}$

B In Problems 21–30, perform the indicated operation.

21. -4^2 -16

22. $(-4)^2$ 16

23. $(-5)^2$ 25

24. -5^2 -25

25. -5^3 -125

26. $(-5)^3$ -125

27. $(-6)^4$ 1296

28. -6^4 -1296

29. $-\left(\frac{1}{2}\right)^5$ $-\frac{1}{32}$

30. $\left(-\frac{1}{2}\right)^5$ $-\frac{1}{32}$

C In Problems 31–60, perform the indicated operation.

31. $\frac{14}{2}$ 7

32. $10 \div 2$ 5

33. $-50 \div 10$ -5

34. $\frac{-20}{5}$ -4

35. $\frac{-30}{10}$ -3

36. $-40 \div 8$ -5

37. $\frac{-0}{3}$ 0

38. $-0 \div 8$ 0

39. $-5 \div 0$ Undefined

40. $\frac{-8}{0}$ Undefined

41. $\frac{0}{7}$ 0

42. $0 \div (-7)$ 0

43. $-15 \div (-3)$ 5

44. $-20 \div (-4)$ 5

45. $\frac{-25}{-5}$ 5

46. $\frac{-16}{-2}$ 8

47. $\frac{18}{-9}$ -2

48. $\frac{35}{-7}$ -5

49. $30 \div (-5)$ -6

50. $80 \div (-10)$ -8

51. $\frac{3}{5} \div \left(-\frac{4}{7}\right)$ $-\frac{21}{20}$

52. $\frac{4}{9} \div \left(-\frac{1}{7}\right)$ $-\frac{28}{9}$

53. $-\frac{2}{3} \div \left(-\frac{7}{6}\right)$ $\frac{4}{7}$

54. $-\frac{5}{6} \div \left(-\frac{25}{18}\right)$ $\frac{3}{5}$

55. $-\frac{5}{8} \div \frac{7}{8}$ $-\frac{5}{7}$

56. $-\frac{4}{5} \div \frac{8}{15}$ $-\frac{3}{2}$

57. $\frac{-3.1}{6.2}$ -0.5 or $-\frac{1}{2}$

58. $\frac{1.2}{-4.8}$ -0.25 or $-\frac{1}{4}$

59. $\frac{-1.6}{-9.6}$ $0.1\overline{6}$ or $\frac{1}{6}$

60. $\frac{-9.8}{-1.4}$ 7

APPLICATIONS

Use the following information in Problems 61–65:

1 all-beef frank	+45 calories
1 slice of bread	+65 calories
Running (1 min)	−15 calories
Swimming (1 min)	−7 calories

61. If a person eats 2 beef franks and runs for 5 min, what is the caloric gain or loss? 15 calories; gain

62. If a person eats 2 beef franks and runs for 30 min, what is the caloric gain or loss? 360 calories; loss

63. If a person eats 2 beef franks with 2 slices of bread and then runs for 15 min, what is the caloric gain or loss? 5 calories; loss

64. If a person eats 2 beef franks with 2 slices of bread and then runs for 15 min and swims for 30 min, what is the caloric gain or loss? 215 calories; loss

65. *"Burning off" calories* If a person eats 2 beef franks, how many minutes does the person have to run to "burn off" the calories? (*Hint:* You must *spend* the calories contained in the 2 beef franks.) 6 minutes

66. *Automobile acceleration* The highest road-tested acceleration for a standard production car is from 0 to 60 miles per hour in 3.275 seconds for a Ford RS 200 Evolution. What was the acceleration of this car? Give your answer to one decimal place. 18.3 mi/hr each second

67. *Automobile acceleration* The highest road-tested acceleration for a street-legal car is from 0 to 60 miles per hour in 3.89 seconds for a Jankel Tempest. What was the acceleration of this car? Give your answer to one decimal place. 15.4 mi/hr each second

68. *Cost of saffron* Which food do you think is the most expensive? It is saffron, which comes from Spain. It costs $472.50 to buy 3.5 ounces at Harrods, a store in Great Britain. What is the cost of 1 ounce of saffron? $135

69. *Long-distance telephone charges* The price of a long-distance call from Tampa to New York is $3.05 for the first 3 minutes and $0.70 for each additional minute or fraction thereof. What is the cost of a 5-minute long-distance call from Tampa to New York? $4.45

70. *Long-distance telephone charges* The price of a long-distance call from Tampa to New York using a different phone company is $3 for the first 3 minutes and $0.75 for each additional minute or fraction thereof. What is the cost of a 5-minute long-distance call from Tampa to New York using this phone company? $4.50

SKILL CHECKER

Try the Skill Checker Exercises so you'll be ready for the next section.

Find:

71. $3 \cdot \frac{1}{3}$ 1

72. $9 \cdot \frac{1}{9}$ 1

73. $7 + (-7)$ 0

74. $-9 + 9$ 0

USING YOUR KNOWLEDGE

Have a Decidedly Lovable Day!

Have you met anybody *nice* today or did you have an *unpleasant* experience? Perhaps the person you met was *very nice* or your experience *very unpleasant.* Psychologists and linguists have a numerical way to indicate the difference between *nice* and *very nice* or between *unpleasant* and *very unpleasant.* Suppose you assign a positive number (+2, for example) to the adjective *nice,* a negative number (say, −2) to *unpleasant,* and a positive number greater than 1 (say +1.75) to *very.* Then, *very nice* means

Very nice

$$(1.75) \cdot (2) = 3.50$$

and *very unpleasant* means

Very unpleasant

$$(1.75) \cdot (-2) = -3.50$$

Here are some adverbs and adjectives and their average numerical values, as rated by a panel of college students. (Values differ from one panel to another.)

Adverbs		Adjectives	
Slightly	0.54	Wicked	−2.5
Rather	0.84	Disgusting	−2.1
Decidedly	0.16	Average	−0.8
Very	1.25	Good	3.1
Extremely	1.45	Lovable	2.4

Find the value of each.

75. Slightly wicked −1.35

76. Decidedly average −0.128

77. Extremely disgusting −3.045

78. Rather lovable 2.016

79. Very good 3.875

By the way, if you got all the answers correct, you are 4.495!

WRITE ON

80. Why is $-a^2$ always negative for any nonzero value of a? Answers may vary.

81. Why is $(-a)^2$ always positive for any nonzero value of a? Answers may vary.

82. Is $(-1)^{100}$ positive or negative? What about $(-1)^{99}$? Positive; negative

83. Explain why $\frac{a}{0}$ is not defined. Answers may vary.

MASTERY TEST

If you know how to do these problems, you have learned your lesson!

Find:

84. $\frac{-3.6}{1.2}$ -3

85. $\frac{-3.1}{-12.4}$ 0.25 or $\frac{1}{4}$

86. $\frac{6.5}{-1.3}$ -5

87. $-3 \cdot 11$ -33

88. $9 \cdot (-10)$ -90

89. $-5 \cdot (-11)$ 55

90. -9^2 -81

91. $(-8)^2$ 64

92. $\left(-\frac{1}{5}\right)^2$ $\frac{1}{25}$

93. $-2.2(3.2)$ -7.04

94. $-1.3(-4.1)$ 5.33

95. $-\frac{3}{7}\left(-\frac{4}{5}\right)$ $\frac{12}{35}$

96. $\frac{6}{7}\left(-\frac{2}{3}\right)$ $-\frac{4}{7}$

97. $\frac{3}{5} \div \left(-\frac{4}{7}\right)$ $-\frac{21}{20}$

98. $-\frac{6}{7} \div \left(-\frac{3}{5}\right)$ $\frac{10}{7}$

99. $-\frac{4}{5} \div \frac{8}{5}$ $-\frac{1}{2}$

100. The driver of a car traveling at 50 miles per hour slams on the brakes and slows down to 10 miles per hour in 5 seconds. What is the acceleration of the car?
-8 mi/hr each second

1.5 ORDER OF OPERATIONS

To Succeed, Review How To . . .

1. Add, subtract, multiply, and divide real numbers (pp. 61–66, 70–72, 73–76).
2. Evaluate expressions containing exponents (pp. 72–73).

Objectives

A Evaluate expressions using the correct order of operations.

B Evaluate expressions with more than one grouping symbol.

C Solve an application.

GETTING STARTED

Collecting the Rent

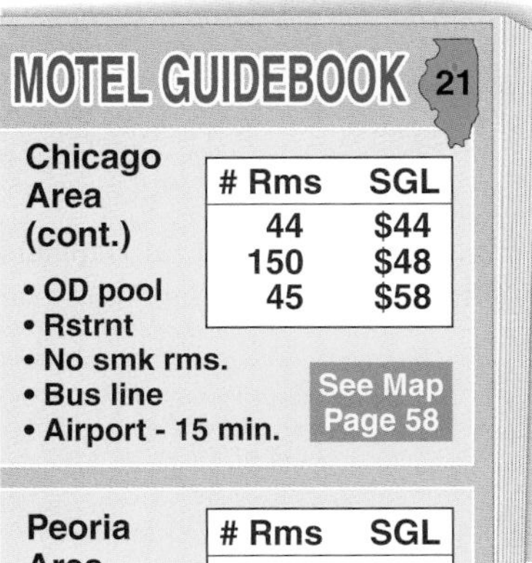

MOTEL GUIDEBOOK
21
Chicago Area (cont.)
Rms SGL
44 $44
150 $48
45 $58
• OD pool
• Rstrnt
• No smk rms.
• Bus line
• Airport - 15 min.
See Map Page 58
Peoria
Rms SGL

Now that we know how to do the fundamental operations with real numbers, we need to know *in what order* we should do them. Let's suppose all rooms in a motel are taken. (For simplicity, we won't include extra persons in the rooms.) How can we figure out how much money we should collect? To do this, we first multiply the price of each room by the number of rooms available at that price. Next, we add all these figures to get the final answer. The calculations look like this:

$$44 \cdot 44 = \$1936$$

$$150 \cdot 48 = \$7200$$

$$45 \cdot 58 = \$2610$$

$$\$1936 + \$7200 + \$2610 = \$11{,}746 \quad \text{Total}$$

Note that we *multiplied* before we *added.* This is the correct order of operations. As you will see, changing the order of operations can change the answer!

Using the Correct Order of Operations

If we want to find the answer to $3 \cdot 4 + 5$, do we

(1) add 4 and 5 first and then multiply by 3? That is, $3 \cdot 9 = 27$?

or

(2) multiply 3 by 4 first and then add 5? That is, $12 + 5 = 17$?

In (1), the answer is 27. In (2), the answer is 17. What if we write $3 \cdot (4 + 5)$ or $(3 \cdot 4) + 5$? What do the parentheses mean? To obtain an answer we can agree upon, we need the following rules.

RULES

Order of Operations

Operations are always performed in the following order:

1. Do all calculations inside **grouping symbols** such as parentheses ( ) or brackets [ ].
2. Evaluate all exponents.
3. Do multiplications and divisions as they occur from *left to right.*
4. Do additions and subtractions as they occur from *left to right.*

Web It

How do you remember the order of operations?

Parentheses	***P**lease*
Exponents	***E**xcuse*
Multiplication	***M**y*
Division	***D**ear*
Addition	***A**unt*
Subtraction	***S**ally*

If you remember **P**lease **E**xcuse **M**y **D**ear **A**unt **S**ally, you should remember **P**arentheses **E**xponents **M**ultiplication **D**ivision **A**ddition **S**ubtraction! **PEMDAS** for short. If you want to see the rules using a flowchart go to the Bello Website at mhhe.com/bello and try link 1-5-1.

With these conventions

$$\begin{aligned} & 3 \cdot 4 + 5 \\ &= 12 + 5 && \text{First multiply.} \\ &= 17 && \text{Then add.} \end{aligned}$$

Similarly,

$$\begin{aligned} & 3 \cdot (4 + 5) \\ &= 3 \cdot 9 && \text{First add inside parentheses.} \\ &= 27 && \text{Then multiply.} \end{aligned}$$

But

$$\begin{aligned} & (3 \cdot 4) + 5 \\ &= 12 + 5 && \text{First multiply inside parentheses.} \\ &= 17 && \text{Then add.} \end{aligned}$$

Note that multiplications and divisions are done *in order,* from left to right. This means that sometimes you do multiplications first, sometimes you do divisions first, depending on the order in which they occur. Thus,

$$\begin{aligned} & 12 \div 4 \cdot 2 \\ &= 3 \cdot 2 && \text{First divide.} \\ &= 6 && \text{Then multiply.} \end{aligned}$$

But

$$\begin{aligned} & 12 \cdot 4 \div 2 \\ =\ & 48 \div 2 \quad \text{First multiply.} \\ =\ & 24 \quad \text{Then divide.} \end{aligned}$$

EXAMPLE 1 **Evaluating expressions**

Find the value of:

a. $8 \cdot 9 - 3$ **b.** $27 + 3 \cdot 5$

SOLUTION

a.

$$\begin{aligned} & 8 \cdot 9 - 3 \\ =\ & 72 - 3 \\ =\ & 69 \end{aligned}$$

Do multiplications and divisions in order from left to right ($8 \cdot 9 = 72$).

Then do additions and subtractions in order from left to right ($72 - 3 = 69$).

b.

$$\begin{aligned} & 27 + 3 \cdot 5 \\ =\ & 27 + 15 \\ =\ & 42 \end{aligned}$$

Do multiplications and divisions in order from left to right ($3 \cdot 5 = 15$).

Then do additions and subtractions in order from left to right ($27 + 15 = 42$).

PROBLEM 1

Find the value of:

a. $3 \cdot 8 - 9$ **b.** $22 + 4 \cdot 9$

EXAMPLE 2 **Expressions with grouping symbols and exponents**

Find the value of:

a. $63 \div 7 - (2 + 3)$ **b.** $8 \div 2^3 + 3 - 1$

SOLUTION

a.

$$\begin{aligned} & 63 \div 7 - (2 + 3) \\ =\ & 63 \div 7 - 5 \\ =\ & 9 - 5 \\ =\ & 4 \end{aligned}$$

First do the operation inside the parentheses.

Next do the division.

Then do the subtraction.

b.

$$\begin{aligned} & 8 \div 2^3 + 3 - 1 \\ =\ & 8 \div 8 + 3 - 1 \\ =\ & 1 + 3 - 1 \\ =\ & 4 - 1 \\ =\ & 3 \end{aligned}$$

First do the exponentation.

Next do the division.

Then do the addition.

Do the final subtraction.

PROBLEM 2

Find the value of:

a. $56 \div 8 - (3 + 1)$

b. $54 \div 3^3 + 5 - 2$

Answers

1. a. 15 **b.** 58 **2. a.** 3 **b.** 5

EXAMPLE 3 Expression with grouping symbols

Find the value of: $8 \div 4 \cdot 2 + 3(5 - 2) - 3 \cdot 2$

SOLUTION

$8 \div 4 \cdot 2 + 3(5 - 2) - 3 \cdot 2$

$= 8 \div 4 \cdot 2 + 3\ (3) - 3 \cdot 2$ First do the operations inside the parentheses. Now do the multiplications and divisions in order from left to right:

$= 2 \cdot 2 + 3(3) - 3 \cdot 2$ This means do $8 \div 4 = 2$ first.

$= 4 + 3(3) - 3 \cdot 2$ Then do $2 \cdot 2 = 4$.

$= 4 + 9 - 3 \cdot 2$ Next do $3(3) = 9$.

$= 4 + 9 - 6$ And finally, do $3 \cdot 2 = 6$.

$= 13 - 6$ We are through with multiplications and divisions. Now do the addition.

$= 7$ The final operation is a subtraction.

PROBLEM 3

Find the value of:

$10 \div 5 \cdot 2 + 4(5 - 3) - 4 \cdot 2$

B Evaluating Expressions with More than One Grouping Symbol

Suppose a local department store is having a great sale. The advertised items are such good deals that you decide to buy two bedspreads and two mattresses. The price of one bedspread and one mattress is \$14 + \$88. Thus the price of two of each is $2 \cdot (14 + 88)$. If you then decide to buy a lamp, the total price is

$$[2 \cdot (14 + 88)] + 12$$

We have used two types of grouping symbols in $[2 \cdot (14 + 88)] + 12$, parentheses ( ) and brackets [ ]. There is one more grouping symbol, braces { }. Typically, the order of these grouping symbols is {[( )]}. Here is the rule we use to handle grouping symbols.

SUPER SATURDAY SALE
KING or QUEEN BEDSPREADS
$29.95 $14.00
QUEEN-SIZE MATTRESS
$88.00
VELVET SWAG LAMPS
$39.00 $12.00

RULE

Grouping Symbols

When grouping symbols occur within other grouping symbols, perform the computations in the *innermost* grouping symbols first.

Answer

3. 4

Thus to find the value of $[2 \cdot (14 + 88)] + 12$, we first add 14 and 88 (the operation inside parentheses, the innermost grouping symbols), then multiply by 2, and finally add 12. Here is the procedure:

$[2 \cdot (14 + 88)] + 12$	Given.
$= [2 \cdot (102)] + 12$	Add 14 and 88 inside the parentheses.
$= 204 + 12$	Multiply 2 by 102 inside the brackets.
$= 216$	Do the final addition.

EXAMPLE 4 **Expressions with three grouping symbols**

Find the value of: $20 \div 4 + \{2 \cdot 9 - [3 + (6 - 2)]\}$

SOLUTION The innermost grouping symbols are the parentheses, so we do the operations inside the parentheses, then inside the brackets, and, finally, inside the braces. Here are the details.

$20 \div 4 + \{2 \cdot 9 - [3 + (6 - 2)]\}$	Given.
$= 20 \div 4 + \{2 \cdot 9 - [3 + 4]\}$	Subtract inside the parentheses $(6 - 2 = 4)$.
$= 20 \div 4 + \{2 \cdot 9 - 7\}$	Add inside the brackets $(3 + 4 = 7)$.
$= 20 \div 4 + \{18 - 7\}$	Multiply inside the braces $(2 \cdot 9 = 18)$.
$= 20 \div 4 + 11$	Subtract inside the braces $(18 - 7 = 11)$.
$= 5 + 11$	Divide $(20 \div 4 = 5)$.
$= 16$	Do the final addition.

PROBLEM 4

Find the value of:

$50 \div 5 + \{2 \cdot 7 - [4 + (5 - 2)]\}$

Fraction bars are sometimes used as grouping symbols to indicate an expression representing a single number. To find the value of such expressions, simplify above and below the fraction bars following the order of operations. Thus

$\dfrac{2(3 + 8) + 4}{2(4) - 10}$	
$= \dfrac{2(11) + 4}{2(4) - 10}$	Add inside the parentheses in the numerator $(3 + 8 = 11)$.
$= \dfrac{22 + 4}{8 - 10}$	Multiply in the numerator $[2(11) = 22]$ and in the denominator $[2(4) = 8]$.
$= \dfrac{26}{-2}$	Add in the numerator $(22 + 4 = 26)$, subtract in the denominator $(8 - 10 = -2)$.
$= -13$	Do the final division. (Remember to use the rules of signs.)

Answer

4. 17

EXAMPLE 5 Using a fraction bar as a grouping symbol

Find the value of:

$$-5^2 + \frac{3(4-8)}{2} + 10 \div 5$$

SOLUTION

$-5^2 + \frac{3(4-8)}{2} + 10 \div 5$	Given.
$= -25 + \frac{3(4-8)}{2} + 10 \div 5$	Do the exponentiation ($5^2 = 25$, so $-5^2 = -25$).
$= -25 + \frac{3(-4)}{2} + 10 \div 5$	Subtract inside the parentheses ($4 - 8 = -4$).
$= -25 + \frac{-12}{2} + 10 \div 5$	Multiply above the division bar [$3(-4) = -12$].
$= -25 + (-6) + 10 \div 5$	Divide $\left(\frac{-12}{2} = -6\right)$.
$= -25 + (-6) + 2$	Divide ($10 \div 5 = 2$).
$= -31 + 2$	Add [$-25 + (-6) = -31$].
$= -29$	Do the final addition.

PROBLEM 5

Find the value of:

$$-3^2 + \frac{2(3-6)}{3} + 20 \div 5$$

C Solving an Application

EXAMPLE 6 Your heart and grouping symbols

Your ideal heart rate while exercising is found by subtracting your age A from 205, multiplying the result by 7, and dividing the answer by 10. In symbols, this is

$$\text{Ideal rate} = [(205 - A) \cdot 7] \div 10$$

Suppose you are 25 years old ($A = 25$). What is your ideal heart rate?

SOLUTION

Ideal rate $= [(205 - A) \cdot 7] \div 10$	
$= [(205 - 25) \cdot 7] \div 10$	Substitute 25 for A.
$= [180 \cdot 7] \div 10$	Subtract ($205 - 25 = 180$).
$= 1260 \div 10$	Multiply inside brackets ($180 \cdot 7 = 1260$).
$= 126$	Do the final division.

PROBLEM 6

Use the formula to find your ideal heart rate assuming you are 20 years old. Round the answer to the nearest whole number.

Answers

5. -7
6. 130 (rounded from 129.5)

Web It

To read and review order of operations, go to the Bello Website at mhhe.com/bello and try link 1-5-2.

To see whether you are really ready for the exercises, try link 1-5-3. It has a tutorial and a pretest.

Finally, for a tutorial following the indicated steps, go to the Bello Website and try link 1-5-4.

Calculate It Order of Operations

How does a calculator handle the order of operations? It does it *automatically*. For example, we've learned that $3 + 4 \cdot 5 = 3 + 20 = 23$; to do this with a calculator, we enter the expression by pressing 3 + 4 × 5 ENTER. The result is shown in Window 1.

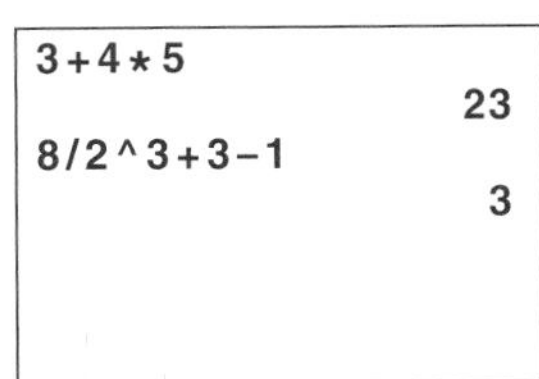

Window 1

To do Example 2(b), you have to know how to enter exponents in your calculator. Many calculators use the ^ key followed by the exponent. Thus, to enter $8 \div 2^3 + 3 - 1$, press 8 ÷ 2 ^ 3 + 3 − 1 ENTER. The result, 3, is shown in Window 1. Note that the calculator follows the order of operations *automatically*.

Finally, you must be extremely careful when you evaluate expressions with division bars such as

$$\frac{2(3 + 8) + 4}{2(4) - 10}$$

When fraction bars are used as grouping symbols, they must be entered as sets of parentheses. Thus you must enter

to obtain −13 (see Window 2). Note that we didn't use any extra parentheses to enter 2(4).

```
(2(3+8)+4)/(2*4-
10)
                -13
```

Window 2

Exercises 1.5

Boost *your* GRADE at mathzone.com!

MathZone

- Practice Problems
- Self-Tests
- Videos
- NetTutor
- e-Professors

A In Problems 1–20, find the value of the given expression.

1. $4 \cdot 5 + 6$ 26

2. $3 \cdot 4 + 6$ 18

3. $7 + 3 \cdot 2$ 13

4. $6 + 9 \cdot 2$ 24

5. $7 \cdot 8 - 3$ 53

6. $6 \cdot 4 - 9$ 15

7. $20 - 3 \cdot 5$ 5

8. $30 - 6 \cdot 5$ 0

9. $48 \div 6 - (3 + 2)$ 3

10. $81 \div 9 - (4 + 5)$ 0

11. $3 \cdot 4 \div 2 + (6 - 2)$ 10

12. $3 \cdot 6 \div 2 + (5 - 2)$ 12

13. $36 \div 3^2 + 4 - 1$ 7

14. $16 \div 2^3 + 3 - 2$ 3

15. $8 \div 2^3 - 3 + 5$ 3

16. $9 \div 3^2 - 8 + 5$ −2

17. $10 \div 5 \cdot 2 + 8 \cdot (6 - 4) - 3 \cdot 4$ 8

18. $15 \div 3 \cdot 3 + 2 \cdot (5 - 2) + 8 \div 4$ 23

19. $4 \cdot 8 \div 2 - 3(4 - 1) + 9 \div 3$ 10

20. $6 \cdot 3 \div 3 - 2(3 - 2) - 8 \div 2$ 0

B In Problems 21–40, find the value of the given expression.

21. $20 \div 5 + \{3 \cdot 4 - [4 + (5 - 3)]\}$ 10

22. $30 \div 6 + \{4 \div 2 \cdot 3 - [3 + (5 - 4)]\}$ 7

23. $(20 - 15) \cdot [20 \div 2 - (2 \cdot 2 + 2)]$ 20

24. $(30 - 10) \cdot [52 \div 4 - (3 \cdot 3 + 3)]$ 20

25. $\{4 \div 2 \cdot 6 - (3 + 2 \cdot 3) + [5(3 + 2) - 1]\}$ 27

26. $-6^2 + \dfrac{4(6-8)}{2} + 15 \div 3$ −35

27. $-7^2 + \dfrac{3(8-4)}{4} + 10 \div 2 \cdot 3$ −31

28. $-5^2 + \dfrac{6(3-7)}{4} + 9 \div 3 \cdot 2$ −25

29. $(-6)^2 \cdot 4 \div 4 - \dfrac{3(7-9)}{2} - 4 \cdot 3 \div 2^2$ 36

30. $(-4)^2 \cdot 3 \div 8 - \dfrac{4(6-10)}{2} - 3 \cdot 8 \div 2^3$ 11

31. $\left[\dfrac{7-(-3)}{8-6}\right]\left[\dfrac{3+(-8)}{7-2}\right]$ −5

32. $\left[\dfrac{4-(-3)}{8-1}\right]\left[\dfrac{-4-(-9)}{8-3}\right]$ 1

33. $\dfrac{(-3)(-2)(-4)}{3(-2)} - (-3)^2$ −5

34. $\dfrac{(-2)(-4)(-5)}{10(-2)} - (-4)^2$ −14

35. $\dfrac{(-10)(-6)}{(-2)(-3)(-5)} - (-3)^3$ 25

36. $\dfrac{(-8)(-10)}{(-3)(-4)(-2)} - (-2)^3$ $\frac{14}{3}$

37. $\dfrac{(-2)^3(-3)(-9)}{(-2)^2(-3)^2} - 3^2$ −15

38. $\dfrac{(-2)(-3)^2(-10)}{(-2)^2(-3)} - 3^2$ −24

39. $-5^2 - \dfrac{(-2)(-3)(-4)}{(-12)(-2)}$ −24

40. $-6^2 - \dfrac{(-3)(-5)(-8)}{(-3)(-4)}$ −26

APPLICATIONS

41. *Gasoline octane rating* Have you noticed the octane rating of gasoline at the gas pump? This octane rating is given by

$$\frac{R + M}{2}$$

where R is a number measuring the performance of gasoline using the Research Method and M is a number measuring the performance of gasoline using the Motor Method. If a certain gasoline has $R = 92$ and $M = 82$, what is its octane rating? 87

42. *Gasoline octane rating* If a gasoline has $R = 97$ and $M = 89$, what is its octane rating? 93

43. *Exercise pulse rate* If A is your age, the minimum pulse rate you should maintain during aerobic activities is $0.72(220 - A)$. What is the minimum pulse rate you should maintain if you are

a. 20 years old? 144 **b.** 45 years old? 126

44. *Exercise pulse rate* If A is your age, the maximum pulse rate you should maintain during aerobic activities is $0.88(220 - A)$. What is the maximum pulse rate you should maintain if you are

a. 20 years old? 176 **b.** 45 years old? 154

SKILL CHECKER

Try the Skill Checker Exercises so you'll be ready for the next section.

Find:

45. $11 \cdot \dfrac{1}{11}$ 1

46. $17 \cdot \dfrac{1}{17}$ 1

47. $-2.3 + 2.3$ 0

48. $-1.7 + 1.7$ 0

49. $-2.5 \cdot \dfrac{1}{-2.5}$ 1

50. $-3.7 \cdot \dfrac{1}{-3.7}$ 1

USING YOUR KNOWLEDGE

Children's Dosages

What is the corresponding dose (amount) of medication for children when the adult dosage is known? There are several formulas that tell us.

51. *Friend's rule* (for children under 2 years):

(Age in months · Adult dose) ÷ 150 = Child's dose

Suppose a child is 10 months old and the adult dose of aspirin is a 75-milligram tablet. What is the child's dose? [*Hint:* Simplify (10 · 75) ÷ 150.] 5 milligrams

52. *Clarke's rule* (for children over 2 years):

(Weight of child · Adult dose) ÷ 150 = Child's dose

If a 7-year-old child weighs 75 pounds and the adult dose is 4 tablets a day, what is the child's dose? [*Hint:* Simplify (75 · 4) ÷ 150.] 2 tablets a day

53. *Young's rule* (for children between 3 and 12):

(Age · Adult dose) ÷ (Age + 12) = Child's dose

Suppose a child is 6 years old and the adult dose of an antibiotic is 4 tablets every 12 hours. What is the child's dose? [*Hint:* Simplify (6 · 4) ÷ (6 + 12).] $1\frac{1}{3}$ tablets every 12 hours

WRITE ON

54. State the rules for the order of operations, and explain why they are needed. Answers may vary.

55. Explain whether the parentheses are needed when finding

a. $2 + (3 \cdot 4)$ Answers may vary.

b. $2 \cdot (3 + 4)$ Answers may vary.

56. When evaluating an expression, do you *always* have to

a. do multiplications before divisions? Give examples to support your answer. No; answers may vary.

b. do additions before subtractions? Give examples to support your answer. No; answers may vary.

MASTERY TEST

If you know how to do these problems, you have learned your lesson!

Find the value of:

57. $63 \div 9 - (2 + 5)$ 0

58. $-64 \div 8 - (6 - 2)$ −12

59. $16 + 2^3 + 3 - 9$ 18

60. $3 \cdot 4 - 18$ −6

61. $18 + 4 \cdot 5$ 38

62. $12 \div 4 \cdot 2 + 2(5 - 3) - 3 \cdot 4$ −2

63. $15 \div 3 + \{2 \cdot 4 - [6 + (2 - 5)]\}$ 10

64. $-6^2 + \frac{4(6 - 12)}{3} + 8 \div 2$ −40

65. The ideal heart rate while exercising for a person A years old is $[(205 - A) \cdot 7] \div 10$. What is the ideal heart rate for a 35-year-old person? 119

1.6 PROPERTIES OF THE REAL NUMBERS

To Succeed, Review How To . . .

1. Add, subtract, multiply, and divide real numbers (pp. 61–66, 70–72, 73–76).

Objectives

A Identify which of the properties (associative or commutative) are used in a statement.

B Use the commutative and associative properties to simplify expressions.

C Identify which of the properties (identity or inverse) are used in a statement.

D Use the properties to simplify expressions.

E Use the distributive property to remove parentheses in an expression.

GETTING STARTED

Clarifying Statements

Look at the sign we found hanging outside a lawn mower repair shop. What does it mean? Do the owners want

a small (engine mechanic)?
or
a (small engine) mechanic?

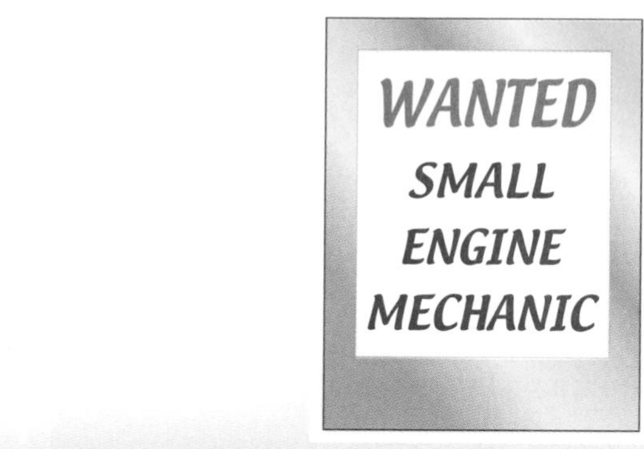

The second meaning is the intended one. Do you see why? The manner in which you *associate* the words makes a difference. Now use parentheses to show how you think the following words should be associated:

Guaranteed used cars

Huge tire sale

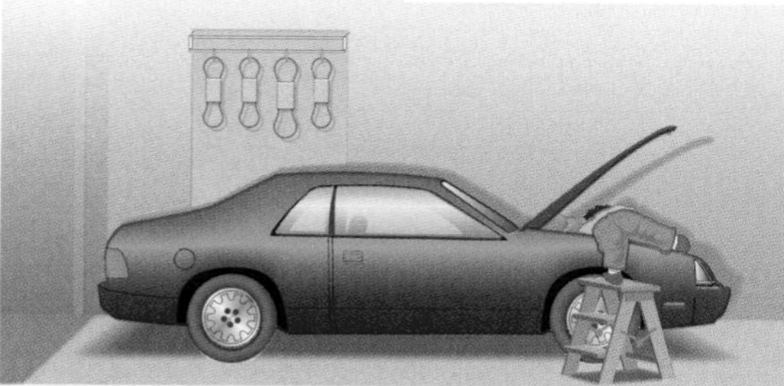

If you wrote

Guaranteed (used cars) and Huge (tire sale)

you are well on your way to understanding the associative property. In algebra, if we are multiplying or adding, the way in which we associate (combine) the numbers makes *no* difference in the answer. This is the associative property, and we shall study it and several other properties of real numbers in this section.

A Identifying the Associative and Commutative Properties

How would you add $17 + 98 + 2$? You would probably add 98 and 2, get 100, and then add 17 to obtain 117. Even though the order of operations tells us to add from left to right, we can group the *addends* (the numbers we are adding) any way we want without changing the sum. Thus $(17 + 98) + 2 = 17 + (98 + 2)$. This fact can be stated as follows.

Associative Property of Addition $a + (b + c) = (a + b) + c$ for any numbers a, b, and c

The associative property of addition tells us that the *grouping* does not matter in addition.

What about multiplication? Does the grouping matter? For example, is $2 \cdot (3 \cdot 4)$ the same as $(2 \cdot 3) \cdot 4$? Since both calculations give 24, the manner in which we group these numbers doesn't matter either. This fact can be stated as follows.

Associative Property of Multiplication $a \cdot (b \cdot c) = (a \cdot b) \cdot c$ for any numbers a, b, and c

Teaching Tip

Have students give examples that illustrate why subtraction and division are not associative or commutative.

The associative property of multiplication tells us that the *grouping* does not matter in multiplication.

We have now seen that grouping numbers differently in addition and multiplication yields the same answer. What about order? As it turns out, the order in which we do additions or multiplications doesn't matter either. For example, $2 + 3 = 3 + 2$ and $5 \cdot 4 = 4 \cdot 5$. In general, we have the following properties.

Commutative Property of Addition $a + b = b + a$ for any numbers a and b

Commutative Property of Multiplication $a \cdot b = b \cdot a$ for any numbers a and b

The commutative properties of addition and multiplication tell us that *order* doesn't matter in addition or multiplication.

EXAMPLE 1 Identifying properties

Name the property illustrated in each of the following statements:

a. $a \cdot (b \cdot c) = (b \cdot c) \cdot a$

b. $(5 + 9) + 2 = 5 + (9 + 2)$

c. $(5 + 9) + 2 = (9 + 5) + 2$

d. $(6 \cdot 2) \cdot 3 = 6 \cdot (2 \cdot 3)$

SOLUTION

a. We changed the *order* of multiplication. The commutative property of multiplication was used.

b. We changed the *grouping* of the numbers. The associative property of addition was used.

c. We changed the *order* of the 5 and the 9 within the parentheses. The commutative property of addition was used.

d. Here we changed the *grouping* of the numbers. We used the associative property of multiplication.

PROBLEM 1

Name the property illustrated in each of the following statements:

a. $5 \cdot (4 \cdot 6) = (5 \cdot 4) \cdot 6$

b. $(x \cdot y) \cdot z = z \cdot (x \cdot y)$

c. $(4 + 8) + 7 = 4 + (8 + 7)$

d. $(4 + 3) + 5 = (3 + 4) + 5$

Answers

1. a. Associative prop. of mult. **b.** Commutative prop. of mult. **c.** Associative prop. of addition **d.** Commutative prop. of addition

EXAMPLE 2 **Using the properties**

Use the correct property to complete the following statements:

a. $-7 + (3 + 2.5) = (-7 + \_\_\_\_) + 2.5$ **b.** $4 \cdot \frac{1}{8} = \frac{1}{8} \cdot \_\_\_\_$

c. $\_\_\_\_ + \frac{1}{4} = \frac{1}{4} + 1.5$ **d.** $\left(\frac{1}{7} \cdot 3\right) \cdot 2 = \frac{1}{7} \cdot (3 \cdot \_\_\_\_)$

SOLUTION

a. $-7 + (3 + 2.5) = (-7 + \underline{3}) + 2.5$ Associative property of addition

b. $4 \cdot \frac{1}{8} = \frac{1}{8} \cdot \underline{4}$ Commutative property of multiplication

c. $\underline{1.5} + \frac{1}{4} = \frac{1}{4} + 1.5$ Commutative property of addition

d. $\left(\frac{1}{7} \cdot 3\right) \cdot 2 = \frac{1}{7} \cdot (3 \cdot \underline{2})$ Associative property of multiplication

PROBLEM 2

Use the correct property to complete the statement:

a. $5 \cdot \frac{1}{3} = \frac{1}{3} \cdot \_\_\_$

b. $\frac{1}{3} \cdot (4 \cdot \_\_\_) = \left(\frac{1}{3} \cdot 4\right) \cdot 7$

c. $\_\_\_ + \frac{1}{5} = \frac{1}{5} + \frac{1}{3}$

d. $(-8 + \_\_\_) + 3.1 = -8 + (4 + 3.1)$

B Using the Associative and Commutative Properties to Simplify Expressions

The associative and commutative properties are also used to simplify expressions. For example, suppose x is a number: $(7 + x) + 8$ can be simplified by adding the numbers together as follows.

$(7 + x) + 8 = 8 + (7 + x)$ Commutative property of addition
$= (8 + 7) + x$ Associative property of addition
$= 15 + x$ You can also write the answer as $x + 15$ using the commutative property of addition.

Of course, you normally just add the 7 and 8 without going through all these steps, but this example shows *why* and *how* your shortcut can be done.

EXAMPLE 3 **Using the properties to simplify expressions**

Simplify: $6 + 4x + 8$

SOLUTION

$6 + 4x + 8 = (6 + 4x) + 8$ Order of operations
$= 8 + (6 + 4x)$ Commutative property of addition
$= (8 + 6) + 4x$ Associative property of addition
$= 14 + 4x$ Add (parentheses aren't needed).

PROBLEM 3

Simplify: $3 + 5x + 8$

You can also write the answer as $4x + 14$ using the commutative property of addition.

C Identifying Identities and Inverses

The properties we've just mentioned are applicable to all real numbers. We now want to discuss two special numbers that have unique properties, the numbers 0 and 1. If we add 0 to a number, the number is unchanged; that is, the number 0 preserves the **identity** of all numbers under addition. Thus, 0 is called the *identity element for addition*. This definition can be stated as follows.

Answers

2. a. 5 **b.** 7 **c.** $\frac{1}{3}$ **d.** 4
3. $11 + 5x$ or $5x + 11$.

Web It

If you want to see a humorous presentation of the zero concept, go to link 1-6-1 at the Bello Website, mhhe.com/bello.

IDENTITY ELEMENT FOR ADDITION

Zero is the **identity element for addition;** that is,

$$a + 0 = 0 + a = a \quad \text{for any number } a$$

The number 1 preserves the **identity** of all numbers under multiplication; that is, if we multiply a number by 1, the number remains unchanged. Thus, 1 is called the *identity element for multiplication,* as stated here.

IDENTITY ELEMENT FOR MULTIPLICATION

The number 1 is the **identity element for multiplication;** that is,

$$a \cdot 1 = 1 \cdot a = a \quad \text{for any number } a$$

We complete our list of properties by stating two ideas that we will discuss fully later.

Teaching Tip

Zero is its own additive inverse but has no reciprocal.

ADDITIVE INVERSE (OPPOSITE)

For every number a, there exists another number $-a$ called its **additive inverse** (or **opposite**) such that $a + (-a) = 0$.

When dealing with multiplication, the *multiplicative inverse* is called the *reciprocal.*

MULTIPLICATIVE INVERSE (RECIPROCAL)

For every number a (except 0), there exists another number $\frac{1}{a}$ called the **multiplicative inverse** (or **reciprocal**) such that

$$a \cdot \frac{1}{a} = 1$$

EXAMPLE 4 Identifying the property used

Name the property illustrated in each of the following statements:

a. $5 + (-5) = 0$ **b.** $\frac{8}{3} \cdot \frac{3}{8} = 1$

c. $0 + 9 = 9$ **d.** $7 \cdot 1 = 7$

SOLUTION

a. $5 + (-5) = 0$ Additive inverse

b. $\frac{8}{3} \cdot \frac{3}{8} = 1$ Multiplicative inverse

c. $0 + 9 = 9$ Identity element for addition

d. $7 \cdot 1 = 7$ Identity element for multiplication

PROBLEM 4

Name the property illustrated:

a. $8 \cdot 1 = 8$ **b.** $\frac{7}{5} \cdot \frac{5}{7} = 1$

c. $3 + 0 = 3$ **d.** $(-4) + 4 = 0$

Answers

4. a. Identity element for mult.
b. Multiplicative inverse
c. Identity element for add.
d. Additive inverse

EXAMPLE 5 Using the properties

Fill in the blank so that the result is a true statement:

a. ___ $\cdot$ 0.5 = 0.5 **b.** $\frac{3}{4}$ + ___ = 0 **c.** ___ + 2.5 = 0

d. ___ $\cdot \frac{3}{5} = 1$ **e.** 1.8 $\cdot$ ___ = 1.8 **f.** $\frac{3}{4}$ + ___ = $\frac{3}{4}$

SOLUTION

a. $\underline{1} \cdot 0.5 = 0.5$ — Identity element for multiplication

b. $\frac{3}{4} + \underline{\left(-\frac{3}{4}\right)} = 0$ — Additive inverse property

c. $\underline{-2.5} + 2.5 = 0$ — Additive inverse property

d. $\underline{\frac{5}{3}} \cdot \frac{3}{5} = 1$ — Multiplicative inverse property

e. $1.8 \cdot \underline{1} = 1.8$ — Identity element for multiplication

f. $\frac{3}{4} + \underline{0} = \frac{3}{4}$ — Identity element for addition

PROBLEM 5

Fill in the blank so that the result is a true statement:

a. $\frac{7}{3}$ + ___ = $\frac{7}{3}$

b. $\frac{4}{9} \cdot$ ___ = 1

c. $\frac{5}{8}$ + ___ = 0

d. 2.4 $\cdot$ ___ = 2.4

e. 9.1 + ___ = 0

f. 0.2 $\cdot$ ___ = 0.2

D Using the Identity and Inverse Properties to Simplify Expressions

EXAMPLE 6 Using the properties of addition

Use the properties of addition to simplify: $-3x + 5 + 3x$

SOLUTION

$-3x + 5 + 3x = (-3x + 5) + 3x$	Order of operations
$= 3x + (-3x + 5)$	Commutative property of addition
$= [3x + (-3x)] + 5$	Associative property of addition
$= 0 + 5$	Additive inverse property
$= 5$	Identity element for addition

PROBLEM 6

Use the addition properties to simplify: $-5x + 7 + 5x$

Keep in mind that after you get enough practice, you won't need to go through all the steps we show here. We have given you all the steps to make sure you understand *why* and *how* we simplify expressions.

E Using the Distributive Property to Remove Parentheses

The properties we've discussed all contain a single operation. Now suppose you wish to multiply a number, say 7, by the sum of 4 and 5. As it turns out, $7 \cdot (4 + 5)$ can be obtained in two ways:

$$7 \cdot (4 + 5) \rightarrow 7 \cdot 9 \rightarrow 63$$

Add within the parentheses first.

Answers

5. a. 0 **b.** $\frac{9}{4}$ **c.** $-\frac{5}{8}$ **d.** 1 **e.** -9.1 **f.** 1 **6.** 7

or

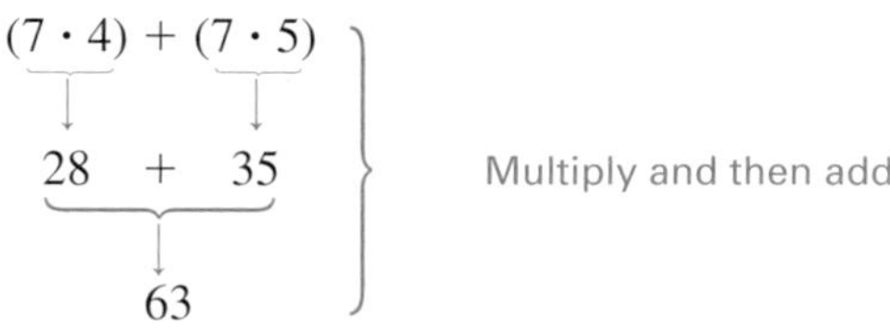

Thus

$$7 \cdot (4 + 5) = (7 \cdot 4) + (7 \cdot 5)$$ *First* multiply 4 by 7 and *then* 5 by 7.

The parentheses in $(7 \cdot 4) + (7 \cdot 5)$ can be omitted since, by the order of operations, multiplications must be done first. Thus, multiplication distributes over addition as follows:

Distributive Property $a(b + c) = ab + ac$ for any numbers a, b, and c

Note that $a(b + c)$ means $a \cdot (b + c)$, ab means $a \cdot b$, and ac means $a \cdot c$. The distributive property can also be extended to more than two numbers inside the parentheses. Thus $a(b + c + d) = ab + ac + ad$. Moreover, $(b + c)a = ba + ca$.

EXAMPLE 7 Using the distributive property

Use the distributive property to multiply the following (x, y, and z are real numbers):

a. $8(2 + 4)$ **b.** $3(x + 5)$ **c.** $3(x + y + z)$ **d.** $(x + y)z$

SOLUTION

a. $8(2 + 4) = 8 \cdot 2 + 8 \cdot 4$

$= 16 + 32$ Multiply first.

$= 48$ Add next.

b. $3(x + 5) = 3x + 3 \cdot 5$

$= 3x + 15$ This expression cannot be simplified further since we don't know the value of x.

c. $3(x + y + z) = 3x + 3y + 3z$

d. $(x + y)z = xz + yz$

PROBLEM 7

Use the distributive property to multiply:

a. $(a + b)c$ **b.** $5(a + b + c)$

c. $4(a + 7)$ **d.** $3(2 + 9)$

EXAMPLE 8 Using the distributive property

Use the distributive property to multiply:

a. $-2(a + 7)$ **b.** $-3(x - 2)$ **c.** $-(a - 2)$

SOLUTION

a. $-2(a + 7) = -2a + (-2)7$

$= -2a - 14$

b. $-3(x - 2) = -3x + (-3)(-2)$

$= -3x + 6$

c. What does $-(a - 2)$ mean? Since

$$-a = -1 \cdot a$$

PROBLEM 8

Use the distributive property to multiply:

a. $-3(x + 5)$

b. $-5(a - 2)$

c. $-(x - 4)$

Answers

7. a. $ac + bc$ **b.** $5a + 5b + 5c$ **c.** $4a + 28$ **d.** 33
8. a. $-3x - 15$ **b.** $-5a + 10$ **c.** $-x + 4$

we have

$$\begin{aligned}-(a-2) &= -1\cdot(a-2)\\ &= -1\cdot a + (-1)\cdot(-2)\\ &= -a+2\end{aligned}$$

Finally, for easy reference, here is a list of all the properties we've studied.

Properties of the Real Numbers

If a, b, and c are real numbers, then the following properties hold.

Addition	Multiplication	Property
$a+b=b+a$	$a\cdot b=b\cdot a$	Commutative property
$a+(b+c)=(a+b)+c$	$a\cdot(b\cdot c)=(a\cdot b)\cdot c$	Associative property
$a+0=0+a=a$	$a\cdot 1=1\cdot a=a$	Identity property
$a+(-a)=0$	$a\cdot\frac{1}{a}=1\quad(a\neq 0)$	Inverse property
$a(b+c)=ab+ac$		Distributive property

CAUTION

The commutative and associative properties apply to addition and multiplication but **not** to subtraction and division. For example,

$$3-5\neq 5-3 \quad\text{and}\quad 6\div 2\neq 2\div 6$$

Also,

$$5-(4-2)\neq(5-4)-2 \quad\text{and}\quad 6\div(3\div 3)\neq(6\div 3)\div 3$$

Web It

If you want to see a complete lesson dealing with the properties of the real numbers, go to the Bello Website at mhhe.com/bello and try link 1-6-2.

For a lighter presentation with humorous situations, go to link 1-6-3 and select the topics you want.

Finally, do you want to know what properties the whole numbers satisfy? Go to link 1-6-4 at the Bello Website and find out!

Exercises 1.6

A In Problems 1–10, name the property illustrated in each statement.

1. $9+8=8+9$
Commutative property of addition

2. $b\cdot a=a\cdot b$ Commutative property of multiplication

3. $4\cdot 3=3\cdot 4$
Commutative property of multiplication

4. $(a+4)+b=a+(4+b)$
Associative property of addition

5. $3(x+6)=3x+3\cdot 6$
Distributive property

6. $8(2+x)=8\cdot 2+8x$
Distributive property

7. $a\cdot(b\cdot c)=a\cdot(c\cdot b)$
Commutative property of multiplication

8. $a\cdot(b\cdot c)=(a\cdot b)\cdot c$
Associative property of multiplication

9. $a+(b+3)=(a+b)+3$
Associative property of addition

10. $(a+3)+b=(3+a)+b$
Commutative property of addition

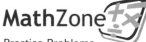

B In Problems 11–18, name the property illustrated in each statement, then find the missing number that makes the statement correct.

11. $-3 + (5 + 1.5) = (-3 + \_\_\_) + 1.5$
Associative property of addition; 5

12. $(-4 + \_\_\_) + \frac{1}{5} = -4 + \left(3 + \frac{1}{5}\right)$
Associative property of addition; 3

13. $7 \cdot \frac{1}{8} = \frac{1}{8} \cdot \_\_\_$
Commutative property of multiplication; 7

14. $\frac{9}{4} \cdot \_\_\_ = 3 \cdot \frac{9}{4}$
Commutative property of multiplication; 3

15. $\_\_\_ + \frac{3}{4} = \frac{3}{4} + 6.5$
Commutative property of addition; 6.5

16. $7.5 + \_\_\_ = \frac{5}{8} + 7.5$
Commutative property of addition; $\frac{5}{8}$

17. $\left(\frac{3}{7} \cdot 5\right) \cdot \_\_\_ = \frac{3}{7} \cdot (5 \cdot 2)$
Associative property of multiplication; 2

18. $\left(\frac{2}{9} \cdot \_\_\_\right) \cdot 4 = \frac{2}{9} \cdot (2 \cdot 4)$
Associative property of multiplication; 2

B D In Problems 19–24, simplify.

19. $5 + 2x + 8$ $13 + 2x$

20. $-10 + 2y + 12$ $2 + 2y$

21. $\frac{1}{5}a - 3 + 3 - \frac{1}{5}a$ 0

22. $\frac{2}{3}b + 4 - 4 - \frac{2}{3}b$ 0

23. $5c + (-5c) + \frac{1}{5} \cdot 5$ 1

24. $-\frac{1}{7} \cdot 7 - 4x + 4x$ -1

C In Problems 25–28, name the property illustrated in each statement.

25. $9 \cdot \frac{1}{9} = 1$ Multiplicative inverse property

26. $10 \cdot 1 = 10$ Identity property for multiplication

27. $8 + 0 = 8$ Identity property for addition.

28. $-6 + 6 = 0$ Additive inverse property.

E In Problems 29–38, use the distributive property to fill in the blank.

29. $8(9 + \_\_\_) = 8 \cdot 9 + 8 \cdot 3$ 3

30. $7(\_\_\_ + 10) = 7 \cdot 4 + 7 \cdot 10$ 4

31. $\_\_\_(3 + b) = 15 + 5b$ 5

32. $3(\_\_\_ + b) = 3 \cdot 8 + 3b$ 8

33. $\_\_\_(b + c) = ab + ac$ a

34. $3(x + 4) = 3x + 3 \cdot \_\_\_$ 4

35. $-4(x + \_\_\_) = -4x + 8$ -2

36. $-3(x + 5) = -3x + \_\_\_$ -15

37. $-2(5 + c) = -2 \cdot 5 + \_\_\_ \cdot c$ -2

38. $-3(\_\_\_ + a) = 6 + (-3)a$ -2

In Problems 39–78, use the distributive property to multiply.

39. $6(4 + x)$
$24 + 6x$

40. $3(2 + x)$
$6 + 3x$

41. $8(x + y + z)$
$8x + 8y + 8z$

42. $5(x + y + z)$
$5x + 5y + 5z$

43. $6(x + 7)$
$6x + 42$

44. $7(x + 2)$
$7x + 14$

45. $(a + 5)b$
$ab + 5b$

46. $(a + 2)c$
$ac + 2c$

47. $6(5 + b)$
$30 + 6b$

48. $3(7 + b)$
$21 + 3b$

49. $-4(x + y)$
$-4x - 4y$

50. $-3(a + b)$
$-3a - 3b$

51. $-9(a + b)$
$-9a - 9b$

52. $-6(x + y)$
$-6x - 6y$

53. $-3(4x + 2)$
$-12x - 6$

54. $-2(3a + 9)$
$-6a - 18$

55. $-\left(\frac{3a}{2} - \frac{6}{7}\right)$ $-\frac{3a}{2} + \frac{6}{7}$

56. $-\left(\frac{2x}{3} - \frac{1}{5}\right)$ $-\frac{2x}{3} + \frac{1}{5}$

57. $-(2x - 6y)$ $-2x + 6y$

58. $-(3a - 6b)$ $-3a + 6b$

59. $-(2.1 + 3y)$ $-2.1 - 3y$

60. $-(5.4 + 4b)$ $-5.4 - 4b$

61. $-4(a + 5)$ $-4a - 20$

62. $-6(x + 8)$ $-6x - 48$

63. $-x(6 + y)$ $-6x - xy$

64. $-y(2x + 3)$ $-2xy - 3y$

65. $-8(x - y)$ $-8x + 8y$

66. $-9(a - b)$ $-9a + 9b$

67. $-3(2a - 7b)$ $-6a + 21b$

68. $-4(3x - 9y)$ $-12x + 36y$

69. $0.5(x + y - 2)$ $0.5x + 0.5y - 1.0$

70. $0.8(a + b - 6)$ $0.8a + 0.8b - 4.8$

71. $\frac{6}{5}(a - b + 5)$ $\frac{6}{5}a - \frac{6}{5}b + 6$

72. $\frac{2}{3}(x - y + 4)$ $\frac{2}{3}x - \frac{2}{3}y + \frac{8}{3}$

73. $-2(x - y + 4)$ $-2x + 2y - 8$

74. $-4(a - b + 8)$ $-4a + 4b - 32$

75. $-0.3(x + y - 6)$ $-0.3x - 0.3y + 1.8$

76. $-0.2(a + b - 3)$ $-0.2a - 0.2b + 0.6$

77. $-\frac{5}{2}(a - 2b + c - 1)$ $-\frac{5}{2}a + 5b - \frac{5}{2}c + \frac{5}{2}$

78. $-\frac{4}{7}(2a - b + 3c - 5)$ $-\frac{8}{7}a + \frac{4}{7}b - \frac{12}{7}c + \frac{20}{7}$

SKILL CHECKER

Try the Skill Checker Exercises so you'll be ready for the next section.

Find:

79. 3^4 81

80. 2^3 8

81. $(-3)^4$ 81

82. $(-2)^4$ 16

83. $(-3)^3$ -27

84. $(-2)^3$ -8

85. $(-2)^6$ 64

86. $(-3)^6$ 729

USING YOUR KNOWLEDGE

Multiplication Made Easy

The distributive property can be used to simplify certain multiplications. For example, to multiply 8 by 43, we can write

$$\begin{aligned} 8(43) &= 8(40 + 3) \\ &= 320 + 24 \\ &= 344 \end{aligned}$$

Use this idea to multiply:

87. 7(38) 266

88. 8(23) 184

89. 6(46) 276

90. 9(52) 468

WRITE ON

91. Explain why zero has no reciprocal. Answers may vary.

92. We have seen that addition is commutative since $a + b = b + a$. Is subtraction commutative? Explain. No; answers may vary.

93. The associative property of addition states that

$$a + (b + c) = (a + b) + c.$$

Is $a - (b - c) = (a - b) - c$? Explain. No; answers may vary.

94. Is $a + (b \cdot c) = (a + b)(a + c)$? Give examples to support your conclusion. No; answers may vary.

95. Multiplication is distributive over addition since $a \cdot (b + c) = a \cdot b + a \cdot c$. Is multiplication distributive over subtraction? Explain. Yes; answers may vary.

MASTERY TEST

If you know how to do these problems, you have learned your lesson!

Use the distributive property to multiply:

96. $-3(a + 4)$ $-3a - 12$

97. $-4(a - 5)$ $-4a + 20$

98. $-(a - 2)$ $-a + 2$

99. $4(a + 6)$ $4a + 24$

100. $2(x + y + z)$ $2x + 2y + 2z$

Simplify:

101. $-2x + 7 + 2x$ 7

102. $-4 - 2x + 2x + 4$ 0

103. $7 + 4x + 9$ $16 + 4x$

104. $-3 - 4x + 2$ $-1 - 4x$

Fill in the blank so that the result is a true statement:

105. $\frac{2}{5} + \_\_\_ = \frac{2}{5}$ 0

106. $\frac{1}{4} + \_\_\_ = 0$ $-\frac{1}{4}$

107. $-3.2 \cdot \_\_\_ = -3.2$ 1

108. $\_\_\_ + 4.2 = 4.2$ 0

109. $3(x + \_\_\_) = 3x + 6$ 2

110. $-2(x + 5) = -2x + \_\_\_$ -10

Name the property illustrated in each of the following statements:

111. $3 \cdot 1 = 3$
Identity property for multiplication

112. $\frac{5}{6} \cdot \frac{6}{5} = 1$
Multiplicative inverse

113. $2 + 0 = 2$
Identity property for addition

114. $-3 + 3 = 0$
Additive inverse

115. $(6 \cdot x) \cdot 2 = 6 \cdot (x \cdot 2)$
Associative property for multiplication

116. $3 \cdot (4 \cdot 5) = 3 \cdot (5 \cdot 4)$
Commutative property for multiplication

117. $(x + 2) + 5 = x + (2 + 5)$
Associative property for addition

118. $-3 \cdot 1 = -3$
Identity property for multiplication

Name the property illustrated in each statement, then find the missing number that makes the statement correct:

119. $\left(\frac{1}{5} \cdot 4\right) \cdot 2 = \frac{1}{5} \cdot (4 \cdot \_\_\_)$
Associative property for multiplication; 2

120. $\frac{1}{3} + 2.4 = \_\_\_ + \frac{1}{3}$
Commutative property for addition; 2.4

121. $\frac{2}{5} \cdot \_\_\_ = 1 \cdot \frac{2}{5}$
Commutative property for multiplication; 1

122. $\frac{2}{11} + \_\_\_ = 0$
Additive inverse; $-\frac{2}{11}$

123. $-2 + (3 + 1.4) = (-2 + \_\_\_) + 1.4$
Associative property for addition; 3

1.7 SIMPLIFYING EXPRESSIONS

To Succeed, Review How To . . .

1. Add and subtract real numbers (pp. 61–66).
2. Use the distributive property to simplify expressions (pp. 92–94).

Objectives

A Add and subtract like terms.

B Use the distributive property to remove parentheses and then combine like terms.

C Translate words into algebraic expressions and solve applications.

GETTING STARTED

Combining Like Terms

If 3 tacos cost \$1.39 and 6 tacos cost \$2.39, then 3 tacos + 6 tacos will cost \$1.39 + \$2.39; that is, 9 tacos will cost \$3.78. Note that

$$3 \text{ tacos} + 6 \text{ tacos} = 9 \text{ tacos} \quad \text{and}$$

$$\$1.39 + \$2.39 = \$3.78$$

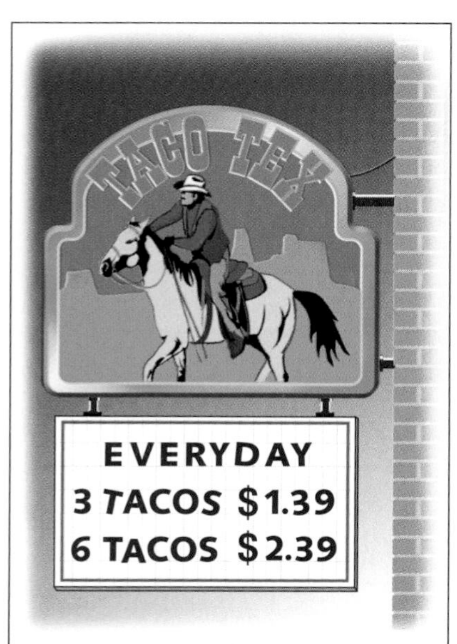

In algebra, an **expression** is a collection of numbers and letters representing numbers (**variables**) connected by operation signs. The parts to be added or subtracted in these expressions are called **terms.** Thus xy^2 is an expression with one term; $x + y$ is an expression with two terms, x and y; and $3x^2 - 2y + z$ has three terms, $3x^2$, $-2y$, and z.

The term "3 tacos" uses the number 3 to tell "how many." The number 3 is called the **numerical coefficient** (or simply the **coefficient**) of the term. Similarly, the terms $5x$, y, and $-8xy$ have numerical coefficients of 5, 1, and -8, respectively.

When two or more terms are *exactly alike* (except possibly for their coefficients or the order in which the factors are multiplied), they are called **like terms.** Thus like terms contain the *same variables* with the *same exponents,* but their coefficients may be different. So 3 tacos and 6 tacos are like terms, $3x$ and $-5x$ are like terms, and $-xy^2$ and $7xy^2$ are like terms. On the other hand, $2x$ and $2x^2$ or $2xy^2$ and $2x^2y$ are *not* like terms. In this section we learn how to simplify expressions using like terms.

Adding and Subtracting Like Terms

In an algebraic expression, like terms can be combined into a single term just as 3 tacos and 6 tacos can be combined into the single term 9 tacos. To combine like terms, make sure that the variable parts of the terms to be combined are identical (only the coefficients may be different), and then add (or subtract) the coefficients and keep the variables. This can be done using the distributive property:

$$\underbrace{3x}_{(x+x+x)} + \underbrace{5x}_{(x+x+x+x+x)} = \underbrace{(3+5)x}_{x+x+x+x+x+x+x+x} = 8x$$

$$\underbrace{2x^2}_{(x^2+x^2)} + \underbrace{4x^2}_{(x^2+x^2+x^2+x^2)} = \underbrace{(2+4)x^2}_{x^2+x^2+x^2+x^2+x^2+x^2} = 6x^2$$

$$\underbrace{3ab}_{(ab+ab+ab)} + \underbrace{2ab}_{(ab+ab)} = \underbrace{(3+2)ab}_{ab+ab+ab+ab+ab} = 5ab$$

But what about another situation, such as combining $-3x$ and $-2x$? We first write

$$\underbrace{-3x}_{(-x)+(-x)+(-x)} + \underbrace{(-2x)}_{(-x)+(-x)}$$

Thus

$$-3x + (-2x) = [-3 + (-2)]x = -5x$$

Note that we write the addition of $-3x$ and $-2x$ as $-3x + (-2x)$, using parentheses around the $-3x$ and $-2x$. We do this to avoid the confusion of writing

$$-3x + -2x$$

Never use two signs of operation together without parentheses.

As you can see, if both quantities to be added are preceded by minus signs, the result is preceded by a minus sign. If they are both preceded by plus signs, the result is preceded by a plus sign. But what about this expression, $-5x + (3x)$? You can visualize this expression as

$$\underbrace{-5x}_{(-x)+(-x)+(-x)+(-x)+(-x)} + \underbrace{3x}_{x+x+x}$$

Teaching Tip

Using the textbook's model for $-5x + 3x = -2x$, have students write a similar model for $-5x + 5x = 0$.

Since we have two more negative x's than positive x's, the result is $-2x$. Thus

$$-5x + (3x) = (-5 + 3)x = -2x \quad \text{Use the distributive property.}$$

On the other hand, $5x + (-3x)$ can be visualized as

$$\underbrace{5x}_{x+x+x+x+x} + \underbrace{(-3x)}_{(-x)+(-x)+(-x)}$$

and the answer is $2x$; that is,

$$5x + (-3x) = [5 + (-3)]x = 2x \quad \text{Use the distributive property.}$$

Now here is an easy one: What is $x + x$? Since $1 \cdot x = x$, the coefficient of x is assumed to be 1. Thus

$$\begin{aligned} x + x &= 1 \cdot x + 1 \cdot x \\ &= (1 + 1)x \\ &= 2x \end{aligned}$$

In all the examples that follow, make sure you use the fact that

$$ac + bc = (a + b)c$$

to combine the numerical coefficients.

EXAMPLE 1 **Combining like terms using addition**

Combine like terms:

a. $-7x + 2x$ **b.** $-4x + 6x$

c. $-2x + (-5x)$ **d.** $x + (-5x)$

SOLUTION

a. $-7x + 2x = (-7 + 2)x = -5x$

b. $-4x + 6x = (-4 + 6)x = 2x$

c. $-2x + (-5x) = [-2 + (-5)]x = -7x$

d. First, recall that $x = 1 \cdot x$. Thus

$$x + (-5x) = 1x + (-5x) = [1 + (-5)]x = -4x$$

PROBLEM 1

Combine like terms:

a. $-9x + 3x$ **b.** $-6x + 8x$

c. $-3x + (-7x)$ **d.** $x + (-6x)$

Subtraction of like terms is defined in terms of addition, as stated here.

SUBTRACTION

To subtract a number b from another number a, add the *additive inverse* (*opposite*) of b to a; that is,

$$a - b = a + (-b)$$

Web It

If you want a computer to do the work for you automatically, step by step, go to the Bello Website at mhhe.com/bello and try link 1-7-1.

Select the topic Combining Like Terms.

As before, to subtract like terms, we use the fact that $ac + bc = (a + b)c$. Thus

Additive inverse

$$3x - 5x = 3x + (-5x) = [3 + (-5)]x = -2x$$

$$-3x - 5x = -3x + (-5x) = [(-3) + (-5)]x = -8x$$

$$3x - (-5x) = 3x + (5x) = (3 + 5)x = 8x$$

$$-3x - (-5x) = -3x + (5x) = (-3 + 5)x = 2x$$

Note that

$$-3x - (-5x) = -3x + 5x$$

In general, we have the following.

SUBTRACTING $-b$

$$a - (-b) = a + b$$ $-(-b)$ is replaced by $+b$, since $-(-b) = +b$.

Answers

1. a. $-6x$ **b.** $2x$ **c.** $-10x$ **d.** $-5x$

We can now combine like terms involving subtraction.

EXAMPLE 2 **Combining like terms using subtraction**

Combine like terms:

a. $7ab - 9ab$

b. $8x^2 - 3x^2$

c. $-5ab^2 - (-8ab^2)$

d. $-6a^2b - (-2a^2b)$

SOLUTION

a. $7ab - 9ab = 7ab + (-9ab) = [7 + (-9)]ab = -2ab$

b. $8x^2 - 3x^2 = 8x^2 + (-3x^2) = [8 + (-3)]x^2 = 5x^2$

c. $-5ab^2 - (-8ab^2) = -5ab^2 + 8ab^2 = (-5 + 8)ab^2 = 3ab^2$

d. $-6a^2b - (-2a^2b) = -6a^2b + 2a^2b = (-6 + 2)a^2b = -4a^2b$

PROBLEM 2

Combine like terms:

a. $3xy - 8xy$

b. $7x^2 - 5x^2$

c. $-3xy^2 - (-7xy^2)$

d. $-5x^2y - (-4x^2y)$

Teaching Tip

In general, the phrase "combining like terms" applies to addition and subtraction.

B Removing Parentheses

Sometimes it's necessary to remove parentheses before combining like terms. For example, to combine like terms in

$$(3x + 5) + (2x - 2)$$

we have to remove the parentheses *first*. If there is a plus sign (or no sign) in front of the parentheses, we can simply remove the parentheses; that is,

$$+(a + b) = a + b \quad \text{and} \quad +(a - b) = a - b$$

With this in mind, we have

$$\begin{aligned}(3x + 5) + (2x - 2) &= 3x + 5 + 2x - 2 \\ &= 3x + 2x + 5 - 2 && \text{Use the commutative property.} \\ &= (3x + 2x) + (5 - 2) && \text{Use the associative property.} \\ &= 5x + 3 && \text{Simplify.}\end{aligned}$$

Note that we used the properties we studied to group like terms together.

Once you understand the use of these properties, you can then see that the simplification of $(3x + 5) + (2x - 2)$ consists of just adding $3x$ to $2x$ and 5 to -2. The computation can then be shown like this:

Like terms

$$\begin{aligned}(3x + 5) + (2x - 2) &= 3x + 5 + 2x - 2 \\ &= 5x + 3\end{aligned}$$

Like terms

We use this idea in the next example.

Web It

For a lesson on how to simplify expressions, go to the Bello Website mhhe.com/bello and try link 1-7-2.

EXAMPLE 3 **Removing parentheses preceded by a plus sign**

Remove parentheses and combine like terms:

a. $(4x - 5) + (7x - 3)$

b. $(3a + 5b) + (4a - 9b)$

SOLUTION We first remove parentheses; then we add like terms.

a. $(4x - 5) + (7x - 3) = 4x - 5 + 7x - 3$
$= 11x - 8$

b. $(3a + 5b) + (4a - 9b) = 3a + 5b + 4a - 9b$
$= 7a - 4b$

PROBLEM 3

Remove parentheses and combine like terms:

a. $(7a - 5) + (8a - 2)$

b. $(4x + 6y) + (3x - 8y)$

Answers

2. a. $-5xy$ **b.** $2x^2$ **c.** $4xy^2$ **d.** $-x^2y$ **3. a.** $15a - 7$ **b.** $7x - 2y$

Teaching Tip

Summary for removing parentheses or brackets.

Expression	Procedure	Result
$+(x+2)$ $(x+2)$	drop parentheses	$x+2$
$-(x+2)$	drop parentheses *and* change all signs inside	$-x-2$
$3(x+2)$ $-4(x+2)$	distribute number outside times *all* terms inside	$3x+6$ $-4x-8$

To simplify

$$4x + 3(x - 2) - (x + 5)$$

recall that $-(x+5) = -1 \cdot (x+5) = -x - 5$. We proceed as follows:

$4x + \underbrace{3(x-2)} - (x+5)$	Given.
$= 4x + 3x - 6 - \underbrace{(x+5)}$	Use the distributive property.
$= 4x + 3x - 6 - x - 5$	Remove parentheses.
$= (4x + 3x - x) + (-6 - 5)$	Combine like terms.
$= 6x - 11$	Simplify.

EXAMPLE 4 **Removing parentheses preceded by a minus sign**

Remove parentheses and combine like terms:

$$8x - 2(x - 1) - (x + 3)$$

SOLUTION

$8x \underbrace{- 2(x-1)} - (x+3)$	
$= 8x - 2x + 2 \underbrace{- (x+3)}$	Use the distributive property.
$= 8x - 2x + 2 - x - 3$	Remove parentheses.
$= (8x - 2x - x) + (2 - 3)$	Combine like terms.
$= 5x + (-1)$ or $5x - 1$	Simplify.

PROBLEM 4

Remove parentheses and combine like terms:

$$9a - 3(a - 2) - (a + 4)$$

Sometimes parentheses occur within other parentheses. To avoid confusion, we use different grouping symbols. Thus we usually don't write $((x + 5) + 3)$. Instead, we write $[(x + 5) + 3]$. To combine like terms in such expressions, remove the *innermost grouping symbols first,* as we illustrate in the next example.

EXAMPLE 5 **Removing grouping symbols**

Remove grouping symbols and simplify:

$$[(x^2 - 1) + (2x + 5)] + [(x - 2) - (3x^2 + 3)]$$

SOLUTION We first remove the innermost parentheses and then combine like terms. Thus

$[(x^2 - 1) + (2x + 5)] + [(x - 2) - (3x^2 + 3)]$	
$= [x^2 - 1 + 2x + 5] + [x - 2 - 3x^2 - 3]$	Remove parentheses. Note that $-(3x^2 + 3) = -3x^2 - 3$.
$= [x^2 + 2x + 4] + [-3x^2 + x - 5]$	Combine like terms inside the brackets.
$= x^2 + 2x + 4 - 3x^2 + x - 5$	Remove brackets.
$= -2x^2 + 3x - 1$	Combine like terms.

PROBLEM 5

Remove grouping symbols and simplify:

$$[(a^2 - 2) + (3a + 4)] + [(a - 5) - (4a^2 + 2)]$$

Teaching Tip

When combining like terms, we usually write terms in order from the term with highest exponent on the variable to the one with the lowest.

C Translating Words into Algebraic Expressions: Applications

As we mentioned at the beginning of this section, we express ideas in algebra using expressions. In most applications, problems are first stated in words and have to be translated into algebraic expressions using mathematical symbols. Here's a short mathematics dictionary that will help you translate word problems.

Answers

4. $5a + 2$ **5.** $-3a^2 + 4a - 5$

Mathematics Dictionary

Addition (+)	Subtraction (−)
Write: $a + b$ (read "a plus b")	Write: $a - b$ (read "a minus b")
Words: Plus, sum, increase, more than, more, added to	Words: Minus, difference, decrease, less than, less, subtracted from
Examples: a plus b The sum of a and b a increased by b b more than a b added to a	Examples: a minus b The difference of a and b a decreased by b b less than a b subtracted from a

Multiplication (× or ·)	Division (÷ or the bar —)
Write: $a \cdot b$, ab, $(a)b$, $a(b)$, or $(a)(b)$ (read "a times b" or simply ab)	Write: $a \div b$ or $\frac{a}{b}$ (read "a divided by b")
Words: Times, of, product	Words: Divided by, quotient
Examples: a times b The product of a and b	Examples: a divided by b The quotient of a and b

Teaching Tip

Remind students that "of" usually indicates multiplication. Thus,

$$\frac{1}{2} \text{ of } b \text{ means } \frac{1}{2} \times b = \frac{b}{2}$$

The words and phrases contained in our mathematics dictionary involve the four fundamental operations of arithmetic. We use these words and phrases to translate sentences into *equations*. An **equation** is a sentence stating that two expressions are equal. Here are some words that in mathematics mean "equals":

Equals (=) means is, the same as, yields, gives, is obtained by

Teaching Tip

Avoid using "×" for a multiplication symbol. Also, the more commonly used symbol for division is the fraction bar.

We use these ideas in Example 6.

EXAMPLE 6 **Translating and finding the number of terms**

Write in symbols and indicate the number of terms to the right of the equals sign:

a. The area A of a circle is obtained by multiplying π by the square of the radius r.

b. The perimeter P of a rectangle is obtained by adding twice the length L to twice the width W.

c. The current I in a circuit is given by the quotient of the voltage V and the resistance R.

SOLUTION

a. The area A of a circle / is obtained by / multiplying π by the square of the radius r.

$$A \quad = \quad \pi r^2$$

There is one term, πr^2, to the right of the equals sign.

PROBLEM 6

Write in symbols and indicate the number of terms to the right of the equals sign:

a. The surface area S of a sphere of radius r is obtained by multiplying 4π by the square of the radius r.

b. The perimeter P of a triangle is obtained by adding the lengths a, b, and c of each of the sides.

c. If P dollars are invested for one year at a rate r, the principal P is the quotient of the interest I and the rate r.

Answers

6. a. $S = 4\pi r^2$; one term
b. $P = a + b + c$; three terms
c. $P = \frac{I}{r}$; one term

b. The perimeter P of a rectangle | is obtained by | adding twice the length L to twice the width W.

$$P = 2L + 2W$$

There are two terms, $2L$ and $2W$, to the right of the equals sign.

c. The current I in a circuit | is given by | the quotient of the voltage V and the resistance R.

$$I = \frac{V}{R}$$

There is one term to the right of the equals sign.

EXAMPLE 7 Number game

Ready for some fun? Let's play a game. Pick a number. Add 2 to it. Triple the result. Subtract 6. Divide by 3. What do you get? The original number! How do we know? Translate and simplify!

SOLUTION The steps are on the left, and the translation is on the right.

Pick a number.	n
Add 2 to it.	$n + 2$
Triple the result.	$3(n + 2)$
Subtract 6.	$3(n + 2) - 6$
Divide by 3.	$\frac{3(n + 2) - 6}{3}$
What is the answer? Simplify!	
Distributive law.	$\frac{3n + 6 - 6}{3}$
Simplify numerator.	$\frac{3n}{3}$
Reduce.	n

Thus, you always get the original number n. Try it using 4, -2, and $\frac{1}{3}$.

PROBLEM 7

You try this one and find out what the answer will be: Pick a number. Add 3 to it. Double the result. Subtract twice the original number. Divide the result by 2. What is the answer?

Calculate It

You can follow the steps in the example with a calculator. Let's do it for the number 4. Here are the keystrokes:

The answer is the original number, 4. Your window should look like this:

```
4+2
               6
Ans*3-6
              12
Ans/3
               4
```

If you have a calculator, try it for -2. The keystrokes are identical and your final answer should be -2.

Answer

7. $\frac{2(n + 3) - 2n}{2} = 3$, so you always get 3.

Now that you know how to translate sentences into equations, let's translate symbols into words. For example, in the equation

$$W_T = W_b + W_s + W_e$$ Read "W sub T equals W sub b plus W sub s plus W sub e."

the letters T, b, s, and e are called **subscripts.** They help us represent the total weight W_T of a McDonald's® biscuit, which is composed of the weight of the biscuit W_b, the weight of the sausage W_s, and the weight of the eggs W_e. With this information, how would you translate the equation $W_T = W_b + W_s + W_e$? Here is one way: The total weight W_T of a McDonald's biscuit with sausage and eggs equals the weight W_b of the biscuit plus the weight W_s of the sausage plus the weight W_e of the eggs. (If you look at *The Fast-Food Guide,* you will see that $W_b = 75$ grams, $W_s = 43$ grams, and $W_e = 57$ grams.) Now try Example 8.

Teaching Tip

Remember, terms are separated by plus or minus signs.

EXAMPLE 8 **Translating symbols to words**

Translate into words: $P_T = P_A + P_B + P_C$, where P_T is the total pressure exerted by a gas, P_A is the partial pressure exerted by gas A, P_B is the partial pressure exerted by gas B, and P_C is the partial pressure exerted by gas C.

SOLUTION The total pressure P_T exerted by a gas is the sum of the partial pressures P_A, P_B, and P_C exerted by gases A, B, and C, respectively.

PROBLEM 8

Translate into words: $T_K = T_C + 273.15$, where T_K is the temperature in degrees Kelvin and T_C is the temperature in degrees Celsius.

You will find some problems like this in the *Using Your Knowledge.*

A NOTE ABOUT SUBSCRIPTS

Subscripts are used to distinguish individual variables that together represent a whole set of quantities. For example, W_1, W_2, and W_3 could be used to represent different weights; t_1, t_2, t_3, and t_4 to represent different times; and d_1 and d_2 to represent different distances.

Web It

If you want a tutorial that will show you how to simplify expressions and even has a calculator to do it for you, go to the Bello Website at mhhe.com/bello and try link 1-7-3.

For a lesson on translating words into algebraic equations with an assignment and answers, go to link 1-7-4.

Finally, for a colorful lesson with many examples, practice problems, and answers, go to link 1-7-5.

Answer

8. The temperature T_K in degrees Kelvin is the sum of the temperature T_C in degrees Celsius and 273.15.

Exercises 1.7

A In Problems 1–30, combine like terms (simplify).

1. $19a + (-8a)$
$11a$

2. $-2b + 5b$
$3b$

3. $-8c + 3c$
$-5c$

4. $-5d + (-7d)$
$-12d$

5. $4n^2 + 8n^2$
$12n^2$

6. $3x^2 + (-9x^2)$
$-6x^2$

7. $-3ab^2 + (-4ab^2)$
$-7ab^2$

8. $9ab^2 + (-3ab^2)$
$6ab^2$

9. $-4abc + 7abc$
$3abc$

10. $6xyz + (-9xyz)$
$-3xyz$

11. $0.7ab + (-0.3ab) + 0.9ab$ $1.3ab$

12. $-0.5x^2y + 0.8x^2y + 0.3x^2y$ $0.6x^2y$

13. $-0.3xy^2 + 0.2x^2y + (-0.6xy^2)$ $0.2x^2y - 0.9xy^2$

14. $0.2x^2y + (-0.3xy^2) + 0.4xy^2$ $0.2x^2y + 0.1xy^2$

15. $8abc^2 + 3ab^2c + (-8abc^2)$ $3ab^2c$

16. $3xy + 5ab + 2xy + (-3ab)$ $5xy + 2ab$

17. $-8ab + 9xy + 2ab + (-2xy)$ $-6ab + 7xy$

18. $7 + \frac{1}{2}x + 3 + \frac{1}{2}x$ $10 + x$

19. $\frac{1}{5}a + \frac{3}{7}a^2b + \frac{2}{5}a + \frac{1}{7}a^2b$ $\frac{3}{5}a + \frac{4}{7}a^2b$

20. $\frac{4}{9}ab + \frac{4}{5}ab^2 + \left(-\frac{1}{9}ab\right) + \left(-\frac{1}{5}a^2b\right)$ $\frac{1}{3}ab + \frac{4}{5}ab^2 - \frac{1}{5}a^2b$

21. $13x - 2x$ $11x$

22. $8x - (-2x)$ $10x$

23. $6ab - (-2ab)$ $8ab$

24. $4.2xy - (-3.7xy)$ $7.9xy$

25. $-4a^2b - (3a^2b)$ $-7a^2b$

26. $-8ab^2 - 4a^2b$ $-8ab^2 - 4a^2b$

27. $3.1t^2 - 3.1t^2$ 0

28. $-4.2ab - 3.8ab$ $-8.0ab$

29. $0.3x^2 - 0.3x^2$ 0

30. $0 - (-0.8xy^2)$ $0.8xy^2$

B In Problems 31–55, remove parentheses and combine like terms.

31. $(3xy + 5) + (7xy - 9)$ $10xy - 4$

32. $(8ab - 9) + (7 - 2ab)$ $6ab - 2$

33. $(7R - 2) + (8 - 9R)$ $-2R + 6$

34. $(5xy - 3ab) + (9ab - 8xy)$ $-3xy + 6ab$

35. $(5L - 3W) + (W - 6L)$ $-L - 2W$

36. $(2ab - 2ac) + (ab - 4ac)$ $3ab - 6ac$

37. $5x - (8x + 1)$ $-3x - 1$

38. $3x - (7x + 2)$ $-4x - 2$

39. $\frac{2x}{9} - \left(\frac{x}{9} - 2\right)$ $\frac{x}{9} + 2$

40. $\frac{5x}{7} - \left(\frac{2x}{7} - 3\right)$ $\frac{3x}{7} + 3$

41. $4a - (a + b) + 3(b + a)$ $6a + 2b$

42. $8x - 3(x + y) - (x - y)$ $4x - 2y$

43. $7x - 3(x + y) - (x + y)$ $3x - 4y$

44. $4(b - a) + 3(b + a) - 2(a + b)$ $5b - 3a$

45. $-(x + y - 2) + 3(x - y + 6) - (x + y - 16)$ $x - 5y + 36$

46. $[(a^2 - 4) + (2a^3 - 5)] + [(4a^3 + a) + (a^2 + 9)]$ $6a^3 + 2a^2 + a$

47. $(x^2 + 7 - x) + [-2x^3 + (8x^2 - 2x) + 5]$ $-2x^3 + 9x^2 - 3x + 12$

48. $[(0.4x - 7) + 0.2x^2] + [(0.3x^2 - 2) - 0.8x]$ $0.5x^2 - 0.4x - 9$

49. $\left[\left(\frac{1}{4}x^2 + \frac{1}{5}x\right) - \frac{1}{8}\right] + \left[\left(\frac{3}{4}x^2 - \frac{3}{5}x\right) + \frac{5}{8}\right]$ $x^2 - \frac{2}{5}x + \frac{1}{2}$

50. $3[3(x + 2) - 10] + [5 + 2(5 + x)]$ $11x + 3$

51. $2[3(2a - 4) + 5] - [2(a - 1) + 6]$ $10a - 18$

52. $-2[6(a - b) + 2a] - [3b - 4(a - b)]$ $-12a + 5b$

53. $-3[4a - (3 + 2b)] - [6(a - 2b) + 5a]$ $-23a + 18b + 9$

54. $-[-(x + y) + 3(x - y)] - [4(x + y) - (3x - 5y)]$ $-3x - 5y$

55. $-[-(0.2x + y) + 3(x - y)] - [2(x + 0.3y) - 5]$ $-4.8x + 3.4y + 5$

C In Problems 56–70, translate the sentences into equations and indicate the number of terms to the right of the equals sign.

56. The number of hours H a growing child should sleep is the difference between 17 and one-half the child's age A. $H = 17 - \frac{1}{2}A$; 2 terms

57. The reciprocal of f is the sum of the reciprocal of u and the reciprocal of v. $\frac{1}{f} = \frac{1}{u} + \frac{1}{v}$; 2 terms

58. The height h attained by a rocket is the sum of its height a at burnout and

$$\frac{r^2}{20}$$

where r is the speed at burnout. $h = a + \frac{r^2}{20}$; 2 terms

59. The Fahrenheit temperature F can be obtained by adding 40 to the quotient of n and 4, where n is the number of cricket chirps in 1 min. $F = \frac{n}{4} + 40$; 2 terms

60. The radius r of a circle is the quotient of the circumference C and 2π. $r = \frac{C}{2\pi}$; 1 term

61. The interest received I is the product of the principal P, the rate r, and the time t. $I = Prt$; 1 term

62. The total profit P_T equals the total revenue R_T minus the total cost C_T. $P_T = R_T - C_T$; 2 terms

63. The perimeter P is the sum of the lengths of the sides a, b, and c. $P = a + b + c$; 3 terms

64. The volume V of a certain gas is the pressure P divided by the temperature T. $V = \frac{P}{T}$; 1 term

65. The area A of a triangle is the product of the length of the base b and the height h, divided by 2. $A = \frac{bh}{2}$; 1 term

66. The area A of a circle is the product of π and the square of the radius r. $A = \pi r^2$; 1 term

67. The kinetic energy K of a moving object is the product of a constant C, the mass m of the object, and the square of its velocity v. $K = Cmv^2$; 1 term

68. The arithmetic mean m of two numbers is found by dividing their sum by 2. $m = \frac{a+b}{2}$; 1 term

69. The distance in feet d that an object falls is the product of 16 and the square of the time in second t that the object has been falling. $d = 16t^2$; 1 term

70. The horse power (hp) that a shaft can safely transmit is the product of a constant C, the speed s of the shaft, and the cube of the shaft's diameter d. hp $= Csd^3$; 1 term

SKILL CHECKER

Try the Skill Checker Exercises so you'll be ready for the next section.

Find:

71. $\frac{1}{4} + \frac{3}{8}$ $\frac{5}{8}$

72. $\frac{3}{5} + \frac{1}{3}$ $\frac{14}{15}$

73. $\frac{3}{4} - \frac{1}{5}$ $\frac{11}{20}$

74. $\frac{5}{6} - \frac{2}{9}$ $\frac{11}{18}$

USING YOUR KNOWLEDGE

Geometric Formulas

The ideas in this section can be used to simplify formulas. For example, the **perimeter** (distance around) of the given rectangle is found by following the blue arrows. The perimeter is

$$P = W + L + W + L$$
$$= 2W + 2L$$

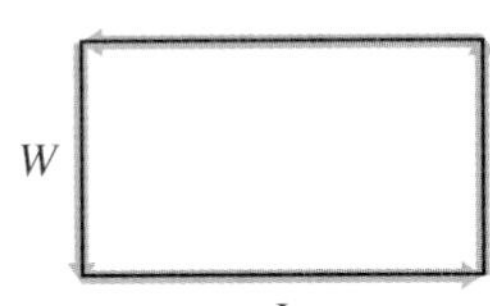

Use this idea to find the perimeter P of the given figure; then write a formula for P in symbols and in words.

75. The square of side S
$4S$; $P = 4S$

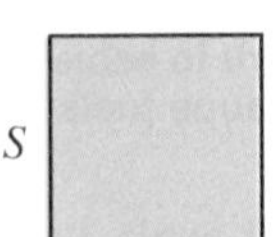

76. The parallelogram of base b and side s
$2b + 2s$; $P = 2b + 2s$

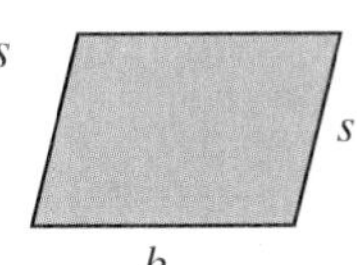

To obtain the actual measurement of certain perimeters, we have to add like terms if the measurements are given in feet and inches. For example, the perimeter of the rectangle shown here is

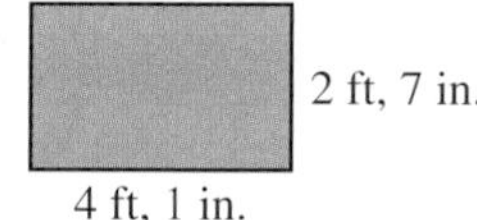

$$P = (2 \text{ ft} + 7 \text{ in.}) + (4 \text{ ft} + 1 \text{ in.}) + (2 \text{ ft} + 7 \text{ in.}) + (4 \text{ ft} + 1 \text{ in.})$$
$$= 12 \text{ ft} + 16 \text{ in.}$$

Since 16 inches = 1 foot + 4 inches,

$$P = 12 \text{ ft} + (1 \text{ ft} + 4 \text{ in.})$$
$$= 13 \text{ ft} + 4 \text{ in.}$$

Use these ideas to obtain the perimeter of the given rectangles.

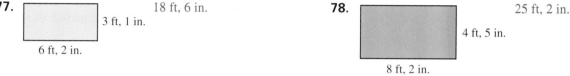

77. 18 ft, 6 in.

3 ft, 1 in.

6 ft, 2 in.

78. 25 ft, 2 in.

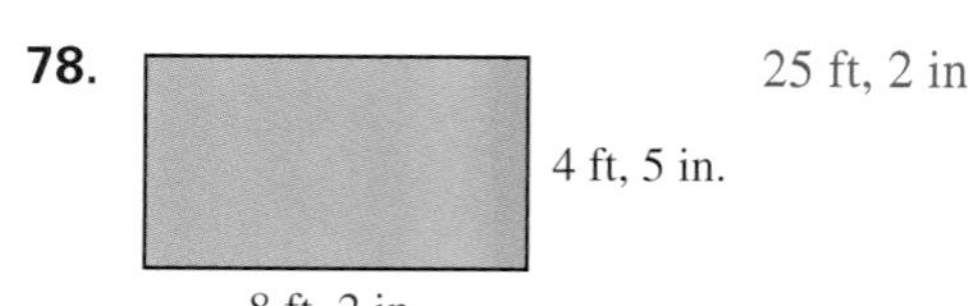

79. The U.S. Postal Service has a regulation stating that "the sum of the length and girth of a package may be no more than 108 inches." What is the sum of the length and girth of the rectangular package below? (*Hint:* The girth of the package is obtained by measuring the length of the red line.) $2h + 2w + L$

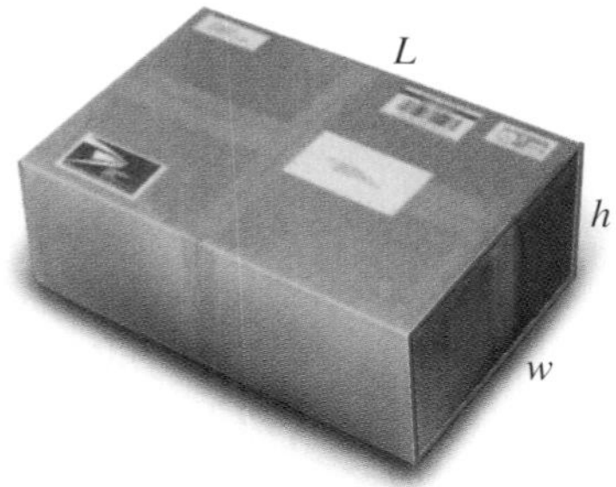

80. Write in simplified form the height of the step block shown. $6x$

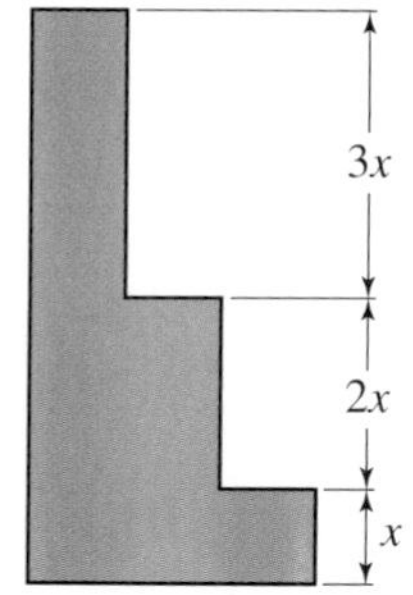

81. Write in simplified form the length of the metal plate shown. $(7x + 2)$ in.

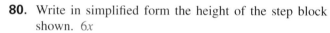

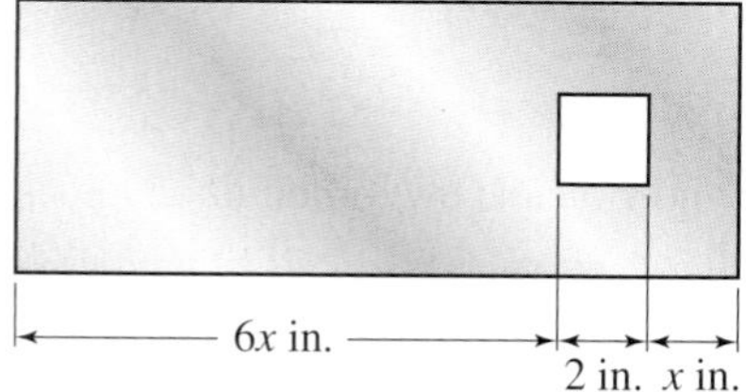

WRITE ON

82. Explain the difference between a factor and a term. Answers may vary.

83. Write the procedure you use to combine like terms. Answers may vary.

84. Explain how to remove parentheses when no sign or a plus sign precedes an expression within parentheses. Answers may vary.

85. Explain how to remove parentheses when a minus sign precedes an expression within parentheses. Answers may vary.

MASTERY TEST

If you know how to do these problems, you have learned your lesson!

Write in symbols and indicate the number of terms to the right of the equals sign:

86. When P dollars are invested at r percent, the amount of money A in an account at the end of 1 year is the sum of P and the product of P and r. $A = P + Pr$; 2 terms

87. The area A of a rectangle is obtained by multiplying the length L by the width W. $A = LW$; 1 term

Remove parentheses and combine like terms:

88. $9x - 3(x - 2) - (2x + 7)$ $4x - 1$

89. $(8y - 7) + 2(3y - 2)$ $14y - 11$

90. $(4t + u) + 4(5t - 7u)$ $24t - 27u$

91. $-6ab^3 - (-3ab^3)$ $-3ab^3$

Remove grouping symbols and simplify:

92. $[(2x^2 - 3) + (3x + 1)] + 2[(x - 1) - (x^2 + 2)]$
$5x - 8$

93. $3[(5 - 2x^2) + (2x - 1)] - 3[(5 - 2x) - (3 + x^2)]$
$-3x^2 + 12x + 6$

COLLABORATIVE LEARNING

Form three groups. The steps to a number trick are shown. Group 1 starts with number 8, Group 2 with -3, and Group 3 with -1.2.

Steps	Group 1	Group 2	Group 3
Pick a number:	8	−3	−1.2
1. ________	10	−1	0.8
2. ________	30	−3	2.4
3. ________	24	−9	−3.6
4. ________	8	−3	−1.2
5. ________	0	0	0

Each group should fill in the blanks in the first column and find out how the trick works. The first group that discovers what the steps are wins! Of course, the other groups can challenge the results. Compare the five steps. Do all groups get the same steps? What are the steps?

The steps given here will give you a different final answer depending on the number you start with.

Number + 2

Ans · 3

Ans − 9

Ans/6

Are you ready for the race? All groups should start with the number 5. What answers do they get? The group that finishes first wins! Then do it again. Start with the number 7 this time. What answer do the groups get now?

Here is the challenge: Write an algebraic expression representing the steps in the table. Again, the group that finishes first wins.

Ultimate challenge: The groups write down the steps that describe the expression $3(\frac{x+2}{3} + 4) - x + 5$.

Research Questions

1. In the *Human Side of Algebra* at the beginning of this chapter, we mentioned the Hindu numeration system. The Egyptians and Babylonians also developed numeration systems. Write a report about each of these numeration systems, detailing the symbols used for the digits 1–9, the base used, and the manner in which fractions were written.
2. Write a report on the life and works of Mohammed al-Khowarizmi, with special emphasis on the books he wrote.
3. We have now studied the four fundamental operations. But do you know where the symbols used to indicate these operations originated?
   a. Write a report about Johann Widmann's *Mercantile Arithmetic* (1489), indicating which symbols of operation were found in the book for the first time and the manner in which they were used.
   b. Introduced in 1557, the original equals sign used longer lines to indicate equality. Why were the two lines used to denote equality, what was the name of the person who introduced the symbol, and in what book did the notation first appear?
4. In this chapter we discussed mathematical expressions. Two of the signs of operation were used for the first time in the earliest-known treatise on algebra to write an algebraic expression. What is the name of this first treatise on algebra, and what is the name of the Dutch mathematician who used these two symbols in 1514?
5. The *Codex Vigilanus* written in Spain in 976 contains the Hindu-Arabic numerals 1–9. Write a report about the notation used for these numbers and the evolution of this notation.

Summary

SECTION	ITEM	MEANING	EXAMPLE
1.1A	Arithmetic expressions	Expressions containing numbers and operation signs	$3 + 4 - 8, 9 \cdot 4 \div 6$ are arithmetic expressions.
1.1A	Algebraic expressions	Expressions containing numbers, operation signs, and variables	$3x - 4y + 3z$, $2x \div y + 9z$, and $7x - 9y \div 3z$ are algebraic expressions.
1.1A	Factor	The items to be multiplied	In the expression $4xy$, the factors are 4, x, and y.
1.1B	Evaluate	To substitute a value for one or more of the variables in an expression	Evaluating $3x - 4y + 3z$ when $x = 1$, $y = 2$, and $z = 3$ yields $3 \cdot 1 - 4 \cdot 2 + 3 \cdot 3$ or 4.
1.2A	Additive inverse (opposite)	The additive inverse of any integer a is $-a$.	The additive inverse of 5 is -5, and the additive inverse of -8 is 8.
1.2B	Absolute value of n, denoted by $\|n\|$	The absolute value of a number n is the distance from n to 0.	$\|-7\| = 7$, $\|13\| = 13$, $\|0.2\| = 0.2$, and $\|-\frac{1}{4}\| = \frac{1}{4}$.

SECTION	ITEM	MEANING	EXAMPLE
1.2C	Set of natural numbers	$\{1, 2, 3, 4, 5, \ldots\}$	4, 13, and 497 are natural numbers.
	Set of whole numbers	$\{0, 1, 2, 3, 4, \ldots\}$	0, 92, and 384 are whole numbers.
	Set of integers	$\{\ldots, -2, -1, 0, 1, 2, \ldots\}$	-98, 0, and 459 are integers.
	Rational numbers	Numbers that can be written in the form $\frac{a}{b}$, a and b integers and b not 0	$\frac{1}{5}$, $\frac{-3}{4}$, and $\frac{0}{5}$ are rational numbers.
	Irrational numbers	Numbers that cannot be written as the ratio of two integers	$\sqrt{2}$, $-\sqrt[3]{2}$, and π are irrational numbers.
	Real numbers	The rationals and the irrationals	3, 0, -9, $\sqrt{2}$, and $\sqrt[3]{5}$ are real numbers.
1.3A	Addition of real numbers	If both numbers have the *same* sign, add their absolute values and give the sum the common sign. If the numbers have *different* signs, subtract their absolute values and give the difference the sign of the number with the larger absolute value.	$-3 + (-5) = -8$ $-3 + 5 = +2$ $-3 + 1 = -2$
1.3B	Subtraction of real numbers	$a - b = a + (-b)$	$3 - 5 = 3 + (-5) = -2$ and $4 - (-2) = 4 + 2 = 6$
1.4A, C	Multiplication and division of real numbers	When multiplying or dividing two numbers with the *same* sign, the answer is *positive;* with *different* signs, the answer is *negative.*	$(-3)(-5) = 15$ $\frac{-15}{-3} = 5$ $(-3)(5) = -15$ $\frac{-15}{3} = -5$
1.4B	Exponents Base	In the expression 3^2, 2 is the exponent. In the expression 3^2, 3 is the base.	3^2 means $3 \cdot 3$
1.5	Order of operations	1. Operations inside grouping symbols	$4^2 \div 2 \cdot 3 - 4 + [2(6 + 1) - 4]$ $= 4^2 \div 2 \cdot 3 - 4 + [2(7) - 4]$ $= 4^2 \div 2 \cdot 3 - 4 + [14 - 4]$
		2. Exponents	$= 4^2 \div 2 \cdot 3 - 4 + [10]$ $= 16 \div 2 \cdot 3 - 4 + [10]$
		3. Multiplications and divisions	$= 8 \cdot 3 - 4 + 10$ $= 24 - 4 + 10$
		4. Additions and subtractions	$= 20 + 10$ $= 30$

(*Continued*)

SECTION	ITEM	MEANING	EXAMPLE
1.6A	Associative property of addition	For any numbers a, b, c, $a + (b + c) = (a + b) + c$.	$3 + (4 + 9) = (3 + 4) + 9$
	Associative property of multiplication	For any numbers a, b, c, $a \cdot (b \cdot c) = (a \cdot b) \cdot c$.	$5 \cdot (2 \cdot 8) = (5 \cdot 2) \cdot 8$
	Commutative property of addition	For any numbers a and b, $a + b = b + a$.	$2 + 9 = 9 + 2$
	Commutative property of multiplication	For any numbers a and b, $a \cdot b = b \cdot a$.	$18 \cdot 5 = 5 \cdot 18$
1.6C	Identity element for addition	0 is the identity element for addition.	$3 + 0 = 0 + 3 = 3$
	Identity element for multiplication	1 is the identity element for multiplication.	$1 \cdot 7 = 7 \cdot 1 = 7$
	Additive inverse (opposite)	For any number a, its additive inverse is $-a$.	The additive inverse of 5 is -5 and that of -8 is 8.
	Multiplicative inverse (reciprocal)	The multiplicative inverse of a is $\frac{1}{a}$ if a is not 0 (0 has no reciprocal).	The multiplicative inverse of $\frac{3}{4}$ is $\frac{4}{3}$.
1.6E	Distributive property	For any numbers a, b, c, $a(b + c) = ab + ac$.	$3(4 + x) = 3 \cdot 4 + 3 \cdot x$
1.7	Expression	A collection of numbers and letters connected by operation signs.	xy^3, $x + y$, $x + y - z$, and $xy^3 - y$ are expressions.
	Terms	The parts that are to be added or subtracted in an expression.	In the expression $3x^2 - 4x + 5$, the terms are $3x^2$, $-4x$, and 5.
	Numerical coefficient or coefficient	The part of the term indicating "how many."	In the term $3x^2$, 3 is the coefficient. In the term $-4x$, -4 is the coefficient. The coefficient of x is 1.
	Like terms	Terms with the same variables and exponents.	$3x$ and $-4x$ are like terms. $-5x^2$ and $8x^2$ are like terms. $-xy^2$ and $3xy^2$ are like terms. *Note:* $-x^2y$ and $-xy^2$ **are not** like terms.
1.7C	Sums and differences	The sum of a and b is $a + b$. The difference of a and b is $a - b$.	The sum of 3 and x is $3 + x$. The difference of 6 and x is $6 - x$.
	Product	The product of a and b is $a \cdot b$, $(a)(b)$, $a(b)$, $(a)b$, or ab.	The product of 8 and x is $8x$.
	Quotient	The quotient of a and b is $\frac{a}{b}$.	The quotient of 7 and x is $\frac{7}{x}$.

Review Exercises

(If you need help with these exercises, look in the section indicated in brackets.)

1. [1.1A] Write in symbols:

a. The sum of a and b $a + b$

b. a minus b $a - b$

c. $7a$ plus $2b$ minus 8 $7a + 2b - 8$

2. [1.1A] Write using juxtaposition:

a. 3 times m $3m$

b. $-m$ times n times r $-mnr$

c. $\frac{1}{7}$ of m $\frac{1}{7}m$

d. The product of 8 and m $8m$

3. [1.1A] Write in symbols:

a. The quotient of m and 9 $\frac{m}{9}$

b. The quotient of 9 and n $\frac{9}{n}$

4. [1.1A] Write in symbols:

a. The quotient of $(m + n)$ and r $\frac{m+n}{r}$

b. The sum of m and n, divided by the difference of m and n $\frac{m+n}{m-n}$

5. [1.1B] For $m = 9$ and $n = 3$, evaluate:

a. $m + n$ 12 **b.** $m - n$ 6 **c.** $4m$ 36

6. [1.1B] For $m = 9$ and $n = 3$, evaluate:

a. $\frac{m}{n}$ 3 **b.** $2m - 3n$ 9 **c.** $\frac{2m + n}{n}$ 7

7. [1.2A] Find the additive inverse (opposite) of:

a. -5 5 **b.** $\frac{2}{3}$ $-\frac{2}{3}$ **c.** 0.37 -0.37

8. [1.2B] Find:

a. $|-8|$ 8 **b.** $\left|3\frac{1}{2}\right|$ $3\frac{1}{2}$ **c.** $-|0.76|$ -0.76

9. [1.2C] Classify each of the numbers as natural, whole, integer, rational, irrational, or real. (*Note:* More than one category may apply.)

a. -7 Integer, rational, real **b.** 0 Whole, integer, rational, real

c. 0.666. . . Rational, real **d.** 0.606006000. . . Irrational, real

e. $\sqrt{41}$ Irrational, real **f.** $-1\frac{1}{8}$ Rational, real

10. [1.3A] Find:

a. $7 + (-5)$ 2 **b.** $(-0.3) + (-0.5)$ -0.8

c. $-\frac{3}{4} + \frac{1}{5}$ $-\frac{11}{20}$ **d.** $3.6 + (-5.8)$ -2.2

e. $\frac{3}{4} + \left(-\frac{1}{2}\right)$ $\frac{1}{4}$

11. [1.3B] Find:

a. $-16 - 4$ -20 **b.** $-7.6 - (-5.2)$ -2.4

c. $\frac{5}{6} - \frac{9}{4}$ $-\frac{17}{12}$

12. [1.3C] Find:

a. $20 - (-12) + 15 - 12 - 5$ 30

b. $-17 + (-7) + 10 - (-7) - 8$ -15

13. [1.4A] Find:

a. $-5 \cdot 7$ -35 **b.** $8(-2.3)$ -18.4

c. $-6(3.2)$ -19.2 **d.** $-\frac{3}{7}\left(-\frac{4}{5}\right)$ $\frac{12}{35}$

14. [1.4B] Find:

a. $(-4)^2$ 16 **b.** -3^2 -9

c. $\left(-\frac{1}{3}\right)^3$ $-\frac{1}{27}$ **d.** $-\left(-\frac{1}{3}\right)^3$ $\frac{1}{27}$

15. [1.4C] Find:

a. $\frac{-40}{10}$ -4 **b.** $-8 \div (-4)$ 2

c. $\frac{3}{8} \div \left(-\frac{9}{16}\right)$ $-\frac{2}{3}$ **d.** $-\frac{3}{4} \div \left(-\frac{1}{2}\right)$ $\frac{3}{2}$

16. [1.5A] Find:

a. $64 \div 8 - (3 - 5)$ 10

b. $27 \div 3^2 + 5 - 8$ 0

17. [1.5B] Find:

a. $20 \div 5 + \{2 \cdot 3 - [4 + (7 - 9)]\}$ 8

b. $-6^2 + \frac{2(3-6)}{2} + 8 \div (-4)$ -41

18. [1.5C] The maximum pulse rate you should maintain during aerobic activities is $0.80(220 - A)$, where A is your age. What is the maximum pulse rate you should maintain if you are 30 years old? 152

19. [1.6A] Name the property illustrated in each statement.

a. $x + (y + z) = x + (z + y)$ Commutative property of addition

b. $6 \cdot (8 \cdot 7) = (6 \cdot 8) \cdot 7$ Associative property of multiplication

c. $x + (y + z) = (x + y) + z$ Associative property of addition

20. [1.6B] Simplify:

a. $6 + 4x - 10$ $4x - 4$

b. $-8 + 7x + 10 - 15x$ $-8x + 2$

21. [1.6C, E] Fill in the blank so that the result is a true statement.

a. $\underline{1} \cdot \frac{1}{5} = \frac{1}{5}$

b. $\frac{3}{4} + \underline{-\frac{3}{4}} = 0$

c. $\underline{-3.7} + 3.7 = 0$

d. $\underline{\frac{2}{3}} \cdot \frac{3}{2} = 1$

e. $\underline{0} + \frac{2}{7} = \frac{2}{7}$

f. $3(x + \underline{-5}) = 3x + (-15)$

22. [1.6E] Use the distributive property to multiply.

a. $-3(a + 8)$ $-3a - 24$

b. $-4(x - 5)$ $-4x + 20$

c. $-(x - 4)$ $-x + 4$

23. [1.7A] Combine like terms.

a. $-7x + 2x$ $-5x$

b. $-2x + (-8x)$ $-10x$

c. $x + (-9x)$ $-8x$

d. $-6a^2b - (-9a^2b)$ $3a^2b$

24. [1.7B] Remove parentheses and combine like terms.

a. $(4a - 7) + (8a + 2)$ $12a - 5$

b. $9x - 3(x + 2) - (x + 3)$ $5x - 9$

25. [1.7C] Write in symbols and indicate the number of terms to the right of the equals sign.

a. The number of minutes m you will wait in line at the bank is equal to the number of people p ahead of you divided by the number n of tellers, times 2.75. $m = \frac{p}{n}(2.75)$; one term

b. The normal weight W of an adult (in pounds) can be estimated by subtracting 220 from the product of $\frac{11}{2}$ and h, where h is the person's height (in inches). $W = \frac{11h}{2} - 220$; two terms

Practice Test 1

(Answers on pages 116–117)

1. Write in symbols:

a. The sum of g and h b. g minus h

c. $6g$ plus $3h$ minus 8

2. Write using juxtaposition:

a. 3 times g b. $-g$ times h times r

c. $\frac{1}{5}$ of g d. The product of 9 and g

3. Write in symbols:

a. The quotient of g and 8

b. The quotient of 8 and h

4. Write in symbols:

a. The quotient of $(g + h)$ and r

b. The sum of g and h, divided by the difference of g and h

5. For $g = 4$ and $h = 3$, evaluate:

a. $g + h$ b. $g - h$ c. $5g$

6. For $g = 8$ and $h = 4$, evaluate:

a. $\frac{g}{h}$ b. $2g - 3h$ c. $\frac{2g + h}{h}$

7. Find the additive inverse (opposite) of:

a. -9 b. $\frac{3}{5}$ c. $0.222\ldots$

8. Find:

a. $\left|-\frac{1}{4}\right|$ b. $|13|$ c. $-|0.92|$

9. Consider the set $\{-8, \frac{5}{3}, \sqrt{2}, 0, 3.4, 0.333\ldots, -3\frac{1}{2}, 7, 0.123\ldots\}$. List the numbers in the set that are:

a. Natural numbers b. Whole numbers

c. Integers d. Rational numbers

e. Irrational numbers f. Real numbers

10. Find:

a. $9 + (-7)$ b. $(-0.9) + (-0.8)$

c. $-\frac{3}{4} + \frac{2}{5}$ d. $1.7 + (-3.8)$

e. $\frac{4}{5} + \left(-\frac{1}{2}\right)$

11. Find:

a. $-18 - 6$ **b.** $-9.2 - (-3.2)$

c. $\frac{1}{6} - \frac{5}{4}$

12. Find:

a. $10 - (-15) + 12 - 15 - 6$

b. $-15 + (-8) + 12 - (-8) - 9$

13. Find:

a. $-6 \cdot 8$ **b.** $9(-2.3)$

c. $-7(8.2)$ **d.** $-\frac{2}{7}\left(-\frac{3}{5}\right)$

14. Find:

a. $(-7)^2$ **b.** -9^2

c. $\left(-\frac{2}{3}\right)^3$ **d.** $-\left(-\frac{2}{3}\right)^3$

15. Find:

a. $\frac{-50}{10}$ **b.** $-14 \div (-7)$

c. $\frac{5}{8} \div \left(-\frac{11}{16}\right)$ **d.** $-\frac{3}{4} \div \left(-\frac{7}{9}\right)$

16. Find:

a. $56 \div 7 - (4 - 9)$

b. $36 \div 2^2 + 7$

17. Find the value of:

a. $30 \div 6 + \{3 \cdot 4 - [2 + (8 - 10)]\}$

b. $-7^2 + \frac{4(3 - 9)}{2} + 12 \div (-4)$

18. The handicap H of a bowler with average A is $H = 0.80(200 - A)$. What is the handicap of a bowler whose average is 180?

19. Name the property illustrated in each statement.

a. $x + (y + z) = (x + y) + z$

b. $6 \cdot (8 \cdot 7) = 6 \cdot (7 \cdot 8)$

c. $x + (y + z) = (y + z) + x$

20. Simplify:

a. $7 - 4x - 10$

b. $-9 + 6x + 12 - 13x$

21. Fill in the blank so that the result is a true statement.

a. $9.2 + \_\_\_ = 0$ **b.** $\_\_\_ \cdot \frac{5}{2} = 1$

c. $\_\_\_ \cdot 0.3 = 0.3$ **d.** $\frac{1}{4} + \_\_\_ = 0$

e. $\_\_\_ + \frac{3}{7} = \frac{3}{7}$

f. $-3(x + \_\_\_) = -3x + (-15)$

22. Use the distributive property to multiply.

a. $-3(x + 9)$ **b.** $-5(a - 4)$

c. $-(x - 8)$

23. Combine like terms.

a. $-3x + (-9x)$

b. $-9ab^2 - (-6ab^2)$

24. Remove parentheses and combine like terms.

a. $(5a - 6) - (7a + 2)$

b. $8x - 4(x - 2) - (x + 2)$

25. Write in symbols and indicate the number of terms to the right of the equals sign:

The temperature F (in degrees Fahrenheit) can be found by finding the sum of 37 and one-quarter the number of chirps C a cricket makes in 1 min.

Answers to Practice Test

ANSWER	IF YOU MISSED	REVIEW		
	QUESTION	SECTION	EXAMPLES	PAGE
1. **a.** $g + h$ **b.** $g - h$ **c.** $6g + 3h - 8$	1	1.1	1	46
2. **a.** $3g$ **b.** $-ghr$ **c.** $\frac{1}{5}g$ **d.** $9g$	2	1.1	2	46
3. **a.** $\frac{g}{8}$ **b.** $\frac{8}{h}$	3	1.1	3a, b	47
4. **a.** $\frac{g+h}{r}$ **b.** $\frac{g+h}{g-h}$	4	1.1	3c, d	47
5. **a.** 7 **b.** 1 **c.** 20	5	1.1	4a, b, c	47
6. **a.** 2 **b.** 4 **c.** 5	6	1.1	4d, e	47
7. **a.** 9 **b.** $-\frac{3}{5}$ **c.** $-0.222\ldots$	7	1.2	1, 2	52–53
8. **a.** $\frac{1}{4}$ **b.** 13 **c.** -0.92	8	1.2	3, 4	54
9. **a.** 7 **b.** 0, 7 **c.** $-8, 0, 7$ **d.** $-8, \frac{5}{3}, 0, 3.4, 0.333\ldots, -3\frac{1}{2}, 7$ **e.** $\sqrt{2}, 0.123\ldots$ **f.** All	9	1.2	5	56
10. **a.** 2 **b.** -1.7 **c.** $-\frac{7}{20}$ **d.** -2.1 **e.** $\frac{3}{10}$	10	1.3	1, 2, 3, 4, 5	62–64
11. **a.** -24 **b.** -6 **c.** $-\frac{13}{12}$	11	1.3	6, 7	65
12. **a.** 16 **b.** -12	12	1.3	8	66
13. **a.** -48 **b.** -20.7 **c.** -57.4 **d.** $\frac{6}{35}$	13	1.4	1, 2	71–72
14. **a.** 49 **b.** -81 **c.** $-\frac{8}{27}$ **d.** $\frac{8}{27}$	14	1.4	3, 4	73
15. **a.** -5 **b.** 2 **c.** $-\frac{10}{11}$ **d.** $\frac{27}{28}$	15	1.4	5, 6	74–75
16. **a.** 13 **b.** 16	16	1.5	1, 2	81
17. **a.** 17 **b.** -64	17	1.5	3, 4, 5	82–84
18. 16	18	1.5	6	84

ANSWER	IF YOU MISSED	REVIEW		
	QUESTION	SECTION	EXAMPLES	PAGE
19. **a.** Associative property of addition **b.** Commutative property of multiplication **c.** Commutative property of addition	19	1.6	1, 2	89–90
20. **a.** $-4x - 3$ **b.** $3 - 7x$	20	1.6	3	90
21. **a.** -9.2 **b.** $\frac{2}{5}$ **c.** 1 **d.** $-\frac{1}{4}$ **e.** 0 **f.** 5	21	1.6	4, 5, 7, 8	91–94
22. **a.** $-3x - 27$ **b.** $-5a + 20$ **c.** $-x + 8$	22	1.6	7, 8	93–94
23. **a.** $-12x$ **b.** $-3ab^2$	23	1.7	1, 2	100–101
24. **a.** $-2a - 8$ **b.** $3x + 6$	24	1.7	3, 4, 5	101–102
25. $F = 37 + \frac{1}{4}C$; two terms	25	1.7	6, 8	103–105

Equations, Problem Solving, and Inequalities

2

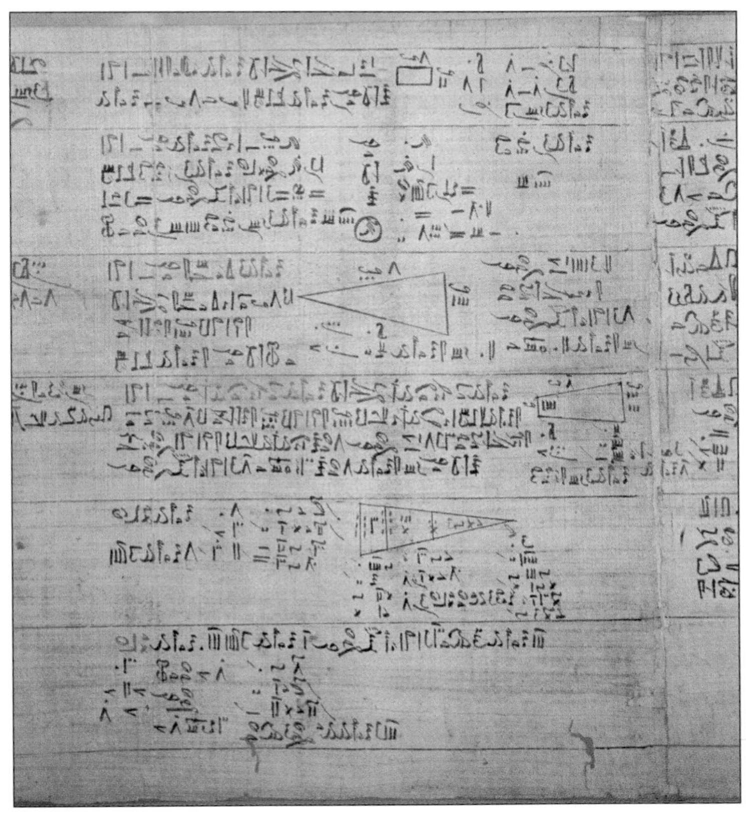

The Human Side of Algebra

Most of the mathematics used in ancient Egypt is preserved in the Rhind papyrus, a document bought in 1858 in Luxor, Egypt, by Henry Rhind. Problem 24 in this document reads:

> A quantity and its $\frac{1}{7}$ added become 19. What is the quantity?

If we let q represent the quantity, we can translate the problem as

$$q + \frac{1}{7}q = 19$$

Unfortunately, the Egyptians were unable to simplify $q + \frac{1}{7}q$ because the sum is $\frac{8}{7}q$, and they didn't have a notation for the fraction $\frac{8}{7}$. How did they attempt to find the answer? They used the method of "false position." That is, they *assumed* the answer was 7, which yields $7 + \frac{1}{7} \cdot 7$ or 8. But the sum as stated in the problem is not 8, it's 19. How can you make the 8 into a 19? By multiplying by $\frac{19}{8}$, that is, by finding $8 \cdot \frac{19}{8} = 19$. But if you multiply the 8 by $\frac{19}{8}$, you should also multiply the assumed answer, 7, by $\frac{19}{8}$ to obtain the true answer, $7 \cdot \frac{19}{8}$. If you solve the equation $\frac{8}{7}q = 19$, you will see that $7 \cdot \frac{19}{8}$ is indeed the correct answer!

Pretest for Chapter 2

(Answers on page 121)

1. Does the number 8 satisfy the equation $1 = 9 - x$?

2. Solve $x - \frac{1}{7} = \frac{3}{7}$.

3. Solve $-3x + \frac{5}{8} + 4x - \frac{3}{8} = \frac{5}{8}$.

4. Solve $3 = 3(x - 1) + 6 - 2x$.

5. Solve $3 + 5(x + 1) = 9 + 5x$.

6. Solve $-5 - 2(x - 1) = -3 - 2x$.

7. Solve $\frac{2}{5}x = -4$.

8. Solve $-\frac{2}{5}x = -6$.

9. Solve $\frac{x}{4} + \frac{2x}{5} = 13$.

10. Solve $\frac{x}{3} - \frac{x}{7} = 4$.

11. Solve $\frac{x - 2}{4} - \frac{x + 1}{7} = 0$.

12. What percent of 44 is 11?

13. Nine is 45% of what number?

14. Solve $\frac{1}{5} - \frac{x}{3} = \frac{28(x + 6)}{15}$.

15. Solve for h in $V = \frac{4}{3}\pi r^2 h$.

16. The sum of two numbers is 85. If one of the numbers is 15 more than the other, what are the numbers?

17. A man has invested a certain amount of money in stocks and bonds. His annual return from these investments is \$780. If the stocks produce \$230 more in returns than the bonds, how much money does he receive annually from each investment?

18. Find the measure of an angle whose supplement is 30° less than 4 times its complement.

19. A freight train leaves a station traveling at 30 miles per hour. Two hours later, a passenger train leaves the same station, traveling in the same direction at 40 miles per hour. How long does it take for the passenger train to catch the freight train?

20. How many pounds of coffee selling for \$1.20 per pound should be mixed with 20 pounds of coffee selling for \$1.80 per pound to obtain a mixture that sells for \$1.60 per pound?

21. An investor bought some municipal bonds yielding 5% annually and some certificates of deposit yielding 7% annually. If her total investment amounts to \$20,000 and her annual return is \$1100, how much money is invested in bonds and how much in certificates of deposit?

22. The cost C of riding a taxi is $C = 1.95 + 0.95m$, where m is the number of miles (or fraction) you travel.

a. Solve for m.

b. How many miles did you travel if the cost of the ride was \$24.75?

23. Find x and the measures of the marked angles.

a.

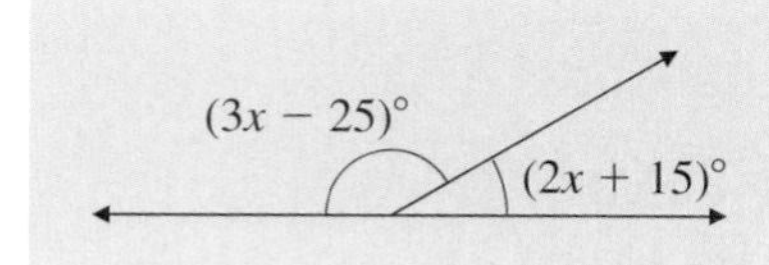

b.

(3x − 20)°
(5x − 40)°

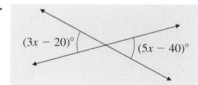

c.

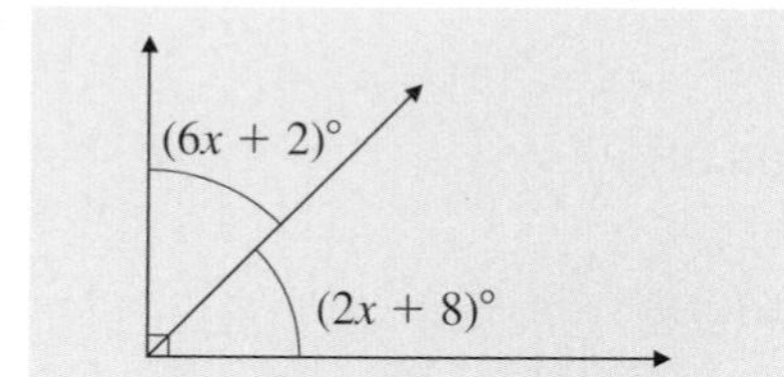

24. Fill in the blank with < or > to make the resulting statement true.

a. -3 _______ -6

b. $-\frac{1}{5}$ _______ 5

25. Solve and graph the inequality.

a. $-\frac{x}{3} + \frac{x}{5} \leq \frac{x - 5}{5}$

b. $x + 2 \leq 4$ and $-2x < 6$

Answers to Pretest

ANSWER	IF YOU MISSED	REVIEW		
	QUESTION	SECTION	EXAMPLES	PAGE
1. Yes	1	2.1	1	123
2. $x = \frac{4}{7}$	2	2.1	2	124
3. $x = \frac{3}{8}$	3	2.1	3	125–126
4. $x = 0$	4	2.1	4, 5	127–129
5. No solution	5	2.1	6	129
6. All real numbers	6	2.1	7	130
7. $x = -10$	7	2.2	1, 2, 3	136–140
8. $x = 15$	8	2.2	1, 2, 3	136–140
9. $x = 20$	9	2.2	4	142–143
10. $x = 21$	10	2.2	5	143
11. $x = 6$	11	2.2	5	143
12. 25%	12	2.2	7	145
13. 20	13	2.2	8	145
14. -5	14	2.3	1, 2, 3	151–155
15. $h = \frac{3V}{4\pi r^2}$	15	2.3	4, 5, 6	156–158
16. 35 and 50	16	2.4	1, 2	164–165
17. \$275 from bonds and \$505 from stocks	17	2.4	3	165–166
18. 50°	18	2.4	4	167
19. 6 hours	19	2.5	1, 2, 3	172–175
20. 10 pounds	20	2.5	4	176–177
21. \$15,000 in bonds, \$5000 in certificates	21	2.5	5	178–179
22. **a.** $m = \frac{C - 1.95}{0.95}$ **b.** 24 miles	22	2.6	1, 2	184–185
23. **a.** 89° and 91° **b.** both are 10° **c.** 62° and 28°	23	2.6	7	190–191
24. **a.** $>$ **b.** $<$	24	2.7	1	198
25. **a.** 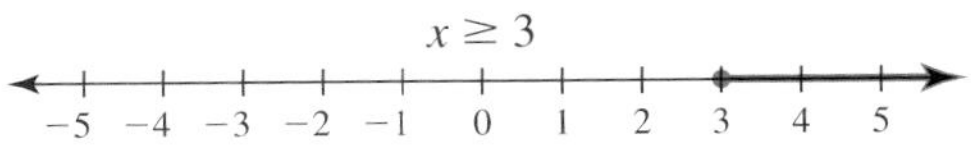 **b.** 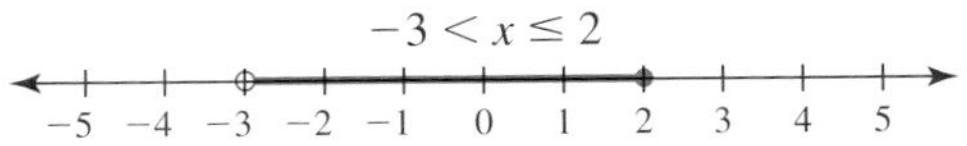	25	2.7	2, 3, 4, 5, 6	199–206

2.1 THE ADDITION AND SUBTRACTION PROPERTIES OF EQUALITY

To Succeed, Review How To . . .

1. Add and subtract real numbers (pp. 61–66).
2. Follow the correct order of operations (p. 80).
3. Simplify expressions (pp. 98–105).

Objectives

A Determine whether a number satisfies an equation.

B Use the addition and subtraction properties of equality to solve equations.

C Use both properties together to solve an equation.

GETTING STARTED

A Lot of Garbage!

In this section we study some ideas that will enable us to solve equations. Do you know what an equation is? Here is an example. How much waste do you generate every day? According to Franklin Associates, LTD, the average American generates about 2.7 pounds of waste daily if we *exclude* paper products! If w and p represent the total amount of waste and paper products generated daily by the average American, $w - p = 2.7$. Further research indicates that $p = 1.7$ pounds; thus $w - 1.7 = 2.7$. The statement

$$w - 1.7 = 2.7$$

is an *equation,* a statement indicating that two expressions are equal. Some equations are *true* $(1 + 1 = 2)$, some are *false* $(2 - 5 = 3)$, and some $(w - 1.7 = 2.7)$ are neither true nor false. The equation $w - 1.7 = 2.7$ is a *conditional* equation in which the *variable* or unknown is w. To find the total amount of waste generated daily (w), we have to *solve* $w - 1.7 = 2.7$; that is, we must find the value of the variable that makes the equation a true statement. We learn how to do this next.

A Finding Solutions

In the equation $w - 1.7 = 2.7$ in the *Getting Started,* the variable w can be replaced by many numbers, but only *one* number will make the resulting statement *true.* This number is called the *solution* of the equation. Can you find the solution of $w - 1.7 = 2.7$? Since w is the total amount of waste, $w = 1.7 + 2.7 = 4.4$, and 4.4 is the solution of the equation.

Web It

To find out more about recycling, go to the Bello Website at mhhe.com/bello and try link 2-1-1.

SOLUTIONS

The **solutions** of an equation are the replacements of the variable that make the equation a *true* statement. When we find the solution of an equation, we say that we have **solved** the equation.

How do we know whether a given number solves (or *satisfies*) an equation? We write the number in place of the variable in the given equation and see whether the result is true. For example, to decide whether 4 is a solution of the equation

$$x + 1 = 5$$

we replace x by 4. This gives the true statement:

$$4 + 1 = 5$$

so 4 is a solution of the equation $x + 1 = 5$.

EXAMPLE 1 Verifying a solution

Determine whether the given number is a solution of the equation:

a. 9; $x - 4 = 5$ **b.** 8; $5 = 3 - y$ **c.** 10; $\frac{1}{2}z - 5 = 0$

SOLUTION

a. If x is 9, $x - 4 = 5$ becomes $9 - 4 = 5$, which is a *true* statement. Thus 9 is a solution of the equation.

b. If y is 8, $5 = 3 - y$ becomes $5 = 3 - 8$, which is a *false* statement. Hence, 8 is not a solution of the equation.

c. If z is 10, $\frac{1}{2}z - 5 = 0$ becomes $\frac{1}{2}(10) - 5 = 0$, which is a *true* statement. Thus 10 is a solution of the equation.

PROBLEM 1

Determine whether the given number is a solution of the equation:

a. 7; $x - 5 = 3$

b. 4; $1 = 5 - y$

c. 6; $\frac{1}{3}z - 2 = 0$

We have learned how to determine whether a number satisfies an equation. Now to find such a number, we must find an *equivalent* equation whose solution is obvious.

EQUIVALENT EQUATIONS

Two equations are **equivalent** if their solutions are the same.

How do we find these equivalent equations? We use the properties of equality.

B Using the Addition and Subtraction Properties of Equality

Look at the ad in the illustration. It says that $5 has been cut from the price of a gallon of paint and that the sale price is $6.69. What was the old price p of the paint? Since the old price p was cut by $5, the new price is $p - 5$. Since the new price is $6.69,

$$p - 5 = 6.69$$

To find the old price p, we add back the $5 that was cut. That is,

$$p - 5 + 5 = 6.69 + 5$$ We add 5 to both sides of the equation to obtain an equivalent equation.

$$p = 11.69$$

Thus the old price was $11.69; this can be verified, since $11.69 - 5 = 6.69$. Note that by adding 5 to both sides of the equation $p - 5 = 6.69$, we produced an equivalent equation, $p = 11.69$, whose solution is obvious. This example illustrates the fact that we can *add* the same number on both sides of an equation and produce an *equivalent* equation—that is, an equation whose solution is identical to the solution of the original one. Here is the property.

Answers

1. a. No **b.** Yes **c.** Yes

The Addition Property of Equality The equation $a = b$ is equivalent to

$$a + c = b + c \quad \text{for any number } c$$

We use this property in the next example.

EXAMPLE 2 **Using the addition property**

Solve:

a. $x - 3 = 9$ **b.** $x - \frac{1}{7} = \frac{5}{7}$

SOLUTION

a. This problem is similar to our example of paint prices. To solve the equation, we need x by itself on one side of the equation. We can achieve this by adding 3 (the additive inverse of -3) on both sides of the equation.

$$x - 3 = 9$$

$$x - 3 + 3 = 9 + 3 \quad \text{Add 3 to both sides.}$$

$$x = 12$$

Thus 12 is the solution of $x - 3 = 9$.

CHECK Substituting 12 for x in the original equation, we have $12 - 3 = 9$, a true statement.

b.

$$x - \frac{1}{7} = \frac{5}{7}$$

$$x - \frac{1}{7} + \frac{1}{7} = \frac{5}{7} + \frac{1}{7} \quad \text{Add } \tfrac{1}{7} \text{ to both sides.}$$

$$x = \frac{6}{7} \quad \text{Simplify.}$$

Thus $\frac{6}{7}$ is the solution of $x - \frac{1}{7} = \frac{5}{7}$.

CHECK $\frac{6}{7} - \frac{1}{7} = \frac{5}{7}$ is a true statement.

PROBLEM 2

Solve:

a. $x - 5 = 7$ **b.** $x - \frac{1}{5} = \frac{3}{5}$

Sometimes it's necessary to simplify an equation before we isolate x on one side. For example, to solve the equation

$$3x + 5 - 2x - 9 = 6x + 5 - 6x$$

we first simplify both sides of the equation by collecting like terms:

$$3x + 5 - 2x - 9 = 6x + 5 - 6x$$

$$(3x - 2x) + (5 - 9) = (6x - 6x) + 5 \quad \text{Group like terms.}$$

$$x + (-4) = 0 + 5 \quad \text{Combine like terms.}$$

$$x - 4 = 5 \quad \text{Rewrite } x + (-4) \text{ as } x - 4.$$

$$x - 4 + 4 = 5 + 4 \quad \text{Now add 4 to both sides.}$$

$$x = 9$$

Thus 9 is the solution of the equation.

Answers

2. a. 12 **b.** $\frac{4}{5}$

CHECK We substitute 9 for x in the original equation. To save time, we use the following diagram:

$3x + 5 - 2x - 9 \stackrel{?}{=}$	$6x + 5 - 6x$
$3(9) + 5 - 2(9) - 9$	$6(9) + 5 - 6(9)$
$27 + 5 - 18 - 9$	$54 + 5 - 54$
$32 - 18 - 9$	5
$32 - 27$	
5	

Teaching Tip

Note that $\stackrel{?}{=}$ means, "are they equal?" We use a 'T' diagram, substituting the solution for the variable in each side of the original equation. After performing the operations, the same number should be on the left and the right.

Since both sides yield 5, our result is correct.

Now suppose that the price of an article is increased by \$3 and the article currently sells for \$8. What was its old price, p? The equation here is

$$\underbrace{p}_{\text{Old price}} \quad \underbrace{+}_{\text{went up}} \quad \underbrace{3}_{\$3} \quad \underbrace{=}_{\text{and is now}} \quad \underbrace{8}_{\$8.}$$

To solve this equation, we have to bring the price down; that is, we need to subtract 3 on both sides of the equation:

$$p + 3 - 3 = 8 - 3$$
$$p = 5$$

Thus the old price was \$5. We have *subtracted* 3 on both sides of the equation. Here is the property that allows us to do this.

> **The Subtraction Property of Equality**
>
> The equation $a = b$ is equivalent to
>
> $a - c = b - c$ for any number c

This property tells us that we can *subtract* the same number on both sides of an equation to produce an *equivalent* equation. Note that since $a - c = a + (-c)$, you can think of *subtracting* c as *adding* $(-c)$.

EXAMPLE 3 Using the subtraction property

Solve:

a. $2x + 4 - x + 2 = 10$

b. $-3x + \frac{5}{7} + 4x - \frac{3}{7} = \frac{6}{7}$

SOLUTION

a. To solve the equation, we need to get x by itself on the left; that is, we want $x = \square$, where $\square$ is a number. We proceed as follows:

$$2x + 4 - x + 2 = 10$$
$$x + 6 = 10 \quad \text{Simplify.}$$
$$x + 6 - 6 = 10 - 6 \quad \text{Subtract 6 from both sides.}$$
$$x = 4$$

Thus 4 is the solution of the equation.

PROBLEM 3

Solve:

a. $5y + 2 - 4y + 3 = 17$

b. $-5z + \frac{3}{8} + 6z - \frac{1}{8} = \frac{5}{8}$

Answers

3. a. 12 **b.** $\frac{3}{8}$

CHECK

$2x + 4 - x + 2 \stackrel{?}{=} 10$	
$2(4) + 4 - 4 + 2$	10
$8 + 2$	
10	

b. $-3x + \frac{5}{7} + 4x - \frac{3}{7} = \frac{6}{7}$

$$x + \frac{2}{7} = \frac{6}{7} \qquad \text{Simplify.}$$

$$x + \frac{2}{7} - \frac{2}{7} = \frac{6}{7} - \frac{2}{7} \qquad \text{Subtract } \tfrac{2}{7} \text{ from both sides.}$$

$$x = \frac{4}{7}$$

Thus $\frac{4}{7}$ is the solution of the equation.

CHECK

$-3x + \frac{5}{7} + 4x - \frac{3}{7} \stackrel{?}{=} \frac{6}{7}$	
$-3\left(\frac{4}{7}\right) + \frac{5}{7} + 4\left(\frac{4}{7}\right) - \frac{3}{7}$	$\frac{6}{7}$
$\underbrace{-\frac{12}{7} + \frac{5}{7}} + \underbrace{\frac{16}{7} - \frac{3}{7}}$	
$-\frac{7}{7} + \frac{13}{7}$	
$\frac{6}{7}$	

Teaching Tip

Make sure students *check* their work. There is *no* reason to get the wrong answer, since you can always CHECK!

C Using Both Properties Together

Can we solve the equation $2x - 7 = x + 2$? Let's try. If we add 7 to both sides, we obtain

$$2x - 7 + 7 = x + 2 + 7$$
$$2x = x + 9$$

But this is not yet a solution. To solve this equation, we must get x by itself on the left—that is, $x = \square$, where $\square$ is a number. How do we do this? We want variables on one side of the equation (and we have them: $2x$) but only specific numbers on the other (here we are in trouble because we have an x on the right). To "get rid of" this x, we subtract x from both sides:

$$2x - x = x - x + 9 \qquad \text{Remember, } x = 1x.$$
$$x = 9$$

Thus 9 is the solution of the equation.

CHECK

$2x - 7 \stackrel{?}{=} x + 2$	
$2(9) - 7$	$9 + 2$
11	11

By the way, you do not have to have the variable by itself on the *left* side of the equation. You may have the variables on the *right* side of the equation and your solution may be of the form $\square = x$, where $\square$ is a number.

Teaching Tip

How do we decide on which side, right or left, we want to have the variable?

The answer is: on whichever side has the "most" variables.

For $2x - 7 = x + 2$, variables
↑ two x's ↑ one x
should be on the left.

A recommended procedure is to isolate the variable on the side of the equation that contains the highest coefficient of variables after simplification. Thus, when solving $7x + 3 = 4x + 6$, isolate the variables on the left side of the equation. On the other hand, when solving $4x + 6 = 7x + 3$, isolate the variables on the right side of the equation. Obviously, your solution would be the same in either case. [See Example 4(c).]

Now let's review our procedure for solving equations by adding or subtracting.

PROCEDURE

Solving Equations by Adding or Subtracting

1. Simplify both sides if necessary.
2. Add or subtract the same numbers on both sides of the equation so that one side contains only variables.
3. Add or subtract the same expressions on both sides of the equation so that the other side contains only numbers.

We use these three steps to solve the next example.

EXAMPLE 4 **Solving equations by adding or subtracting**

Solve:

a. $3 = 8 + x$

b. $4y - 3 = 3y + 8$

c. $0 = 3(z - 2) + 4 - 2z$

d. $2(x + 1) = 3x + 5$

SOLUTION

a.

$3 = 8 + x$ Given.

1. Both sides of the equation are already simplified. $3 = 8 + x$
2. Subtract 8 on both sides. $3 - 8 = 8 - 8 + x$

$-5 = x$

Step 3 is not necessary here, and the solution is -5. $x = -5$

CHECK $3 \stackrel{?}{=} 8 + x$

3	$8 + (-5)$
	3

b.

$4y - 3 = 3y + 8$ Given.

1. Both sides of the equation are already simplified. $4y - 3 = 3y + 8$
2. Add 3 on both sides. $4y - 3 + 3 = 3y + 8 + 3$

$4y = 3y + 11$

3. Subtract $3y$ on both sides. $4y - 3y = 3y - 3y + 11$

$y = 11$

The solution is 11.

PROBLEM 4

Solve:

a. $5 = 7 + x$

b. $5x - 2 = 4x + 3$

c. $0 = 3(y - 3) + 7 - 2y$

d. $3(z + 1) = 4z + 8$

Answers

4. a. -2 **b.** 5 **c.** 2 **d.** -5

CHECK $4y - 3 \stackrel{?}{=} 3y + 8$

$4(11) - 3$	$3(11) + 8$
$44 - 3$	$33 + 8$
41	41

c. $0 = 3(z - 2) + 4 - 2z$ Given.

1. Simplify by using the distributive property and combining like terms.

$$0 = 3z - 6 + 4 - 2z$$
$$0 = z - 2$$

2. Add 2 on both sides.

$$0 + 2 = z - 2 + 2$$
$$2 = z$$

Step 3 is not necessary, and the solution is 2.

$$z = 2$$

CHECK $0 \stackrel{?}{=} 3(z - 2) + 4 - 2z$

0	$3(2 - 2) + 4 - 2(2)$
	$3(0) + 4 - 4$
	$0 + 4 - 4$
	0

d. $2(x + 1) = 3x + 5$ Given.

1. Simplify. $2x + 2 = 3x + 5$
2. Subtract 2 on both sides. $2x + 2 - 2 = 3x + 5 - 2$

$$2x = 3x + 3$$

3. Subtract $3x$ on both sides so all the variables are on the left.

$$2x - 3x = 3x - 3x + 3$$
$$-x = 3$$

If $-x = 3$, then $x = -3$ because the opposite of a number is the number with its sign changed; that is, if the opposite of x is 3, then x itself must be -3. Thus the solution is -3.

CHECK $2(x + 1) \stackrel{?}{=} 3x + 5$

$2(-3 + 1)$	$3(-3) + 5$
$2(-2)$	$-9 + 5$
-4	-4

Keep in mind the following rule.

SOLVING $-x = a$

If a is a real number and $-x = a$, then $x = -a$.

Teaching Tip

You could avoid some work if you look at Step 1, $2x + 2 = 3x + 5$ and note that there are "more" x's on the right. If you isolate the x's on the right by subtracting $2x$ and then subtracting 5, you obtain $-3 = x$ directly.

Teaching Tip

If "$-x$" is read "the opposite of," it will be easier to understand.

EXAMPLE 5 More practice solving equations by adding or subtracting

Solve: $8x + 7 = 9x + 3$

SOLUTION

1. The equation is already simplified.	$8x + 7 = 9x + 3$	Given.
2. Subtract 7 on both sides.	$8x + 7 - 7 = 9x + 3 - 7$	
	$8x = 9x - 4$	
3. Subtract $9x$ on both sides.	$8x - 9x = 9x - 9x - 4$	
	$-x = -4$	

Since $-x = -4$, then $x = -(-4) = 4$, and the solution is 4.

CHECK $8x + 7 \stackrel{?}{=} 9x + 3$

$8(4) + 7$	$9(4) + 3$
$32 + 7$	$36 + 3$
39	39

PROBLEM 5

Solve: $6y + 5 = 7y + 2$

Teaching Tip

You can avoid some work if you isolate the x's on the right. Remember, there are nine x's on the right and only eight x's on the left.

The equations in Examples 2–5 each had exactly one solution. For an equation that can be written as $ax + b = c$, there are three possibilities for the solution:

1. The equation has **one** solution. This is a **conditional** equation.
2. The equation has **no** solution. This is a **contradictory** equation.
3. The equation has **infinitely many** solutions. This is an **identity.**

EXAMPLE 6 Solving a contradictory equation

Solve: $3 + 8(x + 1) = 5 + 8x$

SOLUTION

	$3 + 8(x + 1) = 5 + 8x$	Given.
1. Simplify by using the distributive property and combining like terms.	$3 + 8x + 8 = 5 + 8x$	
	$11 + 8x = 5 + 8x$	
2. Subtract 5 on both sides.	$11 - 5 + 8x = 5 - 5 + 8x$	
	$6 + 8x = 8x$	
3. Subtract $8x$ on both sides.	$6 + 8x - 8x = 8x - 8x$	
	$6 = 0$	

The statement "$6 = 0$" is a *false* statement. When this happens, it indicates that the equation has *no* solution—that is, it's a contradictory equation and we write "no solution."

PROBLEM 6

Solve: $5 + 3(z - 1) = 4 + 3z$

Teaching Tip

If you are solving an equation and the variable term cancels out resulting in

1. a **false** statement, the equation has **no** solution (see Example 6).
2. a **true** statement, the equation has **infinitely many** solutions (see Example 7).

Answers

5. 3 **6.** No solution

EXAMPLE 7 **Solving an equation with infinitely many solutions**

Solve: $7 + 2(x + 1) = 9 + 2x$

SOLUTION

$$7 + 2(x + 1) = 9 + 2x \quad \text{Given.}$$

1. Simplify by using the distributive property and combining like terms.

$$7 + 2x + 2 = 9 + 2x$$
$$9 + 2x = 9 + 2x$$

You could stop here. Since both sides are *identical,* this equation is an identity. Every real number is a solution. But what happens if you go on? Let's see.

2. Subtract 9 on both sides.

$$9 - 9 + 2x = 9 - 9 + 2x$$
$$2x = 2x$$

3. Subtract $2x$ on both sides.

$$2x - 2x = 2x - 2x$$
$$0 = 0$$

The statement "$0 = 0$" is a true statement. When this happens, it indicates that *any* real number is a solution. (Try $x = -1$ or $x = 0$ in the original equation.) The equation has infinitely many solutions, and we write "all real numbers" for the solution.

PROBLEM 7

Solve: $3 + 4(y + 2) = 11 + 4y$

EXAMPLE 8 **Application: College tuition and fees**

How much is your tuition this year? It is estimated that annual tuition and required fees for in-state students at 2-year colleges is given by the expression $400x + 7400$, where x is the number of years after 1998.

a. Determine the estimated tuition and fees for 1998 ($x = 0$).

b. In how many years will tuition and fees reach \$11,000?

SOLUTION

a. When $x = 0$,

$$400x + 7400 = 400(0) + 7400$$
$$= 7400$$

Thus, estimated tuition and fees for 1998 ($x = 0$) amounted to \$7400.

b. The expression representing tuition and fees is $400x + 7400$. We want to estimate when this expression will reach \$11,000; that is, we want to solve the equation

$$400x + 7400 = 11{,}000$$
$$400x = 11{,}000 - 7400 \quad \text{Subtract 7400.}$$
$$400x = 3600 \quad \text{Simplify.}$$
$$x = 9 \quad \text{Since } 400 \cdot 9 = 3600.$$

In the next section, we will show you that the equation $400x = 3600$ can be solved by *dividing* both sides of the equation by 400. The answer is the same, 9. What does this mean? It means that 9 years after 1998, that is, in $1998 + 9 = 2007$, annual tuition and fees at 2-year colleges will reach \$11,000. If you want to confirm this, see the *Calculate It* section that follows.

PROBLEM 8

a. What will the estimated annual tuition and fees be in the year 2003?

b. When will tuition and fees be estimated to be \$9000?

Web It

For a lesson dealing with solving equations by steps, try link 2-1-2 at the Bello Website at mhhe.com/bello. If you prefer, you can enter your own problem and have it explained at link 2-1-3.

Answers

7. All real numbers

8. a. 2003 is 5 years after 1998. The fees then will be $400(5) + 7400 = \$9400$.

b. Solve $400x + 7400 = 9000$ The answer is 4, so in $1998 + 4 = 2002$, tuition and fees will be estimated to be \$9000.

Calculate It Checking and Getting Results

Can you use a calculator to check your results? You can, but you will be a little bit ahead of the game because you first need to know how to graph the equations. First, the equations we are solving are called **linear** equations. Pictures of linear equations, which are called graphs, are lines. Let's graph the equation from Example 5. The equation $8x + 7 = 9x + 3$ has two sides: the left side, $8x + 7$, and the right side, $9x + 3$. If we make a graph of $8x + 7$ and $9x + 3$, we will get two lines. The point at which the lines intersect is the point at which $8x + 7 = 9x + 3$; that is, the value of x at the intersection is the solution of the equation $8x + 7 = 9x + 3$.

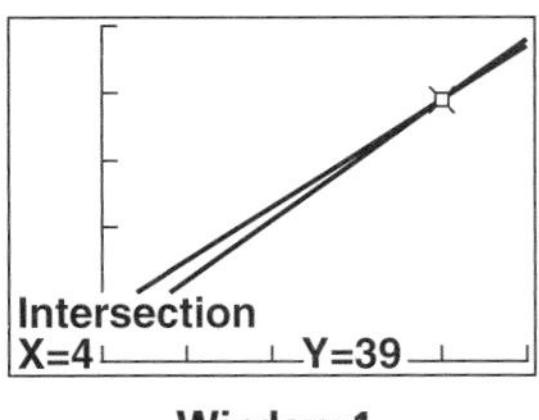

Window 1

But how do you graph $8x + 7$ and $9x + 3$? First, you have to give $8x + 7$ and $9x + 3$ names. With a TI-83 Plus, this is done by pressing Y= and entering $8x + 7$ for Y_1 and $9x + 3$ for Y_2. Next, you have to tell your calculator which part of the lines you want to see. As you recall, when we checked the solution $x = 4$ in Example 5, the evaluation of $8x + 7$ yielded 39. This means that your window has to be made large enough for you to see $Y_1 = 8(4) + 7 = 39$ when $x = 4$. We will use a $[-1, 5]$ display with a scale of 1 unit for the x's and a $[-1, 50]$ window with a scale of 10 units for the y's. (Press WINDOW and select Xmin $= -1$, Xmax $= 5$, Xscl $= 1$, Ymin $= -1$, Ymax $= 50$, and Yscl $= 10$.)

Now, tell your calculator to show the graph by pressing GRAPH. The display will look like Window 1. To find the point of intersection (where the lines meet), use your TRACE key or your intersect feature (2nd TRACE 5 and ENTER three times). At the point of intersection, $x = 4$, so 4 is the solution of the equation $8x + 7 = 9x + 3$. If you do the same for Example 6, the lines never meet. (See Window 2.) They are *parallel* lines, and that's why there's no solution! Now see what you and your calculator can do with Example 7.

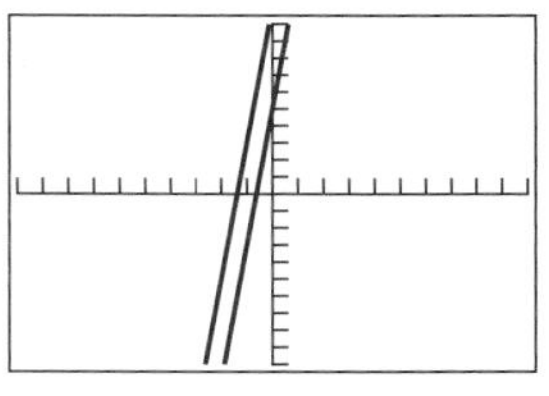

Window 2

Can we actually *solve* equations using a calculator? Most calculators have an EQUATION SOLVER feature. First, tell the calculator that you want to do some MATH. Press MATH then ▲ followed by 0. For the equation of Example 8(b), $400x + 7400 = 11{,}000$, you have to do some algebra before the calculator solves it for you! Why? Because the calculator accepts only equations with a 0 on one side, so subtract 11,000 from both sides of the equation to get $0 = 400x - 3600$. Now, enter $400x - 3600$. (Enter the x by pressing ALPHA STO▸. Do you see the x above the STO▸?) Next you have to tell the calculator which variable you want to solve for, so move the cursor next to the x and press ALPHA SOLVE (above the ENTER key). The solver tells you that $x = 9$ as shown. Do not worry about the rest of the information at this time! Think about this: isn't it easier to just solve the equation?

```
400X-3600=0
▪X=9
 bound={-1E99, 1...
▪left-rt=0
```

Window 3

Exercises 2.1

A In Problems 1–10, determine whether the given number is a solution of the equation. (Do not solve.)

1. $x = 3; x - 1 = 2$ Yes

2. $x = 4; 6 = x - 10$ No

3. $y = -2; 3y + 6 = 0$ Yes

4. $z = -3; -3z + 9 = 0$ No

5. $n = 2; 12 - 3n = 6$ Yes

6. $m = 3\frac{1}{2}; 3\frac{1}{2} + m = 7$ Yes

7. $d = 10; \frac{2}{5}d + 1 = 3$ No

8. $c = 2.3; 3.4 = 2c - 1.4$ No

9. $a = 2.1; 4.6 = 11.9 - 3a$ No

10. $x = \frac{1}{10}; 0.2 = \frac{7}{10} - 5x$ Yes

B In Problems 11–30, solve and check the given equations.

11. $x - 5 = 9$ $x = 14$

12. $y - 3 = 6$ $y = 9$

13. $11 = m - 8$ $m = 19$

14. $6 = n - 2$ $n = 8$

15. $y - \frac{2}{3} = \frac{8}{3}$ $y = \frac{10}{3}$

16. $R - \frac{4}{3} = \frac{35}{3}$ $R = 13$

17. $2k - 6 - k - 10 = 5$ $k = 21$

18. $3n + 4 - 2n - 6 = 7$ $n = 9$

19. $\frac{1}{4} = 2z - \frac{2}{3} - z$ $z = \frac{11}{12}$

20. $\frac{7}{2} = 3v - \frac{1}{5} - 2v$ $v = \frac{37}{10}$

21. $0 = 2x - \frac{3}{2} - x - 2$ $x = \frac{7}{2}$

22. $0 = 3y - \frac{1}{4} - 2y - \frac{1}{2}$ $y = \frac{3}{4}$

23. $\frac{1}{5} = 4c + \frac{1}{5} - 3c$ $c = 0$

24. $0 = 6b + \frac{19}{2} - 5b - \frac{1}{2}$ $b = -9$

25. $-3x + 3 + 4x = 0$ $x = -3$

26. $-5y + 4 + 6y = 0$ $y = -4$

27. $\frac{3}{4}y + \frac{1}{4} + \frac{1}{4}y = \frac{1}{4}$ $y = 0$

28. $\frac{1}{2}y + \frac{2}{3} + \frac{1}{2}y = \frac{1}{3}$ $y = -\frac{1}{3}$

29. $3.4 = -3c + 0.8 + 2c + 0.1$
$c = -2.5$

30. $1.7 = -3c + 0.3 + 4c + 0.4$
$c = 1$

C In Problems 31–55, solve and check the given equations.

31. $6p + 9 = 5p$ $p = -9$

32. $7q + 4 = 6q$ $q = -4$

33. $3x + 3 + 2x = 4x$ $x = -3$

34. $2y + 4 + 6y = 7y$ $y = -4$

35. $4(m - 2) + 2 - 3m = 0$ $m = 6$

36. $3(n + 4) + 2 = 2n$ $n = -14$

37. $5(y - 2) = 4y + 8$ $y = 18$

38. $3(z - 1) = 4z + 1$ $z = -4$

39. $3a - 1 = 2(a - 4)$ $a = -7$

40. $4(b + 1) = 5b - 3$ $b = 7$

41. $5(c - 2) = 6c - 2$ $c = -8$

42. $-4R + 6 = 5 - 3R + 8$ $R = -7$

43. $3x + 5 - 2x + 1 = 6x + 4 - 6x$
$x = -2$

44. $6f - 2 - 4f = -2f + 5 + 3f$
$f = 7$

45. $-2g + 4 - 5g = 6g + 1 - 14g$
$g = -3$

46. $-2x + 3 + 9x = 6x - 1$
$x = -4$

47. $6(x + 4) + 4 - 2x = 4x$
No solution

48. $6(y - 1) - 2 + 2y = 8y + 4$
No solution

49. $10(z - 2) + 10 - 2z = 8(z + 1) - 18$
All real numbers

50. $7(a + 1) - 1 - a = 6(a + 1)$
All real numbers

51. $3b + 6 - 2b = 2(b - 2) + 4$
$b = 6$

52. $3b + 2 - b = 3(b - 2) + 5$
$b = 3$

53. $2p + \frac{2}{3} - 5p = -4p + 7\frac{1}{3}$
$p = \frac{20}{3}$

54. $4q + \frac{2}{7} - 6q = -3q + 2\frac{2}{7}$
$q = 2$

55. $5r + \frac{3}{8} - 9r = -5r + 1\frac{1}{2}$
$r = \frac{9}{8}$

APPLICATIONS

56. *Price increases* The price of an item is increased by \$7; it now sells for \$23. What was the old price of the item? \$16

57. *Average hourly earnings* In a certain year, the average hourly earnings were \$9.81, an increase of 40¢ over the previous year. What were the average hourly earnings the previous year? \$9.41

58. *Consumer Price Index* The Consumer Price Index for housing in a recent year was 169.6, a 5.7-point increase over the previous year. What was the Consumer Price Index for housing the previous year? 163.9

59. *Medical costs* The cost of medical care increased 142.2 points in a 6-year period. If the cost of medical care reached the 326.9 mark, what was it 6 years ago? 184.7

60. *SAT scores* In the last 10 years, mathematics scores in the Scholastic Aptitude Test (SAT) have declined 16 points, to 476. What was the mathematics score 10 years ago? 492

61. *Waste generation* From 1960 to 1990, the amount of waste generated each year increased by a whopping 107.9 million tons, ultimately reaching 195.7 million tons! How much waste was generated in 1960? The figure reached 229.9 million tons in 1999. What was the increase (in million tons) from 1960? 87.8 million tons; 142.1 million tons

62. *Materials recovery* From 1960 to 1990, the amount of materials recovered each year increased by 27.5 million tons to 33.4 million tons. How much waste was recovered in 1960? By 1999, 30.3 million more tons were recovered than in 1990. How many million tons is that? 5.9 million tons; 63.7 million tons

63. *Popularity of exercise-walking* What is the most popular sports activity in the United States? The answer is exercise-walking! According to the Census Bureau, 38% of the females 7 years and older engage in this activity. If this percent is 17% more than the percent of males who exercise-walk, what percent of males engage in exercise-walking? 21%

64. *Popularity of swimming* The most popular sports activity among males 7 years and older is swimming, which is enjoyed by 26% of males. If this is 1% more than the percent of females swimming, what percent of females participate in swimming? 25%

65. *Popularity of cultural activities* According to the Census Bureau, among people 18 years old and older, 8% more attended historic parks than art museums. If 35% of the population attended historic parks, what percent attended art museums? 27%

66. *Red meat consumption* According to the U.S. Department of Agriculture, from 1970 to 1992, the per capita consumption of red meat declined by 17.6 pounds to 114.1 pounds. What was the per capita consumption of red meat in 1970? In 1999, consumption went up 3.6 pounds from the 1992 level. What was the per capita consumption in 1999? 131.7 pounds; 117.7 pounds

SKILL CHECKER

Try the Skill Checker Exercises so you'll be ready for the next section.

Find:

67. $4(-5)$ -20

68. $6(-3)$ -18

69. $-\frac{2}{3}\left(\frac{3}{4}\right)$ $-\frac{1}{2}$

70. $-\frac{5}{7}\left(\frac{7}{10}\right)$ $-\frac{1}{2}$

Find the reciprocal of each number:

71. $\frac{3}{2}$ $\frac{2}{3}$

72. $\frac{2}{5}$ $\frac{5}{2}$

Find the LCM of each pair of numbers:

73. 6 and 16 48

74. 9 and 12 36

75. 10 and 8 40

76. 30 and 18 90

USING YOUR KNOWLEDGE

Some Detective Work

In this section we learned how to determine whether a given number *satisfies* an equation. Let's use this idea to do some detective work!

77. Suppose the police department finds a femur bone from a human female. The relationship between the length f of the femur and the height H of a female (in centimeters) is given by

$$H = 1.95f + 72.85$$

If the length of the femur is 40 centimeters and a missing female is known to be 120 centimeters tall, can the bone belong to the missing female? No

78. If the length of the femur in Problem 77 is 40 centimeters and a missing female is known to be 150.85 centimeters tall, can the bone belong to the missing female? Yes

79. Have you seen police officers measuring the length of a skid mark after an accident? There's a formula for this. It relates the velocity V_a at the time of an accident and the length L_a of the skid mark at the time of the accident to the velocity and length of a test skid mark. The test skid mark is obtained by driving a car at a predetermined speed V_t, skidding to a stop, and then measuring the length of the skid L_t. The formula is

$$V_a^2 = \frac{L_a V_t^2}{L_t}$$

If $L_t = 36$, $L_a = 144$, $V_t = 30$, and the driver claims that at the time of the accident his velocity V_a was 50 miles per hour, can you believe him? No

80. Would you believe the driver in Problem 79 if he says he was going 90 miles per hour? No

WRITE ON

81. Explain what is meant by the solution of an equation. Answers may vary.

82. Explain what is meant by equivalent equations. Answers may vary.

83. Make up an equation that has no solution and one that has infinitely many solutions. Answers may vary.

84. If the next-to-last step in solving an equation is $-x = -5$, what is the solution of the equation? Explain. $x = 5$; answers may vary.

MASTERY TEST

If you know how to do these problems, you have learned your lesson!

Solve:

85. $5 + 4(x + 1) = 3 + 4x$ No solution

86. $2 + 4(x + 1) = 5x + 3$ $x = 3$

87. $x - 5 = 4$ $x = 9$

88. $x - \frac{1}{5} = \frac{3}{5}$ $x = \frac{4}{5}$

89. $x - 2.3 = 3.4$ $x = 5.7$

90. $x - \frac{1}{7} = \frac{1}{4}$ $x = \frac{11}{28}$

91. $2x + 6 - x + 2 = 12$ $x = 4$

92. $3x + 5 - 2x + 3 = 7$ $x = -1$

93. $-5x + \frac{2}{9} + 6x - \frac{4}{9} = \frac{5}{9}$ $x = \frac{7}{9}$

94. $5y - 2 = 4y + 1$ $y = 3$

95. $0 = 4(z - 3) + 5 - 3z$ $z = 7$

96. $2 - (4x + 1) = 1 - 4x$ All real numbers

97. $3(x + 2) + 3 = 2 - (1 - 3x)$ No solution

98. $3(x + 1) - 3 = 2x - 5$ $x = -5$

Determine whether the given number is a solution of the equation:

99. $-7; \frac{1}{7}z - 1 = 0$ No

100. $-\frac{3}{5}; \frac{5}{3}x + 1 = 0$ Yes

2.2 THE MULTIPLICATION AND DIVISION PROPERTIES OF EQUALITY

To Succeed, Review How To . . .

1. Multiply and divide signed numbers (pp. 70–76).
2. Find the reciprocal of a number (p. 14).
3. Find the LCM of two or more numbers (p. 17).
4. Write a fraction as a percent, and vice versa (pp. 32–34).

Objectives

A Use the multiplication and division properties of equality to solve equations.

B Multiply by reciprocals to solve equations.

C Multiply by LCMs to solve equations.

D Solve applications involving percents.

GETTING STARTED

How Good a Deal Is This?

The tire in the ad is on sale at half price. It now costs \$28. What was its old price, p? Since you are paying half price for the tire, the new price is $\frac{1}{2}$ of p—that is, $\frac{1}{2}p$ or $\frac{p}{2}$. Since this price is \$28, we have

$$\frac{p}{2} = 28$$

But what was the old price? Twice as much, of course. Thus to obtain the old price p, we multiply both sides of the equation by 2, the reciprocal of $\frac{1}{2}$, to obtain

$$2 \cdot \frac{p}{2} = 2 \cdot 28 \qquad \text{Note that } \not{2} \cdot \frac{p}{\not{2}} = 1p \text{ or } p.$$

$$p = 56$$

Hence the old price was \$56, as can be easily checked, since $\frac{56}{2} = 28$.

This example shows how you can *multiply* both sides of an equation by a nonzero number and obtain an *equivalent* equation—that is, an equation whose solution is the same as the original one. This is the multiplication property, one of the properties we will study in this section.

A Using the Multiplication and Division Properties of Equality

The ideas discussed in the *Getting Started* can be generalized as the following property.

The Multiplication Property of Equality

The equation $a = b$ is equivalent to

$$ac = bc \quad \text{for any nonzero number } c$$

This means that we can multiply both sides of an equation by the same nonzero number and obtain an equivalent equation. We use this property next.

EXAMPLE 1 **Using the multiplication property**

Solve:

a. $\frac{x}{3} = 2$ **b.** $\frac{y}{5} = -3$

SOLUTION

a. We multiply both sides of $\frac{x}{3} = 2$ by 3, the reciprocal of $\frac{1}{3}$.

$$3 \cdot \frac{x}{3} = 3 \cdot 2$$

$$\overset{1}{\cancel{3}} \cdot \frac{x}{\cancel{3}} = 6 \qquad \text{Note that } 3 \cdot \tfrac{1}{3} = 1\text{, since 3 and } \tfrac{1}{3} \text{ are reciprocals.}$$

$$x = 6$$

Thus the solution is 6.

CHECK $\frac{x}{3} \overset{?}{=} 2$

$\frac{6}{3}$	2
2	

b. We multiply both sides of $\frac{y}{5} = -3$ by 5, the reciprocal of $\frac{1}{5}$.

$$5 \cdot \frac{y}{5} = 5(-3)$$

$$\overset{1}{\cancel{5}} \cdot \frac{y}{\cancel{5}} = -15 \qquad \text{Recall that } 5 \cdot (-3) = -15.$$

$$y = -15$$

Thus the solution is -15.

CHECK $\frac{y}{5} \overset{?}{=} -3$

$\frac{-15}{5}$	-3
-3	

PROBLEM 1

Solve:

a. $\frac{x}{5} = 3$ **b.** $\frac{y}{4} = -5$

Teaching Tip

In Example 1a, we multiply both sides by 3 since multiplication by 3 "undoes" division by 3.

Suppose the price of an article is doubled, and it now sells for \$50. What was its original price, p? Half as much, right? Here is the equation:

$$2p = 50$$

We solve it by dividing both sides by 2 (to find half as much):

$$\frac{2p}{2} = \frac{50}{2}$$

$$\frac{\overset{1}{\cancel{2}}p}{\cancel{2}} = 25$$

Thus the original price p is \$25, as you can check:

$$2 \cdot 25 = 50$$

Answers

1. a. 15 **b.** -20

Web It

If you need a lesson showing you how to solve equations by multiplication and/or division, go to link 2-2-1 at the Bello Website at mhhe.com/bello. As before, if you want to enter your own problem and see an explanation, try link 2-2-2.

Note that dividing both sides by 2 (the *coefficient* of p) is the same as *multiplying* by $\frac{1}{2}$. Thus you can also solve $2p = 50$ by multiplying by $\frac{1}{2}$ (the reciprocal of 2) to obtain

$$\frac{1}{2} \cdot 2p = \frac{1}{2} \cdot 50$$
$$p = 25$$

This example suggests that, just as the addition property of equality lets us *add* a number on each side of an equation, the division property lets us *divide* each side of an equation by a (nonzero) number to obtain an *equivalent* equation. We now state this property and use it in the next example.

The Division Property of Equality

The equation $a = b$ is equivalent to

$$\frac{a}{c} = \frac{b}{c} \quad \text{for any nonzero number } c$$

This means that we can divide both sides of an equation by the same nonzero number and obtain an equivalent equation. Note that we can also *multiply* both sides of $a = b$ by the *reciprocal* of c—that is, by $\frac{1}{c}$—to obtain

$$\frac{1}{c} \cdot a = \frac{1}{c} \cdot b \quad \text{or} \quad \frac{a}{c} = \frac{b}{c} \qquad \text{Same result!}$$

We shall solve equations by multiplying by reciprocals in the next section.

EXAMPLE 2 Using the division property

Solve:

a. $8x = 24$ **b.** $5x = -20$ **c.** $-3x = 7$

SOLUTION

a. We need to get x by itself on the left. That is, we need $x = \square$, where $\square$ is a number. Thus we divide both sides of the equation by 8 (the coefficient of x) to obtain

$$\frac{8x}{8} = \frac{24}{8}$$
$$\frac{\overset{1}{\cancel{8}}x}{\cancel{8}} = 3$$
$$x = 3$$

The solution is 3.

CHECK

$8x \stackrel{?}{=} 24$	
$8 \cdot 3$	24
24	

You can also solve this problem by *multiplying* both sides of $8x = 24$ by $\frac{1}{8}$, the reciprocal of 8.

$$\frac{1}{8} \cdot 8x = \frac{1}{8} \cdot 24$$
$$1x = \frac{1}{\cancel{8}} \cdot \frac{\overset{3}{\cancel{24}}}{1}$$
$$x = 3$$

PROBLEM 2

Solve:

a. $3x = 12$

b. $7x = -21$

c. $-5x = 20$

Teaching Tip

In Example 2a, we divide both sides by 8 since division by 8 "undoes" multiplication by 8.

Answers

2. a. 4 **b.** -3 **c.** -4

b. Here, we divide both sides of $5x = -20$ by 5 (the coefficient of x) so that we have x by itself on the left. Thus

$$\frac{5x}{5} = \frac{-20}{5}$$

$$x = -4$$

The solution is -4.

CHECK

$$5x \stackrel{?}{=} -20$$

$5 \cdot (-4)$	-20
-20	

Of course, you can also solve this problem by *multiplying* both sides by $\frac{1}{5}$, the reciprocal of 5. When solving these types of equations, you always have the option of dividing both sides of the equation by a specific number or multiplying both sides of the equation by the reciprocal of the number.

$$\frac{1}{5} \cdot 5x = \frac{1}{5} \cdot (-20)$$

$$1x = \frac{1}{\cancel{5}_1} \cdot \frac{\cancel{-20}^{-4}}{1}$$

$$x = -4$$

c. In this case we divide both sides of $-3x = 7$ by -3 (the coefficient of x). We then have

$$\frac{\cancel{-3}^{1}x}{\cancel{-3}} = \frac{7}{-3}$$

$$x = -\frac{7}{3}$$

Recall that the quotient of two numbers with different signs is negative.

The solution is $-\frac{7}{3}$.

CHECK

$$-3x \stackrel{?}{=} 7$$

$-3\left(-\frac{7}{3}\right)$	7
7	

As you know, this problem can also be solved by multiplying both sides by $-\frac{1}{3}$, the reciprocal of -3.

$$-\frac{1}{3} \cdot (-3x) = -\frac{1}{3} \cdot 7$$

$$1x = -\frac{1}{3} \cdot \frac{7}{1}$$

$$x = -\frac{7}{3}$$

Teaching Tip

Have students do Example 2c by dividing by $+3$ to see what happens to the variable.

$$\frac{-3x}{+3} = \frac{7}{+3}$$

$$-x = \frac{7}{3}$$

What other step is needed to solve for x? Point out that dividing by -3 eliminates the extra step.

B Multiplying by Reciprocals

In Example 2, the coefficients of the variables were integers. In such cases it's easy to divide each side of the equation by this coefficient. When the coefficient of the variable is a fraction, it's easier to multiply each side of the equation by the reciprocal of the coefficient. Thus to solve $-3x = 7$, divide each side by -3, but to solve $\frac{3}{4}x = 18$, multiply each side by the reciprocal of the coefficient of x, that is, by $\frac{4}{3}$, as shown next.

EXAMPLE 3 **Solving equations by multiplying by reciprocals**

Solve:

a. $\frac{3}{4}x = 18$ **b.** $-\frac{2}{5}x = 8$ **c.** $-\frac{3}{8}x = -15$

SOLUTION

a. Multiply both sides of $\frac{3}{4}x = 18$ by the reciprocal of $\frac{3}{4}$, that is, by $\frac{4}{3}$. We then get

$$\frac{4}{3}\left(\frac{3}{4}x\right) = \frac{4}{3}(18)$$

$$1 \cdot x = \frac{4}{\cancel{3}_1} \cdot \frac{\cancel{18}^6}{1} = \frac{24}{1}$$

$$x = 24$$

Hence the solution is 24.

CHECK

$$\frac{3}{4}x \stackrel{?}{=} 18$$

$\frac{3}{4}(24)$	18
$\frac{3}{\cancel{4}_1} \cdot \frac{\cancel{24}^6}{1}$	
18	

b. Multiplying both sides of $-\frac{2}{5}x = 8$ by the reciprocal of $-\frac{2}{5}$, that is, by $-\frac{5}{2}$, we have

$$-\frac{5}{2}\left(-\frac{2}{5}x\right) = -\frac{5}{2}(8)$$

$$1 \cdot x = \frac{-5}{\cancel{2}_1} \cdot \frac{\cancel{8}^4}{1} = -\frac{20}{1}$$

$$x = -20$$

The solution is -20.

PROBLEM 3

Solve:

a. $\frac{3}{5}x = 12$

b. $-\frac{2}{5}x = 6$

c. $-\frac{4}{5}x = -8$

Answers

3. a. 20 **b.** -15 **c.** 10

CHECK $-\frac{2}{5}x \stackrel{?}{=} 8$

$-\frac{2}{5}(-20)$	8
$-\frac{\overset{1}{2}}{\underset{1}{5}}\left(\frac{\overset{-4}{\cancel{-20}}}{1}\right)$	
8	

c. Here we multiply both sides of $-\frac{3}{8}x = -15$ by $-\frac{8}{3}$, the reciprocal of $-\frac{3}{8}$, to obtain

$$-\frac{8}{3}\left(-\frac{3}{8}x\right) = -\frac{8}{3}(-15)$$

$$1 \cdot x = \frac{-8(-15)}{3} = \frac{-8(\overset{-5}{\cancel{-15}})}{\underset{1}{\cancel{3}}}$$

$$= 40$$

The solution is 40.

CHECK $-\frac{3}{8}x \stackrel{?}{=} -15$

$-\frac{3}{8}(40)$	-15
$-\frac{3}{\underset{1}{\cancel{8}}}(\overset{5}{\cancel{40}})$	
-15	

C Multiplying by the LCM

Finally, if the equation we are solving contains sums or differences of fractions, we first eliminate these fractions by multiplying each term in the equation by the smallest number that is a multiple of each of the denominators. This number is called the *least common multiple* (or LCM for short) of the denominators. Let's see why. Which equation would you rather solve?

$$3x + 2x = 1 \quad \text{or} \quad \frac{x}{2} + \frac{x}{3} = \frac{1}{6}$$

Probably the first! But if you multiply each term in the second equation by the LCM of 2, 3, and 6 (which is 6), you obtain the first equation!

Do you remember how to find the LCM of two numbers? If you don't, here's a quick way to do it. Suppose you wish to solve the equation

$$\frac{x}{6} + \frac{x}{16} = 22$$

To find the LCM of 6 and 16, write the denominators in a horizontal row (see step 1 to the right) and divide each of them by the largest number that will divide *both* of them. In this case, the number is 2. The quotients are 3 and 8, as shown in step 3. Since there are no numbers other than 1 that will divide both 3 and 8, the LCM is the product of 2 and the final quotients 3 and 8, as indicated in step 4.

1. 6 16

2. $2\underline{|6 \quad 16}$

3. $2\underline{|6 \quad 16}$
3 8

4. $2\underline{|6 \quad 16}$
3—8 → $2 \cdot 3 \cdot 8 = 48$ is the LCM.

Now we can multiply each side of the equation by this LCM (48):

$$48\left(\frac{x}{6} + \frac{x}{16}\right) = 48 \cdot 22$$ To avoid confusion, place parentheses around $\frac{x}{6} + \frac{x}{16}$.

$$\overset{8}{\cancel{48}} \cdot \frac{x}{\underset{1}{\cancel{6}}} + \overset{3}{\cancel{48}} \cdot \frac{x}{\underset{1}{\cancel{16}}} = 48 \cdot 22$$ Use the distributive property. Don't multiply $48 \cdot 22$ yet; we'll simplify this later.

$$8x + 3x = 48 \cdot 22$$ Simplify the left side.

$$11x = 48 \cdot 22$$ Combine like terms.

$$\frac{\cancel{11}x}{\cancel{11}} = \frac{48 \cdot \overset{2}{\cancel{22}}}{\underset{1}{\cancel{11}}}$$ Divide both sides by 11. Do you see why we waited to multiply $48 \cdot 22$?

$$x = 96$$

The solution is 96.

Teaching Tip

Tell students to avoid the temptation of multiplying 48 by 22. There may be some simplification later!

CHECK $\frac{x}{6} + \frac{x}{16} \stackrel{?}{=} 22$

$\frac{96}{6} + \frac{96}{16}$	22
$16 + 6$	
22	

Note that if we wish to clear fractions in

$$\frac{a}{b} + \frac{c}{d} = \frac{e}{f}$$

we can use the following procedure.

PROCEDURE

Clearing Fractions

To clear fractions in an equation, multiply both sides of the equation by the LCM of the denominators, or, equivalently, multiply each term by the LCM.

Thus if we multiply both sides of $\frac{a}{b} + \frac{c}{d} = \frac{e}{f}$ by the LCM of b, d, and f (which we shall call L), we get

$$L\left(\frac{a}{b} + \frac{c}{d}\right) = L\frac{e}{f}$$ Note the added parentheses.

or

$$\frac{La}{b} + \frac{Lc}{d} = \frac{Le}{f}$$

Thus to clear the fractions here, we multiply each term by L (using the distributive property).

EXAMPLE 4 **Solving equations by multiplying by the LCM**

Solve:

a. $\frac{x}{10} + \frac{x}{8} = 9$ **b.** $\frac{x}{3} - \frac{x}{8} = 10$

SOLUTION

a. The LCM of 10 and 8 is 40 (since the first four multiples of 10 are 10, 20, 30, and 40, and 8 divides 40). You can also find the LCM by writing

$$\begin{array}{r|ll} 2 & 10 & 8 \\ \hline & 5 & 4 \end{array} \rightarrow 2 \cdot 5 \cdot 4 = 40$$

Multiplying each term by 40, we have

$$40 \cdot \frac{x}{10} + 40 \cdot \frac{x}{8} = 40 \cdot 9$$

$$\begin{aligned} 4x + 5x &= 40 \cdot 9 && \text{Simplify.} \\ 9x &= 40 \cdot 9 && \text{Combine like terms.} \\ x &= 40 && \text{Divide by 9.} \end{aligned}$$

The solution is 40.

CHECK

$$\begin{array}{c|c} \frac{x}{10} + \frac{x}{8} & \stackrel{?}{=} 9 \\ \hline \frac{40}{10} + \frac{40}{8} & 9 \\ 4 + 5 & \\ 9 & \end{array}$$

b. The LCM of 3 and 8 is $3 \cdot 8 = 24$, since the largest number that divides 3 and 8 is 1.

$$\begin{array}{r|ll} 1 & 3 & 8 \\ \hline & 3 & 8 \end{array} \rightarrow 1 \cdot 3 \cdot 8 = 24$$

Multiplying each term by 24 yields

$$24 \cdot \frac{x}{3} - 24 \cdot \frac{x}{8} = 24 \cdot 10$$

$$\begin{aligned} 8x - 3x &= 24 \cdot 10 && \text{Simplify.} \\ 5x &= 24 \cdot 10 && \text{Combine like terms.} \\ \frac{\cancel{5}x}{\cancel{5}} &= \frac{24 \cdot \overset{2}{\cancel{10}}}{\underset{1}{\cancel{5}}} && \text{Divide by 5.} \\ x &= 48 \end{aligned}$$

The solution is 48.

PROBLEM 4

Solve:

a. $\frac{x}{10} + \frac{x}{6} = 8$ **b.** $\frac{x}{4} - \frac{x}{5} = 1$

Answers

4. a. 30 **b.** 20

CHECK

$$\frac{x}{3} - \frac{x}{8} \stackrel{?}{=} 10$$

$\frac{48}{3} - \frac{48}{8}$	10
$16 - 6$	
10	

In some cases, the numerators of the fractions involved contain more than one term. However, the procedure for solving the equation is still the same, as we illustrate in Example 5.

EXAMPLE 5 **More practice clearing fractions**

Solve:

a. $\frac{x+1}{3} + \frac{x-1}{10} = 5$

b. $\frac{x+1}{3} - \frac{x-1}{8} = 4$

SOLUTION

a. The LCM of 3 and 10 is $3 \cdot 10 = 30$, since 3 and 10 don't have any common factors. Multiplying each term by 30, we have

$$\overset{10}{\cancel{30}}\left(\frac{x+1}{\cancel{3}}\right) + \overset{3}{\cancel{30}}\left(\frac{x-1}{\cancel{10}}\right) = 30 \cdot 5 \quad \text{Note the parentheses!}$$

$$10(x+1) + 3(x-1) = 150$$

$$10x + 10 + 3x - 3 = 150 \quad \text{Use the distributive property.}$$

$$13x + 7 = 150 \quad \text{Combine like terms.}$$

$$13x = 143 \quad \text{Subtract 7.}$$

$$x = 11 \quad \text{Divide by 13.}$$

We leave the check to you.

b. Here the LCM is $3 \cdot 8 = 24$. Multiplying each term by 24, we obtain

$$\overset{8}{\cancel{24}}\left(\frac{x+1}{\cancel{3}}\right) - \overset{3}{\cancel{24}}\left(\frac{x-1}{\cancel{8}}\right) = 24 \cdot 4 \quad \text{Note the parentheses.}$$

$$8(x+1) - 3(x-1) = 96$$

$$8x + 8 - 3x + 3 = 96 \quad \text{Use the distributive property.}$$

$$5x + 11 = 96 \quad \text{Combine like terms.}$$

$$5x = 85 \quad \text{Subtract 11.}$$

$$x = 17 \quad \text{Divide by 5.}$$

Be sure you check this answer in the original equation.

PROBLEM 5

Solve:

a. $\frac{x+2}{3} + \frac{x-2}{4} = 6$

b. $\frac{x+2}{5} - \frac{x-2}{3} = 0$

Teaching Tip

Remind students to place parentheses around fractions when multiplying by the LCM.

Answers

5. a. 10 **b.** 8

D Solving Applications: Percent Problems

Percent problems are among the most common types of problems, not only in mathematics, but also in many other fields. Basically, there are three types of percent problems.

Type 1 asks you to find a number that is a given percent of a specific number.

Example: 20% (read "20 percent") of 80 is what number?

Type 2 asks you what percent of a number is another given number.

Example: What percent of 20 is 5?

Type 3 asks you to find a number when it's known that a given number equals a percent of the unknown number.

Example: 10 is 40% of what number?

To do these problems, you need only recall how to translate words into equations and how to write percents as fractions (Sections R.3 and 1.7).

Now do you remember what 20% means? The symbol % is read as "percent," which means "per hundred." Recall that

$$20\% = \frac{20}{100} = \frac{\overset{1}{\cancel{20}}}{\underset{5}{\cancel{100}}} = \frac{1}{5}$$

Similarly,

$$60\% = \frac{60}{100} = \frac{\overset{3}{\cancel{60}}}{\underset{5}{\cancel{100}}} = \frac{3}{5}$$

$$17\% = \frac{17}{100}$$

We are now ready to solve some percent problems.

EXAMPLE 6 Finding a percent of a number (type 1)

Twenty percent of 80 is what number?

SOLUTION

Let's translate this.

20%	of	80	is	what number?
↓	↓	↓	↓	↓
$\frac{20}{100}$	$\cdot$	80	$=$	n

$$\frac{1}{5} \cdot 80 = n \qquad \text{Since } \tfrac{20}{100} = \tfrac{1}{5}$$

$$\frac{80}{5} = n \qquad \text{Multiply } \tfrac{1}{5} \text{ by } \tfrac{80}{1}.$$

$$n = 16 \qquad \text{Reduce } \tfrac{80}{5}.$$

Thus 20% of 80 is 16.

PROBLEM 6

Forty percent of 30 is what number?

Answer

6. 12

EXAMPLE 7 **Finding a percent (type 2)**

What percent of 20 is 5?

SOLUTION

Let's translate this.

What percent of 20 is 5?

$$x \cdot 20 = 5$$

$$\frac{x \cdot 20}{20} = \frac{5}{20} \quad \text{Divide by 20.}$$

$$x = \frac{1}{4} \quad \text{Reduce } \tfrac{5}{20}.$$

But x represents a percent, so we must change $\frac{1}{4}$ to a percent.

$$x = \frac{1}{4} = \frac{25}{100} \quad \text{or} \quad 25\%$$

Thus 5 is 25% of 20.

PROBLEM 7

What percent of 30 is 6?

Teaching Tip

Some students may already know how to do these by the proportion:

$$\frac{\text{is}}{\text{of}} = \frac{\text{percent}}{100}$$

Example 6 $\frac{n}{80} = \frac{20}{100}$

Example 7 $\frac{5}{20} = \frac{p}{100}$

Example 8 $\frac{10}{n} = \frac{40}{100}$

To solve the proportion they can set cross products equal or multiply both sides by the LCM.

Teaching Tip

Point out that $\frac{1}{4}$ is *not* the correct answer. Since the problem asks "what percent," the answer must be a percent.

EXAMPLE 8 **Finding a number (type 3)**

Ten is 40% of what number?

SOLUTION

First we translate:

10 is 40% of what number?

$$10 = \frac{40}{100} \cdot n$$

$$\frac{40}{100} \cdot n = 10 \quad \text{Rearrange.}$$

$$\frac{2}{5} \cdot n = 10 \quad \text{Reduce } \tfrac{40}{100}.$$

$$\frac{5}{2} \cdot \frac{2}{5} \cdot n = \frac{5}{2} \cdot 10 \quad \text{Multiply by } \tfrac{5}{2}.$$

$$n = 25$$

Thus ten is 40% of 25.

PROBLEM 8

Twenty is 40% of what number?

Web It

If you want to practice with all three types of percent problems, go to link 2-2-3 at the Bello Website at mhhe.com/bello.

Answers

7. 20% **8.** 50

EXAMPLE 9 **Application: Angioplasty versus TPA**

The study cited in the illustration claims that you can save more lives with angioplasty (a procedure in which a balloon-tipped instrument is inserted in your arteries, the balloon is inflated, and the artery is unclogged!) than with a blood-clot-breaking drug called TPA. Of the 451 patients studied, 226 were randomly assigned for TPA and 225 for angioplasty. After 6 months the results were as follows:

a. 6.2% of the angioplasty patients died. To the nearest whole number, how many patients is that?

b. 7.1% of the drug therapy patients died. To the nearest whole number, how many patients is that?

c. 2.2% of the angioplasty patients had strokes. To the nearest whole number, how many patients is that?

d. 4% of the drug therapy group patients had strokes. To the nearest whole number, how many patients is that?

SOLUTION

a. We need to take 6.2% of 225 (the number of angioplasty patients).

$$6.2\% \text{ of } 225 \quad \text{Means} \quad 0.062 \cdot 225 = 13.95$$

or 14 when rounded to the nearest whole number.

b. Here, we need 7.1% of 226 = $0.071 \cdot 226 = 16.046$, or 16 when rounded to the nearest whole number.

c. 2.2% of 225 = $0.022 \cdot 225 = 4.95$, or 5 when rounded to the nearest whole number.

d. 4% of 226 = $0.04 \cdot 226 = 9.04$, or 9 when rounded to the nearest whole number.

Now, look at the answers 14, 16, 5, and 9 for parts **a–d,** respectively. Are those answers faithfully depicted in the graphs? What do you think happened? (You will revisit this study in Exercises 71–80.)

PROBLEM 9

The study also said that after six months 5.3% of the angioplasty patients had a heart attack.

a. To the nearest whole number, how many patients is that?

b. In addition, 10.6% of the drug therapy group had a heart attack. To the nearest whole number, how many patients is that?

Are the answers for **a** and **b** faithfully depicted in the graph?

Saving Lives

Community hospitals without on-site cardiac units can save more lives with angioplasty than with drug treatment, a new study of 451 heart attack victims shows.

Clot-breaking drug
Angioplasty

Number of patients with the following outcome, six weeks after treatment:

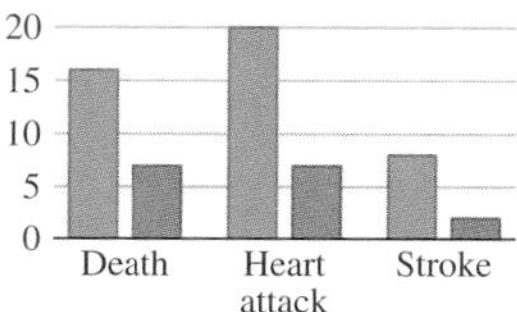

Number of patients with the following outcome, six months after treatment:

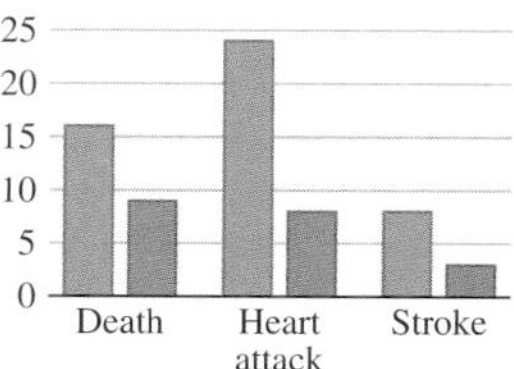

Note: The study was conducted at 11 community hospitals without on-site cardiac surgery units.

Answers

9. a. 12 **b.** 24

Calculate It Verifying Solutions

Can we verify the solutions in this section with a calculator? Of course, and the procedure is still the same. To check Example 5(a), graph

$$Y_1 = \frac{x+1}{3} + \frac{x-1}{10} \quad \text{and} \quad Y_2 = 5$$

(If you've forgotten how to do this, go back to the *Calculate It* in Section 2.1.) Make sure the numerators $x + 1$ and $x - 1$ are in parentheses, as shown in Window 1. Also, make sure your x-values go far enough to the right so you can see the intersection of the two lines (see Window 2).

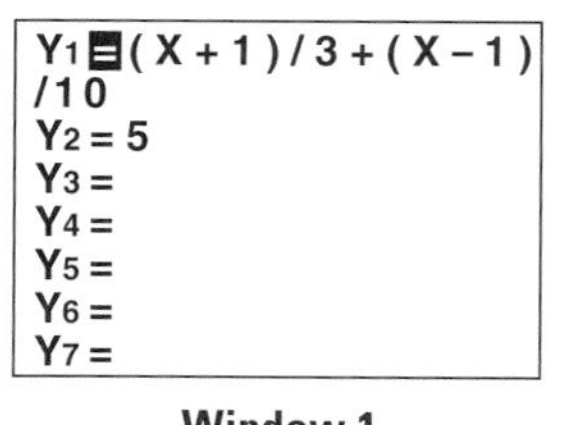

Window 1

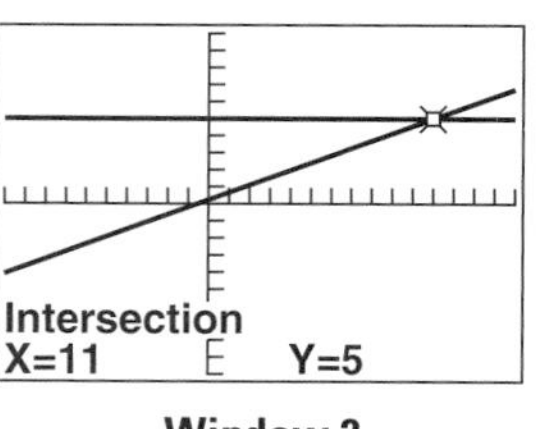

Window 2

Now use your calculator to do Example 5(b) and also to check the solutions you obtain in the exercises set.

Exercises 2.2

A In Problems 1–26, solve the equations.

1. $\frac{x}{7} = 5$ $x = 35$

2. $\frac{y}{2} = 9$ $y = 18$

3. $-4 = \frac{x}{2}$ $x = -8$

4. $\frac{a}{5} = -6$ $a = -30$

5. $\frac{b}{-3} = 5$ $b = -15$

6. $7 = \frac{c}{-4}$ $c = -28$

7. $-3 = \frac{f}{-2}$ $f = 6$

8. $\frac{g}{-4} = -6$ $g = 24$

9. $\frac{v}{4} = \frac{1}{3}$ $v = \frac{4}{3}$

10. $\frac{w}{3} = \frac{2}{7}$ $w = \frac{6}{7}$

11. $\frac{x}{5} = \frac{-3}{4}$ $x = -\frac{15}{4}$

12. $\frac{-8}{9} = \frac{y}{2}$ $y = -\frac{16}{9}$

13. $3z = 33$ $z = 11$

14. $4y = 32$ $y = 8$

15. $-42 = 6x$ $x = -7$

16. $7b = -49$ $b = -7$

17. $-8c = 56$ $c = -7$

18. $-5d = 45$ $d = -9$

19. $-5x = -35$ $x = 7$

20. $-12 = -3x$ $x = 4$

21. $-3y = 11$ $y = -\frac{11}{3}$

22. $-5z = 17$ $z = -\frac{17}{5}$

23. $-2a = 1.2$ $a = -0.6$

24. $-3b = 1.5$ $b = -0.5$

25. $3t = 4\frac{1}{2}$ $t = \frac{3}{2}$

26. $4r = 6\frac{2}{3}$ $r = \frac{5}{3}$

B In Problems 27–40, solve the equations.

27. $\frac{1}{3}x = -0.75$ $x = -2.25$

28. $\frac{1}{4}y = 0.25$ $y = 1.00$

29. $-6 = \frac{3}{4}C$ $C = -8$

30. $-2 = \frac{2}{9}F$ $F = -9$

31. $\frac{5}{6}a = 10$ $a = 12$

32. $24 = \frac{2}{7}z$ $z = 84$

33. $-\frac{4}{5}y = 0.4$ $y = -0.5$

34. $0.5x = \frac{-1}{4}$ $x = -\frac{1}{2}$

35. $\frac{-2}{11}p = 0$ $p = 0$

36. $\frac{-4}{9}q = 0$ $q = 0$

37. $-18 = \frac{3}{5}t$ $t = -30$

38. $-8 = \frac{2}{7}R$ $R = -28$

39. $\frac{7x}{0.02} = -7$ $x = -0.02$

40. $-6 = \frac{6y}{0.03}$ $y = -0.03$

C In Problems 41–60, solve the equations.

41. $\frac{y}{2} + \frac{y}{3} = 10$ $y = 12$

42. $\frac{a}{4} + \frac{a}{3} = 14$ $a = 24$

43. $\frac{x}{7} + \frac{x}{3} = 10$ $x = 21$

44. $\frac{z}{6} + \frac{z}{4} = 20$ $z = 48$

45. $\frac{x}{5} + \frac{x}{10} = 6$ $x = 20$

46. $\frac{r}{2} + \frac{r}{6} = 8$ $r = 12$

47. $\frac{t}{6} + \frac{t}{8} = 7$ $t = 24$

48. $\frac{f}{9} + \frac{f}{12} = 14$ $f = 72$

49. $\frac{x}{2} + \frac{x}{5} = \frac{7}{10}$ $x = 1$

50. $\frac{a}{3} + \frac{a}{7} = \frac{20}{21}$ $a = 2$

51. $\frac{c}{3} - \frac{c}{5} = 2$ $c = 15$

52. $\frac{F}{4} - \frac{F}{7} = 3$ $F = 28$

53. $\frac{W}{6} - \frac{W}{8} = \frac{5}{12}$ $W = 10$

54. $\frac{m}{6} - \frac{m}{10} = \frac{4}{3}$ $m = 20$

55. $\frac{x}{5} - \frac{3}{10} = \frac{1}{2}$ $x = 4$

56. $\frac{3y}{7} - \frac{1}{14} = \frac{1}{14}$ $y = \frac{1}{3}$

57. $\frac{x + 4}{4} - \frac{x + 2}{3} = -\frac{1}{2}$ $x = 10$

58. $\frac{w - 1}{2} + \frac{w}{8} = \frac{7w + 1}{16}$ $w = 3$

59. $\frac{x}{6} + \frac{3}{4} = x - \frac{7}{4}$ $x = 3$

60. $\frac{x}{6} + \frac{4}{3} = x - \frac{1}{3}$ $x = 2$

D Translate into an equation and solve.

61. 30% of 40 is what number? 12

62. 17% of 80 is what number? 13.6

63. 40% of 70 is what number? 28

64. What percent of 40 is 8? 20%

65. What percent of 30 is 15? 50%

66. What percent of 40 is 4? 10%

67. 30 is 20% of what number? 150

68. 20 is 40% of what number? 50

69. 12 is 60% of what number? 20

70. 24 is 50% of what number? 48

Refer to the study described in Example 9. It was determined that not all 451 patients actually received treatment. As a matter of fact, only 211 patients received the clot-breaking drug (TPA) and 171 received angioplasty. Exercises 71–76 refer to the chart shown here. Give your answers to one decimal place.

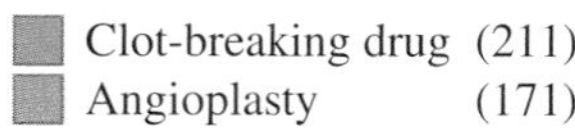

Number of patients with the following outcome, six weeks after treatment:

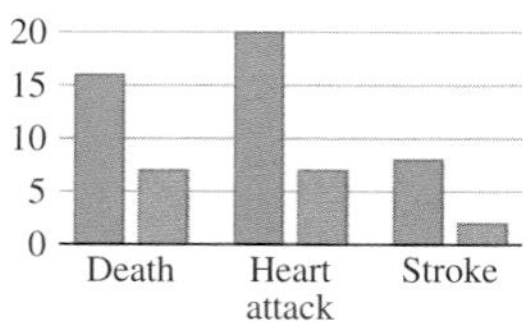

Source: Data from *Journal of the American Medical Association.*

71. As you can see, 16 of the patients receiving the clot-breaking drug died. What percent of the 211 patients is that? 7.6%

72. What percent of the 171 patients receiving angioplasty died? 4.1%

73. What percent of the 211 patients receiving the clot-breaking drug had heart attacks? 9.5%

74. What percent of the 171 patients receiving angioplasty had heart attacks? 4.1%

75. What percent of the 211 patients receiving the clot-breaking drug had a stroke? 3.8%

76. What percent of the 171 patients receiving angioplasty had a stroke? 1.2%

Exercises 77–80 refer to the chart shown here. Give your answers to one decimal place.

Clot-breaking drug (211)
Angioplasty (171)

Number of patients with the following outcome, six months after treatment:

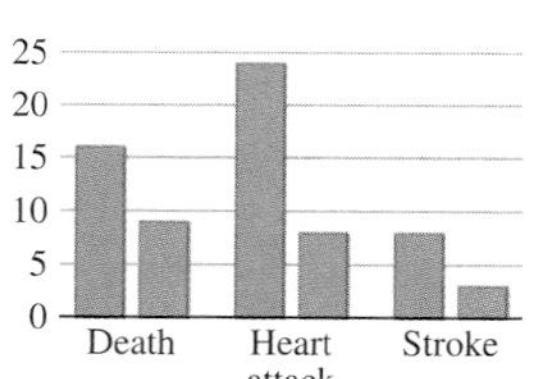

Source: Data from *Journal of the American Medical Association.*

77. As you can see, 9 of the patients receiving angioplasty died. What percent of the 171 patients is that? 5.3%

78. What percent of the 171 patients receiving angioplasty had a heart attack? 4.7%

79. What percent of the 211 patients receiving the clot-breaking drug had a stroke? 3.8%

80. What percent of the 171 patients receiving angioplasty had a stroke? 1.8%

81. An item is selling for half its original price. It now sells for $12. What was the original price? $24

82. The price of an item has tripled and it now sells for $36. What was the original price? $12

83. An item is on sale at one-third off. It now costs $8. What was the original price? $12

84. An item can be bought for one third of its original price. If it's now selling for $17, what was the original price? $51

85. A brewer advertises that its popular light beer has $\frac{1}{3}$ fewer calories than its regular beer. If this light beer has 100 calories, how many calories does the regular beer have? 150 calories

---

SKILL CHECKER

Try the Skill Checker Exercises so you'll be ready for the next section.

Use the distributive property to multiply:

86. $3(6 - x)$ $18 - 3x$

87. $5(8 - y)$ $40 - 5y$

88. $6(8 - 2y)$ $48 - 12y$

89. $9(6 - 3y)$ $54 - 27y$

90. $-3(4x - 2)$ $-12x + 6$

91. $-5(3x - 4)$ $-15x + 20$

Find:

92. $20 \cdot \frac{3}{4}$ 15

93. $24 \cdot \frac{1}{6}$ 4

94. $-5 \cdot \left(-\frac{4}{5}\right)$ 4

95. $-7 \cdot \left(-\frac{3}{7}\right)$ 3

USING YOUR KNOWLEDGE

The Cost of a Strike

Have you ever been on strike? If so, after negotiating a salary increase, did you come out ahead? How can you find out? Let's take an example. If an employee earns \$10 an hour and works 250 8-hour days a year, her salary would be $10 \cdot 8 \cdot 250 =$ \$20,000. What percent increase in salary is needed to make up for the lost time during the strike? First, we must know how many days' wages were lost (the average strike in the United States lasts about 100 days). Let's assume that 100 days were lost. The wages for 100 days would be $10 \cdot 8 \cdot 100 =$ \$8000. The employee needs \$20,000 plus \$8000 or \$28,000 to make up the lost wages. Let x be the multiple of the old wages needed to make up the loss within 1 year. We then have

$$20{,}000 \cdot x = 28{,}000$$

$$\frac{20{,}000 \cdot x}{20{,}000} = \frac{28{,}000}{20{,}000} \qquad \text{Divide by 20,000.}$$

$$x = 1.40 \quad \text{or} \quad 140\%$$

Thus the employee needs a 40% increase to recuperate her lost wages within 1 year!

96. Assume that an employee works 8 hours a day for 250 days. If the employee makes \$20 an hour, what is her yearly salary? \$40,000

97. If the employee goes on strike for 100 days, find the amount of her lost wages. \$16,000

98. What percent increase will be necessary to make up for the lost wages within 1 year? 40%

WRITE ON

99. What is the difference between

a. an expression and an equation? Answers may vary.

b. simplifying an expression and solving an equation? Answers may vary.

100. When solving the equation $-3x = 18$, would it be easier to divide by -3 or to multiply by the reciprocal of -3? Explain your answer and solve the equation. Divide by -3; answers may vary.

101. When solving the equation $-\frac{3}{4}x = 15$, would it be easier to divide by $-\frac{3}{4}$ or to multiply by the reciprocal of $-\frac{3}{4}$? Explain your answer and solve the equation. Multiply by the reciprocal of $-\frac{3}{4}$; answers may vary.

MASTERY TEST

If you know how to do these problems, you have learned your lesson!

102. 15 is 30% of what number? 50

103. What percent of 45 is 9? 20%

104. 40% of 60 is what number? 24

Solve:

105. $\frac{x+2}{4} + \frac{x-1}{5} = 3$ $x = 6$

106. $\frac{x+3}{2} - \frac{x-2}{3} = 5$ $x = 17$

107. $\frac{y}{6} + \frac{y}{10} = 8$ $y = 30$

108. $\frac{y}{5} - \frac{y}{8} = 3$ $y = 40$

109. $\frac{4}{5}y = 8$ $y = 10$

110. $-\frac{3}{4}y = 6$ $y = -8$

111. $-\frac{2}{7}y = -4$ $y = 14$

112. $-7y = 16$ $y = -\frac{16}{7}$

113. $\frac{x}{2} = -7$ $x = -14$

114. $-\frac{y}{4} = -3$ $y = 12$

2.3 LINEAR EQUATIONS

To Succeed, Review How To . . .

1. Find the LCM of three numbers (p. 17).
2. Add, subtract, multiply, and divide real numbers (pp. 61–66, 70–76).
3. Use the distributive property to remove parentheses (pp. 92–94).

Objectives

A Solve linear equations in one variable.

B Solve a literal equation for one of the unknowns.

GETTING STARTED

Getting the Best Deal

Suppose you want to rent a midsize car. Which is the better deal, paying \$39.99 per day *plus* \$0.20 per mile or paying \$49.99 per day with unlimited mileage? Well, it depends on how many miles you travel! First, let's see how many miles you have to travel before the costs are the same. The total daily cost, C, based on traveling m miles consists of two parts: \$39.99 plus the additional charge at \$0.20 per mile. Thus

$$\underbrace{C}_{\text{The total daily cost } C} \quad \underbrace{=}_{\text{consists of}} \quad \underbrace{\$39.99}_{\text{fixed cost}} + \underbrace{0.20m}_{\text{mileage.}}$$

If the costs are identical, C must be \$49.99, that is,

$$\underbrace{39.99 + 0.20m}_{\text{The total daily cost } C} \quad \underbrace{=}_{\text{is the same as}} \quad \underbrace{\$49.99}_{\text{the flat rate.}}$$

Solving for m will tell us how many miles we must go for the mileage rate and the flat rate to be the same.

$$39.99 + 0.20m = 49.99 \qquad \text{Given.}$$

$$39.99 - 39.99 + 0.20m = 49.99 - 39.99 \qquad \text{Subtract 39.99.}$$

$$0.20m = 10$$

$$\frac{0.20m}{0.20} = \frac{10}{0.20} \qquad \text{Divide by 0.20.}$$

$$m = 50$$

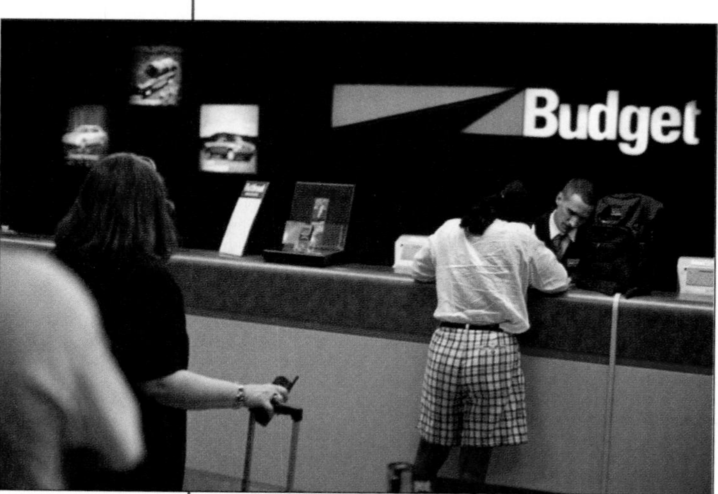

Thus if you plan to travel fewer than 50 miles, the mileage rate is better. If you travel 50 miles, the cost is the same for both, and if you travel more than 50 miles the flat rate is better. In this section we shall learn how to solve equations such as $39.99 + 0.20m = 49.99$, a type of equation called a *linear equation in one variable*.

Solving Linear Equations in One Variable

Let's concentrate on the idea we used to solve $39.99 + 0.20m = 49.99$. Our main objective is for our solution to be in the form of $m = \square$, where $\square$ is a number. Because of this, we first subtracted 39.99 and then divided by 0.20. This technique works for *linear equations.* Here is the definition.

LINEAR EQUATION An equation that can be written in the form

$$ax + b = c$$

where a is not 0 ($a \neq 0$) and a, b, and c are real numbers, is a **linear equation.**

How do we solve linear equations? The same way we solved for m in the *Getting Started.* We use the properties of equality that we studied in Sections 2.1 and 2.2. Remember the steps?

1. The equation is already simplified. $ax + b = c$
2. Subtract b. $ax + b - b = c - b$

   $ax = c - b$
3. Divide by a. $\dfrac{ax}{a} = \dfrac{c - b}{a}$

   $x = \dfrac{c - b}{a}$

Linear equations may have

a. *One* solution (as in Examples 1 and 2)
b. *No* solution ($x + 1 = x + 2$ has no solution), or
c. *Infinitely many* solutions [$2x + 2 = 2(x + 1)$ has infinitely many solutions]

EXAMPLE 1 Solving linear equations

Solve:

a. $3x + 7 = 13$

b. $-5x - 3 = 1$

SOLUTION

a. 1. The equation is already simplified. $3x + 7 = 13$
   2. Subtract 7. $3x + 7 - 7 = 13 - 7$

      $3x = 6$
   3. Divide by 3 (or multiply by the reciprocal of 3). $\dfrac{3x}{3} = \dfrac{6}{3}$

      $x = 2$

   The solution is 2.

   CHECK

$3x + 7 \stackrel{?}{=} 13$	
$3(2) + 7$	13
$6 + 7$	
13	

PROBLEM 1

Solve:

a. $4x + 5 = 17$

b. $-3x - 5 = 2$

Answers

1. a. 3 **b.** $-\frac{7}{3}$

b. **1.** The equation is already simplified. $-5x - 3 = 1$

2. Add 3. $-5x - 3 + 3 = 1 + 3$

$-5x = 4$

3. Divide by -5 (or multiply by the reciprocal of -5). $\frac{-5x}{-5} = \frac{4}{-5}$

$x = -\frac{4}{5}$

The solution is $-\frac{4}{5}$.

CHECK

$-5x - 3 \stackrel{?}{=} 1$	
$-5\left(-\frac{4}{5}\right) - 3$	1
$4 - 3$	
1	

Note that the main idea is to place the variables on one side of the equation so you can write the solution in the form $x = \square$ (or $\square = x$), where $\square$ is a number (a constant). Can we solve the equation $5(x + 2) = 3(x + 1) + 9$? This equation is *not* of the form $ax + b = c$, but we can write it in this form if we use the distributive property to remove parentheses, as shown in Example 2.

EXAMPLE 2 **Using the distributive property to solve a linear equation**

Solve: $5(x + 2) = 3(x + 1) + 9$

SOLUTION

1. There are no fractions to clear. $5(x + 2) = 3(x + 1) + 9$

2. Use the distributive property. $5x + 10 = 3x + 3 + 9$

$5x + 10 = 3x + 12$ Add 3 and 9.

3. Subtract 10 on both sides. $5x + 10 - 10 = 3x + 12 - 10$

$5x = 3x + 2$

4. Subtract $3x$ on both sides. $5x - 3x = 3x - 3x + 2$

$2x = 2$

5. Divide both sides by 2 (or multiply by the reciprocal of 2). $\frac{2x}{2} = \frac{2}{2}$

$x = 1$

The solution is 1.

6. CHECK

$5(x + 2) \stackrel{?}{=}$	$3(x + 1) + 9$
$5(1 + 2)$	$3(1 + 1) + 9$
$5(3)$	$3(2) + 9$
15	$6 + 9$
	15

PROBLEM 2

Solve: $7(x + 1) = 4(x + 2) + 5$

Answer

2. 2

What if we have fractions in the equation? No problem. We can clear them by multiplying by the LCM, as we did in Section 2.2. For example, to solve

$$\frac{3}{4} + \frac{x}{10} = 1$$

we first multiply each term by 20, the LCM of 4 and 10. We can obtain the LCM by noting that 20 is the first multiple of 10 that is divisible by 4, or by writing

$$\begin{array}{l|ll} 2 & 4 & 10 \\ \hline & 2 & 5 \end{array} \rightarrow 2 \cdot 2 \cdot 5 = 20 \quad \text{is the LCM}$$

Now, solve $\frac{3}{4} + \frac{x}{10} = 1$ as follows:

1. Clear the fractions by multiplying by 20, the LCM of 4 and 10. $\quad \overset{5}{\cancel{20}} \cdot \frac{3}{\cancel{4}} + \overset{2}{\cancel{20}} \cdot \frac{x}{\cancel{10}} = 20 \cdot 1$
2. Simplify. $\quad 15 + 2x = 20$
3. Subtract 15. $\quad 15 - 15 + 2x = 20 - 15$
4. The right side has numbers only. $\quad 2x = 5$
5. Divide by 2 (or multiply by the reciprocal of 2). $\quad \frac{2x}{2} = \frac{5}{2}$, $\quad x = \frac{5}{2}$

The solution is $\frac{5}{2}$.

6. **CHECK**

$\frac{3}{4} + \frac{x}{10} \stackrel{?}{=} 1$	
$\frac{3}{4} + \frac{\frac{5}{2}}{10}$	1
$\frac{3}{4} + \frac{5}{2} \cdot \frac{1}{10}$	
$\frac{3}{4} + \frac{5}{20}$	
$\frac{3}{4} + \frac{1}{4}$	
1	

There is a shortcut for finding the LCM of two or more numbers:

1. Check if one is a multiple of another.
2. Select the largest denominator (10).
3. Double it (20), triple it (30), and so on until the other number (4) exactly divides into the doubled or tripled quantity. Since 4 exactly divides into 20, the LCM is 20.

Note: If you were looking for the LCM of 15 and 25, select the 25. Double it (50) but 15 does not exactly divide into 50. Triple the 25 (75). Now 15 exactly divides into 75, so the LCM of 15 and 25 is 75.

The procedure we have used to solve the preceding examples can be used to solve any linear equation. As before, what we need to do is isolate the variables on one side of the equation and the numbers on the other so that we can write the solution in the form $x = \square$ or $\square = x$. Here's how we accomplish this, with a step-by-step example shown on the right. (If you have forgotten how to find the LCM for three numbers, see Section R.2.)

Teaching Tip

Show students that this procedure is the same as multiplying both sides of the equation by 20, like this:

$$20\left(\frac{3}{4} + \frac{x}{10}\right) = 20 \cdot 1$$

This new procedure saves one step (using the distributive property).

Web It

If you want a tutorial with example problems that will tell you how to solve linear equations, try link 2-3-1 on the Bello Website, mhhe.com/bello.

You can also try link 2-3-2.

If you want the computer to solve linear equations for you, go to link 2-3-3.

PROCEDURE

Solving Linear Equations

$$\frac{x}{4} - \frac{1}{6} = \frac{7}{12}(x - 2) \quad \text{Given.}$$

1. Clear any fractions by multiplying each term on both sides of the equation by the LCM of the denominators, 12.

   $12 \cdot \frac{x}{4} - 12 \cdot \frac{1}{6} = 12 \cdot \underbrace{\left[\frac{7}{12}(x - 2)\right]}_{\text{This is one term.}}$

2. Remove parentheses and collect like terms (simplify) if necessary.

   $3x - 2 = 7(x - 2)$

   $3x - 2 = 7x - 14$

3. Add or subtract the same number on both sides of the equation so that the numbers are isolated on one side.

   $3x - 2 + 2 = 7x - 14 + 2$

   $3x = 7x - 12$

4. Add or subtract the same term or expression on both sides of the equation so that the variables are isolated on the other side.

   $3x - 7x = 7x - 7x - 12$

   $-4x = -12$

5. If the coefficient of the variable is not 1, divide both sides of the equation by the coefficient (or, equivalently, multiply by the reciprocal of the coefficient of the variable).

   $\frac{-4x}{-4} = \frac{-12}{-4}$

   $x = 3$ The solution

6. Be sure to check your answer in the original equation.

   CHECK $\frac{x}{4} - \frac{1}{6} \stackrel{?}{=} \frac{7}{12}(x - 2)$

$\frac{3}{4} - \frac{1}{6}$	$\frac{7}{12}(3 - 2)$
$\frac{9}{12} - \frac{2}{12}$	$\frac{7}{12} \cdot 1$
$\frac{7}{12}$	$\frac{7}{12}$

EXAMPLE 3 More practice solving linear equations

Solve:

a. $\frac{7}{24} = \frac{x}{8} + \frac{1}{6}$

b. $\frac{1}{5} - \frac{x}{4} = \frac{7(x + 3)}{10}$

SOLUTION We use the six-step procedure.

a. $\frac{7}{24} = \frac{x}{8} + \frac{1}{6}$ Given.

1. Clear the fractions; the LCM is 24. $\;\; \cancel{24} \cdot \frac{7}{\cancel{24}} = \overset{3}{\cancel{24}} \cdot \frac{x}{\cancel{8}} + \overset{4}{\cancel{24}} \cdot \frac{1}{\cancel{6}}$
2. Simplify. $\;\; 7 = 3x + 4$

PROBLEM 3

Solve:

a. $\frac{20}{21} = \frac{x}{7} + \frac{x}{3}$

b. $\frac{1}{4} - \frac{x}{5} = \frac{17(x + 4)}{20}$

Answers

3. a. 2 **b.** -3

3. Subtract 4. $7 - 4 = 3x + 4 - 4$

4. The left side has numbers only. $3 = 3x$

5. Divide by 3 (or multiply by the reciprocal of 3). $\dfrac{3}{3} = \dfrac{3x}{3}$

$1 = x$

$x = 1$ The solution

6. CHECK $\dfrac{7}{24} \overset{?}{=} \dfrac{x}{8} + \dfrac{1}{6}$

$\dfrac{7}{24}$	$\dfrac{1}{8} + \dfrac{1}{6}$
	$\dfrac{3}{24} + \dfrac{4}{24}$
	$\dfrac{7}{24}$

b. $\dfrac{1}{5} - \dfrac{x}{4} = \dfrac{7(x+3)}{10}$ Given.

1. Clear the fractions; the LCM is 20. Remember, you can get the LCM by selecting 10 (the largest of 5, 4, and 10) and doubling it, which gives 20. Since 5 and 4 exactly divide into 20, 20 is the LCM of 5, 4, and 10. $\overset{4}{\cancel{20}} \cdot \dfrac{1}{\cancel{5}} - \overset{5}{\cancel{20}} \cdot \dfrac{x}{\cancel{4}} = \overset{2}{\cancel{20}} \cdot \dfrac{7(x+3)}{\cancel{10}}$

2. Simplify and use the distributive law. $4 - 5x = 14(x + 3)$

$4 - 5x = 14x + 42$

3. Subtract 4. $4 - 4 - 5x = 14x + 42 - 4$

$-5x = 14x + 38$

4. Subtract $14x$. $-5x - 14x = 14x - 14x + 38$

$-19x = 38$

5. Divide by -19 (or multiply by the reciprocal of -19). $\dfrac{-19x}{-19} = \dfrac{38}{-19}$

$x = -2$ The solution

6. CHECK $\dfrac{1}{5} - \dfrac{x}{4} \overset{?}{=} \dfrac{7(x+3)}{10}$

$\dfrac{1}{5} - \dfrac{-2}{4}$	$\dfrac{7(-2+3)}{10}$
$\dfrac{1}{5} + \dfrac{1}{2}$	$\dfrac{7(1)}{10}$
$\dfrac{2}{10} + \dfrac{5}{10}$	$\dfrac{7}{10}$
$\dfrac{7}{10}$	

B Solving Literal Equations

The procedures for solving linear equations that we've just described can also be used to solve some *literal equations*. A **literal equation** is an equation that contains several variables. In business, science, and engineering, literal equations are usually given as formulas such as the area A of a circle of radius r ($A = \pi r^2$), the interest earned on a principal P at a given rate r for a given time t ($I = Prt$), and so on. Unfortunately, these formulas are not always in the form we need to solve the problem at hand. However, as it turns out, we can use the same methods we've just learned to solve for a particular variable in such a formula. For example, let's solve for P in the formula $I = Prt$. To keep track of the variable P, we first circle it in color:

Web It

For a lesson on solving literal equations, go to link 2-3-4 on the Bello Website at mhhe.com/bello.

For a video on solving literal equations, try link 2-3-5.

$$I = \textcircled{P}rt$$

$$\frac{I}{rt} = \frac{\textcircled{P}rt}{rt} \quad \text{Divide by } rt.$$

$$\frac{I}{rt} = \textcircled{P} \quad \text{Simplify.}$$

$$\textcircled{P} = \frac{I}{rt} \quad \text{Rewrite.}$$

Now let's look at another example.

EXAMPLE 4 Solving a literal equation

A trapezoid is a four-sided figure in which only two of the sides are parallel. The area of the trapezoid shown here is

$$A = \frac{h}{2}(b_1 + b_2)$$

where h is the altitude and b_1 and b_2 are the bases. Solve for b_2.

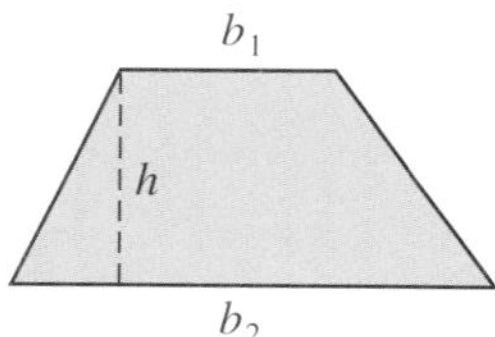

SOLUTION We will circle b_2 to remember that we are solving for it. Now we use our six-step procedure.

$$A = \frac{h}{2}(b_1 + \textcircled{b_2}) \quad \text{Given.}$$

1. Clear the fraction; the LCM is 2.

$$2 \cdot A = \cancel{2} \cdot \frac{h}{\cancel{2}}(b_1 + \textcircled{b_2})$$

$$2A = h(b_1 + \textcircled{b_2})$$

2. Remove parentheses.

$$2A = hb_1 + h\textcircled{b_2}$$

3. There are no numbers to isolate.
4. Subtract the same term, hb_1, on both sides.

$$2A - hb_1 = hb_1 - hb_1 + h\textcircled{b_2}$$

$$2A - hb_1 = h\textcircled{b_2}$$

5. Divide both sides by the coefficient of b_2, h.

$$\frac{2A - hb_1}{h} = \frac{h\textcircled{b_2}}{h}$$

$$\frac{2A - hb_1}{h} = \textcircled{b_2}$$

$$b_2 = \frac{2A - hb_1}{h} \quad \text{The solution}$$

6. No check is necessary.

PROBLEM 4

The speed S of an ant in centimeters per second is $S = \frac{1}{6}(C - 4)$, where C is the temperature in degrees Celsius. Solve for C.

Answer

4. $C = 6S + 4$

EXAMPLE 5 Using a pattern to solve a literal equation

The cost of renting a car is \$40 per day, plus 20¢ per mile m *after* the first 150 miles. (The first 150 miles are free.)

a. What is the formula for the total daily cost C based on the miles m traveled?

b. Solve for m in the equation.

SOLUTION

a. Let's try to find a pattern to the general formula.

Miles Traveled	Daily Cost	Per-Mile Cost	Total Cost
50	\$40	0	\$40
100	\$40	0	\$40
150	\$40	0	\$40
200	\$40	0.20(200 − 150) = 10	\$40 + \$10 = \$50
300	\$40	0.20(300 − 150) = 30	\$40 + \$30 = \$70

Yes, there is a pattern; can you see that the daily cost when traveling more than 150 miles is $C = 40 + 0.20(m - 150)$?

b. Again, we circle the variable we want to isolate.

$$C = 40 + 0.20(m - 150) \quad \text{Given.}$$

$$C = 40 + 0.20m - 30 \quad \text{Use the distributive property.}$$

$$C = 10 + 0.20m \quad \text{Simplify.}$$

$$C - 10 = 10 - 10 + 0.20m \quad \text{Subtract 10.}$$

$$C - 10 = 0.20m$$

$$\frac{C - 10}{0.20} = \frac{0.20m}{0.20} \quad \text{Divide by 0.20.}$$

$$\frac{C - 10}{0.20} = m$$

PROBLEM 5

Some people claim that the relationship between shoe size S and foot length L is $S = 3L - 24$, where L is the length of the foot in inches.

a. If a person wears size 12 shoes, what is the length L of their foot?

b. Solve for L.

Web It

For a website dealing with problems similar to the one in Example 5, go to link 2-3-5 on the Bello Website at mhhe.com/bello.

As you know, you can use your calculator to graph certain equations. Let's look at Example 5 again. Suppose you want to find the number m of miles you can travel for different costs (say, when C = \$50, \$100, \$150, and so on). You are given $C = 40 + 0.20(m - 150)$ and you solve for m, as before, to obtain

$$\frac{C - 10}{0.20} = m$$

Using x instead of C for the cost and y instead of m for the miles (it's easier to work with x's and y's on the calculator), you graph

$$y = \frac{x - 10}{0.20}$$

The graph is shown in Figure 1. Note that the cost x starting at \$50 (shown on the horizontal axis) occurs in \$50 increments, while the number of miles traveled (vertical scale) is in 100-mile increments. You can also use the trace feature offered by most calculators to automatically find the number of miles y you can travel when the cost x is \$50. Do you see the answer in Figure 2? It says you can travel 200 miles when you pay \$50.

Answers

5. a. 12 in. **b.** $L = \frac{S + 24}{3}$

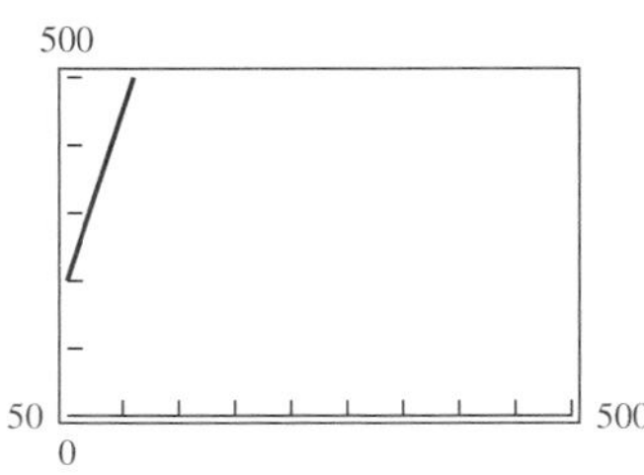

Figure 1
The x-scale (horizontal) shows the cost in \$50 increments and the y-scale (vertical) shows the miles in 100-mile increments.

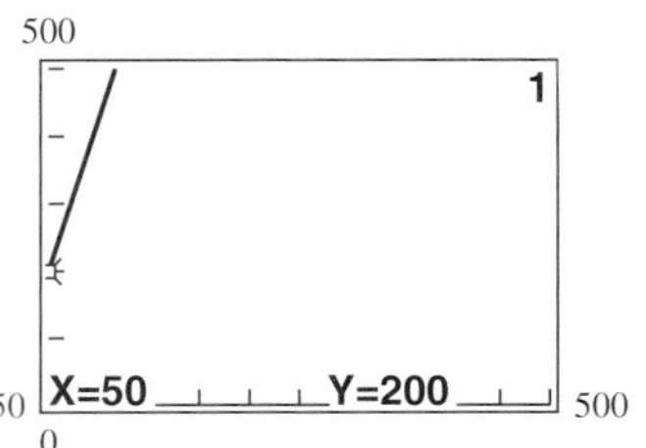

Figure 2
This shows that when the cost is \$50, you can travel 200 miles.

But what does this have to do with solving literal equations? The point is that if you want to find the graph of $2x + 3y = 6$ using a calculator, you have to **know how to solve for y.** Let's do that next.

EXAMPLE 6 **Solving for a specified variable**

Solve for y: $2x + 3y = 6$

SOLUTION Remember, we want to isolate the y.

$$2x + 3y = 6 \quad \text{Given.}$$
$$2x - 2x + 3y = 6 - 2x \quad \text{Subtract } 2x.$$
$$\frac{3y}{3} = \frac{6 - 2x}{3} \quad \text{Divide by 3.}$$
$$y = \frac{6 - 2x}{3} \quad \text{The solution}$$

The result (done with a calculator) is shown in Figure 3.

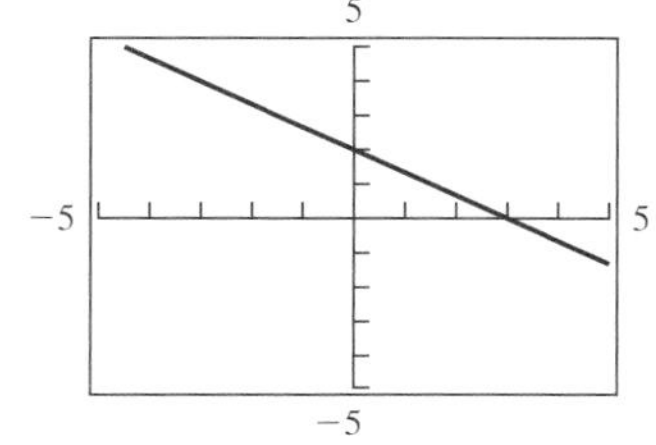

Figure 3
Note that this time both the x- and y-scales contain positive and negative integers. Why do you think the graph of the equation in Example 5 didn't have any negative numbers on the scale?

PROBLEM 6

Solve for x: $3x + 4y = 7$

Answer

6. $x = \frac{7 - 4y}{3}$

Exercises 2.3

Boost *your* GRADE at mathzone.com!

- Practice Problems
- Self-Tests
- Videos
- NetTutor
- e-Professors

A In Problems 1–50, solve the equation.

1. $3x - 12 = 0$ $x = 4$

2. $5a + 10 = 0$ $a = -2$

3. $2y + 6 = 8$ $y = 1$

4. $4b - 5 = 3$ $b = 2$

5. $-3z - 4 = -10$ $z = 2$

6. $-4r - 2 = 6$ $r = -2$

7. $-5y + 1 = -13$ $y = \frac{14}{5}$

8. $-3x + 1 = -9$ $x = \frac{10}{3}$

9. $3x + 4 = x + 10$ $x = 3$

10. $4x + 4 = x + 7$ $x = 1$

11. $5x - 12 = 6x - 8$ $x = -4$

12. $5x + 7 = 7x + 19$ $x = -6$

13. $4v - 7 = 6v + 9$ $v = -8$

14. $8t + 4 = 15t - 10$ $t = 2$

15. $6m - 3m + 12 = 0$ $m = -4$

16. $10k + 15 - 5k = 25$ $k = 2$

17. $10 - 3z = 8 - 6z$ $z = -\frac{2}{3}$

18. $8 - 4y = 10 + 6y$ $y = -\frac{1}{5}$

19. $5(x+2)=3(x+3)+1$ $x=0$

20. $y-(4-2y)=7(y-1)$ $y=\frac{3}{4}$

21. $5(4-3a)=7(3-4a)$ $a=\frac{1}{13}$

22. $\frac{3}{4}y-4.5=\frac{1}{4}y+1.3$ $y=11.6$

23. $-\frac{7}{8}c+5.6=-\frac{5}{8}c-3.3$ $c=35.6$

24. $x+\frac{2}{3}x=10$ $x=6$

25. $-2x+\frac{1}{4}=2x+\frac{4}{5}$ $x=-\frac{11}{80}$

26. $6x+\frac{1}{7}=2x-\frac{2}{7}$ $x=-\frac{3}{28}$

27. $\frac{x-1}{2}+\frac{x-2}{2}=3$ $x=\frac{9}{2}$

28. $\frac{3x+5}{3}+\frac{x+3}{3}=12$ $x=7$

29. $\frac{x}{5}-\frac{x}{4}=1$ $x=-20$

30. $\frac{x}{3}-\frac{x}{2}=1$ $x=-6$

31. $\frac{x+1}{4}-\frac{2x-2}{3}=3$ $x=-5$

32. $\frac{z+4}{3}=\frac{z+6}{4}$ $z=2$

33. $\frac{2h-1}{3}=\frac{h-4}{12}$ $h=0$

34. $\frac{5-6y}{7}-\frac{-7-4y}{3}=2$ $y=-\frac{11}{5}$

35. $\frac{2w+3}{2}-\frac{3w+1}{4}=1$ $w=-1$

36. $\frac{7r+2}{6}+\frac{1}{2}=\frac{r}{4}$ $r=-\frac{10}{11}$

37. $\frac{8x-23}{6}+\frac{1}{3}=\frac{5}{2}x$ $x=-3$

38. $\frac{x+1}{2}+\frac{x+2}{3}+\frac{x+4}{4}=-8$ $x=-\frac{122}{13}$

39. $\frac{x-5}{2}-\frac{x-4}{3}=\frac{x-3}{2}-(x-2)$ $x=\frac{5}{2}$

40. $\frac{x+1}{2}+\frac{x+2}{3}+\frac{x+3}{4}=16$ $x=13$

41. $-4x+\frac{1}{2}=4\left(\frac{1}{8}-x\right)$ All real numbers

42. $-6x+\frac{2}{3}=4\left(\frac{1}{5}-x\right)$ $x=-\frac{1}{15}$

43. $\frac{1}{2}(8x+4)-5=\frac{1}{4}(4x+8)+1$ $x=2$

44. $\frac{1}{3}(3x+9)+2=\frac{1}{9}(9x+18)+3$ All real numbers

45. $x+\frac{x}{2}-\frac{3x}{5}=9$ $x=10$

46. $\frac{5x}{3}-\frac{3x}{4}+\frac{11}{6}=0$ $x=-2$

47. $\frac{4x}{9}-\frac{3}{2}=\frac{5x}{6}-\frac{3x}{2}$ $x=\frac{27}{20}$

48. $\frac{7x}{2}-\frac{4x}{3}+\frac{2x}{5}=-\frac{11}{6}$ $x=-\frac{5}{7}$

49. $\frac{3x+4}{2}-\frac{1}{8}(19x-3)=1-\frac{7x+18}{12}$ $x=\frac{69}{7}$

50. $\frac{11x-2}{3}-\frac{1}{2}(3x-1)=\frac{17x+7}{6}-\frac{2}{9}(7x-2)$ $x=2$

B In Problems 51–60, solve each equation for the indicated variable.

51. $C=2\pi r$; solve for r. $r=\frac{C}{2\pi}$

52. $A=bh$; solve for h. $h=\frac{A}{b}$

53. $3x+2y=6$; solve for y. $y=\frac{6-3x}{2}$

54. $5x-6y=30$; solve for y. $y=\frac{5x-30}{6}$

55. $A=\pi(r^2+rs)$; solve for s. $s=\frac{A-\pi r^2}{\pi r}$

56. $T=2\pi(r^2+rh)$; solve for h. $h=\frac{T-2\pi r^2}{2\pi r}$

57. $\frac{V_2}{V_1}=\frac{P_1}{P_2}$; solve for V_2. $V_2=\frac{P_1V_1}{P_2}$

58. $\frac{a}{b}=\frac{c}{d}$; solve for a. $a=\frac{bc}{d}$

59. $S=\frac{f}{H-h}$; solve for H. $H=\frac{f+Sh}{S}$

60. $I=\frac{E}{R+nr}$; solve for R. $R=\frac{E-Inr}{I}$

APPLICATIONS

61. *Shoe size and length of foot* The relationship between a person's shoe size S and the length of the person's foot L (in inches) is given by (a) $S=3L-22$ for a man and (b) $S=3L-21$ for a woman. Solve for L for parts (a) and (b). (a) $L=\frac{S+22}{3}$; (b) $L=\frac{S+21}{3}$

62. *Man's weight and height* The relationship between a man's weight W (in pounds) and his height H (in inches) is given by $W=5H-190$. Solve for H. $H=\frac{W+190}{5}$

63. *Woman's weight and height* The relationship between a woman's weight W (in pounds) and her height H (in inches) is given by $W = 5H - 200$. Solve for H. $H = \frac{W + 200}{5}$

64. *Sleep hours and child's age* The number H of hours a growing child A years old should sleep is $H = 17 - \frac{A}{2}$. Solve for A. $A = 34 - 2H$

Recall from Exercise 61 that the relationship between shoe size S and the length of a person's foot L (in inches) is given by

$$S = 3L - 22 \quad \text{for men}$$
$$S = 3L - 21 \quad \text{for women}$$

65. If Tyrone wears size 11 shoes, what is the length L of his foot? 11 inches

66. If Maria wears size 7 shoes, what is the length L of her foot? $\frac{28}{3} = 9\frac{1}{3}$ inches

67. Sam's size 7 tennis shoes fit Sue perfectly! What size women's tennis shoe does Sue wear? Size 8

68. *Package delivery charges* The cost of first-class mail is 37 cents for the first ounce and 23 cents for each additional ounce. A delivery company will charge $7.27 for delivering a package weighing up to 2 lb (32 oz). When would the U.S. Postal Service price, $P = 0.37 + 0.23(x - 1)$, where x is the weight of the package in ounces, be the same as the delivery company's price? 31 oz

69. *Cost of parking* The parking cost at a garage is $C = 1 + 0.75(h - 1)$, where h is the number of hours you park and C is the cost in dollars. When is the cost C equal to $11.50? 15 hr

70. *Baseball run production* Do you follow major-league baseball? Did you know that the average number of runs per game was highest in 1996? The average number of runs scored per game for the National League can be approximated by $N = 0.165x + 4.68$, where x is the number of years after 1996. For the American League, the approximation is $A = -0.185x + 5.38$. When will $N = A$? That is, when will the National League run production be the same as that of the American League? When $x = 2$; in 1998

SKILL CHECKER

Try the Skill Checker Exercises so you'll be ready for the next section.

Write in symbols:

71. The quotient of $(a + b)$ and c $\frac{a + b}{c}$

72. The quotient of a and $(b - c)$ $\frac{a}{b - c}$

73. The product of a and the sum of b and c $a(b + c)$

74. The difference of a and b, times c $(a - b)c$

75. The difference of a and the product of b and c $a - bc$

USING YOUR KNOWLEDGE

More Car Deals!

In the *Getting Started* at the beginning of this section, we discussed a method that would tell us which was the better deal: using the mileage rate or the flat rate to rent a car. In Problems 76–78, we ask similar questions based on the rates given.

76. Suppose you wish to rent a subcompact that costs $30.00 per day, plus $0.15 per mile. Write a formula for the cost C based on traveling m miles. $C = 30 + 0.15m$

77. How many miles do you have to travel so that the mileage rate, cost C in Problem 76, is the same as the flat rate, which is $40 per day? $66\frac{2}{3}$ miles

78. If you were planning to travel 300 miles during the week, would you use the mileage rate or the flat rate given in Problems 76 and 77? the mileage rate

79. In a free-market economy, merchants sell goods to make a profit. These profits are obtained by selling goods at a *markup* in price. This markup M can be based on the cost C or on the selling price S. The formula relating the selling price S, the cost C, and the markup M is

$$S = C + M$$

A merchant plans to have a 20% markup on cost.

a. What would the selling price S be? $S = 1.2C$

b. Use the formula obtained in part **a** to find the selling price of an article that cost the merchant $8. $9.60

80. a. If the markup on an article is 25% of the selling price, use the formula $S = C + M$ to find the cost C of the article. $C = 0.75S$

b. Use the formula obtained in part **a** to find the cost of an article that sells for $80. $60

WRITE ON

81. The definition of a linear equation in one variable states that a cannot be zero. What happens if $a = 0$? Answers may vary.

82. In this section we asked you to solve a formula for a specified variable. Write a paragraph explaining what that means. Answers may vary.

83. The simplification of a linear equation led to the following step:

$$3x = 2x$$

If you divide both sides by x, you get $3 = 2$, which indicates that the equation has no solution. What's wrong with this reasoning? $x = 0$ and division by 0 is undefined.

MASTERY TEST

If you know how to do these problems, you have learned your lesson!

84. Solve for b_1 in the equation $A = \frac{h}{2}(b_1 + b_2)$. $b_1 = \frac{2A - hb_2}{h}$

85. Solve for m in the equation $50 = 40 + 0.20(m - 100)$. $m = 150$

Solve:

86. $\frac{7}{12} = \frac{x}{4} + \frac{x}{3}$ $x = 1$

87. $\frac{1}{3} - \frac{x}{5} = \frac{8(x + 2)}{15}$ $x = -1$

88. $10(x + 2) = 6(x + 1) + 18$ $x = 1$

89. $-5(x + 2) = -3(x + 1) - 9$ $x = 1$

90. $-4x - 5 = 2$ $x = -\frac{7}{4}$

91. $3x + 8 = 11$ $x = 1$

2.4 PROBLEM SOLVING: INTEGER, GENERAL, AND GEOMETRY PROBLEMS

To Succeed, Review How To . . .

1. Translate sentences into equations (pp. 102–104).
2. Solve linear equations (pp. 150–155).

Objectives

Use the RSTUV method to solve:

A Integer problems

B General word problems

C Geometry word problems

GETTING STARTED

Twin Problem Solving

Now that you've learned how to solve equations, you are ready to apply this knowledge to solve real-world problems. These problems are usually stated in words and consequently are called **word** or **story problems.** This is an area in which many students encounter difficulties, but don't panic; we are about to give you a surefire method of tackling word problems.

To start, let's look at a problem that may be familiar to you. Look at the photo of the twins, Mary and Margaret. At birth, Margaret was 3 ounces heavier than Mary, and together they weighed 35 ounces. Can you find their weights? There you have it, a word problem! In this section we shall study an effective way to solve these problems.

Here's how we solve word problems. It's as easy as 1-2-3-4-5.

PROCEDURE

RSTUV Method for Solving Word Problems

1. **R**ead the problem carefully and decide what is asked for (the unknown).
2. **S**elect a variable to represent this unknown.
3. **T**hink of a plan to help you write an equation.
4. **U**se algebra to solve the resulting equation.
5. **V**erify the answer.

Web It

For an introduction to problem solving with many types of problems, go to link 2-4-1 at the Bello Website at mhhe.com/bello.

You can click on the type of problem you want and see its solution.

If you really want to learn how to do word problems, this is the method you have to master. Study it carefully, and then use it. *It works!* How do we remember all of these steps? Easy. Look at the first letter in each sentence. We call this the **RSTUV method.**

NOTE

We will present problem solving in a 5-step (RSTUV) format. Follow the instructions at each step until you understand them. Now, cover the Example you are studying (a 3 by 5 index card will do) and work the corresponding margin problem. Check the answer and see if you are correct; if not, study the steps in the Example again.

The mathematics dictionary in Section 1.7 also plays an important role in the solution of word problems. The words contained in the dictionary are often the **key** to translating the problem. But there are other details that may help you. Here's a restatement of the RSTUV method that offers hints and tips.

HINTS AND TIPS

Our problem-solving procedure (RSTUV) contains five steps. The numbered steps are listed, with hints and tips following each.

1. Read the problem.
Mathematics is a language. As such, you have to learn how to read it. You may not understand or even get through reading the problem the first time. That's OK. Read it again and as you do, pay attention to key words or instructions such as *compute, draw, write, construct, make, show, identify, state, simplify, solve,* and *graph.* (Can you think of others?)

2. Select the unknown.
How can you answer a question if you don't know what the question is? One good way to find the unknown (variable) is to look for the question mark "?" and read the material to its left. Try to determine what is given and what is missing.

3. Think of a plan.
Problem solving requires many skills and strategies. Some of them are *look for a pattern; examine a related problem; use a formula; make tables, pictures, or diagrams; write an equation; work backward; and make a guess.* In algebra, the plan should lead to writing an equation or an inequality.

4. Use algebra to solve the resulting equation.
If you are studying a mathematical technique, it's almost certain that you will have to use it in solving the given problem. Look for ways the technique you're studying could be used to solve the problem.

5. Verify the answer.
Look back and check the results of the original problem. Is the answer reasonable? Can you find it some other way?

Are you ready to solve the problem in the *Getting Started* now? Let's restate it for you:

At birth, Mary and Margaret (the Stimson twins) together weighed 35 ounces. Margaret was 3 ounces heavier than Mary. Can you find their weights?

Basically, this problem is about sums. Let's put our RSTUV method to use.

SUMMING THE WEIGHTS OF MARY AND MARGARET

1. Read the problem.
Read the problem slowly—not once, but two or three times. (Reading algebra is *not* like reading a magazine; you may have to read algebra problems several times before you understand them.)

2. Select the unknown. (*Hint:* Let the unknown be the quantity you know nothing about.)
Let w (in ounces) represent the weight for Mary (which makes Margaret $w + 3$ ounces, since she was 3 ounces heavier).

3. Think of a plan.
Translate the sentence into an equation: Together

$$\underbrace{\text{Margaret}}_{(w+3)} \quad \underbrace{\text{and}}_{+} \quad \underbrace{\text{Mary}}_{w} \quad \underbrace{\text{weighed}}_{=} \quad \underbrace{\text{35 oz.}}_{35}$$

4. Use algebra to solve the resulting equation.

$(w + 3) + w = 35$	Given.
$w + 3 + w = 35$	Remove parentheses.
$2w + 3 = 35$	Combine like terms.
$2w + 3 - 3 = 35 - 3$	Subtract 3.
$2w = 32$	Simplify.
$\frac{2w}{2} = \frac{32}{2}$	Divide by 2.
$w = 16$	Mary's weight

Thus Mary weighed 16 ounces, and Margaret weighed $16 + 3 = 19$ ounces.

5. Verify the solution.
Do they weigh 35 ounces together? Yes, $16 + 19$ is 35.

Solving Integer Problems

Sometimes a problem uses words that may be new to you. For example, a popular type of word problem in algebra is the **integer** problem. You remember the integers—they are the positive integers 1, 2, 3, and so on; the negative integers -1, -2, -3, and so on; and 0. Now, if you are given any integer, can you find the integer that comes right after it? Of course, you simply add 1 to the given integer and you have the answer. For example, the integer that comes after 4 is $4 + 1 = 5$ and the one that comes after -4 is $-4 + 1 = -3$. In general, if n is any integer, the integer that follows n is $n + 1$. We usually say that n and $n + 1$ are **consecutive integers.** We have illustrated this idea here.

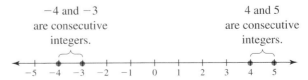

-4 and -3 are consecutive integers.

4 and 5 are consecutive integers.

-5 -4 -3 -2 -1 0 1 2 3 4 5

Teaching Tip

Students may use one of the following to select the unknown for consecutive integer problems:

Consecutive Integer	Consecutive Even/Odd
$n = 1\text{st}$	$n = 1\text{st (even/odd)}$
$n + 1 = 2\text{nd}$	$n + 2 = 2\text{nd (even/odd)}$
$n + 2 = 3\text{rd}$	$n + 4 = 3\text{rd (even/odd)}$
and so on.	and so on.

Web It

To look at an extensive set of varied practice problems dealing mostly with integers, go to link 2-4-2 at the Bello Website at mhhe.com/bello.

Now suppose you are given an *even* integer (an integer divisible by 2) such as 8 and you are asked to find the next even integer (which is 10). This time, you must add 2 to 8 to obtain the answer. What is the next even integer after 24? $24 + 2 = 26$. If n is even, the next even integer after n is $n + 2$. What about the next *odd* integer after 5? First, recall that the odd integers are $\ldots, -3, -1, 1, 3, \ldots$. The next odd integer after 5 is $5 + 2 = 7$. Similarly, the next odd integer after 21 is $21 + 2 = 23$. In general, if n is odd, the next odd integer after n is $n + 2$. Thus if you need two consecutive even or odd integers and you call the first one n, the next one will be $n + 2$. We shall use all these ideas as well as the RSTUV method in Example 1.

EXAMPLE 1 A consecutive integer problem

The sum of three consecutive even integers is 126. Find the integers.

SOLUTION We use the RSTUV method.

1. Read the problem.
We are asked to find three consecutive *even* integers.

2. Select the unknown.
Let n be the first of the integers. Since we want three consecutive even integers, we need to find the next two consecutive even integers. The next even integer after n is $n + 2$, and the one after $n + 2$ is $(n + 2) + 2 = n + 4$. Thus the three consecutive even integers are n, $n + 2$, and $n + 4$.

3. Think of a plan.
Translate the sentence into an equation.

$$\underbrace{\text{The sum of 3 consecutive even integers}}_{n + (n + 2) + (n + 4)} \quad \underbrace{\text{is}}_{=} \quad \underbrace{126.}_{126}$$

4. Use algebra to solve the resulting equation.

$$n + (n + 2) + (n + 4) = 126 \quad \text{Given.}$$
$$n + n + 2 + n + 4 = 126 \quad \text{Remove parentheses.}$$
$$3n + 6 = 126 \quad \text{Combine like terms.}$$
$$3n + 6 - 6 = 126 - 6 \quad \text{Subtract 6.}$$
$$3n = 120 \quad \text{Simplify.}$$
$$\frac{3n}{3} = \frac{120}{3} \quad \text{Divide by 3.}$$
$$n = 40$$

Thus the three consecutive even integers are 40, $40 + 2 = 42$, and $40 + 4 = 44$.

5. Verify the solution.
Since $40 + 42 + 44 = 126$, our result is correct.

PROBLEM 1

The sum of three consecutive odd integers is 129. Find the integers.

Answer

1. 41, 43, 45

The RSTUV method works for numbers other than consecutive integers. We solve a different type of integer problem in the next example.

EXAMPLE 2 **An integer problem**

Twelve less than 5 times a number is the same as the number increased by 4. Find the number.

SOLUTION We use the RSTUV method.

1. Read the problem.
We are asked to find a certain number.

2. Select the unknown.
Let n represent the number.

3. Think of a plan.
Translate the sentence into an equation.

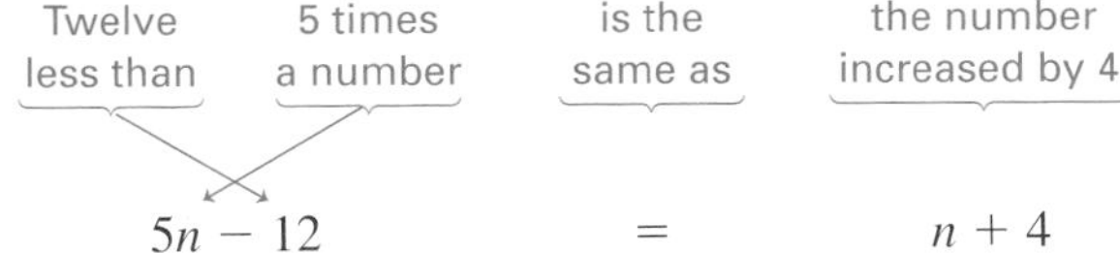

$$5n - 12 \quad = \quad n + 4$$

4. Use algebra to solve the resulting equation.

$$\begin{aligned} 5n - 12 &= n + 4 && \text{Given.} \\ 5n - 12 + 12 &= n + 4 + 12 && \text{Add 12.} \\ 5n &= n + 16 && \text{Simplify.} \\ 5n - n &= n - n + 16 && \text{Subtract } n. \\ 4n &= 16 && \text{Simplify.} \\ \frac{4n}{4} &= \frac{16}{4} && \text{Divide by 4.} \\ n &= 4 \end{aligned}$$

Thus the number is 4.

5. Verify the solution.
Is 12 less than 5 times 4 the same as 4 increased by 4? That is, is $5(4) - 12 = 4 + 4$?

$$\begin{aligned} 5(4) - 12 &= 4 + 4 \\ 20 - 12 &= 8 \end{aligned}$$

Yes, this sentence is true.

PROBLEM 2

Fourteen less than 3 times a number is the same as the number increased by 2. Find the number.

B General Word Problems

Many interesting problems can be solved using the RSTUV method. The next example is typical.

EXAMPLE 3 **How's your diet?**

Have you eaten at a fast-food restaurant lately? If you eat a cheeseburger and fries, you would consume 1070 calories. As a matter of fact, the fries contain 30 more calories than the cheeseburger. How many calories are there in each food?

SOLUTION Again, we use the RSTUV method.

1. Read the problem.
We are asked to find the number of calories in the cheeseburger and in the fries.

PROBLEM 3

A McDonald's® cheeseburger and small fries together contain 540 calories. The cheeseburger has 120 more calories than the fries. How many calories are in each food?

Answers

2. 8 **3.** Cheeseburger, 330; fries, 210

2. Select the unknown.
Let c represent the number of calories in the cheeseburger. This makes the number of calories in the fries $c + 30$, that is, 30 more calories. (We could instead let f be the number of calories in the fries; then $f - 30$ would be the number of calories in the cheeseburger.)

3. Think of a plan.
Translate the problem.

The calories in fries	and	the calories in cheeseburger	total	1070.
$(c + 30)$	$+$	c	$=$	1070

4. Use algebra to solve the equation.

$(c + 30) + c = 1070$	Given.
$c + 30 + c = 1070$	Remove parentheses.
$2c + 30 = 1070$	Combine like terms.
$2c + 30 - 30 = 1070 - 30$	Subtract 30.
$2c = 1040$	Simplify.
$\frac{2c}{2} = \frac{1040}{2}$	Divide by 2.
$c = 520$	

Thus the cheeseburger has 520 calories. Since the fries have 30 calories more than the cheeseburger, the fries have $520 + 30 = 550$ calories.

5. Verify the solution.
Does the total number of calories—that is, $520 + 550$—equal 1070? Since $520 + 550 = 1070$, our results are correct.

C Geometry Word Problems

The last type of problem we discuss here deals with an important concept in geometry: angle measure. Angles are measured using a unit called a **degree** (symbolized by °).

Angle of Measure 1°

One complete revolution around a circle is 360° and $\frac{1}{360}$ of a complete revolution is 1°.

Complementary Angles

Two angles whose sum is 90° are called **complementary angles.** Example: a 30° angle and a 60° angle are complementary since $30° + 60° = 90°$.

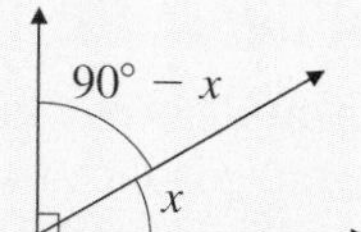

Supplementary Angles

Two angles whose sum is 180° are called **supplementary angles.** Example: a 50° angle and a 130° angle are supplementary since $50° + 130° = 180°$.

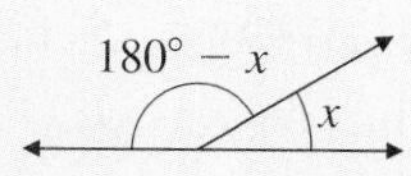

As you can see from the table, if x represents the measure of an angle,

$90 - x$ is the measure of its **complement**

and

$180 - x$ is the measure of its **supplement**

Let's use this information to solve a problem.

EXAMPLE 4 Complementary and supplementary angles

Find the measure of an angle whose supplement is 30° less than 3 times its complement.

SOLUTION As usual, we use the RSTUV method.

1. Read the problem.
We are asked to find the measure of an angle. Make sure you understand the ideas of complement and supplement. If you don't understand them, review those concepts.

2. Select the unknown.
Let m be the measure of the angle. By definition, we know that

$90 - m$ is its complement

$180 - m$ is its supplement

3. Think of a plan.
We translate the problem statement into an equation:

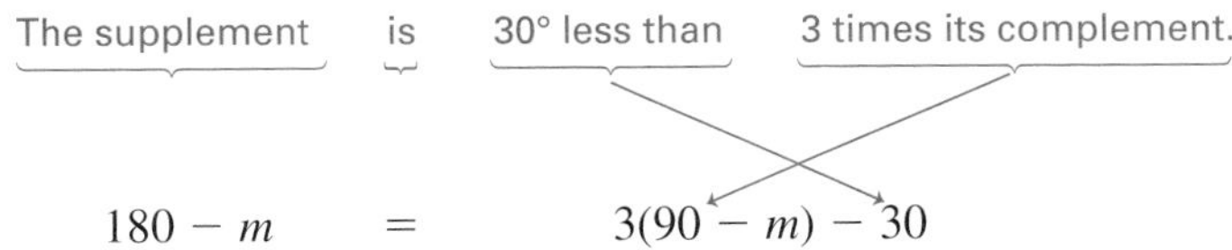
The supplement
is
30° less than
3 times its complement.

$$180 - m \quad = \quad 3(90 - m) - 30$$

4. Use algebra to solve the equation.

$180 - m = 3(90 - m) - 30$	Given.
$180 - m = 270 - 3m - 30$	Remove parentheses.
$180 - m = 240 - 3m$	Combine like terms.
$180 - 180 - m = 240 - 180 - 3m$	Subtract 180.
$-m = 60 - 3m$	Simplify.
$+3m - m = 60 - 3m + 3m$	Add $3m$.
$2m = 60$	Simplify.
$m = 30$	Divide by 2.

Thus the measure of the angle is 30°, which makes its complement $90° - 30° = 60°$ and its supplement $180° - 30° = 150°$.

5. Verify the solution.
Is the supplement (150°) 30° less than 3 times the complement; that is, is $150 = 3 \cdot 60 - 30$? Yes! Our answer is correct.

PROBLEM 4

Find the measure of an angle whose supplement is 45° less than 4 times its complement.

Answer

4. 45°

Web It

To find some geometry word problems, which may introduce additional concepts, go to the Bello Website at mhhe.com/bello and try link 2-4-3.

To get a worksheet with geometry problems, go to link 2-4-4.

Exercises 2.4

Boost ***your*** **GRADE at mathzone.com!**

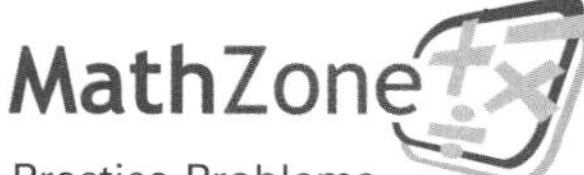

- Practice Problems
- Self-Tests
- Videos
- NetTutor
- e-Professors

A In Problems 1–20, solve the given problem.

1. The sum of three consecutive even integers is 138. Find the integers. 44, 46, 48

2. The sum of three consecutive odd integers is 135. Find the integers. 43, 45, 47

3. The sum of three consecutive even integers is -24. Find the integers. $-10, -8, -6$

4. The sum of three consecutive odd integers is -27. Find the integers. $-11, -9, -7$

5. The sum of two consecutive integers is -25. Find the integers. $-13, -12$

6. The sum of two consecutive integers is -9. Find the integers. $-5, -4$

7. Find three consecutive integers (n, $n + 1$, and $n + 2$) such that the last added to twice the first is 23. 7, 8, 9

8. Maria, Latasha, and Kim live in apartments numbered consecutively (n, $n + 1$, $n + 2$). Kim's apartment number added to twice Maria's apartment number gives 47. What are their apartment numbers? 15, 16, 17

9. Three lockers are numbered consecutively (n, $n + 1$, $n + 2$) in such a way that the sum of the first and last lockers is the same as twice the middle locker number. What are the locker numbers? Any consecutive integers

10. Pedro spent 27 more dollars for his math book than for his English book. If his total bill was \$141, what was the price of each of the books? English: \$57, Math: \$84

11. Tyrone bought two used books. One book was \$24 more than the other. If his total purchase was \$64, what was the price of each of the books? \$20, \$44

12. The total number of points in a basketball game was 179. The winning team scored 5 more points than the losing team. What was the score? 87 to 92

13. Another basketball game turned out to be a rout in which the winning team scored 55 more points than the losing team. If the total number of points in the game was 133, what was the final score? 39 to 94

14. Sandra and Mida went shopping. Their total bill was \$210, but Sandra spent only two-fifths as much as Mida. How much did each one spend? Sandra: \$60, Mida: \$150

15. The total of credit charges on one card is \$4 more than three-eighths the charges on the other. If the total charges are \$147, how much was charged on each card? \$104, \$43

16. Marcus, Liang, and Mourad went to dinner. Liang's bill was \$2 more than Marcus's bill, and Mourad's bill was \$2 more than Liang's bill. If the total bill was \$261 before tip, find the amount for each individual bill. Marcus: \$85, Liang: \$87, Mourad: \$89

17. The sum of three numbers is 254. The second is 3 times the first, and the third is 5 less than the second. Find the numbers. 37, 111, 106

18. The sum of three consecutive integers is 17 less than 4 times the smallest of the three integers. Find the integers. 20, 21, 22

19. The larger of two numbers is 6 times the smaller. Their sum is 147. Find the numbers. 21, 126

20. Five times a certain fraction yields the same as 3 times 1 more than the fraction. Find the fraction. $\frac{3}{2}$

B In Problems 21–34, solve the following problems.

21. Do you have Internet access? Polls indicate that 66% of all adults do. About 25% more online adults access the Internet from home than from work. If 15% go online from other locations (not work or home), what is the percent of online adults accessing the Internet from home? What is the percent of online adults accessing the Internet from work? 38% from home, 13% from work

Source: Harris Interactive poll.

22. The Internet market audience (in thousands) in New York, the leading market, exceeds that in Los Angeles by 727. If their combined market audience is 7955, what are the Internet market audiences (in thousands) in New York and in Los Angeles? New York: 4341, Los Angeles: 3614

23. How much money do you make per week? The salary difference between men over 25 years old and men 16–24 years old is \$330. If men over 25 make \$722 per week, how much do men between 16 and 24 make per week? \$392 per week

24. For women, the difference in weekly salary is \$188 but women over 25 make only \$542 per week. (Compare this with the figure for men in Problem 23.) How much do women between 16 and 24 make per week? \$354 per week

Source: U.S. Bureau of Labor Statistics Household Data.

25. The height of the Empire State Building including its antenna is 1472 feet. The building is 1250 feet high. How high is the antenna? 222 feet

26. The cost C for renting a car is given by $C = 0.10m + 10$, where m is the number of miles traveled. If the total cost amounted to \$17.50, how many miles were traveled? 75

27. A toy rocket goes vertically upward with an initial velocity of 96 feet per second. After t seconds, the velocity of the rocket is given by the formula $v = 96 - 32t$, neglecting air resistance. In how many seconds will the rocket reach its highest point? (*Hint:* At the highest point, $v = 0$.) 3 seconds

28. Refer to Problem 27 and find the number of seconds t that must elapse before the velocity decreases to 16 feet per second. 2.5

29. In a recent school election, 980 votes were cast. The winner received 372 votes more than the loser. How many votes did each of the two candidates receive? 304 and 676

30. The combined annual cost of the U.S. and Russian space programs has been estimated at \$71 billion. The U.S. program is cheaper; it costs \$19 billion less than the Russian program. What is the cost of each program? Russian: \$45 billion, U.S.: \$26 billion

31. A direct-dialed telephone call from Tampa to New York is \$3.05 for the first 3 minutes and \$0.70 for each additional minute or fraction thereof. If Juan's call cost him \$7.95, how many minutes was the call? 10 minutes

32. A parking garage charges \$1.25 for the first hour and \$0.75 for each additional hour or fraction thereof.

a. If Sally paid \$7.25 for parking, for how many hours was she charged? 9

b. If Sally has only \$10 with her, what is the maximum number of hours she can park in the garage? (*Hint:* Check your answer!) 12 hr

33. The cost of a taxicab is \$0.95 plus \$1.25 per mile.

a. If the fare was \$28.45, how long was the ride? 22 miles

b. If a limo service charges \$15 for a 12-mile ride to the airport, which one is the better deal, the cab or the limo? The limo is cheaper.

34. According to the Department of Health and Human Services, the number of heavy alcohol users in the 35-and-older category exceeds the number of heavy alcohol users in the 18–25 category by 844,000. If the total number of persons in these two categories is 7,072,000, how many persons are there in each of these categories? 3,114,000 in 18–25 group, 3,958,000 in the 35-and-older group

C In Problems 35–40, find the measure of the given angles.

35. An angle whose supplement is 20° more than twice its complement 20°

36. An angle whose supplement is 10° more than three times its complement 50°

37. An angle whose supplement is 3 times the measure of its complement 45°

38. An angle whose supplement is 4 times the measure of its complement 60°

39. An angle whose supplement is 54° less than 4 times its complement 42°

40. An angle whose supplement is 41° less than 3 times its complement 24.5°

SKILL CHECKER

Try the Skill Checker Exercises so you'll be ready for the next section.

Solve:

41. $55T = 100$ $T = \dfrac{20}{11}$

42. $88R = 3240$ $R = \dfrac{405}{11}$

43. $15T = 120$ $T = 8$

44. $81T = 3240$ $T = 40$

45. $-75x = -600$ $x = 8$

46. $-45x = -900$ $x = 20$

47. $-0.02P = -70$ $P = 3500$

48. $-0.05R = -100$ $R = 2000$

49. $-0.04x = -40$ $x = 1000$

50. $-0.03x = 30$ $x = 1000$

USING YOUR KNOWLEDGE

Diophantus's Equation

If you are planning to become an algebraist (an expert in algebra), you may not enjoy much fame. As a matter of fact, very little is known about one of the best algebraists of all time, the Greek Diophantus. According to a legend, the following problem is in the inscription on his tomb:

One-sixth of his life God granted him youth. After a twelfth more, he grew a beard. After an additional seventh, he married, and 5 years later, he had a son. Alas, the unfortunate son's life span was only one-half that of his father, who consoled his grief in the remaining 4 years of his life.

51. Use your knowledge to find how many years Diophantus lived. (*Hint:* Let x be the number of years Diophantus lived.) 84

WRITE ON

52. When reading a word problem, what is the first thing you try to determine? Answers may vary.

53. How do you verify your answer in a word problem? Answers may vary.

54. The "T" in the RSTUV method means that you must "Think of a plan." Some strategies you can use in this plan include *look for a pattern* and *make a picture.* Can you think of three other strategies? Answers may vary.

MASTERY TEST

If you know how to do these problems, you have learned your lesson!

55. The sum of three consecutive odd integers is 249. Find the integers. 81, 83, 85

56. Three less than 4 times a number is the same as the number increased by 9. Find the number. 4

57. If you eat a single slice of a 16-inch mushroom pizza and a 10-ounce chocolate shake, you have consumed 530 calories. If the shake has 70 more calories than the pizza, how many calories are there in each? Pizza: 230 calories, Shake: 300 calories

58. Find the measure of an angle whose supplement is 47° less than 3 times its complement. 21.5°

2.5 PROBLEM SOLVING: MOTION, MIXTURE, AND INVESTMENT PROBLEMS

To Succeed, Review How To . . .

1. Translate sentences into equations (pp. 102–104).
2. Solve linear equations (pp. 150–155).

Objectives

Use the RSTUV method to solve:

A Motion problems

B Mixture problems

C Investment problems

GETTING STARTED

Birds in Motion

As we have mentioned previously, some of the strategies we can use to help us think of a plan include *use a formula, make a table,* and *draw a diagram.* We will use these strategies as we solve problems in this section. For example, in the cartoon, the bird is trying an impossible task, unless he turns around! If he does, how does Curly know that it will take him less than 2 hours to fly 100 miles? Because there's a formula to figure this out! If an object moves at a constant rate R for a time T, the distance D traveled by the object is given by

$$D = RT$$

By Permission of John L. Hart FLP, and Creators Syndicate, Inc.

The object could be your car moving at a constant rate R of 55 miles per hour for $T = 2$ hours. In this case, you would have traveled a distance $D = 55 \times 2 = 110$ miles. Here the rate is in miles per *hour* and the time is in *hours.* [Units have to be consistent!] Similarly, if you jog at a constant rate of 5 miles per hour for 2 hours, you would travel $5 \times 2 = 10$ miles. Here the units are in miles per hour, so the time must be in hours. In working with the formula for distance, and especially in more complicated problems, it is often helpful to write the formula in a chart. For example, to figure out how far you jogged, you would write:

	R	$\times$	T	$=$	D
Jogger	5 mi/hr		2 hr		10 mi

As you can see, we used the formula $D = RT$ to solve this problem and organized our information in a chart. We shall continue to use formulas and charts when solving the rest of the problems in this section.

A Motion Problems

Let's go back to the bird problem, which is a **motion problem.** If the bird turns around and is flying at 5 miles per hour with a tail wind of 50 miles per hour, its rate R would be $50 + 5 = 55$ miles per hour. The wind is helping the bird, so the wind speed must be added to his rate. Curly wants to know how long it would take the bird to fly a distance of

100 miles; that is, he wants to find the time T. We can write this information in a table, substituting 55 for R and 100 for D.

	R	$\times$	T	$=$	D
Bird	55		T		100

Since

$$R \times T = D$$

we have

$$55T = 100$$

$$T = \frac{100}{55} = \frac{20}{11} = 1\frac{9}{11}\text{ hr} \quad \text{Divide by 55.}$$

Curly was right; it does take less than 2 hours!

Now let's try some other motion problems.

EXAMPLE 1 Finding the speed

The longest regularly scheduled bus route is Greyhound's "Supercruiser" Miami-to-San Francisco route, a distance of 3240 miles. If this distance is covered in about 82 hours, what is the average speed R of the bus?

SOLUTION We use our RSTUV method.

1. Read the problem.
We are asked to find the rate of the bus.

2. Select the unknown.
Let R represent this rate.

3. Think of a plan. What type of information do you need? How can you enter it so that $R \times T = D$?
Translate the problem and enter the information in a chart.

	R	$\times$	T	$=$	D
Supercruiser	R		82		3240

The equation is

$$R \times 82 = 3240 \quad \text{or} \quad 82R = 3240$$

4. Use algebra to solve the equation.

$$82R = 3240 \quad \text{Given.}$$

$$\frac{82R}{82} = \frac{3240}{82} \quad \text{Divide by 82.}$$

$$R = 39.5 \quad \text{By long division (to the nearest tenth)}$$

```
      39.51
82)3240.0
   246
    780
    738
     42 0
     41 0
      1 00
        82
        18
```

The bus's average speed is 39.5 miles per hour.

PROBLEM 1

Example 1 is about bus routes in the United States. There is a 6000-mile bus trip from Caracas to Buenos Aires that takes 214 hours, including a 12-hour stop in Santiago and a 24-hour stop in Lima. What is the average speed of the bus? *Hint:* Do not count rest time.

Answer

1. 33.7 mi/hr

5. Verify the solution.
Check the arithmetic by substituting 39.5 into the formula $R \times T = D$.

$$39.5 \times 82 = 3239$$

Although the actual distance is 3240, this difference is acceptable because we rounded our answer to the nearest tenth.

Some motion problems depend on the relationship between the distances traveled by the objects involved. When you think of a plan for these types of problems, a diagram can be very useful.

EXAMPLE 2 Finding the time it takes to overtake an object

The Supercruiser bus leaves Miami traveling at an average rate of 40 miles per hour. Three hours later, a car leaves Miami for San Francisco traveling on the same route at 55 miles per hour. How long does it take for the car to overtake the bus?

SOLUTION Again, we use the RSTUV method.

1. Read the problem.
We are asked to find how many hours it takes the car to overtake the bus.

2. Select the unknown.
Let T represent this number of hours.

3. Think of a plan.
Translate the given information and enter it in a chart. Note that if the car goes for T hours, the bus goes for $(T + 3)$ hours (since it left 3 hours earlier).

	R	$\times$	T	$=$	D
Car	55		T		$55T$
Bus	40		$T + 3$		$40(T + 3)$

Since the vehicles are traveling in the same direction, the diagram of this problem looks like this:

Miami —Car→ / —Bus→ San Francisco

When the car overtakes the bus, they will have traveled the *same* distance. According to the chart, the car has traveled $55T$ miles and the bus $40(T + 3)$ miles. Thus

$$\underbrace{55T}_{\text{Distance traveled by car}} \underbrace{=}_{\text{overtakes}} \underbrace{40(T + 3)}_{\text{distance traveled by Supercruiser}}$$

PROBLEM 2

How long does it take if the car in Example 2 is traveling at 60 miles per hour?

Answer

2. 6 hr

4. Use algebra to solve the equation.

$$55T = 40(T + 3) \quad \text{Given.}$$
$$55T = 40T + 120 \quad \text{Simplify.}$$
$$55T - 40T = 40T - 40T + 120 \quad \text{Subtract } 40T.$$
$$15T = 120 \quad \text{Simplify.}$$
$$\frac{15T}{15} = \frac{120}{15} \quad \text{Divide by 15.}$$
$$T = 8$$

It takes the car 8 hours to overtake the bus.

5. Verify the solution.
The car travels for 8 hours at 55 miles per hour; thus it travels $55 \times 8 = 440$ miles, whereas the bus travels at 40 miles per hour for 11 hours, a total of $40 \times 11 = 440$ miles. Since the car traveled the same distance, it overtook the bus in 8 hours.

Web It

For motion problems, go to link 2-5-1 or link 2-5-2 on the Bello Website at mhhe.com/bello.

In Example 2, the two vehicles moved in the *same* direction. A variation of this type of problem involves motion toward each other, as shown in Example 3.

EXAMPLE 3 Two objects moving toward each other

The Supercruiser leaves Miami for San Francisco 3240 miles away, traveling at an average speed of 40 miles per hour. At the same time, a slightly faster bus leaves San Francisco for Miami traveling at 41 miles per hour. How many hours will it take for the buses to meet?

SOLUTION We use the RSTUV method.

1. Read the problem.
We are asked to find how many hours it takes for the buses to meet.

2. Select the unknown.
Let T represent the hours each bus travels before they meet.

3. Think of a plan.
Translate the information and enter it in a chart.

	R	$\times$	T	$=$	D
Supercruiser	40		T		$40T$
Bus	41		T		$41T$

PROBLEM 3

How long does it take if the faster bus travels at 50 miles per hour?

Answer

3. 36 hr

This time the objects are moving toward each other. The distance the Supercruiser travels is $40T$ miles, whereas the bus travels $41T$ miles, as shown here:

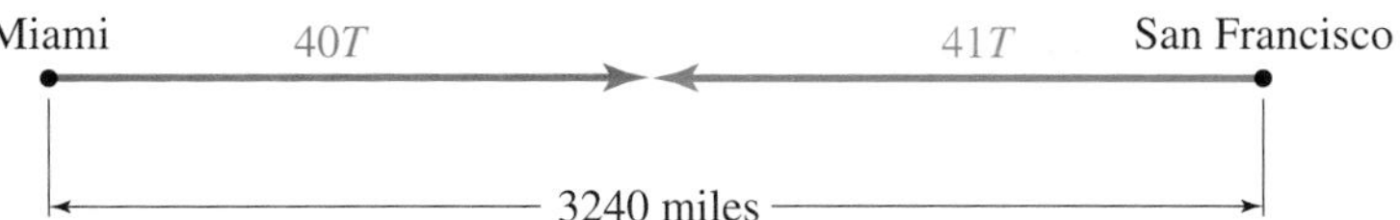

When they meet, the combined distance traveled by *both* buses is 3240 miles. This distance is also $40T + 41T$. Thus we have

$$\underbrace{40T}_{\text{Distance traveled by Supercruiser}} \underbrace{+}_{\text{and}} \underbrace{41T}_{\text{distance traveled by other bus}} \underbrace{=}_{\text{is}} \underbrace{3240}_{\text{total distance.}}$$

4. Use algebra to solve the equation.

$40T + 41T = 3240$ Given.

$81T = 3240$ Combine like terms.

$\frac{81T}{81} = \frac{3240}{81}$ Divide by 81.

$T = 40$

Thus each bus traveled 40 hours before they met.

5. Verify the solution.
The Supercruiser travels $40 \times 40 = 1600$ miles in 40 hours, whereas the other bus travels $40 \times 41 = 1640$ miles. Clearly, the total distance traveled is $1600 + 1640 = 3240$ miles.

B Mixture Problems

Web It

For mixture problems, go to link 2-5-3 on the Bello Website at mhhe.com/bello.

Another type of problem that can be solved using a chart is the **mixture problem.** In a mixture problem, two or more things are combined to form a mixture. For example, dental-supply houses mix pure gold and platinum to make white gold for dental fillings. Suppose one of these houses wishes to make 10 troy ounces of white gold to sell for \$415 per ounce. If pure gold sells for \$400 per ounce and platinum sells for \$475 per ounce, how much of each should the supplier mix?

We are looking for the amount of each material needed to make 10 ounces of a mixture selling for \$415 per ounce. If we let x be the total number of ounces of gold, then $10 - x$ (the balance) must be the number of ounces of platinum. Note that if a quantity T is split into two parts, one part may be x and the other $T - x$. This can be checked by adding $T - x$ and x to obtain $T - x + x = T$. We then enter all the information in a chart. The top line of the chart tells us that if we multiply the price of the item by the ounces used, we will get the total price:

Teaching Tip

Help students see that the last column in these mixture boxes is used to obtain the equation needed to solve the problem. Often, the first terms are added and set equal to the last term in that column.

	Price/Ounce	×	Ounces	=	Total Price
Gold	400		x		$400x$
Platinum	475		$10 - x$		$475(10 - x)$
Mixture	415		10		4150

The first line (following the word *gold*) tells us that the price of gold (\$400) times the amount being used (x ounces) gives us the total price (\$$400x$). In the second line, the price of platinum ($475) times the amount being used ($10 - x$ ounces) gives us the total price of $475(10 − x). Finally, the third line tells us that the price of the mixture is $415, that we need 10 ounces of it, and that its total price will be $4150.

Since the sum of the total prices of gold and platinum in the last column must be equal to the total price of the mixture, it follows that

$$400x + 475(10 - x) = 4150$$

$$400x + 4750 - 475x = 4150 \quad \text{Simplify.}$$

$$4750 - 75x = 4150$$

$$4750 - 4750 - 75x = 4150 - 4750 \quad \text{Subtract 4750.}$$

$$-75x = -600 \quad \text{Simplify.}$$

$$\frac{-75x}{-75} = \frac{-600}{-75} \quad \text{Divide by } -75.$$

$$x = 8$$

Thus the supplier must use 8 ounces of gold and $10 - 8 = 2$ ounces of platinum. You can verify that this is correct! (8 ounces of gold at $400 per ounce and 2 ounces of platinum at $475 per ounce make 10 ounces of the mixture that costs $4150.)

EXAMPLE 4 A mixture for the darkroom

How many ounces of a 50% acetic acid solution should a photographer add to 32 ounces of a 5% acetic acid solution to obtain a 10% acetic acid solution?

SOLUTION We use the RSTUV method.

1. Read the problem.
We are asked to find the number of ounces of the 50% solution (a solution consisting of 50% acetic acid) that should be added to make a 10% solution.

2. Select the unknown.
Let x stand for the number of ounces of 50% solution to be added.

3. Think of a plan. Remember to write the percents as decimals.
To translate the problem, we first use a chart. In this case, the headings should include the percent of acetic acid and the amount to be mixed. The product of these two numbers will then give us the amount of pure acetic acid. Note that the percents have been converted to decimals.

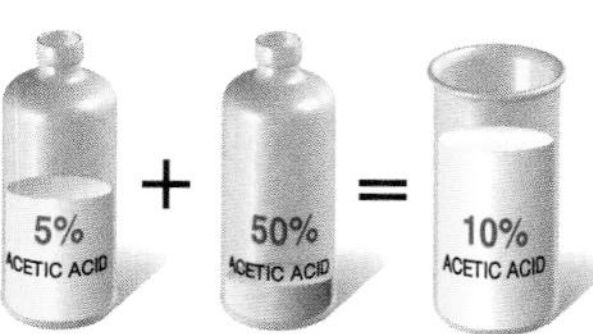

	%	×	Ounces	=	Amount of Pure Acid in Final Mixture
50% solution	0.50		x		$0.50x$
5% solution	0.05		32		1.60
10% solution	0.10		$x + 32$		$0.10(x + 32)$

Since we have x ounces of one solution and 32 ounces of the other, we have $(x + 32)$ ounces of the final mixture.

PROBLEM 4

What if we want to obtain a 30% acetic acid solution?

Answer

4. 40 oz

Since the sum of the amounts of pure acetic acid should be the same as the total amount of pure acetic acid in the final mixture, we have

$$0.50x + 1.60 = 0.10(x + 32) \qquad \text{Given.}$$

4. Use algebra to solve the equation.

$$\begin{aligned}
10 \cdot 0.50x + 10 \cdot 1.60 &= 10 \cdot [0.10(x + 32)] && \text{Multiply by 10 to clear the decimals.}\\
5x + 16 &= 1(x + 32)\\
5x + 16 &= x + 32\\
5x + 16 - 16 &= x + 32 - 16 && \text{Subtract 16.}\\
5x &= x + 16 && \text{Simplify.}\\
5x - x &= x - x + 16 && \text{Subtract } x.\\
4x &= 16 && \text{Simplify.}\\
\frac{4x}{4} &= \frac{16}{4} && \text{Divide by 4.}\\
x &= 4
\end{aligned}$$

Thus the photographer must add 4 ounces of the 50% solution.

5. Verify the solution.

We leave the verification to you.

Investment Problems

Finally, there's another problem that's similar to the mixture problem—the **investment problem.** Investment problems depend on the fact that the simple interest I you can earn (or pay) on principal P invested at rate r for 1 year is given by the formula

$$I = Pr$$

Now suppose you have a total of \$10,000 invested. Part of the money is invested at 6% and the rest at 8%. If the bank tells you that you have earned \$730 interest for 1 year, how much do you have invested at each rate?

In this case, we need to know how much is invested at each rate. Thus if we say that we have invested P dollars at 6%, the rest of the money, that is, $10,000 - P$, would be invested at 8%. Note that for this to work, the sum of the amount invested at 6%, P dollars, plus the rest, $10,000 - P$, must equal \$10,000. Since $P + 10,000 - P = 10,000$, we have done it correctly. This information is entered in a chart, as shown here:

	P	×	r	=	I
6% investment	P		0.06		$0.06P$
8% investment	$10,000 - P$		0.08		$0.08(10,000 - P)$

This column must add to 10,000.

The total interest is the sum of the two expressions.

From the chart we can see that the total interest is the sum of the expressions in the last column, $0.06P + 0.08(10,000 - P)$. Recall that the bank told us that our total interest is \$730. Thus

$$0.06P + 0.08(10,000 - P) = 730$$

Web It

For some worksheets dealing with investment problems, go to link 2-5-4 on the Bello Website at mhhe.com/bello.

Teaching Tip

Students should know that in investment problems $I = Pr$ (interest for one year) is sometimes called the annual income resulting from some investment.

Teaching Tip

Point out to students that multiplying by 100 clears the equation of decimals. Just be careful with the multiplication!

You can solve this equation by first multiplying each term by 100 to clear the decimals:

$$100 \cdot 0.06P + 100 \cdot 0.08(10{,}000 - P) = 100 \cdot 730$$
$$6P + 8(10{,}000 - P) = 73{,}000$$
$$6P + 80{,}000 - 8P = 73{,}000$$
$$-2P = -7000$$
$$P = 3500$$

You could also solve

$$0.06P + 0.08(10{,}000 - P) = 730$$

as follows:

$$0.06P + 800 - 0.08P = 730 \quad \text{Simplify.}$$
$$800 - 0.02P = 730$$
$$800 - 800 - 0.02P = 730 - 800 \quad \text{Subtract 800.}$$
$$-0.02P = -70 \quad \text{Simplify.}$$
$$\frac{-0.02P}{-0.02} = \frac{-70}{-0.02} \quad \text{Divide by } -0.02.$$
$$P = 3500$$

$$\begin{array}{r} 35\,00 \\ 0.02\overline{)70.00} \\ \underline{6} \\ 10 \\ \underline{10} \\ 0 \end{array}$$

Thus \$3500 is invested at 6% and the rest, 10,000 − 3500, or \$6500, is invested at 8%. You can verify that 6% of \$3500 added to 8% of \$6500 yields \$730.

EXAMPLE 5 **An investment portfolio**

A woman has some stocks that yield 5% annually and some bonds that yield 10%. If her investment totals \$6000 and her annual income from the investments is \$500, how much does she have invested in stocks and how much in bonds?

SOLUTION As usual, we use the RSTUV method.

1. Read the problem.
We are asked to find how much is invested in stocks and how much in bonds.

2. Select the unknown.
Let s be the amount invested in stocks. This makes the amount invested in bonds $(6000 - s)$.

3. Think of a plan.
A chart is a good way to visualize this problem. Remember that the headings will be the formula $P \times r = I$ and that the percents must be written as decimals. Now we enter the information:

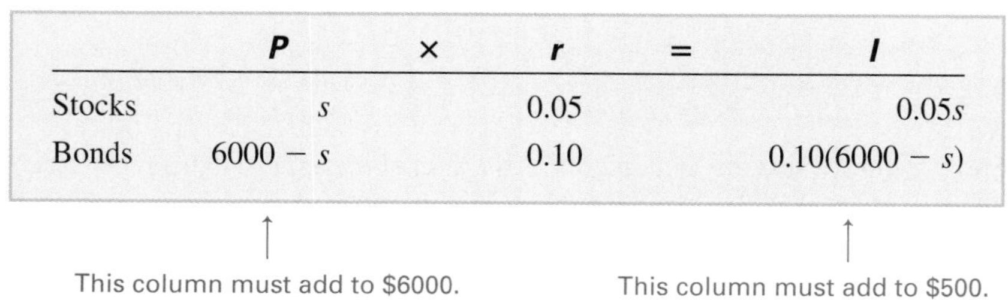

	P	×	r	=	I
Stocks	s		0.05		$0.05s$
Bonds	$6000 - s$		0.10		$0.10(6000 - s)$

This column must add to \$6000.
This column must add to \$500.

PROBLEM 5

What if her income is only \$400?

Answer

5. Stocks, \$4000; bonds, \$2000

The total interest is the sum of the entries in the last column—that is, $0.05s$ and $0.10(6000 - s)$. This amount must be $500:

$$0.05s + 0.10(6000 - s) = 500$$

4. Use algebra to solve the equation.

$0.05s + 0.10(6000 - s) = 500$	Given.
$0.05s + 600 - 0.10s = 500$	Simplify.
$600 - 0.05s = 500$	
$600 - 600 - 0.05s = 500 - 600$	Subtract 600.
$-0.05s = -100$	Simplify.
$\frac{-0.05s}{-0.05} = \frac{-100}{-0.05}$	Divide by −0.05.
$s = 2000$	

Thus the woman has $2000 in stocks and $4000 in bonds.

5. Verify the solution.
To verify the answer, note that 5% of 2000 is 100 and 10% of 4000 is 400, so the total interest is indeed $500.

Teaching Tip

Have students multiply both sides of the equation by 100 to clear decimals. Thus,

$$5s + 10(6000 - s) = 50{,}000$$

Now, simplify and solve.

Exercises 2.5

A In Problems 1–16, use the RSTUV method to solve the motion problems.

1. The distance from Los Angeles to Sacramento is about 400 miles. A bus covers this distance in about 8 hours. What is the average speed of the bus? 50 mi/hr

2. The distance from Boston to New Haven is 260 miles. A car leaves Boston at 7 A.M. and gets to New Haven just in time for lunch, at exactly 12 noon. What is the speed of the car? 52 mi/hr

3. A 120-VHS Memorex videotape contains 246 meters of tape. When played at standard speed (SP), it will play for 120 minutes. What is the rate of play of the tape? Answer to the nearest whole number. 2 meters/min

4. A Laser Writer Select prints one 12-inch page in 6 seconds.

a. What is the rate of output for this printer? 2 in./sec or $\frac{1}{6}$ page/sec

b. How long would it take to print 60 pages at this rate? 360 sec or 6 min

5. The air distance from Miami to Tampa is about 200 miles. If a jet flies at an average speed of 400 miles per hour, how long does it take to go from Tampa to Miami? $\frac{1}{2}$ hr

6. A freight train leaves the station traveling at 30 miles per hour. One hour later, a passenger train leaves the same station traveling at 60 miles per hour in the same direction. How long does it take for the passenger train to overtake the freight train? 1 hr

7. A bus leaves the station traveling at 60 kilometers per hour. Two hours later, a student shows up at the station with a briefcase belonging to her absent-minded professor who is riding the bus. If she immediately starts after the bus at 90 kilometers per hour, how long will it be before she reunites the briefcase with her professor? 4 hr

8. An accountant catches a train that travels at 50 miles per hour, whereas his boss leaves 1 hour later in a car traveling at 60 miles per hour. They had decided to meet at the train station in the next town and, strangely enough, they get there at exactly the same time! If the train and the car traveled in a straight line on parallel paths, how far is it from one town to the other? 300 mi

9. The basketball coach at a local high school left for work on his bicycle traveling at 15 miles per hour. Half an hour later, his wife noticed that he had forgotten his lunch. She got in her car and took his lunch to him. Luckily, she got to school at exactly the same time as her husband. If she made her trip traveling at 60 miles per hour, how far is it from their house to the school? 10 mi

10. A jet traveling 480 miles per hour leaves San Antonio for San Francisco, a distance of 1632 miles. An hour later another plane, going at the same speed, leaves San Francisco for San Antonio. How long will it be before the planes pass each other? 2.2 hr or 2 hr 12 min

11. A car leaves town A going toward B at 50 miles per hour. At the same time, another car leaves B going toward A at 55 miles per hour. How long will it be before the two cars meet if the distance from A to B is 630 miles? 6 hr

12. A contractor has two jobs that are 275 kilometers apart. Her headquarters, by sheer luck, happen to be on a straight road between the two construction sites. Her first crew left headquarters for one job traveling at 70 kilometers per hour. Two hours later, she left headquarters for the other job, traveling at 65 kilometers per hour. If the contractor and her first crew arrived at their job sites simultaneously, how far did the first crew have to drive? 210 km

13. A plane has 7 hours to reach a target and come back to base. It flies out to the target at 480 miles per hour and returns on the same route at 640 miles per hour. How many miles from the base is the target? 1920 mi

14. A man left home driving at 40 miles per hour. When his car broke down, he walked home at a rate of 5 miles per hour; the entire trip (driving and walking) took him $2\frac{1}{4}$ hours. How far from his house did his car break down? 10 mi

15. The space shuttle *Discovery* made a historical approach to the Russian space station *Mir*, coming to a point 366 feet (4392 inches) away from the station. For the first 180 seconds of the final 366-foot approach, the *Discovery* traveled 20 times as fast as during the last 60 seconds of the approach. How fast was the shuttle approaching during the first 180 seconds and during the last 60 seconds? 24 in./sec first 180 sec, 1.2 in./sec last 60 sec

16. If, in Problem 15, the commander decided to approach the *Mir* traveling only 13 times faster for the first 180 seconds as for the last 60 seconds, what would be the shuttle's final rate of approach? 1.83 in./sec

B In Problems 17–27, use the RSTUV method to solve these mixture problems.

17. How many liters of a 40% glycerin solution must be mixed with 10 liters of an 80% glycerin solution to obtain a 65% solution? 6 liters

18. How many parts of glacial acetic acid (99.5% acetic acid) must be added to 100 parts of a 10% solution of acetic acid to give a 28% solution? (Round your answer to the nearest whole part.) 25 parts

19. If the price of copper is 65¢ per pound and the price of zinc is 30¢ per pound, how many pounds of copper and zinc should be mixed to make 70 pounds of brass selling for 45¢ per pound? 30 lb copper, 40 lb zinc

20. Oolong tea sells for $19 per pound. How many pounds of Oolong should be mixed with another tea selling at $4 per pound to produce 50 pounds of tea selling for $7 per pound? 10 pounds

21. How many pounds of Blue Jamaican coffee selling at $5 per pound should be mixed with 80 pounds of regular coffee selling at $2 per pound to make a mixture selling for $2.60 per pound? (The merchant cleverly advertises this mixture as "Containing the incomparable Blue Jamaican coffee"!) 20 pounds

22. How many ounces of vermouth containing 10% alcohol should be added to 20 ounces of gin containing 60% alcohol to make a pitcher of martinis that contains 30% alcohol? 30 ounces

23. Do you know how to make manhattans? They are mixed by combining bourbon and sweet vermouth. How many ounces of manhattans containing 40% vermouth should a bartender mix with manhattans containing 20% vermouth so that she can obtain a half gallon (64 ounces) of manhattans containing 30% vermouth? 32 ounces of each

24. A car radiator contains 30 quarts of 50% antifreeze solution. How many quarts of this solution should be drained and replaced with pure antifreeze so that the new solution is 70% antifreeze? 12 quarts

25. A car radiator contains 30 quarts of 50% antifreeze solution. How many quarts of this solution should be drained and replaced with water so that the new solution is 30% antifreeze? 12 quarts

26. A 12-ounce can of frozen orange juice concentrate is mixed with 3 cans of cold water to obtain a mixture that is 10% juice. What is the percent of pure juice in the concentrate? 40% pure juice

27. The instructions on a 12-ounce can of Welch's Orchard fruit juice state: "Mix with 3 cans cold water" and it will "contain 30% juice when properly reconstituted." What is the percent of pure juice in the concentrate? What does this mean to you? Impossible; it would need 120% of concentrate.

C In Problems 28–36, use the RSTUV method to solve these investment problems.

28. Two sums of money totaling \$15,000 earn, respectively, 5% and 7% annual interest. If the total interest from both investments amounts to \$870, how much is invested at each rate? \$9000 at 5%, \$6000 at 7%

29. An investor invested \$20,000, part at 6% and the rest at 8%. Find the amount invested at each rate if the annual income from the two investments is \$1500. \$5000 at 6%, \$15,000 at 8%

30. A woman invested \$25,000, part at 7.5% and the rest at 6%. If her annual interest from these two investments amounted to \$1620, how much money did she invest at each rate? \$8000 at 7.5%, \$17,000 at 6%

31. A man has a savings account that pays 5% annual interest and some certificates of deposit paying 7% annually. His total interest from the two investments is \$1100, and the total amount of money in the two investments is \$18,000. How much money does he have in the savings account? \$8000

32. A woman invested \$20,000 at 8%. What additional amount must she invest at 6% so that her annual income is \$2200? \$10,000

33. A sum of \$10,000 is split, and the two parts are invested at 5% and 6%, respectively. If the interest from the 5% investment exceeds the interest from the 6% investment by \$60, how much is invested at each rate? \$6000 at 5%, \$4000 at 6%

34. An investor receives \$600 annually from two investments. He has \$500 more invested at 8% than at 6%. Find the amount invested at each rate. \$4000 at 6%, \$4500 at 8%

35. How can you invest \$40,000, part at 6% and the remainder at 10%, so that the interest earned in each of the accounts is the same? \$25,000 at 6%, \$15,000 at 10%

36. An inheritance of \$10,000 was split into two parts and invested. One part lost 5% and the second gained 11%. If the 5% loss was the same as the 11% gain, how much was each investment? \$6875 at 5%, \$3125 at 11%

SKILL CHECKER

Try the Skill Checker Exercises so you'll be ready for the next section.

Find:

37. $-16(2)^2 + 118$ 54 **38.** $-8(3)^2 + 80$ 8 **39.** $-4 \cdot 8 \div 2 + 20$ 4 **40.** $-5 \cdot 6 \div 2 + 25$ 10

USING YOUR KNOWLEDGE

Gesselmann's Guessing

Some of the mixture problems given in this section can be solved using the *guess-and-correct* procedure developed by Dr. Harrison A. Gesselmann of Cornell University. The procedure depends on taking a guess at the answer and then using the calculator to correct this guess. For example, to solve the very first mixture problem presented in this section (p. 175), we have to mix gold and platinum to obtain 10 ounces of a mixture selling for \$415 per ounce. Our first guess is to use *equal amounts* (5 ounces each) of gold and platinum. This gives a mixture with a price per pound equal to the average price of gold (\$400) and platinum (\$475), that is,

$$\frac{400 + 475}{2} = \$437.50 \text{ per pound}$$

As you can see from the following figure, more gold must be used in order to bring the $437.50 average down to the desired $415:

5 oz	Desired	Average	5 oz
$400	$415	$437.50	$475

22.50

37.50

Thus the correction for the additional amount of gold that must be used is

$$\frac{22.50}{37.50} \times 5 \text{ oz}$$

This expression can be obtained by the keystroke sequence

22.50 ÷ 37.50 × 5 ENTER

which gives the correction 3. The correct amount is

First guess	5 oz of gold
+Correction	3 oz of gold
Total =	8 oz of gold

and the remaining 2 ounces is platinum.

If your instructor permits, use this method to work Problems 19 and 23.

WRITE ON

41. Ask your pharmacist or your chemistry instructor if they mix products of different concentrations to make new mixtures. Write a paragraph on your findings. Answers may vary.

42. Most of the problems involved have precisely the information you need to solve them. In real life, however, irrelevant information (called *red herrings*) may be present. Find some problems with red herrings and point them out. Answers may vary.

43. The *guess-and-correct* method explained in the *Using Your Knowledge* also works for investment problems. Write the procedure you would use to solve investment problems using this method. Answers may vary.

MASTERY TEST

If you know how to do these problems, you have learned your lesson!

44. Billy has two investments totaling $8000. One investment yields 5% and the other 10%. If the total annual interest is $650, how much money is invested at each rate? $3000 at 5%, $5000 at 10%

45. How many gallons of a 10% salt solution should be added to 15 gallons of a 20% salt solution to obtain a 16% solution? 10 gallons

46. Two trains are 300 miles apart, traveling toward each other on adjacent tracks. One is traveling at 40 miles per hour and the other at 35 miles per hour. After how many hours do they meet? 4 hr

47. A bus leaves Los Angeles traveling at 50 miles per hour. An hour later a car leaves at 60 miles per hour to try to catch the bus. How long does it take the car to overtake the bus? 5 hr

48. The distance from South Miami to Tampa is 250 miles. This distance can be covered in 5 hours by car. What is the average speed on the trip? 50 mi/hr

2.6 FORMULAS AND GEOMETRY APPLICATIONS

To Succeed, Review How To . . .

1. Reduce fractions (pp. 7–9).
2. Perform the fundamental operations using decimals (pp. 26–29).
3. Evaluate expressions using the correct order of operations (p. 80).

Objectives

A Solve a formula for one variable and use the result to solve a problem.

B Solve problems involving geometric formulas.

C Solve geometric problems involving angle measurement.

D Solve an application.

GETTING STARTED

Do You Want Fries with That?

One way to solve word problems is to use a formula. But which formula? Here's an example in which an incorrect formula has been used. The ad claims that the new Monster Burger has 50% more beef than the Lite Burger because its diameter, 6, is 50% more than 4. But is that the right way to measure them?

Monster Burger 6 inches

Lite Burger 4 inches

"The new **Monster Burger** has **50%** more beef than the **Lite Burger.**"

We should compare the two hamburgers by comparing the *volume* of the beef in each burger. The volume V of a burger is $V = Ah$, where A is the area and h is the height. The area of the Lite Burger is $L = \pi r^2$, where r is the radius (half the distance across the middle) of the burger. For the Lite Burger, $r = 2$ inches, so its area is

$$L = \pi(2 \text{ in.})^2 = 4\pi \text{ in.}^2$$

For the Monster Burger, $r = 3$ inches, so its area M is given by

$$M = \pi(3 \text{ in.})^2 = 9\pi \text{ in.}^2$$

The volume of the Lite Burger is $V = Ah = 4\pi h$, and for the Monster Burger the volume is $9\pi h$. If the height is the same for both burgers, we can simplify the volume calculation. The difference in volumes is $9\pi h - 4\pi h = 5\pi h$, and the *percent* increase for the Monster Burger is given by

$$\text{Percent increase} = \frac{\text{increase}}{\text{base}} = \frac{5\pi h}{4\pi h} = 1.25 \text{ or } 125\%$$

Thus, based on the discussion, the Monster Burgers actually have 125% more beef than the Lite Burgers. What other assumptions must you make for this to be so? (You'll have an opportunity in the *Write On* to give your opinion.) In this section we shall use formulas (like that for the volume or area of a burger) to solve problems.

Problems in many fields of endeavor can be solved if the proper formula is used. For example, why aren't you electrocuted when you hold the two terminals of your car battery? You may know the answer. It's because the voltage V in the battery (12 volts) is too small. We know this from a formula in physics that tells us that the voltage V is the product of the current I and the resistance R; that is, $V = IR$. To find the current I going through your body when you touch both terminals, solve for I by dividing both sides by R to obtain

$$I = \frac{V}{R}$$

A car battery carries 12 volts and $R = 20{,}000$ ohms, so the current is

$$I = \frac{12}{20{,}000} = \frac{6}{10{,}000} = 0.0006 \text{ amp} \qquad \text{Volts divided by ohms yields amperes (amp).}$$

Since it takes about 0.001 ampere to give you a slight shock, your car battery should pose no threat.

A Using Formulas

EXAMPLE 1 Solving problems in anthropology

Anthropologists know how to estimate the height of a man (in centimeters, cm) by using a bone as a clue. To do this, they use the formula

$$H = 2.89h + 70.64$$

where H is the height of the man and h is the length of his humerus.

a. Estimate the height of a man whose humerus bone is 30 centimeters long.

b. Solve for h.

c. If a man is 163.12 centimeters tall, how long is his humerus?

SOLUTION

a. We write 30 in place of h (in parentheses to indicate multiplication):

$$\begin{aligned} H &= 2.89(30) + 70.64 \\ &= 86.70 + 70.64 \\ &= 157.34 \text{ cm} \end{aligned}$$

Thus a man with a 30-centimeter humerus should be about 157 centimeters tall.

b. We circle h to track the variable we are solving for.

$$H = 2.89\textcircled{h} + 70.64 \qquad \text{Given.}$$

$$H - 70.64 = 2.89\textcircled{h} \qquad \text{Subtract 70.64.}$$

$$\frac{H - 70.64}{2.89} = \frac{2.89\textcircled{h}}{2.89} \qquad \text{Divide by 2.89.}$$

Thus

$$\textcircled{h} = \frac{H - 70.64}{2.89}$$

c. This time, we substitute 163.12 for H in the preceding formula to obtain

$$h = \frac{163.12 - 70.64}{2.89} = 32$$

Thus the length of the humerus of a 163.12-centimeter-tall man is 32 centimeters.

PROBLEM 1

The estimated height of a woman (in inches) is given by $H = 2.8h + 28.1$, where h is the length of her humerus (in inches).

a. Estimate the height of a woman whose humerus bone is 15 inches long.

b. Solve for h.

c. If a woman is 61.7 inches tall, how long is her humerus?

Answers

1. a. 70.1 inches **b.** $h = \frac{H - 28.1}{2.8}$ **c.** 12 inches

EXAMPLE 2 Comparing temperatures

The formula for converting degrees Fahrenheit (°F) to degrees Celsius (°C) is

$$C = \frac{5}{9}(F - 32)$$

a. On a hot summer day, the temperature is 95°F. How many degrees Celsius is that?

b. Solve for F.

c. What is F when $C = 10$?

SOLUTION

a. In this case, $F = 95$. So

$$C = \frac{5}{9}(F - 32) = \frac{5}{9}(95 - 32)$$

$$= \frac{5}{9}(63) \qquad \text{Subtract inside the parentheses first.}$$

$$= \frac{5 \cdot 63}{9}$$

$$= 35$$

Thus the temperature is 35°C.

b. We circle F to track the variable we are solving for.

$$C = \frac{5}{9}(\textcircled{F} - 32) \qquad \text{Given.}$$

$$9 \cdot C = 9 \cdot \frac{5}{9}(\textcircled{F} - 32) \qquad \text{Multiply by 9.}$$

$$9C = 5(\textcircled{F} - 32) \qquad \text{Simplify.}$$

$$9C = 5\textcircled{F} - 160 \qquad \text{Use the distributive property.}$$

$$9C + 160 = 5\textcircled{F} - 160 + 160 \qquad \text{Add 160.}$$

$$\frac{9C + 160}{5} = \frac{5\textcircled{F}}{5} \qquad \text{Divide by 5.}$$

Thus

$$\textcircled{F} = \frac{9C + 160}{5}$$

c. Substitute 10 for C:

$$F = \frac{9 \cdot 10 + 160}{5}$$

$$= \frac{250}{5}$$

$$= 50$$

Thus, $F = 50$.

PROBLEM 2

The formula for the speed of an ant is $S = \frac{1}{6}(C - 4)$ centimeters per second, where C is the temperature in degrees Celsius.

a. If the temperature is 22°C, how fast is the ant moving?

b. Solve for C.

c. What is C when $S = 2$ cm/sec?

Teaching Tip

Point out to students that they can simplify $\frac{5 \cdot 63}{9}$ like this

$$\frac{5 \cdot \cancel{63}^{7}}{\cancel{9}_{1}} = 35$$

Do *not* multiply $5 \cdot 63$ first!

Answers

2. a. 3 cm/sec **b.** $C = 6S + 4$ **c.** 16°C

Web It

To learn more about literal equations, go to link 2-6-1 on the Bello Website at mhhe.com/bello.

EXAMPLE 3 **Solving retail problems**

The retail selling price R of an item is obtained by adding the original cost C and the markup M on the item.

a. Write a formula for the retail selling price.

b. Find the markup M of an item that originally cost \$50.

SOLUTION

a. Write the problem in words and then translate it.

The retail selling price	is obtained	by adding the original cost C and the markup M.
R	$=$	$C + M$

b. Here $C = \$50$ and we must solve for M. Substituting 50 for C in $R = C + M$, we have

$$R = 50 + M$$

$$R - 50 = M \qquad \text{Subtract 50.}$$

Thus $M = R - 50$.

PROBLEM 3

The final cost F of an item is obtained by adding the original cost C and the tax T on the item.

a. Write a formula for the final cost F of the item.

b. Find the tax T on an item that originally cost \$10.

B Using Geometric Formulas

Many of the formulas we encounter in algebra come from geometry. For example, to find the *area* of a figure, we must find the number of square units contained in the figure. Unit squares look like these:

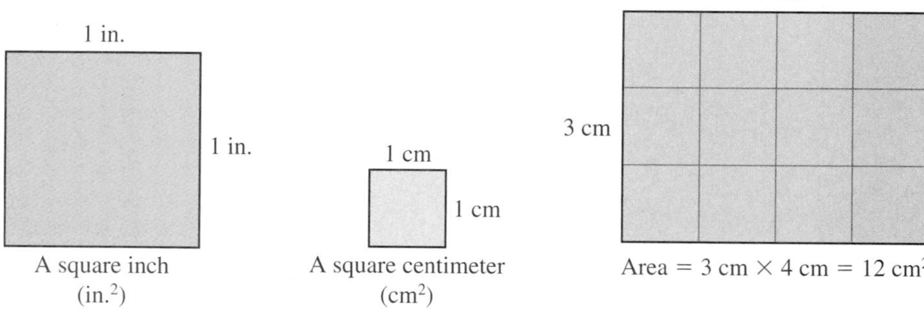

A square inch (in.²)   A square centimeter (cm²)   Area = 3 cm × 4 cm = 12 cm²

Web It

To look at many geometric formulas, go to link 2-6-2 at the Bello Website, mhhe.com/bello.

If you want the computer to calculate areas and volumes, try link 2-6-3.

Now, to find the **area** of a figure, say a rectangle, we must find the number of square units it contains. For example, the area of a rectangle 3 cm by 4 cm is 3 cm × 4 cm = 12 cm² (read "12 square centimeters"), as shown in the diagram.

In general, we can find the area A of a rectangle by multiplying its length L by its width W, as given here.

Answers

3. a. $F = C + T$
b. $T = F - 10$

Area of a Rectangle

The area A of a rectangle	is found	by multiplying its length L by its width W.
A	$=$	LW

What about a rectangle's **perimeter** (distance around)? We can find the perimeter by adding the lengths of the four sides. Since we have two sides of length L and two sides of length W, the perimeter of the rectangle is

$$W + L + W + L = 2L + 2W$$

W

L

In general, we have the following formula.

Perimeter of a Rectangle

The perimeter P of a rectangle of length L and width W is

$$P = 2L + 2W$$

EXAMPLE 4 Finding areas and perimeters

Find:

a. The area of the rectangle shown in the figure.

b. The perimeter of the rectangle shown in the figure.

c. If the perimeter of a rectangle 30 inches long is 110 inches, what is the width of the rectangle?

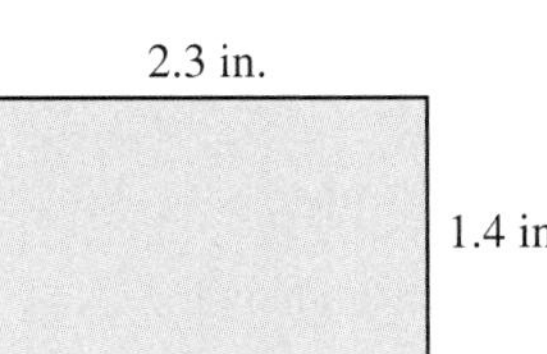

SOLUTION

a. The area:

$$A = LW$$
$$= (2.3\text{ in.}) \cdot (1.4\text{ in.})$$
$$= 3.22\text{ in.}^2$$

b. The perimeter:

$$P = 2L + 2W$$
$$= 2(2.3\text{ in.}) + 2(1.4\text{ in.})$$
$$= 4.6\text{ in.} + 2.8\text{ in.}$$
$$= 7.4\text{ in.}$$

c. The perimeter:

$P = 2L + 2W$

$110\text{ in.} = 2 \cdot (30\text{ in.}) + 2W$ Substitute 30 for L and 110 for P.

$110\text{ in.} = 60\text{ in.} + 2W$ Simplify.

$110\text{ in.} - 60\text{ in.} = 2W$ Subtract 60.

$50\text{ in.} = 2W$

$25\text{ in.} = W$ Divide by 2.

Thus the width of the rectangle is 25 inches.

Note that the area is given in square units, whereas the perimeter is a length and is given in linear units.

PROBLEM 4

a. What is the area of a rectangle 2.4 by 1.2 inches?

b. What is the perimeter of the rectangle in part **a**?

c. If the perimeter of a rectangle 20 inches long is 100 inches, what is the width of the rectangle?

Teaching Tip

Have students solve $P = 2L + 2W$ for W, and then find the value of W by substituting for L and P in the formula.

If we know the area of a rectangle, we can always calculate the area of the shaded triangle:

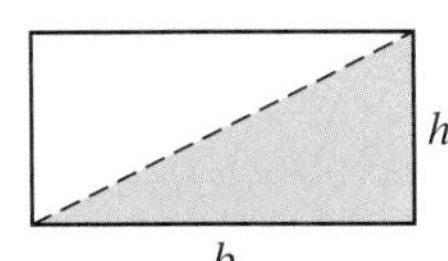

Answers

4. a. 2.88 in.^{2} **b.** 7.2 in.
c. 30 in.

The area of the triangle is $\frac{1}{2}$ the area of the rectangle, which is bh. Thus we have the following formula.

Area of a Triangle

If a triangle has base b and perpendicular height h, its area A is

$$A = \frac{1}{2}bh$$

Note that this time we used b and h instead of L and W. This formula holds true for any type of triangle.

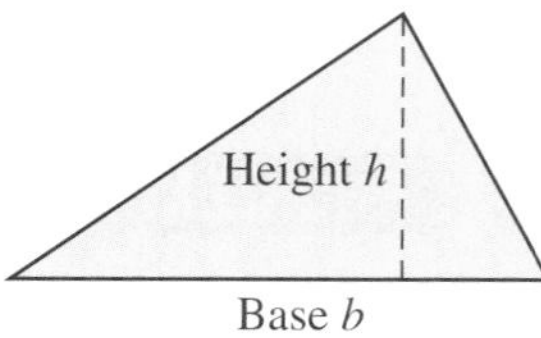

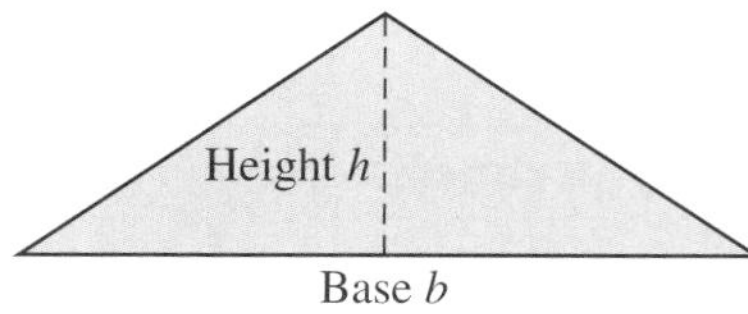

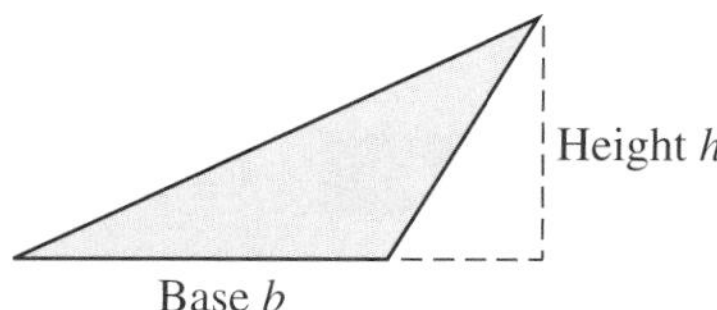

EXAMPLE 5 **Finding the area of a triangle**

a. Find the area of a triangular piece of cloth 20 centimeters long and 10 centimeters high.

b. The area of the triangular sail on a toy boat is 250 square centimeters. If the base of the sail is 20 centimeters long, how high is the sail?

SOLUTION

a. $A = \frac{1}{2}bh$

$= \frac{1}{2}(20\text{ cm}) \cdot (10\text{ cm})$

$= 100\text{ cm}^2$

b. Substituting 250 for A and 20 for b, we obtain

$A = \frac{1}{2}bh$ becomes

$250 = \frac{1}{2} \cdot 20 \cdot h$

$250 = 10h$ Simplify.

$25 = h$ Divide by 10.

Thus the height of the sail is 25 centimeters.

PROBLEM 5

a. Find the area of a triangle 30 inches long and 15 inches high.

b. The area of the sail on a boat is 300 square feet. If the base of the sail is 20 ft long, how high is the sail?

Teaching Tip

Have students solve

$$A = \frac{1}{2}bh$$

for h, and then find the value of h by substituting for A and b in the formula.

The area of a circle can easily be found if we know the **radius** r of the circle, which is the distance from the center of the circle to its edge. Here is the formula.

Area of a Circle

The area A of a circle of radius r is

$$A = \pi \cdot r \cdot r$$
$$= \pi r^2$$

Answers

5. a. 225 in.^{2} **b.** 30 ft

As you can see, the formula for finding the area of a circle involves the number π (read "pie"). The number π is irrational; it cannot be written as a terminating or repeating decimal or a fraction, but it can be *approximated*. In most of our work we shall say that π is about 3.14 or $\frac{22}{7}$; that is, $\pi \approx 3.14$ or $\pi \approx \frac{22}{7}$. The number π is also used in finding the perimeter (distance around) a circle. This perimeter of the circle is called the **circumference** C and is found by using the following formula.

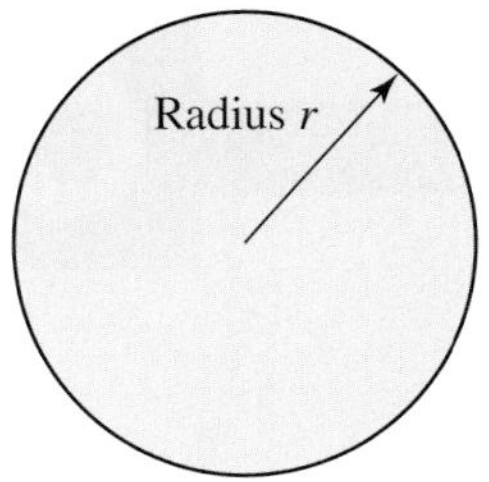

The area A of a circle of radius r is
$A = \pi \cdot r \cdot r = \pi r^2$

Circumference of a Circle

The circumference C of a circle of radius r is

$$C = 2\pi r$$

Since the radius of a circle is half the diameter, another formula for the circumference of a circle is $C = \pi d$, where d is the diameter of the circle.

EXAMPLE 6 A CD's recorded area and circumference

The circumference C of a CD is 12π centimeters.

a. What is the radius r of the CD?

b. Every CD has a circular region in the center with no grooves, as can be seen in the figure. If the radius of this circular region is 2 centimeters and the rest of the CD has grooves, what is the area of the grooved (recorded) region?

SOLUTION

a. The circumference of a circle is

$C = 2\pi r$

$12\pi = 2\pi r$ Substitute 12π for C.

$6 = r$ Divide by 2π.

Thus the radius of the CD is 6 centimeters.

b. To find the grooved (shaded) area, we find the area of the entire CD and subtract the area of the ungrooved (white) region. The area of the entire CD is

$A = \pi r^2$

$A = \pi(6)^2 = 36\pi$ Substitute $r = 6$.

The area of the white region is

$A = \pi(2)^2 = 4\pi$

Thus, the area of the shaded region is $36\pi - 4\pi = 32\pi$ square centimeters.

PROBLEM 6

The circumference C of a CD is 9π inches.

a. What is the radius r of the CD?

b. The radius of the nonrecorded part of this CD is 0.75 in. What is the area of the recorded region?

Answers

6. a. 4.5 in. **b.** 19.6875π in.^{2}

C Solving for Angle Measurements

We've already mentioned complementary angles (two angles whose sum is 90°) and supplementary angles (two angles whose sum is 180°). Now we introduce the idea of **vertical angles.** The figure shows two intersecting lines with angles numbered ①, ❷, ③, and ❹.

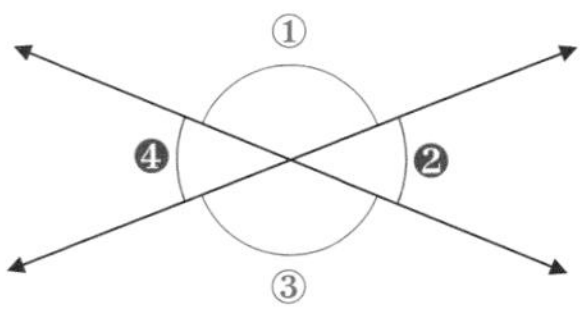

Angles ① and ③ are placed "vertically"; they are called *vertical angles.* Another pair of vertical angles is ❷ and ❹.

VERTICAL ANGLES

Vertical angles have equal measures.

Note that if you add the measures of angles ① and ❷, you get 180°, a **straight angle.** Similarly, the sums of the measures of angles ❷ and ③, ③ and ❹, and ❹ and ① yield straight angles.

EXAMPLE 7 Finding the measures of angles

Find the measures of the marked angles in each figure:

a.

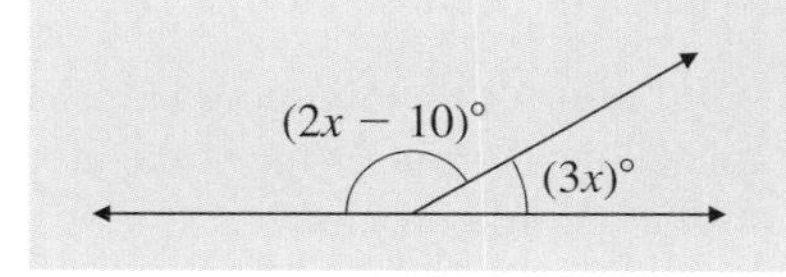

b.

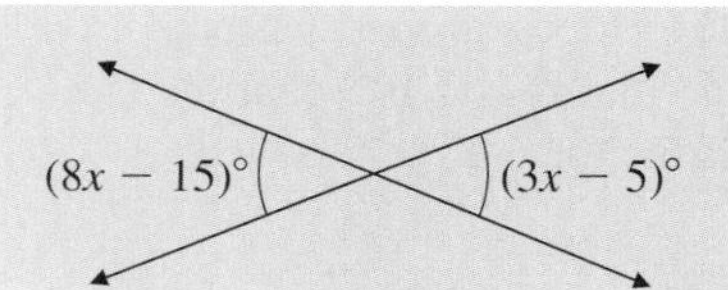

c.

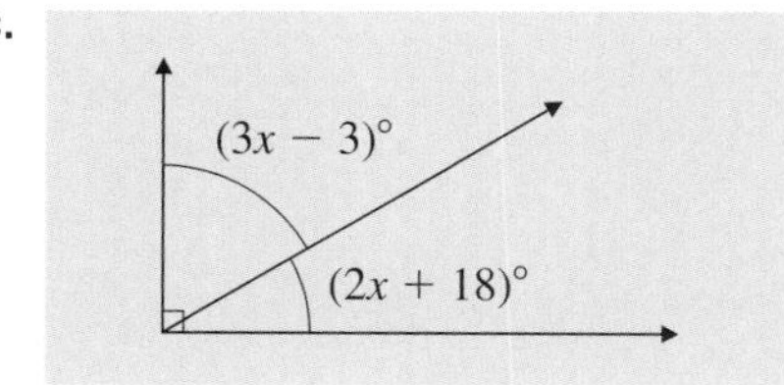

SOLUTION

a. The measure of the sum of the two marked angles must be 180°, since their sum is a straight angle. Thus

$$\begin{aligned} (2x - 10) + 3x &= 180 \\ 5x - 10 &= 180 && \text{Simplify.} \\ 5x &= 190 && \text{Add 10.} \\ x &= 38 && \text{Divide by 5.} \end{aligned}$$

To find the measure of each of the angles, replace x with 38 in $2x - 10$ and in $3x$ to obtain

$$2(38) - 10 = 66 \quad \text{and} \quad 3(38) = 114$$

Thus the measures are 66° and 114°, respectively. Note that $114 + 66 = 180$, so our result is correct.

PROBLEM 7

a. Find the measures if the angle in the diagram for part **a** is $(8x)°$ instead of $(3x)°$.

b. Find the measures if the angle in the diagram for part **b** is $(4x - 3)°$ instead of $(3x - 5)°$.

c. Find the measures if the angle in the diagram for part **c** is $(x + 17)°$ instead of $(2x + 18)°$.

Answers

7. a. 28° and 152°
b. Both are 9°. **c.** 54° and 36°

b. The two marked angles are vertical angles, so their measures must be equal. Thus

$$\begin{aligned} 8x - 15 &= 3x - 5 \\ 8x &= 3x + 10 && \text{Add 15.} \\ 5x &= 10 && \text{Subtract } 3x. \\ x &= 2 && \text{Divide by 5.} \end{aligned}$$

Now, replace x with 2 in $8x - 15$ to obtain $8 \cdot 2 - 15 = 1$. Since the angles are vertical, their measures are both 1°.

c. The measure of the sum of the two marked angles must be 90°, since the angles are complementary. Thus

$$\begin{aligned} (3x - 3) + (2x + 18) &= 90 \\ 5x + 15 &= 90 && \text{Simplify.} \\ 5x &= 75 && \text{Subtract 15.} \\ x &= 15 && \text{Divide by 5.} \end{aligned}$$

Replacing x with 15 in $(3x - 3)$ and $(2x + 18)$, we obtain $3 \cdot 15 - 3 = 42$ and $2 \cdot 15 + 18 = 48$. Thus the measures of the angles are 42° and 48°, respectively. Note that the sum of the measures of the two angles is 90°, as expected.

Solving Applications

Many formulas find uses in our daily lives. For example, do you know the *exact* relationship between your shoe size S and the length L of your foot in inches? Here are the formulas used in the United States:

$$L = \frac{22 + S}{3} \qquad \text{For men}$$

$$L = \frac{21 + S}{3} \qquad \text{For women}$$

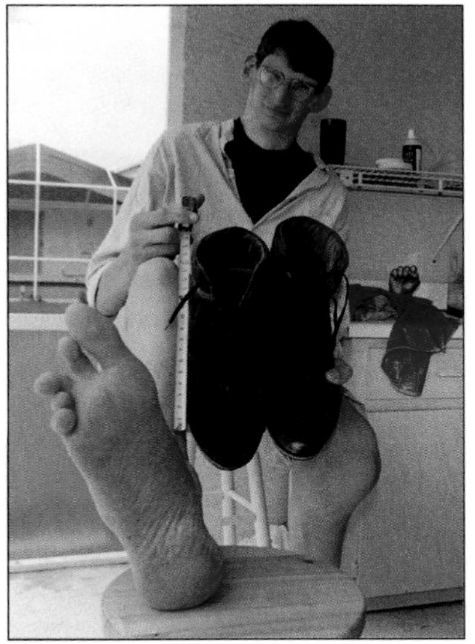

Teaching Tip

Have the students use the formula to estimate the length of their feet when they substitute their shoe size S in the formula.

EXAMPLE 8 Sizing shoes

Use the shoe sizing formulas to find:

a. The length of the foot corresponding to a size 1 shoe for men.

b. The length of the foot corresponding to a size 1 shoe for women.

c. Solve for S and determine the size shoe needed by Matthew McGrory (the man with the largest feet), whose left foot is 18 inches long.

PROBLEM 8

a. Find the length of the foot corresponding to a size 2 shoe for men.

b. Find the length of the foot corresponding to a size 3 shoe for women.

c. McGrory's "smaller" right foot is 17 inches long. What shoe size fits his right foot?

Answers

8. a. 8 in. **b.** 8 in. **c.** 29

SOLUTION

a. To find the length of the foot corresponding to a size 1 shoe for men, substitute 1 for S in

$$L = \frac{22 + S}{3}$$
$$= \frac{22 + 1}{3}$$
$$= \frac{23}{3} = 7\frac{2}{3}\text{ in.}$$

b. This time we substitute 1 for S in

$$L = \frac{21 + S}{3}$$
$$L = \frac{21 + 1}{3} = \frac{22}{3} = 7\frac{1}{3}\text{ in.}$$

c.

$L = \frac{22 + S}{3}$	Given.
$3L = 3 \cdot \frac{22 + S}{3}$	Multiply by the LCM 3.
$3L = 22 + S$	Simplify.
$3L - 22 = S$	Subtract 22.

Since the length of McGrory's foot is 18 in., substitute 18 for L to obtain

$$S = 3 \cdot 18 - 22 = 32$$

Thus McGrory needs a size 32 shoe!

Exercises 2.6

A In Problems 1–10, use the formulas to find the solution.

1. The number of miles D traveled in T hours by an object moving at a rate R (in miles per hour) is given by $D = RT$.

a. Find D when $R = 30$ and $T = 4$. 120

b. Find the distance traveled by a car going 55 miles per hour for 5 hours. 275 mi

c. Solve for R in $D = RT$. $R = \frac{D}{T}$

d. If you travel 180 miles in 3 hours, what is R? 60 mi/hr

2. The rate of travel R of an object moving a distance D in time T is given by

$$R = \frac{D}{T}$$

a. Find R when $D = 240$ miles and $T = 4$ hours. 60 mi/hr

b. Find the rate of travel of a train that traveled 140 miles in 4 hours. 35 mi/hr

c. Solve for T in $R = \frac{D}{T}$. $T = \frac{D}{R}$

d. How long would it take the train of part **b** to travel 105 miles? 3 hr

3. The height H of a man (in inches) is related to his weight W (in pounds) by the formula $W = 5H - 190$.

a. If a man is 60 inches tall, what should his weight be? 110 pounds

b. Solve for H. $H = \frac{W + 190}{5}$

c. If a man weighs 200 pounds, how tall should he be? 78 inches tall (6 ft, 6 in.)

4. The number of hours H a growing child should sleep is

$$H = 17 - \frac{A}{2}$$

where A is the age of the child in years.

a. How many hours should a 6-year-old sleep? 14 hr

b. Solve for A in $H = 17 - \frac{A}{2}$. $A = 34 - 2H$

c. At what age would you expect a child to sleep 11 hours? 12 years of age

5. The formula for converting degrees Celsius to degrees Fahrenheit is

$$F = \frac{9}{5}C + 32$$

a. If the temperature is 15°C, what is the corresponding Fahrenheit temperature? 59°F

b. Solve for C. $C = \frac{5}{9}(F - 32)$

c. What is the corresponding Celsius temperature when $F = 50$? 10°C

6. The profit P a business makes is obtained by subtracting the expenses E from the income I.

a. Write a formula for the profit P. $P = I - E$

b. Find the profit of a business with \$2700 in income and \$347 in expenses. \$2353

c. Solve for E in the formula you wrote in part **a**. $E = I - P$

d. If the profit is \$750 and the income is \$1300, what are the expenses? \$550

7. The energy efficiency ratio (EER) for an air conditioner is obtained by dividing the British thermal units (Btu) per hour by the watts w.

a. Write a formula that will give the EER of an air conditioner. $\text{EER} = \frac{\text{Btu}}{w}$

b. Find the EER of an air conditioner with a capacity of 9000 British thermal units per hour and a rating of 1000 watts. 9

c. Solve for the British thermal units in your formula for EER. $\text{Btu} = (\text{EER})(w)$

d. How many British thermal units does a 2000-watt air conditioner produce if its EER is 10? 20,000

8. The capital C of a business is the difference between the assets A and the liabilities L.

a. Write a formula that will give the capital of a business. $C = A - L$

b. If a business has \$4800 in assets and \$2300 in liabilities, what is the capital of the business? \$2500

c. Solve for L in your formula for C. $L = A - C$

d. If a business has \$18,200 in capital and \$30,000 in assets, what are its liabilities? \$11,800

9. The selling price S of an item is the sum of the cost C and the margin (markup) M.

a. Write a formula for the selling price of a given item. $S = C + M$

b. A merchant wishes to have a \$15 margin (markup) on an item costing \$52. What should be the selling price of this item? \$67

c. Solve for M in your formula for S. $M = S - C$

d. If the selling price of an item is \$18.75 and its cost is \$10.50, what is the markup? \$8.25

10. The tip speed S_T of a propeller is equal to π times the diameter d of the propeller times the number N of revolutions per second.

a. Write a formula for S_T. $S_T = \pi dN$

b. If a propeller has a 2-meter diameter and it is turning at 100 revolutions per second, find S_T. (Use $\pi \approx 3.14$.) 628 m/sec

c. Solve for N in your formula for S_T. $N = \frac{S_T}{\pi d}$

d. What is N when $S_T = 275\pi$ and $d = 2$? 137.5 rev/sec

B In Problems 11–15 use the geometric formulas to find the solution.

11. The perimeter P of a rectangle of length L and width W is $P = 2L + 2W$.

a. Find the perimeter of a rectangle 10 centimeters by 20 centimeters. 60 cm

b. What is the length of a rectangle with a perimeter of 220 centimeters and a width of 20 centimeters? 90 cm

12. The perimeter P of a rectangle of length L and width W is $P = 2L + 2W$.

a. Find the perimeter of a rectangle 15 centimeters by 30 centimeters. 90 cm

b. What is the width of a rectangle with a perimeter of 180 centimeters and a length of 60 centimeters? 30 cm

13. The circumference C of a circle of radius r is $C = 2\pi r$.

a. Find the circumference of a circle with a radius of 10 inches. (Use $\pi \approx 3.14$.) 62.8 in.

b. Solve for r in $C = 2\pi r$. $r = \frac{C}{2\pi}$

c. What is the radius of a circle whose circumference is 20π inches? 10 in.

14. If the circumference C of a circle of radius r is $C = 2\pi r$, what is the radius of a tire whose circumference is 26π inches? 13 in.

15. The area A of a rectangle of length L and width W is $A = LW$.

a. Find the area of a rectangle 4.2 meters by 3.1 meters. 13.02 m^2

b. Solve for W in the formula $A = LW$. $W = \frac{A}{L}$

c. If the area of a rectangle is 60 square meters and its length is 10 meters, what is the width of the rectangle? 6 m

C In Problems 16–29, find the measure of each marked angle.

16. 85° each

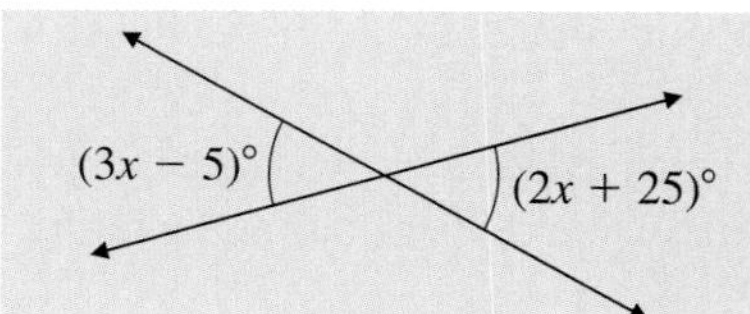

17. 35° each

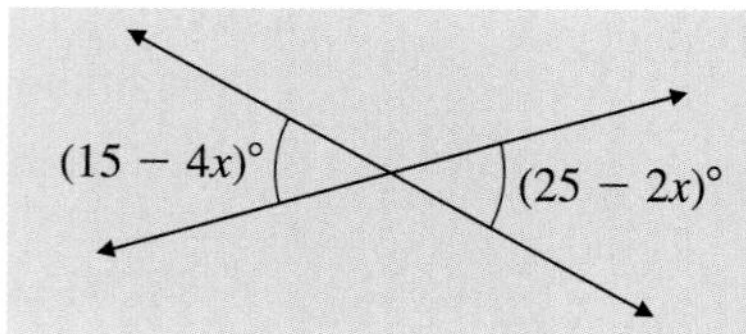

18. 140° each

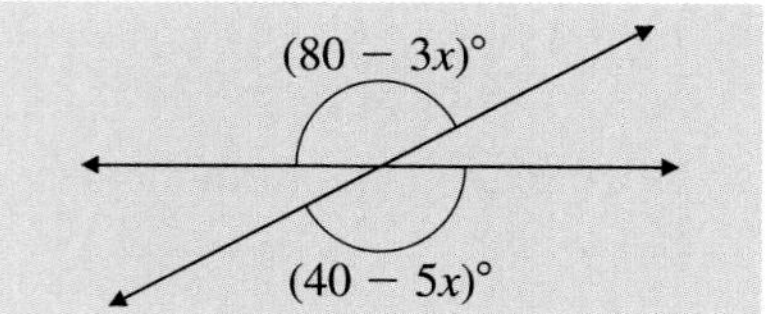

19. 140° each

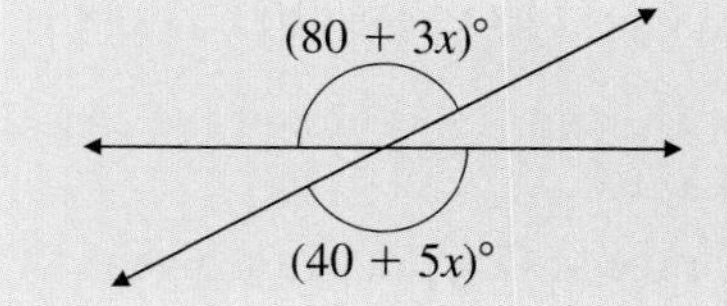

20. 37° each

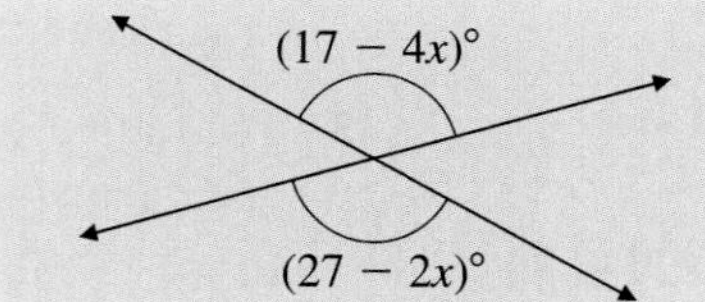

21. 175° each

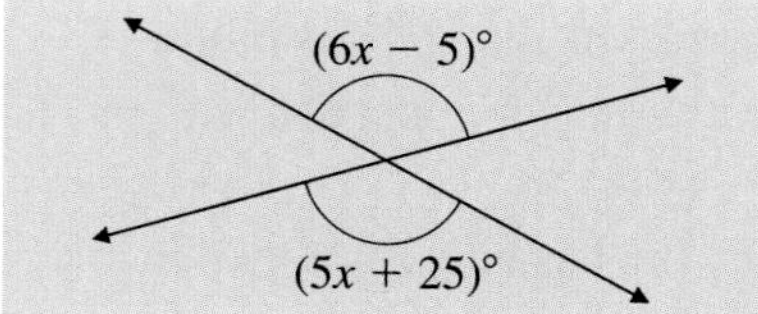

22. 40°, 50°

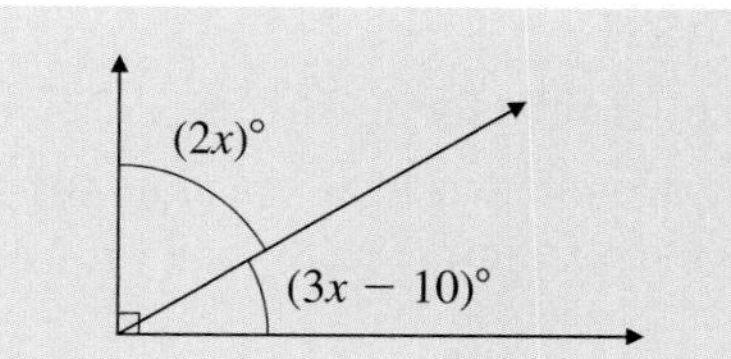

23. 70°, 20°

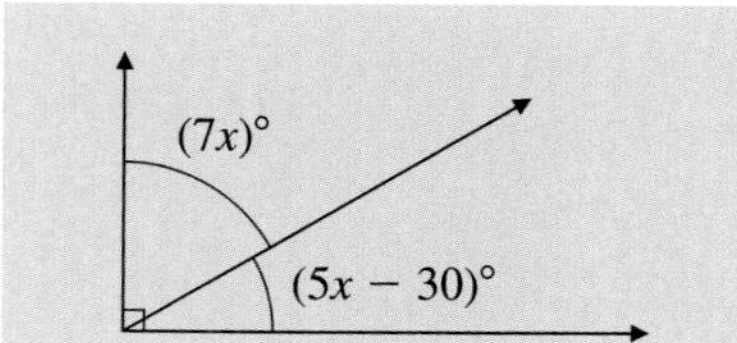

24. 27°, 63°

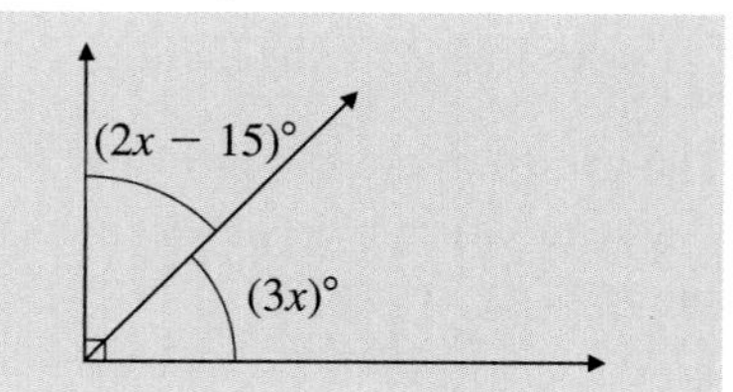

25. 23°, 67°

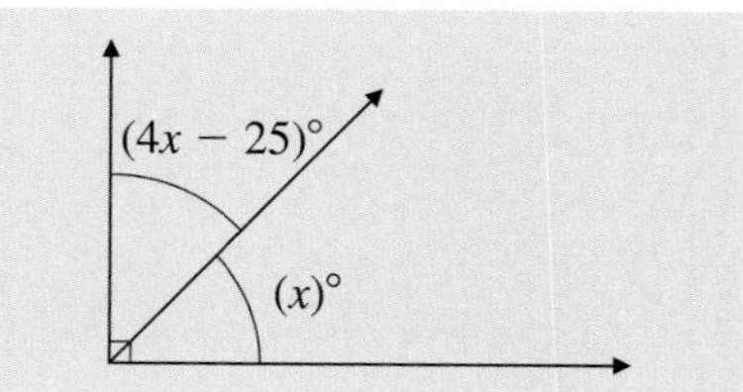

26. 68°, 112°

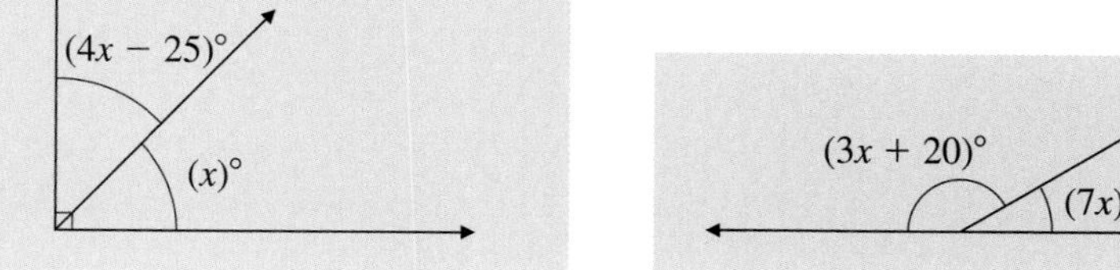

27. 142°, 38°

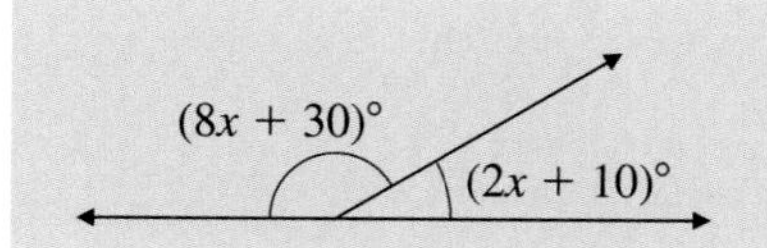

28. 69°, 111°

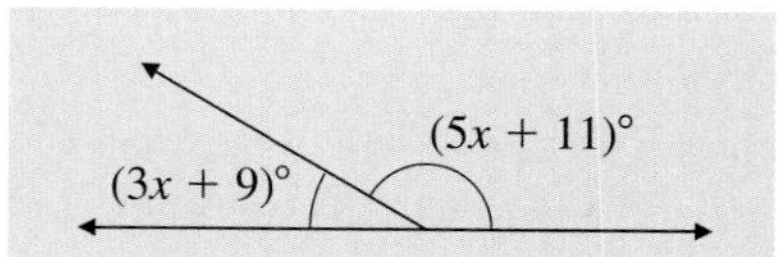

29. 69°, 111°

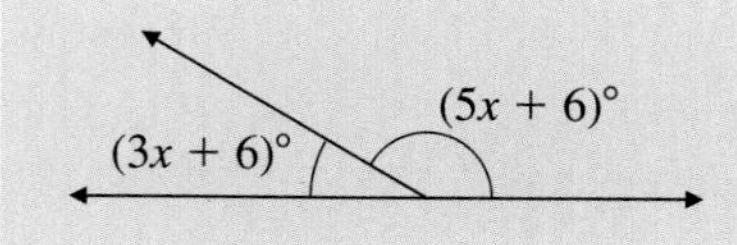

APPLICATIONS

30. *Supersized omelet* One of the largest rectangular omelets ever cooked was 30 feet long and had an 80-foot perimeter. How wide was it? 10 ft

31. *Largest pool* If you were to walk around the largest rectangular pool in the world, in Casablanca, Morocco, you would walk more than 1 kilometer. To be exact, you would walk 1110 meters. If the pool is 480 meters long, how wide is it? 75 m

32. *Football field dimensions* The playing surface of a football field is 120 yards long. A player jogging around the perimeter of this surface jogs 346 yards. How wide is the playing surface of a football field? 53 yd

33. *CD diameter* A point on the rim of a CD record travels 14.13 inches each revolution. What is the diameter of this CD? (Use $\pi \approx 3.14$.) 4.5 in.

34. *Gigantic pizza* One of the largest pizzas ever made had a 251.2-foot circumference! What was its diameter? (Use $\pi \approx 3.14$.) 80 ft

35. *Continental suit sizes* Did you know that clothes are sized differently in different countries? If you want to buy a suit in Europe and you wear a size A in America, your continental (European) size C will be $C = A + 10$.

a. Solve for A. $A = C - 10$

b. If you wear a continental size 50 suit, what would be your American size? 40

36. *Continental dress sizes* Your continental dress size C is given by $C = A + 30$, where A is your American dress size.

a. Solve for A. $A = C - 30$

b. If you wear a continental size 42 dress, what is your corresponding American size? 12

37. *Recreational boats data* According to the National Marine Manufacturers Association, the number N of recreational boats (in millions) has been steadily increasing since 1975 and is given by $N = 9.74 + 0.40t$, where t is the number of years after 1975.

a. What was the number of recreational boats in 1985? 13.74 million

b. Solve for t in $N = 9.74 + 0.40t$. $t = \frac{N - 9.74}{0.40}$ or $t = \frac{5(N - 9.74)}{2}$

c. In what year would you expect the number of recreational boats to reach 17.74 million? 1995

38. *NCAA men's basketball teams* The number N of NCAA men's college basketball teams has been increasing since 1980 according to the formula $N = 720 + 5t$, where t is the number of years after 1980.

a. How many teams would you expect in the year 2000? 820

b. Solve for t. $t = \frac{N - 720}{5}$

c. In what year would you expect the number of teams to reach 795? 1995

SKILL CHECKER

Try the Skill Checker Exercises so you'll be ready for the next section.

Solve:

39. $2x - 1 = x + 3$ $x = 4$

40. $3x - 2 = 2(x - 1)$ $x = 0$

41. $4(x + 1) = 3x + 7$ $x = 3$

42. $\frac{-x}{4} + \frac{x}{6} = \frac{x - 3}{6}$ $x = 2$

43. $\frac{x}{3} - \frac{x}{2} = 1$ $x = -6$

USING YOUR KNOWLEDGE

Living with Algebra

Many practical problems around the house require some knowledge of the formulas we've studied. For example, let's say that you wish to carpet your living room. You need to use the formula for the area A of a rectangle of length L and width W, which is $A = LW$.

44. Carpet sells for \$8 per square yard. If the cost of carpeting a room was \$320 and the room is 20 feet long, how wide is it? (Note that one square yard is nine square feet.) 18 ft

45. If you wish to plant new grass in your yard, you can buy sod squares of grass that can simply be laid on the ground. Each sod square is approximately 1 square foot. If you have 5400 sod squares and you wish to sod an area that is 60 feet wide, how long can the area be? 90 ft

46. If you wish to fence your yard, you need to know its perimeter (the distance around the yard). If the yard is W feet by L feet, the perimeter P is given by $P = 2W + 2L$. If your rectangular yard needs 240 feet of fencing and your yard is 70 feet long, how wide is it? 50 ft

WRITE ON

47. Write an explanation of what is meant by the *perimeter* of a geometric figure. Answers may vary.

48. Write an explanation of what is meant by the *area* of a geometric figure. Answers may vary.

49. Remember the hamburgers in the *Getting Started?* Write two explanations of how the Monster burger can have 75% more beef than the Lite burger and still look like the one in the picture. Answers may vary.

50. To make a fair comparison of the amount of beef in two hamburgers, should you compare the circumferences, areas, or volumes? Explain. Volumes; answers may vary.

51. The Lite burger has a 4-inch diameter whereas the Monster burger has a 6-inch diameter. How much bigger (in percent) is the circumference of the Monster burger? Can you now explain the claim in the ad? Is the claim correct? Explain. 50%; answers may vary.

MASTERY TEST

If you know how to do these problems, you have learned your lesson!

Find the measures of the marked angles:

52. 48°, 42°

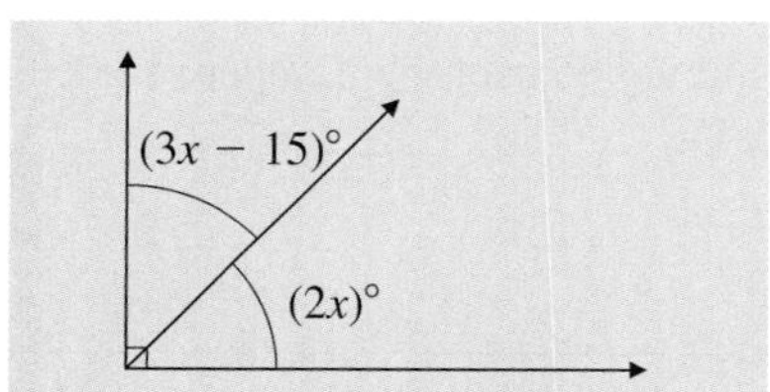

53. 44°, 46°

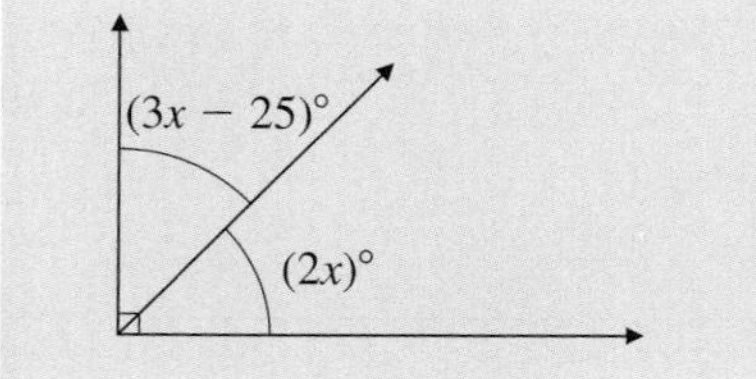

54. 96°, 84°

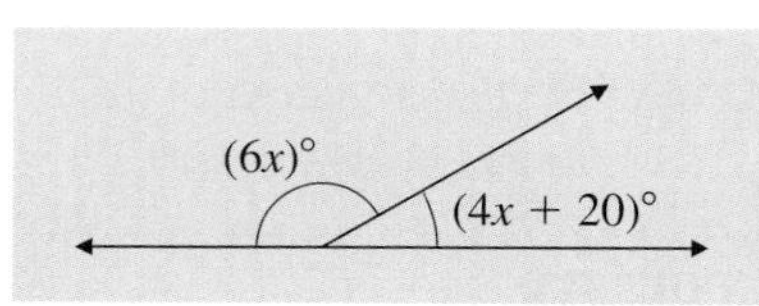

55. 128°, 52°

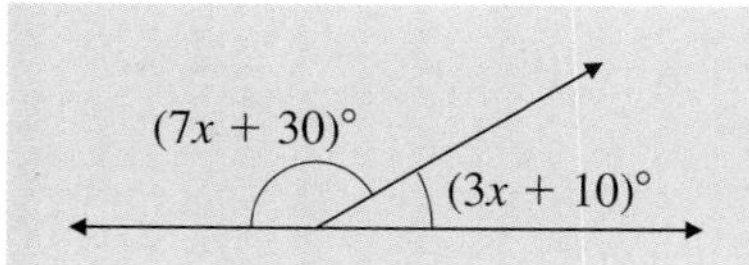

56. 89°, 91°

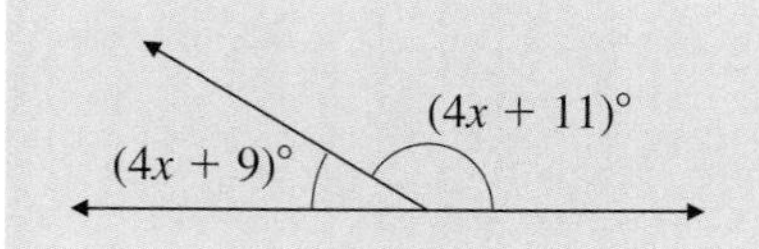

57. 46°, 134°

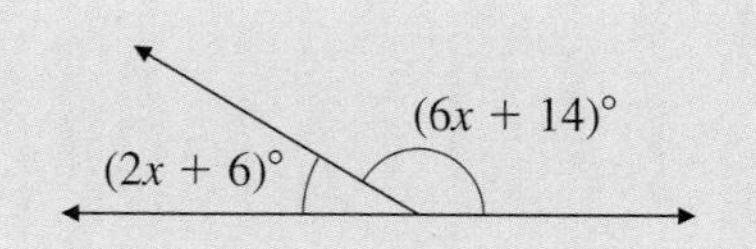

58. A circle has a radius of 10 inches.

a. Find its area and its circumference. $A = 100\pi$ in.^{2} or 314 in.^{2}, $C = 20\pi$ in. or 62.8 in.

b. If the circumference of a circle is 40π inches, what is its radius? 20 in.

59. The formula for estimating the height H of a woman using the length h of her humerus as a clue is $H = 2.75h + 71.48$.

a. Estimate the height of a woman whose humerus bone is 20 centimeters long. 126.48 cm

b. Solve for h. $h = \frac{H - 71.48}{2.75}$

c. If a woman is 140.23 centimeters tall, how long is her humerus? 25 cm

60. The total cost T of an item is obtained by adding its cost C and the tax t on the item.

a. Write a formula for the total cost T. $T = C + t$

b. Find the tax t on an item that cost \$8 if the total after adding the tax is \$8.48. \$0.48

61. According to the Motion Picture Association of America, the number N of motion picture theaters (in thousands) has been growing according to the formula $N = 15 + 0.60t$, where t is the number of years after 1975.

a. How many theaters were there in 1985? (*Hint:* $t = 10$.) 21 thousand

b. Solve for t in $N = 15 + 0.60t$. $t = \frac{N - 15}{0.60}$ or $t = \frac{5(N - 15)}{3}$

c. In what year did the number of theaters total 27,000? 1995

62. The area A of a triangle is $A = \frac{1}{2}bh$, where b is the base of the triangle and h is its height.

a. What is the area of a triangle 15 inches long and with a 10-inch base? 75 in.^{2}

b. Solve for b in $A = \frac{1}{2}bh$. $b = \frac{2A}{h}$

c. The area of a triangle is 18 square inches, and its height is 9 inches. How long is the base of the triangle? 4 in.

63. The formula for converting degrees Fahrenheit F to degrees Celsius C is

$$C = \frac{5}{9}F - \frac{160}{9}$$

a. Find the Celsius temperature on a day in which the thermometer reads 41°F. 5°C

b. Solve for F. $F = \frac{9}{5}C + 32$

c. What is F when $C = 20$? 68°F

2.7 PROPERTIES OF INEQUALITIES

To Succeed, Review How To . . .

1. Add, subtract, multiply, and divide real numbers (pp. 61–66, 70–76).
2. Solve linear equations (pp. 150–155).

Objectives

A Determine which of two numbers is greater.

B Solve and graph linear inequalities.

C Write, solve, and graph compound inequalities.

D Solve an application.

GETTING STARTED

Savings on Sandals

After learning to solve linear equations, we need to learn to solve *linear inequalities.* Fortunately, the rules are very similar, but the notation is a little different. For example, the ad says that the price you will pay for these sandals will be cut \$1 to \$3. That is, you will save \$1 to \$3. If x is the amount of money you can save, what can x be? Well, it is at least \$1 and can be as much as \$3; that is, $x = 1$ or $x = 3$ or x is between 1 and 3. In the language of algebra, we write this fact as

$$1 \le x \le 3$$ Read "1 is less than or equal to x and x is less than or equal to 3" or "x is between 1 and 3, inclusive."

Women's Sandals on SALE!
CUT $1 to $3
Macrame & Canvas

The statement $1 \le x \le 3$ is made up of two parts:

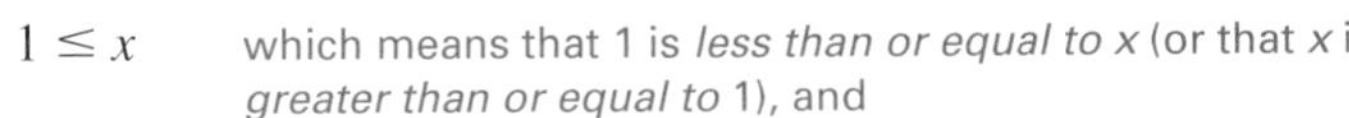

$1 \le x$ which means that 1 is *less than or equal to* x (or that x is *greater than or equal to* 1), and

$x \le 3$ which means that x is *less than or equal to* 3 (or that 3 is *greater than or equal to* x).

These statements are examples of *inequalities,* which we will learn how to solve in this section.

In algebra, an **inequality** is a statement with $>$, $<$, $\geq$, or $\leq$ as its verb. Inequalities can be represented on a number line. Here's how we do it. As you recall, a number line is constructed by drawing a line, selecting a point on this line, and calling it zero (the origin):

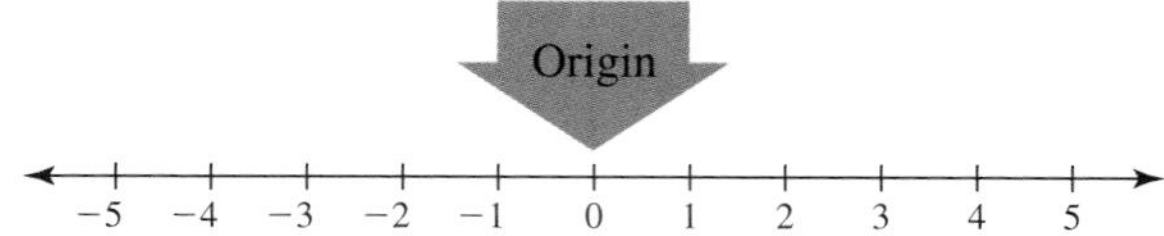

We then locate equally spaced points to the right of the origin on the line and label them with the *positive* integers 1, 2, 3, and so on. The corresponding points to the left of zero are labeled -1, -2, -3, and so on (the *negative* integers). This construction allows the *association of numbers with points on the line*. The number associated with a point is called the **coordinate** of that point. For example, there are points associated with the numbers -2.5, $-1\frac{1}{2}$, $\frac{3}{4}$, and 2.5:

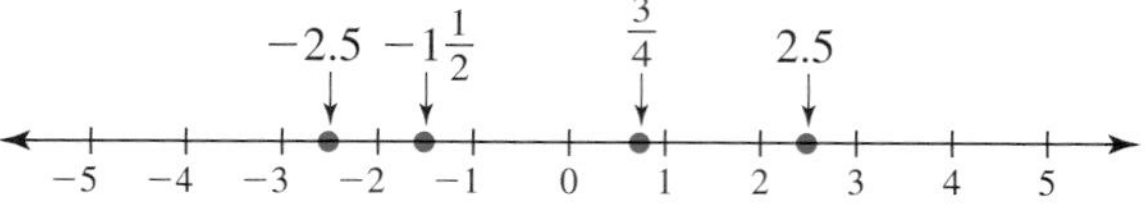

All these numbers are real numbers. As we have mentioned, the real numbers include natural (counting) numbers, whole numbers, integers, fractions, and decimals as well as the irrational numbers (which we discuss in more detail later). Thus the real numbers can all be represented on the number line.

A Order

As you can see, numbers are placed in order on the number line. *Greater* numbers are always to the *right* of *smaller* ones. (The farther to the *right*, the *greater* the number.) Thus any number to the *right* of a second number is said to be **greater than** ($>$) the second number. We also say that the second number is **less than** ($<$) the first number. For example, since 3 is to the right of 1, we write

$$\underbrace{3}_{3}\ \underbrace{\text{is greater than}}_{>}\ \underbrace{1.}_{1} \quad \text{or} \quad \underbrace{1}_{1}\ \underbrace{\text{is less than}}_{<}\ \underbrace{3.}_{3}$$

Similarly,

$$-1 > -3 \quad \text{or} \quad -3 < -1$$
$$0 > -2 \quad \text{or} \quad -2 < 0$$
$$3 > -1 \quad \text{or} \quad -1 < 3$$

Note that the inequality signs $>$ and $<$ always point to the smaller number.

Web It

To learn how to write inequalities, go to link 2-7-1 at the Bello Website at mhhe.com/bello.

Teaching Tip

Students should know that when comparing 3 to 1 there are two appropriate inequality symbols:

$3 > 1$ 3 is greater than 1
$3 \geq 1$ 3 is greater than or equal to 1

EXAMPLE 1 Writing inequalities

Fill in the blank with $>$ or $<$ so that the resulting statement is true.

a. 3 ______ 4 **b.** -4 ______ -3 **c.** -2 ______ -3

SOLUTION We first construct a number line containing these numbers. (Of course, we could just think about the number line without actually drawing one.)

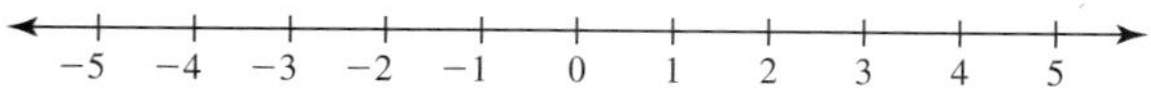

a. Since 3 is to the left of 4, $3 < 4$.

b. Since -4 is to the left of -3, $-4 < -3$.

c. Since -2 is to the right of -3, $-2 > -3$.

PROBLEM 1

Fill in the blank with $>$ or $<$ so that the resulting statement is true.

a. 5 ______ 3 **b.** -1 ______ -4

c. -5 ______ -4

Answers

1. a. $>$ **b.** $>$ **c.** $<$

B Solving and Graphing Inequalities

Just as we solved equations, we can also solve inequalities. We do this by extending the addition and multiplication properties of equality (Sections 2.1 and 2.2) to include inequalities. We say that we have *solved* a given inequality when we obtain an inequality equivalent to the one given and in the form $x < \square$ or $x > \square$. For example, consider the inequality

$$x < 3$$

$x < 3$
$2 < 3$
$1\frac{1}{2} < 3$
$1 < 3$
$0 < 3$
$-\frac{1}{2} < 3$

There are many real numbers that will make this inequality a true statement. A few of them are shown in the accompanying table. Thus 2 is a solution of $x < 3$ because $2 < 3$. Similarly, $-\frac{1}{2}$ is a solution of $x < 3$ because $-\frac{1}{2} < 3$.

As you can see from this table, 2, $1\frac{1}{2}$, 1, 0, and $-\frac{1}{2}$ are *solutions* of the inequality $x < 3$. Of course, we can't list all the real numbers that satisfy the inequality $x < 3$ because there are infinitely many of them, but we can certainly show all the solutions of $x < 3$ *graphically* by using the number line:

$x < 3$

−5 −4 −3 −2 −1 0 1 2 3 4 5

Teaching Tip

The graph of a solution set is a "*picture*" of the solutions to a mathematical statement.

> **NOTE**
>
> $x < 3$ tells you to draw your heavy line to the *left* of 3 because the symbol $<$ points left.

This representation is called the **graph** of the solutions of $x < 3$, which are indicated by the heavy line. Note that there is an open circle at $x = 3$ to indicate that 3 is *not* part of the graph of $x < 3$ (since 3 is not less than 3). Also, the colored arrowhead points to the left (just as the $<$ in $x < 3$ points to the left) to indicate that the heavy line continues to the left without end. On the other hand, the graph of $x \geq 2$ should continue to the right without end:

Teaching Tip

Explain to students that sometimes (or) are used instead of ○. Also, [or] are sometimes used instead of ●.

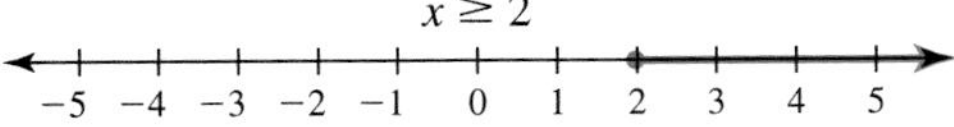

> **NOTE**
>
> $x \geq 2$ tells you to draw your heavy line to the *right* of 2 because the symbol $\geq$ points right.

Moreover, since $x = 2$ is included in the graph, a solid dot appears at the point $x = 2$.

EXAMPLE 2 Graphing inequalities

Graph the inequality on a number line.

a. $x \geq -1$ **b.** $x < -2$

SOLUTION

a. The numbers that satisfy the inequality $x \geq -1$ are the numbers that are *greater than or equal to* -1, that is, the number -1 and all the numbers to

PROBLEM 2

Graph the inequality on a number line.

a. $x \leq -2$ **b.** $x > -3$

Answers

2. a.

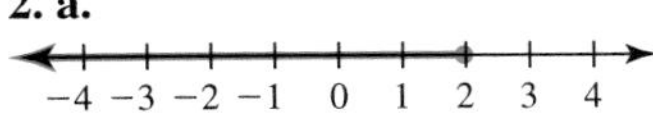

b.

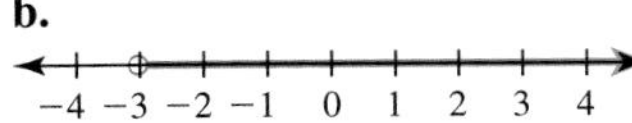

the right of -1 (remember, $\geq$ points to the right and the dot must be solid). The graph is shown here:

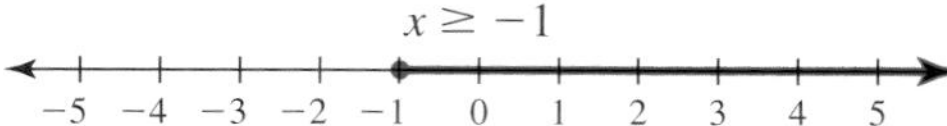

b. The numbers that satisfy the inequality $x < -2$ are the numbers that are *less than* -2, that is, the numbers to the *left of but not including* -2 (note that $<$ points to the left and that the dot is open). The graph of these points is shown here:

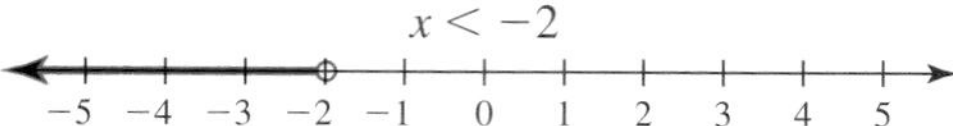

We solve more complicated inequalities just as we solve equations, by finding an equivalent inequality whose solution is obvious. Remember, we have solved a given inequality when we obtain an inequality in the form $x < \square$ or $x > \square$ which is equivalent to the one given. Thus to solve the inequality $2x - 1 < x + 3$, we try to find an equivalent inequality of the form $x < \square$ or $x > \square$. As before, we need some properties. The first of these are the addition and subtraction properties.

If $3 < 4$, then

$$3 + 5 < 4 + 5 \qquad \text{Add 5.}$$

$$8 < 9 \qquad \text{True.}$$

Similarly, if $3 > -2$, then

$$3 + 7 > -2 + 7 \qquad \text{Add 7.}$$

$$10 > 5 \qquad \text{True.}$$

Also, if $3 < 4$, then

$$3 - 1 < 4 - 1 \qquad \text{Subtract 1.}$$

$$2 < 3 \qquad \text{True.}$$

Similarly, if $3 > -2$, then

$$3 - 5 > -2 - 5 \qquad \text{Subtract 5.}$$

$$-2 > -7 \qquad \text{True because } -2 \text{ is to the right of } -7.$$

In general, we have the following properties.

Web It

To study the addition and subtraction properties, as well as the multiplication and division properties, with examples, try link 2-7-2 on the Bello Website at mhhe.com/bello.

Addition and Subtraction Properties of Inequalities

You can *add* or *subtract* the same number a on both sides of an inequality and obtain an equivalent inequality. In symbols,

If	$x < y$	If	$x > y$
then	$x + a < y + a$	then	$x + a > y + a$
or	$x - b < y - b$	or	$x - b > y - b$

Although we use $<$ and $>$ to present the properties in this section, you should understand that the properties also hold when the symbols $\leq$ and $\geq$ are used.

NOTE

Since $x - b = x + (-b)$, subtracting b from both sides is the same as adding the inverse of b, $(-b)$, so you can think of subtracting b as adding $(-b)$.

Now let's return to the inequality $2x - 1 < x + 3$. To solve this inequality, we need the variables by themselves (isolated) on one side, so we proceed as follows:

$2x - 1 < x + 3$	Given.
$2x - 1 + 1 < x + 3 + 1$	Add 1.
$2x < x + 4$	Simplify.
$2x - x < x - x + 4$	Subtract x.
$x < 4$	Simplify.

Any number less than 4 is a solution. The graph of this inequality is as follows:

$2x - 1 < x + 3$ or, equivalently, $x < 4$

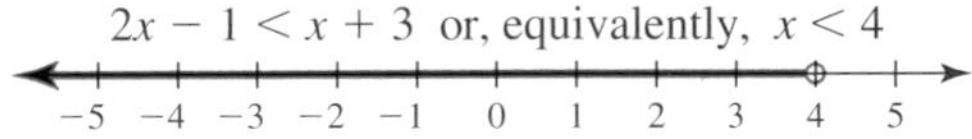

You can check that this solution is correct by selecting any number from the graph (say 0) and replacing x with that number in the original inequality. For $x = 0$, we have $2(0) - 1 < 0 + 3$, or $-1 < 3$, a true statement. Of course, this is only a "partial" check, since we didn't try *all* the numbers in the graph. You can check a little further by selecting a number *not* on the graph to make sure the result is false. For example, when $x = 5$,

$$2x - 1 < x + 3 \quad \text{becomes} \quad 2(5) - 1 < 5 + 3$$
$$10 - 1 < 8$$
$$9 < 8 \quad \text{False.}$$

EXAMPLE 3 **Using the addition and subtraction properties to solve and graph inequalities**

Solve and graph the inequality on a number line:

a. $3x - 2 < 2(x - 2)$ **b.** $4(x + 1) \geq 3x + 7$

SOLUTION

a.

$3x - 2 < 2(x - 2)$	Given.
$3x - 2 < 2x - 4$	Simplify.
$3x - 2 + 2 < 2x - 4 + 2$	Add 2.
$3x < 2x - 2$	Simplify.
$3x - 2x < 2x - 2x - 2$	Subtract $2x$ (or add $-2x$).
$x < -2$	Simplify.

Any number less than -2 is a solution. The graph of this inequality is as follows:

$3x - 2 < 2(x - 2)$ or, equivalently, $x < -2$

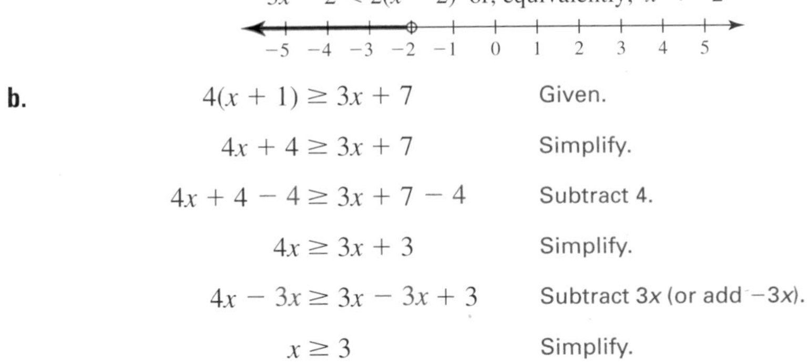

b.

$4(x + 1) \geq 3x + 7$	Given.
$4x + 4 \geq 3x + 7$	Simplify.
$4x + 4 - 4 \geq 3x + 7 - 4$	Subtract 4.
$4x \geq 3x + 3$	Simplify.
$4x - 3x \geq 3x - 3x + 3$	Subtract $3x$ (or add $-3x$).
$x \geq 3$	Simplify.

PROBLEM 3

Solve and graph:

a. $4x - 3 < 3(x - 2)$

b. $3(x + 2) \geq 2x + 5$

Answers

3. a. $x < -3$

b. $x \geq -1$

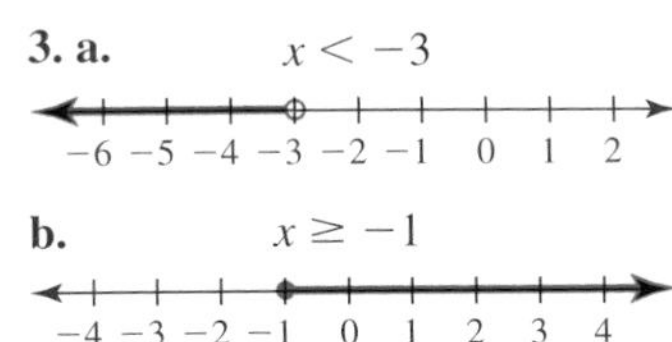

Any number greater than or equal to 3 is a solution. The graph of this inequality is as follows:

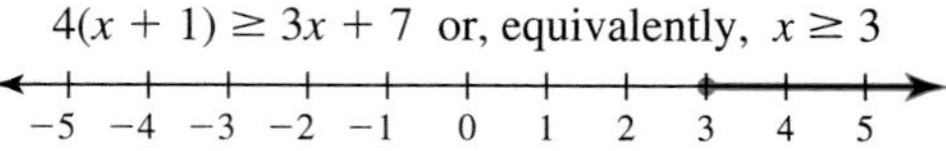

How do we solve an inequality such as $\frac{x}{2} < 3$? If half a number is less than 3, the number must be less than 6. This suggests that you can *multiply* (or *divide*) both sides of an inequality by a *positive* number and obtain an equivalent inequality.

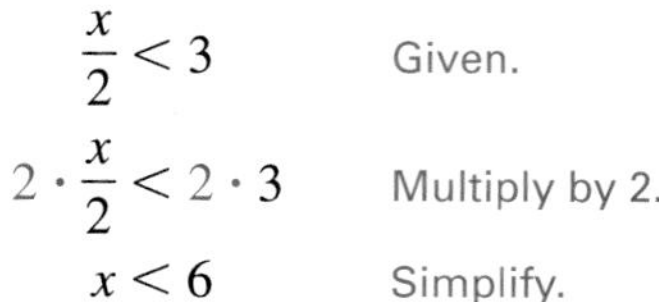

$$\frac{x}{2} < 3 \qquad \text{Given.}$$

$$2 \cdot \frac{x}{2} < 2 \cdot 3 \qquad \text{Multiply by 2.}$$

$$x < 6 \qquad \text{Simplify.}$$

And to solve

$$2x < 8$$

$$\frac{2x}{2} < \frac{8}{2} \qquad \text{Divide by 2 (or multiply by the reciprocal of 2).}$$

$$x < 4 \qquad \text{Simplify.}$$

Any number less than 4 is a solution. Let's try some more examples.

If $3 < 4$, then

$$5 \cdot 3 < 5 \cdot 4$$

$$15 < 20 \qquad \text{True.}$$

Note that we are multiplying or dividing *both* sides of the inequality by a **positive** number. In such cases, we do not change the inequality symbol.

If $-2 > -10$, then

$$5 \cdot (-2) > 5 \cdot (-10)$$

$$-10 > -50 \qquad \text{True.}$$

Also, if $6 < 8$, then

$$\frac{6}{2} < \frac{8}{2}$$

$$3 < 4 \qquad \text{True.}$$

Similarly, if $-6 > -10$, then

$$-\frac{6}{2} > -\frac{10}{2}$$

$$-3 > -5 \qquad \text{True.}$$

Here are the properties we've just used.

Multiplication and Division Properties of Inequalities for Positive Numbers

You can *multiply* or *divide* both sides of an inequality by any *positive* number a and obtain an equivalent inequality. In symbols,

If	$x < y$	(and a is positive)
then	$ax < ay$	
or	$\frac{x}{a} < \frac{y}{a}$	

If	$x > y$	(and a is positive)
then	$ax > ay$	
or	$\frac{x}{a} > \frac{y}{a}$	

NOTE

Since dividing x by a is the same as multiplying x by the reciprocal of a, you can think of dividing by a as multiplying by the reciprocal of a.

EXAMPLE 4 **Using the multiplication and division properties with positive numbers**

Solve and graph the inequality on a number line:

a. $5x + 3 \le 2x + 9$ **b.** $4(x - 1) > 2x + 6$

SOLUTION

a.

$5x + 3 \le 2x + 9$	Given.
$5x + 3 - 3 \le 2x + 9 - 3$	Subtract 3 (or add −3).
$5x \le 2x + 6$	Simplify.
$5x - 2x \le 2x - 2x + 6$	Subtract 2*x* (or add −2*x*).
$3x \le 6$	Simplify.
$\frac{3x}{3} \le \frac{6}{3}$	Divide by 3 (or multiply by the reciprocal of 3).
$x \le 2$	Simplify.

Any number less than or equal to 2 is a solution. The graph is as follows:

$5x + 3 \le 2x + 9$ or, equivalently, $x \le 2$

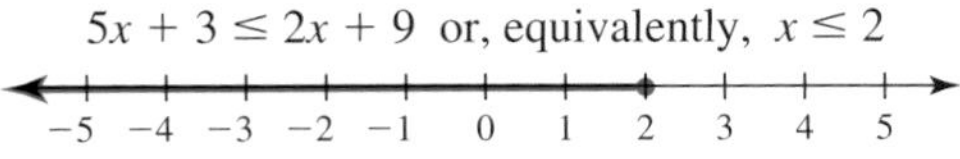

b.

$4(x - 1) > 2x + 6$	Given.
$4x - 4 > 2x + 6$	Simplify.
$4x - 4 + 4 > 2x + 6 + 4$	Add 4.
$4x > 2x + 10$	Simplify.
$4x - 2x > 2x - 2x + 10$	Subtract 2*x* (or add −2*x*).
$2x > 10$	Simplify.
$\frac{2x}{2} > \frac{10}{2}$	Divide by 2 (or multiply by the reciprocal of 2).
$x > 5$	Simplify.

Any number greater than 5 is a solution. The graph is as follows:

$4(x - 1) > 2x + 6$ or, equivalently, $x > 5$

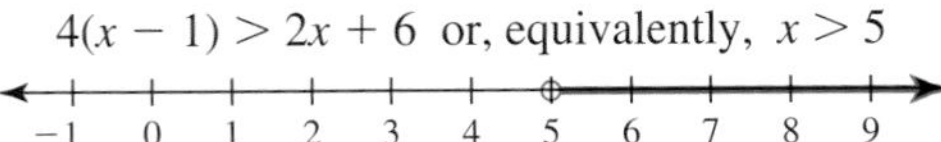

PROBLEM 4

Solve and graph:

a. $4x + 3 \le 2x + 5$

b. $5(x - 1) > 3x + 1$

You may have noticed that the multiplication (or division) property allows us to multiply or divide only by a *positive* number. However, to solve the inequality $-2x < 4$, we need to divide by -2 or multiply by $-\frac{1}{2}$, a number that is *not* positive.

Let's first see what happens when we divide both sides of an inequality by a *negative* number. Consider the inequality

$$2 < 4$$

If we divide both sides of this inequality by -2, we get

$$\frac{2}{-2} < \frac{4}{-2} \quad \text{or} \quad -1 < -2$$

which is *not* true. To obtain a true statement, we must *reverse the inequality sign* and write:

$$-1 > -2$$

Answers

4. a. $x \le 1$

−4 −3 −2 −1 0 1 2 3 4

b. $x > 3$

−2 −1 0 1 2 3 4 5 6

Similarly, consider

$$-6 > -8$$
$$\frac{-6}{-2} > \frac{-8}{-2}$$
$$3 > 4$$

which again is not true. However, the statement becomes true when we reverse the inequality sign and write $3 < 4$. So if we *divide* both sides of an inequality by a *negative* number, we must reverse the inequality sign to obtain an equivalent inequality. Similarly, if we *multiply* both sides of an inequality by a *negative* number, we must reverse the inequality sign to obtain an equivalent inequality:

$2 < 4$ Given.

$-3 \cdot 2 > -3 \cdot 4$ If we multiply by −3, we reverse the inequality sign.

$-6 > -12$

Note that now we are multiplying or dividing *both* sides of the inequality by a **negative** number. In such cases, we reverse the inequality symbol.

Here are some more examples.

$3 < 12$

$-2 \cdot 3 > -2 \cdot 12$ Reverse the sign.

$-6 > -24$

$8 > 4$

$\frac{8}{-2} < \frac{4}{-2}$ Reverse the sign.

$-4 < -2$

These properties are stated here.

Multiplication and Division Properties of Inequalities for Negative Numbers

You can *multiply* or *divide* both sides of an inequality by a *negative* number a and obtain an equivalent inequality provided you *reverse* the inequality sign. In symbols,

If	$x < y$	(and a is negative)	If	$x > y$	(and a is negative)
then	$ax > ay$	Reverse the sign.	then	$ax < ay$	Reverse the sign.
or	$\frac{x}{a} > \frac{y}{a}$	Reverse the sign.	or	$\frac{x}{a} < \frac{y}{a}$	Reverse the sign.

We use these properties in the next example.

EXAMPLE 5 Using the multiplication and division properties with negative numbers

Solve:

a. $-3x < 15$ **b.** $\frac{-x}{4} > 2$ **c.** $3(x - 2) \le 5x + 2$

SOLUTION

a. To solve this inequality, we need the x by itself on the left; that is, we have to divide both sides by −3. Of course, when we do this, we must reverse the inequality sign.

$-3x < 15$ Given.

$\frac{-3x}{-3} > \frac{15}{-3}$ Divide by −3 and reverse the sign.

$x > -5$ Simplify.

Any number greater than −5 is a solution.

PROBLEM 5

Solve:

a. $-4x < 20$

b. $\frac{-x}{3} > 2$

c. $2(x - 1) \le 4x + 1$

Answers

5. a. $x > -5$ **b.** $x < -6$ **c.** $x \ge -\frac{3}{2}$

b. Here we multiply both sides by -4 and reverse the inequality sign.

$$\frac{-x}{4} > 2 \qquad \text{Given.}$$

$$-4\left(\frac{-x}{4}\right) < -4 \cdot 2 \qquad \text{Multiply by } -4 \text{ and reverse the inequality sign.}$$

$$x < -8 \qquad \text{Simplify.}$$

Any number less than -8 is a solution.

c.

$$3(x - 2) \le 5x + 2 \qquad \text{Given.}$$

$$3x - 6 \le 5x + 2 \qquad \text{Simplify.}$$

$$3x - 6 + 6 \le 5x + 2 + 6 \qquad \text{Add 6.}$$

$$3x \le 5x + 8 \qquad \text{Simplify.}$$

$$3x - 5x \le 5x - 5x + 8 \qquad \text{Subtract } 5x \text{ (or add } -5x\text{).}$$

$$-2x \le 8 \qquad \text{Simplify.}$$

$$\frac{-2x}{-2} \ge \frac{8}{-2} \qquad \text{Divide by } -2 \text{ (or multiply by the reciprocal of } -2\text{) and reverse the inequality sign.}$$

$$x \ge -4 \qquad \text{Simplify.}$$

Thus any number greater than or equal to -4 is a solution.

Teaching Tip

Remind students of the mnemonic for the six-step method: "**C**harles **R**aves **C**onstantly **A**bout **A**lgebra's **M**agnificence." Make sure students remember that on the "**M**" step if they are multiplying or dividing by a *negative* number, the inequality symbol has to be *reversed.*

Of course, to solve more complicated inequalities (such as those involving fractions), we simply follow the six-step procedure we use for solving linear equations (p. 154).

EXAMPLE 6 Using the six-step procedure to solve an inequality

Solve:

$$\frac{-x}{4} + \frac{x}{6} < \frac{x - 3}{6}$$

SOLUTION We follow the six-step procedure for linear equations.

$$\frac{-x}{4} + \frac{x}{6} < \frac{x - 3}{6} \qquad \text{Given.}$$

1. Clear the fractions; the LCM is 12. $12 \cdot \left(\frac{-x}{4}\right) + 12 \cdot \frac{x}{6} < 12\left(\frac{x - 3}{6}\right)$

2. Remove parentheses (use the distributive property). $-3x + 2x < 2(x - 3)$

Collect like terms. $-x < 2x - 6$

3. There are no numbers on the left, only the variable $-x$.

4. Subtract $2x$. $-x - 2x < 2x - 2x - 6$

$$-3x < -6$$

PROBLEM 6

Solve:

$$\frac{-x}{3} + \frac{x}{4} < \frac{x - 8}{4}$$

Answer

6. $x > 6$

5. Divide by the coefficient of x, -3, and reverse the inequality sign.

$$\frac{-3x}{-3} > \frac{-6}{-3}$$

Remember that when you divide both sides of the equation by -3, which is negative, you have to reverse the inequality sign from $<$ to $>$.

$$x > 2$$

Thus any number greater than 2 is a solution.

6. **CHECK** Try $x = 12$. (It's a good idea to try the LCM. Do you see why?)

$\frac{-x}{4} + \frac{x}{6} \overset{?}{<} \frac{x-3}{6}$	
$\frac{-12}{4} + \frac{12}{6}$	$\frac{12-3}{6}$
$-3 + 2$	$\frac{9}{6}$
-1	$\frac{3}{2}$

Since $-1 < \frac{3}{2}$, the inequality is true. Of course, this only partially *verifies* the answer, since we're unable to try *every* solution.

C Solving and Graphing Compound Inequalities

What about inequalities like the one at the beginning of this section? Inequalities such as

$$1 \le x \le 3$$

are called **compound inequalities** because they are equivalent to two other inequalities; that is, $1 \le x \le 3$ means $1 \le x$ *and* $x \le 3$. Thus if we are asked to solve the inequalities

$$2 \le x \quad \text{and} \quad x \le 4$$

we write

$$2 \le x \le 4$$

$2 \le x \le 4$ is a compound inequality.

The graph of this inequality consists of all the points between 2 and 4, inclusive, as shown here:

$2 \le x \le 4$

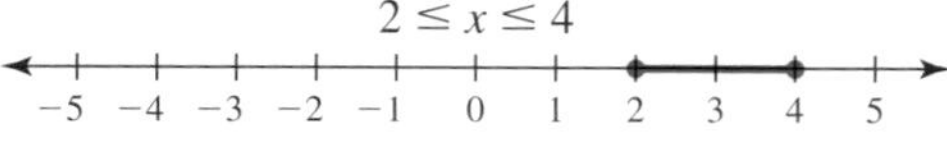

Web It

For a lesson on compound inequalities, go to the Bello Website at mhhe.com/bello and try link 2-7-3.

Teaching Tip

Compound inequalities such as $1 \le x \le 3$ are easy to remember as "between" statements since x is "between" 1 and 3. Also, point out that since 1 is the smaller number both arrows point toward it.

NOTE

To graph $2 \le x \le 4$, place a solid dot at 2, a solid dot at 4, and draw the line segment *between* 2 and 4.

The key to solving compound inequalities is to try writing the inequality in the form $a \le x \le b$ (or $a < x < b$), where a and b are real numbers. This form is called the *required* solution. Of course, if we try to write $x < 3$ and $x > 7$ in this form, we get $7 < x < 3$, which is not true.

EXAMPLE 7 Solving compound inequalities

Solve and graph on a number line:

a. $1 < x$ and $x < 3$

b. $5 \geq -x$ and $x \leq -3$

c. $x + 1 \leq 5$ and $-2x < 6$

SOLUTION

a. The inequalities $1 < x$ *and* $x < 3$ are written as $1 < x < 3$. Thus the solution consists of the numbers *between* 1 and 3, as shown here.

Draw an open circle at 1, an open circle at 3, and draw the line segment between 1 and 3.

$1 < x < 3$

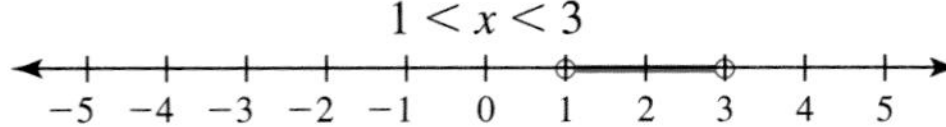

b. Since we wish to write $5 \geq -x$ and $x \leq -3$ in the form $a \leq x \leq b$, we multiply both sides of $5 \geq -x$ by -1 to obtain

$$-1 \cdot 5 \leq -1 \cdot (-x)$$
$$-5 \leq x$$

We now have

$$-5 \leq x \quad \text{and} \quad x \leq -3$$

that is,

$$-5 \leq x \leq -3$$

Thus the solution consists of all the numbers between -5 and -3 *inclusive*.

Note that -5 and -3 have solid dots.

$-5 \leq x \leq -3$

−5 −4 −3 −2 −1 0 1 2 3 4 5

c. We solve $x + 1 \leq 5$ by subtracting 1 from both sides to obtain

$$x + 1 - 1 \leq 5 - 1$$
$$x \leq 4$$

We now have

$$x \leq 4 \quad \text{and} \quad -2x < 6$$

We then divide both sides of $-2x < 6$ by -2 to obtain $x > -3$. We now have

$$x \leq 4 \quad \text{and} \quad x > -3$$

Rearranging these inequalities, we write

$$-3 < x \quad \text{and} \quad x \leq 4$$

that is,

$$-3 < x \leq 4$$

PROBLEM 7

Solve and graph:

a. $2 < x$ and $x < 4$

b. $3 \geq -x$ and $x \leq -1$

c. $x + 2 \leq 6$ and $-3x < 6$

Teaching Tip

When compound statements are written using the word *and,* we solve each of the statements separately and then graph the numbers they have in common (the intersection).

Answers

7. a. **b.**

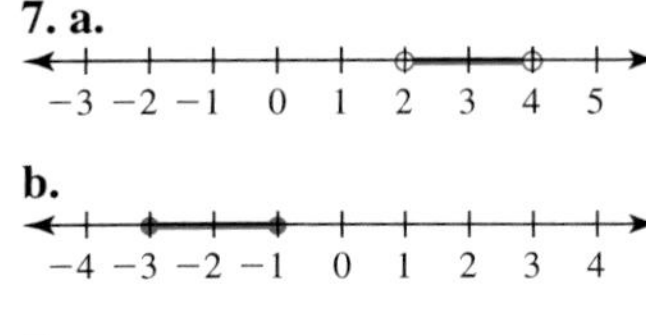

c.

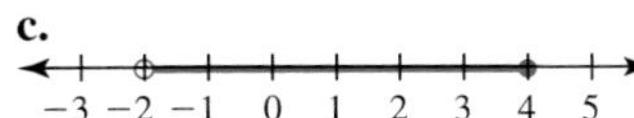

Here the solution consists of all numbers between -3 and 4 and the number 4 itself:

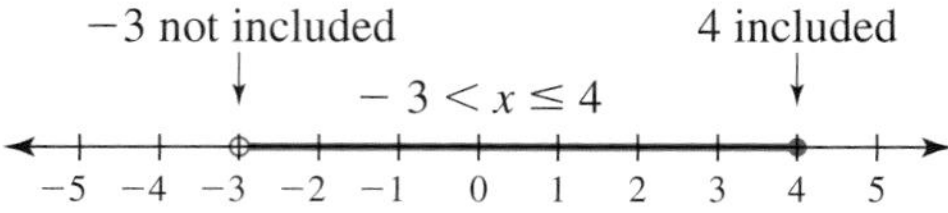

Teaching Tip

Warn students that the variable (x in this case) is usually on the left side of the inequality.

CAUTION

When graphing **linear inequalities** (inequalities that can be written in the form $ax + b \le c$, where a, b, and c are real numbers and a is not 0), the graph is usually a *ray* as shown here:

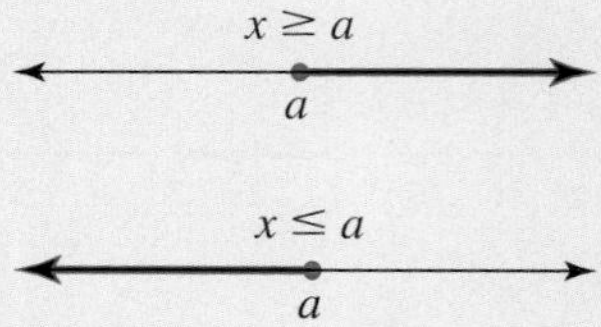

On the other hand, when graphing a **compound inequality,** the graph is usually a *line segment:*

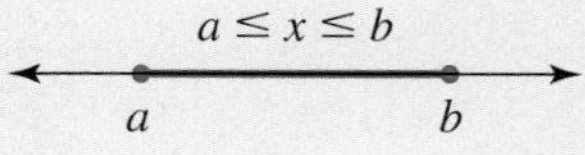

D Solving an Application

EXAMPLE 8 Is smoking decreasing?

Do you think the number of smokers is increasing or decreasing? According to the National Health Institute Survey, the percent of smokers t years after 1965 is given by $P = 43 - 0.64t$. This means that in 1965 ($t = 0$), the percent P of smokers was $P = 43 - 0.64(0) = 43\%$. If the trend continues, in what year would you expect the percent of smokers to be less than 23%? (Assume t, the number of years after 1965, is an integer.)

SOLUTION We want to find the values of t for which $P = 43 - 0.64t < 23$.

$43 - 0.64t < 23$ Given.

$43 - 43 - 0.64t < 23 - 43$ Subtract 43.

$-0.64t < -20$ Simplify.

$\frac{-0.64t}{-0.64} > \frac{-20}{-0.64}$ Divide by -0.64. (Don't forget to reverse the inequality!)

$t > 31.25$ Simplify.

This means that 31.25 or 32 years after 1965—that is, by $1965 + 32 = 1997$—the number of smokers should be less than 23%.

PROBLEM 8

Tobacco use is decreasing in the 12-to-17-year age group. The percent of smokers in that group is $P = 20 - 2x$, where x is the number of years after 1997 ($x = 0$). Using this information, when would the percent of smokers in this group be less than 10%?

Answer

8. $x > 5$, or more than 5 years after 1997, that is, in 2003

Calculate It Checking Linear Inequalities

Can we check the solutions of linear inequalities with a calculator? Again (as with linear equations), the answer is yes! Let's check Example 6; that is, let's find the values of x for which

$$\frac{-x}{4} + \frac{x}{6} < \frac{x-3}{6}$$

First, use a decimal window and let

$$Y_1 = \frac{-x}{4} + \frac{x}{6} \quad \text{and} \quad Y_2 = \frac{x-3}{6}$$

We want to find when $Y_1 < Y_2$, so we graph them and find that the lines intersect when $x = 2$. You can find that $x = 2$ is the intersection by using the TRACE and ZOOM keys or by using the intersection feature of a TI-83 Plus. To the left of the point at which $x = 2$, the line representing Y_2 is below the line representing Y_1; thus $Y_2 < Y_1$. To the right of $x = 2$ (that is, when $x > 2$), the line Y_2 is above the line Y_1, so $Y_1 < Y_2$, as desired. Thus any number greater than 2 ($x > 2$) is a solution, as shown in Window 1. To see this better, you can adjust your window to cover y-values ranging from -1 to 1. Press WINDOW and adjust the values (Window 2). Then graph Y_1 and Y_2 again, as shown in Window 3. You can see the graph much better now! Use this idea to verify your answers in the exercise set.

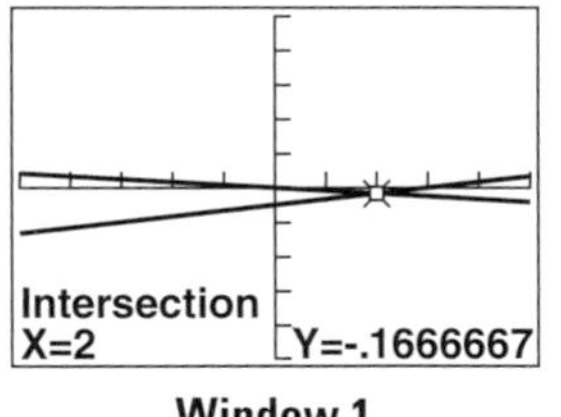

Window 1

WINDOW FORMAT
Xmin=-5
Xmax=5
Xscl=1
Ymin=-1
Ymax=1
Yscl=1

Window 2

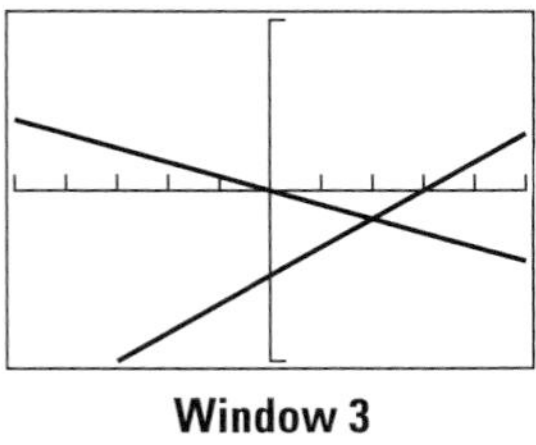
Window 3

Exercises 2.7

A In Problems 1–10, fill in the blank with $>$ or $<$ so that the resulting statement is true.

1. $8 \underline{\;<\;} 9$

2. $-8 \underline{\;>\;} -9$

3. $-4 \underline{\;>\;} -9$

4. $7 \underline{\;>\;} 3$

5. $\frac{1}{4} \underline{\;<\;} \frac{1}{3}$

6. $\frac{1}{5} \underline{\;<\;} \frac{1}{2}$

7. $-\frac{2}{3} \underline{\;>\;} -1$

8. $-\frac{1}{5} \underline{\;>\;} -1$

9. $-3\frac{1}{4} \underline{\;<\;} -3$

10. $-4\frac{1}{5} \underline{\;<\;} -4$

B In Problems 11–30, solve and graph the inequalities on a number line.

11. $2x + 6 \le 8$ $x \le 1$

12. $4y - 5 \le 3$ $y \le 2$

13. $-3y - 4 \ge -10$ $y \le 2$

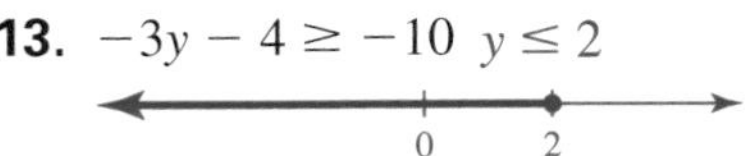

14. $-4z - 2 \ge 6$ $z \le -2$

15. $-5x + 1 < -14$ $x > 3$

16. $-3x + 1 < -8$ $x > 3$

17. $3a + 4 \le a + 10$ $a \le 3$

18. $4b + 4 \le b + 7$ $b \le 1$

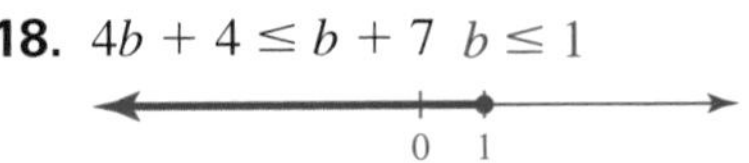

19. $5z - 12 \ge 6z - 8$ $z \le -4$

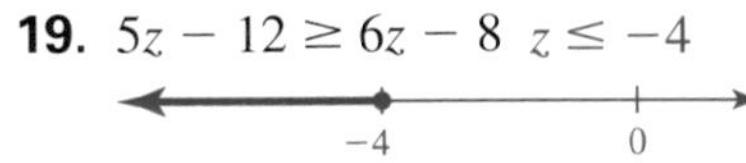

20. $5z + 7 \ge 7z + 19$ $z \le -6$

21. $10 - 3x \le 7 - 6x$ $x \le -1$

22. $8 - 4y \le -12 + 6y$ $y \ge 2$

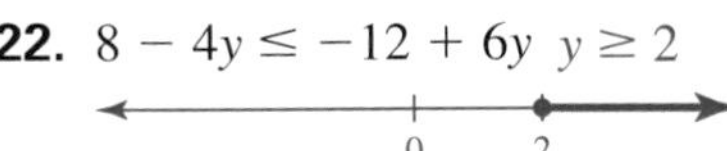

23. $5(x + 2) < 3(x + 3) + 1$ $x < 0$

24. $5(4 - 3x) < 7(3 - 4x) + 12$ $x < 1$

25. $-2x + \frac{1}{4} \ge 2x + \frac{4}{5}$ $x \le -\frac{11}{80}$

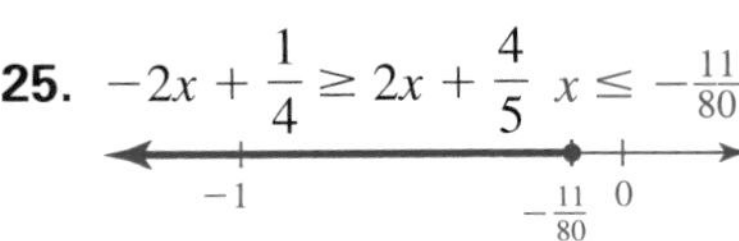

26. $6x + \frac{1}{7} \geq 2x - \frac{2}{7}$ $x \geq -\frac{3}{28}$

27. $\frac{x}{5} - \frac{x}{4} \leq 1$ $x \geq -20$

28. $\frac{x}{3} - \frac{x}{2} \leq 1$ $x \geq -6$

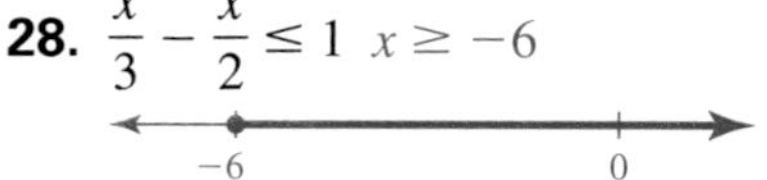

29. $\frac{7x+2}{6} + \frac{1}{2} \geq \frac{3}{4}x$ $x \geq -2$

30. $\frac{8x-23}{6} + \frac{1}{3} \geq \frac{5}{2}x$ $x \leq -3$

C In Problems 31–40, solve and graph the inequalities on a number line.

31. $x < 3$ and $-x < -2$ $2 < x < 3$

32. $-x < 5$ and $x < 2$ $-5 < x < 2$

33. $x + 1 < 4$ and $-x < -1$
$1 < x < 3$

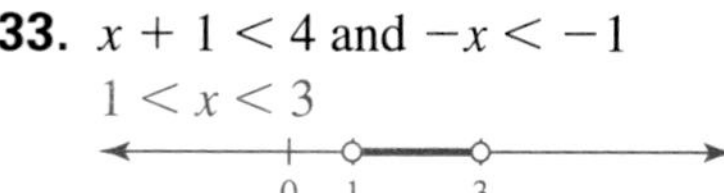

34. $x - 2 < 1$ and $-x < 2$
$-2 < x < 3$

35. $x - 2 < 3$ and $2 > -x$
$-2 < x < 5$

36. $x - 3 < 1$ and $1 > -x$
$-1 < x < 4$

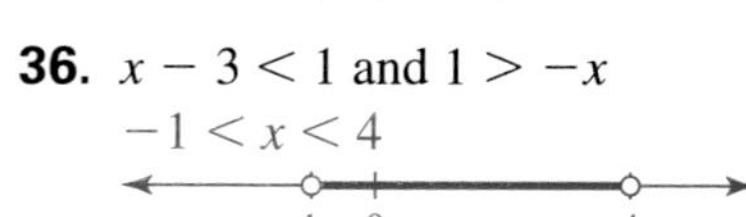

37. $x + 2 < 3$ and $-4 < x + 1$
$-5 < x < 1$

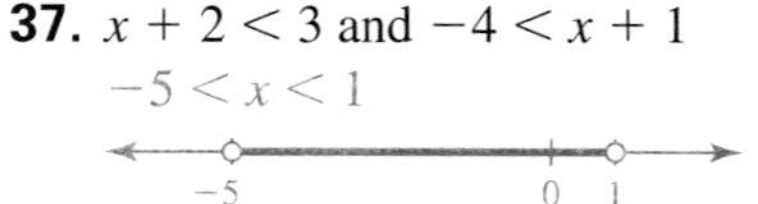

38. $x + 4 < 5$ and $-1 < x + 2$
$-3 < x < 1$

39. $x - 1 > 2$ and $x + 7 < 12$
$3 < x < 5$

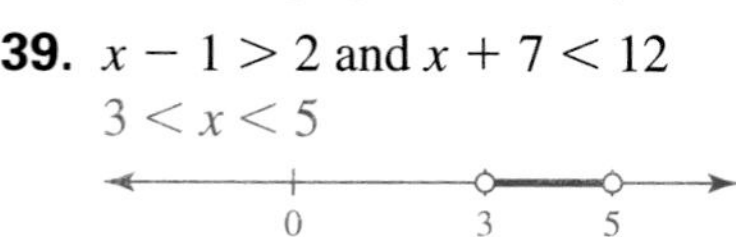

40. $x - 2 > 1$ and $-x > -5$ $3 < x < 5$

In Problems 41–50, write the given information as an inequality.

41. The temperature t in your refrigerator is between 20°F and 40°F. $20°\text{F} < t < 40°\text{F}$

42. The height h (in feet) of any mountain is always less than or equal to that of Mount Everest, 29,029 feet. $h \leq 29{,}029$ ft

43. Joe's salary s for this year will be between \$12,000 and \$13,000. $\$12{,}000 < s < \$13{,}000$

44. Your gas mileage m (in miles per gallon) is between 18 and 22, depending on your driving. $18 < m < 22$

45. The number of possible eclipses e in a year varies from 2 to 7, inclusive. $2 \leq e \leq 7$

46. The assets a of the Du Pont family are in excess of \$150 billion. $a >$ \$150 billion

47. The cost c of ordinary hardware (tools, mowers, and so on) is between \$3.50 and \$4.00 per pound. $\$3.50 < c < \$4.00$

48. The range r (in miles) of a rocket is always less than 19,000 miles. $r < 19{,}000$ mi

49. The altitude a (in feet) attained by the first liquid-fueled rocket was less than 41 feet. $a < 41$ ft

50. The number of days d a person remained in the weightlessness of space before 1988 did not exceed 370. $d \leq 370$

APPLICATIONS

51. *NCAA men's basketball* The number N of NCAA men's college basketball teams t years after 1980 is given by $N = 720 + 5t$. When would you expect that the NCAA would have more than 800 teams? After 1996

52. *How is your heart?* According to the U.S. National Center for Health Statistics, the number D of deaths per 100,000 population attributed to heart disease has been decreasing for t years after 1999 and is given by $D = -10t + 260$. After what year would you expect the number of deaths per 100,000 to be less than 200? 2005

53. *Cost of a hospital room* From 1980 on, the daily room charge C of a hospital room has been increasing and is given by $C = 127 + 17t$, where t is the number of years after 1980. After what year would you expect the average hospital room rate to surpass the \$300 per day mark? 1991

54. *Poultry consumption* The annual per capita consumption of poultry products (in pounds) t years after 1985 is given by $C = 45 + 2t$. When would you expect the per capita consumption of poultry products to exceed 60 pounds per year? (See the Collaborative Learning at the end of the chapter.) 1993

55. *Tuna consumption* The annual per capita consumption C of canned tuna (in pounds) t years after 1989 has been decreasing and is given by $C = 3.91 - 0.13t$. In what year would the per capita consumption fall below 3 pounds per year? 1997

SKILL CHECKER

Try the Skill Checker Exercises so you'll be ready for the next section.

Find:

56. $-4 \cdot 9$ -36

57. $-5 \cdot 8$ -40

58. $-2(-6)$ 12

59. $-5(-8)$ 40

60. $\frac{-6}{2}$ -3

61. $\frac{-18}{9}$ -2

62. $\frac{-21}{-3}$ 7

63. $\frac{-35}{-5}$ 7

64. $\frac{40}{-8}$ -5

USING YOUR KNOWLEDGE

A Question of Inequality

Can you solve the problem in the cartoon? Let

J = Joe's height
B = Bill's height
F = Frank's height
S = Sam's height

Translate each statement into an equation or an inequality.

65. Joe is 5 feet (60 inches) tall. $J = 5$ ft $= 60$ in.

66. Bill is taller than Frank. $B > F$

67. Frank is 3 inches shorter than Sam. $F = S - 3$

68. Frank is taller than Joe. $F > J$

69. Sam is 6 feet 5 inches (77 inches) tall. $S = 6$ ft 5 in. $= 77$ in.

70. According to the statement in Problem 66, Bill is taller than Frank, and according to the statement in Problem 68, Frank is taller than Joe. Write these two statements as an inequality of the form $a > b > c$. $B > F > J$

71. Based on the answer to Problem 70 and the fact that you can obtain Frank's height by using the results of Problems 67 and 69, what can you really say about Bill's height? $B > 74$ in. or $B > 6$ ft 2 in.

WRITE ON

72. Write the similarities and differences in the procedures used to solve equations and inequalities. Answers may vary.

73. As you solve an inequality, when do you have to change the direction of the inequality? When multiplying or dividing by a negative number.

74. A student wrote "$2 < x < -5$" to indicate that x was between 2 and -5. Why is this wrong? Answers may vary.

75. Write the steps you would use to solve the inequality $-3x < 15$. Answers may vary.

MASTERY TEST

If you know how to do these problems, you have learned your lesson!

Solve and graph on a number line:

76. $x + 2 \le 6$ and $-3x \le 9$ $\quad -3 \le x \le 4$

77. $3 \ge -x$ and $x \le -1$ $\quad -3 \le x \le -1$

78. $2 < x$ and $x < 4$ $\quad 2 < x < 4$

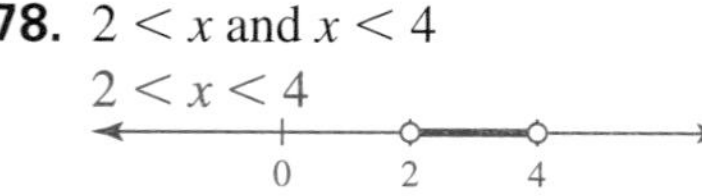

79. $\frac{-x}{3} + \frac{x}{4} < \frac{x-4}{4}$ $\quad x > 3$

80. $4x + 5 < x + 11$ $\quad x < 2$

81. $3(x - 1) > x + 3$ $\quad x > 3$

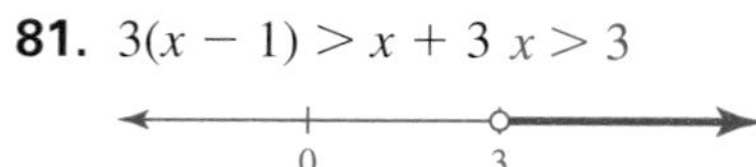

82. $4x - 7 < 3(x - 2)$ $\quad x < 1$

83. $3(x + 1) \ge 2x + 5$ $\quad x \ge 2$

84. $x \ge -2$ $\quad x \ge -2$

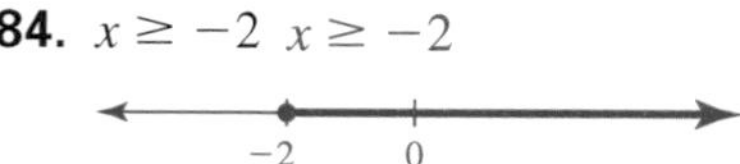

85. $x < 1$ $\quad x < 1$

Fill in the blank with $>$ or $<$ so that the resulting statement is true:

86. 5 __<__ 7

87. -2 __<__ -1

88. -4 __>__ -5

89. $\frac{1}{3}$ __>__ -3

90. According to the U.S. Department of Agriculture, the total daily grams of fat F consumed per person is $F = 181.5 + 0.8t$, where t is the number of years after 2000. After what year would you expect the daily consumption of grams of fat to exceed 189.5? After 2010

COLLABORATIVE LEARNING

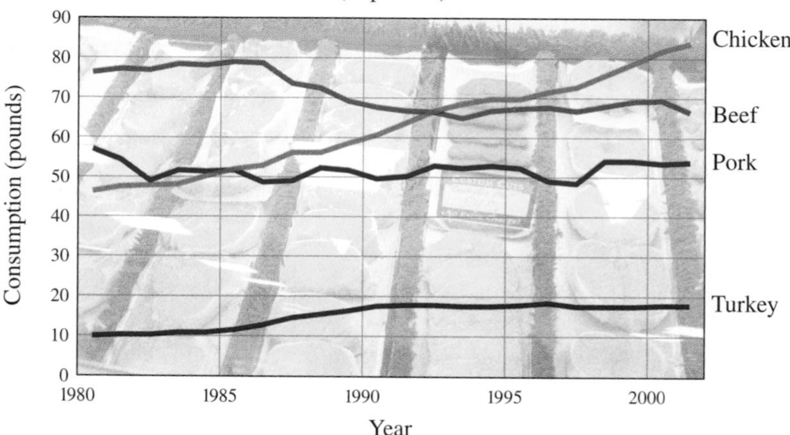

Source: American Meat Institute.

Form four groups: Beefies, porkies, chickens, and turkeys.

1. In Problem 54 of Section 2.7, we gave a formula for the consumption C of poultry in the United States. Use the formula to estimate the U.S. consumption of poultry in 2001. How close is the estimate to the one given in the table? (2001 is the last year shown.)
2. From 1990 on, the consumption of beef (B), pork (P), and turkey (T) has been rather steady. Write a consumption equation that approximates the annual per capita consumption of *your product*.
3. The consumption of chicken (C) has been increasing about two pounds per year since 1980 so that $C = 45 + 2t$. In what year is C highest? In what year is it lowest?
4. Write an inequality comparing the annual consumption of *your product* and C.
5. In what year(s) was the annual consumption of *your product* less than C, equal to C, and greater than C?

Research Questions

1. What does the word *papyrus* mean? Explain how the Rhind papyrus got its name.
2. Write a report on the contents and origins of the Rhind papyrus.
3. The Rhind papyrus is one of two documents detailing Egyptian mathematics. What is the name of the other document, and what type of material does it contain?
4. Write a report about the rule of false position and the rule of double false position.
5. Find out who invented the symbols for greater than (>) and less than (<).
6. What is the meaning of the word *geometry?* Give an account of the origin of the subject.
7. Problem 50 in the Rhind papyrus gives the method for finding the area of a circle. Write a description of the problem and the method used.

Summary

SECTION	ITEM	MEANING	EXAMPLE
2.1	Equation	A statement indicating that two expressions are **equal**	$x - 8 = 9$, $\frac{1}{2} - 2x = \frac{3}{4}$, and $0.2x + 8.9 = \frac{1}{2} + 6x$ are equations.
2.1A	Solutions	The solutions of an equation are the replacements of the variable that make the equation a **true** statement.	4 is a solution of $x + 1 = 5$.
	Equivalent equations	Two equations are **equivalent** if their solutions are the **same.**	$x + 1 = 4$ and $x = 3$ are equivalent.
2.1B	The addition property of equality	$a = b$ is equivalent to $a + c = b + c$.	$x - 1 = 2$ is equivalent to $x - 1 + 1 = 2 + 1$.
	The subtraction property of equality	$a = b$ is equivalent to $a - c = b - c$.	$x + 1 = 2$ is equivalent to $x + 1 - 1 = 2 - 1$.
2.1C	Conditional equation	An equation with **one** solution	$x + 7 = 9$ is a conditional equation whose solution is 2.
	Contradictory equation	An equation with **no** solution	$x + 1 = x + 2$ is a contradictory equation.
	Identity	An equation with **infinitely** many solutions	$2(x + 1) - 5 = 2x - 3$ is an identity. Any real number is a solution.
2.2A	The multiplication property of equality	$a = b$ is equivalent to $ac = bc$ if c is not 0.	$\frac{x}{2} = 3$ is equivalent to $2 \cdot \frac{x}{2} = 2 \cdot 3$.
	The division property of equality	$a = b$ is equivalent to $\frac{a}{c} = \frac{b}{c}$ if c is not 0.	$2x = 6$ is equivalent to $\frac{2x}{2} = \frac{6}{2}$.
2.2B	Reciprocal	The reciprocal of $\frac{a}{b}$ is $\frac{b}{a}$.	The reciprocal of $\frac{5}{2}$ is $\frac{2}{5}$.
2.2C	LCM (least common multiple)	The smallest number that is a multiple of each of the given numbers	The LCM of 3, 8, and 9 is 72.
2.3A	Linear equation	An equation that can be written in the form $ax + b = c$	$5x + 5 = 2x + 6$ is a linear equation (it can be written as $3x + 5 = 6$).
2.3B	Literal equation	An equation that contains letters other than the variable for which we wish to solve	$I = Prt$ and $C = 2\pi r$ are literal equations.
2.4	RSTUV method	To solve word problems, **R**ead, **S**elect a variable, **T**ranslate, **U**se algebra, and **V**erify your answer.	
2.4A	Consecutive integers	If n is an integer, the next consecutive integer is $n + 1$. If n is an even (odd) integer, the next consecutive even (odd) integer is $n + 2$.	4, 5, and 6 are three consecutive integers. 2, 4, and 6 are three consecutive even integers.

SECTION	ITEM	MEANING	EXAMPLE
2.4C	Complementary angles	Two angles whose sum is 90°	Two angles with measures 50° and 40° are complementary.
	Supplementary angles	Two angles whose sum is 180°	Two angles with measures 35° and 145° are supplementary.
2.6B	Perimeter	The distance around a geometric figure	The perimeter P of a rectangle with length L and width W is $P = 2L + 2W$.
	Area of a rectangle	The area A of a rectangle of length L and width W is $A = LW$.	The area of a rectangle 8 inches long and 4 inches wide is $A = 8 \text{ in.} \cdot 4 \text{ in.} = 32 \text{ in.}^2$
	Area of a triangle	The area A of a triangle with base b and height h is $A = \frac{1}{2}bh$.	The area of a triangle with base 5 cm and height 10 cm is $A = \frac{1}{2} \cdot 5 \text{ cm} \cdot 10 \text{ cm} = 25 \text{ cm}^2$.
	Area of a circle	The area A of a circle of radius r is $A = \pi r^2$.	The area of a circle whose radius is 5 inches is $A = \pi(5)^2$, that is, $25\pi \text{ in.}^2$
	Circumference of a circle	The circumference of a circle with radius r is $C = 2\pi r$.	The circumference of a circle whose radius is 10 inches is $C = 2\pi(10)$, that is, 20π in.
2.6C	Vertical angles	Angles ① and ② are vertical angles.	1 2
2.7	Inequality	A statement with $>$, $<$, $\geq$, or $\leq$ for its verb	$2x + 1 > 5$ and $3x - 5 \leq 7 - x$ are inequalities.
2.7B	The addition property of inequalities	$a < b$ is equivalent to $a + c < b + c$.	$x - 1 < 2$ is equivalent to $x - 1 + 1 < 2 + 1$.
	The subtraction property of inequalities	$a < b$ is equivalent to $a - c < b - c$.	$x + 1 < 2$ is equivalent to $x + 1 - 1 < 2 - 1$.
	The multiplication property of inequalities	$a < b$ is equivalent to $ac < bc$ if $c > 0$ or $ac > bc$ if $c < 0$	$\frac{x}{2} < 3$ is equivalent to $2 \cdot \frac{x}{2} < 2 \cdot 3$.
	The division property of inequalities	$a < b$ is equivalent to $\frac{a}{c} < \frac{b}{c}$, if $c > 0$ or $\frac{a}{c} > \frac{b}{c}$, if $c < 0$	$-2x < 6$ is equivalent to $\frac{-2x}{-2} > \frac{6}{-2}$.

Review Exercises

(If you need help with these exercises, look in the section indicated in brackets.)

1. [2.1A] Determine whether the given number satisfies the equation.

a. $5; 7 = 14 - x$ No **b.** $4; 13 = 17 - x$ Yes

c. $-2; 8 = 6 - x$ Yes

2. [2.1B] Solve the given equation.

a. $x - \frac{1}{3} = \frac{1}{3}$ $x = \frac{2}{3}$ **b.** $x - \frac{5}{7} = \frac{2}{7}$ $x = 1$

c. $x - \frac{5}{9} = \frac{1}{9}$ $x = \frac{2}{3}$

3. [2.1B] Solve the given equation.

a. $-3x + \frac{5}{9} + 4x - \frac{2}{9} = \frac{5}{9}$ $x = \frac{2}{9}$

b. $-2x + \frac{4}{7} + 3x - \frac{2}{7} = \frac{6}{7}$ $x = \frac{4}{7}$

c. $-4x + \frac{5}{6} + 5x - \frac{1}{6} = \frac{5}{6}$ $x = \frac{1}{6}$

4. [2.1C] Solve the given equation.

a. $3 = 4(x - 1) + 2 - 3x$ $x = 5$

b. $4 = 5(x - 1) + 9 - 4x$ $x = 0$

c. $5 = 6(x - 1) + 8 - 5x$ $x = 3$

5. [2.1C] Solve the given equation.

a. $6 + 3(x + 1) = 2 + 3x$ No solution

b. $-2 + 4(x - 1) = -7 - 4x$ $x = -\frac{1}{8}$

c. $-1 - 2(x + 1) = 3 - 2x$ No solution

6. [2.1C] Solve the given equation.

a. $5 + 2(x + 1) = 2x + 7$ All real numbers

b. $-2 + 3(x - 1) = -5 + 3x$ All real numbers

c. $-3 - 4(x - 1) = 1 - 4x$ All real numbers

7. [2.2A] Solve the given equation.

a. $\frac{1}{5}x = -3$ $x = -15$ **b.** $\frac{1}{7}x = -2$ $x = -14$

c. $5x = -10$ $x = -2$

8. [2.2B] Solve the given equation.

a. $-\frac{3}{4}x = -9$ $x = 12$ **b.** $-\frac{3}{5}x = -9$ $x = 15$

c. $-\frac{2}{3}x = -6$ $x = 9$

9. [2.2C] Solve the given equation.

a. $\frac{x}{3} + \frac{2x}{4} = 5$ $x = 6$ **b.** $\frac{x}{4} + \frac{3x}{2} = 6$ $x = \frac{24}{7}$

c. $\frac{x}{5} + \frac{3x}{10} = 10$ $x = 20$

10. [2.2C] Solve the given equation.

a. $\frac{x}{3} - \frac{x}{4} = 1$ $x = 12$ **b.** $\frac{x}{2} - \frac{x}{7} = 10$ $x = 28$

c. $\frac{x}{4} - \frac{x}{5} = 2$ $x = 40$

11. [2.2C] Solve the given equation.

a. $\frac{x - 1}{4} - \frac{x + 1}{6} = 1$ $x = 17$

b. $\frac{x - 1}{6} - \frac{x + 1}{8} = 0$ $x = 7$

c. $\frac{x - 1}{8} - \frac{x + 1}{10} = 0$ $x = 9$

12. [2.2D] Solve.

a. What percent of 30 is 6? 20%

b. What percent of 40 is 4? 10%

c. What percent of 50 is 10? 20%

13. [2.2D] Solve.

a. 20 is 40% of what number? 50

b. 30 is 90% of what number? $33\frac{1}{3}$

c. 25 is 75% of what number? $33\frac{1}{3}$

14. [2.3A] Solve.

a. $\frac{1}{5} - \frac{x}{4} = \frac{19(x + 4)}{20}$ $x = -3$

b. $\frac{1}{5} - \frac{x}{4} = \frac{6(x + 5)}{5}$ $x = -4$

c. $\frac{1}{5} - \frac{x}{4} = \frac{29(x + 6)}{20}$ $x = -5$

15. [2.3B] Solve.

a. $A = \frac{1}{2}bh$; solve for h. $h = \frac{2A}{b}$

b. $C = 2\pi r$; solve for r. $r = \frac{C}{2\pi}$

c. $V = \frac{bh}{3}$; solve for b. $b = \frac{3V}{h}$

16. [2.4A] Find the numbers described.

a. The sum of two numbers is 84 and one of the numbers is 20 more than the other. 32, 52

b. The sum of two numbers is 47 and one of the numbers is 19 more than the other. 14, 33

c. The sum of two numbers is 81 and one of the numbers is 23 more than the other. 29, 52

17. [2.4B] If you eat a fried chicken breast and a 3-ounce piece of apple pie, you have consumed 578 calories.

a. If the pie has 22 more calories than the chicken breast, how many calories are in each? Chicken: 278, pie: 300

b. Repeat part **a** where the number of calories consumed is 620 and the pie has 38 more calories than the chicken breast. Chicken: 291, pie: 329

c. Repeat part **a** where the number of calories consumed is 650 and the pie has 42 more calories than the chicken breast. Chicken: 304, pie: 346

18. [2.4C] Find the measure of an angle whose supplement is

a. 20 degrees less than 3 times its complement. 35°

b. 30 degrees less than 3 times its complement. 30°

c. 40 degrees less than 3 times its complement. 25°

19. [2.5A] Solve.

a. A car leaves a town traveling at 40 miles per hour. An hour later, another car leaves the same town traveling at 50 miles per hour in the same direction. How long does it take the second car to overtake the first one? 4 hours

b. Repeat part **a** where the first car travels at 30 miles per hour and the second one at 50 miles per hour. $1\frac{1}{2}$ hours

c. Repeat part **a** where the first car travels at 40 miles per hour and the second one at 60 miles per hour. 2 hours

20. [2.5B] Solve.

a. How many pounds of a product selling at \$1.50 per pound should be mixed with 15 pounds of another product selling at \$3 per pound to obtain a mixture selling at \$2.40 per pound? 10 pounds

b. Repeat part **a** where the products sell for \$2, \$3, and \$2.50, respectively. 15 pounds

c. Repeat part **a** where the products sell for \$6, \$2, and \$4.50, respectively. 25 pounds

21. [2.5C] Solve.

a. A woman invests \$30,000, part at 5% and part at 6%. Her annual interest amounts to \$1600. How much does she have invested at each rate? \$10,000 at 6%, \$20,000 at 5%

b. Repeat part **a** where the rates are 7% and 9%, respectively, and her annual return amounts to \$2300. \$20,000 at 7%, \$10,000 at 9%

c. Repeat part **a** where the rates are 6% and 10%, respectively, and her annual return amounts to \$2000. \$25,000 at 6%, \$5000 at 10%

22. [2.6A] The cost C of a long-distance call is $C = 3.05m + 3$, where m is the number of minutes the call lasts.

a. Solve for m and then find the length of a call that cost \$27.40. $m = \frac{C-3}{3.05}$; 8 minutes

b. Repeat part **a** where $C = 3.15m + 3$ and the call cost \$34.50. $m = \frac{C-3}{3.15}$; 10 minutes

c. Repeat part **a** where $C = 3.25m + 2$ and the call cost \$21.50. $m = \frac{C-2}{3.25}$; 6 minutes

23. [2.6C] Find the measures of the marked angles.

a. 100°, 80°

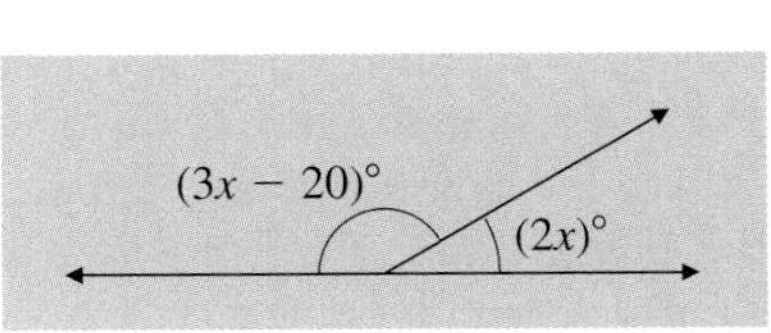

b. Both 60°

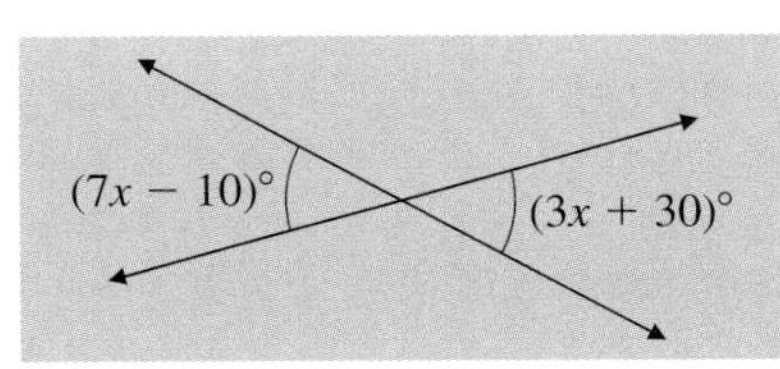

c. 65°, 25°

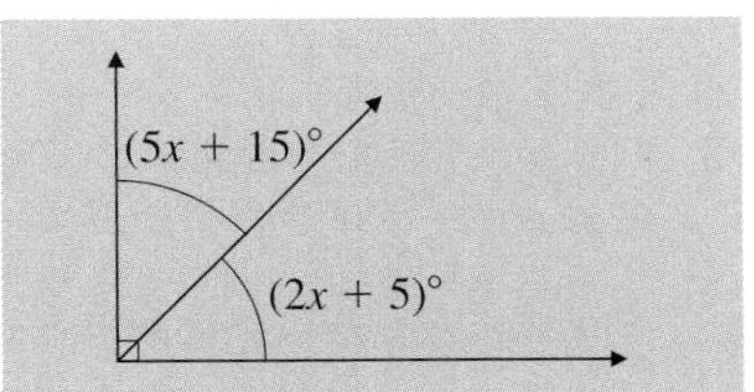

24. [2.7A] Fill in the blank with the symbol $<$ or $>$ to make the resulting statement true.

a. -8 ___$<$___ -7

b. $\frac{1}{2}$ ___$>$___ -3

c. 4 ___$<$___ $4\frac{1}{3}$

25. [2.7B] Solve and graph the given inequality.

a. $4x - 2 < 2(x + 2)$ $x < 3$

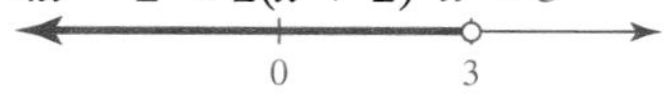

b. $5x - 4 < 2(x + 1)$ $x < 2$

c. $7x - 1 < 3(x + 1)$ $x < 1$

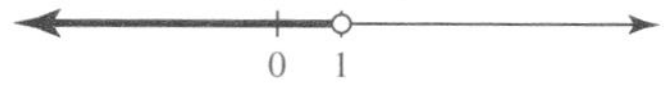

26. [2.7B] Solve and graph the given inequality.

a. $6(x - 1) \geq 4x + 2$ $x \geq 4$

b. $5(x - 1) \geq 2x + 1$ $x \geq 2$

c. $4(x - 2) \geq 2x + 2$ $x \geq 5$

27. [2.7B] Solve and graph the given inequality.

a. $\frac{x}{3} + \frac{x}{6} \leq \frac{x-1}{6}$ $x \geq \frac{1}{2}$

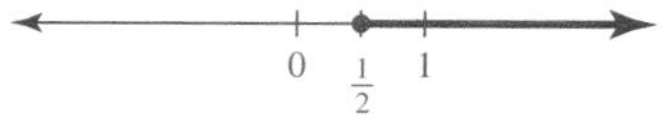

b. $\frac{x}{4} + \frac{x}{7} \leq \frac{x-1}{7}$ $x \geq \frac{4}{7}$

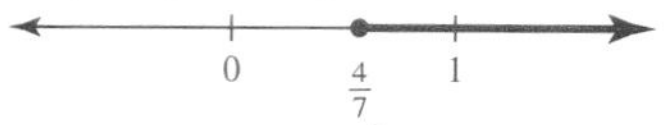

c. $\frac{x}{5} + \frac{x}{3} \leq \frac{x-1}{3}$ $x \geq \frac{5}{3}$

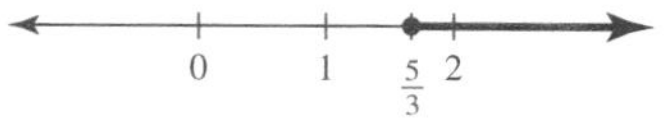

28. [2.7C] Solve and graph the compound inequality.

a. $x + 2 \leq 4$ and $-2x < 6$ $-3 < x \leq 2$

b. $x + 3 \leq 5$ and $-3x < 9$ $-3 < x \leq 2$

c. $x + 1 \leq 2$ and $-4x < 8$ $-2 < x \leq 1$

Practice Test 2

(Answers on pages 219–220)

1. Does the number 3 satisfy the equation $6 = 9 - x$?

2. Solve $x - \frac{2}{7} = \frac{3}{7}$.

3. Solve $-2x + \frac{7}{8} + 3x - \frac{5}{8} = \frac{5}{8}$.

4. Solve $2 = 3(x - 1) + 5 - 2x$.

5. Solve $2 + 5(x + 1) = 8 + 5x$.

6. Solve $-3 - 2(x - 1) = -1 - 2x$.

7. Solve $\frac{2}{3}x = -4$.

8. Solve $-\frac{2}{3}x = -6$.

9. Solve $\frac{x}{4} + \frac{2x}{3} = 11$.

10. Solve $\frac{x}{3} - \frac{x}{5} = 2$.

11. Solve $\frac{x-2}{5} - \frac{x+1}{8} = 0$.

12. What percent of 55 is 11?

13. Nine is 36% of what number?

14. Solve $\frac{1}{5} - \frac{x}{3} = \frac{23(x+5)}{15}$.

15. Solve for h in $S = \frac{1}{3}\pi r^2 h$.

16. The sum of two numbers is 75. If one of the numbers is 15 more than the other, what are the numbers?

17. A man has invested a certain amount of money in stocks and bonds. His annual return from these investments is \$840. If the stocks produce \$230 more in returns than the bonds, how much money does he receive annually from each investment?

18. Find the measure of an angle whose supplement is 50° less than 3 times its complement.

19. A freight train leaves a station traveling at 30 miles per hour. Two hours later, a passenger train leaves the same station traveling in the same direction at 42 miles per hour. How long does it take for the passenger train to catch the freight train?

20. How many pounds of coffee selling for \$1.10 per pound should be mixed with 30 pounds of coffee selling for \$1.70 per pound to obtain a mixture that sells for \$1.50 per pound?

21. An investor bought some municipal bonds yielding 5% annually and some certificates of deposit yielding 7% annually. If her total investment amounts to \$20,000 and her annual return is \$1160, how much money is invested in bonds and how much in certificates of deposit?

22. The cost C of riding a taxi is $C = 1.95 + 0.85m$, where m is the number of miles (or fraction) you travel.

a. Solve for m.

b. How many miles did you travel if the cost of the ride was \$20.65?

23. Find x and the measures of the marked angles.

a.

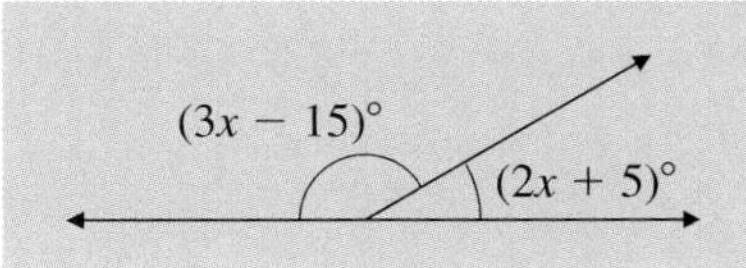

b.

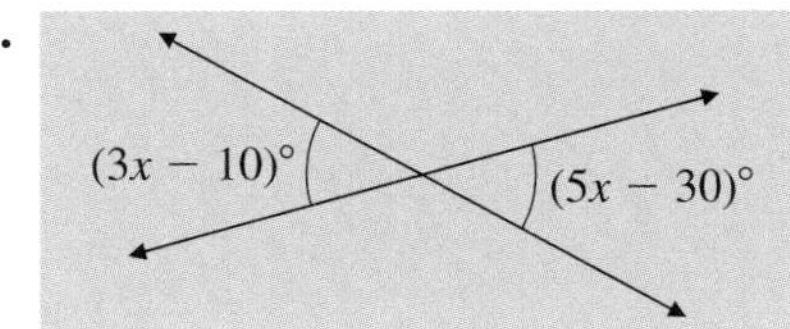

c.

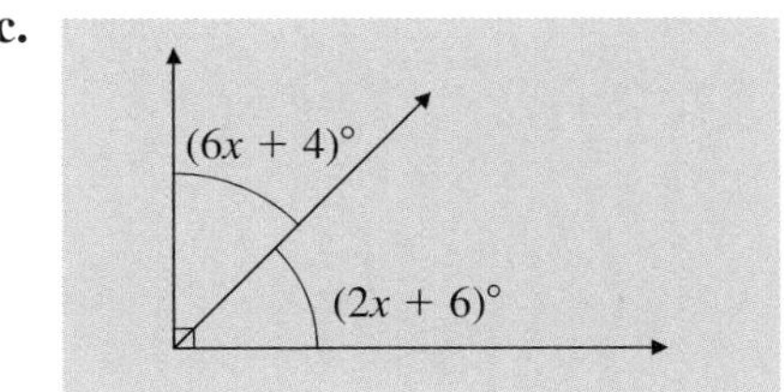

24. Fill in the blank with < or > to make the resulting statement true.

a. -3 ______ -5

b. $-\frac{1}{3}$ ______ 3

25. Solve and graph the inequality.

a. $-\frac{x}{2} + \frac{x}{4} \le \frac{x+2}{4}$

b. $x + 1 \le 3$ and $-2x < 6$

Answers to Practice Test

ANSWER	IF YOU MISSED	REVIEW		
	QUESTION	SECTION	EXAMPLES	PAGE
1. Yes	1	2.1	1	123
2. $x = \frac{5}{7}$	2	2.1	2	124
3. $x = \frac{3}{8}$	3	2.1	3	125–126
4. $x = 0$	4	2.1	4, 5	127–129
5. No solution	5	2.1	6	129
6. All real numbers	6	2.1	7	130

ANSWER	IF YOU MISSED	REVIEW		
	QUESTION	**SECTION**	**EXAMPLES**	**PAGE**
7. $x = -6$	7	2.2	1, 2, 3	136–140
8. $x = 9$	8	2.2	1, 2, 3	136–140
9. $x = 12$	9	2.2	4	142–143
10. $x = 15$	10	2.2	5	143
11. $x = 7$	11	2.2	5	143
12. 20%	12	2.2	7	145
13. 25	13	2.2	8	145
14. $x = -4$	14	2.3	1, 2, 3	151–155
15. $h = \frac{3S}{\pi r^2}$	15	2.3	4, 5, 6	156–158
16. 30 and 45	16	2.4	1, 2	164–165
17. \$305 from bonds and \$535 from stocks	17	2.4	3	165–166
18. 20°	18	2.4	4	167
19. 5 hours	19	2.5	1, 2, 3	172–175
20. 15 pounds	20	2.5	4	176–177
21. \$12,000 in bonds, \$8000 in certificates	21	2.5	5	178–179
22. a. $m = \frac{C - 1.95}{0.85}$ **b.** 22 miles	22	2.6	1, 2	184–185
23. a. $x = 38$; 99° and 81° **b.** $x = 10$; both are 20° **c.** $x = 10$; 64° and 26°	23	2.6	3, 4, 5, 6, 7	186–191
24. a. $>$ **b.** $<$	24	2.7	1	198
25. a. 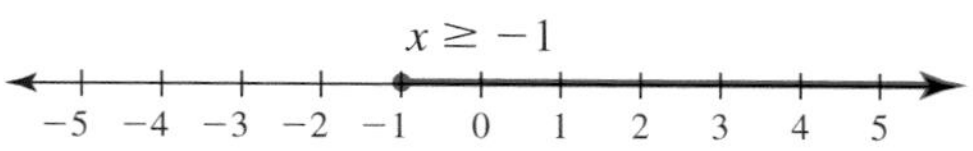 **b.** 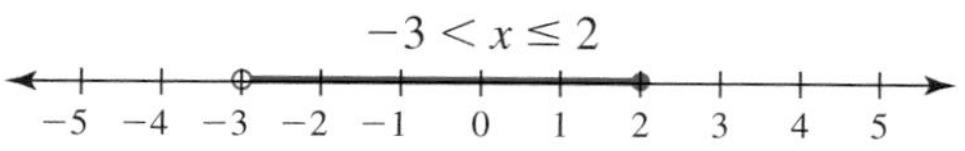	25	2.7	2, 3, 4, 5, 6	199–206

Cumulative Review Chapters 1–2

1. Find the additive inverse (opposite) of −7. 7

2. Find: $\left|-9\frac{9}{10}\right|$ $9\frac{9}{10}$

3. Find: $-\frac{2}{7}+\left(-\frac{2}{9}\right)$ $-\frac{32}{63}$

4. Find: $-0.7-(-8.9)$ 8.2

5. Find: $(-2.4)(3.6)$ −8.64

6. Find: $-(2^4)$ −16

7. Find: $-\frac{7}{8}\div\left(-\frac{5}{24}\right)$ $\frac{21}{5}$

8. Evaluate $y\div 5\cdot x-z$ for $x=6$, $y=60$, $z=3$. 69

9. Which law is illustrated by the following statement?

   $9\cdot(8\cdot 5)=9\cdot(5\cdot 8)$ Commutative law of multiplication

10. Multiply: $6(5x+7)$ $30x+42$

11. Combine like terms: $-5cd-(-6cd)$ cd

12. Simplify: $2x-2(x+4)-3(x+1)$ $-3x-11$

13. Write in symbols: The quotient of $(a-4b)$ and c $\frac{a-4b}{c}$

14. Does the number 4 satisfy the equation $11=15-x$? Yes

15. Solve for x: $5=4(x-3)+4-3x$ $x=13$

16. Solve for x: $-\frac{7}{3}x=-21$ $x=9$

17. Solve for x: $\frac{x}{3}-\frac{x}{5}=2$ $x=15$

18. Solve for x: $4-\frac{x}{4}=\frac{2(x+1)}{9}$ $x=8$

19. Solve for b in the equation $S=6a^2b$. $b=\frac{S}{6a^2}$

20. The sum of two numbers is 155. If one of the numbers is 35 more than the other, what are the numbers? 60 and 95

21. Maria has invested a certain amount of money in stocks and bonds. The annual return from these investments is $595. If the stocks produce $105 more in returns than the bonds, how much money does Maria receive annually from each type of investment? $350 from stocks and $245 from bonds

22. Train A leaves a station traveling at 40 mph. Six hours later, train B leaves the same station traveling in the same direction at 50 mph. How long does it take for train B to catch up to train A? 24 hr

23. Arlene purchased some municipal bonds yielding 12% annually and some certificates of deposit yielding 14% annually. If Arlene's total investment amounts to $5000 and the annual income is $660, how much money is invested in bonds and how much is invested in certificates of deposit? $2000 in bonds and $3000 in certificates of deposit

24. Solve and graph: $-\frac{x}{6}+\frac{x}{5}\le\frac{x-5}{5}$

    $x\ge 6$

    0 6

Graphs of Linear Equations

3

The Human Side of Algebra

René Descartes, "the reputed founder of modern philosophy," was born March 31, 1596, near Tours, France. His frail health caused his formal education to be delayed until he was 8, when his father enrolled him at the Royal College at La Fleche. It was soon noticed that the boy needed more than normal rest, and he was advised to stay in bed as long as he liked in the morning. Descartes followed this advice and made a lifelong habit of staying in bed late whenever he could.

The idea of analytic geometry came to Descartes while he watched a fly crawl along the ceiling near a corner of his room. Descartes described the path of the fly in terms of its distance from the adjacent walls by developing the *Cartesian coordinate system,* which we study in Section 3.1. Impressed by his knowledge of philosophy and mathematics, Queen Christine of Sweden engaged Descartes as a private tutor. He arrived in Sweden to discover that she expected him to teach her philosophy at 5 o'clock in the morning in the ice-cold library of her palace. Deprived of his beloved morning rest, Descartes caught "inflammation of the lungs," from which he died on February 11, 1650, at age 53.

Pretest for Chapter 3

(Answers on pages 225–226)

1. Find the coordinates of each point and determine the quadrant in which it is located:

a. A

b. B

c. C

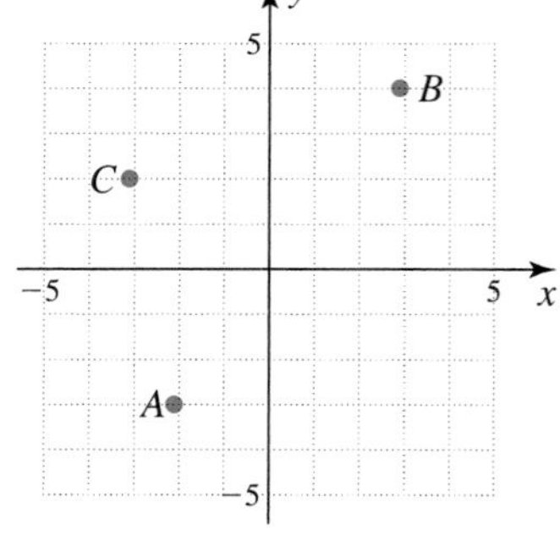

2. The ordered pairs represent the old wind chill temperature in degrees Fahrenheit when the wind speed in miles per hour is as indicated.

a. Graph the ordered pairs.

Wind Speed	Wind Chill
10	−15
20	−31
30	−41

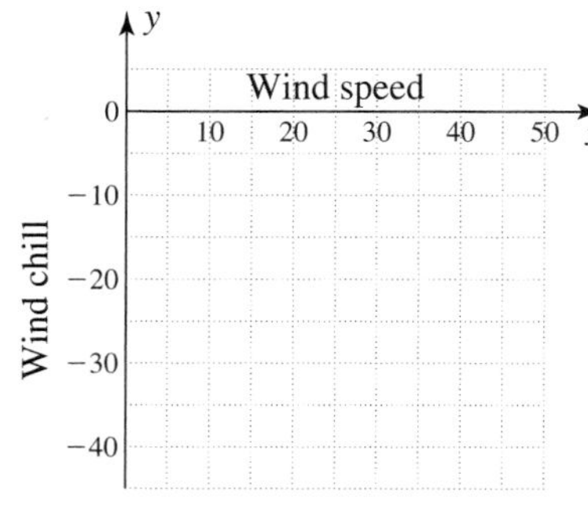

b. What does $(20, -31)$ mean?

c. What is the number s in $(s, -15)$?

3. Determine whether the ordered pair $(1, -2)$ is a solution of $4x - y = -2$.

4. Find x in the ordered pair $(x, 6)$ so that the ordered pair satisfies the equation $4x - y = 10$.

5. Graph $y = -\frac{2}{3}x - 2$.

6. The average daily consumption g of protein (in grams) per person can be approximated by $g = 100 + 0.8t$, where t is the number of years after 2000. Graph $g = 100 + 0.8t$.

7. Graph $5x - 2y - 10 = 0$.

8. Graph $4x - 6y = 0$.

9. Graph $4x - 8 = 0$.

10. Find the slope of the line going through:

a. $(3, -7)$ and $(-3, -1)$

b. $(4, -6)$ and $(-2, 4)$

11. Find the slope of the line going through:

a. $(-3, -2)$ and $(-3, 6)$

b. $(3, 4)$ and $(4, 4)$

12. Find the slope of the line $y - 2x = 8$.

13. Decide whether the lines are parallel, perpendicular, or neither.

a. $3y = x + 5$
$2x - 6y = 6$

b. $3y = -x + 5$
$9x - 3y = 6$

14. The number N of stores in a city t years after 2000 can be approximated by

$$N = 0.6t + 15 \text{ (hundred)}$$

a. What is the slope of this line?

b. What does the slope represent?

c. How many stores were added each year?

Answers to Pretest

<table>
<tr><th>ANSWER</th><th>IF YOU MISSED</th><th colspan="3">REVIEW</th></tr>
<tr><th></th><th>QUESTION</th><th>SECTION</th><th>EXAMPLES</th><th>PAGE</th></tr>
<tr><td>1. a. $(-2, -3)$; Quadrant III</td><td>1</td><td>3.1</td><td>2</td><td>229</td></tr>
<tr><td>b. $(3, 4)$; Quadrant I</td><td></td><td>3.1</td><td>5</td><td>232</td></tr>
<tr><td>c. $(-3, 2)$; Quadrant II</td><td></td><td></td><td></td><td></td></tr>
<tr><td>2. a. 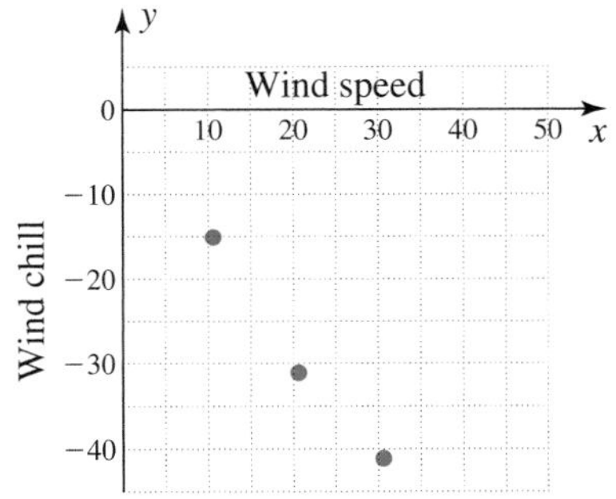
b. When the wind speed is 20 mi/hr, the wind chill is −31°F.
c. 10</td><td>2</td><td>3.1</td><td>6</td><td>233</td></tr>
<tr><td>3. No</td><td>3</td><td>3.2</td><td>1</td><td>242</td></tr>
<tr><td>4. $x = 4$</td><td>4</td><td>3.2</td><td>2</td><td>244</td></tr>
<tr><td>5. 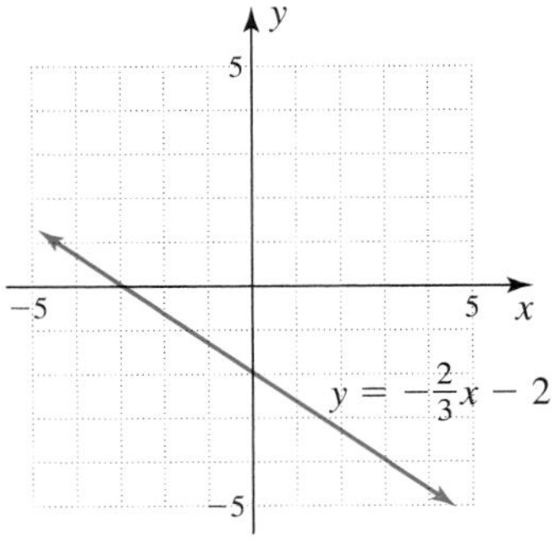</td><td>5</td><td>3.2</td><td>5</td><td>247</td></tr>
<tr><td>6. 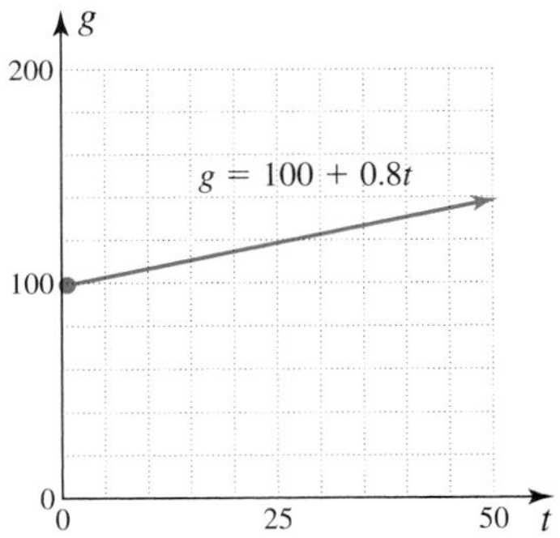</td><td>6</td><td>3.2</td><td>6</td><td>248</td></tr>
</table>

ANSWER	IF YOU MISSED	REVIEW		
	QUESTION	SECTION	EXAMPLES	PAGE
7. 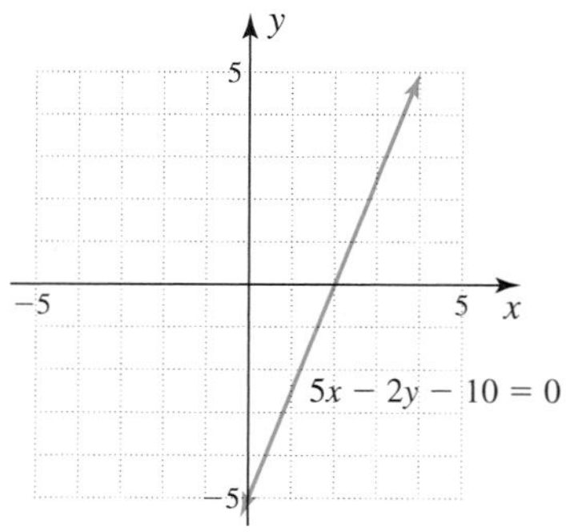	7	3.3	1, 2	255–256
8. 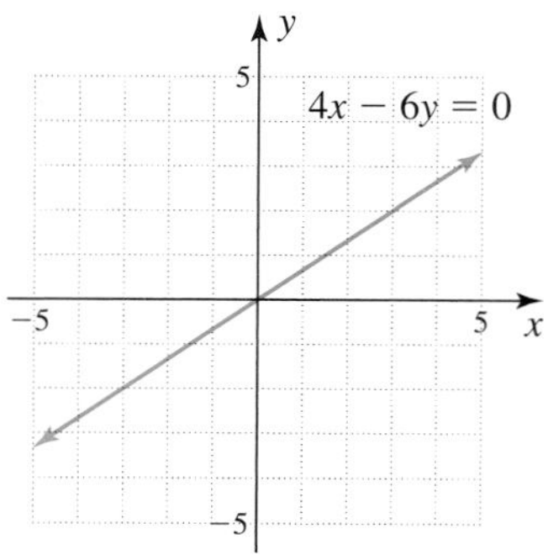	8	3.3	3	258
9. 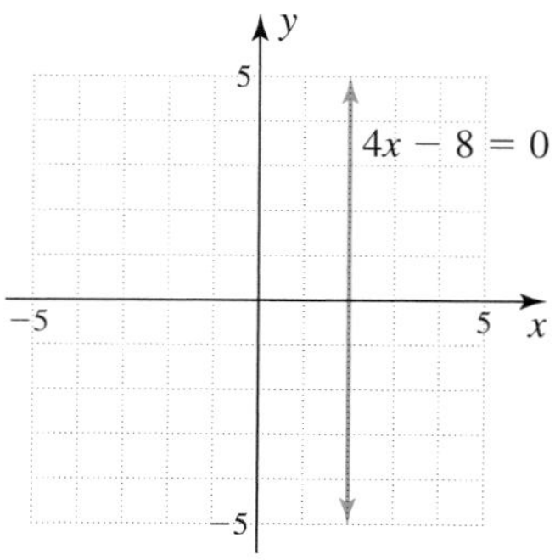	9	3.3	4a	260
10. a. -1 **b.** $-\frac{5}{3}$	10	3.4	1, 2	269
11. a. Undefined **b.** 0	11	3.4	3	270
12. 2	12	3.4	4	271
13. a. Parallel **b.** Perpendicular	13	3.4	5	273
14. a. 0.6 **b.** Annual increase in the number of stores **c.** 60	14	3.4	6	274

3.1 GRAPHS AND APPLICATIONS

To Succeed, Review How To . . .

1. Evaluate an expression (p. 47).
2. Solve linear equations (p. 151).

Objectives

A Graph (plot) ordered pairs of numbers.

B Determine the coordinates of a point in the plane.

C Read and interpret ordered pairs on a line graph.

D Read and interpret ordered pairs on a bar graph.

E Find the quadrant in which a point lies.

F Given a chart or ordered pairs, make a line graph.

GETTING STARTED

Hurricanes and Graphs

The map shows the position of Hurricane Desi. The hurricane is near the intersection of the vertical line indicating 90° longitude and the horizontal line indicating 25° latitude. This point can be identified by assigning to it an **ordered pair** of numbers, called **coordinates,** showing the longitude first and the latitude second. Thus the hurricane would have the coordinates

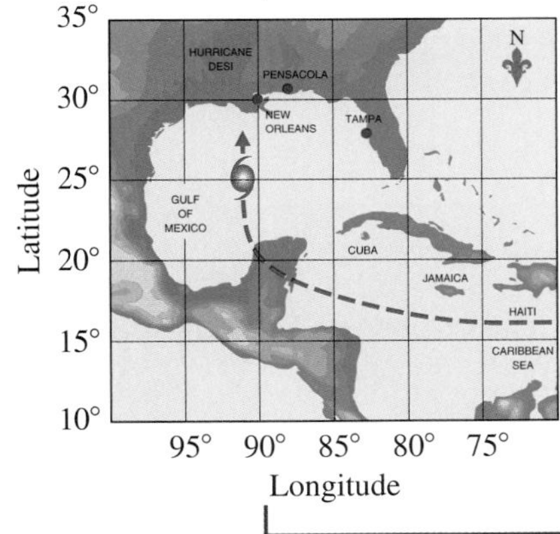

(91, 25)

This is the longitude (units right or left). This is the latitude (units up or down).

This method is used to give the position of cities, islands, ships, airplanes, and so on. For example, the coordinates of New Orleans on the map are (90, 30), whereas those of Pensacola are approximately (88, 31). In mathematics we use a system very similar to this one to locate points in a plane. In this section we learn how to graph points in a *Cartesian plane* and then examine the relationship of these points to linear equations.

−4 −3 −2 −1 0 1 2 3 4

Figure 1

Here is the way we construct a **Cartesian coordinate system:**

1. Draw a number line (Figure 1).
2. Draw another number line perpendicular to the first one and crossing it at 0 (see Figure 2).

On the number line, each point on the graph is a **number.** On a coordinate plane, each point is the **graph** of an ordered pair. The individual numbers in an ordered pair are called **coordinates.** For example, the point P in Figure 3 is associated with the ordered pair (2, 3). The first coordinate in P is 2 and the second coordinate is 3. The point $Q(-1, 2)$ has a first coordinate of -1 and a second coordinate of 2. We call the horizontal number line the ***x*-axis** and label it with the letter x; the vertical number line is called the ***y*-axis** and is labeled with the letter y. We can now say that the point $P(2, 3)$ has x-coordinate (**abcissa**) 2, and y-coordinate (**ordinate**) 3.

Sometimes locating or graphing a point in the coordinate plane is called "plotting the point."

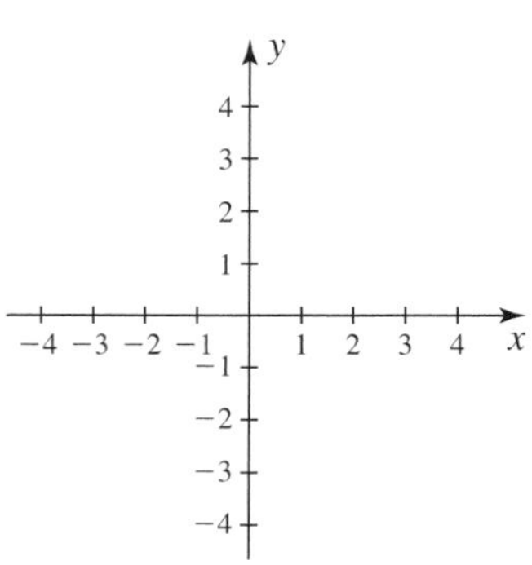

Figure 2

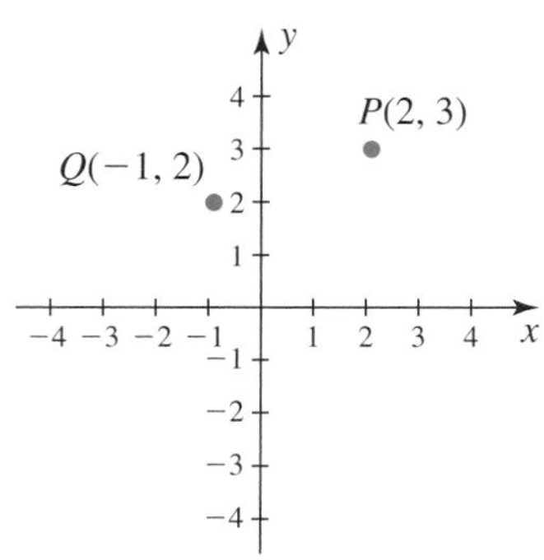

Figure 3

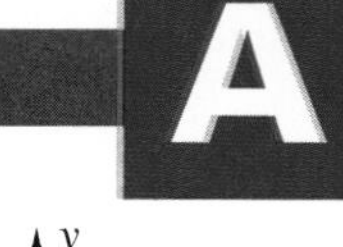

Graphing Ordered Pairs

In general, if a point P has coordinates (x, y), we can always locate or graph the point in the coordinate plane. We start at the origin and go x units to the *right* if x is *positive;* we go to the *left* if x is *negative*. We then go y units *up* if y is *positive, down* if y is *negative*.

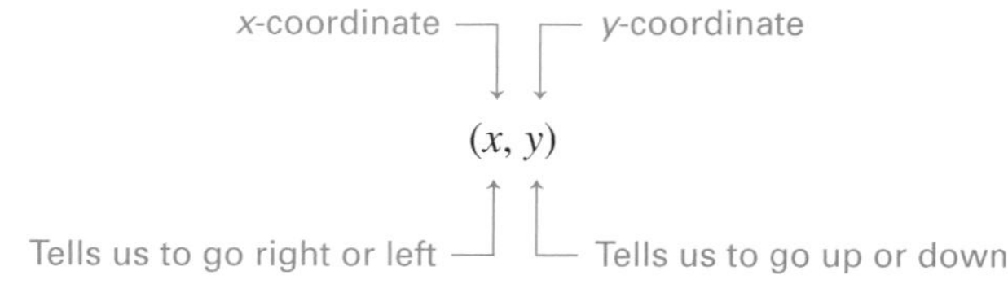

Figure 4

For example, to graph the point $R(3, -2)$, we start at the origin and go 3 units right (since the x-coordinate 3 is positive) and 2 units down (since the y-coordinate -2 is negative). The point is graphed in Figure 4.

Web It

For interactive graphing, try link 3-1-1 at the Bello Website at mhhe.com/bello.

NOTE

All points on the x-axis have y-coordinate 0 (zero units up or down); all points on the y-axis have x-coordinate 0 (zero units right or left).

EXAMPLE 1 **Graphing points in the coordinate plane**

Graph the points:

a. $A(1, 3)$ **b.** $B(2, -1)$
c. $C(-4, 1)$ **d.** $D(-2, -4)$

SOLUTION

a. We start at the origin. To reach point $(1, 3)$, we go 1 unit to the right and 3 units up. The graph of A is shown in Figure 5.

b. To graph $(2, -1)$, we start at the origin, go 2 units right and 1 unit down. The graph of B is shown in Figure 5.

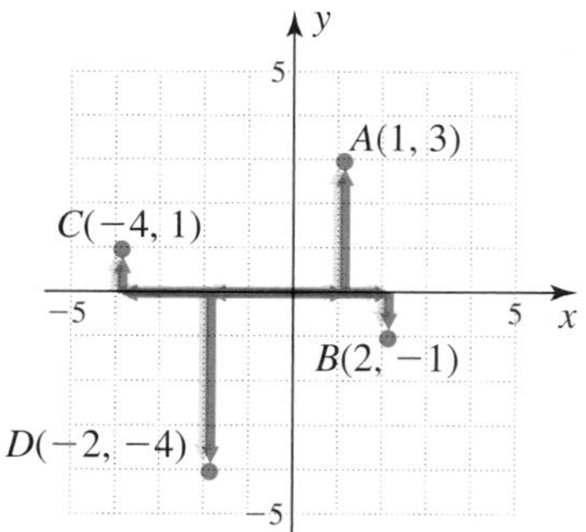

Figure 5

PROBLEM 1

Graph:

a. $A(2, 4)$ **b.** $B(4, -2)$
c. $C(-3, 2)$ **d.** $D(-4, -4)$

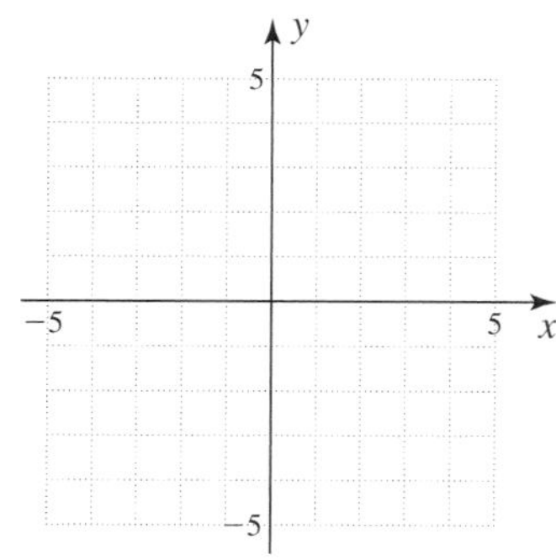

c. As usual, we start at the origin: $(-4, 1)$ means to go 4 units left and 1 unit up, as shown in Figure 5.

d. The point $D(-2, -4)$ has both coordinates negative. Thus from the origin we go 2 units left and 4 units down; see Figure 5.

Finding Coordinates

Since every point P in the plane is associated with an ordered pair (x, y), we should be able to find the coordinates of any point as shown in Example 2.

EXAMPLE 2 Finding coordinates

Determine the coordinates of each of the points in Figure 6.

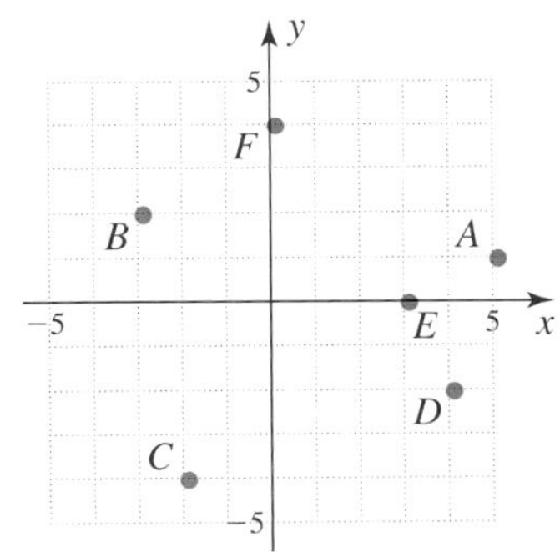

Figure 6

SOLUTION

Point A is 5 units to the right of the origin and 1 unit above the horizontal axis. The ordered pair corresponding to A is $(5, 1)$. The coordinates of the other four points can be found in a similar manner. Here is the summary.

Point	Start at the origin, move:	Coordinates
A	5 units *right,* 1 unit *up*	$(5, 1)$
B	3 units *left,* 2 units *up*	$(-3, 2)$
C	2 units *left,* 4 units *down*	$(-2, -4)$
D	4 units *right,* 2 units *down*	$(4, -2)$
E	3 units *right,* 0 units *up*	$(3, 0)$
F	0 units *right,* 4 units *up*	$(0, 4)$

PROBLEM 2

Determine the coordinates of each of the points.

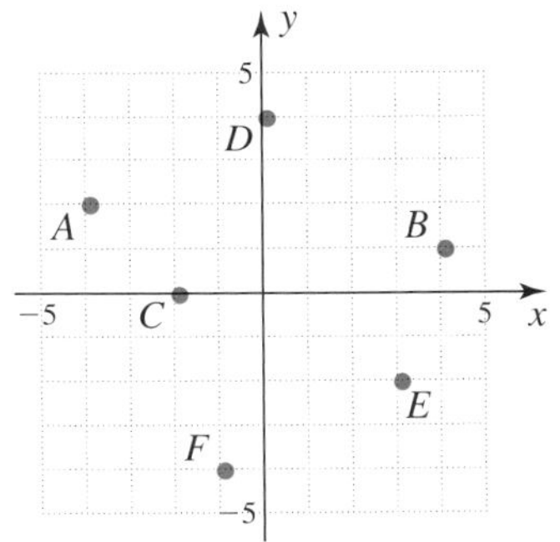

Answers

1.

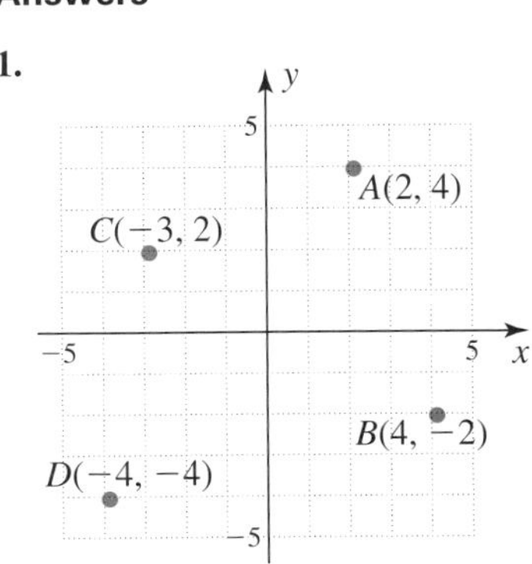

2. $A(-4, 2)$; $B(4, 1)$; $C(-2, 0)$; $D(0, 4)$; $E(3, -2)$; $F(-1, -4)$

Applications: Line Graphs

Now that you know how to find the coordinates of a point, we learn how to read and interpret line graphs.

EXAMPLE 3 Going to the dogs

Figure 7 gives the age coversion from human years to dog years. The ordered pair (1, 12) means that 1 human year is equivalent to about 12 dog years.

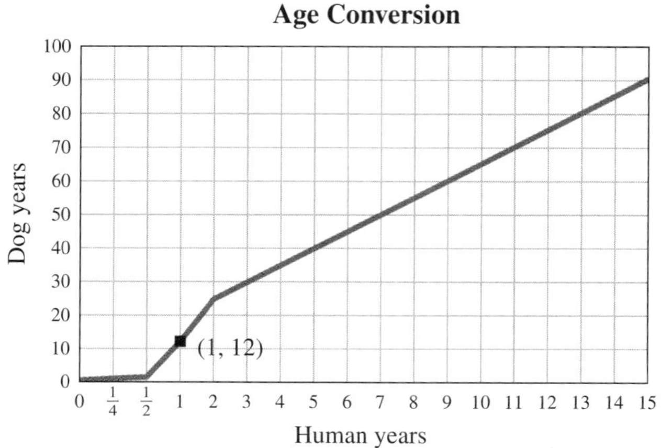

Figure 7
Source: Data from Cindy's K9 Clips.

a. What does the ordered pair (3, 30) represent?

b. If a dog is 9 years old in human years, how old is it in dog years? Write the ordered pair corresponding to this situation.

c. If retirement age is 65, what is the retirement age for dogs in human years? That is, how many human years correspond to 65 dog years? Write the ordered pair corresponding to this situation.

SOLUTION

a. Three human years is equivalent to 30 dog years.

b. Start at (0, 0) and move right to 9 on the horizontal axis. (See Figure 8.) Now, go up until you reach the graph as shown. The point is 60 units high (the y-coordinate is 60). Thus, 9 years old in human years is equivalent to 60 years old in dog years. The ordered pair corresponding to this situation is (9, 60).

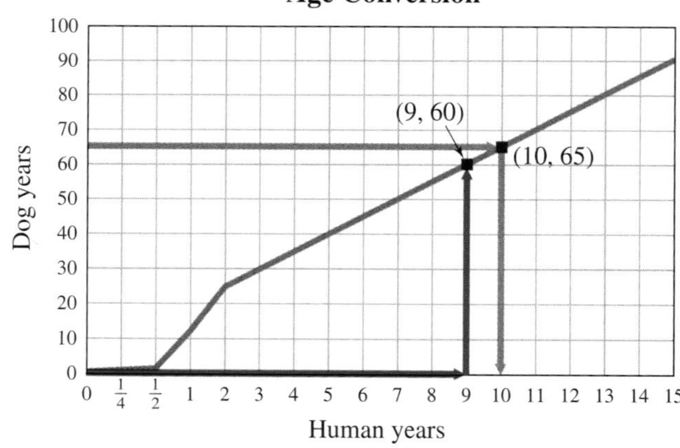

Figure 8

c. We have to find how many human years are represented by 65 dog years. This time, we go to the point at which y is 65 units high, then move right until we reach the graph. At the point for which y is 65 on the graph, x is 10. Thus, the equivalent retirement age for dogs is 10 human years. At that age, they are entitled to Canine Security benefits! The ordered pair corresponding to this situation is (10, 65).

PROBLEM 3

a. What does the ordered pair (5, 40) in Figure 7 represent?

b. If a dog is 11 human years old, how old is it in dog years?

c. If the drinking age for humans is 21, what is the equivalent drinking age for dogs in human years? (Answer to the nearest whole number.)

Answers

3. a. 5 human years are equivalent to 40 dog years. **b.** 70 **c.** 2

Applications: Bar Graphs

Another popular use of ordered pairs is **bar graphs,** in which certain *categories* are paired with certain numbers. For example, if you have a \$1000 balance at 18% interest and are making the minimum \$25 payment each month on your credit card, it will take you forever (actually 5 years) to pay it off. If you decide to pay it off in 12 months, how much would your payment be? The bar graph in Figure 9 tells you, provided you know how to read it! First, start at the 0 point and move right horizontally until you get to the category labeled 12 months (blue arrow), then go up vertically to the end of the bar (red arrow). According to the vertical scale labeled Monthly payment, the arrow is 92 units long, meaning that the monthly payment will be \$92 per month. The ordered pair corresponding to this situation is (12, 92).

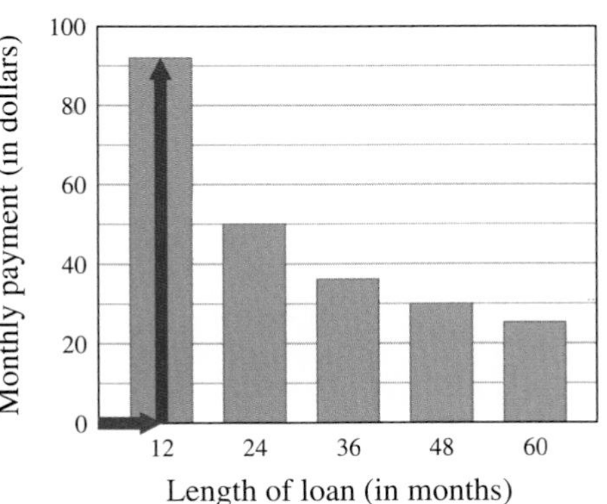

Figure 9
Source: Data from KJE Computer Solutions, LLC.

EXAMPLE 4 **Paying off your debt**

a. Referring to the graph, what would your payment be if you decide to pay off the \$1000 in 24 months?

b. How much would you save if you do so?

SOLUTION

a. To find the payment corresponding to 24 months, move right on the horizontal axis to the category labeled 24 months and then vertically to the end of the bar in Figure 10. According to the vertical scale, the monthly payment will be \$50.

b. If you pay the minimum \$25 payment for 60 months, you would pay $60 \times \$25 = \$1500$. If you pay \$50 for 24 months, you would pay $24 \times \$50 = \$1200$ and would save \$300 ($\$1500 - \$1200$).

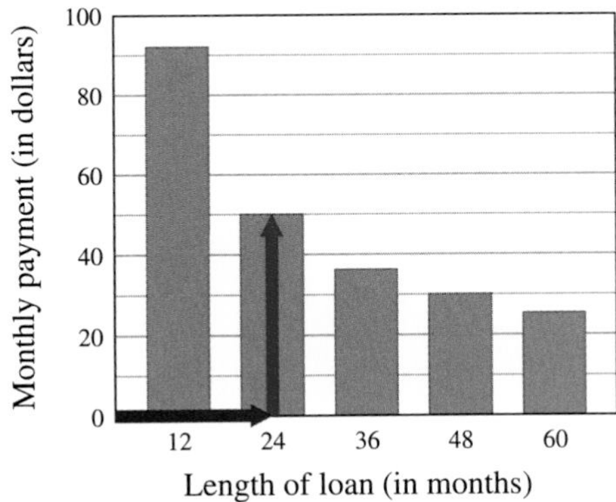

Figure 10
Source: Data from KJE Computer Solutions, LLC.

PROBLEM 4

a. What would your payment be if you decide to pay off the \$1000 in 48 months?

b. How much would you save if you do so?

Answers

4. a. \$30 per month
b. \$60 ($\$1500 - \$1440$)

E Quadrants

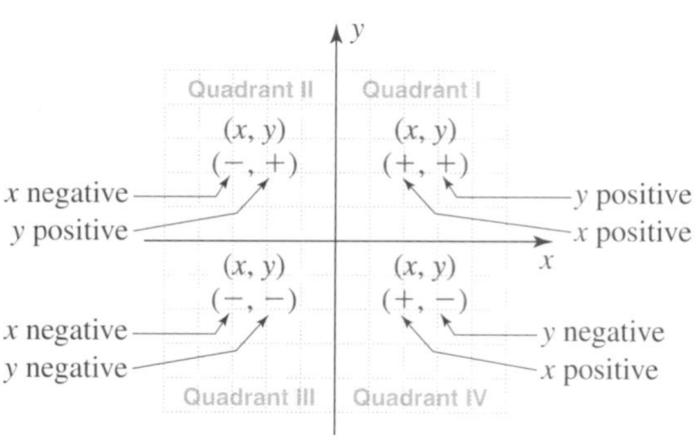

Figure 11

Figure 11 shows that the x- and y-axes divide the plane into four regions called **quadrants.** These quadrants are numbered in counterclockwise order using Roman numerals and starting in the upper right-hand region. What can we say about the coordinates of the points in each quadrant?

In Quadrant I: Both coordinates positive

In Quadrant II: First coordinate negative, second positive

In Quadrant III: Both coordinates negative

In Quadrant IV: First coordinate positive, second negative

EXAMPLE 5 Find the quadrant

In which quadrant, if any, are the points located?

a. $A(-3, 2)$ **b.** $B(1, -2)$ **c.** $C(3, 4)$

d. $D(-4, -1)$ **e.** $E(0, 3)$ **f.** $F(2, 0)$

SOLUTION The points are shown in the graph.

a. The point $A(-3, 2)$ is in the second quadrant.

b. The point $B(1, -2)$ is in the fourth quadrant.

c. The point $C(3, 4)$ is in the first quadrant.

d. The point $D(-4, -1)$ is in the third quadrant.

e. The point $E(0, 3)$ is on the y-axis (*no quadrant*).

f. The point $F(2, 0)$ is on the x-axis (*no quadrant*).

At this point, you may be wondering: Why do we need to know about quadrants? There are two reasons:

1. If you are using a graphing calculator, you have to specify the WINDOW (basically, the quadrants) you want to display.
2. If you are graphing ordered pairs, you need to adjust your graph to show the quadrants in which your ordered pairs lie. We will illustrate how this works in the next section.

PROBLEM 5

In which quadrant, if any, are the points located?

a. $A(1, 3)$ **b.** $B(-2, -2)$

c. $C(-4, 2)$ **d.** $D(2, -1)$

e. $E(0, 1)$ **f.** $F(3, 0)$

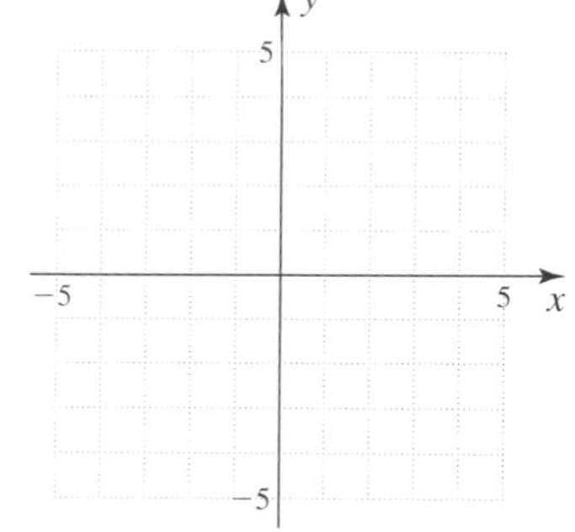

Answers

5. a. QI **b.** QIII **c.** QII
d. QIV **e.** y-axis **f.** x-axis

Applications: Making Line Graphs

Suppose you want to graph the ordered pairs in the table (we give equal time to cats!). The ordered pair (c, h) would pair the actual age c of a cat (in "regular" time) to the equivalent human age h. Since both c and h are positive, the graph would have to be in Quadrant I. Values of c would be from 0 to 21 and values of h from 10 to 100, suggesting a scale of 10 units for each equivalent human year. Let us do all this in Example 6.

Cat's Actual Age	Equivalent Human Age	Cat's Actual Age	Equivalent Human Age
6 months	10 years	10 years	56 years
8 months	13 years	12 years	64 years
1 year	15 years	14 years	72 years
2 years	24 years	16 years	80 years
4 years	32 years	18 years	88 years
6 years	40 years	20 years	96 years
8 years	48 years	21 years	100 years

It was once thought that 1 year in the life of a cat was equivalent to 7 years of a human life. Recently, a new scale has been accepted: after the first 2 years, the cat's life proceeds more slowly in relation to human life and each feline year is approximately 4 human years. The general consensus is that at about age 7 a cat can be considered "middle-aged," and age 10 and beyond "old."

Source: Data from X Mission Internet.

EXAMPLE 6 Equal time for cats

a. Make a coordinate grid to graph the ordered pairs in the table.

b. Graph the ordered pairs starting with (1, 15).

c. What does (21, 100) mean?

d. Study the pattern (2, 24), (4, 32), (6, 40). What would be the number h in $(8, h)$?

e. What is the number c in $(c, 56)$?

f. The oldest cat, Spike, lived to an equivalent human age of 140 years. What was Spike's actual age?

SOLUTION Parts **a** and **b** are shown in the graph.

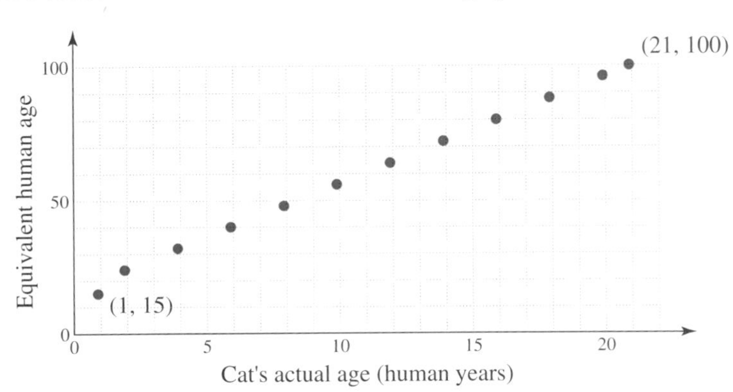

PROBLEM 6

a. What does the ordered pair (16, 80) in the table mean?

b. What is h in $(12, h)$?

c. What is c in $(c, 72)$?

d. CNN reports that there is a cat living in Texas whose equivalent human age is 148 years. What is the cat's actual age?

Source: CNN.

Answers

6. a. (16, 80) means that if a cat's actual age is 16, its equivalent human age is 80. **b.** 64 **c.** 14 **d.** 33

c. (21, 100) means that if a cat's actual age is 21, its equivalent human age is 100 years.

d. The numbers for the second coordinate are 24, 32, and 40 (increasing by 8). The next number would be 48, so the next ordered pair would be (8, 48). You can verify this from the table or the graph.

e. From the table or the graph, the ordered pair whose second coordinate is 56 is (10, 56), so $c = 10$.

f. Neither the table nor the graph goes to 140 (the graph stops at 100). But we can look at the last two ordered pairs in the table and follow the pattern, or we can extend the graph and find the answer. Here are the last two ordered pairs in the table:

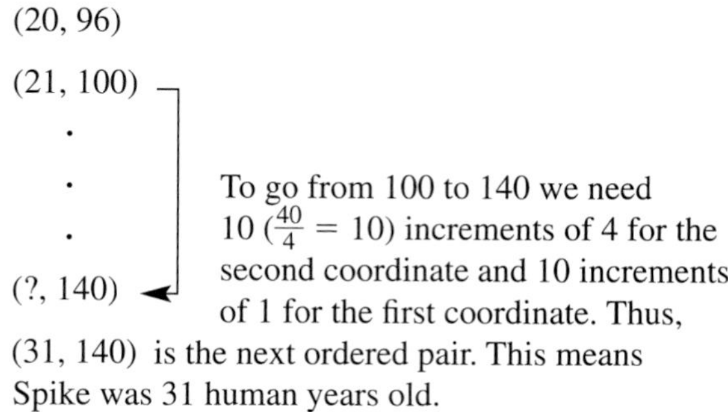

(20, 96)

(21, 100)

.

.

.

(?, 140)

To go from 100 to 140 we need 10 ($\frac{40}{4} = 10$) increments of 4 for the second coordinate and 10 increments of 1 for the first coordinate. Thus, (31, 140) is the next ordered pair. This means Spike was 31 human years old.

As you can see, *line graphs* can be used to show a certain change over a period of time. If we were to connect the points in the graph of Example 6 starting with (2, 24), we would have a *straight line.* We shall study linear equations and their graphs in the next section.

Calculate It Graphing Points

We can do most of the work in this section with our calculators. To do Example 1, first adjust the **viewing screen** or **window** of the calculator. Since the values of x range from -5 to 5, denoted by $[-5, 5]$, and the values of y also range from -5 to 5, we have a $[-5, 5]$ by $[-5, 5]$ viewing rectangle, or window, as shown in Window 1.

```
WINDOW
Xmin=-5
Xmax=5
Xscl=1
Ymin=-5
Ymax=5
Yscl=1
```

Window 1

To graph the points A, B, C, and D, set the calculator on the statistical graph mode (2nd Y= 1), turn the plot on (ENTER), select the type of plot, and the list of numbers you are going to use for the x-coordinates (L_1) and the y-coordinates (L_2) as well as the type of mark you want the calculator to make (Window 2). Now, press STAT 1 and enter the x-coordinates of A, B, C, and D under L_1 and the y-coordinates of A, B, C, and D under L_2. Finally, press GRAPH to obtain the points A, B, C, and D in Window 3. Use these ideas to do Problems 1–5 in the exercise set.

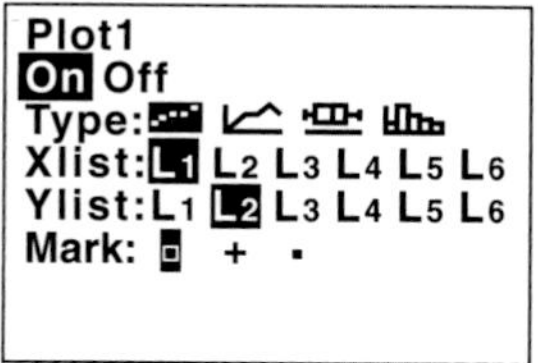

Window 2

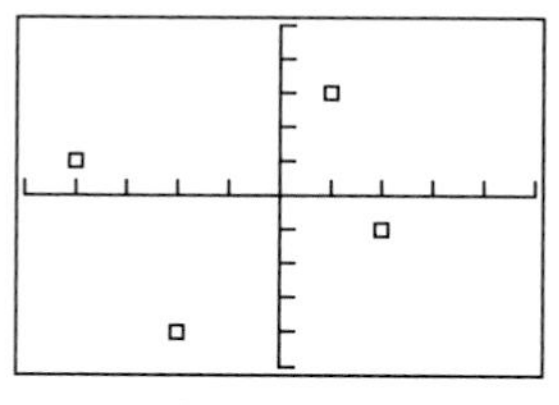

Window 3

Exercises 3.1

A In Problems 1–5, graph the points.

1. **a.** $A(1, 2)$ **b.** $B(-2, 3)$ **c.** $C(-3, 1)$ **d.** $D(-4, -1)$

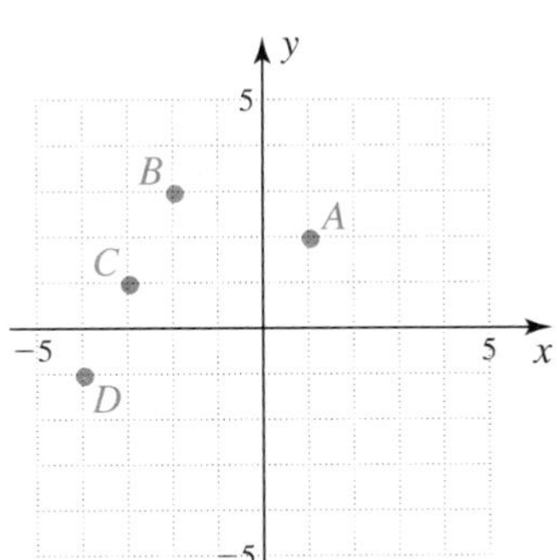

2. **a.** $A\left(-2\frac{1}{2}, 3\right)$ **b.** $B\left(-1, 3\frac{1}{2}\right)$ **c.** $C\left(-\frac{1}{2}, -4\frac{1}{2}\right)$ **d.** $D\left(\frac{1}{3}, 4\right)$

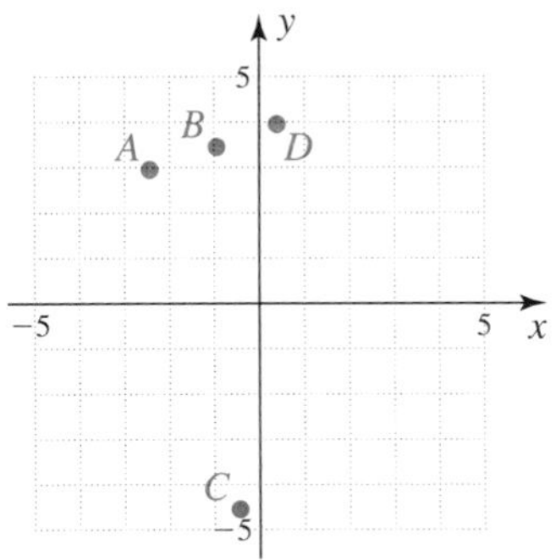

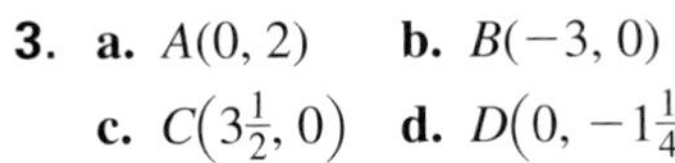

3. **a.** $A(0, 2)$ **b.** $B(-3, 0)$ **c.** $C\left(3\frac{1}{2}, 0\right)$ **d.** $D\left(0, -1\frac{1}{4}\right)$

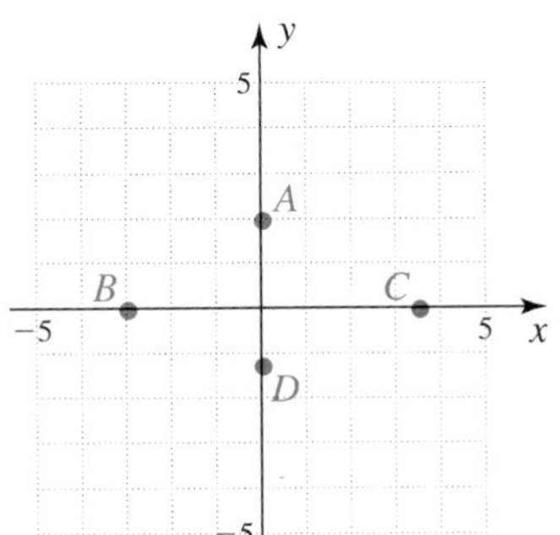

4. **a.** $A(20, 20)$ **b.** $B(-10, 20)$ **c.** $C(35, -15)$ **d.** $D(-25, -45)$

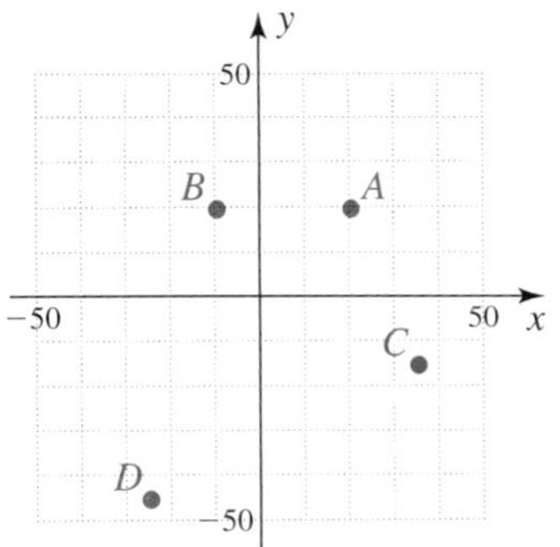

5. **a.** $A(0, 40)$ **b.** $B(-35, 0)$ **c.** $C(-40, -15)$ **d.** $D(0, -25)$

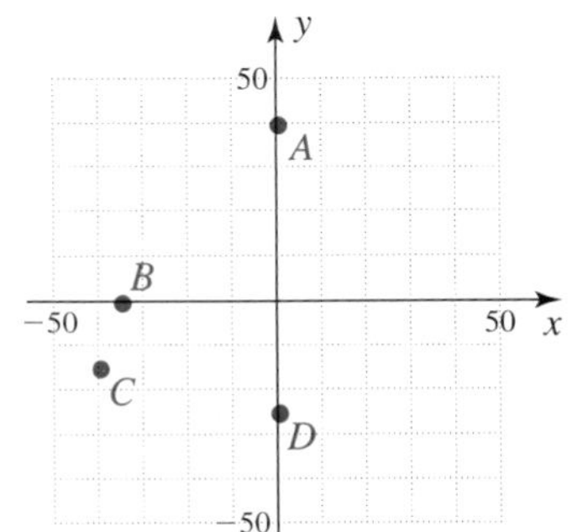

B–E In Problems 6–10, give the coordinates of the points and the quadrant in which each point lies.

6.

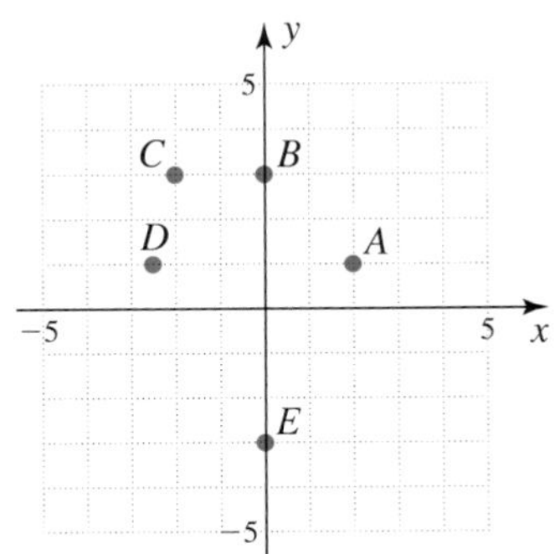

$A(2, 1)$; $B(0, 3)$; $C(-2, 3)$; $D\left(-2\frac{1}{2}, 1\right)$; $E(0, -3)$; Quadrant I: A; Quadrant II: C, D; y-axis: B, E

7.

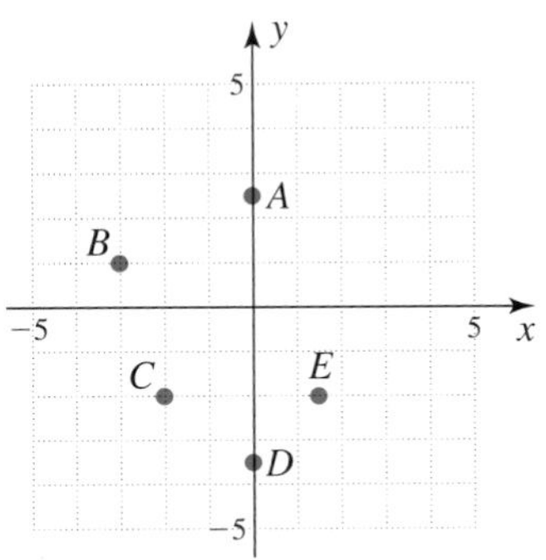

$A\left(0, 2\frac{1}{2}\right)$; $B(-3, 1)$; $C(-2, -2)$; $D\left(0, -3\frac{1}{2}\right)$; $E\left(1\frac{1}{2}, -2\right)$; Quadrant II: B; Quadrant III: C; Quadrant IV; E; y-axis: A, D

8.

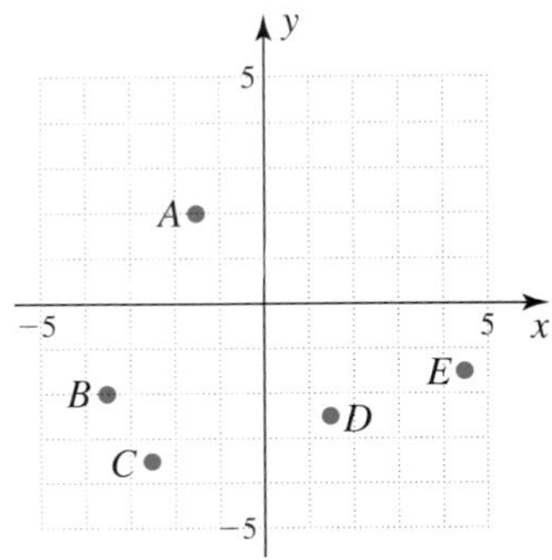

$A\left(-\frac{1}{2}, 2\right)$; $B\left(-3\frac{1}{2}, -2\right)$; $C\left(-2\frac{1}{2}, -3\frac{1}{2}\right)$; $D\left(1\frac{1}{2}, -2\frac{1}{2}\right)$; $E\left(4\frac{1}{2}, -1\frac{1}{2}\right)$; Quadrant II: A; Quadrant III: B, C; Quadrant IV; D, E

9.

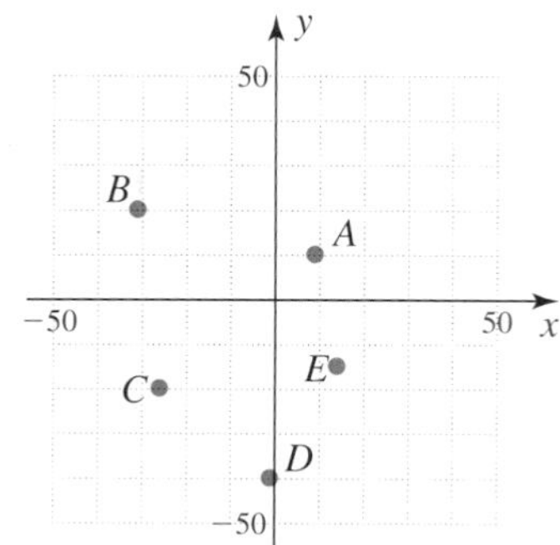

$A(10, 10)$; $B(-30, 20)$; $C(-25, -20)$; $D(0, -40)$; $E(15, -15)$; Quadrant I: A; Quadrant II: B; Quadrant III; C; Quadrant IV: E; y-axis: D

10.

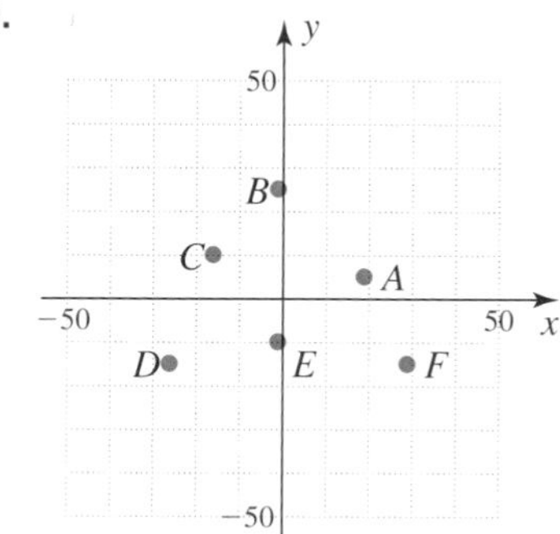

$A(20, 5)$; $B(0, 25)$; $C(-15, 10)$; $D(-25, -15)$; $E(0, -10)$; $F(30, -15)$; Quadrant I: A; Quadrant II: C; Quadrant III; D; Quadrant IV; F; y-axis: B, E

Use the following graph to work Problems 11–14.

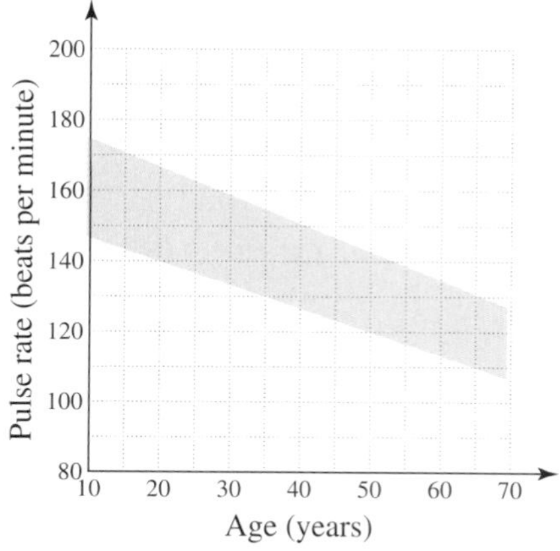

11. What is the lower limit pulse rate for a 20-year-old? Write the answer as an ordered pair. (20, 140)

12. What is the upper limit pulse rate for a 20-year-old? Write the answer as an ordered pair. (20, 170)

13. What is the upper limit pulse rate for a 45-year-old? Write the answer as an ordered pair. (45, 150)

14. What is the lower limit pulse rate for a 50-year-old? Write the answer as an ordered pair. (50, 120)

C Use the following graph to work Problems 15–24.

The green graph shows the pounds of waste generated per capita (per person) each day and the blue graph shows the total waste generated in millions of tons.

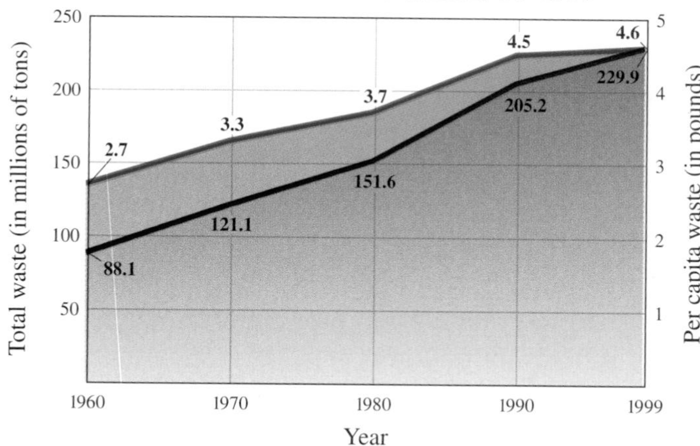

Source: Data from U.S. Environmental Protection Agency.

15. How many pounds of waste were generated per person each day in 1960? 2.7

16. How many pounds of waste were generated per person each day in 1999? 4.6

17. How many more pounds were generated per person each day in 1999 than in 1960? 1.9

18. How many more pounds were generated per person each day in 1999 than in 1990? 0.1

19. Based on your answer to Problem 18, how many pounds of waste would you expect to be generated per person each day in 2008? 4.7

20. What was the total waste (in millions of tons) generated in 1960? 88.1

21. What was the total waste (in millions of tons) generated in 1999? 229.9

22. How many more million tons of waste were generated in 1999 than in 1960? 141.8

23. How many more million tons of waste were generated in 1999 than in 1990? 24.7

24. Based on your answer to Problem 23, how many million tons of waste would you expect to be generated in 2008? 254.6

Use the following graph to work Problems 25–28.

The graph shows the amount owed on a $1000 debt at an 18% interest rate when the minimum $25 payment is made or when a new monthly payment of $92 is made. Thus, at the current monthly payment of $25, it will take 60 months to pay the $1000 balance (you got to $0!). On the other hand, with a new $92 monthly payment, you pay off the $1000 balance in 12 months.

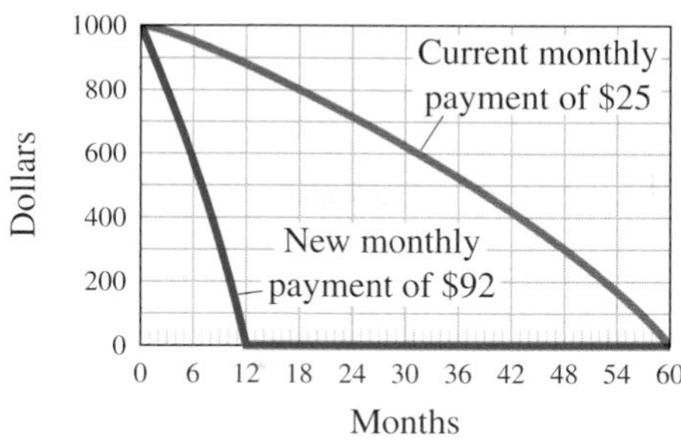

Source: Data from KJE Computer Solutions, LLC.

25. What is your balance after 6 months if you are paying $25 a month? $950

26. What is your balance after 6 months if you are paying $92 per month? $575

27. What is your balance after 18 months if you are paying $25 a month? $800

28. What is your balance after 48 months if you are paying $25 a month? $300

Use the following graph to work Problems 29–33.

Reduction in Young Americans Returning Through El Paso Border Crossing After Juarez Bars Closing Time Change

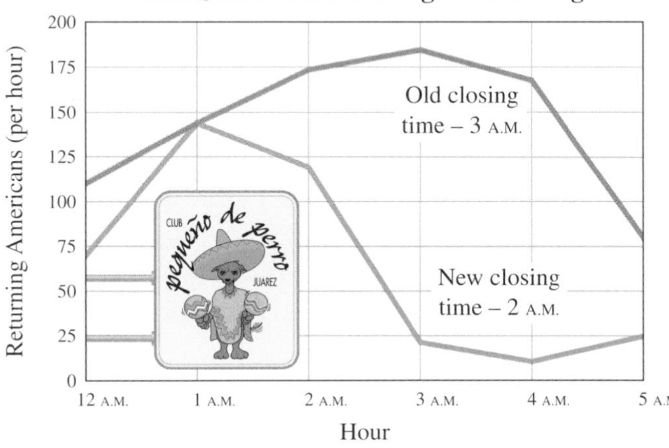

Source: Data from Institute for Public Strategies.

The graph shows the number of young Americans crossing the border at El Paso after the bar closing hour in Juarez, Mexico, changed from 3 A.M. to 2 A.M.

29. About how many people crossed the border at 12 A.M. when the new bar closing time was 2 A.M.? 70

30. About how many people crossed the border at 12 A.M. when the old bar closing time was 3 A.M.? 110

31. At what time was the difference in the number of border crossers greatest? Can you suggest an explanation for this? 3 A.M.

32. At what time was the number of border crossers the same? 1 A.M.

33. What was the difference in the number of border crossers at 5 A.M.? 55

D Use the following graph to work Problems 34–38.

Change in Number of Crossers with BAC* Over 0.08 per Weekend Night from the Juarez Bar Closing Time Shift

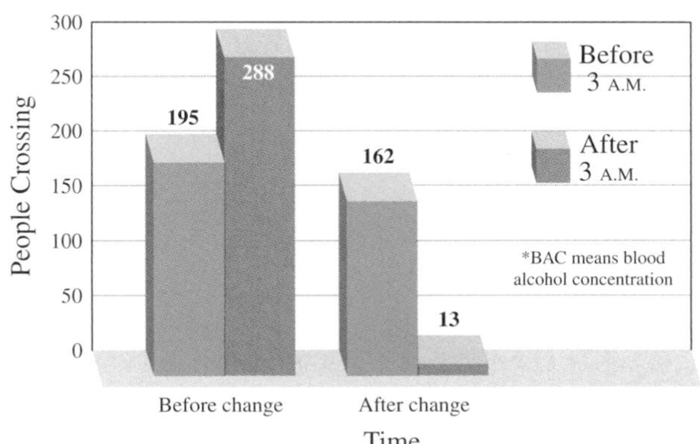

Source: Data from Institute for Public Strategies.

The numbers of crossers before 3 A.M. and after 3 A.M. are shown in the graph.

34. What was the total number of crossers before the change? 483

35. How many of those were before 3 A.M. and how many after 3 A.M.? Before 3 A.M.: 195; after 3 A.M.: 288

36. How many crossers were there after the change? 175

37. How many of those were before and how many after 3 A.M.? Before 3 A.M.: 162; after 3 A.M.: 13

38. Why do you think there were so many crossers before the change? Answers will vary.

Use the following graph to work Problems 39–42.

The graph shows the percent of selected materials and the number of tons recycled in each category in a recent year. The total amount of recycled materials was 70 million tons.

Recycling Rates of Selected Materials
(70 million tons total)

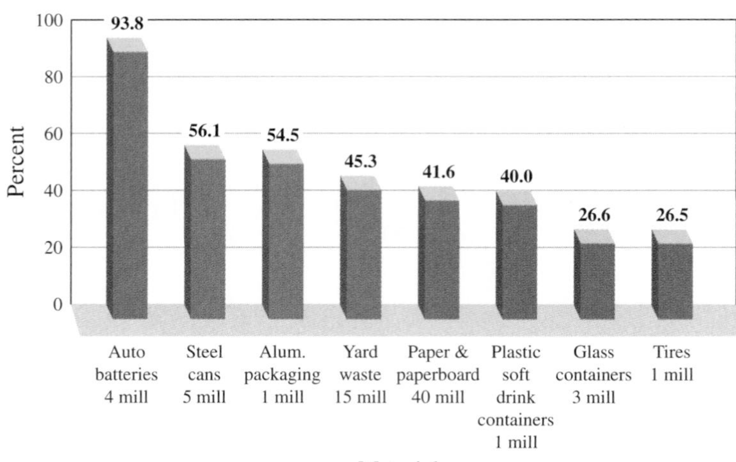

Source: Data from U.S. Environmental Protection Agency.

39. Which material was:

a. The most recycled? Auto batteries

b. The least recycled? Tires

40. What percent of the auto batteries were recycled? What percent of the 70 million tons total were auto batteries? 93.8, 5.71%

41. What percent of the tires were recycled? What percent of the 70 million tons total were tires? 26.5; 1.43%

42. How many more tons of glass containers than aluminum packaging were recycled? 2 million tons

Use the following graph to work Problems 43–46.

How does the down payment affect the monthly payment? The graph tells you!

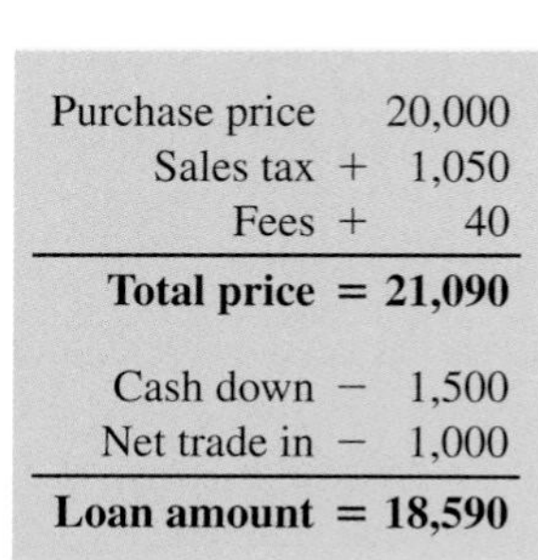

Purchase price	20,000
Sales tax +	1,050
Fees +	40
Total price =	**21,090**
Cash down −	1,500
Net trade in −	1,000
Loan amount =	**18,590**

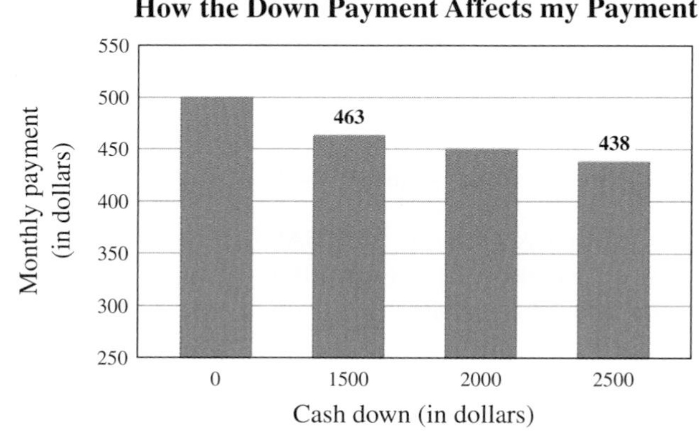

Source: Data from KJE Computer Solutions, LLC.

43. What is the monthly payment if the down payment is $0? $500

44. What is the monthly payment if the down payment is $2000? $450

45. What is the difference in the monthly payment if you increase the down payment from $1500 to $2500? $25

46. What is the difference in the annual amount paid if you increase the down payment from $1500 to $2500? $300

F The chart shows the dog's actual age and the equivalent human age for dogs of different weights and will be used in Problems 47–54.

Using different colors, graph the equivalent human age for a dog weighing:

Dog's age	WEIGHT (POUNDS)			
	15–30	30–49	50–74	75–100
1	12	12	14	16
2	19	21	23	23
3	25	25	26	29
4	32	32	35	35
5	36	36	37	39
6	40	42	43	45
7	44	45	46	48
8	48	50	52	52
9	52	53	54	56
10	56	56	58	61
11	60	60	62	65
12	62	63	66	70
13	66	67	70	75
14	70	70	75	80
15	72	76	80	85

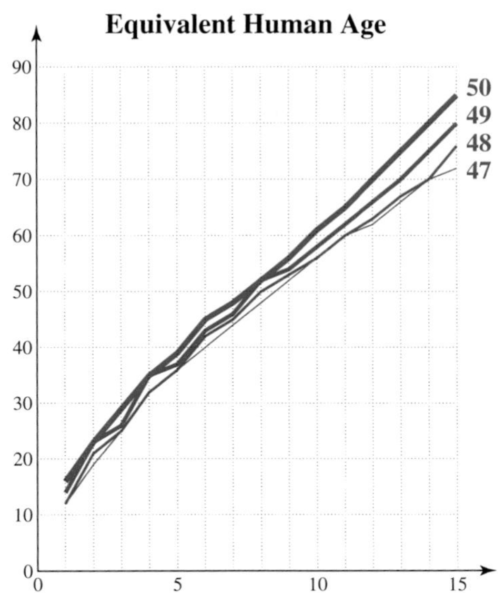

47. Between 15 and 30 pounds

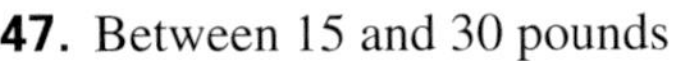

48. Between 30 and 49 pounds

49. Between 50 and 74 pounds

50. Between 75 and 100 pounds

51. If a dog's age is 15, which of the weight scales will make it appear oldest? 75–100 pounds

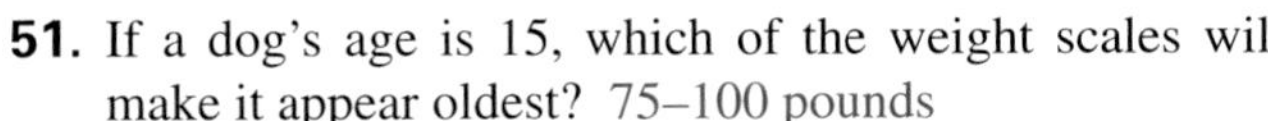

52. If a dog's age is 15, which of the weight scales will make it appear youngest? 15–30 pounds

53. Based on your answers to Problems 51 and 52, the _____ (less, more) a dog weighs, the _____ (younger, older) it appears on the human age scale. Less/younger, or more/older

54. If a dog is 6 years old, what is its equivalent human age? *Hint:* There are four answers! 15–30 pounds: 40; 30–49 pounds: 42; 50–74 pounds: 43; 75–100 pounds: 45

55. Based on the answers to Problem 54, at which dog's age(s) is there complete disagreement as to what the equivalent human age should be? Ages 6, 7, 9, 12, 13, and 15

SKILL CHECKER

Try the Skill Checker Exercises so you'll be ready for the next section.

Solve:

56. $C = 0.10m + 10$ when $m = 30$ $C = 13$

57. $3x + y = 9$ when $x = 0$ $y = 9$

58. $y = 50x + 450$ when $x = 2$ $y = 550$

59. $3x + y = 6$ when $x = 1$ $y = 3$

60. $3x + y = 6$ when $y = 0$ $x = 2$

USING YOUR KNOWLEDGE

Graphing the Risks of "Hot" Exercise

The ideas we presented in this section are vital for understanding graphs. For example, do you exercise in the summer? To determine the risk of exercising in the heat, you must know how to read the following graph.

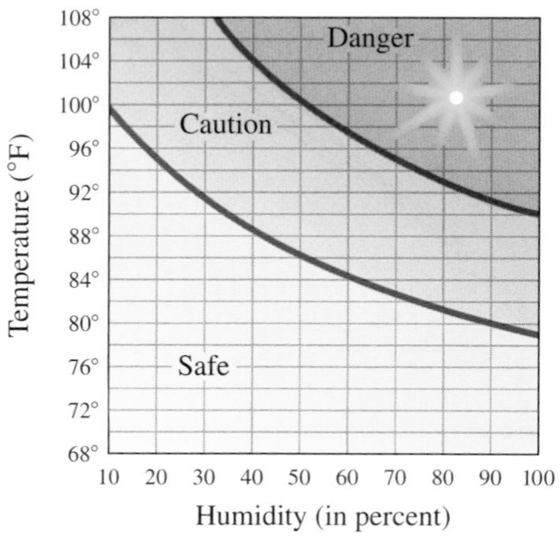

This can be done by first finding the temperature (the y-axis) and then reading across from it to the right, stopping at the vertical line representing the relative humidity. Thus on a 90°F day, if the humidity is less than 30%, the weather is in the safe zone.

Use your knowledge to answer the following questions about exercising in the heat.

61. If the humidity is 50%, how high can the temperature go and still be in the safe zone for exercising? (Answer to the nearest degree.) 86°F

62. If the humidity is 70%, at what temperature will the danger zone start? 95°F

63. If the temperature is 100°F, what does the humidity have to be so that it is safe to exercise? Less than 10%

64. Between what temperatures should you use caution when exercising if the humidity is 80%? Between 81°F and 93°F

65. Suppose the temperature is 86°F and the humidity is 60%. How many degrees can the temperature rise before you get to the danger zone? 12°F

WRITE ON

66. Suppose the point (a, b) lies in the first quadrant. What can you say about a and b? a and b are both positive

67. Suppose the point (a, b) lies in the second quadrant. What can you say about a and b? Is it possible that $a = b$? a is negative, b is positive; no

68. In what quadrant(s) can you have points (a, b) such that $a = b$? Explain. Quadrants I and III; answers may vary

69. Suppose the ordered pairs (a, b) and (c, d) have the same point as their graphs—that is, $(a, b) = (c, d)$. What is the relationship between a, b, c, and d? $a = c$ and $b = d$

70. Now suppose $(a + 5, b - 7)$ and $(3, 5)$ have the same point as their graphs. What are the values of a and b? $a = -2$; $b = 12$

MASTERY TEST

If you know how to do these problems, you have learned your lesson!

Graph the points:

71. $A(4, 2)$

72. $B(3, -2)$

73. $C(-2, 1)$

74. $D(0, -3)$

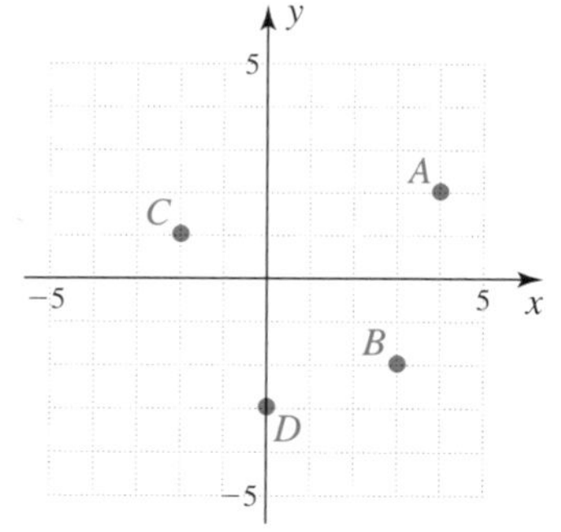

75. The projected cost (in cents per minute) for wireless phone use for the next four years is given in the table. Graph these points.

Year	Cost (cents/min)
1	25
2	23
3	22
4	20

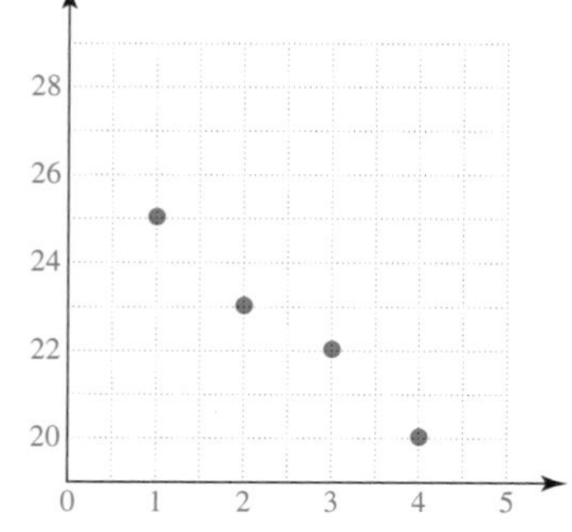

76. Referring to the graph, find an ordered pair giving the equivalent dog age (dog years) for a dog that is actually 7 years old (human years). (7, 50)

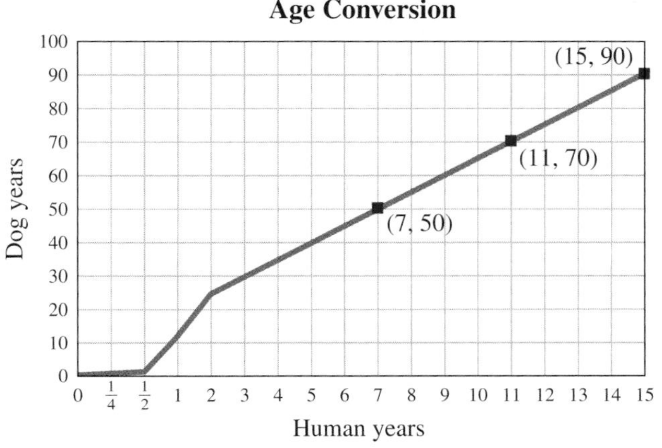

77. Referring to the graph in Problem 76, find the age (human years) of a dog that is 70 years old (dog years). 11 human years

78. What does the ordered pair (15, 90) in the graph in Problem 76 mean? 15 human years is equivalent to 90 dog years.

79. In which quadrant are the following points?

a. $(-2, 3)$ Quadrant II **b.** $(3, -2)$ Quadrant IV

c. $(2, 2)$ Quadrant I **d.** $(-1, -1)$ Quadrant III

80. Give the coordinates of each of the points shown in the graph. $A(-3, 3)$; $B(2, 0)$; $C(-1, -3)$; $D(4, -5)$

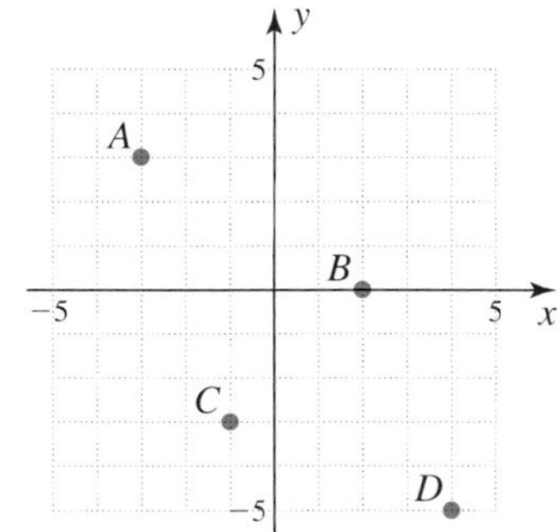

3.2 GRAPHING LINEAR EQUATIONS IN TWO VARIABLES

To Succeed, Review How To . . .

1. Graph ordered pairs (p. 228).
2. Read and interpret ordered pairs on graphs (p. 230).

Objectives

A Determine whether a given ordered pair is a solution of an equation.

B Find ordered pairs that are solutions of a given equation.

C Graph linear equations of the form $y = mx + b$ and $Ax + By = C$.

D Solve applications involving linear equations.

GETTING STARTED

Old Type Cordless Phone Sales

Figure 12 shows the sales of old-fashioned 900 MHz cordless phones (in millions of units). The x-coordinate gives the *year,* whereas the y-coordinate gives the *number* of units sold.

(x, y)

x is the year — y is units sold

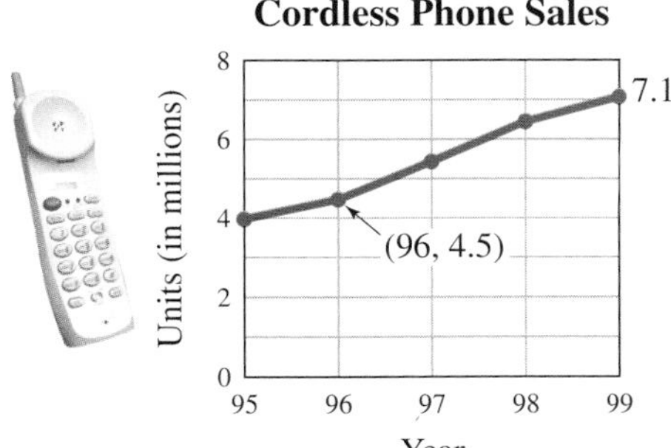

Figure 12

Thus in 1995, the number of units sold was 4 million, and the ordered pair giving this information was (1995, 4). In 1996, the number of units sold was about 4.5 million, corresponding to the ordered pair (1996, 4.5). What ordered pair corresponds to sales in 1999?

Solutions of Equations

Web It

For a lesson on verifying whether an ordered pair is a solution of (satisfies) an equation, go to link 3-2-1 on the Bello Website at mhhe.com/bello.

For a review of graphing ordered pairs and a tutorial dealing with ordered pairs as solutions of equations, try link 3-2-2.

How do we construct the graph showing the cordless phone sales? First, we label the x-axis with the years from 1995 to 1999, and the y-axis with the number of units sold (in millions). We then graph the ordered pairs (x, y) representing the year and the sales and join the resulting points with a line. Does this line have any relation to algebra? The answer is *yes,* especially if the graph is a straight line. Thus if we label the x-axis with the years 0, 1, 2, 3, and 4 (corresponding to the years 1995–1999, respectively), the line representing the number of millions of units sold can be approximated by the equation

$$y = \frac{3}{4}x + 4$$

Note that in 1995 (when $x = 0$),

$$y = \frac{3}{4} \cdot 0 + 4 = 4 \text{ (million)}$$

If we write $x = 0$ and $y = 4$ as the ordered pair (0, 4), then we say that the ordered pair (0, 4) **satisfies,** or is a **solution** of, the equation $y = \frac{3}{4}x + 4$.

What about 1997? Since 1997 is 2 years after 1995,

$$x = 2 \quad \text{and} \quad y = \frac{3}{4} \cdot 2 + 4 = \frac{6}{4} + 4 = \frac{3}{2} + 4 = 5\frac{1}{2} = 5.5$$

Hence the ordered pair (2, 5.5) also satisfies the equation $y = \frac{3}{4}x + 4$, since $5.5 = \frac{3}{4} \cdot 2 + 4$. Note that the equation $y = \frac{3}{4}x + 4$ is different from all other equations we've studied because it has two variables, x and y. Because of this, such equations are called **equations in two variables.**

As you can see, to determine whether an ordered pair is a solution of an equation in two variables, we substitute the x-coordinate for x and the y-coordinate for y in the given equation. If the resulting statement is *true,* then the ordered pair is a solution of, or *satisfies,* the equation. Thus (2, 5) is a solution of $y = x + 3$, since in the ordered pair (2, 5), $x = 2$, $y = 5$, and

$$y = x + 3$$

becomes

$$5 = 2 + 3$$

which is a true statement.

EXAMPLE 1 **Determining whether an ordered pair is a solution**

Determine whether the given ordered pairs are solutions of $2x + 3y = 10$:

a. (2, 2) **b.** (−3, 4) **c.** (−4, 6)

SOLUTION

a. In the ordered pair (2, 2), $x = 2$ and $y = 2$. Substituting in

$$2x + 3y = 10$$

we get

$$2(2) + 3(2) = 10 \quad \text{or} \quad 4 + 6 = 10$$

PROBLEM 1

Determine whether the ordered pairs are solutions of $3x + 2y = 10$.

a. (2, 2)

b. (−3, 4)

c. (−4, 11)

Answers

1. a. Yes; $3(2) + 2(2) = 10$
b. No; $3(-3) + 2(4) \neq 10$
c. Yes; $3(-4) + 2(11) = 10$

which is true. Thus (2, 2) is a solution of

$$2x + 3y = 10$$

You can summarize your work as shown here:

$2x + 3y$	$= 10$
$2(2) + 3(2)$	10
$4 + 6$	
10	True

b. In the ordered pair $(-3, 4)$, $x = -3$ and $y = 4$. Substituting in

$$2x + 3y = 10$$

we get

$$2(-3) + 3(4) = 10$$

$$-6 + 12 = 10$$

which is *not* true. Thus $(-3, 4)$ is not a solution of the given equation. You can summarize your work as shown here:

$2x + 3y$	$= 10$
$2(-3) + 3(4)$	10
$-6 + 12$	
6	False

c. In the ordered pair $(-4, 6)$, the x-coordinate is -4 and the y-coordinate is 6. Substituting these numbers for x and y, respectively, we have

$$2x + 3y = 10$$

$$2(-4) + 3(6) = 10$$

or

$$-8 + 18 = 10$$

which is a true statement. Thus $(-4, 6)$ satisfies, or is a solution of, the given equation. You can summarize your work as shown here.

$2x + 3y$	$= 10$
$2(-4) + 3(6)$	10
$-8 + 18$	
10	True

B Finding Missing Coordinates

In some cases, rather than verifying that a certain ordered pair *satisfies* an equation, we actually have to *find* ordered pairs that are solutions of the given equation. For example, we might have the equation

$$y = 2x + 5$$

and would like to find several ordered pairs that satisfy this equation. To do this we substitute any number for x in the equation and then find the corresponding y-value. A good number to choose is $x = 0$. In this case,

$$y = 2 \cdot 0 + 5 = 5$$

Zero is easy to work with because the value of y is easily found when 0 is substituted for x.

Teaching Tip

In Problems 7–16, many of the equations are in $ax + by = c$ form. To prepare the students, do the following example.
Given:

$$-3x - y = 9$$

Find:

$$(x, -3) \text{ and } (2, y)$$

$-3x - (-3) = 9$	$-3(2) - y = 9$
$-3x + 3 = 9$	$-6 - y = 9$
$-3x = 6$	$-y = 15$
$x = -2$	$y = -15$

Thus the ordered pair (0, 5) satisfies the equation $y = 2x + 5$. For $x = 1$, $y = 2 \cdot 1 + 5 = 7$; hence (1, 7) also satisfies the equation.

We can let x be any number in an equation and then find the corresponding y-value. Conversely, we can let y be any number in the given equation and then find the x-value. For example, if we are given the equation

$$y = 3x - 2$$

and we are asked to find the value of x in the ordered pair $(x, 7)$, we simply let y be 7 and obtain

$$7 = 3x - 2$$

We then solve for x by rewriting the equation as

$$3x - 2 = 7$$
$$3x = 9 \quad \text{Add 2.}$$
$$x = 3 \quad \text{Divide by 3.}$$

Thus $x = 3$ and the ordered pair satisfying $y = 3x - 2$ is (3, 7), as can be verified since $7 = 3(3) - 2$.

EXAMPLE 2 Finding the missing variable

Complete the given ordered pairs so that they satisfy the equation $y = 4x + 3$:

a. $(x, 11)$ **b.** $(-2, y)$

SOLUTION

a. In the ordered pair $(x, 11)$, y is 11. Substituting 11 for y in the given equation, we have

$$11 = 4x + 3$$
$$4x + 3 = 11 \quad \text{Rewrite with } 4x + 3 \text{ on the left.}$$
$$4x = 8 \quad \text{Subtract 3.}$$
$$x = 2 \quad \text{Divide by 4.}$$

Thus $x = 2$ and the ordered pair is (2, 11).

b. Here $x = -2$. Substituting this value in $y = 4x + 3$ yields

$$y = 4(-2) + 3 = -8 + 3 = -5$$

Thus $y = -5$ and the ordered pair is $(-2, -5)$.

PROBLEM 2

Complete the ordered pairs so that they satisfy the equation $y = 3x + 4$.

a. $(x, 7)$ **b.** $(-2, y)$

EXAMPLE 3 Approximating cholesterol levels

Figure 13 shows the decrease in cholesterol with exercise over a 12-week period. If C is the cholesterol level and w is the number of weeks elapsed, the line shown can be approximated by $C = -3w + 215$. What is the cholesterol level at the end of 1 week and at the end of 12 weeks:

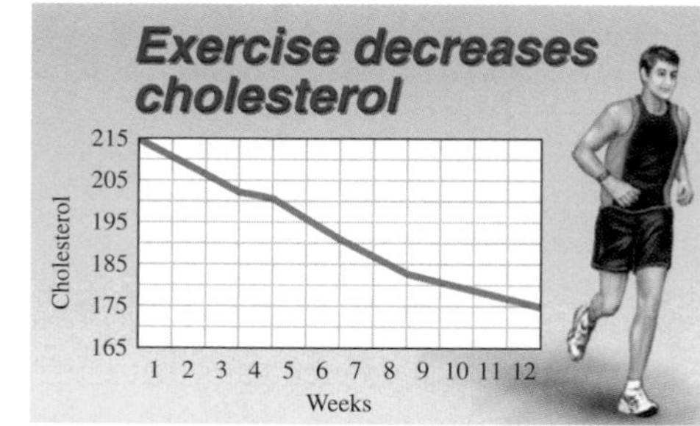

Figure 13

a. According to the graph?

b. Using the equation $C = -3w + 215$?

PROBLEM 3

What is the cholesterol level at the end of 4 weeks:

a. According to the graph?

b. Using the equation in Example 3?

Answers

2. a. (1, 7) **b.** $(-2, -2)$
3. a. About 202 **b.** 203

SOLUTION

a. The cholesterol level at the end of 1 week corresponds to the point (1, 211) on the graph (the point of intersection of the vertical line to the *right* of 1 on the graph). Thus the cholesterol level at the end of 1 week is 211. Similarly, the cholesterol level at the end of 12 weeks corresponds to the point (12, 175). Thus the cholesterol level at the end of 12 weeks is 175.

b. From the equation $C = -3w + 215$, the cholesterol level at the end of 1 week ($w = 1$) is given by $C = -3(1) + 215 = 212$ (close to the 211 on the graph!) and the cholesterol level at the end of 12 weeks ($w = 12$) is $C = -3(12) + 215 = -36 + 215$, or 179.

Teaching Tip

After Example 3, have students discuss the method that would give the most accurate ordered pair. Is it the graph or the equation? Again, remind them that the graph is the *picture* of all solutions.

C Graphing Linear Equations

Suppose you want to rent a car that costs \$30 per day plus \$0.20 per mile traveled. If we have the equation for the daily cost C based on the number m of miles traveled, we can graph this equation. The equation is

$$C = \underbrace{0.20m}_{\text{20¢ per mile}} + \underbrace{30}_{\text{\$30 each day}}$$

Now remember that a solution of this equation must be an ordered pair of numbers of the form (m, C). For example, if you travel 10 miles, $m = 10$ and the cost is

$$C = 0.20(10) + 30 = 2 + 30 = \$32$$

Thus (10, 32) is an ordered pair satisfying the equation; that is, (10, 32) is a *solution* of the equation. If we go 20 miles, $m = 20$ and

$$C = 0.20(20) + 30 = 4 + 30 = \$34$$

Hence (20, 34) is also a solution.

As you see, we can go on forever finding solutions. It's much better to organize our work, list these two solutions and some others, and then graph the points obtained in the Cartesian coordinate system. In this system the number of miles m will appear on the horizontal axis, and the cost C on the vertical axis. The corresponding points given in the table appear in the accompanying figure.

m	C
0	30
10	32
20	34
30	36

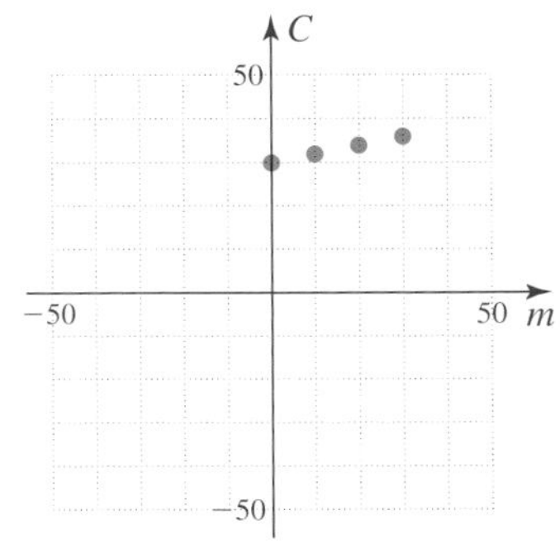

Note that m must be *positive* or *zero*. We have selected values for m that make the cost easy to compute—namely, 0, 10, 20, and 30.

It seems that if we join the points appearing in the graph, we obtain a straight line. Of course, if we knew this for sure, we could have saved time! Why? Because if we know that the graph of an equation is a straight line, we simply find *two* solutions of the equation, graph the two points, and then join them with a straight line. As it turns out, the graph of $C = 0.20m + 30$ *is* a straight line.

The procedure used to graph $C = 0.20m + 30$ can be generalized to graph other lines. Here are the steps.

PROCEDURE

Graphing Lines

1. Choose a value for one variable, calculate the value of the other variable, and graph the resulting ordered pair.
2. Repeat step 1 to obtain at least two ordered pairs.
3. Graph the ordered pairs and draw a line passing through the points. (You can use a third ordered pair as a check.)

This procedure is called **graphing,** and the following rule lets us know that the graph is indeed a *straight line.*

RULE

Straight-Line Graphs

The graph of a linear equation of the form

$$Ax + By = C \quad \text{or} \quad y = mx + b$$

where A, B, C, m, and b are constants (A and B not both 0) is a **straight line,** and every straight line has an equation that can be written in one of these forms.

Thus, the graph of $C = 0.20m + 30$ is a straight line and the graph of $3x + y = 6$ is also a straight line, as we show next.

EXAMPLE 4 Graphing lines of the form $Ax + By = C$

Graph: $3x + y = 6$

SOLUTION The equation is of the form $Ax + By = C$, and thus the graph is a straight line. Since two points determine a line, we shall graph two points and join them with a straight line, the graph of the equation. Two easy points to use occur when we let $x = 0$ and find y, and let $y = 0$ and find x. For $x = 0$,

$$3x + y = 6$$

becomes

$$3 \cdot 0 + y = 6 \quad \text{or} \quad y = 6$$

Thus (0, 6) is on the graph.

PROBLEM 4

Graph: $2x + y = 4$

When $y = 0$, $3x + y = 6$ becomes

$$3x + 0 = 6$$
$$3x = 6$$
$$x = 2$$

Hence (2, 0) is also on the graph.

We chose $x = 0$ and $y = 0$ because the calculations are easy. There are infinitely many points that satisfy the equation $3x + y = 6$, but the points (0, 6) and (2, 0)—the *y*- and *x-intercepts*—are easy to find.

It's a good idea to pick a *third* point as a *check.* For example, if we let $x = 1$, $3x + y = 6$ becomes

$$3 \cdot 1 + y = 6$$
$$3 + y = 6$$
$$y = 3$$

Answer

4.

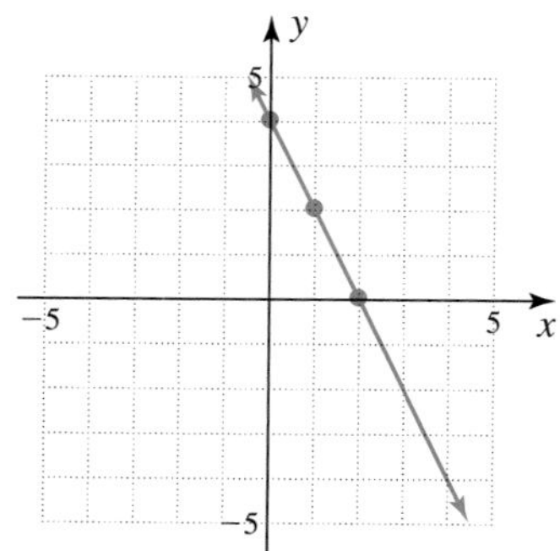

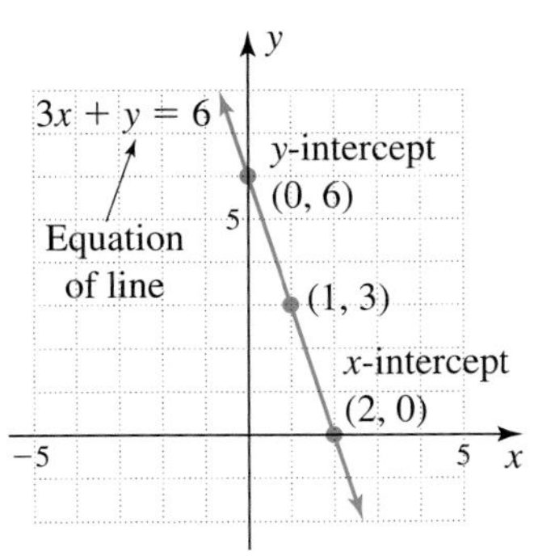

Figure 14

Now we have our third point, (1, 3), as shown in the following table. The points (0, 6), (2, 0), and (1, 3), as well as the completed graph of the line, are shown in Figure 14.

x	y	
0	6	← y-intercept
2	0	← x-intercept
1	3	

Note that the point (0, 6) where the line crosses the y-axis is called the ***y*-intercept,** and the point (2, 0) where the line crosses the x-axis is called the ***x*-intercept.**

The graph in Figure 14 cannot show the entire line, which extends indefinitely in both directions; that is, the graph shown is a part of a line that continues without end in both directions, as indicated by the arrows.

EXAMPLE 5 Graphing lines of the form $y = mx + b$

Graph: $y = -2x + 6$

SOLUTION The equation is of the form $y = mx + b$, which is a straight line. Thus, we follow the procedure for graphing lines.

1. Let $x = 0$.

$$y = -2x + 6 \quad \text{becomes}$$
$$y = -2(0) + 6 = 6$$

Thus, the point (0, 6) is on the graph. (See Figure 15.)

2. Let $y = 0$.

$$y = -2x + 6 \quad \text{becomes}$$
$$0 = -2x + 6$$
$$2x = 6 \quad \text{Add } 2x.$$
$$x = 3 \quad \text{Divide by 2.}$$

Hence, (3, 0) is also on the line.

3. Graph (0, 6) and (3, 0), and draw a line passing through both points. To make sure, we will use a third point. Let $x = 1$.

$$y = -2x + 6 \quad \text{becomes}$$
$$y = -2(1) + 6 = 4$$

Thus, the point (1, 4) is also on the line, as shown in Figure 15.

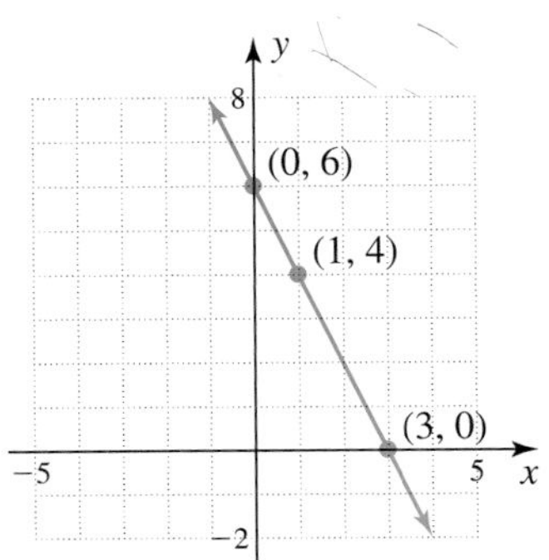

Figure 15

PROBLEM 5

Graph: $y = -3x + 6$

Web It

For a lesson on how to graph lines, go to link 3-2-3 on the Bello Website at mhhe.com/bello.

Answer

5.

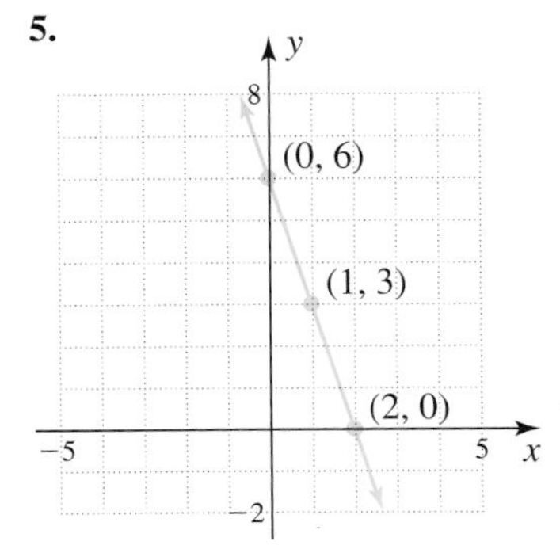

Applications

Have you heard the term "wind chill factor"? It is the temperature you actually *feel* because of the wind. Thus, for example, if the temperature is 10 degrees Fahrenheit (10°F) and the wind is blowing at 15 miles per hour, you will *feel* as if the temperature is −7°F. If the wind is blowing at a *constant* 15 miles per hour, the wind chill factor W can be roughly approximated by

$$W = 1.3t - 20, \quad \text{where } t \text{ is the temperature in °F}$$

EXAMPLE 6 **Very cool graphing**

Graph: $W = 1.3t - 20$

SOLUTION First, note that we have to label the axes differently. Instead of y we have W, and instead of x we have t. Next, note that the wind chill factors we will get seem to be negative (try $t = -5$, $t = 0$, and $t = 5$). Thus, we will concentrate on points in the third and fourth quadrants, as shown in the grid. Now let's obtain some points.

For $t = -5$, $W = 1.3(-5) - 20 = -26.5$. Graph $(-5, -26.5)$.

For $t = 0$, $W = 1.3(0) - 20 = -20$. Graph $(0, -20)$.

For $t = 5$, $W = 1.3(5) - 20 = -13.5$. Graph $(5, -13.5)$.

Note that the equation $W = 1.3t - 20$ applies only when the wind is blowing at 15 miles per hour! Moreover, you should recognize that the graph is a line (it is of the form $y = mx + b$ with W instead of y and t instead of x) and that the graph of the line belongs mostly in the third and fourth quadrants.

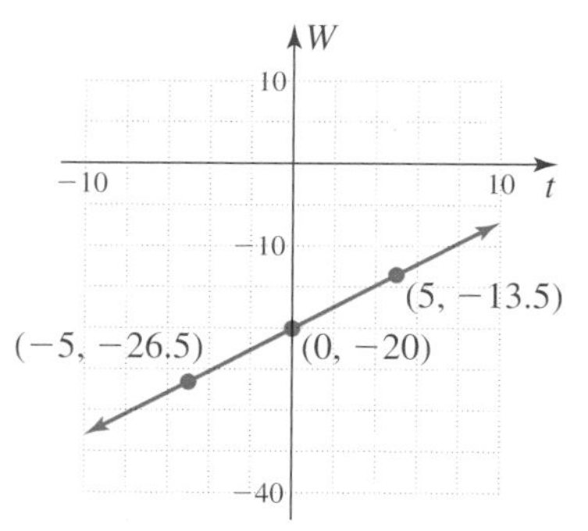

PROBLEM 6

Graph: $W = 1.2t - 20$

Answer

6.

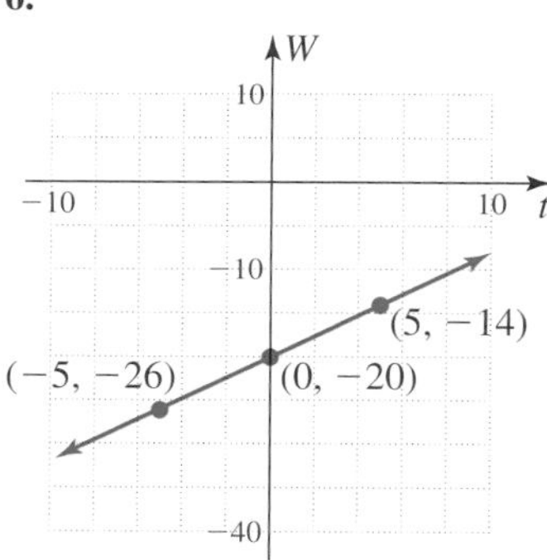

Calculate It Graphing Lines

You can satisfy three of the objectives of this section using a calculator. We will show you how by graphing the equation $2x + 3y = 10$, determining whether the point (2, 2) satisfies the equation, and finding the x-coordinate in the ordered pair $(x, -6)$ using a calculator.

The graph of $2x + 3y = 10$ consists of all points satisfying $2x + 3y = 10$. To graph this equation, first solve for y obtaining

$$y = \frac{10}{3} - \frac{2x}{3}$$

If you want to determine whether (2, 2) is a solution of the equation, turn the statistical plot off (press 2nd Y= 1 and select OFF), then set the window for integers (ZOOM 8 ENTER), and enter

$$Y_1 = \frac{10}{3} - \frac{2x}{3}$$

Press GRAPH.

Use TRACE to move the cursor around. In Window 1, $x = 2$ and $y = 2$ are shown; thus (2, 2) satisfies the equation. To find x in $(x, -6)$, use TRACE until you are at the point where $y = -6$; then read the value of x, which is 14. Try it! You can use these ideas to do Problems 1–16.

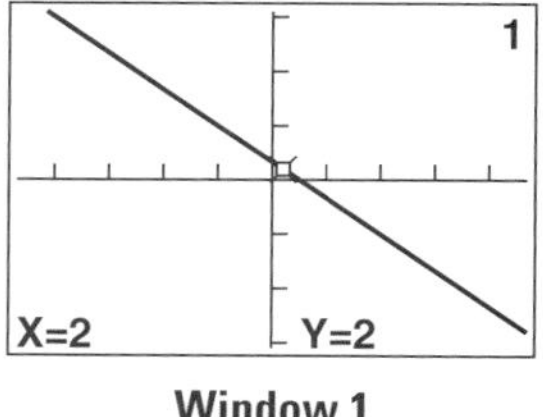

Window 1

We do not want to leave you with the idea that all equations you will encounter are linear equations. Later in the book we shall study:

1. Absolute-value equations of the form $y = |x|$
2. Quadratic equations of the form $y = x^2$
3. Cubic equations of the form $y = x^3$

These types of equations can be graphed using the same procedure as that for graphing lines, with one notable exception: in step (3), you draw a smooth curve passing through the points. The results are shown here.

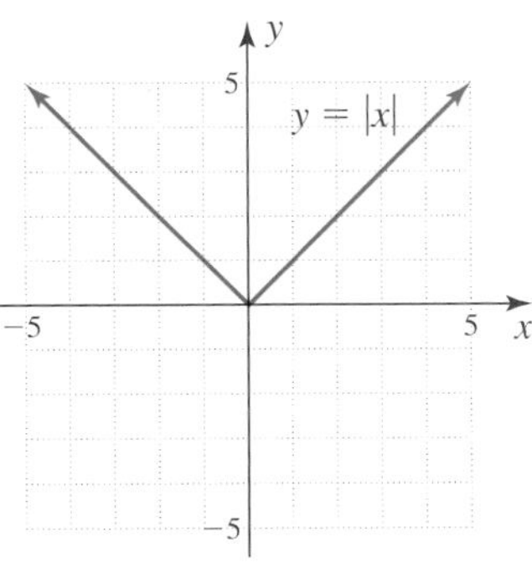

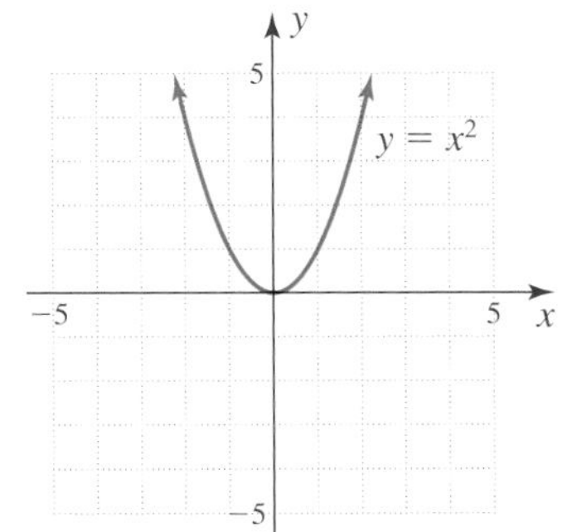

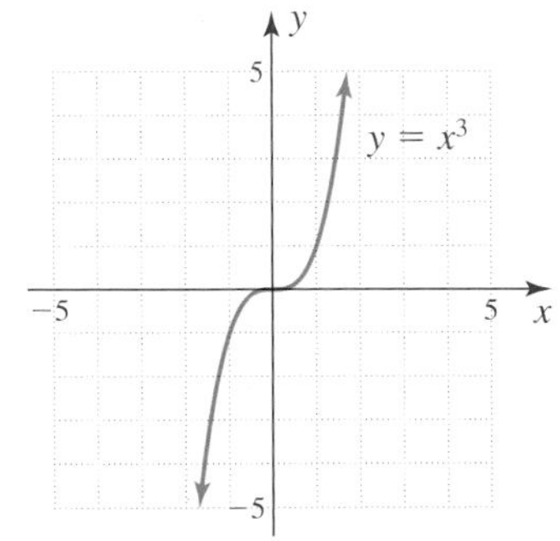

Exercises 3.2

A In Problems 1–6, determine whether the ordered pair is a solution of the equation.

1. $(3, 2)$; $x + 2y = 7$ Yes

2. $(4, 2)$; $x - 3y = 2$ No

3. $(5, 3)$; $2x - 5y = -5$ Yes

4. $(-2, 1)$; $-3x = 5y + 1$ Yes

5. $(2, 3)$; $-5x = 2y + 4$ No

6. $(-1, 1)$; $4y = -2x + 2$ Yes

B In Problems 7–16, find the missing coordinate.

7. $(3, \_\_)$ is a solution of $2x - y = 6$. 0

8. $(-2, \_\_)$ is a solution of $-3x + y = 8$. 2

9. $(\_\_, 2)$ is a solution of $3x + 2y = -2$. -2

10. $(\_\_, -5)$ is a solution of $x - y = 0$. -5

11. $(0, \_\_)$ is a solution of $3x - y = 3$. -3

12. $(0, \_\_)$ is a solution of $x - 2y = 8$. -4

13. $(\_\_, 0)$ is a solution of $2x - y = 6$. 3

14. $(\_\_, 0)$ is a solution of $-2x - y = 10$. -5

15. $(-3, \_\_)$ is a solution of $-2x + y = 8$. 2

16. $(-5, \_\_)$ is a solution of $-3x - 2y = 9$. 3

C In Problems 17–40, graph the equation.

17. $2x + y = 4$

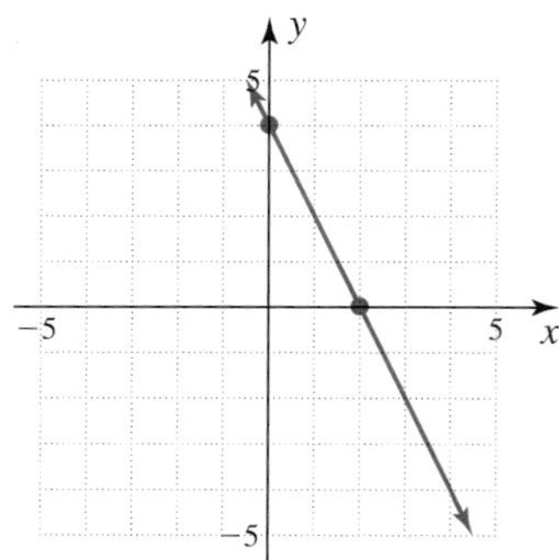

18. $y + 3x = 3$

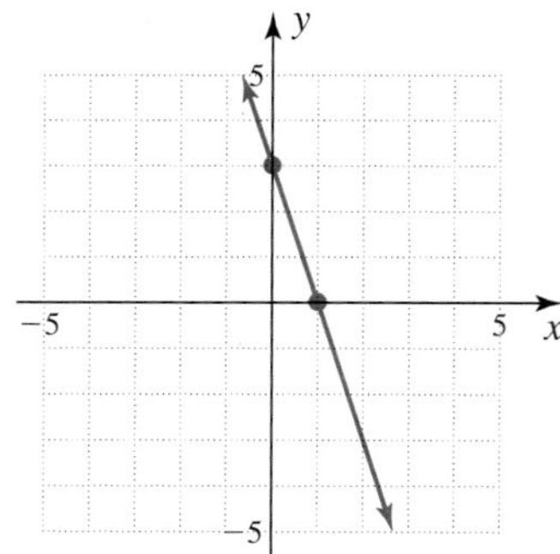

19. $-2x - 5y = -10$

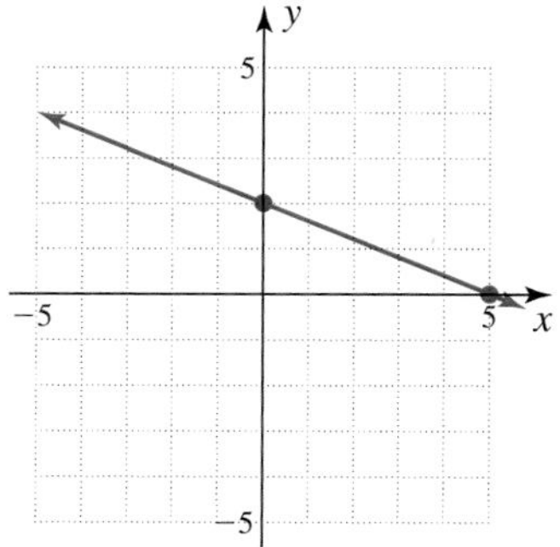

20. $-3x - 2y = -6$

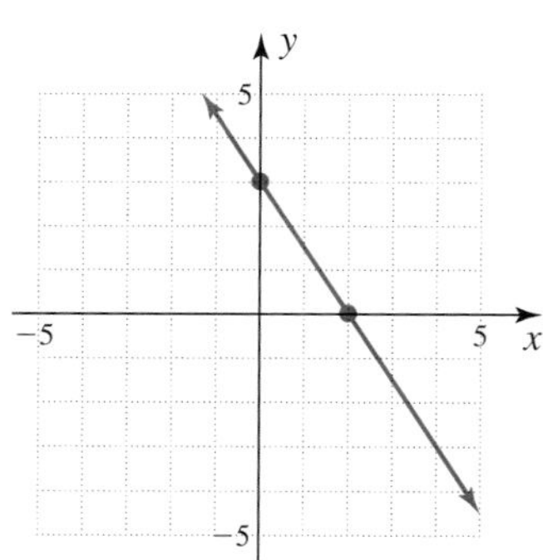

21. $y + 3 = 3x$

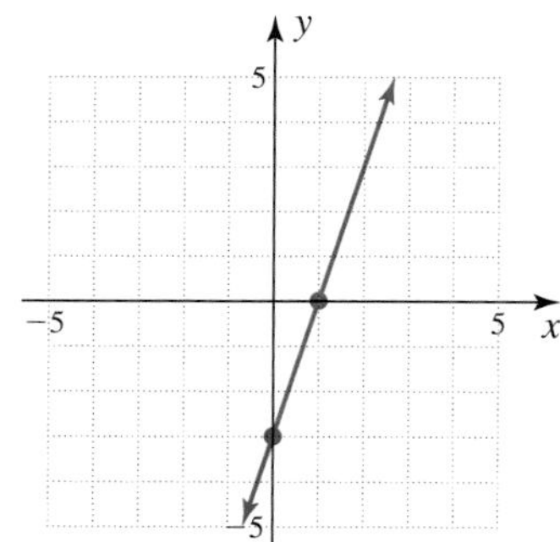

22. $y - 4 = -2x$

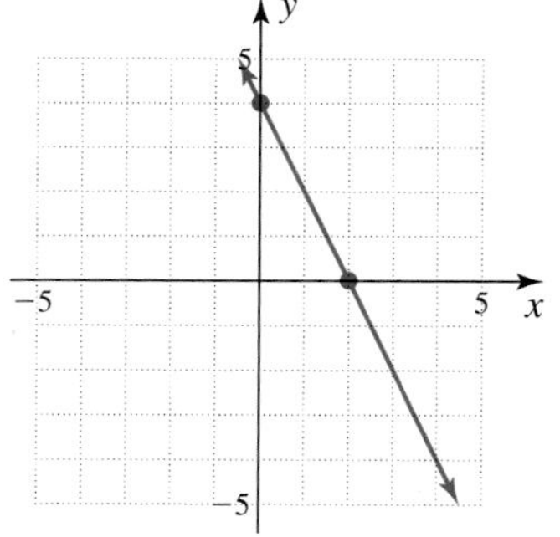

23. $6 = 3x - 6y$

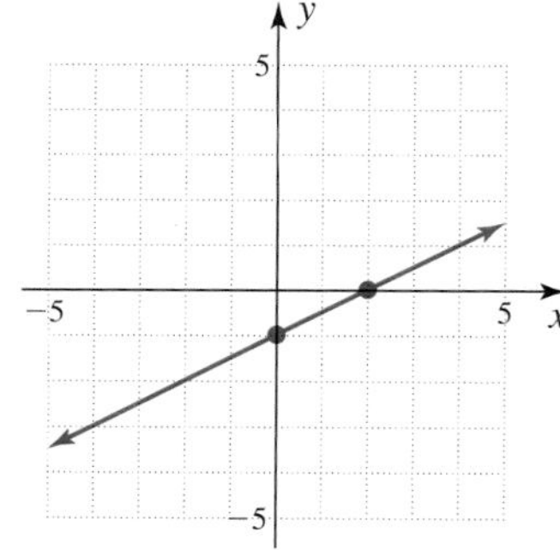

24. $6 = 2x - 3y$

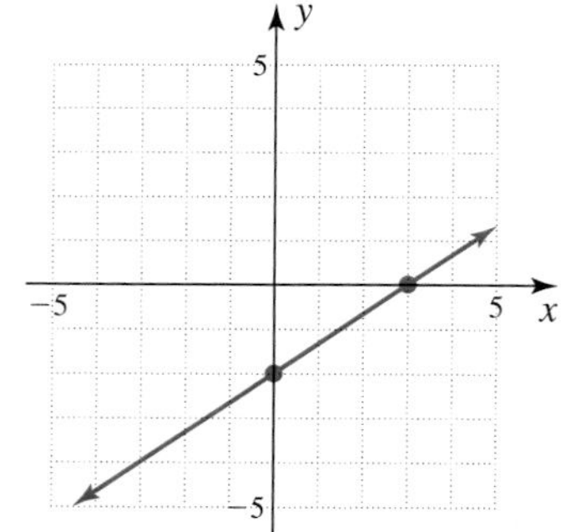

25. $-3y = 4x + 12$

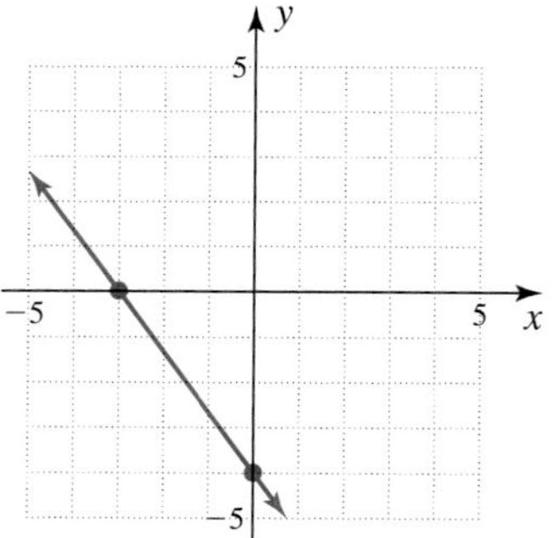

26. $-2x = 5y + 10$

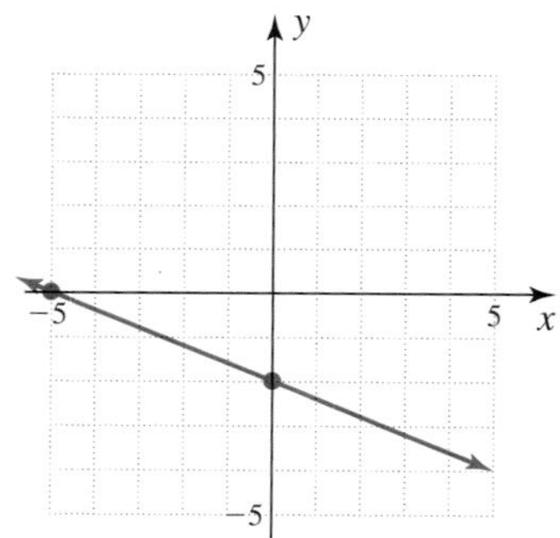

27. $-2y = -x + 4$

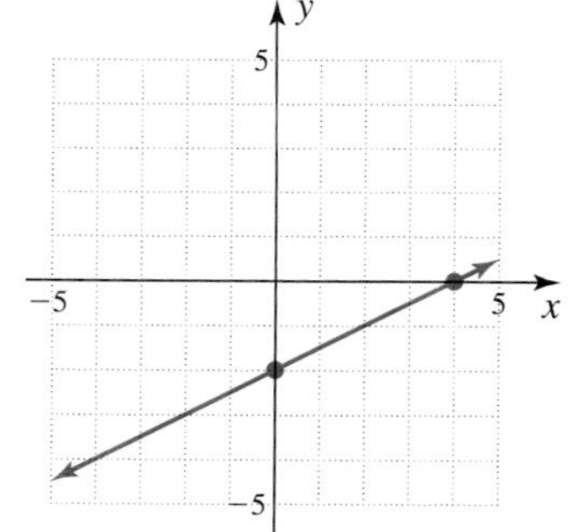

28. $-3y = -x + 6$

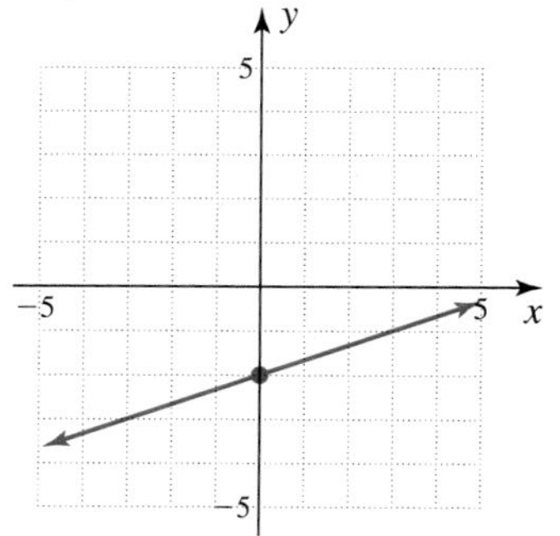

29. $-3y = -6x + 3$

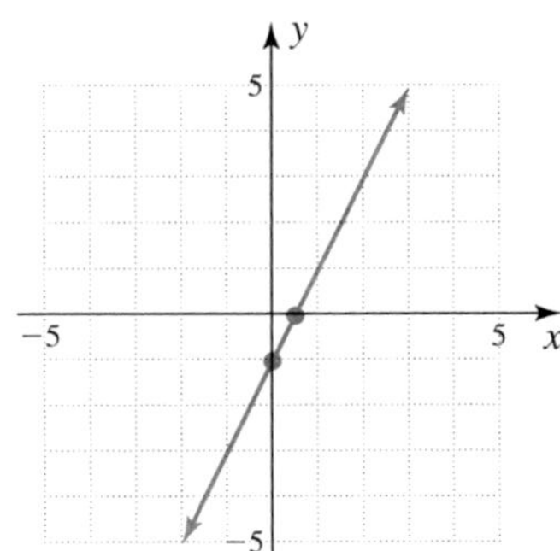

30. $-4y = -2x + 4$

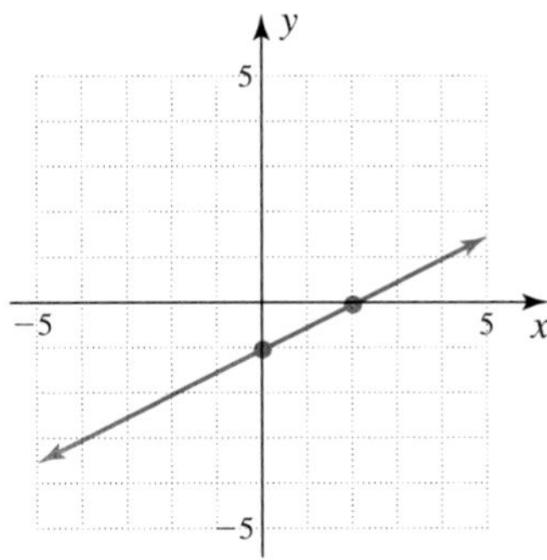

31. $y = 2x + 4$

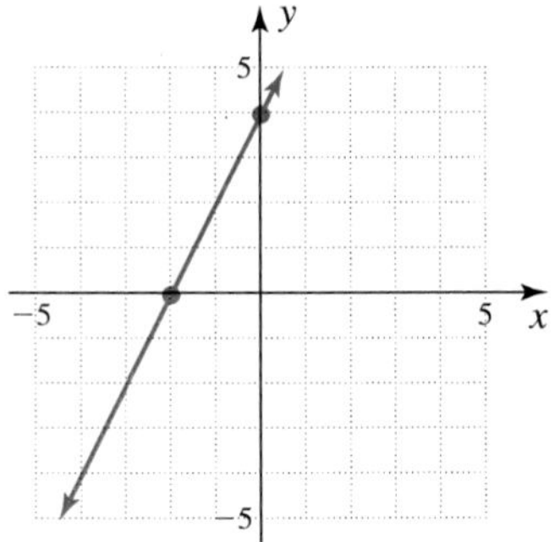

32. $y = 3x + 6$

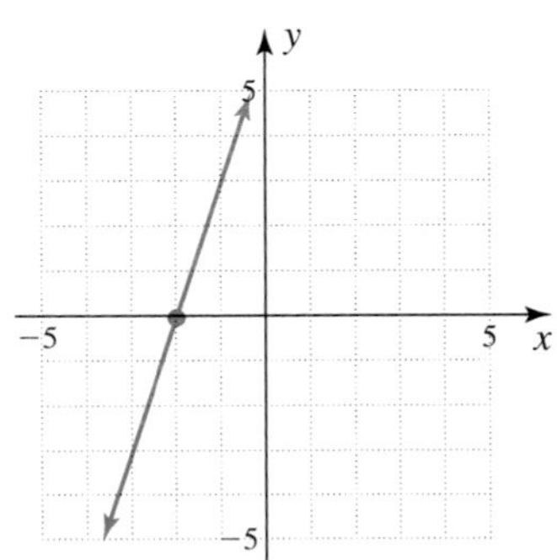

33. $y = -2x + 4$

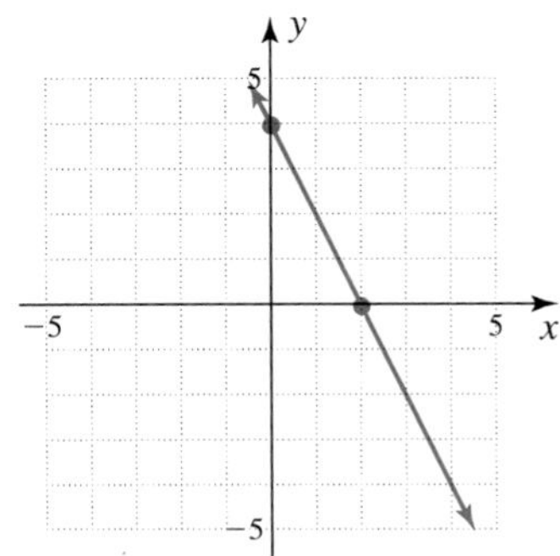

34. $y = -3x + 6$

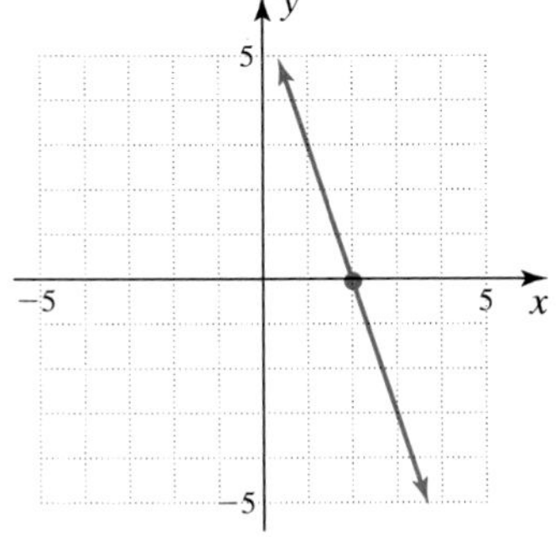

35. $y = -3x - 6$

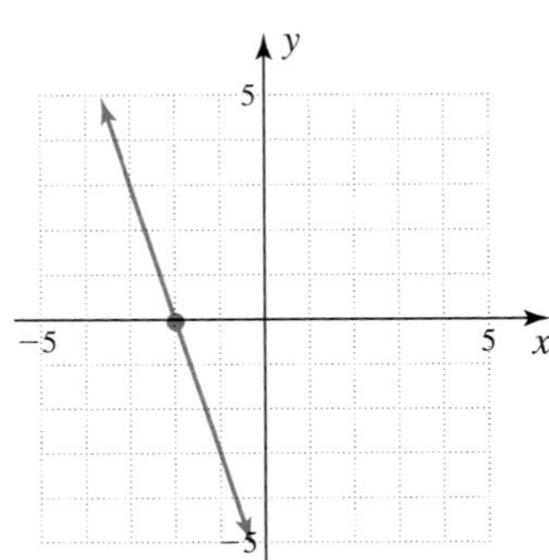

36. $y = -2x - 4$

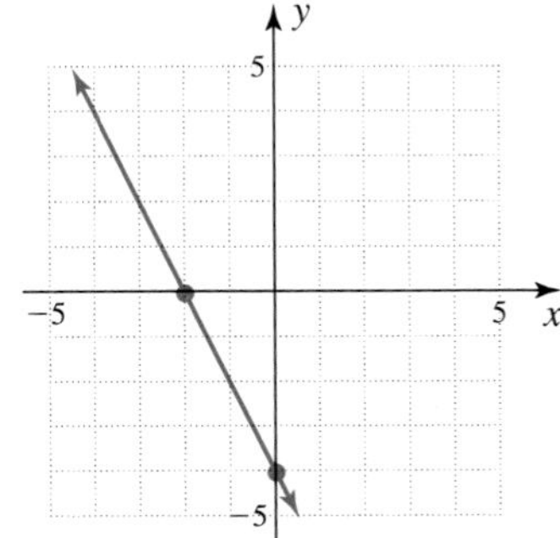

37. $y = \frac{1}{2}x - 2$

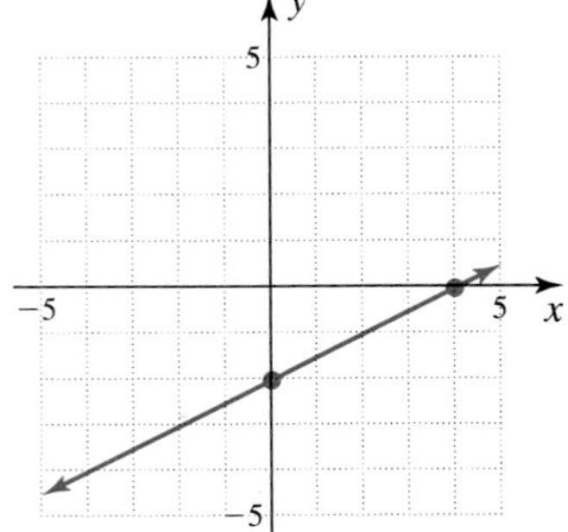

38. $y = \frac{1}{3}x - 1$

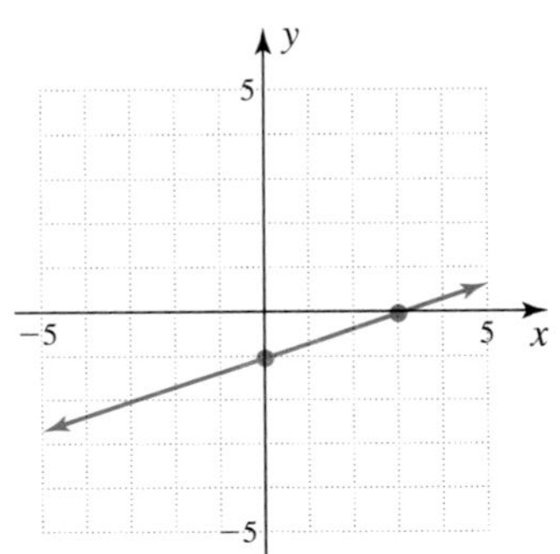

39. $y = -\frac{1}{2}x - 2$

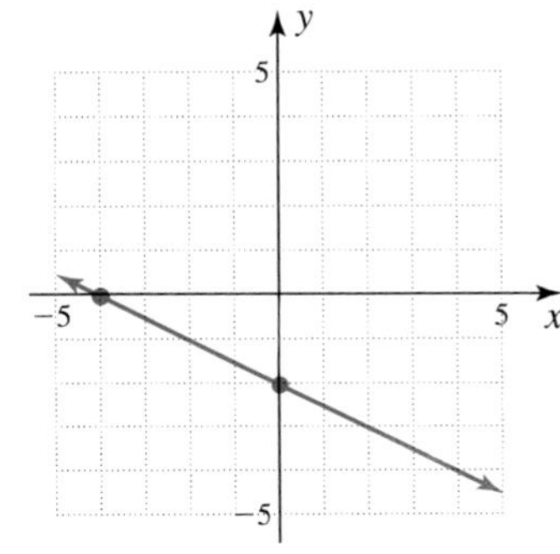

40. $y = -\frac{1}{3}x - 1$

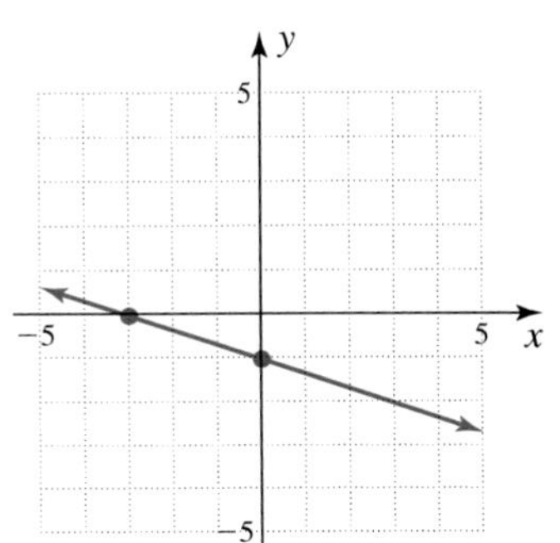

APPLICATIONS

41. *Wind chill factor* When the wind is blowing at a constant 5 miles per hour, the wind chill factor W can be approximated by $W = 1.1t - 9$, where t is the temperature in °F and W is the wind chill factor.

a. Find W when $t = 0$. -9

b. Find W when $t = 10$. 2

c. Graph W.

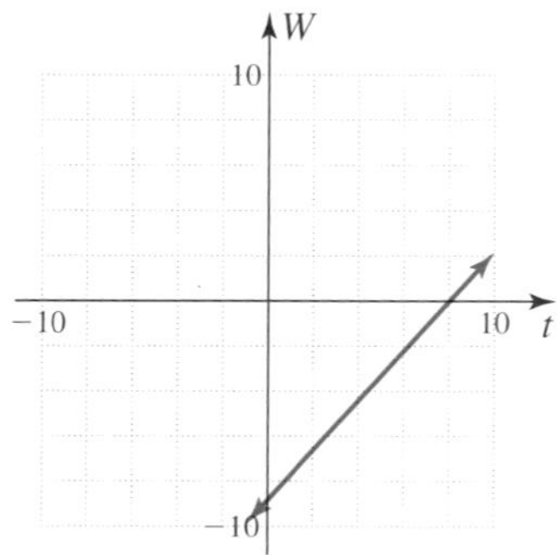

42. *Wind chill factor* When the wind is blowing at a constant 20 miles per hour, the wind chill factor W can be approximated by $W = 1.3t - 21$, where t is the temperature in °F and W is the wind chill factor.

a. Find W when $t = 0$. -21

b. Find W when $t = 10$. -8

c. Graph W.

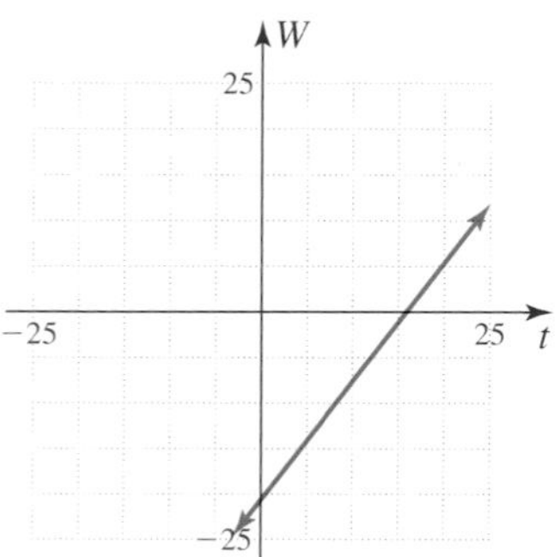

43. *College tuition and fees* Do you want a rough approximation of the annual cost C (in dollars) of tuition and required fees at a two-year college? This cost can be approximated by $C = 200t + 7000$, where t the number of years after 1996.

a. What was the cost of tuition in 1996? $7000

b. What was it in 2001? $8000

c. What would it be in 2006? $9000

d. Make a graph of the line representing C.

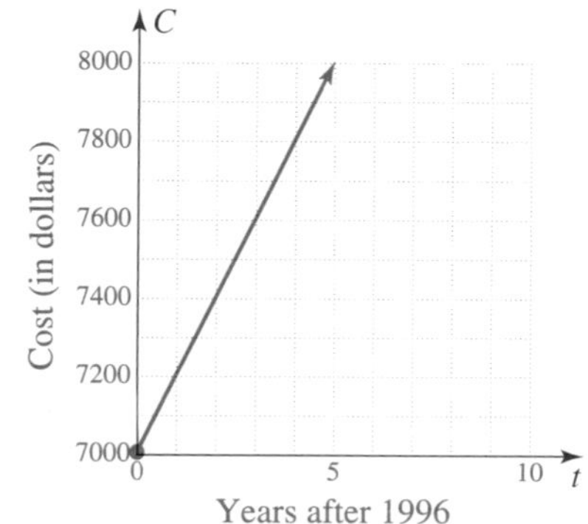

44. *Cost of textbooks* The annual average cost C (in dollars) of textbooks at a two-year college is approximately $C = 30t + 680$, where t is the number of years after 2000.

a. What was the cost of textbooks in 2000? $680

b. What was it in 2001? $710

c. What would it be in 2005? $830

d. Make a graph of the line representing C.

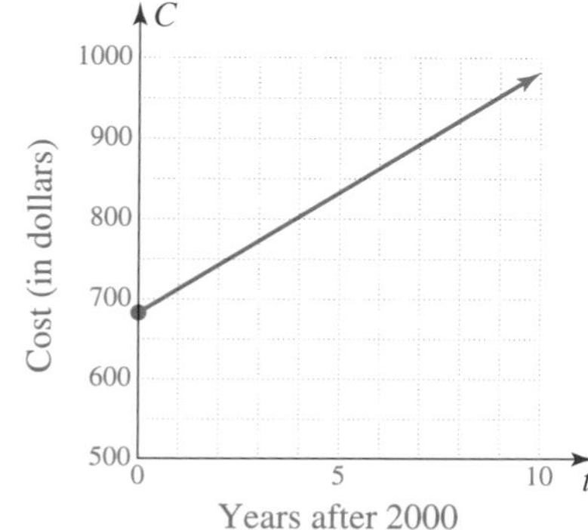

SKILL CHECKER

Try the Skill Checker Exercises so you'll be ready for the next section.

45. Solve: $2x + 4 = 0$ $x = -2$

46. Solve: $3 + 3y = 0$ $y = -1$

47. Solve: $0 = 0 - 0.1t$ $t = 0$

48. Solve: $0 = 8 - 0.1t$ $t = 80$

USING YOUR KNOWLEDGE

You already know how to determine whether an ordered pair satisfies an equation. Anthropological detectives use this knowledge to estimate the living height of its owner using one dried bone as a clue. Suppose a detective finds a 17.9-inch femur bone from a male. To find the height H (in inches) of its owner, use the formula $H = 1.88f + 32.010$ and determine that the owner's height H must have been $H = 1.88(17.9) + 32.010 \approx 66$ inches. Of course, the ordered pair (17.9, 66) satisfies the equation. Now, for the fun part. Suppose you find a 17.9-inch femur bone, but this time you are looking for a missing female 66 inches tall. Can this femur belong to her? No! How do we know? The ordered pair (17.9, 66) *does not* satisfy the equation for female femur bones, which is $H = 1.945f + 28$. Try it! Now, use the formulas to work Problems 49–54.

Living Height (inches)

Male	Female
$H = 1.880f + 32.010$	$H = 1.945f + 28.379$
$H = 2.894h + 27.811$	$H = 2.754h + 28.140$
$H = 3.271r + 33.829$	$H = 3.343r + 31.978$
$H = 2.376t + 30.970$	$H = 2.352t + 29.439$

where f, h, r, and t represent the length of the femur, humerus, radius, and tibia bones, respectively.

49. An 18-inch humerus bone from a male subject has been found. Can the bone belong to a 6' 8" basketball player missing for several weeks? Yes

50. A 16-inch humerus bone from a female subject has been found. Can the bone belong to a missing 5'2" female student? No

51. A radius bone measuring 14 inches is found. Can it belong to Sandy Allen, the world's tallest living woman? (She is 7 feet tall!) How long to the nearest inch should her radius bone be? No; 16 in.

Source: Guinness World Records.

52. The tallest woman in medical history was Zeng Jinlian. A tibia bone measuring 29 inches is claimed to be hers. If the claim proves to be true, how tall was she, and what ordered pair would satisfy the equation relating the length of the tibia and the height of a woman? 97.6" or 8'2" tall; $(t, 2.352t + 29.439)$

53. A 10-inch radius of a man is found. Can it belong to a man 66.5 inches tall? Yes

54. The longest recorded bone—an amazing 29.9 inches long—is the femur of the German giant Constantine who died in 1902. How tall was he, and what ordered pair would satisfy the equation relating the length of his femur and his height? 88.2" or 7'4" tall; (29.9, 88.2)

WRITE ON

55. Write the procedure you use to show that an ordered pair (a, b) satisfies an equation. Answers may vary.

56. Write the procedure you use to determine whether the graph of an equation is a straight line. Answers may vary.

57. Write the procedure you use to graph a line. Answers may vary.

58. In your own words, write what "wind chill factor" means. Answers may vary.

MASTERY TEST

If you know how to do these problems, you have learned your lesson!

59. Determine whether the ordered pair is a solution of $3x + 2y = 10$.

a. (1, 2) No **b.** (2, 1) No

60. Graph: $2x + y = 6$

61. Graph: $y = -3x + 6$

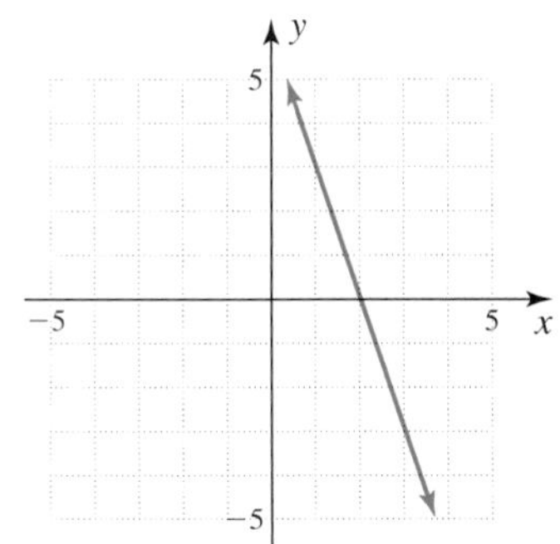

Web It

To find out where these bones are and how the owners' heights are determined (in centimeters this time), go to link 3-2-4 on the Bello Website at mhhe.com/bello.

62. Complete the ordered pairs so they satisfy the equation $y = 3x + 2$.

a. $(x, 5)$ 1

b. $(-3, y)$ -7

63. When the wind is blowing at a constant 10 miles per hour, the wind chill factor W can be approximated by $W = 1.3t - 18$, where t is the temperature in °F and W is the wind chill factor.

a. Find W when $t = 0$. -18

b. Find W when $t = 10$. -5

c. Graph W.

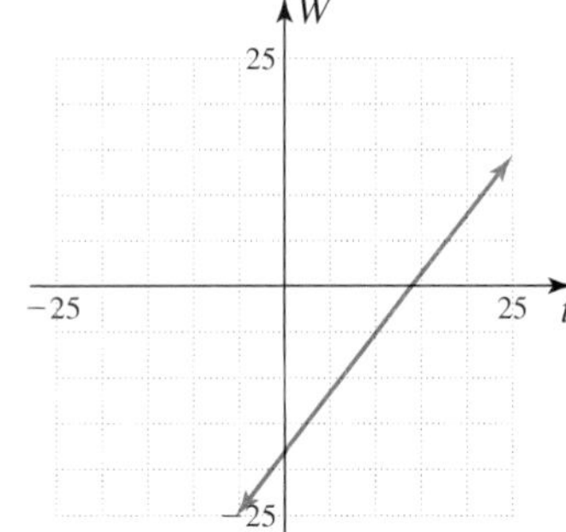

3.3 GRAPHING LINES USING INTERCEPTS

To Succeed, Review How To . . .

1. Solve linear equations (p. 151).
2. Graph ordered pairs of numbers (p. 228).

Objectives

A Graph lines using intercepts.

B Graphs lines passing through the origin.

C Graph horizontal and vertical lines.

D Solve applications involving graphs of lines.

GETTING STARTED How Much Money Do You Have?

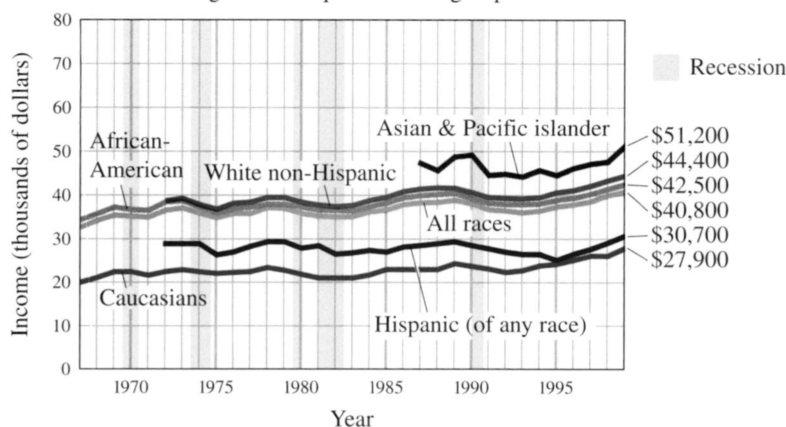

Source: Data from U.S. Census Bureau, Current Population Survey, March 1968 to 2000.

What is your household income? As you can see from the graphs, some of the median household incomes were slowly increasing, whereas others were almost flat over the time period shown. For example, the median household income h for Hispanics was slightly under $30,000 in 1972 and slightly over $30,000 in 1999. The graph of h can be approximated by the horizontal line $h = 30{,}000$. In this section we learn how to graph horizontal and vertical lines, and lines whose graphs go through the origin. How do we know which lines are horizontal, which ones are vertical, and which ones pass through the origin? We will tell you in this section!

Graphing Lines Using Intercepts

In the preceding section we mentioned that the graph of a linear equation of the form $Ax + By = C$ is a *straight line,* and we graphed such lines by finding ordered pairs that satisfied the equation of the line. But there is a quicker and simpler way to graph these equations by using the idea of *intercepts.* Here are the definitions we need.

Web It

To learn to graph equations using intercepts, go to link 3-3-1 at the Bello Website at mhhe.com/bello.

PROCEDURE

1. The ***x*-intercept** $(a, 0)$ of a line is the point at which the line crosses the x-axis. To find the x-intercept, let $y = 0$ and solve for x.
2. The ***y*-intercept** $(0, b)$ of a line is the point at which the line crosses the y-axis. To find the y-intercept, let $x = 0$ and solve for y.

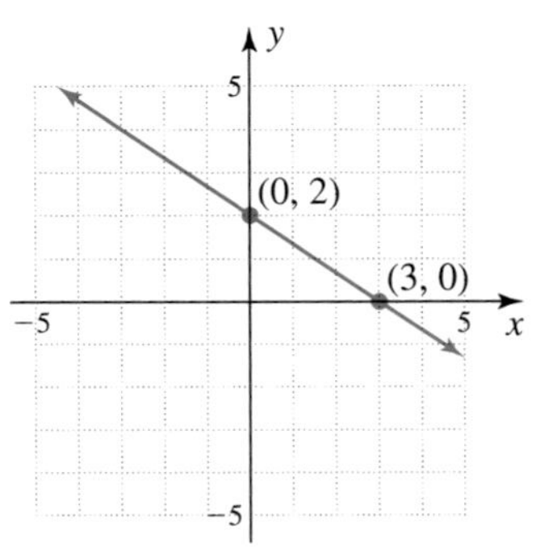

Figure 16

The graph in Figure 16 shows a line with y-intercept $(0, 2)$ and x-intercept $(3, 0)$. How do we use these ideas to graph lines? Here is the procedure.

PROCEDURE

To Graph a Line Using the Intercepts

1. Find the x-intercept $(a, 0)$.
2. Find the y-intercept $(0, b)$.
3. Graph the points $(a, 0)$ and $(0, b)$, and connect them with a line.
4. Find a third point to use as a check (make sure the point is on the line!).

EXAMPLE 1 **Graphing lines using intercepts**

Graph: $5x + 2y = 10$

SOLUTION First we find the x- and y-intercepts.

1. Let $x = 0$ in $5x + 2y = 10$. Then

$$5(0) + 2y = 10$$
$$2y = 10$$
$$y = 5$$

Hence $(0, 5)$ is the y-intercept.

2. Let $y = 0$ in $5x + 2y = 10$. Then

$$5x + 2(0) = 10$$
$$5x = 10$$
$$x = 2$$

Hence, $(2, 0)$ is the x-intercept.

PROBLEM 1

Graph: $2x + 5y = 10$

Answer

1.

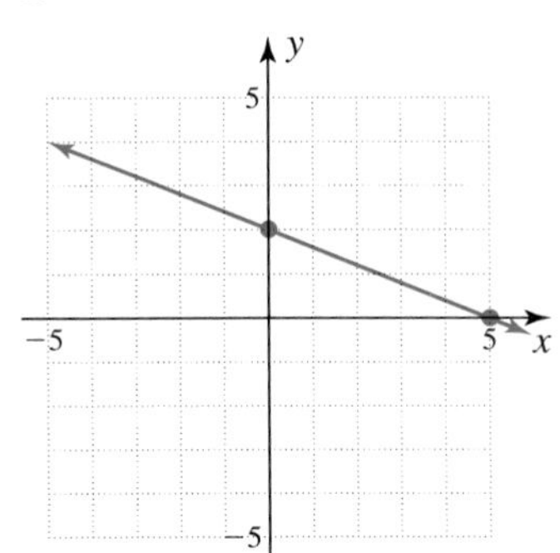

3. Now we graph the points (0, 5) and (2, 0) and connect them with a line as shown in Figure 17.

4. We need to find a *third* point to use as a *check:* we let $x = 4$ and replace x with 4 in

$$5x + 2y = 10$$
$$5(4) + 2y = 10$$
$$20 + 2y = 10$$
$$2y = -10 \quad \text{Subtract 20.}$$
$$y = -5$$

The three points (0, 5), (2, 0), and (4, −5) as well as the completed graph are shown in Figure 17.

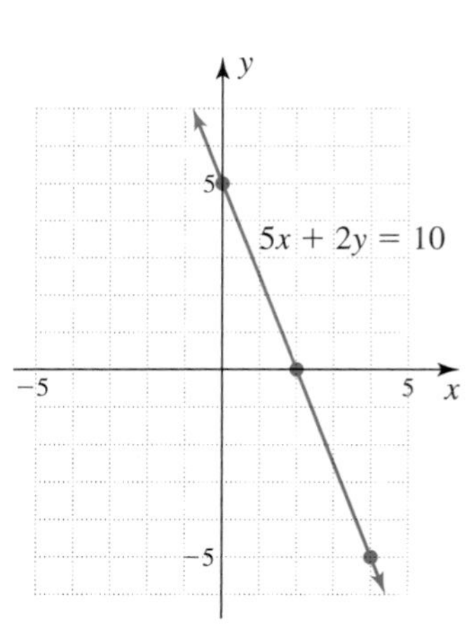

Figure 17

The procedure used to graph a line using the intercepts works even when the line is *not* written in the form $Ax + By = C$, as we will show in Example 2.

EXAMPLE 2 **Graphing lines using intercepts**

Graph: $2x - 3y - 6 = 0$

SOLUTION

1. Let $y = 0$. Then

$$2x - 3(0) - 6 = 0$$
$$2x - 6 = 0$$
$$2x = 6 \quad \text{Add 6.}$$
$$x = 3 \quad \text{Divide by 2.}$$

(3, 0) is the x-intercept.

2. Let $x = 0$. Then

$$2(0) - 3y - 6 = 0$$
$$-3y - 6 = 0$$
$$-3y = 6 \quad \text{Add 6.}$$
$$y = -2 \quad \text{Divide by } -3.$$

(0, −2) is the y-intercept.

3. Connect (0, −2) and (3, 0) with a line, the graph of $2x - 3y - 6 = 0$, see Figure 18.

4. Use a third point as a check. Examine the line. It seems that for $x = -3$, y is −4. Let us check.

If $x = -3$, we have

$$2(-3) - 3y - 6 = 0$$
$$-6 - 3y - 6 = 0$$
$$-3y - 12 = 0 \quad \text{Simplify.}$$
$$-3y = 12 \quad \text{Add 12.}$$
$$y = -4 \quad \text{Divide by } -3.$$

PROBLEM 2

Graph: $3x - 2y - 6 = 0$

Teaching Tip

The procedure used in Example 2 is sometimes called the *intercept* method because you have to find both *intercepts*. Have students discuss whether or not lines have to have both intercepts. If not, have them sketch other possibilities.

Answer

2.

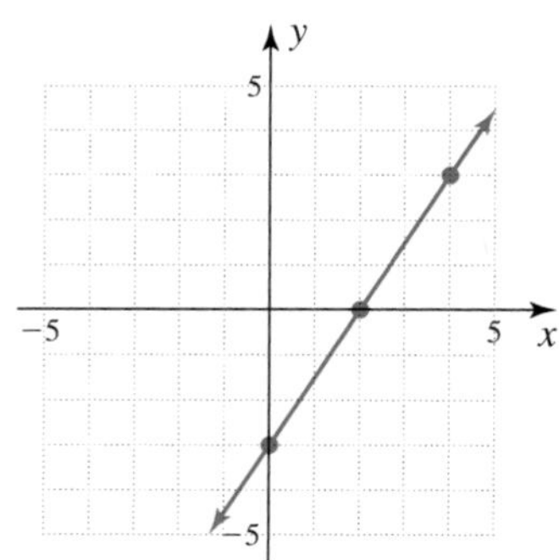

As we suspected, the resulting point $(-3, -4)$ is on the line. Do you see why we picked $x = -3$? We did so because -3 seemed to be the only x-coordinate that yielded a y-value that was an integer.

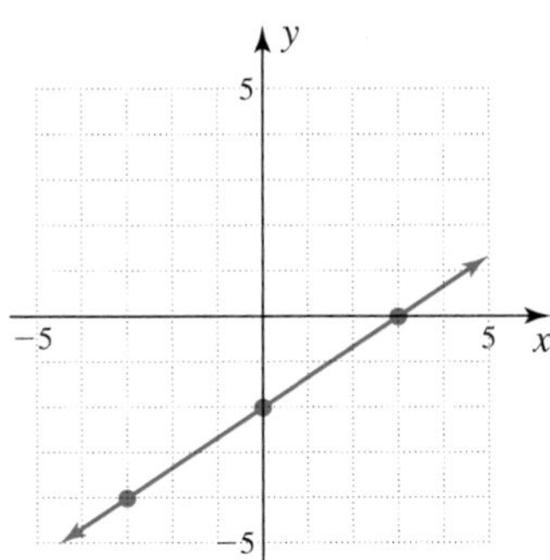

Figure 18

B Graphing Lines Through the Origin

Sometimes it isn't possible to use the x- and y-intercepts to graph an equation. For example, to graph $x + 5y = 0$, we can start by letting $x = 0$ to obtain $0 + 5y = 0$ or $y = 0$. This means that $(0, 0)$ is part of the graph of the line $x + 5y = 0$. If we now let $y = 0$, we get $x = 0$, which is the same ordered pair. The line $x + 5y = 0$ goes through the origin $(0, 0)$ as seen in Figure 19, so we need to find another point. An easy one is $x = 5$. When 5 is substituted for x in $x + 5y = 0$, we obtain

$$5 + 5y = 0 \quad \text{or} \quad y = -1$$

Thus a second point on the graph is $(5, -1)$. We join the points $(0, 0)$ and $(5, -1)$ to get the graph of the line shown in Figure 19. To use a third point as a check, let $x = -5$, which gives $y = 1$ and the point $(-5, 1)$, also shown on the line.

Here is the procedure for recognizing and graphing lines that go through the origin.

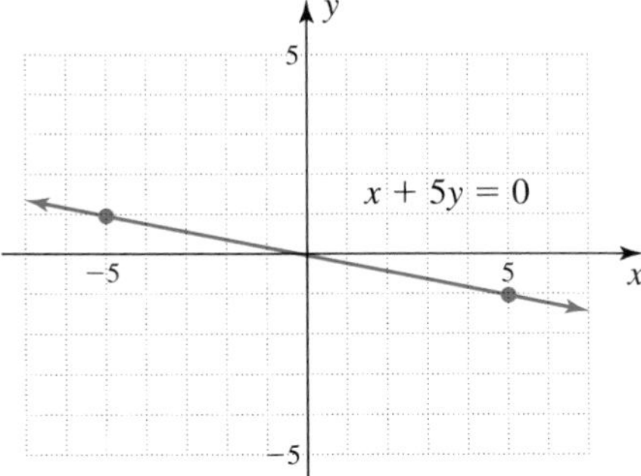

Figure 19

STRAIGHT-LINE GRAPH THROUGH THE ORIGIN

The graph of an equation of the form $Ax + By = 0$, A and B constants and not both zero, is a straight line that goes through the origin.

PROCEDURE

Graphing Lines Through the Origin

To graph a line through the origin, use the point $(0, 0)$, find another point, and draw the line passing through $(0, 0)$ and this other point. Find a third point and verify that it is on the graph of the line.

EXAMPLE 3 Graphing lines going through the origin

Graph: $2x + y = 0$

SOLUTION The line $2x + y = 0$ is of the form $Ax + By = 0$ so it goes through the origin (0, 0). To find another point on the line, we let $y = 4$ in $2x + y = 0$:

$$2x + 4 = 0$$
$$x = -2$$

Now we join the points (0, 0) and (−2, 4) with a line, the graph of $2x + y = 0$, as shown in Figure 20. Note that the check point (2, −4) is on the graph.

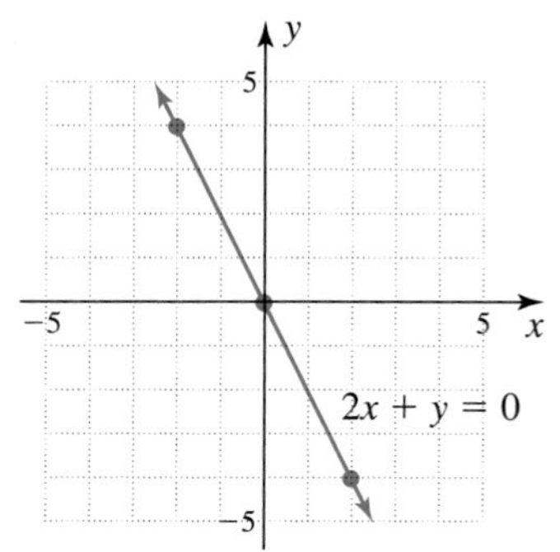

Figure 20

PROBLEM 3

Graph: $3x + y = 0$

Here is a summary of the techniques used to graph lines that are *not* vertical or horizontal.

To graph $Ax + By = C$ (A, B, and C not 0)	To graph $Ax + By = 0$ (A and B not 0)
Find two points on the graph (preferably the x- and y-intercepts) and join them with a line. The result is the graph of $Ax + By = C$.	The graph goes through (0, 0). Find another point and join (0, 0) and the point with a line. The result is the graph of $Ax + By = 0$.
To graph $2x + y = -4$: Let $x = 0$, find $y = -4$, and graph (0, −4). Let $y = 0$, find $x = -2$, and graph (−2, 0).	To graph $x + 4y = 0$, let $x = -4$ in $x + 4y = 0$ $-4 + 4y = 0$ $y = 1$
Join (0, −4) and (−2, 0) to obtain the graph.	Join (0, 0) and (−4, 1) to obtain the graph.
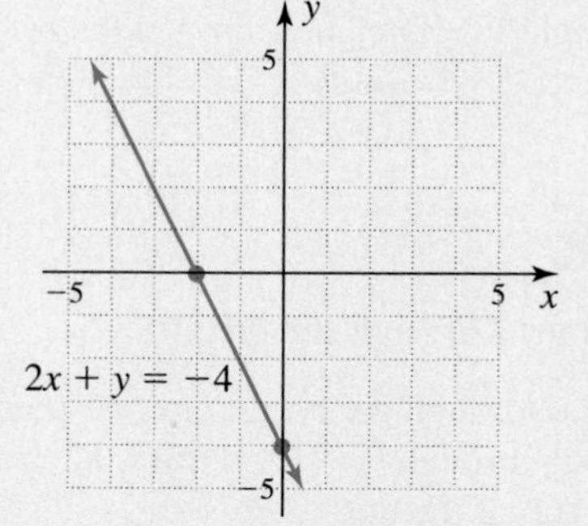	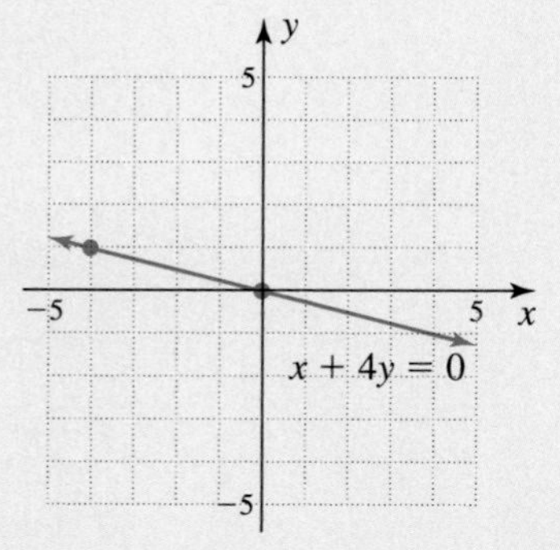

Answer

3.

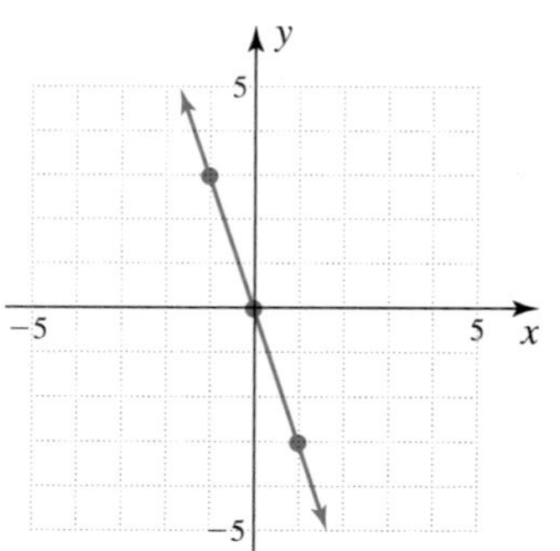

Graphing Horizontal and Vertical Lines

Not all linear equations are written in the form $Ax + By = C$. For example, consider the equation $y = 3$. It may seem that this equation is not in the form $Ax + By = C$. However, we can write the equation $y = 3$ as

$$0 \cdot x + y = 3$$

which is an equation written in the desired form. How do we graph the equation $y = 3$? Since it doesn't matter what value we give x, the result is always $y = 3$, as can be seen in the following table:

x	y
0	3
1	3
2	3

In the equation $0 \cdot x + y = 3$, x can be any number; you always get $y = 3$.

These ordered pairs, as well as the graph of $y = 3$, appear in Figure 21.

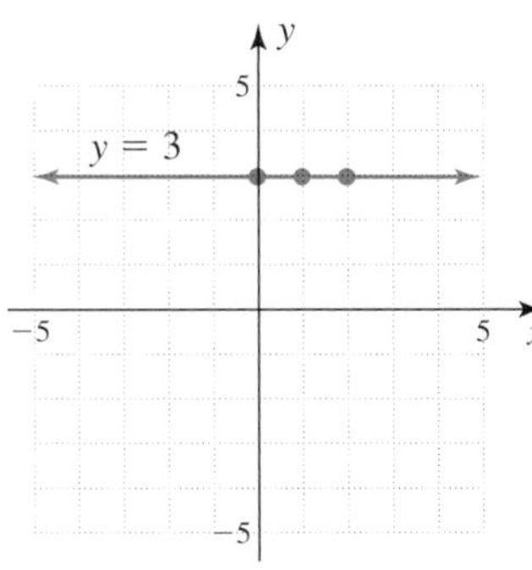

Figure 21

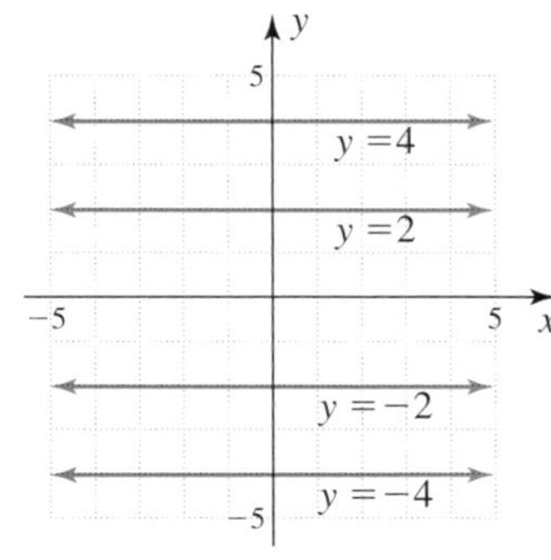

Figure 22

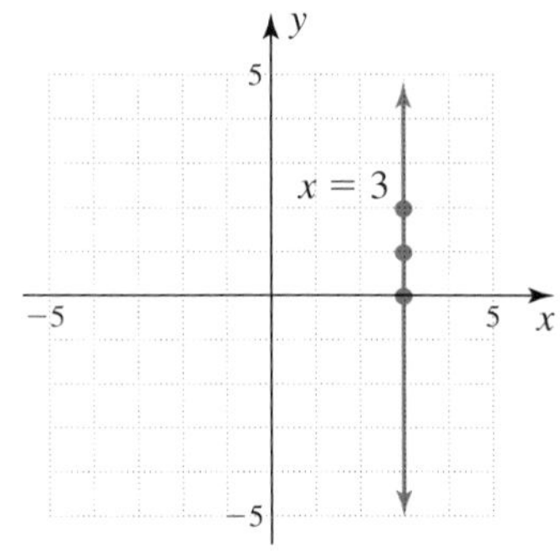

Figure 23

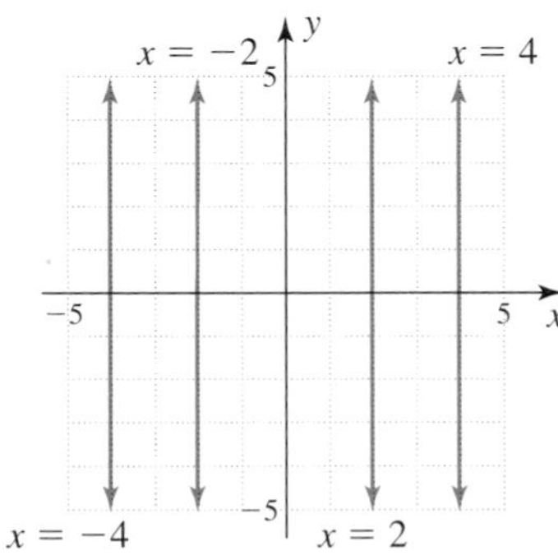

Figure 24

Note that the equation $y = 3$ has for its graph a *horizontal* line crossing the y-axis at $y = 3$. The graphs of some other horizontal lines, all of which have equations of the form $y = k$ (k a constant), appear in Figure 22.

If the graph of any equation $y = k$ is a horizontal line, what would the graph of the equation $x = 3$ be? A vertical line, of course! We first note that the equation $x = 3$ can be written as

$$x + 0 \cdot y = 3$$

Thus the equation is of the form $Ax + By = C$ so that its graph is a straight line. Now for any value of y, the value of x remains 3. Three values of x and the corresponding y-values appear in the following table. These three points, as well as the completed graph, are shown in Figure 23. The graphs of other vertical lines $x = -4$, $x = -2$, $x = 2$, and $x = 4$ are given in Figure 24.

x	y
3	0
3	1
3	2

Here are the formal definitions.

Web It

Want to try a video? Go to link 3-3-2 at the Bello Website at mhhe.com/bello.

Teaching Tip

The key to recognizing that the equation's graph is horizontal or vertical is that one variable is missing. As in Example 4, solving for the only variable will give $y = k$ or $x = k$.

GRAPH OF $y = k$

The graph of any equation of the form

$y = k$ where k is a constant

is a **horizontal line** crossing the y-axis at k.

GRAPH OF $x = k$

The graph of any equation of the form

$x = k$ where k is a constant

is a **vertical line** crossing the x-axis at k.

EXAMPLE 4 Graphing vertical and horizontal lines

Graph: **a.** $2x - 4 = 0$ **b.** $3 + 3y = 0$

SOLUTION

a. We first solve for x.

$2x - 4 = 0$ Given.

$2x = 4$ Add 4.

$x = 2$ Divide by 2.

The graph of $x = 2$ is a vertical line crossing the x-axis at 2, as shown in Figure 25.

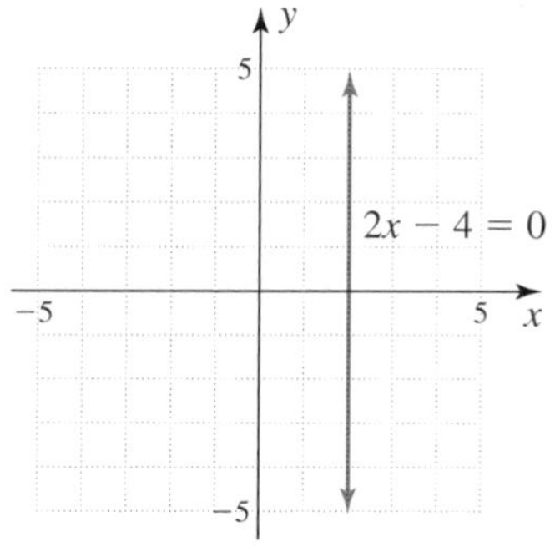

Figure 25

b. In this case, we solve for y.

$3 + 3y = 0$ Given.

$3y = -3$ Subtract 3.

$y = -1$ Divide by 3.

The graph of $y = -1$ is a horizontal line crossing the y-axis at -1, as shown in Figure 26.

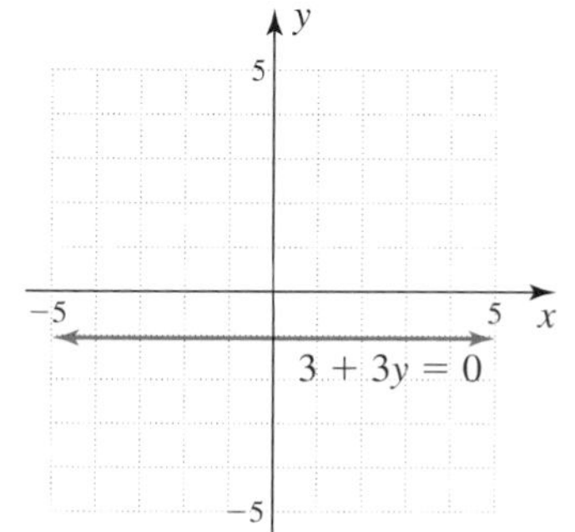

Figure 26

PROBLEM 4

Graph:

a. $3x - 3 = 0$

b. $4 + 2y = 0$

Answer

4.

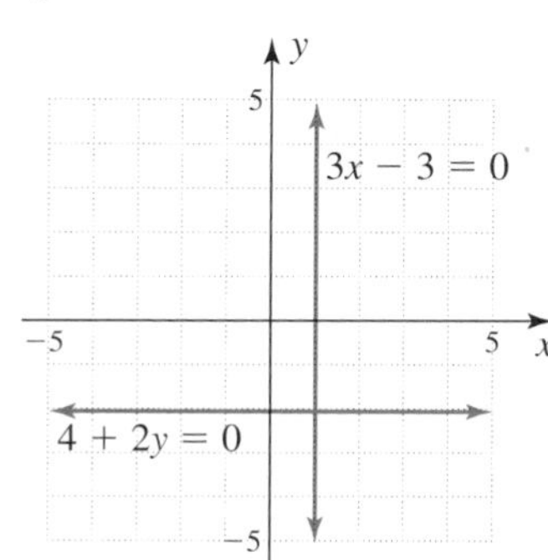

D Solving an Application

As we saw in the *Getting Started,* graphing a linear equation in two variables can be very useful in solving real-world problems. Here is another example.

EXAMPLE 5 Gender differences in hospital stays

Figure 27 shows the average stay in the hospital (in days) for males and females using data from the U.S. National Center for Health Statistics. If t represents the number of years after 1970 and d represents the number of days, the equations can be approximated as follows:

For males: $d = 9 - 0.1t$
For females: $d = 8 - 0.1t$

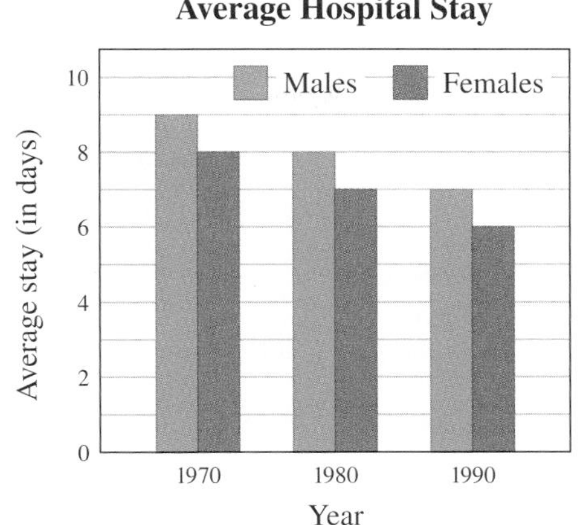

Figure 27

a. Graph these two equations.

b. Can t be negative?

c. Can d be negative?

d. Based on your answers to parts **b** and **c**, in what quadrant do these two graphs make sense?

SOLUTION

a. We find the t- and d-intercepts for males and females, graph them, and connect them with a line.

For males, when $t = 0$,

$$d = 9 - 0.1(0) = 9$$

Thus (0, 9) is the d-intercept.
To find the t-intercept, let $d = 0$:

$$0 = 9 - 0.1t$$
$$0.1t = 9$$
$$t = 90$$

Thus (90, 0) is the t-intercept. Graph the intercepts (0, 9) and (90, 0) for males and join them with a line:

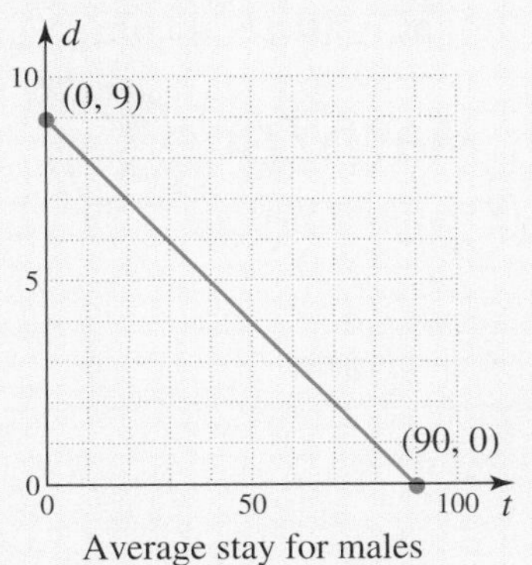

For females, when $t = 0$,

$$d = 8 - 0.1(0) = 8$$

Thus (0, 8) is the d-intercept. To find the t-intercept, let $d = 0$:

$$0 = 8 - 0.1t$$
$$0.1t = 8$$
$$t = 80$$

Thus (80, 0) is the t-intercept. Graph the intercepts (0, 8) and (80, 0) for females and join them with a line:

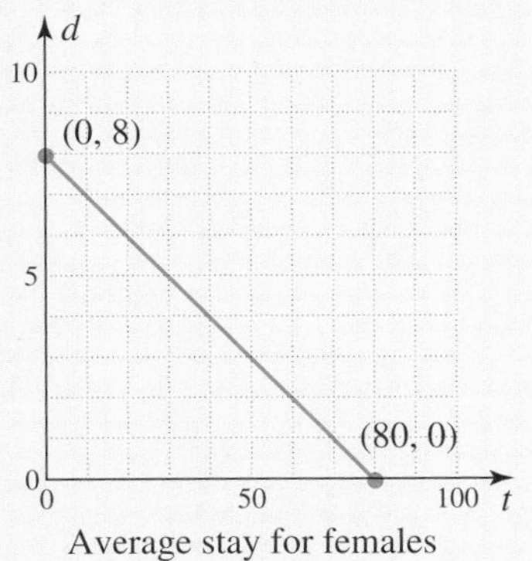

Note that the scale on the t-axis is not the same as that on the d-axis.

b. Because t represents the number of years after 1970, it cannot be negative.

c. Because d represents the number of days, it cannot be negative.

d. Since t and d cannot be negative, the graphs make sense only in quadrant I.

PROBLEM 5

Graph:

a. $d = 9 - 0.2t$

b. $d = 8 - 0.2t$

Answer

5.

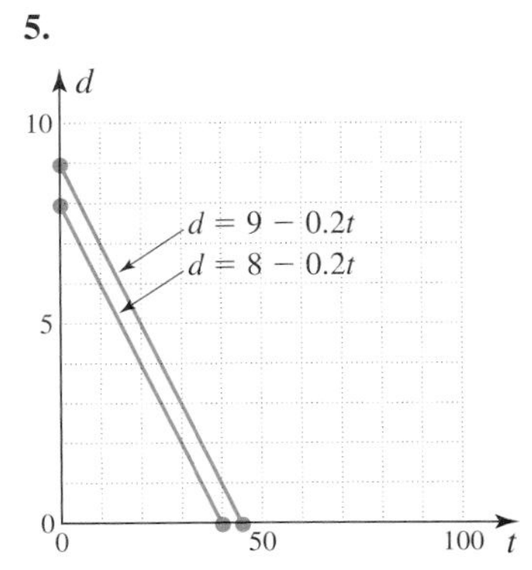

Calculate It **Graphing Lines**

Your calculator can easily do the examples in this section, but you have to *know* how to solve for y to enter the given equations. (Even with a calculator, you have to know algebra!) Then to graph $3x + y = 6$ (Example 4 in Section 3.2), we first solve for y to obtain $y = 6 - 3x$; thus we enter $Y_1 = 6 - 3x$ and press [GRAPH] to obtain the result shown in Window 1. We used the *default* or *standard* window, which is a $[-10, 10]$ by $[-10, 10]$ rectangle. If you have a calculator, do Examples 2 and 3 of Section 3.3 now. Note that you can graph $3 + 3y = 0$ (Example 4b) by solving for y to obtain $y = -1$, but you *cannot* graph $2x - 4 = 0$ with most calculators. Can you see why?

To graph the equations in Example 5, you have to select the appropriate window. Let's see why. Suppose you simply enter $d = 9 - 0.1t$, by using y instead of d and x instead of t. The result is shown in Window 2. To see more of the graph, you have to adjust the window. Algebraically, we know that the t-intercept is $(90, 0)$ and the d-intercept is $(0, 9)$; thus an appropriate window might be $[0, 90]$ by $[0, 10]$. Now you only have to select the scales for x and y. Since the x's go from 0 to 90, make Xscl = 10. The y's go from 0 to 10, so we can let Yscl = 1. (See Window 3.) The completed graph appears with a scale that allows us to see most of the graph in Window 4. Now do you see how the equation $d = 9 - 0.1t$ can be graphed and why it's graphed in quadrant I?

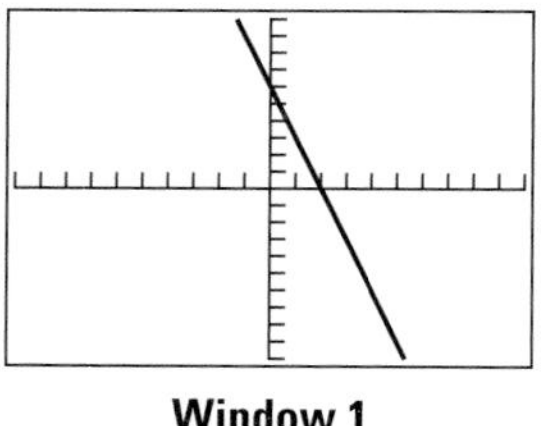

Window 1

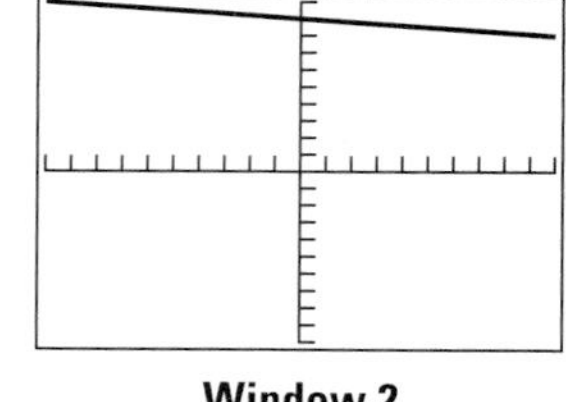

Window 2

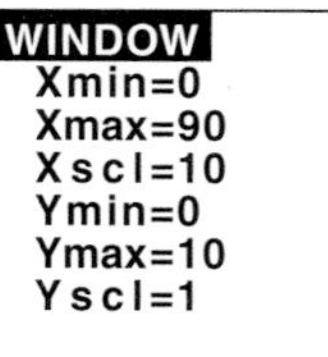

Window 3

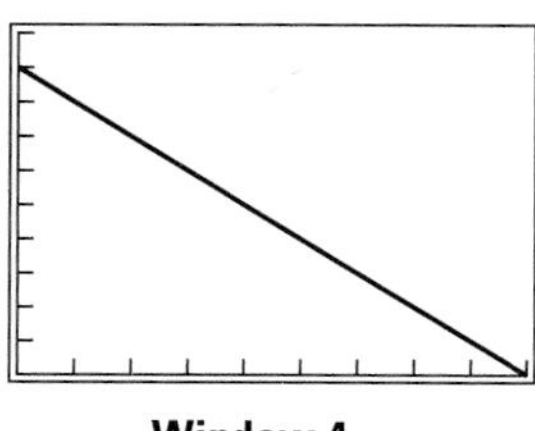

Window 4

You've gotten this far, so you should be rewarded! Suppose you want to find out how long the projected average hospital stay would be for males in the year 2000. Some calculators find the value of y when given x. Press [2nd] [Y=] 30 [ENTER] to find the value of $y = 0.1x - 9$ when $X = 30$ $(2000 - 1970 = 30)$. (If your calculator doesn't have this feature, you can use [ZOOM] and [TRACE] to find the answer.) Window 5 shows the answer. Can you see what it is? Can you do it without your calculator? Now see if you can find the projected average hospital stay for females in the year 2000.

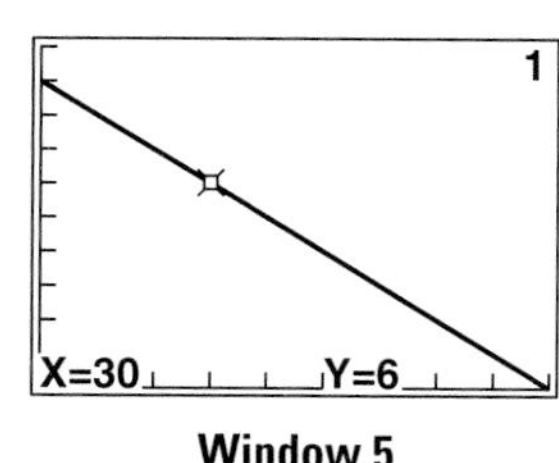

Window 5

Exercises 3.3

A In Problems 1–10, graph the equations.

1. $x + 2y = 4$

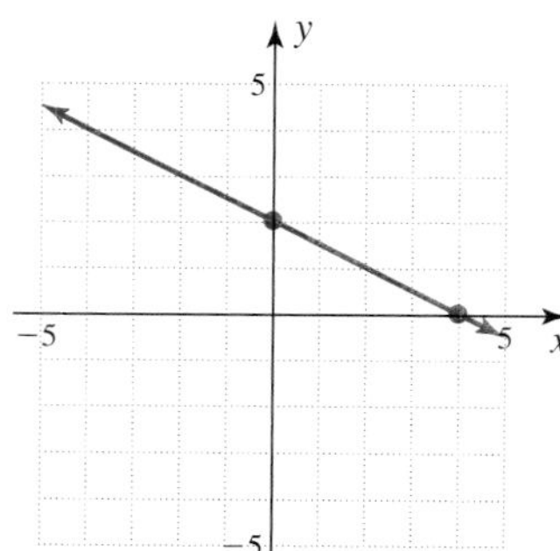

2. $y + 2x = 2$

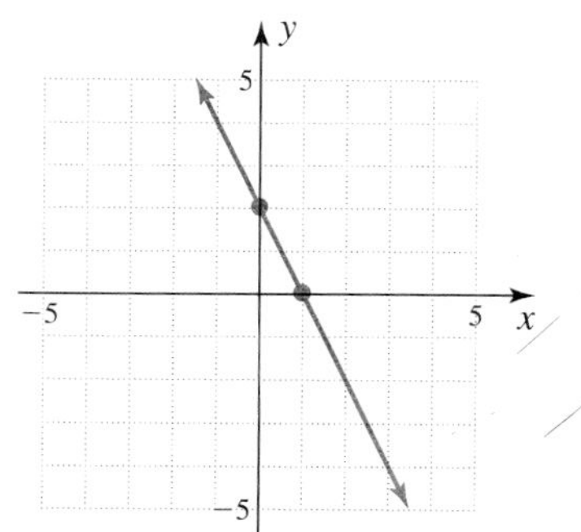

3. $-5x - 2y = -10$

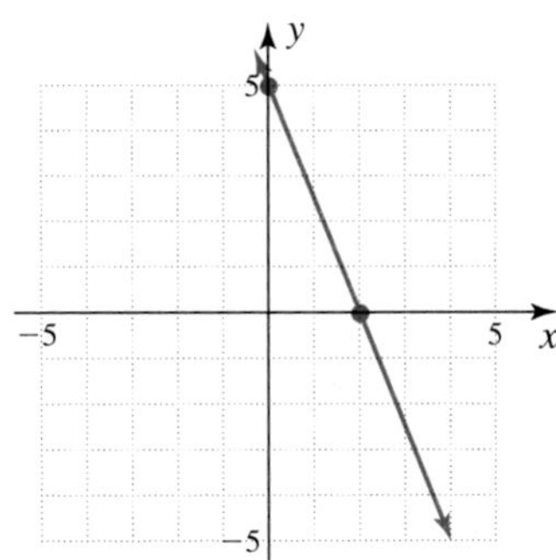

4. $-2x - 3y = -6$

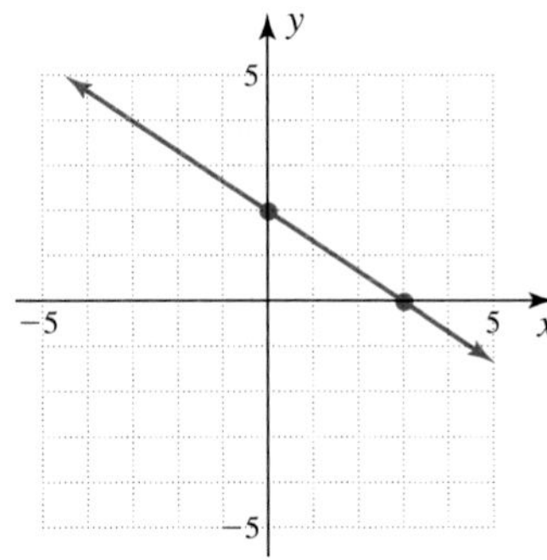

5. $y - 3x - 3 = 0$

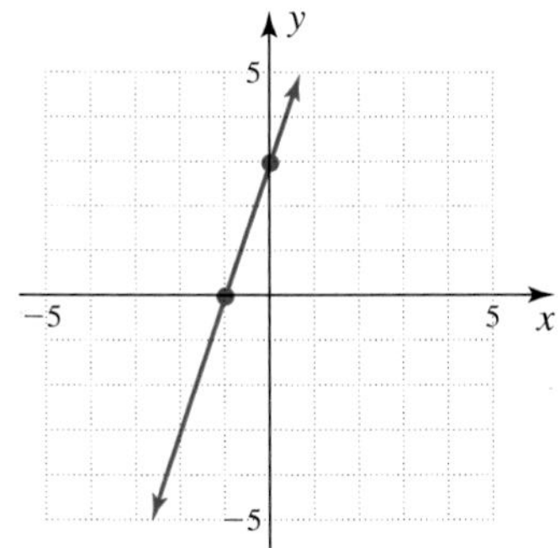

6. $y + 2x - 4 = 0$

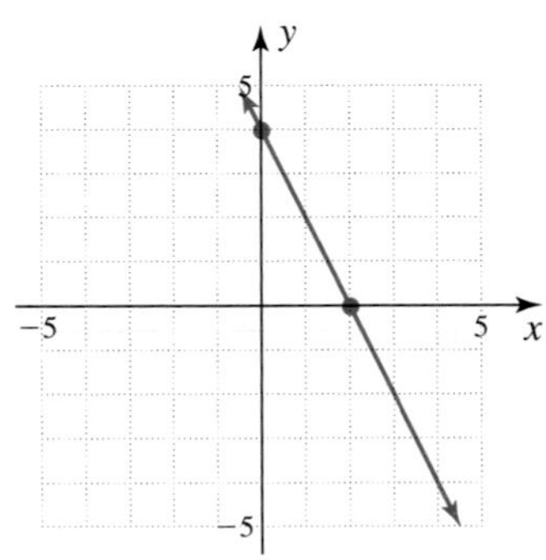

7. $6 = 6x - 3y$

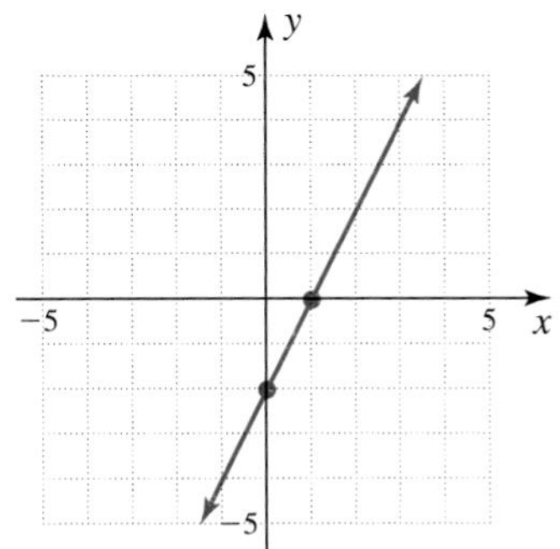

8. $6 = 3y - 2x$

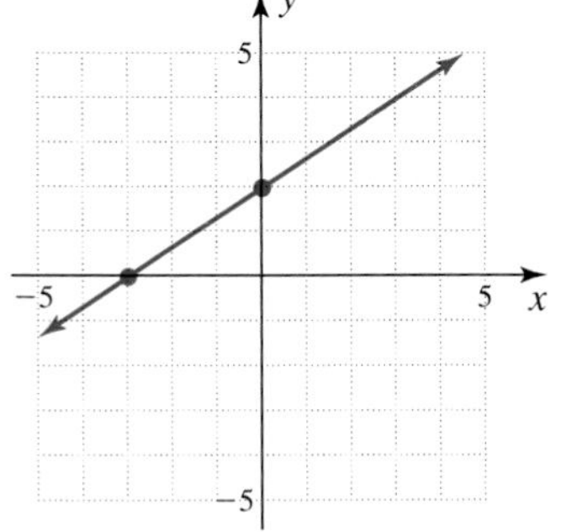

9. $3x + 4y + 12 = 0$

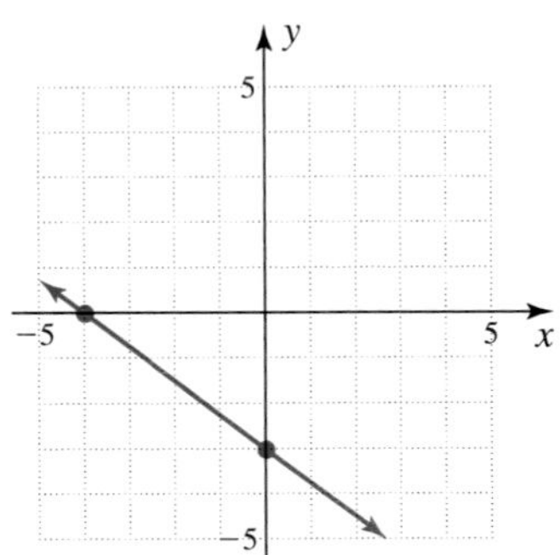

10. $5x + 2y + 10 = 0$

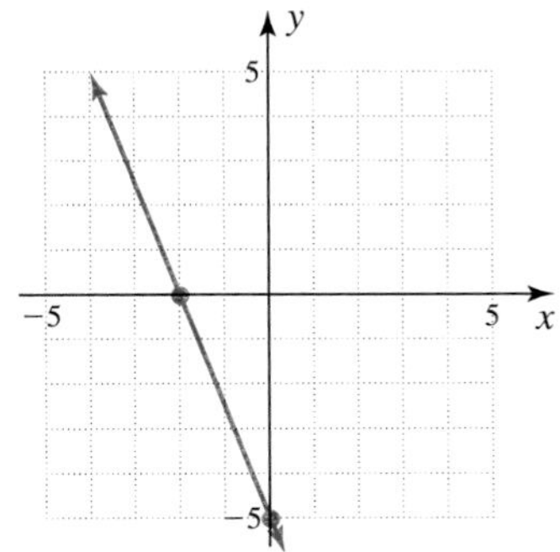

B In Problems 11–20, graph the equations.

11. $3x + y = 0$

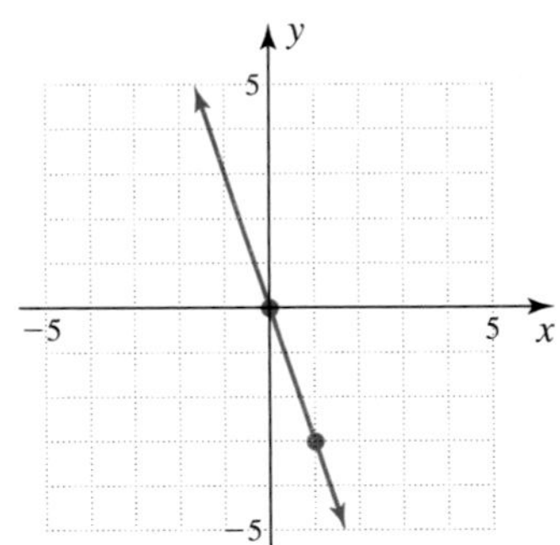

12. $4x + y = 0$

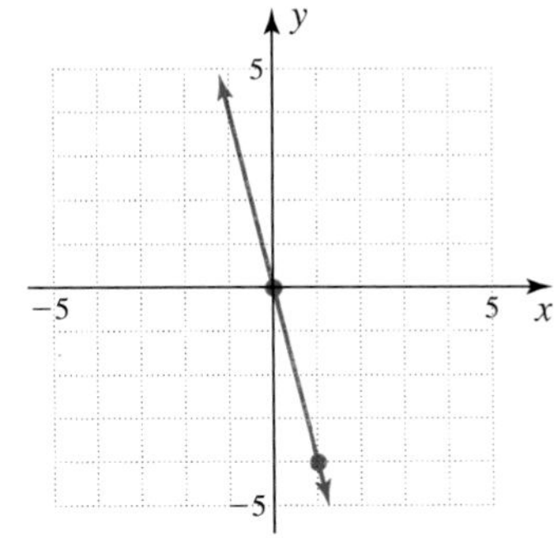

13. $2x + 3y = 0$

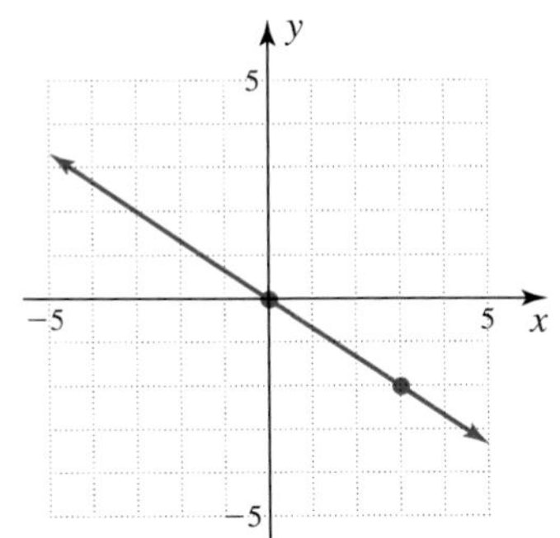

14. $3x + 2y = 0$

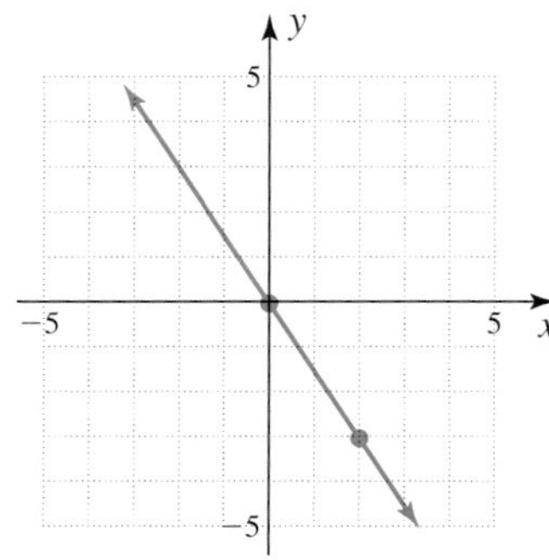

15. $-2x + y = 0$

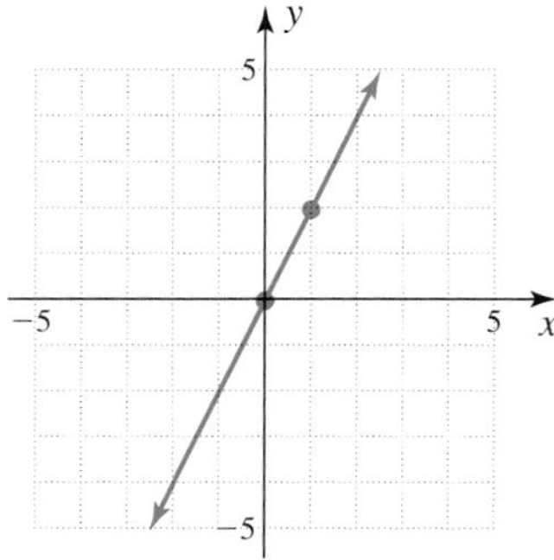

16. $-3x + y = 0$

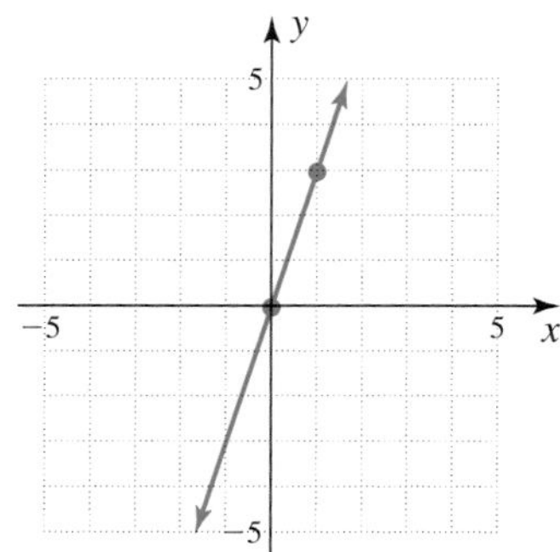

17. $2x - 3y = 0$

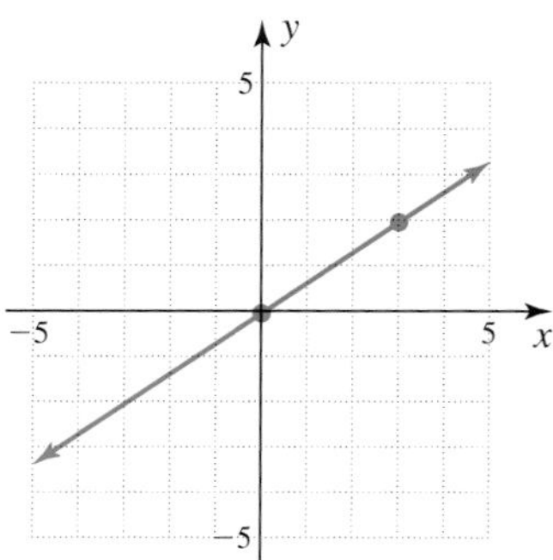

18. $3x - 2y = 0$

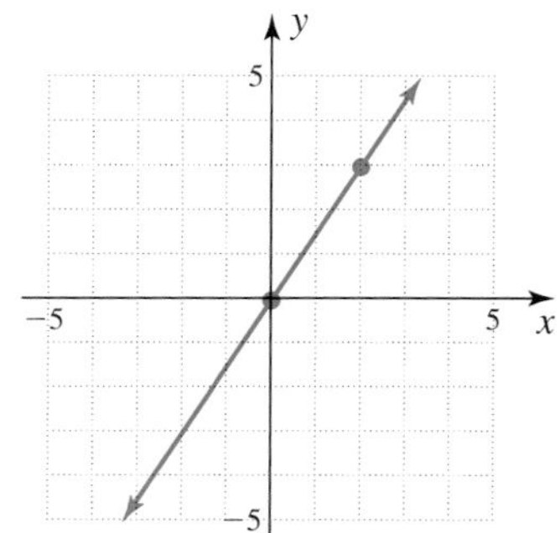

19. $-3x = -2y$

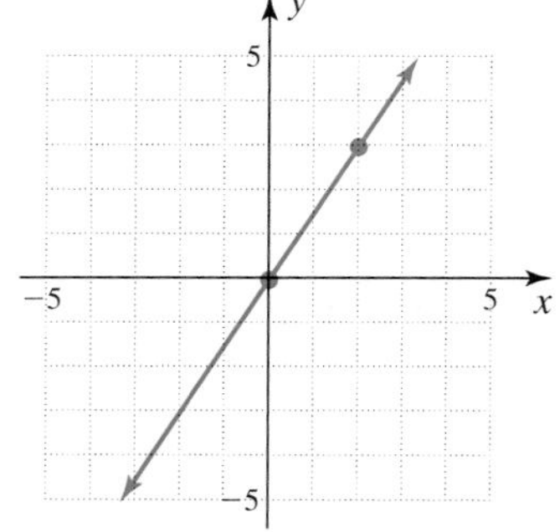

20. $-2x = -3y$

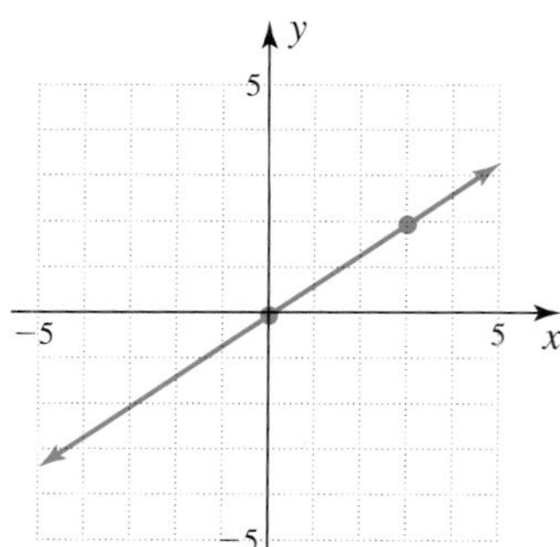

C In Problems 21–30, graph the equations.

21. $y = -4$

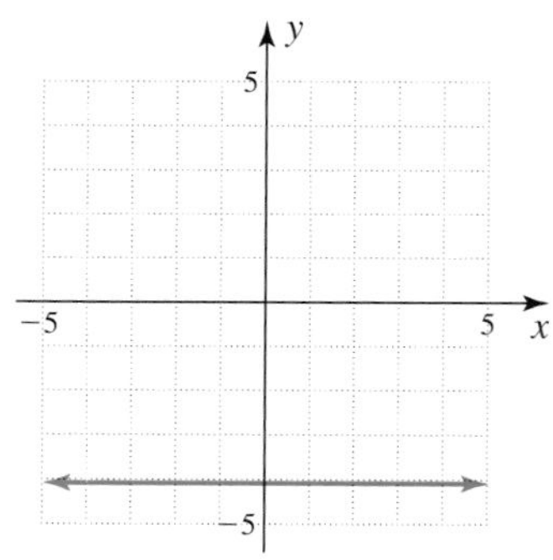

22. $y = -\frac{3}{2}$

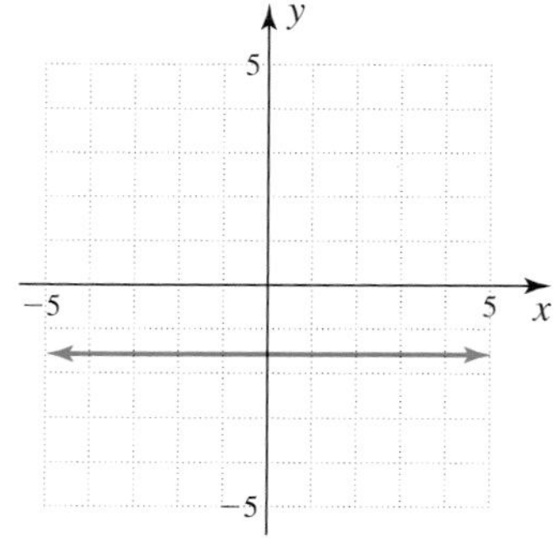

23. $2y + 6 = 0$

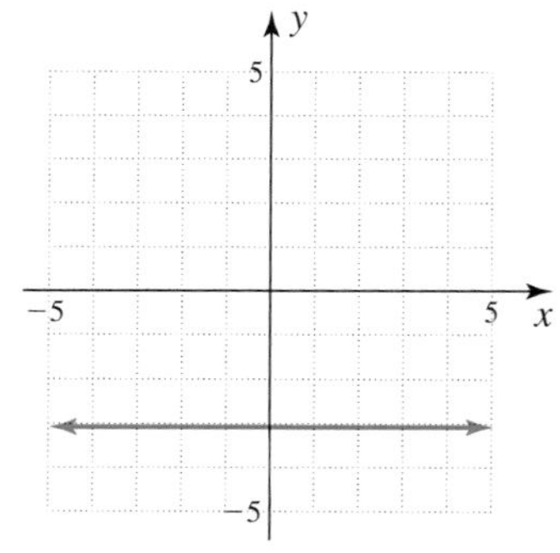

24. $-3y + 9 = 0$

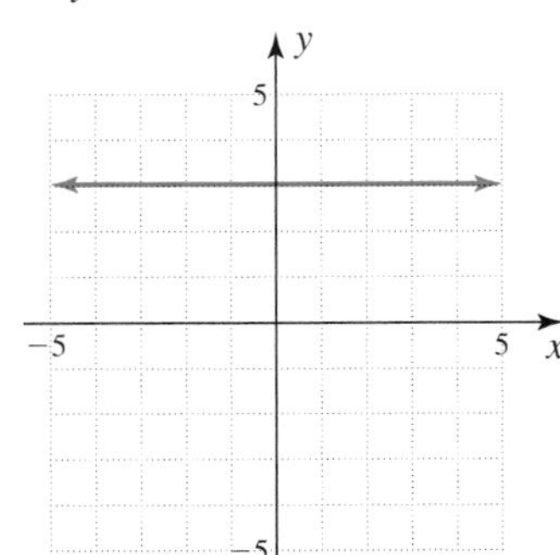

25. $x = -\frac{5}{2}$

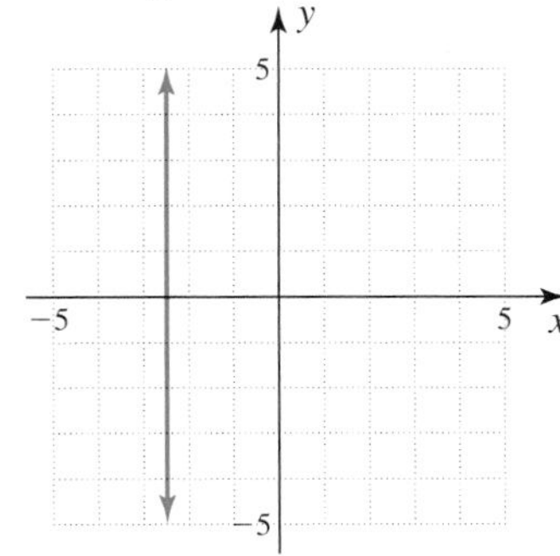

26. $x = \frac{7}{2}$

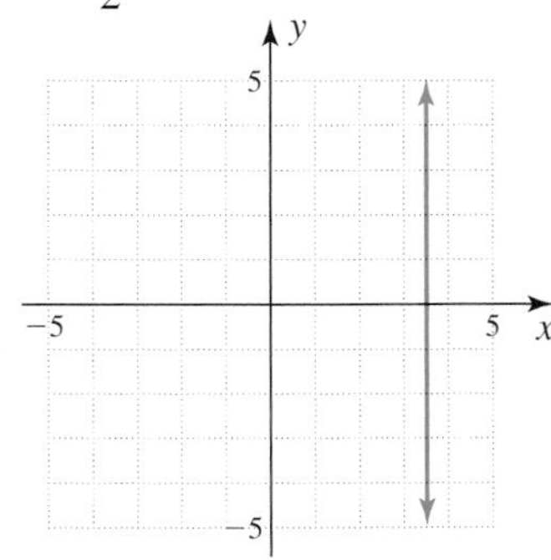

27. $2x + 4 = 0$

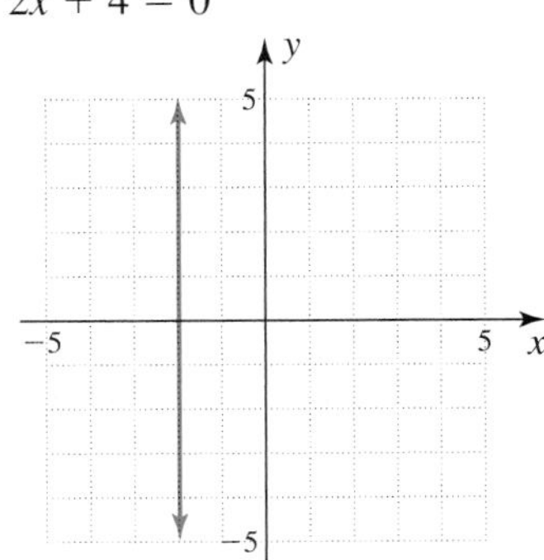

28. $3x - 12 = 0$

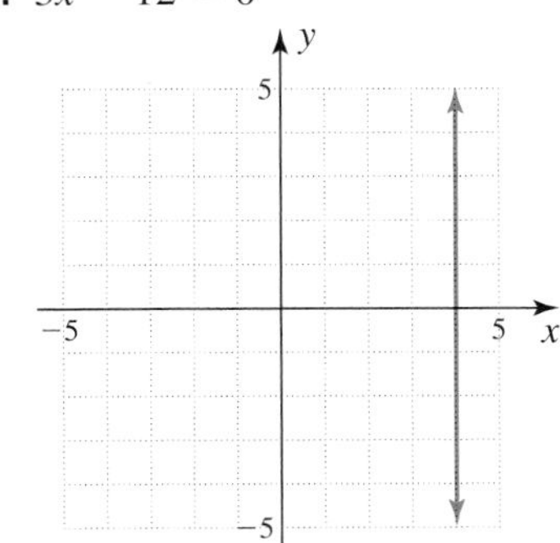

29. $2x - 9 = 0$

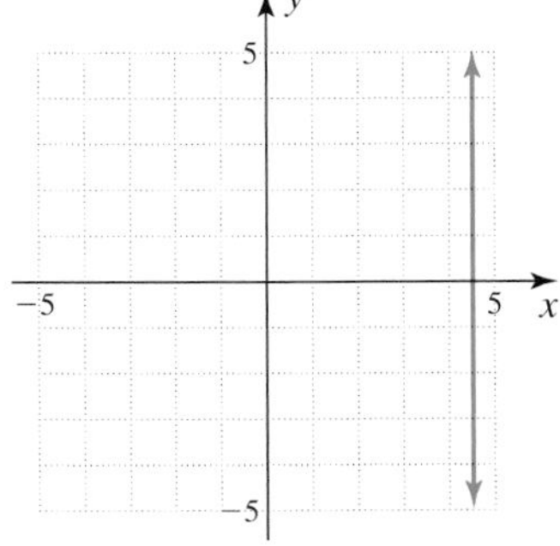

30. $-2x + 7 = 0$

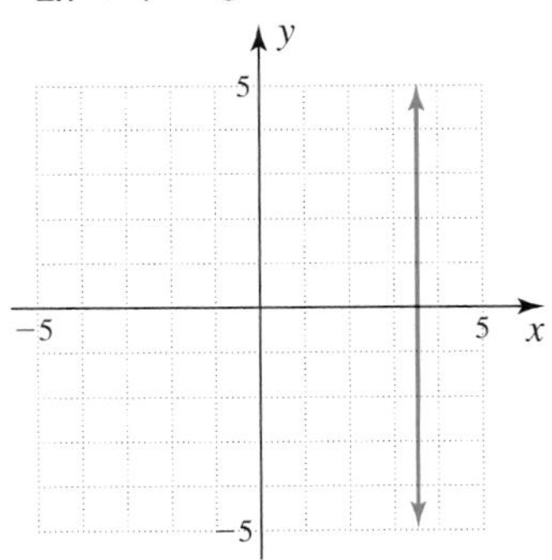

APPLICATIONS

31. *Average stay in hospital* According to the U.S. National Center for Health Statistics, the total average stay in the hospital, d (in days), can be approximated by $d = 5.6 - 0.08t$, where t is the number of years after 2000.

a. What is the d-intercept? 5.6

b. What is the t-intercept? 70

c. Graph the equation.

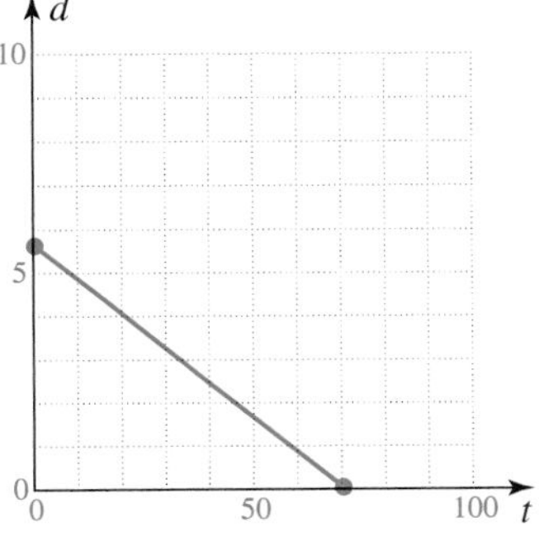

32. *Sales of jazz recordings* According to the Recording Industry Association, the percent p of U.S. dollar sales for jazz recordings can be approximated by $p = 6 - 0.6t$, where t is the number of years after 2000.

a. Find the t- and p-intercepts for this equation. t: 10, p: 6

b. Graph the equation.

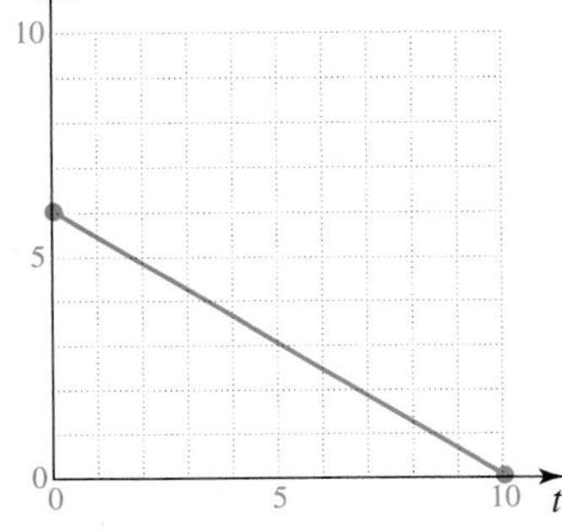

33. *Daily fat intake* According to the U.S. Department of Agriculture, the total daily fat intake g (in grams) per person can be approximated by $g = 140 + t$, where t is the number of years after 1950.

a. What was the daily fat intake per person in 1950? 140 g

b. What was the daily fat intake per person in 1990? 180 g

c. What would you project the daily fat intake to be in the year 2000? 190 g

d. Use the information from parts **a–c** to graph $g = 140 + t$.

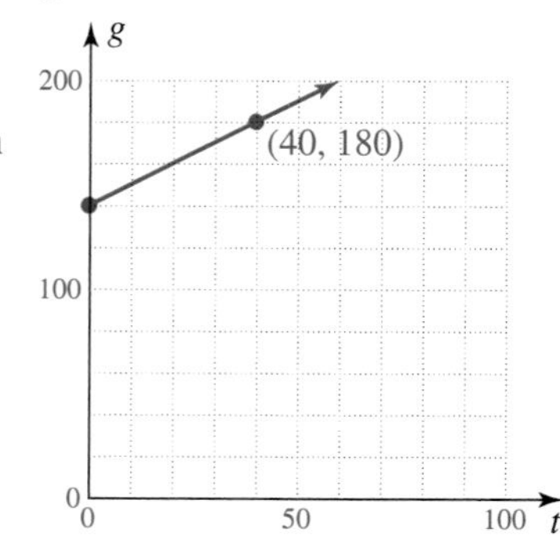

34. *Death rates* According to the U.S. Health and Human Services, the number D of deaths from heart disease per 100,000 population can be approximated by $D = -5t + 290$, where t is the number of years after 1998.

a. How many deaths per 100,000 population were there in 1998? 290

b. How many deaths per 100,000 population would you expect in 2008? 240

c. Find the t- and D-intercepts for $D = -5t + 290$ and graph the equation.
t: 58, D: 290,

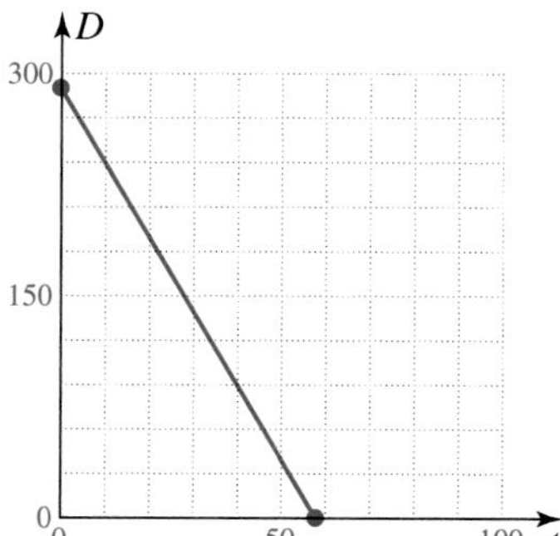

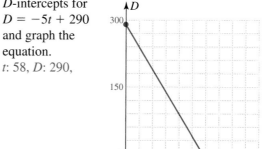

SKILL CHECKER

Try the Skill Checker Exercises so you'll be ready for the next section.

Find:

35. $3 - (-6)$ 9 **36.** $5 - (-7)$ 12 **37.** $-6 - 3$ -9 **38.** $-5 - 7$ -12

USING YOUR KNOWLEDGE

And the Grease Goes On!

In Example 3 of Section 3.2 (p. 244), we mentioned that the decrease in the cholesterol level C after w weeks elapsed could be approximated by $C = -3w + 215$.

39. According to this equation, what was the cholesterol level initially? 215

40. What was the cholesterol level at the end of 12 weeks? 179

41. Use the results of Problems 39 and 40 to graph the equation $C = -3w + 215$.

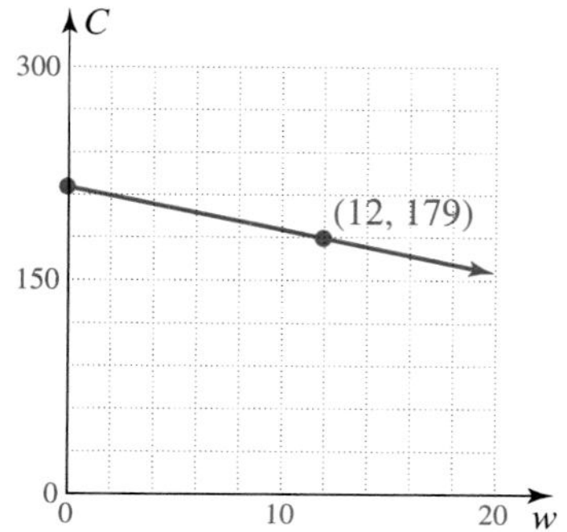

42. According to the graph shown in Example 3 (p. 244), the initial cholesterol level was 215. If we assume that the initial cholesterol level is 230, we can approximate the reduction in cholesterol by $C = -3w + 230$.

a. Find the intercepts and graph the equation on the same coordinate axes as $C = -3w + 215$. C: 230, w: $76\frac{2}{3}$,

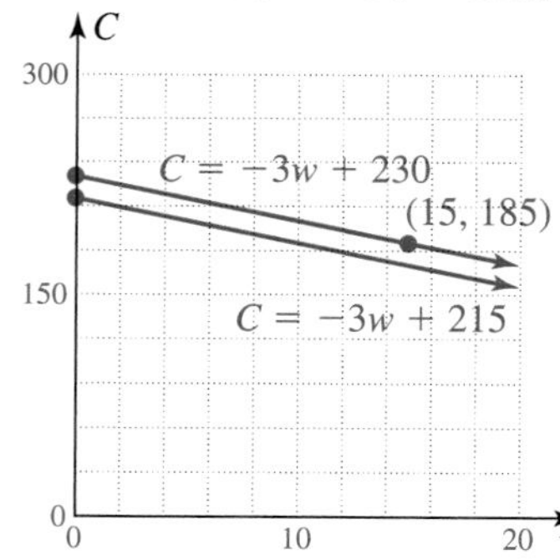

b. How many weeks does it take for a person with an initial cholesterol level of 230 to reduce it to 175? $18\frac{1}{3}$

WRITE ON

43. If in the equation $Ax + By = C, A = 0$ and B and C are not zero, what type of graph will result? a horizontal line

44. If in the equation $Ax + By = C, B = 0$ and A and C are not zero, what type of graph will result? a vertical line

45. If in the equation $Ax + By = C, C = 0$ and A and B are not zero, what type of graph will result? a line through the origin

46. What are the intercepts of the line $Ax + By = C$, and how would you find them? Answers may vary.

47. How many points are needed to graph a straight line? two

MASTERY TEST

If you know how to do these problems, you have learned your lesson!

Graph:

48. $x + 2y = 4$

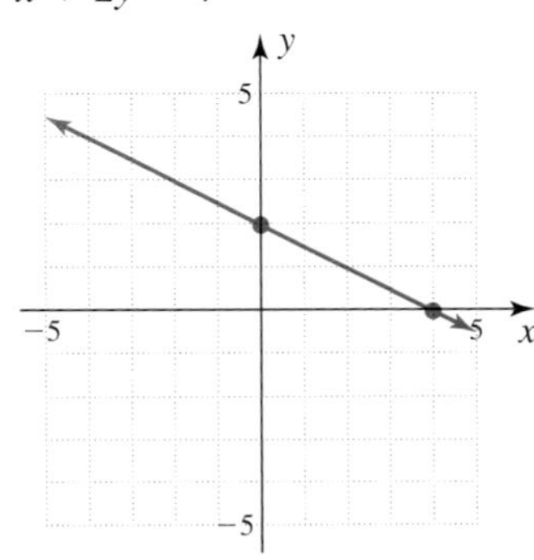

49. $-2x + y = 4$

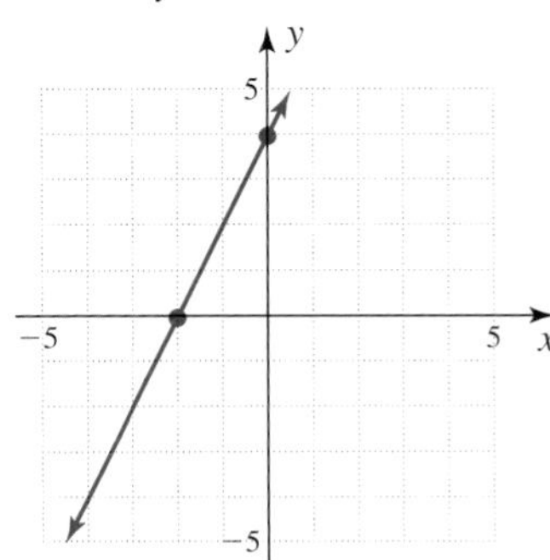

50. $3x - 6 = 0$

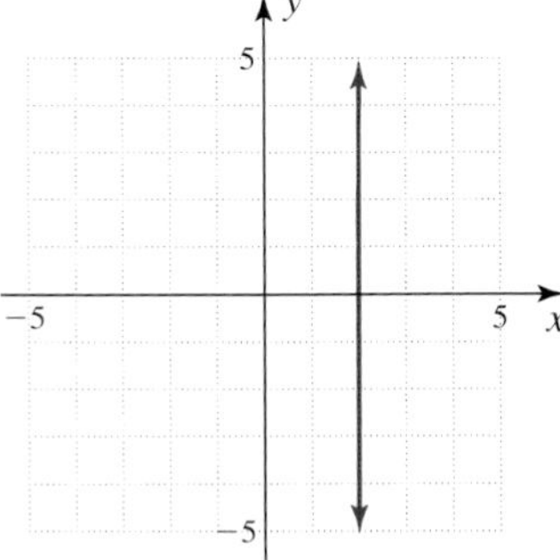

51. $4 + 2y = 0$

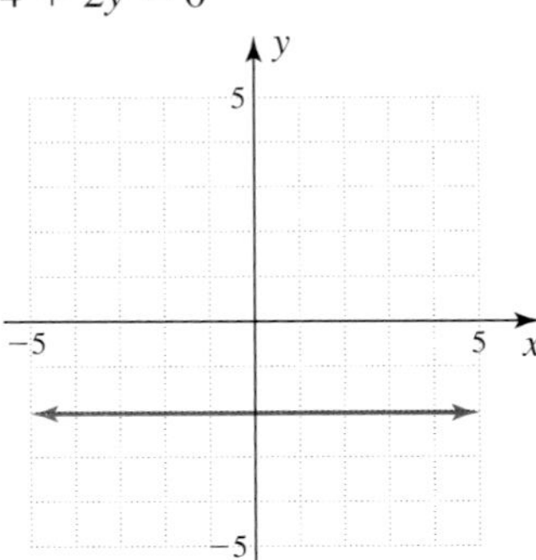

52. $-4x + y = 0$

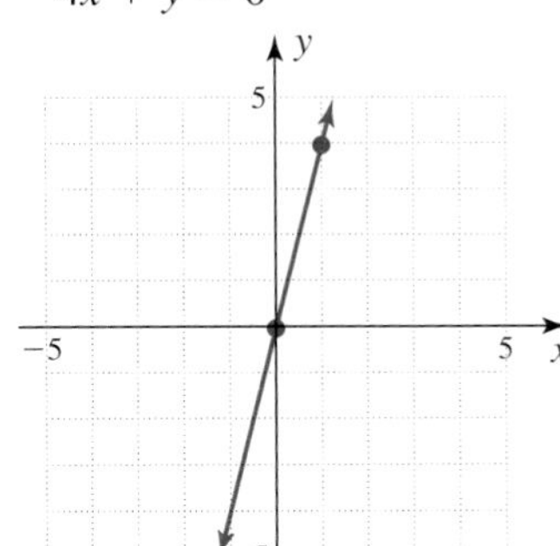

53. $-x + 4y = 0$

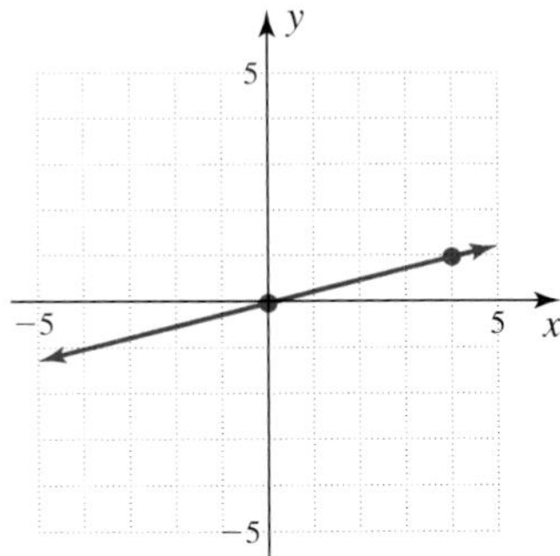

54. The number N (in millions) of recreational boats in the United States can be approximated by $N = 10 + 0.4t$, where t is the number of years after 1975.

a. How many recreational boats were there in 1975? 10 million

b. How many would you expect in 1995? 18 million

c. Find the t- and N-intercepts of $N = 10 + 0.4t$ and graph the equation. t: -25, N: 10

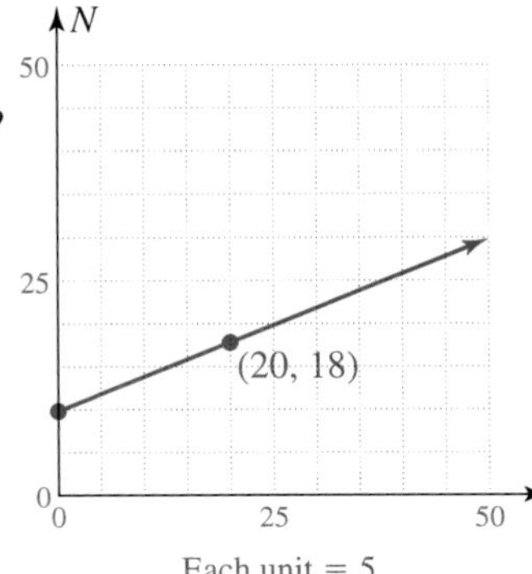

Each unit = 5

55. The relationship between the Continental dress size C and the American dress size A is $C = A + 30$.

a. Find the intercepts for $C = A + 30$ and graph the equation. C: 30, A: -30,

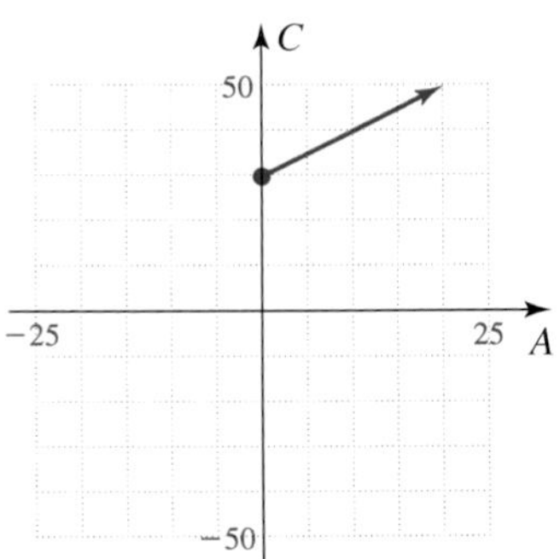

Each vertical unit = 10; each horizontal unit = 5

b. In what quadrant should the graph be? Quadrant I

3.4 THE SLOPE OF A LINE

To Succeed, Review How To . . .

1. Add, subtract, multiply, and divide signed numbers (pp. 61, 70).
2. Solve an equation for a specified variable (p. 156).

Objectives

A Find the slope of a line given two points.

B Find the slope of a line given the equation of the line.

C Determine whether two lines are parallel, perpendicular, or neither.

D Solve an application.

GETTING STARTED

Online Services Are Sloping Upward

Can you tell from the graph when the number of subscribers to commercial online services soared? After 1993! The annual increase in subscribers in 1993 was

$$\frac{\text{Difference in subscribers}}{\text{Difference in years}} = \frac{3.5 - 3.0}{93 - 92} = \frac{0.5}{1} = 0.5 \text{ (million)}$$

The annual increase in subscribers from 1993 to 1997 was

$$\frac{13.5 - 3.5}{97 - 93} = \frac{10}{4} = 2.5 \text{ (million)}$$

Source: ClickZ Network.

As you can see from the graph, the line gets "steeper" from 1993 to 1997 than from 1992 to 1993. The "steepness" of the lines ('92 to '93 and '93 to '97) was calculated by comparing the *vertical* change of the line (the **rise**) to the *horizontal* change (the **run**). This measure of steepness is called the **slope** of the line. In this section we shall learn how to find the slope of a line when two points are given or when the equation of the line is given and also how to determine when lines are parallel, perpendicular, or neither. (By the way, the "soaring" did not continue for long. As a matter of fact, the number of online subscribers declined by 2.7 million at the end of the third quarter of 2001!)

Finding Slopes from Two Points

The steepness of a line can be measured by using the ratio of the **vertical rise** (or **fall**) to the corresponding **horizontal run.** This ratio is called the *slope*. For example, a staircase that rises 3 feet in a horizontal distance of 4 feet is said to have a slope of $\frac{3}{4}$. The definition of slope is as follows.

Web It

For an interesting lesson on finding slopes, go to link 3-4-1 on the Bello Website at mhhe.com/bello.

SLOPE

The **slope** m of the line going through the points (x_1, y_1) and (x_2, y_2), where $x_1 \neq x_2$, is given by

$$m = \frac{y_2 - y_1}{x_2 - x_1} = \frac{\text{rise} \uparrow}{\text{run} \rightarrow}$$

Teaching Tip

The slope of a line is a numerical way of describing steepness. The formula

$$m = \frac{y_2 - y_1}{x_2 - x_1}$$

is the way to assign the number. Have students discuss other real-life examples of slope.
Examples:
1. the pitch of a roof
2. the incline of a mountain, etc.

The slope for a vertical line such as $x = 3$ is not defined because all points on this line have the same x-value, 3, and hence

$$m = \frac{y_2 - y_1}{3 - 3} = \frac{y_2 - y_1}{0}$$

which is undefined (see Example 3). The slope of a horizontal line is zero because all points on such a line have the same y-values. Thus for the line $y = 7$,

$$m = \frac{7 - 7}{x_2 - x_1} = \frac{0}{x_2 - x_1} = 0$$ See Example 3.

EXAMPLE 1 Finding the slope given two points: Positive slope

Find the slope of the line going through the points $(0, -6)$ and $(3, 3)$ in Figure 28.

SOLUTION Suppose we choose $(x_1, y_1) = (0, -6)$ and $(x_2, y_2) = (3, 3)$. Then we use the equation for slope to obtain

$$m = \frac{3 - (-6)}{3 - 0} = \frac{9}{3} = 3$$

If we choose $(x_1, y_1) = (3, 3)$ and $(x_2, y_2) = (0, -6)$, then

$$m = \frac{-6 - 3}{0 - 3} = \frac{-9}{-3} = 3$$

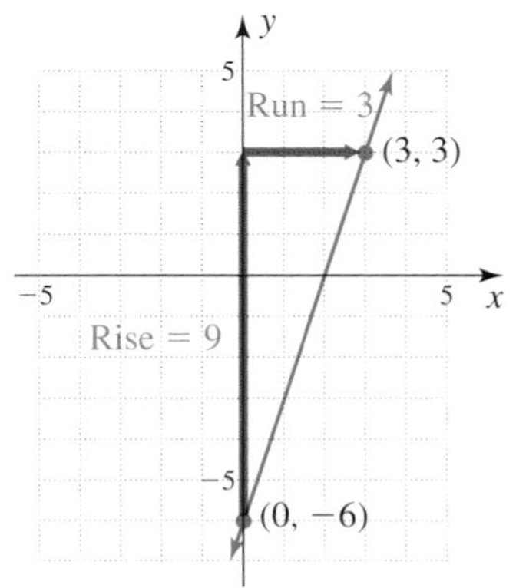

Figure 28
Line with positive slope: $\frac{\text{Rise}}{\text{Run}} = \frac{9}{3} = 3$

As you can see, it makes no difference which point is labeled (x_1, y_1) and which is labeled (x_2, y_2). Since an interchange of the two points simply changes the sign of both the numerator and the denominator in the slope formula, the result is the same in both cases.

PROBLEM 1

Find the slope of the line going through the points $(0, -6)$ and $(2, 2)$.

EXAMPLE 2 Finding the slope given two points: Negative slope

Find the slope of the line that goes through the points $(3, -4)$ and $(-2, 3)$ in Figure 29.

SOLUTION We take $(x_1, y_1) = (-2, 3)$ so that $(x_2, y_2) = (3, -4)$. Then

$$m = \frac{-4 - 3}{3 - (-2)} = -\frac{7}{5}$$

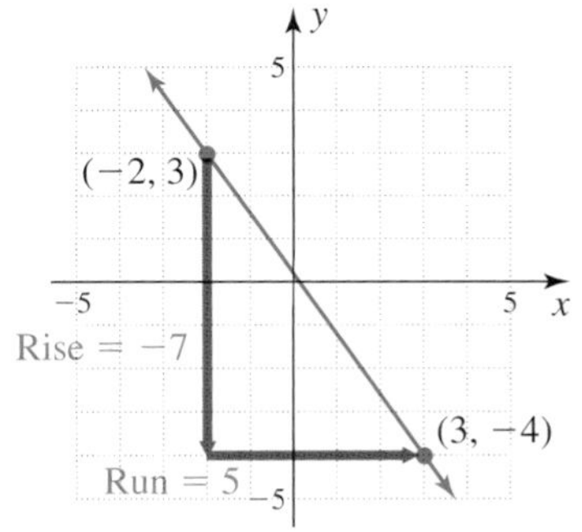

Figure 29
Line with negative slope: $\frac{\text{Rise}}{\text{Run}} = \frac{-7}{5}$

PROBLEM 2

Find the slope of the line going through the points $(2, -4)$ and $(-3, 2)$.

Examples 1 and 2 are illustrations of the fact that a line that rises from left to right has a *positive slope* and one that falls from left to right has a *negative slope*. What about vertical and horizontal lines? We shall discuss them in the next example.

Answers

1. 4 **2.** $-\frac{6}{5}$

EXAMPLE 3 Finding the slopes of vertical and horizontal lines

Find the slope of the line passing through:

a. $(-4, -2)$ and $(-4, 3)$ **b.** $(1, 4)$ and $(4, 4)$

SOLUTION

a. Substituting $(-4, -2)$ for (x_1, y_1) and $(-4, 3)$ for (x_2, y_2) in the equation for slope, we obtain

$$m = \frac{3 - (-2)}{-4 - (-4)} = \frac{3 + 2}{-4 + 4} = \frac{5}{0}$$

which is undefined. Thus the slope of a vertical line is *undefined* (see Figure 30).

b. This time $(x_1, y_1) = (1, 4)$ and $(x_2, y_2) = (4, 4)$, so

$$m = \frac{4 - 4}{4 - 1} = \frac{0}{3} = 0$$

Thus the slope of a horizontal line is zero (see Figure 30).

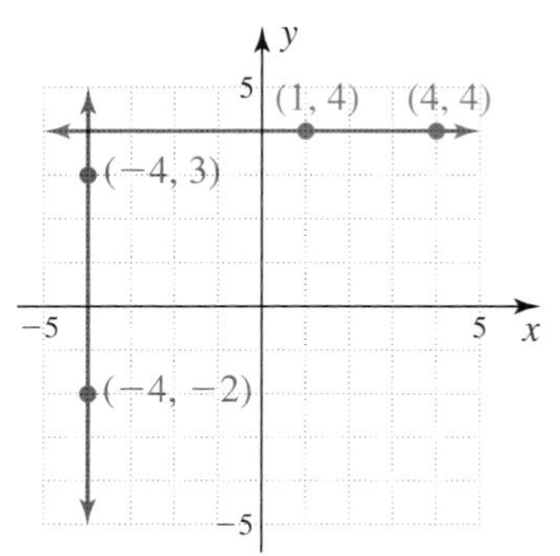

Figure 30

PROBLEM 3

Find the slope of the line passing through:

a. $(4, 1)$ and $(-3, 1)$

b. $(-2, 4)$ and $(-2, 1)$

Web It

To practice finding the slope of vertical and horizontal lines, try link 3-4-2 on the Bello Website at mhhe.com/bello.

Teaching Tip

Students should note that undefined slope is not the same as zero slope. It might be helpful if they note that a horizontal line has not begun to rise, therefore the best number to describe its steepness is 0. However, as a line gets steeper the number describing its steepness gets larger so a vertical line, being the steepest, should have the "largest number" to describe its steepness. Since there is no "largest number," the steepness of a vertical line cannot be defined.

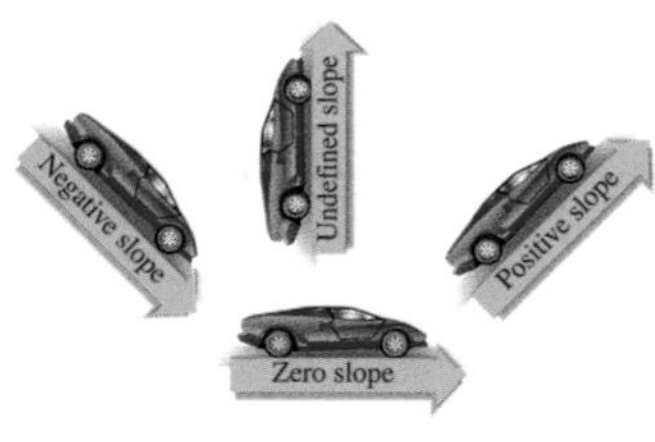

Answers

3. a. 0 **b.** Undefined

Let's summarize our work with slopes so far.

A line with **positive** slope *rises* from left to right.

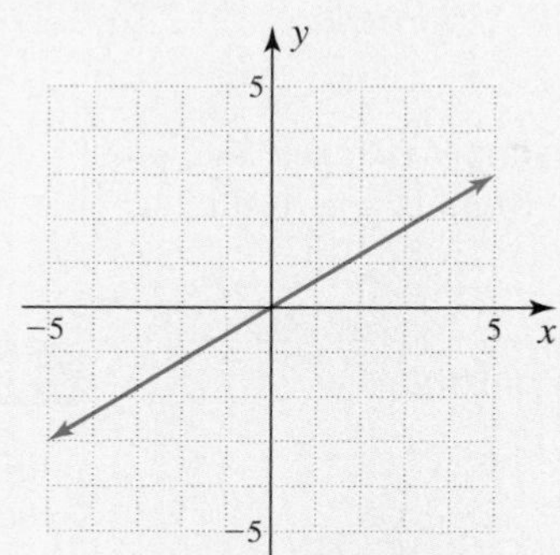

Positive slope

A line with **negative** slope *falls* from left to right.

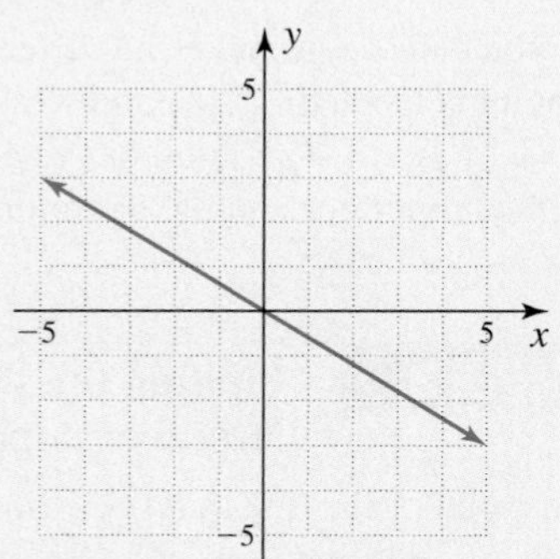

Negative slope

A *horizontal* line with an equation of the form $y = k$ has **zero** slope.

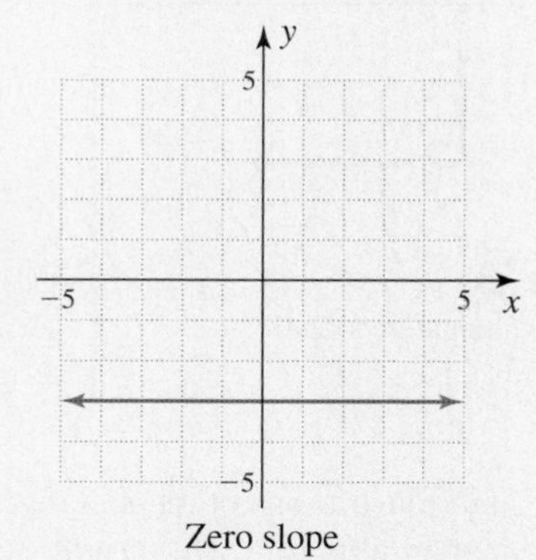

Zero slope

A *vertical* line with an equation of the form $x = k$ has an **undefined** slope.

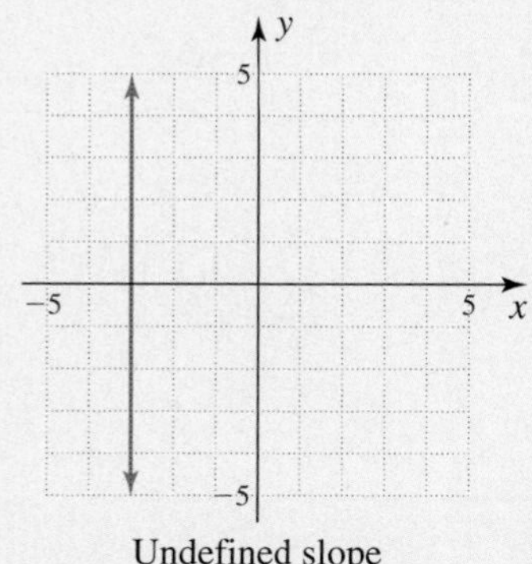

Undefined slope

B Finding Slopes from Equations

In the *Getting Started,* we found the slope of the line by using a ratio. The slope of a line can also be found from its equation. Thus if we approximate the number of subscribers to commercial online services between 1993 and 1997 by the equation

$$y = 2.5x + 1 \quad \text{(millions)}$$

where x is the number of years after 1992 and y is the number of subscribers, we can find the slope of $y = 2.5x + 1$ by using two values for x and then finding the corresponding y-values.

For $x = 1$, $y = 2.5(1) + 1 = 3.5$, so $(1, 3.5)$ is on the line.

For $x = 2$, $y = 2.5(2) + 1 = 6$, so $(2, 6)$ is on the line.

The slope of $y = 2.5x + 1$ passing through $(1, 3.5)$ and $(2, 6)$ is

$$m = \frac{6 - 3.5}{2 - 1} = 2.5$$

which is simply the **coefficient** of x in the equation $y = 2.5x + 1$. This idea can be generalized as follows.

Web It

For an interactive site dealing with the equation $y = mx + b$, go to link 3-4-3 on the Bello Website at mhhe.com/bello.

Teaching Tip

After doing this example and finding the slope to be the same as the coefficient of x, have students predict the slope for $2x + 3y = 6$. They will probably guess 2. Then have them find the slope by using two ordered pairs. Have them discuss the differences between two equations and why the slope is the coefficient of x in one equation and not the other.

SLOPE OF $y = mx + b$

The slope of the line defined by the equation $y = mx + b$ is m.

Thus if you are given the equation of a line and you want to find its slope, you would use this procedure.

PROCEDURE

Finding a Slope

1. Solve the equation for y.
2. The slope is m, the coefficient of x.

EXAMPLE 4 Finding the slope given an equation

Find the slopes of the following lines:

a. $2x + 3y = 6$ **b.** $3x - 2y = 4$

SOLUTION

a. We follow the two-step procedure.

1. Solve $2x + 3y = 6$ for y.

$$2x + 3y = 6 \quad \text{Given.}$$

$$3y = -2x + 6 \quad \text{Subtract } 2x.$$

$$y = -\frac{2}{3}x + \frac{6}{3} \quad \text{Divide each term by 3.}$$

$$y = -\frac{2}{3}x + 2$$

2. Since the coefficient of x is $-\frac{2}{3}$, the slope is $-\frac{2}{3}$.

PROBLEM 4

Find the slope of the line:

a. $3x + 2y = 6$

b. $2x - 3y = 9$

Answers

4. a. $-\frac{3}{2}$ **b.** $\frac{2}{3}$

b. We follow the steps.

1. Solve for y.

$$3x - 2y = 4 \qquad \text{Given.}$$

$$-2y = -3x + 4 \qquad \text{Subtract } 3x.$$

$$y = \frac{-3}{-2}x + \frac{4}{-2} \qquad \text{Divide each term by } -2.$$

$$y = \frac{3}{2}x - 2$$

2. The slope is the coefficient of x, so the slope is $\frac{3}{2}$.

C Finding Parallel and Perpendicular Lines

Parallel lines are lines in the plane that never intersect. In Figure 31, the lines $x + 3y = 6$ and $x + 3y = -3$ appear to be parallel lines. How can we be sure? By solving each equation for y and determining whether both lines have the same slope.

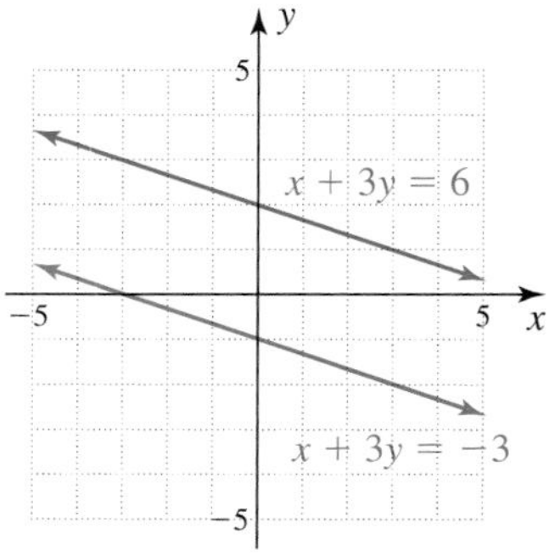

Figure 31

For $x + 3y = 6$: $\quad y = -\frac{1}{3}x + 2 \qquad$ The slope is $-\frac{1}{3}$.

For $x + 3y = -3$: $\quad y = -\frac{1}{3}x - 1 \qquad$ The slope is $-\frac{1}{3}$.

Since the two lines have the same slope but different y-intercepts, they are parallel lines.

The two lines in Figure 32 appear to be **perpendicular;** that is, they meet at a 90° angle. Note that their slopes are *negative reciprocals* and that the product of their slopes is

$$-\frac{a}{b} \cdot \frac{b}{a} = -1$$

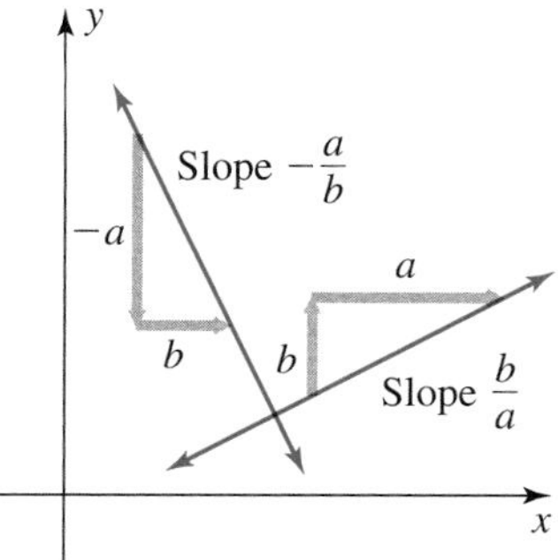

Figure 32

The lines $2x + y = 2$ and $x - 2y = -4$ have graphs that also appear to be perpendicular (see Figure 33). To show that this is the case, we check their slopes.

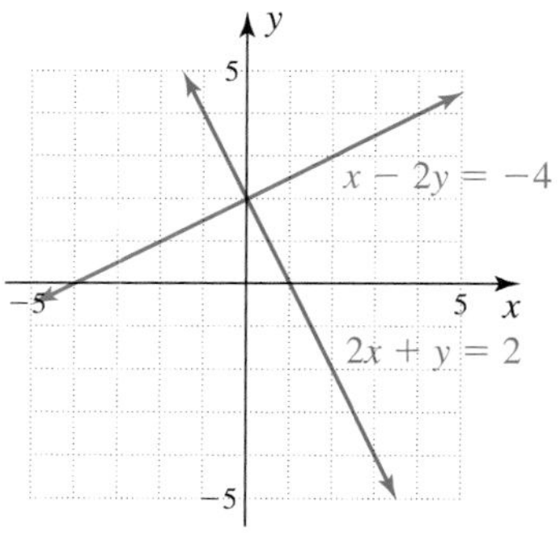

Figure 33

For $2x + y = 2$: $\quad y = -2x + 2 \qquad$ The slope is -2.

For $x - 2y = -4$: $\quad y = \frac{1}{2}x + 2 \qquad$ The slope is $\frac{1}{2}$.

Since the product of the slopes is $-2 \cdot \frac{1}{2} = -1$, the lines are perpendicular.

Web It

For a well illustrated lesson dealing with parallel and perpendicular lines, go to link 3-4-4 on the Bello Website at mhhe.com/bello.

SLOPES OF PARALLEL AND PERPENDICULAR LINES

Two lines with the **same** slope but different y-intercepts are **parallel.**

Two lines whose slopes have a product of **−1** are **perpendicular.**

EXAMPLE 5 Finding whether two given lines are parallel, perpendicular, or neither

Decide whether the pair of lines is parallel, perpendicular, or neither:

a. $x - 3y = 6$
$2x - 6y = -12$

b. $2x + y = 6$
$x + y = 4$

c. $2x + y = 5$
$x - 2y = 4$

SOLUTION

a. We find the slope of each line by solving the equations for y.

$x - 3y = 6$ Given. $-3y = -x + 6$ Subtract x. $y = \frac{1}{3}x - 2$ Divide by -3. The slope is $\frac{1}{3}$.	$2x - 6y = -12$ Given. $-6y = -2x - 12$ Subtract $2x$. $y = \frac{1}{3}x + 2$ Divide by -6. The slope is $\frac{1}{3}$.

Since the slopes are equal and the y-intercepts are different, the lines are parallel, as shown in Figure 34.

b. We find the slope of each line by solving the equations for y.

$2x + y = 6$ Given. $y = -2x + 6$ Subtract $2x$. The slope is -2.	$x + y = 4$ Given. $y = -x + 4$ Subtract x. The slope is -1.

Since the slopes are -2 and -1, their product is not -1, so the lines are neither parallel nor perpendicular, as shown in Figure 35.

c. We again solve both equations for y to find their slopes.

$2x + y = 5$ Given. $y = -2x + 5$ Subtract $2x$. The slope is -2.	$x - 2y = 4$ Given. $-2y = -x + 4$ Subtract x. $y = \frac{1}{2}x - 2$ Divide by -2. The slope is $\frac{1}{2}$.

Since the product of the slopes is $-2 \cdot \frac{1}{2} = -1$, the lines are perpendicular, as shown in Figure 36.

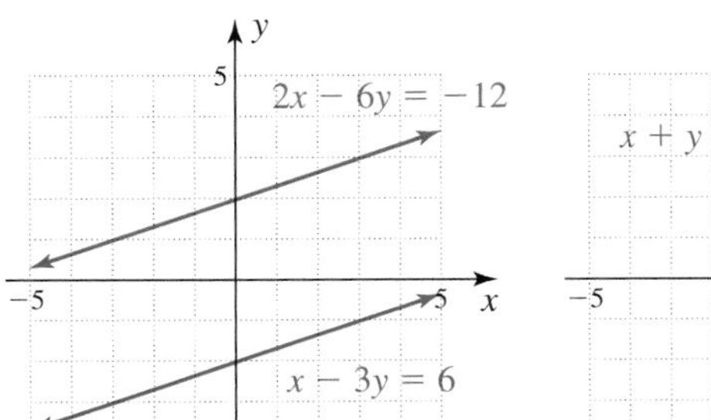

Figure 34

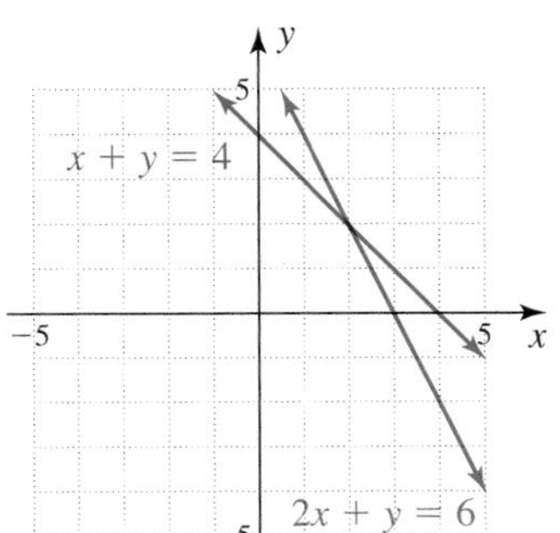

Figure 35

Figure 36

PROBLEM 5

Decide whether the pair of lines is parallel, perpendicular, or neither.

a. $x - 2y = 3$
$2x - 4y = 6$

b. $3x + y = 6$
$x + y = 2$

c. $3x + y = 6$
$x - 3y = 5$

Answers

5. a. Parallel **b.** Neither **c.** Perpendicular

Solving an Application

In the *Getting Started,* we saw that the number of commercial online subscribers *increased* at a rate of 2.5 million per year. The slope of a line can be used to describe a rate of change. For example, the number N of deaths (per 100,000 population) due to heart disease can be approximated by $N = -5t + 300$, where t is the number of years after 1960. Since the slope of the line $N = -5t + 300$ is -5, this means that the number of deaths per 100,000 population due to heart disease is *decreasing* by 5 every year.

EXAMPLE 6 Escalating health care costs

As you have probably heard, health care costs have been soaring for several years. According to the Health Insurance Association of America, the average daily hospital room charge C can be approximated by $C = 17t + 127$ (dollars), where t is the number of years after 1980.

a. What is the slope of $C = 17t + 127$?

b. What does the slope represent?

SOLUTION

a. The slope of the line $C = 17t + 127$ is 17.

b. The slope 17 represents the annual *increase* (in dollars) for the average daily room charge.

PROBLEM 6

The decrease in cholesterol level C over w weeks is given by $C = -3w + 215$.

a. What is the slope of $C = -3w + 215$?

b. What does the slope represent?

Calculate It Using STAT to find $y = ax + b$

In this section we learned how to find the slope of a line given two points. Your calculator can do better! Let's try Example 1, where we are given the points $(0, -6)$ and $(3, 3)$ and asked to find the slope. Press STAT 1 and enter the two x-coordinates (0 and 3) under L_1 and the two y-coordinates (-6 and 3) under L_2. Your calculator is internally programmed to use a process called *regression* to find the equation of the line passing through these two points. To do so, press STAT, move the cursor right to reach CALC, enter 4 for LinReg (Linear Regression), and then press ENTER. The resulting line shown in Window 1 is written in the slope-intercept form $y = ax + b$. Clearly, the slope is 3 and the y-intercept is -6. See what happens when you use your calculator to do Example 2.

What happens when you do Example 3a? You get the warning shown in Window 2, indicating that the slope is *undefined.* Here you have to *know* some algebra to conclude that the resulting line is a *vertical* line. (Your calculator can't do that for you!)

```
LinReg
 y=ax+b
 a=3
 b=-6
 r=1
```

Window 1

```
ERR:DOMAIN
1:Quit
2:Goto
```

Window 2

Teaching Tip

For $N = -5t + 300$, have students complete and discuss the following chart to emphasize that -5 as a slope means that the number of deaths each year is decreasing by 5.

t (number of years)	N (number of deaths)
0	?
1	?
2	?
3	?

Answers

6. a. -3 **b.** The cholesterol is dropping 3 points each week.

Exercises 3.4

A In Problems 1–14, find the slope of the line that goes through the two given points.

1. (1, 2) and (3, 4) $m = 1$

2. (1, −2) and (−3, −4) $m = \frac{1}{2}$

3. (0, 5) and (5, 0) $m = -1$

4. (3, −6) and (5, −6) $m = 0$

5. (−1, −3) and (7, −4) $m = -\frac{1}{8}$

6. (−2, −5) and (−1, −6) $m = -1$

7. (0, 0) and (12, 3) $m = \frac{1}{4}$

8. (−1, −1) and (−10, −10) $m = 1$

9. (3, 5) and (−2, 5) $m = 0$

10. (4, −3) and (2, −3) $m = 0$

11. (4, 7) and (−5, 7) $m = 0$

12. (−3, −5) and (−2, −5) $m = 0$

13. $\left(-\frac{1}{2}, -\frac{1}{3}\right)$ and $\left(2, -\frac{1}{3}\right)$ $m = 0$

14. $\left(-\frac{1}{5}, 2\right)$ and $\left(-\frac{1}{5}, 1\right)$ Undefined

B In Problems 15–26, find the slope of the given line.

15. $y = 3x + 7$ $m = 3$

16. $y = -4x + 6$ $m = -4$

17. $-3y = 2x - 4$ $m = -\frac{2}{3}$

18. $4y = 6x + 3$ $m = \frac{3}{2}$

19. $x + 3y = 6$ $m = -\frac{1}{3}$

20. $-x + 2y = 3$ $m = \frac{1}{2}$

21. $-2x + 5y = 5$ $m = \frac{2}{5}$

22. $3x - y = 6$ $m = 3$

23. $y = 6$ $m = 0$

24. $x = 7$ Undefined

25. $2x - 4 = 0$ Undefined

26. $2y - 3 = 0$ $m = 0$

C In Problems 27–37, determine whether the given lines are parallel, perpendicular, or neither.

27. $y = 2x + 5$ and $4x - 2y = 7$ Parallel

28. $y = 4 - 5x$ and $15x + 3y = 3$ Parallel

29. $2x + 5y = 8$ and $5x - 2y = -9$ Perpendicular

30. $3x + 4y = 4$ and $2x - 6y = 7$ Neither

31. $x + 7y = 7$ and $2x + 14y = 21$ Parallel

32. $y - 5x = 12$ and $y - 3x = 8$ Neither

33. $2x + y = 7$ and $-2x - y = 9$ Parallel

34. $2x - 4 = 0$ and $x - 1 = 0$ Parallel

35. $2y - 4 = 0$ and $3y - 6 = 0$ Neither; the lines coincide

36. $3y = 6$ and $2x = 6$ Perpendicular

37. $3x = 7$ and $2y = 7$ Perpendicular

APPLICATIONS

38. *Daily fat consumption* According to the U.S. Department of Agriculture, the daily fat consumption F, per person can be approximated by

$$F = 190 + 0.8t \text{ (grams)}$$

where t is the number of years after 2000.

a. What is the slope of this line? $m = 0.8$

b. What does the slope represent? The rate at which the daily fat consumption per person changes.

39. *Average hospital stay* According to the National Center for Health Statistics, the average hospital stay S can be approximated by

$$S = 5 - 0.1t \text{ (days)}$$

where t is the number of years after 2000.

a. What is the slope of this line? $m = -0.1$

b. Is the average hospital stay increasing or decreasing? Decreasing

c. What does the slope represent? The rate at which the average hospital stay is changing.

40. *Daily seafood consumption* According to the U.S. Department of Agriculture, the daily consumption T of tuna can be approximated by

$$T = 3.87 - 0.13t \text{ (pounds)}$$

and the daily consumption F of fish and shellfish can be approximated by

$$F = -0.29t + 15.46$$

a. Is the consumption of tuna increasing or decreasing? Decreasing

b. Is the consumption of fish and shellfish increasing or decreasing? Decreasing

c. Which consumption (tuna or fish and shellfish) is decreasing faster? Fish and shellfish

41. *Life expectancy of women* The average life span (life expectancy) y of an American woman is given by $y = 0.15t + 80$, where t is the number of years after 2000.

a. What is the slope of this line? 0.15

b. Is the life span of American women increasing or decreasing? Increasing

c. What does the slope represent? The slope 0.15 represents the annual increase in the life span of American women.

Source: U.S. National Center for Health Statistics, *Statistical Abstract of the United States.*

42. *Life expectancy of men* The average life span y of an American man is given by $y = 0.15t + 74$, where t is the number of years after 2000.

a. What is the slope of this line? 0.15

b. Is the life span of American men increasing or decreasing? Increasing

c. What does the slope represent? The slope 0.15 represents the annual increase in the life span of American men.

Source: U.S. National Center for Health Statistics, *Statistical Abstract of the United States.*

43. *Velocity of a thrown ball* The speed v of a ball thrown up with an initial velocity of 128 feet per second is given by $v = 128 - 32t$, where v is the velocity (in feet per second) and t is the number of seconds after the ball is thrown.

a. What is the slope of this line? -32

b. Is the velocity of the ball increasing or decreasing? Decreasing

c. What does the slope represent? The slope -32 represents the decrease in the velocity of the ball.

44. *Velocity of a thrown ball* The speed v of a ball thrown up with an initial velocity of 15 meters per second is given by $v = 15 - 5t$, where v is the velocity (in meters per second) and t is the number of seconds after the ball is thrown.

a. What is the slope of this line? -5

b. Is the velocity of the ball increasing or decreasing? Decreasing

c. What does the slope represent? The slope -5 represents the decrease in the velocity of the ball.

45. *Daily fat consumption* The number of fat grams f consumed daily by the average American can be approximated by $f = 165 + 0.4t$, where t is the number of years after 2000.

a. What is the slope of this line? 0.4

b. Is the consumption of fat increasing or decreasing? Increasing

c. What does the slope represent? The slope 0.4 represents the annual increase in the consumption of fat.

Source: U.S. Dept. of Agriculture, *Statistical Abstract of the United States.*

46. *Milk products consumption* The number of gallons of milk products g consumed annually by the average American can be approximated by $g = 24 - 0.2t$, where t is the number of years after 2000.

a. What is the slope of this line? -0.2

b. Is the consumption of milk products increasing or decreasing? Decreasing

c. What does the slope represent? The slope -0.2 represents the annual decrease in the consumption of milk products.

Source: U.S. Dept. of Agriculture, *Statistical Abstract of the United States.*

SKILL CHECKER

Try the Skill Checker Exercises so you'll be ready for the next section.

Simplify:

47. $2[x - (-4)]$ $2x + 8$

48. $3[x - (-6)]$ $3x + 18$

49. $-2[x - (-1)]$ $-2x - 2$

USING YOUR KNOWLEDGE

Up, Down, or Away!

The slope of a line can be positive, negative, zero, or undefined. Use your knowledge to sketch the following lines.

50. A line that has a negative slope

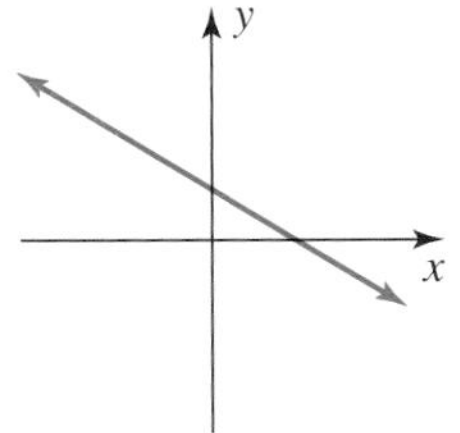

51. A line that has a positive slope

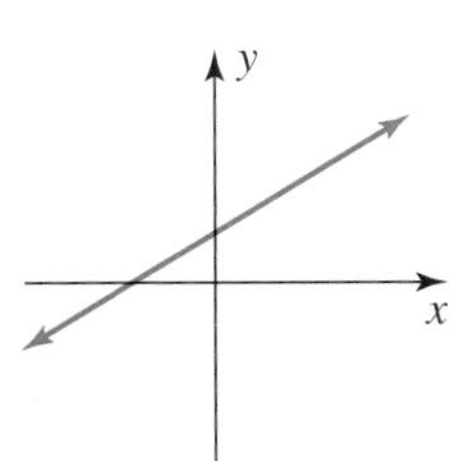

52. A line with zero slope

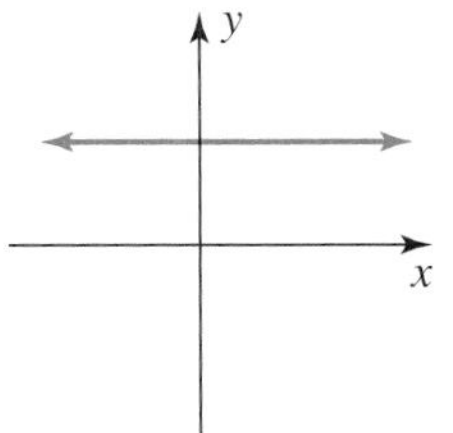

53. A line with an undefined slope

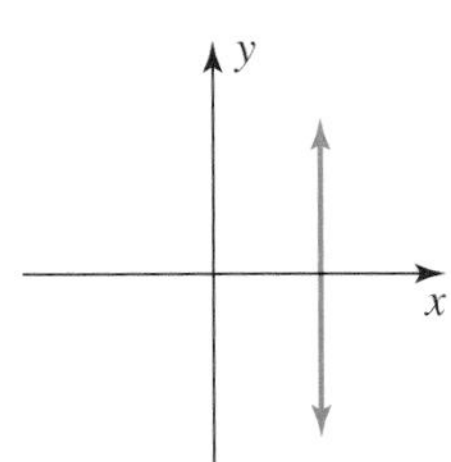

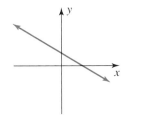

WRITE ON

54. Write in your own words what is meant by the *slope* of a line.
Answers may vary.

55. Explain why the slope of a horizontal line is zero.
Answers may vary.

56. Explain why the slope of a vertical line is undefined.
Answers may vary.

MASTERY TEST

If you know how to do these problems, you have learned your lesson!

Find the slope of the line going through:

57. $(2, -3)$ and $(4, -5)$ $m = -1$

58. $(-1, 2)$ and $(4, -2)$ $m = -\frac{4}{5}$

59. $(-4, 2)$ and $(-4, 5)$ Undefined

60. $(-3, 4)$ and $(-5, 4)$ $m = 0$

61. Find the slope of the line $3x + 2y = 6$. $m = -\frac{3}{2}$

62. Find the slope of the line $-3x + 4y = 12$. $m = \frac{3}{4}$

Decide whether the lines are parallel, perpendicular, or neither:

63. $-x + 3y = -6$ and $2x - 6y = -7$ Parallel

64. $2x + 3y = 5$ and $3x - 2y = 5$ Perpendicular

65. $3x - 2y = 6$ and $-2x - 3y = 6$ Perpendicular

66. The U.S. population P (in millions) can be approximated by $P = 2.2t + 180$, where t is the number of years after 1960.

a. What is the slope of this line? $m = 2.2$

b. How fast is the U.S. population growing each year? (State your answer in millions.) 2.2 million

COLLABORATIVE LEARNING

Hourly Earnings in Selected Industries

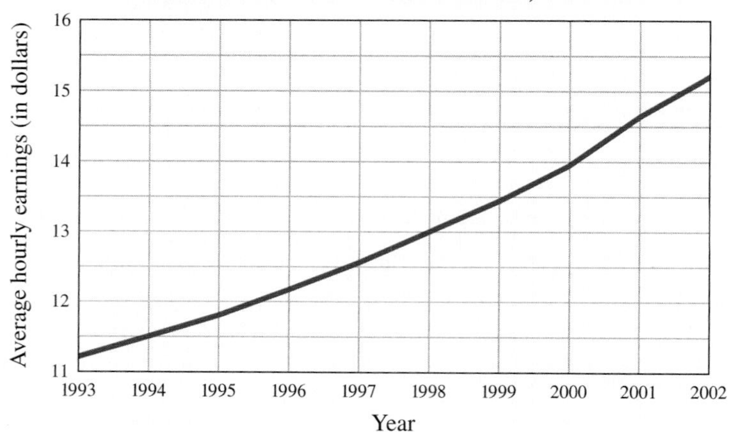

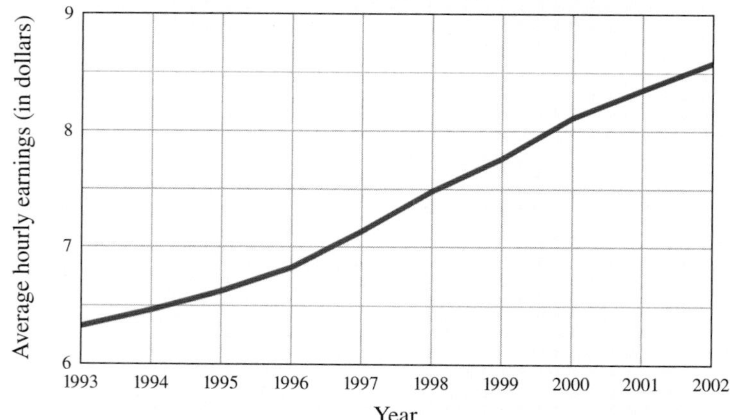

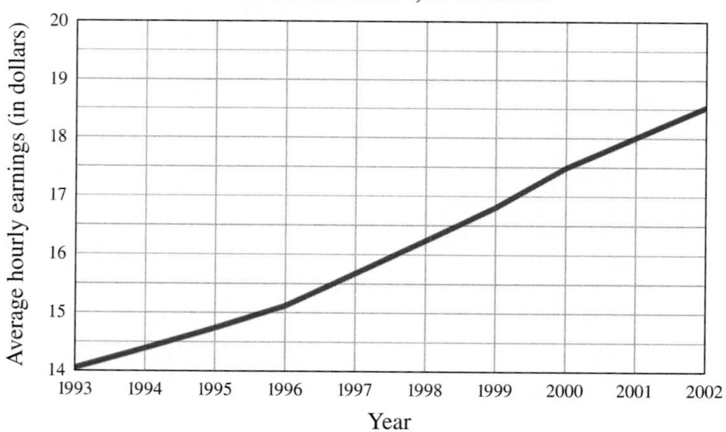

Source: U.S. Department of Labor, Bureau of Labor Statistics. Select the desired industry.

The three charts show the average hourly earnings (H) of production workers in Education and Health Services (E), Leisure and Hospitality (L), and Construction (C). Form three groups, E, L, and C.

1. What was your group salary (to the nearest dollar) in 1993?
2. What was your group salary (to the nearest dollar) in 2002?
3. Use 1993 = 0. What is (0, H), where H is the hourly earnings to the nearest dollar for each of the groups?
4. What is (9, H), where H is the hourly earnings to the nearest dollar for each of the groups?
5. What is the slope of the line for each of the groups?
6. What is the equation of the line for the hourly earnings H for each of the groups?
7. What would be the predicted hourly earnings for each of the groups in 2010?
8. If you were making a career decision based on the equations obtained in Question 6, which career would you choose?

Research Questions

1. Some historians claim that the official birthday of analytic geometry is November 10, 1619. Investigate and write a report on why this is so and on the event that led Descartes to the discovery of analytic geometry.
2. Find out what led Descartes to make his famous pronouncement "*Je pense, donc je suis*" (I think, therefore, I am) and write a report about the contents of one of his works, *La Geometrie.*
3. From 1629 to 1633, Descartes "was occupied with building up a cosmological theory of vortices to explain all natural phenomena." Find out the name of the treatise in which these theories were explained and why it wasn't published until 1654, after his death.

Summary

SECTION	ITEM	MEANING	EXAMPLE
3.1	x-axis	A horizontal number line on a coordinate plane	y Q II Q I y-axis x x-axis Q III Q IV
	y-axis	A vertical number line on a coordinate plane	
	Abscissa	The first coordinate in an ordered pair	The abscissa in the ordered pair (3, 4) is 3.
	Ordinate	The second coordinate in an ordered pair	The ordinate in the ordered pair (3, 4) is 4.
	Quadrant	One of the four regions into which the axes divide the plane	
3.2	Solution of an equation	The ordered pair (a, b) is a solution of an equation if when the values for a and b are substituted for the variables in the equation, the result is a true statement.	(4, 5) is a solution of the equation $2x + 3y = 23$ because if 4 and 5 are substituted for x and y in $2x + 3y = 23$, the result $2(4) + 3(5) = 23$ is true.
3.2C	Graph of a linear equation	The graph of a linear equation of the form $y = mx + b$ is a straight line.	The graph of the equation $y = 3x + 6$ is a straight line.
	Graph of a linear equation	The graph of a linear equation of the form $Ax + By = C$ is a straight line, and every straight line has an equation of this form.	The linear equation $3x + 6y = 12$ has a straight line for its graph.
3.3A	x-intercept	The point at which a line crosses the x-axis	The x-intercept of the line $3x + 6y = 12$ is (4, 0).
	y-intercept	The point at which a line crosses the y-axis	The y-intercept of the line $3x + 6y = 12$ is (0, 2).
3.3C	Horizontal line	A line whose equation can be written in the form $y = k$	The line $y = 3$ is a horizontal line.
	Vertical line	A line whose equation can be written in the form $x = k$	The line $x = -5$ is a vertical line.
3.4A	Slope of a line	The ratio of the vertical change to the horizontal change of a line	
	Slope of a line through (x_1, y_1) and (x_2, y_2), $x_1 \neq x_2$	$m = \dfrac{y_2 - y_1}{x_2 - x_1}$	The slope of the line through (3, 5) and (6, 8) is $m = \dfrac{8 - 5}{6 - 3} = 1$.

SECTION	ITEM	MEANING	EXAMPLE
3.4B	Slope of the line $y = mx + b$	The slope of the line $y = mx + b$ is m.	The slope of the line $y = \frac{3}{5}x + 7$ is $\frac{3}{5}$.
3.4C	Parallel lines	Two lines are parallel if their slopes are equal and they have different y-intercepts.	The lines $y = 2x + 5$ and $3y - 6x = 8$ are parallel (both have a slope of 2).
	Perpendicular lines	Two lines are perpendicular if the product of their slopes is -1.	The lines defined by $y = 2x + 1$ and $y = -\frac{1}{2}x + 2$ are perpendicular since $2 \cdot (-\frac{1}{2}) = -1$.

Review Exercises

(If you need help with these exercises, look in the section indicated in brackets.)

1. [3.1A] Graph the points.

a. A: $(-1, 2)$ **b.** B: $(-2, 1)$

c. C: $(-3, 3)$

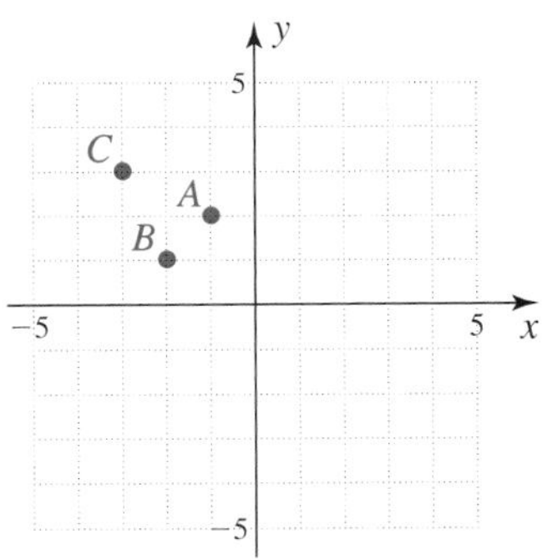

2. [3.1B] Find the coordinates of the points shown on the graph. A: (1, 1); B: (−1, −2); C: (3, −3)

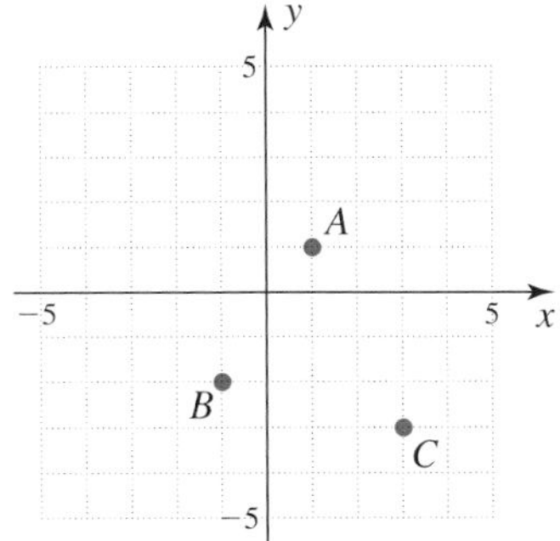

3. [3.1C] The graph shows the new wind chill temperatures (top) and the old wind chill temperatures (bottom) for different wind speeds.

a. On the top graph, what does the ordered pair $(10, -10)$ represent? For a wind speed of 10 mi/hr, the wind chill temperature is −10°F.

b. If the wind speed is 40 miles per hour, what is the approximate new wind chill temperature? −22°F

c. If the wind speed is 40 miles per hour, what is the approximate old wind chill temperature? −45°F

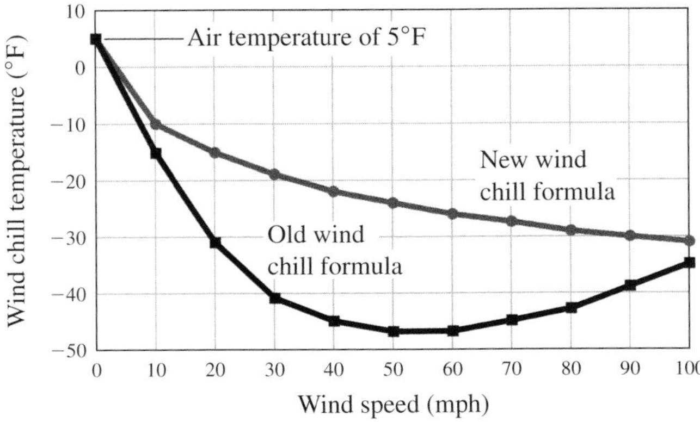

Source: Data from National Oceanic and Atmospheric Administration.

4. [3.1D] The bar graph indicates the number of months and monthly payment needed to pay off a $500 loan at 18% annual interest.

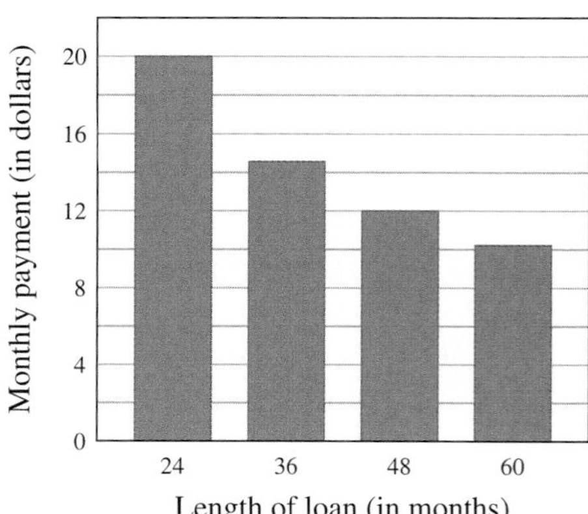

Source: Data from KJE Computer Solutions, LLC.

To the nearest dollar, what is the monthly payment if you want to pay off the loan in:

a. 60 months? $10

b. 48 months? $12

c. 24 months? $20

5. [3.1E] Determine the quadrant in which each of the points is located:

a. A Quadrant II

b. B Quadrant III

c. C Quadrant IV

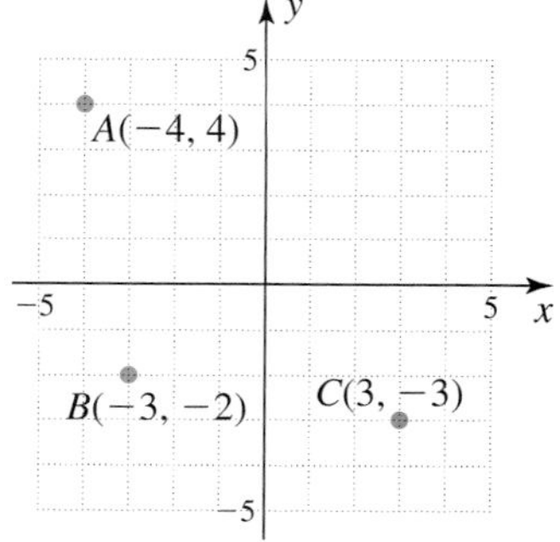

6. [3.1F] The ordered pairs represent the wind chill temperature in degrees Fahrenheit when the wind speed in miles per hour is as indicated.

Wind Speed	Wind Chill
10	−10
20	−15
30	−18

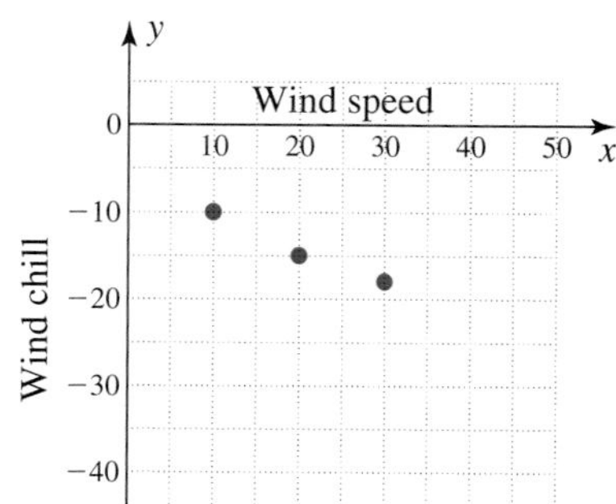

a. Graph the ordered pairs.

b. What does $(10, -10)$ mean? For a wind speed of 10 mi/hr, the wind chill temperature is −10°F.

c. What is the number s in $(s, -15)$? $s = 20$

7. [3.2A] Determine whether the given point is a solution of $x - 2y = -3$.

a. $(1, -2)$ No

b. $(2, -1)$ No

c. $(-1, 1)$ Yes

8. [3.2B] Find x in the given ordered pair so that the pair satisfies the equation $2x - y = 4$.

a. $(x, 2)$ $x = 3$

b. $(x, 4)$ $x = 4$

c. $(x, 0)$ $x = 2$

9. [3.2C] Graph:

a. $x + y = 4$

b. $x + y = 2$

c. $x + 2y = 2$

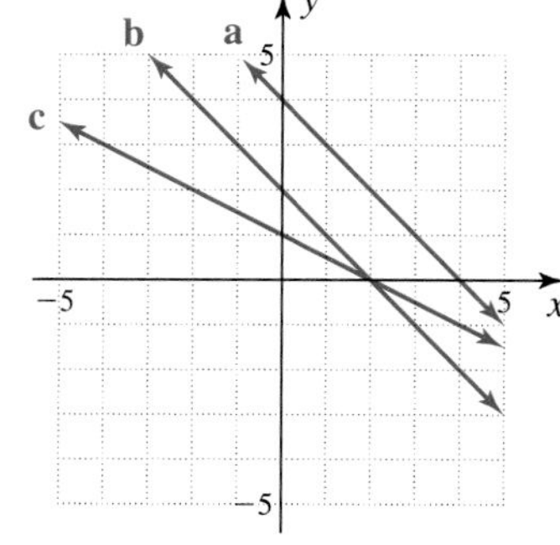

10. [3.2C] Graph:

a. $y = \frac{3}{2}x + 3$

b. $y = -\frac{3}{2}x + 3$

c. $y = \frac{3}{4}x + 4$

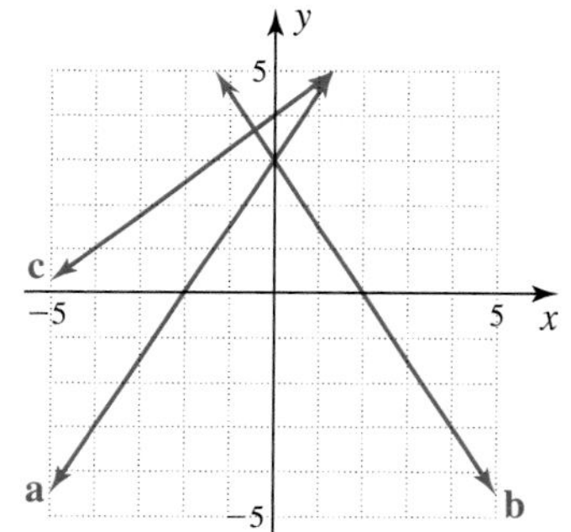

11. [3.2D] The average annual consumption g of milk products (in gallons) per person can be approximated by $g = 30 - 0.2t$, where t is the number of years after 1980. Graph:

a. $g = 30 - 0.2t$

b. $g = 20 - 0.2t$

c. $g = 10 - 0.2t$

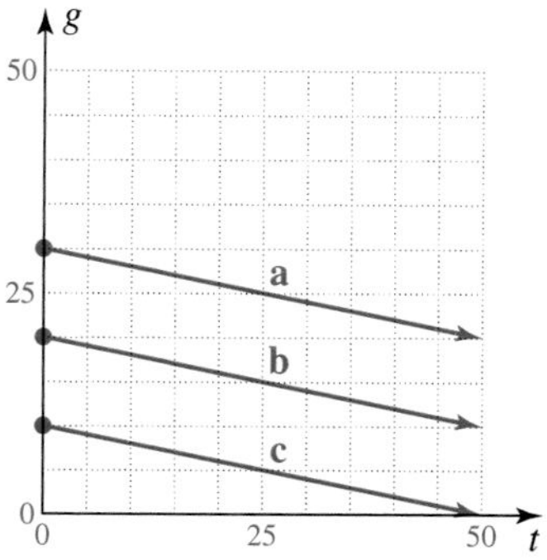

12. [3.3A] Graph:

a. $2x - 3y - 12 = 0$

b. $3x - 2y - 12 = 0$

c. $2x + 3y + 12 = 0$

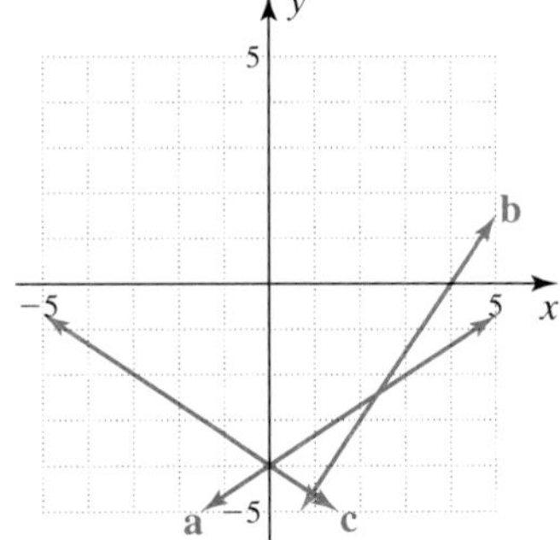

13. [3.3B] Graph:

a. $3x + y = 0$

b. $-2x + 3y = 0$

c. $-3x + 2y = 0$

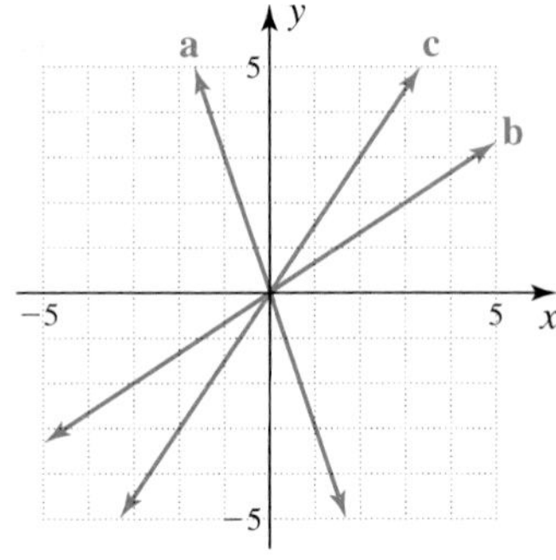

14. [3.3C] Graph:

a. $2x - 6 = 0$

b. $2x - 2 = 0$

c. $2x - 4 = 0$

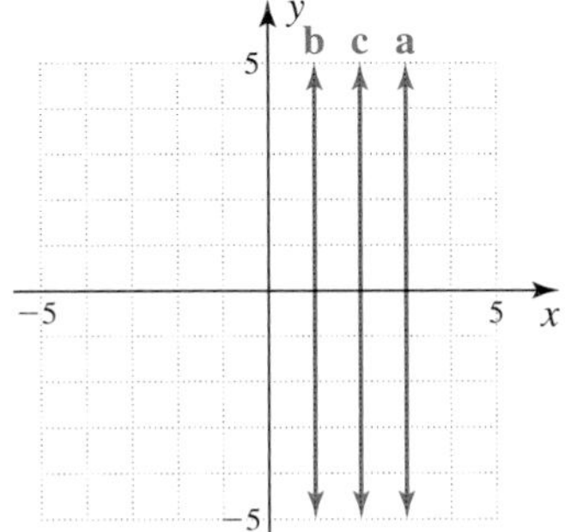

15. [3.3C] Graph:

a. $y = -1$

b. $y = -3$

c. $y = -4$

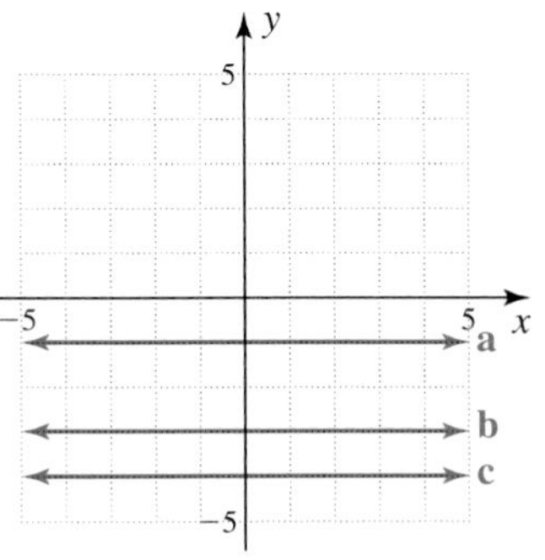

16. [3.4A] Find the slope of the line going through the given points.

a. $(1, -4)$ and $(2, -3)$ $m = 1$

b. $(5, -2)$ and $(8, 5)$ $m = \frac{7}{3}$

c. $(3, -4)$ and $(4, -8)$ $m = -4$

17. [3.4A] Find the slope of the line passing through:

a. $(-5, 2)$ and $(-5, 4)$ Undefined

b. $(-3, 5)$ and $(3, 5)$ 0

c. $(-2, -1)$ and $(-2, -5)$ Undefined

18. [3.4B] Find the slope of the line.

a. $3x + 2y = 6$ $m = -\frac{3}{2}$

b. $x + 4y = 4$ $m = -\frac{1}{4}$

c. $-2x + 3y = 6$ $m = \frac{2}{3}$

19. [3.4C] Decide whether the lines are parallel, perpendicular, or neither.

a. $2x + 3y = 6$
$6x = 6 - 4y$ Neither

b. $3x + 2y = 4$
$-2x + 3y = 4$ Perpendicular

c. $2x + 3y = 6$
$-2x - 3y = 6$ Parallel

20. [3.4D] The number N of theaters t years after 1975 can be approximated by

$$N = 0.6t + 15 \text{ (thousand)}$$

a. What is the slope of this line? $m = 0.6$

b. What does the slope represent? The change (increase) in the number of theaters per year

c. How many theaters were added each year? 600

Practice Test 3

(Answers on pages 286–288)

1. Graph the point $(-2, 3)$.

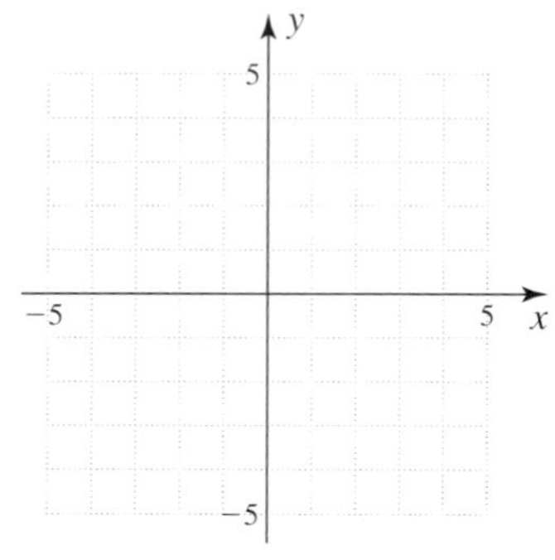

2. Find the coordinates of point A shown on the graph.

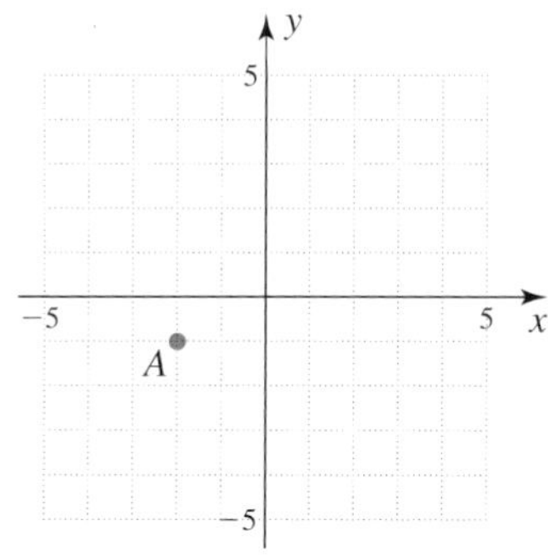

3. The graph shows the new wind chill temperatures (top) and the old wind chill temperatures (bottom) for different wind speeds.

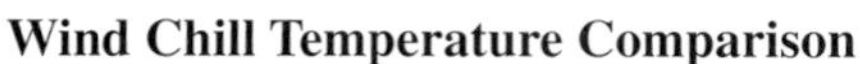

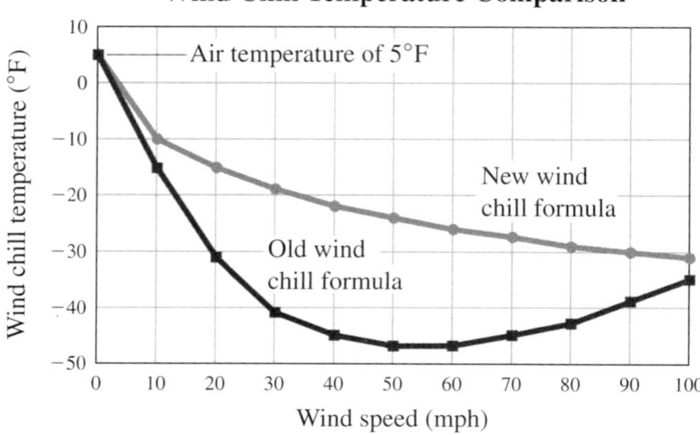

Source: Data from National Oceanic and Atmospheric Administration.

a. On the top graph, what does the ordered pair $(20, -15)$ represent?

b. If the wind speed is 90 miles per hour, what is the approximate new wind chill temperature?

c. If the wind speed is 90 miles per hour, what is the approximate old wind chill temperature?

4. The bar graph indicates the number of months and monthly payment needed to pay off a \$1000 loan at 8% annual interest.

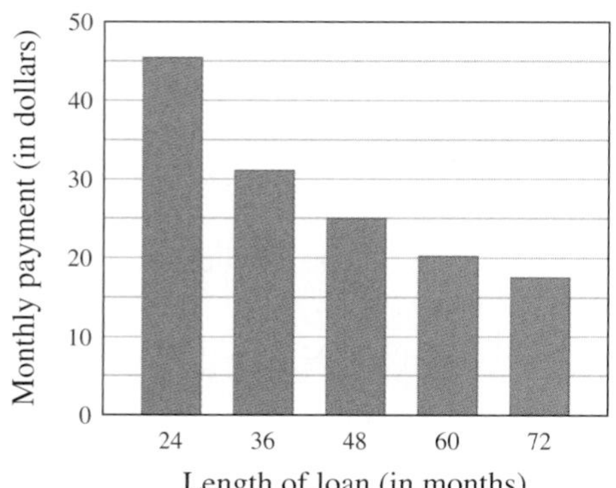

Source: Data from KJE Computer Solutions, LLC.

What is the monthly payment if you want to pay off the loan in:

a. 60 months? **b.** 48 months? **c.** 24 months?

5. Determine the quadrant in which each of the points is located:

a. A

b. B

c. C

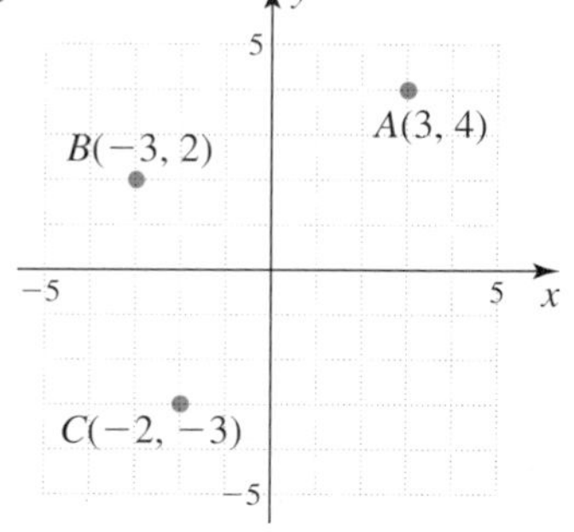

6. The ordered pairs represent the old wind chill temperature in degrees Fahrenheit when the wind speed in miles per hour is as indicated.

Wind Speed	Wind Chill
10	−15
20	−31
30	−41

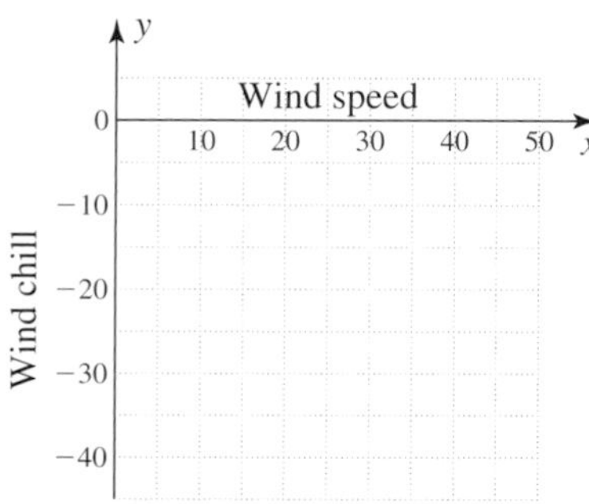

a. Graph the ordered pairs.

b. What does $(10, -15)$ mean?

c. What is the number s in $(s, -15)$?

7. Determine whether the ordered pair $(1, -2)$ is a solution of $2x - y = -2$.

8. Find x in the ordered pair $(x, 2)$ so that the ordered pair satisfies the equation $3x - y = 10$.

9. Graph $x + 2y = 4$.

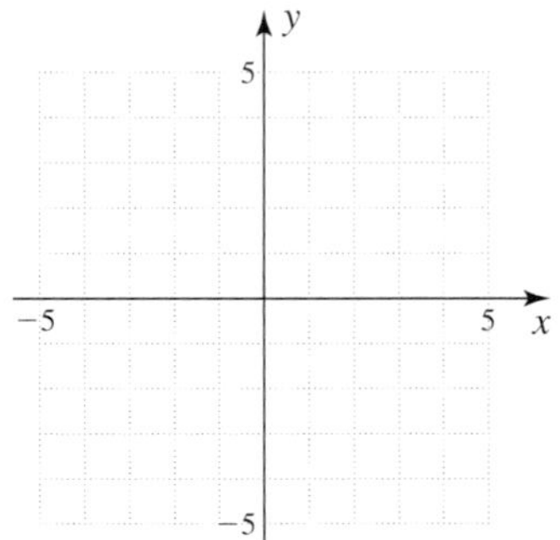

10. Graph $y = -\frac{3}{2}x - 3$.

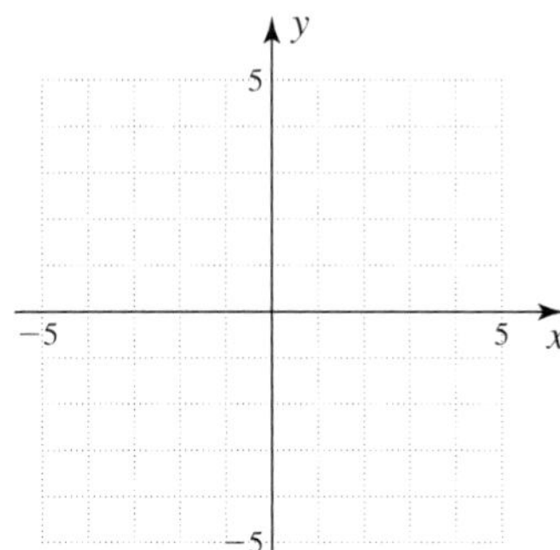

11. The average daily consumption g of protein (in grams) per person can be approximated by $g = 100 + 0.7t$, where t is the number of years after 2000. Graph $g = 100 + 0.7t$.

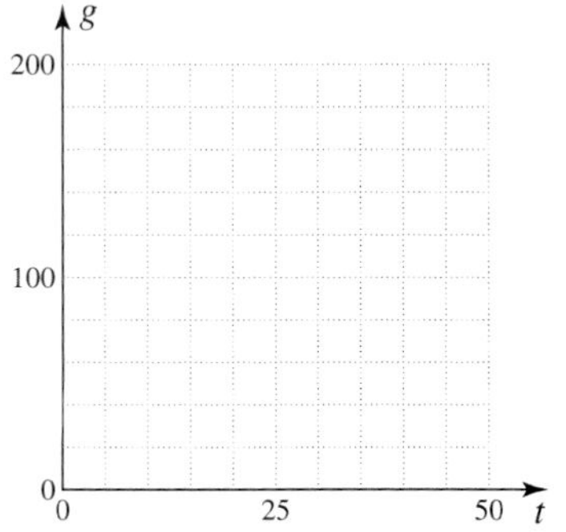

12. Graph $2x - 5y - 10 = 0$.

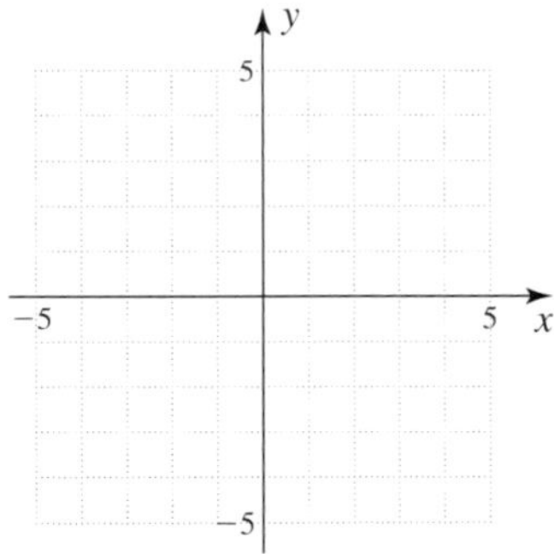

13. Graph $2x - 3y = 0$.

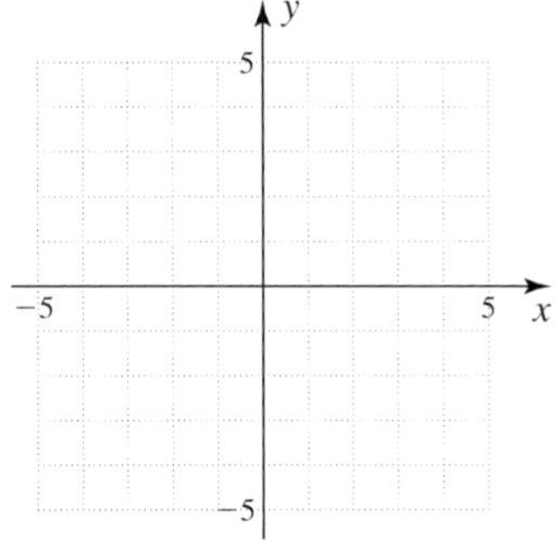

14. Graph $3x - 6 = 0$.

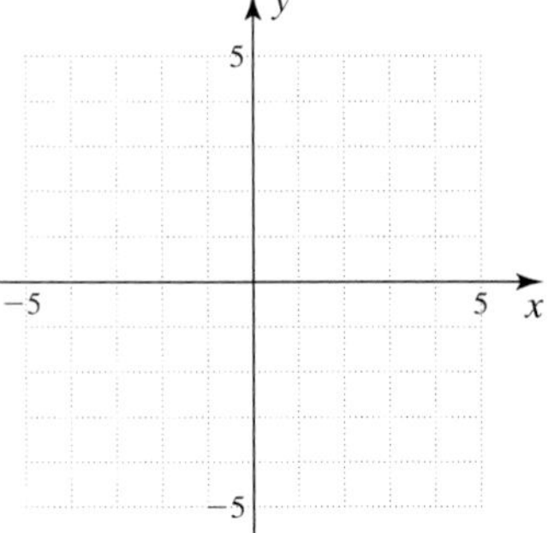

15. Graph $y = -4$.

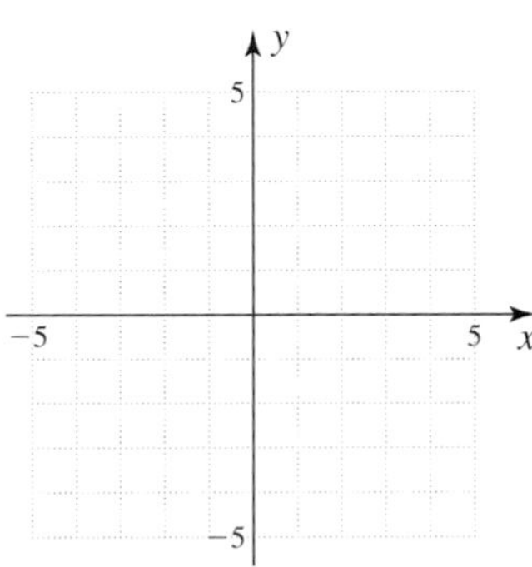

16. Find the slope of the line going through:

a. $(2, -8)$ and $(-4, -2)$

b. $(6, -8)$ and $(-4, 6)$

17. Find the slope of the line going through:

a. $(-3, -2)$ and $(-3, 4)$

b. $(2, 4)$ and $(4, 4)$

18. Find the slope of the line $y - 2x = 6$.

19. Decide whether the lines are parallel, perpendicular, or neither.

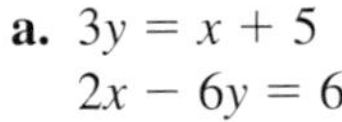

a. $3y = x + 5$
$2x - 6y = 6$

b. $3y = -x + 5$
$9x - 3y = 6$

20. The number N of stores in a city t years after 2000 can be approximated by

$$N = 0.8t + 15 \text{ (hundred)}$$

a. What is the slope of this line?

b. What does the slope represent?

c. How many stores were added each year?

Answers to Practice Test

<table>
<tr><th>ANSWER</th><th>IF YOU MISSED</th><th colspan="3">REVIEW</th></tr>
<tr><th></th><th>QUESTION</th><th>SECTION</th><th>EXAMPLES</th><th>PAGE</th></tr>
<tr><td>

1.

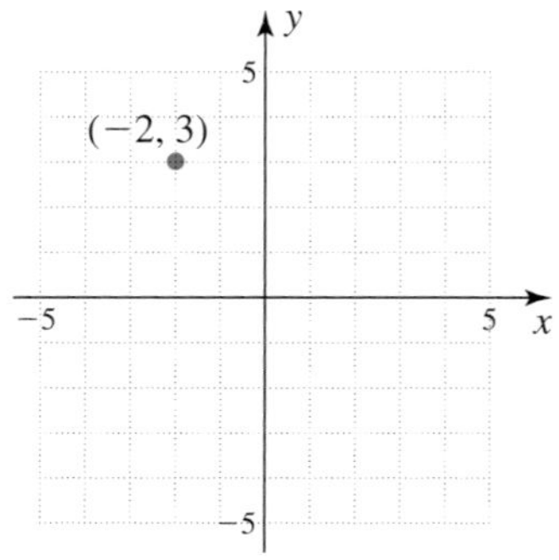

</td><td>1</td><td>3.1</td><td>1</td><td>228</td></tr>
<tr><td>

2. $(-2, -1)$

</td><td>2</td><td>3.1</td><td>2</td><td>229</td></tr>
<tr><td>

3. a. When the wind speed is 20 miles per hour, the wind chill temperature is -15 degrees Fahrenheit.
b. -30 degrees Fahrenheit
c. -40 degrees Fahrenheit

</td><td>3</td><td>3.1</td><td>3</td><td>230</td></tr>
<tr><td>

4. a. \$20
b. \$25
c. \$45

</td><td>4</td><td>3.1</td><td>4</td><td>231</td></tr>
<tr><td>

5. a. Quadrant I
b. Quadrant II
c. Quadrant III

</td><td>5</td><td>3.1</td><td>5</td><td>232</td></tr>
<tr><td>

6. a.

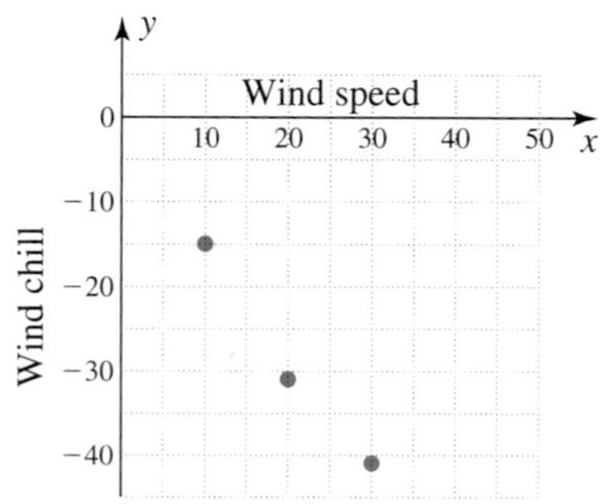

b. When the wind speed is 10 mi/hr, the wind chill temperature is -15°F.
c. 10

</td><td>6</td><td>3.1</td><td>6</td><td>233–234</td></tr>
<tr><td>

7. No

</td><td>7</td><td>3.2</td><td>1</td><td>242–243</td></tr>
<tr><td>

8. $x = 4$

</td><td>8</td><td>3.2</td><td>2</td><td>244</td></tr>
</table>

ANSWER	IF YOU MISSED	REVIEW		
	QUESTION	SECTION	EXAMPLES	PAGE
9. $x + 2y = 4$	9	3.2	4	246
10. $y = -\frac{3}{2}x - 3$	10	3.2	5	247
11. $g = 100 + 0.7t$	11	3.2	6	248
12. $2x - 5y - 10 = 0$	12	3.3	1, 2	255–256

ANSWER	IF YOU MISSED	REVIEW		
	QUESTION	SECTION	EXAMPLES	PAGE
13. 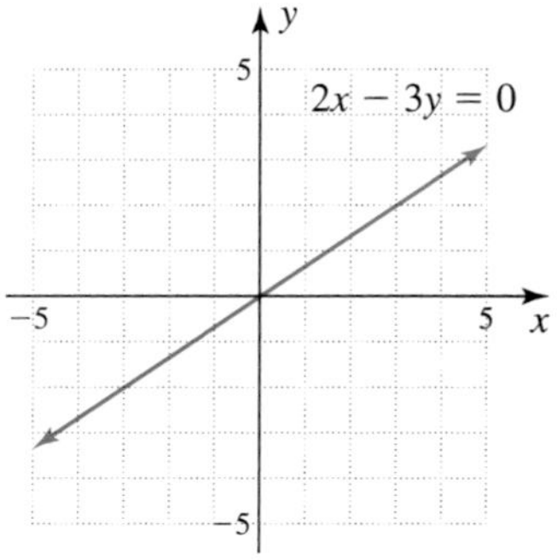	13	3.3	3	258
14. 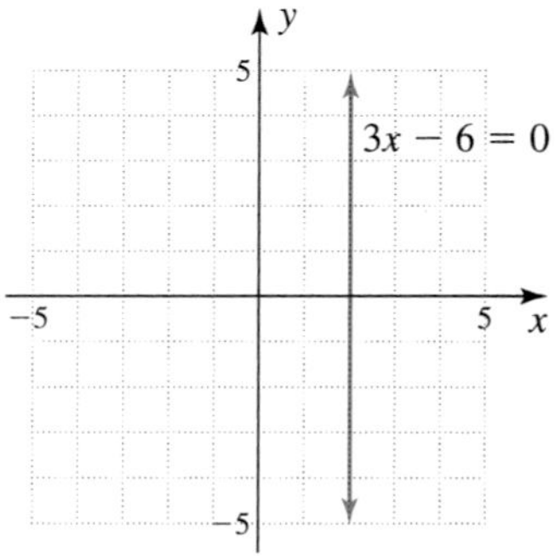	14	3.3	4a	260
15. 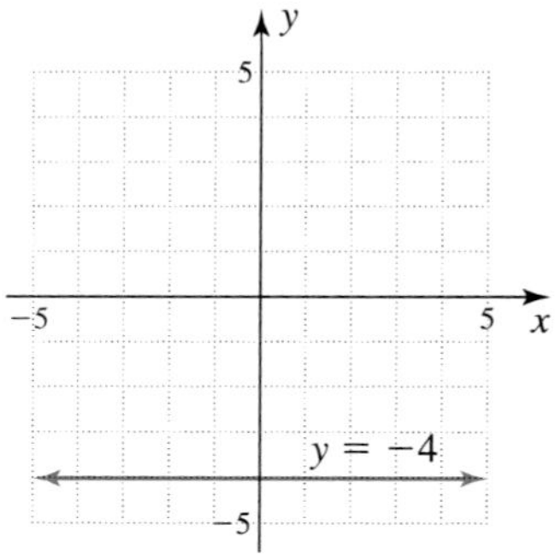	15	3.3	4b	260
16. a. -1 **b.** $-\frac{7}{5}$	16	3.4	1, 2	269
17. a. Undefined **b.** 0	17	3.4	3	270
18. 2	18	3.4	4	271
19. a. Parallel **b.** Perpendicular	19	3.4	5	273
20. a. 0.8 **b.** Annual increase in the number of stores **c.** 80	20	3.4	6	274

Cumulative Review Chapters 1–3

1. Find the additive inverse (opposite) of -1. 1

2. Find: $\left|-3\frac{1}{7}\right|$ $3\frac{1}{7}$

3. Find: $-\frac{1}{6}+\left(-\frac{3}{8}\right)$ $-\frac{13}{24}$

4. Find: $9.7-(-3.3)$ 13.0

5. Find: $(-2.6)(7.6)$ -19.76

6. Find: $(-6)^4$ 1296

7. Find: $-\frac{6}{7}\div\left(-\frac{1}{14}\right)$ 12

8. Evaluate $y \div 2 \cdot x - z$ for $x = 2, y = 8, z = 3$. 5

9. Which law is illustrated by the following statement?

$$(4 + 3) + 9 = (3 + 4) + 9$$

Commutative law of addition

10. Multiply: $3(6x - 7)$ $18x - 21$

11. Combine like terms: $-4cd^2 - (-5cd^2)$ cd^2

12. Simplify: $3x - 3(x + 4) - (x + 2)$ $-x - 14$

13. Write in symbols: The quotient of $(m + n)$ and p. $\frac{m+n}{p}$

14. Does the number -14 satisfy the equation $1 = 15 - x$? No

15. Solve for x: $1 = 5(x - 2) + 5 - 4x$ $x = 6$

16. Solve for x: $-\frac{8}{3}x = -24$ $x = 9$

17. Solve for x: $\frac{x}{6}-\frac{x}{9}=3$ $x = 54$

18. Solve for x: $4-\frac{x}{5}=\frac{2(x+1)}{11}$ $x = 10$

19. Solve for d in the equation $S = 7c^2d$. $d=\frac{S}{7c^2}$

20. The sum of two numbers is 170. If one of the numbers is 40 more than the other, what are the numbers? 65 and 105

21. Dave has invested a certain amount of money in stocks and bonds. The total annual return from these investments is \$625. If the stocks produce \$245 more in returns than the bonds, how much money does Dave receive annually from each type of investment? \$435 from stocks and \$190 from bonds

22. Train A leaves a station traveling at 20 mi/hr. Two hours later, train B leaves the same station traveling in the same direction at 40 mi/hr. How long does it take for train B to catch up to train A? 2 hr

23. Martin purchased some municipal bonds yielding 10% annually and some certificates of deposit yielding 11% annually. If Martin's total investment amounts to \$25,000 and the annual income is \$2630, how much money is invested in bonds and how much is invested in certificates of deposit? \$12,000 in bonds, \$13,000 in certificates of deposit

24. Graph: $-\frac{x}{7} + \frac{x}{6} \le \frac{x-6}{6}$

25. Graph the point $C(-3, 4)$.

26. What are the coordinates of point A? $(2, -4)$

27. Determine whether the ordered pair $(-5, 1)$ is a solution of $4x + y = -21$. No

28. Find x in the ordered pair $(x, 2)$ so that the ordered pair satisfies the equation $3x - y = -8$. $x = -2$

29. Graph: $2x + y = 8$

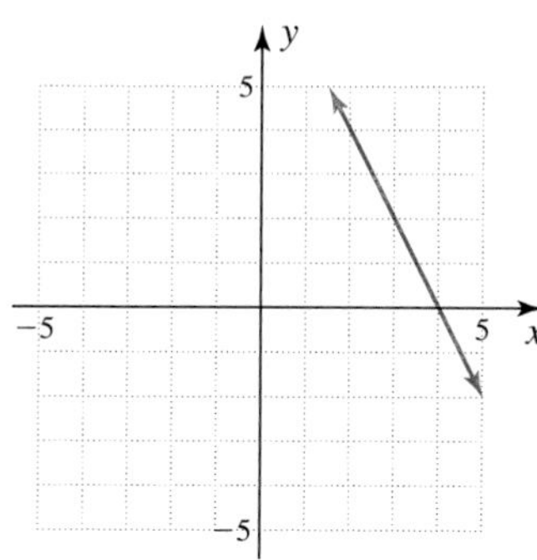

30. Graph: $4x - 8 = 0$

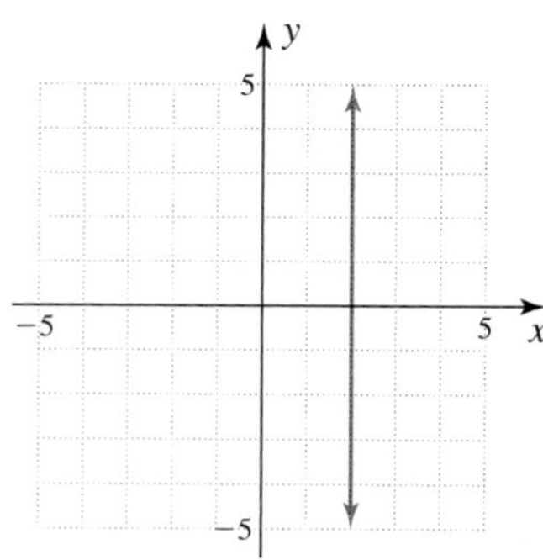

31. Find the slope of the line going through the points $(-5, 3)$ and $(-6, -6)$. 9

32. What is the slope of the line $4x - 2y = 18$? 2

33. Find the pair of parallel lines. (1) and (3)

(1) $-5y = 4x + 7$

(2) $-15y + 12x = 7$

(3) $12x + 15y = 7$

Exponents and Polynomials

The Human Side of Algebra

In the "Golden Age" of Greek mathematics, 300–200 B.C., three mathematicians "stood head and shoulders above all the others of the time." One of them was Apollonius of Perga in Southern Asia Minor. Around 262–190 B.C., Apollonius developed a method of "tetrads" for expressing large numbers, using an equivalent of exponents of the single myriad (10,000). It was not until about the year 250 that the *Arithmetica* of Diophantus advanced the idea of exponents by denoting the square of the unknown as Δ^{Y}, the first two letters of the word *dunamis,* meaning "power." Similarly, K^{Y} represented the cube of the unknown quantity. It was not until 1360 that Nicole Oresme of France gave rules equivalent to the product and power rules of exponents that we study in this chapter. Finally, around 1484, a manuscript written by the French mathematician Nicholas Chuquet contained the *denominacion* (or power) of the unknown quantity, so that our algebraic expressions $3x$, $7x^2$, and $10x^3$ were written as .3. and $.7.^2$ and $.10.^3$. What about zero and negative exponents? $8x^0$ became $.8.^0$ and $8x^{-2}$ was written as $.8.^{2.m}$, meaning ".8. *seconds moins,*" or 8 to the negative two power. Some things do change!

Pretest for Chapter 4

(Answers on page 293)

1. Find.

a. $(2a^4b)(-6ab^4)$

b. $(-2x^4yz)(-6xy^5z^6)$

c. $\dfrac{18x^7y^8}{-9xy^3}$

2. Find.

a. $(2x^3y^4)^3$

b. $(-3x^5y^3)^2$

3. Find.

a. $\left(\dfrac{3}{5}\right)^3$

b. $\left(\dfrac{3x^2}{y^5}\right)^3$

4. Simplify and write the answer without negative exponents.

a. $\left(\dfrac{1}{x}\right)^{-8}$ **b.** $\dfrac{x^{-8}}{x^{-8}}$

c. $\dfrac{x^{-7}}{x^{-8}}$

5. Simplify.

a. $3x^2x^{-6}$

b. $\left(\dfrac{2x^{-3}y^6}{3x^2y^5}\right)^{-2}$

6. Write in scientific notation.

a. 58,000,000

b. 0.00000087

7. Perform the indicated operations.

a. $(3 \times 10^5) \times (7.1 \times 10^6)$

b. $\dfrac{2.84 \times 10^{-3}}{7.1 \times 10^{-4}}$

8. Classify as a monomial (M), binomial (B), or trinomial (T):

a. $8x + 5$

b. $6x^3$

c. $9x^2 - 2 + 6x$

9. Write the polynomial $-3x + 7 + 9x^2$ in descending order and find its degree.

10. Find the value of $-16t^2 + 100$ when $t = 3$.

11. Add $-6x + 3x^2 - 8$ and $-6x^2 - 4 + 3x$.

12. Subtract $6x - 2 + 8x^2$ from $2x^2 - 3x$.

13. Remove parentheses (simplify): $-6x^2(x + 8y)$

14. Find $(x + 6)(x - 3)$.

15. Find $(x + 4)(x - 8)$.

16. Find $(5x - 3y)(4x - 3y)$.

17. Expand $(2x + 5y)^2$.

18. Expand $(3x - 7y)^2$.

19. Find $(3x - 5y)(3x + 5y)$.

20. Find $(x + 2)(x^2 + 3x + 2)$.

21. Find $2x(x + 2)(x + 5)$.

22. Expand $(x + 6)^3$.

23. Expand $\left(2x^2 - \dfrac{1}{2}\right)^2$.

24. Find $(2x^2 + 7)(2x^2 - 7)$.

25. Divide $2x^3 - 19x + 5$ by $x - 3$.

Answers to Pretest

ANSWER	IF YOU MISSED	REVIEW		
	QUESTION	SECTION	EXAMPLES	PAGE
1. a. $-12a^5b^5$ **b.** $12x^5y^6z^7$ **c.** $-2x^6y^5$	1	4.1	1, 2, 3	296–298
2. a. $8x^9y^{12}$ **b.** $9x^{10}y^6$	2	4.1	4, 5	299–300
3. a. $\frac{27}{125}$ **b.** $\frac{27x^6}{y^{15}}$	3	4.1	6	300
4. a. x^8 **b.** 1 **c.** x	4	4.2	1, 2, 5	309, 311
5. a. $\frac{3}{x^4}$ **b.** $\frac{9x^{10}}{4y^2}$	5	4.2	4, 6	311, 312
6. a. 5.8×10^7 **b.** 8.7×10^{-7}	6	4.3	1, 2	320
7. a. 2.13×10^{12} **b.** 4	7	4.3	3, 4	320–321
8. a. B **b.** M **c.** T	8	4.4	1	327
9. $9x^2 - 3x + 7$; degree 2	9	4.4	2, 3	328
10. -44	10	4.4	4, 5, 6	329–330
11. $-3x^2 - 3x - 12$	11	4.5	1, 2	337–338
12. $-6x^2 - 9x + 2$	12	4.5	3	338–339
13. $-6x^3 - 48x^2y$	13	4.6	1, 2	347–348
14. $x^2 + 3x - 18$	14	4.6	3a	350
15. $x^2 - 4x - 32$	15	4.6	3b	350
16. $20x^2 - 27xy + 9y^2$	16	4.6	4b	350
17. $4x^2 + 20xy + 25y^2$	17	4.7	1	358
18. $9x^2 - 42xy + 49y^2$	18	4.7	2	359
19. $9x^2 - 25y^2$	19	4.7	3	360
20. $x^3 + 5x^2 + 8x + 4$	20	4.7	4	361–362
21. $2x^3 + 14x^2 + 20x$	21	4.7	5	362–363
22. $x^3 + 18x^2 + 108x + 216$	22	4.7	6	363
23. $4x^4 - 2x^2 + \frac{1}{4}$	23	4.7	7	363
24. $4x^4 - 49$	24	4.7	8	364
25. $(2x^2 + 6x - 1)$ R 2	25	4.8	1, 2, 3, 4	370–372

4.1 THE PRODUCT, QUOTIENT, AND POWER RULES FOR EXPONENTS

To Succeed, Review How To...

1. Multiply and divide integers (pp. 70–75).
2. Use the commutative and associative properties (p. 90).

Objectives

A Multiply expressions using the product rule for exponents.

B Divide expressions using the quotient rule for exponents.

C Use the power rules to simplify expressions.

GETTING STARTED

Squares, Cubes, and Exponents

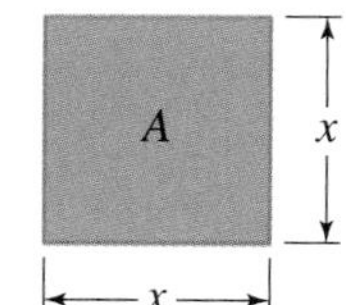

Exponential notation is used to indicate how many times a quantity is to be used as a *factor*. For example, the area A of the *square* in the figure can be written as

$A = x \cdot x = x^2$ Read "x squared."

The exponent 2 indicates that the x is used as a factor twice. Similarly, the volume V of the *cube* is

$V = x \cdot x \cdot x = x^3$ Read "x cubed."

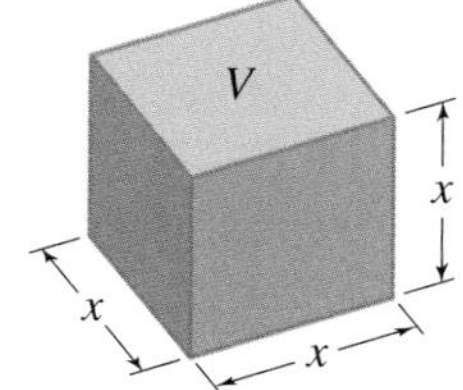

This time, the exponent 3 indicates that the x is used as a factor three times.

If a variable x (called the **base**) is to be used n times as a factor, we use the following definition:

$$\underbrace{x \cdot x \cdot x \cdot \cdots \cdot x}_{n \text{ factors}} = \underset{\text{Base}}{x}^{n} \longleftarrow \text{Exponent}$$

When n is a natural number, some of the powers of x are

$x \cdot x \cdot x \cdot x \cdot x = x^5$ x to the fifth power

$x \cdot x \cdot x \cdot x = x^4$ x to the fourth power

$x \cdot x \cdot x = x^3$ x to the third power (also read "x cubed")

$x \cdot x = x^2$ x to the second power (also read "x squared")

$x = x^1$ x to the first power (also read as x)

Note that if the base carries no exponent, the exponent is assumed to be 1, that is,

$$a = a^1, \quad b = b^1, \quad \text{and} \quad c = c^1$$

Moreover, for a not 0, a^0 is defined to be 1.

In this section we shall learn how to multiply and divide expressions containing exponents by using the product and quotient rules.

A Multiplying Expressions

We are now ready to multiply expressions involving exponents. For example, to multiply x^2 by x^3, we first write

$$\underbrace{\overbrace{x \cdot x}^{x^2} \cdot \overbrace{x \cdot x \cdot x}^{x^3}}_{x^5}$$

or

Here we've just added the exponents of x^2 and x^3, 2 and 3, to find the exponent of the result, 5. Similarly, $2^2 \cdot 2^3 = 2^{2+3} = 2^5$ and $3^2 \cdot 3^3 = 3^{2+3} = 3^5$. Thus

$$a^3 \cdot a^4 = a^{3+4} = a^7$$

and

$$b^2 \cdot b^4 = b^{2+4} = b^6$$

From these and similar examples, we have the following product rule (sometimes called the product law) for exponents.

PRODUCT RULE FOR EXPONENTS

If m and n are positive integers,

$$x^m \cdot x^n = x^{m+n}$$

NOTE

Note that $x^m \cdot y^n \neq (x \cdot y)^{m+n}$ because the bases x and y are not equal.

This rule means that to multiply expressions with the *same base,* we keep the base and *add* the exponents.

NOTE

Before you apply the product rule, make sure that the *bases* are the same.

Of course, some expressions may have numerical coefficients other than 1. For example, the expression $3x^2$ has the numerical coefficient 3. Similarly, the numerical coefficient of $5x^3$ is 5. If we decide to multiply $3x^2$ by $5x^3$, we just multiply numbers by numbers (coefficients) and letters by letters. This procedure is possible because of the *commutative and associative properties of multiplication* we've studied. Using these two properties, we then write

$$\begin{aligned}(3x^2)(5x^3) &= (3 \cdot 5)(x^2 \cdot x^3) \quad \text{We use parentheses to indicate multiplication.}\\ &= 15x^{2+3}\\ &= 15x^5\end{aligned}$$

and

$$\begin{aligned}&(8x^2y)(4xy^2)(2x^5y^3)\\ &= (8 \cdot 4 \cdot 2) \cdot (x^2 \cdot x^1 \cdot x^5)(y^1 \cdot y^2 \cdot y^3) \quad \text{Note that } x = x^1 \text{ and } y = y^1.\\ &= 64x^8y^6\end{aligned}$$

NOTE

Be sure you understand the difference between adding and multiplying expressions. Thus

$$5x^2 + 7x^2 = 12x^2$$

but

$$(5x^2)(7x^2) = 35x^{2+2} = 35x^4$$

EXAMPLE 1 **Multiplying expressions with positive coefficients**

Multiply:

a. $(5x^4)(3x^7)$

b. $(3ab^2c^3)(4a^2b)(2bc)$

SOLUTION

a. We use the commutative and associative properties to write the coefficients and the letters together:

$$\begin{aligned}(5x^4)(3x^7) &= (5 \cdot 3)(x^4 \cdot x^7)\\ &= 15x^{4+7}\\ &= 15x^{11}\end{aligned}$$

b. Using the commutative and associative properties, and the fact that $a = a^1$, $b = b^1$, and $c = c^1$, we write

$$\begin{aligned}(3ab^2c^3)(4a^2b)(2bc) &= (3 \cdot 4 \cdot 2)(a^1 \cdot a^2)(b^2 \cdot b^1 \cdot b^1)(c^3 \cdot c^1)\\ &= 24a^{1+2}b^{2+1+1}c^{3+1}\\ &= 24a^3b^4c^4\end{aligned}$$

PROBLEM 1

Multiply:

a. $(3x^5)(4x^7)$

b. $(4ab^3c^2)(3ab^3)(2bc)$

To multiply expressions involving signed coefficients, we recall the rule of signs for multiplication:

RULES

Signs for Multiplication

1. When multiplying two numbers with the *same* (like) sign, the product is *positive* (+).
2. When multiplying two numbers with *different* (unlike) signs, the product is *negative* (−).

Thus to multiply $(-3x^5)$ by $(8x^2)$, we first note that the expressions have *different* signs. The product should have a *negative* coefficient; that is,

$$\begin{aligned}(-3x^5)(8x^2) &= (-3 \cdot 8)(x^5 \cdot x^2)\\ &= -24x^{5+2}\\ &= -24x^7\end{aligned}$$

Of course, if the expressions have the *same* sign, the product should have a *positive* coefficient. Thus

$$(-2x^3)(-7x^5) = (-2)(-7)(x^3 \cdot x^5)$$

Note that the product of −2 and −7 was written as (−2)(−7) and *not* as (−2 · −7) to avoid confusion.

$$= +14x^{3+5}$$

$$= 14x^8$$

Recall that +14 = 14.

Web It

To study, practice, and see a lesson dealing with the laws of exponents, go to link 4-1-1 at the Bello Website at mhhe.com/bello.

For another lesson with exercises and solutions, go to link 4-1-2.

Finally, to see some practice problems involving coefficients, go to link 4-1-3.

Answers

1. a. $12x^{12}$ **b.** $24a^2b^7c^3$

EXAMPLE 2 **Multiplying expressions with negative coefficients**

Multiply:

a. $(7a^2bc)(-3ac^4)$ **b.** $(-2xyz^3)(-4x^3yz^2)$

SOLUTION

a. Since the coefficients have *different* signs, the result must have a *negative* coefficient. Hence

$$\begin{aligned}(7a^2bc)(-3ac^4) &= (7)(-3)(a^2 \cdot a^1)(b^1)(c^1 \cdot c^4)\\ &= -21a^{2+1}b^1c^{1+4}\\ &= -21a^3bc^5\end{aligned}$$

b. The expressions have the *same* sign, so the result must have a *positive* coefficient. That is,

$$\begin{aligned}(-2xyz^3)(-4x^3yz^2) &= [(-2)(-4)](x^1 \cdot x^3)(y^1 \cdot y^1)(z^3 \cdot z^2)\\ &= +8x^{1+3}y^{1+1}z^{3+2}\\ &= 8x^4y^2z^5\end{aligned}$$

PROBLEM 2

Multiply:

a. $(2a^3bc)(-5ac^5)$

b. $(-3x^3yz^4)(-2xyz^4)$

Teaching Tip

Have students discover the rule for signs when multiplying more than two terms. That is, an even number of negative signs gives a positive product and an odd number of negative signs gives a negative product. This hint will help students with Problems 11–16.

B Dividing Expressions

We are now ready to discuss the division of one expression by another. As you recall, the same rule of signs that applies to the multiplication of integers applies to the division of integers. We write this rule for easy reference.

RULES

Signs for Division

1. When dividing two numbers with the *same* (like) sign, the quotient is *positive* (+).
2. When dividing two numbers with *different* (unlike) signs, the quotient is *negative* (−).

Web It

For a question generator that will simplify almost any expression and explain the steps, go to link 4-1-4 on the Bello Website at mhhe.com/bello.

Read the examples that are given before you practice with the generator.

Now we know what to do with the numerical coefficients when we divide one expression by another. But what about the exponents? To divide expressions involving exponents, we need another rule. So to divide x^5 by x^3, we first use the definition of exponent and write

$$\frac{x^5}{x^3} = \frac{x \cdot x \cdot x \cdot x \cdot x}{x \cdot x \cdot x} \quad (x \neq 0)$$

Remember, division by zero is not allowed!

Since $(x \cdot x \cdot x)$ is common to the numerator and denominator, we have

$$\frac{x^5}{x^3} = \frac{\cancel{(x \cdot x \cdot x)} \cdot x \cdot x}{\cancel{(x \cdot x \cdot x)}} = x \cdot x = x^2$$

Answers

2. a. $-10a^4bc^6$ **b.** $6x^4y^2z^8$

Here the colored x's mean that we divided the numerator and denominator by the common factor $(x \cdot x \cdot x)$. Of course, you can immediately see that the exponent 2 in the answer is simply the difference of the original two exponents, 5 and 3; that is,

$$\frac{x^5}{x^3} = x^{5-3} = x^2 \quad (x \neq 0)$$

Similarly,

$$\frac{x^7}{x^4} = x^{7-4} = x^3 \quad (x \neq 0)$$

and

$$\frac{y^4}{y^1} = y^{4-1} = y^3 \quad (y \neq 0)$$

We can now state the rule for dividing expressions involving exponents.

Teaching Tip

Students should do some division with numerical bases.

Examples:

$$\frac{3^7}{3^2} = 3^{7-2} = 3^5$$

$$\frac{5^3}{5} = 5^{3-1} = 5^2$$

Remind them that the bases do not divide out. That is, we *keep the base* and subtract the exponents just as we do with the variable bases.

QUOTIENT RULE FOR EXPONENTS

If m and n are positive integers,

$$\frac{x^m}{x^n} = x^{m-n} \quad (x \neq 0)$$

where m is greater than n.

NOTE

Before you apply the quotient rule, make sure that the bases are the same.

EXAMPLE 3 Dividing expressions with negative coefficients

Find the quotient:

$$\frac{24x^2y^6}{-6xy^4}$$

SOLUTION We could write

$$\frac{24x^2y^6}{-6xy^4} = \frac{2 \cdot 2 \cdot (2 \cdot 3 \cdot x) \cdot x \cdot (y \cdot y \cdot y \cdot y) \cdot y \cdot y}{-(2 \cdot 3 \cdot x) \cdot (y \cdot y \cdot y \cdot y)}$$

$$= -2 \cdot 2 \cdot x \cdot y \cdot y$$

$$= -4xy^2$$

but to save time, it's easier to divide 24 by -6, x^2 by x, and y^6 by y^4, like this:

$$\frac{24x^2y^6}{-6xy^4} = \frac{24}{-6} \cdot \frac{x^2}{x} \cdot \frac{y^6}{y^4}$$

$$= -4 \cdot x^{2-1} \cdot y^{6-4}$$

$$= -4xy^2$$

PROBLEM 3

Find the quotient:

$$\frac{25a^3b^7}{-5ab^3}$$

Answer

3. $-5a^2b^4$

C Simplifying Expressions Using the Power Rules

Web It

Take advantage of the question generator on link 4-1-4 on the Bello Website at mhhe.com/bello. It does powers, too!

Suppose we wish to find $(5^3)^2$. By definition, squaring a quantity means that we multiply the quantity by itself. Thus

$$(5^3)^2 = 5^3 \cdot 5^3 = 5^{3+3} \quad \text{or} \quad 5^6$$

We could get this answer by multiplying exponents in $(5^3)^2$ to obtain $5^{3\cdot 2} = 5^6$. Similarly,

$$(x^2)^3 = x^2 \cdot x^2 \cdot x^2 = x^{2\cdot 3} = x^6$$

Again, we multiplied exponents in $(x^2)^3$ to get x^6. We use these ideas to state the following power rule for exponents.

POWER RULE FOR EXPONENTS

If m and n are positive integers,

$$(x^m)^n = x^{mn}$$

This rule says that when raising a *power* to a *power,* we keep the base and *multiply* the exponents.

EXAMPLE 4 **Raising a power to a power**

Simplify:

a. $(2^3)^4$ **b.** $(x^2)^3$ **c.** $(y^4)^5$

SOLUTION

a. $(2^3)^4 = 2^{3\cdot 4} = 2^{12}$ **b.** $(x^2)^3 = x^{2\cdot 3} = x^6$ **c.** $(y^4)^5 = y^{4\cdot 5} = y^{20}$

PROBLEM 4

Simplify:

a. $(3^2)^4$ **b.** $(a^3)^5$ **c.** $(b^5)^3$

Sometimes we need to raise several factors inside parentheses to a power, such as in $(x^2y^3)^3$. We use the definition of cubing (see the *Getting Started*) and write:

$$\begin{aligned}(x^2y^3)^3 &= x^2y^3 \cdot x^2y^3 \cdot x^2y^3 \\ &= (x^2 \cdot x^2 \cdot x^2)(y^3 \cdot y^3 \cdot y^3) \\ &= (x^2)^3(y^3)^3 \\ &= x^6y^9\end{aligned}$$

Since we are cubing x^2y^3, we could get the same answer by multiplying each of the exponents in x^2y^3 by 3 to obtain $x^{2\cdot 3}y^{3\cdot 3} = x^6y^9$. Thus to raise several factors inside parentheses to a power, we raise each factor to the given power, as stated in the following rule.

POWER RULE FOR PRODUCTS

If m, n, and k are positive integers,

$$(x^my^n)^k = (x^m)^k(y^n)^k = x^{mk}y^{nk}$$

NOTE

The power rule applies to products only:

$$(x + y)^n \neq x^n + y^n$$

Answers

4. a. 3^8 **b.** a^{15} **c.** b^{15}

EXAMPLE 5 **Using the power rule for products**

Simplify:

a. $(3x^2y^2)^3$ **b.** $(-2x^2y^3)^3$

SOLUTION

a. $(3x^2y^2)^3 = 3^3(x^2)^3(y^2)^3$ Note that since 3 is a *factor* in $3x^2y^2$, 3 is also raised to the third power.

$= 27x^6y^6$

b. $(-2x^2y^3)^3 = (-2)^3(x^2)^3(y^3)^3$

$= -8x^6y^9$

PROBLEM 5

Simplify:

a. $(2a^3b^2)^4$ **b.** $(-3a^3b^2)^3$

Teaching Tip

Do a numerical example to demonstrate the power rule for products.

Example:

Sum: $(2 + 4)^3 \stackrel{?}{=} (2)^3 + (4)^3$

$(6)^3 \stackrel{?}{=} 8 + 64$

$216 \neq 72$

However, product:

$(2 \cdot 4)^3 = (2)^3 \cdot (4)^3$

$(8)^3 = 8 \cdot 64$

$512 = 512$

Just as we have a product and a quotient rule for exponents, we also have a power rule for products and for quotients. Since the quotient

$$\frac{a}{b} = a \cdot \frac{1}{b},$$

we can use the power rule for products and some of the real-number properties to obtain the following.

THE POWER RULE FOR QUOTIENTS

If m is a positive integer,

$$\left(\frac{x}{y}\right)^m = \frac{x^m}{y^m} \quad (y \neq 0)$$

This means that to raise a quotient to a power, we raise the numerator and denominator in the quotient to that power.

EXAMPLE 6 **Using the power rule for quotients**

Simplify:

a. $\left(\frac{4}{5}\right)^3$ **b.** $\left(\frac{2x^2}{y^3}\right)^4$

SOLUTION

a. $\left(\frac{4}{5}\right)^3 = \frac{4^3}{5^3} = \frac{64}{125}$

b. $\frac{(2x^2)^4}{(y^3)^4} = \frac{2^4 \cdot (x^2)^4}{(y^3)^4}$

$= \frac{16x^{2 \cdot 4}}{y^{3 \cdot 4}}$

$= \frac{16x^8}{y^{12}}$

PROBLEM 6

Simplify:

a. $\left(\frac{2}{3}\right)^3$ **b.** $\left(\frac{3a^2}{b^4}\right)^3$

Answers

5. a. $16a^{12}b^8$ **b.** $-27a^9b^6$

6. a. $\frac{8}{27}$ **b.** $\frac{27a^6}{b^{12}}$

Here is a summary of the rules we've just studied.

RULES FOR EXPONENTS

If m, n, and k are positive integers, the following rules apply:

	Rule	Example
1.	Product rule for exponents: $x^m x^n = x^{m+n}$	$x^5 \cdot x^6 = x^{5+6} = x^{11}$
2.	Quotient rule for exponents: $\frac{x^m}{x^n} = x^{m-n}$ $(m > n, x \neq 0)$	$\frac{p^8}{p^3} = p^{8-3} = p^5$
3.	Power rule for products: $(x^m y^n)^k = x^{mk}y^{nk}$	$(x^4y^3)^4 = x^{4\cdot4}y^{3\cdot4} = x^{16}y^{12}$
4.	Power rule for quotients: $\left(\frac{x}{y}\right)^m = \frac{x^m}{y^m}$ $(y \neq 0)$	$\left(\frac{a^3}{b^4}\right)^6 = \frac{a^{3\cdot6}}{b^{4\cdot6}} = \frac{a^{18}}{b^{24}}$

Teaching Tip

When several rules are used, try to apply them in the following order:

1. Power rule
2. Product rule
3. Quotient rule

Now let's look at an example where several of these rules are used.

EXAMPLE 7 Using the power rule for products and quotients

Simplify:

a. $(3x^4)^3(-2y^3)^2$ **b.** $\left(\frac{3}{5}\right)^3 \cdot 5^4$

SOLUTION

a. $(3x^4)^3(-2y^3)^2 = (3)^3(x^4)^3(-2)^2(y^3)^2$ Use Rule 3.

$= 27x^{4\cdot3}(4)y^{3\cdot2}$ Use Rule 3.

$= (27 \cdot 4)x^{12}y^6$

$= 108x^{12}y^6$

b. $\left(\frac{3}{5}\right)^3 \cdot 5^4 = \frac{3^3}{5^3} \cdot 5^4$ Use Rule 4.

$= \frac{3^3}{5^3} \cdot \frac{5^4}{1}$ Since $5^4 = \frac{5^4}{1}$

$= \frac{3^3 \cdot 5^4}{5^3}$ Multiply.

$= 3^3 \cdot 5$ Since $\frac{5^4}{5^3} = 5^{4-3} = 5$

$= 27 \cdot 5$ Since $3^3 = 27$

$= 135$ Multiply.

PROBLEM 7

Simplify:

a. $(2a^3)^4(-3b^2)^3$ **b.** $\left(\frac{2}{3}\right)^3 \cdot 3^5$

So far, we have discussed only the theory and rules for exponents. Does anybody need or use exponents? We will show one application in the next example.

Answers

7. a. $-432a^{12}b^6$ **b.** 72

EXAMPLE 8 Application: Volume of an ice cream cone

The volume V of a cone is $V = \frac{1}{3}\pi r^2 h$, where r is the radius of the base (opening) and h is the height of the cone. The volume S of a hemisphere is $S = \frac{2}{3}\pi r^3$. The ice cream cone in the photo is 10 cm high and has a radius of 2 cm; the mound of ice cream is 2 cm high. To make computation easier, assume that π is about 3; that is, $\pi \approx 3$.

a. How much ice cream does it take to fill the cone?

b. What is the volume of the mound of ice cream?

c. What is the total volume of ice cream (cone plus mound) in the photo?

SOLUTION

a. The volume V of the cone is $V = \frac{1}{3}\pi r^2 h$, where $r = 2$ and $h = 10$. Thus, $V = \frac{1}{3}\pi(2)^2(10) = \frac{40\pi}{3}$ cm^3 or approximately 40 cm^3 (cubic centimeters).

b. The volume of the mound of ice cream is the volume of the hemisphere:

$$S = \frac{2}{3}\pi r^3 = \frac{2}{3}\pi(2)^3 = \frac{16\pi}{3}\text{ cm}^3 \approx 16\text{ cm}^3$$

c. The total amount of ice cream is $\frac{40\pi}{3} + \frac{16\pi}{3}$ cm^3 $= \frac{56\pi}{3}$ cm^3 or about 56 cm^3.

PROBLEM 8

Answer the same questions as in Example 8 if the cone is 4 inches high and has a 1-inch radius. Can you see that the hemisphere must have a 1-inch radius?

Calculate It Checking Exponents

Can we check the rules of exponents using a calculator? We can do it if we have only one variable and we agree that **two expressions are equivalent if their graphs are identical.** Thus, to check Example 1(a) we have to check that $(5x^4)(3x^7) = 15x^{11}$. We will consider $(5x^4)(3x^7)$ and $15x^{11}$ separately. As a matter of fact, let $Y_1 = (5x^4)(3x^7)$ and $Y_2 = 15x^{11}$; enter Y_1 in your calculator by pressing Y= and entering $(5x^4)(3x^7)$. (Remember that exponents are entered by pressing ^.) Now, press GRAPH and the graph appears, as shown at the left. Now enter $Y_2 = 15x^{11}$. You get the same graph, indicating that the graphs actually coincide and thus are equivalent. Now, for a numerical quick surprise, press 2nd GRAPH. The screen shows the information on the right. What does it mean? It means that for the values of x in the table (0, 1, 2, 3, etc.), Y_1 and Y_2 have the same values.

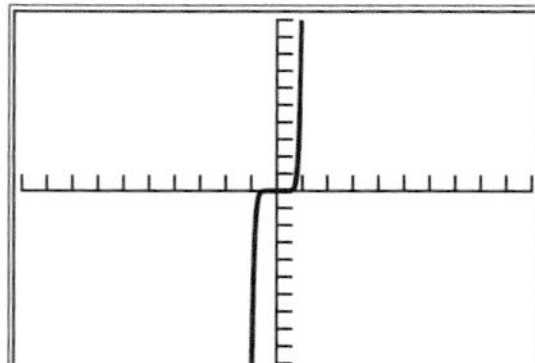

X	Y1	Y2
0	0	0
1	15	15
2	30720	30720
3	2.66E6	2.66E6
4	6.29E7	6.29E7
5	7.32E8	7.32E8
6	5.44E9	5.44E9

X=0

What if you had done the original computation incorrectly? Say you had decided that $(5x^4)(3x^7) = 15x^{28}$. Then the graphs would be different, as shown here. Try the table and confirm the different results!

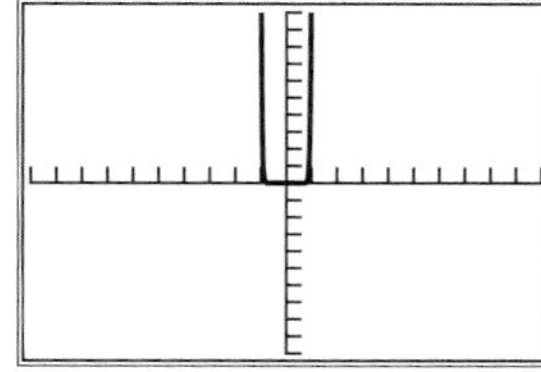

Answers

8. a. $\frac{4\pi}{3}$ in.^{3} $\approx$ 4 in.^{3}
b. $\frac{2\pi}{3}$ in.^{3} $\approx$ 2 in.^{3}
c. 2π in.^{3} $\approx$ 6 in.^{3}

Exercises 4.1

A In Problems 1–16, find the product.

1. $(4x)(6x^2)$ $24x^3$

2. $(2a^2)(3a^3)$ $6a^5$

3. $(5ab^2)(6a^3b)$ $30a^4b^3$

4. $(-2xy)(x^2y)$ $-2x^3y^2$

5. $(-xy^2)(-3x^2y)$ $3x^3y^3$

6. $(x^2y)(-5xy)$ $-5x^3y^2$

7. $b^3\left(\frac{-b^2c}{5}\right)$ $-\frac{b^5c}{5}$

8. $-a^3b\left(\frac{ab^4c}{3}\right)$ $-\frac{a^4b^5c}{3}$

9. $\left(\frac{-5xy^2z}{2}\right)\left(\frac{-3x^2yz^5}{5}\right)$ $\frac{3x^3y^3z^6}{2}$

10. $(-2x^2yz^3)(4xyz)$ $-8x^3y^2z^4$

11. $(-2xyz)(3x^2yz^3)(5x^2yz^4)$ $-30x^5y^3z^8$

12. $(-a^2b)(-0.4b^2c^3)(1.5abc)$ $0.6a^3b^4c^4$

13. $(a^2c^3)(-3b^2c)(-5a^2b)$ $15a^4b^3c^4$

14. $(ab^2c)(-ac^2)(-0.3bc^3)(-2.5a^3c^2)$ $-0.75a^5b^3c^8$

15. $(-2abc)(-3a^2b^2c^2)(-4c)(-b^2c)$ $24a^3b^5c^5$

16. $(xy)(yz)(xz)(yz)$ $x^2y^3z^3$

B In Problems 17–34, find the quotient.

17. $\frac{x^7}{x^3}$ x^4

18. $\frac{8a^3}{4a^2}$ $2a$

19. $\frac{-8a^4}{16a^2}$ $-\frac{a^2}{2}$

20. $\frac{9y^5}{6y^2}$ $\frac{3y^3}{2}$

21. $\frac{12x^5y^3}{6x^2y}$ $2x^3y^2$

22. $\frac{18x^6y^2}{9xy}$ $2x^5y$

23. $\frac{-6x^6y^3}{12x^3y}$ $-\frac{x^3y^2}{2}$

24. $\frac{8x^8y^4}{4x^5y^2}$ $2x^3y^2$

25. $\frac{-14a^8y^6}{-21a^5y^2}$ $\frac{2a^3y^4}{3}$

26. $\frac{-2a^5y^8}{-6a^2y^3}$ $\frac{a^3y^5}{3}$

27. $\frac{-27a^2b^8c^3}{-36ab^5c^2}$ $\frac{3ab^3c}{4}$

28. $\frac{-5x^6y^8z^5}{10x^2y^5z^2}$ $-\frac{x^4y^3z^3}{2}$

29. $\frac{3a^3 \cdot a^5}{2a^4}$ $\frac{3a^4}{2}$

30. $\frac{y^2 \cdot y^8}{y \cdot y^3}$ y^6

31. $\frac{(2x^2y^3)(-3x^5y)}{6xy^3}$ $-x^6y$

32. $\frac{(-3x^3y^2z)(4xy^3z)}{6xy^2z}$ $-2x^3y^3z$

33. $\frac{(-x^2y)(x^3y^2)}{x^3y}$ $-x^2y^2$

34. $\frac{(-8x^2y)(-7x^5y^3)}{-2x^2y^3}$ $-28x^5y$

C In Problems 35–70, simplify.

35. $(2^2)^3$ $2^6 = 64$

36. $(3^1)^2$ $3^2 = 9$

37. $(3^2)^1$ $3^2 = 9$

38. $(2^3)^2$ $2^6 = 64$

39. $(x^3)^3$ x^9

40. $(y^2)^4$ y^8

41. $(y^3)^2$ y^6

42. $(x^4)^3$ x^{12}

43. $(-a^2)^3$ $-a^6$

44. $(-b^3)^5$ $-b^{15}$

45. $(2x^3y^2)^3$ $8x^9y^6$

46. $(3x^2y^3)^2$ $9x^4y^6$

47. $(2x^2y^3)^2$ $4x^4y^6$

48. $(3x^4y^4)^3$ $27x^{12}y^{12}$

49. $(-3x^3y^2)^3$ $-27x^9y^6$

50. $(-2x^5y^4)^4$ $16x^{20}y^{16}$

51. $(-3x^6y^3)^2$ $9x^{12}y^6$

52. $(y^4z^3)^5$ $y^{20}z^{15}$

53. $(-2x^4y^4)^3$ $-8x^{12}y^{12}$

54. $(-3y^5z^3)^4$ $81y^{20}z^{12}$

55. $\left(\frac{2}{3}\right)^4$ $\frac{16}{81}$

56. $\left(\frac{-3}{4}\right)^3$ $-\frac{27}{64}$

57. $\left(\frac{3x^2}{2y^3}\right)^3$ $\frac{27x^6}{8y^9}$

58. $\left(\frac{3a^2}{2b^3}\right)^4$ $\frac{81a^8}{16b^{12}}$

59. $\left(\frac{-2x^2}{3y^3}\right)^4$ $\frac{16x^8}{81y^{12}}$

60. $\left(\frac{-3x^2}{2y^3}\right)^4$ $\frac{81x^8}{16y^{12}}$

61. $(2x^3)^2(3y^3)$ $12x^6y^3$

62. $(3a)^3(2b)^2$ $108a^3b^2$

63. $(-3a)^2(-4b)^3$ $-576a^2b^3$

64. $(-2x^2)^3(-3y^3)^2$ $-72x^6y^6$

65. $-(4a^2)^2(-3b^3)^2$ $-144a^4b^6$

66. $-(3x^3)^3(-2y^3)^3$ $216x^9y^9$

67. $\left(\frac{2}{3}\right)^5 \cdot 3^6$ 96

68. $\left(\frac{4}{5}\right)^3 \cdot 5^4$ 320

69. $\left(\frac{x}{y}\right)^5 \cdot y^7$ x^5y^2

70. $\left(\frac{2a^2}{3b}\right)^5 \cdot 3^4b^7$ $\frac{32a^{10}b^2}{3}$

APPLICATIONS

Compound Interest

If you invest P dollars at rate r compounded annually, the compound amount A in your account at the end of n years will be

$$A = P(1 + r)^n$$

Use this formula in Problems 71–74.

71. If you invest \$1000 at 8% compounded annually, how much will be in your account at the end of 3 years? \$1259.71

72. If you invest \$500 at 10% compounded annually, how much will be in your account at the end of 3 years? \$665.50

73. Suppose you invest \$1000 at 10% compounded annually. How much will be in your account at the end of 4 years? \$1464.10

74. Which investment will produce more money: \$500 invested at 8% compounded annually for 3 years or \$600 invested at 7% compounded annually for 3 years? How much more money? \$600 at 7% for 3 years; \$105.17

Problems 75–80 refer to the photo.

75. A standard pickle container measures x by x by $\frac{2}{3}x$ inches. What is the volume V of the container? $V = \frac{2}{3}x^3$

76. If $x = 6$ inches, what is the volume of the container? 144 in.^{3}

77. A Cuban sandwich takes about one cubic inch of pickles. How many sandwiches can be made with the contents of the container? 144 sandwiches

78. A container of pickles holds about 24 ounces. If pickles cost \$0.045 per ounce, what is the cost of the pickles in the container? \$1.08

79. What is the cost of the pickles in a Cuban sandwich (see Problem 77)? \$0.0075 or 0.75¢

80. How many ounces of pickles does a Cuban sandwich take (see Problem 77)? $\frac{1}{6}$ oz

Problems 81–85 refer to the photo.

81. The dimensions of the green bean container are x by $2x$ by $\frac{1}{2}x$. What is the volume V of the container? $V = x^3$

82. If $x = 6$ inches, what is the volume of the container? 216 in.^{3}

83. Green beans are served in a small bowl that has the approximate shape of a hemisphere with a radius of 2 inches. If the volume of a hemisphere is $S = \frac{2}{3}\pi r^3$, what is the volume of one serving of green beans? Use $\pi \approx 3$. 16 in.^{3}

84. How many servings of green beans does the whole container hold? $13\frac{1}{2}$ or 13.5 servings

85. If each serving of green beans costs \$0.24 and sells for \$1.99, how much does the restaurant makes per container? $\approx$ \$23.63

Problems 86–90 refer to the photo.

86. The volume V of a cylinder is $V = \pi r^2 h$, where r is the radius and h is the height. If the pot is x inches high and its radius is $\frac{1}{2}x$, what is the volume of the pot? Use $\pi \approx 3$. $V = \frac{3}{4}x^3$ in.^{3}

87. When the pot is used for cooking beans, it is only $\frac{3}{4}$ full. What is the volume of the beans in the pot? $V = \frac{9}{16}x^3$ in.^{3}

88. If the height of the pot is 10 inches, what is the volume of the beans? 562.5 in.^{3}

89. A serving of beans is about 1 cup, or 15 in.^{3}, of beans. How many servings of beans are in a pot of beans? 37.5 servings

90. How many pots of beans are needed to serve 150 people? 4 pots

SKILL CHECKER

Try the Skill Checker Exercises so you'll be ready for the next section.

Find the product:

91. $8.1 \cdot 10$ 81

92. $3.42 \cdot 10^2$ 342

93. $9.142 \cdot 10^3$ 9142

94. $3.2 \cdot 4$ 12.8

USING YOUR KNOWLEDGE

Follow That Pattern

Many interesting patterns involve exponents. Use your knowledge to find the answers to the following problems.

95.
$1^2 = 1$
$(11)^2 = 121$
$(111)^2 = 12{,}321$
$(1111)^2 = 1{,}234{,}321$

a. Find $(11{,}111)^2$. 123,454,321
b. Find $(111{,}111)^2$. 12,345,654,321

96.
$1^2 = 1$
$2^2 = 1 + 2 + 1$
$3^2 = 1 + 2 + 3 + 2 + 1$
$4^2 = 1 + 2 + 3 + 4 + 3 + 2 + 1$

a. Use this pattern to write 5^2.
$5^2 = 1 + 2 + 3 + 4 + 5 + 4 + 3 + 2 + 1$
b. Use this pattern to write 6^2.
$6^2 = 1 + 2 + 3 + 4 + 5 + 6 + 5 + 4 + 3 + 2 + 1$

97.
$1 + 3 = 2^2$
$1 + 3 + 5 = 3^2$
$1 + 3 + 5 + 7 = 4^2$

a. Find $1 + 3 + 5 + 7 + 9$. $5^2 = 25$
b. Find $1 + 3 + 5 + 7 + 9 + 11 + 13$. $7^2 = 49$

98. Can you discover your own pattern? What is the largest number you can construct by using the number 9 three times? (It is not 999!) $(9^9)^9$

WRITE ON

99. Explain why the product rule does not apply to the expression $x^2 \cdot y^3$. The bases are different.

100. Explain why the product rule does not apply to the expression $x^2 + y^3$. This is a sum, not a product.

101. What is the difference between the product rule and the power rule? Answers may vary.

102. Explain why you cannot use the quotient rule stated in the text to conclude that

$$\frac{x^m}{x^m} = 1 = x^0, \quad x \neq 0$$

Answers may vary.

MASTERY TEST

If you know how to do these problems, you have learned your lesson!

Simplify:

103. $(5a^3)(6a^8)$ $30a^{11}$

104. $(3x^2yz)(2xy^2)(8xz^3)$ $48x^4y^3z^4$

105. $(-3x^3yz)(4xz^4)$ $-12x^4yz^5$

106. $(-3ab^2c^4)(-5a^4bc^3)$ $15a^5b^3c^7$

107. $\dfrac{15x^4y^7}{-3x^2y}$ $-5x^2y^6$

108. $\dfrac{-5x^2y^6z^4}{15xy^4z^2}$ $-\dfrac{xy^2z^2}{3}$

109. $\dfrac{-3a^2b^9c^2}{-12ab^2c}$ $\dfrac{ab^7c}{4}$

110. $(3^2)^2$ 81

111. $(y^5)^4$ y^{20}

112. $(3x^2y^3)^2$ $9x^4y^6$

113. $\left(\dfrac{3y^2}{x^5}\right)^3$ $\dfrac{27y^6}{x^{15}}$

114. $(-2x^2y^3)^2(3x^4y^5)^3$ $108x^{16}y^{21}$

115. $\left(\dfrac{1}{2^5}\right) \cdot 2^7$ 4

116. $\left(\dfrac{x^4}{y^7}\right) \cdot y^9$ x^4y^2

4.2 INTEGER EXPONENTS

To Succeed, Review How To...

1. Raise numbers to a power (p. 72).
2. Use the rules of exponents (p. 301).

Objectives

A Write an expression with negative exponents as an equivalent one with positive exponents.

B Write a fraction involving exponents as a number with a negative power.

C Multiply and divide expressions involving negative exponents.

GETTING STARTED

Negative Exponents in Science

In science and technology, negative numbers are used as exponents. For example, the diameter of the DNA molecule pictured is 10^{-8} meter, and the time it takes for an electron to go from source to screen in a TV tube is 10^{-6} second. So what do the numbers 10^{-8} and 10^{-6} mean? Look at the pattern obtained by *dividing by 10* in each step:

Exponents decrease by 1.

$$10^3 = 1000$$
$$10^2 = 100$$
$$10^1 = 10$$
$$10^0 = 1$$

Divide by 10 at each step.

Hence the following also holds true:

Exponents decrease by 1.

$$10^{-1} = \frac{1}{10} = \frac{1}{10}$$
$$10^{-2} = \frac{1}{100} = \frac{1}{10^2}$$
$$10^{-3} = \frac{1}{1000} = \frac{1}{10^3}$$

Divide by 10 at each step.

Thus the diameter of a DNA molecule is

$$10^{-8} = \frac{1}{10^8} = 0.00000001 \text{ meter}$$

and the time elapsed between source and screen in a TV tube is

$$10^{-6} = \frac{1}{10^6} = 0.000001 \text{ second}$$

These are very small numbers and very cumbersome to write, so, for convenience, we use negative exponents to represent them. In this section we shall study expressions involving negative and zero exponents.

In the pattern used in the *Getting Started,* $10^0 = 1$. In general, we make the following definition.

ZERO EXPONENT

$$\text{For } x \neq 0, \quad x^0 = 1$$

Thus $5^0 = 1$, $8^0 = 1$, $6^0 = 1$, $(3y)^0 = 1$, and $(-2x^2)^0 = 1$.

A Negative Exponents

Let us look again at the pattern in the *Getting Started;* as you can see,

$$10^{-1} = \frac{1}{10^1}, \quad 10^{-2} = \frac{1}{10^2}, \quad 10^{-3} = \frac{1}{10^3}, \quad \text{and} \quad 10^{-n} = \frac{1}{10^n}$$

We also have the following definition.

Web It

For a review of exponents and a lesson on negative exponents, go to link 4-2-1 on the Bello Website at mhhe.com/bello.

Teaching Tip

Another way to introduce b^0 and b^{-n} is to use the quotient rule.

Example:

If $\frac{x^3}{x^5} = x^{3-5} = x^{-2}$

and $\frac{x^3}{x^5} = \frac{\cancel{x} \cdot \cancel{x} \cdot \cancel{x}}{\cancel{x} \cdot \cancel{x} \cdot \cancel{x} \cdot x \cdot x} = \frac{1}{x^2}$

then $x^{-2} = \frac{1}{x^2}$

Also, if $\frac{x^2}{x^2} = x^{2-2} = x^0$

and $\frac{x^2}{x^2} = \frac{\cancel{x} \cdot \cancel{x}}{\cancel{x} \cdot \cancel{x}} = 1$

then $x^0 = 1$

NEGATIVE EXPONENT

If n is a positive integer,

$$x^{-n} = \frac{1}{x^n} \quad (x \neq 0)$$

This definition says that x^{-n} and x^n are reciprocals, since

$$x^{-n} \cdot x^n = \frac{1}{x^n} \cdot x^n = 1$$

By definition, then,

$$5^{-2} = \frac{1}{5^2} = \frac{1}{5 \cdot 5} = \frac{1}{25} \quad \text{and} \quad 2^{-3} = \frac{1}{2^3} = \frac{1}{2 \cdot 2 \cdot 2} = \frac{1}{8}$$

When n is positive, we obtain this result:

$$\left(\frac{1}{x}\right)^{-n} = \frac{1}{\left(\frac{1}{x}\right)^n} = \frac{1}{\frac{1^n}{x^n}} = x^n$$

which can be stated as follows.

***n*TH POWER OF A QUOTIENT**

If n is a positive integer,

$$\left(\frac{1}{x}\right)^{-n} = x^n$$

Thus

$$\left(\frac{1}{4}\right)^{-3} = 4^3 = 64 \quad \text{and} \quad \left(\frac{1}{5}\right)^{-2} = 5^2 = 25$$

EXAMPLE 1 **Rewriting with positive exponents**

Use positive exponents to rewrite and simplify:

a. 6^{-2} **b.** 4^{-3} **c.** $\left(\frac{1}{7}\right)^{-3}$ **d.** $\left(\frac{1}{x}\right)^{-5}$

SOLUTION

a. $6^{-2} = \frac{1}{6^2} = \frac{1}{6 \cdot 6} = \frac{1}{36}$

b. $4^{-3} = \frac{1}{4^3} = \frac{1}{4 \cdot 4 \cdot 4} = \frac{1}{64}$

c. $\left(\frac{1}{7}\right)^{-3} = 7^3 = 343$

d. $\left(\frac{1}{x}\right)^{-5} = x^5$

PROBLEM 1

Rewrite and simplify:

a. 2^{-3} **b.** 3^{-3} **c.** $\left(\frac{1}{2}\right)^{-3}$ **d.** $\left(\frac{1}{a}\right)^{-4}$

If we want to rewrite

$$\frac{x^{-2}}{y^{-3}}$$

without negative exponents, we use the definition of negative exponents to rewrite x^{-2} and y^{-3}:

$$\frac{x^{-2}}{y^{-3}} = \frac{\frac{1}{x^2}}{\frac{1}{y^3}} = \frac{1}{x^2} \cdot \frac{y^3}{1} = \frac{y^3}{x^2}$$

so that the expression is in the simpler form,

$$\frac{x^{-2}}{y^{-3}} = \frac{y^3}{x^2}$$

Here is the rule for changing fractions with negative exponents to equivalent ones with positive exponents.

Web It

To see the rules of exponents, enter your problems, and see the solutions, go to link 4-2-2 on the Bello Website at mhhe.com/bello.

SIMPLIFYING FRACTIONS WITH NEGATIVE EXPONENTS

For any nonzero numbers x and y and any positive integers m and n,

$$\frac{x^{-m}}{y^{-n}} = \frac{y^n}{x^m}$$

This means that to write an equivalent fraction without negative exponents, we interchange numerators and denominators and make the exponents positive.

EXAMPLE 2 **Writing equivalent fractions with positive exponents**

Write as an equivalent fraction without negative exponents and simplify:

a. $\frac{3^{-2}}{4^{-3}}$ **b.** $\frac{a^{-7}}{b^{-3}}$ **c.** $\frac{x^2}{y^{-4}}$

SOLUTION

a. $\frac{3^{-2}}{4^{-3}} = \frac{4^3}{3^2} = \frac{64}{9}$

b. $\frac{a^{-7}}{b^{-3}} = \frac{b^3}{a^7}$

c. $\frac{x^2}{y^{-4}} = x^2y^4$

PROBLEM 2

Write as an equivalent expression without negative exponents and simplify:

a. $\frac{2^{-4}}{3^{-3}}$ **b.** $\frac{x^{-9}}{y^{-4}}$ **c.** $\frac{a^5}{b^{-8}}$

Answers

1. a. $\frac{1}{8}$ **b.** $\frac{1}{27}$ **c.** 8 **d.** a^4

2. a. $\frac{3^3}{2^4} = \frac{27}{16}$ **b.** $\frac{y^4}{x^9}$ **c.** a^5b^8

B Writing Fractions Using Negative Exponents

As we saw in the *Getting Started,* we can write fractions involving powers in the denominator using negative exponents.

EXAMPLE 3 Writing equivalent fractions with negative exponents

Write using negative exponents:

a. $\frac{1}{5^4}$ **b.** $\frac{1}{7^5}$ **c.** $\frac{1}{x^5}$ **d.** $\frac{3}{x^4}$

SOLUTION We use the definition of negative exponents.

a. $\frac{1}{5^4} = 5^{-4}$ **b.** $\frac{1}{7^5} = 7^{-5}$

c. $\frac{1}{x^5} = x^{-5}$ **d.** $\frac{3}{x^4} = 3 \cdot \frac{1}{x^4} = 3x^{-4}$

PROBLEM 3

Write using negative exponents:

a. $\frac{1}{7^6}$ **b.** $\frac{1}{8^5}$

c. $\frac{1}{a^9}$ **d.** $\frac{7}{a^6}$

C Multiplying and Dividing Expressions with Negative Exponents

In Section 4.1, we multiplied expressions that contained positive exponents. For example,

$$x^5 \cdot x^8 = x^{5+8} = x^{13} \quad \text{and} \quad y^2 \cdot y^3 = y^{2+3} = y^5$$

Can we multiply expressions involving negative exponents using the same idea? Let's see.

$$x^5 \cdot x^{-2} = x^5 \cdot \frac{1}{x^2} = \frac{x^5}{x^2} = x^3$$

Adding exponents,

$$x^5 \cdot x^{-2} = x^{5+(-2)} = x^3 \quad \text{Same answer!}$$

Similarly,

$$x^{-3} \cdot x^{-2} = \frac{1}{x^3} \cdot \frac{1}{x^2} = \frac{1}{x^{3+2}} = \frac{1}{x^5} = x^{-5}$$

Adding exponents,

$$x^{-3} \cdot x^{-2} = x^{-3+(-2)} = x^{-5} \quad \text{Same answer again!}$$

So, we have the following rule.

PRODUCT RULE FOR EXPONENTS

If m and n are integers,

$$x^m \cdot x^n = x^{m+n}$$

This rule says that when we *multiply* expressions with the same base, we keep the base and *add* the exponents. Note that the rule does *not* apply to $x^7 \cdot y^6$ because the bases are different.

Answers

3. a. 7^{-6} **b.** 8^{-5} **c.** a^{-9} **d.** $7a^{-6}$

EXAMPLE 4 **Using the product rule**

Multiply and simplify (that is, write the answer without negative exponents):

a. $2^6 \cdot 2^{-4}$ **b.** $4^3 \cdot 4^{-5}$ **c.** $y^{-2} \cdot y^{-3}$ **d.** $a^{-5} \cdot a^5$

SOLUTION

a. $2^6 \cdot 2^{-4} = 2^{6+(-4)} = 2^2 = 4$

b. $4^3 \cdot 4^{-5} = 4^{3+(-5)} = 4^{-2} = \frac{1}{4^2} = \frac{1}{16}$

c. $y^{-2} \cdot y^{-3} = y^{-2+(-3)} = y^{-5} = \frac{1}{y^5}$

d. $a^{-5} \cdot a^5 = a^{-5+5} = a^0 = 1$

PROBLEM 4

Multiply and simplify:

a. $3^{-4} \cdot 3^6$ **b.** $2^4 \cdot 2^{-6}$

c. $b^{-3} \cdot b^{-5}$ **d.** $x^{-7} \cdot x^7$

Teaching Tip

Students should be reminded to use the product rule *before* applying the definition for negative exponents.

NOTE

We wrote the answers in Example 4 *without* using negative exponents. In algebra, it's customary to write answers *without* negative exponents.

In Section 4.1, we divided expressions with the same base. Thus

$$\frac{7^5}{7^2} = 7^{5-2} = 7^3 \quad \text{and} \quad \frac{8^3}{8} = 8^{3-1} = 8^2$$

The rule used there can be extended to any exponents that are integers.

QUOTIENT RULE FOR EXPONENTS

If m and n are integers,

$$\frac{x^m}{x^n} = x^{m-n}$$

This rule says that when we *divide* expressions with the *same* base, we keep the base and *subtract* the exponents. Note that

$$\frac{x^m}{x^n} = x^m \cdot \frac{1}{x^n} = x^m \cdot x^{-n} = x^{m-n}$$

EXAMPLE 5 **Using the quotient rule**

Simplify:

a. $\frac{6^5}{6^{-2}}$ **b.** $\frac{x}{x^5}$ **c.** $\frac{y^{-2}}{y^{-2}}$ **d.** $\frac{z^{-3}}{z^{-4}}$

SOLUTION

a. $\frac{6^5}{6^{-2}} = 6^{5-(-2)} = 6^{5+2} = 6^7$

b. $\frac{x}{x^5} = x^{1-5} = x^{-4} = \frac{1}{x^4}$

c. $\frac{y^{-2}}{y^{-2}} = y^{-2-(-2)} = y^{-2+2} = y^0 = 1$ Recall the definition for the zero exponent.

d. $\frac{z^{-3}}{z^{-4}} = z^{-3-(-4)} = z^{-3+4} = z^1 = z$

PROBLEM 5

Simplify:

a. $\frac{3^5}{3^{-3}}$ **b.** $\frac{a}{a^{-6}}$

c. $\frac{b^{-5}}{b^{-5}}$ **d.** $\frac{a^{-4}}{a^6}$

Answers

4. a. 9 **b.** $\frac{1}{4}$ **c.** $\frac{1}{b^8}$ **d.** 1

5. a. 3^8 **b.** a^7 **c.** 1 **d.** $\frac{1}{a^{10}}$

Here is a summary of the definitions and rules of exponents we've just discussed.

RULES FOR EXPONENTS

Read and memorize the rules and see how they are applied before you go on.

If m, n, and k are integers, the following rules apply:

Rule		Example
Product rule for exponents:	$x^m x^n = x^{m+n}$	$x^{-2} \cdot x^6 = x^{-2+6} = x^4$
Zero exponent:	$x^0 = 1 \quad (x \neq 0)$	$9^0 = 1$, $y^0 = 1$, and $(3a)^0 = 1$
Negative exponent:	$x^{-n} = \frac{1}{x^n} \quad (x \neq 0)$	$3^{-4} = \frac{1}{3^4} = \frac{1}{81}$, $y^{-7} = \frac{1}{y^7}$
Quotient rule for exponents:	$\frac{x^m}{x^n} = x^{m-n} \quad (x \neq 0)$	$\frac{p^8}{p^3} = p^{8-3} = p^5$
Power rule for exponents:	$(x^m)^n = x^{mn}$	$(a^3)^9 = a^{3 \cdot 9} = a^{27}$
Power rule for products:	$(x^m y^n)^k = x^{mk} y^{nk}$	$(x^4 y^3)^4 = x^{4 \cdot 4} y^{3 \cdot 4} = x^{16} y^{12}$
Power rule for quotients:	$\left(\frac{x}{y}\right)^m = \frac{x^m}{y^m} \quad (y \neq 0)$	$\left(\frac{a^3}{b^4}\right)^8 = \frac{a^{3 \cdot 8}}{b^{4 \cdot 8}} = \frac{a^{24}}{b^{32}}$
Negative to positive exponent:	$\left(\frac{1}{x}\right)^{-n} = x^n$	$\left(\frac{1}{a^2}\right)^{-n} = (a^2)^n = a^{2n}$
Negative to positive exponent:	$\frac{x^{-m}}{y^{-n}} = \frac{y^n}{x^m}$	$\frac{x^{-7}}{y^{-9}} = \frac{y^9}{x^7}$

Now let's do an example that requires us to use several of these rules.

EXAMPLE 6 Using the rules to simplify a quotient

Simplify:

$$\left(\frac{2x^2y^3}{3x^3y^{-2}}\right)^{-4}$$

SOLUTION

$$\left(\frac{2x^2y^3}{3x^3y^{-2}}\right)^{-4} = \left(\frac{2x^{2-3}y^{3-(-2)}}{3}\right)^{-4} \quad \text{Use the quotient rule.}$$

$$= \left(\frac{2x^{-1}y^5}{3}\right)^{-4} \quad \text{Simplify.}$$

$$= \frac{2^{-4}x^4y^{-20}}{3^{-4}} \quad \text{Use the power rule.}$$

$$= \frac{3^4x^4y^{-20}}{2^4} \quad \text{Negative to positive}$$

$$= \frac{3^4x^4}{2^4y^{20}} \quad \text{Definition of negative exponent}$$

$$= \frac{81x^4}{16y^{20}} \quad \text{Simplify.}$$

PROBLEM 6

Simplify:

$$\left(\frac{3a^5b^4}{4a^6b^{-3}}\right)^{-3}$$

Answer

6. $\frac{64a^3}{27b^{21}}$

Let's use exponents in a practical way. What are the costs involved when you are driving a car? Gas, repairs, the car loan payment. What else? Depreciation! The depreciation on a car is the difference between what you paid for the car and what the car is worth now. When you pay P dollars for your car it is then worth 100% of P, but here is what happens to the value as years go by:

Year	Value
0	100% of P
1	90% of $P = (1 - 0.10)P$
2	90% of $(1 - 0.10)P = (1 - 0.10)(1 - 0.10)P = (1 - 0.10)^2P$
3	90% of $(1 - 0.10)^2P = (1 - 0.10)(1 - 0.10)^2P = (1 - 0.10)^3P$

After n years, the value will be

$$(1 - 0.10)^nP$$

But what about x years *ago*? x years ago means $-x$, so the value C of the car x years ago would have been

$$C = (1 - 0.10)^{-x}P$$

There are more examples using exponents in the Using Your Knowledge Exercises.

EXAMPLE 7 **Car depreciation**

Lashonda bought a 3-year-old car for \$10,000. If the car depreciates 10% each year:

a. What will be the value of the car in 2 years?

b. What was the price of the car when it was new?

SOLUTION

a. The value in 2 years will be

$$10{,}000(1 - 0.10)^2 = 10{,}000(0.90)^2 = \$8100$$

b. The car was new 3 years ago and it was then worth

$$10{,}000(1 - 0.10)^{-3} = 10{,}000(0.90)^{-3} = \$13{,}717.42$$

PROBLEM 7

a. Find the value of the car in 3 years.

b. Find the value of the car 2 years ago.

Answers

7. a. \$7290 **b.** \$12,345.68

Exercises 4.2

A In Problems 1–16, write using positive exponents and then simplify.

1. 4^{-2} $\frac{1}{4^2} = \frac{1}{16}$

2. 2^{-3} $\frac{1}{2^3} = \frac{1}{8}$

3. 5^{-3} $\frac{1}{5^3} = \frac{1}{125}$

4. 7^{-2} $\frac{1}{7^2} = \frac{1}{49}$

5. $\left(\frac{1}{8}\right)^{-2}$ $8^2 = 64$

6. $\left(\frac{1}{6}\right)^{-3}$ $6^3 = 216$

7. $\left(\frac{1}{x}\right)^{-7}$ x^7

8. $\left(\frac{1}{y}\right)^{-6}$ y^6

9. $\frac{4^{-2}}{3^{-3}}$ $\frac{3^3}{4^2} = \frac{27}{16}$

10. $\frac{2^{-4}}{3^{-2}}$ $\frac{3^2}{2^4} = \frac{9}{16}$

11. $\frac{5^{-2}}{3^{-4}}$ $\frac{3^4}{5^2} = \frac{81}{25}$

12. $\frac{6^{-2}}{5^{-3}}$ $\frac{5^3}{6^2} = \frac{125}{36}$

13. $\frac{a^{-5}}{b^{-6}}$ $\frac{b^6}{a^5}$

14. $\frac{p^{-6}}{q^{-5}}$ $\frac{q^5}{p^6}$

15. $\frac{x^{-9}}{y^{-9}}$ $\frac{y^9}{x^9}$

16. $\frac{t^{-3}}{s^{-3}}$ $\frac{s^3}{t^3}$

B In Problems 17–22, write using negative exponents.

17. $\frac{1}{2^3}$ 2^{-3}

18. $\frac{1}{3^4}$ 3^{-4}

19. $\frac{1}{y^5}$ y^{-5}

20. $\frac{1}{b^6}$ b^{-6}

21. $\frac{1}{q^5}$ q^{-5}

22. $\frac{1}{t^4}$ t^{-4}

C In Problems 23–46, multiply and simplify. (Remember to write your answers without negative exponents.)

23. $3^5 \cdot 3^{-4}$ 3

24. $4^{-6} \cdot 4^8$ 16

25. $2^{-5} \cdot 2^7$ 4

26. $3^8 \cdot 3^{-5}$ 27

27. $4^{-6} \cdot 4^4$ $\frac{1}{16}$

28. $5^{-4} \cdot 5^2$ $\frac{1}{25}$

29. $6^{-1} \cdot 6^{-2}$ $\frac{1}{216}$

30. $3^{-2} \cdot 3^{-1}$ $\frac{1}{27}$

31. $2^{-4} \cdot 2^{-2}$ $\frac{1}{64}$

32. $4^{-1} \cdot 4^{-2}$ $\frac{1}{64}$

33. $x^6 \cdot x^{-4}$ x^2

34. $y^7 \cdot y^{-2}$ y^5

35. $y^{-3} \cdot y^5$ y^2

36. $x^{-7} \cdot x^8$ x

37. $a^3 \cdot a^{-8}$ $\frac{1}{a^5}$

38. $b^4 \cdot b^{-7}$ $\frac{1}{b^3}$

39. $x^{-5} \cdot x^3$ $\frac{1}{x^2}$

40. $y^{-6} \cdot y^2$ $\frac{1}{y^4}$

41. $x \cdot x^{-3}$ $\frac{1}{x^2}$

42. $y \cdot y^{-5}$ $\frac{1}{y^4}$

43. $a^{-2} \cdot a^{-3}$ $\frac{1}{a^5}$

44. $b^{-5} \cdot b^{-2}$ $\frac{1}{b^7}$

45. $b^{-3} \cdot b^3$ 1

46. $a^6 \cdot a^{-6}$ 1

In Problems 47–60, divide and simplify.

47. $\frac{3^4}{3^{-1}}$ $3^5 = 243$

48. $\frac{2^2}{2^{-2}}$ $2^4 = 16$

49. $\frac{4^{-1}}{4^2}$ $\frac{1}{4^3} = \frac{1}{64}$

50. $\frac{3^{-2}}{3^3}$ $\frac{1}{3^5} = \frac{1}{243}$

51. $\frac{y}{y^3}$ $\frac{1}{y^2}$

52. $\frac{x}{x^4}$ $\frac{1}{x^3}$

53. $\frac{x}{x^{-2}}$ x^3

54. $\frac{y}{y^{-3}}$ y^4

55. $\frac{x^{-3}}{x^{-1}}$ $\frac{1}{x^2}$

56. $\frac{x^{-4}}{x^{-2}}$ $\frac{1}{x^2}$

57. $\frac{x^{-3}}{x^4}$ $\frac{1}{x^7}$

58. $\frac{y^{-4}}{y^5}$ $\frac{1}{y^9}$

59. $\frac{x^{-2}}{x^{-5}}$ x^3

60. $\frac{y^{-3}}{y^{-6}}$ y^3

In Problems 61–70, simplify.

61. $\left(\frac{a}{b^3}\right)^2$ $\frac{a^2}{b^6}$

62. $\left(\frac{a^2}{b}\right)^3$ $\frac{a^6}{b^3}$

63. $\left(\frac{-3a}{2b^2}\right)^{-3}$ $-\frac{8b^6}{27a^3}$

64. $\left(\frac{-2a^2}{3b^0}\right)^{-2}$ $\frac{9}{4a^4}$

65. $\left(\frac{a^{-4}}{b^2}\right)^{-2}$ a^8b^4

66. $\left(\frac{a^{-2}}{b^3}\right)^{-3}$ a^6b^9

67. $\left(\frac{x^5}{y^{-2}}\right)^{-3}$ $\frac{1}{x^{15}y^6}$

68. $\left(\frac{x^6}{y^{-3}}\right)^{-2}$ $\frac{1}{x^{12}y^6}$

69. $\left(\frac{x^{-4}y^3}{x^5y^5}\right)^{-3}$ $x^{27}y^6$

70. $\left(\frac{x^{-2}y^0}{x^7y^2}\right)^{-2}$ $x^{18}y^4$

SKILL CHECKER

Try the Skill Checker Exercises so you'll be ready for the next section.

Find:

71. 8.39×10^2 839

72. 7.314×10^3 7314

73. 8.16×10^{-2} 0.0816

74. 3.15×10^{-3} 0.00315

USING YOUR KNOWLEDGE

Exponential Population Growth

The idea of exponents can be used to measure population growth. Thus if we assume that the world population is increasing about 2% each year (experts say the rate is between 1% and 3%), we can predict the population of the world next year by *multiplying* the present world population by 1.02 (100% + 2% = 102% = 1.02). If we let the world population be P, we have

$$\text{Population in 1 year} = 1.02P$$

$$\text{Population in 2 years} = 1.02(1.02P) = (1.02)^2P$$

$$\text{Population in 3 years} = 1.02(1.02)^2P = (1.02)^3P$$

Web It

You can check the world population at www.census.gov/ipc/www/world.html.

75. If the population P in 2000 was 6 billion people, what would it be in 2 years—that is, in 2002? (Give your answer to three decimal places.) 6.242 billion

76. What will the population be 5 years after 2000? (Give your answer to three decimal places.) 6.624 billion

To find the population 1 year from now, we multiply by 1.02. What should we do to find the population 1 year *ago?* We divide by 1.02. Thus if the population today is P,

$$\text{Population 1 year ago} = \frac{P}{1.02} = P \cdot 1.02^{-1}$$

$$\text{Population 2 years ago} = \frac{P \cdot 1.02^{-1}}{1.02} = P \cdot 1.02^{-2}$$

$$\text{Population 3 years ago} = \frac{P \cdot 1.02^{-2}}{1.02} = P \cdot 1.02^{-3}$$

77. If the population P in 2000 was 6 billion people, what was it in 1990? (Give your answer to three decimal places.) 4.922 billion

78. What was the population in 1985? (Give your answer to three decimal places.) 4.458 billion

Do you know what inflation is? It is the tendency for prices and wages to become more expensive, making your money worth less today than in the future. The formula for the cost C of an item n years from now if the present value (value now) is P dollars and the inflation rate is $r\%$ is

$$C = P(1 + r)^n$$

79. In 2002, the cheapest Super Bowl tickets were \$400. If we assume a 3% (0.03) inflation rate, how much would tickets cost:

a. Next year \$412

b. In 5 years \$463.71

80. In 1967 a Super Bowl ticket was \$12. Assuming a 3% inflation rate:

a. How much should the tickets cost 35 years later? Is the cost more or less than the actual \$400 price? \$33.77; less

b. If a ticket is \$400 now, how much should it have been 35 years ago? \$142.15

81. In 1967 the Consumer Price Index (CPI) was 35. Thirty-five years later, it was 180, so ticket prices should be scaled up by a factor of $\frac{180}{35} \approx 5.14$. Using this factor, how much should a \$12 ticket price be 35 years later? \$61.68

82. The average annual cost for tuition and fees at a 4-year public college is about \$20,000 and rising at a 3.5% rate. How much would you have to pay in tuition and fees 4 years from now? \$22,950.46

83. The average price of a house is \$170,000. If the inflation rate is 2%, what would the average price be in 3 years? \$180,405.36

84. If the average price of a house now is \$170,000 and the inflation rate is 2%, what was the average price 3 years ago? \$160,194.80

85. A credit card account is a loan for which you pay at least 9% annually. If you purchase \$100 with your card, make no payments for 4 months, and your rate is 1.5% a month, how much would you owe at the end of the 4 months? \$106.14

86. The number of bacteria in a sample after n hours growing at a 50% rate is

$$100(1 + 0.50)^n$$

a. What would the number of bacteria be after 6 hours? 1139

b. What was the number of bacteria 2 hours ago? 44

87. On a Monday morning, you count 40 ants in your room. On Tuesday, you find 60. By Wednesday, they number 90. If the population continues to grow at this rate:

a. What is the percent rate of growth? 50% per day

b. Write an equation of the form $P(1 + r)^n$ that models the number A of ants after n days. $A = 40(1 + 0.50)^n$

c. How many ants would you expect on Friday? 203

d. How many ants would you expect in 2 weeks (14 days)? 11,677

e. Name at least one factor that would cause the ant population to slow down its growth. Answers may vary.

88. Referring to Problem 87, how many ants (to the nearest whole number) were there on the previous Sunday? On the previous Saturday? Sunday: 27; Saturday: 18

WRITE ON

89. Give three different explanations for why $x^0 = 1$ $(x \neq 0)$. Answers may vary.

90. By definition, if n is a positive integer,

$$x^{-n} = \frac{1}{x^n} \quad (x \neq 0)$$

a. Does this rule hold if n is *any* integer? Explain and give some examples. Yes; answers may vary.

b. Why do we have to state $x \neq 0$ in this definition? Division by zero is not allowed.

91. Does $x^{-2} + y^{-2} = (x + y)^{-2}$? Explain why or why not. No; answers may vary.

92. Does

$$x^{-1} + y^{-1} = \frac{1}{x + y}?$$

Explain why or why not. No; answers may vary.

MASTERY TEST

If you know how to do these problems, you have learned your lesson!

Write using positive exponents and simplify:

93. $\frac{7^{-2}}{6^{-2}}$ $\frac{6^2}{7^2} = \frac{36}{49}$

94. $\frac{p^{-3}}{q^{-7}}$ $\frac{q^7}{p^3}$

95. $\left(\frac{1}{9}\right)^{-2}$ $9^2 = 81$

96. $\left(\frac{1}{r}\right)^{-4}$ r^4

Write using negative exponents:

97. $\frac{1}{7^6}$ 7^{-6}

98. $\frac{1}{w^5}$ w^{-5}

Simplify and write the answer with positive exponents:

99. $5^6 \cdot 5^{-4}$ $5^2 = 25$

100. $x^{-3} \cdot x^{-7}$ $x^{-10} = \frac{1}{x^{10}}$

101. $\frac{z^{-5}}{z^{-7}}$ z^2

102. $\frac{r}{r^8}$ $\frac{1}{r^7}$

103. $\left(\frac{2xy^3}{3x^3y^{-3}}\right)^{-4}$ $\frac{81x^8}{16y^{24}}$

104. $\left(\frac{3x^{-2}y^{-3}}{2x^3y^{-2}}\right)^{-5}$ $\frac{32x^{25}y^5}{243}$

105. The price of a used car is \$5000. If the car depreciates (loses its value) by 10% each year:

a. What will the value of the car be in 3 years? \$3645.00

b. What was the value of the car 2 years ago. \$6172.84

4.3 APPLICATION OF EXPONENTS: SCIENTIFIC NOTATION

To Succeed, Review How To . . .

1. Use the rules of exponents (p. 312).
2. Multiply and divide real numbers (pp. 70–75).

Objectives

A Write numbers in scientific notation.

B Multiply and divide numbers in scientific notation.

C Solve applications.

GETTING STARTED Sun Facts

How many facts do you know about the sun? Here is some information taken from an encyclopedia.

Mass: 2.19×10^{27} tons

Temperature: 1.8×10^{6} degrees Fahrenheit

Energy generated per minute: 2.4×10^{4} horsepower

All the numbers here are written as products of a number between 1 and 10 and an appropriate power of 10. This is called *scientific notation.* When written in standard notation, these numbers are

2,190,000,000,000,000,000,000,000,000

1,800,000

24,000

It's easy to see why so many technical fields use scientific notation. In this section we shall learn how to write numbers in scientific notation and how to perform multiplications and divisions using these numbers.

A Scientific Notation

We define scientific notation as follows:

SCIENTIFIC NOTATION

A number in **scientific notation** is written as

$$M \times 10^n$$

where M is a number between 1 and 10 and n is an integer.

How do we write a number in scientific notation? First, recall that when we *multiply* a number by a power of 10 ($10^1 = 10$, $10^2 = 100$, and so on), we simply move the decimal point as many places to the *right* as indicated by the exponent of 10. Thus

$7.31 \times 10^1 = 7.31 = 73.1$ Exponent 1; move decimal point 1 place right.

$$72.813 \times 10^2 = 72.813 = 7281.3$$ Exponent 2; move decimal point 2 places right.

$$160.7234 \times 10^3 = 160.7234 = 160{,}723.4$$ Exponent 3; move decimal point 3 places right.

On the other hand, if we *divide* a number by a power of 10, we move the decimal point as many places to the *left* as indicated by the exponent of 10. Thus

$$\frac{7}{10} = 0.7 = 7 \times 10^{-1}$$

$$\frac{8}{100} = 0.08 = 8 \times 10^{-2}$$

and

$$\frac{4.7}{100{,}000} = 0.000047 = 4.7 \times 10^{-5}$$

Remembering the following procedure makes it easy to write a number in scientific notation.

Teaching Tip

This is one place where negative exponents are not simplified further. That is, *do not* write

$$3.5 \times 10^{-2} \text{ as } 3.5 \times \frac{1}{10^2}.$$

PROCEDURE

Writing a Number in Scientific Notation ($M \times 10^n$)

1. Move the decimal point in the given number so that there is only one nonzero digit to its left. The resulting number is M.
2. Count the number of places you moved the decimal point in step 1. If the decimal point was moved to the *left*, n is *positive;* if it was moved to the *right*, n is *negative.*
3. Write $M \times 10^n$.

For example,

$$5.3 = 5.3 \times 10^0$$ The decimal point in 5.3 must be moved 0 places to get 5.3.

1 $$87 = 8.7 \times 10^1 = 8.7 \times 10$$ The decimal point in 87 must be moved 1 place *left* to get 8.7.

4 $$68{,}000 = 6.8 \times 10^4$$ The decimal point in 68,000 must be moved 4 places *left* to get 6.8.

−1 $$0.49 = 4.9 \times 10^{-1}$$ The decimal point in 0.49 must be moved 1 place *right* to get 4.9.

−2 $$0.072 = 7.2 \times 10^{-2}$$ The decimal point in 0.072 must be moved 2 places *right* to get 7.2.

NOTE

After completing step 1 in the procedure, decide whether you should make the number obtained *larger* (n positive) or *smaller* (n negative).

EXAMPLE 1 **Writing a number in scientific notation**

The approximate distance to the sun is 93,000,000 miles, and the wavelength of its ultraviolet light is 0.000035 centimeter. Write 93,000,000 and 0.000035 in scientific notation.

SOLUTION

$93{,}000{,}000 = 9.3 \times 10^7$

$0.000035 = 3.5 \times 10^{-5}$

Teaching Tip

In Example 1, we want to make 9.3 *larger* (equal to 93,000,000), so n is *positive* (+7). Then we want to make 3.5 *smaller* (equal to 0.000035), so n is *negative* (−5).

PROBLEM 1

The approximate distance to the moon is 239,000 miles and its mass is 0.012 that of the earth. Write 239,000 and 0.012 in scientific notation.

EXAMPLE 2 **Changing scientific notation to standard notation**

A jumbo jet weighs 7.75×10^5 pounds, whereas a house spider weighs 2.2×10^{-4} pound. Write these weights in standard notation.

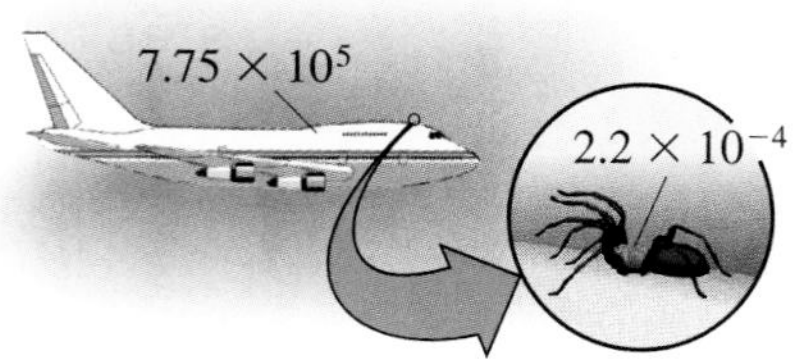

SOLUTION

$7.75 \times 10^5 = 775{,}000$ To multiply by 10^5, move the decimal point 5 places right.

$2.2 \times 10^{-4} = 0.00022$ To multiply by 10^{-4}, move the decimal point 4 places left.

PROBLEM 2

In an average year, Carnival Cruise™ line puts more than 10.1×10^6 mints, each weighing 5.25×10^{-2} ounce, on guest's pillows. Write 10.1×10^6 and 5.25×10^{-2} in standard notation.

B Multiplying and Dividing Using Scientific Notation

Consider the product $300 \cdot 2000 = 600{,}000$. In scientific notation, we would write

$$(3 \times 10^2) \cdot (2 \times 10^3) = 6 \times 10^5$$

To find the answer, we can multiply 3 by 2 to obtain 6 and multiply 10^2 by 10^3, obtaining 10^5. To multiply numbers in scientific notation, we proceed in a similar manner; here's the procedure.

PROCEDURE

Multiplying Using Scientific Notation

1. Multiply the decimal parts first and write the result in scientific notation.
2. Multiply the powers of 10 using the product rule.
3. The answer is the product obtained in steps 1 and 2 after simplification.

EXAMPLE 3 **Multiplying numbers in scientific notation**

Multiply:

a. $(5 \times 10^3) \times (8.1 \times 10^4)$ **b.** $(3.2 \times 10^2) \times (4 \times 10^{-5})$

SOLUTION

a. We multiply the decimal parts first, then write the result in scientific notation.

$$5 \times 8.1 = 40.5 = 4.05 \times 10$$

Next we multiply the powers of 10.

$10^3 \times 10^4 = 10^7$ Add exponents 3 and 4 to obtain 7.

The answer is $(4.05 \times 10) \times 10^7$, or 4.05×10^8.

PROBLEM 3

Multiply:

a. $(6 \times 10^4) \times (5.2 \times 10^5)$

b. $(3.1 \times 10^3) \times (5 \times 10^{-6})$

Answers

1. 2.39×10^5; 1.2×10^{-2}
2. 10,100,000; 0.0525
3. a. 3.12×10^{10}
b. 1.55×10^{-2}

b. Multiply the decimals and write the result in scientific notation.

$$3.2 \times 4 = 12.8 = 1.28 \times 10$$

Multiply the powers of 10.

$$10^2 \times 10^{-5} = 10^{2-5} = 10^{-3}$$

The answer is $(1.28 \times 10) \times 10^{-3}$, or $1.28 \times 10^{1+(-3)} = 1.28 \times 10^{-2}$.

Division is done in a similar manner. For example,

$$\frac{3.2 \times 10^5}{1.6 \times 10^2}$$

is found by dividing 3.2 by 1.6 (yielding 2) and 10^5 by 10^2, which is 10^3. The answer is 2×10^3.

EXAMPLE 4 **Dividing numbers in scientific notation**

Find: $(1.24 \times 10^{-2}) \div (3.1 \times 10^{-3})$

SOLUTION

First divide 1.24 by 3.1 to obtain $0.4 = 4 \times 10^{-1}$. Now divide powers of 10:

$$\begin{aligned} 10^{-2} \div 10^{-3} &= 10^{-2-(-3)} \\ &= 10^{-2+3} \\ &= 10^1 \end{aligned}$$

The answer is $(4 \times 10^{-1}) \times 10^1 = 4 \times 10^0 = 4$.

PROBLEM 4

Find: $(1.23 \times 10^{-3}) \div (4.1 \times 10^{-4})$

C Solving Applications

Since so many fields need to handle large numbers, applications using scientific notation are not hard to come by. Take, for example, the field of astronomy.

EXAMPLE 5 **The energizer**

The total energy received from the sun each minute is 1.02×10^{19} calories. Since the area of the earth is 5.1×10^{18} square centimeters, the amount of energy received per square centimeter of the earth's surface every minute (the solar constant) is

$$\frac{1.02 \times 10^{19}}{5.1 \times 10^{18}}$$

Simplify this expression.

SOLUTION Dividing 1.02 by 5.1, we obtain $0.2 = 2 \times 10^{-1}$. Now, $10^{19} \div 10^{18} = 10^{19-18} = 10^1$. Thus the final answer is

$$(2 \times 10^{-1}) \times 10^1 = 2 \times 10^0 = 2$$

This means that the earth receives about 2 calories of energy per square centimeter each minute.

PROBLEM 5

The population density for a country is the number of people per square mile. Monaco's population density is

$$\frac{3.3 \times 10^4}{7.5 \times 10^{-1}}$$

Simplify this expression.

Answers

4. 3 **5.** 44,000/mi^2

Now, let's talk about more "earthly" matters.

EXAMPLE 6 **Money, money, money!**

In a recent year, the Treasury Department reported printing the following amounts of money in the specified denominations:

\$3,500,000,000 in	\$1 bills	\$2,160,000,000 in	\$20 bills
\$1,120,000,000 in	\$5 bills	\$250,000,000 in	\$50 bills
\$640,000,000 in	\$10 bills	\$320,000,000 in	\$100 bills

a. Write these numbers in scientific notation.

b. Determine how much money was printed (in billions).

SOLUTION

a. $3{,}500{,}000{,}000 = 3.5 \times 10^9$
$1{,}120{,}000{,}000 = 1.12 \times 10^9$
$640{,}000{,}000 = 6.4 \times 10^8$
$2{,}160{,}000{,}000 = 2.16 \times 10^9$
$250{,}000{,}000 = 2.5 \times 10^8$
$320{,}000{,}000 = 3.2 \times 10^8$

b. Since we have to write all the quantities in billions (a billion is 10^9), we have to write all numbers using 9 as the exponent. First, let's consider 6.4×10^8. To write this number with an exponent of 9, we write

$$6.4 \times 10^8 = (0.64 \times 10) \times 10^8 = 0.64 \times 10^9$$

Similarly,

$$2.5 \times 10^8 = (0.25 \times 10) \times 10^8 = 0.25 \times 10^9$$

and

$$3.2 \times 10^8 = (0.32 \times 10) \times 10^8 = 0.32 \times 10^9$$

Writing the other numbers, we get

$$\begin{array}{r} 3.5 \times 10^9 \\ 1.12 \times 10^9 \\ \underline{2.16 \times 10^9} \\ 7.99 \times 10^9 \end{array}$$

Add the entire column.

Thus 7.99 billion dollars were printed.

PROBLEM 6

How much money was minted in coins? The amounts of money minted in the specified denominations are as follows:

Pennies:	\$1,025,740,000
Nickels:	\$66,183,600
Dimes:	\$233,530,000
Quarters:	\$466,850,000
Half-dollars:	\$15,355,000

a. Write these numbers in scientific notation.

b. Determine how much money was minted (in billions).

Teaching Tip

Remember, you can only add apples to apples, so we have to write all the numbers in billions (10^9).

Answers

6. a. Pennies: 1.02574×10^9
Nickels: 6.61836×10^7
Dimes: 2.3353×10^8
Quarters: 4.6685×10^8
Half-dollars: 1.5355×10^7
b. \$1.8076586 billion

Web It

For a lesson with practice problems, go to link 4-3-1 at the Bello Website at mhhe.com/bello.

Another lesson that deals with operations in scientific notation and provides practice problems is at link 4-3-2.

Exercises 4.3

Boost *your* GRADE at mathzone.com!

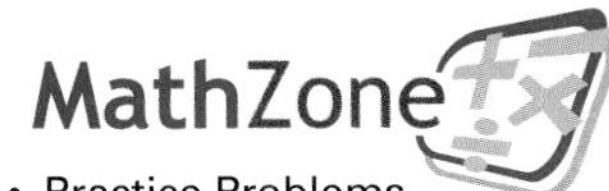

- Practice Problems
- Self-Tests
- Videos
- NetTutor
- e-Professors

A In Problems 1–10, write the given number in scientific notation.

1. 54,000,000 (working women in the United States) 5.4×10^7

2. 68,000,000 (working men in the United States) 6.8×10^7

3. 287,190,520 (U.S. population now) 2.8719052×10^8

4. 299,000,000 (estimated U.S. population in the year 2010) 2.99×10^8

5. 1,900,000,000 (dollars spent on water beds and accessories in 1 year) 1.9×10^9

6. 0.035 (ounces in a gram) 3.5×10^{-2}

7. 0.00024 (probability of four-of-a-kind in poker) 2.4×10^{-4}

8. 0.000005 (the gram-weight of an amoeba) 5×10^{-6}

9. 0.000000002 (the gram-weight of one liver cell) 2×10^{-9}

10. 0.00000009 (wavelength of an X ray in centimeters) 9×10^{-8}

In Problems 11–20, write the given number in standard notation.

11. 1.53×10^2 (pounds of meat consumed per person per year in the United States) 153

12. 5.96×10^2 (pounds of dairy products consumed per person per year in the United States) 596

13. 8×10^6 (bagels eaten per day in the United States) 8,000,000

14. 2.01×10^6 (estimated number of jobs created in service industries between now and the year 2010) 2,010,000

15. 6.85×10^9 (estimated worth, in dollars, of the five wealthiest women) 6,850,000,000

16. 1.962×10^{10} (estimated worth, in dollars, of the five wealthiest men) 19,620,000,000

17. 2.3×10^{-1} (kilowatts per hour used by your TV) 0.23

18. 4×10^{-2} (inches in 1 millimeter) 0.04

19. 2.5×10^{-4} (thermal conductivity of glass) 0.00025

20. 4×10^{-11} (energy, in joules, released by splitting one uranium atom) 0.00000000004

B In Problems 21–30, perform the indicated operations (give your answer in scientific notation).

21. $(3 \times 10^4) \times (5 \times 10^5)$ 1.5×10^{10}

22. $(5 \times 10^2) \times (3.5 \times 10^3)$ 1.75×10^6

23. $(6 \times 10^{-3}) \times (5.1 \times 10^6)$ 3.06×10^4

24. $(3 \times 10^{-2}) \times (8.2 \times 10^5)$ 2.46×10^4

25. $(4 \times 10^{-2}) \times (3.1 \times 10^{-3})$ 1.24×10^{-4}

26. $(3.1 \times 10^{-3}) \times (4.2 \times 10^{-2})$ 1.302×10^{-4}

27. $\dfrac{4.2 \times 10^5}{2.1 \times 10^2}$ 2×10^3

28. $\dfrac{5 \times 10^6}{2 \times 10^3}$ 2.5×10^3

29. $\dfrac{2.2 \times 10^4}{8.8 \times 10^6}$ 2.5×10^{-3}

30. $\dfrac{2.1 \times 10^3}{8.4 \times 10^5}$ 2.5×10^{-3}

APPLICATIONS

31. *Vegetable consumption* The average American eats 80 pounds of vegetables each year. Since there are about 290 million Americans, the number of pounds of vegetables consumed each year should be $(8 \times 10^1) \times (2.9 \times 10^8)$.

a. Write this number in scientific notation. 2.32×10^{10}

b. Write this number in standard notation. 23,200,000,000

32. *Soft drink consumption* The average American drinks 44.8 gallons of soft drinks each year. Since there are about 290 million Americans, the number of gallons of soft drinks consumed each year should be $(4.48 \times 10^1) \times (2.9 \times 10^8)$.

a. Write this number in scientific notation. 1.2992×10^{10}

b. Write this number in standard notation. 12,992,000,000

33. *Garbage production* America produces 148.5 million tons of garbage each year. Since a ton is 2000 pounds, and there are about 360 days in a year and 290 million Americans, the number of pounds of garbage produced each day of the year for each man, woman, and child in America is

$$\frac{(1.485 \times 10^8) \times (2 \times 10^3)}{(2.9 \times 10^8) \times (3.6 \times 10^2)}$$

Write this number in standard notation. 2.844827586

34. *Velocity of light* The velocity of light can be measured by dividing the distance from the sun to the earth (1.47×10^{11} meters) by the time it takes for sunlight to reach the earth (4.9×10^2 seconds). Thus the velocity of light is

$$\frac{1.47 \times 10^{11}}{4.9 \times 10^2}$$

How many meters per second is that? 3×10^8

35. *Nuclear fission* Nuclear fission is used as an energy source. Do you know how much energy a gram of uranium-235 gives? The answer is

$$\frac{4.7 \times 10^9}{235} \text{ kilocalories}$$

Write this number in scientific notation. 2×10^7

36. *How much do you owe?* The national debt of the United States is about $\$6 \times 10^{12}$ (about \$6 trillion). If we assume that the U.S. population is 288 million, each citizen actually owes $\frac{6 \times 10^{12}}{2.88 \times 10^8}$ dollars. How much money is that? Approximate the answer (to the nearest cent). You can see the answer for this very minute at http://www.brillig.com/debt_clock/. \$20,833.33

37. *Internet e-mail projections* An Internet travel site called Travelocity® sends about 730 million e-mails a year, and that number is projected to grow to 4.38 billion per year. If a year has 365 days, how many e-mails per day is Travelocity sending now, and how many are they projecting to send later? Now: 2,000,000; later: 12,000,000

Source: Iconocast.com.

38. *Product information e-mails* Every year, 18.25 billion e-mails requesting product information or service inquiries are sent. If a year has 365 days, how many e-mails a day is that? 50,000,000

Source: Warp 9, Inc.

39. *E-mail proliferation* It is predicted that in 2005, about 288 million people in the United States will be sending 36 million e-mails every day. How many e-mails will each person be sending per year? Round your answer to three decimal places. 45.625

SKILL CHECKER

Try the Skill Checker Exercises so you'll be ready for the next section.

Find:

40. $-16(2)^2 + 118$ 54

41. $-8(3)^2 + 80$ 8

42. $3(2^2) - 5(3) + 8$ 5

43. $-4 \cdot 8 \div 2 + 20$ 4

44. $-5 \cdot 6 \div 2 + 25$ 10

USING YOUR KNOWLEDGE

Astronomical Quantities

As we have seen, scientific notation is especially useful when very large quantities are involved. Here's another example. In astronomy we find that the speed of light is 299,792,458 meters per second.

45. Write 299,792,458 in scientific notation. 2.99792458×10^8

Astronomical distances are so large that they are measured in astronomical units (AU). An astronomical unit is defined as the average separation (distance) of the earth and the sun—that is, 150,000,000 kilometers.

46. Write 150,000,000 in scientific notation. 1.5×10^8

47. Distances in astronomy are also measured in *parsecs:* 1 parsec $= 2.06 \times 10^5$ AU. Thus 1 parsec $= (2.06 \times 10^5) \times (1.5 \times 10^8)$ kilometers. Written in scientific notation, how many kilometers is that? 3.09×10^{13}

48. Astronomers also measure distances in *light-years,* the distance light travels in 1 year: 1 light-year $= 9.46 \times 10^{12}$ kilometers. The closest star, Proxima Centauri, is 4.22 light-years away. In scientific notation using two decimal places, how many kilometers is that? 3.99×10^{13}

49. Since 1 parsec $= 3.09 \times 10^{13}$ kilometers (see Problem 47) and 1 light-year $= 9.46 \times 10^{12}$ kilometers, the number of light-years in a parsec is

$$\frac{3.09 \times 10^{13}}{9.46 \times 10^{12}}$$

Write this number in standard notation using two decimal places. 3.27

WRITE ON

50. Explain why the procedure used to write numbers in scientific notation works. Answers may vary.

51. What are the advantages and disadvantages of writing numbers in scientific notation? Answers may vary.

MASTERY TEST

If you know how to do these problems, you have learned your lesson!

52. The width of the asteroid belt is 2.8×10^8 kilometers. The speed of *Pioneer 10,* a U.S. space vehicle, in passing through this belt was 1.14×10^5 kilometers per hour. Thus *Pioneer 10* took

$$\frac{2.8 \times 10^8}{1.4 \times 10^5}$$

hours to go through the belt. How many hours is that? 2×10^3

53. The distance to the moon is about 289,000 miles, and its mass is 0.12456 that of the earth. Write these numbers in scientific notation. Distance: 2.89×10^5; mass: 1.2456×10^{-1}

54. The Concorde (a supersonic passenger plane) weighs 4.08×10^5 pounds, and a cricket weighs 3.125×10^{-4} pound. Write these weights in standard notation. Concorde: 408,000; cricket: 0.0003125

Find. Write the answer in scientific and standard notation.

55. $(2.52 \times 10^{-2}) \div (4.2 \times 10^{-3})$ $6 \times 10^0 = 6$

56. $(4.1 \times 10^2) \times (3 \times 10^{-5})$ $1.23 \times 10^{-2} = 0.0123$

57. $(6 \times 10^4) \times (2.2 \times 10^3)$ $1.32 \times 10^8 = 132{,}000{,}000$

58. $(3.2 \times 10^{-2}) \div (1.6 \times 10^{-5})$ $2 \times 10^3 = 2000$

4.4 POLYNOMIALS: AN INTRODUCTION

To Succeed, Review How To . . .

1. Evaluate expressions (pp. 72, 81).
2. Add, subtract, and multiply expressions (pp. 92–94, 98).

Objectives

A Classify polynomials.

B Find the degree of a polynomial.

C Write a polynomial in descending order.

D Evaluate polynomials.

GETTING STARTED A Diving Polynomial

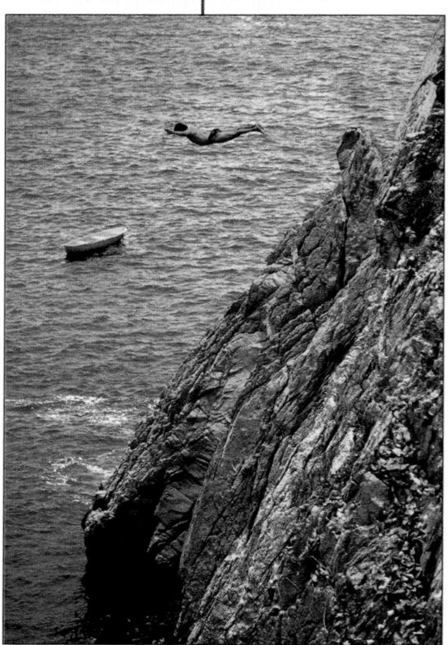

The diver in the photo jumped from a height of 118 feet. Do you know how many feet above the water the diver will be after t seconds? Scientists have determined a formula for finding the answer:

$$-16t^2 + 118 \quad \text{(feet above the water)}$$

The expression $-16t^2 + 118$ is an example of a *polynomial.* Here are some other polynomials:

$$5x, \quad 9x - 2, \quad -5t^2 + 18t - 4, \quad \text{and} \quad y^5 - 2y^2 + \frac{4}{5}y - 6$$

We construct these polynomials by adding or subtracting products of numbers and variables raised to whole-number exponents. Of course, if we use any other operations, the result may not be a polynomial. For example,

$$x^2 - \frac{3}{x} \quad \text{and} \quad x^{-7} + 4x$$

are *not* polynomials (we divided by the variable x in the first one and used negative exponents in the second one). In this section we shall learn how to classify polynomials, find their degrees, write them in descending order, and evaluate them.

A Classifying Polynomials

Polynomials can be used to track and predict the amount of waste (in millions of tons) generated annually in the United States. The polynomial approximating this amount is

$$-0.001t^3 + 0.06t^2 + 2.6t + 88.6$$

where t is the number of years after 1960. But how do we *predict* how much waste will be generated in the year 2010 using this polynomial? We will show you in Example 6! First, let's look at the definition of polynomials.

POLYNOMIAL

A **polynomial** is an algebraic expression formed by using the operations of addition and subtraction on products of numbers and variables raised to whole-number exponents.

The parts of a polynomial separated by plus signs are called the **terms** of the polynomial. If there are subtraction signs, we can rewrite the polynomial using addition signs, since we know that $a - b = a + (-b)$. Thus

$5x$ has one term:	$5x$	
$9x - 2$ has two terms:	$9x$ and -2	Recall that $9x - 2 = 9x + (-2)$.
$-5t^2 + 18t - 4$ has three terms:	$-5t^2$, $18t$, and -4	$-5t^2 + 18t - 4 = -5t^2 + 18t + (-4)$

Polynomials are classified according to the number of terms they have. Thus

$5x$ has *one* term; it is called a *monomial.*	*mono* means one.
$9x - 2$ has *two* terms; it is called a *binomial.*	*bi* means two.
$-5t^2 + 18t - 4$ has *three* terms; it is called a *trinomial.*	*tri* means three.

Clearly, a polynomial of one term is a **monomial;** two terms, a **binomial;** and three, a **trinomial.** If a polynomial has more than three terms, we simply refer to it as a polynomial.

EXAMPLE 1 Classifying polynomials

Classify each of the following polynomials as a monomial, binomial, or trinomial:

a. $6x - 1$ **b.** -8 **c.** $-4 + 3y - y^2$ **d.** $5(x + 2) - 3$

SOLUTION

a. $6x - 1$ has two terms; it is a binomial.

b. -8 has only one term; it is a monomial.

c. $-4 + 3y - y^2$ has three terms; it is a trinomial.

d. $5(x + 2) - 3$ is a binomial. Note that $5(x + 2)$ is *one* term.

PROBLEM 1

Classify as monomial, binomial, or trinomial:

a. -5

b. $-3 + 4y + 6y^2$

c. $8x - 3$

d. $8(x + 9) - 3(x - 1)$

B Finding the Degree of a Polynomial

All the polynomials we have seen contain only one variable and are called *polynomials in one variable.* Polynomials in one variable, such as $x^2 + 3x - 7$, can also be classified according to the *highest* exponent of the variable. The highest exponent of the variable is called the **degree** of the polynomial. To find the degree of a polynomial, you simply examine each term and find the highest exponent of the variable. Thus the degree of $3x^2 + 5x^4 - 2$ is found by looking at the exponent of the variable in each of the terms.

The exponent in $3x^2$ is 2.

The exponent in $5x^4$ is 4.

The exponent in -2 is 0 because $-2 = -2x^0$. (Recall that $x^0 = 1$.)

Thus the degree of $3x^2 + 5x^4 - 2$ is 4, the highest exponent of the variable in the polynomial. Similarly, the degree of $4y^3 - 3y^5 + 9y^2$ is 5, since 5 is the highest exponent of the variable present in the polynomial. By convention, a number such as -4 or 7 is called a **polynomial of degree 0,** because if $a \neq 0$, $a = ax^0$. Thus $-4 = -4x^0$ and $7 = 7x^0$ are polynomials of degree 0. The number 0 itself is called the **zero polynomial** and is *not* assigned a degree. (Note that $0 \cdot x^1 = 0$; $0 \cdot x^2 = 0$, $0 \cdot x^3 = 0$, and so on, so the zero polynomial cannot have a degree.)

Web It

To learn more about the terminology associated with polynomials and for an interactive feature dealing with the degree of a polynomial, go to link 4-4-1 on the Bello Website at mhhe.com/bello.

Answers

1. a. Monomial **b.** Trinomial **c.** Binomial **d.** Binomial

EXAMPLE 2 **Finding the degree of a polynomial**

Find the degree:

a. $-2t^2 + 7t - 2 + 9t^3$

b. 8

c. $-3x + 7$

d. 0

SOLUTION

a. The highest exponent of the variable t in the polynomial $-2t^2 + 7t - 2 + 9t^3$ is 3; thus the degree of the polynomial is 3.

b. The degree of 8 is, by convention, 0.

c. Since $x = x^1$, $-3x + 7$ can be written as $-3x^1 + 7$, making the degree of the polynomial 1.

d. 0 is the zero polynomial; it does not have a degree.

PROBLEM 2

Find the degree:

a. 9

b. $-5z^2 + 2z - 8$

c. 0

d. $-8y + 1$

C Writing a Polynomial in Descending Order

Teaching Tip

Have students think of polynomial examples that may have a real-life application.

Example:

50 mi/hr (1 hr) = 50 miles

50 mi/hr (2 hr) = 100 miles

In general, $50t$ represents the distance traveled.

The degree of a polynomial is easier to find if we agree to write the polynomial in **descending order;** that is, the term with the *highest* exponent is written *first,* the *second* highest is *next,* and so on. Fortunately, the associative and commutative properties of addition permit us to do this rearranging! Thus instead of writing $3x^2 - 5x^3 + 4x - 2$, we rearrange the terms and write $-5x^3 + 3x^2 + 4x - 2$ with exponents in the terms arranged in *descending* order. Similarly, to write $-3x^3 + 7 + 5x^4 - 2x$ in descending order, we use the associative and commutative properties and write $5x^4 - 3x^3 - 2x + 7$. Of course, it would not be incorrect to write this polynomial in ascending order (or with no order at all); it is just that we *agree* to write polynomials in descending order for uniformity and convenience.

EXAMPLE 3 **Writing polynomials in descending order**

Write in descending order:

a. $-9x + x^2 - 17$

b. $-5x^3 + 3x - 4x^2 + 8$

SOLUTION

a. $-9x + x^2 - 17 = x^2 - 9x - 17$

b. $-5x^3 + 3x - 4x^2 + 8 = -5x^3 - 4x^2 + 3x + 8$

PROBLEM 3

Write in descending order:

a. $-4x^2 + 3x^3 - 8 + 2x$

b. $-3y + y^2 - 1$

D Evaluating Polynomials

Now, let's return to the diver in the *Getting Started.* You may be wondering why his height above the water after t seconds was $-16t^2 + 118$ feet. This expression doesn't even look like a number! But polynomials represent numbers when they are *evaluated.* So, if our diver is $-16t^2 + 118$ feet above the water after t seconds, then after 1 second (that is, when $t = 1$), our diver will be

$$-16(1)^2 + 118 = -16 + 118 = 102 \text{ ft}$$

above the water.

Answers

2. a. 0 **b.** 2 **c.** No degree **d.** 1

3. a. $3x^3 - 4x^2 + 2x - 8$

b. $y^2 - 3y - 1$

Web It

For some examples and practice problems on evaluating polynomials, go to link 4-4-2 on the Bello Website at mhhe.com/bello.

Teaching Tip

Have students take their examples from the previous *Teaching Tip* and create a problem to evaluate using this new solution.

Example:

If $D(t) = 50t$, find $D(3)$.

After 2 seconds (that is, when $t = 2$), his height will be

$$-16(2)^2 + 118 = -16 \cdot 4 + 118 = 54 \text{ ft}$$

above the water.

We sometimes say that

$$\text{At } t = 1, \quad -16t^2 + 118 = 102$$
$$\text{At } t = 2, \quad -16t^2 + 118 = 54$$

and so on.

In algebra, polynomials in one variable can be represented by using symbols such as $P(t)$ (read "P of t"), $Q(x)$, and $D(y)$, where the symbol in parentheses indicates the variable being used. Thus $P(t) = -16t^2 + 118$ is the polynomial representing the height of the diver above the water and $G(t)$ is the polynomial representing the amount of waste generated annually in the United States. With this notation, $P(1)$ represents the value of the polynomial $P(t)$ when 1 is substituted for t in the polynomial; that is,

$$P(1) = -16(1)^2 + 118 = 102$$

and

$$P(2) = -16(2)^2 + 118 = 54$$

and so on.

EXAMPLE 4 Evaluating a polynomial

When $t = 3$, what is the value of $P(t) = -16t^2 + 118$?

SOLUTION When $t = 3$,

$$P(t) = -16t^2 + 118$$

becomes

$$\begin{aligned} P(3) &= -16(3)^2 + 118 \\ &= -16(9) + 118 \\ &= -144 + 118 \\ &= -26 \end{aligned}$$

PROBLEM 4

Find the value of $P(x) = -16t^2 + 90$ when $t = 2$.

Note that in this case, the answer is *negative*, which means that the diver should be *below* the water's surface. However, since he can't continue to free-fall after hitting the water, we conclude that it took him between 2 and 3 seconds to hit the water.

EXAMPLE 5 Evaluating polynomials

Evaluate $Q(x) = 3x^2 - 5x + 8$ when $x = 2$.

SOLUTION When $x = 2$,

$$Q(x) = 3x^2 - 5x + 8$$

becomes

$$\begin{aligned} Q(2) &= 3(2^2) - 5(2) + 8 & \\ &= 3(4) - 5(2) + 8 & \text{Multiply } 2 \cdot 2 = 2^2. \\ &= 12 - 10 + 8 & \text{Multiply } 3 \cdot 4 \text{ and } 5 \cdot 2. \\ &= 2 + 8 & \text{Subtract } 12 - 10. \\ &= 10 & \text{Add } 2 + 8. \end{aligned}$$

Note that to evaluate this polynomial, we followed the order of operations studied in Section 1.5.

PROBLEM 5

Evaluate $R(x) = 5x^2 - 3x + 9$ when $x = 3$.

Answers

4. 26 **5.** 45

EXAMPLE 6 A lot of waste

a. If $G(t) = -0.001t^3 + 0.06t^2 + 2.6t + 88.6$ is the amount of waste (in millions of tons) generated annually in the United States and t is the number of years *after* 1960, how much waste was generated in 1960 ($t = 0$)?

b. How much waste would be generated in the year 2010?

SOLUTION

a. At $t = 0$, $G(t) = -0.001t^3 + 0.06t^2 + 2.6t + 88.6$ becomes

$G(0) = -0.001(0)^{-3} + 0.06(0)^2 + 2.6(0) + 88.6 = 88.6$ (million tons)

Thus 88.6 million tons were generated in 1960.

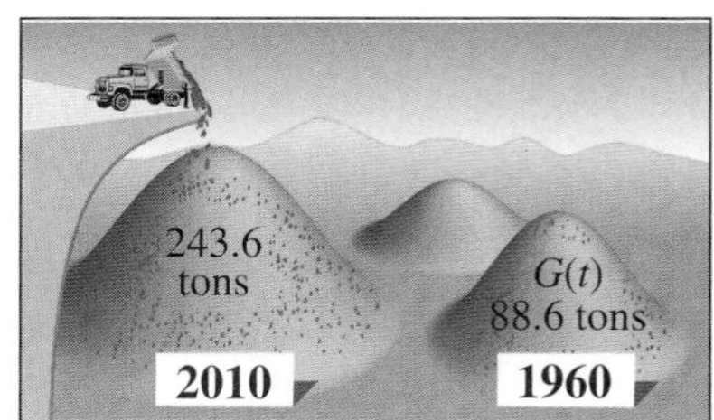

b. The year 2010 is 2010 − 1960 = 50 years *after* 1960. This means that $t = 50$ and

$$\begin{aligned} G(50) &= -0.001(50)^3 + 0.06(50)^2 + 2.6(50) + 88.6 \\ &= -125 + 0.06(2500) + 2.6(50) + 88.6 \\ &= 243.6 \quad \text{(million tons)} \end{aligned}$$

The prediction is that 243.6 million tons of waste will be generated in the year 2010.

PROBLEM 6

a. How much waste was generated in 1961?

b. How much waste was generated in 2000?

EXAMPLE 7 Don't drink and drive!

Do you know how many drinks it takes before you are considered legally drunk? In many states you are drunk if you have a blood alcohol level (BAL) of 0.10 or even lower (0.08). The accompanying chart shows your BAL after consuming 3 ounces of alcohol (6 beers with 4% alcohol or 30 ounces of 10% wine or 7.5 ounces of vodka or whiskey) in the time period shown. The polynomial equation $y = -0.0226x + 0.1509$ approximates the BAL for a 150-pound male and $y = -0.0257x + 0.1663$ the BAL of a 150-pound female.

a. Use the graph to find the BAL for a male after 0.5 hour.

b. Use the graph to find the BAL for a female after 1 hour.

c. Evaluate $y = -0.0226x + 0.1509$ for $x = 0.5$. Does your answer coincide with the answer to part **a**?

d. Evaluate $y = -0.0257x + 0.1663$ for $x = 1$. Does your answer coincide with the answer to part **b**?

3 Ounces of Alcohol Consumed in Given Time

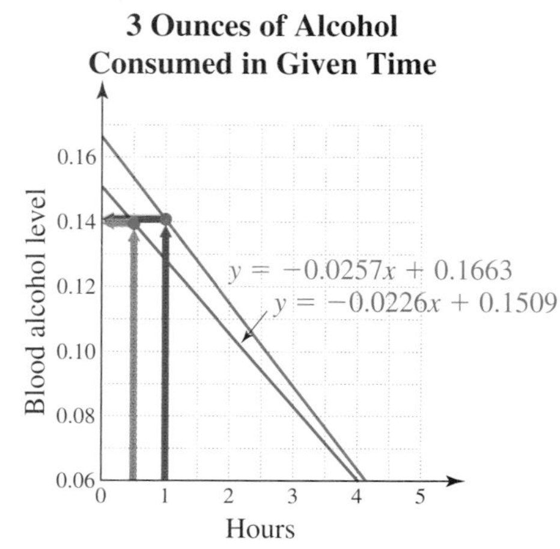

PROBLEM 7

a. Use the graph to find the BAL for a male after 3 hours.

b. Use the graph to find the BAL for a female after 3 hours.

c. Evaluate $y = -0.0226x + 0.1509$ for $x = 3$.

d. Evaluate $y = -0.0257x + 0.1663$ for $x = 3$.

Answers

6. a. 91.259 million tons **b.** 224.60 million tons
7. a. 0.082 **b.** 0.09 **c.** 0.0831 **d.** 0.0892

SOLUTION

a. First, locate 0.5 on the x-axis. Move vertically until you reach the blue line and then horizontally (left) to the y-axis. The y-value at that point is approximately 0.14. This means that the BAL of a male 0.5 hour after consuming 3 ounces of alcohol is about 0.14 (legally drunk!).

b. This time, locate 1 on the x-axis, move vertically until you reach the red line, and then move horizontally to the y-axis. The y-value at that point is a little more than 0.14, so we estimate the answer to be 0.141 (legally drunk!).

c. When $x = 0.5$,

$$y = -0.0226x + 0.1509$$
$$= -0.0226(0.5) + 0.1509 = 0.1396$$

which is very close to the 0.14 from part **a**.

d. When $x = 1$,

$$y = -0.0257x + 0.1663$$
$$= -0.0257(1) + 0.1663 = 0.1406$$

which is also very close to the 0.141 from part **b**.

Calculate It Evaluating Polynomials

If you have a calculator, you can evaluate polynomials in several ways. One way is to make a picture (graph) of the polynomial and use the TRACE and ZOOM keys. Or, better yet, if your calculator has a "value" feature, it will automatically find the value of a polynomial for a given number. Thus to find the value of $G(t) = -0.001t^3 + 0.06t^2 + 2.6t + 88.6$ when $t = 50$ in Example 6, first graph the polynomial. With a TI-83 Plus, press Y= and enter $-0.001x^3 + 0.06x^2 + 2.6x + 88.6$ for Y_1. (Note that we used X's instead of t's because X's are easier to enter.) If you then press GRAPH, nothing will show in your window! Why? Because a standard window gives values of X only between -10 and 10 and corresponding $Y_1 = G(x)$ values between -10 and 10. Adjust the X- and Y-values to those shown in Window 1 and press GRAPH again. To evaluate $G(X)$ at $X = 50$ with a TI-83 Plus, press 2nd TRACE 1. When the calculator prompts you by showing $X =$, enter 50 and press ENTER. The result is shown in Window 2 as $Y = 243.6$. This means that 50 years after 1960—that is, in the year 2010—243.6 million tons of waste will be generated.

```
WINDOW FORMAT
 Xmin=0
 Xmax=50
 Xscl=10
 Ymin=0
 Ymax=200
 Yscl=50
```

Window 1

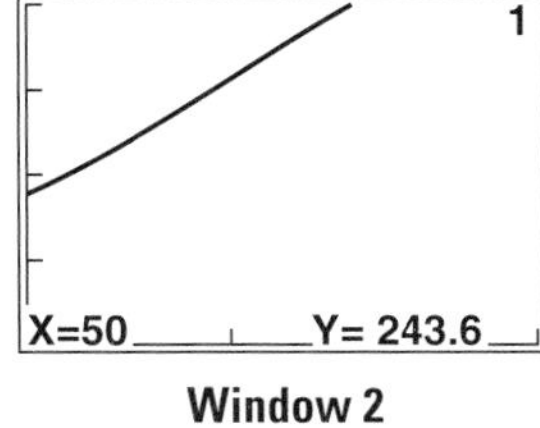

Window 2

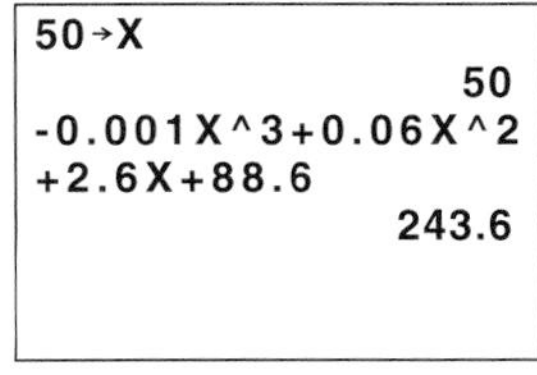

Window 3

You can also evaluate $G(X)$ by first storing the value you wish ($X = 50$) by pressing 50 STO→ X,T,θ,n ENTER, then entering $-0.001x^3 + 0.06x^2 + 2.6x + 88.6$, and finally pressing ENTER again. The result is shown in Window 3.

The beauty of the first method is that now you can evaluate $G(20)$, $G(50)$, or $G(a)$ for any number a by simply entering the value of a and pressing ENTER. You don't have to reenter $G(X)$ or adjust the window again!

Are you ready for a secret? You can get any value of $G(t)$ by pressing 2nd GRAPH. Do it now so you can see what happens!

By the way, if you look in the *Statistical Abstract of the United States* and use $G(X)$ to estimate the amount of waste generated in 1995 ($X = 35$), your answer will be a very close approximation to the actual 211.4 million tons generated in that year.

Exercises 4.4

A B In Problems 1–10, classify as a monomial (M), binomial (B), trinomial (T), or polynomial (P) and give the degree.

1. $-5x + 7$ B; 1

2. $8 + 9x^3$ B; 3

3. $7x$ M; 1

4. $-3x^4$ M; 4

5. $-2x + 7x^2 + 9$ T; 2

6. $-x + x^3 - 2x^2$ T; 3

7. 18 M; 0

8. 0 Zero polynomial; no degree

9. $9x^3 - 2x$ B; 3

10. $-7x + 8x^6 + 3x^5 + 9$ P; 6

B C In Problems 11–20, write in descending order and give the degree of each polynomial.

11. $-3x + 8x^3$ $8x^3 - 3x$; 3

12. $7 - 2x^3$ $-2x^3 + 7$; 3

13. $4x - 7 + 8x^2$ $8x^2 + 4x - 7$; 2

14. $9 - 3x + x^3$ $x^3 - 3x + 9$; 3

15. $5x + x^2$ $x^2 + 5x$; 2

16. $-3x - 7x^3$ $-7x^3 - 3x$; 3

17. $3 + x^3 - x^2$ $x^3 - x^2 + 3$; 3

18. $-3x^2 + 8 - 2x$ $-3x^2 - 2x + 8$; 2

19. $4x^5 + 2x^2 - 3x^3$ $4x^5 - 3x^3 + 2x^2$; 5

20. $4 - 3x^3 + 2x^2 + x$ $-3x^3 + 2x^2 + x + 4$; 3

D In Problems 21–24, find the value of the polynomial when (a) $x = 2$ and (b) $x = -2$.

21. $3x - 2$ **a.** 4 **b.** -8

22. $x^2 - 3$ **a.** 1 **b.** 1

23. $2x^2 - 1$ **a.** 7 **b.** 7

24. $x^3 - 1$ **a.** 7 **b.** -9

25. If $P(x) = 3x^2 - x - 1$, find
a. $P(2)$ 9 **b.** $P(-2)$ 13

26. If $Q(x) = 2x^2 + 2x + 1$, find
a. $Q(2)$ 13 **b.** $Q(-2)$ 5

27. If $R(x) = 3x - 1 + x^2$, find
a. $R(2)$ 9 **b.** $R(-2)$ -3

28. If $S(x) = 2x - 3 - x^2$, find
a. $S(2)$ -3 **b.** $S(-2)$ -11

29. If $T(y) = -3 + y + y^2$, find
a. $T(2)$ 3 **b.** $T(-2)$ -1

30. If $U(r) = -r - 4 - r^2$, find
a. $U(2)$ -10 **b.** $U(-2)$ -6

APPLICATIONS

31. *Height of dropped object* If an object drops from an altitude of k feet, its height above ground after t seconds is given by $-16t^2 + k$ feet. If the object is dropped from an altitude of 150 feet, what would be the height of the object after

a. t seconds? $(-16t^2 + 150)$ ft

b. 1 second? 134 ft **c.** 2 seconds? 86 ft

32. *Velocity of dropped object* After t seconds have passed, the velocity of an object dropped from a height of 96 feet is $-32t$ feet per second. What would be the velocity of the object after

a. 1 second? -32 ft/sec

b. 2 seconds? -64 ft/sec

33. *Height of dropped object* If an object drops from an altitude of k meters, its height above the ground after t seconds is given by $-4.9t^2 + k$ meters. If the object is dropped from an altitude of 200 meters, what would be the height of the object after

a. t seconds? $(-4.9t^2 + 200)$ m

b. 1 second? 195.1 m **c.** 2 seconds? 180.4 m

34. *Velocity of dropped object* After t seconds have passed, the velocity of an object dropped from a height of 300 meters is $-9.8t$ meters per second. What would be the velocity of the object after

a. 1 second? -9.8 m/sec

b. 2 seconds? -19.6 m/sec

35. *Annual number of robberies* According to FBI data, the annual number of robberies (per 100,000 population) can be approximated by

$$R(t) = 1.76t^2 - 17.24t + 251$$

where t is the number of years after 1980.

a. What was the number of robberies (per 100,000) in 1980 ($t = 0$)? 251

b. How many robberies per 100,000 would you predict in the year 2000? In 2010? About 610; about 1318

36. *Annual number of assaults* The number of aggravated assaults (per 100,000) can be approximated by

$$A(t) = -0.2t^3 + 4.7t^2 - 15t + 300$$

where t is the number of years after 2000.

a. What was the number of aggravated assaults (per 100,000) in 2000 ($t = 0$)? 300

b. How many aggravated assaults per 100,000 would you predict for the year 2020? 280

37. *Rock music sales* What type of music do you buy? According to the Recording Industry Association of America, the percent of U.S. dollar music sales spent annually on rock music is given by

$$R(t) = 0.07t^3 - 1.15t^2 + 4t + 30$$

where t is the number of years after 1992. (Give answers to two decimals.)

a. What percent of music sales were spent on rock music during 1992? 30%

b. What percent of music sales were spent on rock music during 2002? 25%

c. Were rock music sales increasing or decreasing from 1992 to 2002? Decreasing

d. During 1992, the value of all music sold amounted to $9.02 billion. How much was spent on rock music? $2.71 billion

e. During 2002, the value of all music sold amounted to $13.0 billion. How much was spent on rock music? $3.25 billion

38. *Country music sales* The percent of U.S. dollar music sales spent annually on country music is given by

$$C(t) = -0.04t^2 - 0.6t + 18$$

where t is the number of years after 1992.

a. What percent of music sales were spent on country music during 1992? 18%

b. What percent of music sales were spent on country music during 2002? 8%

c. Were country music sales increasing or decreasing from 1992 to 2002? Decreasing

d. During 1992, the value of all music sold amounted to $9.02 billion. How much was spent on country music? $1.62 billion

e. During 2002, the value of all music sold amounted to $13.0 billion. How much was spent on country music? $1.04 billion

39. *Record low temperatures* According to the *USA Today Weather Almanac,* the coldest city in the United States (based on average annual temperature) is International Falls, Minnesota. Record low temperatures (in °F) there can be approximated by $L(m) = -4m^2 + 57m - 175$, where m is the number of the month starting with March ($m = 3$) and ending with December ($m = 12$).

a. Find the record low during July. 28°

b. If $m = 1$ were allowed, what would be the record low in January? Does the answer seem reasonable? Do you see why $m = 1$ is not one of the choices? −122°; 122° below 0 is unreasonable

40. *Record low temperatures* According to the *USA Today Weather Almanac,* the hottest city in the United States (based on average annual temperature) is Key West, Florida. Record low temperatures there can be approximated by $L(m) = -0.12m^2 + 2.9m + 77$, where m is the number of the month starting with January ($m = 1$) and ending with December ($m = 12$).

a. Find the record low during January. (Answer to the nearest whole number.) 80°

b. In what two months would you expect the highest-ever temperature to have occurred? What is m for each of the two months? July, August; $m = 7, m = 8$

c. Find $L(m)$ for each of the two months of part **b**. Which is higher? $L(7) = 91; L(8) = 93$

d. The highest temperature ever recorded in Key West was 95°F and occurred in August 1957. How close was your approximation? 2°

41. *Internet use in China* The number of Internet users in China (in millions) is shown in the figure and can be approximated by $N(t) = 0.5t^2 + 4t + 2.1$, where t is the number of years after 1998.

a. Use the graph to find the number of users in 2003. 34 million

b. Evaluate $N(t)$ for $t = 5$. Is the result close to the approximation in part **a**? $N(5) = 34.6$; the result is close

Internet Use in China

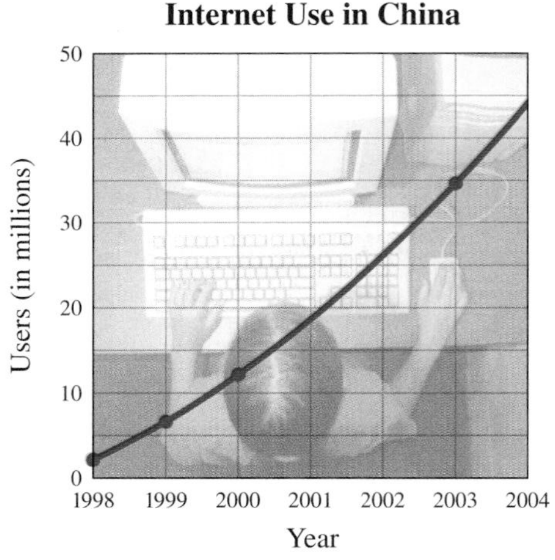

Source: Data from *USA Today*, May 9, 2000.

42. *Stopping distances* The stopping distance needed for a 3000-pound car to come to a complete stop when traveling at the indicated speeds is shown in the figure.

a. Use the graph to esimate the number of feet it takes the car to stop when traveling at 85 miles per hour. 550 ft

b. Use the quadratic polynomial $D(t) = 0.05s^2 + 2.2s + 0.75$, where s is speed in mi/hr, to approximate the number of feet it takes for the car to stop when traveling at 85 mi/hr. $D(85) = 549$ ft

Auto Stopping Distance

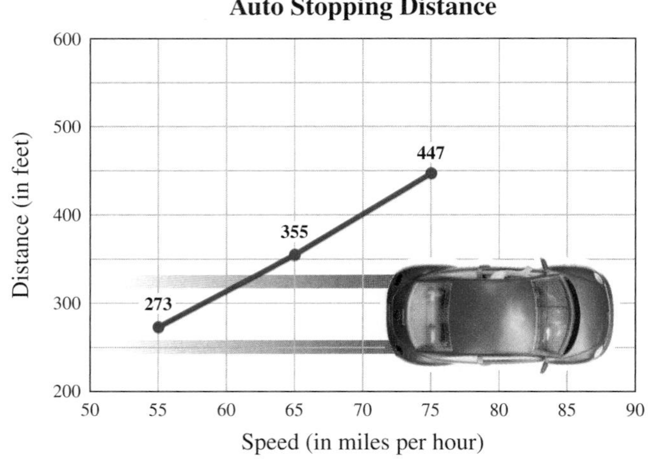

Source: Data from *USA Today*/Foundation for Traffic Safety.

SKILL CHECKER

Try the Skill Checker Exercises so you'll be ready for the next section.

Find:

43. $-3ab + (-4ab)$ $-7ab$

44. $-8a^2b + (-5a^2b)$ $-13a^2b$

45. $-3x^2y + 8x^2y - 2x^2y$ $3x^2y$

46. $-2xy^2 + 7xy^2 - 9xy^2$ $-4xy^2$

47. $5xy^2 - (-3xy^2)$ $8xy^2$

48. $7x^2y - (-8x^2y)$ $15x^2y$

USING YOUR KNOWLEDGE

Faster and Faster Polynomials

We've already stated that if an object is simply *dropped* from a certain height, its velocity after t seconds is given by $-32t$ feet per second. What will happen if we actually *throw* the object down with an initial velocity, say v_0? Since the velocity $-32t$ is being helped by the velocity v_0, the new final velocity will be given by $-32t + v_0$ (v_0 is *negative* if the object is thrown *downward*).

49. Find the velocity after t seconds have elapsed of a ball thrown downward with an initial velocity of 10 feet per second. $(-32t - 10)$ ft/sec

50. What will be the velocity of the ball in Problem 49 after

a. 1 second? -42 ft/sec **b.** 2 seconds? -74 ft/sec

51. In the metric system, the velocity after t seconds of an object thrown downward with an initial velocity v_0 is given by

$$-9.8t + v_0 \quad \text{(meters)}$$

What would be the velocity of a ball thrown downward with an initial velocity of 2 meters per second after

a. 1 second? -11.8 m/sec

b. 2 seconds? -21.6 m/sec

52. The height of an object after t seconds have elapsed depends on two factors: the initial velocity v_0 and the height s_0 from which the object is thrown. The polynomial giving this height is

$$-16t^2 + v_0t + s_0 \quad \text{(feet)}$$

where v_0 is the initial velocity and s_0 is the height from which the object is thrown. What would be the height of a ball thrown downward from a 300-foot tower with an initial velocity of 10 feet per second after

a. 1 second? 274 ft **b.** 2 seconds? 216 ft

WRITE ON

53. Write your own definition of a polynomial. Answers may vary.

54. Is $x^2 + \frac{1}{x} + 2$ a polynomial? Why or why not? No; division by x.

55. Is $x^{-2} + x + 3$ a polynomial? Why or why not? No; negative exponent on x.

56. Explain how to find the degree of a polynomial in one variable. Answers may vary.

57. The degree of x^4 is 4. What is the degree of 7^4? Why? 0; $7^4 = 2401x^0$

58. What does "evaluate a polynomial" mean? To find the value of the polynomial for a given value of the variable.

MASTERY TEST

If you know how to do these problems, you have learned your lesson!

59. Evaluate $2x^2 - 3x + 10$ when $x = 2$. 12

60. If $P(x) = 3x^3 - 7x + 9$, find $P(3)$. 69

61. When $t = 2.5$, what is the value of $-16t^2 + 118$? 18

Find the degree of:

62. $-5y - 3$ 1

63. $4x^2 - 5x^3 + x^8$ 8

64. -9 0

65. 0 No degree

Write in descending order:

66. $-2x^4 + 5x - 3x^2 + 9$ $-2x^4 - 3x^2 + 5x + 9$

67. $-8 + 5x^2 - 3x$ $5x^2 - 3x - 8$

Classify as a monomial, binomial, trinomial, or polynomial:

68. $-4t + t^2 - 8$ Trinomial

69. $-5y$ Monomial

70. $278 + 6x$ Binomial

71. $2x^3 - x^2 + x - 1$ Polynomial

72. The amount of waste recovered (in millions of tons) in the United States can be approximated by $R(t) = 0.04t^2 - 0.59t + 7.42$, where t is the number of years after 1960.

a. How many million tons were recovered in 1960? 7.42 million tons

b. How many million tons would you predict will be recovered in the year 2010? 77.92 million tons

73. Refer to Example 7.

a. Use the graph to find the BAL for a female after 2 hours. 0.115

b. Use the graph to find the BAL for a male after 2 hours. 0.105

c. Evaluate $y = -0.0257x + 0.1663$ for $x = 2$. Is the answer close to that of part **a**? 0.1149; very close

d. Evaluate $y = -0.0226x + 0.1509$ for $x = 2$. Is the answer close to that of part **b**? 0.1057; very close

4.5 ADDITION AND SUBTRACTION OF POLYNOMIALS

To Succeed, Review How To . . .

1. Add and subtract like terms (pp. 98–101).
2. Remove parentheses in expressions preceded by a minus sign (p. 101).

Objectives

A Add polynomials.

B Subtract polynomials.

C Find areas by adding polynomials.

D Solve applications.

GETTING STARTED

Wasted Waste

The annual amount of waste (in millions of tons) generated in the United States is approximated by $G(t) = -0.001t^3 + 0.06t^2 + 2.6t + 88.6$, where t is the number of years after 1960. How much of this waste is recovered? That amount can be approximated by $R(t) = 0.06t^2 - 0.59t + 6.4$. From these two approximations, we can estimate that the amount of waste actually "wasted" (not recovered) is $G(t) - R(t)$. To find this difference, we simply subtract like terms. To make the procedure more familiar, we write it in columns:

Waste Generated

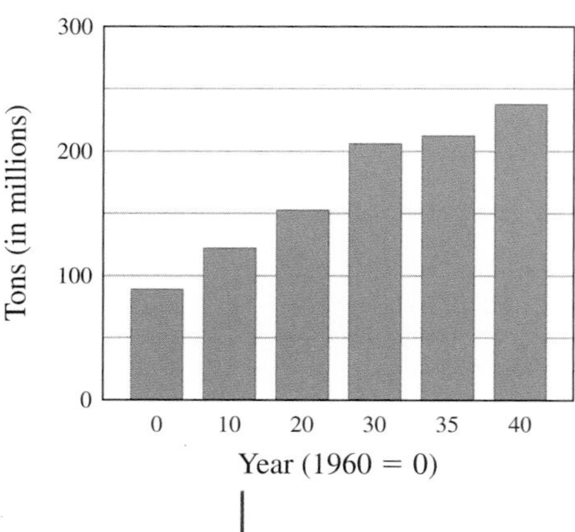

Waste Recovered

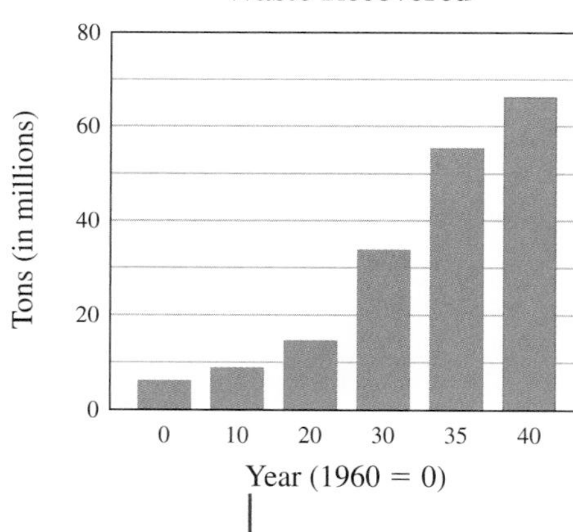

$$\begin{array}{rrrrr} G(t) = & -0.001t^3 & + 0.06t^2 & + 2.6t & + 88.6 \\ \underline{(-)\ R(t)} = (-) & & 0.06t^2 & - 0.59t & + 6.4 \\ \hline & -0.001t^3 & + & + 3.19t & + 82.2 \end{array}$$

Thus the amount of waste generated and *not* recovered is

$$G(t) - R(t) = -0.001t^3 + 3.19t + 82.2.$$

Let's see what this means in millions of tons. Since t is the number of years after 1960, $t = 0$ in 1960, and the amount of waste generated, the amount of waste recycled, and the amount of waste not recovered are as follows:

$$G(0) = -0.001(0)^3 + 0.06(0)^2 + 2.6(0) + 88.6 = 88.6 \quad \text{(million tons)}$$

$$R(0) = 0.06(0)^2 - 0.59(0) + 6.4 = 6.4 \quad \text{(million tons)}$$

$$G(0) - R(0) = 88.6 - 6.4 = 82.2 \quad \text{(million tons)}$$

As you can see, there is much more material *not* recovered than material recovered. How can we find out whether the situation is changing? One way is to predict how much waste will be produced and how much recovered, say, in the year 2010. The amount can be approximated by $G(50) - R(50)$. Then we find out if, percentagewise, the situation is getting better. In 1960, the percent of materials recovered was 6.4/88.6, or about 7.2%. What percent would it be in the year 2010? In this section we will learn how to add and subtract polynomials and use these ideas to solve applications.

A Adding Polynomials

The addition of monomials and polynomials is a matter of combining like terms. For example, suppose we wish to add $3x^2 + 7x - 3$ and $5x^2 - 2x + 9$; that is, we wish to find

$$(3x^2 + 7x - 3) + (5x^2 - 2x + 9)$$

Using the commutative, associative, and distributive properties, we write

$$\begin{aligned}(3x^2 + 7x - 3) + (5x^2 - 2x + 9) &= (3x^2 + 5x^2) + (7x - 2x) + (-3 + 9)\\ &= (3 + 5)x^2 + (7 - 2)x + (-3 + 9)\\ &= 8x^2 + 5x + 6\end{aligned}$$

Similarly, the sum of $4x^3 + \frac{3}{7}x^2 - 2x + 3$ and $6x^3 - \frac{1}{7}x^2 + 9$ is written as

$$\begin{aligned}&\left(4x^3 + \frac{3}{7}x^2 - 2x + 3\right) + \left(6x^3 - \frac{1}{7}x^2 + 9\right)\\ &\quad= (4x^3 + 6x^3) + \left(\frac{3}{7}x^2 - \frac{1}{7}x^2\right) + (-2x) + (3 + 9)\\ &\quad= 10x^3 + \frac{2}{7}x^2 - 2x + 12\end{aligned}$$

In both examples, the polynomials have been written in descending order for convenience in combining like terms.

Web It

To practice combining like terms, go to link 4-5-1 on the Bello Website at mhhe.com/bello.

EXAMPLE 1 Adding polynomials

Add: $3x + 7x^2 - 7$ and $-4x^2 + 9 - 3x$

SOLUTION We first write both polynomials in descending order and then combine like terms to obtain

$$\begin{aligned}(7x^2 + 3x - 7) + (-4x^2 - 3x + 9) &= (7x^2 - 4x^2) + (3x - 3x) + (-7 + 9)\\ &= 3x^2 + 0 + 2\\ &= 3x^2 + 2\end{aligned}$$

PROBLEM 1

Add: $5x + 8x^2 - 3$ and $-3x^2 + 8 - 5x$

As in arithmetic, the addition of polynomials can be done by writing the polynomials in descending order and then placing like terms in columns. In arithmetic, you add 345 and 678 by writing the numbers in a column:

$$\begin{array}{r} +345\\ \underline{+678} \end{array}$$

Units
Tens
Hundreds

Thus to add $4x^3 + 3x - 7$ and $7x - 3x^3 + x^2 + 9$, we first write both polynomials in descending order with like terms in the same column, leaving space for any missing terms. We then add the terms in each of the columns:

The x^2 term is missing in $4x^3 + 3x - 7$.

$$\begin{array}{rrrr} 4x^3 & & +\ 3x & -\ 7\\ -3x^3 & +\ x^2 & +\ 7x & +\ 9\\ \hline x^3 & +\ x^2 & +\ 10x & +\ 2 \end{array}$$

Answer

1. $5x^2 + 5$

EXAMPLE 2 **More practice adding polynomials**

Add: $-3x + 7x^2 - 2$ and $-4x^2 - 3 + 5x$

SOLUTION We first write both polynomials in descending order, place like terms in a column, and then add as shown:

$$\begin{array}{r} 7x^2 - 3x - 2 \\ \underline{-4x^2 + 5x - 3} \\ 3x^2 + 2x - 5 \end{array}$$

Horizontally, we write:

$$\begin{aligned} &(7x^2 - 3x - 2) + (-4x^2 + 5x - 3) \\ &= \underbrace{(7x^2 - 4x^2)} + \underbrace{(-3x + 5x)} + \underbrace{(-2 - 3)} \\ &= 3x^2 + 2x + (-5) \\ &= 3x^2 + 2x - 5 \end{aligned}$$

PROBLEM 2

Add: $-5y + 8y^2 - 3$ and $-5y^2 - 4 + 6y$

B Subtracting Polynomials

To subtract polynomials, we first recall that

$$a - (b + c) = a - b - c$$

To remove the parentheses from an expression preceded by a minus sign, we must change the sign of each term *inside* the parentheses. This is the same as multiplying each term inside the parentheses by -1. Thus

$$\begin{aligned} (3x^2 - 2x + 1) - (4x^2 + 5x + 2) &= 3x^2 - 2x + 1 - 4x^2 - 5x - 2 \\ &= (3x^2 - 4x^2) + (-2x - 5x) + (1 - 2) \\ &= -x^2 - 7x + (-1) \\ &= -x^2 - 7x - 1 \end{aligned}$$

Here's how we do it using columns:

$$\begin{array}{r} 3x^2 - 2x + 1 \\ \underline{(-)\ 4x^2 + 5x + 2} \end{array} \xrightarrow{\text{is written}} \begin{array}{r} 3x^2 - 2x + 1 \\ \underline{(+)\ -4x^2 - 5x - 2} \\ -x^2 - 7x - 1 \end{array}$$

Note that we changed the sign of *every* term in $4x^2 + 5x + 2$ and wrote $-4x^2 - 5x - 2$.

Web It

To practice with the subtraction of polynomials, go to link 4-5-2 on the Bello Website at mhhe.com/bello and enter your problem. Be careful with parentheses!

For a more conventional approach and different strategies, try link 4-5-3.

Teaching Tip

The minus sign in $a - (b + c)$ is used to mean "take the opposite sign" of all the terms inside the parentheses.

NOTE

"Subtract b from a" means to find $a - b$.

EXAMPLE 3 **Subtracting polynomials**

Subtract $4x - 3 + 7x^2$ from $5x^2 - 3x$.

SOLUTION We first write the problem in columns, then change the signs and add:

$$\begin{array}{r} 5x^2 - 3x \\ \underline{(-)7x^2 + 4x - 3} \end{array} \xrightarrow{\text{is written}} \begin{array}{r} 5x^2 - 3x \\ \underline{(+)-7x^2 - 4x + 3} \\ -2x^2 - 7x + 3 \end{array}$$

Thus the answer is $-2x^2 - 7x + 3$.

PROBLEM 3

Subtract $5y - 4 + 8y^2$ from $6y^2 - 4y$.

Answers

2. $3y^2 + y - 7$
3. $-2y^2 - 9y + 4$

To do it horizontally, we write

$$(5x^2 - 3x) - (7x^2 + 4x - 3)$$

$$= 5x^2 - 3x - 7x^2 - 4x + 3$$ Change the sign of every term in $7x^2 + 4x - 3$.

$$= (5x^2 - 7x^2) + (-3x - 4x) + 3$$ Use the commutative and associative properties.

$$= -2x^2 - 7x + 3$$

Teaching Tip

Do a numerical example to illustrate the phrase "subtract from."

Example:

Subtract 7 from 9 means $9 - 7$.

Just as in arithmetic, we can add or subtract more than two polynomials. For example, to add the polynomials $-7x + x^2 - 3$, $6x^2 - 8 + 2x$, and $3x - x^2 + 5$, we simply write each of the polynomials in descending order with like terms in the same column and add:

$$\begin{array}{r} x^2 - 7x - 3 \\ 6x^2 + 2x - 8 \\ -x^2 + 3x + 5 \\ \hline 6x^2 - 2x - 6 \end{array}$$

Or, horizontally, we write

$$(x^2 - 7x - 3) + (6x^2 + 2x - 8) + (-x^2 + 3x + 5)$$
$$= (x^2 + 6x^2 - x^2) + (-7x + 2x + 3x) + (-3 - 8 + 5)$$
$$= 6x^2 + (-2x) + (-6)$$
$$= 6x^2 - 2x - 6$$

EXAMPLE 4 **More practice adding polynomials**

Add: $x^3 + 2x - 3x^2 - 5$, $-8 + 2x - 5x^2$, and $7x^3 - 4x + 9$

SOLUTION We first write all the polynomials in descending order with like terms in the same column and then add:

$$\begin{array}{rrrr} x^3 & - 3x^2 & + 2x & - 5 \\ & - 5x^2 & + 2x & - 8 \\ 7x^3 & & - 4x & + 9 \\ \hline 8x^3 & - 8x^2 & & - 4 \end{array}$$

Horizontally, we have

$$(x^3 - 3x^2 + 2x - 5) + (-5x^2 + 2x - 8) + (7x^3 - 4x + 9)$$
$$= (x^3 + 7x^3) + (-3x^2 - 5x^2) + (2x + 2x - 4x) + (-5 - 8 + 9)$$
$$= 8x^3 + (-8x^2) + 0x + (-4)$$
$$= 8x^3 - 8x^2 - 4$$

PROBLEM 4

Add: $y^3 + 3y - 4y^2 - 6$, $-9 + 3y - 6y^2$, and $6y^3 - 5y + 8$

C Finding Areas

Addition of polynomials can be used to find the sum of the areas of several rectangles. To find the total area of the shaded rectangles, add the individual areas.

2 A 5 | 3 B 2 | 2 C 3

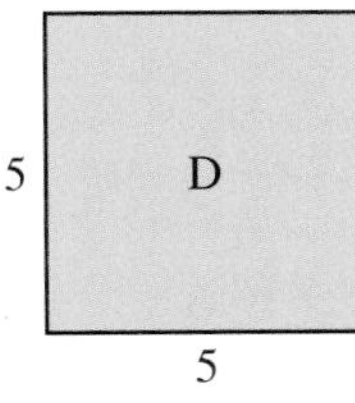

Answer

4. $7y^3 - 10y^2 + y - 7$

Web It

To practice adding polynomials using sums of areas, go back to link 4-5-3 on the Bello Website at mhhe.com/bello.

Since the area of a rectangle is the product of its length and its width, we have

$$\underbrace{5 \cdot 2}_{\text{Area of A}} + \underbrace{2 \cdot 3}_{\text{Area of B}} + \underbrace{3 \cdot 2}_{\text{Area of C}} + \underbrace{5 \cdot 5}_{\text{Area of D}}$$

$$= 10 + 6 + 6 + 25$$

Thus

$$10 + 6 + 6 + 25 = 47 \quad \text{(square units)}$$

This same procedure can be used when some of the lengths are represented by variables, as shown in the next example.

EXAMPLE 5 Finding sums of areas

Find the sum of the areas of the shaded rectangles:

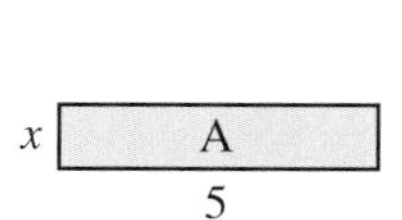

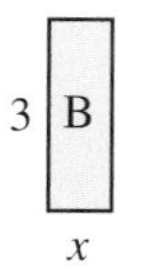

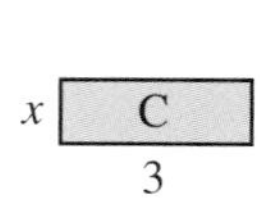

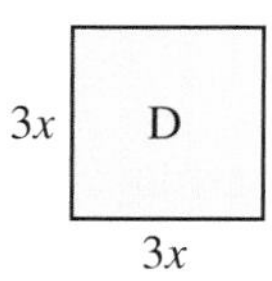

SOLUTION The total area in square units is

$$\underbrace{5x}_{\text{Area of A}} + \underbrace{3x}_{\text{Area of B}} + \underbrace{3x}_{\text{Area of C}} + \underbrace{(3x)^2}_{\text{Area of D}}$$

$$11x + 9x^2$$

or

$$9x^2 + 11x \quad \text{In descending order}$$

PROBLEM 5

Find the sum of the areas of the shaded rectangles:

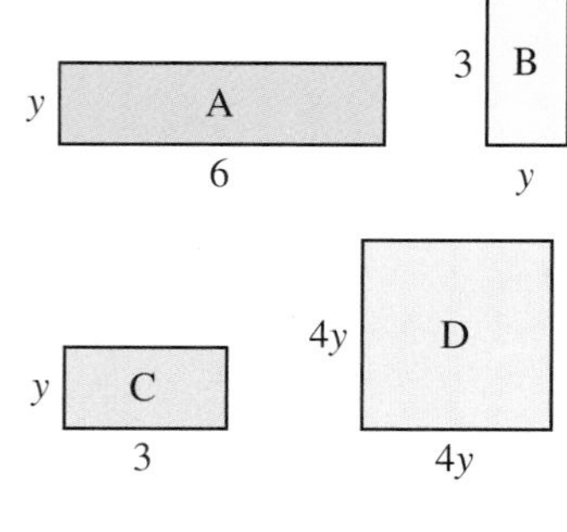

D Solving Applications

The concepts we've just studied can be used to examine important issues in American life. Let's see how we can use these concepts to study health care.

EXAMPLE 6 Paying to get well

Based on Social Security Administration statistics, the annual amount of money spent on hospital care by the average American can be approximated by

$$H(t) = 6.6t^2 + 30t + 682 \quad \text{(dollars)}$$

while the annual amount spent for doctors' services is approximated by

$$D(t) = 41t + 293$$

where t is the number of years after 1985.

a. Find the annual amount of money spent for doctors' services and hospital care.

b. How much was spent for doctors' services in 1985?

c. How much was spent on hospital care in 1985?

d. How much would you predict will be spent on hospital care and doctors' services in the year 2005?

PROBLEM 6

How much was spent on hospital care and doctors' services in 2000?

Answers

5. $12y + 16y^2$ **6.** \$3525

SOLUTION

a. To find the annual amount spent for doctors' services and hospital care, we find $D(t) + H(t)$:

$$\begin{aligned} D(t) &= \phantom{6.6t^2 + {}} 41t + 293 \\ H(t) &= 6.6t^2 + 30t + 682 \\ \hline & 6.6t^2 + 71t + 975 \end{aligned}$$

Thus the annual amount spent for doctors' services and hospital care is $6.6t^2 + 71t + 975$.

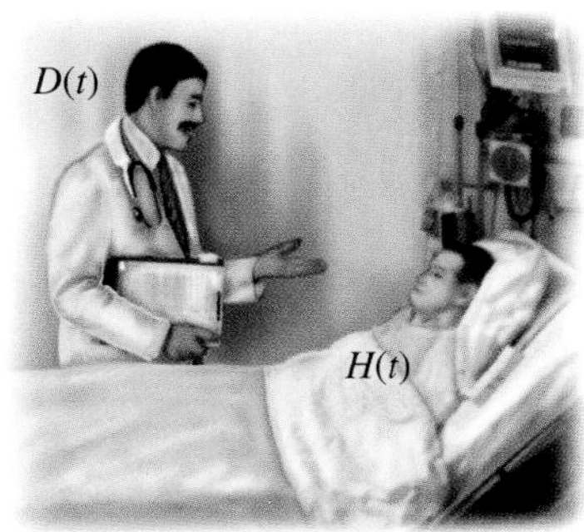

b. To find how much was spent for doctors' services in 1985 ($t = 0$), find $D(0) = 41(0) + 293$. Thus \$293 was spent for doctors' services in 1985.

c. The amount spent on hospital care in 1985 ($t = 0$) was $H(0) = 6.6(0)^2 + 30(0) + 682 = 682$. Thus \$682 was spent for hospital care in 1985.

d. The year 2005 corresponds to $t = 2005 - 1985 = 20$. To predict the amount to be spent on hospital care and doctors' services in the year 2005, we find $6.6(20)^2 + 71(20) + 975 = \$5035$. Note that it's easier to evaluate $D(t) + H(t) = 6.6t^2 + 71t + 975$ for $t = 20$ than it is to evaluate $D(20)$ and $H(20)$ and then add.

EXAMPLE 7 More drinking

In Example 7 of Section 4.4, we introduced the polynomial equations $y = -0.0226x + 0.1509$ and $y = -0.0257x + 0.1663$ approximating the blood alcohol level (BAL) for a 150-pound male or female, respectively. In these equations x represents the time since consuming 3 ounces of alcohol.

It is known that the burn-off rate of alcohol is 0.015 per hour (that is, the BAL is reduced by 0.015 per hour if no additional alcohol is consumed). Find a polynomial that would approximate the BAL for a male x hours after consuming the 3 ounces of alcohol.

SOLUTION The initial BAL is $-0.0226x + 0.1509$, but this level is *decreased* by 0.015 each hour. Thus, the actual BAL after x hours is $-0.0226x + 0.1509 - 0.015x$, or $-0.0376x + 0.1509$. (Check with the calculator at link 4-4-5 on the Bello Website at mhhe.com/bello.)

PROBLEM 7

Find a polynomial that would approximate the BAL for a female x hours after consuming 3 ounces of alcohol.

Answer

7. $-0.0407x + 0.1663$

Exercises 4.5

A In Problems 1–30, add as indicated.

1. $(5x^2 + 2x + 5) + (7x^2 + 3x + 1)$ $12x^2 + 5x + 6$

2. $(3x^2 - 5x - 5) + (9x^2 + 2x + 1)$ $12x^2 - 3x - 4$

3. $(-3x + 5x^2 - 1) + (-7 + 2x - 7x^2)$
$-2x^2 - x - 8$

4. $(3 - 2x^2 + 7x) + (-6 + 2x^2 - 5x)$
$2x - 3$

5. $(2x + 5x^2 - 2) + (-3 + 5x - 8x^2)$
$-3x^2 + 7x - 5$

6. $-3x - 2 + 3x^2$ and $-4 + 5x - 6x^2$
$-3x^2 + 2x - 6$

7. $-2 + 5x$ and $-3 - x^2 - 5x$
$-x^2 - 5$

8. $-4x + 2 - 6x^2$ and $2 + 5x$
$-6x^2 + x + 4$

9. $x^3 - 2x + 3$ and $-2x^2 + x - 5$
$x^3 - 2x^2 - x - 2$

10. $x^4 - 3 + 2x - 3x^3$ and $3x^4 - 2x^2 + 5 - x$
$4x^4 - 3x^3 - 2x^2 + x + 2$

11. $-6x^3 + 2x^4 - x$ and $2x^2 + 5 + 2x - 2x^3$
$2x^4 - 8x^3 + 2x^2 + x + 5$

12. $\frac{1}{2}x^3 + x^2 - \frac{1}{5}x$ and $\frac{3}{5}x + \frac{1}{2}x^3 - 3x^2$
$x^3 - 2x^2 + \frac{2}{5}x$

13. $\frac{1}{3} - \frac{2}{5}x^2 + \frac{3}{4}x$ and $\frac{1}{4}x - \frac{1}{5}x^2 + \frac{2}{3}$
$-\frac{3}{5}x^2 + x + 1$

14. $0.3x - 0.1 - 0.4x^2$ and $0.1x^2 - 0.1x + 0.6$
$-0.3x^2 + 0.2x + 0.5$

15. $0.2x - 0.3 + 0.5x^2$ and $-\frac{1}{10} + \frac{1}{10}x - \frac{1}{10}x^2$ $0.4x^2 + 0.3x - 0.4$

16.
$$\begin{array}{rr} & -x^2 + 5x + 2 \\ (+) & 3x^2 - 7x - 2 \\ \hline & 2x^2 - 2x \end{array}$$

17.
$$\begin{array}{rr} & -3x^2 + 2x - 4 \\ (+) & x^2 - 4x + 7 \\ \hline & -2x^2 - 2x + 3 \end{array}$$

18.
$$\begin{array}{rr} & -2x^4 \qquad\quad + 2x - 1 \\ (+) & - x^3 - 3x + 5 \\ \hline & -2x^4 - x^3 - \ x + 4 \end{array}$$

19.
$$\begin{array}{rr} & 3x^4 \qquad - 3x + 4 \\ (+) & x^3 - 2x - 5 \\ \hline & 3x^4 + x^3 - 5x - 1 \end{array}$$

20.
$$\begin{array}{rr} & -3x^4 \qquad\quad + 2x^2 - \ x + 5 \\ (+) & - 2x^3 \qquad\quad + 5x - 7 \\ \hline & -3x^4 - 2x^3 + 2x^2 + 4x - 2 \end{array}$$

21.
$$\begin{array}{rr} & -5x^4 \qquad\quad - 5x^2 + 3 \\ (+) & 5x^3 + 3x^2 - 5 \\ \hline & -5x^4 + 5x^3 - 2x^2 - 2 \end{array}$$

22.
$$\begin{array}{rr} & 3x^3 \qquad\quad + \ x - 1 \\ & x^2 - 2x + 5 \\ (+) & 5x^3 \qquad\quad - \ x \\ \hline & 8x^3 + x^2 - 2x + 4 \end{array}$$

23.
$$\begin{array}{rr} & 5x^3 - \ x^2 \qquad\quad - 3 \\ & 5x + 9 \\ (+) & - 3x^2 \qquad\quad - 7 \\ \hline & 5x^3 - 4x^2 + 5x - 1 \end{array}$$

24.
$$\begin{array}{l} -\frac{1}{3}x^3 \qquad\quad -\frac{1}{2}x + 5 \\ \qquad -\frac{1}{5}x^2 + \frac{1}{2}x - 1 \\ (+)\ \frac{2}{3}x^3 \qquad\quad +\ x - 2 \\ \hline \frac{1}{3}x^3 - \frac{1}{5}x^2 + x + 2 \end{array}$$

25.
$$\begin{array}{l} -\frac{2}{7}x^3 + \frac{1}{6}x^2 \qquad + 2 \\ \frac{1}{7}x^3 \qquad\qquad + 5x - 3 \\ (+) \qquad -\frac{5}{6}x^2 \qquad + 1 \\ \hline -\frac{1}{7}x^3 - \frac{2}{3}x^2 + 5x \end{array}$$

26.
$$\begin{array}{l} \qquad -\frac{1}{8}x^2 - \frac{1}{3}x + \frac{1}{5} \\ -x^3 + \frac{3}{8}x^2 \qquad -\frac{2}{5} \\ (+)\ -3x^3 \qquad -\frac{2}{3}x + \frac{4}{5} \\ \hline -4x^3 + \frac{1}{4}x^2 - x + \frac{3}{5} \end{array}$$

27.
$$\begin{array}{l} -\frac{1}{7}x^3 \qquad\qquad\qquad + 2 \\ \qquad -\frac{1}{9}x^2 - x - 3 \\ (+)\ -\frac{2}{7}x^3 + \frac{2}{9}x^2 + 2x - 5 \\ \hline -\frac{3}{7}x^3 + \frac{1}{9}x^2 + x - 6 \end{array}$$

28.
$$\begin{array}{l} -2x^4 + 5x^3 - 2x^2 + 3x - 5 \\ \qquad 8x^3 \qquad - 2x + 5 \\ -x^4 \qquad + 3x^2 - x - 2 \\ (+) \qquad 6x^3 \qquad + 2x + 5 \\ \hline -3x^4 + 19x^3 + x^2 + 2x + 3 \end{array}$$

29.
$$\begin{array}{l} \qquad -6x^3 + 2x^2 \qquad + 1 \\ -x^4 + 3x^3 - 5x^2 + 3x \\ \qquad -x^3 \qquad - 7x + 2 \\ (+)\ -3x^4 \qquad\qquad + 3x - 1 \\ \hline -4x^4 - 4x^3 - 3x^2 - x + 2 \end{array}$$

30.
$$\begin{array}{l} \qquad -3x^4 \qquad + 2x^2 \qquad - 5 \\ x^5 + x^4 - 2x^3 + 7x^2 + 5x \\ \qquad 2x^4 \qquad - 2x^2 \qquad + 7 \\ (+)\ 7x^5 \qquad + 2x^3 \qquad - 2x \\ \hline 8x^5 + \qquad\qquad 7x^2 + 3x + 2 \end{array}$$

B In Problems 31–50, subtract as indicated.

31. $(7x^2 + 2) - (3x^2 - 5)$
$4x^2 + 7$

32. $(8x^2 - x) - (7x^2 + 3x)$
$x^2 - 4x$

33. $(3x^2 - 2x - 1) - (4x^2 + 2x + 5)$
$-x^2 - 4x - 6$

34. $(-3x + x^2 - 1) - (5x + 1 - 3x^2)$
$4x^2 - 8x - 2$

35. $(-1 + 7x^2 - 2x) - (5x + 3x^2 - 7)$
$4x^2 - 7x + 6$

36. $(7x^3 - x^2 + x - 1) - (2x^2 + 3x + 6)$
$7x^3 - 3x^2 - 2x - 7$

37. $(5x^2 - 2x + 5) - (3x^3 - x^2 + 5)$
$-3x^3 + 6x^2 - 2x$

38. $(3x^2 - x - 7) - (5x^3 + 5 - x^2 + 2x)$
$-5x^3 + 4x^2 - 3x - 12$

39. $(6x^3 - 2x^2 - 3x + 1) - (-x^3 - x^2 - 5x + 7)$
$7x^3 - x^2 + 2x - 6$

40. $(x - 3x^2 + x^3 + 9) - (-8 + 7x - x^2 + x^3)$
$-2x^2 - 6x + 17$

41.
$$\begin{array}{r} 6x^2 - 3x + 5 \\ (-)\ 3x^2 + 4x - 2 \\ \hline 3x^2 - 7x + 7 \end{array}$$

42.
$$\begin{array}{r} 7x^2 + 4x - 5 \\ (-)\ 9x^2 - 2x + 5 \\ \hline -2x^2 + 6x - 10 \end{array}$$

43.
$$\begin{array}{r} 3x^2 - 2x - 1 \\ (-)\ 3x^2 - 2x - 1 \\ \hline 0 \end{array}$$

44.
$$\begin{array}{r} 5x^2 \qquad\ \ - 1 \\ (-)\ 3x^2 - 2x + 1 \\ \hline 2x^2 + 2x - 2 \end{array}$$

45.
$$\begin{array}{r} 4x^3 \qquad\quad - 2x + 5 \\ (-) \qquad 3x^2 + 5x - 1 \\ \hline 4x^3 - 3x^2 - 7x + 6 \end{array}$$

46.
$$\begin{array}{r} -3x^2 + 5x - 2 \\ (-)\ x^3 - 2x^2 \qquad\ + 5 \\ \hline -x^3 - x^2 + 5x - 7 \end{array}$$

47.
$$\begin{array}{r} 3x^3 \qquad\qquad - 2 \\ (-) \qquad 2x^2 - x + 6 \\ \hline 3x^3 - 2x^2 + x - 8 \end{array}$$

48.
$$\begin{array}{r} x^2 - 2x + 1 \\ (-)\ -3x^3 + x^2 + 5x - 2 \\ \hline 3x^3 \qquad\quad - 7x + 3 \end{array}$$

49.
$$\begin{array}{r} -5x^3 \qquad\quad + x - 2 \\ (-) \qquad 5x^2 - 3x + 7 \\ \hline -5x^3 - 5x^2 + 4x - 9 \end{array}$$

50.
$$\begin{array}{r} 6x^3 \qquad\quad + 2x - 5 \\ (-) \qquad -3x^2 - x \qquad \\ \hline 6x^3 + 3x^2 + 3x - 5 \end{array}$$

C In Problems 51–55, find the sum of the areas of the shaded rectangles.

51.

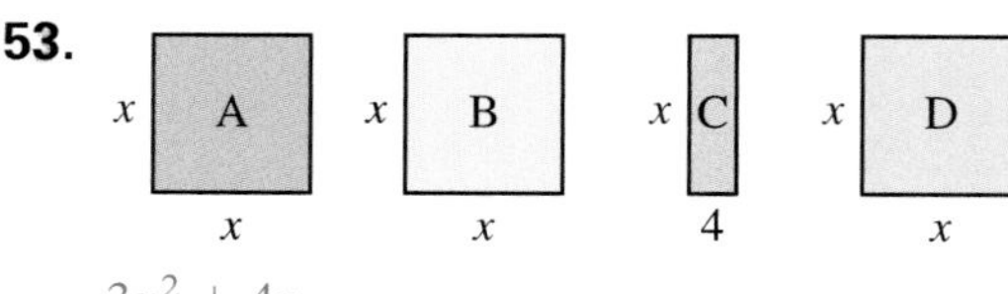

$2x^2 + 6x$

52.

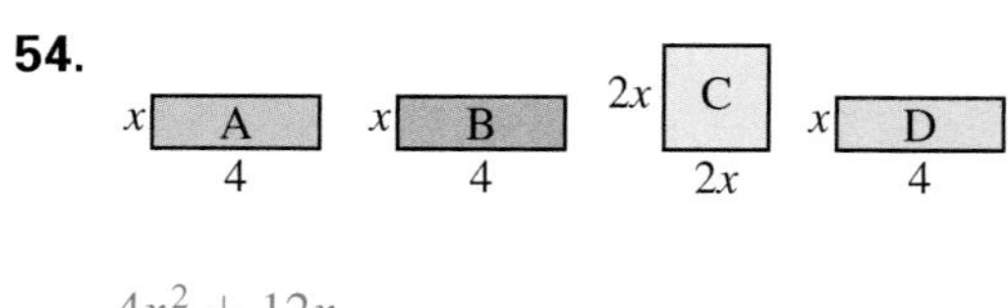

$x^2 + 15x$

53.

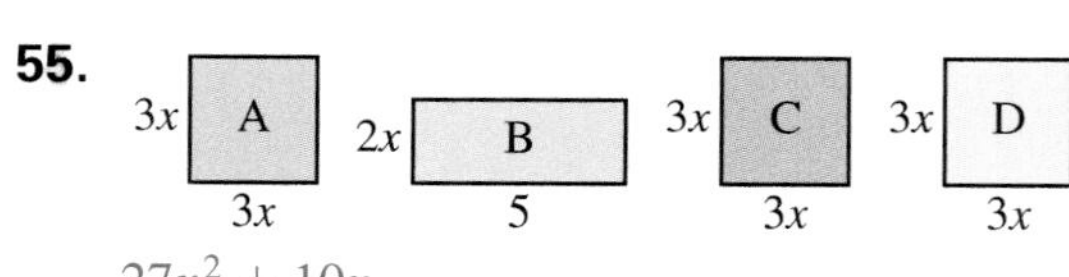

$3x^2 + 4x$

54.

$4x^2 + 12x$

55.

$27x^2 + 10x$

APPLICATIONS

56. *Annual tuna consumption* Based on Census Bureau estimates, the annual consumption of canned tuna (in pounds) per person is $C(t) = 3.87 - 0.13t$, while the consumption of turkey (in pounds) is $T(t) = -0.15t^2 + 0.8t + 13$, where t is the number of years after 1989.

a. How many pounds of canned tuna and how many pounds of turkey were consumed per person in 1989? Tuna: 3.87 lb; Turkey: 13 lb

b. Find the difference between the amount of turkey consumed and the amount of canned tuna consumed. $T(t) - C(t) = -0.15t^2 + 0.93t + 9.13$

c. Use your answer to part **b** to find the difference in the amount of turkey and the amount of canned tuna you would expect to be consumed in the year 2000. 1.21 lb

57. *Annual poultry consumption* Poultry products consist of turkey and chicken. Annual chicken consumption per person (in pounds) is $C(t) = 0.05t^3 - 0.7t^2 - t + 36$ and annual turkey consumption (in pounds) is $T(t) = -0.15t^2 + 1.6t + 9$, where t is the number of years after 1985.

a. Find the total annual poultry consumption per person. $P(t) = 0.05t^3 - 0.85t^2 + 0.6t + 45$

b. How much poultry was consumed per person in 1995? How much poultry do you predict would be consumed in 2005? $P(10) = 16$ lb; $P(20) = 117$ lb

58. *College costs* How much are you paying for tuition and fees? In a four-year public institution, the amount $T(t)$ you pay for tuition and fees (in dollars) can be approximated by $T(t) = 45t^2 + 110t + 3356$, where t is the number of years after 2000 ($2000 = 0$).

a. What would you predict tuition and fees to be in 2005? \$5031

b. The cost of books t years after 2000 can be approximated by $B(t) = 27.5t + 680$. What would be the cost of books in 2005? \$817.50

c. What polynomial would represent the cost of tuition and fees and books t years after 2000? $T(t) + B(t) = 45t^2 + 137.5t + 4036$

d. What would you predict the cost of tuition and fees and books would be in 2005? \$5848.50

59. *College expenses* The three major college expenses are: tuition and fees, books, and room and board. They can be approximated, respectively, by:

$$T(t) = 45t^2 + 110t + 3356$$
$$B(t) = 27.5t + 680$$
$$R(t) = 32t^2 + 200t + 4730$$

where t is the number of years after 2000 ($2000 = 0$).

a. Write a polynomial representing the total cost of tuition and fees, books, and room and board t years after 2000. $T(t) + B(t) + R(t) = 77t^2 + 337.5t + 8766$

b. What was the cost of tuition and fees, books, and room and board in 2000? \$8766

c. What would you predict the cost of tuition and fees, books, and room and board would be in 2005? \$12,378.50

60. *Student loans* If you are an undergraduate dependent student you can apply for a Stafford Loan. The amount of these loans can be approximated by $S(t) = 562.5t^2 + 312.5t + 2625$, where t is between 0 and 2 inclusive. If t is between 3 and 5 inclusive, then $S(t) =$ \$5500.

a. How much money can you get from a Stafford Loan the first year? $S(1) = \$3500$

b. What about the second year? $S(2) = \$5500$

c. What about the fifth year? $S(5) = \$5500$

61. *College financial aid* Assume that you have to pay tuition and fees, books, and room and board in 2003 but have your Stafford Loan to decrease expenses. (See Problems 59 and 60.) Write a polynomial that would approximate how much you would have to pay t years after 2000 (t between 0 and 2).
$T(t) + B(t) + R(t) - S(t) = -485.5t^2 + 25t + 6141$

SKILL CHECKER

Try the Skill Checker Exercises so you'll be ready for the next section.

Simplify:

62. $(-5x^3) \cdot (2x^4)$ $-10x^7$

63. $(-2x^4) \cdot (3x^5)$ $-6x^9$

64. $5(x - 3)$ $5x - 15$

65. $6(y - 4)$ $6y - 24$

66. $-3(2y - 3)$ $-6y + 9$

USING YOUR KNOWLEDGE

Business Polynomials

Polynomials are also used in business and economics. For example, the revenue R may be obtained by subtracting the cost C of the merchandise from its selling price S. In symbols, this is

$$R = S - C$$

Now the cost C of the merchandise is made up of two parts: the *variable cost* per item and the *fixed cost.* For example, if you decide to manufacture Frisbees™, you might spend \$2 per Frisbee in materials, labor, and so forth. In addition, you might have \$100 of fixed expenses. Then the cost for manufacturing x Frisbees is

Cost C of merchandise	is	cost per Frisbee	and	fixed expenses.
C	$=$	$2x$	$+$	100

If x Frisbees are then sold for \$3 each, the total selling price S is $3x$, and the revenue R would be

$$\begin{aligned} R &= S - C \\ &= 3x - (2x + 100) \\ &= 3x - 2x - 100 \\ &= x - 100 \end{aligned}$$

Thus if the selling price S is \$3 per Frisbee, the variable costs are \$2 per Frisbee, and the fixed expenses are \$100, the revenue after manufacturing x Frisbees is given by

$$R = x - 100$$

In Problems 67–69, find the revenue R for the given cost C and selling price S.

67. $C = 3x + 50$; $S = 4x$ $R = x - 50$

68. $C = 6x + 100$; $S = 8x$ $R = 2x - 100$

69. $C = 7x$; $S = 9x$ $R = 2x$

70. In Problem 68, how many items were manufactured if the revenue was zero? 50

71. If the merchant of Problem 68 suffered a \$40 loss (−\$40 revenue), how many items were produced? 30

WRITE ON

72. Write the procedure you use to add polynomials. Answers may vary.

73. Write the procedure you use to subtract polynomials. Answers may vary.

74. Explain the difference between "subtract $x^2 + 3x - 5$ from $7x^2 - 2x + 9$" and "subtract $7x^2 - 2x + 9$ from $x^2 + 3x - 5$." What is the answer in each case? Answers may vary.

75. List the advantages and disadvantages of adding (or subtracting) polynomials horizontally or in columns. Answers may vary.

MASTERY TEST

If you know how to do these problems, you have learned your lesson!

Add:

76. $3x + 3x^2 - 6$ and $-5x^2 + 10 - x$ $-2x^2 + 2x + 4$

77. $-5x + 8x^2 - 3$ and $-3x^2 + 4 + 8x$ $5x^2 + 3x + 1$

Subtract:

78. $3 - 4x^2 + 5x$ from $9x^2 - 2x$ $13x^2 - 7x - 3$

79. $9 + x^3 - 3x^2$ from $10 + 7x^2 + 5x^3$ $4x^3 + 10x^2 + 1$

80. Add $2x^3 + 3x - 5x^2 - 2$, $-6 + 5x - 2x^2$, and $6x^3 - 2x + 8$. $8x^3 - 7x^2 + 6x$

81. Find the sum of the areas of the shaded rectangles:

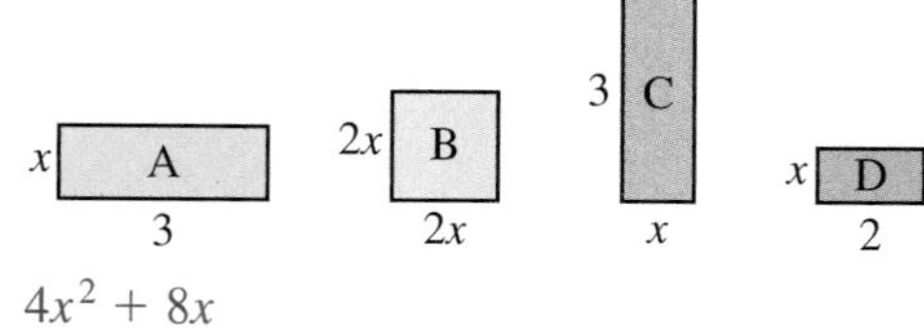

$4x^2 + 8x$

82. The number of robberies (per 100,000 population) can be approximated by $R(t) = 1.85t^2 - 19.14t + 262$, while the number of aggravated assaults is approximated by $A(t) = -0.2t^3 + 4.7t^2 - 15t + 300$, where t is the number of years after 1960.

a. Were there more aggravated assaults or more robberies per 100,000 in 1960? More aggravated assaults

b. Find the difference between the number of aggravated assaults and the number of robberies per 100,000.
$A(t) - R(t) = -0.2t^3 + 2.85t^2 + 4.14t + 38$

c. What would this difference be in the year 2000? In 2010? $A(40) - R(40) = -8036.4$; $A(50) - R(50) = -17{,}630$

4.6 MULTIPLICATION OF POLYNOMIALS

To Succeed, Review How To . . .

1. Multiply expressions (pp. 92, 98, 101, 294, 301).
2. Use the distributive property to remove parentheses in an expression (p. 101).

Objectives

A Multiply two monomials.

B Multiply a monomial and a binomial.

C Multiply two binomials using the FOIL method.

D Solve an application.

GETTING STARTED

Deflections on a Bridge

How much does the beam bend (deflect) when a car or truck goes over the bridge? There's a formula that can tell us. For a certain beam of length L, the deflection at a distance x from one end is given by

$$(x - L)(x - 2L)$$

To multiply these two binomials, we must first learn how to do several related types of multiplication.

A Multiplying Two Monomials

We've already multiplied two monomials in Section 4.1. The idea is to use the associative and commutative properties and the rules of exponents, as shown in the next example.

Web It

For practice problems and a tutorial on multiplying monomials, go to link 4-6-1 on the Bello Website at mhhe.com/bello.

EXAMPLE 1 **Multiplying two monomials**

Multiply: $(-3x^2)$ by $(2x^3)$

SOLUTION

$(-3x^2)(2x^3) = (-3 \cdot 2)(x^2 \cdot x^3)$ Use the associative and commutative properties.

$= -6x^{2+3}$ Use the rules of exponents.

$= -6x^5$

PROBLEM 1

Multiply: $(-4y^3)$ by $(5y^4)$

Answer

1. $-20y^7$

B Multiplying a Monomial and a Binomial

In Sections 1.6 and 1.7, we also multiplied a monomial and a binomial. The procedure was based on the distributive property, as shown next.

EXAMPLE 2 **Multiplying a monomial by a binomial**

Remove parentheses (simplify):

a. $5(x - 2y)$ **b.** $(x^2 + 2x)3x^4$

SOLUTION

a. $5(x - 2y) = 5x - 5 \cdot 2y$
$= 5x - 10y$

b. $(x^2 + 2x)3x^4 = x^2 \cdot 3x^4 + 2x \cdot 3x^4$ Since $(a + b)c = ac + bc$
$= 3 \cdot x^2 \cdot x^4 + 2 \cdot 3 \cdot x \cdot x^4$
$= 3x^6 + 6x^5$

PROBLEM 2

Simplify:

a. $4(a - 3b)$ **b.** $(a^2 + 3a)4a^5$

NOTE

You can use the commutative property first and write

$$(x^2 + 2x)3x^4 = 3x^4(x^2 + 2x)$$
$$= 3x^6 + 6x^5 \quad \text{(Same answer!)}$$

C Multiplying Two Binomials Using the FOIL Method

Another way to multiply $(x + 2)(x + 3)$ is to use the distributive property $a(b + c) = ab + ac$. Think of $x + 2$ as a, which makes x like b and 3 like c. Here's how it's done.

$$\underbrace{(x+2)}_{a}\,(\underbrace{x}_{b} + \underbrace{3}_{c}) = \underbrace{(x+2)}_{a}\underbrace{x}_{b} + \underbrace{(x+2)}_{a}\underbrace{3}_{c}$$
$$= x \cdot x + 2 \cdot x + x \cdot 3 + 2 \cdot 3$$
$$= x^2 + 2x + 3x + 6$$
$$= x^2 + 5x + 6$$

Similarly,

$$(x - 3)(x + 5) = (x - 3)x + (x - 3)5$$
$$= x \cdot x + (-3) \cdot x + x \cdot 5 + (-3) \cdot 5$$
$$= x^2 - 3x + 5x - 15$$
$$= x^2 + 2x - 15$$

Can you see a pattern developing? Look at the answers:

$$(x + 2)(x + 3) = x^2 + 5x + 6$$
$$(x - 3)(x + 5) = x^2 + 2x - 15$$

Web It

To learn about four different methods of multiplying binomials, go to link 4-6-2 on the Bello Website at mhhe.com/bello.

Teaching Tip

See if students can recognize that both these products of binomials lead to a trinomial result. Give them a hint that in a later chapter they will be given the trinomial and asked to find the two binomial factors.

Answers

2. a. $4a - 12b$ **b.** $4a^7 + 12a^6$

It seems that the *first* term in each answer (x^2) is obtained by multiplying the *first* terms in the factors (x and x). Similarly, the *last* terms (6 and -15) are obtained by multiplying the *last* terms ($2 \cdot 3$ and $-3 \cdot 5$). Here's how it works so far:

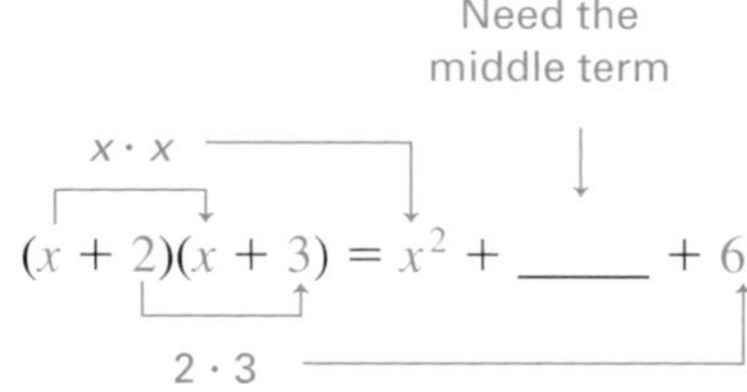

Web It

To practice FOIL with your very own polynomials, go to link 4-6-3 on the Bello Website at mhhe.com/bello.

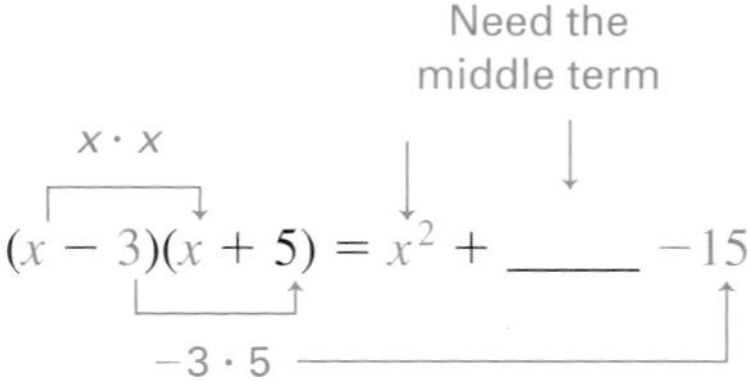

But what about the *middle* terms? In $(x + 2)(x + 3)$, the middle term is obtained by adding $3x$ and $2x$, which is the same as the result we got when we multiplied the *outer* terms (x and 3) and added the product of the inner terms (2 and x). Here's a diagram that shows how the middle term is obtained:

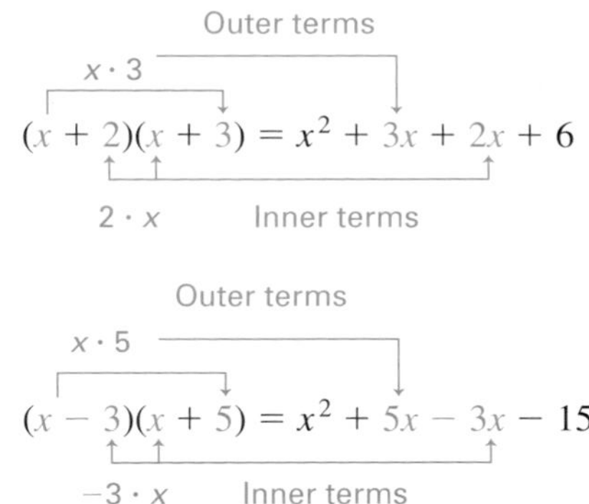

Do you see how it works now? Here is a summary of this method.

PROCEDURE

FOIL Method for Multiplying Binomials

First terms are multiplied first.
Outer terms are multiplied second.
Inner terms are multiplied third.
Last terms are multiplied last.

Of course, we call this method the **FOIL method.** We shall do one more example, step by step, to give you additional practice.

F	$(x + 7)(x - 4) \rightarrow x^2$	First: $x \cdot x$
O	$(x + 7)(x - 4) \rightarrow x^2 - 4x$	Outer: $-4 \cdot x$
I	$(x + 7)(x - 4) \rightarrow x^2 - 4x + 7x$	Inner: $7 \cdot x$
L	$(x + 7)(x - 4) = x^2 - 4x + 7x - 28$	Last: $7 \cdot (-4)$
	$= x^2 + 3x - 28$	

EXAMPLE 3 Using FOIL to multiply two binomials

Find:

a. $(x + 5)(x - 2)$ **b.** $(x - 4)(x + 3)$

SOLUTION

(First) F, (Outer) O, (Inner) I, (Last) L

$$\textbf{a. } (x + 5)(x - 2) = \underbrace{x \cdot x}_{} - \underbrace{2x + 5x}_{} - \underbrace{5 \cdot 2}_{}$$
$$= x^2 + 3x - 10$$

$$\textbf{b. } (x - 4)(x + 3) = \underbrace{x \cdot x}_{} + \underbrace{3x - 4x}_{} - \underbrace{4 \cdot 3}_{}$$
$$= x^2 - x - 12$$

PROBLEM 3

Find:

a. $(a + 4)(a - 3)$

b. $(a - 5)(a + 4)$

As in the case of arithmetic, we can use the ideas we've just discussed to do more complicated problems. Thus we can use the FOIL method to multiply expressions such as $(2x + 5)$ and $(3x - 4)$. We proceed as before; just remember your laws of exponents and the FOIL sequence.

EXAMPLE 4 Using FOIL to multiply two binomials

Find:

a. $(2x + 5)(3x - 4)$ **b.** $(3x - 2)(5x - 1)$

SOLUTION

(First) F, (Outer) O, (Inner) I, (Last) L

$$\textbf{a. } (2x + 5)(3x - 4) = \underbrace{(2x)(3x)}_{} + \underbrace{(2x)(-4)}_{} + \underbrace{5(3x)}_{} + \underbrace{(5)(-4)}_{}$$
$$= 6x^2 - \underbrace{8x + 15x}_{} - 20$$
$$= 6x^2 + 7x - 20$$

F, O, I, L

$$\textbf{b. } (3x - 2)(5x - 1) = \underbrace{(3x)(5x)}_{} + \underbrace{3x(-1)}_{} - \underbrace{2(5x)}_{} - \underbrace{2(-1)}_{}$$
$$= 15x^2 - \underbrace{3x - 10x}_{} + 2$$
$$= 15x^2 - 13x + 2$$

PROBLEM 4

Find:

a. $(3a + 5)(2a - 3)$

b. $(2a - 3)(4a - 1)$

Does FOIL work when the binomials to be multiplied contain more than one variable? Fortunately, yes. Again, just remember the sequence and the laws of exponents. For example, to multiply $(3x + 2y)$ by $(2x + 5y)$, we proceed as follows:

F, O, I, L

$$(2x + 5y)(3x + 2y) = \underbrace{(2x)(3x)}_{} + \underbrace{(2x)(2y)}_{} + \underbrace{(5y)(3x)}_{} + \underbrace{(5y)(2y)}_{}$$
$$= 6x^2 + \underbrace{4xy + 15xy}_{} + 10y^2$$
$$= 6x^2 + 19xy + 10y^2$$

Answers

3. a. $a^2 + a - 12$
b. $a^2 - a - 20$
4. a. $6a^2 + a - 15$
b. $8a^2 - 14a + 3$

EXAMPLE 5 Multiplying binomials involving two variables

Find:

a. $(5x + 2y)(2x + 3y)$ **b.** $(3x - y)(4x - 3y)$

SOLUTION

a.
$$\begin{aligned}(5x + 2y)(2x + 3y) &= \overset{F}{(5x)(2x)} + \overset{O}{(5x)(3y)} + \overset{I}{(2y)(2x)} + \overset{L}{(2y)(3y)}\\ &= 10x^2 + 15xy + 4xy + 6y^2\\ &= 10x^2 + 19xy + 6y^2\end{aligned}$$

b.
$$\begin{aligned}(3x - y)(4x - 3y) &= \overset{F}{(3x)(4x)} + \overset{O}{(3x)(-3y)} + \overset{I}{(-y)(4x)} + \overset{L}{(-y)(-3y)}\\ &= 12x^2 - 9xy - 4xy + 3y^2\\ &= 12x^2 - 13xy + 3y^2\end{aligned}$$

PROBLEM 5

Find:

a. $(4a + 3b)(3a + 5b)$

b. $(2a - b)(3a - 4b)$

Teaching Tip

In this chapter the outer and inner terms will combine but in some cases they don't.

Example:

$$(x + 2b)(x + y)$$
$$= x^2 + xy + 2bx + 2by$$

Now one more thing. How do we multiply the expression in the *Getting Started?*

$$(x - L)(x - 2L)$$

We do it in the next example.

EXAMPLE 6 Multiplying binomials involving two variables

Perform the indicated operation:

$$(x - L)(x - 2L)$$

SOLUTION

$$\begin{aligned}(x - L)(x - 2L) &= \overset{F}{x \cdot x} + \overset{O}{(x)(-2L)} + \overset{I}{(-L)(x)} + \overset{L}{(-L)(-2L)}\\ &= x^2 - 2xL - xL + 2L^2\\ &= x^2 - 3xL + 2L^2\end{aligned}$$

PROBLEM 6

Perform the indicated operation:

$$(y - 2L)(y - 3L)$$

D Solving an Application

Suppose we wish to find out how much is spent annually on hospital care. According to the American Hospital Association, average daily room charges can be approximated by $C(t) = 160 + 14t$ (dollars), where t is the number of years after 1990. On the other hand, the U.S. National Health Center for Health Statistics indicated that the average stay (in days) in the hospital can be approximated by $D(t) = 7 - 0.2t$, where t is the number of years after 1990. The amount spent annually on hospital care is given by

$$\text{Cost per day} \times \text{Number of days} = C(t) \times D(t)$$

We find this product next.

Answers

5. a. $12a^2 + 29ab + 15b^2$
b. $6a^2 - 11ab + 4b^2$
6. $y^2 - 5Ly + 6L^2$

EXAMPLE 7 **Getting well in U.S. hospitals**

Find: $C(t) \times D(t) = (160 + 14t)(7 - 0.2t)$

SOLUTION We use the FOIL method:

$$\begin{aligned}(160 + 14t)(7 - 0.2t) &= \overset{F}{\underbrace{160 \cdot 7}} + \overset{O}{\underbrace{160 \cdot (-0.2t)}} + \overset{I}{\underbrace{14t \cdot 7}} + \overset{L}{\underbrace{14t \cdot (-0.2t)}} \\ &= 1120 - \underbrace{32t + 98t} - 2.8t^2 \\ &= 1120 + 66t - 2.8t^2\end{aligned}$$

Thus the total amount spent annually on hospital care is $1120 + 66t - 2.8t^2$. Can you calculate what this amount was for 2000? What will it be for the year 2010?

PROBLEM 7

Suppose that at a certain hospital, the cost per day is $(200 + 15t)$ and the average stay (in days) is $(10 - 0.1t)$. What is the amount spent annually at this hospital?

Some students prefer a **grid method** to multiply polynomials. Thus, to do Example 4(a), $(2x + 5)(3x - 4)$, create a grid separated into four compartments. Place the term $(2x + 5)$ at the top and the term $(3x - 4)$ on the side of the grid. Multiply the rows and columns of the grid as shown. (After you get some practice, you can skip the initial step and write $6x^2$, $15x$, $-8x$, and -20 in the grid.)

	$2x$ +	5
$3x$	$3x \cdot 2x$ $6x^2$	$3x \cdot 5$ $15x$
-4	$-4 \cdot 2x$ $-8x$	$-4 \cdot 5$ -20

Finish by writing the results of each of the grid boxes: $6x^2 + 15x - 8x - 20$

And combining like terms: $6x^2 + 7x - 20$

You can try using this technique in the margin problems or in the exercises!

Calculate It Checking Equivalency

In the Section 4.1 *Calculate It,* we agreed that **two expressions are equivalent if their graphs are identical.** Thus, to check Example 1 we have to check that $(-3x^2)(2x^3) = -6x^5$. Let $Y_1 = (-3x^2)(2x^3)$ and $Y_2 = -6x^5$. Press GRAPH and the graph shown here will appear. To confirm the result numerically, press 2nd GRAPH and you get the result in the table.

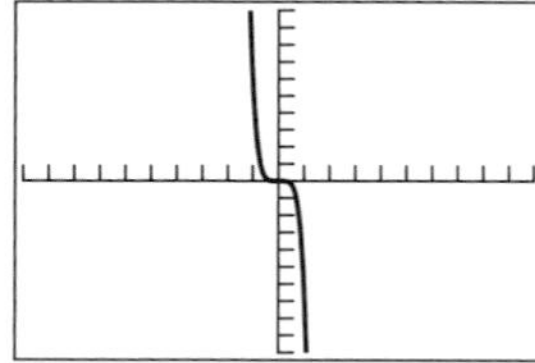

X	Y1	Y2
0	0	0
1	-6	-6
2	-192	-192
3	-1458	-1458
4	-6144	-6144
5	-18750	-18750
6	-46656	-46656

X=0

You can check the rest of the examples except Examples 5 and 6. Why?

Answer

7. $-1.5t^2 + 130t + 2000$

Exercises 4.6

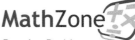

A In Problems 1–6, find the product.

1. $(5x^3)(9x^2)$ $45x^5$

2. $(8x^4)(9x^3)$ $72x^7$

3. $(-2x)(5x^2)$ $-10x^3$

4. $(-3y^2)(4y^3)$ $-12y^5$

5. $(-2y^2)(-3y)$ $6y^3$

6. $(-5z)(-3z)$ $15z^2$

B In Problems 7–20, remove parentheses (simplify).

7. $3(x + y)$ $3x + 3y$

8. $5(2x + y)$ $10x + 5y$

9. $5(2x - y)$ $10x - 5y$

10. $4(3x - 4y)$ $12x - 16y$

11. $-4x(2x - 3)$ $-8x^2 + 12x$

12. $-6x(5x - 3)$ $-30x^2 + 18x$

13. $(x^2 + 4x)x^3$ $x^5 + 4x^4$

14. $(x^2 + 2x)x^2$ $x^4 + 2x^3$

15. $(x - x^2)4x$ $4x^2 - 4x^3$

16. $(x - 3x^2)5x$ $5x^2 - 15x^3$

17. $(x + y)3x$ $3x^2 + 3xy$

18. $(x + 2y)5x^2$ $5x^3 + 10x^2y$

19. $(2x - 3y)(-4y^2)$ $-8xy^2 + 12y^3$

20. $(3x^2 - 4y)(-5y^3)$ $-15x^2y^3 + 20y^4$

C In Problems 21–56, use the FOIL method to perform the indicated operation.

21. $(x + 1)(x + 2)$ $x^2 + 3x + 2$

22. $(y + 3)(y + 8)$ $y^2 + 11y + 24$

23. $(y + 4)(y - 9)$ $y^2 - 5y - 36$

24. $(y + 6)(y - 5)$ $y^2 + y - 30$

25. $(x - 7)(x + 2)$ $x^2 - 5x - 14$

26. $(z - 2)(z + 9)$ $z^2 + 7z - 18$

27. $(x - 3)(x - 9)$ $x^2 - 12x + 27$

28. $(x - 2)(x - 11)$ $x^2 - 13x + 22$

29. $(y - 3)(y - 3)$ $y^2 - 6y + 9$

30. $(y + 4)(y + 4)$ $y^2 + 8y + 16$

31. $(2x + 1)(3x + 2)$ $6x^2 + 7x + 2$

32. $(4x + 3)(3x + 5)$ $12x^2 + 29x + 15$

33. $(3y + 5)(2y - 3)$ $6y^2 + y - 15$

34. $(4y - 1)(3y + 4)$ $12y^2 + 13y - 4$

35. $(5z - 1)(2z + 9)$ $10z^2 + 43z - 9$

36. $(2z - 7)(3z + 1)$ $6z^2 - 19z - 7$

37. $(2x - 4)(3x - 11)$ $6x^2 - 34x + 44$

38. $(5x - 1)(2x - 1)$ $10x^2 - 7x + 1$

39. $(4z + 1)(4z + 1)$ $16z^2 + 8z + 1$

40. $(3z - 2)(3z - 2)$ $9z^2 - 12z + 4$

41. $(3x + y)(2x + 3y)$ $6x^2 + 11xy + 3y^2$

42. $(4x + z)(3x + 2z)$ $12x^2 + 11xz + 2z^2$

43. $(2x + 3y)(x - y)$ $2x^2 + xy - 3y^2$

44. $(3x + 2y)(x - 5y)$ $3x^2 - 13xy - 10y^2$

45. $(5z - y)(2z + 3y)$ $10z^2 + 13yz - 3y^2$

46. $(2z - 5y)(3z + 2y)$ $6z^2 - 11yz - 10y^2$

47. $(3x - 2z)(4x - z)$ $12x^2 - 11xz + 2z^2$

48. $(2x - 3z)(5x - z)$ $10x^2 - 17xz + 3z^2$

49. $(2x - 3y)(2x - 3y)$ $4x^2 - 12xy + 9y^2$

50. $(3x + 5y)(3x + 5y)$ $9x^2 + 30xy + 25y^2$

51. $(3 + 4x)(2 + 3x)$ $6 + 17x + 12x^2$

52. $(2 + 3x)(3 + 2x)$ $6 + 13x + 6x^2$

53. $(2 - 3x)(3 + x)$ $6 - 7x - 3x^2$

54. $(3 - 2x)(2 + x)$ $6 - x - 2x^2$

55. $(2 - 5x)(4 + 2x)$ $8 - 16x - 10x^2$

56. $(3 - 5x)(2 + 3x)$ $6 - x - 15x^2$

APPLICATIONS

57. *Area of a rectangle* The area A of a rectangle is obtained by multiplying its length L by its width W; that is, $A = LW$. Find the area of the rectangle shown in the figure. $(x^2 + 7x + 10)$ square units

$x + 2$ $x + 5$

58. *Area of a rectangle* Use the formula in Problem 57 to find the area of a rectangle of width $x - 4$ and length $x + 3$. $(x^2 - x - 12)$ square units

59. *Height of a thrown object* The height reached by an object t seconds after being thrown upward with a velocity of 96 feet per second is given by $16t(6 - t)$. Use the distributive property to simplify this expression. $(96t - 16t^2)$ ft

60. *Resistance* The resistance R of a resistor varies with the temperature T according to the equation $R = (T + 100)(T + 20)$. Use the distributive property to simplify this expression. $R = T^2 + 120T + 2000$

61. *Gas property expression* In chemistry, when V is the volume and P is the pressure of a certain gas, we find the expression $(V_2 - V_1)(CP + PR)$, where C and R are constants. Use the distributive property to simplify this expression. $V_2CP + V_2PR - V_1CP - V_1PR$

The garage shown is 40 feet by 20 feet. You want to convert it to a bigger garage with two storage areas, S_1 and S_2.

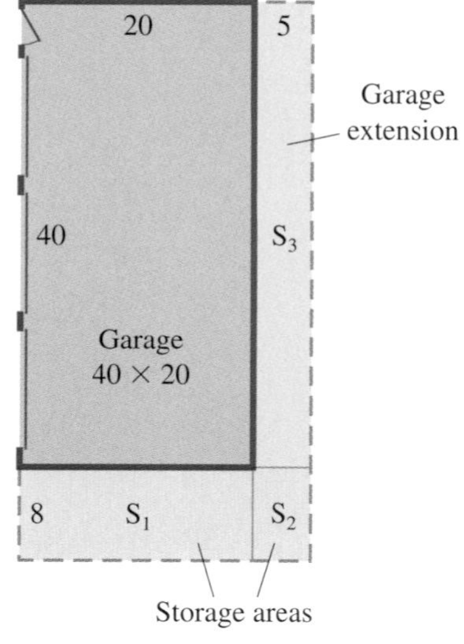

62. What is the area of the current garage? 800 ft^2

63. If you extend the long side by 8 feet and the short side by 5 feet, what is the area of the new garage? 1200 ft^2

64. Calculate the areas of S_1, S_2, and S_3, and write your answers in the appropriate places in the diagram. $S_1 = 160$ ft^2; $S_2 = 40$ ft^2; $S_3 = 200$ ft^2

65. Determine the total area of the new garage by adding the area of the original garage to the areas of S_1, S_2, and S_3; that is, add the answers you obtained in Problems 62 and 64. $800 + 160 + 40 + 200 = 1200$ ft^2

66. Is the area of the new garage (Problem 63) the same as the answer in Problem 65? Yes

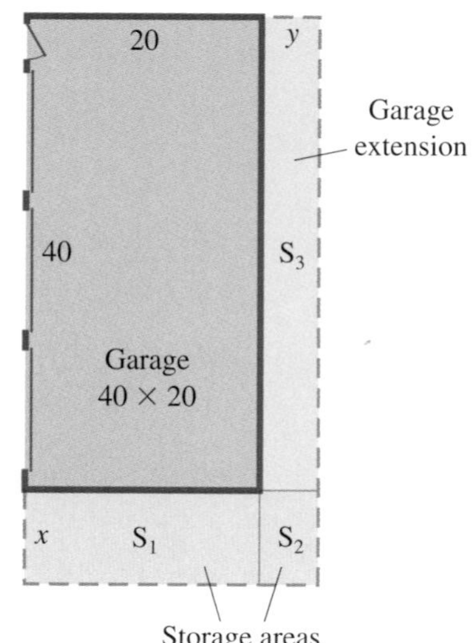

67. If you are not sure how big you want the storage rooms, extend the long side of the garage by x feet and the short side by y feet.

a. Find the area of S_1. $S_1 = 20x$ ft^2

b. Find the area of S_2. $S_2 = xy$ ft^2

c. Find the area of S_3. $S_3 = 40y$ ft^2

68. The area of the new garage is $(40 + x)(20 + y)$. Simplify this expression. $800 + 40y + 20x + xy$

69. Add the areas of S_1, S_2, S_3, and the area of the original garage. Is the answer the same as the one you obtained in Problem 68? $800 + 20x + xy + 40y$; yes

SKILL CHECKER

Try the Skill Checker Exercises so you'll be ready for the next section.

Find:

70. $(4y)^2$ $16y^2$

71. $(3x)^2$ $9x^2$

72. $(-A)^2$ A^2

73. $(-3A)^2$ $9A^2$

74. $A(-A)$ $-A^2$

USING YOUR KNOWLEDGE

Profitable Polynomials

Do you know how to find the profit made in a certain business? The profit P is the difference between the revenue R and the expense E of doing business. In symbols,

$$P = R - E$$

Of course, R depends on the number n of items sold and their price p (in dollars); that is,

$$R = np$$

Thus

$$P = np - E$$

75. If $n = -3p + 60$ and $E = -5p + 100$, find P.
$P = -3p^2 + 65p - 100$

76. If $n = -2p + 50$ and $E = -3p + 300$, find P.
$P = -2p^2 + 53p - 300$

77. In Problem 75, if the price was \$2, what was the profit P? \$18

78. In Problem 76, if the price was \$10, what was the profit P? \$30

WRITE ON

79. Will the product of two monomials always be a monomial? Explain. Yes; answers may vary.

80. If you multiply a monomial and a binomial, will you ever get a trinomial? Explain. No; answers may vary.

81. Will the product of two binomials (after combining like terms) always be a trinomial? Explain.
No; consider the results in Problem 82.

82. Multiply:

$$(x + 1)(x - 1) =$$

$$(y + 2)(y - 2) =$$

$$(z + 3)(z - 3) =$$

What is the pattern? $(X + A)(X - A) = X^2 - A^2$

MASTERY TEST

If you know how to do these problems, you have learned your lesson!

Multiply:

83. $(-7x^4)(5x^2)$
$-35x^6$

84. $(-8a^3)(-5a^5)$
$40a^8$

85. $(x + 7)(x - 3)$
$x^2 + 4x - 21$

86. $(x - 2)(x + 8)$
$x^2 + 6x - 16$

87. $(3x + 4)(3x - 1)$
$9x^2 + 9x - 4$

88. $(4x + 3y)(3x + 2y)$
$12x^2 + 17xy + 6y^2$

89. $(5x - 2y)(2x - 3y)$
$10x^2 - 19xy + 6y^2$

90. $(x - L)(x - 3L)$
$x^2 - 4xL + 3L^2$

91. Simplify $6(x - 3y)$.
$6x - 18y$

92. Simplify $(x^3 + 5x)(-4x^5)$.
$-4x^8 - 20x^6$

4.7 SPECIAL PRODUCTS OF POLYNOMIALS

To Succeed, Review How To . . .

1. Use the FOIL method to multiply polynomials (pp. 348–351).
2. Multiply expressions (pp. 92, 98, 101, 294, 301).
3. Use the distributive property to simplify expressions (p. 101).

Objectives

Expand (simplify) binomials of the form

A $(X + A)^2$

B $(X - A)^2$

C $(X + A)(X - A)$

D Multiply a binomial by a trinomial.

E Multiply any two polynomials.

GETTING STARTED

Expanding Your Property

In Section 4.6, we learned how to use the distributive property and the FOIL method to multiply two binomials. In this section we shall develop patterns that will help us in multiplying certain binomial products that occur frequently. For example, do you know how to find the area of this property lot? Since the land is a square, the area is

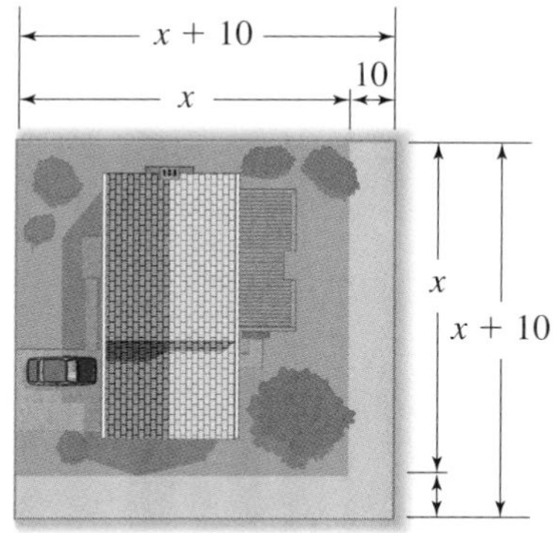

$$(X + 10)(X + 10) = (X + 10)^2$$

The expression $(X + 10)^2$ is the square of the binomial $X + 10$. You can use the FOIL method, of course, but this type of expression is so common in algebra that we have *special products* or formulas that we use to multiply (or expand) them. We are now ready to study several of these special products. You will soon find that we've already studied the first of these special products: the FOIL method is actually special product 1!

As we implied in the *Getting Started,* there's another way to look at the FOIL method: we can format it as a special product. Do you recall the FOIL method?

$$\begin{aligned}(X + A)(X + B) &= \overset{\text{F}}{X^2} + \overset{\text{O}}{XB} + \overset{\text{I}}{AX} + \overset{\text{L}}{AB} \\ &= X^2 + (B + A)X + AB \\ &= X^2 + (A + B)X + AB\end{aligned}$$

Thus we have our first special product.

Web It

To see videos dealing with the multiplication of polynomials, go to link 4-7-1 on the Bello Website at mhhe.com/bello.

For a lesson on multiplying binomials, go to link 4-7-2.

For a tutorial on multiplying polynomials, including special products, try link 4-7-3.

PRODUCT OF TWO BINOMIALS

Special Product 1

$$(X + A)(X + B) = X^2 + (A + B)X + AB \qquad \textbf{(SP1)}$$

Teaching Tip

Be sure students notice that SP1 does not work for binomials with leading coefficients that are not the same.

Example:

$(2x + 3)(5x - 4)$

For example, $(x - 3)(x + 5)$ can be multiplied using SP1 with $X = x$, $A = -3$, and $B = 5$. Thus

$$\begin{aligned}(x - 3)(x + 5) &= x^2 + (-3 + 5)x + (-3)(5)\\ &= x^2 + 2x - 15\end{aligned}$$

Similarly, to expand $(2x - 5)(2x + 1)$, we can let $X = 2x$, $A = -5$, and $B = 1$ in SP1. Hence

$$\begin{aligned}(2x - 5)(2x + 1) &= (2x)^2 + (-5 + 1)2x + (-5)(1)\\ &= 4x^2 - 8x - 5\end{aligned}$$

Why have we gone to the trouble of reformatting FOIL? Because, as you will see, three very predictable and very handy special products can be derived from SP1.

A Squaring Sums

Our second special product (SP2) deals with squaring polynomial sums. Let's see how it is developed by expanding $(X + 10)^2$. First, we start with SP1; we let $A = B$ and note that

$$(X + A)(X + A) = (X + A)^2$$

Foiled again

$(X + A)^2 = X^2 + 2AX + A^2$

Now we let $A = 10$, so in our expansion of $(X + 10)^2$ [or $(X + A)^2$], we have

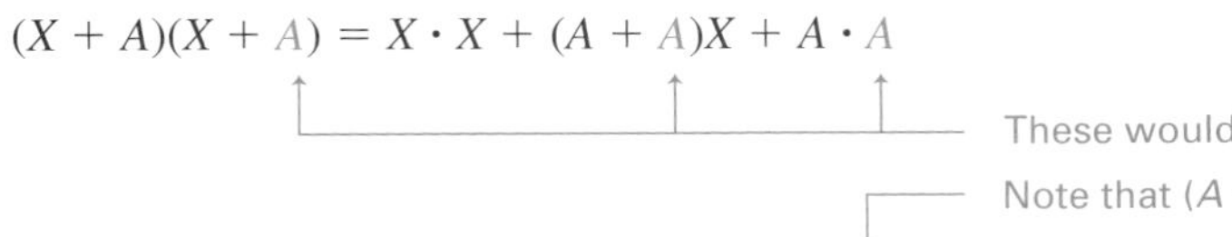

$(X + A)(X + A) = X \cdot X + (A + A)X + A \cdot A$
These would be B in SP1.
Note that $(A + A)X = 2AX$.

Thus

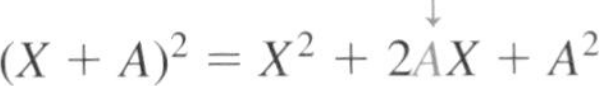

Now we have our second special product, SP2:

THE SQUARE OF A BINOMIAL SUM

Special Product 2

$(X + A)^2 = X^2 + 2AX + A^2$ **(SP2)** Note that $(X + A)^2 \neq X^2 + A^2$. (See the *Using Your Knowledge* in Exercises 4.7.)

Here is the pattern used in SP2:

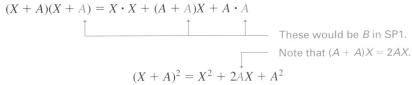

$$\underbrace{(X}_{\text{First term}} + \underbrace{A)^2}_{\text{Second term}} = \underbrace{X^2}_{\text{Square the first term.}} + \underbrace{2AX}_{\text{Multiply the terms and double.}} + \underbrace{A^2}_{\text{Square the last term.}}$$

Teaching Tip

Again, encourage students to use the special product because soon they will be given the resulting trinomial and asked to factor it into the appropriate binomial squared. These steps will be used in reverse.

If you examine this pattern carefully, you can see that SP2 is a result of the FOIL method when a binomial sum is squared. Can we obtain the same product with FOIL?

$$\begin{aligned}(X + A)(X + A) &= \overset{F}{X \cdot X} + \overset{O}{AX} + \overset{I}{AX} + \overset{L}{A \cdot A}\\ &= X^2 + 2AX + A^2\end{aligned}$$

The answer is yes! We are now ready to expand $(X + 10)^2$ using SP2:

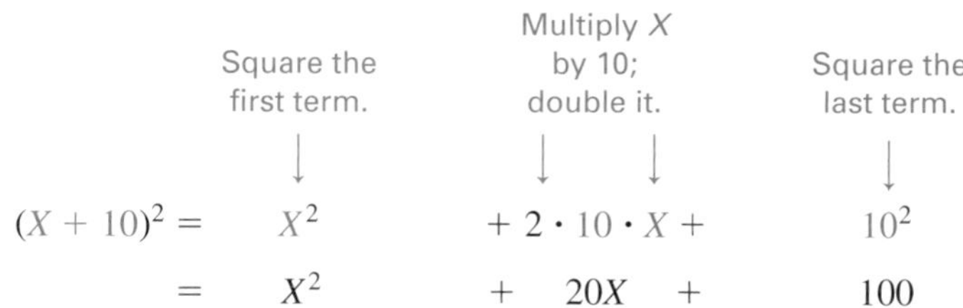

$$(X + 10)^2 = X^2 + 2 \cdot 10 \cdot X + 10^2$$
$$= X^2 + 20X + 100$$

Similarly,

$$(x + 7)^2 = x^2 + 2 \cdot 7 \cdot x + 7^2$$
$$= x^2 + 14x + 49$$

EXAMPLE 1 **Squaring binomial sums**

Find (expand):

a. $(x + 9)^2$ **b.** $(2x + 3)^2$ **c.** $(3x + 4y)^2$

SOLUTION

a. $(x + 9)^2 = x^2 + 2 \cdot 9 \cdot x + 9^2$ Let $A = 9$ in SP2.
$= x^2 + 18x + 81$

b. $(2x + 3)^2 = (2x)^2 + 2 \cdot 3 \cdot 2x + 3^2$ Let $X = 2x$ and $A = 3$ in SP2.
$= 4x^2 + 12x + 9$

c. $(3x + 4y)^2 = (3x)^2 + 2(4y)(3x) + (4y)^2$ Let $X = 3x$ and $A = 4y$ in SP2.
$= 9x^2 + 24xy + 16y^2$

PROBLEM 1

Find (expand):

a. $(y + 6)^2$ **b.** $(2y + 1)^2$

c. $(2x + 3y)^2$

Teaching Tip

When squaring a number such as 5 the result, 25, is called a perfect square number. Hence, when squaring a binomial such as $(x + 9)$ the result, $x^2 + 18x + 81$, is called a perfect square trinomial.

But remember, you do not need to use SP2, you can *always* use FOIL; SP2 simply saves you time and work.

B Squaring Differences

Can we also expand $(X - 10)^2$? Of course! But we first have to learn how to expand $(X - A)^2$. To do this, we simply write $-A$ instead of A in SP2 to obtain

$$(X - A)^2 = X^2 + 2(-A)X + (-A)^2$$
$$= X^2 - 2AX + A^2$$

This is the special product we need, and we call it SP3, the square of a binomial difference.

THE SQUARE OF A BINOMIAL DIFFERENCE

Special Product 3

$$(X - A)^2 = X^2 - 2AX + A^2 \qquad \textbf{(SP3)}$$

Note that $(X - A)^2 \neq X^2 - A^2$, so the only difference between the square of a sum and the square of a difference is the sign preceding $2AX$. Here is the comparison:

$$(X + A)^2 = X^2 + 2AX + A^2 \qquad \textbf{(SP2)}$$
$$(X - A)^2 = X^2 - 2AX + A^2 \qquad \textbf{(SP3)}$$

Answers

1. a. $y^2 + 12y + 36$
b. $4y^2 + 4y + 1$
c. $4x^2 + 12xy + 9y^2$

Keeping this in mind,

$$(X - 10)^2 = X^2 - 2 \cdot 10 \cdot X + 10^2$$
$$= X^2 - 20X + 100$$

Similarly,

$$(x - 3)^2 = x^2 - 2 \cdot 3 \cdot x + 3^2$$
$$= x^2 - 6x + 9$$

EXAMPLE 2 **Squaring binomial differences**

Find (expand):

a. $(x - 5)^2$ **b.** $(3x - 2)^2$ **c.** $(2x - 3y)^2$

SOLUTION

a. $(x - 5)^2 = x^2 - 2 \cdot 5 \cdot x + 5^2$ Let $A = 5$ in SP3.
$= x^2 - 10x + 25$

b. $(3x - 2)^2 = (3x)^2 - 2 \cdot 2 \cdot 3x + 2^2$ Let $X = 3x$ and $A = 2$ in SP3.
$= 9x^2 - 12x + 4$

c. $(2x - 3y)^2 = (2x)^2 - 2 \cdot 3y \cdot 2x + (3y)^2$ Let $X = 2x$ and $A = 3y$ in SP3.
$= 4x^2 - 12xy + 9y^2$

PROBLEM 2

Find (expand):

a. $(y - 4)^2$

b. $(2y - 3)^2$

c. $(3x - 2y)^2$

C Multiplying Sums and Differences

We have one more special product, and this one is especially clever! Suppose we multiply the sum of two terms by the difference of the same two terms; that is, suppose we wish to multiply

$$(X + A)(X - A)$$

If we substitute $-A$ for B in SP1, then

$$(X + A)(X + B) = X^2 + (A + B)X + AB$$

becomes

$$(X + A)(X - A) = X^2 + (A - A)X + A(-A)$$
$$= X^2 + 0X - A^2$$
$$= X^2 - A^2$$

This gives us our last very special product, SP4.

THE PRODUCT OF THE SUM AND DIFFERENCE OF TWO TERMS

Special Product 4

$$(X + A)(X - A) = X^2 - A^2 \quad \textbf{(SP4)}$$

Note that $(X + A)(X - A) = (X - A)(X + A)$; so $(X - A)(X + A) = X^2 - A^2$. Thus to multiply the sum and difference of two terms, we simply square the first term and then subtract from this the square of the last term. Checking this result using the FOIL method,

Teaching Tip

When multiplying the sum and differences of two terms, the result is called "the difference of two perfect squares."

Web It

For an audiovisual tutorial and practice, try link 4-7-4 on the Bello Website at mhhe.com/bello.

Answers

2. a. $y^2 - 8y + 16$
b. $4y^2 - 12y + 9$
c. $9x^2 - 12xy + 4y^2$

we have

$$\overset{\text{F}}{} \quad \overset{\text{O}}{} \quad \overset{\text{I}}{} \quad \overset{\text{L}}{}$$

$$(X + A)(X - A) = X^2 \underbrace{- AX + AX}_{0} - A^2$$

$$= X^2 - A^2$$

Since the middle term is *always* zero, we have

$$(x + 3)(x - 3) = x^2 - 3^2 = x^2 - 9$$

$$(x + 6)(x - 6) = x^2 - 6^2 = x^2 - 36$$

Similarly, by the commutative property, we also have

$$(x - 3)(x + 3) = x^2 - 3^2 = x^2 - 9$$

$$(x - 6)(x + 6) = x^2 - 6^2 = x^2 - 36$$

EXAMPLE 3 **Finding the product of the sum and difference of two terms**

Find:

a. $(x + 10)(x - 10)$ **b.** $(2x + y)(2x - y)$

c. $(3x - 5y)(3x + 5y)$

SOLUTION

a. $(x + 10)(x - 10) = x^2 - 10^2$ Let $A = 10$ in SP4.

$= x^2 - 100$

b. $(2x + y)(2x - y) = (2x)^2 - y^2$ Let $X = 2x$ and $A = y$ in SP4.

$= 4x^2 - y^2$

c. $(3x - 5y)(3x + 5y) = (3x)^2 - (5y)^2$ Let $X = 3x$ and $A = 5y$ in SP4.

$= 9x^2 - 25y^2$

Note that

$$(3x - 5y)(3x + 5y) = (3x + 5y)(3x - 5y)$$

by the commutative property, so SP4 still applies.

PROBLEM 3

Find:

a. $(y + 9)(y - 9)$

b. $(3x + y)(3x - y)$

c. $(3x - 2y)(3x + 2y)$

D Multiplying a Binomial by a Trinomial

Can we use the FOIL method to multiply any two polynomials? Unfortunately, no. But wait; if algebra is a generalized arithmetic, we should be able to multiply a binomial by a trinomial using the same techniques we employ to multiply, say, 23 by 342.

Answers

3. a. $y^2 - 81$ **b.** $9x^2 - y^2$ **c.** $9x^2 - 4y^2$

Web It

To learn about multiplying a binomial by a trinomial, try link 4-7-5 on the Bello Website at mhhe.com/bello.

First, let's review how we multiply 23 by 342. Here are the steps.

Step 1	Step 2	Step 3
$\begin{array}{r} 342 \\ \times\ 23 \\ \hline 1026 \end{array}$	$\begin{array}{r} 342 \\ \times\ 23 \\ \hline 1026 \\ 6840 \\ \hline \end{array}$	$\begin{array}{r} 342 \\ \times\ 23 \\ \hline 1026 \\ 6840 \\ \hline 7866 \end{array}$
$3 \times 342 = 1026$	$20 \times 342 = 6840$	$1026 + 6840 = 7866$

Now let's use this same technique to multiply two polynomials, say $(x + 5)$ and $(x^2 + x - 2)$.

Step 1	Step 2	Step 3
$\begin{array}{r} x^2 + x - 2 \\ x + 5 \\ \hline 5x^2 + 5x - 10 \end{array}$	$\begin{array}{r} x^2 + x - 2 \\ x + 5 \\ \hline 5x^2 + 5x - 10 \\ x^3 + x^2 - 2x \\ \hline \end{array}$	$\begin{array}{r} x^2 + x - 2 \\ x + 5 \\ \hline 5x^2 + 5x - 10 \\ x^3 + x^2 - 2x \\ \hline x^3 + 6x^2 + 3x - 10 \end{array}$
$5(x^2 + x - 2)$ $= 5x^2 + 5x - 10$	$x(x^2 + x - 2)$ $= x^3 + x^2 - 2x$	$\begin{array}{r} 5x^2 + 5x - 10 \\ x^3 + x^2 - 2x \\ \hline x^3 + 6x^2 + 3x - 10 \end{array}$

Note that in step 3, all *like* terms are placed in the same column so they can be combined.

For obvious reasons, this method is called the **vertical scheme** and can be used when one of the polynomials to be multiplied has *three or more terms.* Of course, we could have obtained the same result by using the distributive property:

$$(a + b)c = ac + bc$$

The procedure would look like this:

$$\begin{aligned} \underbrace{(x + 5)}_{(a+b)} \cdot \underbrace{(x^2 + x - 2)}_{c} &= \underbrace{x}_{a} \cdot \underbrace{(x^2 + x - 2)}_{c} + \underbrace{5}_{b} \cdot \underbrace{(x^2 + x - 2)}_{c} \\ &= x^3 + x^2 - 2x + 5x^2 + 5x - 10 \\ &= x^3 + (x^2 + 5x^2) + (-2x + 5x) - 10 \\ &= x^3 + 6x^2 + 3x - 10 \end{aligned}$$

EXAMPLE 4 Multiplying a binomial by a trinomial

Find: $(x - 3)(x^2 - 2x - 4)$

SOLUTION Using the vertical scheme, we proceed as follows:

$$\begin{array}{rl} x^2 - 2x - 4 & \\ x - 3 & \\ \hline -3x^2 + 6x + 12 & \text{Multiply } x^2 - 2x - 4 \text{ by } -3. \\ x^3 - 2x^2 - 4x & \text{Multiply } x^2 - 2x - 4 \text{ by } x. \\ \hline x^3 - 5x^2 + 2x + 12 & \text{Add like terms.} \end{array}$$

PROBLEM 4

Find: $(y - 2)(y^2 - y - 3)$

Answer

4. $y^3 - 3y^2 - y + 6$

Thus the result is $x^3 - 5x^2 + 2x + 12$. You can also do this problem by using the distributive property $(a + b)c = ac + bc$. The procedure would look like this:

$$\begin{aligned}
\underbrace{(x - 3)}_{(a - b)} \cdot \underbrace{(x^2 - 2x - 4)}_{c} &= \underbrace{x}_{a} \cdot \underbrace{(x^2 - 2x - 4)}_{c} - \underbrace{3}_{b} \cdot \underbrace{(x^2 - 2x - 4)}_{c} \\
&= x^3 - 2x^2 - 4x - 3x^2 + 6x + 12 \\
&= x^3 + (-2x^2 - 3x^2) + (-4x + 6x) + 12 \\
&= x^3 - 5x^2 + 2x + 12
\end{aligned}$$

Note that the same result is obtained in both cases.

E Multiplying Any Two Polynomials

Here is the idea we used in Example 4.

PROCEDURE

Multiplying *Any* Two Polynomials (Term-By-Term Multiplication)

To multiply two polynomials, multiply each term of one by every term of the other and add the results.

Teaching Tip

Because SP1–SP4 are so important for factoring, see if students can name the type of result for the following examples.

$(x + 2)(x - 3) = ?$ trinomial

$(x + 4)^2 = ?$ perfect square binomial

$(x + 2)(x - 2) = ?$ difference of two squares

$(x - 3)^2 = ?$ perfect square trinomial

Then reverse the procedure and see if they can write an example of the two binomials that would produce the following products.

a. trinomial

b. perfect square trinomial

c. difference of two perfect squares

Now that you've learned all the basic techniques used to multiply polynomials, you should be able to tackle any polynomial multiplication. To do this, you must *first* decide what special product (if any) is involved. Here are the special products we've studied.

SPECIAL PRODUCTS

$(X + A)(X + B) = X^2 + (A + B)X + AB$ **(SP1 or FOIL)**

$(X + A)(X + A) = (X + A)^2 = X^2 + 2AX + A^2$ **(SP2)**

$(X - A)(X - A) = (X - A)^2 = X^2 - 2AX + A^2$ **(SP3)**

$(X + A)(X - A) = X^2 - A^2$ **(SP4)**

Of course, the FOIL method always works for the last three types of equations, but since it's more laborious, learning to recognize special products 2–4 is definitely worth the effort!

EXAMPLE 5 **Using SP1 (FOIL) and the distributive property**

Find: $3x(x + 5)(x + 6)$

SOLUTION One way to approach this problem is to save the $3x$ multiplication until last. First, use FOIL (SP1):

$$\begin{aligned}
& \qquad\quad \text{F} \quad \text{O} \quad \text{I} \quad \text{L} \\
3x(x + 5)(x + 6) &= 3x(x^2 + 6x + 5x + 30) \\
&= 3x(x^2 + 11x + 30) \\
&= 3x^3 + 33x^2 + 90x \qquad \text{Use the distributive property.}
\end{aligned}$$

PROBLEM 5

Find: $2y(y + 2)(y + 3)$

Answer

5. $2y^3 + 10y^2 + 12y$

Alternatively, you can use the distributive property to multiply $(x + 5)$ by $3x$ and then proceed with FOIL, as shown here:

$$3x(x + 5)(x + 6) = (3x \cdot x + 3x \cdot 5)(x + 6) \quad \text{Multiply } 3x(x + 5).$$
$$= (3x^2 + 15x)(x + 6) \quad \text{Simplify.}$$
$$= \overset{F}{3x^2 \cdot x} + \overset{O}{3x^2 \cdot 6} + \overset{I}{15x \cdot x} + \overset{L}{15x \cdot 6} \quad \text{Use FOIL.}$$
$$= 3x^3 + 18x^2 + 15x^2 + 90x \quad \text{Simplify.}$$
$$= 3x^3 + 33x^2 + 90x \quad \text{Collect like terms.}$$

Of course, both methods produce the same result.

Teaching Tip

Remind students that $3x$ gets distributed only once. For example, in multiplying 3(5)(2), 3 gets multiplied to only one of the two numbers (5 or 2) and then the result is multiplied times the third number.

EXAMPLE 6 Cubing a binomial

Find: $(x + 3)^3$

SOLUTION Recall that $(x + 3)^3 = (x + 3)(x + 3)(x + 3)$. Thus we can square the sum $(x + 3)$ first and then use the distributive property. It goes like this:

Square of sum

$$(x + 3)(x + 3)(x + 3) = (x + 3)(x^2 + 6x + 9)$$
$$= x(x^2 + 6x + 9) + 3(x^2 + 6x + 9) \quad \text{Use the distributive property.}$$
$$= x \cdot x^2 + x \cdot 6x + x \cdot 9 + 3 \cdot x^2 + 3 \cdot 6x + 3 \cdot 9$$
$$= x^3 + 9x^2 + 27x + 27$$

PROBLEM 6

Find: $(y + 2)^3$

Teaching Tip

For Example 6,

$$(x + 3)^3 \neq x^3 + 3^3$$

Illustrate with a numerical example, such as $(2 + 3)^3$.

$$(2 + 3)^3 \stackrel{?}{=} 2^3 + 3^3$$
$$(5)^3 \stackrel{?}{=} 8 + 27$$
$$125 \neq 35$$

EXAMPLE 7 Squaring a binomial difference involving fractions

Find:

$$\left(5t^2 - \frac{1}{2}\right)^2$$

SOLUTION This is the square of a difference. Using SP3, we obtain

$$\left(5t^2 - \frac{1}{2}\right)^2 = (5t^2)^2 - 2 \cdot \frac{1}{2} \cdot (5t^2) + \left(\frac{1}{2}\right)^2$$
$$= 25t^4 - 5t^2 + \frac{1}{4}$$

PROBLEM 7

Find:

$$\left(3y^2 - \frac{1}{3}\right)^2$$

EXAMPLE 8 Finding the product of the sum and difference of two terms

Find: $(2x^2 + 5)(2x^2 - 5)$

SOLUTION This time, we use SP4 because we have the product of the sum and difference of two terms.

$$(2x^2 + 5)(2x^2 - 5) = (2x^2)^2 - (5)^2$$
$$= 4x^4 - 25$$

PROBLEM 8

Find: $(3y^2 + 2)(3y^2 - 2)$

Answers

6. $y^3 + 6y^2 + 12y + 8$

7. $9y^4 - 2y^2 + \frac{1}{9}$ **8.** $9y^4 - 4$

EXAMPLE 9 Finding Areas

In Example 5 of Section 4.5, we used the addition of polynomials to find the total area of several rectangles. As you recall, the area A of a rectangle is the product of its length L and its width W; that is, $A = L \times W$. If the length of the rectangle is $(x + 5)$ and its width is $(x - 3)$, what is its area?

SOLUTION Since the area is $L \times W$, where $L = (x + 5)$ and $W = (x - 3)$, we have

$$\begin{aligned} A &= (x + 5)(x - 3) \\ &= x^2 - 3x + 5x - 15 \\ &= x^2 + 2x - 15 \end{aligned}$$

(x − 3) A (x + 5)

PROBLEM 9

The area A of a circle of radius r is $A = \pi r^2$. What is the area of a circle with radius $(x - 2)$?

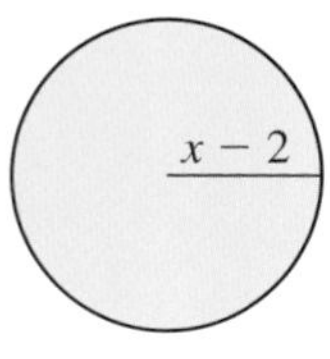

How do you know which of the special products you should use?

The trick to using these special products effectively, of course, is being able to *recognize* situations where they apply. Here are some tips.

PROCEDURE

Choosing the Appropriate Method for Multiplying Two Polynomials

1. Is the product the square of a binomial? If so, use SP2 or SP3:

$$(X + A)^2 = (X + A)(X + A) = X^2 + 2AX + A^2 \quad \textbf{(SP2)}$$

$$(X - A)^2 = (X - A)(X - A) = X^2 - 2AX + A^2 \quad \textbf{(SP3)}$$

Note that both answers have *three* terms.

2. Are the two binomials in the product the sum and difference of the same two terms? If so, use SP4:

$$(X + A)(X - A) = X^2 - A^2 \quad \textbf{(SP4)}$$

The answer has *two* terms.

3. Is the binomial product different from those in 1 and 2? If so, use FOIL. The answer will have *three* or *four* terms.
4. Is the product different from all the ones mentioned in 1, 2, and 3? If so, multiply every term of the first polynomial by every term of the second and collect like terms.

Calculate It Checking Results

As we have done previously, we can use a calculator to check the results of the examples in this section. You know the procedure: let the original expression be Y_1 and the answer be Y_2. If the graphs are identical, the answer is correct. Thus, to check Example 5, let $Y_1 = 3x(x + 5)(x + 6)$ and $Y_2 = 3x^3 + 33x^2 + 90x$. Press GRAPH and you get only a partial graph, as shown here on the left. Since we are missing the bottom part of the graph, we need more values of y that are negative, so we have to fix the window we use for the graph. Press WINDOW and let Ymin $= -20$. Still, we do not see the whole graph, so try again with Ymin $= -90$.

Press GRAPH and this time we see the complete graph as shown on the right.

You should now check Example 7, but use x instead of t. Can you check Example 1(c)? Why not?

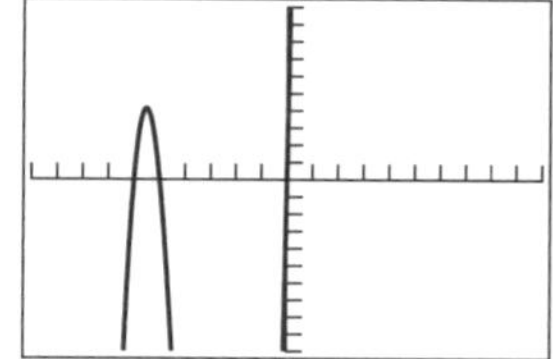

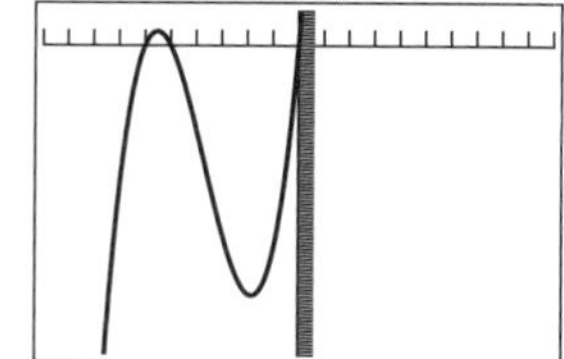

A final word of advice before you do Exercises 4.7: SP1 (the FOIL method) is very important and should be thoroughly understood before you attempt the problems. This is because *all* the special products are *derived* from this formula. As a matter of fact, you can successfully complete most of the problems in Exercises 4.7 if you fully understand the result given in SP1.

Answer

9. $\pi(x - 2)^2 = \pi x^2 - 4\pi x + 4\pi$

Exercises 4.7

A In Problems 1–6, expand the given binomial sum.

1. $(x + 1)^2$ $x^2 + 2x + 1$
2. $(x + 6)^2$ $x^2 + 12x + 36$
3. $(2x + 1)^2$ $4x^2 + 4x + 1$
4. $(2x + 3)^2$ $4x^2 + 12x + 9$
5. $(3x + 2y)^2$ $9x^2 + 12xy + 4y^2$
6. $(4x + 5y)^2$ $16x^2 + 40xy + 25y^2$

B In Problems 7–16, expand the given binomial difference.

7. $(x - 1)^2$ $x^2 - 2x + 1$
8. $(x - 2)^2$ $x^2 - 4x + 4$
9. $(2x - 1)^2$ $4x^2 - 4x + 1$
10. $(3x - 4)^2$ $9x^2 - 24x + 16$
11. $(3x - y)^2$ $9x^2 - 6xy + y^2$
12. $(4x - y)^2$ $16x^2 - 8xy + y^2$
13. $(6x - 5y)^2$ $36x^2 - 60xy + 25y^2$
14. $(4x - 3y)^2$ $16x^2 - 24xy + 9y^2$
15. $(2x - 7y)^2$ $4x^2 - 28xy + 49y^2$
16. $(3x - 5y)^2$ $9x^2 - 30xy + 25y^2$

C In Problems 17–30, find the product.

17. $(x + 2)(x - 2)$ $x^2 - 4$
18. $(x + 1)(x - 1)$ $x^2 - 1$
19. $(x + 4)(x - 4)$ $x^2 - 16$
20. $(2x + y)(2x - y)$ $4x^2 - y^2$
21. $(3x + 2y)(3x - 2y)$ $9x^2 - 4y^2$
22. $(2x + 5y)(2x - 5y)$ $4x^2 - 25y^2$
23. $(x - 6)(x + 6)$ $x^2 - 36$
24. $(x - 11)(x + 11)$ $x^2 - 121$
25. $(x - 12)(x + 12)$ $x^2 - 144$
26. $(x - 9)(x + 9)$ $x^2 - 81$
27. $(3x - y)(3x + y)$ $9x^2 - y^2$
28. $(5x - 6y)(5x + 6y)$ $25x^2 - 36y^2$
29. $(2x - 7y)(2x + 7y)$ $4x^2 - 49y^2$
30. $(5x - 8y)(5x + 8y)$ $25x^2 - 64y^2$

In Problems 31–40, use the special products to multiply the given expressions.

31. $(x^2 + 2)(x^2 + 5)$ $x^4 + 7x^2 + 10$
32. $(x^2 - 3)(x^2 + 2)$ $x^4 - x^2 - 6$
33. $(x^2 + y)^2$ $x^4 + 2x^2y + y^2$
34. $(2x^2 + y)^2$ $4x^4 + 4x^2y + y^2$
35. $(3x^2 - 2y^2)^2$ $9x^4 - 12x^2y^2 + 4y^4$
36. $(4x^3 - 5y^3)^2$ $16x^6 - 40x^3y^3 + 25y^6$
37. $(x^2 - 2y^2)(x^2 + 2y^2)$ $x^4 - 4y^4$
38. $(x^2 - 3y^2)(x^2 + 3y^2)$ $x^4 - 9y^4$
39. $(2x + 4y^2)(2x - 4y^2)$ $4x^2 - 16y^4$
40. $(5x^2 + 2y)(5x^2 - 2y)$ $25x^4 - 4y^2$

D In Problems 41–52, find the product.

41. $(x + 3)(x^2 + x + 5)$ $x^3 + 4x^2 + 8x + 15$
42. $(x + 2)(x^2 + 5x + 6)$ $x^3 + 7x^2 + 16x + 12$
43. $(x + 4)(x^2 - x + 3)$ $x^3 + 3x^2 - x + 12$
44. $(x + 5)(x^2 - x + 2)$ $x^3 + 4x^2 - 3x + 10$
45. $(x + 3)(x^2 - x - 2)$ $x^3 + 2x^2 - 5x - 6$
46. $(x + 4)(x^2 - x - 3)$ $x^3 + 3x^2 - 7x - 12$
47. $(x - 2)(x^2 + 2x + 4)$ $x^3 - 8$
48. $(x - 3)(x^2 + x + 1)$ $x^3 - 2x^2 - 2x - 3$
49. $-(x - 1)(x^2 - x + 2)$ $-x^3 + 2x^2 - 3x + 2$
50. $-(x - 2)(x^2 - 2x + 1)$ $-x^3 + 4x^2 - 5x + 2$
51. $-(x - 4)(x^2 - 4x - 1)$ $-x^3 + 8x^2 - 15x - 4$
52. $-(x - 3)(x^2 - 2x - 2)$ $-x^3 + 5x^2 - 4x - 6$

E In Problems 53–80, find the product.

53. $2x(x + 1)(x + 2)$ $2x^3 + 6x^2 + 4x$
54. $3x(x + 2)(x + 5)$ $3x^3 + 21x^2 + 30x$
55. $3x(x - 1)(x + 2)$ $3x^3 + 3x^2 - 6x$
56. $4x(x - 2)(x + 3)$ $4x^3 + 4x^2 - 24x$
57. $4x(x - 1)(x - 2)$ $4x^3 - 12x^2 + 8x$
58. $2x(x - 3)(x - 1)$ $2x^3 - 8x^2 + 6x$

59. $5x(x + 1)(x - 5)$ $5x^3 - 20x^2 - 25x$

60. $6x(x + 2)(x - 4)$ $6x^3 - 12x^2 - 48x$

61. $(x + 5)^3$ $x^3 + 15x^2 + 75x + 125$

62. $(x + 4)^3$ $x^3 + 12x^2 + 48x + 64$

63. $(2x + 3)^3$ $8x^3 + 36x^2 + 54x + 27$

64. $(3x + 2)^3$ $27x^3 + 54x^2 + 36x + 8$

65. $(2x + 3y)^3$ $8x^3 + 36x^2y + 54xy^2 + 27y^3$

66. $(3x + 2y)^3$ $27x^3 + 54x^2y + 36xy^2 + 8y^3$

67. $(4t^2 + 3)^2$ $16t^4 + 24t^2 + 9$

68. $(5t^2 + 1)^2$ $25t^4 + 10t^2 + 1$

69. $(4t^2 + 3u)^2$ $16t^4 + 24t^2u + 9u^2$

70. $(5t^2 + u)^2$ $25t^4 + 10t^2u + u^2$

71. $\left(3t^2 - \frac{1}{3}\right)^2$ $9t^4 - 2t^2 + \frac{1}{9}$

72. $\left(4t^2 - \frac{1}{2}\right)^2$ $16t^4 - 4t^2 + \frac{1}{4}$

73. $\left(3t^2 - \frac{1}{3}u\right)^2$ $9t^4 - 2t^2u + \frac{1}{9}u^2$

74. $\left(4t^2 - \frac{1}{2}u\right)^2$ $16t^4 - 4t^2u + \frac{1}{4}u^2$

75. $(3x^2 + 5)(3x^2 - 5)$ $9x^4 - 25$

76. $(4x^2 + 3)(4x^2 - 3)$ $16x^4 - 9$

77. $(3x^2 + 5y^2)(3x^2 - 5y^2)$ $9x^4 - 25y^4$

78. $(4x^2 + 3y^2)(4x^2 - 3y^2)$ $16x^4 - 9y^4$

79. $(4x^3 - 5y^3)(4x^3 + 5y^3)$ $16x^6 - 25y^6$

80. $(2x^4 - 3y^3)(2x^4 + 3y^3)$ $4x^8 - 9y^6$

81. Find the volume V of the cone when the height is $h = x + 4$ and the radius is $r = x + 2$.

Cone

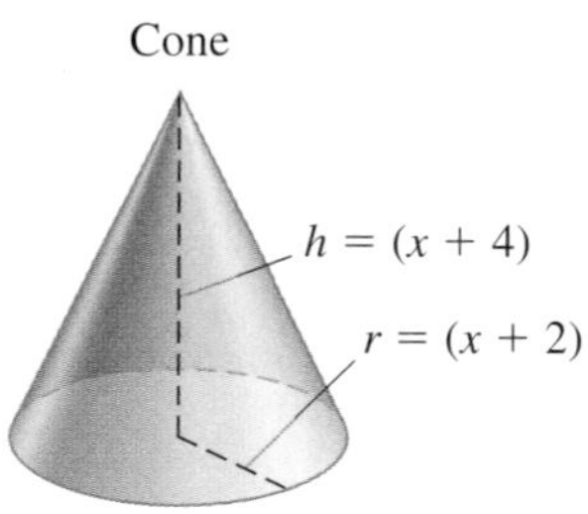

$V = \frac{1}{3}\pi r^2h$

$\frac{(x^3 + 8x^2 + 20x + 16)\pi}{3} = \frac{1}{3}\pi x^3 + \frac{8}{3}\pi x^2 + \frac{20}{3}\pi x + \frac{16}{3}\pi$

82. Find the volume V of the parallelepiped (box) when the width is $a = x$, the length is $c = x + 5$, and the height is $b = x + 4$.

Parallelepiped

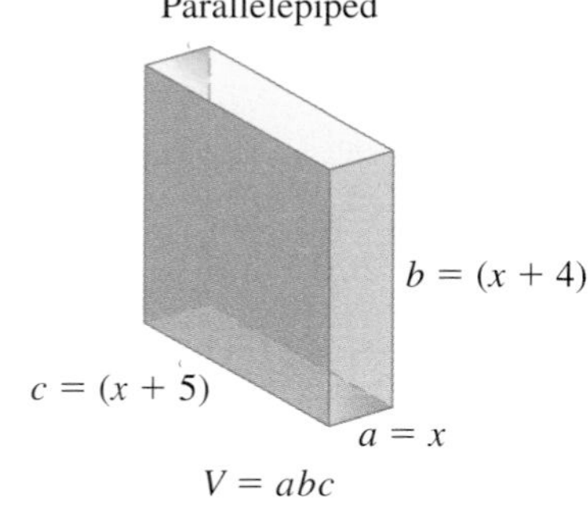

$V = abc$

$x^3 + 9x^2 + 20x$

Sphere

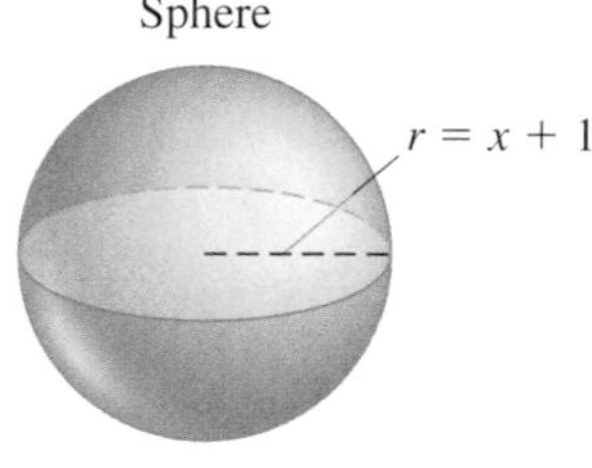

$V = \frac{4}{3}\pi r^3$

83. Find the volume V of the sphere when the radius is $r = x + 1$.
$\frac{4(x^3 + 3x^2 + 3x + 1)\pi}{3} = \frac{4}{3}\pi x^3 + 4\pi x^2 + 4\pi x + \frac{4}{3}\pi$

84. Find the volume V of the hemisphere (half of the sphere) when the radius is $r = x + 2$.
$\frac{2(x^3 + 6x^2 + 12x + 8)\pi}{3} = \frac{2}{3}\pi x^3 + 4\pi x^2 + 8\pi x + \frac{16}{3}\pi$

85. Find the volume V of the cylinder when the height is $h = x + 2$ and the radius is $r = x + 1$.

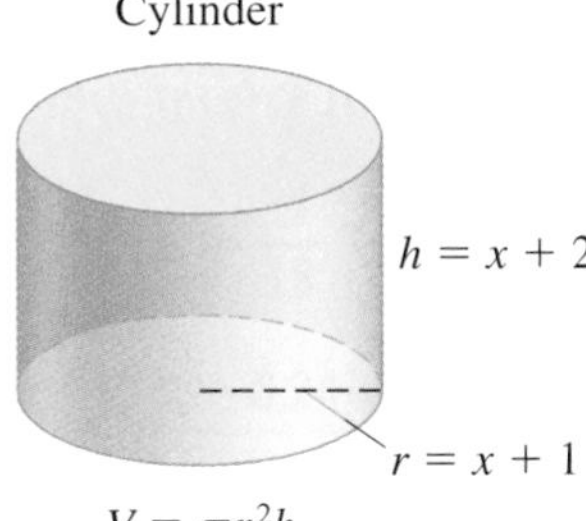

$V = \pi r^2 h$

$(x^3 + 4x^2 + 5x + 2)\pi = \pi x^3 + 4\pi x^2 + 5\pi x + 2\pi$

86. Find the volume V of the cube when the length of the side is $a = x + 3$.

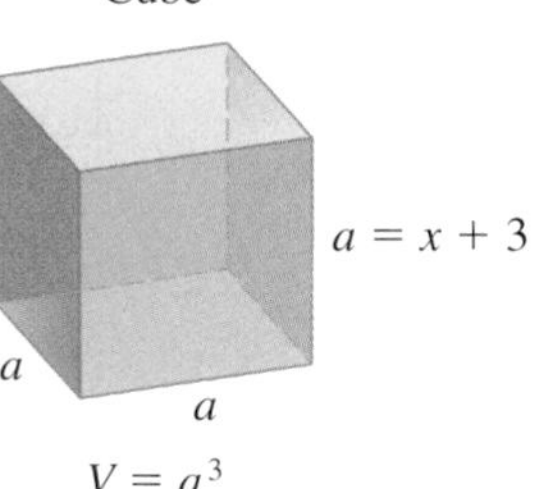

$V = a^3$

$x^3 + 9x^2 + 27x + 27$

APPLICATIONS

87. *Drain current* The drain current I_D for a certain gate source is

$$I_D = 0.004\left(1 + \frac{V}{2}\right)^2$$

where V is the voltage at the gate source. Expand this expression. $I_D = 0.004 + 0.004V + 0.001V^2$

88. *Transconductance* The equation of a transconductance curve is

$$I_D = I_{DSS}\left[1 - \frac{V_1}{V_2}\right]^2$$

Expand this expression. (Here I_{DSS} is the current from drain to source with shorted gate, V_1 is the voltage at the gate source, and V_2 is the voltage when the gate source is off.) $I_D = I_{DSS} - 2I_{DSS}\frac{V_1}{V_2} + I_{DSS}\frac{V_1^2}{V_2^2}$

89. *Heat transfer* The heat transmission between two objects of temperatures T_2 and T_1 involves the expression

$$(T_1^2 + T_2^2)(T_1^2 - T_2^2)$$

Multiply this expression. $T_1^4 - T_2^4$

90. *Deflection of a beam* The deflection of a certain beam involves the expression $w(l^2 - x^2)^2$. Expand this expression. $wl^4 - 2wl^2x^2 + wx^4$

91. *Heat from convection* The heat output from a natural draught convector is given by $K(t_n - t_a)^2$. Expand this expression. $Kt_n^2 - 2Kt_nt_a + Kt_a^2$

SKILL CHECKER

Try the Skill Checker Exercises so you'll be ready for the next section.

Find:

92. $\frac{28x^4}{4x^2}$ $7x^2$

93. $\frac{-5x^2}{10x^2}$ $-\frac{1}{2}$

94. $\frac{10x^2}{5x}$ $2x$

USING YOUR KNOWLEDGE

Binomial Fallacies

A common fallacy (mistake) when multiplying binomials is to assume that

$$(x + y)^2 = x^2 + y^2$$

Here are some arguments that should convince you this is *not* true.

95. Let $x = 1$, $y = 2$.

a. What is $(x + y)^2$? 9

b. What is $x^2 + y^2$? 5

c. Is $(x + y)^2 = x^2 + y^2$? No

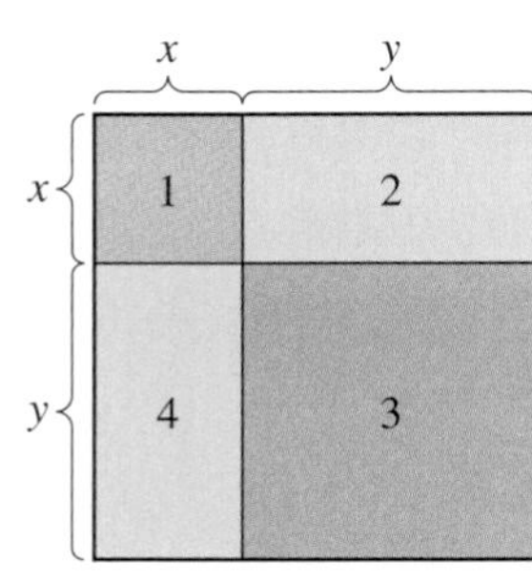

96. Let $x = 2$, $y = 1$.

a. What is $(x - y)^2$? 1

b. What is $x^2 - y^2$? 3

c. Is $(x - y)^2 = x^2 - y^2$? No

97. Look at the large square. Its area is $(x + y)^2$. The square is divided into four smaller areas numbered 1, 2, 3, and 4. What is the area of

a. square 1? x^2 **b.** rectangle 2? xy

c. square 3? y^2 **d.** rectangle 4? xy

98. The total area of the square is $(x + y)^2$. It's also the sum of the four areas numbered 1, 2, 3, and 4. What is the sum of these four areas? (Simplify your answer.) $x^2 + 2xy + y^2$

99. From your answer to Problem 98, what can you say about $x^2 + 2xy + y^2$ and $(x + y)^2$? They are equal.

100. The sum of the areas of the squares numbered 1 and 3 is $x^2 + y^2$. Is it true that $x^2 + y^2 = (x + y)^2$? No

WRITE ON

101. Write the procedure you use to square a binomial sum. Answers may vary.

102. Write the procedure you use to square a binomial difference. Answers may vary.

103. If you multiply two binomials, you usually get a trinomial. What is the exception? Explain. When the binomials are the sum and difference of the same two terms.

104. Describe the procedure you use to multiply two polynomials. Answers may vary.

MASTERY TEST

If you know how to do these problems, you have learned your lesson!

Expand and simplify:

105. $(x + 7)^2$ $x^2 + 14x + 49$

106. $(x - 2y)^2$ $x^2 - 4xy + 4y^2$

107. $(2x + 5y)^2$ $4x^2 + 20xy + 25y^2$

108. $(x - 12)^2$ $x^2 - 24x + 144$

109. $(x^2 + 2y)^2$ $x^4 + 4x^2y + 4y^2$

110. $(x - 3y)^2$ $x^2 - 6xy + 9y^2$

111. $(3x - 2y)(3x + 2y)$ $9x^2 - 4y^2$

112. $(3x^2 + 2y)(3x^2 - 2y)$ $9x^4 - 4y^2$

113. $(3x + 4)(5x - 6)$ $15x^2 + 2x - 24$

114. $(2x + 5)(5x - 2)$ $10x^2 + 21x - 10$

115. $(7x - 2y)(3x - 4y)$ $21x^2 - 34xy + 8y^2$

116. $(3y - 4x)(4y - 3x)$ $12y^2 - 25xy + 12x^2$

117. $(x + 2)(x^2 + x + 1)$ $x^3 + 3x^2 + 3x + 2$

118. $(x - 3)(x^2 - 3x - 1)$ $x^3 - 6x^2 + 8x + 3$

119. $(x^2 - 2)(x^2 + x + 1)$ $x^4 + x^3 - x^2 - 2x - 2$

120. $(x + 3)^3$ $x^3 + 9x^2 + 27x + 27$

121. $(x + 2y)^3$ $x^3 + 6x^2y + 12xy^2 + 8y^3$

122. $5x(x + 2)(x - 4)$ $5x^3 - 10x^2 - 40x$

123. $3x(x^2 + 1)(x^2 - 1)$ $3x^5 - 3x$

124. $-4x(x^2 + 1)^2$ $-4x^5 - 8x^3 - 4x$

4.8 DIVISION OF POLYNOMIALS

To Succeed, Review How To . . .

1. Use the distributive property to simplify expressions (p. 101).
2. Multiply polynomials (pp. 347–351).
3. Use the quotient rule of exponents (p. 298).

Objectives

A Divide a polynomial by a monomial.

B Divide one polynomial by another polynomial.

GETTING STARTED

Refrigerator Efficiency

In the preceding sections, we learned to add, subtract, and multiply polynomials. We are now ready for division. For example, do you know how efficient your refrigerator or your heat pump is? Their efficiency E is given by the quotient

$$E = \frac{T_1 - T_2}{T_1}$$

where T_1 and T_2 are the initial and final temperatures between which they operate. Can you do this division?

$$\frac{T_1 - T_2}{T_1}$$

Yes, you can, if you follow the steps:

$$\frac{T_1 - T_2}{T_1} = (T_1 - T_2) \div T_1$$

$$= (T_1 - T_2)\left(\frac{1}{T_1}\right)$$ Division by T_1 is the same as multiplication by $\frac{1}{T_1}$.

$$= T_1\left(\frac{1}{T_1}\right) - T_2\left(\frac{1}{T_1}\right)$$ Use the distributive law.

$$= \frac{T_1}{T_1} - \frac{T_2}{T_1}$$ Multiply.

$$= 1 - \frac{T_2}{T_1}$$ Since $\frac{T_1}{T_1} = 1$

In this section we will learn how to divide a polynomial by a monomial, and then we will generalize this idea, which will enable us to divide one polynomial by another.

A Dividing a Polynomial by a Monomial

To divide the binomial $4x^3 - 8x^2$ by the monomial $2x$, we proceed as we did in the *Getting Started.*

$$\frac{4x^3 - 8x^2}{2x} = (4x^3 - 8x^2) \div 2x$$

$$= (4x^3 - 8x^2)\left(\frac{1}{2x}\right)$$ Remember that to *divide* by $2x$, you multiply by the reciprocal $\frac{1}{2x}$.

$$= 4x^3\left(\frac{1}{2x}\right) - 8x^2\left(\frac{1}{2x}\right)$$ Use the distributive property.

$$= \frac{4x^3}{2x} - \frac{8x^2}{2x}$$ Multiply.

Teaching Tip

Show a numerical example first:

$$\frac{12-6}{2} \stackrel{?}{=} \frac{12}{2} - \frac{6}{2}$$

$$\frac{6}{2} \stackrel{?}{=} 6 - 3$$

$$3 = 3$$

Web It

For a calculator that will do the division for you and then explain it, try link 4-8-1 on the Bello Website at mhhe.com/bello.

$$= \frac{\overset{2x^2}{\cancel{4x^3}}}{\cancel{2x}} - \frac{\overset{4x}{\cancel{8x^2}}}{\cancel{2x}}$$ Use the quotient rule.

$$= 2x^2 - 4x$$ Simplify.

We then have

$$\frac{4x^3 - 8x^2}{2x} = \frac{4x^3}{2x} - \frac{8x^2}{2x} = 2x^2 - 4x$$

This example suggests the following rule.

RULE TO DIVIDE A POLYNOMIAL BY A MONOMIAL

To divide a polynomial by a monomial, divide *each* term in the polynomial by the *monomial.*

EXAMPLE 1 **Dividing a polynomial by a monomial**

Find:

a. $\frac{28x^4 - 14x^3}{7x^2}$ b. $\frac{20x^3 - 5x^2 + 10x}{10x^2}$

SOLUTION

a. $\frac{28x^4 - 14x^3}{7x^2} = \frac{28x^4}{7x^2} - \frac{14x^3}{7x^2}$

$= 4x^2 - 2x$

b. $\frac{20x^3 - 5x^2 + 10x}{10x^2} = \frac{20x^3}{10x^2} - \frac{5x^2}{10x^2} + \frac{10x}{10x^2}$

$= 2x - \frac{1}{2} + \frac{1}{x}$

PROBLEM 1

Find:

a. $\frac{27y^4 - 18y^3}{9y^2}$

b. $\frac{8y^3 - 4y^2 + 12y}{4y^2}$

B Dividing One Polynomial by Another Polynomial

Web It

To learn about the division of polynomials in which there is a remainder, try link 4-8-2 on the Bello Website at mhhe.com/bello.

If we wish to divide a polynomial, called the **dividend,** by another polynomial, called the **divisor,** we proceed very much as we did in long division in arithmetic. To show you that this is so, let's go through the division of 337 by 16 and $(x^2 + 3x + 3)$ (dividend) by $(x + 1)$ (divisor) side by side.

1. $16\overline{)337}$ with 2 above. Divide 33 by 16. It goes twice. Write 2 over the 33.

$x + 1\overline{)x^2 + 3x + 3}$ with x above. Divide x^2 by x $\left(\frac{x^2}{x} = x\right)$ It goes x times. Write x over the $3x$.

Answers

1. a. $3y^2 - 2y$ **b.** $2y - 1 + \frac{3}{y}$

Teaching Tip

Be sure students recognize that long division is used when the divisors are *other than* monomials.

2. $16\overline{)337}$ with quotient 2; -32; remainder 1.

Multiply 2 by 16 and subtract the product 32 from 33 to obtain 1.

$x+1\overline{)x^2+3x+3}$ with quotient x; $(-)\,x^2 + x$; $0 + 2x$

Multiply x by $x + 1$ and subtract the product $x^2 + x$ from $x^2 + 3x$ to obtain $0 + 2x$.

3. $16\overline{)337}$ with quotient 21; -32; 17

Bring down the 7. Now, divide 17 by 16. It goes once. Write 1 after the 2.

$x+1\overline{)x^2+3x+3}$ with quotient $x + 2$; $(-)\,x^2 + x$; $0 + 2x + 3$

Bring down the 3. Now, divide $2x$ by x $\left(\frac{2x}{x} = 2\right)$ It goes 2 times. Write $+2$ after the x.

4. $16\overline{)337}$ with quotient 21; -32; 17; -16; 1

Multiply 1 by 16 and subtract the result from 17. The remainder is 1.

$x+1\overline{)x^2+3x+3}$ with quotient $x + 2$; $(-)\,x^2 + x$; $0 + 2x + 3$; $(-)2x + 2$; 1

Multiply 2 by $x + 1$ to obtain $2x + 2$. Subtract this result from $2x + 3$. The remainder is 1.

5. The answer **(quotient)** can be written as 21 R 1 (read "21 remainder 1") or as

$$21 + \frac{1}{16}, \text{ which is } 21\tfrac{1}{16}$$

The answer **(quotient)** can be written as $(x + 2)$ R 1 (read "$x + 2$ remainder 1") or as

$$x + 2 + \frac{1}{x+1}$$

6. You can check this answer by multiplying 21 by 16 (336) and adding the remainder 1 to obtain 337, the dividend.

You can check the answer by multiplying $(x + 2)(x + 1)$, obtaining $x^2 + 3x + 2$, and then adding the remainder 1 to get $x^2 + 3x + 3$, the dividend.

EXAMPLE 2 **Dividing a polynomial by a binomial**

Divide: $x^2 + 2x - 17$ by $x - 3$

SOLUTION

x^2 divided by x is x.

$5x$ divided by x is 5.

$$\begin{array}{r} x + 5 \\ x - 3\overline{)x^2 + 2x - 17} \\ (-)\,x^2 - 3x \quad\quad \\ 5x - 17 \\ (-)\,5x - 15 \\ -\;2 \end{array}$$

$\frac{x^2}{x} = x$; $x(x - 3) = x^2 - 3x$

$\frac{5x}{x} = 5$; $5(x - 3) = 5x - 15$

Remainder

Thus $(x^2 + 2x - 17) \div (x - 3) = (x + 5)$ R -2 or

$$x + 5 + \frac{-2}{x - 3}$$

PROBLEM 2

Divide: $y^2 + y - 7$ by $y - 2$

Teaching Tip

Suggest that students use the mnemonic in the right-hand column to remember the four steps.

1. Divide.	1. Daddy
2. Multiply.	2. Mother
3. Subtract.	3. Sister
4. Bring down.	4. Brother

Answer

2. $y + 3 + \frac{-1}{y-2}$

If there are missing terms in the polynomial being divided, we insert zero coefficients, as shown in the next example.

EXAMPLE 3 **Dividing a third-degree polynomial by a binomial**

Divide: $2x^3 - 2 - 4x$ by $2 + 2x$

SOLUTION We write the polynomials in *descending* order, inserting $0x^2$ in the dividend, since the x^2 term is missing. We then have

$$\begin{array}{r|l}
x^2 - x - 1 & \\
2x + 2\overline{)2x^3 + 0x^2 - 4x - 2} & \left\{\frac{2x^3}{2x} = x^2\right. \\
(-)\ 2x^3 + 2x^2 \qquad\qquad & \quad x^2(2x + 2) = 2x^3 + 2x^2 \\
\hline
0 - 2x^2 - 4x \quad & \left\{-2x^2 \text{ divided by } 2x \text{ is } -x.\right. \\
(-) -2x^2 - 2x \quad & \quad -x(2x + 2) = -2x^2 - 2x. \\
\hline
-2x - 2 & \left\{-2x \text{ divided by } 2x \text{ is } -1.\right. \\
(-) -2x - 2 & \quad -1(2x + 2) = -2x - 2 \\
\hline
0 & \text{There is no remainder.}
\end{array}$$

Thus $(2x^3 - 4x - 2) \div (2x + 2) = x^2 - x - 1$.

PROBLEM 3

Divide: $y^3 + y^2 - 7y - 3$ by $3 + y$

Web It

To learn about the division of polynomials when there is no remainder, try link 4-8-3 on the Bello Website at mhhe.com/bello.

EXAMPLE 4 **Dividing a fourth-degree polynomial by a binomial**

Divide: $x^4 + x^3 - 3x^2 + 1$ by $x^2 - 3$

SOLUTION We write the polynomials in *descending* order, inserting $0x$ for the missing term in the dividend. We then have

$$\begin{array}{r|l}
x^2 + x & \\
x^2 - 3\overline{)x^4 + x^3 - 3x^2 + 0x + 1} & \left\{x^4 \text{ divided by } x^2 \text{ is } x^2.\right. \\
(-)\ x^4 \qquad - 3x^2 \qquad\qquad & \quad x^2(x^2 - 3) = x^4 - 3x^2 \\
\hline
0 + x^3 \qquad\qquad + 1 & \left\{x^3 \text{ divided by } x^2 \text{ is } x.\right. \\
(-)x^3 \qquad\quad - 3x \quad & \quad x(x^2 - 3) = x^3 - 3x \\
\hline
3x + 1 & \text{We cannot divide } 3x + 1 \text{ by } x^2\text{, so we stop. The remainder is } 3x + 1.
\end{array}$$

In general, we stop the division when the degree of the remainder is less than the degree of the divisor or when the remainder is zero. Thus $(x^4 + x^3 - 3x^2 + 1) \div (x^2 - 3) = (x^2 + x)$ R $(3x + 1)$. You can also write the answer as

$$x^2 + x + \frac{3x + 1}{x^2 - 3}$$

PROBLEM 4

Divide: $y^4 + y^3 - 2y^2 + y + 1$ by $y^2 - 2$

Answers

3. $y^2 - 2y - 1$

4. $y^2 + y + \frac{3y + 1}{y^2 - 2}$

Web It

For a complete lesson on division of polynomials, go to link 4-8-4 on the Bello Website at mhhe.com/bello.

PROCEDURE

Checking the Result

1. Multiply the divisor $x^2 - 3$ by the quotient $x^2 + x$. (Use FOIL.)
2. Add the remainder $3x + 1$.
3. The result must be the dividend

$$x^4 + x^3 - 3x^2 + 1 \quad \text{(and it is!)}$$

$$\underbrace{(x^2 - 3)}_{\text{Divisor}}\underbrace{(x^2 + x)}_{\text{Quotient}} = \begin{array}{r} x^4 + x^3 - 3x^2 - 3x \\ (+) \qquad\qquad\qquad 3x + 1 \\ \hline x^4 + x^3 - 3x^2 \qquad + 1 \end{array}$$

Calculate It Checking Quotients

Now, how would you check Example 4? The original problem is $Y_1 = (x^4 + x^3 - 3x^2 + 1)/(x^2 - 3)$ and the answer is $Y_2 = x^2 + x + (3x + 1)/(x^2 - 3)$. Be very careful when entering the expressions Y_1 and Y_2. Note the parentheses! Press GRAPH. Did you get the same graph for Y_1 and Y_2? If you are not sure, let us check the numerical values by pressing 2nd GRAPH. Y_1 and Y_2 seem to have the same values, so we are probably correct. Why "probably"? Because we are not able to check that for every value of x (column 1), Y_1 and Y_2 (columns 2 and 3) have the same value. However, the lists give plausible evidence that they do.

X	Y1	Y2
0	-.3333	-.3333
1	0	0
2	16	16
3	13.667	13.667
4	21	21
5	30.727	30.727
6	42.576	42.576
X=0		

Exercises 4.8

Boost your GRADE at mathzone.com!

MathZone

- Practice Problems
- Self-Tests
- Videos
- NetTutor
- e-Professors

A In Problems 1–10, divide.

1. $\dfrac{3x + 9y}{3}$ $x + 3y$
2. $\dfrac{6x + 8y}{2}$ $3x + 4y$
3. $\dfrac{10x - 5y}{5}$ $2x - y$
4. $\dfrac{24x - 12y}{6}$ $4x - 2y$
5. $\dfrac{8y^3 - 32y^2 + 16y}{-4y^2}$ $-2y + 8 - \dfrac{4}{y}$
6. $\dfrac{9y^3 - 45y^2 + 9}{-3y^2}$ $-3y + 15 - \dfrac{3}{y^2}$
7. $10x^2 + 8x$ by x $10x + 8$
8. $12x^2 + 18x$ by x $12x + 18$
9. $15x^3 - 10x^2$ by $5x^2$ $3x - 2$
10. $18x^4 - 24x^2$ by $3x^2$ $6x^2 - 8$

B In Problems 11–40, divide.

11. $x^2 + 5x + 6$ by $x + 3$ $x + 2$
12. $x^2 + 9x + 20$ by $x + 4$ $x + 5$
13. $y^2 + 3y - 11$ by $y + 5$ $(y - 2)$ R -1
14. $y^2 + 2y - 16$ by $y + 5$ $(y - 3)$ R -1
15. $2x + x^2 - 24$ by $x - 4$ $x + 6$
16. $4x + x^2 - 21$ by $x - 3$ $x + 7$

17. $-8 + 2x + 3x^2$ by $2 + x$
$3x - 4$

18. $-6 + x + 2x^2$ by $2 + x$
$2x - 3$

19. $2y^2 + 9y - 36$ by $7 + y$
$(2y - 5)$ R -1

20. $3y^2 + 13y - 32$ by $6 + y$
$(3y - 5)$ R -2

21. $2x^3 - 4x - 2$ by $2x + 2$
$x^2 - x - 1$

22. $3x^3 - 9x - 6$ by $3x + 3$
$x^2 - x - 2$

23. $y^4 - y^2 - 2y - 1$ by $y^2 + y + 1$
$y^2 - y - 1$

24. $y^4 - y^2 - 4y - 4$ by $y^2 + y + 2$
$y^2 - y - 2$

25. $8x^3 - 6x^2 + 5x - 9$ by $2x - 3$
$(4x^2 + 3x + 7)$ R 12

26. $2x^4 - x^3 + 7x - 2$ by $2x + 3$
$(x^3 - 2x^2 + 3x - 1)$ R 1

27. $x^3 - 8$ by $x - 2$
$x^2 + 2x + 4$

28. $x^3 + 64$ by $x + 4$
$x^2 - 4x + 16$

29. $8y^3 - 64$ by $2y - 4$
$4y^2 + 8y + 16$

30. $27x^3 - 8$ by $3x - 2$
$9x^2 + 6x + 4$

31. $x^4 - x^2 - 2x + 2$ by $x^2 - x - 1$
$(x^2 + x + 1)$ R 3

32. $y^4 - y^3 - 3y - 9$ by $y^2 - y - 3$ $y^2 + 3$

33. $x^5 - x^4 + 6x^2 - 5x + 3$ by $x^2 - 2x + 3$ $x^3 + x^2 - x + 1$

34. $y^6 - y^5 + 6y^3 - 5y^2 + 3y$ by $y^2 - 2y + 3$
$y^4 + y^3 - y^2 + y$

35. $m^4 - 11m^2 + 34$ by $m^2 - 3$
$(m^2 - 8)$ R 10

36. $n^3 - n^2 - 6n$ by $n^2 + 3n$
$(n - 4)$ R 6

37. $\dfrac{x^3 - y^3}{x - y}$
$x^2 + xy + y^2$

38. $\dfrac{x^3 + 8}{x + 2}$
$x^2 - 2x + 4$

39. $\dfrac{x^3 + 8}{x - 2}$ $(x^2 + 2x + 4)$ R 16

40. $\dfrac{x^5 + 32}{x - 2}$ $(x^4 + 2x^3 + 4x^2 + 8x + 16)$ R 64

SKILL CHECKER

Try the Skill Checker Exercises so you'll be ready for the next section.

Find the LCM:

41. 20 and 18 180

42. 30 and 16 240

43. 40 and 12 120

44. 10, 18, and 12 180

45. 20, 30, and 18 180

46. 40, 15, and 10 120

USING YOUR KNOWLEDGE

Is It Profitable?

In business, the average cost per unit of a product, denoted by $\overline{C(x)}$, is defined by

$$\overline{C(x)} = \frac{C(x)}{x}$$

where $C(x)$ is a polynomial in the variable x, and x is the number of units produced.

47. If $C(x) = 3x^2 + 5x$, find $\overline{C(x)}$. $\overline{C(x)} = 3x + 5$

48. If $C(x) = 30 + 3x^2$, find $\overline{C(x)}$. $\overline{C(x)} = \frac{30}{x} + 3x$

The average profit $\overline{P(x)}$ is

$$\overline{P(x)} = \frac{P(x)}{x}$$

where $P(x)$ is a polynomial in the variable x, and x is the number of units sold in a certain period of time.

49. If $P(x) = 50x + x^2 - 7000$ (dollars), find the average profit. $\overline{P(x)} = 50 + x - \frac{7000}{x}$

50. If in Problem 49, 100 units are sold in a period of 1 week, what is the average profit? \$80

WRITE ON

51. How can you check that your answer is correct when you divide one polynomial by another? Explain. Answers may vary.

52. A problem in a recent test stated: "Find the quotient of $x^2 + 5x + 6$ and $x + 2$." Do you have to divide $x + 2$ by $x^2 + 5x + 6$ or do you have to divide $x^2 + 5x + 6$ by $x + 2$? Explain. Answers may vary.

53. When you are dividing one polynomial by another, when do you stop the division process? Answers may vary.

MASTERY TEST

If you know how to do these problems, you have learned your lesson!

Divide:

54. $x^4 + x^3 - 2x^2 + 1$ by $x^2 - 2$
$(x^2 + x)$ R $(2x + 1)$

55. $2x^3 + x - 3$ by $x - 1$
$2x^2 + 2x + 3$

56. $x^2 + 4x - 15$ by $x - 2$
$(x + 6)$ R -3

57. $\frac{24x^4 - 18x^3}{6x^2}$ $4x^2 - 3x$

58. $\frac{16x^4 - 4x^2 + 8x}{8x^2}$ $2x^2 - \frac{1}{2} + \frac{1}{x}$

59. $\frac{-6y^3 + 12y^2 + 3}{-3y^2}$ $2y - 4 - \frac{1}{y^2}$

COLLABORATIVE LEARNING

How fast can you go?

How fast can you obtain information to solve a problem? Form three groups: library, the Web, and bookstore (where you can look at books, papers, and so on for free). Each group is going to research car prices. Select a car model that has been on the market for at least 5 years. Each of the groups should find:

1. The new car value and the value of a 3-year-old car of the same model
2. The estimated depreciation rate for the car
3. The estimated value of the car in 3 years
4. A graph comparing age and value of the car for the next 5 years
5. An equation of the form $C = P(1 - r)^n$ or $C = rn + b$, where n is the number of years after purchase and r is the depreciation rate

Which group finished first? Share the procedure used to obtain your information so the most efficient research method can be established.

Research Questions

1. In the *Human Side of Algebra* at the beginning of this chapter, we mentioned that in the Golden Age of Greek mathematics, roughly from 300 to 200 B.C., three mathematicians "stood head and shoulders above all the others of the time." Apollonius was one. Who were the other two, and what were their contributions to mathematics?
2. Write a short paragraph about Diophantus and his contributions to mathematics.
3. Write a paper about the *Arithmetica* written by Diophantus. How many books were in this *Arithmetica?* What is its relationship to algebra?
4. It is said that "one of Diophantus' main contributions was the 'syncopation' of algebra." Explain what this means.
5. Write a short paragraph about Nicole Oresme and his mathematical achievements.
6. Expand on the discussion of the works of Nicholas Chuquet described in the *Human Side of Algebra.*
7. Write a short paper about the contributions of François Viète regarding the development of algebraic notation.

Summary

SECTION	ITEM	MEANING	EXAMPLE
4.1A	Product rule for exponents	$x^m \cdot x^n = x^{m+n}$	$x^3 \cdot x^5 = x^{3+5} = x^8$
4.1B	Quotient rule for exponents	$\frac{x^m}{x^n} = x^{m-n}$	$\frac{x^8}{x^3} = x^{8-3} = x^5$
4.1C	Power rule for products Power rule for quotients	$(x^m y^n)^k = x^{mk} y^{nk}$ $\left(\frac{x}{y}\right)^m = \frac{x^m}{y^m}$	$(x^3 y^2)^4 = x^{12} y^8$ $\left(\frac{x^2}{y^3}\right)^4 = \frac{x^8}{y^{12}}$
4.2	Zero exponent	For $x \neq 0$, $x^0 = 1$	$3^0 = 1$, $(-8)^0 = 1$, $(3x)^0 = 1$
4.2A	x^{-n}, n a positive integer	$x^{-n} = \frac{1}{x^n}$	$2^{-3} = \frac{1}{2^3} = \frac{1}{8}$
4.3A	Scientific notation	A number is in scientific notation when written in the form $M \times 10^n$, where M is a number between 1 and 10 and n is an integer.	3×10^{-3} and 2.7×10^5 are in scientific notation.
4.4A	Polynomial	An algebraic expression formed by using the operations of addition and subtraction on products of numbers and a variable raised to whole-number exponents	$x^2 + 3x - 5$, $2x + 8 - x^3$, and $9x^7 - 3x^3 + 4x^8 - 10$ are polynomials but $\sqrt{x} - 3$, $\frac{x^2 - 2x + 3}{x}$, and $x^{3/2} - x$ are not.
	Terms	The parts of a polynomial separated by plus signs are the terms.	The terms of $x^2 - 2x + 3$ are x^2, $-2x$, and 3.
	Monomial	A polynomial with one term	$3x$, $7x^2$, and $-3x^{10}$ are monomials.
	Binomial	A polynomial with two terms	$3x + x^2$, $7x - 8$, and $x^3 - 8x^7$ are binomials.
	Trinomial	A polynomial with three terms	$-8 + 3x + x^2$, $7x - 8 + x^4$, and $x^3 - 8x^7 + 9$ are trinomials.
4.4B	Degree	The degree of a polynomial is the highest exponent of the variable.	The degree of $8 + 3x + x^2$ is 2, and the degree of $7x - 8 + x^4$ is 4.
4.6C	FOIL method (SP1)	To multiply two binomials such as $(x + a)(x + b)$, multiply the **F**irst terms, the **O**uter terms, the **I**nner terms, and the **L**ast terms and add.	F O I L $(x + 2)(x + 3) = x^2 + 3x + 2x + 6$ $= x^2 + 5x + 6$
4.7A	The square of a binomial sum (SP2)	$(X + A)^2 = X^2 + 2AX + A^2$	$(x + 5)^2 = x^2 + 2 \cdot 5 \cdot x + 5^2$ $= x^2 + 10x + 25$

(Continued)

SECTION	ITEM	MEANING	EXAMPLE
4.7B	The square of a binomial difference (SP3)	$(X - A)^2 = X^2 - 2AX + A^2$	$(x - 5)^2 = x^2 - 2 \cdot 5 \cdot x + 5^2$ $= x^2 - 10x + 25$
4.7C	The product of the sum and difference of two terms (SP4)	$(X + A)(X - A) = X^2 - A^2$	$(x + 7)(x - 7) = x^2 - 7^2$ $= x^2 - 49$
4.8A	Dividing a polynomial by a monomial	To divide a polynomial by a monomial, divide each term in the polynomial by the monomial.	$\frac{35x^5 - 21x^3}{7x^2} = \frac{35x^5}{7x^2} - \frac{21x^3}{7x^2}$ $= 5x^3 - 3x$
4.8B	Dividing one polynomial by another	To divide one polynomial by another, use polynomial long division.	Divide $6x^2 + 11x - 35$ by $3x - 5$. $\begin{array}{r} 2x + 7 \\ 3x - 5\overline{)6x^2 + 11x - 35} \\ (-)6x^2 - 10x \\ \hline 21x - 35 \\ (-)21x - 35 \\ \hline 0 \end{array}$ Thus, $(6x^2 + 11x - 35) \div (3x - 5)$ is $2x + 7$.

Review Exercises

(If you need help with these questions, look in the section indicated in brackets.)

1. [4.1A] Find the product.

a. (i) $(3a^2b)(-5ab^3)$ $-15a^3b^4$
(ii) $(4a^2b)(-6ab^4)$ $-24a^3b^5$
(iii) $(5a^2b)(-7ab^3)$ $-35a^3b^4$

b. (i) $(-2xy^2z)(-3x^2yz^4)$ $6x^3y^3z^5$
(ii) $(-3x^2yz^2)(-4xy^3z)$ $12x^3y^4z^3$
(iii) $(-4xyz)(-5xy^2z^3)$ $20x^2y^3z^4$

2. [4.1B] Find the quotient.

a. (i) $\frac{16x^6y^8}{-8xy^4}$ $-2x^5y^4$
(ii) $\frac{24x^7y^6}{-4xy^3}$ $-6x^6y^3$
(iii) $\frac{-18x^8y^7}{9xy^4}$ $-2x^7y^3$

b. (i) $\frac{-8x^9y^7}{-16x^4y}$ $\frac{x^5y^6}{2}$
(ii) $\frac{-5x^7y^8}{-10x^6y}$ $\frac{xy^7}{2}$
(iii) $\frac{-3x^9y^7}{-9x^8y}$ $\frac{xy^6}{3}$

3. [4.1C] Simplify.

a. $(2^2)^3$ $2^6 = 64$ **b.** $(2^2)^2$ $2^4 = 16$
c. $(3^2)^2$ $3^4 = 81$

4. [4.1C] Simplify.

a. $(y^3)^2$ y^6 **b.** $(x^2)^3$ x^6
c. $(a^4)^5$ a^{20}

5. [4.1C] Simplify.

a. $(4xy^3)^2$ $16x^2y^6$ **b.** $(2x^2y)^3$ $8x^6y^3$
c. $(3x^2y^2)^3$ $27x^6y^6$

6. [4.1C] Simplify.

a. $(-2xy^3)^3$ $-8x^3y^9$ **b.** $(-3x^2y^3)^2$ $9x^4y^6$
c. $(-2x^2y^2)^3$ $-8x^6y^6$

7. [4.1C] Simplify.

a. $\left(\frac{2y^2}{x^4}\right)^2$ $\frac{4y^4}{x^8}$ **b.** $\left(\frac{3x}{y^3}\right)^3$ $\frac{27x^3}{y^9}$

c. $\left(\frac{2x^2}{y^4}\right)^4$ $\frac{16x^8}{y^{16}}$

8. [4.1C] Simplify.

a. $(2x^4)^3(-2y^2)^2$ $32x^{12}y^4$

b. $(3x^2)^2(-2y^3)^3$ $-72x^4y^9$

c. $(4x^3)^2(-2y^4)^4$ $256x^6y^{16}$

9. [4.2A] Write using positive exponents and then simplify.

a. 2^{-3} $\frac{1}{2^3} = \frac{1}{8}$ **b.** 3^{-4} $\frac{1}{3^4} = \frac{1}{81}$

c. 5^{-2} $\frac{1}{5^2} = \frac{1}{25}$

10. [4.2A] Write using positive exponents.

a. $\left(\frac{1}{x}\right)^{-4}$ x^4 **b.** $\left(\frac{1}{y}\right)^{-3}$ y^3

c. $\left(\frac{1}{z}\right)^{-5}$ z^5

11. [4.2B] Write using negative exponents.

a. $\frac{1}{x^5}$ x^{-5} **b.** $\frac{1}{y^7}$ y^{-7}

c. $\frac{1}{z^8}$ z^{-8}

12. [4.2C] Multiply and simplify.

a. (i) $2^8 \cdot 2^{-5}$ 8 **b. (i)** $y^{-3} \cdot y^{-5}$ $\frac{1}{y^8}$

(ii) $2^6 \cdot 2^{-4}$ 4 **(ii)** $y^{-2} \cdot y^{-3}$ $\frac{1}{y^5}$

(iii) $3^6 \cdot 3^{-3}$ 27 **(iii)** $y^{-4} \cdot y^{-2}$ $\frac{1}{y^6}$

13. [4.2C] Divide and simplify.

a. (i) $\frac{x}{x^5}$ $\frac{1}{x^4}$ **b. (i)** $\frac{a^{-2}}{a^{-2}}$ 1 **c. (i)** $\frac{x^{-2}}{x^{-3}}$ x

(ii) $\frac{x}{x^7}$ $\frac{1}{x^6}$ **(ii)** $\frac{a^{-4}}{a^{-4}}$ 1 **(ii)** $\frac{x^{-5}}{x^{-8}}$ x^3

(iii) $\frac{x}{x^9}$ $\frac{1}{x^8}$ **(iii)** $\frac{a^{-10}}{a^{-10}}$ 1 **(iii)** $\frac{x^{-7}}{x^{-9}}$ x^2

14. [4.2C] Simplify.

a. $\left(\frac{2x^3y^4}{3x^4y^{-3}}\right)^{-2}$ $\frac{9x^2}{4y^{14}}$ **b.** $\left(\frac{3x^5y^{-3}}{2x^7y^4}\right)^{-3}$ $\frac{8x^6y^{21}}{27}$

c. $\left(\frac{3x^{-5}y^{-4}}{2x^{-6}y^{-8}}\right)^{-2}$ $\frac{4}{9x^2y^8}$

15. [4.3A] Write in scientific notation.

a. (i) 44,000,000 4.4×10^7

(ii) 4,500,000 4.5×10^6

(iii) 460,000 4.6×10^5

b. (i) 0.0014 1.4×10^{-3}

(ii) 0.00015 1.5×10^{-4}

(iii) 0.000016 1.6×10^{-5}

16. [4.3B] Perform the indicated operations and write the answer in scientific notation.

a. (i) $(2 \times 10^2) \times (1.1 \times 10^3)$ 2.2×10^5

(ii) $(3 \times 10^2) \times (3.1 \times 10^4)$ 9.3×10^6

(iii) $(4 \times 10^2) \times (3.1 \times 10^5)$ 1.24×10^8

b. (i) $\frac{1.15 \times 10^{-3}}{2.3 \times 10^{-4}}$ 5

(ii) $\frac{1.38 \times 10^{-3}}{2.3 \times 10^{-4}}$ 6

(iii) $\frac{1.61 \times 10^{-3}}{2.3 \times 10^{-4}}$ 7

17. [4.4A] Classify as a monomial (M), binomial (B), or trinomial (T).

a. $9x^2 - 9 + 7x$ T **b.** $7x^2$ M **c.** $3x - 1$ B

18. [4.4B] Find the degree of the given polynomial.

a. $3x^2 - 7x + 8x^4$ 4 **b.** $-4x + 2x^2 - 3$ 2

c. $8 + 3x - 4x^2$ 2

19. [4.4C] Write the given polynomial in descending order.

a. $4x^2 - 8x + 9x^4$ $9x^4 + 4x^2 - 8x$

b. $-3x + 4x^2 - 3$ $4x^2 - 3x - 3$

c. $8 + 3x - 4x^2$ $-4x^2 + 3x + 8$

20. [4.4D] Find the value of $-16t^2 + 300$ for each value of t.

a. $t = 1$ 284 **b.** $t = 3$ 156 **c.** $t = 5$ -100

21. [4.5A] Add the given polynomials.

a. $-5x + 7x^2 - 3$ and $-2x^2 - 7 + 4x$ $5x^2 - x - 10$
b. $-3x^2 + 8x - 1$ and $3 + 7x - 2x^2$ $-5x^2 + 15x + 2$
c. $-4 + 3x^2 - 5x$ and $6x^2 - 2x + 5$ $9x^2 - 7x + 1$

22. [4.5B] Subtract the first polynomial from the second.

a. $3x - 4 + 7x^2$ from $6x^2 - 4x$ $-x^2 - 7x + 4$
b. $5x - 3 + 2x^2$ from $9x^2 - 2x$ $7x^2 - 7x + 3$
c. $6 - 2x + 5x^2$ from $2x - 5$ $-5x^2 + 4x - 11$

23. [4.6A] Find.

a. $(-6x^2)(3x^5)$ $-18x^7$
b. $(-8x^3)(5x^6)$ $-40x^9$
c. $(-9x^4)(3x^7)$ $-27x^{11}$

24. [4.6B] Remove parentheses (simplify).

a. $-2x^2(x + 2y)$ $-2x^3 - 4x^2y$
b. $-3x^3(2x + 3y)$ $-6x^4 - 9x^3y$
c. $-4x^3(5x + 7y)$ $-20x^4 - 28x^3y$

25. [4.6C] Find.

a. $(x + 6)(x + 9)$ $x^2 + 15x + 54$
b. $(x + 2)(x + 3)$ $x^2 + 5x + 6$
c. $(x + 7)(x + 9)$ $x^2 + 16x + 63$

26. [4.6C] Find.

a. $(x + 7)(x - 3)$ $x^2 + 4x - 21$
b. $(x + 6)(x - 2)$ $x^2 + 4x - 12$
c. $(x + 5)(x - 1)$ $x^2 + 4x - 5$

27. [4.6C] Find.

a. $(x + 3)(x - 7)$ $x^2 - 4x - 21$
b. $(x + 2)(x - 6)$ $x^2 - 4x - 12$
c. $(x + 1)(x - 5)$ $x^2 - 4x - 5$

28. [4.6C] Find.

a. $(3x - 2y)(2x - 3y)$ $6x^2 - 13xy + 6y^2$
b. $(5x - 3y)(4x - 3y)$ $20x^2 - 27xy + 9y^2$
c. $(4x - 3y)(2x - 5y)$ $8x^2 - 26xy + 15y^2$

29. [4.7A] Expand.

a. $(2x + 3y)^2$ $4x^2 + 12xy + 9y^2$
b. $(3x + 4y)^2$ $9x^2 + 24xy + 16y^2$
c. $(4x + 5y)^2$ $16x^2 + 40xy + 25y^2$

30. [4.7B] Expand.

a. $(2x - 3y)^2$ $4x^2 - 12xy + 9y^2$
b. $(3x - 2y)^2$ $9x^2 - 12xy + 4y^2$
c. $(5x - 2y)^2$ $25x^2 - 20xy + 4y^2$

31. [4.7C] Find.

a. $(3x - 5y)(3x + 5y)$ $9x^2 - 25y^2$
b. $(3x - 2y)(3x + 2y)$ $9x^2 - 4y^2$
c. $(3x - 4y)(3x + 4y)$ $9x^2 - 16y^2$

32. [4.7D] Find.

a. $(x + 1)(x^2 + 3x + 2)$ $x^3 + 4x^2 + 5x + 2$
b. $(x + 2)(x^2 + 3x + 2)$ $x^3 + 5x^2 + 8x + 4$
c. $(x + 3)(x^2 + 3x + 2)$ $x^3 + 6x^2 + 11x + 6$

33. [4.7E] Find.

a. $3x(x + 1)(x + 2)$ $3x^3 + 9x^2 + 6x$
b. $4x(x + 1)(x + 2)$ $4x^3 + 12x^2 + 8x$
c. $5x(x + 1)(x + 2)$ $5x^3 + 15x^2 + 10x$

34. [4.7E] Expand.

a. $(x + 2)^3$ $x^3 + 6x^2 + 12x + 8$
b. $(x + 3)^3$ $x^3 + 9x^2 + 27x + 27$
c. $(x + 4)^3$ $x^3 + 12x^2 + 48x + 64$

35. [4.7E] Expand.

a. $\left(5x^2 - \frac{1}{2}\right)^2$ $25x^4 - 5x^2 + \frac{1}{4}$
b. $\left(7x^2 - \frac{1}{2}\right)^2$ $49x^4 - 7x^2 + \frac{1}{4}$
c. $\left(9x^2 - \frac{1}{2}\right)^2$ $81x^4 - 9x^2 + \frac{1}{4}$

36. [4.7E] Expand.

a. $(3x^2 + 2)(3x^2 - 2)$ $9x^4 - 4$
b. $(3x^2 + 4)(3x^2 - 4)$ $9x^4 - 16$
c. $(2x^2 + 5)(2x^2 - 5)$ $4x^4 - 25$

37. [4.8A] Find.

a. $\dfrac{18x^3 - 9x^2}{9x}$ $2x^2 - x$
b. $\dfrac{20x^3 - 10x^2}{5x}$ $4x^2 - 2x$
c. $\dfrac{24x^3 - 12x^2}{6x}$ $4x^2 - 2x$

38. [4.8B] Divide.

a. $x^2 + 4x - 12$ by $x - 2$ $x + 6$
b. $x^2 + 4x - 21$ by $x - 3$ $x + 7$
c. $x^2 + 4x - 32$ by $x - 4$ $x + 8$

39. [4.8B] Divide.

a. $8x^3 - 16x - 8$ by $2 + 2x$ $4x^2 - 4x - 4$
b. $12x^3 - 24x - 12$ by $2 + 2x$ $6x^2 - 6x - 6$
c. $4x^3 - 8x - 4$ by $2 + 2x$ $2x^2 - 2x - 2$

40. [4.8B] Divide.

a. $2x^3 - 20x + 8$ by $x - 3$ $(2x^2 + 6x - 2)$ R 2
b. $2x^3 - 21x + 12$ by $x - 3$ $(2x^2 + 6x - 3)$ R 3
c. $3x^3 - 4x + 5$ by $x - 1$ $(3x^2 + 3x - 1)$ R 4

41. [4.8B] Divide.

a. $x^4 + x^3 - 4x^2 + 1$ by $x^2 - 4$ $(x^2 + x)$ R $(4x + 1)$
b. $x^4 + x^3 - 5x^2 + 1$ by $x^2 - 5$ $(x^2 + x)$ R $(5x + 1)$
c. $x^4 + x^3 - 6x^2 + 1$ by $x^2 - 6$ $(x^2 + x)$ R $(6x + 1)$

Practice Test 4

(Answers on page 382)

1. Find.

a. $(2a^3b)(-6ab^3)$ **b.** $(-2x^2yz)(-6xy^3z^4)$
c. $\dfrac{18x^5y^7}{-9xy^3}$

2. Find.

a. $(2x^3y^2)^3$
b. $(-3x^2y^3)^2$

3. Find.

a. $\left(\dfrac{3}{4}\right)^3$ **b.** $\left(\dfrac{3x^3}{y^4}\right)^3$

4. Simplify and write the answer without negative exponents.

a. $\left(\dfrac{1}{x}\right)^{-7}$ **b.** $\dfrac{x^{-6}}{x^{-6}}$ **c.** $\dfrac{x^{-6}}{x^{-7}}$

5. Simplify.

a. $3x^2x^{-4}$ **b.** $\left(\dfrac{2x^{-3}y^4}{3x^2y^3}\right)^{-2}$

6. Write in scientific notation.

a. 48,000,000 **b.** 0.00000037

7. Perform the indicated operations.

a. $(3 \times 10^4) \times (7.1 \times 10^6)$
b. $\dfrac{2.84 \times 10^{-2}}{7.1 \times 10^{-3}}$

8. Classify as a monomial (M), binomial (B), or trinomial (T).

a. $3x - 5$ **b.** $5x^3$ **c.** $8x^2 - 2 + 5x$

9. Write the polynomial $-3x + 7 + 8x^2$ in descending order and find its degree.

10. Find the value of $-16t^2 + 100$ when $t = 2$.

11. Add $-4x + 8x^2 - 3$ and $-5x^2 - 4 + 2x$.

12. Subtract $5x - 2 + 8x^2$ from $3x^2 - 2x$.

13. Remove parentheses (simplify): $-2x^2(x + 3y)$.

14. Find $(x + 8)(x - 3)$.

15. Find $(x + 4)(x - 6)$.

16. Find $(5x - 2y)(4x - 3y)$.

17. Expand $(3x + 5y)^2$.

18. Expand $(2x - 7y)^2$.

19. Find $(2x - 5y)(2x + 5y)$.

20. Find $(x + 2)(x^2 + 5x + 3)$.

21. Find $3x(x + 2)(x + 5)$.

22. Expand $(x + 7)^3$.

23. Expand $\left(3x^2 - \dfrac{1}{2}\right)^2$.

24. Find $(3x^2 + 7)(3x^2 - 7)$.

25. Divide $2x^3 - 9x + 5$ by $x - 2$.

Answers to Practice Test

ANSWER	IF YOU MISSED	REVIEW		
	QUESTION	SECTION	EXAMPLES	PAGE
1. a. $-12a^4b^4$ **b.** $12x^3y^4z^5$ **c.** $-2x^4y^4$	1	4.1	1, 2, 3	296–298
2. a. $8x^9y^6$ **b.** $9x^4y^6$	2	4.1	4, 5	299–300
3. a. $\frac{27}{64}$ **b.** $\frac{27x^9}{y^{12}}$	3	4.1	6	300
4. a. x^7 **b.** 1 **c.** x	4	4.2	1, 2, 5	309, 311
5. a. $\frac{3}{x^2}$ **b.** $\frac{9x^{10}}{4y^2}$	5	4.2	4, 6	311, 312
6. a. 4.8×10^7 **b.** 3.7×10^{-7}	6	4.3	1, 2	320
7. a. 2.13×10^{11} **b.** 4	7	4.3	3, 4	320–321
8. a. B **b.** M **c.** T	8	4.4	1	327
9. $8x^2 - 3x + 7$; 2	9	4.4	2, 3	328
10. 36	10	4.4	4, 5, 6	329–330
11. $3x^2 - 2x - 7$	11	4.5	1, 2	337–338
12. $-5x^2 - 7x + 2$	12	4.5	3	338–339
13. $-2x^3 - 6x^2y$	13	4.6	1, 2	347–348
14. $x^2 + 5x - 24$	14	4.6	3a	350
15. $x^2 - 2x - 24$	15	4.6	3b	350
16. $20x^2 - 23xy + 6y^2$	16	4.6	4b	350
17. $9x^2 + 30xy + 25y^2$	17	4.7	1	358
18. $4x^2 - 28xy + 49y^2$	18	4.7	2	359
19. $4x^2 - 25y^2$	19	4.7	3	360
20. $x^3 + 7x^2 + 13x + 6$	20	4.7	4	361–362
21. $3x^3 + 21x^2 + 30x$	21	4.7	5	362–363
22. $x^3 + 21x^2 + 147x + 343$	22	4.7	6	363
23. $9x^4 - 3x^2 + \frac{1}{4}$	23	4.7	7	363
24. $9x^4 - 49$	24	4.7	8	364
25. $(2x^2 + 4x - 1)$ R 3	25	4.8	1, 2, 3, 4	370–372

Cumulative Review Chapters 1–4

1. Find the additive inverse (opposite) of -7. 7

2. Find: $\left|-9\frac{9}{10}\right|$ $9\frac{9}{10}$

3. Find: $-\frac{1}{9} + \left(-\frac{1}{6}\right)$ $-\frac{5}{18}$

4. Find: $7.2 - (-6.4)$ 13.6

5. Find: $(-2.4)(2.6)$ -6.24

6. Find: $-(2^4)$ -16

7. Find: $-\frac{3}{4} \div \left(-\frac{7}{8}\right)$ $\frac{6}{7}$

8. Evaluate $y \div 5 \cdot x - z$ for $x = 5$, $y = 50$, $z = 3$. 47

9. Which law is illustrated by the following statement?

$$6 \cdot (5 \cdot 2) = (6 \cdot 5) \cdot 2$$

Associative law of multiplication

10. Multiply: $3(2x - 8)$ $6x - 24$

11. Combine like terms: $-8xy^3 - (-xy^3)$ $-7xy^3$

12. Simplify: $5x + (x + 4) - 2(x - 3)$ $4x + 10$

13. Write in symbols: The quotient of $(m + 3n)$ and p $\frac{m + 3n}{p}$

14. Does the number -3 satisfy the equation $12 = 15 - x$? No

15. Solve for x: $5 = 5(x - 3) + 5 - 4x$ $x = 15$

16. Solve for x: $-\frac{8}{7}x = -56$ $x = 49$

17. Solve for x: $\frac{x}{4} - \frac{x}{9} = 5$ $x = 36$

18. Solve for x: $5 - \frac{x}{3} = \frac{3(x + 1)}{7}$ $x = 6$

19. Solve for b in the equation $S = 6a^2b$. $b = \frac{S}{6a^2}$

20. The sum of two numbers is 100. If one of the numbers is 20 more than the other, what are the numbers? 40 and 60

21. Dave has invested a certain amount of money in stocks and bonds. The total annual return from these investments is \$615. If the stocks produce \$245 more in returns than the bonds, how much money does Dave receive annually from each type of investment? \$430 from stocks and \$185 from bonds

22. Train A leaves a station traveling at 30 mph. Six hours later, train B leaves the same station traveling in the same direction at 40 mph. How long does it take for train B to catch up to train A? 18 hr

23. Susan purchased some municipal bonds yielding 8% annually and some certificates of deposit yielding 11% annually. If Susan's total investment amounts to \$17,000 and the annual income is \$1630, how much money is invested in bonds and how much is invested in certificates of deposit? \$8000 in bonds; \$9000 in certificates of deposit

24. Graph: $-\frac{x}{2} + \frac{x}{9} \geq \frac{x - 9}{9}$

0 2

25. Graph the point $C(4, -4)$.

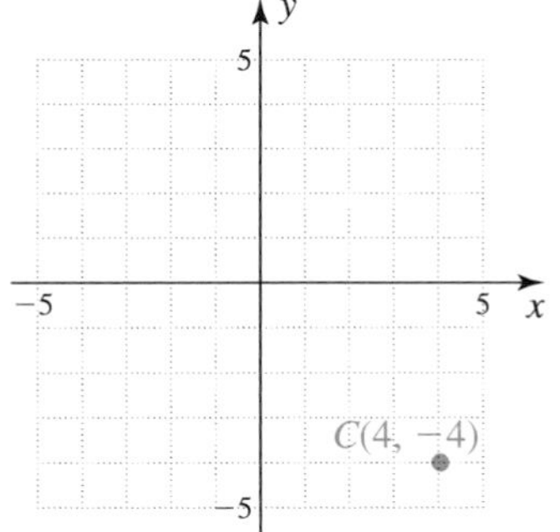

26. What are the coordinates of point A? $(3, -2)$

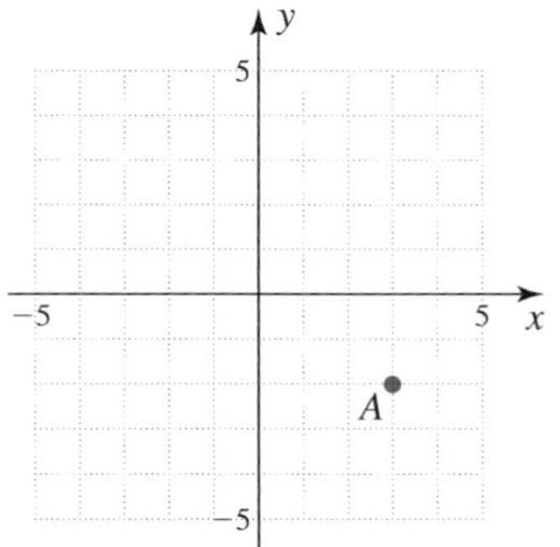

27. Determine whether the ordered pair $(-5, 1)$ is a solution of $5x + 3y = -22$. Yes

28. Find x in the ordered pair $(x, -2)$ so that the ordered pair satisfies the equation $5x - 4y = 13$. $x = 1$

29. Graph: $5x + y = 5$

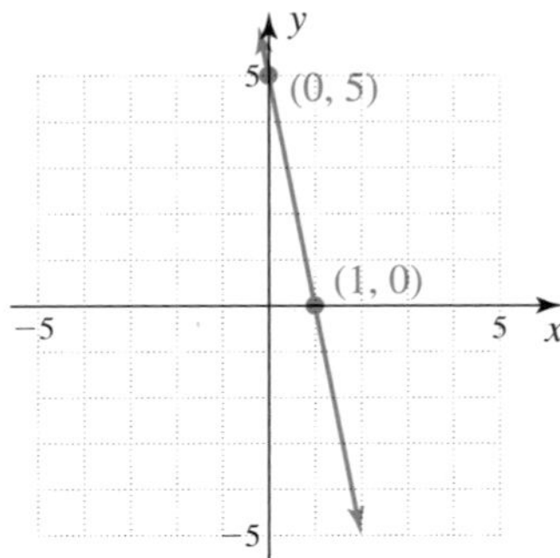

30. Graph: $3y - 15 = 0$

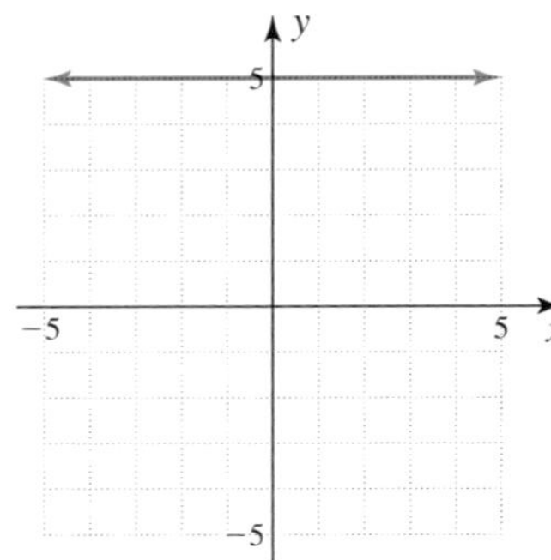

31. Find the slope of the line going through the points $(-7, -1)$ and $(-2, 6)$. $\frac{7}{5}$

32. What is the slope of the line $8x - 4y = 14$? 2

33. Find the pair of parallel lines.

(1) $18y + 24x = 8$

(2) $3y = 4x + 8$

(3) $24x - 18y = 8$ (2) and (3)

34. Find: $(-3x^4y)(-6x^3y^2)$ $18x^7y^3$

35. Find: $\frac{25x^2y^5}{-5xy^7}$ $-\frac{5x}{y^2}$

36. Find: $\frac{x^{-6}}{x^{-9}}$ x^3

37. Multiply and simplify: $x^8 \cdot x^{-3}$ x^5

38. Simplify: $(4x^4y^{-4})^3$ $\frac{64x^{12}}{y^{12}}$

39. Write in scientific notation: 8,000,000 8.0×10^6

40. Divide and express the answer in scientific notation: $(26.04 \times 10^{-5}) \div (6.2 \times 10^3)$ 4.2×10^{-8}

41. Classify as a monomial, binomial, or trinomial: $2x^2$ Monomial

42. Find the degree of the polynomial: $3x^2 - 3x + 1$ 2

43. Write the polynomial in descending order: $-3x^2 - x - 3x^3 + 4$ $-3x^3 - 3x^2 - x + 4$

44. Find the value of $3x^3 + 2x^2$ when $x = 2$. 32

45. Add $(x + 6x^3 - 2)$ and $(-2x^3 - 1 + 8x)$. $4x^3 + 9x - 3$

46. Remove parentheses (simplify): $-3x(3x^2 + 4y)$ $-9x^3 - 12xy$

47. Find (expand): $(3x + 2y)^2$ $9x^2 + 12xy + 4y^2$

48. Find: $(4x - 3y)(4x + 3y)$ $16x^2 - 9y^2$

49. Find (expand): $\left(5x^2 - \frac{1}{2}\right)^2$ $25x^4 - 5x^2 + \frac{1}{4}$

50. Find: $(4x^2 + 9)(4x^2 - 9)$ $16x^4 - 81$

51. Divide $(3x^3 - 20x^2 + 29x - 17)$ by $(x - 5)$. $(3x^2 - 5x + 4)$ R 3

Factoring

5

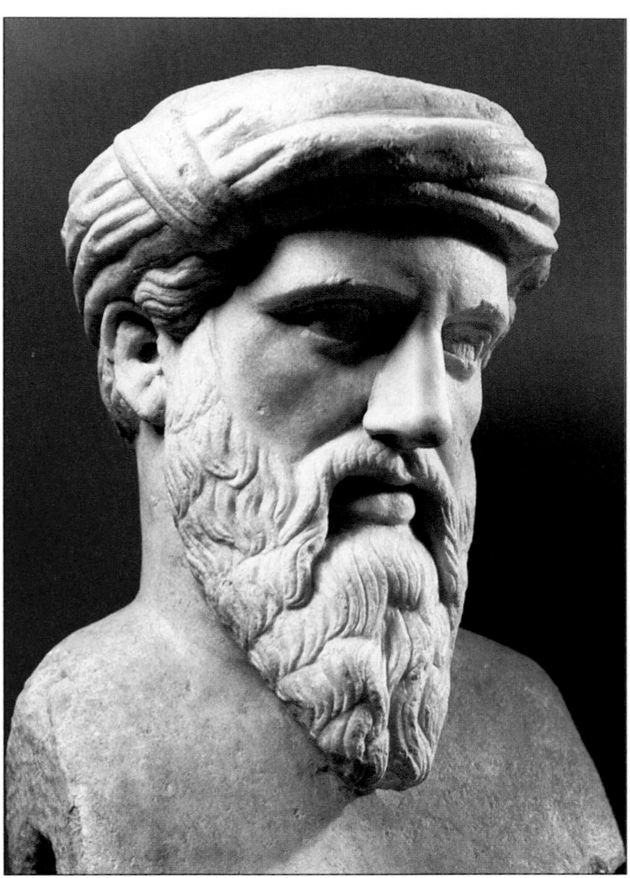

The Human Side of Algebra

One of the most famous mathematicians of antiquity is Pythagoras, to whom has been ascribed the theorem that bears his name (see Section 5.7). It's believed that he was born between 580 and 569 B.C. on the Aegean island of Samos, from which he was later banned by the powerful tyrant Polycrates. When he was about 50, Pythagoras moved to Croton, a colony in southern Italy, where he founded a secret society of 300 young aristocrats called the Pythagoreans. Four subjects were studied: arithmetic, music, geometry, and astronomy. Students attending lectures were divided into two groups: *acoustici* (listeners) and *mathematici*. How did one get to join the *mathematici* in those days? One listened to the master's voice (*acoustici*) from behind a curtain for a period of 3 years! According to Burton's *History of Mathematics,* the Pythagoreans had "strange initiations, rites, and prohibitions." Among them was their refusal "to eat beans, drink wine, pick up anything that had fallen or stir a fire with an iron," but what set them apart from other sects was their philosophy that "knowledge is the greatest purification," and to them, knowledge meant mathematics. You can now gain some of this knowledge by studying Pythagoras' theorem yourself, and evaluating its greatness.

Pretest for Chapter 5

(Answers on page 387)

1. Find the GCF of 25 and 30.

2. Find the GCF of $18x^4y^6$ and $30x^5y^7$.

3. Factor $12x^3 - 24x^5$.

4. Factor $\frac{3}{7}x^6 - \frac{4}{7}x^5 + \frac{2}{7}x^4 + \frac{1}{7}x^2$.

5. Factor $3x^3 + 6x^2y + x + 2y$.

6. Factor $x^2 - 9x + 18$.

7. Factor $6x^2 - 5 + 13x$.

8. Factor $6x^2 - 17xy + 5y^2$.

9. Factor $9x^2 + 12xy + 4y^2$.

10. Factor $x^2 - 18x + 81$.

11. Factor $4x^2 - 12xy + 9y^2$.

12. Factor $x^2 - 81$.

13. Factor $9x^2 - 25y^2$.

14. Factor $8x^3 + 27y^3$.

15. Factor $8x^3 - 27y^3$.

16. Factor $2x^3 - 2x^2 - 8x$.

17. Factor $3x^3 - 12x^2 - 15x$.

18. Factor $2x^2 + 4x + x + 2$.

19. Factor $9kx^2 + 12kx + 4k$.

20. Factor $-9x^4 + 9x^2$.

21. Factor $-4x^2 + 12xy - 9y^2$.

22. Solve $x^2 - 2x - 15 = 0$.

23. Solve $3x^2 - 4x = 4$.

24. Solve $(2x - 3)(x - 4) = 2(x - 2) + 1$.

25. Solve $y(3y + 7) = -2$.

26. The product of two consecutive odd integers is 45 more than 10 times the larger of the two integers. Find the integers.

27. The product of two consecutive integers is 20 less than 10 times the smaller of the two integers. What are the integers?

28. The area of the rectangle is numerically 12 more than its perimeter. If the length of the rectangle is 6 inches more than its width, what are the dimensions of the rectangle?

29. A rectangular 10-inch television screen (measured diagonally) is 2 inches longer than it is high. What are the dimensions of the screen?

30. A car traveled 54 feet after the brakes were applied. How fast was the car going when the brakes were applied? (Use the formula $b = 0.06v^2$, where b is the braking distance and v is the speed of the car when the brakes are applied.)

Answers to Pretest

ANSWERS	IF YOU MISSED	REVIEW		
	QUESTION	SECTION	EXAMPLES	PAGE
1. 5	1	5.1	1	389–390
2. $6x^4y^6$	2	5.1	2	391
3. $12x^3(1 - 2x^2)$	3	5.1	3, 4	393
4. $\frac{1}{7}x^2(3x^4 - 4x^3 + 2x^2 + 1)$	4	5.1	5	393
5. $(x + 2y)(3x^2 + 1)$	5	5.1	6–8	394–395
6. $(x - 3)(x - 6)$	6	5.2	1–4	401–403
7. $(3x - 1)(2x + 5)$	7	5.3	2, 3	409–410
8. $(3x - y)(2x - 5y)$	8	5.3	4–7	410, 412–414
9. $(3x + 2y)^2$	9	5.4	2	419
10. $(x - 9)^2$	10	5.4	3a, b	419
11. $(2x - 3y)^2$	11	5.4	3c	419
12. $(x + 9)(x - 9)$	12	5.4	4a, b	420
13. $(3x + 5y)(3x - 5y)$	13	5.4	4c	420
14. $(2x + 3y)(4x^2 - 6xy + 9y^2)$	14	5.5	1a, b	425
15. $(2x - 3y)(4x^2 + 6xy + 9y^2)$	15	5.5	1c, d	425
16. $2x(x^2 - x - 4)$	16	5.5	2	426
17. $3x(x + 1)(x - 5)$	17	5.5	2	426
18. $(x + 2)(2x + 1)$	18	5.5	3	427
19. $k(3x + 2)^2$	19	5.5	4	427
20. $-9x^2(x + 1)(x - 1)$	20	5.5	7	428–429
21. $-(2x - 3y)^2$	21	5.5	7	428–429
22. $x = 5$ or $x = -3$	22	5.6	1	433–434
23. $x = 2$ or $x = -\frac{2}{3}$	23	5.6	2	434–435
24. $x = 5$ or $x = \frac{3}{2}$	24	5.6	3, 4	435–436
25. $y = -2$ or $y = -\frac{1}{3}$	25	5.6	5	436–437
26. 13 and 15, or −5 and −3	26	5.7	1	441
27. 5 and 6, or 4 and 5	27	5.7	1	441
28. 4 inches by 10 inches	28	5.7	2, 3	442–443
29. 6 inches by 8 inches	29	5.7	4	444–445
30. 30 mi/hr	30	5.7	5	446

5.1 COMMON FACTORS AND GROUPING

To Succeed, Review How To . . .

1. Write a number as a product of primes (pp. 8–9).
2. Write a polynomial in descending order (p. 328).
3. Use the distributive property (pp. 92–94).

Objectives

A Find the greatest common factor (GCF) of numbers.

B Find the GCF of terms.

C Factor out the GCF.

D Factor a four-term expression by grouping.

GETTING STARTED

Expansion Joints and Factoring

Have you ever noticed as you were going over a bridge that there is always a piece of metal at about the midpoint? This piece of metal is called an *expansion joint,* and it prevents the bridge from cracking when it expands or contracts. We can describe this expansion algebraically. In general, if α is the coefficient of linear expansion, L is the length of the material, and t_2 and t_1 are the high and low temperatures in degrees Celsius, then the linear expansion of a solid is

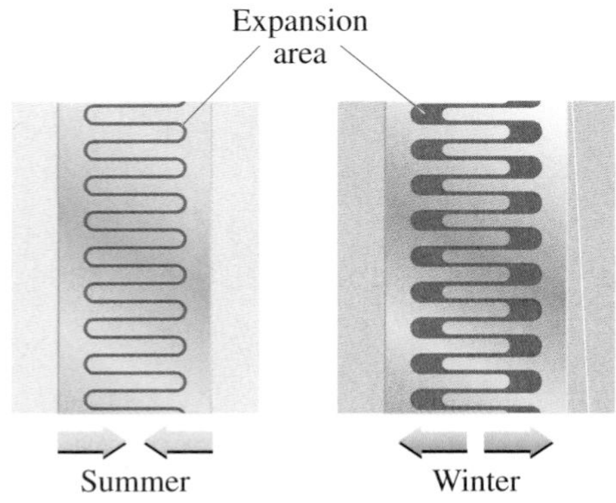

$$e = \alpha L t_2 - \alpha L t_1$$

The expression on the right-hand side of the equation can be written in a simpler way if we *factor* it. You will get a chance to factor this expression in Exercises 5.1, Problem 81! In this section we learn how to factor polynomials by finding a common factor and by grouping.

To *factor* an expression, we write the expression as a product of its factors. Does this sound complicated? It isn't! Suppose we give you the numbers 3 and 5 and tell you to multiply them. You will probably write

$$3 \times 5 = 15$$

Web It

To review how to factor a number as a product of primes, go to link 5-1-1 on the Bello Website at mhhe.com/bello.

In the reverse process, we give you the product 15 and tell you to **factor** it! (Factoring is the *reverse* of multiplying.) You will then write

$$\underset{\text{Product}}{15} = \underbrace{3 \times 5}_{\text{Factors}}$$

Why? Because you know that multiplying 3 by 5 gives you 15. What about factoring the number 20? Here you can write

$$20 = 4 \times 5 \quad \text{or} \quad 20 = 2 \times 10$$

Note that $20 = 4 \times 5$ and $20 = 2 \times 10$ are not **completely** factored. They contain factors that are *not* prime numbers, numbers divisible only by themselves and 1. The first few prime numbers are 2, 3, 5, 7, 11, and so on. Since neither of these factorizations is *complete,* we note that in the first factorization $4 = (2 \times 2)$, so

$$20 = (2 \times 2) \times 5$$

whereas in the second factorization, $10 = (2 \times 5)$. Thus

$$20 = 2 \times (2 \times 5)$$

In either case, the complete factorization for 20 is

$$20 = 2 \times 2 \times 5$$

From now on, when we say factor a *number,* we mean factor it *completely*. Moreover, the numerical factors are assumed to be prime numbers unless noted otherwise. Thus we don't factor 20 as

$$20 = \frac{1}{4} \times 80$$

With these preliminaries out of the way, we can now factor some algebraic expressions. There are different factoring techniques for different situations. We will study most of them in this chapter.

Finding the Greatest Common Factor (GCF) of Numbers

When the numbers 15 and 20 are completely factored, we have

$$15 = 3 \cdot 5 \quad \text{and} \quad 20 = 2 \cdot 2 \cdot 5$$

As you can see, 5 is a common factor of 15 and 20. Similarly, 3 is a common factor of 18 and 30, because 3 is a factor of 18 and 3 is a factor of 30. Other common factors of 18 and 30 are 1, 2, and 6. The greatest (largest) of the common factors of 18 and 30 is 6, so 6 is the *greatest common factor* (GCF) of 18 and 30. Note that the GCF is the same as the GCD we already studied in Section R.1. In general, we have:

Web It

For a lesson on how to find the GCF of several numbers, go to link 5-1-2 on the Bello Website at mhhe.com/bello.

GREATEST COMMON FACTOR OF A LIST OF INTEGERS

The **greatest common factor (GCF)** of a list of integers is the *largest* common factor of the integers in the list.

EXAMPLE 1 Finding the GCF of a List of Numbers

Find the greatest common factor (GCF) of:

a. 45 and 60 **b.** 36, 60, and 108 **c.** 10, 17, and 12

SOLUTION

a. To discover the common factors, write each of the numbers as a product of primes using exponents:

$$45 = 3 \cdot 3 \cdot 5 \quad = 3^2 \cdot 5$$

$$60 = 2 \cdot 2 \cdot 3 \cdot 5 = 2^2 \cdot 3 \cdot 5$$

The common factors are 3 and 5; thus the GCF is $3 \cdot 5 = 15$. Note that to find the GCF, we simply choose the prime factors with the *smallest* exponents and find their product.

PROBLEM 1

Find the GCF of:

a. 30 and 45

b. 45, 60, and 108

c. 20, 13, and 18

Answers

1. a. 15 **b.** 3 **c.** 1

Here is a shortcut: Write

$$\underline{|45 \quad 60}$$

Divide each number by a prime divisor common to all numbers. The product of all divisors is the GCF.

$$\begin{array}{r|rr} 3 & 45 & 60 \\ \hline 5 & 15 & 20 \\ \hline & 3 & 4 \end{array}$$

GCF $= 3 \cdot 5 = 15$

b. We write each of the integers in factored form using exponents and then select the prime factors with the *smallest* exponents.

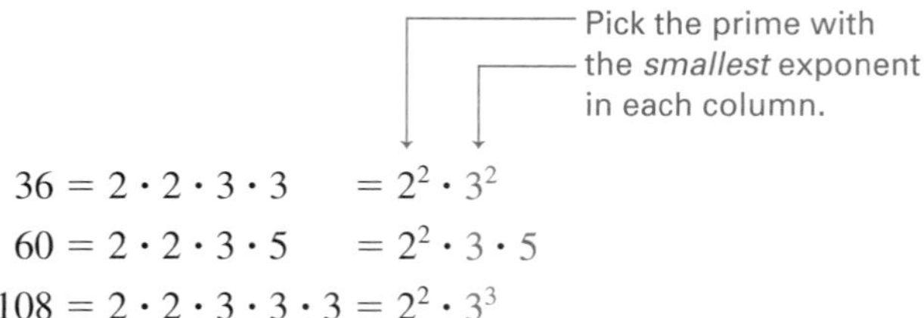
Pick the prime with the *smallest* exponent in each column.

$$\begin{aligned} 36 &= 2 \cdot 2 \cdot 3 \cdot 3 &&= 2^2 \cdot 3^2 \\ 60 &= 2 \cdot 2 \cdot 3 \cdot 5 &&= 2^2 \cdot 3 \cdot 5 \\ 108 &= 2 \cdot 2 \cdot 3 \cdot 3 \cdot 3 &&= 2^2 \cdot 3^3 \end{aligned}$$

We select 2^2 from the first column and 3 from the second. Thus, the GCF is $2^2 \cdot 3 = 12$.

Shortcut: Write

$$\underline{|36 \quad 60 \quad 108}$$

Divide by a prime divisor common to all numbers:

$$\begin{array}{r|rrr} 2 & 36 & 60 & 108 \\ \hline 2 & 18 & 30 & 54 \\ \hline 3 & 9 & 15 & 27 \\ \hline & 3 & 5 & 9 \end{array}$$

GCF $= 2 \cdot 2 \cdot 3 = 12$ as before.

c. First, write

$$\begin{aligned} 10 &= 2 \cdot 5 &&= 2 \cdot 5 \\ 17 &= 17 &&= 17 \\ 12 &= 2 \cdot 2 \cdot 3 &&= 2^2 \cdot 3 \end{aligned}$$

There are no primes common to all three numbers, so the GCF is 1.

Shortcut: Write

$$\underline{|10 \quad 17 \quad 12}$$

There is no prime divisor common to 10, 17, and 12, so the GCF is 1.

B Finding the Greatest Common Factor (GCF) of Terms

We can also find the GCF for a list of terms. Simply write the terms in factored form using exponents, and then select the factors with the *smallest* exponents. For example, the GCF of x^5, x^3, and x^6 is x^3 because x^3 is the factor with the *smallest* exponent, 3. Similarly, to find the GCF of $a^3b^4c^2$ and $a^4b^2c^3$ write

$$a^3b^4c^2$$
$$a^4b^2c^3$$

We select the factors with the *smallest* exponents in each column: a^3 from the a column, b^2 from the b column, and c^2 from the c column. Thus, the GCF is $a^3 \cdot b^2 \cdot c^2$.

EXAMPLE 2 **Finding the GCF of a List of Terms**

Find the GCF of:

a. $24a^6, 18a^4, -30a^9$ **b.** $a^5b^3, b^6a^4, a^2b^5, a^8$ **c.** $-x^3y, -xy^3$

SOLUTION

a. Write:

$$24a^6 = 2^3 \cdot 3a^6$$
$$18a^4 = 2 \cdot 3^2a^4$$
$$-30a^9 = -1 \cdot 2 \cdot 3 \cdot 5a^9$$

The common factors with the smallest exponents are 2, 3, and a^4, so the GCF $= 2 \cdot 3 \cdot a^4 = 6a^4$.

b. Rewrite b^6a^4 as a^4b^6 so we can compare exponents, place all terms in a column, and select the factor with the lowest exponent from each column:

$$a^5b^3$$
$$a^4b^6$$
$$a^2b^5$$
$$a^8$$

There are no factors containing b in a^8, so the GCF is a^2.

c. Write:

$$-x^3y = -1 \cdot 1x^3y$$
$$-xy^3 = -1 \cdot 1xy^3$$

The factors of -1 are -1 and 1, but the *greatest* (largest) common factor is 1, so the GCF $= 1 \cdot xy = xy$.

Note that in a list containing negative terms, sometimes a negative common factor is preferred. Thus, in Example 2(c) we may prefer $-xy$ as the common factor. Both answers, xy and $-xy$, are correct.

PROBLEM 2

Find the GCF of:

a. $12x^8$, $9x^5$, and $-30x^8$

b. x^4y^3, y^4x^6, x^3y^8, and y^9

C Factoring Out the Greatest Common Factor

We can use the ideas we have learned about GCF to write polynomials in factored form. To start, let's compare the multiplications with the factors and see whether we can discover a pattern.

Finding the Product	Finding the Factors
$4(x + y) = 4x + 4y$	$4x + 4y = 4(x + y)$
$5(a - 2b) = 5a - 10b$	$5a - 10b = 5(a - 2b)$
$2x(x + 3) = 2x^2 + 6x$	$2x^2 + 6x = 2x(x + 3)$

Answers

2. a. $3x^5$ **b.** y^3

What do all these operations have in common? They use the *distributive property.* When multiplying, we have

$$a(b + c) = ab + ac$$

When factoring, we have

$ab + ac = a(b + c)$ We are factoring a monomial (a) from a binomial ($ab + ac$).

Of course, we are now more interested in the latter operation. It tells us that to factor a binomial, we must find a factor (a in this case) common to all terms. The first step in having a completely factored expression is to select the *greatest common factor,* ax^n. Here's how we do it.

GREATEST COMMON FACTOR OF A POLYNOMIAL

The term ax^n is the **greatest common factor (GCF)** of a polynomial if

1. a is the *greatest integer* that divides each of the coefficients of the polynomial, and
2. n is the *smallest exponent* of x in all the terms of the polynomial.

Teaching Tip

Remind students that in this definition "divides" means exactly divides; that is, the remainder is zero.

Thus to factor $6x^3 + 18x^2$, we could write

$$6x^3 + 18x^2 = 3x(2x^2 + 6x)$$

but this is not completely factored because $2x^2 + 6x$ can be factored further. Here the *greatest* integer dividing 6 and 18 is 6, and the smallest exponent of x in all terms is x^2. Thus the complete factorization is

$6x^3 + 18x^2 = 6x^2(x + 3)$ Where $6x^2$ is the GCF

Note that if you write $6x^3 + 18x^2$ as the sum $2 \cdot 3 \cdot x^3 + 2 \cdot 3 \cdot 3x^2$, you have *not* factored the polynomial; you have only factored the coefficients of the terms. The complete factorization is the product $6x^2(x + 3)$.

You can check this by multiplying $6x^2(x + 3)$. Of course, it might help your accuracy and understanding if you were to write an intermediate step indicating the common factor present in each term. Thus to factor out the GFC of $6x^3 + 18x^2$, you would write

$$6x^3 + 18x^2 = 6x^2 \cdot x + 6x^2 \cdot 3$$
$$= 6x^2(x + 3)$$

Note that since the 6 is a coefficient, it is *not* written in factored form; that is, we write $6x^2(x + 3)$ and *not* $2 \cdot 3x^2(x + 3)$.

Similarly, to factor $4x - 28$, you would write

$$4x - 28 = 4 \cdot x - 4 \cdot 7$$
$$= 4(x - 7)$$

Remember, if you write $4x - 28$ as $2 \cdot 2 \cdot x - 2 \cdot 2 \cdot 7 \cdot x$, you have factored the terms and *not* the binomial. The factorization of $4x - 28$ is the *product* $4(x - 7)$.

One more thing. When an expression such as $-3x + 12$ is to be factored, we have two possible factorizations:

$$-3(x - 4) \quad \text{or} \quad 3(-x + 4)$$

The first one, $-3(x - 4)$, is the *preferred* one, since in that case the first term of the binomial, $x - 4$ has a positive sign.

EXAMPLE 3 Factoring out a common factor from a binomial

Factor completely:

a. $8x + 24$ **b.** $-6y + 12$ **c.** $10x^2 - 25x^3$

SOLUTION

a. $8x + 24 = 8 \cdot x + 8 \cdot 3$
$= 8(x + 3)$ 8 is the GCF.

b. $-6y + 12 = -6 \cdot y - 6(-2)$
$= -6(y - 2)$ -6 is the GCF.

c. $10x^2 - 25x^3 = 5x^2 \cdot 2 - 5x^2 \cdot 5x$ $5x^2$ is the GCF.
$= 5x^2(2 - 5x)$ or, better yet, $-5x^2(5x - 2)$

Check your results by multiplying the factors obtained, $-5x^2$ and $(5x - 2)$.

PROBLEM 3

Factor completely:

a. $6a + 18$ **b.** $-9y + 27$

c. $15a^2 - 45a^3$

Teaching Tip

Here is an alternate plan for factoring out the greatest common factor.

1. Write the GCF down with a set of parentheses next to it.
2. Divide each term by the GCF and place results in the parentheses.
3. Check by multiplication.

Example:

$$\begin{array}{lccc} & 10x^2 & - & 25x^3 \\ & \frac{10x^2}{5x^2} & & \frac{-25x^3}{5x^2} \\ \text{GCF} & \downarrow & & \downarrow \\ & 5x^2(2 & - & 5x) \end{array}$$

We can also factor polynomials with more than two terms, as shown next.

EXAMPLE 4 Factoring out a common factor from a polynomial

Factor completely:

a. $6x^3 + 12x^2 + 18x$ **b.** $10x^6 - 15x^5 + 20x^4 + 30x^2$

c. $2x^3 + 4x^4 + 8x^5$

SOLUTION

a. $6x^3 + 12x^2 + 18x = 6x \cdot x^2 + 6x \cdot 2x + 6x \cdot 3$ $6x$ is the GCF.
$= 6x(x^2 + 2x + 3)$

b. $10x^6 - 15x^5 + 20x^4 + 30x^2 = 5x^2 \cdot 2x^4 - 5x^2 \cdot 3x^3 + 5x^2 \cdot 4x^2 + 5x^2 \cdot 6$
$= 5x^2(2x^4 - 3x^3 + 4x^2 + 6)$ $5x^2$ is the GCF.

c. $2x^3 + 4x^4 + 8x^5 = 2x^3 \cdot 1 + 2x^3 \cdot 2x + 2x^3 \cdot 4x^2$
$= 2x^3(1 + 2x + 4x^2)$ $2x^3$ is the GCF.

PROBLEM 4

Factor completely:

a. $5x^3 + 15x^2 + 20x$

b. $15x^7 - 20x^6 + 10x^5 + 25x^3$

c. $3a^3 + 9a^4 + 12a^5$

EXAMPLE 5 Factoring out a common factor that is a fraction

Factor completely: $\frac{3}{4}x^2 - \frac{1}{4}x + \frac{5}{4}$

SOLUTION As you can see, we are not working with integers here. This is a very special situation, but this expression can still be factored. Here's how to find the common factor:

$$\frac{3}{4}x^2 - \frac{1}{4}x + \frac{5}{4} = \frac{1}{4} \cdot 3x^2 - \frac{1}{4} \cdot x + \frac{1}{4} \cdot 5$$
$$= \frac{1}{4}(3x^2 - x + 5)$$

PROBLEM 5

Factor completely: $\frac{4}{5}a^2 - \frac{1}{5}a + \frac{2}{5}$

Answers

3. a. $6(a + 3)$ **b.** $-9(y - 3)$
c. $15a^2(1 - 3a)$ or $-15a^2(3a - 1)$
4. a. $5x(x^2 + 3x + 4)$
b. $5x^3(3x^4 - 4x^3 + 2x^2 + 5)$
c. $3a^3(1 + 3a + 4a^2)$
5. $\frac{1}{5}(4a^2 - a + 2)$

D Factoring by Grouping

Teaching Tip

Point out to students that the GCF may be a binomial, such as $(x + 2)$.

Can we factor $x^3 + 2x^2 + 3x + 6$? It seems that there's no common factor here except 1. However, we can group and factor the first two terms and also the last two terms, and then use the distributive property. Here are the steps we use.

1. Group terms with common factors using the associative property. $x^3 + 2x^2 + 3x + 6 = (x^3 + 2x^2) + (3x + 6)$
2. Factor each resulting binomial. (Same) $= x^2(x + 2) + 3(x + 2)$
3. Factor out the GCF, $(x + 2)$, using the distributive property. $= (x + 2)(x^2 + 3)$

Web It

To learn more about factoring by grouping and to see some examples and explanations, go to link 5-1-3 on the Bello Website at mhhe.com/bello.

Thus

$$x^3 + 2x^2 + 3x + 6 = (x + 2)(x^2 + 3)$$

Note that $x^2(x + 2) + 3(x + 2)$ can also be written as $(x^2 + 3)(x + 2)$, since $ac + bc = (a + b)c$. Hence

$$x^3 + 2x^2 + 3x + 6 = (x^2 + 3)(x + 2)$$

Either factorization is correct. You can check this by multiplying $(x + 2)(x^2 + 3)$ or $(x^2 + 3)(x + 2)$.

EXAMPLE 6 **Factor by grouping**

Factor completely:

a. $3x^3 + 6x^2 + 2x + 4$ **b.** $6x^3 - 3x^2 - 4x + 2$

SOLUTION

a. We proceed by steps, as before.

1. Group terms with common factors using the associative property. $3x^3 + 6x^2 + 2x + 4 = (3x^3 + 6x^2) + (2x + 4)$
2. Factor each resulting binomial. (Same) $= 3x^2(x + 2) + 2(x + 2)$
3. Factor out the GCF, $(x + 2)$, using the distributive property. $= (x + 2)(3x^2 + 2)$

Note that if you write $3x^3 + 2x + 6x^2 + 4$ in step 1, your answer would be $(3x^2 + 2)(x + 2)$. Since by the commutative property $(3x^2 + 2)(x + 2) = (x + 2)(3x^2 + 2)$, both answers are correct. We will factor our polynomials by first writing them in *descending* order.

b. Again, we proceed by steps.

1. Group terms with common factors using the associative property. $6x^3 - 3x^2 - 4x + 2 = (6x^3 - 3x^2) - (4x - 2)$ Note that $-(4x - 2) = -4x + 2$.
2. Factor each resulting binomial. (Same) $= 3x^2(2x - 1) - 2(2x - 1)$
3. Factor out the GCF, $(2x - 1)$. $= (2x - 1)(3x^2 - 2)$

Thus $6x^3 - 3x^2 - 4x + 2 = (2x - 1)(3x^2 - 2)$. Note that $2x - 1$ and $3x^2 - 2$ *cannot* be factored any further, so the polynomial is completely factored.

PROBLEM 6

Factor completely:

a. $2a^3 + 6a^2 + 5a + 15$

b. $6a^3 - 2a^2 - 3a + 1$

Answers

6. a. $(a + 3)(2a^2 + 5)$
b. $(3a - 1)(2a^2 - 1)$

EXAMPLE 7 **Factor by grouping**

Factor completely:

a. $2x^3 - 4x^2 - x + 2$ **b.** $6x^4 - 9x^2 + 4x^2 - 6$

SOLUTION

a. 1. Group terms with common factors using the associative property. $2x^3 - 4x^2 - x + 2 = (2x^3 - 4x^2) - (x - 2)$

Same

2. Factor each resulting binomial. $= 2x^2(x - 2) - 1(x - 2)$

3. Factor out the GCF, $(x - 2)$. $= (x - 2)(2x^2 - 1)$

b. We proceed as usual.

1. Group terms with common factors using the associative property. $6x^4 - 9x^2 + 4x^2 - 6 = (6x^4 - 9x^2) + (4x^2 - 6)$

Same

2. Factor each resulting binomial. $= 3x^2(2x^2 - 3) + 2(2x^2 - 3)$

3. Factor out the GCF, $(2x^2 - 3)$. $= (2x^2 - 3)(3x^2 + 2)$

PROBLEM 7

Factor completely:

a. $3a^3 - 9a^2 - a + 3$

b. $6a^4 - 4a^2 + 9a^2 - 6$

Teaching Tip

Make sure students write $-(x - 2)$ as $-1(x - 2)$ so they can reach the last step, $(x - 2)(2x^2 - 1)$, correctly.

EXAMPLE 8 **Factor a polynomial with two variables by grouping**

Factor completely: $6x^2 + 2xy - 9xy - 3y^2$

SOLUTION Our steps serve us well in this situation also.

1. Group terms with common factors using the associative property. $6x^2 + 2xy - 9xy - 3y^2 = (6x^2 + 2xy) - (9xy + 3y^2)$

Note that $-(9xy + 3y^2) = -9xy - 3y^2$.

Same

2. Factor each resulting binomial. $= 2x(3x + y) - 3y(3x + y)$

3. Factor out the GCF, $(3x + y)$. $= (3x + y)(2x - 3y)$

PROBLEM 8

Factor completely:

$6a^2 + 3ab - 4ab - 2b^2$

Calculate It Checking Factorization

If you believe a picture is worth a thousand words, you can use your calculator to check factorization problems. The idea is this: If in Example 6(a) you get the same picture (graph) for $3x^3 + 6x^2 + 2x + 4$ and for $(x + 2)(3x^2 + 2)$, these two polynomials must be equal; that is, $3x^3 + 6x^2 + 2x + 4 = (x + 2)(3x^2 + 2)$. Now, use your calculator to show this. Graph the polynomials by pressing Y= and entering $Y_1 = 3x^3 + 6x^2 + 2x + 4$ and $Y_2 = (x + 2)(3x^2 + 2)$. Then press GRAPH. If you get the same graph for Y_1 and Y_2, which means you see only one picture as shown in the window, you probably have the correct factorization. But really, wouldn't you rather do it algebraically by multiplying $(x + 2)$ and $(3x^2 + 2)$?

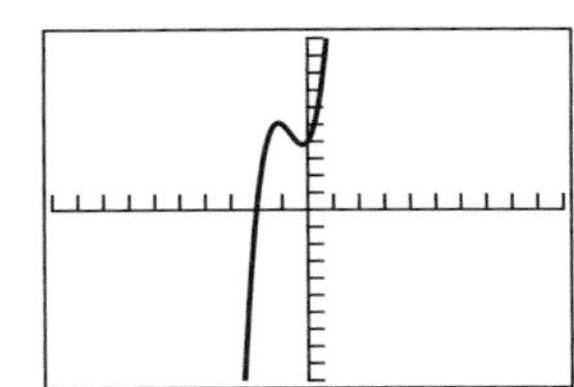

Answers

7. a. $(a - 3)(3a^2 - 1)$
b. $(3a^2 - 2)(2a^2 + 3)$
8. $(2a + b)(3a - 2b)$

Exercises 5.1

Boost *your* GRADE at mathzone.com!

MathZone

- Practice Problems
- Self-Tests
- Videos
- NetTutor
- e-Professors

A In Problems 1–6, find the GCF.

1. 20, 24 4
2. 30, 70 10
3. 16, 48, 88 8
4. 52, 26, 130 26
5. 8, 19, 12 1
6. 10, 41, 18 1

B In Problems 7–20, find the GCF.

7. a^3, a^8 a^3
8. y^7, y^9 y^7
9. x^3, x^6, x^{10} x^3
10. b^6, b^8, b^{10} b^6
11. $5y^6, 10y^7$ $5y^6$
12. $12y^9, 24y^4$ $12y^4$
13. $8x^3, 6x^7, 10x^9$ $2x^3$
14. $9a^3, 6a^7, 12a^5$ $3a^3$
15. $9b^2c, 12bc^2, 15b^2c^2$ $3bc$
16. $12x^4y^3, 18x^3y^4, 6x^5y^5$ $6x^3y^3$
17. $9y^2, 6x^3, 3x^3y$ 3
18. $15ab^3,\ 25b^3c^4,\ 10a^3bc$ $5b$
19. $18a^4b^3z^4, 27a^5b^3z^4, 81ab^3z$ $9ab^3z$
20. $6x^2y^2z^2, 9xy^2z^2, 15x^2y^2z^2$ $3xy^2z^2$

C In Problems 21–56, factor completely.

21. $3x + 15$ $3(x + 5)$
22. $5x + 45$ $5(x + 9)$
23. $9y - 18$ $9(y - 2)$
24. $11y - 33$ $11(y - 3)$
25. $-5y + 20$ $-5(y - 4)$
26. $-4y + 28$ $-4(y - 7)$
27. $-3x - 27$ $-3(x + 9)$
28. $-6x - 36$ $-6(x + 6)$
29. $4x^2 + 32x$ $4x(x + 8)$
30. $5x^3 + 20x$ $5x(x^2 + 4)$
31. $6x - 42x^2$ $6x(1 - 7x)$
32. $7x - 14x^3$ $7x(1 - 2x^2)$
33. $-5x^2 - 25x^4$ $-5x^2(1 + 5x^2)$
34. $-3x^3 - 18x^6$ $-3x^3(1 + 6x^3)$
35. $3x^3 + 6x^2 + 9x$ $3x(x^2 + 2x + 3)$
36. $8x^3 + 4x^2 - 16x$ $4x(2x^2 + x - 4)$
37. $9y^3 - 18y^2 + 27y$ $9y(y^2 - 2y + 3)$
38. $10y^3 - 5y^2 + 10y$ $5y(2y^2 - y + 2)$
39. $6x^6 + 12x^5 - 18x^4 + 30x^2$ $6x^2(x^4 + 2x^3 - 3x^2 + 5)$
40. $5x^7 - 15x^6 + 10x^3 - 20x^2$ $5x^2(x^5 - 3x^4 + 2x - 4)$
41. $8y^8 + 16y^5 - 24y^4 + 8y^3$ $8y^3(y^5 + 2y^2 - 3y + 1)$
42. $12y^9 - 4y^6 + 6y^5 + 8y^4$ $2y^4(6y^5 - 2y^2 + 3y + 4)$
43. $\frac{4}{7}x^3 + \frac{3}{7}x^2 - \frac{9}{7}x + \frac{3}{7}$ $\frac{1}{7}(4x^3 + 3x^2 - 9x + 3)$
44. $\frac{2}{5}x^3 + \frac{3}{5}x^2 - \frac{2}{5}x + \frac{4}{5}$ $\frac{1}{5}(2x^3 + 3x^2 - 2x + 4)$
45. $\frac{7}{8}y^9 + \frac{3}{8}y^6 - \frac{5}{8}y^4 + \frac{5}{8}y^2$ $\frac{1}{8}y^2(7y^7 + 3y^4 - 5y^2 + 5)$
46. $\frac{4}{3}y^7 - \frac{1}{3}y^5 + \frac{2}{3}y^4 - \frac{5}{3}y^3$ $\frac{1}{3}y^3(4y^4 - y^2 + 2y - 5)$
47. $3(x + 4) - y(x + 4)$ $(x + 4)(3 - y)$
48. $5(y - 2) - x(y - 2)$ $(y - 2)(5 - x)$
49. $x(y - 2) - (y - 2)$ $(y - 2)(x - 1)$
50. $y(x + 3) - (x + 3)$ $(x + 3)(y - 1)$
51. $c(t + s) - (t + s)$ $(t + s)(c - 1)$
52. $p(x - q) - (x - q)$ $(x - q)(p - 1)$
53. $4x^3 + 4x^4 - 12x^5$ $4x^3(1 + x - 3x^2)$

54. $5x^6 + 10x^7 - 5x^8$
$5x^6(1 + 2x - x^2)$

55. $6y^7 - 12y^9 - 6y^{11}$
$6y^7(1 - 2y^2 - y^4)$

56. $7x^9 - 7x^{13} - 14x^{15}$
$7x^9(1 - x^4 - 2x^6)$

D In Problems 57–80, factor completely by grouping.

57. $x^3 + 2x^2 + x + 2$
$(x + 2)(x^2 + 1)$

58. $x^3 + 3x^2 + x + 3$
$(x + 3)(x^2 + 1)$

59. $y^3 - 3y^2 + y - 3$
$(y - 3)(y^2 + 1)$

60. $y^3 - 5y^2 + y - 5$
$(y - 5)(y^2 + 1)$

61. $4x^3 + 6x^2 + 2x + 3$
$(2x + 3)(2x^2 + 1)$

62. $6x^3 + 3x^2 + 2x + 1$
$(2x + 1)(3x^2 + 1)$

63. $6x^3 - 2x^2 + 3x - 1$
$(3x - 1)(2x^2 + 1)$

64. $6x^3 - 9x^2 + 2x - 3$
$(2x - 3)(3x^2 + 1)$

65. $4y^3 + 8y^2 + y + 2$
$(y + 2)(4y^2 + 1)$

66. $2y^3 - 6y^2 - y + 3$
$(y - 3)(2y^2 - 1)$

67. $2a^3 + 3a^2 + 2a + 3$
$(2a + 3)(a^2 + 1)$

68. $3a^3 + 2a^2 + 3a + 2$
$(3a + 2)(a^2 + 1)$

69. $3x^4 + 12x^2 + x^2 + 4$
$(x^2 + 4)(3x^2 + 1)$

70. $2x^4 + 2x^2 + x^2 + 1$
$(x^2 + 1)(2x^2 + 1)$

71. $6y^4 + 9y^2 + 2y^2 + 3$
$(2y^2 + 3)(3y^2 + 1)$

72. $12y^4 + 8y^2 + 3y^2 + 2$
$(3y^2 + 2)(4y^2 + 1)$

73. $4y^4 + 12y^2 + y^2 + 3$
$(y^2 + 3)(4y^2 + 1)$

74. $2y^4 + 2y^2 + y^2 + 1$
$(y^2 + 1)(2y^2 + 1)$

75. $3a^4 - 6a^2 - 2a^2 + 4$
$(a^2 - 2)(3a^2 - 2)$

76. $4a^4 - 12a^2 - 3a^2 + 9$
$(a^2 - 3)(4a^2 - 3)$

77. $6a - 5b + 12ad - 10bd$
$(6a - 5b)(1 + 2d)$

78. $3x - 2y + 15xz - 10yz$
$(3x - 2y)(1 + 5z)$

79. $x^2 - y - 3x^2z + 3yz$
$(x^2 - y)(1 - 3z)$

80. $x^2 + 2y - 2x^2 - 4y$
$-(x^2 + 2y)$

APPLICATIONS

81. *Linear expansion* Factor $\alpha Lt_2 - \alpha Lt_1$, where α is the coefficient of linear expansion, L the length of the material, and t_2 and t_1 the high and low temperatures in degrees Celsius. $\alpha L(t_2 - t_1)$

82. *Restoring force of spring* Factor the expression $-kx - k\ell$ (k is a constant), which represents the restoring force of a spring stretched an amount ℓ from its equilibrium position and then an additional x units. $-k(x + \ell)$

83. *Equivalent resistance of circuits* When solving for the equivalent resistance of two circuits, we need to factor the expression $R^2 - R - R + 1$. Factor this expression by grouping. $(R - 1)(R - 1)$

84. *Bending moment of cantilever beam* The bending moment of a cantilever beam of length L, at x inches from its support, involves the expression $L^2 - Lx - Lx + x^2$. Factor this expression by grouping. $(L - x)(L - x)$

85. *Surface area of cylinder* The surface area A of a right circular cylinder is given by $A = 2\pi rh + 2\pi r^2$, where h is the height of the cylinder and r is its radius. Factor $2\pi rh + 2\pi r^2$. $2\pi r(h + r)$

86. *Area of trapezoid* The area A of a trapezoid is given by $A = \frac{1}{2}b_1h + \frac{1}{2}b_2h$, where h is the altitude of the trapezoid and b_1 and b_2 are the lengths of the bases. Factor $\frac{1}{2}b_1h + \frac{1}{2}b_2h$. $\frac{1}{2}h(b_1 + b_2)$

SKILL CHECKER

Try the Skill Checker Exercises so you'll be ready for the next section.

Multiply:

87. $(x + 5)(x + 3)$ $x^2 + 8x + 15$

88. $(x - 5)(x + 2)$ $x^2 - 3x - 10$

89. $(x - 1)(x - 4)$ $x^2 - 5x + 4$

90. $(x + 2)(x + 1)$ $x^2 + 3x + 2$

91. $(x + 1)(2x - 3)$ $2x^2 - x - 3$

92. $(x - 2)(x + 4)$ $x^2 + 2x - 8$

USING YOUR KNOWLEDGE

Factoring Formulas

Many formulas can be simplified by factoring. Here are a few; factor the expressions given in each problem.

93. The vertical shear at any section of a cantilever beam of uniform cross section is

$$-w\ell + wz$$

Factor this expression. $-w(\ell - z)$

94. The bending moment of any section of a cantilever beam of uniform cross section is

$$-P\ell + Px$$

Factor this expression. $-P(\ell - x)$

95. The surface area of a square pyramid is

$$a^2 + 2as$$

Factor this expression. $a(a + 2s)$

96. The energy of a moving object is given by

$$800m - mv^2$$

Factor this expression. $m(800 - v^2)$

97. The height of a rock thrown from the roof of a certain building is given by

$$-16t^2 + 80t + 240$$

Factor this expression. (*Hint:* -16 is a common factor.) $-16(t^2 - 5t - 15)$

WRITE ON

98. What do we mean by a factored expression, and how can you check whether the result is correct? Answers may vary.

99. Explain the procedure you use to factor a monomial from a polynomial. Can you use the definition of GCF to factor $\frac{1}{2}x + \frac{1}{2}$? Explain. Answers may vary.

MASTERY TEST

If you know how to do these problems, you have learned your lesson!

Factor:

100. $3x^3 - 6x^2 - x + 2$
$(x - 2)(3x^2 - 1)$

101. $6x^4 + 2x^2 - 9x^2 - 3$
$(3x^2 + 1)(2x^2 - 3)$

102. $2x^3 + 2x^2 + 3x + 3$
$(x + 1)(2x^2 + 3)$

103. $6x^3 - 9x^2 - 2x + 3$
$(2x - 3)(3x^2 - 1)$

104. $\frac{2}{5}x^2 - \frac{3}{5}x - \frac{1}{5}$
$\frac{1}{5}(2x^2 - 3x - 1)$

105. $3x^6 - 6x^5 + 12x^4 + 27x^2$
$3x^2(x^4 - 2x^3 + 4x^2 + 9)$

106. $7x^3 + 14x^2 - 49x$
$7x(x^2 + 2x - 7)$

107. $12x^2 + 6xy - 10xy - 5y^2$
$(2x + y)(6x - 5y)$

108. $6x + 48$
$6(x + 8)$

109. $-3y + 21$
$-3(y - 7)$

110. $4x^2 - 32x^3$
$4x^2(1 - 8x)$

111. $5x(x + b) + 6y(x + b)$
$(x + b)(5x + 6y)$

112. $3x + 7y - 12x^2 - 28xy$
$(3x + 7y)(1 - 4x)$

113. Find the GCF of 14 and 24. 2

114. Find the GCF of $20x^4$, $35x^5y$, and $40xy^7$. $5x$

5.2 FACTORING $x^2 + bx + c$

To Succeed, Review How To . . .

1. Expand $(X + A)(X + B)$ (pp. 356–357).
2. Multiply integers (p. 71).
3. Know the definition of a prime number (p. 8).

Objectives

A Factor trinomials of the form $x^2 + bx + c$.

GETTING STARTED

Supply and Demand

Why do prices go up? One reason is related to the supply of the product. Large supply, low prices; small supply, higher prices. The supply function for a certain item can be stated as

$$p^2 + 3p - 70$$

where p is the price of the product. This trinomial is factorable by using *reverse* multiplication (factoring) as we did in Section 5.1. But why do we need to know how to factor? If we know how to factor $p^2 + 3p - 70$, we can solve the equation $p^2 + 3p - 70 = 0$ by rewriting the left side in factored form as

Supply Affects Price

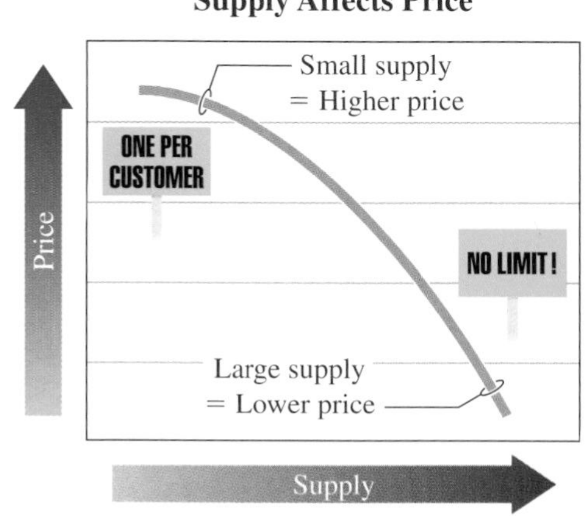

$$(p + 10)(p - 7) = 0$$

Do you see the solutions $p = -10$ and $p = 7$? We will learn how to factor trinomials in this section and then use this knowledge in Sections 5.6 and 5.7 to solve equations and applications.

Factoring Trinomials of the Form $x^2 + bx + c$

Since factoring is the reverse of multiplying, we can use the special products that we derived from the FOIL method (Section 4.7), reverse them, and have some equally useful factoring rules. The basis for the factoring rules is special product 1 (SP1), which you also know as the FOIL method. We now rewrite this product as a factoring rule.

Teaching Tip

Have students multiply the following and compare the results:

$(x + 2)(x + 3) = x^2 + 5x + 6$
$(x - 4)(x + 5) = x^2 + x - 20$
$(x - 1)(x - 7) = x^2 - 8x + 7$

Each of these results is called a trinomial of the form $x^2 + bx + c$.

FACTORING RULE 1

$$X^2 + (A + B)X + AB = (X + A)(X + B) \quad \textbf{(F1)}$$

Thus to factor $x^2 + bx + c$, we need to find two binomials whose product is $x^2 + bx + c$. Now suppose we wish to factor the polynomial

$$x^2 + 8x + 15$$

Web It

For a short and simple explanation of how to factor $x^2 + bx + c$, go to link 5-2-1 on the Bello Website at mhhe.com/bello.

To do this, we use F1:

$$\underbrace{X^2} + \underbrace{(A + B)X} + \underbrace{AB}$$
$$x^2 + \quad 8x \quad + 15$$

As you can see, 15 is used instead of AB and 8 instead of $A + B$; that is, we have two numbers A and B such that $AB = 15$ and $A + B = 8$. We write the possible factors of $AB = 15$ and their sums in a table.

Factors	Sum
15, 1	16
5, 3	8

The correct numbers are $A = 5$ and $B = 3$. Now

$$X^2 + \underbrace{(A + B)}X + \underbrace{AB} = (X + A)(X + B)$$
$$x^2 + \underbrace{(5 + 3)x} + \underbrace{5 \cdot 3} = (x + 5)(x + 3)$$
$$x^2 + \quad 8x \quad + \quad 15 = (x + 5)(x + 3)$$

Remember, to factor $x^2 + 8x + 15$, we need two integers whose product is 15 and whose sum is 8. The integers 5 and 3 will do! So the commutative property allows us to write

$$x^2 + 8x + 15 = (x + 3)(x + 5)$$

as our answer.

What about factoring $x^2 - 8x + 15$? We still need two numbers A and B whose product is 15, but this time their sum must be -8; that is, we need $AB = 15$ and $A + B = -8$. Letting $A = -5$ and $B = -3$ will do it. Check it out:

$$(x - 5)(x - 3) = x^2 - 3x - 5x + (-5)(-3)$$
$$= x^2 - 8x + 15$$

Do you see how this works? To factor a trinomial of the form $x^2 + bx + c$, we must find two numbers A and B so that $A + B = b$ and $AB = c$. Then

$$x^2 + bx + c = x^2 + (A + B)x + AB = (x + A)(x + B)$$

For example, to factor

$$x^2 \overset{b}{\boxed{-3}}x \overset{c}{\boxed{-10}}$$

we need two numbers whose product is -10 (c) and whose sum is -3 (b). Here's a table showing the possibilities for the factors and their sum.

Factors	Sum
−10, 1	−9
10, −1	9
5, −2	3
−5, 2	−3

← This is the only one in which the sum is −3.

The numbers are -5 and $+2$. Thus

$$x^2 - 3x - 10 = (x - 5)(x + 2)$$

Note that the answer $(x + 2)(x - 5)$ is also correct by the commutative property. You can check this by multiplying $(x - 5)(x + 2)$ or $(x + 2)(x - 5)$.

Similarly, to factor $x^2 + (5)x (-14)$, we need two numbers whose product is -14 (so the numbers have different signs) and whose sum is 5 (so the larger one is positive). The numbers are -2 and $+7$ ($-2 \cdot 7 = -14$ and $-2 + 7 = 5$). Thus

$$x^2 + 5x - 14 = (x - 2)(x + 7).$$

We can also write $x^2 + 5x - 14 = (x + 7)(x - 2)$.

Here are the factorizations we have obtained:

$$x^2 + 8x + 15 = (x + 5)(x + 3)$$
$$x^2 - 8x + 15 = (x - 5)(x - 3)$$
$$x^2 - 3x - 10 = (x - 5)(x + 2)$$
$$x^2 + 5x - 14 = (x - 2)(x + 7)$$

All of them involve factoring a trinomial of the form $x^2 + bx + c$; in each case, we need to find two integers whose product is b and whose sum is c. In general, here is the procedure you need:

PROCEDURE

Factoring $x^2 + bx + c$

Find *two* integers whose product is c and whose sum is b.

1. If b and c are positive, *both* integers must be positive.
2. If c is positive and b is negative, *both* integers must be negative.
3. If c is negative, *one* integer must be positive and *one* negative.

EXAMPLE 1 Factoring a trinomial with all positive terms

Factor completely: $6 + 5x + x^2$

SOLUTION Write $6 + 5x + x^2$ in decreasing order as $x^2 + 5x + 6$. To factor $x^2 + 5x + 6$ we need two numbers with product 6 and sum 5. The possibilities are listed in the table. The two factors whose sum is 5 are 3 and 2. Thus,

Factors	Sum
6, 1	7
3, 2	5

$$x^2 + 5x + 6 = (x + 3)(x + 2)$$

CHECK $(x + 3)(x + 2) = x^2 + 2x + 3x + 3 \cdot 2$
$= x^2 + 5x + 6$

PROBLEM 1

Factor completely: $8 + 9x + x^2$

EXAMPLE 2 Factoring a trinomial with a negative middle term

Factor completely: $x^2 - 6x + 5$

SOLUTION To factor $x^2 - 6x + 5$ we need two numbers with product 5 and sum -6. In order to obtain the positive product 5, both factors must be negative. There is only one possibility, so the desired numbers are -5 and -1. Thus,

Factors	Sum
$-5, -1$	-6

$$x^2 - 6x + 5 = (x - 5)(x - 1)$$

CHECK $(x - 5)(x - 1) = x^2 - 1x - 5x + (-5) \cdot (-1)$
$= x^2 - 6x + 5$

PROBLEM 2

Factor completely: $x^2 - 8x + 7$

Answers

1. $(x + 8)(x + 1)$
2. $(x - 7)(x - 1)$

EXAMPLE 3 Factoring a trinomial with two negative terms

Factor completely: $x^2 - 3x - 4$

SOLUTION To factor $x^2 - 3x - 4$ we need two numbers with product -4 and sum -3. To obtain the negative product -4, one number must be positive and one negative. The possibilities are shown in the table. The desired numbers are -4 and 1, so

Factors	Sum
4, −1	3
−4, 1	−3
−2, 2	0

$$x^2 - 3x - 4 = (x - 4)(x + 1)$$

CHECK $(x - 4)(x + 1) = x^2 + 1x - 4x + (-4) \cdot (1)$
$= x^2 - 3x - 4$

PROBLEM 3

Factor completely: $x^2 - 2x - 8$

As you recall from Section R.1, a *prime number* is a number whose only factors are itself and 1. Thus, if we want to factor the number 15 as a product of primes, we write $15 = 3 \cdot 5$. On the other hand, if we want to factor the number 17, we are unable to do it because 17 is prime. If a polynomial cannot be factored using only integers, we say that the polynomial is *prime* and call it a **prime polynomial.** Look at the next example and discover which of the two polynomials is prime.

EXAMPLE 4 Factoring a trinomial with a positive middle term and a negative last term

Factor if possible:

a. $p^2 + 3p - 70$

b. $p^2 + 4p - 15$

SOLUTION

a. We need two numbers with product -70 and sum 3. Since -70 is negative, the numbers must have different signs. Moreover, since the sum of the two numbers is 3, the *larger* number must be *positive*. Here are the possibilities:

Factors	Sum	
70, −1	69	
35, −2	33	
14, −5	9	
10, −7	3	⟵ The only ones with sum 3

The numbers we need then are 10 and -7. Thus

$$p^2 + 3p - 70 = (p + 10)(p - 7)$$

b. This time we need two numbers with product -15 and sum 4. Here are the possibilities:

Factors	Sum
15, −1	14
−15, 1	−14
5, −3	2
−5, 3	−2

PROBLEM 4

Factor if possible:

a. $y^2 + 4y - 14$

b. $y^2 + 3y - 40$

Answers

3. $(x - 4)(x + 2)$
4. a. $y^2 + 4y - 14$ is prime
b. $(y + 8)(y - 5)$

None of the pairs of factors has a sum of 4. Thus, the trinomial $p^2 + 4p - 15$ cannot be factored using only integers. The polynomial $p^2 + 4p - 15$ is **prime.**

EXAMPLE 5 **Factoring a trinomial with two variables**

Factor completely: $x^2 + 5ax + 6a^2$

SOLUTION This problem is very similar to Example 1, except here we have two variables, x and a. The procedure, however, is the same. We need two expressions whose product is $6a^2$ and whose sum is $5a$. The possible factors are $6a$ and a, or $3a$ and $2a$. Since $3a + 2a = 5a$, the appropriate factors are $3a$ and $2a$. Thus,

$$x^2 + 5ax + 6a^2 = (x + 3a)(x + 2a)$$

CHECK $(x + 3a)(x + 2a) = x^2 + 2ax + 3ax + 6a^2$
$= x^2 + 5ax + 6a^2$

PROBLEM 5

Factor completely: $x^2 + 6ax + 8a^2$

Now, do you remember GCFs? Sometimes we have to factor out the GCF in an expression so that we can completely factor it. How? We will show you in Example 6.

EXAMPLE 6 **Factoring a trinomial with a GCF**

Factor completely:

a. $x^5 - 6x^4 + 9x^3$ **b.** $ax^6 - 3ax^5 - 18ax^4$

SOLUTION

a. The GCF of $x^5 - 6x^4 + 9x^3$ is the term with the smallest exponent, that is, x^3. Thus,

$$x^5 - 6x^4 + 9x^3 = x^3(x^2 - 6x + 9)$$

Now, we have to factor $x^2 - 6x + 9$ by finding two numbers whose product is 9 and whose sum is -6. To obtain the positive product 9, both factors must be negative; -3 and -3 will do, so

$$x^2 - 6x + 9 = (x - 3)(x - 3) = (x - 3)^2$$

Here is the complete procedure for you to follow:

$x^5 - 6x^4 + 9x^3 = x^3(x^2 - 6x + 9)$ Factor out the GCF x^3.
$= x^3(x - 3)(x - 3)$ Factor $x^2 - 6x + 9$.
$= x^3(x - 3)^2$ Write $(x - 3)(x - 3)$ as $(x - 3)^2$.

You should check the result!

b. $ax^6 - 3ax^5 - 18ax^4 = ax^4(x^2 - 3x - 18)$ Factor out the GCF, ax^4.
$= ax^4(x + 3)(x - 6)$ Factor $x^2 - 3x - 18$ using 3 and -6, whose product is -18 and whose sum is -3.

We leave the check for you.

PROBLEM 6

Factor completely:

a. $y^5 - 4y^4 + 4y^3$

b. $b^2y^6 - b^2y^5 - 20b^2y^4$

Answers

5. $(x + 4a)(x + 2a)$
6. a. $y^3(y - 2)^2$
b. $b^2y^4(y + 4)(y - 5)$

Exercises 5.2

A In Problems 1–40, factor completely.

1. $y^2 + 6y + 8$ $(y + 2)(y + 4)$

2. $y^2 + 10y + 21$ $(y + 7)(y + 3)$

3. $x^2 + 7x + 10$ $(x + 2)(x + 5)$

4. $x^2 + 13x + 22$ $(x + 11)(x + 2)$

5. $y^2 + 3y - 10$ $(y + 5)(y - 2)$

6. $y^2 + 5y - 24$ $(y + 8)(y - 3)$

7. $x^2 + 5x - 14$ $(x + 7)(x - 2)$

8. $x^2 + 5x - 36$ $(x + 9)(x - 4)$

9. $x^2 - 6x - 7$ $(x + 1)(x - 7)$

10. $x^2 - 7x - 8$ $(x - 8)(x + 1)$

11. $y^2 - 5y - 14$ $(y + 2)(y - 7)$

12. $y^2 - 4y - 12$ $(y - 6)(y + 2)$

13. $y^2 - 3y + 2$ $(y - 2)(y - 1)$

14. $y^2 - 11y + 30$ $(y - 6)(y - 5)$

15. $x^2 - 5x + 4$ $(x - 4)(x - 1)$

16. $x^2 - 12x + 27$ $(x - 9)(x - 3)$

17. $x^2 + 3x + 4$ Not factorable

18. $x^2 - 5x + 6$ $(x - 3)(x - 2)$

19. $-7 - 7x + x^2$ Not factorable

20. $-5 - 5x + x^2$ Not factorable

21. $x^2 + 3ax + 2a^2$ $(x + a)(x + 2a)$

22. $x^2 + 12ax + 35a^2$ $(x + 7a)(x + 5a)$

23. $z^2 + 6bz + 9b^2$ $(z + 3b)(z + 3b)$

24. $z^2 + 4bz + 16b^2$ Not factorable

25. $r^2 + ar - 12a^2$ $(r + 4a)(r - 3a)$

26. $r^2 + 3ar - 10a^2$ $(r + 5a)(r - 2a)$

27. $x^2 + 9ax - 10a^2$ $(x + 10a)(x - a)$

28. $x^2 + 2ax - 8a^2$ $(x + 4a)(x - 2a)$

29. $-b^2 - 2by + 3y^2$ $-(b - y)(b + 3y)$

30. $-b^2 + by + 110y^2$ $-(b - 11y)(b + 10y)$

31. $m^2 - am - 2a^2$ $(m - 2a)(m + a)$

32. $m^2 - 2bm - 15b^2$ $(m - 5b)(m + 3b)$

33. $2t^3 + 10t^2 + 8t$ $2t(t + 1)(t + 4)$

34. $3t^3 + 12t^2 + 9t$ $3t(t + 1)(t + 3)$

35. $a^2x^3 + 3a^2x^2 + 2ax^2$ $ax^2(ax + 3a + 2)$

36. $b^3x^5 + 4b^3x^4 + 2b^3x^3$ $b^3x^3(x^2 + 4x + 2)$

37. $b^3x^7 + b^3x^6 - 12b^3x^5$ $b^3x^5(x - 3)(x + 4)$

38. $b^5y^8 + 3b^5y^7 - 10b^5y^6$ $b^5y^6(y - 2)(y + 5)$

39. $2c^5z^6 + 4c^5z^5 - 30c^5z^4$ $2c^5z^4(z - 3)(z + 5)$

40. $3z^8 - 12z^7y - 63z^6y$ $3z^6(z^2 - 4yz - 21y)$

APPLICATIONS

In Problems 41–42, the expression

$$-5t^2 + V_0t + h$$

represents the altitude of an object t seconds after being thrown from a height of h meters with an initial velocity of V_0 meters per second.

41. *Altitude of a thrown object*

a. Find the expression for the altitude of an object thrown upward with an initial velocity V_0 of 5 meters per second from a building 10 meters high. $-5t^2 + 5t + 10$

b. Factor the expression. $-5(t - 2)(t + 1)$

c. When the product obtained in part **b** is 0, the altitude of the object is 0 and the object is on the ground. What values of t will make the product 0? $t = 2$ and $t = -1$

d. Based on the answer to part **c**, how long does it take the object to return to the ground? 2 seconds

42. *Altitude of a thrown object*

a. Find the expression for the altitude of an object thrown upward with an initial velocity V_0 of 10 meters per second from a building 40 meters high. $-5t^2 + 10t + 40$

b. Factor the expression. $-5(t - 4)(t + 2)$

c. When the product obtained in part **b** is 0, the altitude of the object is 0 and the object is on the ground. What values of t will make the product 0? $t = 4$ and $t = -2$

d. Based on the answer to part **c**, how long does it take the object to return to the ground? 4 seconds

43. *Descent of a rock* The height (in feet) of a rock thrown from the roof of a building after t seconds is given by

$$-16t^2 + 32t + 240$$

a. Factor this expression. $-16(t - 5)(t + 3)$

b. When the product obtained in part **a** is 0, the altitude of the rock is 0 and the rock is on the ground. What values of t will make the product 0? $t = 5$ and $t = -3$

c. Based on the answer to part **b**, how long does it take the rock to return to the ground? 5 seconds

44. *Chlorofluorocarbon production* Do you know what freon is? It is a gas containing chlorofluorocarbons (CFC) used in old air conditioners and linked to the depletion of the ozone layer. Their production (in thousands of tons) can be represented by the expression

$$-0.04t^2 + 2.8t + 120$$

where t is the number of years after 1960.

a. Factor $-0.04t^2 + 2.8t + 120$ using -0.04 as the GCF. $-0.04(t - 100)(t + 30)$

b. When will the product obtained in part **a** be 0? $t = 100$ and $t = -30$

c. Based on the answer to part **b**, when will the production of CFCs be 0? (The Montreal Protocol called for a 50% decrease by the year 2000.) 2060

SKILL CHECKER

Try the Skill Checker Exercises so you'll be ready for the next section.

Find (expand):

45. $(x + 4)^2$ $x^2 + 8x + 16$

46. $(x + 6)^2$ $x^2 + 12x + 36$

47. $(x - 3)^2$ $x^2 - 6x + 9$

48. $(x - 5)^2$ $x^2 - 10x + 25$

49. $(3x + 2y)^2$ $9x^2 + 12xy + 4y^2$

50. $(4x + 3y)^2$ $16x^2 + 24xy + 9y^2$

51. $(5x - 2y)^2$ $25x^2 - 20xy + 4y^2$

52. $(3x - 4y)^2$ $9x^2 - 24xy + 16y^2$

53. $(3x + 4y)(3x - 4y)$ $9x^2 - 16y^2$

54. $(5x + 2y)(5x - 2y)$ $25x^2 - 4y^2$

WRITE ON

55. When factoring any trinomial, what should your first step be?

56. Ana says that a polynomial can be factored as $(x - 3)(x - 2)$. Tyrone insists that the answer is really $(x - 2)(x - 3)$. Who is correct?

57. Now, Tyrone says he reworked the problem and his new answer is $(2 - x)(3 - x)$. Ana still says she got $(x - 2)(x - 3)$. Who is correct?

58. Explain in your own words what a *prime polynomial* is.

59. When factoring $x^2 + bx + c$ (b and c positive) as $(x + A)(x + B)$, what can you say about A and B?

60. When factoring $x^2 + bx + c$ (b negative, c positive) as $(x + A)(x + B)$, what can you say about A and B?

61. When factoring $x^2 + bx + c$ (c negative) as $(x + A)(x + B)$, what can you say about A and B?

MASTERY TEST

If you know how to do these problems, you have learned your lesson!

In Problems 62–73, factor completely.

62. $8x^2 - 12xy + 2xy - 3y^2$ $(2x - 3y)(4x + y)$

63. $3y^3 - 6y^2 - y + 2$ $(y - 2)(3y^2 - 1)$

64. $3y^3 + 9y^2 + 2y + 6$ $(y + 3)(3y^2 + 2)$

65. $2y^3 - 6y^2 - y + 3$ $(y - 3)(2y^2 - 1)$

66. $y^2 + 5y + 6$ $(y + 2)(y + 3)$

67. $x^2 + 7xy + 12y^2$ $(x + 4y)(x + 3y)$

68. $x^2 + 5x - 14$ $(x + 7)(x - 2)$

69. $x^2 - 7xy + 10y^2$ $(x - 2y)(x - 5y)$

70. $3y^4 + 6y^5 + 9y^6$ $3y^4(1 + 2y + 3y^2)$

71. $z^2 - 10z - 25$ Not factorable

72. $kx^3 - 2kx^2 - 15kx$ $kx(x - 5)(x + 3)$

73. $2y^3 + 6y^5 + 10y^7$ $2y^3(1 + 3y^2 + 5y^4)$

5.3 FACTORING $ax^2 + bx + c, a \neq 1$

To Succeed, Review How To . . .

1. Expand $(X + A)(X + B)$ (pp. 356–357).
2. Use FOIL to expand polynomials (pp. 348–351).
3. Multiply integers (p. 71).

Objectives

A Use the *ac* test to determine whether $ax^2 + bx + c$ is factorable.

B Factor $ax^2 + bx + c$ by grouping.

C Factor $ax^2 + bx + c$ using FOIL.

GETTING STARTED

When There Is Smoke We Need Water

How much water is the fire truck pumping? It depends on many factors, but one of them is the friction loss inside the hose. If the friction loss is 36 lb/in.^{2}, we have to know how to factor $2g^2 + g - 36$ to find the answer.

This expression is of the form $ax^2 + bx + c, a \neq 1$, and we can factor it two ways: by **grouping** or by using **FOIL.** How do we know whether $2g^2 + g - 36$ is even factorable? We will show you how to tell and ask you to factor $2g^2 + g - 36$ in Problem 61.

A The *ac* Test for $ax^2 + bx + c, a \neq 1$

ac TEST FOR $ax^2 + bx + c$

A trinomial of the form $ax^2 + bx + c$ is factorable if there are two integers with product *ac* and sum *b*.

Note that a and c are the first and last numbers in $ax^2 + bx + c$ (hence the name *ac* test), and b is the coefficient of x. A diagram may help you visualize this test.

ac TEST

We need two numbers whose product is *ac*.

$$ax^2 + bx + c$$

The sum of the numbers must be b.

Note: Before you use the *ac* test, factor the GCF and write the polynomial in descending order.

Thus, to determine whether $2g^2 + g - 36$ is factorable, we need two numbers whose product is $a \cdot c = 2 \cdot (-36) = -72$ and whose sum is $b = 1$. Since $9 \cdot (-8) = -72$ and $9 + (-8) = 1$, $2g^2 + g - 36$ is factorable.

EXAMPLE 1 **Using the *ac* test to find whether a polynomial is factorable**

Determine whether the given polynomial is factorable:

a. $6x^2 + 7x + 2$ **b.** $2x^2 + 5x + 4$

SOLUTION

a. To find out whether $6x^2 + 7x + 2$ is factorable, we use these three steps:

1. Multiply $a = 6$ by $c = 2$ $(6 \times 2 = 12)$.
2. Find two integers whose product is $ac = 12$ and whose sum is $b = 7$.

We need two numbers whose product is $6 \cdot 2 = 12$.

$$6x^2 + 7x + 2$$

The sum of the two numbers must be 7.

3. A little searching will produce 4 and 3.

CHECK $\underbrace{4 \times 3 = 12}_{\text{Product}}$, $\underbrace{4 + 3 = 7}_{\text{Sum}}$

Thus $6x^2 + 7x + 2$ is factorable.

b. Consider the trinomial $2x^2 + 5x + 4$. Here is the *ac* test for this trinomial.

1. Multiply 2 by 4 $(2 \times 4 = 8)$.
2. Find two integers whose product is 8 and whose sum is 5.
3. The factors of 8 are 4 and 2, and 8 and 1. Neither pair adds up to 5 $(4 + 2 = 6, 8 + 1 = 9)$. Thus the trinomial $2x^2 + 5x + 4$ is *not* factorable using factors containing only integer coefficients. A polynomial that cannot be factored using only factors with integer coefficients is called a **prime polynomial.**

PROBLEM 1

Determine whether the polynomial is factorable:

a. $4y^2 + 3y + 2$

b. $3y^2 + 5y + 2$

B Factoring Trinomials of the Form $ax^2 + bx + c$ by Grouping

At this point, you should be convinced that the *ac* test really tells you whether a trinomial of the form $ax^2 + bx + c$ is factorable; however, we still don't know *how* to do the actual factorization. But we are in luck; the number *ac* still plays an important part in factoring this trinomial. In fact, the number *ac* is so important that we shall call it the **key number** in the factorization of $ax^2 + bx + c$. To get a little practice, we have found and circled the key numbers of a few trinomials.

	a	*c*	*ac*
$6x^2 + 8x + 5$	6	5	(30)
$2x^2 - 7x - 4$	2	−4	(−8)
$-3x^2 + 2x + 5$	−3	5	(−15)

Answers

1. a. Not factorable (prime)
b. Factorable ($ac = 6$; factors are 3 and 2; sum $b = 3 + 2 = 5$)

Web It

To practice some more with the *ac* test, go to the Bello Website at mhhe.com/bello and try link 5-3-1.

Teaching Tip

Have students multiply the following and compare results:

$(x + 3)(2x - 5) = 2x^2 + x - 15$

$(3x + 1)(2x + 7) = 6x^2 + 23x + 7$

$(5x - 2)(3x - 1) = 15x^2 - 11x + 2$

Each of these results is called a trinomial of the form $ax^2 + bx + c$, where $a \neq 1$. Now, have students find ac and b for the three given trinomials. All of them are factorable, of course!

As before, by examining the key number and the coefficient of the middle term, you can determine whether a trinomial is factorable. For example, the key number of the trinomial $6x^2 + 8x + 5$ is 30. But since there are *no* integers with sum 8 whose product is 30, this trinomial is *not* factorable. (The factors of 30 are 6 and 5, 10 and 3, 15 and 2, and 30 and 1. None of these pairs has a sum of 8.)

On the other hand, the key number for $2x^2 - 7x - 4$ is -8, and -8 has two factors (-8 and 1) whose product is -8 and whose sum is the coefficient of the middle term, -7. Thus $2x^2 - 7x - 4$ is factorable; here are the steps:

1. Find the key number $[2 \cdot (-4) = -8]$.	$2x^2 - 7x - 4$	(-8)
2. Find the factors of the key number and use the appropriate ones to rewrite the middle term.	$2x^2 - 8x + 1x - 4$	$-8, 1$ $-8(1) = -8$ $-8 + 1 = -7$
3. Group the terms into pairs (as we did in Section 5.1).	$(2x^2 - 8x) + (1x - 4)$	
4. Factor each pair.	$2x(x - 4) + 1(x - 4)$	
5. Note that $(x - 4)$ is the GCF.	$(x - 4)(2x + 1)$	

Thus $2x^2 - 7x - 4 = (x - 4)(2x + 1)$. You can check to see that this is the correct factorization by multiplying $(x - 4)$ by $(2x + 1)$.

Now a word of warning: You can write the factorization of $ax^2 + bx + c$ in *two* ways. Suppose you wish to factor the trinomial $5x^2 + 7x + 2$. Here's one way:

1. Find the key number $[5 \cdot 2 = 10]$.	$5x^2 + 7x + 2$	(10)
2. Find the factors of the key number; use them to rewrite the middle term.	$5x^2 + 5x + 2x + 2$	5, 2
3. Group the terms into pairs.	$(5x^2 + 5x) + (2x + 2)$	
4. Factor each pair.	$5x(x + 1) + 2(x + 1)$	
5. Note that $(x + 1)$ is the GCF.	$(x + 1)(5x + 2)$	

Thus $5x^2 + 7x + 2 = (x + 1)(5x + 2)$. But there's another way:

1. Find the key number $[5 \cdot 2 = 10]$.	$5x^2 + 7x + 2$	
2. Find the factors of the key number and use them to rewrite the middle term.	$5x^2 + 2x + 5x + 2$	2, 5
3. Group the terms into pairs.	$(5x^2 + 2x) + (5x + 2)$	
4. Factor each pair.	$x(5x + 2) + 1(5x + 2)$	
5. Note that $(5x + 2)$ is the GCF.	$(5x + 2)(x + 1)$	

In this case, we found that

$$5x^2 + 7x + 2 = (5x + 2)(x + 1)$$

Is the correct factorization $(x + 1)(5x + 2)$ or $(5x + 2)(x + 1)$? The answer is that *both* factorizations are correct! This is because the multiplication of real numbers is commutative and the variable x, as well as the trinomials involved, also represent real numbers; thus the *order* in which the product is written (according to the commutative property of multiplication) *makes no difference in the final answer.*

When factoring a trinomial by grouping, just remember to write the polynomial in descending order and factor out the GCF (if any).

EXAMPLE 2 Factoring trinomials of the form $ax^2 + bx + c$ by grouping

Factor:

a. $4 - 3x + 6x^2$ **b.** $4x^2 - 3 - 4x$

SOLUTION

a. We first write the polynomial in descending order as $6x^2 - 3x + 4$, then proceed by steps:

1. Find the key number $[6 \cdot 4 = 24]$. $6x^2 - 3x + 4$ (24)
2. Find the factors of the key number and use them to rewrite the middle term. Unfortunately, it's impossible to find two numbers with product 24 and sum -3. This trinomial is *not* factorable.

b. We first rewrite the polynomial (*in descending order*) as $4x^2 - 4x - 3$, and then proceed by steps.

Step		
1. Find the key number $[4 \cdot (-3) = -12]$.	$4x^2 - 4x - 3$	(−12)
2. Find the factors of the key number and use them to rewrite the middle term.	$4x^2 - 6x + 2x - 3$	−6, 2
3. Group the terms into pairs.	$(4x^2 - 6x) + (2x - 3)$	
4. Factor each pair.	$2x(2x - 3) + 1(2x - 3)$	
5. Note that $(2x - 3)$ is the GCF.	$(2x - 3)(2x + 1)$	

Thus $4x^2 - 4x - 3 = (2x - 3)(2x + 1)$, as can easily be verified by multiplication.

PROBLEM 2

Factor:

a. $9x^2 - 2 - 3x$

b. $3 - 4x + 2x^2$

Teaching Tip

Make sure students write the polynomial in descending order *before* using the *ac* test. Otherwise they may incorrectly find the *ac* number for $4x^2 - 3 - 4x$ to be -16!

Factoring problems in which the third term in step 2 contains a negative number as a coefficient requires that special care be taken with the signs. Thus to factor the trinomial $4x^2 - 5x + 1$, we proceed as follows:

Step		
1. Find the key number $[4 \cdot 1 = 4]$.	$4x^2 - 5x + 1$	(4)
2. Find the factors of the key number and use them to rewrite the middle term.	$4x^2 - 4x - 1x + 1$	−4, −1

Note that the third term has a negative coefficient, −1.

Answers

2. a. $(3x - 2)(3x + 1)$
b. Not factorable (prime)

Teaching Tip

Assure students that the *ac* test is infallible! If $(x - 1)$ is a factor of the first pair, it *will always* be a factor of the second pair.

3. Group the terms into pairs. $(4x^2 - 4x) + (-1x + 1)$
4. Factor each pair. $4x(x - 1) - 1(x - 1)$ Recall that $-1(x - 1) = -x + 1$.
5. Note that $(x - 1)$ is the GCF. $(x - 1)(4x - 1)$ If the first pair has $(x - 1)$ as a factor, the second pair will also have $(x - 1)$ as a factor.

Thus $4x^2 - 5x + 1 = (x - 1)(4x - 1)$.

EXAMPLE 3 Factoring trinomials of the form $ax^2 + bx + c$ by grouping

Factor: $5x^2 - 11x + 2$

SOLUTION

1. Find the key number $[5 \cdot 2 = 10]$. $5x^2 \underline{- 11x} + 2$ (10)
2. Find the factors of the key number and use them to rewrite the middle term. $5x^2 \underline{- 10x - 1x} + 2$ $-10, -1$
3. Group the terms into pairs. $(5x^2 - 10x) + (-1x + 2)$
4. Factor each pair. $5x(x - 2) - 1(x - 2)$
5. Note that $(x - 2)$ is the GCF. $(x - 2)(5x - 1)$

Thus the factorization of $5x^2 - 11x + 2$ is $(x - 2)(5x - 1)$.

PROBLEM 3

Factor: $4x^2 - 13x + 3$

Teaching Tip

Have students multiply the following to get familiar with a trinomial in two variables:

$(x + 3y)(x - y) = x^2 + 2xy - 3y^2$

$(2x + y)(x + 4y) = 2x^2 + 9xy + 4y^2$

So far, we have factored trinomials in one variable only. A procedure similar to the one used for factoring a trinomial of the form $ax^2 + bx + c$ can be used to factor certain trinomials in two variables. We illustrate the procedure in the next example.

EXAMPLE 4 Factoring trinomials with two variables by grouping

Factor: $6x^2 - xy - 2y^2$

SOLUTION

1. Find the key number $[6 \cdot (-2) = -12]$. $6x^2 \underline{- xy} - 2y^2$ (−12)
2. Find the factors of the key number and use them to rewrite the middle term. $6x^2 \underline{- 4xy + 3xy} - 2y^2$ $-4, 3$
3. Group the terms into pairs. $(6x^2 - 4xy) + (3xy - 2y^2)$
4. Factor each pair. $2x(3x - 2y) + y(3x - 2y)$
5. Note that $(3x - 2y)$ is the GCF. $(3x - 2y)(2x + y)$

Thus $6x^2 - xy - 2y^2 = (3x - 2y)(2x + y)$.

PROBLEM 4

Factor: $4x^2 - 4xy - 3y^2$

Answers

3. $(x - 3)(4x - 1)$
4. $(2x - 3y)(2x + y)$

C Factoring $ax^2 + bx + c$ by FOIL (Trial and Error)

Sometimes, it's easier to factor a polynomial of the form $ax^2 + bx + c$ by using a FOIL process (*trial and error*). This is especially so when a or c is a prime number such as 2, 3, 5, 7, 11, and so on. Here's how we do this.

Web It

For a video dealing with factoring, go to link 5-3-2 on the Bello Website at mhhe.com/bello.

You need RealPlayer® to play the video.

For several examples of factorization using FOIL (trial and error), see link 5-3-3.

PROCEDURE

Factoring by FOIL (Trial and Error)

Product must be c.

FOIL

$$ax^2 + bx + c = (\_\_x + \_\_)(\_\_x + \_\_)$$

FOIL

Product must be a.

Now,

1. The product of the numbers in the *first* (F) blanks must be a.
2. The coefficients of the *outside* (O) products and the *inside* (I) products must add up to b.
3. The product of the numbers in the *last* (L) blanks must be c.

For example, to factor $2x^2 + 5x + 3$, we write:

3

$$2x^2 + 5x + 3 = (\_\_x + \_\_)(\_\_x + \_\_)$$

2

We first look for two numbers whose product is 2. These numbers are 2 and 1, or -2 and -1. We have these possibilities:

$$(2x + \_\_)(x + \_\_) \quad \text{or} \quad (-2x + \_\_)(-x + \_\_)$$

Let's agree that we want the first coefficients inside the parentheses to be *positive*. This eliminates products involving $(-2x + \_\_)$. Now we look for numbers whose product is 3. These numbers are 3 and 1, or -3 and -1, which we substitute into the blanks, to obtain:

$$(2x + 3)(x + 1)$$
$$(2x + 1)(x + 3)$$
$$(2x - 3)(x - 1)$$
$$(2x - 1)(x - 3)$$

Since the final result must be $2x^2 + 5x + 3$, the first expression (shaded) yields the desired factorization:

$$2x^2 + 5x + 3 = (2x + 3)(x + 1)$$

You can save some time if you notice that all coefficients are positive, so the trial numbers must be positive. That leaves 2, 1 and 3, 1 as the only possibilities.

EXAMPLE 5 Factoring by FOIL (trial and error): All terms positive

Factor: $3x^2 + 7x + 2$

SOLUTION Since we want the first coefficients in the factorization to be positive, the only two factors of 3 we consider are 3 and 1. We then look for the numbers whose product will equal 2:

$$3x^2 + 7x + 2 = (3x + \underbrace{___)(x + ___}_{2})$$

These factors are 2 and 1, and the possibilities are

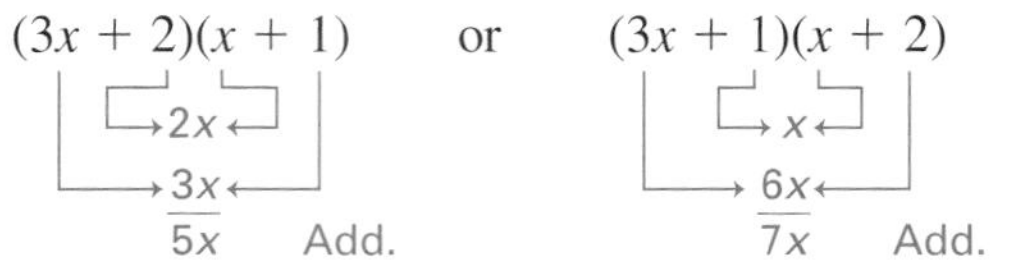
(3x + 2)(x + 1)
2x
3x
5x Add.
or
(3x + 1)(x + 2)
x
6x
7x Add.

Since the second product, $(3x + 1)(x + 2)$, yields the correct middle term, $7x$, we have

$$3x^2 + 7x + 2 = (3x + 1)(x + 2)$$

PROBLEM 5

Factor using FOIL (trial and error): $3x^2 + 5x + 2$

Note that the trial-and-error method is based on FOIL (Section 4.6). Thus to *multiply* $(2x + 3)(3x + 4)$ using FOIL, we write

$$(2x + 3)(3x + 4) = \overset{\text{F}}{6x^2} + \overset{\text{O}}{8x} + \overset{\text{I}}{9x} + \overset{\text{L}}{12}$$
$$= \underbrace{6x^2}_{\substack{\text{F} \\ 2\cdot 3}} + \underbrace{17x}_{\substack{\text{O}+\text{I} \\ 2\cdot 4 + 3\cdot 3}} + \underbrace{12}_{\substack{\text{L} \\ 3\cdot 4}}$$

Now to *factor* $6x^2 + 17x + 12$, we do the reverse, using trial and error. Since the factors of 6 are 6 and 1, or 3 and 2 (we won't use -6 and -1, or -3 and -2, because then the first coefficients will be negative), the possible combinations are

$$(6x + ___)(x + ___) \qquad (3x + ___)(2x + ___)$$

Since the product of the last two numbers is 12, the possible factors are 12, 1; 6, 2; and 3, 4. The possibilities are

$(6x + 12)(x + 1)^\star$	$(6x + 1)(x + 12)$
$(6x + 6)(x + 2)^\star$	$(6x + 2)(x + 6)^\star$
$(6x + 3)(x + 4)^\star$	$(6x + 4)(x + 3)^\star$
$(3x + 12)(2x + 1)^\star$	$(3x + 1)(2x + 12)^\star$
$(3x + 6)(2x + 2)^\star$	$(3x + 2)(2x + 6)^\star$
$(3x + 3)(2x + 4)^\star$	$(3x + 4)(2x + 3)$

Note that a term in the starred items have a common factor, but $6x^2 + 17x + 12$ has *no* common factor other than 1. We can thus eliminate all starred products.

Thus $6x^2 + 17x + 12 = (3x + 4)(2x + 3)$.

Note that if there is a common factor, we must factor it out first. Thus to factor $12x^2 + 2x - 2$, we must *first* factor out the common factor 2, as illustrated in Example 6.

Answer

5. $(3x + 2)(x + 1)$

EXAMPLE 6 **Factoring by FOIL (trial and error): Last term negative**

Factor: $12x^2 + 2x - 2$

SOLUTION Since 2 is a common factor, we first factor it out to obtain

$$12x^2 + 2x - 2 = 2 \cdot 6x^2 + 2 \cdot x - 2 \cdot 1$$
$$= 2(6x^2 + x - 1)$$

Now we factor $6x^2 + x - 1$. The factors of 6 are 6, 1 or 3, 2. Thus

$$6x^2 + x - 1 = (6x + \_\_\_)(x + \_\_\_)$$

or

$$6x^2 + x - 1 = (3x + \_\_\_)(2x + \_\_\_)$$

The product of the last two terms must be -1. The possible factors are -1, 1. The possibilities are

$$(6x - 1)(x + 1) \qquad (6x + 1)(x - 1)$$
$$(3x - 1)(2x + 1) \qquad (3x + 1)(2x - 1)$$

The only product that yields $6x^2 + x - 1$ is $(3x - 1)(2x + 1)$. Try it! This example shows why this method is sometimes called *trial* and error. Thus

$$12x^2 + 2x - 2 = 2(6x^2 + x - 1)$$
$$= 2(3x - 1)(2x + 1)$$

↑ Remember to write the common factor.

PROBLEM 6

Factor using FOIL (trial and error): $18x^2 + 3x - 6$

EXAMPLE 7 **Factoring by FOIL (trial and error): Two variables**

Factor: $6x^2 - 11xy - 10y^2$

SOLUTION Since there are no common factors, look for the factors of 6: 6 and 1, or 3 and 2. The possibilities for our factorization are

$$(6x + \_\_\_)(x + \_\_\_) \quad \text{or} \quad (3x + \_\_\_)(2x + \_\_\_)$$

The last term, $-10y^2$, has the following possible factors:

$$10y, -y \qquad -10y, y \qquad 2y, -5y \qquad -2y, 5y$$
$$-y, 10y \qquad y, -10y \qquad -5y, 2y \qquad 5y, -2y$$

Look daunting? Don't despair; just look more closely. Can you see that some trials like $(6x + 10y)(3x - y)$, which correspond to the factors $10y, -y$, or $(3x - y)(2x + 10y)$, which correspond to $-y, 10y$, can't be correct because they contain 2 as a common factor? As you can see, $6x^2 - 11xy - 10y^2$ has no common factors (other than 1), so let's try

$$(3x + 10y)(2x - y)$$

$20xy$
$(+) -3xy$
$17xy$ Add.

PROBLEM 7

Factor using FOIL (trial and error): $6x^2 - 5xy - 6y^2$

Answers

6. $3(2x - 1)(3x + 2)$
7. $(3x + 2y)(2x - 3y)$

$17xy$ is not the correct middle term. Next we try

$$(3x - 2y)(2x + 5y)$$

$$\begin{array}{r} -4xy \\ (+)\ 15xy \\ \hline 11xy \end{array} \quad \text{Add.}$$

Again, $11xy$ is not the correct middle term, but it's close! The trial is incorrect but only because of the sign of the middle term. The correct factorization is found by interchanging the signs in the binomials. Thus

$$(3x + 2y)(2x - 5y) = 6x^2 - 11xy - 10y^2$$

$$\begin{array}{r} 4xy \\ (+)\ -15xy \\ \hline -11xy \end{array} \quad \text{Add.}$$

CAUTION

Some students and some instructors prefer the FOIL (trial-and-error) method over the grouping method. You can use either method and your answer will be the same. Which one should you use? The one you understand better or the one your instructor asks you to use!

Exercises 5.3

A In Problems 1–50, determine whether the polynomial is factorable.

B If the polynomial is factorable, factor it.

C Use the FOIL method to factor if you wish.

1. $2x^2 + 5x + 3$ $(2x + 3)(x + 1)$

2. $2x^2 + 7x + 3$ $(2x + 1)(x + 3)$

3. $6x^2 + 11x + 3$ $(2x + 3)(3x + 1)$

4. $6x^2 + 17x + 5$ $(3x + 1)(2x + 5)$

5. $6x^2 + 11x + 4$ $(2x + 1)(3x + 4)$

6. $5x^2 + 2x + 1$ Not factorable

7. $2x^2 + 3x - 2$ $(x + 2)(2x - 1)$

8. $2x^2 + x - 3$ $(2x + 3)(x - 1)$

9. $3x^2 + 16x - 12$ $(x + 6)(3x - 2)$

10. $6x^2 + x - 12$ $(3x - 4)(2x + 3)$

11. $4y^2 - 11y + 6$ $(4y - 3)(y - 2)$

12. $3y^2 - 17y + 10$ $(3y - 2)(y - 5)$

13. $4y^2 - 8y + 6$ $2(2y^2 - 4y + 3)$

14. $3y^2 - 11y + 6$ $(3y - 2)(y - 3)$

15. $6y^2 - 10y - 4$ $2(3y + 1)(y - 2)$

16. $12y^2 - 10y - 12$ $2(3y + 2)(2y - 3)$

17. $12y^2 - y - 6$ $(3y + 2)(4y - 3)$

18. $3y^2 - y - 1$ Not factorable

19. $18y^2 - 21y - 9$ $3(3y + 1)(2y - 3)$

20. $36y^2 - 12y - 15$ $3(6y - 5)(2y + 1)$

21. $3x^2 + 2 + 7x$ $(3x + 1)(x + 2)$

22. $2x^2 + 2 + 5x$ $(2x + 1)(x + 2)$

23. $5x^2 + 2 + 11x$ $(5x + 1)(x + 2)$

24. $5x^2 + 3 + 12x$ Not factorable

25. $6x^2 - 5 + 15x$ Not factorable

26. $5x^2 - 8 + 6x$ $(5x - 4)(x + 2)$

27. $3x^2 - 2 - 5x$ $(3x + 1)(x - 2)$

28. $5x^2 - 8 - 6x$ $(5x + 4)(x - 2)$

29. $15x^2 - 2 + x$ $(5x + 2)(3x - 1)$

30. $8x^2 + 15 - 14x$ Not factorable

31. $8x^2 + 20xy + 8y^2$ $4(2x + y)(x + 2y)$

32. $12x^2 + 28xy + 8y^2$ $4(3x + y)(x + 2y)$

33. $6x^2 + 7xy - 3y^2$ $(2x + 3y)(3x - y)$

34. $3x^2 + 13xy - 10y^2$ $(3x - 2y)(x + 5y)$

35. $7x^2 - 10xy + 3y^2$ $(7x - 3y)(x - y)$

36. $6x^2 - 17xy + 5y^2$ $(3x - y)(2x - 5y)$

37. $15x^2 - xy - 2y^2$ $(3x + y)(5x - 2y)$

38. $5x^2 - 6xy - 8y^2$ $(5x + 4y)(x - 2y)$

39. $15x^2 - 2xy - 2y^2$ Not factorable

40. $4x^2 - 13xy - 3y^2$ Not factorable

41. $12r^2 + 17r - 5$ $(3r + 5)(4r - 1)$

42. $20s^2 + 7s - 6$ $(5s - 2)(4s + 3)$

43. $22t^2 - 29t - 6$ $(11t + 2)(2t - 3)$

44. $39u^2 - 23u - 6$ Not factorable

45. $18x^2 - 21x + 6$ $3(3x - 2)(2x - 1)$

46. $12x^2 - 22x + 6$ $2(3x - 1)(2x - 3)$

47. $6ab^2 + 5ab + a$ $a(3b + 1)(2b + 1)$

48. $6bc^2 + 13bc + 6b$ $b(3c + 2)(2c + 3)$

49. $6x^5y + 25x^4y^2 + 4x^3y^3$ $x^3y(6x + y)(x + 4y)$

50. $12p^4q^3 + 11p^3q^4 + 2p^2q^5$ $p^2q^3(4p + q)(3p + 2q)$

In Problems 51–60, first factor out -1. [*Hint:* To factor $-6x^2 + 7x + 2$, the first step will be

$$-6x^2 + 7x + 2 = -1(6x^2 - 7x - 2) = -(6x^2 - 7x - 2)$$

then factor inside the parentheses.]

51. $-6x^2 - 7x - 2$ $-(3x + 2)(2x + 1)$

52. $-12y^2 - 11y - 2$ $-(4y + 1)(3y + 2)$

53. $-9x^2 - 3x + 2$ $-(3x + 2)(3x - 1)$

54. $-6y^2 - 5y + 6$ $-(3y - 2)(2y + 3)$

55. $-8m^2 + 10mn + 3n^2$ $-(4m + n)(2m - 3n)$

56. $-6s^2 + st + 2t^2$ $-(3s - 2t)(2s + t)$

57. $-8x^2 + 9xy - y^2$ $-(8x - y)(x - y)$

58. $-6y^2 + 3xy + 2x^2$ $-(6y^2 - 3xy - 2x^2)$

59. $-x^3 - 5x^2 - 6x$ $-x(x + 3)(x + 2)$

60. $-y^3 + 3y^2 - 2y$ $-y(y - 2)(y - 1)$

APPLICATIONS

61. *Flow rate* To find the flow g (in hundreds of gallons per minute) in 100 feet of $2\frac{1}{2}$-inch rubber-lined hose when the friction loss is 36 pounds per square inch, we need to factor the expression

$$2g^2 + g - 36$$

Factor this expression. $(2g + 9)(g - 4)$

62. *Flow rate* To find the flow g (in hundreds of gallons per minute) in 100 feet of $2\frac{1}{2}$-inch rubber-lined hose when the friction loss is 55 pounds per square inch, we must factor the expression

$$2g^2 + g - 55$$

Factor this expression. $(2g + 11)(g - 5)$

63. *Equivalent resistance* When solving for the equivalent resistance R of two electric circuits, we find the expression

$$2R^2 - 3R + 1$$

Factor this expression. $(2R - 1)(R - 1)$

64. *Rate of ascent* To find the time t at which an object thrown upward at 12 meters per second will be 4 meters above the ground, we must evaluate the expression

$$5t^2 - 12t + 4$$

Factor this expression. $(5t - 2)(t - 2)$

65. *Box office receipts* Have you been to the movies lately? How much was your ticket? According to the U.S. Department of Commerce, box office receipts from 1980 to 1990 can be approximated by

$$B(t) \cdot P(t) = 0.004t^2 + 0.2544t + 2.72$$

where $B(t)$ is the number of people buying tickets (in thousands), $P(t)$ is the average ticket price (in dollars), and t is the number of years after 1980.

a. Factor $0.004t^2 + 0.2544t + 2.72$.
[*Hint:* Try $B(t) \cdot P(t) = (\_\_\_ + 1)(\_\_\_ + 2.72)$.]
$(0.02t + 1)(0.2t + 2.72)$

b. Based on your answer to part **a**, what was the average price of a ticket in 1980? (The actual average price for 1980 was \$2.72.) \$2.72

c. Based on your answer to part **a**, what was the average price of a ticket in 1985? (The actual average price for 1985 was \$3.55.) \$3.72

66. *Entertainment and reading expenditures* According to the annual *Consumer Educational Survey* of the Bureau of Labor Statistics, the annual average amount of money spent for entertainment and reading can be approximated by $1300 + 4.16t + 3.12t^2$ (dollars), where t is the number of years after 1985. If you want to approximate how much is spent *weekly* on entertainment and reading, write

$$1300 + 4.16t + 3.12t^2 = 52 \cdot E(t)$$

a. Where does the 52 come from? Number of weeks in a year

b. What does $E(t)$ represent? Amount spent weekly on entertainment and reading

c. How much was spent weekly for entertainment and reading in 1985? (The actual amount was \$25.21.) \$25

d. How much was spent weekly on entertainment and reading in 1995? In 2005? \$31.80; \$50.60

SKILL CHECKER

Try the Skill Checker Exercises so you'll be ready for the next section.

Expand:

67. $(x + 8)^2$ $x^2 + 16x + 64$

68. $(x - 7)^2$ $x^2 - 14x + 49$

69. $(3x - 2)^2$ $9x^2 - 12x + 4$

70. $(2x + 3)^2$ $4x^2 + 12x + 9$

71. $(2x + 3y)^2$ $4x^2 + 12xy + 9y^2$

72. $(2x - 3y)^2$ $4x^2 - 12xy + 9y^2$

73. $(3x + 5y)(3x - 5y)$ $9x^2 - 25y^2$

74. $(2x - 5y)(2x + 5y)$ $4x^2 - 25y^2$

75. $(x^2 + 4)(x^2 - 4)$ $x^4 - 16$

76. $(x^2 - 3)(x^2 + 3)$ $x^4 - 9$

USING YOUR KNOWLEDGE

Factoring Applications

The ideas presented in this section are important in many fields. Use your knowledge to factor the given expressions.

77. To find the deflection of a beam of length L at a distance of 3 feet from its end, we must evaluate the expression

$$2L^2 - 9L + 9$$

Factor this expression. $(2L - 3)(L - 3)$

78. In Problem 77, if the distance from the end is x feet, then we must use the expression

$$2L^2 - 3xL + x^2$$

Factor this expression. $(2L - x)(L - x)$

79. The height after t seconds of an object thrown upward at 12 meters per second is

$$-5t^2 + 12t$$

To determine the time at which the object will be 7 meters above ground, we must solve the equation

$$5t^2 - 12t + 7 = 0$$

Factor this trinomial. $(5t - 7)(t - 1)$

WRITE ON

80. Mourad says that the ac key number for $2x^2 + 1 + 3x$ is 2 and hence $2x^2 + 1 + 3x$ is factorable. Bill says ac is 6. Who is right? Mourad

81. A student gives $(3x - 1)(x - 2)$ as the answer to a factoring problem. Another student gets $(2 - x)(1 - 3x)$. Which student is correct? Explain.
Both; answers may vary.

82. When factoring $6x^2 + 11x + 3$ by grouping, student A writes

$$6x^2 + 11x + 3 = 6x^2 + 9x + 2x + 3$$

Student B writes

$$6x^2 + 11x + 3 = 6x^2 + 2x + 9x + 3$$

a. What will student A's answer be? $(2x + 3)(3x + 1)$

b. What will student B's answer be? $(3x + 1)(2x + 3)$

c. Who is right, student A or student B? Explain.
Both; answers may vary.

MASTERY TEST

If you know how to do these problems, you have learned your lesson!

Factor:

83. $3x^2 - 4 - 4x$
$(3x + 2)(x - 2)$

84. $2x^2 - 11x + 5$
$(2x - 1)(x - 5)$

85. $2x^2 - xy - 6y^2$
$(2x + 3y)(x - 2y)$

86. $3x^2 + 5x + 2$
$(3x + 2)(x + 1)$

87. $16x^2 + 4x - 2$
$2(4x - 1)(2x + 1)$

88. $3x^3 + 7x^2 + 2x$
$x(3x + 1)(x + 2)$

89. $3x^4 + 5x^3 - 3x^2$
$x^2(3x^2 + 5x - 3)$

90. $5x^2 - 2x + 2$
Not factorable

5.4 FACTORING SQUARES OF BINOMIALS

To Succeed, Review How To . . .

1. Expand the square of a binomial sum or difference (pp. 357–359).
2. Find the product of the sum and difference of two terms (pp. 359–360).

Objectives

A Recognize the square of a binomial (a perfect square trinomial).

B Factor a perfect square trinomial.

C Factor the difference of two squares.

GETTING STARTED

A Moment for a Crane

What is the moment (the product of a quantity and the distance from a perpendicular axis) on the crane? At x feet from its support, the moment involves the expression

$$\frac{w}{2}(x^2 - 20x + 100)$$

where w is the weight of the crane in pounds per foot. The expression $x^2 - 20x + 100$ is the result of *squaring a binomial* and is called a **perfect square trinomial** because $x^2 - 20x + 100 = (x - 10)^2$. Similarly, $x^2 + 12x + 36 = (x + 6)^2$ is the square of a binomial. We can factor these two expressions by using the special products studied in Section 4.7 in reverse. For example, $x^2 - 20x + 100$ has the same form as SP3, whereas $x^2 + 12x + 36$ looks like SP2. In this section we continue to study the *reverse* process of multiplying binomials, that of *factoring* perfect square trinomials.

Recognizing Squares of Binomials

We start by rewriting the products in SP2 and SP3 so you can use them for factoring.

Web It

To look at the procedure used to recognize binomial squares, go to link 5-4-1 on the Bello Website at mhhe.com/bello.

FACTORING RULES 2 AND 3: PERFECT SQUARE TRINOMIALS

$X^2 + 2AX + A^2 = (X + A)^2$ Note that $X^2 + A^2 \neq (X + A)^2$ **(F2)**

$X^2 - 2AX + A^2 = (X - A)^2$ Note that $X^2 - A^2 \neq (X - A)^2$ **(F3)**

Teaching Tip

Have students square the following binomials and compare the results with these three criteria for a perfect square trinomial.

$(x + 5)^2 = x^2 + 10x + 25$
$(3x - 2)^2 = 9x^2 - 12x + 4$

Note that to be the **square of a binomial** (a *perfect square trinomial*), a trinomial must satisfy three conditions:

1. The first and last terms (X^2 and A^2) must be perfect squares.
2. There must be no minus signs before A^2 or X^2.
3. The middle term is twice the product of the expressions being squared in step 1 ($2AX$) or its additive inverse ($-2AX$). Note the X and A are the terms of the binomial being squared to obtain the perfect square trinomial.

EXAMPLE 1 Deciding whether an expression is the square of a binomial

Determine whether the given expression is the square of a binomial:

a. $x^2 + 8x + 16$ **b.** $x^2 + 6x - 9$

c. $x^2 + 4x + 16$ **d.** $4x^2 - 12xy + 9y^2$

SOLUTION In each case, we check the three conditions necessary for having a perfect square trinomial.

a. **1.** x^2 and $16 = 4^2$ are perfect squares.

2. There are no minus signs before x^2 or 16.

3. The middle term is twice the product of the expressions being squared in step 1, x and 4; that is, the middle term is $2 \cdot (x \cdot 4) = 8x$. Thus $x^2 + 8x + 16$ *is* a perfect square trinomial (the square of a binomial).

b. **1.** x^2 and $9 = 3^2$ are perfect squares.

2. However, there's a minus sign before the 9. Thus $x^2 + 6x - 9$ is *not* the square of a binomial.

c. **1.** x^2 and $16 = 4^2$ are perfect squares.

2. There are no minus signs before x^2 or 16.

3. The middle term should be $2 \cdot (x \cdot 4) = 8x$, but instead it's $4x$. Thus $x^2 + 4x + 16$ is *not* a perfect square trinomial.

d. **1.** $4x^2 = (2x)^2$ and $9y^2 = (3y)^2$ are perfect squares.

2. There are no minus signs before $4x^2$ or $9y^2$.

3. The middle term is the additive inverse of twice the product of the expressions being squared in step 1; that is, $-2 \cdot (2x \cdot 3y) = -12xy$. Thus $4x^2 - 12xy + 9y^2$ *is* a perfect square trinomial.

PROBLEM 1

Determine whether the expression is the square of a binomial:

a. $y^2 + 9y + 9$

b. $y^2 + 4y + 4$

c. $y^2 + 8y - 16$

d. $9x^2 - 12xy + 4y^2$

Teaching Tip

A reminder about perfect squares:

These are perfect squares:		Their square roots are:	
1	x^2	$\sqrt{1} = 1$	$\sqrt{x^2} = x$
4	x^4	$\sqrt{4} = 2$	$\sqrt{x^4} = x^2$
9	x^6	$\sqrt{9} = 3$	$\sqrt{x^6} = x^3$
16	x^8	$\sqrt{16} = 4$	$\sqrt{x^8} = x^4$
etc.		etc.	

Web It

For practice problems, with answers, dealing with perfect square trinomials, go to the Bello Website at mhhe.com/bello and try link 5-4-2.

B Factoring Perfect Square Trinomials

The formula given in F2 can be used to factor any trinomials that are perfect squares. For example, the trinomial $9x^2 + 12x + 4$ can be factored using F2 if we first notice that

1. $9x^2$ and 4 are perfect squares, since $9x^2 = (3x)^2$ and $4 = 2^2$.
2. There are no minus signs before $9x^2$ or 4.
3. $12x = 2 \cdot (2 \cdot 3x)$. (The middle term is twice the product of 2 and $3x$, the expressions being squared in step 1.)

Answers

1. a. No **b.** Yes **c.** No **d.** Yes

Teaching Tip

Here is an alternate method for factoring perfect square trinomials:

1. If the first and last terms of the trinomial are both positive and perfect squares, write their square roots in parentheses to the second power using the middle sign of the trinomial as the sign inside the binomial.
2. Verify the middle term by multiplying the two terms of the binomial and doubling the result.

Example:

$x^2 + 6x + 9$

$\sqrt{x^2}$ $\sqrt{9}$

$(x + 3)^2$

$2 \cdot x \cdot 3 = 6x$

We then write

$$\begin{aligned} 9x^2 + 12x + 4 &= \overset{X^2}{(3x)^2} + \overset{+2}{2} \cdot \overset{AX}{(2 \cdot 3x)} + \overset{+A^2}{2^2} \\ &= (3x + 2)^2 \end{aligned}$$

We are letting $X = 3x$, $A = 2$ in F2.

Here are some other examples of this form; study these examples carefully before you continue.

$$9x^2 + 6x + 1 = \overset{X^2}{(3x)^2} + \overset{+2}{2} \cdot \overset{AX}{(1 \cdot 3x)} + \overset{+A^2}{1^2} = (3x + 1)^2$$

Letting $X = 3x$, $A = 1$ in F2.

$$16x^2 + 24x + 9 = (4x)^2 + 2 \cdot (3 \cdot 4x) + 3^2 = (4x + 3)^2$$

Here $X = 4x$, $A = 3$.

$$4x^2 + 12xy + 9y^2 = (2x)^2 + 2 \cdot (3y \cdot 2x) + (3y)^2 = (2x + 3y)^2$$

Here $X = 2x$, $A = 3y$.

> **NOTE**
>
> The key for factoring these trinomials is to recognize that the first and last terms are *perfect squares.* (Of course, you have to check the middle term also.)

EXAMPLE 2 Factoring perfect square trinomials

Factor:

a. $x^2 + 16x + 64$ **b.** $25x^2 + 20x + 4$ **c.** $9x^2 + 12xy + 4y^2$

SOLUTION

a. We first write the trinomial in the form $X^2 + 2AX + A^2$. Thus

$$x^2 + 16x + 64 = x^2 + 2 \cdot (8 \cdot x) + 8^2 = (x + 8)^2$$

b. $25x^2 + 20x + 4 = (5x)^2 + 2 \cdot (2 \cdot 5x) + 2^2 = (5x + 2)^2$

c. $9x^2 + 12xy + 4y^2 = (3x)^2 + 2 \cdot (2y \cdot 3x) + (2y)^2 = (3x + 2y)^2$

PROBLEM 2

Factor:

a. $y^2 + 6y + 9$

b. $4x^2 + 28xy + 49y^2$

c. $9y^2 + 12y + 4$

Of course, we use the same technique (but with F3) to factor $x^2 - 16x + 64$ or $25x^2 - 20x + 4$. Do you recall F3?

$$\underbrace{x^2}_{X^2} \; \underbrace{- 16x}_{-2AX} \; \underbrace{+ 64}_{+A^2} = x^2 - 2 \cdot (8 \cdot x) + 8^2 = \underbrace{(x - 8)^2}_{(X - A)^2}$$

Similarly,

$$25x^2 - 20x + 4 = (5x)^2 - 2 \cdot (2 \cdot 5x) + 2^2 = (5x - 2)^2$$

EXAMPLE 3 Factoring perfect square trinomials

Factor:

a. $x^2 - 10x + 25$ **b.** $4x^2 - 12x + 9$ **c.** $4x^2 - 20xy + 25y^2$

SOLUTION

a. $x^2 - 10x + 25 = x^2 - 2 \cdot (5 \cdot x) + 5^2 = (x - 5)^2$

b. $4x^2 - 12x + 9 = (2x)^2 - 2 \cdot (3 \cdot 2x) + 3^2 = (2x - 3)^2$

c. $4x^2 - 20xy + 25y^2 = (2x)^2 - 2 \cdot (5y \cdot 2x) + (5y)^2 = (2x - 5y)^2$

PROBLEM 3

Factor:

a. $y^2 - 4y + 4$ **b.** $9y^2 - 12y + 4$

c. $9x^2 - 30xy + 25y^2$

Answers

2. a. $(y + 3)^2$ **b.** $(2x + 7y)^2$ **c.** $(3y + 2)^2$ **3. a.** $(y - 2)^2$ **b.** $(3y - 2)^2$ **c.** $(3x - 5y)^2$

C Factoring the Difference of Two Squares

Web It

For a site where you can enter your own problem and get the solution explained, go to link 5-4-3 on the Bello Website at mhhe.com/bello.

Can we factor $x^2 - 9$ as a product of two binomials? Note that $x^2 - 9$ has no middle term. The only special product with no middle term we've studied is the product of the sum and the difference of two terms (SP4). Here is the corresponding factoring rule:

FACTORING RULE 4: THE DIFFERENCE OF TWO SQUARES

$$X^2 - A^2 = (X + A)(X - A) \qquad \textbf{(F4)}$$

We can now factor binomials of the form $x^2 - 16$ and $9x^2 - 25y^2$. To do this, we proceed as follows:

$$\begin{array}{c}X^2 \quad - \quad A^2 \quad = (X + A)\,(X - A)\\ x^2 - 16 = (x)^2 - (4)^2 = (x + 4)(x - 4)\end{array}$$

Check your answers by using FOIL.

and

$$\begin{array}{c}X^2 \quad - \quad A^2 \quad = (X + A)\quad(X - A)\\ 9x^2 - 25y^2 = (3x)^2 - (5y)^2 = (3x + 5y)(3x - 5y)\end{array}$$

Web It

For a site reviewing most of the factoring we have done, go to link 5-4-4 on the Bello Website at mhhe.com/bello.

NOTE

$x^2 + A^2$ *cannot* be factored!

EXAMPLE 4 **Factoring the difference of two squares**

Factor:

a. $x^2 - 4$ **b.** $25x^2 - 9$

c. $16x^2 - 9y^2$ **d.** $x^4 - 16$

e. $\frac{1}{4}x^2 - \frac{1}{9}$ **f.** $7x^3 - 28x$

SOLUTION

a. $\begin{array}{l}\quad\; X^2 \;\; - A^2 \quad = (X + A)\,(X - A)\\ x^2 - 4 = (x)^2 - (2)^2 = (x + 2)(x - 2)\end{array}$

b. $25x^2 - 9 = (5x)^2 - (3)^2 = (5x + 3)(5x - 3)$

c. $16x^2 - 9y^2 = (4x)^2 - (3y)^2 = (4x + 3y)(4x - 3y)$

d. $x^4 - 16 = (x^2)^2 - (4)^2 = (x^2 + 4)(x^2 - 4)$

But $(x^2 - 4)$ itself is factorable, so $(x^2 + 4)(x^2 - 4) = (x^2 + 4)(x + 2)(x - 2)$. Thus

$$x^4 - 16 = \underbrace{(x^2 + 4)}_{\text{Not factorable}}(x + 2)(x - 2)$$

PROBLEM 4

Factor:

a. $y^2 - 1$ **b.** $9y^2 - 25$

c. $9y^2 - 25x^2$ **d.** $y^4 - 81$

e. $\frac{1}{9}y^2 - \frac{1}{4}$ **f.** $5y^3 - 45y$

Answers

4. a. $(y + 1)(y - 1)$
b. $(3y + 5)(3y - 5)$
c. $(3y + 5x)(3y - 5x)$
d. $(y^2 + 9)(y + 3)(y - 3)$
e. $(\frac{1}{3}y + \frac{1}{2})(\frac{1}{3}y - \frac{1}{2})$
f. $5y(y + 3)(y - 3)$

e. We start by writing $\frac{1}{4}x^2 - \frac{1}{9}$ as the difference of two squares. To obtain $\frac{1}{4}x^2$, we must square $\frac{1}{2}x$ and to obtain $\frac{1}{9}$, we must square $\frac{1}{3}$. Thus

$$\frac{1}{4}x^2 - \frac{1}{9} = \left(\frac{1}{2}x\right)^2 - \left(\frac{1}{3}\right)^2$$
$$= \left(\frac{1}{2}x + \frac{1}{3}\right)\left(\frac{1}{2}x - \frac{1}{3}\right)$$

f. We start by finding the GCF of $7x^3 - 28x$, which is $7x$. We then write

$$7x^3 - 28x = 7x(x^2 - 4)$$
$$= 7x(x + 2)(x - 2) \quad \text{Factor } x^2 - 4 \text{ as } (x + 2)(x - 2).$$

Teaching Tip

Have students multiply the following and compare results:

$$(x + 3)(x - 3) = x^2 - 9$$
$$(2x + 1)(2x - 1) = 4x^2 - 1$$
$$(x - 5y)(x + 5y) = x^2 - 25y^2$$

Each of these results is called the difference of two squares.

Calculate It More Factoring Checking

Can you use a calculator to check factoring? Yes, but you still have to accept the following: If the graphs of two polynomials are identical, the polynomials are identical. Consider Example 2b, $25x^2 + 20x + 4 = (5x + 2)^2$. Graph $Y_1 = 25x^2 + 20x + 4$ and $Y_2 = (5x + 2)^2$. The graphs are the same! How do you know there are two graphs? Press TRACE and ▲, then ▼. Do you see the little number at the top right of the screen? It tells you which curve is showing. The bottom of the screen shows that for both Y_1 and Y_2 when $x = 0$, $y = 4$. Graph $Y_3 = (5x + 2)^2 + 10$. Press TRACE. Are the values for Y_1, Y_2, and Y_3 the same? They shouldn't be! Do you see why? Use this technique to check other factoring problems.

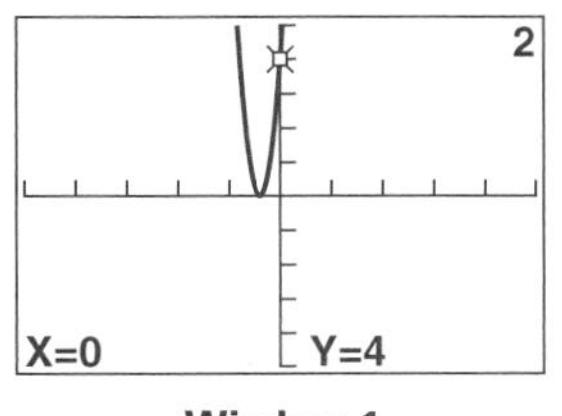

Window 1

Exercises 5.4

A In Problems 1–10, determine whether the given expression is a perfect square trinomial (the square of a binomial).

1. $x^2 + 14x + 49$ Yes
2. $x^2 + 18x + 81$ Yes
3. $25x^2 + 10x - 1$ No
4. $9x^2 + 12x - 4$ No
5. $25x^2 + 10x + 1$ Yes
6. $9x^2 + 12x + 4$ Yes
7. $y^2 - 4y - 4$ No
8. $y^2 - 20y - 100$ No
9. $16y^2 - 40yz + 25z^2$ Yes
10. $49y^2 - 56yz + 16z^2$ Yes

B In Problems 11–34, factor completely.

11. $x^2 + 2x + 1$ $(x + 1)^2$
12. $x^2 + 6x + 9$ $(x + 3)^2$
13. $3x^2 + 30x + 75$ $3(x + 5)^2$
14. $2x^2 + 28x + 98$ $2(x + 7)^2$
15. $9x^2 + 6x + 1$ $(3x + 1)^2$
16. $16x^2 + 8x + 1$ $(4x + 1)^2$
17. $9x^2 + 12x + 4$ $(3x + 2)^2$
18. $25x^2 + 10x + 1$ $(5x + 1)^2$
19. $16x^2 + 40xy + 25y^2$ $(4x + 5y)^2$
20. $9x^2 + 30xy + 25y^2$ $(3x + 5y)^2$
21. $25x^2 + 20xy + 4y^2$ $(5x + 2y)^2$
22. $36x^2 + 60xy + 25y^2$ $(6x + 5y)^2$
23. $y^2 - 2y + 1$ $(y - 1)^2$
24. $y^2 - 4y + 4$ $(y - 2)^2$
25. $3y^2 - 24y + 48$ $3(y - 4)^2$

26. $2y^2 - 40y + 200$ $2(y - 10)^2$ **27.** $9x^2 - 6x + 1$ $(3x - 1)^2$ **28.** $4x^2 - 20x + 25$ $(2x - 5)^2$

29. $16x^2 - 56x + 49$ $(4x - 7)^2$ **30.** $25x^2 - 30x + 9$ $(5x - 3)^2$ **31.** $9x^2 - 12xy + 4y^2$ $(3x - 2y)^2$

32. $16x^2 - 40xy + 25y^2$ $(4x - 5y)^2$ **33.** $25x^2 - 10xy + y^2$ $(5x - y)^2$ **34.** $49x^2 - 56xy + 16y^2$ $(7x - 4y)^2$

C In Problems 35–60, factor completely.

35. $x^2 - 49$ $(x + 7)(x - 7)$ **36.** $x^2 - 121$ $(x + 11)(x - 11)$ **37.** $9x^2 - 49$ $(3x + 7)(3x - 7)$

38. $16x^2 - 81$ $(4x + 9)(4x - 9)$ **39.** $25x^2 - 81y^2$ $(5x + 9y)(5x - 9y)$ **40.** $81x^2 - 25y^2$ $(9x + 5y)(9x - 5y)$

41. $x^4 - 1$ $(x^2 + 1)(x + 1)(x - 1)$ **42.** $x^4 - 256$ $(x^2 + 16)(x + 4)(x - 4)$ **43.** $16x^4 - 1$ $(4x^2 + 1)(2x + 1)(2x - 1)$

44. $16x^4 - 81$ $(4x^2 + 9)(2x + 3)(2x - 3)$ **45.** $\frac{1}{9}x^2 - \frac{1}{16}$ $\left(\frac{1}{3}x + \frac{1}{4}\right)\left(\frac{1}{3}x - \frac{1}{4}\right)$ **46.** $\frac{1}{4}y^2 - \frac{1}{25}$ $\left(\frac{1}{2}y + \frac{1}{5}\right)\left(\frac{1}{2}y - \frac{1}{5}\right)$

47. $\frac{1}{4}z^2 - 1$ $\left(\frac{1}{2}z + 1\right)\left(\frac{1}{2}z - 1\right)$ **48.** $\frac{1}{9}r^2 - 1$ $\left(\frac{1}{3}r + 1\right)\left(\frac{1}{3}r - 1\right)$ **49.** $1 - \frac{1}{4}s^2$ $\left(1 + \frac{1}{2}s\right)\left(1 - \frac{1}{2}s\right)$

50. $1 - \frac{1}{9}t^2$ $\left(1 + \frac{1}{3}t\right)\left(1 - \frac{1}{3}t\right)$ **51.** $\frac{1}{4} - \frac{1}{9}y^2$ $\left(\frac{1}{2} + \frac{1}{3}y\right)\left(\frac{1}{2} - \frac{1}{3}y\right)$ **52.** $\frac{1}{9} - \frac{1}{16}u^2$ $\left(\frac{1}{3} + \frac{1}{4}u\right)\left(\frac{1}{3} - \frac{1}{4}u\right)$

53. $\frac{1}{9} + \frac{1}{4}x^2$ Not factorable **54.** $\frac{1}{4} + \frac{1}{25}n^2$ Not factorable **55.** $3x^3 - 12x$ $3x(x + 2)(x - 2)$

56. $4y^3 - 16y$ $4y(y + 2)(y - 2)$ **57.** $5t^3 - 20t$ $5t(t + 2)(t - 2)$ **58.** $7t^3 - 63t$ $7t(t + 3)(t - 3)$

59. $5t - 20t^3$ $5t(1 + 2t)(1 - 2t)$ **60.** $2s - 18s^3$ $2s(1 + 3s)(1 - 3s)$

In Problems 61–74, you have to use a variety of factoring methods.

61. $49x^2 + 28x + 4$ $(7x + 2)^2$ **62.** $49y^2 + 42y + 9$ $(7y + 3)^2$ **63.** $x^2 - 100$ $(x + 10)(x - 10)$

64. $x^2 - 144$ $(x + 12)(x - 12)$ **65.** $x^2 + 20x + 100$ $(x + 10)^2$ **66.** $x^2 + 18x + 81$ $(x + 9)^2$

67. $9 - 16m^2$ $(3 + 4m)(3 - 4m)$ **68.** $25 - 9n^2$ $(5 + 3n)(5 - 3n)$ **69.** $9x^2 - 30xy + 25y^2$ $(3x - 5y)^2$

70. $16x^2 - 40xy + 25y^2$ $(4x - 5y)^2$ **71.** $x^4 - 16$ $(x^2 + 4)(x + 2)(x - 2)$ **72.** $16y^4 - 81$ $(4y^2 + 9)(2y + 3)(2y - 3)$

73. $3x^3 - 75x$ $3x(x + 5)(x - 5)$ **74.** $2y^3 - 72y$ $2y(y + 6)(y - 6)$

SKILL CHECKER

Try the Skill Checker Exercises so you'll be ready for the next section.

Multiply:

75. $(R + r)(R - r)$ $R^2 - r^2$ **76.** $(P + q)(P - q)$ $P^2 - q^2$

Factor completely:

77. $6x^2 - 18x - 24$ $6(x + 1)(x - 4)$ **78.** $4x^4 + 12x^3 + 40x^2$ $4x^2(x^2 + 3x + 10)$

79. $2x^2 - 18$ $2(x + 3)(x - 3)$ **80.** $3x^2 - 27$ $3(x + 3)(x - 3)$

USING YOUR KNOWLEDGE

How Does It Function?

Many business ideas are made precise by using expressions called **functions.** These expressions are often given in unfactored form. Use your knowledge to factor the given expressions (functions).

81. When x units of an item are demanded by consumers, the price per unit is given by the **demand** function $D(x)$ (read "D of x"):

$$D(x) = x^2 - 14x + 49$$

Factor this expression. $D(x) = (x - 7)^2$

82. When x units are supplied by sellers, the price per unit of an item is given by the **supply** function $S(x)$:

$$S(x) = x^2 + 4x + 4$$

Factor this expression. $S(x) = (x + 2)^2$

83. When x units are produced, the cost function $C(x)$ for a certain item is given by

$$C(x) = x^2 + 12x + 36$$

Factor this expression. $C(x) = (x + 6)^2$

84. When the market price is p dollars, the supply function $S(p)$ for a certain commodity is given by

$$S(p) = p^2 - 6p + 9$$

Factor this expression. $S(p) = (p - 3)^2$

WRITE ON

85. The difference of two squares can be factored. Can you factor $a^2 + b^2$? Can you say that the sum of two squares can *never* be factored? Explain.
Answers may vary.

86. Can you factor $4a^2 + 16b^2$? Think of the implications for Problem 85. $4(a^2 + 4b^2)$

87. What binomial multiplied by $(x + 2)$ gives a perfect square trinomial? $(x + 2)$

88. What binomial multiplied by $(2x - 3y)$ gives a perfect square trinomial? $(2x - 3y)$

MASTERY TEST

If you know how to do these problems, you have learned your lesson!

Factor, if possible:

89. $x^2 - 1$ $(x + 1)(x - 1)$

90. $9x^2 - 16$ $(3x + 4)(3x - 4)$

91. $9x^2 - 25y^2$ $(3x + 5y)(3x - 5y)$

92. $x^2 - 6x + 9$ $(x - 3)^2$

93. $9x^2 - 24xy + 16y^2$ $(3x - 4y)^2$

94. $9x^2 - 12x + 4$ $(3x - 2)^2$

95. $16x^2 + 24xy + 9y^2$ $(4x + 3y)^2$

96. $x^2 + 4x + 4$ $(x + 2)^2$

97. $9x^2 + 30x + 25$ $(3x + 5)^2$

98. $4x^2 - 20xy + 25y^2$ $(2x - 5y)^2$

99. $9x^2 + 4$ Not factorable

100. $\frac{1}{36}x^2 - \frac{1}{49}$ $\left(\frac{1}{6}x + \frac{1}{7}\right)\left(\frac{1}{6}x - \frac{1}{7}\right)$

101. $\frac{1}{81} - \frac{1}{4}x^2$ $\left(\frac{1}{9} + \frac{1}{2}x\right)\left(\frac{1}{9} - \frac{1}{2}x\right)$

102. $12m^3 - 3mn^2$
$3m(2m + n)(2m - n)$

103. $18x^3 - 50xy^2$
$2x(3x + 5y)(3x - 5y)$

104. $9x^3 + 25xy^2$ $x(9x^2 + 25y^2)$

Determine whether the expression is the square of a binomial. If it is, factor it.

105. $x^2 + 6x + 9$ Yes; $(x + 3)^2$

106. $x^2 + 8x + 64$ No

107. $x^2 + 6x - 9$ No

108. $x^2 + 8x - 64$ No

109. $4x^2 - 20xy + 25y^2$ Yes; $(2x - 5y)^2$

5.5 A GENERAL FACTORING STRATEGY

To Succeed, Review How To . . .

1. Factor a polynomial using F1–F4 (pp. 399–403; 417–421).

Objectives

A Factor the sum or difference of two cubes.

B Factor a polynomial by using the general factoring strategy.

C Factor expressions whose leading coefficient is -1.

GETTING STARTED

Factoring and Medicine

In an artery (see the cross section in the photo), the speed (in centimeters per second) of the blood is given by

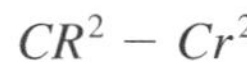

$$CR^2 - Cr^2$$

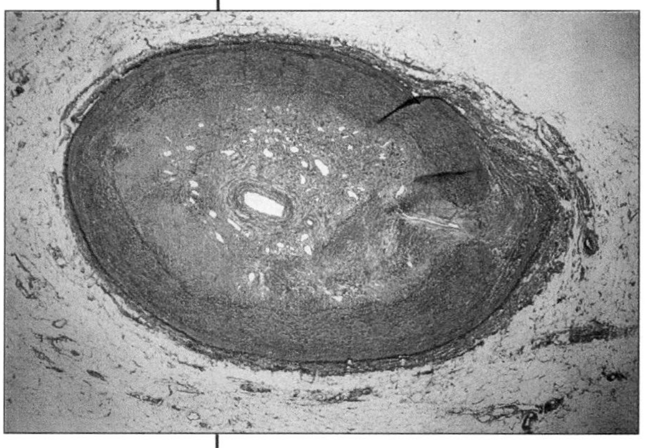

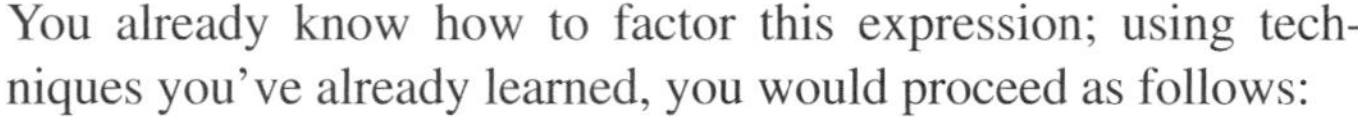

You already know how to factor this expression; using techniques you've already learned, you would proceed as follows:

1. Factor out any common factors (in this case, C). $\quad CR^2 - Cr^2 = C(R^2 - r^2)$
2. Look at the terms inside the parentheses. You have the difference of two square terms in the expression: $R^2 - r^2$, so you factor it. $\quad = C(R + r)(R - r)$
3. Make sure the expression is *completely* factored. Note that $C(R + r)(R - r)$ cannot be factored further.

What we've just used here is a *strategy* for factoring polynomials—a logical way to call up any of the techniques you've studied when they fit the expression you are factoring. In this section we shall study one more type of factoring: sums or differences of cubes. We will then examine in more depth the general factoring strategy for polynomials.

Teaching Tip

Have students multiply $(x + y)(x^2 - xy + y^2)$ and $(x - y)(x^2 + xy + y^2)$ before these two general examples.

A Factoring Sums or Differences of Cubes

We've already factored the difference of two squares $X^2 - A^2$. Can we factor the difference of two cubes $X^3 - A^3$? Not only can we factor $X^3 - A^3$, we can even factor $X^3 + A^3$! Since factoring is "reverse multiplication," let's start with two multiplication problems: $(X + A)(X^2 - AX + A^2)$ and $(X - A)(X^2 + AX + A^2)$.

$$\begin{array}{rrrrl}
 & X^2 & - AX & + A^2 & \\
 & \times & X & + A & \\
\hline
 & AX^2 & - A^2X & + A^3 & \leftarrow \text{Multiply } A(X^2 - AX + A^2). \\
X^3 & - AX^2 & + A^2X & & \leftarrow \text{Multiply } X(X^2 - AX + A^2). \\
\hline
X^3 & & & + A^3 &
\end{array}$$

$$\begin{array}{rrrrl}
 & X^2 & + AX & + A^2 & \\
 & \times & X & - A & \\
\hline
 & -AX^2 & - A^2X & - A^3 & \leftarrow \text{Multiply } -A(X^2 + AX + A^2). \\
X^3 & + AX^2 & + A^2X & & \leftarrow \text{Multiply } X(X^2 + AX + A^2). \\
\hline
X^3 & & & - A^3 &
\end{array}$$

Teaching Tip

Have students try to factor $x^2 + 3x + 9$ and then discuss why it can't be done.

Thus

$$\overbrace{X^3 + A^3 = (X + A)(X^2 - AX + A^2)}^{\text{Same} \quad \text{Different}} \quad \text{and} \quad \overbrace{X^3 - A^3 = (X - A)(X^2 + AX + A^2)}^{\text{Same} \quad \text{Different}}.$$

This gives us our final factoring rules.

FACTORING RULES 5 AND 6: THE SUM AND DIFFERENCE OF TWO CUBES

$$X^3 + A^3 = (X + A)(X^2 - AX + A^2) \quad \textbf{(F5)}$$

$$X^3 - A^3 = (X - A)(X^2 + AX + A^2) \quad \textbf{(F6)}$$

Web It

For more examples and practice problems dealing with factoring sums and differences of cubes, go to link 5-5-1 on the Bello Website at mhhe.com/bello.

NOTE

The trinomials $X^2 - AX + A^2$ and $X^2 + AX + A^2$ cannot be factored further.

EXAMPLE 1 Factoring sums and differences of cubes

Factor completely:

a. $x^3 + 27$ **b.** $8x^3 + y^3$

c. $m^3 - 8n^3$ **d.** $27r^3 - 8s^3$

SOLUTION

a. We rewrite $x^3 + 27$ as the sum of two cubes and then use F5:

$$\begin{aligned} x^3 + 27 &= (x)^3 + (3)^3 \\ &= (x + 3)(x^2 - 3x + 3^2) && \text{Letting } X = x \text{ and } A = 3 \text{ in F5} \\ &= (x + 3)(x^2 - 3x + 9) \end{aligned}$$

b. This is also the sum of two cubes, so we write:

$$\begin{aligned} 8x^3 + y^3 &= (2x)^3 + (y)^3 \\ &= (2x + y)[(2x)^2 - 2xy + y^2] && \text{Letting } X = 2x \text{ and } A = y \text{ in F5} \\ &= (2x + y)(4x^2 - 2xy + y^2) \end{aligned}$$

c. Here we have the difference of two cubes, so we use F6. We start by writing the problem as the difference of two cubes:

$$\begin{aligned} m^3 - 8n^3 &= (m)^3 - (2n)^3 \\ &= (m - 2n)[m^2 + m(2n) + (2n)^2] && \text{Letting } A = m \text{ and } A = 2n \text{ in F6} \\ &= (m - 2n)(m^2 + 2mn + 4n^2) \end{aligned}$$

d. We write the problem as the difference of two cubes and then use F6.

$$\begin{aligned} 27r^3 - 8s^3 &= (3r)^3 - (2s)^3 \\ &= (3r - 2s)[(3r)^2 + (3r)(2s) + (2s)^2] && \text{Letting } X = 3r \text{ and } A = 2s \text{ in F6} \\ &= (3r - 2s)(9r^2 + 6rs + 4s^2) \end{aligned}$$

Note that you can verify all of these results by multiplying the factors in the final answer.

PROBLEM 1

Factor completely:

a. $y^3 + 8$ **b.** $27y^3 + 8$

c. $y^3 - 27z^3$ **d.** $8a^3 - 27b^3$

Teaching Tip

Here's an alternate method for factoring cubes. Since the result is a binomial times a trinomial, place parentheses accordingly.

1. To get the *binomial factor,* take the cube roots using the same sign.
2. To get the *trinomial factor,* square the terms of the binomial factor and use them as the first and last terms of the trinomial. The middle term of the trinomial is the result of multiplying the two terms of the binomial factor and changing the sign.

Example:

$x^3 + 27$

$\sqrt[3]{x^3}\ \sqrt[3]{27}$

$\downarrow \quad \downarrow$

$(x + 3)(x^2 - 3x + 9)$

square square

middle term:

(change sign) $-x \cdot 3 = -3x$

Answers

1. a. $(y + 2)(y^2 - 2y + 4)$
b. $(3y + 2)(9y^2 - 6y + 4)$
c. $(y - 3z)(y^2 + 3yz + 9z^2)$
d. $(2a - 3b)(4a^2 + 6ab + 9b^2)$

B Using a General Factoring Strategy

We have now studied several factoring techniques. How do you know which one to use? Here is a general factoring strategy that can help you answer this question. Remember that when we say *factor,* we mean *factor completely* using integer coefficients.

Web It

For a site with its own general factoring strategy, go to link 5-5-1 on the Bello Website at mhhe.com/bello.

PROCEDURE

A General Factoring Strategy

1. Factor out all common factors (the GCF).
2. Look at the number of terms inside the parentheses (or in the original polynomial). If there are

Four terms:	Factor by grouping.
Three terms:	Check whether the expression is a perfect square trinomial. If so, factor it. Otherwise, use the *ac* test to factor.
Two terms and *squared:*	Look for the difference of two squares $(X^2 - A^2)$ and factor it. Note that $X^2 + A^2$ is not factorable.
Two terms and *cubed:*	Look for the sum of two cubes $(X^3 + A^3)$ or the difference of two cubes $(X^3 - A^3)$ and factor it.

3. Make sure the expression is completely factored.

You can check your results by multiplying the factors you obtain.

EXAMPLE 2 Using the general factoring strategy

Factor completely:

a. $6x^2 - 18x - 24$ **b.** $4x^4 + 12x^3 + 40x^2$

SOLUTION

a. We follow the steps in our general factoring strategy.

1. Factor out the common factor: $6x^2 - 18x - 24 = 6(x^2 - 3x - 4)$
2. $x^2 - 3x - 4$ has three terms, and it is factored by finding two numbers whose product is -4 and whose sum is -3. These numbers are 1 and -4. Thus $x^2 - 3x - 4 = (x + 1)(x - 4)$

   We then have $6x^2 - 18x - 24 = 6(x + 1)(x - 4)$
3. This expression cannot be factored any further.

b.
1. Here the GCF is $4x^2$. Thus $4x^4 + 12x^3 + 40x^2 = 4x^2(x^2 + 3x + 10)$
2. The trinomial $x^2 + 3x + 10$ is *not* factorable since there are no numbers whose product is 10 with a sum of 3.
3. The complete factorization is simply $4x^4 + 12x^3 + 40x^2 = 4x^2(x^2 + 3x + 10)$

PROBLEM 2

Factor completely:

a. $7a^2 - 14a - 21$

b. $5a^4 + 10a^3 + 25a^2$

Answers

2. a. $7(a + 1)(a - 3)$
b. $5a^2(a^2 + 2a + 5)$

EXAMPLE 3 Using the general factoring strategy with four terms

Factor completely: $3x^3 + 9x^2 + x + 3$

SOLUTION

1. There are no common factors.
2. Since the expression has four terms, we factor by grouping:

$$\begin{aligned} 3x^3 + 9x^2 + x + 3 &= (3x^3 + 9x^2) + (x + 3) \\ &= 3x^2(x + 3) + 1 \cdot (x + 3) \\ &= (x + 3)(3x^2 + 1) \end{aligned}$$

3. This result cannot be factored any further, so the factorization is complete.

PROBLEM 3

Factor completely: $3a^3 + 6a^2 + a + 2$

EXAMPLE 4 Using the general factoring strategy with a perfect square trinomial

The heat output from a natural draught convector is $kt_n^2 - 2kt_nt_a + kt_a^2$. ($t_n^2$ is read as "t sub n squared." The "n" is called a **subscript.**) Factor this expression.

SOLUTION As usual, we proceed by steps.

1. The common factor is k. Hence

$$kt_n^2 - 2kt_nt_a + kt_a^2 = k(t_n^2 - 2t_nt_a + t_a^2)$$

2. $t_n^2 - 2t_nt_a + t_a^2$ is a perfect square trinomial, which factors into $(t_n - t_a)^2$. Thus

$$kt_n^2 - 2kt_nt_a + kt_a^2 = k(t_n - t_a)^2$$

3. This expression cannot be factored further.

PROBLEM 4

Factor completely: $kt_1^2 - 2kt_1t_2 + kt_2^2$

Teaching Tip

Point out in Example 4 that the expression uses t_n and t_a. Although these are the same letter, they are not considered like terms since they have different subscripts. Also, in Example 5 the expression uses D^4 and d^4. They are not considered like terms because one is capitalized and the other is lowercase.

EXAMPLE 5 Using the general factoring strategy with the difference of two squares

Factor completely: $D^4 - d^4$

SOLUTION

1. There are no common factors.
2. The expression has *two* squared terms separated by a minus sign, so it's the difference of *two* squares. Thus

$$\begin{aligned} D^4 - d^4 &= (D^2)^2 - (d^2)^2 \\ &= (D^2 + d^2)(D^2 - d^2) \end{aligned}$$

3. The expression $D^2 - d^2$ is also the difference of two squares, which can be factored into $(D + d)(D - d)$. Thus

$$\begin{aligned} D^4 - d^4 &= (D^2 + d^2)(D^2 - d^2) \\ &= (D^2 + d^2)(D + d)(D - d) \end{aligned}$$

Note that $D^2 + d^2$, which is the *sum* of two squares, *cannot* be factored.

PROBLEM 5

Factor completely: $m^4 - n^4$

Answers

3. $(a + 2)(3a^2 + 1)$
4. $k(t_1 - t_2)^2$
5. $(m^2 + n^2)(m + n)(m - n)$

EXAMPLE 6 **Using the general factoring strategy with sums and differences of cubes**

Factor completely:

a. $8x^5 - x^2y^3$ b. $8x^5 + x^3y^2$

SOLUTION

a. We proceed as usual by steps.

1. The GCF is x^2, so we factor it out. $8x^5 - x^2y^3 = x^2(8x^3 - y^3)$
2. $8x^3 - y^3$ is the difference of two cubes with $X = 2x$ and $A = y$. $= x^2(2x - y)[(2x)^2 + (2x)y + y^2]$
3. Note that the expression cannot be factored further. $= x^2(2x - y)(4x^2 + 2xy + y^2)$

b. The GCF is x^3.

1. Factor the GCF. $8x^5 + x^3y^2 = x^3(8x^2 + y^2)$
2. $8x^2 + y^2$ is the sum of two squares and is not factorable. Thus $8x^5 + x^3y^2 = x^3(8x^2 + y^2)$
3. Note that the expression cannot be factored further.

PROBLEM 6

Factor completely:

a. $27a^5 - a^2b^3$

b. $64a^5 + a^3b^2$

C Using −1 as a Factor

In the preceding examples, we did not factor expressions in which the leading coefficient is preceded by a minus sign. Can some of these expressions be factored? The answer is yes, but we must first factor -1 from each term. Thus to factor $-x^2 + 6x - 9$, we first write

$$-x^2 + 6x - 9 = -1 \cdot (x^2 - 6x + 9)$$
$$= -1 \cdot (x - 3)^2 \quad \text{Note that } x^2 - 6x + 9 = (x - 3)^2.$$
$$= -(x - 3)^2 \quad \text{Since } -1 \cdot a = -a,\ -1 \cdot (x - 3)^2 = -(x - 3)^2.$$

Teaching Tip

Be sure students realize that factoring out -1 from each term is like dividing each term by -1.

EXAMPLE 7 **Factoring −1**

Factor, if possible:

a. $-x^2 - 8x - 16$ b. $-4x^2 + 12xy - 9y^2$

c. $-9x^2 - 12xy + 4y^2$ d. $-4x^4 + 25x^2$

SOLUTION

a. We first factor out -1 to obtain

$$-x^2 - 8x - 16 = -1 \cdot (x^2 + 8x + 16)$$
$$= -1 \cdot (x + 4)^2 \quad x^2 + 8x + 16 = (x + 4)^2$$
$$= -(x + 4)^2$$

PROBLEM 7

Factor:

a. $-a^2 - 6a - 9$

b. $-9a^2 + 12ab - 4b^2$

c. $-4a^2 - 12ab + 9b^2$ d. $-4a^4 + 9a^2$

Answers

6. a. $a^2(3a - b)(9a^2 + 3ab + b^2)$
b. $a^3(64a^2 + b^2)$
7. a. $-(a + 3)^2$ **b.** $-(3a - 2b)^2$
c. $-(4a^2 + 12ab - 9b^2)$
d. $-a^2(2a + 3)(2a - 3)$

b. $-4x^2 + 12xy - 9y^2 = -1 \cdot (4x^2 - 12xy + 9y^2)$
$= -1 \cdot (2x - 3y)^2$ $\quad 4x^2 - 12xy + 9y^2 = (2x - 3y)^2$
$= -(2x - 3y)^2$

c. $-9x^2 - 12xy + 4y^2 = -1 \cdot (9x^2 + 12xy - 4y^2)$
$9x^2 = (3x)^2$ and $4y^2 = (2y)^2$, but $9x^2 + 12xy - 4y^2$ has a minus sign before the last term $4y^2$ and is *not* a perfect square trinomial, so $9x^2 + 12xy - 4y^2$ is not factorable; thus $-9x^2 - 12xy + 4y^2 = -(9x^2 + 12xy - 4y^2)$.

d. $-4x^4 + 25x^2 = -x^2 \cdot (4x^2 - 25)$ $\quad$ The GCF is $-x^2$.
$= -x^2 \cdot (2x + 5)(2x - 5)$ $\quad 4x^2 - 25 = (2x + 5)(2x - 5)$
$= -x^2(2x + 5)(2x - 5)$

Exercises 5.5

A In Problems 1–10, factor completely.

1. $x^3 + 8$
$(x + 2)(x^2 - 2x + 4)$

2. $y^3 + 8$
$(y + 2)(y^2 - 2y + 4)$

3. $8m^3 - 27$
$(2m - 3)(4m^2 + 6m + 9)$

4. $8y^3 - 27x^3$
$(2y - 3x)(4y^2 + 6xy + 9x^2)$

5. $27m^3 - 8n^3$
$(3m - 2n)(9m^2 + 6mn + 4n^2)$

6. $27x^2 - x^5$
$x^2(3 - x)(9 + 3x + x^2)$

7. $64s^3 - s^6$
$s^3(4 - s)(16 + 4s + s^2)$

8. $t^7 - 8t^4$
$t^4(t - 2)(t^2 + 2t + 4)$

9. $27x^4 + 8x^7$
$x^4(3 + 2x)(9 - 6x + 4x^2)$

10. $8y^8 + 27y^5$
$y^5(2y + 3)(4y^2 - 6y + 9)$

B In Problems 11–46, factor completely.

11. $3x^2 - 3x - 18$
$3(x - 3)(x + 2)$

12. $4x^2 - 12x - 16$
$4(x - 4)(x + 1)$

13. $5x^2 + 11x + 2$
$(5x + 1)(x + 2)$

14. $6x^2 + 19x + 10$
$(3x + 2)(2x + 5)$

15. $3x^3 + 6x^2 + 21x$
$3x(x^2 + 2x + 7)$

16. $6x^3 + 18x^2 + 12x$
$6x(x + 2)(x + 1)$

17. $2x^4 - 4x^3 - 10x^2$
$2x^2(x^2 - 2x - 5)$

18. $3x^4 - 12x^3 - 9x^2$
$3x^2(x^2 - 4x - 3)$

19. $4x^4 + 12x^3 + 18x^2$
$2x^2(2x^2 + 6x + 9)$

20. $5x^4 + 25x^3 + 30x^2$
$5x^2(x + 3)(x + 2)$

21. $3x^3 + 6x^2 + x + 2$
$(x + 2)(3x^2 + 1)$

22. $2x^3 + 8x^2 + x + 4$
$(x + 4)(2x^2 + 1)$

23. $3x^3 + 3x^2 + 2x + 2$
$(x + 1)(3x^2 + 2)$

24. $4x^3 + 8x^2 + 3x + 6$
$(x + 2)(4x^2 + 3)$

25. $2x^3 + 2x^2 - x - 1$
$(x + 1)(2x^2 - 1)$

26. $3x^3 + 6x^2 - x - 2$
$(x + 2)(3x^2 - 1)$

27. $3x^2 + 24x + 48$
$3(x + 4)^2$

28. $2x^2 + 12x + 18$
$2(x + 3)^2$

29. $kx^2 + 4kx + 4k$
$k(x + 2)^2$

30. $kx^2 + 10kx + 25k$
$k(x + 5)^2$

31. $4x^2 - 24x + 36$
$4(x - 3)^2$

32. $5x^2 - 20x + 20$
$5(x - 2)^2$

33. $kx^2 - 12kx + 36k$
$k(x - 6)^2$

34. $kx^2 - 10kx + 25k$
$k(x - 5)^2$

35. $3x^3 + 12x^2 + 12x$
$3x(x + 2)^2$

36. $2x^3 + 16x^2 + 32x$
$2x(x + 4)^2$

37. $18x^3 + 12x^2 + 2x$
$2x(3x + 1)^2$

38. $12x^3 + 12x^2 + 3x$
$3x(2x + 1)^2$

39. $12x^4 - 36x^3 + 27x^2$
$3x^2(2x - 3)^2$

40. $18x^4 - 24x^3 + 8x^2$
$2x^2(3x - 2)^2$

41. $x^4 - 1$
$(x^2 + 1)(x + 1)(x - 1)$

42. $x^4 - 16$
$(x^2 + 4)(x + 2)(x - 2)$

43. $x^4 - y^4$
$(x^2 + y^2)(x + y)(x - y)$

44. $x^4 - z^4$
$(x^2 + z^2)(x + z)(x - z)$

45. $x^4 - 16y^4$
$(x^2 + 4y^2)(x + 2y)(x - 2y)$

46. $x^4 - 81y^4$
$(x^2 + 9y^2)(x + 3y)(x - 3y)$

C In Problems 47–70, factor.

47. $-x^2 - 6x - 9$
$-(x + 3)^2$

48. $-x^2 - 10x - 25$
$-(x + 5)^2$

49. $-x^2 - 4x - 4$
$-(x + 2)^2$

50. $-x^2 - 12x - 36$
$-(x + 6)^2$

51. $-4x^2 - 4xy - y^2$
$-(2x + y)^2$

52. $-9x^2 - 6xy - y^2$
$-(3x + y)^2$

53. $-9x^2 - 12xy - 4y^2$
$-(3x + 2y)^2$

54. $-4x^2 - 12xy - 9y^2$
$-(2x + 3y)^2$

55. $-4x^2 + 12xy - 9y^2$
$-(2x - 3y)^2$

56. $-9x^2 + 12xy - 4y^2$
$-(3x - 2y)^2$

57. $-18x^3 - 24x^2y - 8xy^2$
$-2x(3x + 2y)^2$

58. $-12x^3 - 36x^2y - 27xy^2$
$-3x(2x + 3y)^2$

59. $-18x^3 - 60x^2y - 50xy^2$
$-2x(3x + 5y)^2$

60. $-12x^3 - 60x^2y - 75xy^2$
$-3x(2x + 5y)^2$

61. $-x^3 + x$
$-x(x + 1)(x - 1)$

62. $-x^3 + 9x$
$-x(x + 3)(x - 3)$

63. $-x^4 + 4x^2$
$-x^2(x + 2)(x - 2)$

64. $-x^4 + 16x^2$
$-x^2(x + 4)(x - 4)$

65. $-4x^4 + 9x^2$
$-x^2(2x + 3)(2x - 3)$

66. $-9x^4 + 4x^2$
$-x^2(3x + 2)(3x - 2)$

67. $-2x^4 + 16x$
$-2x(x - 2)(x^2 + 2x + 4)$

68. $-24x^4 + 3x$
$-3x(2x - 1)(4x^2 + 2x + 1)$

69. $-16x^5 - 2x^2$
$-2x^2(2x + 1)(4x^2 - 2x + 1)$

70. $-3x^5 + 24x^2$
$-3x^2(x - 2)(x^2 + 2x + 4)$

SKILL CHECKER

Try the Skill Checker Exercises so you'll be ready for the next section.

Use the *ac* test to factor (if possible).

71. $10x^2 + 13x - 3$
$(2x + 3)(5x - 1)$

72. $3x^2 - 5x - 1$
Not factorable

73. $2x^2 - 5x - 3$
$(2x + 1)(x - 3)$

74. $2x^2 - 3x - 2$
$(2x + 1)(x - 2)$

USING YOUR KNOWLEDGE

Factoring Engineering Problems

Many of the ideas presented in this section are used by engineers and technicians. Use your knowledge to factor the given expressions.

75. The bend allowance needed to bend a piece of metal of thickness t through an angle A when the inside radius of the bend is R_1 is given by the expression

$$\frac{2\pi A}{360}R_1 + \frac{2\pi A}{360}Kt \quad \frac{2\pi A}{360}(R_1 + Kt)$$

where K is a constant. Factor this expression.

76. The change in kinetic energy of a moving object of mass m with initial velocity v_1 and terminal velocity v_2 is given by

$$\frac{1}{2}mv_1^2 - \frac{1}{2}mv_2^2 \quad \frac{1}{2}m(v_1 + v_2)(v_1 - v_2)$$

Factor this expression.

77. The parabolic distribution of shear stress on the cross section of a certain beam is given by

$$\frac{3Sd^2}{2bd^3} - \frac{12Sz^2}{2bd^3} \quad \frac{3S}{2bd^3}(d + 2z)(d - 2z)$$

Factor this expression.

78. The polar moment of inertia J of a hollow round shaft of inner diameter d_1 and outer diameter d is given by

$$\frac{\pi d^4}{32} - \frac{\pi d_1^4}{32} \quad \frac{\pi}{32}(d^2 + d_1^2)(d + d_1)(d - d_1)$$

Factor this expression.

WRITE ON

79. Write the procedure you use to factor the sum of two cubes. Answers may vary.

80. Write the procedure you use to factor the difference of two cubes. Answers may vary.

81. A student factored $x^4 - 7x^2 - 18$ as $(x^2 + 2)(x^2 - 9)$. The student did not get full credit on the answer. Why?
$x^2 - 9$ can be factored as $(x + 3)(x - 3)$.

MASTERY TEST

If you know how to do these problems, you have learned your lesson!

Factor completely:

82. $8x^2 - 16x - 24$
$8(x - 3)(x + 1)$

83. $5x^4 - 10x^3 + 20x^2$
$5x^2(x^2 - 2x + 4)$

84. $3x^3 + 12x^2 + x + 4$
$(x + 4)(3x^2 + 1)$

85. $6x^2 - x - 35$
$(3x + 7)(2x - 5)$

86. $2x^4 + 7x^3 - 15x^2$
$x^2(2x - 3)(x + 5)$

87. $27t^3 - 64$
$(3t - 4)(9t^2 + 12t + 16)$

88. $kt_n^2 + 2kt_nt_a + kt_a^2$
$k(t_n + t_a)^2$

89. $x^4 - 81$
$(x^2 + 9)(x + 3)(x - 3)$

90. $-x^2 - 10x - 25$
$-(x + 5)^2$

91. $-9x^2 - 30xy - 25y^2$
$-(3x + 5y)^2$

92. $-9x^2 + 30xy - 25y^2$
$-(3x - 5y)^2$

93. $64y^3 + 27x^3$
$(4y + 3x)(16y^2 - 12xy + 9x^2)$

94. $-9x^4 + 4x^2$
$-x^2(3x + 2)(3x - 2)$

95. $-x^5 - x^2y^3$
$-x^2(x + y)(x^2 - xy + y^2)$

5.6 SOLVING QUADRATIC EQUATIONS BY FACTORING

To Succeed, Review How To . . .

1. Factor an expression of the form $ax^2 + bx + c$ (pp. 406–414).
2. Solve a linear equation (pp. 150–155).

Objectives

A Solve quadratic equations by factoring.

GETTING STARTED

Quadratics and Gravity

Let's suppose the girl throws the ball with an initial velocity of 4 meters per second. If she releases the ball 1 meter above the ground, the *height* of the ball after t seconds is

$$-5t^2 + 4t + 1$$

To find out how long it takes the ball to hit the ground, we set this expression equal to zero. The reason for this is that when the ball *is* on the ground, the height is zero. Thus we write

Remember, the girl's hand is 1 meter *above* the ground.

$$-5t^2 + 4t + 1 = 0$$

This is the height after t seconds. This is the height when the ball hits the ground.

The equation is a *quadratic equation,* an equation in which the greatest exponent of the variable is 2 and which can be written as $at^2 + bt + c = 0, a \neq 0$. We learn how to solve these equations next.

Solving Quadratic Equations by Factoring

How can we solve the equation given in the *Getting Started?* For starters, we make the leading coefficient positive by multiplying each side of the equation by -1 to obtain

$$5t^2 - 4t - 1 = 0 \qquad -1(-5t^2 + 4t + 1) = 5t^2 - 4t - 1 \text{ and } -1 \cdot 0 = 0$$

Since $a = 5$ and $c = 1$, the ac number is -5 and we write

$$5t^2 - 5t + 1t - 1 = 0 \qquad \text{Write } -4t \text{ as } -5t + 1t.$$

$$5t(t - 1) + 1(t - 1) = 0 \qquad \text{The GCF is } (t - 1).$$

$$(t - 1)(5t + 1) = 0$$

(You can also use reverse FOIL or trial and error.) At this point, we note that the product of two expressions $(t - 1)$ and $(5t + 1)$ gives us a result of zero. What does this mean?

We know that if we have two numbers and at least one of them is zero, then their product is zero. For example,

$$-5 \cdot 0 = 0 \qquad \frac{3}{2} \cdot 0 = 0 \qquad x \cdot 0 = 0$$

$$0 \cdot 8 = 0 \qquad 0 \cdot x = 0 \qquad 0 \cdot 0 = 0$$

As you can see, in all these cases at least one of the factors is zero. In general, it can be shown that if the product of the two factors is zero, at least one of the factors *must be* zero. We shall call this idea the **principle of zero product.**

PRINCIPLE OF ZERO PRODUCT

If $A \cdot B = 0$, then $A = 0$ or $B = 0$ (or both A and B are equal to 0).

Now, let's go back to our original equation. We can think of $(t - 1)$ as A and $(5t + 1)$ as B. Then our equation

$$(t - 1)(5t + 1) = 0$$

becomes

$$A \cdot B = 0$$

By the principle of zero product, if $A \cdot B = 0$, then

$$A = 0 \quad \text{or} \quad B = 0$$

Thus

$$t - 1 = 0 \quad \text{or} \quad 5t + 1 = 0$$

$$t = 1 \qquad 5t = -1 \qquad \text{We added 1 in the first equation and subtracted 1 in the second.}$$

$$t = 1 \qquad t = \frac{-1}{5}$$

NOTE

When solving quadratic equations, you usually get two answers, but sometimes in application problems one of the answers must be discarded. It is always a good idea to check that the answers you get apply to the original conditions of the problem.

Web It

To practice with the principle of zero product (sometimes called the zero product rule) go to link 5-6-1 on the Bello Website at mhhe.com/bello.

Teaching Tip

Have students practice naming ac.

Examples:

Given	ac
$x^2 + 5x + 6 = 0$	$1(6) = 6$
$3x^2 + 11x - 4 = 0$	$3(-4) = -12$
$5x^2 - x + 2 = 0$	$5(2) = 10$

Thus the ball reaches the ground after 1 second or after $\frac{-1}{5}$ second. The second answer is negative, which is impossible because we start timing when $t = 0$, so we can see that the ball thrown by the girl at 4 meters per second will reach the ground after $t = 1$ second. You can check this by letting $t = 1$ in the original equation:

$$-5t^2 + 4t + 1 \stackrel{?}{=} 0$$
$$-5(1)^2 + 4(1) + 1 \stackrel{?}{=} 0$$
$$-5 + 4 + 1 \stackrel{?}{=} 0$$
$$0 = 0$$

Similarly, if we want to find how long a ball thrown from level ground at 10 meters per second takes to return to the ground, we need to solve the equation

$$5t^2 - 10t = 0$$

Factoring, we obtain

$$5t(t - 2) = 0$$

By the principle of zero product,

$5t = 0$ or $t - 2 = 0$

$t = 0$ or $t = 2$ If $5t = 0$, then $t = 0$.

Thus the ball returns to the ground after 2 seconds. (The other possible answer, $t = 0$, indicates that the ball was on the ground when $t = 0$, which is true.)

EXAMPLE 1 Solving a quadratic equation by factoring

Solve:

a. $3x^2 + 11x - 4 = 0$ **b.** $6x^2 - x - 2 = 0$

SOLUTION

a. To solve this equation, we must first factor the left-hand side. Here's how we do it.

$3x^2 + \underline{11x} - 4 = 0$ The key number is -12 $(= 3(-4))$; then determine $12(-1) = -12$ and $12 + (-1) = 11$.

$3x^2 + \underline{12x - 1x} - 4 = 0$ Rewrite the middle term.

$3x(x + 4) - 1(x + 4) = 0$ Factor each pair.

$(x + 4)(3x - 1) = 0$ Factor out the GCF, $(x + 4)$.

$x + 4 = 0$ or $3x - 1 = 0$ Use the principle of zero product.

$x = -4$ $3x = 1$ Solve each equation.

$x = -4$ $x = \dfrac{1}{3}$

Thus the possible solutions are $x = -4$ and $x = \frac{1}{3}$. To verify that -4 is a correct solution, we substitute -4 in the original equation to obtain

$$3(-4)^2 + 11(-4) - 4 = 3(16) - 44 - 4$$
$$= 48 - 44 - 4 = 0$$

We leave it to you to verify that $\frac{1}{3}$ is also a solution.

PROBLEM 1

Solve:

a. $3x^2 + 8x - 3 = 0$

b. $6x^2 - 7x - 3 = 0$

Answers

1. a. $x = -3$ and $x = \frac{1}{3}$
b. $x = \frac{3}{2}$ and $x = -\frac{1}{3}$

b. As before, we must factor the left-hand side.

$6x^2 - x - 2 = 0$	The key number is -12.
$6x^2 - 4x + 3x - 2 = 0$	Rewrite the middle term.
$2x(3x - 2) + 1(3x - 2) = 0$	Factor each pair.
$(3x - 2)(2x + 1) = 0$	Factor out the GCF, $(3x - 2)$.
$3x - 2 = 0$ or $2x + 1 = 0$	Use the principle of zero product.
$3x = 2$ $\quad$ $2x = -1$	Solve each equation.
$x = \frac{2}{3}$ $\quad$ $x = -\frac{1}{2}$	

Thus the solutions are $\frac{2}{3}$ and $-\frac{1}{2}$. You can check this by substituting these values in the original equation.

Teaching Tip

Before Example 2, have students practice writing quadratic equations in standard form.

Examples:

Given	$ax^2 + bx + c = 0$
$x^2 - 5x = 2$	$x^2 - 5x - 2 = 0$
$4x - 1 = 3x^2$	$3x^2 - 4x + 1 = 0$
$-x^2 + 7 = 6x$	$x^2 + 6x - 7 = 0$

In the preceding discussion, we solved equations in which the highest exponent of the variable was 2. These equations are called *quadratic equations.* A quadratic equation is an equation in which the greatest exponent of the variable is 2. Moreover, to solve these quadratic equations, one of the sides of the equation must equal zero. These two ideas can be summarized as follows.

QUADRATIC EQUATION IN STANDARD FORM

If a, b, and c are real numbers ($a \neq 0$),

$$ax^2 + bx + c = 0$$

is a **quadratic equation in standard form.**

To solve a quadratic equation by the method of factoring, the equation must be in *standard form.*

EXAMPLE 2 Solving a quadratic equation *not* in standard form

Solve: $10x^2 + 13x = 3$

SOLUTION This equation is not in standard form, so we can't solve it as written. However, if we subtract 3 from each side of the equation, we have

$$10x^2 + 13x - 3 = 0$$

which is in standard form. We can now solve by factoring using trial and error or the ac test. To use the ac test, we write

$10x^2 + 13x - 3 = 0$	The key number is -30.
$10x^2 + 15x - 2x - 3 = 0$	Rewrite the middle term.
$5x(2x + 3) - 1(2x + 3) = 0$	Factor each pair.
$(2x + 3)(5x - 1) = 0$	Factor out the GCF, $(2x + 3)$.
$2x + 3 = 0$ or $5x - 1 = 0$	Use the principle of zero product.
$2x = -3$ $\quad$ $5x = 1$	Solve each equation.
$x = -\frac{3}{2}$ $\quad$ $x = \frac{1}{5}$	

PROBLEM 2

Solve: $6x^2 + 13x = 5$

Answer

2. $x = -\frac{5}{2}$ and $x = \frac{1}{3}$

Thus the solutions of the equation are

$$x = -\frac{3}{2} \quad \text{or} \quad x = \frac{1}{5}$$

Check this!

Sometimes we need to simplify before we write an equation in standard form. For example, to solve the equation

$$(3x + 1)(x - 1) = 3(x + 1) - 2$$

we need to remove parentheses by multiplying the factors involved and write the equation in standard form. Here's how we do it:

$$(3x + 1)(x - 1) = 3(x + 1) - 2$$ Given.

$$3x^2 - 2x - 1 = 3x + 3 - 2$$ Multiply.

$$3x^2 - 2x - 1 = 3x + 1$$ Simplify on the right.

$$3x^2 - 5x - 2 = 0$$ Subtract $3x$ and 1 from each side.

Now we factor using the *ac* test. (We could also use trial and error.)

$$3x^2 \underline{- 6x + 1x} - 2 = 0$$ The key number is -6.

$$3x(x - 2) + 1(x - 2) = 0$$ Factor each pair.

$$(x - 2)(3x + 1) = 0$$ Factor out the GCF, $(x - 2)$.

$$x - 2 = 0 \quad \text{or} \quad 3x + 1 = 0$$ Use the principle of zero product.

$$x = 2 \qquad 3x = -1$$

$$x = 2 \quad \text{or} \quad x = -\frac{1}{3}$$ Solve each equation.

Remember to check this by substituting $x = 2$ and then $x = -\frac{1}{3}$ in the original equation.

EXAMPLE 3 **Solving a quadratic equation by simplifying first**

Solve: $(2x + 1)(x - 2) = 2(x - 1) + 3$

SOLUTION

$$2x^2 - 3x - 2 = 2x - 2 + 3$$ Multiply.

$$2x^2 - 3x - 2 = 2x + 1$$ Simplify.

$$2x^2 - 5x - 2 = 1$$ Subtract $2x$.

$$2x^2 - 5x - 3 = 0$$ Subtract 1.

$$2x^2 - 6x + 1x - 3 = 0$$ Rewrite the middle term.

$$2x(x - 3) + 1(x - 3) = 0$$ Factor each pair.

$$(x - 3)(2x + 1) = 0$$ Factor out the GCF, $(x - 3)$.

$$x - 3 = 0 \quad \text{or} \quad 2x + 1 = 0$$ Use the principle of zero product.

$$x = 3 \qquad x = -\frac{1}{2}$$ Solve each equation.

Remember to check these answers.

PROBLEM 3

Solve: $(4n + 1)(n - 2) = 4(n + 1) - 3$

Answer

3. $n = -\frac{1}{4}$ and $n = 3$

Finally, we can always use the general factoring strategy developed in Section 5.5 to solve certain equations. We illustrate this possibility in Example 4.

EXAMPLE 4 **Solving a quadratic equation requiring simplification**

Solve: $(4x - 1)(x - 1) = 2(x + 2) - 3x - 4$

SOLUTION

$4x^2 - 5x + 1 = 2x + 4 - 3x - 4$	Multiply.
$4x^2 - 5x + 1 = -x$	Simplify.
$4x^2 - 4x + 1 = 0$	Add x.
$4x^2 - 2x - 2x + 1 = 0$	Rewrite the middle term.
$2x(2x - 1) - 1(2x - 1) = 0$	Factor each pair.
$(2x - 1)(2x - 1) = 0$	Factor out the GCF, $(2x - 1)$.
$2x - 1 = 0$ or $2x - 1 = 0$	Use the principle of zero product.
$x = \frac{1}{2}$ $\quad x = \frac{1}{2}$	Solve the equations.

Don't forget to check the answer!

PROBLEM 4

Solve:

$(3m + 1)(3m + 2) = 5(m + 1) - 2m - 4$

Note that in Example 4 there is really only *one* solution, $x = \frac{1}{2}$. A lot of work could be avoided if you notice that the expression $4x^2 - 4x + 1 = 0$ is a perfect square trinomial (F3) that can be factored as $(2x - 1)^2$ or, equivalently, $(2x - 1)(2x - 1)$. To avoid extra work, follow the general factoring strategy from Section 5.5.

In Example 5, we will solve several quadratic equations not in standard form. To do so, we make the right-hand side of the equation 0 and use the general factoring strategy to factor the left-hand side.

EXAMPLE 5 **Solving quadratic equations using the general factoring strategy**

Solve:

a. $9z^2 - 16 = 0$ **b.** $y(3y + 7) = -2$ **c.** $m^2 = 3m$

SOLUTION

a. The right-hand side is 0, so we use the general factoring strategy to factor $9z^2 - 16$:

$9z^2 - 16 = 0$	Given.
$(3z + 4)(3z - 4) = 0$	Factor the difference of squares.
$3z + 4 = 0$ or $3z - 4 = 0$	Use the principle of zero product.
$z = -\frac{4}{3}$ $\quad z = \frac{4}{3}$	Solve the equations.

Check that the solutions are $-\frac{4}{3}$ and $\frac{4}{3}$ by substituting each number in the original equation.

b. We start by adding 2 to both sides of the equation so that the right-hand side is 0

$y(3y + 7) + 2 = -2 + 2$	Add 2.
$3y^2 + 7y + 2 = 0$	Simplify.
$(3y + 1)(y + 2) = 0$	Factor.
$3y + 1 = 0$ or $y + 2 = 0$	Use the principle of zero product.
$y = -\frac{1}{3}$ $\quad y = -2$	Solve.

Check to make sure the solutions are $-\frac{1}{3}$ and -2.

PROBLEM 5

Solve:

a. $16x^2 - 9 = 0$

b. $y(2y + 3) = -1$

c. $n^2 = 4n$

Answers

4. $m = -\frac{1}{3}$
5. a. $x = \frac{3}{4}$ and $x = -\frac{3}{4}$
b. $y = -\frac{1}{2}$ and $y = -1$
c. $n = 0$ and $n = 4$

c. Start by subtracting $3m$, then factor the result:

$m^2 = 3m$	Given.
$m^2 - 3m = 0$	Subtract $3m$.
$m(m - 3) = 0$	Factor.
$m = 0$ or $m - 3 = 0$	Use the principle of zero product.
$m = 0$ $\quad m = 3$	Solve.

Check the solutions 0 and 3 in the original equation.

The principle of zero product can be applied to solve certain nonquadratic equations, as long as one side of the equation is 0 and the other side can be written as a product. Example 6 illustrates such a situation.

EXAMPLE 6 **Extending the principle of zero product to solve other types of equations**

Solve: $(v - 2)(v^2 - v - 12) = 0$

SOLUTION The right-hand side of the equation is 0 as needed. Avoid the temptation of multiplying the expressions on the left-hand side! Remember, what we need is a *product* of factors so we can use the principle of zero product. With this in mind, factor $v^2 - v - 12$ as shown:

$(v - 2)(v^2 - v - 12) = 0$	Given.
$(v - 2)(v - 4)(v + 3) = 0$	Factor $v^2 - v - 12$.
$v - 2 = 0$ or $v - 4 = 0$ or $v + 3 = 0$	Use the principle of zero product.
$v = 2$ $\quad v = 4$ $\quad v = -3$	Solve.

The solutions are 2, 4, and -3. Check that this is the case by substituting each number in the original equation.

PROBLEM 6

Solve: $(m - 3)(m^2 - m - 2) = 0$

Before you attempt the exercises, we remind you of the steps used to solve quadratic equations by factoring:

PROCEDURE TO SOLVE QUADRATICS BY FACTORING

1. Perform the necessary operations on both sides of the equation so that the right-hand side is 0.
2. Use the general factoring strategy to factor the left side of the equation, if necessary.
3. Use the principle of zero product and make each factor on the left equal 0.
4. Solve each of the resulting equations.
5. Check the results by substituting the solutions obtained in step 4 in the original equation.

Answer

6. $m = 3$ and $m = 2$ and $m = -1$

Calculate It Solving Quadratics

You can solve quadratics with a calculator, but you still have to know some algebra to write the equations in standard form. To acquaint yourself with graphing quadratics, graph $Y_1 = x^2$, $Y_2 = x^2 + 1$, and $Y_3 = 2x^2 - 1$. All of these graphs are **parabolas** (see Window 1). Now try $Y_4 = -x^2$, $Y_5 = -x^2 + 1$, and $Y_6 = -2x^2 - 3$. The results are still parabolas but they are "upside down" (see Window 2). This happens when the coefficient of the x^2 term is negative. Now we are ready to solve quadratic equations using our calculators.

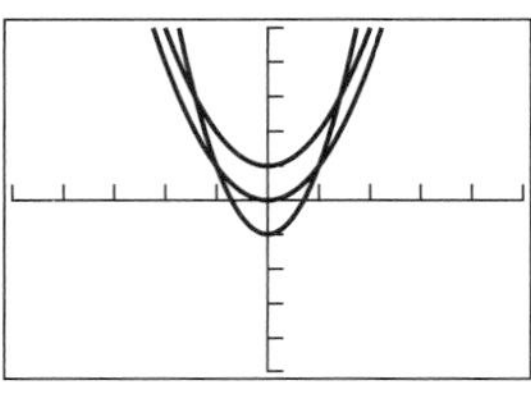

Window 1

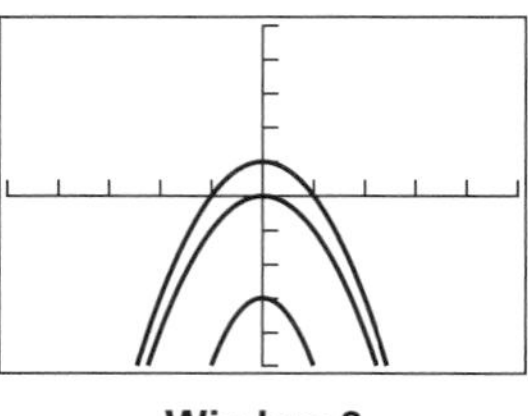

Window 2

Let's look at Example 3, where we have to solve $(2x + 1)(x - 2) = 2(x - 1) + 3$. As we mentioned, *you have to know the algebra* to write the equation in the standard form, $2x^2 - 5x - 3 = 0$. Now graph $Y = 2x^2 - 5x - 3$ (see Window 3). Where are the points where $Y = 0$? They are on the horizontal axis (the x-axis). For any point on the x-axis, its y-value is $y = 0$. The graph appears to have x-values 3 and $-\frac{1}{2}$ at the two points where the graph crosses the x-axis ($y = 0$). You can use your TRACE and ZOOM keys to confirm this.

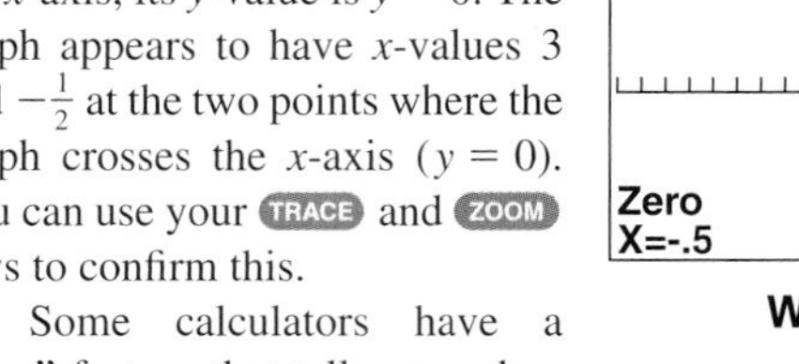

Window 3

Some calculators have a "zero" feature that tells you when $y = 0$. On a TI-83 Plus, enter 2nd TRACE 2 to activate the zero feature. The calculator asks you to select a left bound—that is, a point on the curve to the left of where the curve crosses the horizontal axis. Pick one near $-\frac{1}{2}$ using your ◀ and ▶ keys to move the cursor. Press ENTER. Select a right bound—that is, a point on the curve to the right of where the curve crosses the horizontal axis, and press ENTER again. The calculator then asks you to guess. Use the calculator's guess by pressing ENTER. The zero is given as $-.5$ (see Window 3). Do the same to find the other zero, $x = 3$.

Use these techniques to solve some of the other examples in this section and some of the problems in Exercises 5.6.

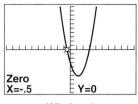

Exercises 5.6

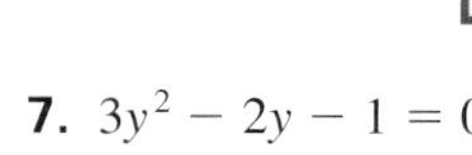

A In Problems 1–20, solve the given equation.

1. $2x^2 + 7x + 3 = 0$ $x = -3$ or $x = -\dfrac{1}{2}$

2. $2x^2 + 5x + 3 = 0$ $x = -\dfrac{3}{2}$ or $x = -1$

3. $2x^2 + x - 3 = 0$ $x = 1$ or $x = -\dfrac{3}{2}$

4. $6x^2 + x - 12 = 0$ $x = \dfrac{4}{3}$ or $x = -\dfrac{3}{2}$

5. $3y^2 - 11y + 6 = 0$ $y = 3$ or $y = \dfrac{2}{3}$

6. $4y^2 - 11y + 6 = 0$ $y = \dfrac{3}{4}$ or $y = 2$

7. $3y^2 - 2y - 1 = 0$ $y = 1$ or $y = -\dfrac{1}{3}$

8. $12y^2 - y - 6 = 0$ $y = \dfrac{3}{4}$ or $y = -\dfrac{2}{3}$

9. $6x^2 + 11x = -4$ $x = -\dfrac{4}{3}$ or $x = -\dfrac{1}{2}$

10. $5x^2 + 6x = -1$ $x = -\dfrac{1}{5}$ or $x = -1$

11. $3x^2 - 5x = 2$ $x = 2$ or $x = -\dfrac{1}{3}$

12. $12x^2 - x = 6$ $x = \dfrac{3}{4}$ or $x = -\dfrac{2}{3}$

13. $5x^2 + 6x = 8$ $x = \dfrac{4}{5}$ or $x = -2$

14. $6x^2 + 13x = 5$ $x = \dfrac{1}{3}$ or $x = -\dfrac{5}{2}$

15. $5x^2 - 13x = -8$ $x = 1$ or $x = \dfrac{8}{5}$

16. $3x^2 + 5x = -2$ $x = -\dfrac{2}{3}$ or $x = -1$

17. $3y^2 = 17y - 10$ $y = 5$ or $y = \dfrac{2}{3}$

18. $3y^2 = 2y + 1$ $y = -\dfrac{1}{3}$ or $y = 1$

19. $2y^2 = -5y - 2$ $y = -2$ or $y = -\dfrac{1}{2}$

20. $5y^2 = -6y + 8$ $y = \dfrac{4}{5}$ or $y = -2$

In Problems 21–36, solve (you may use F2 and F3):

21. $9x^2 + 6x + 1 = 0$ $x = -\dfrac{1}{3}$

22. $x^2 + 14x + 49 = 0$ $x = -7$

23. $y^2 - 8y = -16$ $y = 4$

24. $y^2 - 20y = -100$ $y = 10$

25. $9x^2 + 12x = -4$ $x = -\dfrac{2}{3}$

26. $25x^2 + 10x = -1$ $x = -\dfrac{1}{5}$

27. $4y^2 - 20y = -25$ $y = \dfrac{5}{2}$

28. $16y^2 - 56y = -49$ $y = \dfrac{7}{4}$

29. $x^2 = -10x - 25$ $x = -5$

30. $x^2 = -16x - 64$
$x = -8$

31. $(2x - 1)(x - 3) = 3x - 5$
$x = 4$ or $x = 1$

32. $(3x + 1)(x - 2) = x + 7$
$x = 3$ or $x = -1$

33. $(2x + 3)(x + 4) = 2(x - 1) + 4$
$x = -2$ or $x = -\frac{5}{2}$

34. $(5x - 2)(x + 2) = 3(x + 1) - 7$
$x = 0$ or $x = -1$

35. $(2x - 1)(x - 1) = x - 1$
$x = 1$

36. $(3x - 2)(3x - 1) = 1 - 3x$ $x = \frac{1}{3}$

In Problems 37–60, solve.

37. $4x^2 - 1 = 0$ $x = \frac{1}{2}$ or $x = -\frac{1}{2}$

38. $9x^2 - 1 = 0$ $x = \frac{1}{3}$ or $x = -\frac{1}{3}$

39. $4y^2 - 25 = 0$ $y = \frac{5}{2}$ or $y = -\frac{5}{2}$

40. $25y^2 - 9 = 0$ $y = \frac{3}{5}$ or $y = -\frac{3}{5}$

41. $z^2 = 9$ $z = 3$ or $z = -3$

42. $z^2 = 25$ $z = 5$ or $z = -5$

43. $25x^2 = 49$ $x = \frac{7}{5}$ or $x = -\frac{7}{5}$

44. $9x^2 = 64$ $x = \frac{8}{3}$ or $x = -\frac{8}{3}$

45. $m^2 = 5m$ $m = 0$ or $m = 5$

46. $m^2 = 9m$
$m = 0$ or $m = 9$

47. $2n^2 = 10n$
$n = 0$ or $n = 5$

48. $3n^2 = 12n$
$n = 0$ or $n = 4$

49. $y(y + 11) = -24$
$y = -3$ or $y = -8$

50. $y(y + 15) = -56$
$y = -7$ or $y = -8$

51. $y(y - 16) = -63$
$y = 9$ or $y = 7$

52. $y(y - 19) = -88$
$y = 11$ or $y = 8$

53. $(v - 2)(v^2 + 3v + 2) = 0$
$v = 2$ or $v = -1$ or $v = -2$

54. $(v - 1)(v^2 + 5v + 6) = 0$
$v = 1$ or $v = -2$ or $v = -3$

55. $(m^2 - 3m + 2)(m - 4) = 0$
$m = 2$ or $m = 1$ or $m = 4$

56. $(m^2 - 4m + 3)(m - 6) = 0$
$m = 3$ or $m = 1$ or $m = 6$

57. $(n^2 - 3n - 4)(n + 2) = 0$
$n = 4$ or $n = -1$ or $n = -2$

58. $(n^2 - 4n - 5)(n + 8) = 0$
$n = 5$ or $n = -1$ or $n = -8$

59. $(x^2 + 2x - 3)(x - 1) = 0$
$x = 1$ or $x = -3$

60. $(x^2 + 3x - 4)(x - 1) = 0$
$x = -4$ or $x = 1$

SKILL CHECKER

Try the Skill Checker Exercises so you'll be ready for the next section.

Simplify.

61. $H^2 + (3 + H)^2$ $2H^2 + 6H + 9$

62. $H^2 + (6 + H)^2$ $2H^2 + 12H + 36$

Expand.

63. $(H - 9)(H + 3)$ $H^2 - 6H - 27$

64. $(H - 8)(H + 4)$ $H^2 - 4H - 32$

USING YOUR KNOWLEDGE

The "Function" of Quadratics

The ideas you've studied in this section are useful for solving problems in business.

65. Solve the equation in Problem 81 of the *Using Your Knowledge* in Exercises 5.4 when $D(x) = 0$. $x = 7$

66. Solve the equation in Problem 84 of Exercises 5.4 when $S(p) = 0$. $p = 3$

WRITE ON

67. What is a quadratic equation and what procedure do you use to solve it? Answers may vary.

68. Write the steps you use to solve $x(x - 1) = 0$ and then write the difference between this procedure and the one you would use to solve $3x(x - 1) = 0$.
Answers may vary.

69. The equation $x^2 + 2x + 1 = 0$ can be solved by factoring. How many solutions does it have? The original equation is equivalent to $(x + 1)^2 = 0$. Why do you think -1 is called a *double root* for the equation?
1; answers may vary.

MASTERY TEST

If you know how to do these problems, you have learned your lesson!

Solve:

70. $10x^2 - 13x = 3$ $x = \frac{3}{2}$ or $x = -\frac{1}{5}$

71. $(3x - 2)(x - 1) = 2(x + 3) + 2$ $x = 3$ or $x = -\frac{2}{3}$

72. $(9x - 2)(x - 1) = 2(x + 1) - 7x - 1$ $x = \frac{1}{3}$

73. $5x^2 + 9x - 2 = 0$ $x = -2$ or $x = \frac{1}{5}$

74. $3x^2 - 2x - 5 = 0$ $x = -1$ or $x = \frac{5}{3}$

75. $x(x - 1) = 0$ $x = 0$ or $x = 1$

76. $2x(x + 3) = 0$ $x = 0$ or $x = -3$

77. $(x - 4)(x^2 + 4x - 5) = 0$ $x = 4$ or $x = -5$ or $x = 1$

78. $25y^2 - 36 = 0$ $y = \frac{6}{5}$ or $y = -\frac{6}{5}$

79. $m(3m + 5) = -2$ $m = -\frac{2}{3}$ or $m = -1$

80. $n^2 = 11n$ $n = 0$ or $n = 11$

5.7 APPLICATIONS OF QUADRATICS

To Succeed, Review How To . . .

1. Solve integer problems (pp. 163–165).
2. Use FOIL to expand polynomials (pp. 348–351).
3. Multiply integers (p. 71).

Objectives

Use the RSTUV procedure to solve:

A Integer problems

B Area and perimeter problems

C Problems involving the Pythagorean Theorem

D Motion problems

GETTING STARTED

Stop That Car!

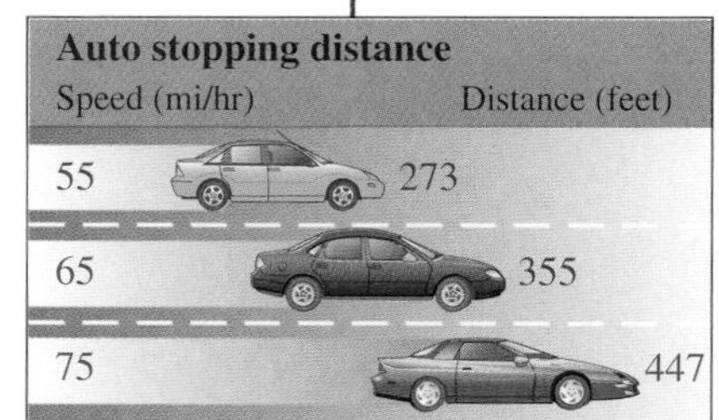

Auto stopping distance
Speed (mi/hr)
Distance (feet)
55
273
65
355
75
447

The diagram shows the distance needed to stop a car traveling at the indicated speeds. At speed m (in miles per hour), that stopping distance $D(m)$ in feet is given by

$$D(m) = 0.05m^2 + 2.2m + 0.75$$

If you are one car length (13 ft) behind a stopped car, how fast can you be traveling and still be able to stop before hitting the car?

In this case, the actual distance is 13 feet, so we have to solve

$$D(m) = 0.05m^2 + 2.2m + 0.75 = 13$$

$$5m^2 + 220m + 75 = 1300 \quad \text{Multiply by 100 (clear the decimals).}$$

$$5m^2 + 220m - 1225 = 0 \quad \text{Subtract 1300 (write in standard form).}$$

$$m^2 + 44m - 245 = 0 \quad \text{Divide by 5.}$$

$$(m - 5)(m + 49) = 0 \quad \text{Factor.}$$

$$m = 5 \quad \text{or} \quad m = -49 \quad \text{Use the principle of zero product and solve.}$$

So, you can be going 5 mph and stop in 13 feet. (Any speed over that and you hit it!)

Consecutive Integer Problems

Do you remember the integer problems of Chapter 2? Here's the terminology we need to solve a problem involving integers and quadratic equations.

Web It

For a lesson dealing with consecutive integer problems, go to link 5-7-1 on the Bello Website at mhhe.com/bello.

Terminology	Notation	Examples
Two consecutive integers	$n, n + 1$	$3, 4; -6, -5$
Three consecutive integers	$n, n + 1, n + 2$	$7, 8, 9; -4, -3, -2$
Two consecutive even integers	$n, n + 2$	$8, 10; -6, -4$
Two consecutive odd integers	$n, n + 2$	$13, 15; -21, -19$

As before, we solve this problem using the RSTUV method.

EXAMPLE 1 A consecutive integer problem

The product of two consecutive even integers is 10 more than 7 times the larger of the two integers. Find the integers.

SOLUTION

1. Read the problem.
We are asked to find two consecutive even integers.

2. Select the unknown.
Let n and $n + 2$ be the integers ($n + 2$ being the larger).

3. Think of a plan.
We first translate the problem:

The product of two consecutive integers	is	10	more than	7 times the larger.
$n(n + 2)$	$=$	10	$+$	$7(n + 2)$

4. Use algebra to solve the equation.

$n^2 + 2n = 10 + 7n + 14$	Use the distributive property.
$n^2 + 2n = 24 + 7n$	Simplify.
$n^2 + 2n - 24 - 7n = 0$	Subtract $24 + 7n$.
$n^2 - 5n - 24 = 0$	Simplify.
$(n - 8)(n + 3) = 0$	Factor $(-8 \cdot 3 = -24, -8 + 3 = -5)$.
$n - 8 = 0$ or $n + 3 = 0$	Use the principle of zero product.
$n = 8$ $\quad$ $n = -3$	Solve each equation.

If $n = 8$ is the first integer, the second is $n + 2 = 8 + 2 = 10$. The solution $n = -3$ is not acceptable because -3 is *not* even. Thus there is only one pair of integers that satisfies the problem: 8 and 10.

5. Verify the solution.

The product of two consecutive integers	is	10	more than	7 times the larger.	
$8 \cdot 10$	$=$	10	$+$	$7 \cdot 10$	True.

PROBLEM 1

The product of two odd integers is 10 more than 5 times the smaller of the two integers. Find the integers.

Answer

1. 5 and 7

B Area and Perimeter

Quadratic equations are also used in geometry. As you recall, if L and W are the length and width of a rectangle, then the perimeter P is $P = 2L + 2W$, and the area A is $A = LW$. Let's use these ideas in the next example.

EXAMPLE 2 Finding the dimensions of a room

A rectangular room is 4 feet longer than it is wide. The area of the room numerically exceeds its perimeter by 92. What are the dimensions of the room?

SOLUTION

1. Read the problem.
We are asked to find the dimensions of the room; we also know it is a rectangle.

2. Select the unknown.
Let W be the width. Since the room is 4 feet longer than it is wide, the length is $W + 4$.

3. Think of a plan.
Two measurements are involved: the area and the perimeter. Let's start with a picture:

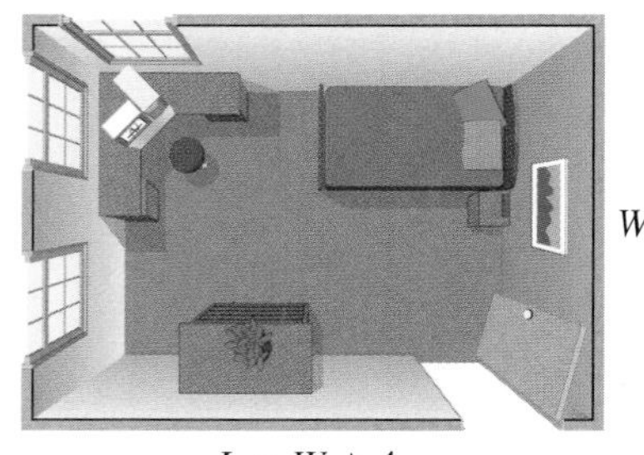

The area of the rectangle is

$$W(W + 4)$$

The perimeter of the rectangle is

$$2(W + 4) + 2W = 4W + 8$$

Look at the wording of the problem. We can restate it as follows:

$$\underbrace{W(W+4)}_{\text{The area of the room}} \quad \underbrace{=}_{\text{is equal to}} \quad \underbrace{(4W+8)+92}_{\text{its perimeter plus 92.}}$$

4. Use algebra to solve the equation.

$$W^2 + 4W = 4W + 100 \qquad \text{Simplify both sides.}$$

$$W^2 - 100 = 0 \qquad \text{Subtract } 4W + 100.$$

$$(W + 10)(W - 10) = 0 \qquad \text{Factor.}$$

$$W + 10 = 0 \quad \text{or} \quad W - 10 = 0 \qquad \text{Use the principle of zero product.}$$

$$W = -10 \qquad W = 10 \qquad \text{Solve each equation.}$$

Since a rectangle can't have a negative width, discard the -10, so the width is 10 feet and the length is 4 more feet or 14 feet. Thus the dimensions of the room are 10 feet by 14 feet.

5. Verify the solution.
The area of the room is $10 \cdot 14 = 140$ and the perimeter is $2 \cdot 10 + 2 \cdot 14 = 20 + 28 = 48$. The area (140) must exceed the perimeter (48) by 92. Does $140 = 48 + 92$? Yes, so our answer is correct.

PROBLEM 2

What are the dimensions of the room in Example 2 if the area exceeds the perimeter by 56?

Answer

2. 8 ft by 12 ft

EXAMPLE 3 Finding the dimensions of a monitor

Denise's monitor is 3 inches wider than it is high and has 130 square inches of viewing area.

a. Find the dimensions of her monitor.

b. Denise wants a Trinitron® monitor, which is 4 inches wider than it is high and has 62 more square inches of viewing area than her current monitor. What are the dimensions of the Trinitron monitor?

SOLUTION

a. 1. Read the problem.
We want the dimensions of the monitor.

2. Select the unknown.
Let H be the height of the monitor.

3. Think of a plan.
Draw a picture. H is the height. The width is 3 inches more than the height; that is, the width is $H + 3$. The area of the rectangle is the height H times the width, $H + 3$. It is also 130 square inches. Thus,

$$H(H + 3) = 130$$

4. Use algebra to solve the equation.

$$H(H + 3) = 130$$

$$H^2 + 3H = 130 \quad \text{Simplify.}$$

$$H^2 + 3H - 130 = 0 \quad \text{Subtract 130.}$$

$$(H + 13)(H - 10) = 0 \quad \text{Factor.}$$

$$H + 13 = 0 \quad \text{or} \quad H - 10 = 0 \quad \text{Use the principle of zero product.}$$

$$H = -13 \quad \text{or} \quad H = 10 \quad \text{Solve.}$$

Discard $H = -13$. (Why?) Thus, the dimensions of the monitor are 10 in. by 13 in.

b. What are the Trinitron dimensions? Let the height be H, so the width is $H + 4$. The area of the Trinitron screen is $H(H + 4)$, which is 62 square inches more than her old screen; that is,

$$H(H + 4) = 130 + 62 = 192$$

4. Use algebra to solve the equation.

$$H(H + 4) = 192$$

$$H^2 + 4H = 192 \quad \text{Simplify.}$$

$$H^2 + 4H - 192 = 0 \quad \text{Subtract 192.}$$

$$(H - 12)(H + 16) = 0 \quad \text{Use the principle of zero product.}$$

$$H = 12 \quad \text{or} \quad H = -16 \quad \text{Solve.}$$

Discard $H = -16$. The dimensions of the Trinitron are 12 in. by 16 in.

5. Verify the solution!

PROBLEM 3

a. What are the dimensions of Denise's current monitor if it is 3 inches wider than it is high and has 108 square inches of viewing area?

b. What are the dimensions of a Cinema monitor, which is 6 inches wider than it is high and has double the viewing area of the monitor in part **a**?

Answers

3. a. 9 in. by 12 in.
b. 12 in. by 18 in.

C The Pythagorean Theorem

Suppose you were to buy a 20-inch television set. What does that 20-inch measurement represent? It's the diagonal length of the rectangular screen. The relationship between the length L, the height H, and the diagonal measurement d of the screen can be found by using the Pythagorean Theorem, which is stated here.

PYTHAGOREAN THEOREM

If the longest side of a right triangle (a triangle with a 90° angle) is of length c and the other two sides are of length a and b, respectively, then

$$a^2 + b^2 = c^2$$

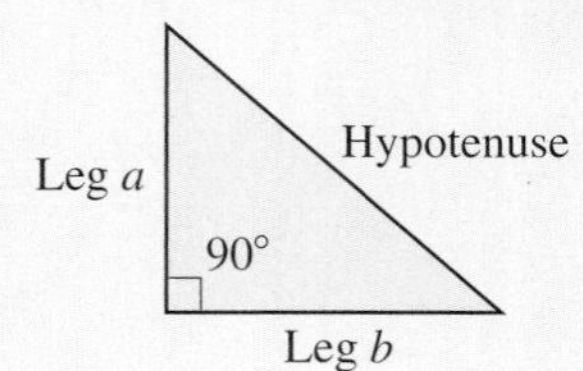

Note that the longest side of the triangle is called the **hypotenuse,** and the two shorter sides are called the **legs** of the triangle.

Web It

To see an animated proof of this theorem, go to link 5-7-2 on the Bello Website at mhhe.com/bello.

You need Java to use this link.

If you want a calculator to find the answer for you, try link 5-7-3.

In the case of the television set, $H^2 + L^2 = 20^2$. If the screen is 16 inches long ($L = 16$), can we find its height H? Substituting $L = 16$ into $H^2 + L^2 = 20^2$, we obtain

$$H^2 + 16^2 = 20^2$$

$$H^2 + 256 = 400 \quad \text{Simplify.}$$

$$H^2 - 144 = 0 \quad \text{Subtract 400.}$$

$$(H - 12)(H + 12) = 0 \quad \text{Factor.}$$

$$H - 12 = 0 \quad \text{or} \quad H + 12 = 0 \quad \text{Use the principle of zero product.}$$

$$H = 12 \qquad H = -12 \quad \text{Solve each equation.}$$

Since H represents the height of the screen, we discard -12 as an answer. (The height cannot be negative.) Thus the height of the screen is 12 inches.

EXAMPLE 4 Computing a computer monitor's dimensions

The screen in a rectangular computer monitor is 3 inches wider than it is high, and its diagonal is 6 inches longer than its height. What are the dimensions of the monitor?

SOLUTION

1. Read the problem.
We are asked to find the dimensions, which involves finding the width, height, and diagonal of the screen.

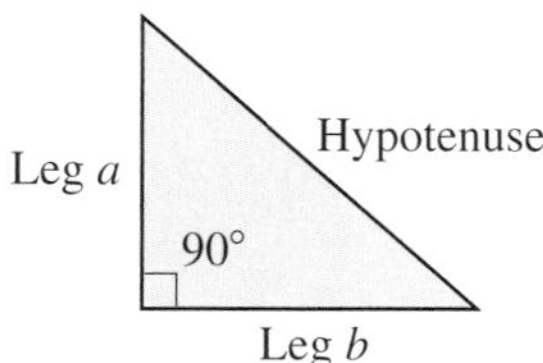

2. Select the unknown.
Since all measurements are given in terms of the height, let H be the height.

PROBLEM 4

A rectangular book is 2 inches higher than it is wide. If the diagonal is 4 inches longer than the width of the book, what are the dimensions of the cover?

Answer

4. 8 in. by 6 in.

3. Think of a plan.

It's a good idea to start with a picture so we can see the relationships among the measurements. Note that the width is $3 + H$ (3 inches wider than the height) and the diagonal is $6 + H$ (6 inches longer than the height). We enter this information in a diagram, as shown here.

4. Use the Pythagorean Theorem to describe the relationship among the measurements and then solve the resulting equation.

According to the Pythagorean Theorem:

$$H^2 + (3 + H)^2 = (6 + H)^2$$

$$H^2 + 9 + 6H + H^2 = 36 + 12H + H^2 \quad \text{Expand } (3 + H)^2 \text{ and } (6 + H)^2.$$

$$H^2 + 9 + 6H + H^2 - 36 - 12H - H^2 = 0 \quad \text{Subtract } 36 + 12H + H^2.$$

$$H^2 - 6H - 27 = 0 \quad \text{Simplify.}$$

$$(H - 9)(H + 3) = 0 \quad \text{Factor } (-9 \cdot 3 = -27;\ -9 + 3 = -6).$$

$$H - 9 = 0 \quad \text{or} \quad H + 3 = 0 \quad \text{Use the principle of zero product.}$$

$$H = 9 \qquad H = -3 \quad \text{Solve each equation.}$$

Since H is the height, we discard -3, so the height of the monitor is 9 inches, the width is 3 more inches, or 12 inches, and the diagonal is 6 more inches than the height, or $9 + 6 = 15$ inches.

5. Verify the solution.

Looking at the diagram and using the Pythagorean Theorem, we see that $9^2 + 12^2$ must be 15^2, that is,

$$9^2 + 12^2 = 15^2$$

Since $81 + 144 = 225$ is a true statement, our dimensions are correct.

By the way, television sets claiming to be 25 inches or 27 inches (meaning the length of the screen measured diagonally is 25 or 27 inches) hardly ever measure 25 or 27 inches. This is easy to confirm using the Pythagorean Theorem. For example, a 27-inch Panasonic® has a screen that is 16 inches high and 21 inches long. Can it really be 27 inches diagonally? If this were the case, $16^2 + 21^2$ would equal 27^2. Is this true? Measure a couple of TV or computer screens and see whether the manufacturers' claims are true!

D Motion Problems

In the *Getting Started,* we gave a formula for the stopping distance for a car traveling at the indicated speeds. That distance involves the **braking distance *b***, the distance it takes to stop a car *after* the brakes are applied. This distance is given by

$$b = 0.06v^2$$

Web It

To learn more about braking distance and reaction time, try links 5-7-4, 5-7-5, and 5-7-6 on the Bello Website at mhhe.com/bello.

where v is the speed of the car when the brakes are applied. It also involves the reaction distance

$$r = 1.5tv$$

where t is the driver's reaction time (in seconds) and v is the speed of the car in miles per hour.

EXAMPLE 5 Finding the speed of a car based on braking distance

A car traveled 96 feet *after* the brakes were applied. How fast was the car going when the brakes were applied?

SOLUTION

1. Read the problem.
We want to find how fast the car was going.

2. Select the unknown.
Let v represent the velocity.

3. Think of a plan.
The braking distance b is 96.
The formula for b is $b = 0.06v^2$.
Thus, $0.06v^2 = 96$.

4. Use algebra to solve the equation.

$$0.06v^2 = 96$$
$$6v^2 = 9600 \quad \text{Multiply by 100.}$$
$$6v^2 - 9600 = 0 \quad \text{Subtract 9600.}$$
$$6(v^2 - 1600) = 0 \quad \text{Factor out the GCF.}$$
$$6(v + 40)(v - 40) = 0 \quad \text{Factor.}$$
$$v + 40 = 0 \quad \text{or} \quad v - 40 = 0 \quad \text{Solve.}$$
$$v = -40 \quad \text{or} \quad v = 40$$

Discard the -40 because the velocity was positive, so the velocity of the car was 40 miles per hour.

5. Verify the solution.
Substitute $v = 40$ in $b = 0.06v^2$:

$$b = 0.06(40)^2 = 0.06(1600) = 96$$

so the answer is correct.

PROBLEM 5

A car traveled 150 feet *after* the brakes were applied. How fast was the car going when the brakes were applied?

Answer

5. 50 mi/hr

Exercises 5.7

A In Problems 1–6, use the RSTUV method to solve the integer problems.

1. The product of two consecutive integers is 8 less than 10 times the smaller of the two integers. Find the integers. 8, 9 or 1, 2

2. The product of two consecutive even integers is 4 more than 5 times the smaller of the two integers. Find the integers. 4, 6

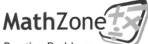

3. The product of two consecutive odd integers is 7 more than their sum. What are the integers? 3, 5 or -3, -1

4. Find three consecutive even integers such that the square of the largest exceeds the sum of the squares of the other two by 12. 0, 2, 4 or 4, 6, 8

5. Find two consecutive even integers such that the square of the larger integer is 4 less than 10 times the smaller integer. 4, 6 or 2, 4

6. The sum of two consecutive integers is 29 less than their product. Find the integers. 6, 7 or -5, -4

Use the information in the table to solve Problems 7–14.

Name	Geometric Shapes	Name	Geometric Shapes
Triangle Area $= \frac{1}{2}bh$	h, b	**Rectangle** Area $= LW$	W, L
Trapezoid Area $= \frac{1}{2}h(b_1 + b_2)$	b_1, h, b_2	**Circle** Area $= \pi r^2$	r
Parallelogram Area $= Lh$	h, W, L		

B In Problems 7–14, use the RSTUV method to solve the area and perimeter problems.

7. The area A of a triangle is 40 square inches. Find its dimensions if the base b is 2 inches more than its height h. $b = 10$ in., $h = 8$ in.

8. The area A of a trapezoid is 105 square centimeters. If the height h is the same length as the smaller side b_1 and 1 inch less than the longer side b_2, find the height h. $h = 10$ cm

9. The area A of a parallelogram is 150 square inches. If the length of the parallelogram is 5 inches more than its height, what are the dimensions? $L = 15$ in., $h = 10$ in.

10. The area of a rectangle is 96 square centimeters. If the width of the rectangle is 4 centimeters less than its length, what are the dimensions of the rectangle? $L = 12$ cm, $W = 8$ cm

11. The area of a circle is 49π square units. If the radius r is $x + 3$ units, find x and then find the radius of the circle. $x = 4$; $r = 7$ units

12. A rectangular room is 5 feet longer than it is wide. The area of the room numerically exceeds its perimeter by 100. What are the dimensions of the room? $L = 15$ ft, $W = 10$ ft

13. The biggest lasagna ever made had an area of 250 square feet. If it was made in the shape of a rectangle 45 feet longer than wide, what were its dimensions? (It was made in Dublin and weighed 3609 pounds, 10 ounces.) $L = 50$ ft, $W = 5$ ft

14. The biggest strawberry shortcake needed 360 square feet of strawberry topping. If this area was numerically 225 more than 3 times its length, what were the dimensions of this rectangular shortcake? 45 ft by 8 ft

C In Problems 15–20, use the RSTUV procedure to solve the problems.

15. The biggest television ever built is the 289 Sony® Jumbo Tron. (This means the screen measured 289 feet diagonally.) If the length of the screen was 150 feet, how high was it? About 247 ft

16. A television screen is 2 inches longer than it is high. If the diagonal length of the screen is 2 inches more than its length, what is the diagonal measurement of this screen? 10 in.

17. The hypotenuse of a right triangle is 4 inches longer than the shortest side and 2 inches longer than the remaining side. Find the dimensions of the triangle. 6 in., 8 in., 10 in.

18. The hypotenuse of a right triangle is 16 inches longer than the shortest side and 2 inches longer than the remaining side. Find the dimensions of the triangle. 10 in., 24 in., 26 in.

19. One of the sides of a right triangle is 3 inches longer than the shortest side. If the hypotenuse is 3 inches longer than the longer side, what are the dimensions of the triangle? 9 in., 12 in., 15 in.

20. The sides of a right triangle are consecutive even integers. Find their lengths. 6, 8, 10

D In Problems 21–26, solve the motion problems. Do not round off answers resulting in decimals. (*Hint:* For Problems 21–24, use the formulas from Example 5.)

21. A car traveled 54 feet after the driver applied the brakes. How fast was the car going when the brakes were applied? 30 mi/hr

22. A car traveled 216 feet after the brakes were applied. How fast was the car going when the brakes were applied? 60 mi/hr

23. A car left a 73.5-foot skid mark on the road.

a. How fast was the car going when the brakes were applied and the skid mark was made? 35 mi/hr

b. If the speed limit was 30 mi/hr, was the car speeding? Yes

24. A police officer measured a skid mark to be 150 feet long. The driver said he was going under the speed limit, which was 45 mi/hr. Was the driver correct? How long would the skid mark have to be if he was indeed going 45 mi/hr when the brakes were applied? No (50 mi/hr); 121.5 ft

25. Pedro reacts very quickly. In fact, his reaction time is 0.4 sec. When driving on a highway, Pedro saw a danger signal ahead and tried to stop. If his car traveled 120 ft before stopping, how fast was he going when he saw the sign? Use

$$d = 1.5tv + 0.06v^2$$

for the stopping distance, where d is in feet, t is in seconds, and v is in mi/hr. 40 mi/hr

26. Vilmos' reaction time is 0.3 sec. He tried to avoid hitting a student walking on the road.

a. If his car traveled 33 ft before stopping, how fast was he going? 20 mi/hr

b. How many feet would he travel at that speed, using the formula given in the *Getting Started?* 64.75 ft

SKILL CHECKER

Try the Skill Checker Exercises so you'll be ready for the next section.

27. Write $\frac{5}{6}$ with a denominator of 18. $\frac{15}{18}$

28. Factor: $6x - 12y$ $6(x - 2y)$

29. Factor: $18x - 36y$ $18(x - 2y)$

30. Factor: $x^2 + 2x - 15$ $(x + 5)(x - 3)$

31. Factor: $x^2 + 5x - 6$ $(x + 6)(x - 1)$

32. Factor: $x^2 - 9$ $(x + 3)(x - 3)$

USING YOUR KNOWLEDGE

More Motion Problems

In Problems 33–36, use the fact that the height $H(t)$ of an object thrown downward from a height of h meters and initial velocity V_0 is

$$H(t) = -5t^2 - V_0t + h$$

33. An object is thrown downward at 5 meters per second from a height of 10 meters. How long does it take the object to hit the ground? 1 second

34. An object is thrown downward from a height of 28 meters with an initial velocity of 4 meters per second. How long does it take the object to hit the ground? 2 seconds

35. An object is thrown downward from a building 15 meters high at 10 meters per second. How long does it take the object to hit the ground? 1 second

36. How long does it take a package thrown downward from a plane at 10 meters per second to hit the ground 175 meters below? 5 seconds

WRITE ON

37. In the *Getting Started*, the stopping distance is given by

$$D(m) = 0.05m^2 + 2.2m + 0.75$$

In Problem 25, it is

$$d = 1.5tv + 0.06v^2$$

If you are traveling at 20 mi/hr, which formula can you use to evaluate your stopping distance? Why?
The first formula; answers may vary.

38. What additional information do you need in Problem 37 to use the second formula? The reaction time t.

39. Name at least three factors that would influence the stopping distance of a car. Explain. Answers may vary.

40. Suppose you are driving at 20 mi/hr. What reaction time will give the same stopping distance for both formulas? About 1.36 seconds.

MASTERY TEST

If you know how to do these problems, you have learned your lesson!

41. The product of two consecutive odd integers is the difference between 23 and the sum of the two integers. What are the integers? 3, 5 or -7, -5

42. A rectangular room is 2 feet longer than it is wide. Its area numerically exceeds 10 times its longer side by 28. What are the dimensions of the room? $L = 14$ ft, $W = 12$ ft

43. The hypotenuse of a right triangle is 8 inches longer than the shortest side and 1 inch longer than the remaining side. What are the dimensions of the triangle? 5 in., 12 in., 13 in.

44. The product of the length of the sides of a right triangle is 325 less than the length of the hypotenuse squared. If the longer side is 5 units longer than the shorter side, what are the dimensions of the triangle? 15 ft, 20 ft, 25 ft

45. The distance b (in feet) it takes to stop a car *after* the brakes are applied is given by

$$b = 0.06v^2$$

A car traveled 150 feet after the driver applied the brakes. How fast was the car going when the brakes were applied? 50 mi/hr

COLLABORATIVE LEARNING

Shortcut patterns for solving quadratic equations

In this chapter we have used several methods to solve quadratic equations. Now we are going to develop a shortcut to solve these quadratic equations by letting you discover some applicable patterns.

Form three groups of students. Each of the groups will solve the equations assigned to them and enter the required information.

Recall that to factor $x^2 + bx + c = 0$, we need two numbers whose product is c and whose sum is b.

Group 1	**Numbers required to factor**	**Solutions**
$x^2 + 3x + 2 = 0$	2, 1	$-2, -1$
$x^2 - 7x + 12 = 0$		
$x^2 + x - 2 = 0$		
$x^2 - x - 2 = 0$		
Group 2		
$x^2 + 5x + 6 = 0$	5, 1	$-5, -1$
$x^2 - 5x - 6 = 0$		
$x^2 + x - 6 = 0$		
$x^2 - x - 6 = 0$		
Group 3		
$x^2 + 7x + 12 = 0$	3, 4	$-3, -4$
$x^2 - 3x + 2 = 0$		
$x^2 + x - 12 = 0$		
$x^2 - x - 12 = 0$		

Based on the patterns you see, have each of the groups complete the following conjecture (guess):

The factorization of $x^2 + bx + c = 0$ requires two factors F_1 and F_2 whose product is ________ and whose sum is ________ .

The solutions of $x^2 + bx + c = 0$ are ________ and ________ .

See whether all groups agree.

Is there a pattern that can be used to solve $ax^2 + bx + c = 0$ when $a \neq 1$? First, recall that to factor $ax^2 + bx + c$ we need two factors whose product is ac and whose sum is b. Here are some examples and some work for all groups.

	ac	**Factors**	**Solutions**
$4x^2 - 4x + 1 = 0$	4	$-2, -2$	$\frac{1}{2}$
$3x^2 - 5x - 2 = 0$	-6	$-6, 1$	$2, -\frac{1}{3}$
$10x^2 + 13x - 3 = 0$	-30	$15, -2$	$-\frac{3}{2}, \frac{1}{5}$

Based on the patterns you see, have each of the groups complete the following conjecture (guess):

The factorization of $ax^2 + bx + c = 0$ requires two factors F_1 and F_2 whose product is ________ and whose sum is ________ .

The solutions of $ax^2 + bx + c = 0$ are ________ and ________ .

See whether all groups agree.

Research Questions

1. Proposition 4 in Book II of Euclid's *Elements* states:

   If a straight line be cut at random, the square on the whole is equal to the squares on the segments and twice the rectangles contained by the segments.

   This statement corresponds to one of the rules of factoring we've studied in this chapter. Which rule is it, and how does it relate?

2. Proposition 5 of Book II of the *Elements* also corresponds to one of the factoring formulas we studied. Which one is it, and what does the proposition say?

3. Many scholars ascribe the Pythagorean Theorem to Pythagoras. However, other versions of the theorem exist:

   a. The ancient Chinese proof

   b. Bhaskara's proof

   c. Euclid's proof

   d. Garfield's proof

   e. Pappus's generalization

   Select three of these versions and write a paper giving details, if possible, telling where they appeared, who authored them, and what they said.

4. There are different versions regarding Pythagoras's death. Write a short paper detailing the circumstances of his death, where it occurred, and how.

5. Write a report about Pythagorean triples.

6. In *The Human Side of Algebra,* we mentioned that the Pythagoreans studied arithmetic, music, geometry, and astronomy. Write a report about the Pythagoreans' theory of music.

7. Write a report about the Pythagoreans' theory of astronomy.

Summary

SECTION	ITEM	MEANING	EXAMPLE
5.1A	Greatest common factor (GCF) of numbers	The GCF of a list of integers is the largest common factor of the integers in the list.	The GCF of 45 and 75 is 15.
5.1B	Greatest common factor (GCF) of variable terms	The GCF of a list of terms is the product of each of the variables raised to the lowest exponent to which they occur.	The GCF of x^2y^3, x^4y^2, and x^5y^4 is x^2y^2.
5.1C	Greatest common factor (GCF) of a polynomial	ax^n is the GCF of a polynomial if a divides each of the coefficients and n is the smallest exponent of x in the polynomial.	$3x^2$ is the GCF of $6x^4 - 9x^3 + 3x^2$.
5.2A	Factoring Rule 1 (F1) Factoring a trinomial of the form $x^2 + bx + c$	$X^2 + (A + B)X + AB = (X + A)(X + B)$ To factor $x^2 + bx + c$, find two numbers whose product is c and whose sum is b.	$x^2 + 8x + 15 = (x + 5)(x + 3)$ To factor $x^2 + 7x + 10$, find two numbers whose product is 10 and whose sum is 7. The numbers are 5 and 2. Thus $x^2 + 7x + 10 = (x + 5)(x + 2)$.
5.3A	*ac* test	$ax^2 + bx + c$ is factorable if there are two integers with product ac and sum b.	$3x^2 + 8x + 5$ is factorable. (There are two integers whose product is 15 and whose sum is 8: 5 and 3.) $2x^2 + x + 3$ is not factorable. (There are no integers whose product is 6 and whose sum is 1.)
5.4A, B	Factoring squares of binomials (F2 and F3)	$X^2 + 2AX + A^2 = (X + A)^2$ $X^2 - 2AX + A^2 = (X - A)^2$	$x^2 + 10x + 25 = (x + 5)^2$ $x^2 - 10x + 25 = (x - 5)^2$
5.4C	Factoring the difference of two squares (F4)	$X^2 - A^2 = (X + A)(X - A)$	$x^2 - 36 = (x + 6)(x - 6)$
5.5A	Factoring the sum (F5) or difference (F6) of cubes	$X^3 + A^3 = (X + A)(X^2 - AX + A^2)$ $X^3 - A^3 = (X - A)(X^2 + AX + A^2)$	$x^3 + 64 = (x + 4)(x^2 - 4x + 16)$ $x^3 - 64 = (x - 4)(x^2 + 4x + 16)$
5.5B	General factoring strategy	1. Factor out the GCF. 2. Look at the number of terms inside the parentheses or in the original polynomial. *Four terms:* Grouping *Three terms:* Perfect square trinomial or *ac* test *Two terms:* Difference of two squares, sum of two cubes, difference of two cubes 3. Make sure the expression is completely factored.	

SECTION	ITEM	MEANING	EXAMPLE
5.6A	Quadratic equation	An equation in which the greatest exponent of the variable is 2. If a, b, and c are real numbers, $ax^2 + bx + c = 0$ is in standard form.	$x^2 + 3x - 7 = 0$ is a quadratic equation in standard form.
	Principle of zero product	If $A \cdot B = 0$, then $A = 0$ or $B = 0$.	If $(x + 1)(x + 2) = 0$, then $x + 1 = 0$ or $x + 2 = 0$.
5.7C	Pythagorean Theorem	If the longest side of a right triangle (a triangle with a 90° angle) is of length c and the two other sides are of length a and b, then $a^2 + b^2 = c^2$.	Leg a, Hypotenuse, 90°, Leg b If the length of leg a is 3 inches and the length of leg b is 4 inches, then the length h of the hypotenuse is $3^2 + 4^2 = h^2$ Thus $9 + 16 = h^2$ $25 = h^2$ $5 = h$
5.7D	Braking distance $b = 0.06v^2$	If a car is moving v miles per hour, b is the distance (in feet) needed to stop the car *after* the brakes are applied.	The braking distance b for a car moving at 20 mi/hr is $b = 0.06(20^2) = 24$ feet.

Review Exercises

(If you need help with these exercises, look in the section indicated in brackets.)

1. [5.1A] Find the GCF of:

a. 60 and 90 30
b. 12 and 18 6
c. 27, 80, and 17 1

2. [5.1B] Find the GCF of:

a. $24x^7, 18x^5, -30x^{10}$ $6x^5$
b. $18x^8, 12x^9, -20x^{10}$ $2x^8$
c. $x^6y^4, y^7x^5, x^3y^6, x^9$ x^3

3. [5.1C] Factor.

a. $20x^3 - 55x^5$ $5x^3(4 - 11x^2)$
b. $14x^4 - 35x^6$ $7x^4(2 - 5x^2)$
c. $16x^7 - 40x^9$ $8x^7(2 - 5x^2)$

4. [5.1C] Factor.

a. $\frac{3}{7}x^6 - \frac{5}{7}x^5 + \frac{2}{7}x^4 - \frac{1}{7}x^2$ $\frac{1}{7}x^2(3x^4 - 5x^3 + 2x^2 - 1)$
b. $\frac{4}{9}x^7 - \frac{2}{9}x^6 + \frac{2}{9}x^5 - \frac{1}{9}x^3$ $\frac{1}{9}x^3(4x^4 - 2x^3 + 2x^2 - 1)$
c. $\frac{3}{8}x^9 - \frac{7}{8}x^8 + \frac{3}{8}x^7 - \frac{1}{8}x^5$ $\frac{1}{8}x^5(3x^4 - 7x^3 + 3x^2 - 1)$

5. [5.1D] Factor.

a. $3x^3 - 21x^2 - x + 7$ $(x - 7)(3x^2 - 1)$
b. $3x^3 + 18x^2 + x + 6$ $(x + 6)(3x^2 + 1)$
c. $4x^3 - 8x^2y + x - 2y$ $(x - 2y)(4x^2 + 1)$

6. [5.2A] Factor.

a. $x^2 + 8x + 7$ $(x + 7)(x + 1)$
b. $x^2 - 8x - 9$ $(x - 9)(x + 1)$
c. $x^2 + 6x + 5$ $(x + 5)(x + 1)$

7. [5.2A] Factor.

a. $x^2 - 7x + 10$ $(x - 5)(x - 2)$
b. $x^2 - 9x + 14$ $(x - 7)(x - 2)$
c. $x^2 + 2x - 8$ $(x + 4)(x - 2)$

8. [5.3B] Factor.

a. $6x^2 - 6 + 5x$ $(2x + 3)(3x - 2)$
b. $6x^2 - 1 + x$ $(2x + 1)(3x - 1)$
c. $6x^2 - 5 + 13x$ $(2x + 5)(3x - 1)$

9. [5.3B] Factor.

a. $6x^2 - 17xy + 5y^2$ $(3x - y)(2x - 5y)$
b. $6x^2 - 7xy + 2y^2$ $(3x - 2y)(2x - y)$
c. $6x^2 - 11xy + 4y^2$ $(3x - 4y)(2x - y)$

10. [5.4B] Factor.

a. $x^2 + 4x + 4$ $(x + 2)^2$
b. $x^2 + 10x + 25$ $(x + 5)^2$
c. $x^2 + 8x + 16$ $(x + 4)^2$

11. [5.4B] Factor.

a. $9x^2 + 12xy + 4y^2$ $(3x + 2y)^2$
b. $9x^2 + 30xy + 25y^2$ $(3x + 5y)^2$
c. $9x^2 + 24xy + 16y^2$ $(3x + 4y)^2$

12. [5.4B] Factor.

a. $x^2 - 4x + 4$ $(x - 2)^2$
b. $x^2 - 6x + 9$ $(x - 3)^2$
c. $x^2 - 12x + 36$ $(x - 6)^2$

13. [5.4B] Factor.

a. $4x^2 - 12xy + 9y^2$ $(2x - 3y)^2$
b. $4x^2 - 20xy + 25y^2$ $(2x - 5y)^2$
c. $4x^2 - 28xy + 49y^2$ $(2x - 7y)^2$

14. [5.4C] Factor.

a. $x^2 - 36$ $(x + 6)(x - 6)$
b. $x^2 - 49$ $(x + 7)(x - 7)$
c. $x^2 - 81$ $(x + 9)(x - 9)$

15. [5.4C] Factor.

a. $16x^2 - 81y^2$ $(4x + 9y)(4x - 9y)$
b. $25x^2 - 64y^2$ $(5x + 8y)(5x - 8y)$
c. $9x^2 - 100y^2$ $(3x + 10y)(3x - 10y)$

16. [5.5A] Factor.

a. $m^3 + 125$ $(m + 5)(m^2 - 5m + 25)$
b. $n^3 + 64$ $(n + 4)(n^2 - 4n + 16)$
c. $y^3 + 8$ $(y + 2)(y^2 - 2y + 4)$

17. [5.5A] Factor.

a. $8y^3 - 27x^3$ $(2y - 3x)(4y^2 + 6xy + 9x^2)$
b. $64y^3 - 125x^3$ $(4y - 5x)(16y^2 + 20xy + 25x^2)$
c. $8m^3 - 125n^3$ $(2m - 5n)(4m^2 + 10mn + 25n^2)$

18. [5.5B] Factor.

a. $3x^3 - 6x^2 + 27x$ $3x(x^2 - 2x + 9)$
b. $3x^3 - 6x^2 + 30x$ $3x(x^2 - 2x + 10)$
c. $4x^3 - 8x^2 + 32x$ $4x(x^2 - 2x + 8)$

19. [5.5B] Factor.

a. $2x^3 - 2x^2 - 4x$ $2x(x - 2)(x + 1)$
b. $3x^3 - 6x^2 - 9x$ $3x(x - 3)(x + 1)$
c. $4x^3 - 12x^2 - 16x$ $4x(x - 4)(x + 1)$

20. [5.5B] Factor.

a. $2x^3 + 8x^2 + x + 4$ $(x + 4)(2x^2 + 1)$
b. $2x^3 + 10x^2 + x + 5$ $(x + 5)(2x^2 + 1)$
c. $2x^3 + 12x^2 + x + 6$ $(x + 6)(2x^2 + 1)$

21. [5.5B] Factor.

a. $9kx^2 + 12kx + 4k$ $k(3x + 2)^2$
b. $9kx^2 + 30kx + 25k$ $k(3x + 5)^2$
c. $4kx^2 + 20kx + 25k$ $k(2x + 5)^2$

22. [5.5B] Factor.

a. $-3x^4 + 27x^2$ $-3x^2(x + 3)(x - 3)$
b. $-4x^4 + 64x^2$ $-4x^2(x + 4)(x - 4)$
c. $-5x^4 + 20x^2$ $-5x^2(x + 2)(x - 2)$

23. [5.5C] Factor.

a. $-x^3 - y^3$ $-(x + y)(x^2 - xy + y^2)$
b. $-8m^3 - 27n^3$ $-(2m + 3n)(4m^2 - 6mn + 9n^2)$
c. $-64n^3 - m^3$ $-(4n + m)(16n^2 - 4mn + m^2)$

24. [5.5C] Factor.

a. $-y^3 + x^3$ $-(y - x)(y^2 + xy + x^2)$
b. $-8m^3 + 27n^3$ $-(2m - 3n)(4m^2 + 6mn + 9n^2)$
c. $-64t^3 + 125s^3$ $-(4t - 5s)(16t^2 + 20st + 25s^2)$

25. [5.5C] Factor.

a. $-4x^2 - 12xy + 9y^2$ $-(4x^2 + 12xy - 9y^2)$

b. $-25x^2 - 30xy + 9y^2$ $-(25x^2 + 30xy - 9y^2)$

c. $-16x^2 - 24xy + 9y^2$ $-(16x^2 + 24xy - 9y^2)$

26. [5.6A] Solve.

a. $x^2 - 4x - 5 = 0$ $x = 5$ or $x = -1$

b. $x^2 - 5x - 6 = 0$ $x = 6$ or $x = -1$

c. $x^2 - 6x - 7 = 0$ $x = 7$ or $x = -1$

27. [5.6A] Solve.

a. $2x^2 + x = 10$ $x = -\frac{5}{2}$ or $x = 2$

b. $2x^2 + 3x = 5$ $x = -\frac{5}{2}$ or $x = 1$

c. $2x^2 + x = 3$ $x = -\frac{3}{2}$ or $x = 1$

28. [5.6A] Solve.

a. $(3x + 1)(x - 2) = 2(x - 1) - 4$ $x = 1$ or $x = \frac{4}{3}$

b. $(2x + 1)(x - 4) = 6(x - 4) - 1$ $x = 3$ or $x = \frac{7}{2}$

c. $(2x + 1)(x - 1) = 3(x + 2) - 1$ $x = 3$ or $x = -1$

29. [5.7A] Find the integers if the product of two consecutive even integers is

a. 4 more than 5 times the smaller of the integers 4, 6

b. 4 more than twice the smaller of the integers 2, 4 or −2, 0

c. 10 more than 11 times the smaller of the integers 10, 12

30. [5.7C] The hypotenuse of a right triangle is 6 inches longer than the longest side of the triangle and 12 inches longer than the remaining side. What are the dimensions of the triangle? 18 in., 24 in., 30 in.

Practice Test 5

(Answers on page 456)

1. Find the GCF of 40 and 60.

2. Find the GCF of $18x^2y^4$ and $30x^3y^5$.

3. Factor $10x^3 - 35x^5$.

4. Factor $\frac{4}{5}x^6 - \frac{3}{5}x^5 + \frac{2}{5}x^4 - \frac{1}{5}x^2$.

5. Factor $2x^3 + 6x^2y + x + 3y$.

6. Factor $x^2 - 8x + 12$.

7. Factor $6x^2 - 3 + 7x$.

8. Factor $6x^2 - 11xy + 3y^2$.

9. Factor $4x^2 + 12xy + 9y^2$.

10. Factor $x^2 - 14x + 49$.

11. Factor $9x^2 - 12xy + 4y^2$.

12. Factor $x^2 - 100$.

13. Factor $16x^2 - 25y^2$.

14. Factor $125t^3 + 27s^3$.

15. Factor $8y^3 - 125x^3$.

16. Factor $3x^3 - 6x^2 + 24x$.

17. Factor $2x^3 - 8x^2 - 10x$.

18. Factor $2x^3 + 6x^2 + x + 3$.

19. Factor $4kx^2 + 12kx + 9k$.

20. Factor $-9x^4 + 36x^2$.

21. Factor $-9x^2 - 24xy - 16y^2$.

22. Solve $x^2 - 3x - 10 = 0$.

23. Solve $2x^2 - x = 15$.

24. Solve $(2x - 3)(x - 4) = 2(x - 1) - 1$.

25. Solve $y(2y + 7) = -3$.

26. The product of two consecutive odd integers is 13 more than 10 times the larger of the two integers. Find the integers.

27. The product of two consecutive integers is 14 less than 10 times the smaller of the two integers. What are the integers?

28. The area of a rectangle is numerically 44 more than its perimeter. If the length of the rectangle is 8 inches more than its width, what are the dimensions of the rectangle?

29. A rectangular 10-inch television screen (measured diagonally) is 2 inches wider than it is high. What are the dimensions of the screen?

30. A car traveled 24 feet after the brakes were applied. How fast was the car going when the brakes were applied? $(b = 0.06v^2)$

Answers to Practice Test

ANSWER	IF YOU MISSED	REVIEW		
	QUESTION	SECTION	EXAMPLES	PAGE
1. 20	1	5.1	1	389–390
2. $6x^2y^4$	2	5.1	2	391
3. $5x^3(2 - 7x^2)$	3	5.1	3, 4	393
4. $\frac{1}{5}x^2(4x^4 - 3x^3 + 2x^2 - 1)$	4	5.1	5	393
5. $(x + 3y)(2x^2 + 1)$	5	5.1	6–8	394–395
6. $(x - 2)(x - 6)$	6	5.2	1–4	401–403
7. $(3x - 1)(2x + 3)$	7	5.3	2, 3	409–410
8. $(3x - y)(2x - 3y)$	8	5.3	4–7	410, 412–414
9. $(2x + 3y)^2$	9	5.4	2	419
10. $(x - 7)^2$	10	5.4	3a, b	419
11. $(3x - 2y)^2$	11	5.4	3c	419
12. $(x + 10)(x - 10)$	12	5.4	4a, b	420
13. $(4x + 5y)(4x - 5y)$	13	5.4	4c	420
14. $(5t + 3s)(25t^2 - 15st + 9s^2)$	14	5.5	1a, b	425
15. $(2y - 5x)(4y^2 + 10xy + 25x^2)$	15	5.5	1c, d	425
16. $3x(x^2 - 2x + 8)$	16	5.5	2	426
17. $2x(x + 1)(x - 5)$	17	5.5	2	426
18. $(x + 3)(2x^2 + 1)$	18	5.5	3	427
19. $k(2x + 3)^2$	19	5.5	4	427
20. $-9x^2(x + 2)(x - 2)$	20	5.5	7	428–429
21. $-(3x + 4y)^2$	21	5.5	7	428–429
22. $x = 5$ or $x = -2$	22	5.6	1	433–434
23. $x = 3$ or $x = -\frac{5}{2}$	23	5.6	2	434–435
24. $x = 5$ or $x = \frac{3}{2}$	24	5.6	3, 4	435–436
25. $y = -3$ or $y = -\frac{1}{2}$	25	5.6	5	436–437
26. 11 and 13 or -3 and -1	26	5.7	1	441
27. 7 and 8 or 2 and 3	27	5.7	1	441
28. 6 inches by 14 inches	28	5.7	2, 3	442–443
29. 6 inches by 8 inches	29	5.7	4	444–445
30. 20 mi/hr	30	5.7	5	446

Cumulative Review Chapters 1–5

1. Find: $-\frac{3}{8} + \left(-\frac{1}{6}\right)$ $-\frac{13}{24}$

2. Find: $6.6 - (-9.8)$ 16.4

3. Find: $(-5.5)(5.7)$ -31.35

4. Find: $-(5^2)$ -25

5. Find: $-\frac{5}{6} \div \left(-\frac{5}{18}\right)$ 3

6. Evaluate $y \div 5 \cdot x - z$ for $x = 6$, $y = 60$, $z = 3$. 69

7. Which law is illustrated by the following statement?
$7 \cdot (6 \cdot 4) = (7 \cdot 6) \cdot 4$ Associative law of multiplication

8. Combine like terms: $-8xy^4 - (-9xy^4)$ xy^4

9. Simplify: $2x - (x + 3) - 2(x + 4)$ $-x - 11$

10. Write in symbols: The quotient of $(d + 5e)$ and f $\frac{d + 5e}{f}$

11. Solve for x: $5 = 3(x - 1) + 5 - 2x$ $x = 3$

12. Solve for x: $\frac{x}{7} - \frac{x}{9} = 2$ $x = 63$

13. Solve for x: $8 - \frac{x}{3} = \frac{6(x + 1)}{7}$ $x = 6$

14. The sum of two numbers is 105. If one of the numbers is 25 more than the other, what are the numbers? 40 and 65

15. Train A leaves a station traveling at 50 mi/hr. Two hours later, train B leaves the same station traveling in the same direction at 60 mi/hr. How long does it take for train B to catch up to train A? 10 hr

16. Susan purchased some municipal bonds yielding 11% annually and some certificates of deposit yielding 12% annually. If Susan's total investment amounts to \$6000 and the annual income is \$680, how much money is invested in bonds and how much is invested in certificates of deposit? \$4000 in bonds; \$2000 in certificates of deposit

17. Graph: $-\frac{x}{6} + \frac{x}{2} \geq \frac{x - 2}{2}$

0 6

18. Graph the point $C(3, -3)$.

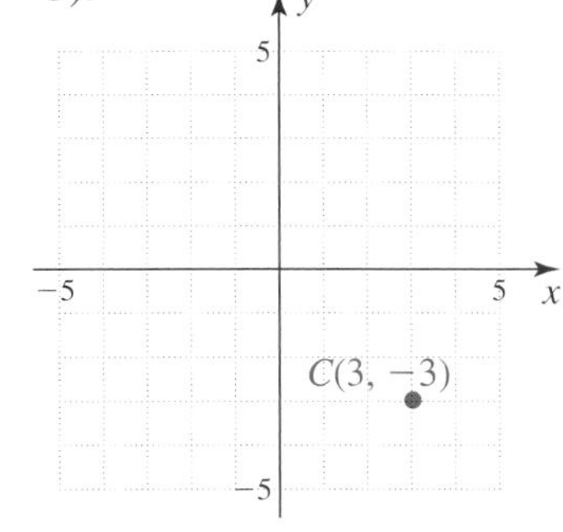

19. Determine whether the ordered pair $(-3, -3)$ is a solution of $5x - y = -18$. No

20. Find x in the ordered pair $(x, 3)$ so that the ordered pair satisfies the equation $2x - 3y = -5$. $x = 2$

21. Graph: $x + y = 4$

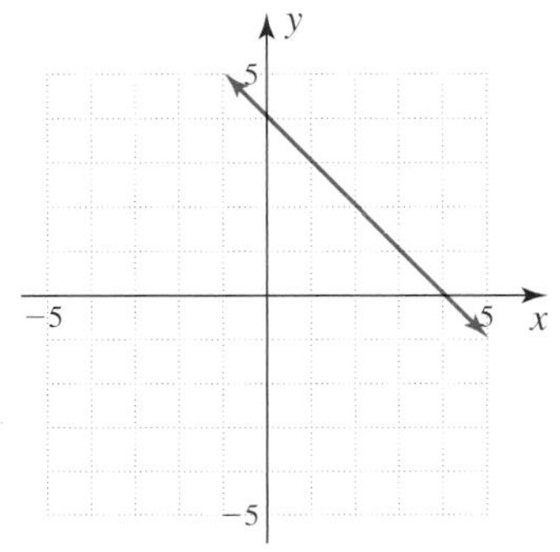

22. Graph: $4y - 20 = 0$

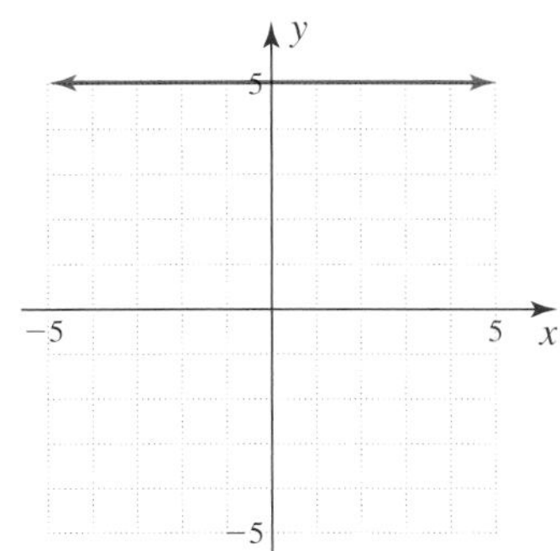

23. Find the slope of the line going through the points (7, 7) and (−4, 5). $\frac{2}{11}$

24. What is the slope of the line $6x - 2y = 15$? 3

25. Find the pair of parallel lines. (1) and (2)

(1) $4y = x - 7$
(2) $7x - 28y = -7$
(3) $28y + 7x = -7$

26. Find: $\dfrac{20x^6y^3}{-5x^4y^7}$ $-\dfrac{4x^2}{y^4}$

27. Find: $\dfrac{x^{-7}}{x^{-9}}$ x^2

28. Multiply and simplify: $x \cdot x^{-5}$ $\dfrac{1}{x^4}$

29. Simplify: $(3x^3y^{-3})^{-3}$ $\dfrac{y^9}{27x^9}$

30. Write in scientific notation: 0.000048 4.8×10^{-5}

31. Divide and express the answer in scientific notation: $(5.98 \times 10^{-3}) \div (1.3 \times 10^2)$ 4.60×10^{-5}

32. Find the degree of the polynomial: $6x^2 + x + 2$. 2

33. Find the value of $x^3 + 2x^2 + 1$ when $x = -2$. 1

34. Add $(-3x^2 - 2x^3 - 6)$ and $(5x^3 - 7 - 4x^2)$. $3x^3 - 7x^2 - 13$

35. Remove parentheses (simplify): $-4x^4(8x^2 + 4y)$ $-32x^6 - 16x^4y$

36. Find (expand): $(3x + 2y)^2$ $9x^2 + 12xy + 4y^2$

37. Find: $(5x - 7y)(5x + 7y)$ $25x^2 - 49y^2$

38. Find (expand): $\left(2x^2 - \dfrac{1}{5}\right)^2$ $4x^4 - \dfrac{4}{5}x^2 + \dfrac{1}{25}$

39. Find: $(5x^2 + 9)(5x^2 - 9)$ $25x^4 - 81$

40. Divide: $(2x^3 + x^2 - 5x - 9)$ by $(x - 2)$. $2x^2 + 5x + 5$ R 1

41. Factor: $12x^6 - 14x^9$ $2x^6(6 - 7x^3)$

42. Factor: $\dfrac{4}{5}x^7 - \dfrac{3}{5}x^6 + \dfrac{4}{5}x^5 - \dfrac{2}{5}x^3$ $\dfrac{1}{5}x^3(4x^4 - 3x^3 + 4x^2 - 2)$

43. Factor: $x^2 - 12x + 27$ $(x - 3)(x - 9)$

44. Factor: $20x^2 - 23xy + 6y^2$ $(5x - 2y)(4x - 3y)$

45. Factor completely: $25x^2 - 49y^2$ $(5x + 7y)(5x - 7y)$

46. Factor completely: $-5x^4 + 80x^2$ $-5x^2(x + 4)(x - 4)$

47. Factor completely: $3x^3 - 6x^2 - 9x$ $3x(x + 1)(x - 3)$

48. Factor completely: $2x^2 + 5x + 6x + 15$ $(2x + 5)(x + 3)$

49. Factor completely: $9kx^2 + 6kx + k$ $k(3x + 1)^2$

50. Solve for x: $4x^2 + 17x = 15$ $x = -5$ or $x = \dfrac{3}{4}$

Rational Expressions

The Human Side of Algebra

The whole-number concept is one of the oldest in mathematics. The concept of rational numbers (so named because they are *ratios* of whole numbers) developed much later because nonliterate tribes had no need for such a concept. Rational numbers evolved over a long period of time, stimulated by the need for certain types of measurement. For example, take a rod of length 1 unit and cut it into two equal pieces. What is the length of each piece? One-half, of course. If the same rod is cut into four equal pieces, then each piece is of length $\frac{1}{4}$. Two of these pieces will have length $\frac{2}{4}$, which tells us that we should have $\frac{2}{4} = \frac{1}{2}$.

It was ideas such as these that led to the development of the arithmetic of the rational numbers.

During the Bronze Age, Egyptian hieroglyphic inscriptions show the reciprocals of integers by using an elongated oval sign. Thus $\frac{1}{8}$ and $\frac{1}{20}$ were respectively written as

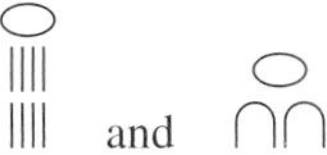

In this chapter, we generalize the concept of a rational number to that of a *rational expression*—that is, the quotient of two polynomials.

Pretest for Chapter 6

(Answers on page 461)

1. Write $\frac{3x}{7y}$ with a denominator of $28y^3$.

2. Reduce $\frac{-9(x^2 - y^2)}{3(x - y)}$ to lowest terms.

3. Simplify $\frac{-x}{x + x^2}$.

4. Reduce $\frac{x^2 + x - 12}{3 - x}$ to lowest terms.

In Problems 5–12, perform the indicated operations and simplify.

5. Multiply $\frac{2y^2}{7} \cdot \frac{21x}{8y}$.

6. Multiply $(x - 1) \cdot \frac{x + 3}{x^2 - 1}$.

7. Divide $\frac{x^2 - 9}{x + 5} \div (x - 3)$.

8. Divide $\frac{x + 2}{x - 2} \div \frac{x^2 - 4}{2 - x}$.

9. Add $\frac{5}{2(x - 2)} + \frac{3}{2(x - 2)}$.

10. Subtract $\frac{10}{3(x + 1)} - \frac{1}{3(x + 1)}$.

11. Add $\frac{3}{x + 1} + \frac{1}{x - 1}$.

12. Subtract $\frac{x + 1}{x^2 + x - 2} - \frac{x + 2}{x^2 - 1}$.

13. Simplify $\dfrac{\frac{3}{4x} - \frac{1}{2x}}{\frac{1}{x} + \frac{2}{3x}}$.

14. Solve $\frac{4x}{x - 2} + 4 = \frac{6x}{x - 2}$.

15. Solve $\frac{x}{x^2 - 9} + \frac{3}{x - 3} = \frac{1}{x + 3}$.

16. Solve $\frac{x}{x + 6} - \frac{1}{6} = \frac{-6}{x + 6}$.

17. Solve $1 + \frac{2}{x - 3} = \frac{12}{x^2 - 9}$.

18. Solve $D = \frac{d_1(1 - d^n)}{(1 + d)^n}$ for d_1.

19. A car travels 150 miles on 9 gallons of gas. How many gallons will it need to travel 300 mi?

20. Solve $\frac{x + 3}{7} = \frac{11}{6}$.

21. A woman can paint a house in 5 hr. Another one can do it in 6 hr. How long would it take both painters working together to finish the job?

22. A boat can travel 10 miles against a current in the same time it takes to travel 30 miles with the current. If the speed of the current is 8 miles per hour, what is the speed of the boat in still water?

23. A baseball player has 20 singles in 80 games. At that rate, how many singles will he have in 160 games?

24. A conversion van company wants to finish 150 vans in 1 year (12 months). To attain this goal, how many vans should be finished by the end of April (the fourth month)?

25. Find the unknown in the given similar triangles.

a.

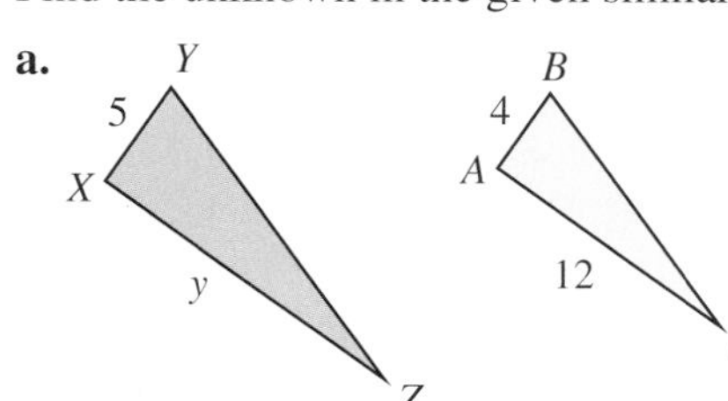

b.

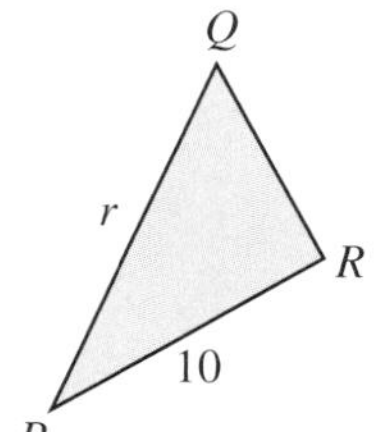

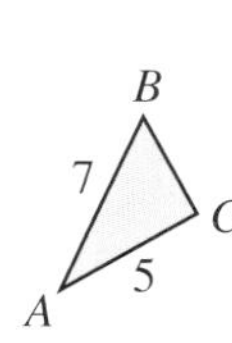

Answers to Pretest

ANSWER	IF YOU MISSED	REVIEW		
	QUESTION	SECTION	EXAMPLES	PAGE
1. $\frac{12xy^2}{28y^3}$	1	6.1	2	466–467
2. $-3(x + y)$ or $-3x - 3y$	2	6.1	3	468–469
3. $\frac{-1}{1 + x}$	3	6.1	4	470
4. $-(x + 4)$ or $-x - 4$	4	6.1	5	471–472
5. $\frac{3xy}{4}$	5	6.2	1, 2	478–479
6. $\frac{x + 3}{x + 1}$	6	6.2	3	479–480
7. $\frac{x + 3}{x + 5}$	7	6.2	4	481
8. $\frac{-1}{x - 2}$ or $\frac{1}{2 - x}$	8	6.2	4, 5	481–482
9. $\frac{4}{x - 2}$	9	6.3	1a	487
10. $\frac{3}{x + 1}$	10	6.3	1b	487
11. $\frac{4x - 2}{x^2 - 1}$	11	6.3	2, 4a	487–490
12. $\frac{-2x - 3}{(x + 2)(x + 1)(x - 1)}$	12	6.3	3, 4b, 5	488–492
13. $\frac{3}{20}$	13	6.4	1, 2	498–499
14. $x = 4$	14	6.5	1, 2	503–504
15. $x = -4$	15	6.5	3	504–505
16. No solution	16	6.5	4	505
17. $x = -5$	17	6.5	5	506–507
18. $d_1 = \frac{D(1 + d)^n}{1 - d^n}$	18	6.5	6	507
19. 18	19	6.6	1	511
20. $x = \frac{59}{6}$	20	6.6	1	511
21. $2\frac{8}{11}$ hr	21	6.6	2	513
22. 16 miles per hour	22	6.6	3	514
23. 40	23	6.6	4	515
24. 50	24	6.6	4	515
25. a. $y = 15$ **b.** $r = 14$	25	6.6	5	516

6.1 BUILDING AND REDUCING RATIONAL EXPRESSIONS

To Succeed, Review How To . . .

1. Factor polynomials (pp. 424–429).
2. Write a fraction with specified denominator (pp. 5–7).
3. Simplify fractions (pp. 7–9).

Objectives

A Determine the values that make a rational expression undefined.

B Build fractions.

C Reduce (simplify) a rational expression to lowest terms.

GETTING STARTED

Recycling Waste and Rational Expressions

We've already mentioned that algebra is generalized arithmetic. In arithmetic we study the natural numbers, the whole numbers, the integers, and the rational numbers. In algebra we've studied expressions and polynomials, and we have also discussed how they follow rules similar to those used with the real numbers. We will find that rational expressions in algebra follow the same rules as rational numbers in arithmetic.

In arithmetic, a **rational number** is a number that can be written in the form $\frac{a}{b}$, where a and b are integers and b is not zero. As usual, a is called the *numerator* and b, the *denominator*. A similar approach is used in algebra.

In algebra, an expression of the form $\frac{A}{B}$, where A and B are polynomials and B is not zero, is called an **algebraic fraction,** or a **rational expression.** So if $G(t) = -0.001t^3 + 0.06t^2 + 2.6t + 88.6$ is a polynomial representing the amount of waste generated in the United States (in millions of tons), $R(t) = 0.06t^2 - 0.59t + 6.4$ is a polynomial representing the amount of waste recovered (in millions of tons), and t is the number of years after 1960, then

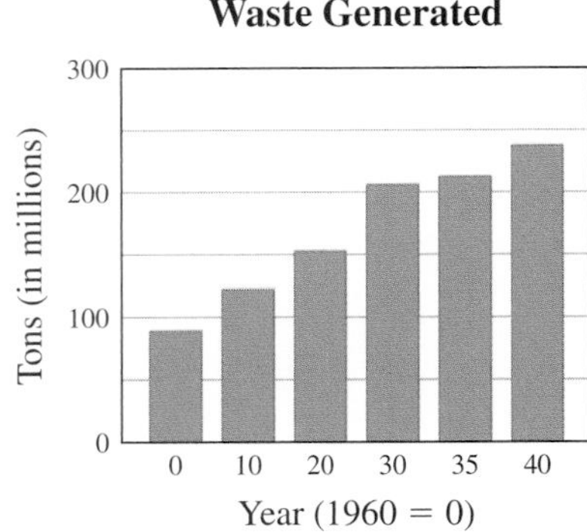

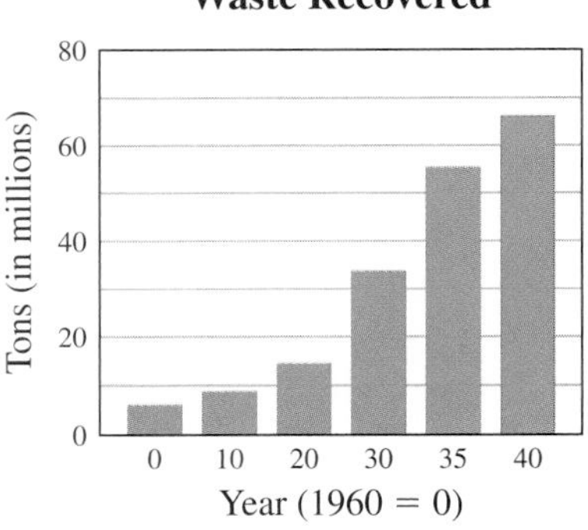

$$\frac{R(t)}{G(t)}$$

is a *rational expression* representing the *fraction* of the waste recovered in those years. Thus in 1960 (when $t = 0$), the fraction is

$$\frac{R(0)}{G(0)} = \frac{0.06(0)^2 - 0.59(0) + 6.4}{-0.001(0)^3 + 0.06(0)^2 + 2.6(0) + 88.6} = \frac{6.4}{88.6}$$

which is about 7%. What percent of the waste will be recovered in the year 2010 when $t = 2010 - 1960$?

In this section we shall learn how to determine when rational expressions are undefined and how to write them with a given denominator and then simplify them.

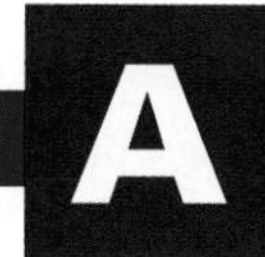

Undefined Rational Expressions

The expressions

$$\frac{8}{x}, \quad \frac{x^2 + 2x + 3}{x + 5}, \quad \frac{y}{y - 1}, \quad \text{and} \quad \frac{x^2 + 3x + 9}{x^2 - 4x + 4}$$

are rational expressions. Of course, since we don't want the denominators of these expressions to be zero, we must place some restrictions on these denominators. Let's see what these restrictions must be.

For $\frac{8}{x}$, x cannot be 0 ($x \neq 0$) because then we would have $\frac{8}{0}$, which is not defined. For

$$\frac{x^2 + 2x + 3}{x + 5}, \quad x \neq -5$$

If x is -5, $\frac{x^2 + 2x + 3}{x + 5} = \frac{25 - 10 + 3}{0}$, which is undefined.

For

$$\frac{y}{y - 1}, \quad y \neq 1$$

If y is 1, $\frac{y}{y - 1} = \frac{1}{0}$, which is undefined.

and for

$$\frac{x^2 + 3x + 9}{x^2 - 4x + 4} = \frac{x^2 + 3x + 9}{(x - 2)^2}, \quad x \neq 2$$

If x is 2, $\frac{x^2 + 3x + 9}{x^2 - 4x + 4} = \frac{4 + 6 + 9}{4 - 8 + 4} = \frac{19}{0}$—again, undefined.

Teaching Tip

Remind students about division by zero using these two examples:

1. Division into zero is OK.

$$\frac{O}{K}$$

2. Division by zero is not possible.

$$\frac{N}{O}$$

To avoid stating repeatedly that the denominators of rational expressions must not be zero, we make the following rule.

RULE

Avoiding Zero Denominators

The variables in a rational expression must not be replaced by numbers that make the denominator zero.

To find the value or values that make the denominator zero: (1) set the denominator equal to zero, (2) solve the resulting equation for the variable.

For example, we found that

$$\frac{x^2 + 2x + 3}{x + 5}$$

is undefined for $x = -5$ by letting $x + 5 = 0$ and solving for x, to obtain $x = -5$. Similarly, we can find the values that make

$$\frac{1}{x^2 + 3x - 4}$$

undefined by letting

$$x^2 + 3x - 4 = 0$$

$$(x + 4)(x - 1) = 0 \quad \text{Factor.}$$

$$x + 4 = 0 \quad \text{or} \quad x - 1 = 0 \quad \text{Solve each equation.}$$

$$x = -4 \qquad x = 1$$

Thus

$$\frac{1}{x^2 + 3x - 4}$$

is undefined when $x = -4$ or $x = 1$. (Check this out!)

EXAMPLE 1 **Finding values that make rational expressions undefined**

Find the values for which the rational expression is undefined:

a. $\dfrac{n}{2n - 3}$ **b.** $\dfrac{x + 1}{x^2 + 5x - 6}$ **c.** $\dfrac{m + 2}{m^2 + 2}$

SOLUTION

a. We set the denominator equal to zero and solve:

$$2n - 3 = 0$$

$$2n = 3 \quad \text{Add 3.}$$

$$n = \frac{3}{2} \quad \text{Divide by 2.}$$

Thus

$$\frac{n}{2n - 3}$$

is undefined for $n = \frac{3}{2}$.

b. We proceed as before and set the denominator equal to zero:

$$x^2 + 5x - 6 = 0$$

$$(x + 6)(x - 1) = 0 \quad \text{Factor.}$$

$$x + 6 = 0 \quad \text{or} \quad x - 1 = 0 \quad \text{Solve each equation.}$$

$$x = -6 \qquad x = 1$$

Thus

$$\frac{x + 1}{x^2 + 5x - 6}$$

is undefined for $x = -6$ or $x = 1$.

c. Setting the denominator equal to zero, we have

$$m^2 + 2 = 0$$

so that $m^2 = -2$. But the square of any real number is not negative, so there are no real numbers m for which the denominator is zero. (Note that m^2 is greater than or equal to zero, so $m^2 + 2$ is always positive). There are *no values* for which

$$\frac{m + 2}{m^2 + 2}$$

is undefined.

PROBLEM 1

Find values for which the rational expression is undefined:

a. $\dfrac{m}{2m + 3}$ **b.** $\dfrac{y + 1}{y^2 + 4y - 5}$

c. $\dfrac{n + 3}{n^2 + 3}$

Web It

To find a discussion of rational expressions avoiding zero denominators (excluded values), go to link 6-1-1 on the Bello Website at mhhe.com/bello.

For a more theoretical approach, check out link 6-1-2.

Teaching Tip

To help students understand that

$$\frac{m + 2}{m^2 + 2}$$

is defined for any values of m, have students evaluate the rational expression for $m = 2$ and $m = -2$.

B Building Fractions

Now that we know that we must avoid zero denominators, recall what we did with fractions in arithmetic. First, we learned how to recognize which ones are equal, and then we used this idea to reduce or build fractions. Let's talk about equality first.

What does this picture tell you (if you write the value of the coins using fractions)? It states that

$$\frac{1}{2} = \frac{2}{4}$$

Answers

1. a. $m = -\frac{3}{2}$ **b.** $y = -5$ or $y = 1$ **c.** No values

As a matter of fact, the following is also true:

$$\frac{1}{2} = \frac{2}{4} = \frac{3}{6} = \frac{4}{8} = \frac{x}{2x} = \frac{x^2}{2x^2}$$

and so on. Do you see the pattern? Here is another way to write it:

$$\frac{1}{2} = \frac{1 \times 2}{2 \times 2} = \frac{2}{4}$$ Note that $\frac{2}{2} = 1$.

$$\frac{1}{2} = \frac{1 \times 3}{2 \times 3} = \frac{3}{6}$$ Here, $\frac{3}{3} = 1$.

$$\frac{1}{2} = \frac{1 \times 4}{2 \times 4} = \frac{4}{8}$$

$$\frac{1}{2} = \frac{1 \times x}{2 \times x} = \frac{x}{2x}$$

$$\frac{1}{2} = \frac{1 \times x^2}{2 \times x^2} = \frac{x^2}{2x^2}$$

Web It

To practice building fractions, go to link 6-1-3 on the Bello Website at mhhe.com/bello. You will find practice problems with answers!

To practice both reducing and building fractions, try link 6-1-4.

We can always obtain a rational expression that is equivalent to another rational expression $\frac{A}{B}$ by multiplying the numerator and denominator of the given rational expression by the same nonzero number or expression, C. Here is this rule stated in symbols.

RULE

Fundamental Rule of Rational Expressions

$$\frac{A}{B} = \frac{A \cdot C}{B \cdot C} \quad (B \neq 0, C \neq 0)$$

Note that

$$\frac{A}{B} = \frac{A \cdot C}{B \cdot C} \quad \text{because} \quad \frac{C}{C} = 1$$

and multiplying by 1 does not change the value of the expression.

This idea is very important for adding or subtracting rational expressions (Section 6.3). For example, to add $\frac{1}{2}$ and $\frac{3}{4}$, we write

$$\frac{1}{2} = \frac{2}{4}$$ Note that $\frac{1}{2} = \frac{1 \times 2}{2 \times 2} = \frac{2}{4}$.

$$+\frac{3}{4} = \frac{3}{4}$$

$$\overline{\qquad \frac{5}{4}}$$

Here we've used the fundamental rule of rational expressions to write the $\frac{1}{2}$ as an equivalent fraction with a denominator of 4, namely, $\frac{2}{4}$. Now, suppose we wish to write $\frac{3}{8}$ with a denominator of 16. How do we do this? First, we write the problem as

$$\frac{3}{8} = \frac{?}{16}$$ Note that $16 = 8 \times 2$.

↑ ↑

Multiply by 2.

and notice that to get 16 as the denominator, we need to multiply the 8 by 2. Of course, we must do the same to the numerator 3; we obtain

Multiply by 2.

$$\frac{3}{8} = \frac{6}{16}$$

By the fundamental rule of rational expressions, if we multiply the denominator by 2, we must multiply the numerator by 2.

Note that $\frac{3 \times 2}{8 \times 2} = \frac{6}{16}$.

Similarly, to write

$$\frac{5x}{3y}$$

with a denominator of $6y^3$, we first write the new equivalent expression

$$\frac{?}{6y^3}$$

with the old denominator $3y$ factored out:

$$\frac{5x}{3y} = \frac{?}{6y^3} = \frac{?}{3y(2y^2)}$$

Write $6y^3$ as $3y(2y^2)$.

Multiply by $2y^2$.

Since the multiplier is $2y^2$, we have

Multiply by $2y^2$.

$$\frac{5x}{3y} = \frac{5x(2y^2)}{3y(2y^2)} = \frac{10xy^2}{6y^3}$$

Thus

$$\frac{5x}{3y} = \frac{10xy^2}{6y^3}$$

We multiplied the denominator by $2y^2$, so we have to multiply the numerator by $2y^2$.

Teaching Tip

Rewriting rational expressions with a specified denominator can also be called "building up" the fraction.

EXAMPLE 2 Rewriting rational expressions with a specified denominator

Write:

a. $\frac{5}{6}$ with a denominator of 18

b. $\frac{2x}{9y^2}$ with a denominator of $18y^3$

c. $\frac{3x}{x - 1}$ with a denominator of $x^2 + 2x - 3$

SOLUTION

a. $\frac{5}{6} = \frac{?}{18}$

Multiply by 3.

Multiply by 3.

$$\frac{5}{6} = \frac{15}{18}$$

Note that $\frac{5 \times 3}{6 \times 3} = \frac{15}{18}$.

PROBLEM 2

Write:

a. $\frac{4}{7}$ with a denominator of 14

b. $\frac{2y}{7x^2}$ with a denominator of $14x^3$

c. $\frac{2x}{x + 2}$ with a denominator of $x^2 + x - 2$

Answers

2. a. $\frac{8}{14}$ **b.** $\frac{4xy}{14x^3}$ **c.** $\frac{2x^2 - 2x}{x^2 + x - 2}$

b. Since $18y^3 = 9y^2(2y)$,

$$\frac{2x}{9y^2} = \frac{?}{9y^2(2y)}$$

↑ Multiply by $2y$. ↑

↓ Multiply by $2y$. ↓

$$\frac{2x}{9y^2} = \frac{2x(2y)}{9y^2(2y)} = \frac{4xy}{18y^3}$$

c. We first note that $x^2 + 2x - 3 = (x - 1)(x + 3)$. Thus

$$\frac{3x}{x-1} = \frac{?}{(x-1)(x+3)}$$

↑ Multiply by $(x + 3)$. ↑

↓ Multiply by $(x + 3)$. ↓

$$\frac{3x}{x-1} = \frac{3x(x+3)}{(x-1)(x+3)} = \frac{3x^2+9x}{x^2+2x-3}$$

Teaching Tip

When building up fractions that have other than monomial denominators, it is important to factor the denominators first.

C Reducing Rational Expressions

Web It

For a lesson and worked examples on reducing rational expressions, try link 6-1-5 on the Bello Website at mhhe.com/bello.

For more practice with reducing rational expressions, try link 6-1-6.

Now that we know how to build up rational expressions, we are now ready to use the reverse process—that is, to **reduce** them. In Example 2(a), we wrote $\frac{5}{6}$ with a denominator of 18; that is, we found out that

$$\frac{5}{6} = \frac{5 \times 3}{6 \times 3} = \frac{15}{18}$$

Of course, you will probably agree that $\frac{5}{6}$ is written in a "simpler" form than $\frac{15}{18}$. Certainly you will eventually agree—though it is hard to see at first glance—that

$$\frac{5}{3} = \frac{5(x+3)(x^2-4)}{3(x+2)(x^2+x-6)}$$

and that $\frac{5}{3}$ is the "simpler" of the two expressions. In algebra, the process of removing all factors common to the numerator and denominator is called **reducing to lowest terms,** or **simplifying** the expression. How do we reduce the rational expression

$$\frac{5(x+3)(x^2-4)}{3(x+2)(x^2+x-6)}$$

to lowest terms? We proceed by steps.

Teaching Tip

Have students review some factoring examples.

$$2x + 4 = 2(x + 2)$$
$$x^2 - 9 = (x + 3)(x - 3)$$
$$x^2 + 5x + 6 = (x + 2)(x + 3)$$

etc.

PROCEDURE

Reducing Rational Expressions to Lowest Terms (Simplifying)

1. Write the numerator and denominator of the expression in factored form.
2. Find the factors that are common to the numerator and denominator.
3. Replace the quotient of the common factors by the number 1, since $\frac{a}{a} = 1$.
4. Rewrite the expression in simplified form.

Let's use this procedure to reduce

$$\frac{5(x+3)(x^2-4)}{3(x+2)(x^2+x-6)}$$

to lowest terms:

1. Write the numerator and denominator in factored form. $\frac{5(x+3)(x+2)(x-2)}{3(x+2)(x+3)(x-2)}$
2. Find the factors that are common to the numerator and denominator (we rearranged them so the common factors are in columns). $\frac{5(x+2)(x+3)(x-2)}{3(x+2)(x+3)(x-2)}$
3. Replace the quotient of the common factors by the number 1. $\frac{5\overset{1}{\cancel{(x+2)}}\overset{1}{\cancel{(x+3)}}\overset{1}{\cancel{(x-2)}}}{3\cancel{(x+2)}\cancel{(x+3)}\cancel{(x-2)}}$
4. Rewrite the expression in simplified form. $\frac{5}{3}$

The whole procedure can be written as

$$\frac{5(x+3)(x^2-4)}{3(x+2)(x^2+x-6)} = \frac{5\overset{1}{\cancel{(x+3)}}\overset{1}{\cancel{(x+2)}}\overset{1}{\cancel{(x-2)}}}{3\cancel{(x+2)}\cancel{(x+3)}\cancel{(x-2)}} = \frac{5}{3}$$

EXAMPLE 3 **Reducing rational expressions to lowest terms**

Simplify:

a. $\frac{-5x^2y}{15xy^3}$

b. $\frac{-8(x^2-y^2)}{-4(x+y)}$

SOLUTION

a. We use our four-step procedure.

1. Write in factored form. $\frac{-5x^2y}{15xy^3} = \frac{(-1)\cdot 5\cdot x\cdot x\cdot y}{3\cdot 5\cdot x\cdot y\cdot y\cdot y}$
2. Find the common factors. $= \frac{5\cdot x\cdot y(-1)\cdot x}{5\cdot x\cdot y\cdot 3\cdot y\cdot y}$
3. Replace the quotient of the common factors by 1. $= \frac{1\cdot(-1)\cdot x}{3\cdot y\cdot y}$
4. Rewrite in simplified form. $= \frac{-x}{3y^2}$

The whole process can be written as

$$\frac{-5x^2y}{15xy^3} = \frac{\overset{-1x}{\cancel{-5x^2y}}}{\underset{3\ \ y^2}{\cancel{15xy^3}}} = \frac{-x}{3y^2}$$

PROBLEM 3

Simplify:

a. $\frac{-3x^2y}{6xy^4}$

b. $\frac{-6(m^2-n^2)}{-3(m-n)}$

Answers

3. a. $\frac{-x}{2y^3}$ **b.** $2(m+n)$

b. We again use our four-step procedure.

1. Write the fractions in factored form. $\frac{-8(x^2 - y^2)}{-4(x + y)} = \frac{(-1) \cdot 2 \cdot 2 \cdot 2(x + y)(x - y)}{(-1) \cdot 2 \cdot 2(x + y)}$

2. Find the common factors. $= \frac{(-1) \cdot 2 \cdot 2 \cdot (x + y)(x - y) \cdot 2}{(-1) \cdot 2 \cdot 2 \cdot (x + y)}$

3. Replace the quotient of the common factors by 1. $= 1 \cdot (x - y) \cdot 2$

4. Rewrite in simplified form. $= 2(x - y)$

The abbreviated form is

$$\frac{-8(x^2 - y^2)}{-4(x + y)} = \frac{\overset{2}{\cancel{-8}}\overset{1}{\cancel{(x + y)}}(x - y)}{\cancel{-4}\cancel{(x + y)}} = 2(x - y)$$

You may have noticed that in Example 3(a) we wrote the answer as

$$\frac{-x}{3y^2}$$

It could be argued that since

$$\frac{-5}{15} = -\frac{1}{3}$$

the answer should be

$$-\frac{x}{3y^2}$$

However, to avoid confusion, we agree to write

$$-\frac{x}{3y^2} \quad \text{as} \quad \frac{-x}{3y^2}$$

with the *negative* sign in the numerator. In general, since a fraction has three possible signs (numerator, denominator, and the sign of the fraction itself), we use the following conventions.

Web It

Want to practice some before you try the next problem? Try the first few problems in link 6-1-6 on the Bello Website at mhhe.com/bello. They will tell you whether you got the answer right or wrong!

STANDARD FORM OF A FRACTION

$-\frac{a}{b}$ is written as $\frac{-a}{b}$ $\qquad$ $\frac{-a}{-b}$ is written as $\frac{a}{b}$

$\frac{a}{-b}$ is written as $\frac{-a}{b}$ $\qquad$ $-\frac{-a}{b}$ is written as $\frac{a}{b}$

$-\frac{-a}{-b}$ is written as $\frac{-a}{b}$ $\qquad$ $-\frac{a}{-b}$ is written as $\frac{a}{b}$

The forms

$$\frac{a}{b} \quad \text{and} \quad \frac{-a}{b}$$

are called the **standard forms** of a fraction and are what we use to write answers involving fractions.

EXAMPLE 4 Reducing to lowest terms

Simplify:

a. $\dfrac{6x - 12y}{18x - 36y}$ b. $-\dfrac{y}{y + xy}$ c. $\dfrac{x + 3}{-(x^2 - 9)}$

SOLUTION

a. $\dfrac{6x - 12y}{18x - 36y} = \dfrac{6(x - 2y)}{18(x - 2y)}$

$= \dfrac{\overset{1}{\cancel{6}}\overset{1}{\cancel{(x - 2y)}}}{\underset{3}{\cancel{18}}\underset{1}{\cancel{(x - 2y)}}}$

$= \dfrac{1}{3}$

b. $-\dfrac{y}{y + xy} = -\dfrac{y}{y(1 + x)}$

$= -\dfrac{\overset{1}{\cancel{y}}}{\cancel{y}(1 + x)}$

$= \dfrac{-1}{1 + x}$

c. $\dfrac{x + 3}{-(x^2 - 9)} = \dfrac{(x + 3)}{-(x + 3)(x - 3)}$

$= \dfrac{\overset{1}{\cancel{(x + 3)}}}{-\cancel{(x + 3)}(x - 3)}$

$= \dfrac{1}{-(x - 3)}$

$= \dfrac{-1}{x - 3}$ In standard form

PROBLEM 4

Simplify:

a. $\dfrac{3x - 12y}{12x - 48y}$

b. $-\dfrac{x}{x + xy}$

c. $\dfrac{y - 3}{-(y^2 - 9)}$

Teaching Tip

Remind students that only factors can be divided, not terms. In Example 4a, if you divide out common terms you won't get the same answer.

$\dfrac{\overset{1}{\cancel{6x}} - \overset{1}{\cancel{12y}}}{\underset{3}{\cancel{18x}} - \underset{3}{\cancel{36y}}} \overset{?}{=} \dfrac{\overset{1}{\cancel{6}}\overset{1}{\cancel{(x - 2y)}}}{\underset{3}{\cancel{18}}\underset{1}{\cancel{(x - 2y)}}}$

$\dfrac{1 - 1}{3 - 3} \overset{?}{=} \dfrac{1 \cdot 1}{3 \cdot 1}$

$\dfrac{0}{0} \neq \dfrac{1}{3}$

Is there a way in which

$$\frac{-1}{x - 3}$$

can be simplified further? The answer is yes. See whether you can see the reasons behind each step:

$$\frac{-1}{x - 3} = \frac{-1}{-(3 - x)} \qquad -(3 - x) = -3 + x = x - 3$$

$$= \frac{1}{3 - x}$$

Thus

$$\frac{-1}{x - 3} = \frac{1}{3 - x}$$

Note that $\frac{-1}{x - 3}$ involves finding the additive inverse of 1 in the numerator and subtracting 3 from x in the denominator, whereas $\frac{1}{3 - x}$ only involves subtracting x from 3 in the denominator. Less work!

Answers

4. a. $\frac{1}{4}$ **b.** $\frac{-1}{1 + y}$ **c.** $\frac{-1}{y + 3}$

And why, you might ask, is this answer simpler than the other? Because

$$\frac{-1}{x-3}$$

is a quotient involving a subtraction $(x - 3)$ and an additive inverse (-1), but

$$\frac{1}{3-x}$$

is a quotient that involves only a subtraction $(3 - x)$. Be aware of these simplifications when writing your answers!

Here is a situation that occurs very frequently in algebra. Can you reduce this expression?

$$\frac{a-b}{b-a}$$

Teaching Tip

Give students a numerical example to demonstrate

$$\frac{a-b}{b-a} = -1.$$

Example:

$$\frac{7-2}{2-7} = \frac{5}{-5} = -1$$

Look at the following steps. Note that the original denominator, $b - a$, can be written as $-(a - b)$ since $-(a - b) = -a + b = b - a$.

$$\frac{a-b}{b-a} = \frac{\overset{1}{\cancel{(a-b)}}}{-\cancel{(a-b)}} \quad \text{Write } b - a \text{ as } -(a - b) \text{ in the denominator.}$$

$$= \frac{1}{-1}$$

$$= -1$$

QUOTIENT OF ADDITIVE INVERSES

$$\frac{a-b}{b-a} = -1$$

This means that a fraction whose numerator and denominator are additive inverses (such as $a - b$ and $b - a$) equals -1. We use this idea in the next example.

EXAMPLE 5 **Reducing rational expressions to lowest terms**

Reduce to lowest terms:

a. $\dfrac{x^2 - 16}{4 + x}$ **b.** $-\dfrac{x^2 - 16}{x + 4}$ **c.** $\dfrac{x^2 + 2x - 15}{3 - x}$

SOLUTION

a. $\dfrac{x^2 - 16}{4 + x} = \dfrac{\cancel{(x+4)}(x-4)}{\cancel{x+4}}$ $\quad$ $4 + x = x + 4$ by the commutative property.

$= x - 4$

b. $-\dfrac{x^2 - 16}{x + 4} = -\dfrac{(x+4)(x-4)}{(x+4)}$

$= -\dfrac{\overset{1}{\cancel{(x+4)}}(x-4)}{\cancel{(x+4)}}$ $\quad$ Note that $\dfrac{x+4}{x+4} = 1$.

$= -(x - 4)$

$= -x + 4$

$= 4 - x$

PROBLEM 5

Reduce to lowest terms:

a. $\dfrac{y^2 - 9}{3 + y}$

b. $\dfrac{y^2 - 9}{y + 3}$

c. $\dfrac{y^2 + y - 12}{3 - y}$

Answers

5. a. $y - 3$ **b.** $y - 3$ **c.** $-(y + 4)$

c. $\dfrac{x^2 + 2x - 15}{3 - x} = \dfrac{(x + 5)(x - 3)}{3 - x}$

$= \dfrac{\overset{-1}{(\cancel{x - 3})}(x + 5)}{(\cancel{3 - x})}$ Note that $\dfrac{x - 3}{3 - x} = -1$.

$= -(x + 5)$

We have used the fact that

$$\frac{a - b}{b - a} = -1$$

in the second step to simplify

$$\frac{(x - 3)}{(3 - x)}$$

Note that the final answer is written as $-(x + 5)$ instead of $-x - 5$, since $-(x + 5)$ is considered to be simpler.

Calculate It **Checking Quotients**

How can you check that your answers are correct? If you have a calculator, you can do it two ways:

1. *By looking at the graph* If the graph of the original problem and the graph of the answer coincide, the answer is probably correct.

Thus, to check the results of Example 5(c), enter

$$Y_1 = \frac{(x^2 + 2x - 15)}{(3 - x)} \quad \text{and} \quad Y_2 = -(x + 5)$$

Note the parentheses in the numerator and denominator of Y_1, and remember that the $-$ in $-(x + 5)$ has to be entered with the (−) key.

2. *By looking at the numerical values of Y_1 and Y_2* (Press 2nd GRAPH.) The graph and the numerical values are as shown.

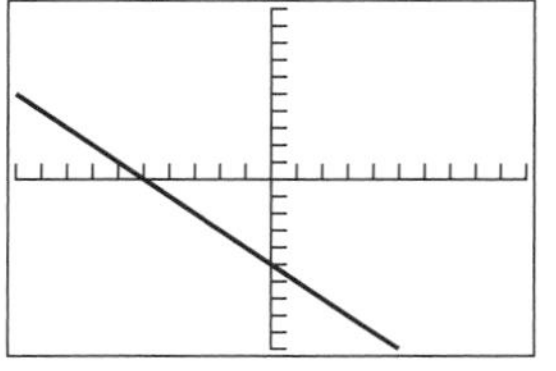

X	Y1	Y2
0	-5	-5
1	-6	-6
2	-7	-7
3	ERROR	-8
4	-9	-9
5	-10	-10
6	-11	-11
X=3		

One more thing: When you are looking at the table of values, what happens when $x = 3$? Why?

Exercises 6.1

A In Problems 1–20, find the value(s) for which the rational expression is undefined.

1. $\dfrac{x}{x - 7}$ $x = 7$

2. $\dfrac{3y}{2y + 9}$ $y = -\dfrac{9}{2}$

3. $\dfrac{y - 5}{2y + 8}$ $y = -4$

4. $\dfrac{y + 4}{3y - 9}$ $y = 3$

5. $\dfrac{x + 9}{x^2 - 9}$ $x = 3$ or $x = -3$

6. $\dfrac{y + 4}{y^2 - 16}$ $y = 4$ or $y = -4$

7. $\dfrac{y + 3}{y^2 + 5}$ None

8. $\dfrac{y + 4}{y^2 + 1}$ None

9. $\dfrac{2y + 9}{y^2 - 6y + 8}$ $y = 2$ or $y = 4$

10. $\dfrac{3x + 5}{x^2 - 7x + 6}$ $x = 1$ or $x = 6$

11. $\dfrac{x^2 - 4}{x^3 + 8}$ $x = -2$

12. $\dfrac{y^2 - 9}{y^3 + 1}$ $y = -1$

13. $\dfrac{x^2 + 4}{x^3 - 27}$ $x = 3$

14. $\dfrac{x^2 + 16}{x^3 - 8}$ $x = 2$

15. $\dfrac{x - 1}{x^2 + 6x + 9}$ $x = -3$

16. $\dfrac{x - 4}{x^2 + 8x + 16}$ $x = -4$

17. $\dfrac{y-3}{y^2-6y-16}$ $y=-2$ or $y=8$

18. $\dfrac{y-5}{y^2-10y+25}$ $y=5$

19. $\dfrac{x+4}{x^3+2x^2+x}$ $x=0$ or $x=-1$

20. $\dfrac{x+5}{x^3+4x^2+4x}$ $x=0$ or $x=-2$

B In Problems 21–32, write the given fraction as an equivalent one with the indicated denominator.

21. $\dfrac{3}{7}$ with a denominator of 21 $\dfrac{9}{21}$

22. $\dfrac{5}{9}$ with a denominator of 36 $\dfrac{20}{36}$

23. $\dfrac{-8}{11}$ with a denominator of 22 $\dfrac{-16}{22}$

24. $\dfrac{-5}{17}$ with a denominator of 51 $\dfrac{-15}{51}$

25. $\dfrac{5x}{6y^2}$ with a denominator of $24y^3$ $\dfrac{20xy}{24y^3}$

26. $\dfrac{7y}{5x^3}$ with a denominator of $10x^4$ $\dfrac{14xy}{10x^4}$

27. $\dfrac{-3x}{7y}$ with a denominator of $21y^4$ $\dfrac{-9xy^3}{21y^4}$

28. $\dfrac{-4y}{7x^2}$ with a denominator of $28x^3$ $\dfrac{-16xy}{28x^3}$

29. $\dfrac{4x}{x+1}$ with a denominator of x^2-x-2 $\dfrac{4x(x-2)}{x^2-x-2}$ or $\dfrac{4x^2-8x}{x^2-x-2}$

30. $\dfrac{5y}{y-1}$ with a denominator of y^2+2y-3 $\dfrac{5y(y+3)}{y^2+2y-3}$ or $\dfrac{5y^2+15y}{y^2+2y-3}$

31. $\dfrac{-5x}{x+3}$ with a denominator of x^2+x-6 $\dfrac{-5x(x-2)}{x^2+x-6}$ or $\dfrac{-5x^2+10x}{x^2+x-6}$

32. $\dfrac{-3y}{y-4}$ with a denominator of y^2-2y-8 $\dfrac{-3y(y+2)}{y^2-2y-8}$ or $\dfrac{-3y^2-6y}{y^2-2y-8}$

C In Problems 33–70, reduce to lowest terms (simplify).

33. $\dfrac{7x^3y}{14xy^4}$ $\dfrac{x^2}{2y^3}$

34. $\dfrac{24xy^3}{6x^3y}$ $\dfrac{4y^2}{x^2}$

35. $\dfrac{-9xy^5}{3x^2y}$ $\dfrac{-3y^4}{x}$

36. $\dfrac{-24x^3y^2}{48xy^4}$ $\dfrac{-x^2}{2y^2}$

37. $\dfrac{-6x^2y}{-12x^3y^4}$ $\dfrac{1}{2xy^3}$

38. $\dfrac{-9xy^4}{-18x^5y}$ $\dfrac{y^3}{2x^4}$

39. $\dfrac{-25x^3y^2}{-5x^2y^4}$ $\dfrac{5x}{y^2}$

40. $\dfrac{-30x^2y^2}{-6x^3y^5}$ $\dfrac{5}{xy^3}$

41. $\dfrac{6(x^2-y^2)}{18(x+y)}$ $\dfrac{x-y}{3}$

42. $\dfrac{12(x^2-y^2)}{48(x+y)}$ $\dfrac{x-y}{4}$

43. $\dfrac{-9(x^2-y^2)}{3(x+y)}$ $-3(x-y)$ or $3y-3x$

44. $\dfrac{-12(x^2-y^2)}{3(x-y)}$ $-4(x+y)$ or $-4x-4y$

45. $\dfrac{-6(x+y)}{24(x^2-y^2)}$ $\dfrac{-1}{4(x-y)}$ or $\dfrac{1}{4y-4x}$

46. $\dfrac{-8(x+3)}{40(x^2-9)}$ $\dfrac{-1}{5(x-3)}$ or $\dfrac{1}{15-5x}$

47. $\dfrac{-5(x-2)}{-10(x^2-4)}$ $\dfrac{1}{2(x+2)}$ or $\dfrac{1}{2x+4}$

48. $\dfrac{-12(x-2)}{-60(x^2-4)}$ $\dfrac{1}{5(x+2)}$ or $\dfrac{1}{5x+10}$

49. $\dfrac{-3(x-y)}{-3(x^2-y^2)}$ $\dfrac{1}{x+y}$

50. $\dfrac{-10(x+y)}{-10(x^2-y^2)}$ $\dfrac{1}{x-y}$

51. $\dfrac{4x-4y}{8x-8y}$ $\dfrac{1}{2}$

52. $\dfrac{6x+6y}{2x+2y}$ 3

53. $\dfrac{4x-8y}{12x-24y}$ $\dfrac{1}{3}$

54. $\dfrac{15y-45x}{5y-15x}$ 3

55. $-\dfrac{6}{6+12y}$ $\dfrac{-1}{1+2y}$

56. $-\dfrac{4}{8+12x}$ $\dfrac{-1}{2+3x}$

57. $-\dfrac{x}{x+2xy}$ $\dfrac{-1}{1+2y}$

58. $-\dfrac{y}{2y+6xy}$ $\dfrac{-1}{2(1+3x)}$ or $\dfrac{-1}{2+6x}$

59. $-\dfrac{6y}{6xy+12y}$ $\dfrac{-1}{x+2}$

60. $-\frac{4x}{8xy+16x}$ $\frac{-1}{2(y+2)}$ or $\frac{-1}{2y+4}$ **61.** $\frac{3x-2y}{2y-3x}$ -1 **62.** $\frac{5y-2x}{2x-5y}$ -1

63. $\frac{x^2+4x-5}{1-x}$ $-(x+5)$ or $-x-5$ **64.** $\frac{x^2-2x-15}{5-x}$ $-(x+3)$ or $-x-3$ **65.** $\frac{x^2-6x+8}{4-x}$ $-(x-2)$ or $2-x$

66. $\frac{x^2-8x+15}{3-x}$ $-(x-5)$ or $5-x$ **67.** $\frac{2-x}{x^2+4x-12}$ $\frac{-1}{x+6}$ **68.** $\frac{3-x}{x^2+3x-18}$ $\frac{-1}{x+6}$

69. $-\frac{3-x}{x^2-5x+6}$ $\frac{1}{x-2}$ **70.** $-\frac{4-x}{x^2-3x-4}$ $\frac{1}{x+1}$

APPLICATIONS

71. *Annual advertising expenditures* How much is spent annually on advertising? The total amount is given by the polynomial $S(t) = -0.3t^2 + 10t + 50$ (millions), where t is the number of years after 1980. The amounts spent on national and local advertisement (in millions) are given by the respective polynomials

$$N(t) = -0.13t^2 + 5t + 30$$

and

$$L(t) = -0.17t^2 + 5t + 20$$

a. What was the total amount spent on advertising in 1980 ($t = 0$)? In 2000? \$50 million; \$130 million

b. What amount was spent on national advertising in 1980? In 2000? \$30 million; \$78 million

c. What amount was spent on local advertising in 1980? In 2000? \$20 million; \$52 million

d. What does the rational expression

$$\frac{N(t)}{S(t)}$$

represent? The fraction of the total advertising that is spent on national advertising

72. *Annual advertising expenditures* Use the information given in Problem 71 to answer these questions:

a. What percent of the total amount spent in 1980 was for national advertising? In 2000? 60%; 60%

b. What percent of the total amount spent in 1980 was for local advertising? In 2000? 40%; 40%

c. What rational expression represents the percent spent for national advertising? $\frac{N(t)}{S(t)} = \frac{-0.13t^2 + 5t + 30}{-0.3t^2 + 10t + 50}$

d. What percent of all advertising would be spent on national advertising in the year 2010? 78.75%

73. *Expenditures for television advertising* The amount spent annually on television advertising can be approximated by the polynomial $T(t) = -0.07t^2 + 2t + 11$ (millions), where t is the number of years after 1980. Use the information in Problem 71 to find what percent of the total amount spent annually on advertising would be

a. Spent on television in the year 2010. 10%

b. Spent on local advertising in the year 2010. 21.25%

c. What does the rational fraction

$$\frac{T(t)}{S(t)}$$

represent? The fraction of the total advertising that is spent on TV advertising

74. *Television spot advertising* The estimated fees for spot advertising on television can be approximated by the polynomial $C(t) = 0.04t^3 - t^2 + 6t + 2.5$ (billions), where t is the number of years after 1980. Of this amount, automotive spot advertising fees can be approximated by the polynomial $A(t) = 0.02t^3 - 0.5t^2 + 3t + 0.3$ (billions).

a. What was the total amount spent on TV spot advertising in 1980? \$2.5 billion

b. What was the amount spent on automotive TV spot advertising in 1980? \$0.3 billion

c. What was the total amount spent on TV spot advertising in 1990? In 2000? \$2.5 billion; \$42.5 billion

d. What was the amount spent on automotive TV spot advertising in 1990? In 2000? \$0.3 billion; \$20.3 billion

e. What does the rational fraction

$$\frac{A(t)}{C(t)}$$

represent? The fraction of the total TV spot advertising that is spent for automotive spot advertising

SKILL CHECKER

Try the Skill Checker Exercises so you'll be ready for the next section.

Multiply:

75. $\frac{3}{2} \cdot \frac{4}{9}$ $\frac{2}{3}$

76. $\frac{3}{5} \cdot \frac{10}{9}$ $\frac{2}{3}$

Factor:

77. $x^2 + 2x - 3$
$(x + 3)(x - 1)$

78. $x^2 + 7x + 12$
$(x + 4)(x + 3)$

79. $x^2 - 7x + 10$
$(x - 5)(x - 2)$

80. $x^2 + 3x - 4$
$(x + 4)(x - 1)$

USING YOUR KNOWLEDGE

Ratios

There is an important relationship between fractions and *ratios.* In general, a **ratio** is a way of comparing two or more numbers. For example, if there are 10 workers in an office, 3 women and 7 men, the ratio of women to men is 3 to 7

$$\frac{3}{7} \quad \begin{matrix} \leftarrow \text{Number of women} \\ \leftarrow \text{Number of men} \end{matrix}$$

On the other hand, if there are 6 men and 4 women in the office, the **reduced ratio** of women to men is

$$\frac{4}{6} = \frac{2}{3} \quad \begin{matrix} \leftarrow \text{Number of women} \\ \leftarrow \text{Number of men} \end{matrix}$$

Use your knowledge to solve the following problems.

81. A class is composed of 40 men and 60 women. Find the reduced ratio of men to women. $\frac{2}{3}$

82. Do you know the teacher-to-student ratio in your school? Suppose your school has 10,000 students and 500 teachers.

a. Find the reduced teacher-to-student ratio. $\frac{1}{20}$

b. If the school wishes to maintain a $\frac{1}{20}$ ratio and the enrollment increases to 12,000 students, how many teachers are needed? 600

83. The transmission ratio in your automobile is defined by

$$\text{Transmission ratio} = \frac{\text{Engine speed}}{\text{Drive shaft speed}}$$

a. If the engine is running at 2000 revolutions per minute and the drive shaft speed is 500 revolutions per minute, what is the reduced transmission ratio? 4 to 1

b. If the transmission ratio of a car is 5 to 1, and the drive shaft speed is 500 revolutions per minute, what is the engine speed? 2500 rpm

We will examine ratios more closely in Section 6.6.

WRITE ON

84. Write the procedure you use to determine the values for which the rational expression

$$\frac{P(x)}{Q(x)}$$

is undefined. Answers may vary.

85. Consider the rational expression

$$\frac{1}{x^2 + a}$$

a. Is this expression always defined when a is positive? Explain. (*Hint:* Let $a = 1, 2$, and so on.)
Yes; answers may vary.

b. Is this expression always defined when a is negative? Explain. (*Hint:* Let $a = -1, -2$, and so on.)
No; answers may vary.

86. Write the procedure you use to reduce a fraction to lowest terms. Answers may vary.

87. If

$$\frac{P(x)}{Q(x)}$$

is equal to -1, what is the relationship between $P(x)$ and $Q(x)$? $P(x) = -Q(x)$

MASTERY TEST

If you know how to do these problems, you have learned your lesson!

Reduce to lowest terms (simplify):

88. $\frac{x^2 - 9}{3 + x}$ $x - 3$

89. $\frac{x^2 - 9}{x + 3}$ $x - 3$

90. $\frac{x^2 - 3x - 10}{5 - x}$ $-(x + 2)$ or $-x - 2$

91. $\frac{10x - 15y}{4x - 6y}$ $\frac{5}{2}$

92. $-\frac{x}{xy + x}$ $\frac{-1}{y + 1}$

93. $\frac{x + 4}{-(x^2 - 16)}$ $\frac{-1}{x - 4}$ or $\frac{1}{4 - x}$

94. $\frac{-3xy^2}{12x^2y}$ $\frac{-y}{4x}$

95. $\frac{-6(x^2 - y^2)}{-3(x - y)}$ $2(x + y)$ or $2x + 2y$

Write:

96. $\frac{7}{8}$ with a denominator of 16 $\frac{14}{16}$

97. $\frac{3x}{8y^2}$ with a denominator of $24y^3$ $\frac{9xy}{24y^3}$

98. $\frac{4x}{x + 2}$ with a denominator of $x^2 - x - 6$ $\frac{4x(x - 3)}{x^2 - x - 6}$ or $\frac{4x^2 - 12x}{x^2 - x - 6}$

99. Find the values for which $\frac{x^2 + 1}{x^2 - 4}$ is undefined. $x = 2$ or $x = -2$

100. Find the values for which $\frac{x + 4}{x^2 - 6x + 8}$ is undefined. $x = 4$ or $x = 2$

6.2 MULTIPLICATION AND DIVISION OF RATIONAL EXPRESSIONS

To Succeed, Review How To . . .

1. Multiply, divide, and reduce fractions (pp. 12–21).
2. Factor trinomials (pp. 399–403, 406–414, 417–419).
3. Factor the difference of two squares (pp. 420–421).

Objectives

A Multiply two rational expressions.

B Divide one rational expression by another.

GETTING STARTED

Gearing for Multiplication

How fast can the last (smallest) gear in this compound gear train go? It depends on the speed of the first gear, and the number of teeth in all the gears! The formula that tells us the number of revolutions per minute (rpm) the last gear can turn is

$$\text{rpm} = \frac{T_1}{t_1} \cdot \frac{T_2}{t_2} \cdot R$$

where T_1 and T_2 are the numbers of teeth in the driving gears, t_1 and t_2 are the numbers of teeth in the driven gears, and R is the number of revolutions per minute the first driving gear is turning. Many useful formulas require that we know how to multiply and divide rational expressions, and we shall learn how to do so in this section.

Multiplying Rational Expressions

Can you simplify this expression?

$$\frac{T_1}{t_1} \cdot \frac{T_2}{t_2} \cdot R$$

Web It

For practice multiplying rational expressions, go to link 6-2-1 on the Bello Website at mhhe.com/bello.

You can even see a video tutorial if you select Audio/Visual Tutorial and Practice.

Of course you can, if you remember how to multiply fractions in arithmetic.* As you recall, in arithmetic the product of two fractions is another fraction whose numerator is the product of the original numerators and whose denominator is the product of the original denominators. Here's how we state this rule in symbols.

RULE

Multiplying Rational Expressions

$$\frac{A}{B} \cdot \frac{C}{D} = \frac{AC}{BD}, \qquad B \neq 0, D \neq 0$$

Thus the formula in the *Getting Started* can be simplified to

$$\text{rpm} = \frac{T_1T_2R}{t_1t_2}$$

Web It

If you need to review fractions, go to link 6-2-2 on the Bello Website at mhhe.com/bello.

If we assume that $R = 12$ and then count the teeth in the gears, we get $T_1 = 48$, $T_2 = 40$, $t_1 = 24$, and $t_2 = 24$. To find the revolutions per minute, we write

$$\text{rpm} = \frac{48}{24} \cdot \frac{40}{24} \cdot \frac{12}{1}$$

Then we have the following:

1. Reduce each fraction. $\quad \frac{\overset{2}{\cancel{48}}}{\underset{1}{\cancel{24}}} \cdot \frac{\overset{5}{\cancel{40}}}{\underset{3}{\cancel{24}}} \cdot \frac{12}{1} = \frac{2}{1} \cdot \frac{5}{3} \cdot \frac{12}{1}$
2. Multiply the numerators. $\quad \frac{120}{1 \cdot 3 \cdot 1}$
3. Multiply the denominators. $\quad \frac{120}{3}$
4. Reduce the answer. $\quad 40$

Thus the speed of the final gear is 40 revolutions per minute. Here is what we have done.

PROCEDURE

Multiplying Rational Expressions

1. Reduce each expression if possible.
2. Multiply the numerators to obtain the new numerator.
3. Multiply the denominators to obtain the new denominator.
4. Reduce the answer if possible.

---

*Multiplication and division of arithmetic fractions is covered in Section R.2.

Note that you could also write

$$\frac{2}{1} \cdot \frac{5}{\not{3}_1} \cdot \frac{\not{12}^4}{1}$$

and obtain $2 \cdot 5 \cdot 4 = 40$ as before.

EXAMPLE 1 **Multiplying rational expressions**

Multiply:

a. $\frac{x}{6} \cdot \frac{7}{y}$ **b.** $\frac{3x^2}{2} \cdot \frac{4y}{9x}$

SOLUTION

a. $\frac{x}{6} \cdot \frac{7}{y} = \frac{7x}{6y}$ ⟵ Multiply numerators. ⟵ Multiply denominators.

b. $\frac{3x^2}{2} \cdot \frac{4y}{9x} = \frac{12x^2y}{18x}$ ⟵ Multiply numerators. ⟵ Multiply denominators.

$= \frac{\overset{2x}{\not{12}\not{x^2}y}}{\underset{3}{\not{18}\not{x}}}$ Reduce the answer.

$= \frac{2xy}{3}$

PROBLEM 1

Multiply:

a. $\frac{m}{4} \cdot \frac{5}{n}$ **b.** $\frac{5y^3}{2} \cdot \frac{4x}{15y}$

The procedure used to find the product in Example 1(b) can be shortened if we do some reduction beforehand. Thus we can write

$$\frac{\overset{1x}{\not{3}\not{x^2}}}{\underset{1}{\not{2}}} \cdot \frac{\overset{2}{\not{4}y}}{\underset{3}{\not{9}\not{x}}} = \frac{2xy}{3}$$

We use this idea in the next example.

EXAMPLE 2 **Multiplying rational expressions involving signed numbers**

Multiply:

a. $\frac{-6x}{7y^2} \cdot \frac{14y}{12x^2}$ **b.** $8y^2 \cdot \frac{9x}{16y^2}$

SOLUTION

a. Since $\frac{14y}{12x^2} = \frac{7y}{6x^2}$, we write $\frac{7y}{6x^2}$ instead of $\frac{14y}{12x^2}$:

$$\frac{-6x}{7y^2} \cdot \frac{14y}{12x^2} = \frac{\overset{-1}{-\not{6}\not{x}}}{\underset{1y}{\not{7}\not{y^2}}} \cdot \frac{\overset{1}{\not{7}\not{y}}}{\underset{1x}{\not{6}\not{x^2}}}$$

$$= \frac{-1}{xy}$$

PROBLEM 2

Multiply:

a. $\frac{-7m}{5n^2} \cdot \frac{15n}{21m^2}$ **b.** $5x^2 \cdot \frac{7y}{10x^2}$

Answers

1. a. $\frac{5m}{4n}$ **b.** $\frac{2xy^2}{3}$ **2. a.** $\frac{-1}{mn}$ **b.** $\frac{7y}{2}$

b. Since $8y^2 = \frac{8y^2}{1}$, we have

$$8y^2 \cdot \frac{9x}{16y^2} = \frac{\overset{1}{\cancel{8y^2}}}{1} \cdot \frac{9x}{\underset{2}{\cancel{16y^2}}} = \frac{9x}{2}$$

Can all problems be done as in these examples? Yes, but note the following.

> **NOTE**
>
> When the numerators and denominators involved are binomials or trinomials, it isn't easy to do the reductions we just did in the examples *unless* the numerators and denominators involved are *factored.*

Thus to multiply

$$\frac{x^2 + 2x - 3}{x^2 + 7x + 12} \cdot \frac{x + 4}{x + 5}$$

we *first* factor and then multiply. Thus

Factor first.

$$\frac{x^2 + 2x - 3}{x^2 + 7x + 12} \cdot \frac{x + 4}{x + 5} = \frac{(x - 1)\overset{1}{\cancel{(x + 3)}}}{\underset{1}{\cancel{(x + 4)}}\underset{1}{\cancel{(x + 3)}}} \cdot \frac{\overset{1}{\cancel{(x + 4)}}}{(x + 5)} = \frac{x - 1}{x + 5}$$

Thus when multiplying fractions involving trinomials, we factor the trinomials and reduce the answer if possible. But note the following.

> **NOTE**
>
> Only **factors** can be divided out (canceled), never terms. Thus
>
> $$\frac{x\overset{1}{\cancel{y}}}{\cancel{y}} = x$$
>
> but
>
> $$\frac{x + y}{y}$$
>
> cannot be reduced (simplified) further.

Teaching Tip

Students often try to reduce

$$\frac{x + y}{y}$$

by canceling terms like this:

$$\frac{x + \cancel{y}}{\cancel{y}} = x$$

Show a numerical example such as

$$\frac{4 + 2}{2} = \frac{6}{2} = 3$$

not

$$\frac{4 + \cancel{2}}{\cancel{2}} = 4$$

EXAMPLE 3 Factoring and multiplying rational expressions

Multiply:

a. $(x - 3) \cdot \frac{x + 5}{x^2 - 9}$

b. $\frac{x^2 - x - 20}{x - 1} \cdot \frac{1 - x}{x + 4}$

PROBLEM 3

Multiply:

a. $(m + 2) \cdot \frac{m + 3}{m^2 - 4}$

b. $\frac{y^2 - y - 12}{y - 2} \cdot \frac{2 - y}{y + 3}$

Answers

3. a. $\frac{m + 3}{m - 2}$ **b.** $4 - y$

SOLUTION

a. Since $(x - 3) = \dfrac{x - 3}{1}$ and $x^2 - 9 = (x + 3)(x - 3)$,

$$(x - 3) \cdot \frac{x + 5}{x^2 - 9} = \frac{\cancel{(x - 3)}}{1} \cdot \frac{x + 5}{(x + 3)\cancel{(x - 3)}}$$

$$= \frac{x + 5}{x + 3}$$

b. Since $x^2 - x - 20 = (x + 4)(x - 5)$ and $\dfrac{1 - x}{x - 1} = -1$,

$$\frac{x^2 - x - 20}{x - 1} \cdot \frac{1 - x}{x + 4} = \frac{\overset{1}{\cancel{(x + 4)}}(x - 5)}{\cancel{(x - 1)}} \cdot \frac{\overset{-1}{\cancel{(1 - x)}}}{\underset{1}{\cancel{x + 4}}}$$

Remember that $\dfrac{a - b}{b - a} = -1$.

$$= -1(x - 5)$$

$$= -x + 5$$

$$= 5 - x$$

Teaching Tip

Remind students to keep the parentheses around $(x - 5)$ when multiplying by the -1.

Yes		No
$-1(x - 5)$		$-1 \cdot x - 5$
$-x + 5$	$\neq$	$-x - 5$

B Dividing Rational Expressions

What about dividing rational expressions? We are in luck! The division of rational expressions uses the same rule as in arithmetic.

Teaching Tip

Remind students that

$$\frac{x - 2}{x - 1} \neq 2$$

Remind them to divide factors, not terms.

RULE

Dividing Rational Expressions

$$\frac{A}{B} \div \frac{C}{D} = \frac{A}{B} \cdot \frac{D}{C} = \frac{AD}{BC}, \quad B, C, \text{and } D \neq 0$$

Web It

For a good lesson covering the division of rational expressions, go to the Bello Website at mhhe.com/bello and try link 6-2-3.

Thus to divide $\frac{A}{B}$ by $\frac{C}{D}$, we simply invert $\frac{C}{D}$ (interchange the numerator and denominator) and multiply. That is, to divide $\frac{A}{B}$ by $\frac{C}{D}$, we multiply $\frac{A}{B}$ by the **reciprocal** (inverse) of $\frac{C}{D}$. For example, to divide

$$\frac{x + 4}{x - 5} \div \frac{x^2 + 3x - 4}{x^2 - 7x + 10}$$

we use the given rule and write

Invert.

$$\frac{x + 4}{x - 5} \div \frac{x^2 + 3x - 4}{x^2 - 7x + 10} = \frac{x + 4}{x - 5} \cdot \frac{x^2 - 7x + 10}{x^2 + 3x - 4}$$

$$= \frac{\overset{1}{\cancel{x + 4}}}{\underset{1}{\cancel{x - 5}}} \cdot \frac{\overset{1}{\cancel{(x - 5)}}(x - 2)}{\underset{1}{\cancel{(x + 4)}}(x - 1)}$$ First factor numerator and denominator.

$$= \frac{x - 2}{x - 1}$$ Reduce.

Here is another example.

EXAMPLE 4 Dividing rational expressions involving the difference of two squares

Divide:

a. $\frac{x^2 - 16}{x + 3} \div (x + 4)$ b. $\frac{x + 5}{x - 5} \div \frac{x^2 - 25}{5 - x}$

SOLUTION

a. Since $(x + 4) = \frac{(x + 4)}{1}$,

Invert.

$$\frac{x^2 - 16}{x + 3} \div \frac{(x + 4)}{1} = \frac{x^2 - 16}{x + 3} \cdot \frac{1}{(x + 4)}$$

$$= \frac{\overset{1}{\cancel{(x + 4)}}(x - 4)}{x + 3} \cdot \frac{1}{\underset{1}{\cancel{(x + 4)}}} \quad \text{Factor } x^2 - 16.$$

$$= \frac{x - 4}{x + 3} \quad \text{Reduce.}$$

Invert.

b. $$\frac{x + 5}{x - 5} \div \frac{x^2 - 25}{5 - x} = \frac{x + 5}{x - 5} \cdot \frac{5 - x}{x^2 - 25}$$

$$= \frac{\overset{1}{\cancel{x + 5}}}{\cancel{x - 5}} \cdot \frac{\overset{-1}{\cancel{5 - x}}}{\underset{1}{\cancel{(x + 5)}}(x - 5)} \quad \text{Factor } x^2 - 25 \text{ and note that } \frac{5 - x}{x - 5} = -1.$$

$$= \frac{-1}{x - 5} \quad \text{Reduce.}$$

Of course, this answer *can* be simplified further (to show fewer negative signs), since

$$\frac{-1}{x - 5} = \frac{(-1)(-1)}{(-1)(x - 5)} \quad \text{Multiply numerator and denominator by } (-1).$$

$$= \frac{1}{-x + 5} = \frac{1}{5 - x}$$

PROBLEM 4

Divide:

a. $\frac{y^2 - 9}{y + 5} \div (y - 3)$

b. $\frac{y + 4}{y - 4} \div \frac{y^2 - 16}{4 - y}$

Here's another example.

EXAMPLE 5 Factoring and dividing rational expressions

Divide:

a. $\frac{x^2 + 5x + 4}{x^2 - 2x - 3} \div \frac{x^2 - 4}{x^2 - 6x + 8}$ b. $\frac{x^2 - 1}{x^2 + x - 6} \div \frac{x^2 - 4x + 3}{x^2 - 4}$

SOLUTION

Invert.

a. $$\frac{x^2 + 5x + 4}{x^2 - 2x - 3} \div \frac{x^2 - 4}{x^2 - 6x + 8} = \frac{x^2 + 5x + 4}{x^2 - 2x - 3} \cdot \frac{x^2 - 6x + 8}{x^2 - 4}$$

$$= \frac{(x + 4)\overset{1}{\cancel{(x + 1)}}}{(x - 3)\underset{1}{\cancel{(x + 1)}}} \cdot \frac{(x - 4)\overset{1}{\cancel{(x - 2)}}}{(x + 2)\underset{1}{\cancel{(x - 2)}}} \quad \text{Factor.}$$

PROBLEM 5

Divide:

a. $\frac{y^2 - 3y + 2}{y^2 - 4y + 3} \div \frac{y^2 - 49}{y^2 - 5y - 14}$

b. $\frac{y^2 - 1}{y^2 - y - 6} \div \frac{y^2 - 3y + 2}{y^2 - 9}$

Answers

4. a. $\frac{y + 3}{y + 5}$ **b.** $\frac{-1}{y - 4} = \frac{1}{4 - y}$

Note: $\frac{1}{4 - y}$ is preferred!

5. a. $\frac{y^2 - 4}{y^2 + 4y - 21}$ **b.** $\frac{y^2 + 4y + 3}{y^2 - 4}$

$$= \frac{(x + 4)(x - 4)}{(x - 3)(x + 2)}$$

$$= \frac{x^2 - 16}{x^2 - x - 6}$$

Invert.

b. $$\frac{x^2 - 1}{x^2 + x - 6} \div \frac{x^2 - 4x + 3}{x^2 - 4} = \frac{x^2 - 1}{x^2 + x - 6} \cdot \frac{x^2 - 4}{x^2 - 4x + 3}$$

$$= \frac{(x + 1)\overset{1}{\cancel{(x - 1)}}}{(x + 3)\underset{1}{\cancel{(x - 2)}}} \cdot \frac{(x + 2)\overset{1}{\cancel{(x - 2)}}}{\underset{1}{\cancel{(x - 1)}}(x - 3)} \quad \text{Factor.}$$

$$= \frac{(x + 1)(x + 2)}{(x + 3)(x - 3)}$$

$$= \frac{x^2 + 3x + 2}{x^2 - 9}$$

Teaching Tip

Note that

$$\frac{(x + 1)(x + 2)}{(x + 3)(x - 3)}$$

is *not* the final answer. You must multiply the factors to obtain the final answer.

A final word of warning! Be very careful when you reduce fractions.

CAUTION

You may cancel factors, but you *must not* cancel terms.

Thus

$$\frac{\cancel{x^2} + 5x + 6}{\cancel{x^2} + 8x + 15} = \frac{5x + 6}{8x + 15} \quad \text{is wrong!}$$

Note that x^2 is a term, *not* a factor. The correct way is to *factor* first and then cancel. Thus

$$\frac{x^2 + 5x + 6}{x^2 + 8x + 15} = \frac{\cancel{(x + 3)}(x + 2)}{\cancel{(x + 3)}(x + 5)} = \frac{x + 2}{x + 5}$$

Of course,

$$\frac{x + 2}{x + 5}$$

cannot be reduced further. To write

$$\frac{x + 2}{x + 5} = \frac{2}{5} \quad \text{is wrong!}$$

Again, x is a term, not a factor. (If you had

$$\frac{2x}{5x} = \frac{2}{5}$$

that would be correct. In the expressions $2x$ and $5x$, x is a *factor* that may be canceled.) Why is

$$\frac{x + 2}{x + 5} \neq \frac{2}{5}?$$

Try it when x is 4.

$$\frac{x + 2}{x + 5} = \frac{4 + 2}{4 + 5} = \frac{6}{9} = \frac{2}{3}$$

Thus the answer *cannot* be $\frac{2}{5}$!

Exercises 6.2

A In Problems 1–30, multiply and simplify.

1. $\frac{x}{3} \cdot \frac{8}{y}$ $\frac{8x}{3y}$

2. $\frac{-x}{4} \cdot \frac{7}{y}$ $\frac{-7x}{4y}$

3. $\frac{-6x^2}{7} \cdot \frac{14y}{9x}$ $\frac{-4xy}{3}$

4. $\frac{-5x^3}{6y^2} \cdot \frac{18y}{-10x}$ $\frac{3x^2}{2y}$

5. $7x^2 \cdot \frac{3y}{14x^2}$ $\frac{3y}{2}$

6. $11y^2 \cdot \frac{4x}{33y}$ $\frac{4xy}{3}$

7. $\frac{-4y}{7x^2} \cdot 14x^3$ $-8xy$

8. $\frac{-3y^3}{8x^2} \cdot -16y$ $\frac{6y^4}{x^2}$

9. $(x - 7) \cdot \frac{x + 1}{x^2 - 49}$ $\frac{x + 1}{x + 7}$

10. $3(x + 1) \cdot \frac{x + 2}{x^2 - 1}$ $\frac{3x + 6}{x - 1}$

11. $-2(x + 2) \cdot \frac{x - 1}{x^2 - 4}$ $\frac{2 - 2x}{x - 2}$

12. $-3(x - 1) \cdot \frac{x - 2}{x^2 - 1}$ $\frac{6 - 3x}{x + 1}$

13. $\frac{3}{x - 5} \cdot \frac{x^2 - 25}{x + 1}$ $\frac{3x + 15}{x + 1}$

14. $\frac{1}{x - 2} \cdot \frac{x^2 - 4}{x - 1}$ $\frac{x + 2}{x - 1}$

15. $\frac{x^2 - x - 6}{x - 2} \cdot \frac{2 - x}{x - 3}$ $-(x + 2)$ or $-x - 2$

16. $\frac{x^2 + 3x - 4}{3 - x} \cdot \frac{x - 3}{x + 4}$ $-(x - 1)$ or $1 - x$

17. $\frac{x - 1}{3 - x} \cdot \frac{x + 3}{1 - x}$ $\frac{x + 3}{x - 3}$

18. $\frac{2x - 1}{5 - 3x} \cdot \frac{3x - 5}{1 - 2x}$ 1

19. $\frac{3(x - 5)}{14(4 - x)} \cdot \frac{7(x - 4)}{6(5 - x)}$ $\frac{1}{4}$

20. $\frac{7(1 - x)}{10(x - 5)} \cdot \frac{5(5 - x)}{14(x - 1)}$ $\frac{1}{4}$

21. $\frac{6x^3}{x^2 - 16} \cdot \frac{x^2 - 5x + 4}{3x^2}$ $\frac{2x^2 - 2x}{x + 4}$

22. $\frac{3a^4}{a^2 - 4} \cdot \frac{a^2 - a - 2}{9a^3}$ $\frac{a^2 + a}{3a + 6}$

23. $\frac{y^2 + 2y - 3}{y - 5} \cdot \frac{y^2 - 3y - 10}{y^2 + 5y - 6}$ $\frac{y^2 + 5y + 6}{y + 6}$

24. $\frac{f^2 + 2f - 8}{f^2 + 7f + 12} \cdot \frac{f^2 + 2f - 3}{f^2 - 3f + 2}$ 1

25. $\frac{2y^2 + y - 3}{6 - 11y - 10y^2} \cdot \frac{5y^3 - 2y^2}{3y^2 - 5y + 2}$ $\frac{y^2}{2 - 3y}$

26. $\frac{3x^2 - x - 2}{2 - x - 6x^2} \cdot \frac{2x^4 - x^3}{3x^2 - 2x - 1}$ $\frac{-x^3}{3x + 1}$

27. $\frac{15x^2 - x - 2}{2x^2 + 5x - 18} \cdot \frac{2x^2 + x - 36}{3x^2 - 11x - 4}$ $\frac{5x - 2}{x - 2}$

28. $\frac{6x^2 + x - 1}{3x^2 + 5x + 2} \cdot \frac{3x^2 - x - 2}{2x^2 - x - 1}$ $\frac{3x - 1}{x + 1}$

29. $\frac{27y^3 + 8}{6y^2 + 19y + 10} \cdot \frac{4y^2 - 25}{9y^2 - 6y + 4}$ $2y - 5$

30. $\frac{8y^3 + 27}{10y^2 + 19y + 6} \cdot \frac{25y^2 - 4}{4y^2 - 6y + 9}$ $5y - 2$

B In Problems 31–64, divide and simplify.

31. $\frac{x^2 - 1}{x + 2} \div (x + 1)$ $\frac{x - 1}{x + 2}$

32. $\frac{x^2 - 4}{x - 3} \div (x + 2)$ $\frac{x - 2}{x - 3}$

33. $\frac{x^2 - 25}{x - 3} \div 5(x + 5)$ $\frac{x - 5}{5x - 15}$

34. $\frac{x^2 - 16}{8(x - 3)} \div 4(x + 4)$ $\frac{x - 4}{32x - 96}$

35. $(x + 3) \div \frac{x^2 - 9}{x + 4}$ $\frac{x + 4}{x - 3}$

36. $4(x - 4) \div \frac{8(x^2 - 16)}{5}$ $\frac{5}{2x + 8}$

37. $\frac{-3}{x-4} \div \frac{6(x+3)}{5(x^2-16)}$ $\frac{-5x-20}{2x+6}$

38. $\frac{-6}{x-2} \div \frac{3(x-1)}{7(x^2-4)}$ $\frac{14x+28}{1-x}$

39. $\frac{-4(x+1)}{3(x+2)} \div \frac{-8(x^2-1)}{6(x^2-4)}$ $\frac{x-2}{x-1}$

40. $\frac{-10(x^2-1)}{6(x^2-4)} \div \frac{5(x+1)}{-3(x+2)}$ $\frac{x-1}{x-2}$

41. $\frac{x+3}{x-3} \div \frac{x^2-1}{3-x}$ $\frac{x+3}{1-x^2}$

42. $\frac{4-x}{x+1} \div \frac{x-4}{x^2-1}$ $1-x$

43. $\frac{x^2-4}{7(x^2-9)} \div \frac{x+2}{14(x+3)}$ $\frac{2x-4}{x-3}$

44. $\frac{x^2-25}{3(x^2-1)} \div \frac{5-x}{6(x+1)}$ $\frac{2x+10}{1-x}$

45. $\frac{3(x^2-36)}{14(5-x)} \div \frac{6(6-x)}{7(x^2-25)}$ $\frac{x^2+11x+30}{4}$

46. $\frac{6(x^2-1)}{35(x^2-4)} \div \frac{12(1-x)}{7(2-x)}$ $\frac{x+1}{10x+20}$

47. $\frac{x+2}{x-1} \div \frac{x^2+5x+6}{x^2-4x+4}$ $\frac{x^2-4x+4}{x^2+2x-3}$

48. $\frac{x-3}{x+2} \div \frac{x^2-4x+3}{x^2-x-6}$ $\frac{x-3}{x-1}$

49. $\frac{x-5}{x+3} \div \frac{5(x-5)}{x^2+9x+18}$ $\frac{x+6}{5}$

50. $\frac{x-3}{x+4} \div \frac{2(x-3)}{x^2+2x-8}$ $\frac{x-2}{2}$

51. $\frac{x^2+2x-3}{x-5} \div \frac{x^2+6x+9}{x^2-2x-15}$ $x-1$

52. $\frac{x^2-3x+2}{x^2-5x+6} \div \frac{x^2-5x+4}{x^2-7x+12}$ 1

53. $\frac{x^2-1}{x^2+3x-10} \div \frac{x^2-3x-4}{x^2-25}$ $\frac{x^2-6x+5}{x^2-6x+8}$

54. $\frac{x^2-4x-21}{x^2-10x+25} \div \frac{x^2+2x-3}{x^2-6x+5}$ $\frac{x-7}{x-5}$

55. $\frac{x^2+3x-4}{x^2+7x+12} \div \frac{x^2+x-2}{x^2+5x+6}$ 1

56. $\frac{x^2+x-2}{x^2+6x-7} \div \frac{x^2-3x-10}{x^2+5x-14}$ $\frac{x-2}{x-5}$

57. $\frac{x^2-y^2}{x^2-2xy} \div \frac{x^2+xy-2y^2}{x^2-4y^2}$ $\frac{x+y}{x}$

58. $\frac{x^2+xy-2y^2}{x^2-4y^2} \div \frac{x^2-y^2}{x^2-2xy}$ $\frac{x}{x+y}$

59. $\frac{x^2+2xy-3y^2}{y^2-7y+10} \div \frac{x^2+5xy-6y^2}{y^2-3y-10}$ $\frac{xy+2x+6y+3y^2}{xy-2x-12y+6y^2}$

60. $\frac{x^2+2xy-8y^2}{x^2+7xy+12y^2} \div \frac{x^2-3xy+2y^2}{x^2+2xy-3y^2}$ 1

61. $\frac{2x^2-x-28}{3x^2-x-2} \div \frac{4x^2+16x+7}{3x^2+11x+6}$ $\frac{x^2-x-12}{2x^2-x-1}$

62. $\frac{15x^2-x-2}{2x^2+5x-18} \div \frac{3x^2-11x-4}{2x^2+x-36}$ $\frac{5x-2}{x-2}$

63. $\frac{(a^3-27)(a^2-9)}{(a-3)^2(a+3)^3} \div \frac{a^2+3a+9}{a^2+3a}$ $\frac{a}{a+3}$

64. $\frac{(y^3-8)(y^2-4)}{(y+2)^2(y-2)^3} \div \frac{y^2+2y+4}{y^2-2y}$ $\frac{y}{y-2}$

SKILL CHECKER

Try the Skill Checker Exercises so you'll be ready for the next section.

Add:

65. $\frac{7}{8} + \frac{2}{5}$ $\frac{51}{40}$

66. $\frac{7}{12} + \frac{1}{18}$ $\frac{23}{36}$

Subtract:

67. $\frac{7}{8} - \frac{2}{5}$ $\frac{19}{40}$

68. $\frac{7}{12} - \frac{1}{18}$ $\frac{19}{36}$

69. $\frac{5}{2} - \frac{1}{6}$ $\frac{7}{3}$

70. $\frac{8}{3} - \frac{3}{4}$ $\frac{23}{12}$

USING YOUR KNOWLEDGE

Resistance, Molecules, and Reordering

71. In the study of parallel resistors, the expression

$$R \cdot \frac{R_T}{R - R_T}$$

occurs, where R is a known resistance and R_T a required one. Do the multiplication. $\frac{RR_T}{R-R_T}$

72. The molecular model predicts that the pressure of a gas is given by

$$\frac{2}{3} \cdot \frac{mv^2}{2} \cdot \frac{N}{v}$$

where m, v, and N represent the mass, velocity, and total number of molecules, respectively. Do the multiplication. $\frac{2mv^2N}{6v} = \frac{mvN}{3}$

73. Suppose a store orders 3000 items each year. If it orders x units at a time, the number N of reorders is

$$N = \frac{3000}{x}$$

If there is a fixed \$20 reorder fee and a \$3 charge per item, the cost of each order is

$$C = 20 + 3x$$

The yearly reorder cost C_R is then given by

$$C_R = N \cdot C$$

Find C_R. $C_R = \frac{60{,}000 + 9000x}{x}$

WRITE ON

74. Write the procedure you use to multiply two rational expressions. Answers may vary.

75. Write the procedure you use to divide one rational expression by another. Answers may vary.

76. Explain why you cannot "cancel" the x's in

$$\frac{x+5}{x+2}$$

to obtain an answer of $\frac{5}{2}$ but you can cancel the x's in $\frac{5x}{2x}$.
The x's are terms in the first case and factors in the second case.

77. Explain what the statement "you can cancel factors but you cannot cancel terms" means and give examples. Answers may vary.

MASTERY TEST

If you know how to do these problems, you have learned your lesson!

Divide and simplify:

78. $\frac{x^2 - 3x + 2}{x^2 - 4x + 3} \div \frac{x^2 - 49}{x^2 + 5x - 14}$ $\quad \frac{x^2 - 4x + 4}{x^2 - 10x + 21}$

79. $\frac{x^2 - 1}{x^2 - x - 6} \div \frac{x^2 - 3x + 2}{x^2 - 9}$ $\quad \frac{x^2 + 4x + 3}{x^2 - 4}$

80. $\frac{x^2 - 25}{x + 1} \div (x + 5)$ $\quad \frac{x - 5}{x + 1}$

81. $\frac{x + 6}{x - 6} \div \frac{x^2 - 36}{6 - x}$ $\quad \frac{1}{6 - x}$

82. $\frac{x^3 + 8}{x - 2} \div \frac{x^2 - 2x + 4}{x^2 - 4}$ $\quad x^2 + 4x + 4$

83. $\frac{x^3 - 27}{x + 3} \div \frac{x^2 + 3x + 9}{x^2 - 9}$ $\quad x^2 - 6x + 9$

Multiply and simplify:

84. $(x - 4) \cdot \frac{x + 8}{x^2 - 16}$ $\quad \frac{x + 8}{x + 4}$

85. $\frac{x^2 + x - 6}{x - 3} \cdot \frac{3 - x}{x + 3}$ $\quad 2 - x$

86. $\frac{-3x}{4y^2} \cdot \frac{18y}{12x^2}$ $\quad \frac{-9}{8xy}$

87. $\frac{6x}{11y^2} \cdot 22y^2$ $\quad 12x$

88. $\frac{x^3 - 64}{x + 1} \cdot \frac{x^2 - 1}{x^2 + 4x + 16}$ $\quad x^2 - 5x + 4$

89. $\frac{2x^2 - x - 28}{3x^2 - x - 2} \cdot \frac{3x^2 + 11x + 6}{4x^2 + 16x + 7}$ $\quad \frac{x^2 - x - 12}{2x^2 - x - 1}$

6.3 ADDITION AND SUBTRACTION OF RATIONAL EXPRESSIONS

To Succeed, Review How To . . .

1. Find the LCD of two or more fractions (pp. 16–17).
2. Add and subtract fractions (pp. 12–21).

Objectives

A Add and subtract rational expressions with the same denominator.

B Add and subtract rational expressions with different denominators.

C Solve an application.

GETTING STARTED

Tennis, Anyone?

The racket hits the ball with such tremendous force that the ball is distorted. Can we find out how much force? The answer is

$$\frac{mv}{t} - \frac{mv_0}{t}$$

where

m = mass of ball
v = velocity of racket
v_0 = "initial" velocity of racket
t = time of contact

Since the expressions involved have the same denominator, subtracting them is easy. As in arithmetic, we simply subtract the numerators and keep the same denominator. Thus

$$\frac{mv}{t} - \frac{mv_0}{t} = \frac{mv - mv_0}{t}$$ ← Subtract numerators. ← Keep the denominator.

In this section we shall learn how to add and subtract rational expressions.

A Adding and Subtracting Expressions with the Same Denominator

As you recall from Section R.2,* $\frac{1}{5} + \frac{2}{5} = \frac{3}{5}$, $\frac{1}{7} + \frac{4}{7} = \frac{5}{7}$, and $\frac{1}{11} + \frac{8}{11} = \frac{9}{11}$. The same procedure works for rational expressions. For example,

$$\frac{3}{x} + \frac{5}{x} = \frac{3+5}{x} = \frac{8}{x}$$ ← Add numerators. ← Keep the denominator.

Similarly,

$$\frac{5}{x+1} + \frac{2}{x+1} = \frac{5+2}{x+1} = \frac{7}{x+1}$$ ← Add numerators. ← Keep the denominator.

and

$$\frac{5}{7(x-1)} + \frac{2}{7(x-1)} = \frac{5+2}{7(x-1)} = \frac{\overset{1}{\cancel{7}}}{\underset{1}{\cancel{7}}(x-1)} = \frac{1}{x-1}$$

Web It

To further review addition and subtraction of fractions, go to links 6-3-1 and 6-3-2 on the Bello Website at mhhe.com/bello.

If you are ready to practice with the addition and subtraction of rational expressions, go to the tutorial on link 6-3-3.

---

*The addition and subtraction of arithmetic fractions is covered in Section R.2.

For subtraction,

$$\frac{8}{x+5} - \frac{2}{x+5} = \frac{8-2}{x+5} = \frac{6}{x+5}$$ ← Subtract numerators. ← Keep the denominator.

and

$$\frac{8}{9(x-3)} - \frac{2}{9(x-3)} = \frac{8-2}{9(x-3)} = \frac{\overset{2}{\cancel{6}}}{\underset{3}{\cancel{9}}(x-3)} = \frac{2}{3(x-3)}$$

EXAMPLE 1 **Adding and subtracting rational expressions: Same denominator**

Find:

a. $\frac{8}{3(x-2)} + \frac{1}{3(x-2)}$ b. $\frac{7}{5(x+4)} - \frac{2}{5(x+4)}$

SOLUTION

a. $\frac{8}{3(x-2)} + \frac{1}{3(x-2)} = \frac{8+1}{3(x-2)} = \frac{\overset{3}{\cancel{9}}}{\underset{1}{\cancel{3}}(x-2)} = \frac{3}{x-2}$ Remember to reduce the answer.

b. $\frac{7}{5(x+4)} - \frac{2}{5(x+4)} = \frac{7-2}{5(x+4)} = \frac{\overset{1}{\cancel{5}}}{\underset{1}{\cancel{5}}(x+4)} = \frac{1}{x+4}$

PROBLEM 1

Find:

a. $\frac{4}{5(y-1)} + \frac{1}{5(y-1)}$

b. $\frac{9}{7(y+2)} - \frac{2}{7(y+2)}$

B Adding and Subtracting Expressions with Different Denominators

Not all rational expressions have the same denominator. To add or subtract rational expressions with different denominators, we again rely on our experiences in arithmetic. Let's practice adding rational numbers before we try rational expressions.

EXAMPLE 2 **Adding fractions: Different denominators**

Add: $\frac{7}{12} + \frac{5}{18}$

SOLUTION We first must find a common denominator—that is, a *multiple* of 12 and 18. Of course, it's more convenient to use the smallest one available. In general, the **lowest common denominator (LCD)** of two fractions is the smallest number that is a multiple of *both* denominators. To find the LCD, we can use successive divisions as in Section R.2.

2	12	18
3	6	9
	2	3 →

PROBLEM 2

Find: $\frac{5}{12} + \frac{7}{18}$

Answers

1. a. $\frac{1}{y-1}$ **b.** $\frac{1}{y+2}$ **2.** $\frac{29}{36}$

The LCD is $2 \times 3 \times 2 \times 3 = 36$. Better yet, we can also factor both numbers and write each factor in a column to obtain

Pick the number with the highest exponent.

$$12 = 2 \cdot 2 \cdot 3 \quad = 2^2 \cdot 3^1$$

$$18 = \quad 2 \cdot 3 \cdot 3 = 2^1 \cdot 3^2$$

Note that all the 2's and all the 3's are written in the *same* column.

Note that since we need a number that is a multiple of 12 and 18, we select the factors raised to the *highest* power in each column—that is, 2^2 and 3^2. The product of these factors is the LCD. Thus the LCD of 12 and 18 is $2^2 \cdot 3^2 = 4 \cdot 9 = 36$, as before. We then write each fraction with a denominator of 36 and add.

$$\frac{7}{12} = \frac{7 \cdot 3}{12 \cdot 3} = \frac{21}{36}$$

We multiply the denominator 12 by 3 (to get 36), so we do the same to the numerator.

$$\frac{5}{18} = \frac{5 \cdot 2}{18 \cdot 2} = \frac{10}{36}$$

Here we multiply the denominator 18 by 2 to get 36, so we do the same to the numerator.

$$\frac{7}{12} + \frac{5}{18} = \frac{21}{36} + \frac{10}{36} = \frac{31}{36}$$

Can you see how this one is done?

The procedure can also be written as

$$\begin{array}{r} \dfrac{7}{12} = \dfrac{21}{36} \\ +\dfrac{5}{18} = \dfrac{10}{36} \\ \hline \dfrac{31}{36} \end{array}$$

Teaching Tip

This vertical method is usually easier for students to use when adding or subtracting rational expressions.

NOTE

Make sure you know how to do this type of problem before you go on. The idea in adding and subtracting rational expressions is the same as the idea used with rational numbers. In fact, every problem in Exercises 6.3 has two parts: one with rational numbers and one with rational expressions. If you know how to do one, you should be able to do the other one.

Now, let's practice subtraction of fractions.

EXAMPLE 3 **Subtracting fractions: Different denominators**

Find: $\frac{11}{15} - \frac{5}{18}$

SOLUTION We can use successive divisions to find the LCD, writing

$$\begin{array}{r|rr} 3 & 15 & 18 \\ \hline & 5 & 6 \end{array}$$

PROBLEM 3

Find: $\frac{13}{15} - \frac{7}{12}$

Answer

3. $\frac{17}{60}$

The LCD is $3 \times 5 \times 6 = 90$. Better yet, we can factor the denominators and write them as follows:

$$15 = 3 \cdot 5 \quad = \quad 3 \cdot 5$$
$$18 = 2 \cdot 3 \cdot 3 = 2 \cdot 3^2$$

Note that all the 2's, all the 3's, and all the 5's are in separate columns.

As before, the LCD is

$$2 \cdot 3^2 \cdot 5 = 2 \cdot 9 \cdot 5 = 90$$

Then we write $\frac{11}{15}$ and $\frac{5}{18}$ as equivalent fractions with a denominator of 90.

$$\frac{11}{15} = \frac{11 \cdot 6}{15 \cdot 6} = \frac{66}{90}$$ Multiply the numerator and denominator by 6.

$$\frac{5}{18} = \frac{5 \cdot 5}{18 \cdot 5} = \frac{25}{90}$$ Multiply the numerator and denominator by 5.

and then subtract

$$\frac{11}{15} - \frac{5}{18} = \frac{66}{90} - \frac{25}{90} = \frac{66 - 25}{90} = \frac{41}{90}$$

Of course, it's possible that the denominators involved have no common factors. In this case, the LCD is the *product* of the denominators. Thus to add $\frac{3}{5}$ and $\frac{4}{7}$, we use $5 \cdot 7 = 35$ as the LCD and write

$$\frac{3}{5} = \frac{3 \cdot 7}{5 \cdot 7} = \frac{21}{35}$$ Multiply the numerator and denominator by 7.

$$\frac{4}{7} = \frac{4 \cdot 5}{7 \cdot 5} = \frac{20}{35}$$ Multiply the numerator and denominator by 5.

Thus

$$\frac{3}{5} + \frac{4}{7} = \frac{21}{35} + \frac{20}{35} = \frac{41}{35}$$

Similarly, the expression

$$\frac{4}{x} + \frac{5}{3}$$

has $3x$ as the LCD. We then write $\frac{4}{x}$ and $\frac{5}{3}$ as equivalent fractions with $3x$ as the denominator and add:

$$\frac{4}{x} = \frac{4 \cdot 3}{x \cdot 3} = \frac{12}{3x}$$ Multiply the numerator and denominator by 3.

$$\frac{5}{3} = \frac{5 \cdot x}{3 \cdot x} = \frac{5x}{3x}$$ Multiply the numerator and denominator by x.

Thus

$$\frac{4}{x} + \frac{5}{3} = \frac{12}{3x} + \frac{5x}{3x} = \frac{12 + 5x}{3x}$$

Teaching Tip

It might be easier for students to draw a line and add once they have built up the fractions rather than rewrite the problem horizontally. Example:

$$\frac{4}{x} = \frac{12}{3x}$$
$$\frac{5}{3} = \frac{5x}{3x}$$
$$\frac{12 + 5x}{3x}$$

EXAMPLE 4 Adding and subtracting rational expressions: Different denominators

Find:

a. $\frac{7}{8} + \frac{2}{x}$

b. $\frac{2}{x-1} - \frac{1}{x+2}$

SOLUTION

a. Since 8 and x don't have any common factors, the LCD is $8x$. We write $\frac{7}{8}$ and $\frac{2}{x}$ as equivalent fractions with $8x$ as denominator and add.

$$\frac{7}{8} = \frac{7 \cdot x}{8 \cdot x} = \frac{7x}{8x}$$

$$\frac{2}{x} = \frac{2 \cdot 8}{x \cdot 8} = \frac{16}{8x}$$

Thus

$$\frac{7}{8} + \frac{2}{x} = \frac{7x}{8x} + \frac{16}{8x}$$

$$= \frac{7x + 16}{8x}$$

Note that when the denominators A and B do not have any common factors, the common denominator becomes AB.

b. Since $(x - 1)$ and $(x + 2)$ don't have any common factors, the LCD of

$$\frac{2}{x-1} - \frac{1}{x+2}$$

is $(x - 1)(x + 2)$. We then write

$$\frac{2}{x-1} \quad \text{and} \quad \frac{1}{x+2}$$

as equivalent fractions with $(x - 1)(x + 2)$ as the denominator.

$$\frac{2}{x-1} = \frac{2 \cdot (x+2)}{(x-1)(x+2)}$$ Multiply numerator and denominator by $(x + 2)$.

$$\frac{1}{x+2} = \frac{1 \cdot (x-1)}{(x+2)(x-1)} = \frac{(x-1)}{(x-1)(x+2)}$$ Multiply numerator and denominator by $(x - 1)$.

Hence

$$\frac{2}{x-1} - \frac{1}{x+2} = \frac{2 \cdot (x+2)}{(x-1)(x+2)} - \frac{(x-1)}{(x-1)(x+2)}$$

$$= \frac{2(x+2) - (x-1)}{(x-1)(x+2)}$$ ← Subtract numerators. ← Keep denominator.

$$= \frac{2x + 4 - x + 1}{(x-1)(x+2)}$$ Remember that $-(x - 1) = -x + 1$.

$$= \frac{x+5}{(x-1)(x+2)}$$ ← Simplify numerator. ← Keep denominator.

PROBLEM 4

Find:

a. $\frac{3}{5} + \frac{2}{y}$

b. $\frac{2}{y-2} - \frac{1}{y+1}$

Answers

4. a. $\frac{3y + 10}{5y}$ **b.** $\frac{y+4}{(y-2)(y+1)}$

Are you ready for a more complicated problem? First, let's state the generalized procedure.

PROCEDURE

Adding (or Subtracting) Fractions with Different Denominators

1. Find the LCD.
2. Write all fractions as equivalent ones with the LCD as the denominator.
3. Add (or subtract) numerators and keep denominators.
4. Reduce if possible.

Teaching Tip

Use the letters in King "LEAR" to help students remember these four steps.

L	LCD	Find LCD.
E	Equivalent	Write equivalent fractions.
A	Add	Add numerators.
R	Reduce	Reduce, if possible.

Let's use these steps to add

$$\frac{x+1}{x^2+x-2}+\frac{x+3}{x^2-1}$$

1. We first find the LCD of the denominators. To do this, we factor the denominators.

$$x^2+x-2=(x+2)(x-1)$$
$$x^2-1=\quad(x-1)(x+1)$$
$$(x+2)(x-1)(x+1)\qquad \text{The LCD}$$

2. We then write

$$\frac{x+1}{x^2+x-2}\quad\text{and}\quad\frac{x+3}{x^2-1}$$

as equivalent fractions with $(x+2)(x-1)(x+1)$ as denominator.

$$\frac{x+1}{x^2+x-2}=\frac{x+1}{(x+2)(x-1)}=\frac{(x+1)(x+1)}{(x+2)(x-1)(x+1)}$$

$$\frac{x+3}{x^2-1}=\frac{x+3}{(x+1)(x-1)}=\frac{(x+3)(x+2)}{(x+1)(x-1)(x+2)}$$

$$=\frac{(x+3)(x+2)}{(x+2)(x-1)(x+1)}$$

3. Add the numerators and keep the denominator.

$$\frac{x+1}{x^2+x-2}+\frac{x+3}{x^2-1}=\frac{(x+1)(x+1)}{(x+2)(x-1)(x+1)}+\frac{(x+3)(x+2)}{(x+2)(x-1)(x+1)}$$

$$=\frac{(x^2+2x+1)+(x^2+5x+6)}{(x+2)(x-1)(x+1)}$$

$$=\frac{2x^2+7x+7}{(x+2)(x-1)(x+1)}$$

Note that the denominator is left as an indicated product.

4. The answer is not reducible, since there are no factors common to the numerator and denominator.

We use this procedure to subtract rational expressions in the next example.

EXAMPLE 5 **Subtracting rational expressions: Different denominators**

Subtract:

$$\frac{x-2}{x^2-x-6} - \frac{x+3}{x^2-9}$$

SOLUTION We use the four-step procedure.

1. To find the LCD, we factor the denominators to obtain

$$x^2-x-6 = (x-3)(x+2)$$
$$x^2-9 = (x+3)(x-3)$$
$$(x+3)(x-3)(x+2) \quad \text{The LCD}$$

2. We write each fraction as an equivalent one with the LCD as the denominator. Hence

$$\frac{x-2}{x^2-x-6} = \frac{(x-2)(x+3)}{(x-3)(x+2)(x+3)}$$

$$\frac{x+3}{x^2-9} = \frac{(x+3)(x+2)}{(x+3)(x-3)(x+2)}$$

3. $$\frac{x-2}{x^2-x-6} - \frac{x+3}{x^2-9} = \frac{(x-2)(x+3)}{(x+3)(x-3)(x+2)} - \frac{(x+3)(x+2)}{(x+3)(x-3)(x+2)}$$

$$= \frac{(x^2+x-6)-(x^2+5x+6)}{(x+3)(x-3)(x+2)}$$

$$= \frac{x^2+x-6-x^2-5x-6}{(x+3)(x-3)(x+2)}$$

Note that $-(x^2+5x+6) = x^2-5x-6$.

$$= \frac{-4x-12}{(x+3)(x-3)(x+2)}$$

$$= \frac{-4(x+3)}{(x+3)(x-3)(x+2)}$$

Factor the numerator and keep the denominator.

4. Reduce.

$$\frac{-4\cancel{(x+3)}}{\cancel{(x+3)}(x-3)(x+2)} = \frac{-4}{(x-3)(x+2)}$$

PROBLEM 5

Subtract:

$$\frac{x-3}{(x+1)(x-2)} - \frac{x+3}{x^2-4}$$

Teaching Tip

Let students know that although

$$\frac{-4}{(x-3)(x+2)} = \frac{-4}{x^2-x-6}$$

the preferred answer is

$$\frac{-4}{(x-3)(x+2)}$$

Answer

5. $\frac{-5x-9}{(x+1)(x+2)(x-2)}$

Solving an Application

EXAMPLE 6 Graduate engineering students

Do you want to be an engineer? The chart shows the total number (T) and number of female (F) graduate science/engineering students (in thousands). These numbers can be approximated by

$$T(t) = -0.4t^2 + 10t + 360$$

$$F(t) = -0.1t^2 + 6t + 125$$

where t is the number of years after 1985.

a. What fraction of the total number is female?

b. What fraction of the total number was female in 1985?

c. What percent of the total was female in 1985? (Answer to the nearest percent.)

d. What fraction of the total would you predict to be female in 2005?

e. Is the percent of female engineering graduates increasing?

SOLUTION

a. The fraction of the total number [$T(t)$] that is female [$F(t)$] is

$$\frac{F(t)}{T(t)} = \frac{-0.1t^2 + 6t + 125}{-0.4t^2 + 10t + 360}$$

b. Since t is the number of years after 1985, $t = 0$ in 1985. Substituting $t = 0$ in

$$\frac{F(t)}{T(t)} = \frac{-0.1t^2 + 6t + 125}{-0.4t^2 + 10t + 360}$$

we obtain

$$\frac{F(0)}{T(0)} = \frac{-0.1(0)^2 + 6(0) + 125}{-0.4(0)^2 + 10(0) + 360} = \frac{125}{360} = \frac{25}{72}$$

c. To convert $\frac{25}{72}$ to a percent, divide 25 by 72 and multiply by 100, obtaining 34.72 or 35% (to the nearest percent).

d. In 2005, $t = 2005 - 1985 = 20$. Substituting 20 in part **a** yields

$$\frac{F(20)}{T(20)} = \frac{-0.1(20)^2 + 6(20) + 125}{-0.4(20)^2 + 10(20) + 360} = \frac{205}{400} = \frac{41}{80} \text{ or about } 51\%$$

e. Yes, it is increasing from 35% in 1985 to 51% in 2005.

PROBLEM 6

a. What is the ratio of total students to female students?

b. According to part **e** of Example 6, the percent of female graduates is increasing, but these figures are only a model. Can this increasing trend continue indefinitely? *Hint:* What would be the theoretical percent of female students in 2020 ($t = 35$)? In 2021?

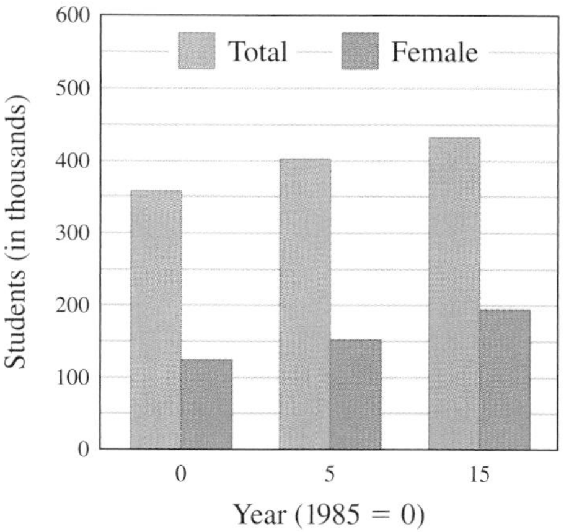

Source: National Science Foundation, *Statistical Abstract of the United States,* Table 784.

And now, a last word before you go on to the exercise set. At this point, you can see that there are great similarities between algebra and arithmetic. In fact, in this very section we use the arithmetic addition of fractions as a model to do the algebraic addition of rational expressions. To show that these similarities are very strong and also to give you more practice, Problems 1–20 in the exercise set consist of two similar problems, an arithmetic one and an algebraic one. Use the practice and experience gained in working one to do the other.

Answers

6. a. $\frac{-0.4t^2 + 10t + 360}{-0.1t^2 + 6t + 125}$

b. No. In 2020 the percent of female students would be 97%, and in 2021 it would be 105%.

Exercises 6.3

Boost *your* GRADE at mathzone.com!

- Practice Problems
- Self-Tests
- Videos
- NetTutor
- e-Professors

A B In Problems 1–40, perform the indicated operations.

1. a. $\frac{2}{7}+\frac{3}{7}$ $\frac{5}{7}$
b. $\frac{3}{x}+\frac{8}{x}$ $\frac{11}{x}$

2. a. $\frac{5}{9}+\frac{2}{9}$ $\frac{7}{9}$
b. $\frac{9}{x-1}+\frac{2}{x-1}$ $\frac{11}{x-1}$

3. a. $\frac{8}{9}-\frac{2}{9}$ $\frac{2}{3}$
b. $\frac{6}{x}-\frac{2}{x}$ $\frac{4}{x}$

4. a. $\frac{4}{7}-\frac{2}{7}$ $\frac{2}{7}$
b. $\frac{6}{x+4}-\frac{2}{x+4}$ $\frac{4}{x+4}$

5. a. $\frac{6}{7}+\frac{8}{7}$ 2
b. $\frac{3}{2x}+\frac{7}{2x}$ $\frac{5}{x}$

6. a. $\frac{1}{9}+\frac{2}{9}$ $\frac{1}{3}$
b. $\frac{3}{8(x-2)}+\frac{1}{8(x-2)}$ $\frac{1}{2(x-2)}$

7. a. $\frac{8}{3}-\frac{2}{3}$ 2
b. $\frac{11}{3(x+1)}-\frac{9}{3(x+1)}$ $\frac{2}{3(x+1)}$

8. a. $\frac{3}{8}-\frac{1}{8}$ $\frac{1}{4}$
b. $\frac{7}{15(x-1)}-\frac{2}{15(x-1)}$ $\frac{1}{3(x-1)}$

9. a. $\frac{8}{9}+\frac{4}{9}$ $\frac{4}{3}$
b. $\frac{7x}{4(x+1)}+\frac{3x}{4(x+1)}$ $\frac{5x}{2(x+1)}$

10. a. $\frac{15}{14}+\frac{3}{14}$ $\frac{9}{7}$
b. $\frac{29x}{15(x-3)}+\frac{4x}{15(x-3)}$ $\frac{11x}{5(x-3)}$

11. a. $\frac{3}{4}-\frac{1}{3}$ $\frac{5}{12}$
b. $\frac{7}{x}-\frac{3}{8}$ $\frac{56-3x}{8x}$

12. a. $\frac{5}{7}-\frac{2}{5}$ $\frac{11}{35}$
b. $\frac{x}{3}-\frac{7}{x}$ $\frac{x^2-21}{3x}$

13. a. $\frac{1}{5}+\frac{1}{7}$ $\frac{12}{35}$
b. $\frac{4}{x}+\frac{x}{9}$ $\frac{x^2+36}{9x}$

14. a. $\frac{1}{3}+\frac{1}{9}$ $\frac{4}{9}$
b. $\frac{5}{x}+\frac{6}{3x}$ $\frac{7}{x}$

15. a. $\frac{2}{5}-\frac{4}{15}$ $\frac{2}{15}$
b. $\frac{4}{7(x-1)}-\frac{3}{14(x-1)}$ $\frac{5}{14(x-1)}$

16. a. $\frac{9}{2}-\frac{5}{8}$ $\frac{31}{8}$
b. $\frac{8}{9(x+3)}-\frac{5}{36(x+3)}$ $\frac{3}{4(x+3)}$

17. a. $\frac{4}{7}+\frac{3}{8}$ $\frac{53}{56}$
b. $\frac{3}{x+1}+\frac{5}{x-2}$ $\frac{8x-1}{(x+1)(x-2)}$

18. a. $\frac{2}{9}+\frac{4}{5}$ $\frac{46}{45}$
b. $\frac{2x}{x+2}+\frac{3x}{x-4}$ $\frac{5x^2-2x}{(x+2)(x-4)}$

19. a. $\frac{7}{8}-\frac{1}{3}$ $\frac{13}{24}$
b. $\frac{6}{x-2}-\frac{3}{x+1}$ $\frac{3x+12}{(x-2)(x+1)}$

20. a. $\frac{6}{7}-\frac{2}{3}$ $\frac{4}{21}$
b. $\frac{4x}{x+1}-\frac{4x}{x+2}$ $\frac{4x}{(x+1)(x+2)}$

21. $\frac{x+1}{x^2+3x-4}+\frac{x+2}{x^2-16}$
$\frac{2x^2-2x-6}{(x+4)(x-4)(x-1)}$

22. $\frac{x-2}{x^2-9}+\frac{x+1}{x^2-x-12}$
$\frac{2x^2-8x+5}{(x+3)(x-3)(x-4)}$

23. $\frac{3x}{x^2+3x-10}+\frac{2x}{x^2+x-6}$
$\frac{5x^2+19x}{(x+5)(x-2)(x+3)}$

24. $\frac{x+3}{x^2-x-2}+\frac{x-1}{x^2+2x+1}$
$\frac{2x^2+x+5}{(x-2)(x+1)^2}$

25. $\frac{1}{x^2-y^2}+\frac{5}{(x+y)^2}$
$\frac{6x-4y}{(x+y)^2(x-y)}$

26. $\frac{3x}{(x+y)^2}+\frac{5x}{(x-y)}$
$\frac{5x^3+10x^2y+5xy^2+3x^2-3xy}{(x+y)^2(x-y)}$

27. $\dfrac{2}{x-5} - \dfrac{3x}{x^2-25}$ $\dfrac{10-x}{(x+5)(x-5)}$

28. $\dfrac{x+3}{x^2-x-2} - \dfrac{x-1}{x^2+2x+1}$ $\dfrac{7x+1}{(x-2)(x+1)^2}$

29. $\dfrac{x-1}{x^2+3x+2} - \dfrac{x+7}{x^2+5x+6}$ $\dfrac{-6x-10}{(x+2)(x+1)(x+3)}$

30. $\dfrac{2}{x^2+3xy+2y^2} - \dfrac{1}{x^2-xy-2y^2}$ $\dfrac{x-6y}{(x+2y)(x-2y)(x+y)}$

31. $\dfrac{y}{y^2-1} + \dfrac{y}{y+1}$ $\dfrac{y^2}{(y+1)(y-1)}$

32. $\dfrac{3y}{y^2-4} - \dfrac{y}{y+2}$ $\dfrac{5y-y^2}{(y+2)(y-2)}$

33. $\dfrac{3y+1}{y^2-16} - \dfrac{2y-1}{y-4}$ $\dfrac{5-4y-2y^2}{(y+4)(y-4)}$

34. $\dfrac{2y+1}{y^2-4} - \dfrac{3y-1}{y+2}$ $\dfrac{-3y^2+9y-1}{(y+2)(y-2)}$

35. $\dfrac{x+1}{x^2-x-2} + \dfrac{x-1}{x^2+2x+1}$ $\dfrac{2x^2-x+3}{(x-2)(x+1)^2}$

36. $\dfrac{y+3}{y^2+y-6} + \dfrac{y-2}{y^2+3y-10}$ $\dfrac{2y+3}{(y-2)(y+5)}$

37. $\dfrac{a}{a-w} - \dfrac{w}{a+w} - \dfrac{a^2+w^2}{a^2-w^2}$ 0

38. $\dfrac{x}{x-y} + \dfrac{y}{x+y} - \dfrac{x^2+y^2}{x^2-y^2}$ $\dfrac{2y}{x+y}$

39. $\dfrac{1}{a^3+8} + \dfrac{a+1}{a^2-2a+4}$ $\dfrac{a^2+3a+3}{(a+2)(a^2-2a+4)}$

40. $\dfrac{c}{c^3-1} + \dfrac{2}{c^2+c+1}$ $\dfrac{3c-2}{(c-1)(c^2+c+1)}$

APPLICATIONS

41. *Medicaid payments* According to data from the Social Security Administration, the number of Medicaid recipients (in millions) from 1980 to 1992 can be approximated by

$$R(t) = 0.01t^3 - 0.12t^2 + 0.28t + 22 \text{ (millions)}$$

The payments they received can be approximated by

$$P(t) = 0.04t^3 - 0.28t^2 + 3.2t + 23 \text{ (billions)}$$

a. If t represents the number of years after 1980, write a rational expression representing the average amount received by each recipient. (Note that the result will be in thousands of dollars.) $\dfrac{P(t)}{R(t)} = \dfrac{(0.04t^3 - 0.28t^2 + 3.2t + 23) \text{ billion}}{(0.01t^3 - 0.12t^2 + 0.28t + 22) \text{ million}}$

b. How much did each recipient get in 1980 ($t = 0$)? \$1045.45

c. What would you project each recipient will be paid in 2000 ($t = 20$)? In 2010 ($t = 30$)? \$4949.66; \$4922.04

42. *AFDC payments* The largest category of Medicaid recipients are AFDC (Aid to Families with Dependent Children) recipients. Their number $A(t)$ and payments received $C(t)$ from 1980 to 1992 can be approximated by

$$A(t) = 0.01t^3 - 0.19t^2 + 0.78t + 14 \text{ (millions)}$$

$$C(t) = 0.1t^3 - 0.1t^2 + 0.4t + 6 \text{ (billions)}$$

a. If t represents the number of years after 1980, write a rational expression representing the average amount received by each AFDC recipient. (The result will be in thousands of dollars.) $\dfrac{C(t)}{A(t)} = \dfrac{(0.1t^3 - 0.1t^2 + 0.4t + 6) \text{ billion}}{(0.01t^3 - 0.19t^2 + 0.78t + 14) \text{ million}}$

b. How much did each recipient get in 1980 ($t = 0$)? \$428.57

c. What would you project each recipient will receive in 2000 ($t = 20$)? In 2010 ($t = 30$)? \$23,035.71; \$19,266.86

43. *Non-AFDC Medicaid payments* If you subtract the average amount paid to each AFDC recipient in 1980 (Problem 42b) from the average amount received by each recipient in 1980 (Problem 41b), you will get the average amount spent for the rest of the Medicaid recipients. What was this average amount in 1980? \$616.88

44. *Non-AFDC Medicaid payments* Use the ideas in Problem 43 to find the average payment for Medicaid recipients who were *not* AFDC recipients in 1990. What do you think is wrong? −\$4873.90

SKILL CHECKER

Try the Skill Checker Exercises so you'll be ready for the next section.

Perform the indicated operations:

45. $1 \div \frac{20}{9}$ $\frac{9}{20}$ **46.** $1 \div \frac{30}{7}$ $\frac{7}{30}$ **47.** $12x\left(\frac{2}{x} + \frac{3}{2x}\right)$ 42 **48.** $12x\left(\frac{4}{3x} - \frac{1}{4x}\right)$ 13 **49.** $x^2\left(1 - \frac{1}{x^2}\right)$ $x^2 - 1$ **50.** $x^2\left(1 + \frac{1}{x}\right)$ $x^2 + x$

USING YOUR KNOWLEDGE

Continuing the Study of Fractions

In Chapter 1, we mentioned that some numbers cannot be written as the ratio of two integers. These numbers are called **irrational numbers.** We can approximate irrational numbers by using a type of fraction called a **continued fraction.** Here's how we do it. From a table of square roots, or a calculator, we find that

$\sqrt{2} \approx 1.4142$ $\approx$ means "approximately equal."

Can we find some continued fraction to approximate $\sqrt{2}$?

51. Try $1 + \frac{1}{2}$ (write it as a decimal). 1.5

52. Try $1 + \cfrac{1}{2 + \frac{1}{2}}$ (write it as a decimal). 1.4

53. Try $1 + \cfrac{1}{2 + \cfrac{1}{2 + \frac{1}{2}}}$ (write it as a decimal). $1.41\overline{6}$

54. Look at the pattern for the approximation of $\sqrt{2}$ given in Problems 51–53. What do you think the next approximation (when written as a continued fraction) will be?

$$\left[1 + \cfrac{1}{2 + \cfrac{1}{2 + \cfrac{1}{2 + \frac{1}{2}}}}\right] \approx 1.4138$$

55. How close is the approximation for $\sqrt{2}$ in Problem 53 to the value $\sqrt{2} \approx 1.4142$? 0.0025

WRITE ON

56. Write the procedure you use to find the LCD of two rational expressions.
Answers may vary.

57. Write the procedure you use to find the sum of two rational expressions
a. with the same denominator.
b. with different denominators.
Answers may vary.

58. Write the procedure you use to find the difference of two rational expressions
a. with the same denominator.
b. with different denominators.
Answers may vary.

MASTERY TEST

If you know how to do these problems, you have learned your lesson!

59. The fraction of the waste recovered in the United States can be approximated by

$$\frac{R(t)}{G(t)}$$

Of this,

$$\frac{P(t)}{G(t)}$$

is paper and paperboard. If

$$P(t) = 0.02t^2 - 0.25t + 6$$
$$G(t) = 0.04t^2 + 2.34t + 90$$
$$R(t) = 0.04t^2 - 0.59t + 7.42$$

and t represents the number of years after 1960, find the fraction of the waste generated that is *not* paper and paperboard. $\frac{R(t) - P(t)}{G(t)} = \frac{0.02t^2 - 0.34t + 1.42}{0.04t^2 + 2.34t + 90}$

Perform the indicated operations:

60. $\frac{x - 3}{x^2 - x - 2} - \frac{x + 3}{x^2 - 4}$ $\frac{-5x - 9}{(x + 2)(x - 2)(x + 1)}$

61. $\frac{5}{x - 1} - \frac{3}{x + 3}$ $\frac{2x + 18}{(x - 1)(x + 3)}$

62. $\frac{4}{5} + \frac{3}{x}$ $\frac{4x + 15}{5x}$

63. $\frac{4}{5(x - 2)} + \frac{6}{5(x - 2)}$ $\frac{2}{x - 2}$

64. $\frac{11}{3(x + 2)} - \frac{2}{3(x + 2)}$ $\frac{3}{x + 2}$

65. $\frac{x + 3}{x^2 - x - 2} - \frac{x - 3}{x^2 - 4}$ $\frac{7x + 9}{(x + 2)(x - 2)(x + 1)}$

6.4 COMPLEX FRACTIONS

To Succeed, Review How To . . .

1. Add, subtract, multiply, and divide fractions (pp. 12–21).

Objective

A Simplify a complex fraction using one of two methods.

GETTING STARTED

Planetary Models and Complex Fractions

We've already learned how to do the four fundamental operations using rational expressions. In some instances, we want to find the quotient of two expressions that contain fractions in the numerator or the denominator or in both. For example, the model of the planets in our solar system shown here is similar to the model designed by the seventeenth-century Dutch mathematician and astronomer Christian Huygens. The gears used in the model were especially difficult to design since they had to make each of the planets revolve around the sun at different rates. For example, Saturn goes around the sun in

$$29 + \cfrac{1}{2 + \cfrac{2}{9}} \text{ yr}$$

The expression

$$\cfrac{1}{2 + \cfrac{2}{9}}$$

is a *complex fraction.* In this section we shall learn how to simplify complex fractions.

A Simplifying Complex Fractions

Before we simplify *complex fractions,* we need a formal definition.

COMPLEX FRACTION

A **complex fraction** is a fraction that has one or more fractions in its numerator, denominator, or both.

Complex fractions can be simplified in either of two ways:

PROCEDURE

Simplifying Complex Fractions

1. Multiply numerator and denominator by the LCD of the fractions involved, or
2. Perform the operations indicated in the numerator and denominator of the complex fraction, and then divide the simplified numerator by the simplified denominator.

Web It

To practice the simplification of complex fractions, go to links 6-4-1 and 6-4-2 on the Bello Website at mhhe.com/bello. You will see some practice problems with solutions and answers.

We illustrate these two methods by simplifying the complex fraction

$$\frac{1}{2+\frac{2}{9}}$$

Method 1. Multiply numerator and denominator by the LCD of the fractions involved (in our case, by 9).

$$\frac{1}{2+\frac{2}{9}} = \frac{9\cdot 1}{9\left(2+\frac{2}{9}\right)}$$

Note that $9\left(2+\frac{2}{9}\right) = 9\cdot 2 + 9\cdot\frac{2}{9} = 18 + 2$.

$$= \frac{9}{18+2}$$

$$= \frac{9}{20}$$

Method 2. Perform the operations indicated in the numerator and denominator of the complex fraction and then divide the numerator by the denominator.

$$\frac{1}{2+\frac{2}{9}} = \frac{1}{\frac{18}{9}+\frac{2}{9}} = \frac{1}{\frac{20}{9}}$$ Add $2+\frac{2}{9}$.

$$= 1 \div \frac{20}{9}$$ Write $\frac{1}{\frac{20}{9}}$ as $1 \div \frac{20}{9}$.

$$= 1\cdot\frac{9}{20}$$ Multiply by the reciprocal of $\frac{20}{9}$.

$$= \frac{9}{20}$$

Either procedure also works for more complicated rational expressions. In the next example we simplify a complex fraction using both methods. Compare the results and see which method you prefer!

EXAMPLE 1 **Simplifying complex fractions: Both methods**

Simplify:

$$\frac{\frac{1}{a}+\frac{2}{b}}{\frac{3}{a}-\frac{1}{b}}$$

SOLUTION

Method 1. The LCD of the fractions involved is ab, so we multiply the numerator and denominator by ab to obtain

$$\frac{ab\cdot\left(\frac{1}{a}+\frac{2}{b}\right)}{ab\cdot\left(\frac{3}{a}-\frac{1}{b}\right)} = \frac{ab\cdot\frac{1}{a}+ab\cdot\frac{2}{b}}{ab\cdot\frac{3}{a}-ab\cdot\frac{1}{b}}$$

Note that $\cancel{a}b\cdot\frac{1}{\cancel{a}} = b$, $a\cancel{b}\cdot\frac{2}{\cancel{b}} = 2a$, $\cancel{a}b\cdot\frac{3}{\cancel{a}} = 3b$, and $a\cancel{b}\cdot\frac{1}{\cancel{b}} = a$.

$$= \frac{b+2a}{3b-a}$$

PROBLEM 1

Simplify:

$$\frac{\frac{2}{a}-\frac{3}{b}}{\frac{1}{a}+\frac{2}{b}}$$

Answer

1. $\frac{2b-3a}{b+2a}$

Method 2. Add the fractions in the numerator and subtract the fractions in the denominator. In both cases, the LCD of the fractions is ab. Hence

$$\frac{\frac{1}{a}+\frac{2}{b}}{\frac{3}{a}-\frac{1}{b}} = \frac{\frac{b}{ab}+\frac{2a}{ab}}{\frac{3b}{ab}-\frac{a}{ab}}$$ Write the fractions with their LCD.

$$= \frac{\frac{b+2a}{ab}}{\frac{3b-a}{ab}}$$ Add in the numerator, subtract in the denominator.

$$= \frac{b+2a}{ab}\cdot\frac{ab}{3b-a}$$ Multiply by the reciprocal of $\frac{3b-a}{ab}$.

$$= \frac{b+2a}{3b-a}$$ Simplify.

Teaching Tip

For Method 2, remind students that division by a fraction means multiplying by its reciprocal.

Example:

$$\frac{\frac{2}{3}}{\frac{5}{7}} = \frac{2}{3} \div \frac{5}{7} = \frac{2}{3}\cdot\frac{7}{5}$$

EXAMPLE 2 Simplifying complex fractions: Method 1

Simplify:

$$\frac{\frac{2}{x}+\frac{3}{2x}}{\frac{4}{3x}-\frac{1}{4x}}$$

SOLUTION We must first find the LCD of x, $2x$, $3x$, and $4x$. Now

$$\begin{aligned} x &= & & & x \\ 2x &= 2 & & & \cdot x \\ 3x &= & & 3 & \cdot x \\ 4x &= 2^2 & & & \cdot x \end{aligned}$$ Write the factors in columns.

The LCD is $2^2 \cdot 3 \cdot x = 12x$. Multiplying numerator and denominator by $12x$, we have

$$\frac{\frac{2}{x}+\frac{3}{2x}}{\frac{4}{3x}-\frac{1}{4x}} = \frac{12x\cdot\left(\frac{2}{x}+\frac{3}{2x}\right)}{12x\cdot\left(\frac{4}{3x}-\frac{1}{4x}\right)}$$

$$= \frac{12x\cdot\frac{2}{x}+12x\cdot\frac{3}{2x}}{12x\cdot\frac{4}{3x}-12x\cdot\frac{1}{4x}}$$ Use the distributive property.

$$= \frac{12\cdot 2+6\cdot 3}{4\cdot 4-3\cdot 1}$$ Simplify.

$$= \frac{24+18}{16-3}$$

$$= \frac{42}{13}$$

PROBLEM 2

Simplify:

$$\frac{\frac{1}{4x}+\frac{2}{3x}}{\frac{3}{2x}-\frac{1}{x}}$$

Answer

2. $\frac{11}{6}$

EXAMPLE 3 Simplifying complex fractions: Method 1

Simplify:

$$\frac{1 - \frac{1}{x^2}}{1 + \frac{1}{x}}$$

SOLUTION Here the LCD of the fractions involved is x^2. Thus

$$\frac{1 - \frac{1}{x^2}}{1 + \frac{1}{x}} = \frac{x^2 \cdot \left(1 - \frac{1}{x^2}\right)}{x^2 \cdot \left(1 + \frac{1}{x}\right)}$$ Multiply numerator and denominator by x^2.

$$= \frac{x^2 \cdot 1 - x^2 \cdot \frac{1}{x^2}}{x^2 \cdot 1 + x^2 \cdot \frac{1}{x}}$$ Use the distributive property.

$$= \frac{x^2 - 1}{x^2 + x}$$ Simplify.

$$= \frac{(x + 1)(x - 1)}{x(x + 1)}$$ Factor.

$$= \frac{x - 1}{x}$$ Simplify.

PROBLEM 3

Simplify:

$$\frac{1 - \frac{1}{x^2}}{1 - \frac{1}{x}}$$

Answer

3. $\frac{x+1}{x}$

Exercises 6.4

A In Problems 1–24, simplify.

1. $\dfrac{\frac{1}{2}}{2 + \frac{1}{2}}$ $\frac{1}{5}$

2. $\dfrac{\frac{1}{4}}{3 + \frac{1}{4}}$ $\frac{1}{13}$

3. $\dfrac{\frac{1}{2}}{2 - \frac{1}{2}}$ $\frac{1}{3}$

4. $\dfrac{\frac{1}{4}}{3 - \frac{1}{4}}$ $\frac{1}{11}$

5. $\dfrac{a - \frac{a}{b}}{1 + \frac{a}{b}}$ $\frac{ab - a}{b + a}$

6. $\dfrac{1 - \frac{1}{a}}{1 + \frac{1}{a}}$ $\frac{a - 1}{a + 1}$

7. $\dfrac{\frac{1}{a} + \frac{1}{b}}{\frac{1}{a} - \frac{1}{b}}$ $\frac{b + a}{b - a}$

8. $\dfrac{\frac{2}{a} + \frac{1}{b}}{\frac{2}{a} - \frac{1}{b}}$ $\frac{2b + a}{2b - a}$

9. $\dfrac{\frac{1}{2a} + \frac{1}{3b}}{\frac{4}{a} - \frac{3}{4b}}$ $\frac{6b + 4a}{48b - 9a}$

10. $\dfrac{\frac{1}{2a} + \frac{1}{4b}}{\frac{2}{a} - \frac{3}{5b}}$ $\frac{10b + 5a}{40b - 12a}$

11. $\dfrac{\frac{1}{3} + \frac{3}{4}}{\frac{3}{8} - \frac{1}{6}}$ $\frac{26}{5}$

12. $\dfrac{\frac{1}{5} + \frac{3}{2}}{\frac{5}{8} - \frac{3}{10}}$ $\frac{68}{13}$

13. $\dfrac{2 + \frac{1}{x}}{4 - \frac{1}{x^2}}$ $\frac{x}{2x - 1}$

14. $\dfrac{3 + \frac{1}{x}}{9 - \frac{1}{x^2}}$ $\frac{x}{3x - 1}$

15. $\dfrac{2 + \frac{2}{x}}{1 + \frac{1}{x}}$ 2

16. $\dfrac{5 + \frac{5}{x^2}}{1 + \frac{1}{x^2}}$ 5

17. $\dfrac{\frac{1}{y} + \frac{1}{x}}{\frac{x}{y} - \frac{y}{x}}$ $\frac{1}{x - y}$

18. $\dfrac{\frac{1}{x} + \frac{1}{y}}{\frac{y}{x} - \frac{x}{y}}$ $\frac{1}{y - x}$

19. $\dfrac{x - 2 - \frac{8}{x}}{x - 3 - \frac{4}{x}}$ $\frac{x + 2}{x + 1}$

20. $\dfrac{x - 2 - \frac{15}{x}}{x - 3 - \frac{10}{x}}$ $\frac{x + 3}{x + 2}$

21. $\dfrac{\frac{1}{x + 5}}{\frac{4}{x^2 - 25}}$ $\frac{x - 5}{4}$

22. $\dfrac{\frac{1}{x - 3}}{\frac{2}{x^2 - 9}}$ $\frac{x + 3}{2}$

23. $\dfrac{\frac{1}{x^2 - 16}}{\frac{2}{x + 4}}$ $\frac{1}{2(x - 4)}$

24. $\dfrac{\frac{3}{x^2 - 64}}{\frac{4}{x + 8}}$ $\frac{3}{4(x - 8)}$

SKILL CHECKER

Try the Skill Checker Exercises so you'll be ready for the next section.

Solve:

25. $19w = 2356$ $w = 124$
26. $18L = 2232$ $L = 124$
27. $9x + 24 = x$ $x = -3$
28. $10x + 36 = x$ $x = -4$
29. $5x = 4x + 3$ $x = 3$
30. $6x = 5x + 5$ $x = 5$

USING YOUR KNOWLEDGE

Around the Sun in Complex Fractions

In the *Getting Started* of this section, we mentioned that Saturn takes

$$29 + \frac{1}{2 + \frac{2}{9}} \text{ yr}$$

to go around the sun. Since we have shown that

$$\frac{1}{2 + \frac{2}{9}} = \frac{9}{20}$$

we know that Saturn takes $29 + \frac{9}{29} = 29\frac{9}{20}$ years to go around the sun.

Use your knowledge to simplify the number of years it takes the following planets to go around the sun.

31. Mercury, $\frac{1}{4 + \frac{1}{6}}$ yr $\frac{6}{25}$ yr

32. Venus, $\frac{1}{1 + \frac{2}{3}}$ yr $\frac{3}{5}$ yr

33. Jupiter, $11 + \frac{1}{1 + \frac{7}{43}}$ yr (Write your answer as a mixed number.) $11\frac{43}{50}$ yr

34. Mars, $1 + \frac{1}{1 + \frac{3}{22}}$ yr (Write your answer as a mixed number.) $1\frac{22}{25}$ yr

WRITE ON

35. What is a complex fraction? A fraction that has one or more fractions in its numerator, denominator, or both.

36. List the advantages and disadvantages of Method 1 when simplifying a complex fraction. Answers may vary.

37. List the advantages and disadvantages of Method 2 when simplifying a complex fraction. Answers may vary.

38. Which method do you prefer to simplify complex fractions? Why? Answers may vary.

39. How do you know which method to use when simplifying complex fractions? Answers may vary.

MASTERY TEST

If you know how to do these problems, you have learned your lesson!

Simplify:

40. $\dfrac{\frac{2}{a} - \frac{3}{b}}{\frac{1}{a} + \frac{2}{b}}$ $\dfrac{2b - 3a}{b + 2a}$

41. $\dfrac{\frac{1}{4x} + \frac{2}{3x}}{\frac{3}{2x} - \frac{1}{x}}$ $\dfrac{11}{6}$

42. $\dfrac{1 - \frac{1}{x^2}}{1 - \frac{1}{x}}$ $\dfrac{x + 1}{x}$

43. $\dfrac{w + 2 - \frac{18}{w - 5}}{w - 1 - \frac{12}{w - 5}}$ $\dfrac{w + 4}{w + 1}$

44. $\dfrac{\frac{8x}{3x + 1} - \frac{3x - 1}{x}}{\frac{4x}{3x + 1} - \frac{2x - 1}{x}}$ $\dfrac{1 + x}{1 + 2x}$

45. $\dfrac{\frac{3}{m - 4} - \frac{16}{m - 3}}{\frac{2}{m - 3} - \frac{15}{m + 5}}$ $\dfrac{m + 5}{m - 4}$

6.5 SOLVING EQUATIONS CONTAINING RATIONAL EXPRESSIONS

To Succeed, Review How To . . .

1. Solve linear equations (pp. 150–155).
2. Find the LCD of two or more rational expressions (pp. 487–492).

Objectives

A Solve equations that contain rational expressions.

B Solve a rational equation for a specified variable.

GETTING STARTED

Salute One of the Largest Flags

This American flag (one of the largest ever) was displayed in J. L. Hudson's store in Detroit. By law, the ratio of length to width of the American flag should be $\frac{19}{10}$. If the length of this flag was 235 feet, what should its width be to conform with the law? To solve this problem, we let W be the width of the flag and set up the equation

$$\frac{19}{10} = \frac{235}{W} \quad \begin{matrix} \leftarrow \text{Length} \\ \leftarrow \text{Width} \end{matrix}$$

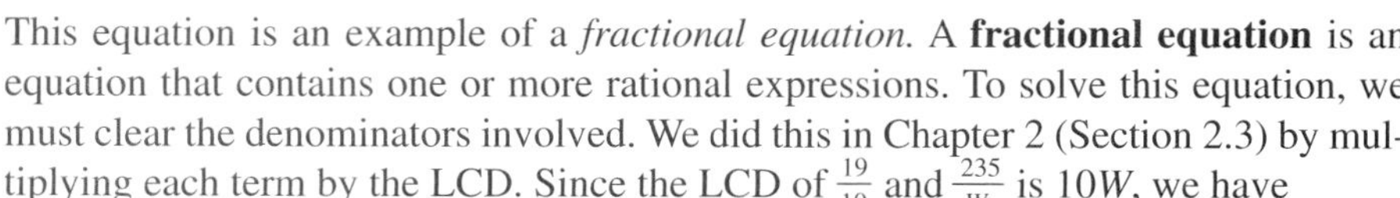

This equation is an example of a *fractional equation.* A **fractional equation** is an equation that contains one or more rational expressions. To solve this equation, we must clear the denominators involved. We did this in Chapter 2 (Section 2.3) by multiplying each term by the LCD. Since the LCD of $\frac{19}{10}$ and $\frac{235}{W}$ is $10W$, we have

$$10W \cdot \frac{19}{10} = \frac{235}{W} \cdot 10W \quad \text{Multiply by the LCD.}$$

$$19W = 2350 \quad \text{Simplify.}$$

$$W = \frac{2350}{19} \quad \text{Divide by 19.}$$

$$W \approx 124 \quad \text{Approximate the answer.}$$

$$\begin{array}{r} 123.6 \\ 19\overline{)2350.0} \\ \underline{19} \\ 45 \\ \underline{38} \\ 70 \\ \underline{57} \\ 130 \\ \underline{114} \\ 16 \end{array}$$

(By the way, the flag was only 104 feet long and weighed 1500 pounds. It was *not* an official flag.) In this section we shall learn how to solve fractional equations.

A Solving Fractional Equations

Web It

For an excellent presentation on solving fractional equations, go to link 6-5-1 on the Bello Website at mhhe.com/bello.

Then, to practice solving the equations (with answers), go to link 6-5-2.

The first step in solving fractional equations is to multiply each side of the equation by the **least common multiple (LCM)** of the denominators present. This is equivalent to multiplying *each term* by the LCD because if you have the equation

$$\frac{a}{b} + \frac{c}{d} = \frac{e}{f}$$

then multiplying each side by L, the LCM of the denominators, gives

$$L \cdot \left(\frac{a}{b} + \frac{c}{d}\right) = L \cdot \frac{e}{f}$$

or, using the distributive property,

$$L \cdot \frac{a}{b} + L \cdot \frac{c}{d} = L \cdot \frac{e}{f}$$

Thus, we have the following result.

Teaching Tip

Remind students that b, d, and f cannot equal zero. Therefore, it will be important to check the potential solutions in the original equation. If any value results in a zero denominator, reject that value.

LEAST COMMON MULTIPLE

Multiplying each side of the equation

$$\frac{a}{b} + \frac{c}{d} = \frac{e}{f}$$

by L is equivalent to multiplying each *term* by L.

The procedure for solving fractional equations is similar to the one used to solve linear equations. Because of this, you may want to review the procedure (see p. 154) before continuing.

EXAMPLE 1 Solving fractional equations

Solve:

$$\frac{3}{4} + \frac{2}{x} = \frac{1}{12}$$

SOLUTION The LCD of $\frac{3}{4}$, $\frac{2}{x}$, and $\frac{1}{12}$ is $12x$.

1. Clear the fractions; the LCD is $12x$. $\quad 12x \cdot \frac{3}{4} + 12x \cdot \frac{2}{x} = 12x \cdot \frac{1}{12}$

2. Simplify. $\quad 9x + 24 = x$

3. Subtract 24 from each side. $\quad 9x = x - 24$

4. Subtract x from each side. $\quad 8x = -24$

5. Divide each side by 8. $\quad x = -3$

The answer is -3.

6. Here is the check:

$$\frac{3}{4} + \frac{2}{x} \stackrel{?}{=} \frac{1}{12}$$

$\frac{3}{4} + \frac{2}{-3}$	$\frac{1}{12}$
$\frac{3 \cdot 3}{4 \cdot 3} + \frac{2 \cdot 4}{-3 \cdot 4}$	
$\frac{9}{12} - \frac{8}{12}$	
$\frac{1}{12}$	

PROBLEM 1

Solve:

$$\frac{4}{5} + \frac{1}{x} = \frac{3}{10}$$

In some cases, the denominators involved may be more complicated. Nevertheless, the procedure used to solve the equation remains the same. Thus we can also use the six-step procedure to solve

$$\frac{2x}{x-1} + 3 = \frac{4x}{x-1}$$

as shown in the next example.

Answer

1. $x = -2$

EXAMPLE 2 Solving fractional equations

Solve:

$$\frac{2x}{x-1} + 3 = \frac{4x}{x-1}$$

SOLUTION Since $x - 1$ is the only denominator, it must be the LCD. We then proceed by steps.

1. Clear the fractions; the LCD is $(x - 1)$.
$$(x-1)\cdot\frac{2x}{x-1} + 3(x-1) = (x-1)\cdot\frac{4x}{x-1}$$
2. Simplify. $2x + 3x - 3 = 4x$
$5x - 3 = 4x$
3. Add 3. $5x = 4x + 3$
4. Subtract $4x$. $x = 3$
5. Division is not necessary.
6. You can easily check this by substitution.

Thus the answer is $x = 3$, as can easily be verified by substituting 3 for x in the original equation.

PROBLEM 2

Solve:

$$\frac{8}{x-1} - 4 = \frac{2x}{x-1}$$

Teaching Tip

For Example 2, when multiplying by the LCD, $(x - 1)$, it might be more obvious to simplify if the LCD is written over 1 for the two fraction multiplications.

Example:

$$\frac{\cancel{(x-1)}}{1}\cdot\frac{2x}{\cancel{x-1}} + 3(x-1) = \frac{\cancel{(x-1)}}{1}\cdot\frac{4x}{\cancel{x-1}}$$

$$2x + 3(x-1) = 4x$$

etc.

So far, the denominators used in the examples have *not* been factorable. In the cases where they are, it's very important that we factor them *before* we find the LCD. For instance, to solve the equation

$$\frac{x}{x^2-16} + \frac{4}{x-4} = \frac{1}{x+4}$$

we first note that

$$x^2 - 16 = (x+4)(x-4)$$

We then write

$$\frac{x}{x^2-16} + \frac{4}{x-4} = \frac{1}{x+4}$$

as

$$\frac{x}{(x+4)(x-4)} + \frac{4}{x-4} = \frac{1}{x+4}$$

The denominator $x^2 - 16$ has been factored as $(x + 4)(x - 4)$.

The solution to this equation is given in the next example.

EXAMPLE 3 Solving fractional equations

Solve:

$$\frac{x}{x^2-16} + \frac{4}{x-4} = \frac{1}{x+4}$$

SOLUTION Since $x^2 - 16 = (x + 4)(x - 4)$, we write the equation with the $x^2 - 16$ factored as

$$\frac{x}{(x+4)(x-4)} + \frac{4}{x-4} = \frac{1}{x+4}$$

PROBLEM 3

Solve:

$$\frac{x}{x^2-9} + \frac{3}{x-3} = \frac{1}{x+3}$$

Answers

2. $x = 2$ **3.** $x = -4$

1. Clear the fractions; the LCD is $(x + 4)(x - 4)$.

$$(x + 4)(x - 4) \cdot \frac{x}{(x + 4)(x - 4)} + (x + 4)(x - 4) \cdot \frac{4}{x - 4} = (x + 4)(x - 4) \cdot \frac{1}{x + 4}$$

2. Simplify.
$$x + 4(x + 4) = x - 4$$
$$x + 4x + 16 = x - 4$$
$$5x + 16 = x - 4$$

3. Subtract 16. $5x = x - 20$
4. Subtract x. $4x = -20$
5. Divide by 4. $x = -5$

Thus the solution is $x = -5$.

6. You can easily check this by substituting -5 for x in the original equation.

By now, you've probably noticed that we always recommend checking the solution by *direct substitution.* The importance of doing this will be made obvious in the next example.

EXAMPLE 4 **Solving fractional equations: No-solution case**

Solve:

$$\frac{x}{x + 4} - \frac{2}{5} = \frac{-4}{x + 4}$$

SOLUTION Here the denominators are 5 and $(x + 4)$.

1. Clear the fractions; the LCD is $5(x + 4)$.

$$5(x + 4) \cdot \frac{x}{x + 4} - \frac{2}{5} \cdot 5(x + 4) = \frac{-4}{x + 4} \cdot 5(x + 4)$$

2. Simplify.
$$5x - 2x - 8 = -20$$
$$3x - 8 = -20$$

3. Add 8. $3x = -12$
4. The variable is already isolated, so step 4 is not necessary.
5. Divide by 3. $x = -4$

Thus the solution seems to be $x = -4$.

6. But now let's do the check. If we substitute -4 for x in the original equation, we have

$$\frac{-4}{-4 + 4} - \frac{2}{5} = \frac{-4}{-4 + 4}$$

or

$$\frac{-4}{0} - \frac{2}{5} = \frac{-4}{0}$$

Division by zero is not defined.

Two of the terms are not defined. Thus this equation has no solution.

PROBLEM 4

Solve:

$$\frac{x}{x - 2} + \frac{3}{4} = \frac{2}{x - 2}$$

Teaching Tip

Instead of waiting to check the solution at the end, students can determine the possible values of x to omit from the solution set before solving by setting each denominator equal to zero and solving. (See Section 6.1.) Thus, for Example 4

$$x + 4 \neq 0$$
$$x \neq -4$$

If done this way, you can stop when you get $x = -4$!

Answer

4. No solution

> **NOTE**
>
> Remember, no matter how careful you are when you get an answer, call it a *possible* or a *proposed* answer. A given number is *not* an answer until you show that, when the number is substituted for the variable in the original equation, the result is a true statement.

Finally, we must point out that the equations resulting when clearing denominators are not *always* linear equations—that is, equations that can be written in the form $ax + b = c$ $(a \neq 0)$. For example, to solve the equation

$$\frac{x^2}{x+2} = \frac{4}{x+2}$$

we first multiply by the LCD $(x + 2)$ to obtain

$$(x+2)\cdot\frac{x^2}{x+2} = (x+2)\cdot\frac{4}{x+2}$$

or

$$x^2 = 4$$

In this equation, the variable x has a 2 as an exponent; thus it is a **quadratic** equation and can be solved when written in standard form—that is, by writing the equation as

$$x^2 - 4 = 0 \quad \text{Recall that a quadratic equation is an equation that can be written in standard form as } ax^2 + bx + c = 0 \quad (a \neq 0).$$

$$(x+2)(x-2) = 0 \quad \text{Factor.}$$

$$x + 2 = 0 \quad \text{or} \quad x - 2 = 0 \quad \text{Use the principle of zero product.}$$

$$x = -2 \quad \text{or} \quad x = 2 \quad \text{Solve each equation.}$$

Teaching Tip

To avoid a zero denominator, $x + 2 \neq 0$; that is, $x \neq -2$. Thus, $x = 2$ is the only solution.

Thus $x = 2$ is a solution since

$$\frac{2^2}{2+2} = \frac{4}{2+2}$$

However, for $x = -2$,

$$\frac{x^2}{x+2} = \frac{2^2}{-2+2} = \frac{4}{0}$$

and the denominator $x + 2$ becomes 0. Thus $x = -2$ is *not* a solution; -2 is called an *extraneous* root. The only solution is $x = 2$.

EXAMPLE 5 **Solving fractional equations: Extraneous roots case**

Solve:

$$1 + \frac{3}{x-2} = \frac{12}{x^2-4}$$

SOLUTION Since $x^2 - 4 = (x + 2)(x - 2)$, the LCD is $(x + 2)(x - 2)$. We then write the equation with the denominator $x^2 - 4$ in factored form and multiply each term by the LCD as before. Here are the steps.

$$(x+2)(x-2)\cdot 1 + (x+2)(x-2)\cdot\frac{3}{x-2} \quad \text{The LCD is } (x+2)(x-2).$$

$$= (x+2)(x-2)\cdot\frac{12}{(x+2)(x-2)}$$

$$(x^2-4) + 3(x+2) = 12 \quad \text{Simplify.}$$

$$x^2 - 4 + 3x + 6 = 12$$

$$x^2 + 3x + 2 = 12$$

PROBLEM 5

Solve:

$$1 - \frac{4}{x^2-1} = \frac{-2}{x-1}$$

Answer

5. $x = -3$

$$x^2 + 3x - 10 = 0$$ Subtract 12 from both sides to write in standard form.

$$(x + 5)(x - 2) = 0$$ Factor.

$$x + 5 = 0 \quad \text{or} \quad x - 2 = 0$$ Use the principle of zero product.

$$x = -5 \qquad x = 2$$ Solve each equation.

Since $x = 2$ makes the denominator $x - 2$ equal to zero, the only possible solution is $x = -5$. This solution can be verified in the original equation.

B Solving Fractional Equations for a Specified Variable

In Section 2.6, we solved a formula for a specified variable. We can also solve fractional equations for a specified variable. Here S_n is the sum of an arithmetic sequence:

$$S_n = \frac{n(a_1 + a_n)}{2}$$

(By the way, don't be intimidated by subscripts such as n in S_n, which is read "S sub n"; they are simply used to distinguish one variable from another—for example, a_1 is different from a_n because the subscripts are different.) So, to solve for n in this equation, we proceed as follows:

$$S_n = \frac{n(a_1 + a_n)}{2}$$ Given. Since the only denominator is 2, the LCD is 2.

1. Clear any fractions; the LCD is 2: $2 \cdot S_n = 2 \cdot \frac{n(a_1 + a_n)}{2}$
2. Simplify. $2S_n = n(a_1 + a_n)$
3. Since we want n by itself, divide by $(a_1 + a_n)$: $\frac{2S_n}{a_1 + a_n} = n$

Thus the solution is $n = \frac{2S_n}{a_1 + a_n}$.

EXAMPLE 6 Solving for a specified variable

The sum S_n of a geometric sequence is

$$S_n = \frac{a_1(1 - r^n)}{1 - r}$$

Solve for a_1.

SOLUTION This time, the only denominator is $1 - r$, so it must be the LCD. As before, we proceed by steps.

$$S_n = \frac{a_1(1 - r^n)}{1 - r}$$ Given.

1. Clear any fractions; the LCD is $1 - r$. $(1 - r)S_n = (1 - r) \cdot \frac{a_1(1 - r^n)}{1 - r}$
2. Simplify. $(1 - r)S_n = a_1(1 - r^n)$
3. Divide both sides by $1 - r^n$ to isolate a_1. $\frac{(1 - r)S_n}{1 - r^n} = a_1$

Thus the solution is $a_1 = \frac{(1 - r)S_n}{1 - r^n}$.

PROBLEM 6

Solve for a in

$$S = \frac{a(r^n - 1)}{r - 1}$$

Answer

6. $a = \frac{S(r - 1)}{r^n - 1}$

You will have more opportunities to practice solving for a specified variable in the *Using Your Knowledge*.

Exercises 6.5

A In Problems 1–50, solve (if possible).

1. $\frac{x}{4} = \frac{3}{2}$ $x = 6$

2. $\frac{x}{8} = \frac{-7}{4}$ $x = -14$

3. $\frac{3}{x} = \frac{3}{4}$ $x = 4$

4. $\frac{6}{x} = \frac{-2}{7}$ $x = -21$

5. $\frac{-8}{3} = \frac{16}{x}$ $x = -6$

6. $\frac{-5}{6} = \frac{10}{x}$ $x = -12$

7. $\frac{4}{3} = \frac{x}{9}$ $x = 12$

8. $\frac{-3}{7} = \frac{x}{14}$ $x = -6$

9. $\frac{2}{5} + \frac{3}{x} = \frac{23}{20}$ $x = 4$

10. $\frac{6}{7} + \frac{2}{x} = \frac{3}{21}$ $x = -\frac{14}{5}$

11. $\frac{3}{x} - \frac{2}{7} = \frac{11}{35}$ $x = 5$

12. $\frac{4}{x} - \frac{2}{9} = \frac{22}{63}$ $x = 7$

13. $\frac{3}{5} + \frac{7x}{10} = 2$ $x = 2$

14. $\frac{2}{7} + \frac{4x}{21} = \frac{2}{3}$ $x = 2$

15. $\frac{3x}{4} - \frac{1}{5} = \frac{13}{10}$ $x = 2$

16. $\frac{2x}{3} - \frac{1}{4} = -1$ $x = \frac{-9}{8}$

17. $\frac{3}{x+2} = \frac{4}{x-1}$ $x = -11$

18. $\frac{2}{x-2} = \frac{5}{x+1}$ $x = 4$

19. $\frac{-1}{x+1} = \frac{3}{x+5}$ $x = -2$

20. $\frac{2}{x-1} = \frac{-3}{x+9}$ $x = -3$

21. $\frac{3x}{x-3} + 2 = \frac{5x}{x-3}$ No solution

22. $\frac{2x}{x-2} + 18 = \frac{8x}{x-2}$ $x = 3$

23. $\frac{5x}{x+1} - 6 = \frac{3x}{x+1}$ $x = \frac{-3}{2}$

24. $\frac{5x}{x+1} - 2 = \frac{2x}{x+1}$ $x = 2$

25. $\frac{x}{x^2-25} + \frac{5}{x-5} = \frac{1}{x+5}$ $x = -6$

26. $\frac{x}{x^2-64} + \frac{8}{x-8} = \frac{1}{x+8}$ $x = -9$

27. $\frac{x}{x^2-49} + \frac{7}{x-7} = \frac{1}{x+7}$ $x = -8$

28. $\frac{x}{x^2-1} + \frac{1}{x-1} = \frac{1}{x+1}$ $x = -2$

29. $\frac{x}{x+3} + \frac{3}{4} = \frac{-3}{x+3}$ No solution

30. $\frac{1}{5} + \frac{x}{x-2} = \frac{2}{x-2}$ No solution

31. $\frac{x}{x-4} - \frac{2}{7} = \frac{4}{x-4}$ No solution

32. $\frac{x}{x-8} - \frac{1}{5} = \frac{8}{x-8}$ No solution

33. $1 + \frac{2}{x-1} = \frac{4}{x^2-1}$ $x = -3$

34. $1 + \frac{2}{x-3} = \frac{5}{x^2-9}$ $x = -4$ or $x = 2$

35. $2 - \frac{6}{x^2-1} = \frac{-3}{x-1}$ $x = \frac{-5}{2}$

36. $2 - \frac{4}{x^2-4} = \frac{-1}{x-2}$ $x = \frac{-5}{2}$

37. $\frac{4}{x-3} - \frac{2}{x-1} = \frac{2}{x+2}$ $x = \frac{1}{7}$

38. $\frac{5}{x+1} - \frac{1}{x+2} = \frac{13}{x+5}$ $x = \frac{-19}{9}$ or $x = 1$

39. $\frac{2x}{x^2-1} + \frac{4}{x-1} = \frac{1}{x-1}$ $x = \frac{-3}{5}$

40. $\frac{3x-2}{x^2-4} + \frac{4}{x+2} = \frac{1}{x-2}$ No solution

41. $\frac{2z+7}{z-3} + 1 = \frac{z+5}{z-6} + 2$ $z = \frac{45}{2}$

42. $\frac{4x-2}{x+4} - 3 = \frac{2-5x}{x-2} + 6$ $x = \frac{34}{5}$

43. $\frac{2y-5}{2} - \frac{1}{y-1} = y + 1$ $y = \frac{5}{7}$

44. $\frac{4}{x+1} - \frac{3}{x} = \frac{1}{x-2}$ $x = 1$

45. $\frac{5}{2v-1} + \frac{2}{v} = \frac{18v}{4v^2-1}$ $v = \frac{2}{5}$

46. $\frac{3y}{y^2-9} - \frac{4}{y+3} = \frac{6-y}{y^2+3y}$ $y = -6$

47. $\frac{z+7}{z-1} - \frac{z+3}{z-2} = \frac{3}{z-6}$ $z = 3$

48. $\frac{x+1}{x-3} - \frac{x+5}{x-2} = -\frac{3}{x-1}$ $x = -5$

49. $\frac{2}{x^2-4x+3} - \frac{5}{x^2-x-6} = \frac{x-7}{(x-1)(x-3)(x+2)}$
$x = 4$

50. $\frac{y-5}{y^2-4} - \frac{1}{y^2+2y-8} = \frac{y^2}{(y^2-4)(y+4)}$
$y = -11$

SKILL CHECKER

Try the Skill Checker Exercises so you'll be ready for the next section.

Solve:

51. $4(x+3) = 45$ $x = \frac{33}{4}$

52. $\frac{d}{3} + \frac{d}{4} = 1$ $d = \frac{12}{7}$

53. $\frac{h}{4} + \frac{h}{6} = 1$ $h = \frac{12}{5}$

54. $60(R-5) = 40(R+5)$ $R = 25$

55. $30(R-5) = 10(R+15)$ $R = 15$

56. $\frac{n}{90} = \frac{1}{15}$ $n = 6$

USING YOUR KNOWLEDGE

Looking for Variables in the Right Places

Use your knowledge of fractional equations to solve the given problem for the indicated variable.

57. The area A of a trapezoid is

$$A = \frac{h(b_1 + b_2)}{2}$$

Solve for h. $h = \frac{2A}{b_1 + b_2}$

58. In an electric circuit, we have

$$\frac{1}{R} = \frac{1}{R_1} + \frac{1}{R_2}$$

Solve for R. $R = \frac{R_1 R_2}{R_1 + R_2}$

59. In refrigeration we find the formula

$$\frac{Q_1}{Q_2 - Q_1} = P$$

Solve for Q_1. $Q_1 = \frac{PQ_2}{P+1}$

60. When studying the expansion of metals, we find the formula

$$\frac{L}{1 + at} = L_0$$

Solve for t. $t = \frac{L - L_0}{aL_0}$

61. Manufacturers of camera lenses use the formula

$$\frac{1}{f} = \frac{1}{a} + \frac{1}{b}$$

Solve for f. $f = \frac{ab}{a+b}$

WRITE ON

62. Explain the difference between adding two rational expressions such as

$$\frac{1}{x} + \frac{1}{2}$$

and solving an equation such as

$$\frac{1}{x} + \frac{1}{2} = 1$$

Answers may vary.

63. Write the procedure you use to solve a fractional equation. Answers may vary.

64. In Section 6.1, we used the fundamental rule of rational expressions to multiply the numerators and denominators of *fractions* so that the resulting fractions have the LCD as their denominator. In this section we multiplied each *term* of an equation by the LCD. Explain the difference. Answers may vary.

MASTERY TEST

If you know how to do these problems, you have learned your lesson!

Solve:

65. $1 - \frac{4}{x^2 - 1} = \frac{-2}{x - 1}$ $x = -3$

66. $\frac{x}{x - 2} + \frac{3}{4} = \frac{2}{x - 2}$ No solution

67. $\frac{x}{x^2 - 9} + \frac{3}{x - 3} = \frac{1}{x + 3}$ $x = -4$

68. $\frac{8}{x - 1} - 4 = \frac{2x}{x - 1}$ $x = 2$

69. $\frac{4}{5} + \frac{1}{x} = \frac{3}{10}$ $x = -2$

70. $\frac{1}{x^2 + 2x - 3} + \frac{1}{x^2 - 9} = \frac{1}{(x - 1)(x^2 - 9)}$ $x = \frac{5}{2}$

Solve for the specified variable:

71. $C = \frac{5}{9}(F - 32)$; F $F = \frac{9}{5}C + 32$

72. $A = P(1 + r)$; r $r = \frac{A - P}{P}$

6.6 RATIO, PROPORTION, AND APPLICATIONS

To Succeed, Review How To . . .

1. Find the LCD of two or more fractions (pp. 16–17).
2. Solve linear equations (pp. 150–155).
3. Use the RSTUV method to solve word problems (p. 162).

Objectives

A Solve proportions.

B Solve applications.

GETTING STARTED

Coffee, Ratio, and Proportion

The manufacturer of this jar of instant coffee claims that 4 ounces of its instant coffee is equivalent to 1 pound (16 ounces) of regular coffee. In mathematics, we say that the ratio of instant coffee used to regular coffee used is 4 to 16. The ratio 4 to 16 can be written as the fraction $\frac{4}{16}$. A **ratio** is a quotient of two numbers. There are *three* ways in which the ratio of a number a to another number b can be written:

1. a to b
2. $a:b$
3. $\frac{a}{b}$

In this section we shall learn how to use ratios to solve proportions and to solve different applications involving these proportions.

Solving Proportions

Let's look more closely at the coffee jar. The label on the back claims that 4 ounces of instant coffee will make 60 cups of coffee. Thus the ratio of ounces to cups, when written as a fraction, is

$$\frac{4}{60} = \frac{1}{15} \quad \begin{matrix} \leftarrow \text{Ounces} \\ \leftarrow \text{Cups} \end{matrix}$$

The fraction $\frac{1}{15}$ is called the *reduced ratio* of ounces of coffee to cups of coffee. It tells us that 1 ounce of coffee will make 15 cups of coffee. Now suppose you want to make 90 cups of coffee. How many ounces do you need? The ratio of ounces to cups is $\frac{1}{15}$, and we need to know how many ounces will make 90 cups. Let n be the number of ounces needed. Then

$$\frac{1}{15} = \frac{n}{90} \quad \begin{matrix} \leftarrow \text{Ounces} \\ \leftarrow \text{Cups} \end{matrix}$$

Note that in both fractions the numerator indicates the number of ounces and the denominator indicates the number of cups. The equation $\frac{1}{15} = \frac{n}{90}$ is an equality between two ratios. In mathematics, an equality between ratios is called a **proportion.** Thus $\frac{1}{15} = \frac{n}{90}$ is a proportion. To solve this proportion, which is simply a fractional equation, we proceed as before. First, since the LCM of 15 and 90 is 90:

1. Multiply by the LCM. $90 \cdot \frac{1}{15} = 90 \cdot \frac{n}{90}$
2. Simplify. $6 = n$

Thus we need 6 ounces of instant coffee to make 90 cups.

We use the same ideas in the next example.

Web It

To view a site dealing with proportions, go to link 6-6-1 on the Bello Website at mhhe.com/bello.

After reading the examples, try the practice problems on link 6-6-2.

Teaching Tip

Since a ratio is a fraction, a proportion is an equality between ratios.

Example:

$$\frac{2}{3} = \frac{4}{6}$$

To read this as a proportion say, "2 compares to 3, the same as 4 compares to 6."

EXAMPLE 1 Traveling ratios

A car travels 140 miles on 8 gallons of gas.

a. What is the reduced ratio of miles to gallons? (Note that another way to state this ratio is to use miles *per* gallon.)

b. How many gallons will be needed to travel 210 miles?

SOLUTION

a. The ratio of miles to gallons is

$$\frac{140}{8} = \frac{35}{2} \quad \begin{matrix} \leftarrow \text{Miles} \\ \leftarrow \text{Gallons} \end{matrix}$$

b. Let g be the gallons needed. The ratio of miles to gallons is $\frac{35}{2}$; it is also $\frac{210}{g}$. Thus

$$\frac{210}{g} = \frac{35}{2}$$

Multiplying by $2g$, the LCD, we have

$$2g \cdot \frac{210}{g} = 2g \cdot \frac{35}{2}$$

$$420 = 35g$$

$$g = \frac{420}{35} = 12$$

Hence 12 gallons of gas will be needed to travel 210 miles.

PROBLEM 1

A car travels 150 miles on 9 gallons of gas. How many gallons does it need to travel 900 miles?

Answer

1. 54 gal

Web It

To practice with this concept, try link 6-2-3 on the Bello Website at mhhe.com/bello.

Teaching Tip

Have students try to solve

$$1 + \frac{2}{3} = \frac{x}{5}$$

by cross products and discuss why it doesn't yield the correct answer.

Example:

No!	Yes
$1 + \frac{2}{3} = \frac{x}{5}$	$1 + \frac{2}{3} = \frac{x}{5}$
$1 + 3x = 10$	$15\left(1 + \frac{2}{3}\right) = \left(\frac{x}{5}\right)15$
$3x = 9$	$15 + 10 = 3x$
$x = 3$	$\frac{25}{3} = x$
Check:	
$1 + \frac{2}{3} \stackrel{?}{=} \frac{3}{5}$	$1 + \frac{2}{3} \stackrel{?}{=} \frac{\frac{25}{3}}{5}$
$\frac{3}{3} + \frac{2}{3} \stackrel{?}{=} \frac{3}{5}$	$\frac{3}{3} + \frac{2}{3} \stackrel{?}{=} \frac{\overset{5}{\cancel{25}}}{3} \cdot \frac{1}{\cancel{5}}$
$\frac{5}{3} \neq \frac{3}{5}$	$\frac{5}{3} = \frac{5}{3}$

Proportions are so common that we use a shortcut method to solve them. The method depends on the fact that if two fractions $\frac{a}{b}$ and $\frac{c}{d}$ are equivalent, their **cross products** are also equal. Here is the rule.

> **RULE**
>
> **Cross Products**
>
> If $\frac{a}{b} = \frac{c}{d}$, then $ad = bc$.
>
> Note that if $\frac{a}{b} = \frac{c}{d}$
>
> then $bd \cdot \frac{a}{b} = bd \cdot \frac{c}{d}$
>
> $ad = bc$

Thus since $\frac{1}{2} = \frac{2}{4}$, $1 \cdot 4 = 2 \cdot 2$, and since $\frac{3}{9} = \frac{1}{3}$, $3 \cdot 3 = 9 \cdot 1$. To solve the proportion

$$\frac{210}{g} = \frac{35}{2}$$

of Example 1(b), we use the cross product and write:

$$210 \cdot 2 = 35g$$

$$\frac{210 \cdot 2}{35} = g \qquad \text{Dividing by 35 yields } g\text{, as before.}$$

$$12 = g$$

> **NOTE**
>
> This technique avoids having to find the LCD first, but it applies only when you have *one* term on each side of the equation!

B Solving Applications

Web It

To review proportions further and look at some applications, try link 6-2-4 on the Bello Website at mhhe.com/bello.

Ratios and proportions can be used to solve work problems. For example, suppose a worker can finish a certain job in 3 days, whereas another worker can do it in 4 days. How many days would it take both workers working together to complete the job? Before we solve this problem, let's see how ratios play a part in the problem itself.

Since the first worker can do the job in 3 days, she does $\frac{1}{3}$ of the job in 1 day.

The second worker can do the job in 4 days, so he does $\frac{1}{4}$ of the job in 1 day.

Working together they do the job in d days.

And in 1 day, they do $\frac{1}{d}$ of the job.

Here's what happens in 1 day:

$$\underbrace{\text{Fraction of the job done by the first person}}_{\frac{1}{3}} + \underbrace{\text{fraction of the job done by the second person}}_{\frac{1}{4}} = \underbrace{\text{fraction of the job done by both persons}}_{\frac{1}{d}}$$

Teaching Tip

Here's an alternate way to organize numbers for a "work" problem.

Workers	Time	Fraction
First	3 days	$\frac{1}{3}$
Second	4 days	$\frac{1}{4}$
Together	x days	$\frac{1}{x}$

Add the last column to get the equation

$$\frac{1}{3} + \frac{1}{4} = \frac{1}{x}$$

This is a fractional equation that can be solved by multiplying each term by the LCD, $3 \cdot 4 \cdot d = 12d$. Note that we can't "cross multiply" here because we have three terms! Thus, we do it like this:

$$\frac{1}{3} + \frac{1}{4} = \frac{1}{d} \qquad \text{Given.}$$

$$12d \cdot \frac{1}{3} + 12d \cdot \frac{1}{4} = 12d \cdot \frac{1}{d} \qquad \text{Multiply each term by the LCD, } 12d.$$

$$4d + 3d = 12 \qquad \text{Simplify.}$$

$$7d = 12 \qquad \text{Combine like terms.}$$

$$d = \frac{12}{7} = 1\frac{5}{7} \qquad \text{Divide by 7.}$$

Thus if they work together they can complete the job in $1\frac{5}{7}$ days.

EXAMPLE 2 Work problems and ratios

A computer can do a job in 4 hours. Computer sharing is arranged with another computer that can finish the job in 6 hours. How long would it take for both computers operating simultaneously to finish the job?

SOLUTION We use the RSTUV method.

1. **Read the problem.**
   We are asked to find the time it will take both computers working together to complete the job.
2. **Select the unknown.**
   Let h be the number of hours it takes to complete the job when both computers are operating simultaneously.
3. **Think of a plan.**
   Translate the problem:
   The first computer does $\frac{1}{4}$ of the job in 1 hour.
   The second computer does $\frac{1}{6}$ of the job in 1 hour.
   When both work together, they do $\frac{1}{h}$ of the job in 1 hour.
   The sum of the fractions of the job done in 1 hour by each computer, $\frac{1}{4} + \frac{1}{6}$, must equal the fraction of the job done each hour when they are operating simultaneously, $\frac{1}{h}$. Thus

$$\frac{1}{4} + \frac{1}{6} = \frac{1}{h}$$

4. **Use algebra to solve the problem.**
   To solve this equation, we proceed as usual.

$$12h \cdot \frac{1}{4} + 12h \cdot \frac{1}{6} = 12h \cdot \frac{1}{h} \qquad \text{Multiply by the LCD, } 12h.$$

$$3h + 2h = 12 \qquad \text{Simplify.}$$

$$5h = 12$$

$$h = \frac{12}{5} = 2\frac{2}{5} \qquad \text{Divide by 5.}$$

$$= 2.4 \text{ hr}$$

   Thus both computers working together take 2.4 hours to do the job.
5. **Verify the solution.**
   We leave the verification to you.

PROBLEM 2

A worker can finish a report in 5 hr. Another worker can do it in 8 hr. How many hours will it take to finish the report if both workers work on it?

Answer

2. $\frac{40}{13} = 3\frac{1}{13}$ hr

Other types of problems often require ratios and proportions for their solution—for example, distance, rate, and time problems. We've already mentioned that the formula relating these three variables is

$$D = RT$$

(Distance ← D; Rate ← R; Time ← T)

We use this formula to solve a motion problem in the next example.

EXAMPLE 3 Going with the flow

Suppose you are cruising down a river one sunny afternoon in your powerboat. Before you know it, you've gone 60 miles. Now, it's time to get back. Perhaps you can make it back in the same time? Wrong! This time you cover only 40 miles in the same time. What happened? You traveled 60 miles downstream in the same time it took to travel 40 miles upstream! Oh yes, the current—it was flowing at 5 miles per hour. What was the speed of your boat in still water?

SOLUTION We use the RSTUV method.

1. **Read the problem.**
   You are asked to find the speed of your boat in still water.
2. **Select the unknown.**
   Let R be the speed of the boat in still water.

   speed downstream: $R + 5$ Current helps.

   speed upstream: $R - 5$ Current hinders.
3. **Think of a plan.**
   Translate the problem. The time taken downstream and upstream must be the same:

$$T_{up} = T_{down}$$

Since $D = RT$, $$T = \frac{D}{R}$$

$$\underbrace{\frac{60}{R+5}}_{T_{up}} = \underbrace{\frac{40}{R-5}}_{T_{down}}$$

4. **Use the cross-product rule to solve the problem.**

$$60(R - 5) = 40(R + 5) \quad \text{Cross multiply.}$$
$$60R - 300 = 40R + 200 \quad \text{Simplify.}$$
$$60R = 40R + 500 \quad \text{Add 300.}$$
$$20R = 500 \quad \text{Subtract } 40R.$$
$$R = 25 \quad \text{Divide by 20.}$$

   Thus the speed of the boat in still water is 25 miles per hour.
5. **Verify the solution.**
   Substitute 25 for R in

$$\frac{60}{R+5} = \frac{40}{R-5}$$
$$\frac{60}{25+5} = \frac{40}{25-5}$$
$$2 = 2 \quad \text{A true statement}$$

PROBLEM 3

A freight train travels 120 miles in the same time a passenger train covers 140 miles. If the passenger train is 5 mi/hr faster, what is the speed of the freight train?

Teaching Tip

Use the following table to organize the numbers.

	Rate	Time	Distance
Downstream	$(R + 5)$	T_{down}	60
Upstream	$(R - 5)$	T_{up}	40

Answer

3. 30 mi/hr

Now, let's use proportion to look at the infamous 1994 Major League baseball season. Remember the 1994 season? It was cut short by a work stoppage (strike); instead of the usual 162 games, each team played only between 110 and 115 games. The abbreviated season deprived several players of an opportunity to break some of baseball's most coveted records. For example, San Francisco's Matt Williams hit 43 home runs in 112 games that year. If he could have maintained this pace for 162 games, how many homers would he have hit? Using proportions, if we assume that he would have hit h home runs in 162 games, this translates as

$$\frac{\text{Homers}}{\text{Games}} \begin{matrix}\rightarrow \\ \rightarrow\end{matrix} \frac{h}{162} = \frac{43}{112}$$

$$112h = 43 \cdot 162 \qquad \text{Cross multiply.}$$

$$h = \frac{43 \cdot 162}{112} = 62 \text{ (to the nearest whole number)} \qquad \text{Divide by 112.}$$

And this *would* have broken Roger Maris's record of 61 home runs in one season!

EXAMPLE 4 Baseball records and proportions

Do you know what a "walk" or a "base on balls" means? This is baseball lingo for when a batter advances to first base after four pitches that are balls; sometimes he's walked intentionally. In the 1994 baseball season, Frank Thomas of the Chicago White Sox walked 121 times in 113 games! The record for "walks" is held by Babe Ruth, who had 170 walks in 152 games in the 1923 season. If Thomas could have maintained his 1994 ratio of walks for 152 games, would he have broken Ruth's record?

SOLUTION As usual, we use our RSTUV method.

1. **Read the problem.**
   We are asked to find out whether the ratio of Thomas's walks would exceed the ratio set by Ruth—that is, would it be more than 170 walks in 152 games?
2. **Select the unknown.**
   Let w be the number of walks Thomas would receive in 152 games. His *hypothetical* ratio of walks to games is then $\frac{w}{152}$.
3. **Think of a plan.**
   We know that he had 121 walks in 113 games, so his *actual* ratio of walks to games was $\frac{121}{113}$. We can now construct a proportion using this information:

$$\frac{w}{152} = \frac{121}{113}$$

4. **Use the cross-product rule to solve the problem.**

$$113w = 121 \cdot 152 \qquad \text{Cross multiply.}$$

$$w = \frac{121 \cdot 152}{113} \qquad \text{Divide by 113.}$$

$$w = 163 \text{ (to the nearest whole number)} \qquad \text{Simplify.}$$

   Ruth's record stands! Since 163 is less than 170, Thomas wouldn't have broken Ruth's record had he continued to walk at the same rate as in the first 113 games. (Some purists say that a modern baseball season consists of 162 games. Would he have broken the record in 162 games? See Problem 4.)
5. **Verify the solution.**
   We leave the verification to you.

PROBLEM 4

Would Thomas have broken Ruth's record in 162 games?

Answer

4. Yes. In 162 games, Thomas would have had 173 walks. However, the record would be for 162 games, not for 152 games!

In everyday life many objects are *similar* but *not* the same; that is, they have the same shape but are not necessarily the same size. Thus a penny and a nickel are similar, and two computer or television screens may be similar. In geometry, figures that have exactly the same shape but not necessarily the same size are called **similar figures.** Look at the two similar triangles:

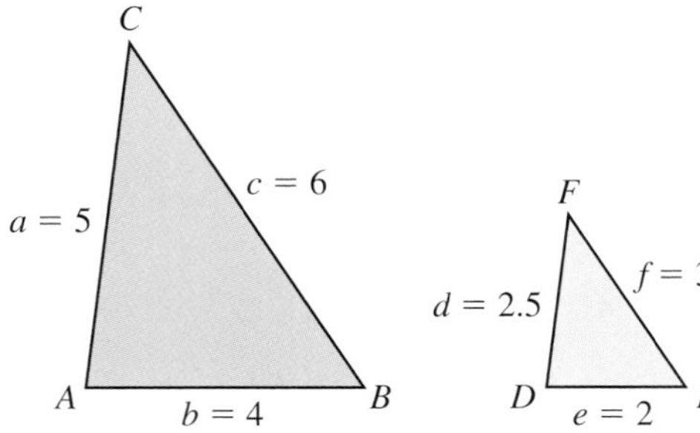

Side *a* corresponds to side *d*.
Side *b* corresponds to side *e*.
Side *c* corresponds to side *f*.

To show that corresponding sides are in proportion, we write the ratios of the corresponding sides:

$$\frac{a}{d} = \frac{5}{2.5} = 2, \quad \frac{b}{e} = \frac{4}{2} = 2, \quad \text{and} \quad \frac{c}{f} = \frac{6}{3} = 2$$

As you can see, corresponding sides are proportional. We use this idea to solve the next example.

EXAMPLE 5 **Similar triangles**

Two similar triangles measured in centimeters (cm) are shown. Find f for the triangle on the right:

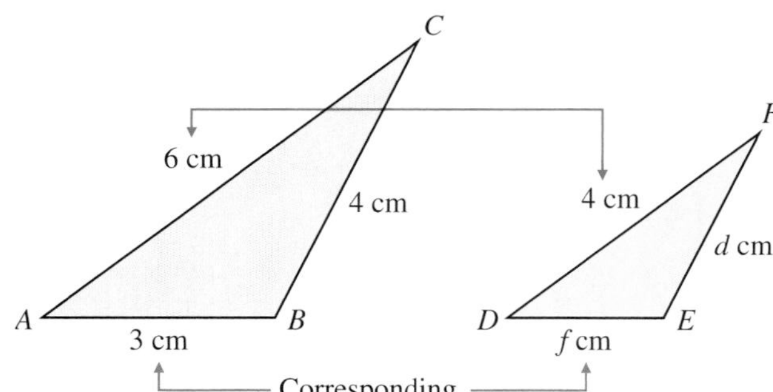

SOLUTION Since the triangles are similar, the corresponding sides must be proportional. Thus

$$\frac{f}{3} = \frac{4}{6}$$

Since $\frac{4}{6} = \frac{2}{3}$, we have

$$\frac{f}{3} = \frac{2}{3}$$

$$3f = 6 \quad \text{Cross multiply.}$$

Solving this equation for f, we get

$$f = \frac{6}{3} = 2 \text{ cm}$$

Thus f is 2 centimeters long.

PROBLEM 5

Find d in the diagram.

Answer

5. $d = \frac{8}{3} = 2\frac{2}{3}$ cm

Exercises 6.6

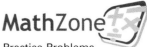

A In Problems 1–13, solve the proportion problems.

1. Do you think your gas station sells a lot of gas? The greatest number of gallons sold through a single pump is claimed by the Downtown Service Station in New Zealand. It sold 7400 Imperial gallons in 24 hours. At that rate, how many Imperial gallons would it sell in 30 hours? 9250

2. Do you like lemons? Bob Blackmore ate 3 whole lemons in 24 seconds. At that rate, how many lemons could he eat in 1 minute (60 seconds)? 7.5

3. Michael Cisneros ate 25 tortillas in 15 minutes. How many could he eat in 1 hour at the same rate? 100

4. If your lawn is yellowing, it may need "essential minor elements" administered at the rate of 2.5 pounds per 100 square feet of lawn. If your lawn covers 150 square feet, how many pounds of essential minor elements do you need? 3.75

5. If you have roses, they may need Ironite® administered at the rate of 4 pounds per 100 square feet. If your rose garden is 15 feet by 10 feet, how many pounds of Ironite do you need? 6

6. The directions for a certain fertilizer recommend that 35 pounds of fertilizer be applied per 1000 square feet of lawn.

a. If your lawn covers 1400 square feet, how many pounds of fertilizer do you need? 49

b. If fertilizer comes in 40-pound bags, how many bags do you need? 2

7. You can fertilize your flowers by using 6-6-6 fertilizer at the rate of 2 pounds per 100 square feet of flower bed. If your flower bed is 12 feet by 15 feet, how many pounds of fertilizer do you need? 3.6

8. Petunias can be planted in a mixture consisting of 2 parts peat moss to 5 parts of potting soil. If you have a 20-pound bag of potting soil, how many pounds of peat moss do you need to make a mixture to plant your petunias? 8

9. According to the U.S. Bureau of the Census, there were 400 homeless persons "visible on the street" for every 1200 homeless in shelters in Dallas.

a. If the homeless persons visible on the street grew to 750, how many would you expect to find in shelters? 2250

b. If each shelter houses 50 homeless persons, how many shelters would be needed when there are 600 homeless persons visible on the street? 36

10. In Minneapolis, there were 30 homeless persons visible on the street for every 1050 homeless in shelters. If the number of homeless visible on the street grew to 75, how many would you expect to find in shelters? 2625

11. What is the ratio of "homeless in shelters" to "homeless visible on the street" in

a. Dallas? (See Problem 9.) 3 to 1

b. Minneapolis? (See Problem 10.) 35 to 1

c. Where is the ratio of "homeless in shelters" to "homeless visible on the street" greater, Dallas or Minneapolis? Why do you think this is so? Minneapolis; climate

12. According to the U.S. Bureau of the Census, about 40,000 persons living in Alabama were born in a foreign country. If Alabama had 4 million people, how many persons born in a foreign country would you expect if the population of Alabama increased to 5 million people? 50,000

13. The District of Columbia had about 60,000 persons born in a foreign country in 1991. If the population of the District of Columbia increased from 600,000 to 700,000, how many foreign-born persons would you expect? 70,000

APPLICATIONS

14. Mr. Gerry Harley, of England, shaved 130 men in 60 minutes. If another barber takes 5 hours to shave the 130 men, how long would it take both men working together to shave the 130 men? 50 min or $\frac{5}{6}$ hr

15. Mr. J. Moir riveted 11,209 rivets in 9 hours (a world record). If it takes another man 12 hours to do this job, how long would it take both men to rivet the 11,209 rivets? $5\frac{1}{7}$ hr

16. It takes a printer 3 hours to print a certain document. If a faster printer can print the document in 2 hours, how long would it take both printers working together to print the document? $1\frac{1}{5}$ hr

17. A computer can send a company's e-mail in 5 minutes. A faster computer does it in 3 minutes. How long would it take both computers operating simultaneously to send out the company's e-mail? $1\frac{7}{8}$ min

18. Two secretaries working together typed the company's annual report in 4 hours. If one of them could type the report by herself in 6 hours, how long would it take the other secretary working alone to type the report? 12 hr

19. Two fax machines can together send all the office mailings in 2 hours. When one of the machines broke down, it took 3 hours to send all the mailings. If the number of faxes sent was the same in both cases, how many hours would it take the remaining fax machine to send all the mailings? 6 hr

20. The train *Grande Vitesse* covers the 264 miles from Paris to Lyons in the same time a regular train covers 124 miles. If the *Grande Vitesse* is 70 miles per hour faster, how fast is it? 132 mi/hr

21. The strongest current on the East Coast of the United States is at St. Johns River in Pablo Creek, Florida, where the current reaches a speed of 6 miles per hour. A motorboat can travel 4 miles downstream on Pablo Creek in the same time it takes to go 16 miles upstream. (No, it is *not* wrong! The St. Johns flows *up*stream.) What is the speed of the boat in still water? 10 mi/hr

22. The strongest current in the United States occurs at Pt. Kootzhahoo in Chatam, Alaska. If a boat that travels at 10 miles per hour in still water takes the same time to travel 2 miles upstream in this area as it takes to travel 18 miles downstream, what is the speed of the current? 8 mi/hr

23. One of the fastest point-to-point trains in the world is the *New Tokaido* from Osaka to Okayama. This train covers 450 miles in the same time a regular train covers 180 miles. If the *Tokaido* is 60 miles per hour faster than the regular train, how fast is it? 100 mi/hr

24. The world's strongest current, reaching 18 miles per hour, is the Saltstraumen in Norway. A motorboat can travel 48 miles downstream in the Saltstraumen in the same time it takes to go 12 miles upstream. What is the speed of the boat in still water? 30 mi/hr

25. In the abbreviated 1994 baseball season, Ken Griffey of the Seattle Mariners hit 40 home runs in 111 games.

a. At this rate, how many home runs would he have hit during a regular season (162 games)? 58

b. Would he have broken Roger Maris's record of 61 home runs in one season? No

26. Frank Thomas gained 291 total bases in 113 games in the 1994 baseball season.

a. At this rate, how many games would he need to reach 457 total bases, Babe Ruth's Major League record? Answer to the nearest whole number. 177

b. A regular season consists of 162 games. Would Thomas have broken Ruth's record? No

27. Frank Thomas scored 106 runs in 113 games in the 1994 baseball season. At this rate, how many games would he need to score 177 runs (Babe Ruth's record)? Could he have broken the record in a regular baseball season consisting of 162 games? 189; no

28. The modern record for the most singles in a season belongs to Lloyd James Waner, who hit 198 singles in 150 games in 1927. To the nearest whole number, how many singles would a player have to hit in a 162-game season to equal Waner's record? 214

In Problems 29–41, the pairs of triangles are similar. Find the lengths of the indicated unknowns.

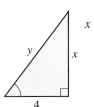

29.

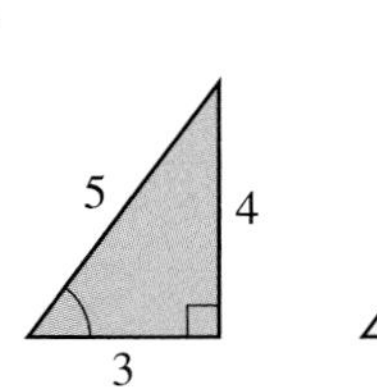

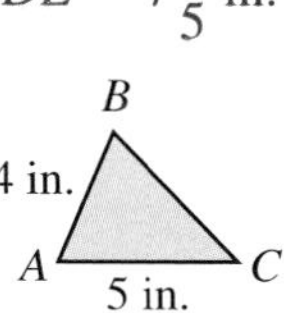

$x = 5\frac{1}{3}$; $y = 6\frac{2}{3}$

30. Side DE $DE = 7\frac{1}{5}$ in.

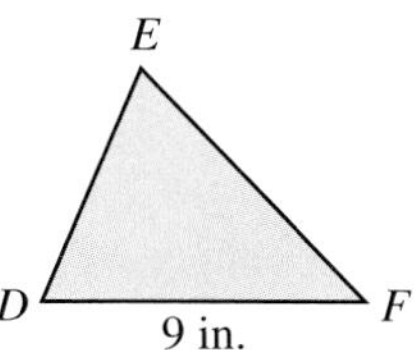

31. Side DE $DE = 13\frac{5}{7}$ in.

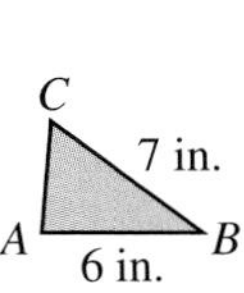

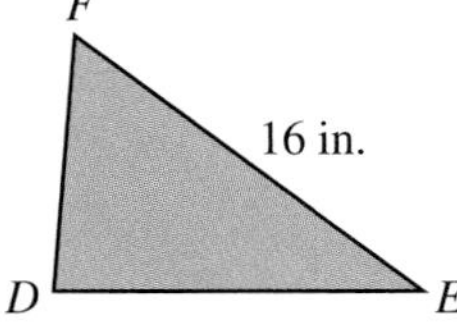

32.

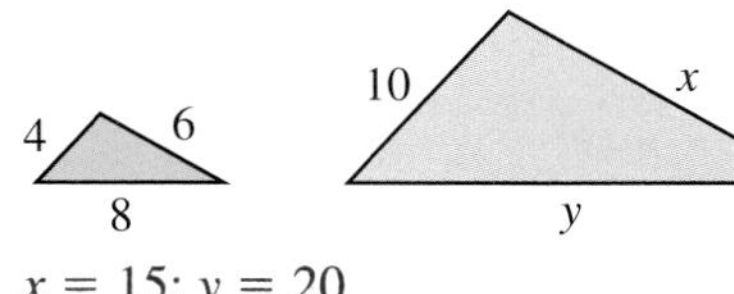

$x = 15$; $y = 20$

33.

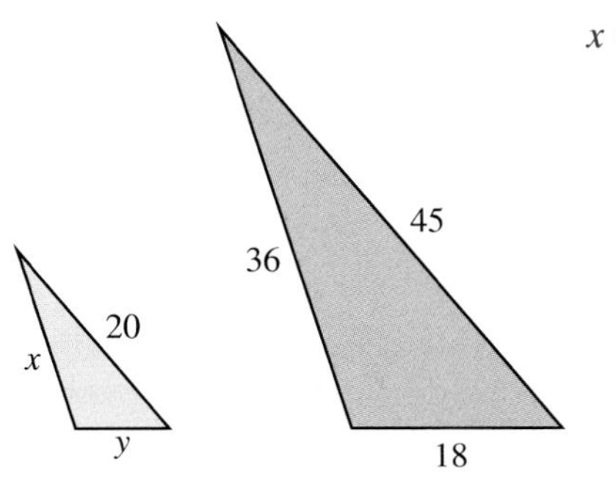

$x = 16$; $y = 8$

34.

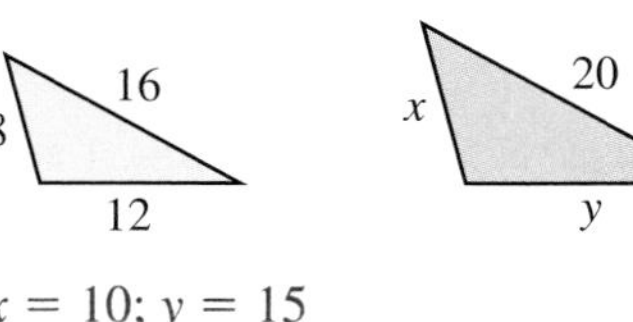

$x = 10$; $y = 15$

35.

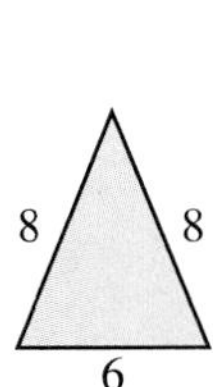

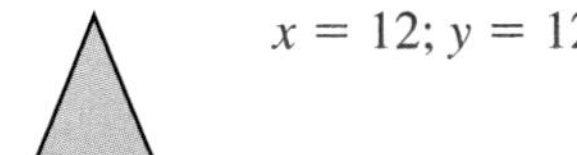

$x = 12$; $y = 12$

36.

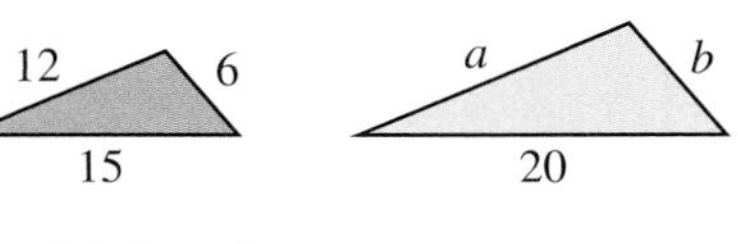

$a = 16$; $b = 8$

37.

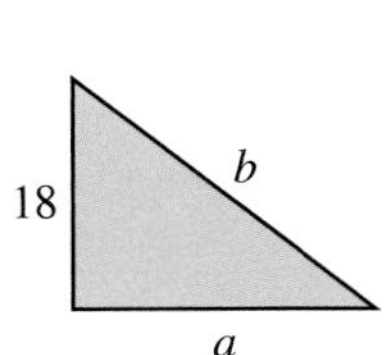

$a = 24$; $b = 30$

38.

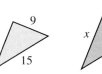

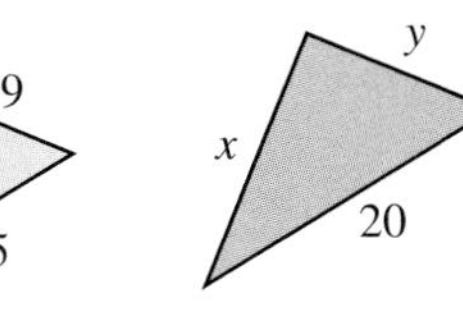

$x = 16$; $y = 12$

39.

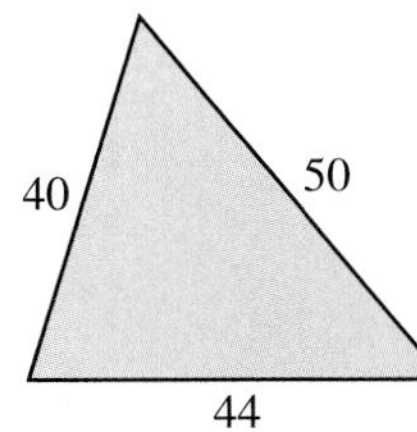

$x = 8$; $y = 8\frac{4}{5}$

40.

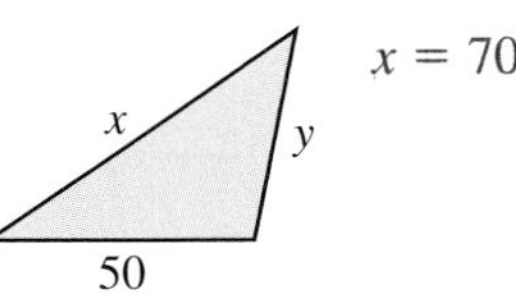

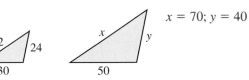

$x = 70$; $y = 40$

41.

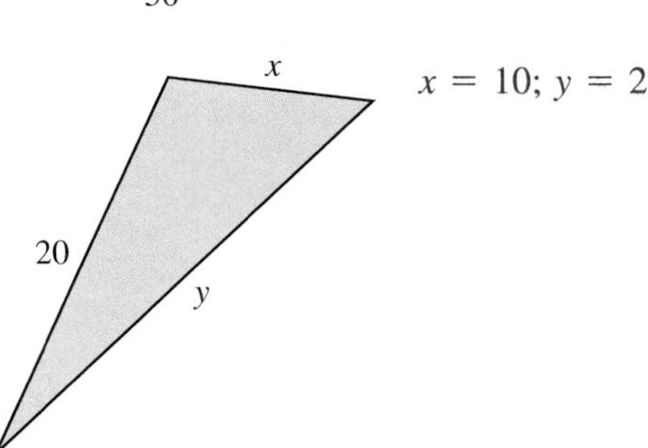

$x = 10$; $y = 25$

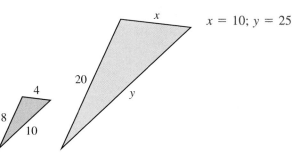

SKILL CHECKER

Try the Skill Checker Exercises so you'll be ready for the next section.

Evaluate:

42. $2a + 3b$ when $a = 2$ and $b = -4$ -8

43. $5x - 2y$ when $x = -2$ and $y = -1$ -8

Solve:

44. $7 = 3x - 2$ $x = 3$

45. $11 = 4x + 3$ $x = 2$

46. $10 = -3x + 2$ $x = \frac{-8}{3}$

USING YOUR KNOWLEDGE

A Matter of Proportion

Proportions are used in many areas other than mathematics. The following problems are typical.

47. In a certain experiment, a physicist stretched a spring 3 inches by applying a 7-pound force to it. How many pounds of force would be required to make the spring stretch 8 inches? $18\frac{2}{3}$ lb

48. Suppose a photographer wishes to enlarge a 2-inch by 3-inch picture so that the longer side is 10 inches. How wide does she have to make it? $6\frac{2}{3}$ in.

49. In carpentry the pitch of a rafter is the ratio of the rise to the run of the rafter. What is the rise of a rafter having a pitch of $\frac{2}{5}$ if the run is 15 feet? 6 ft

50. In an automobile the rear axle ratio is the ratio of the number of teeth in the ring gear to the number of teeth in the pinion gear. A car has a 3-to-1 rear axle ratio, and the ring gear has 60 teeth. How many teeth does the pinion gear have? 20

51. A zoologist took 250 fish from a lake, tagged them, and released them. A few days later 53 fish were taken from the lake, and 5 of them were found to be tagged. Approximately how many fish were in the lake originally? 2650

WRITE ON

52. What is a proportion? Answers may vary.

53. Write the procedure you use to solve a proportion. Answers may vary.

54. We can solve proportions by using their cross products. Can you solve the equation

$$\frac{x}{2} + \frac{x}{4} = 3$$

by using cross products? Explain. Answers may vary.

55. Find some examples of similar figures and then write your own definition of similar figures. Answers may vary.

MASTERY TEST

If you know how to do these problems, you have learned your lesson!

56. A freight train travels 90 miles in the same time a passenger train travels 105 miles. If the passenger train is 5 miles per hour faster, what is the speed of the freight train? 30 mi/hr

57. A typist can finish a report in 5 hours. Another typist can do it in 8 hours. How many hours will it take to finish the report if both work on it? $3\frac{1}{13}$ hr

58. A car travels 150 miles on 9 gallons of gas. How many gallons does the car need to travel 800 miles? 48

59. A baseball player has 160 singles in 120 games. If he continues at this rate, how many singles will he hit in 150 games? 200

60. Find $\frac{d}{2}$ for the triangle in Example 5. $1\frac{1}{3}$ cm

COLLABORATIVE LEARNING

Golden Rectangles

A **golden rectangle** is described as one of the most pleasing shapes to the human eye. Fill in the blanks in the table.

Item	Length	Width	Ratio of length to width	Golden rectangle?
3×5 card				
8.5×11 sheet of paper				
Desktop				
Textbook				
Teacher's desk				
ID card				
Anything else of interest . . .				

Source: Oswego (NY) City School District Regents Exam Prep Center.

1. Which is the ratio of length to width in a golden rectangle?
2. Form several groups. Each group select a topic:
   The golden rectangle in architecture
   The golden section in art
   The golden section in music
   Magical properties of the golden section

Make a report to the rest of the group about your findings.

Research Questions

In this chapter we have written rational numbers either as fractions $(\frac{2}{7})$ or decimals (0.285714285714285714 . . .). The ancient Egyptians used a number system based on *unit fractions*—fractions with a 1 in the numerator. This idea let them represent numbers such as $\frac{1}{7}$ easily enough; other numbers such as $\frac{2}{7}$ were represented as sums of unit fractions (e.g., $\frac{2}{7} = \frac{1}{4} + \frac{1}{28}$). Further, the same fraction could not be used twice (so $\frac{2}{7} = \frac{1}{7} + \frac{1}{7}$ is not allowed). We call a formula representing a sum of distinct unit fractions an *Egyptian fraction.* (*Source:* David Eppstein, ICS, University of California, Irvine.) Do a web search for Egyptian fractions to answer the following questions.

1. Write $\frac{3}{7}$ as an Egyptian fraction.
2. There are many algorithms (procedures) that will show how to write a fraction as an Egyptian fraction. Name three of these algorithms.
3. One of the algorithms uses *continued fractions.* Describe what a *continued fraction* is and how they are used to write fractions as Egyptian fractions.
4. The Indian mathematician Aryabhata (d. 550 A.D.) used a continued fraction to solve a linear *indeterminate equation.* What is an *indeterminate equation?*
5. Write a short paragraph about Aryabhata and his contributions to mathematics.
6. Can you write $\frac{5}{6}$ as an Egyptian fraction? Find two sites that will do it for you.
7. Use the sites you found in question 6 to write $\frac{5}{11}$ as an Egyptian fraction. Do you get the same answer at both sites?

Summary

SECTION	ITEM	MEANING	EXAMPLE
6.1	Rational number	A number that can be written as $\frac{a}{b}$, a and b integers and b not 0	$\frac{3}{4}, -\frac{6}{5}, 0, -8$, and $1\frac{1}{3}$ are rational numbers.
	Rational expression	An expression of the form $\frac{A}{B}$, where A and B are polynomials	$\dfrac{x^2+3}{x-1}$ and $\dfrac{y^2+y-1}{3y^3+7y+8}$ are rational expressions.
6.1B	Fundamental rule of rational expressions	$\dfrac{A}{B}=\dfrac{A\cdot C}{B\cdot C}$ $\quad B\neq 0, C\neq 0$	$\dfrac{3}{4}=\dfrac{3\cdot 5}{4\cdot 5}$ and $\dfrac{x}{x^2+2}=\dfrac{5\cdot x}{5\cdot(x^2+2)}$
6.1C	Reducing a rational expression	The process of removing a common factor from the numerator and denominator of a rational expression	$\dfrac{8}{6}=\dfrac{2\cdot 4}{2\cdot 3}=\dfrac{4}{3}$ is reduced.
	Standard form of a fraction	$\frac{a}{b}$ and $\frac{-a}{b}$ are the standard forms of a fraction.	$-\frac{-a}{-b}$ is written as $\frac{-a}{b}$ and $\frac{a}{-b}$ is written as $\frac{-a}{b}$.
6.2A	Multiplication of rational expressions	$\dfrac{A}{B}\cdot\dfrac{C}{D}=\dfrac{A\cdot C}{B\cdot D}$ $\quad B\neq 0, D\neq 0$	$\dfrac{3}{4}\cdot\dfrac{7}{5}=\dfrac{3\cdot 7}{4\cdot 5}=\dfrac{21}{20}$
6.2B	Division of rational expressions	$\dfrac{A}{B}\div\dfrac{C}{D}=\dfrac{A\cdot D}{B\cdot C}$ $\quad B\neq 0, C\neq 0, D\neq 0$	$\dfrac{3}{7}\div\dfrac{5}{4}=\dfrac{3\cdot 4}{7\cdot 5}=\dfrac{12}{35}$
6.3B	Addition of rational expressions with different denominators	$\dfrac{A}{B}+\dfrac{C}{D}=\dfrac{AD+BC}{BD}$ $\quad B\neq 0, D\neq 0$	$\dfrac{1}{6}+\dfrac{1}{5}=\dfrac{5+6}{6\cdot 5}=\dfrac{11}{30}$
	Subtraction of rational expressions with different denominators	$\dfrac{A}{B}-\dfrac{C}{D}=\dfrac{AD-BC}{BD}$ $\quad B\neq 0, D\neq 0$	$\dfrac{1}{5}-\dfrac{1}{6}=\dfrac{6-5}{5\cdot 6}=\dfrac{1}{30}$
6.4	Complex fraction	A fraction that has other fractions in the numerator, denominator, or both	$\dfrac{\frac{x}{x+2}}{x^2+x+1}$ is a complex fraction.
	Simplifying complex fractions	You can simplify a complex fraction by: 1. Multiplying numerator and denominator by the LCD or 2. Performing the indicated operations in the numerator and denominator and then dividing the numerator by the denominator	Simplify $\dfrac{\frac{1}{x}+1}{\frac{1}{x}-1}$ Method 1: $\dfrac{x\left(\frac{1}{x}+1\right)}{x\left(\frac{1}{x}-1\right)}=\dfrac{1+x}{1-x}$

SECTION	ITEM	MEANING	EXAMPLE
			Method 2: $\dfrac{\frac{1}{x}+\frac{x}{x}}{\frac{1}{x}-\frac{x}{x}} = \dfrac{\frac{1+x}{x}}{\frac{1-x}{x}} = \dfrac{1+x}{x}\cdot\dfrac{x}{1-x}$ $= \dfrac{1+x}{1-x}$
6.5	Fractional equation	An equation containing one or more rational expressions	$\frac{x}{2}+\frac{x}{3}=1$ is a fractional equation.
6.6	Ratio	A quotient of two numbers	3 to 4, 3:4, and $\frac{3}{4}$ are ratios.
6.6A	Proportion Cross products	An equality between ratios If $\frac{a}{b}=\frac{c}{d}$, then $ad = bc$. ad and bc are the cross products.	3:4 as x:6, or $\frac{3}{4}=\frac{x}{6}$ If $\frac{3}{4}=\frac{x}{6}$, then $3\cdot 6 = 4\cdot x$. $3\cdot 6$ and $4\cdot x$ are the cross products.
6.6B	Similar figures	Figures that have the same shape but not necessarily the same size	▭ and ▭ are similar figures.

Review Exercises

(If you need help with these exercises, look in the section indicated in brackets.)

1. [6.1B] Write the given fraction with the indicated denominator.

a. $\frac{5x}{8y}$ with a denominator of $16y^2$ $\frac{10xy}{16y^2}$

b. $\frac{3x}{4y^2}$ with a denominator of $16y^3$ $\frac{12xy}{16y^3}$

c. $\frac{2y}{3x^3}$ with a denominator of $15x^5$ $\frac{10x^2y}{15x^5}$

2. [6.1C] Reduce the given fraction to lowest terms.

a. $\frac{-9(x^2-y^2)}{3(x+y)}$ $-3(x-y)$ or $3y-3x$

b. $\frac{-10(x^2-y^2)}{5(x+y)}$ $-2(x-y)$ or $2y-2x$

c. $\frac{-16(x^2-y^2)}{-4(x+y)}$ $4(x-y)$ or $4x-4y$

3. [6.1A, C] Simplify the given fraction and determine the values for which the expression is undefined.

a. $\frac{-x}{x^2+x}$ $\frac{-1}{x+1}$; 0, -1

b. $\frac{-x}{x^2-x}$ $\frac{-1}{x-1}$ or $\frac{1}{1-x}$; 0, 1

c. $\frac{-x}{x-x^2}$ $\frac{-1}{1-x}$ or $\frac{1}{x-1}$; 0, 1

4. [6.1C] Reduce to lowest terms.

a. $\frac{x^2-3x-18}{6-x}$ $-(x+3)$ or $-x-3$

b. $\frac{x^2-2x-8}{4-x}$ $-(x+2)$ or $-x-2$

c. $\frac{x^2-2x-15}{5-x}$ $-(x+3)$ or $-x-3$

5. [6.2A] Multiply.

a. $\frac{3y^2}{7}\cdot\frac{14x}{9y}$ $\frac{2xy}{3}$

b. $\frac{7y^2}{5}\cdot\frac{15x}{14y}$ $\frac{3xy}{2}$

c. $\frac{6y^3}{7}\cdot\frac{28x}{3y}$ $8xy^2$

6. [6.2A] Multiply.

a. $(x-3)\cdot\frac{x+2}{x^2-9}$ $\frac{x+2}{x+3}$

b. $(x-5)\cdot\frac{x+1}{x^2-25}$ $\frac{x+1}{x+5}$

c. $(x-4)\cdot\frac{x+5}{x^2-16}$ $\frac{x+5}{x+4}$

7. [6.2B] Divide.

a. $\dfrac{x^2-9}{x+2} \div (x+3)$ $\dfrac{x-3}{x+2}$

b. $\dfrac{x^2-16}{x+1} \div (x+4)$ $\dfrac{x-4}{x+1}$

c. $\dfrac{x^2-25}{x+4} \div (x+5)$ $\dfrac{x-5}{x+4}$

8. [6.2B] Divide.

a. $\dfrac{x+5}{x-5} \div \dfrac{x^2-25}{5-x}$ $\dfrac{-1}{x-5}$ or $\dfrac{1}{5-x}$

b. $\dfrac{x+1}{x-1} \div \dfrac{x^2-1}{1-x}$ $\dfrac{-1}{x-1}$ or $\dfrac{1}{1-x}$

c. $\dfrac{x+2}{x-2} \div \dfrac{x^2-4}{2-x}$ $\dfrac{-1}{x-2}$ or $\dfrac{1}{2-x}$

9. [6.3A] Add.

a. $\dfrac{3}{2(x-1)} + \dfrac{1}{2(x-1)}$ $\dfrac{2}{x-1}$

b. $\dfrac{5}{6(x-2)} + \dfrac{7}{6(x-2)}$ $\dfrac{2}{x-2}$

c. $\dfrac{3}{4(x+1)} + \dfrac{1}{4(x+1)}$ $\dfrac{1}{x+1}$

10. [6.3A] Subtract.

a. $\dfrac{7}{2(x+1)} - \dfrac{3}{2(x+1)}$ $\dfrac{2}{x+1}$

b. $\dfrac{11}{5(x+2)} - \dfrac{1}{5(x+2)}$ $\dfrac{2}{x+2}$

c. $\dfrac{17}{7(x+3)} - \dfrac{3}{7(x+3)}$ $\dfrac{2}{x+3}$

11. [6.3B] Add.

a. $\dfrac{2}{x+2} + \dfrac{1}{x-2}$ $\dfrac{3x-2}{(x+2)(x-2)}$

b. $\dfrac{3}{x+1} + \dfrac{1}{x-1}$ $\dfrac{4x-2}{(x+1)(x-1)}$

c. $\dfrac{4}{x+3} + \dfrac{1}{x-3}$ $\dfrac{5x-9}{(x+3)(x-3)}$

12. [6.3B] Subtract.

a. $\dfrac{x-1}{x^2+3x+2} - \dfrac{x+7}{x^2+5x+6}$ $\dfrac{-6x-10}{(x+1)(x+2)(x+3)}$

b. $\dfrac{x+3}{x^2-x-2} - \dfrac{x-1}{x^2+2x+1}$ $\dfrac{7x+1}{(x-2)(x+1)^2}$

c. $\dfrac{x-1}{x^2+3x+2} - \dfrac{x+1}{x^2+x-2}$ $\dfrac{-4x}{(x+2)(x+1)(x-1)}$

13. [6.4A] Simplify.

a. $\dfrac{\dfrac{3}{2x}-\dfrac{1}{x}}{\dfrac{2}{3x}+\dfrac{3}{4x}}$ $\dfrac{6}{17}$

b. $\dfrac{\dfrac{3}{2x}-\dfrac{1}{3x}}{\dfrac{2}{x}+\dfrac{1}{4x}}$ $\dfrac{14}{27}$

c. $\dfrac{\dfrac{3}{2x}-\dfrac{1}{x}}{\dfrac{3}{4x}+\dfrac{4}{3x}}$ $\dfrac{6}{25}$

14. [6.5A] Solve.

a. $\dfrac{2x}{x-1} + 3 = \dfrac{4x}{x-1}$ $x = 3$

b. $\dfrac{6x}{x-5} + 7 = \dfrac{8x}{x-5}$ $x = 7$

c. $\dfrac{5x}{x-4} + 6 = \dfrac{7x}{x-4}$ $x = 6$

15. [6.5A] Solve.

a. $\dfrac{x}{x^2-4} + \dfrac{2}{x-2} = \dfrac{x-3}{x^2-x-6}$ $x = -3$

b. $\dfrac{x}{x^2-16} + \dfrac{4}{x-4} = \dfrac{x-5}{x^2-x-20}$ $x = -5$

c. $\dfrac{x}{x^2-25} + \dfrac{5}{x-5} = \dfrac{x-6}{x^2-x-30}$ $x = -6$

16. [6.5A] Solve.

a. $\dfrac{x}{x+6} - \dfrac{1}{7} = \dfrac{-6}{x+6}$ No solution

b. $\dfrac{x}{x+7} - \dfrac{1}{8} = \dfrac{-7}{x+7}$ No solution

c. $\dfrac{x}{x+8} - \dfrac{1}{9} = \dfrac{-8}{x+8}$ No solution

17. [6.5A] Solve.

a. $3 + \dfrac{5}{x-4} = \dfrac{50}{x^2-16}$ $x = -6$ or $x = \dfrac{13}{3}$

b. $4 + \dfrac{6}{x-5} = \dfrac{84}{x^2-25}$ $x = -7$ or $x = \dfrac{11}{2}$

c. $5 + \dfrac{7}{x-6} = \dfrac{126}{x^2-36}$ $x = -8$ or $x = \dfrac{33}{5}$

18. [6.5B] Solve for the indicated variable.

a. $A = \dfrac{a_1(1-b)}{1-b^n}$; a_1 $a_1 = \dfrac{A(1-b^n)}{1-b}$

b. $B = \dfrac{b_1(1-c^n)}{1-c}$; b_1 $b_1 = \dfrac{B(1-c)}{1-c^n}$

c. $C = \dfrac{c_1(1-d)^n}{d+1}$; c_1 $c_1 = \dfrac{C(d+1)}{(1-d)^n}$

19. [6.6A] A car travels 160 miles on 7 gallons of gas.

a. How many gallons will it need to travel 240 miles? $10\frac{1}{2}$ gal

b. Repeat the problem where the car travels 180 miles on 9 gallons of gas and we wish to go 270 miles. $13\frac{1}{2}$ gal

c. Repeat the problem where the car travels 200 miles on 12 gallons and we wish to go 300 miles. 18 gal

20. [6.6A] Solve using cross products.

a. $\dfrac{x+3}{6} = \dfrac{7}{2}$ $x = 18$

b. $\dfrac{x+4}{8} = \dfrac{9}{5}$ $x = 10\frac{2}{5}$

c. $\dfrac{x+5}{2} = \dfrac{6}{5}$ $x = \dfrac{-13}{5}$

21. [6.6B] A person can do a job in 6 hours. Another person can do it in 8 hours.

a. How long would it take to do the job if both of them work together? $3\frac{3}{7}$ hr

b. Repeat the problem where the first person takes 10 hours and the second person takes 8 hours. $4\frac{4}{9}$ hr

c. Repeat the problem where the first person takes 9 hours and the second person takes 6 hours. $3\frac{3}{5}$ hr

22. [6.6B] A boat can travel 10 miles against a current in the same time it takes to travel 30 miles with the current. What is the speed of the boat in still water if the current flows at

a. 2 miles per hour? 4 mi/hr

b. 4 miles per hour? 8 mi/hr

c. 6 miles per hour? 12 mi/hr

23. [6.6B] A baseball player has 30 home runs in 120 games. At that rate, how many home runs will he have in

a. 128 games? 32

b. 140 games? 35

c. 160 games? 40

24. [6.6B] A company wants to produce 1500 items in 1 year (12 months). To attain this goal, how many items should be produced by the end of

a. September (the 9th month)? 1125

b. October (the 10th month)? 1250

c. November (the 11th month)? 1375

25. [6.6B] Find the unknown in the given similar triangle.

a. Find x. $10\frac{1}{2}$

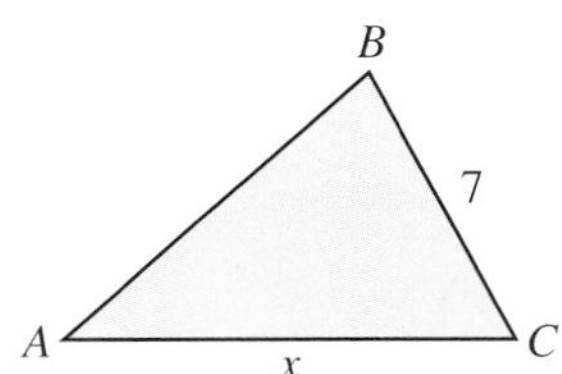

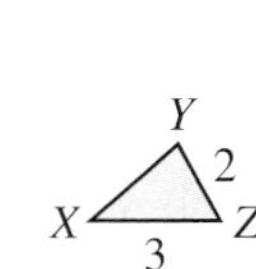

b. Find b. $10\frac{2}{3}$

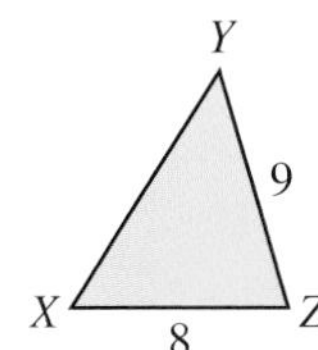

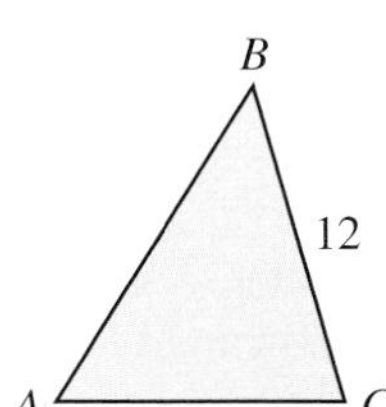

c. Find s. $2\frac{2}{3}$

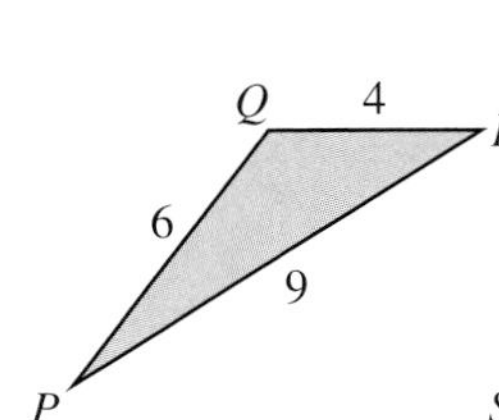

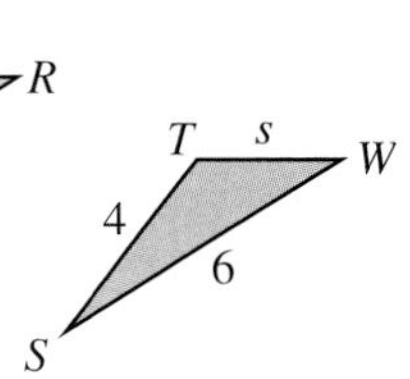

Practice Test 6

(Answers on page 527)

1. Write $\frac{3x}{7y}$ with a denominator of $21y^3$.

2. Reduce $\frac{-6(x^2 - y^2)}{3(x - y)}$ to lowest terms.

3. Determine the values for which the expression $\frac{-x}{x + x^2}$ is undefined and simplify.

4. Reduce to lowest terms $\frac{x^2 + 2x - 8}{2 - x}$.

In Problems 5–12, perform the indicated operations and simplify.

5. Multiply $\frac{2y^2}{7} \cdot \frac{21x}{4y}$.

6. Multiply $(x - 2) \cdot \frac{x + 3}{x^2 - 4}$.

7. Divide $\frac{x^2 - 4}{x + 5} \div (x - 2)$.

8. Divide $\frac{x + 3}{x - 3} \div \frac{x^2 - 9}{3 - x}$.

9. Add $\frac{5}{2(x - 2)} + \frac{1}{2(x - 2)}$.

10. Subtract $\frac{7}{3(x + 1)} - \frac{1}{3(x + 1)}$.

11. Add $\frac{2}{x + 1} + \frac{1}{x - 1}$.

12. Subtract $\frac{x + 1}{x^2 + x - 2} - \frac{x + 2}{x^2 - 1}$.

13. Simplify $\dfrac{\frac{1}{x} - \frac{2}{3x}}{\frac{3}{4x} + \frac{1}{2x}}$.

14. Solve $\frac{3x}{x - 2} + 4 = \frac{5x}{x - 2}$.

15. Solve $\frac{x}{x^2 - 9} + \frac{3}{x - 3} = \frac{1}{x + 3}$.

16. Solve $\frac{x}{x + 5} - \frac{1}{6} = \frac{-5}{x + 5}$.

17. Solve $2 + \frac{4}{x - 3} = \frac{24}{x^2 - 9}$.

18. Solve for d_1 in

$$D = \frac{d_1(1 - d^n)}{(1 + d)^n}$$

19. A car travels 150 miles on 9 gallons of gas. How many gallons will it need to travel 400 miles?

20. Solve $\frac{x + 5}{7} = \frac{11}{6}$.

21. A woman can paint a house in 5 hours. Another one can do it in 8 hours. How long would it take to paint the house if both women work together?

22. A boat can travel 10 miles against a current in the same time it takes to travel 30 miles with the current. If the speed of the current is 8 miles per hour, what is the speed of the boat in still water?

23. A baseball player has 20 singles in 80 games. At that rate, how many singles will he have in 160 games?

24. A conversion van company wants to finish 150 vans in 1 year (12 months). To attain this goal, how many vans should be finished by the end of April (the fourth month)?

25. Find the unknown in the given similar triangles.

a.

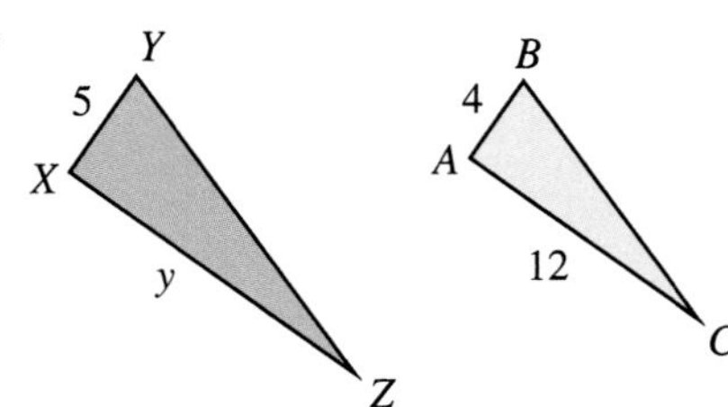

b.

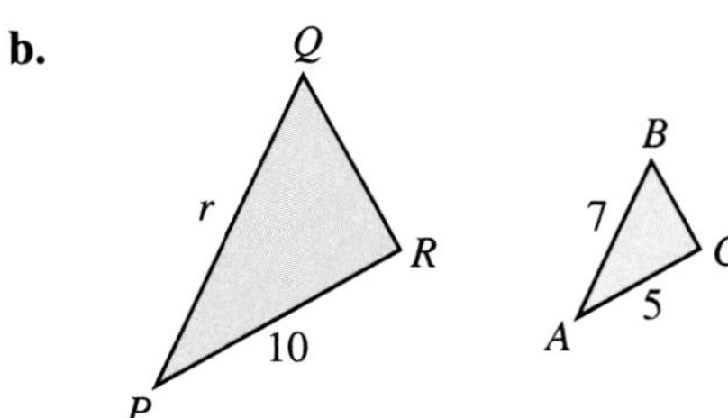

Answers to Practice Test

ANSWER	IF YOU MISSED	REVIEW		
	QUESTION	SECTION	EXAMPLES	PAGE
1. $\frac{9xy^2}{21y^3}$	1	6.1	2	466–467
2. $-2(x + y)$ or $-2x - 2y$	2	6.1	3	468–469
3. $\frac{-1}{1 + x}$; undefined for $x = 0$ and $x = -1$	3	6.1	4, 1	470, 464
4. $-(x + 4)$ or $-x - 4$	4	6.1	5	471–472
5. $\frac{3xy}{2}$	5	6.2	1, 2	478–479
6. $\frac{x + 3}{x + 2}$	6	6.2	3	479–480
7. $\frac{x + 2}{x + 5}$	7	6.2	4	481
8. $\frac{-1}{x - 3}$ or $\frac{1}{3 - x}$	8	6.2	4, 5	481–482
9. $\frac{3}{x - 2}$	9	6.3	1a	487
10. $\frac{2}{x + 1}$	10	6.3	1b	487
11. $\frac{3x - 1}{x^2 - 1}$	11	6.3	2, 4a	487–490
12. $\frac{-2x - 3}{(x + 2)(x + 1)(x - 1)}$	12	6.3	3, 4b, 5	488–492
13. $\frac{4}{15}$	13	6.4	1, 2	498–499
14. $x = 4$	14	6.5	1, 2	503–504
15. $x = -4$	15	6.5	3	504–505
16. No solution	16	6.5	4	505
17. $x = -5$	17	6.5	5	506–507
18. $d_1 = \frac{D(1 + d)^n}{(1 - d^n)}$	18	6.5	6	507
19. 24	19	6.6	1	511
20. $x = \frac{47}{6}$	20	6.6	1	511
21. $3\frac{1}{13}$ hours	21	6.6	2	513
22. 16 miles per hour	22	6.6	3	514
23. 40	23	6.6	4	515
24. 50	24	6.6	4	515
25. a. $y = 15$ **b.** $r = 14$	25	6.6	5	516

Cumulative Review Chapters 1–6

1. Find: $-\frac{1}{6} + \left(-\frac{2}{5}\right)$ $-\frac{17}{30}$

2. Find: $-5.6 - (-8.3)$ 2.7

3. Find: $(-5)^2$ 25

4. Find: $-\frac{7}{8} \div \left(-\frac{7}{16}\right)$ 2

5. Evaluate $y \div 5 \cdot x - z$ for $x = 6$, $y = 60$, $z = 3$. 69

6. Simplify: $x + 4(x - 3) + (x - 2)$ $6x - 14$

7. Write in symbols: The quotient of $(a - b)$ and c. $\frac{a-b}{c}$

8. Solve for x: $5 = 3(x - 1) + 4 - 2x$ $x = 4$

9. Solve for x: $\frac{x}{7} - \frac{x}{9} = 2$ $x = 63$

10. The sum of two numbers is 95. If one of the numbers is 35 more than the other, what are the numbers? 30 and 65

11. Susan purchased some municipal bonds yielding 8% annually and some certificates of deposit yielding 10% annually. If Susan's total investment amounts to \$12,000 and the annual income is \$1020, how much money is invested in bonds and how much is invested in certificates of deposit? \$9000 in bonds; \$3000 in certificates of deposit

12. Graph: $-\frac{x}{2} + \frac{x}{4} \geq \frac{x-4}{4}$

0 2

13. Graph the point $C(-1, -3)$.

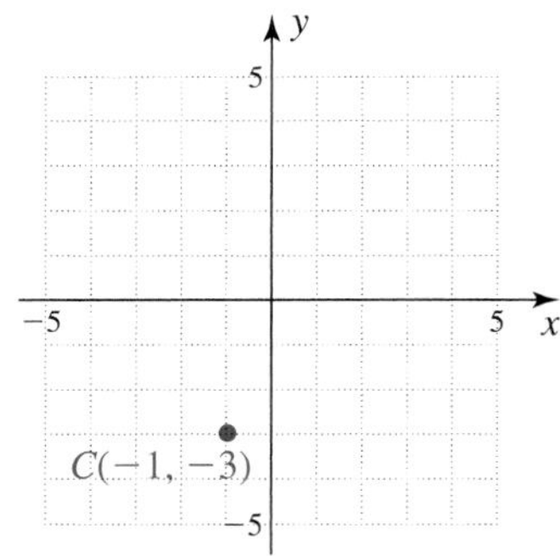

14. Determine whether the ordered pair $(-3, -3)$ is a solution of $5x - y = -12$. Yes

15. Find x in the ordered pair $(x, -2)$ so that the ordered pair satisfies the equation $3x - y = 11$. $x = 3$

16. Graph: $2x + y = 2$

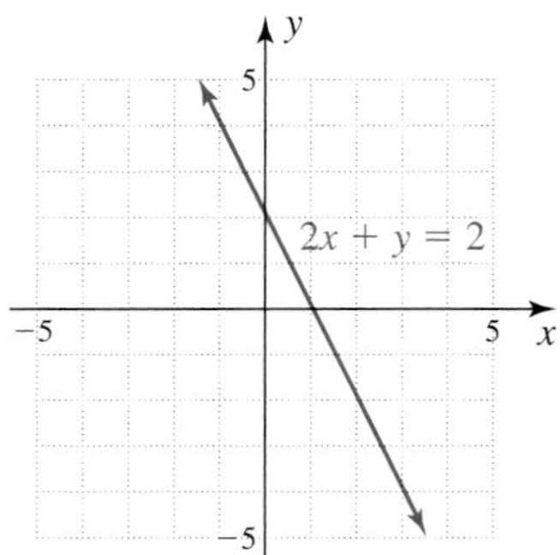

17. Graph: $3y + 6 = 0$

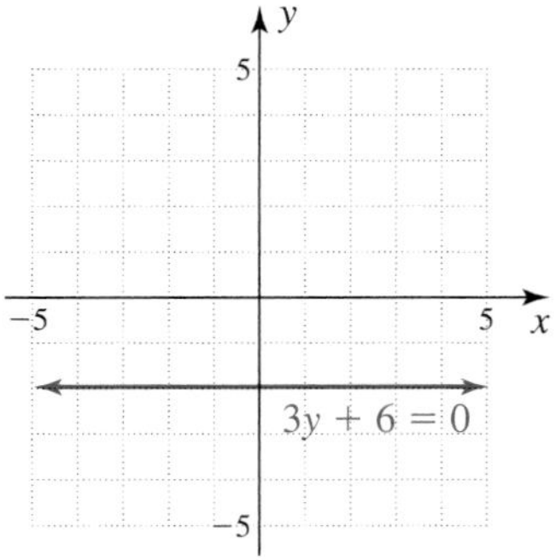

18. Find the slope of the line going through the points $(-7, -9)$ and $(9, -2)$. $\frac{7}{16}$

19. What is the slope of the line $6x - 3y = -10$? 2

20. Find the pair of parallel lines.
(1) $15y + 20x = 8$
(2) $20x - 15y = 8$
(3) $3y = 4x + 8$ (2) and (3)

21. Find: $\frac{x^{-8}}{x^{-9}}$ x

22. Multiply and simplify: $x^5 \cdot x^{-7}$ $\frac{1}{x^2}$

23. Simplify: $(3x^3y^{-2})^{-4}$ $\frac{y^8}{81x^{12}}$

24. Write in scientific notation: 0.00036 3.6×10^{-4}

25. Divide and express the answer in scientific notation: $(14.04 \times 10^{-3}) \div (7.8 \times 10^2)$ 1.80×10^{-5}

26. Find the value of $x^2 + 3x - 2$ when $x = -2$. -4

27. Add $(2x^4 - 6x^6 - 7)$ and $(-6x^6 + 3 + 7x^4)$.
$-12x^6 + 9x^4 - 4$

28. Find (expand): $\left(5x^2 - \frac{1}{5}\right)^2$ $25x^4 - 2x^2 + \frac{1}{25}$

29. Find: $(9x^2 + 2)(9x^2 - 2)$ $81x^4 - 4$

30. Divide $(3x^3 + 11x^2 + 18)$ by $(x + 4)$.
$(3x^2 - x + 4)$ R 2

31. Factor completely: $6x^2 - 9x^4$ $3x^2(2 - 3x^2)$

32. Factor completely: $\frac{2}{5}x^7 - \frac{4}{5}x^6 + \frac{4}{5}x^5 - \frac{1}{5}x^3$
$\frac{1}{5}x^3(2x^4 - 4x^3 + 4x^2 - 1)$

33. Factor completely: $x^2 - 15x + 56$ $(x - 7)(x - 8)$

34. Factor completely: $15x^2 - 37xy + 20y^2$
$(5x - 4y)(3x - 5y)$

35. Factor completely: $16x^2 - 9y^2$ $(4x + 3y)(4x - 3y)$

36. Factor completely: $-4x^4 + 4x^2$ $-4x^2(x + 1)(x - 1)$

37. Factor completely: $3x^3 - 6x^2 - 9x$ $3x(x + 1)(x - 3)$

38. Factor completely: $4x^2 + 4x + 3x + 3$
$(x + 1)(4x + 3)$

39. Factor completely: $16kx^2 - 8kx + k$ $k(4x - 1)^2$

40. Solve for x: $2x^2 + x = 15$ $x = -3, x = \frac{5}{2}$

41. Write $\frac{7x}{6y}$ with a denominator of $12y^3$. $\frac{14xy^2}{12y^3}$

42. Reduce to lowest terms: $\frac{-8(x^2 - y^2)}{4(x - y)}$
$-2(x + y)$ or $-2x - 2y$

43. Reduce to lowest terms: $\frac{x^2 + 4x - 5}{1 - x}$
$-(x + 5)$ or $-x - 5$

44. Multiply: $(x - 7) \cdot \frac{x + 3}{x^2 - 49}$ $\frac{x + 3}{x + 7}$

45. Divide: $\frac{x + 3}{x - 3} \div \frac{x^2 - 9}{3 - x}$ $\frac{-1}{x - 3}$ or $\frac{1}{3 - x}$

46. Add: $\frac{3}{2(x + 8)} + \frac{9}{2(x + 8)}$ $\frac{6}{x + 8}$

47. Subtract: $\frac{x + 5}{x^2 + x - 30} - \frac{x + 6}{x^2 - 25}$
$\frac{-2x - 11}{(x + 6)(x + 5)(x - 5)}$

48. Simplify: $\dfrac{\frac{2}{x} + \frac{3}{2x}}{\frac{1}{3x} - \frac{1}{4x}}$ 42

49. Solve for x: $\frac{3x}{x - 2} + 3 = \frac{4x}{x - 2}$ 3

50. Solve for x: $\frac{x}{x^2 - 49} + \frac{7}{x - 7} = \frac{1}{x + 7}$ $x = -8$

51. Solve for x: $\frac{x}{x + 4} - \frac{1}{5} = \frac{-4}{x + 4}$ No solution

52. Solve for x: $1 + \frac{2}{x - 3} = \frac{12}{x^2 - 9}$ $x = -5$

53. A car travels 120 miles on 6 gallons of gas. How many gallons will it need to travel 460 miles? 23 gal

54. Solve for x: $\frac{x + 8}{9} = \frac{2}{11}$ $-\frac{70}{11}$

55. Maria can paint a kitchen in 6 hours, and James can paint the same kitchen in 7 hours. How long would it take for both working together to paint the kitchen? $3\frac{3}{13}$ hr

Graphs, Slopes, Inequalities, and Applications

7

The Human Side of Algebra

By the middle of the 1600s, the concepts involved in coordinate graphing were becoming well known, and their dissemination was accomplished through math teachers and the textbooks they wrote. One of the notable writers about graphing was an Italian woman named Maria Gaetana Agnesi, who wrote one of the first calculus textbooks. Her motive? To teach her little brother about math! Her book, published in 1748, contained the graph of a curve affectionately called the Witch of Agnesi.

The first person to use the word *graph* was James Joseph Sylvester, a teacher born in London in 1814, who used the word in an article published in 1878. Sylvester passed the knowledge of graphs to his students, writing articles in *Applied Mechanics* and admonishing his students that they would "do well to graph on squared paper some curves like the following." Unfortunately for his students, Sylvester had a dangerous temper and attacked a student with a sword cane at the University of Virginia. The infraction? The student was reading a newspaper during his class. So now you can learn from this lesson: concentrate on your graphs and do not read newspapers in the classroom!

Pretest for Chapter 7

(Answers on pages 533–534)

1. Find an equation of the line going through the point $(2, -5)$ and with slope -4. Write the answer in point-slope form and then graph the line.

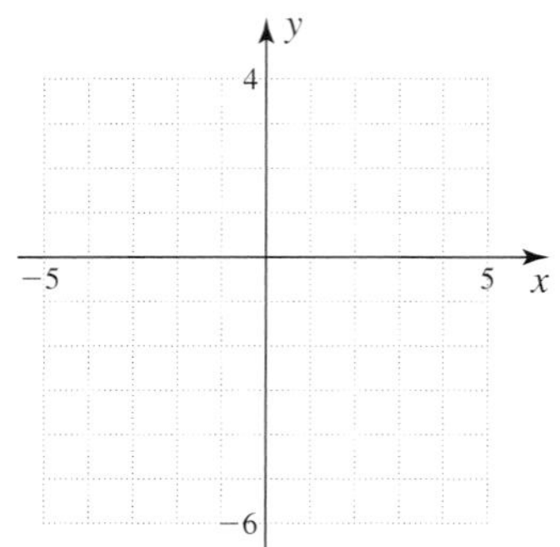

2. Find an equation of the line with slope 6 and y-intercept -4. Write the answer in slope-intercept form and then graph the line.

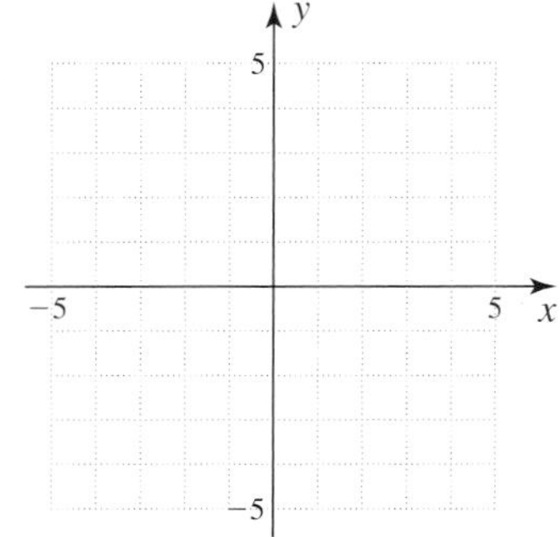

3. Find an equation of the line passing through the points $(2, -8)$ and $(-3, -3)$. Write the equation in standard form.

4. International telephone rates for m minutes are \$10 plus \$0.30 for each minute. If a 10-minute call costs \$13, write an equation for the total cost C and find the cost of a 15-minute call.

5. A cell phone plan costs \$40 per month with 500 free minutes and \$0.60 for each additional minute. Find an equation for the total cost C of the plan when m minutes are used after the first 500. What is the cost when 800 total minutes are used?

6. The bills for two long-distance calls are \$7 for 5 minutes and \$9 for 10 minutes. Find an equation for the total cost C of calls when m minutes are used.

7. Graph $2x - 3y < -6$.

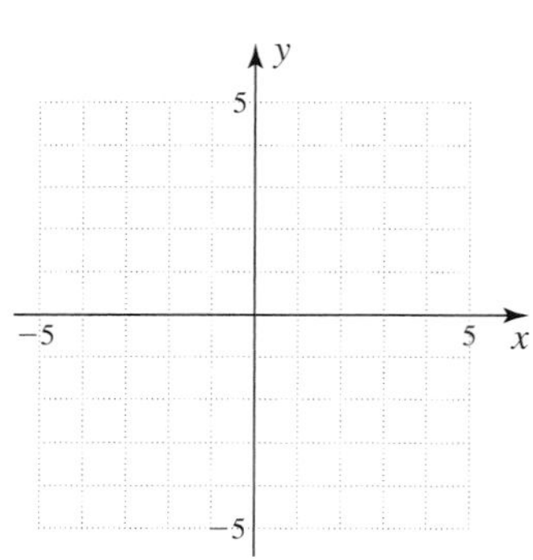

8. Graph $-y \le -2x + 2$.

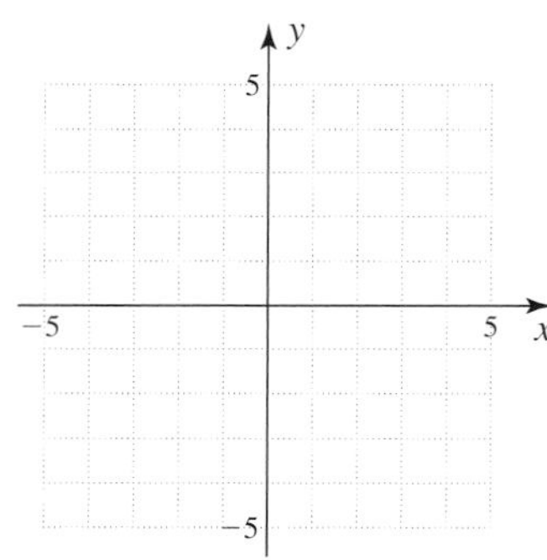

9. Graph $3x - y > 0$.

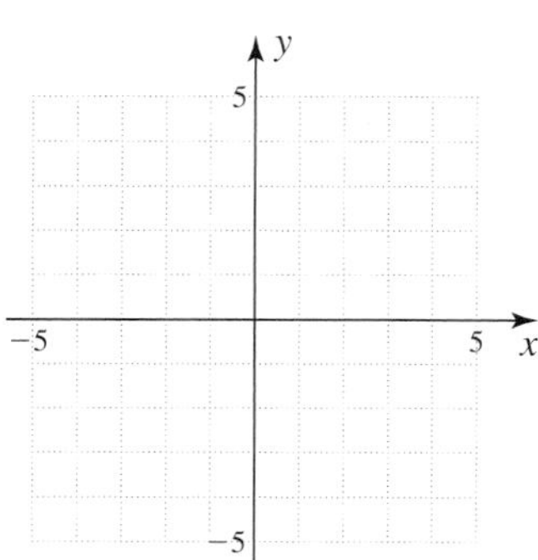

10. Graph $3y - 12 \ge 0$.

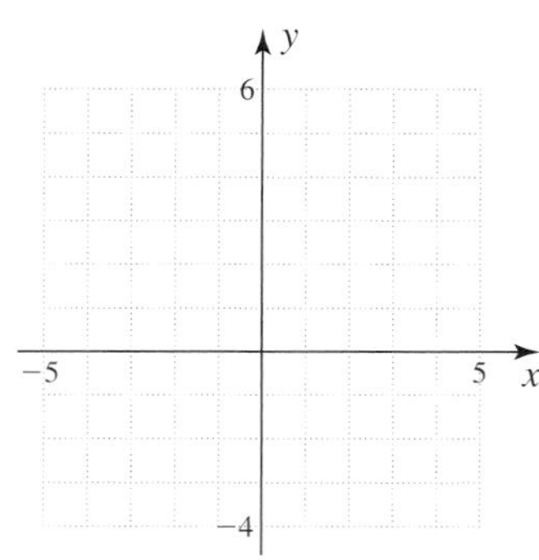

11. Have you raked leaves lately? If you rake leaves for m minutes, the number C of calories used is proportional to the time you rake. If 60 calories are used when you rake for 30 minutes, write an equation of variation and find the number of calories used when you rake for $1\frac{1}{2}$ hours.

12. The maximum weight W that can be supported by a 2-by-4-inch piece of pinewood varies inversely with its length L. If the maximum weight W that can be supported by a 10-foot-long 2-by-4 piece of pine is 600 pounds, find an equation of variation and the maximum weight W that can be supported by a 25-foot length of 2-by-4 pine.

Answers to Pretest

ANSWER	IF YOU MISSED	REVIEW		
	QUESTION	SECTION	EXAMPLES	PAGE
1. $y + 5 = -4(x - 2)$ 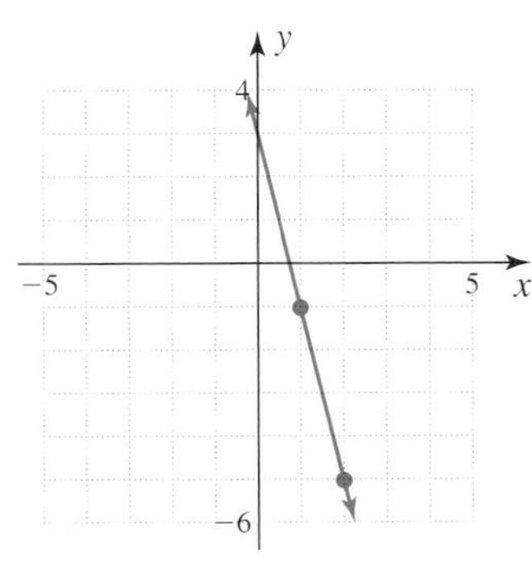	1	7.1	1	536
2. $y = 6x - 4$ 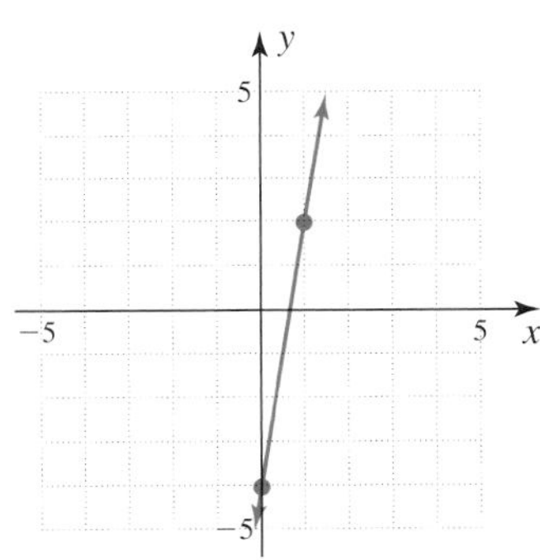	2	7.1	2	537
3. $x + y = -6$	3	7.1	3	538
4. $C = 0.30m + 10$; \$14.50	4	7.2	1	544
5. $C = 0.60m + 40$; \$58	5	7.2	2	544
6. $C = 0.40m + 5$	6	7.2	3	545
7. 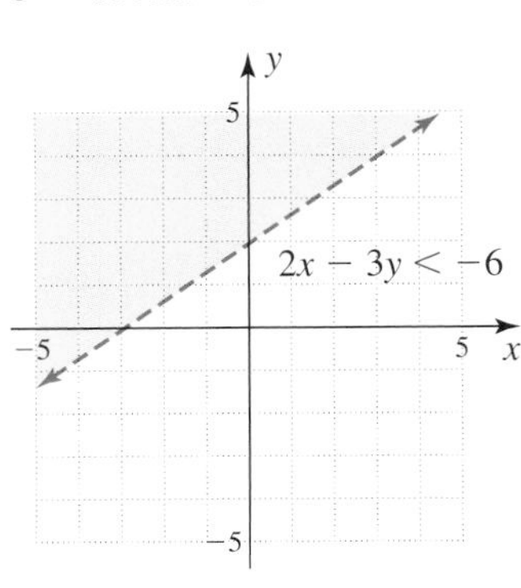	7	7.3	1	551

ANSWER	IF YOU MISSED	REVIEW		
	QUESTION	SECTION	EXAMPLES	PAGE
8. 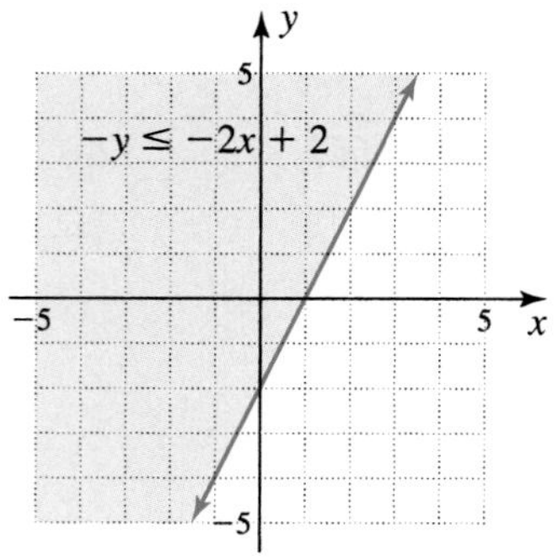 $-y \le -2x + 2$	8	7.3	2	552
9. 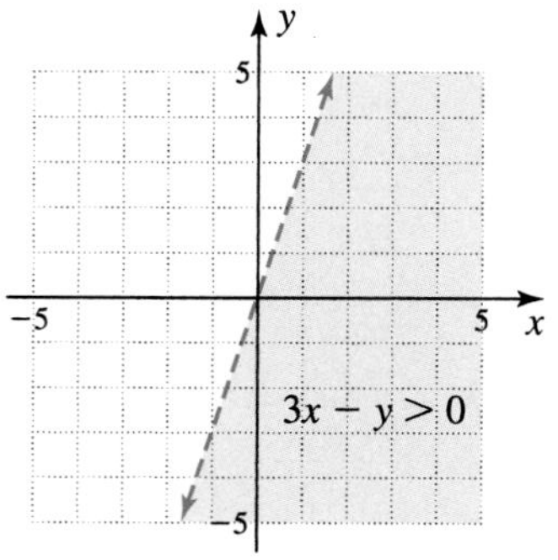 $3x - y > 0$	9	7.3	3	553
10. 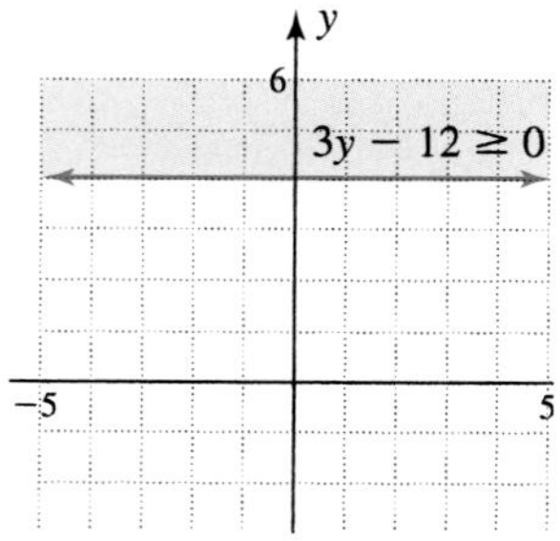 $3y - 12 \ge 0$	10	7.3	4	554
11. $C = 2m$; 180	11	7.4	1, 2	561, 562
12. $W = \frac{6000}{L}$; 240 pounds	12	7.4	3, 4	563

7.1 GRAPHING LINES USING POINTS AND SLOPES

To Succeed, Review How To ...

1. Add, subtract, multiply, and divide signed numbers (pp. 61–66, 70–75).
2. Solve a linear equation for a specified variable (pp. 156–158).

Objectives

Find and graph an equation of a line given:

A Its slope and a point on the line.

B Its slope and y-intercept.

C Two points on the line.

GETTING STARTED

The Formula for Cholesterol Reduction

As we discussed in Example 3, Section 3.2 (p. 244), the cholesterol level C can be approximated by $C = -3w + 215$, where w is the number of weeks elapsed. How did we get this equation? You can see by looking at the graph that the cholesterol level decreases *about* 3 points each week, so the slope of the line is -3. You can make this approximation more exact by using the points (0, 215) and (12, 175) to find the slope

$$m = \frac{215 - 175}{0 - 12} = \frac{-40}{12} \approx -3.3$$

Cholesterol Level Reduction

Since the y-intercept is at 215, you can reason that the cholesterol level starts at 215 and decreases about 3 points each week. Thus the cholesterol level C based on the number of weeks elapsed is

$$C = 215 - 3w$$

or, equivalently,

$$C = -3w + 215$$

If you want a more exact approximation, you can also write

$$C = -3.3w + 215$$

Thus if you are given the slope m and y-intercept b of a line, the equation of the line will be $y = mx + b$.

In this section we shall learn how to find and graph a linear equation when we are given the slope and a point on the line, the slope and the y-intercept, or two points on the line.

Using the Point-Slope Form of a Line

We can use the slope of a line to obtain the equation of the line provided we are given one point on the line. Thus suppose a line has slope m and passes through the point (x_1, y_1). If we let (x, y) be a second point on the line, the slope of the line shown in Figure 1 is given by

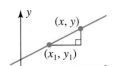

Figure 1

$$\frac{y - y_1}{x - x_1} = m$$

Multiplying both sides by $(x - x_1)$, we get the *point-slope form* of the line.

POINT-SLOPE FORM

The **point-slope form** of the equation of the line going through (x_1, y_1) and having slope m is

$$y - y_1 = m(x - x_1)$$

EXAMPLE 1 **Finding an equation for a line given a point and the slope**

a. Find an equation of the line that goes through the point $(2, -3)$ and has slope $m = -4$.

b. Graph the line.

SOLUTION

a. Using the point-slope form, we get

$$y - (-3) = -4(x - 2)$$
$$y + 3 = -4x + 8$$
$$y = -4x + 5$$

b. To graph this line, we start at the point $(2, -3)$. Since the slope of the line is -4 and, by definition, the slope is

$$\frac{\text{Rise}}{\text{Run}} = \frac{-4}{1}$$

we go 4 units *down* (the rise) and 1 unit *right* (the run), ending at the point $(3, -7)$. We then join the points $(2, -3)$ and $(3, -7)$ with a line, which is the graph of $y = -4x + 5$ (see Figure 2). As a final check, does the point $(3, -7)$ satisfy the equation $y = -4x + 5$? When $x = 3$, $y = -4x + 5$ becomes

$$y = -4(3) + 5 = -7 \qquad \text{(True!)}$$

Thus the point $(3, -7)$ is on the line $y = -4x + 5$. Note that the equation $y + 3 = -4x + 8$ can be written in the standard form $Ax + By = C$.

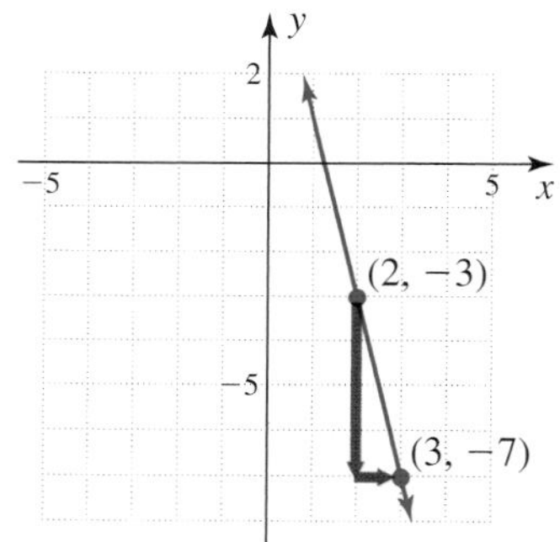

Figure 2

$y + 3 = -4x + 8$	Given.
$4x + y + 3 = 8$	Add $4x$.
$4x + y = 5$	Subtract 3.

PROBLEM 1

a. Find an equation of the line that goes through the point $(2, -4)$ and has slope $m = -3$.

b. Graph the line.

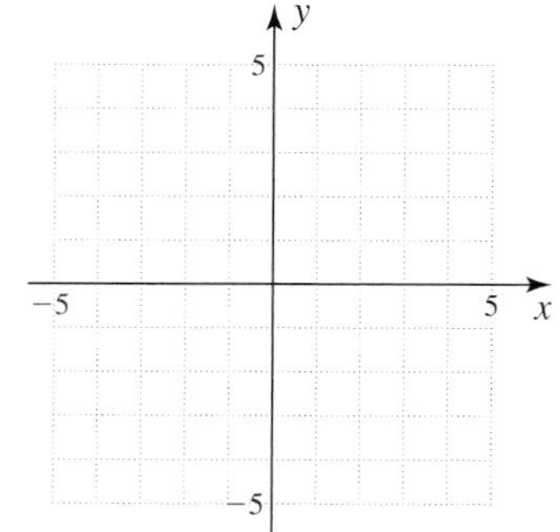

Teaching Tip

In Example 1, if the given slope is a fraction it is easier to use the formula

$$m = \frac{y - y_1}{x - x_1}$$

and cross multiply.

Example:

Write the equation of the line that goes through $(2, 3)$ and has slope $= \frac{3}{4}$.

$$\frac{3}{4} = \frac{y - 3}{x - 2}$$
$$4(y - 3) = 3(x - 2)$$
$$4y - 12 = 3x - 6$$

etc.

Answers

1. a. $y = -3x + 2$

b.

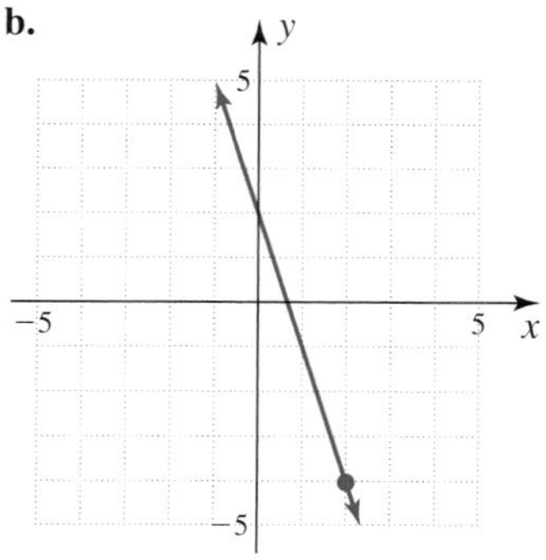

B The Slope-Intercept Form of a Line

An important special case of the point-slope form is that in which the given point is the point where the line intersects the y-axis. Let this point be denoted by $(0, b)$. Then b is called the *y-intercept* of the line. Using the point-slope form, we obtain

$$y - b = m(x - 0)$$

or

$$y - b = mx$$

By adding b to both sides, we get the *slope-intercept form* of the equation of the line.

SLOPE-INTERCEPT FORM

The **slope-intercept form** of the equation of the line having slope m and y-intercept b is

$$y = mx + b$$

EXAMPLE 2 Finding an equation of a line given the slope and the y-intercept

a. Find an equation of the line having slope 5 and y-intercept -4.

b. Graph the line.

SOLUTION

a. In this case $m = 5$ and $b = -4$. Substituting in the slope-intercept form, we obtain

$$y = 5x + (-4) \quad \text{or} \quad y = 5x - 4$$

b. To graph this line, we start at the y-intercept $(0, -4)$. Since the slope of the line is 5 and by definition the slope is

$$\frac{\text{Rise}}{\text{Run}} = \frac{5}{1}$$

go 5 units *up* (the rise) and 1 unit *right* (the run), ending at $(1, 1)$. We join the points $(0, -4)$ and $(1, 1)$ with a line, which is the graph of $y = 5x - 4$ (see Figure 3). Now check that the point $(1, 1)$ is on the line $y = 5x - 4$.

When $x = 1$, $y = 5x - 4$ becomes

$$y = 5(1) - 4 = 1 \quad \text{(True!)}$$

Thus the point $(1, 1)$ is on the line $y = 5x - 4$.

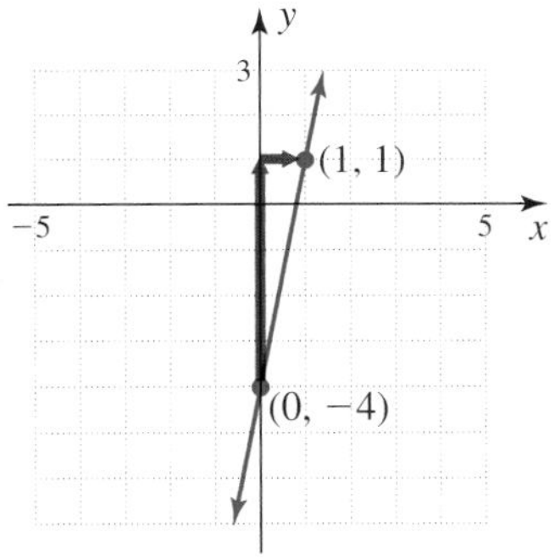

Figure 3

PROBLEM 2

a. Find an equation of the line with slope 3 and y-intercept -6.

b. Graph the line.

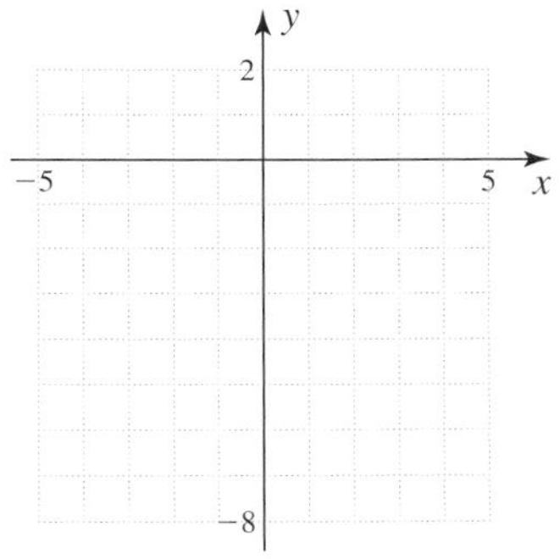

Answers

2. a. $y = 3x - 6$

b.

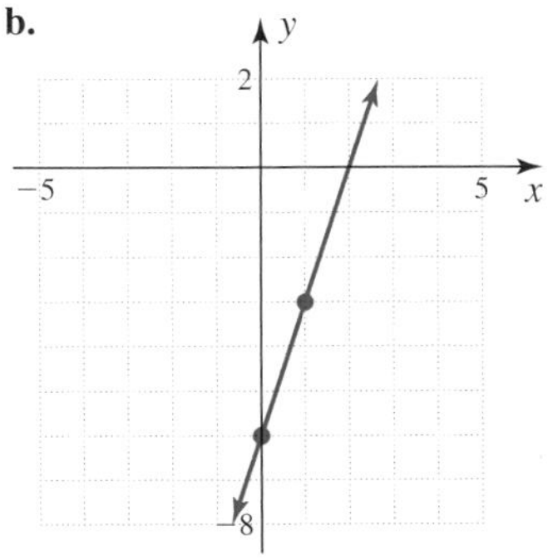

C The Two-Point Form of a Line

If a line passes through two given points, you can find and graph the equation of the line by using the point-slope form as shown in the next example.

EXAMPLE 3 Finding an equation of a line given two points

a. Find an equation of the line passing through the points $(-2, 3)$ and $(1, -3)$.

b. Graph the line.

SOLUTION

a. We first find the slope of the line.

$$m = \frac{3 - (-3)}{-2 - 1} = \frac{6}{-3} = -2$$

Now we can use either the point $(-2, 3)$ or $(1, -3)$ and the point-slope form to find the equation of the line. Using the point $(-2, 3) = (x_1, y_1)$ and $m = -2$,

$$y - y_1 = m(x - x_1)$$

becomes

$$y - 3 = -2[x - (-2)]$$
$$y - 3 = -2(x + 2)$$
$$y - 3 = -2x - 4$$
$$y = -2x - 1$$

or, in standard form, $2x + y = -1$.

b. To graph this line, we simply plot the given points $(-2, 3)$ and $(1, -3)$ and join them with a line, as shown in Figure 4. To check our results, we make sure that both points satisfy the equation $2x + y = -1$. For $(-2, 3)$, let $x = -2$ and $y = 3$ in $2x + y = -1$ to obtain

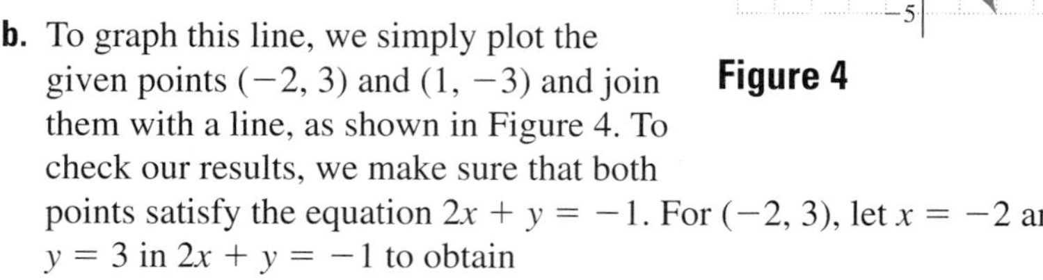

Figure 4

$$2(-2) + 3 = -4 + 3 = -1$$

Thus $(-2, 3)$ satisfies $2x + y = -1$, and our result is correct.

PROBLEM 3

a. Find an equation of the line going through $(5, -6)$ and $(-3, 4)$.

b. Graph the line.

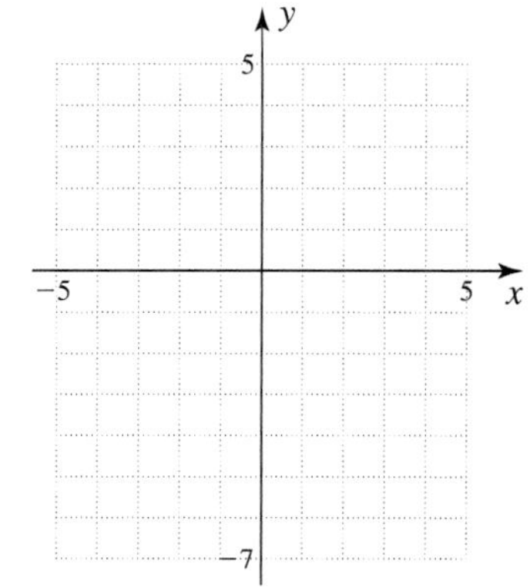

Teaching Tip

Include examples of writing equations of horizontal and vertical lines.

Examples:

1. (2, 3) and (4, 3)

$$m = \frac{3 - 3}{2 - 4} = \frac{0}{-2} = 0$$
$$y - 3 = 0(x - 2)$$
$$y = 3$$

2. (4, 1) and (4, 2)

$$m = \frac{1 - 2}{4 - 4} = \frac{-1}{0} = \text{undefined}$$

Since the slope is undefined, this must be a vertical line and $x = 4$.

Answers

3. a. $y = -\frac{5}{4}x + \frac{1}{4}$

b.

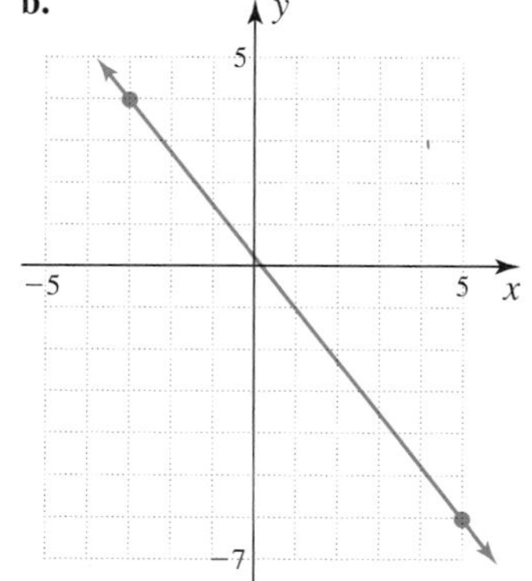

At this point, many students ask, "How do we know which form to use?" The answer depends on what information is given in the problem. The following table helps you make this decision. It's a good idea to examine this table closely before you attempt the problems in Exercises 7.1.

Web It

To see a lesson dealing with lines and slopes, go to link 7-1-1 on the Bello Website at mhhe.com/bello.

To see the derivation of the point-slope formula with animations, try link 7-1-2.

To practice graphing linear equations, try link 7-1-3.

To find the slope-intercept form when two points are given, try link 7-1-4.

Finding the Equation of a Line

Given	Use
A point (x_1, y_1) and the slope m	Point-slope form: $y - y_1 = m(x - x_1)$
The slope m and the y-intercept b	Slope-intercept form: $y = mx + b$
Two points (x_1, y_1) and (x_2, y_2), $x_1 \neq x_2$	Two-point form: $y - y_1 = m(x - x_1)$, where $m = \dfrac{y_2 - y_1}{x_2 - x_1}$

Note that the resulting equation can always be written in the standard form: $Ax + By = C$.

Exercises 7.1

A In Problems 1–6, find the slope-intercept form (if possible) of the equation of the line that has the given properties (m is the slope), then graph the line.

1. Goes through (1, 2); $m = \frac{1}{2}$ $y = \frac{1}{2}x + \frac{3}{2}$

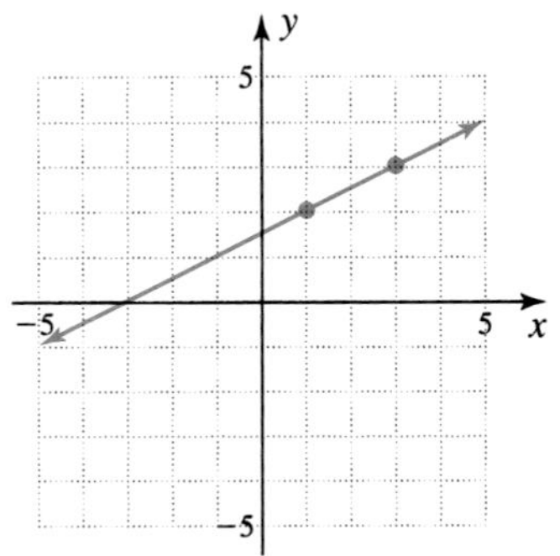

2. Goes through (−1, −2); $m = -2$ $y = -2x - 4$

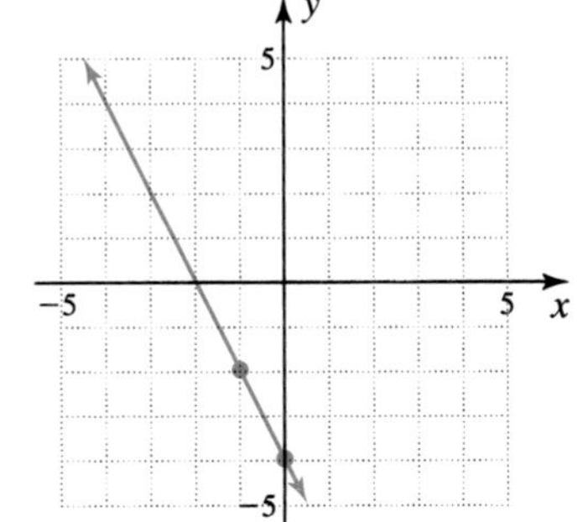

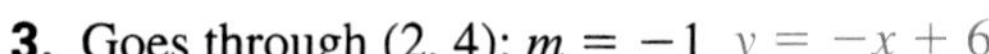

3. Goes through (2, 4); $m = -1$ $y = -x + 6$

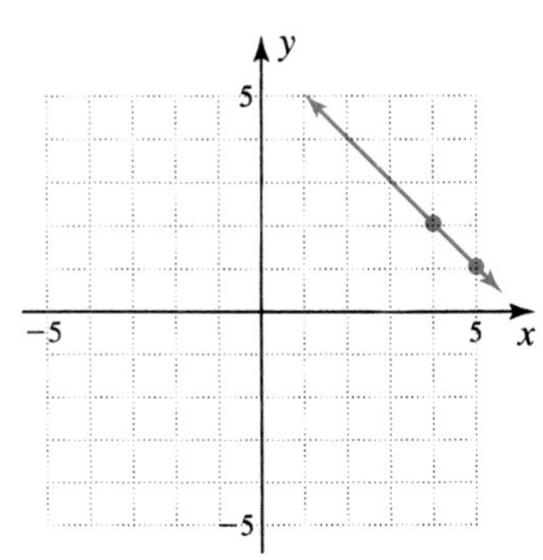

4. Goes through $(-3, 1)$; $m = \frac{3}{2}$
$y = \frac{3}{2}x + \frac{11}{2}$

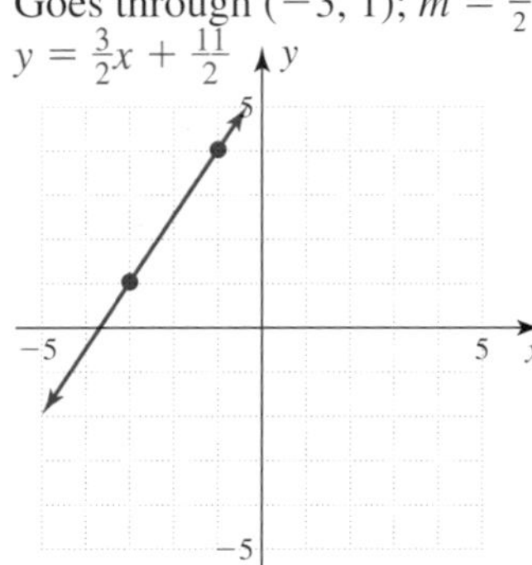

5. Goes through $(4, 5)$; $m = 0$
$y = 5$

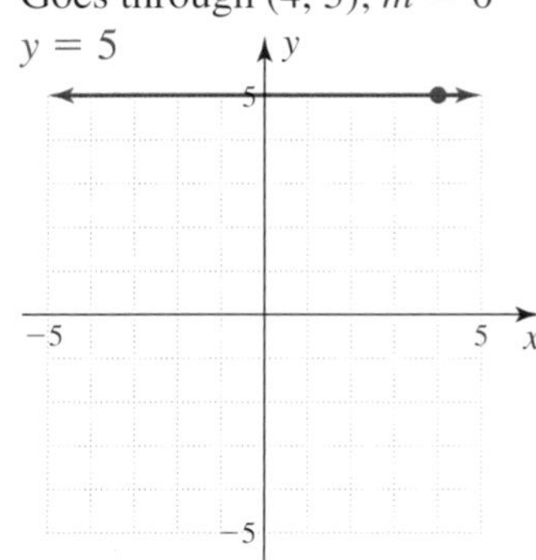

6. Goes through $(3, 2)$; slope is not defined (does not exist) $x = 3$

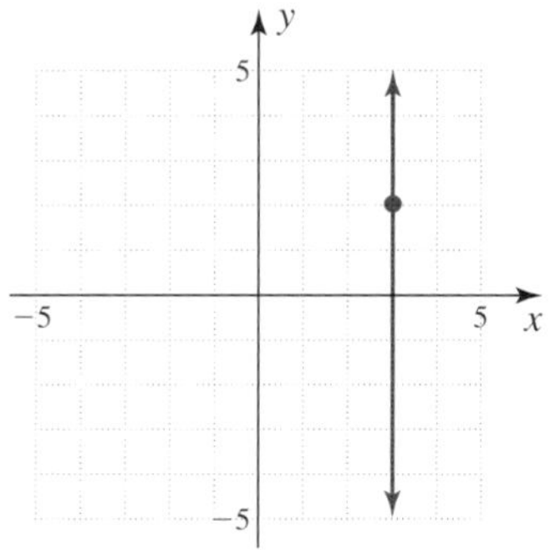

B In Problems 7–16, find an equation of the line with the given slope and intercept.

7. Slope, 2; y-intercept, -3 $y = 2x - 3$

8. Slope, 3; y-intercept, -5 $y = 3x - 5$

9. Slope, -4; y-intercept, 6 $y = -4x + 6$

10. Slope, -6; y-intercept, -7 $y = -6x - 7$

11. Slope, $\frac{3}{4}$; y-intercept, $\frac{7}{8}$ $y = \frac{3}{4}x + \frac{7}{8}$

12. Slope $\frac{7}{8}$; y-intercept, $\frac{3}{8}$ $y = \frac{7}{8}x + \frac{3}{8}$

13. Slope, 2.5; y-intercept, -4.7 $y = 2.5x - 4.7$

14. Slope, 2.8; y-intercept, -3.2 $y = 2.8x - 3.2$

15. Slope, -3.5; y-intercept, 5.9 $y = -3.5x + 5.9$

16. Slope, -2.5; y-intercept, 6.4 $y = -2.5x + 6.4$

In Problems 17–20, find an equation of the line having slope m and y-intercept b, and then graph the line.

17. $m = \frac{1}{4}$, $b = 3$ $y = \frac{1}{4}x + 3$

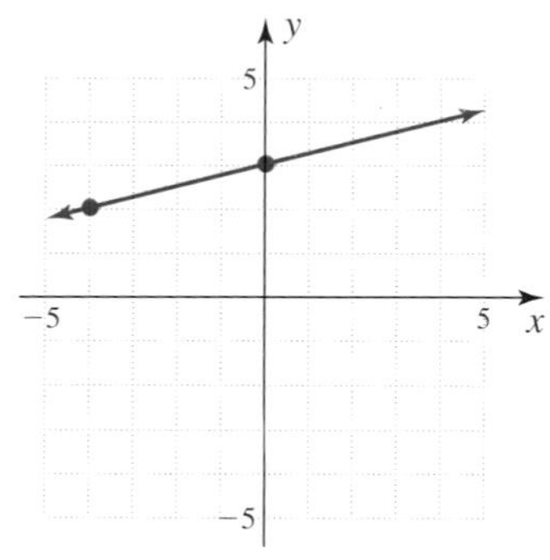

18. $m = -\frac{2}{5}$, $b = 1$ $y = -\frac{2}{5}x + 1$

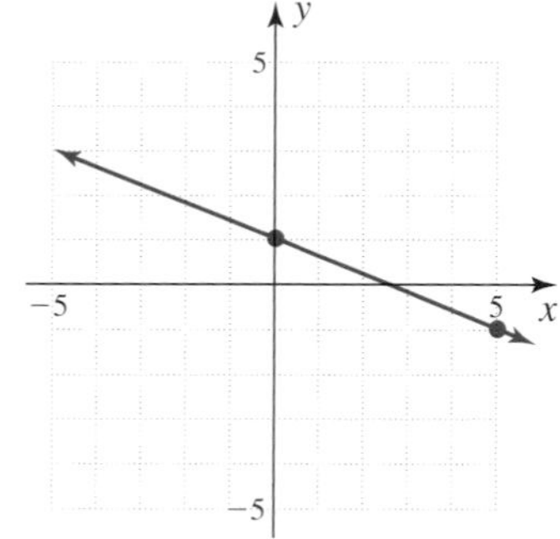

19. $m = -\frac{3}{4}$, $b = -2$ $y = -\frac{3}{4}x - 2$

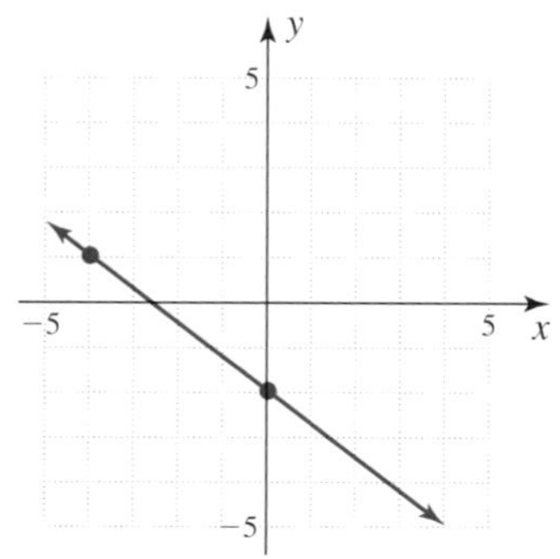

20. $m = -\frac{1}{3}$, $b = -1$ $y = -\frac{1}{3}x - 1$

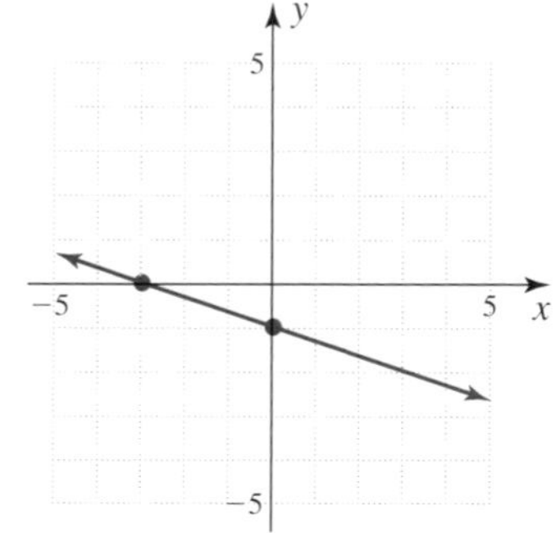

C In Problems 21–30, find an equation of the line passing through the given points, write it in standard form, and graph it.

21. (2, 3) and (7, 8) $x - y = -1$

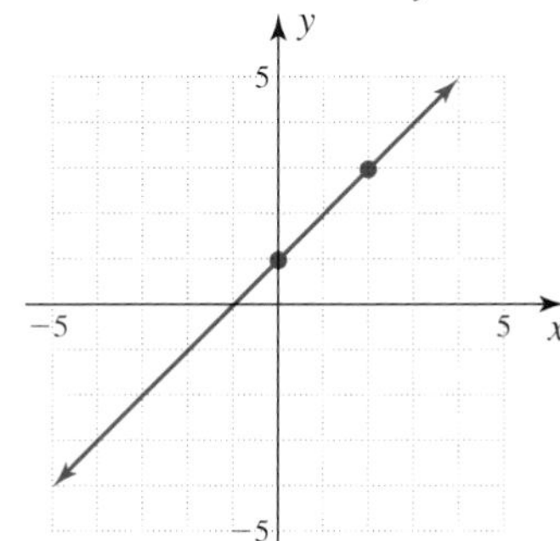

22. (−2, −3) and (1, −6) $x + y = -5$

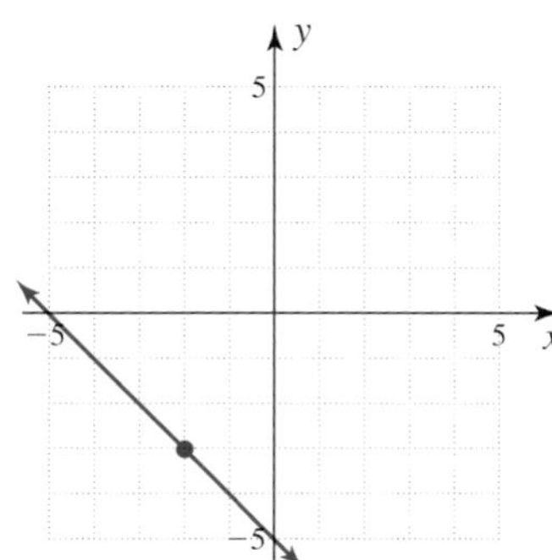

23. (2, 2) and (1, −1) $3x - y = 4$

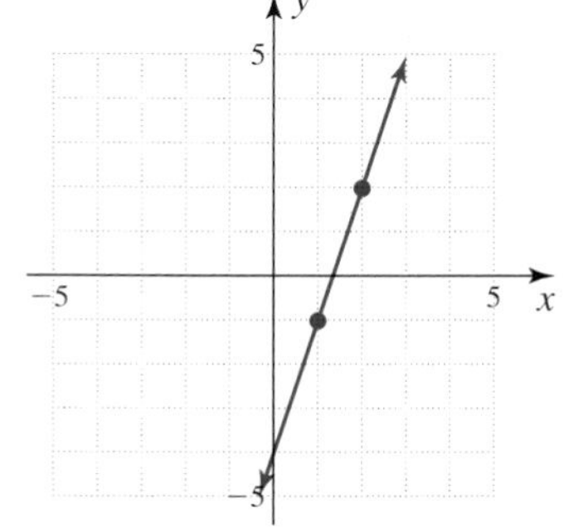

24. (−3, 4) and (−2, 0) $4x + y = -8$

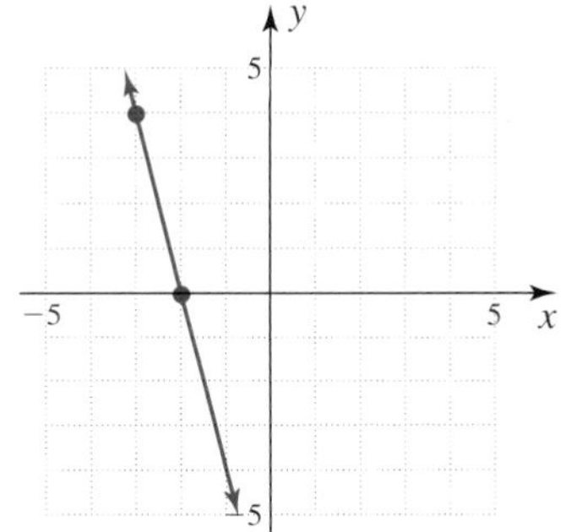

25. (3, 0) and (0, 4) $4x + 3y = 12$

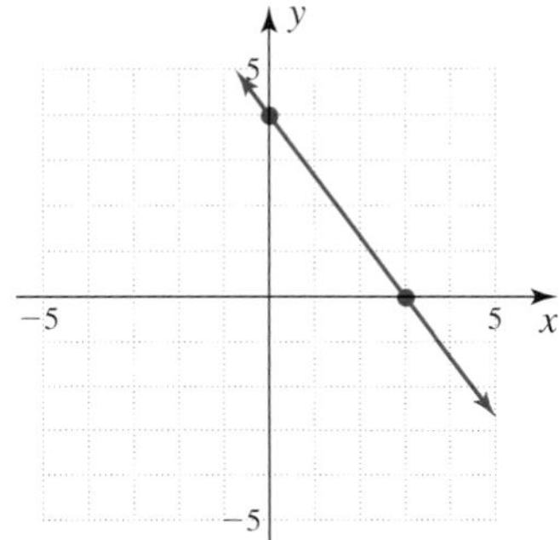

26. (0, −3) and (4, 0) $3x - 4y = 12$

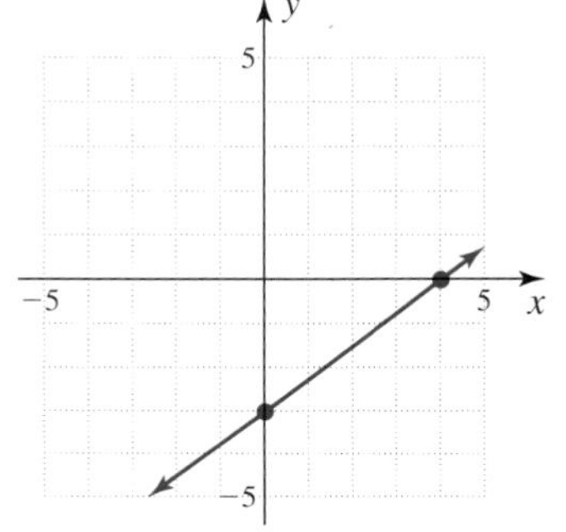

27. (3, 0) and (3, 2) $x = 3$

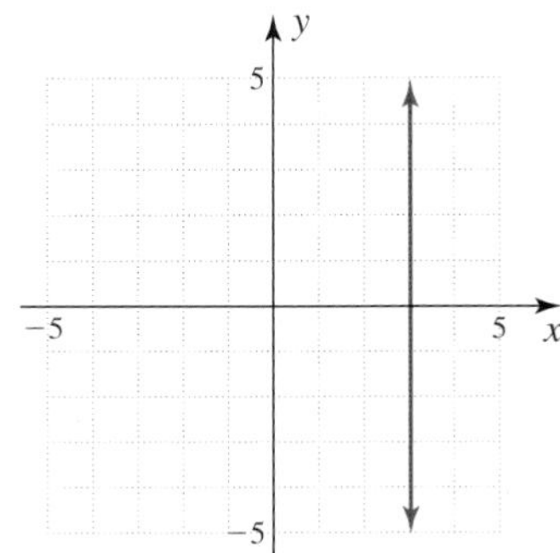

28. (−4, 2) and (−4, 0) $x = -4$

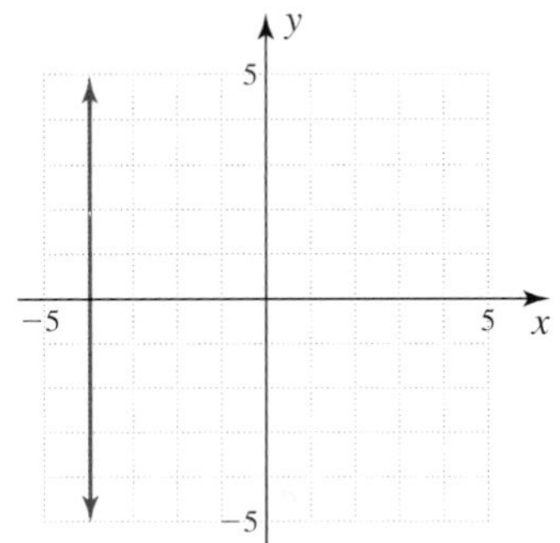

29. (−2, −3) and (1, −3) $y = -3$

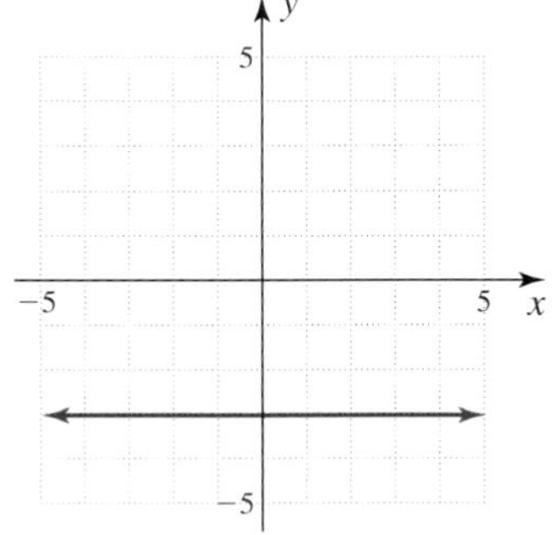

30. (−3, 0) and (−3, 4) $x = -3$

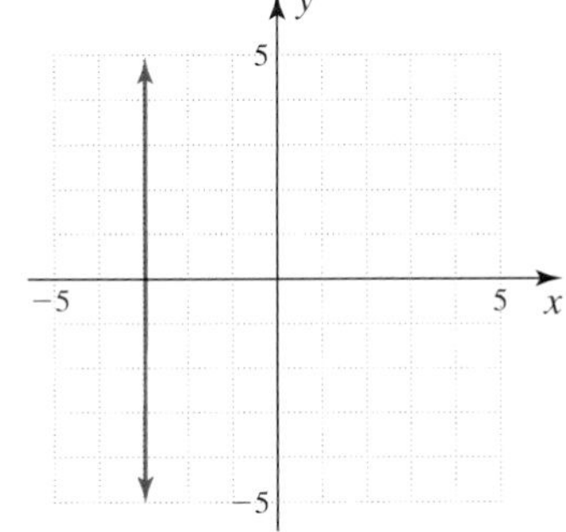

In Problems 31–36, find an equation of the line that satisfies the given conditions.

31. Passes through (0, 3) and (4, 7) $y = x + 3$

32. Passes through (−1, −2) and (3, 6) $y = 2x$

33. Slope 2 and passes through (1, 2) $y = 2x$

34. Slope −3 and passes through (−1, −2) $y = -3x - 5$

35. Slope −2 and y-intercept −3 $y = -2x - 3$

36. Slope 4 and y-intercept −1 $y = 4x - 1$

USING YOUR KNOWLEDGE

The Business of Slopes and Intercepts

The slope m and the y-intercept of an equation play important roles in economics and business. Let's see how.

Suppose you wish to go into the business of manufacturing fancy candles. First, you have to buy some ingredients such as wax, paint, and so on. Assume that all these ingredients cost you \$100. This is the *fixed cost.* Now suppose it costs \$2 to manufacture each candle. This is the *marginal cost.* What would be the total cost y if the marginal cost is \$2, x units are produced, and the fixed cost is \$100? The answer is

$$\underbrace{y}_{\text{Total cost (in dollars)}} = \underbrace{2x}_{\text{Cost for } x \text{ units}} + \underbrace{100}_{\text{Fixed cost}}$$

In general, an equation of the form

$$y = mx + b$$

gives the total cost y of producing x units, when m is the cost of producing 1 unit and b is the fixed cost.

37. Find the total cost y of producing x units of a product costing \$2 per unit if the fixed cost is \$50. $y = 2x + 50$

38. Find the total cost y of producing x units of a product whose production cost is \$7 per unit if the fixed cost is \$300. $y = 7x + 300$

39. The total cost y of producing x units of a certain product is given by

$$y = 2x + 75$$

a. What is the production cost for each unit? \$2

b. What is the fixed cost? \$75

WRITE ON

40. If you are given the equation of a nonvertical line, describe the procedure you would use to find the slope of the line. Answers may vary.

41. If you are given a vertical line whose y-intercept is the origin, what is the name of the line? the y-axis

42. If you are given a horizontal line whose x-intercept is the origin, what is the name of the line? the x-axis

43. How would you write the equation of a vertical line in the standard form $Ax + By = C$? $Ax + 0y = C$

44. How would you write the equation of a horizontal line in the standard form $Ax + By = C$? $0x + By = C$

45. Write an explanation of why a vertical line cannot be written in the form $y = mx + b$. Answers may vary.

MASTERY TEST

If you know how to do these problems, you have learned your lesson!

46. Find an equation of the line with slope 3 and y-intercept -6, and graph the line. $y = 3x - 6$

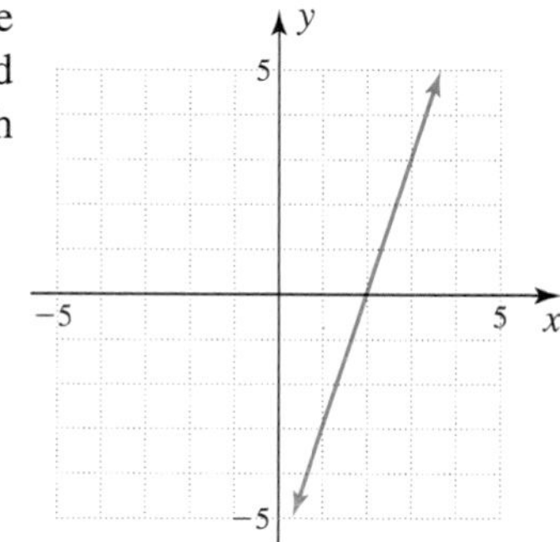

47. Find an equation of the line with slope $-\frac{3}{4}$ and y-intercept 2, and graph the line. $y = -\frac{3}{4}x + 2$

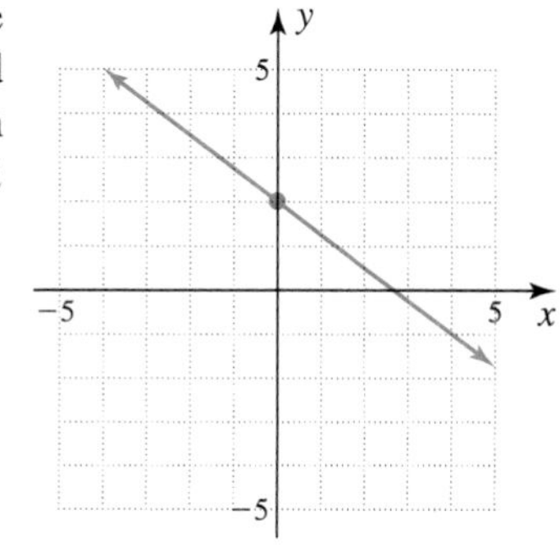

48. Find an equation of the line passing through the point $(-2, 3)$ and with slope -3, and graph the line.
$3x + y = -3$

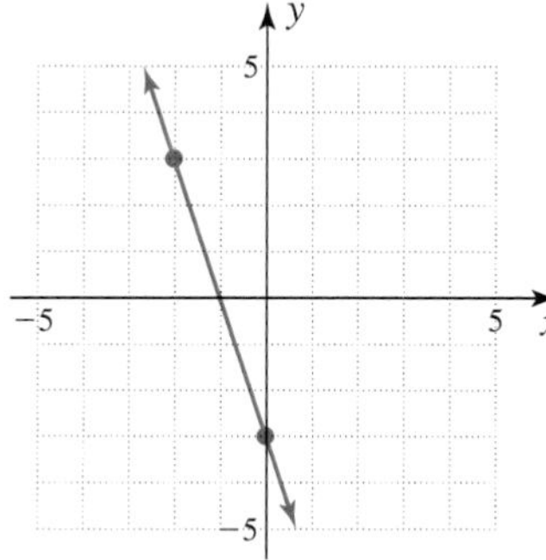

49. Find an equation of the line passing through the point $(3, -1)$ and with slope $\frac{2}{3}$, and graph the line.
$2x - 3y = 9$

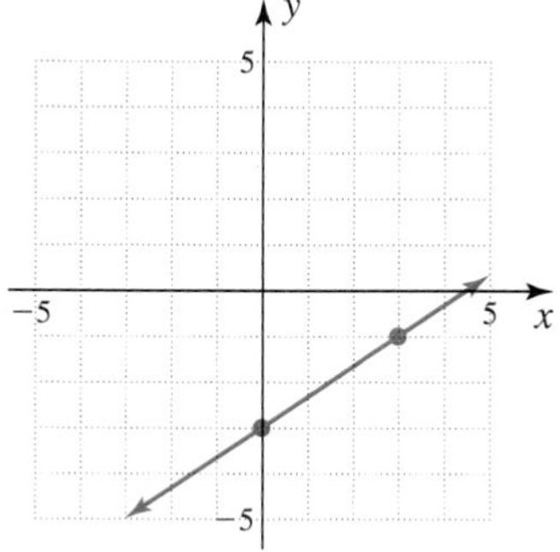

50. Find an equation of the line passing through the points $(-3, 4)$ and $(-2, -6)$, write it in standard form, and graph the line.
$10x + y = -26$

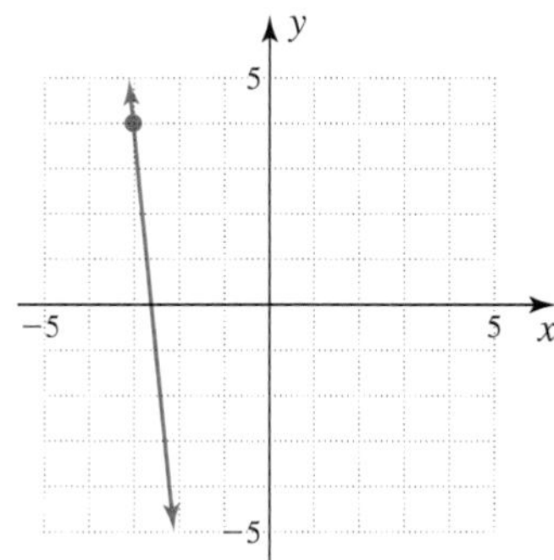

51. Find an equation of the line passing through the points $(-1, -6)$ and $(-3, 4)$, write it in standard form, and graph the line.
$5x + y = -11$

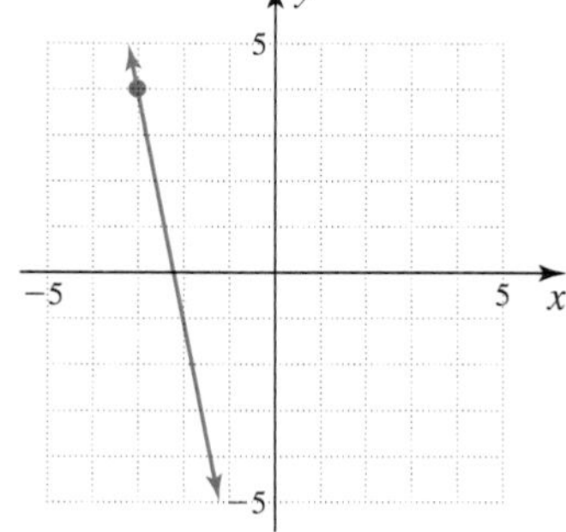

7.2 APPLICATIONS OF EQUATIONS OF LINES

To Succeed, Review How To . . .

1. Find the equation of a line given a point and the slope (pp. 535–536).
2. Find the equation of a line given a point and the y-intercept (p. 537).
3. Find the equation of a line given two points (p. 538).

Objectives

A Solve applications involving the point-slope formula.

B Solve applications involving the slope-intercept formula.

C Solve applications involving the two-point formula.

GETTING STARTED

Taxi! Taxi!

How much is a taxi ride in your city? In Boston, it is \$1.50 for the first $\frac{1}{4}$ mile and \$2 for each additional mile. As a matter of fact, it costs \$21 for a 10-mile ride. Can we find a simple formula based on the number of miles m you ride? If C represents the total cost of the ride and we know that a 10-mile ride costs \$21, we are given one point—namely, (10, 21). We also know that the slope is 2, since the costs are \$2 per additional mile. Using the point-slope form, we have

$$C - 21 = 2(m - 10)$$

or

$$C = 2m - 20 + 21$$

That is,

$$C = 2m + 1$$

Note that this is consistent with the fact that a 10-mile ride costs \$21, since

$$C = 2(10) + 1 = 21$$

In the preceding section, we learned how to find and graph the equation of a line when given a point and the slope, a point and the y-intercept, or two points. In this section we are going to use the appropriate formulas to solve application problems.

A Point-Slope Formula Applications

EXAMPLE 1 **Taxicab fare calculations**

Taxi fares in Key West are \$2.25 for the first $\frac{1}{5}$ mile and then \$2.50 for each additional mile. If a 10-mile ride costs \$26.75, find an equation for the total cost C of an m-mile ride. What will be the cost for a 30-mile ride?

SOLUTION Since we know that a 10-mile ride costs \$26.75, and each additional mile costs \$2.50, we are given the point (10, 26.75) and the slope 2.5.

Using the point-slope form, we have

$$C - 26.75 = 2.5(m - 10)$$

or

$$C = 2.5m - 25 + 26.75$$

That is,

$$C = 2.5m + 1.75$$

For a 30-mile ride, $m = 30$ and $C = 2.5(30) + 1.75 = \$76.75$.

PROBLEM 1

A taxicab company charges \$2.50 for the first $\frac{1}{4}$ mile and \$2.00 for each additional mile. If a 10-mile ride costs \$22, find an equation for the total cost C of an m-mile ride. What is the cost of a 25-mile ride?

Web It

To find the actual way taxi fares are quoted for Boston and Key West, go to links 7-2-1 and 7-2-2 on the Bello Website at mhhe.com/bello.

B Slope-Intercept Formula Applications

Web It

Some other restrictions do apply to the Verizon cell phone plan described here. You can go to link 7-2-3 on the Bello Website at mhhe.com/bello to see!

Do you have a cell phone? Which plan do you have? Most plans have a set number of free minutes for a set fee, after which you pay for additional minutes used. For example, at the present time, Verizon Wireless® has a plan that allows 900 free minutes for \$50 with unlimited weekend minutes. After that, you pay \$0.35 for each additional minute. We will consider such a plan in Example 2.

EXAMPLE 2 **Cell phone plan charges**

Maria subscribed to a cell phone plan with 900 free minutes, a \$50 monthly fee, and \$0.35 for each additional minute. Find an equation for the total cost C of her plan when she uses m minutes after the first 900. What is her cost when she uses 1200 minutes?

SOLUTION Let m be the number of minutes Maria uses after the first 900. If she does not go over the limit, she will pay \$50. The y-intercept will then be 50. Since she pays \$0.35 for each additional minute, the slope is 0.35. Thus, the total cost C when m additional minutes are used is

$$C = 0.35m + 50 \quad \text{for } m > 900$$

If she uses 1200 minutes, she has to pay for 300 of them (1200 − 900 free = 300 paid), and the cost will then be

$$\begin{aligned} C &= 0.35(300) + 50 \\ &= 105 + 50 \\ &= \$155 \end{aligned}$$

Thus, Maria will pay \$155 for her monthly payment.

PROBLEM 2

Find the cost of the plan if the monthly fee is \$40 and each minute after the first 900 costs \$0.50. What is the cost when she uses 1000 minutes?

Answers

1. $C = 2m + 2$; \$52
2. $C = 0.50m + 40$ for $m > 900$; \$90

C Two-Point Formula Applications

Web It

To see the price of more expensive AOL usage plans, go to the Bello Website at mhhe.com/bello and look at link 7-2-4.

To be on the Internet you need an Internet service provider (ISP), a company that provides Internet access to individuals and companies. There are many ISP plans and many have unlimited use; that is, subscribers can stay online as much time as they wish. Economical plans involve a set fee for a set number of hours and an additional charge for extra hours. For example, at the present time, America Online® has a "light usage" plan and a "limited" plan. We discuss the cost for the limited plan in Example 3.

EXAMPLE 3 **AOL subscriber costs**

AOL's limited plan costs \$9.95 for up to 5 hours of use, \$12.90 for 6 hours, and \$18.80 for 8 hours. Find an equation for the total cost C of this plan when h hours are used ($h > 5$). What is the cost for 10 hours?

SOLUTION We will examine ordered pairs of the form (h, C), where C is the cost and h is the number of hours used. The ordered pair corresponding to 6 hours of use costing \$12.90 is (6, 12.90), and the ordered pair corresponding to 8 hours of use costing \$18.80 is (8, 18.80). We now find the slope. Using the slope formula, we have

$$m = \frac{y_2 - y_1}{x_2 - x_1} = \frac{18.80 - 12.90}{8 - 6} = \frac{5.9}{2} = \$2.95$$

To find the equation for C we use the point-slope form with $m = 2.95$ and (6, 12.90) as our point, obtaining

$$C - 12.90 = 2.95(h - 6)$$

or $$C = 2.95h - 17.70 + 12.90$$

or $$C = 2.95h - 4.80$$

For 10 hours of use, $h = 10$ and

$$\begin{aligned} C &= 2.95(10) - 4.80 \\ &= 29.50 - 4.80 \\ &= \$24.70 \end{aligned}$$

PROBLEM 3

There is another plan that costs \$14.95 for up to 5 hours, \$17.45 for 6 hours, and \$24.95 for 9 hours. Find an equation for the total cost C of this plan when h hours are used ($h > 5$). What is the cost when 10 hours are used?

Answer

3. $C = 2.5h + 2.45$; \$27.45

Exercises 7.2

Boost *your* GRADE at mathzone.com!

- Practice Problems
- Self-Tests
- Videos
- NetTutor
- e-Professors

APPLICATIONS

A In Problems 1–10, solve the application.

1. *San Francisco taxi fares* Taxi fares in San Francisco are \$2 for the first mile and \$1.70 for each additional mile. If a 10-mile ride costs \$17.30, find an equation for the total cost C of an m-mile ride. What would the price be for a 30-mile ride? $C = 1.7m + 0.3$; \$51.30

2. *San Francisco taxi fares* A different taxicab company in San Francisco charges \$3 for the first mile and \$1.50 for each additional mile. If a 20-mile ride costs \$31.50, find an equation for the total cost C of an m-mile ride. What would the price of a 10-mile ride be? Which company is cheaper, this one or the company in Problem 1? $C = 1.5m + 1.5$; \$16.50; this one

3. *San Francisco taxi fares* Pedro took a cab in San Francisco and paid the fare quoted in Problem 1. Tyrone paid the fare quoted in Problem 2. Amazingly, they paid the same amount! How far did they ride? 6 miles

4. *New York taxi fares* In New York, a 20-mile cab ride is \$32 and consists of an initial set charge and \$1.50 per mile. Find an equation for the total cost C of an m-mile ride. How much would you have to pay for a 30-mile ride? $C = 1.5m + 2$; \$47

5. *San Francisco taxi fares* The cost C for San Francisco fares (Problem 1) is \$2 for the first mile and \$1.70 for each mile thereafter. We can find C by following these steps:

a. What is the cost of the first mile? \$2

b. If the whole trip is m miles, how many miles do you travel after the first mile? $m - 1$

c. How much do you pay per mile after the first mile? \$1.70

d. What is the cost of all the miles after the first? $1.70(m - 1)$

e. The total cost C is the sum of the cost of the first mile and the cost of all the miles after the first. What is that cost? Is your answer the same as that in Problem 1? $C = 2 + 1.70(m - 1)$ or $C = 1.7m + 0.3$; yes

6. *New York taxi fares* New York fares are easier to compute. They are simply \$2 for the initial set charge and \$1.50 for each mile. Find the total cost C for an m-mile trip. Do you get the same answer as you did in Problem 4? $C = 1.5m + 2$; yes

7. *Cell phone rental overseas* Did you know that you can rent cell phones for your overseas travel? If you are in Paris, the cost for a 1-week rental, including 60 minutes of long-distance calls to New York, is \$175.

a. Find a formula for the total cost C of a rental phone that include m minutes of long-distance calls to New York. $C = 175$, if $m \le 60$

b. What is the weekly charge for the phone? $C = 175$, if $m \le 60$

c. What is the per-minute usage charge? $C = \frac{175}{m}$, if $m \le 60$

8. *International long-distance rates* Long-distance calls from the Hilton Hotel in Paris to New York cost \$7.80 per minute.

a. Find a formula for the cost C of m minutes of long-distance calls from Paris to New York. $C = 7.8m$

b. How many minutes can you use so that the charges are identical to those you would pay when renting the phone of Problem 7? Answer to the nearest minute. 22 minutes

9. *Wind chill temperatures* The table shows the relationship between the actual temperature (x) in degrees Fahrenheit and the wind chill temperature (y) when the wind speed is 5 mi/hr.

Wind Chill (Wind Speed 5 mi/hr)

Temperature	(x)	20	25	30
Wind Chill	(y)	13	19	25

a. Find the slope of the line (the rate of change of the wind chill temperature) using the points (20, 13) and (25, 19). $m = \frac{6}{5}$

b. Find the slope of the line using the points (25, 19) and (30, 25). $m = \frac{6}{5}$

c. Are the two slopes the same? Yes

d. Use the point-slope form to find y using the points (20, 13) and (25, 19). $y = \frac{6}{5}x - 11$

e. Use the point-slope form to find y using the points (25, 19) and (30, 25). Do you get an equation equivalent to the one in part **d**? $y = \frac{6}{5}x - 11$; yes

f. Use your formula to find the wind chill when the temperature is 5°F and the wind speed is 5 mi/hr. -5

10. *Heat index values* The table shows the relationship between the actual temperature (x) and the heat index or apparent temperature (y) when the relative humidity is 50%. *Note:* Temperatures above 105° F can cause severe heat disorders with continued exposure!

Relative Humidity (50%)

Temperature	(x)	100	102	104
Heat Index	(y)	118	124	130

a. Find the slope of the line using the points (100, 118) and (102, 124). $m = 3$

b. Find the slope of the line using the points (102, 124) and (104, 130). $m = 3$

c. Are the two slopes the same? Yes

d. Use the point-slope form to find y using the points (100, 118) and (102, 124). $y = 3x - 182$

e. Use the point-slope form to find y using the points (102, 124) and (104, 130). Do you get an equation equivalent to the one in part **d**? $y = 3x - 182$; yes

f. Use your formula to find the heat index when the temperature is 106° F and the relative humidity is 50%. 136

B In Problems 11–22, solve the application.

11. *Electrician's charges* According to Microsoft Home Advisor®, if you have an electrical failure caused by a blown fuse or circuit breaker you may have to call an electrician. "Plan on spending at least $100 for a service call, plus the electrician's hourly rate." Assume that the service call is $100 and the electrician charges $37.50 an hour.

a. Find an equation for the total cost C of a call lasting h hours. $C = 37.5h + 100$

b. If your bill amounts to $212.50, for how many hours were you charged? 3 hours

12. *Appliance technician's charges* Microsoft Home Advisor suggests that the cost C for fixing a faulty appliance is about $75 for the service call plus the technician's hourly rate. Assume that this rate is $40 per hour.

a. Find an equation for the total cost C of a call lasting h hours. $C = 40h + 75$

b. If your bill amounts to $195, for how many hours were you charged? 3 hours

13. *International phone rental charges* The total cost C for renting a phone for 1 week in London, England, is $40 per week plus $2.30 per minute for long-distance charges when calling the United States.

a. Find a formula for the cost C of renting a phone and using m minutes of long-distance charges. $C = 2.3m + 40$

b. If the total cost C amounted to $201, how many minutes were used? 70 minutes

c. The Dorchester Hotel in London charges $7.20 per minute for long-distance charges to the United States. Find a formula for the cost C of m minutes of long-distance calls from the hotel to the United States. $C = 7.2m$

d. How many minutes can you use so that the charges are identical to those you would pay when renting the phone of part **a** (round to nearest minute)? About 8 minutes

14. *Phone rental charges* The rate for incoming calls for a rental phone is $1.20 per minute. If the rental charge is $50 per week, find a formula for the total cost C of renting a phone for m incoming minutes. If you paid $146 for the rental, for how many incoming minutes were you billed? $C = 1.2m + 50$; 80 minutes

15. *Cell phone costs* Verizon Wireless has a plan that costs $50 per month for 900 anytime minutes and unlimited weekend minutes. The charge for each additional minute is $0.35.

a. Find a formula for the cost C when m additional minutes are used. $C = 0.35m + 50$

b. If your bill was for $88.50, how many additional minutes did you use? 110 minutes

16. *Cell phone costs* A Verizon competitor has a similar plan, but it charges $45 per month and $0.40 for each additional minute.

a. Find a formula for the cost C when m additional minutes are used. $C = 0.4m + 45$

b. If your bill was for $61, how many additional minutes did you use? 40 minutes

17. *Cell phone costs* How many additional minutes do you have to use so that the costs for the plans in Problems 15 and 16 are identical? After how many additional minutes is the plan in Problem 15 cheaper? 100 minutes; after 100 minutes

18. *Estimating a man's height* You can estimate a man's height y (in inches) by multiplying the length x of his femur bone by 1.88 and adding 32 to the result.

a. Write an equation for a man's estimated height y in slope-intercept form. $y = 1.88x + 32$

b. What is the slope? $m = 1.88$

c. What is the y-intercept? $b = 32$

19. *Estimating a female's height* The height y for a female (in inches) can be estimated by multiplying the length x of her femur bone by 1.95 and adding 29 to the result.

a. Write an equation for a woman's height y in slope-intercept form. $y = 1.95x + 29$

b. What is the slope? $m = 1.95$

c. What is the y-intercept? $b = 29$

20. *Estimating height from femur length* Refer to Problems 18 and 19.

a. What would be the length x of a femur that would yield the same height y for a male and a female? About 43 inches

b. What would the estimated height of a person having such a femur bone be? About 113 inches, or 9 ft 5 in.

21. *Average hospital stay* Since 1970, the number y of days the average person stays in the hospital has steadily decreased from 8 days at the rate of 0.1 day per year. If x represents the number of years after 1970, write a slope-intercept equation representing the average number y of days a person stays in the hospital. What would the average stay in the year 2000 be? In 2010? $y = -0.1x + 8$; 5 days; 4 days

22. *Average hospital stay for females* Since 1970, the number y of days the average female stays in the hospital has steadily decreased from 7.5 days at the rate of 0.1 day per year. If x represents the number of years after 1970, write a slope-intercept equation representing the average number y of days a female stays in the hospital. What would the average stay in the year 2000 be? In 2010? $y = -0.1x + 7.5$; 4.5 days; 3.5 days

C In Problems 23–26, solve the application.

23. *How old is your dog in human years?* It is a common belief that 1 human year is equal to 7 dog years. That is not very accurate, since dogs reach adulthood within the first couple of years. A more accurate formula indicates that when a dog is 3 years old in human years, it would be 30 years old in dog years. Moreover, a 9-year-old dog is 60 years old in dog years.

a. Form the ordered pairs (h, d), where h is the age of the dog in human years and d is the age of the dog in dog years. What does (3, 30) mean? What does (9, 60) mean? 3 human years corresponds to 30 dog years; 9 human years corresponds to 60 dog years

b. Find the slope of the line using the points (3, 30) and (9, 60). $m = 5$

c. Find an equation for the age d of a dog based on the dog's age h in human years $(h > 1)$. $d = 5h + 15$

d. If a dog is 4 human years old, how old is it in dog years? 35 dog years

e. If a human could retire at age 65, what is the equivalent retirement age for a dog (in human years)? 10 years

f. The drinking age for humans is usually 21 years. What is the equivalent drinking age for dogs? 1.2 years

24. *How old is your cat in human years?* When a cat is $h = 2$ years old in human years, it is $c = 24$ years old in cat years, and when a cat is $h = 6$ years old in human years, it is $c = 40$ years old in cat years.

a. Form the ordered pairs (h, c), where h is the age of the cat in human years and c is the age of the cat in cat years. What does (2, 24) mean? What does (6, 40) mean? 2 human years corresponds to 24 cat years; 6 human years corresponds to 40 cat years

b. Find the slope of the line using the points (2, 24) and (6, 40). $m = 4$

c. Find an equation for the age c of a cat based on the cat's age h in human years $(h > 1)$. $c = 4h + 16$

d. If a cat is 4 human years old, how old is it in cat years? 32 cat years

e. If a human could retire at age 60, what is the equivalent retirement age for a cat in human years? 11 years

f. The drinking age for humans is usually 21 years. What is the equivalent drinking age for cats? 1.25 years

25. *Blood alcohol concentration* Your blood alcohol level is dependent on the amount of liquor you consume, your weight, and your gender. Suppose you are a 150-pound male and you consume 3 beers (5% alcohol) over a period of 1 hour. Your blood alcohol concentration (BAC) would be 0.052. If you have 5 beers, then your BAC would be 0.103. (You are legally drunk then! Most states regard a BAC of 0.08 as legally drunk.) Consider the ordered pairs (3, 0.052) and (5, 0.103).

a. Find the slope of the line passing through the two points. $m = 0.0255$

b. If b represents the number of beers you had in 1 hour and c represents your BAC, find an equation for c. $c = 0.0255b - 0.0245$

c. Find your BAC when you have had 4 beers in 1 hour. Find your BAC when you have had 6 beers in 1 hour. 0.0775; 0.1285

d. How many beers do you have to drink in 1 hour to be legally drunk (BAC = 0.08)? Answer to the nearest whole number. 4

26. *Blood alcohol concentration* Suppose you are a 125-pound female and you consume 3 beers (5% alcohol) over a period of 1 hour. Your blood alcohol concentration (BAC) would be 0.069. If you have 5 beers, then your BAC would be 0.136. (You are really legally drunk then!) Consider the ordered pairs (3, 0.069) and (5, 0.136).

a. Find the slope of the line passing through the two points. $m = 0.0335$

b. If b represents the number of beers you had in 1 hour and c represents your BAC, find an equation for c. $c = 0.0335b - 0.0315$

c. Find your BAC when you have had 4 beers in 1 hour. Find your BAC when you have had 6 beers in 1 hour. 0.1025; 0.1695

d. How many beers do you have to drink in 1 hour to be legally drunk (BAC = 0.08)? Answer to the nearest whole number. You can check this information by going to link 7-2-5 on the Bello Website at mhhe.com/bello. 3

SKILL CHECKER

Try the Skill Checker Exercises so you'll be ready for the next section.

Fill in the blank with $<$ or $>$ so that the result is a true statement:

27. $0 \underline{>} -8$ **28.** $0 \underline{<} 6$ **29.** $10 \underline{>} 0$

30. $-5 \underline{<} 0$ **31.** $-1 \underline{<} 0$ **32.** $-3 \underline{<} -1$

WRITE ON

33. In general, write the steps you use to find the formula for a linear equation in two variables similar to the ones in Problems 11–16. Answers may vary.

34. What is the minimum number of points you need to find the formula for a linear equation in two variables like the ones in Problems 23–26? Answers may vary.

35. What does the slope you obtained in Problems 25 and 26 mean? Answers may vary.

36. Would the slopes be different if the weights of the male and female are different? Explain. Answers may vary.

USING YOUR KNOWLEDGE

We have so far neglected the graphs of the applications we have studied, but graphs are especially useful when comparing different situations. For example, the America Online "Light Usage" plan costs \$4.95 per month for up to 3 hours and \$2.95 for each additional hour. If C is the cost and h is the number of hours:

37. Graph C when h is between 0 and 3 hours, inclusive.

38. Graph C when h is more than 3 hours.

39. A different AOL plan costs \$14.95 for unlimited hours. Graph the cost for this plan.

40. When is the plan in Problems 37 and 38 cheaper? When is the plan in Problem 39 cheaper? For less than about 6.4 hours; after about 6.4 hours

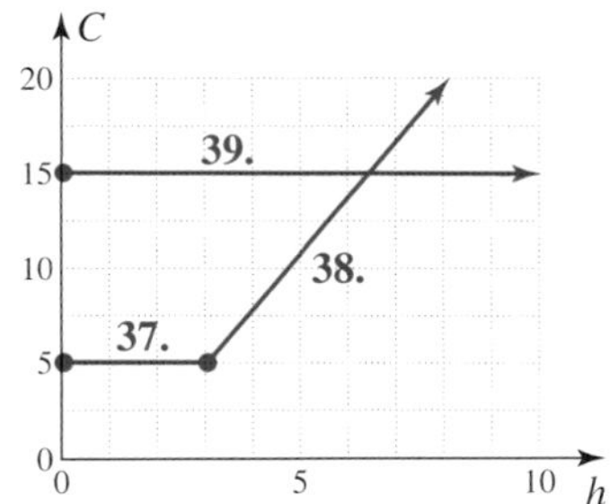

MASTERY TEST

If you know how to do these problems, you have learned your lesson!

41. Irena rented a car and paid \$40 a day and \$0.15 per mile. Find an equation for the total daily cost C when she travels m miles. What is her cost if she travels 450 miles in a day? $C = 0.15m + 40$; \$107.50

42. In 1990 about 186 million tons of trash were produced in the United States. The amount increases by 3.4 million tons each year after 1990. Use the point-slope formula to find an equation for the total amount of trash y (millions of tons) produced x years after 1990:

a. What would be the slope? $m = 3.4$

b. What point would you use to find the equation? (0, 186)

c. What is the equation? $y = 3.4x + 186$

43. Somjit subscribes to an Internet service with a flat monthly rate for up to 20 hours of use. For each hour over this limit, there is an additional per-hour fee. The table shows Somjit's first two bills.

Month	Hours of Use	Monthly Fee
March	30	\$24
April	33	\$26.70

a. What does the point (30, 24) mean? 30 hours cost \$24

b. What does the point (33, 26.70) mean? 33 hours cost \$26.70

c. Find an equation for the cost C of x hours ($x > 20$) of use. $C = 0.9x - 3$

7.3 GRAPHING INEQUALITIES IN TWO VARIABLES

To Succeed, Review How To . . .

1. Use the symbols > and < to compare numbers (p. 198).
2. Graph lines (pp. 245–247).
3. Evaluate an expression (p. 81).

Objective

A Graph linear inequalities in two variables.

GETTING STARTED

Renting Cars and Inequalities

Suppose you want to rent a car costing \$30 a day and \$0.20 per mile. The total cost T depends on the number x of days the car is rented and the number y of miles traveled and is given by

$$\underbrace{T}_{\text{Total cost}} = \underbrace{30x}_{\substack{\text{Cost per day} \times \\ \text{Number of days}}} + \underbrace{0.20y}_{\substack{\text{Cost per mile} \times \\ \text{Number of miles}}}$$

If you want the cost to be *exactly* \$600 ($T = 600$), we graph the equation $600 = 30x + 0.20y$ by finding the intercepts. For $x = 0$,

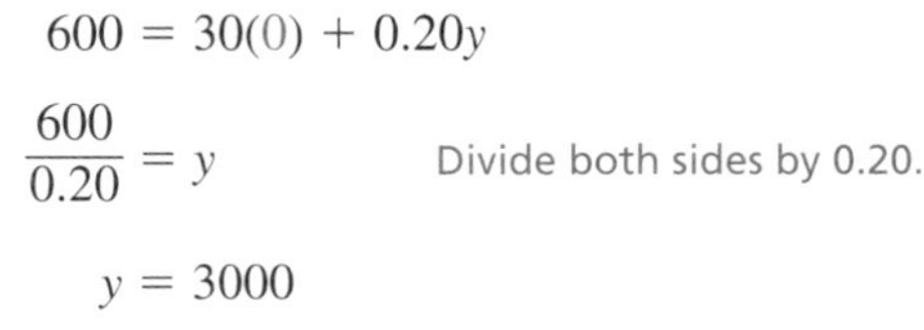

$$600 = 30(0) + 0.20y$$

$$\frac{600}{0.20} = y \qquad \text{Divide both sides by 0.20.}$$

$$y = 3000$$

Thus (0, 3000) is the y-intercept. Now for $y = 0$,

$$600 = 30x + 0.20(0)$$

$$20 = x \qquad \text{Divide both sides by 30.}$$

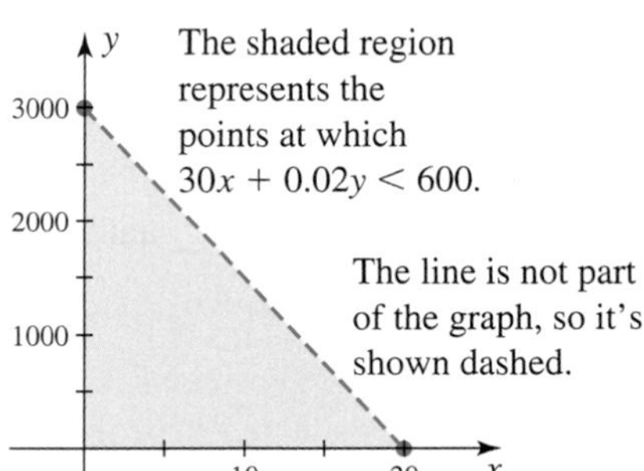

The shaded region represents the points at which 30x + 0.02y < 600.
The line is not part of the graph, so it's shown dashed.

Thus (20, 0) is the x-intercept. Join the two intercepts with a line to obtain the graph shown.

If we want to spend *less* than \$600, the total cost T must satisfy (be a solution of) the inequality

$$30x + 0.20y < 600$$

All points satisfying the equation $30x + 0.20y = 600$ are *on* the line. Where are the points so that $30x + 0.20y < 600$? The line $30x + 0.20y = 600$ divides the plane into three regions:

1. The points *below* the line
2. The points *on* the line
3. The points *above* the line

The test point (0, 0) is *below* the line $30x + 0.20y = 600$ and satisfies $30x + 0.20y < 600$. Thus all the other points *below* the line also satisfy the inequality and are shown shaded in the second graph. The line is *not* part of the answer, so it's shown dashed. In this section we shall learn how to solve linear inequalities in two variables, and we shall also examine why their solutions are regions of the plane.

Graphing Linear Inequalities in Two Variables

The procedure we used in the *Getting Started* can be generalized to graph any linear inequality that can be written in the form $Ax + By < C$. Here are the steps.

Teaching Tip

Remind students that the shading includes all the solutions to the inequality. In other words, we've made a picture of all solutions.

PROCEDURE

Graphing a Linear Inequality

1. Determine the line that is the boundary of the region. If the inequality involves $\le$ or $\ge$, draw the line **solid;** if it involves $<$ or $>$, draw a **dashed** line. The points on the *solid* line are part of the solution set.
2. Use any point (a, b) not on the line as a test point. Substitute the values of a and b for x and y in the inequality. If a true statement results, shade the side of the line containing the test point. If a false statement results, shade the other side.

EXAMPLE 1 Graphing linear inequalities where the line is *not* part of the solution set

Graph: $2x - 4y < -8$

SOLUTION We use the two-step procedure for graphing inequalities.

1. We first graph the boundary line $2x - 4y = -8$.

When $x = 0$, $-4y = -8$, and $y = 2$

When $y = 0$, $2x = -8$, and $x = -4$

x	y
0	2
−4	0

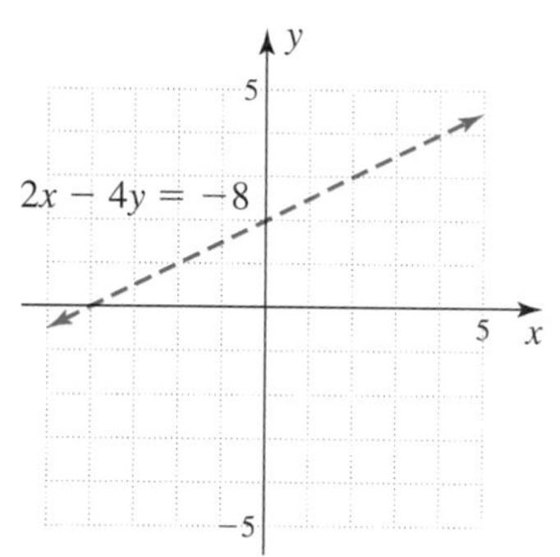

Figure 5

Since the inequality involved is $<$, join the points $(0, 2)$ and $(-4, 0)$ with a dashed line as shown in Figure 5.

2. Select an easy test point and see whether it satisfies the inequality. If it does, the solution lies on the same side of the line as the test point; otherwise, the solution is on the other side of the line.

An easy point is $(0, 0)$, which is *below* the line. If we substitute $x = 0$ and $y = 0$ in the inequality $2x - 4y < -8$, we obtain

$$2 \cdot 0 - 4 \cdot 0 < -8 \quad \text{or} \quad 0 < -8$$

which is *false*. Thus the point $(0, 0)$ is not part of the solution. Because of this, the solution consists of the points *above* (on the other side of) the line $2x - 4y = -8$, as shown shaded in Figure 6. Note that the line itself is shown dashed to indicate that it isn't part of the solution.

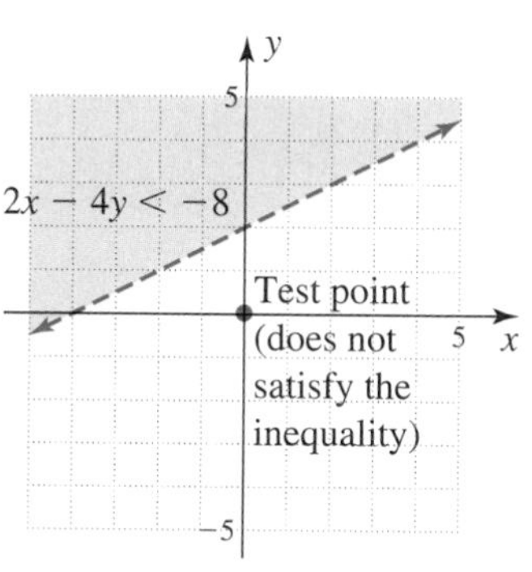

Figure 6

PROBLEM 1

Graph: $3x - 2y < -6$

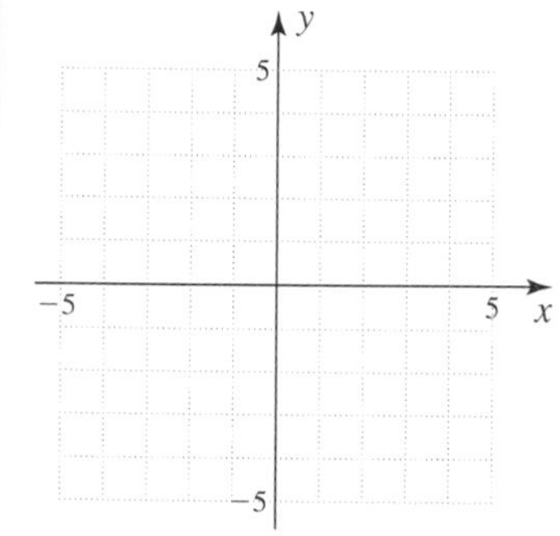

Answer

1.

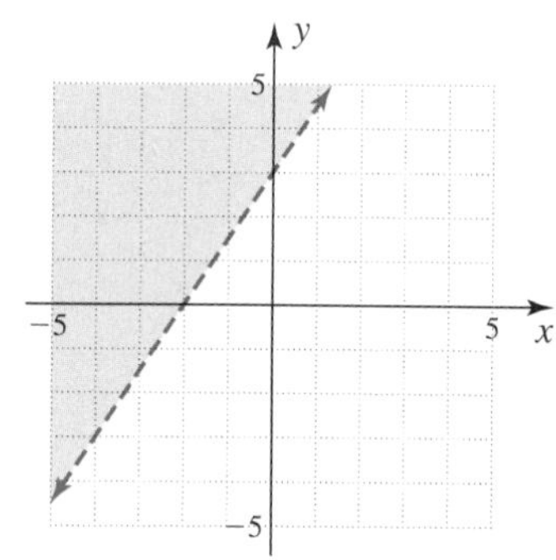

In Example 1, the line was not part of the solution for the given inequality. Next, we give an example in which the line is part of the solution for the inequality.

EXAMPLE 2 **Graphing linear inequalities where the line is part of the solution set**

Graph: $y \le -2x + 6$

SOLUTION As usual, we use the two-step procedure for graphing inequalities.

1. We first graph the line $y = -2x + 6$.

   When $x = 0$, $y = 6$

   When $y = 0$, $0 = -2x + 6$ or $x = 3$

x	y
0	6
3	0

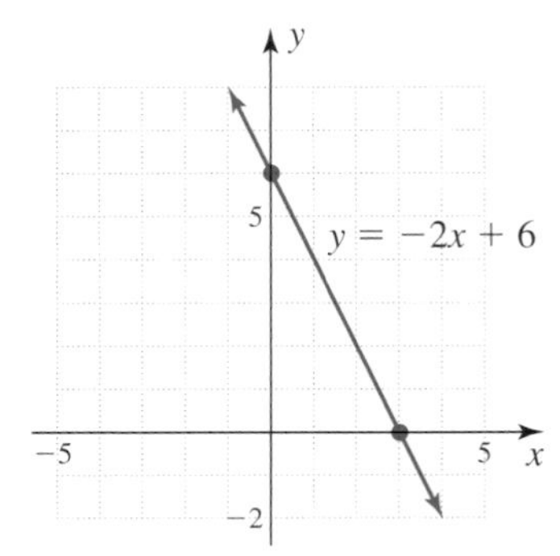

Figure 7

   Since the inequality involved is $\le$, the graph of the line is shown solid in Figure 7.

2. Select the point (0, 0) or any other point below the line as a test point. When $x = 0$ and $y = 0$, the inequality

$$y \le -2x + 6$$

   becomes

$$0 \le -2 \cdot 0 + 6 \quad \text{or} \quad 0 \le 6$$

   which is *true*. Thus all the points on the same side of the line as (0, 0)—that is, the points *below* the line—are solutions of $y \le -2x + 6$. These solutions are shown shaded in Figure 8.

   The line $y = -2x + 6$ is shown solid because it is part of the solution, since $y \le -2x + 6$ allows $y = -2x + 6$. [For example, the point (3, 0) satisfies the inequality $y \le -2x + 6$ because $0 \le -2 \cdot 3 + 6$ yields $0 \le 0$, which is true.]

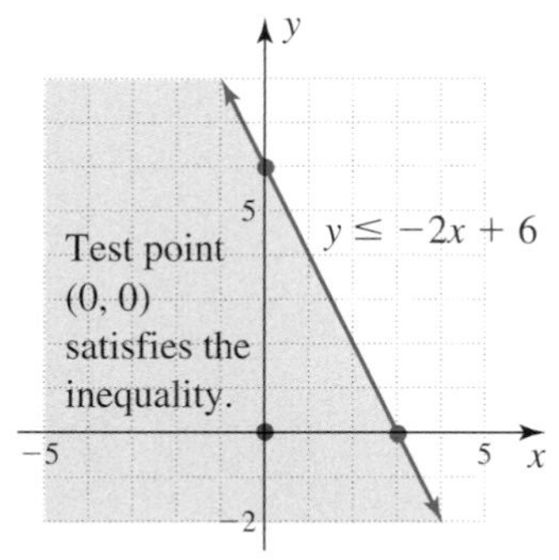

Figure 8

PROBLEM 2

Graph: $y \le -3x + 6$

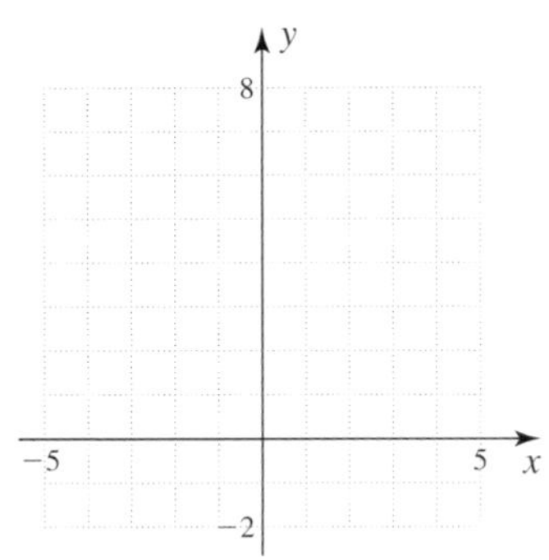

Answer

2.

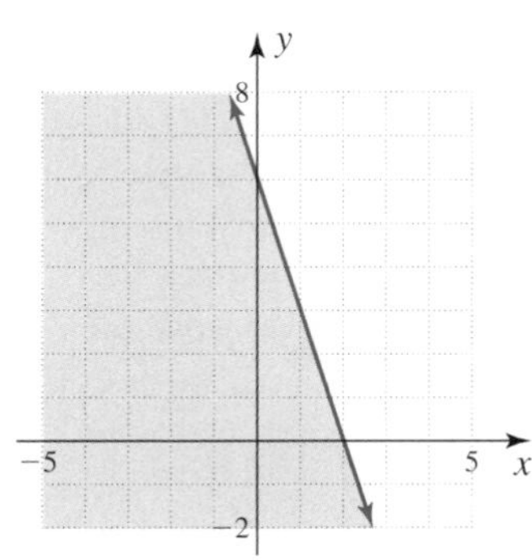

As you recall from Section 3.3, a line with an equation of the form $Ax + By = 0$ passes through the origin, so we cannot use the point (0, 0) as a test point. This is not a problem! Just use any other convenient point, as shown in the next example.

EXAMPLE 3 Graphing linear inequalities where the boundary line goes through the origin

Graph: $y + 2x > 0$

SOLUTION We follow the two-step procedure.

1. We first graph the boundary line $y + 2x = 0$:

When $x = 0$, $y = 0$

When $y = -4$, $-4 + 2x = 0$ or $x = 2$

Join the points (0, 0) and (2, −4) with a dashed line as shown in Figure 9 since the inequality involved is >.

x	*y*
0	0
2	−4

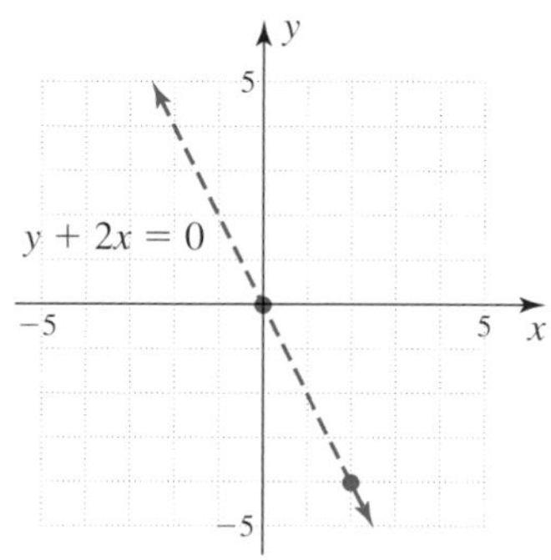

Figure 9

2. Because the line goes through the origin, we cannot use (0, 0) as a test point. A convenient point to use is (1, 1), which is *above* the line. Substituting $x = 1$ and $y = 1$ in $y + 2x > 0$, we obtain

$$1 + 2(1) > 0 \quad \text{or} \quad 3 > 0$$

which is true. Thus we shade the points *above* the line $y + 2x = 0$, as shown in Figure 10.

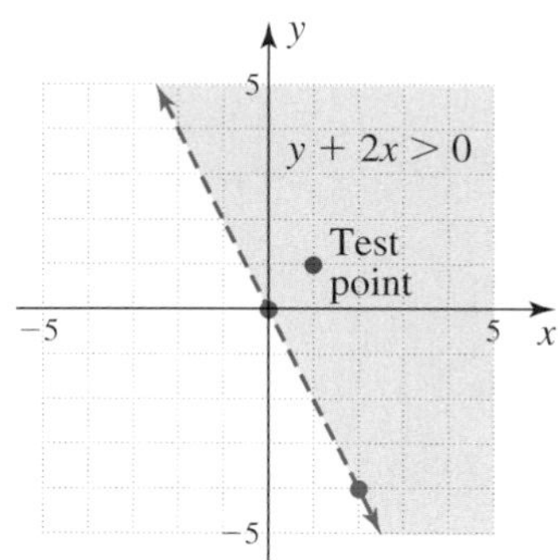

Figure 10

PROBLEM 3

Graph: $y + 3x > 0$

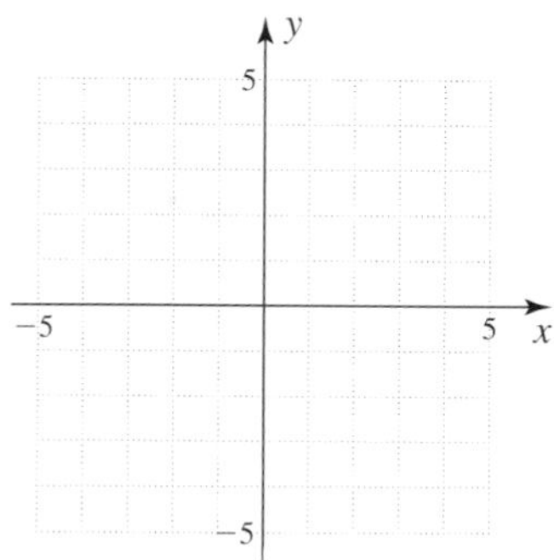

Teaching Tip

In Example 4, have students discuss shading without the use of a test point. That is, once the statement is written with the variable on the left, the arrow indicates the side to shade: > right/above; < left/below.

Finally, if the inequalities involve horizontal or vertical lines, we also proceed in the same manner.

Answer

3.

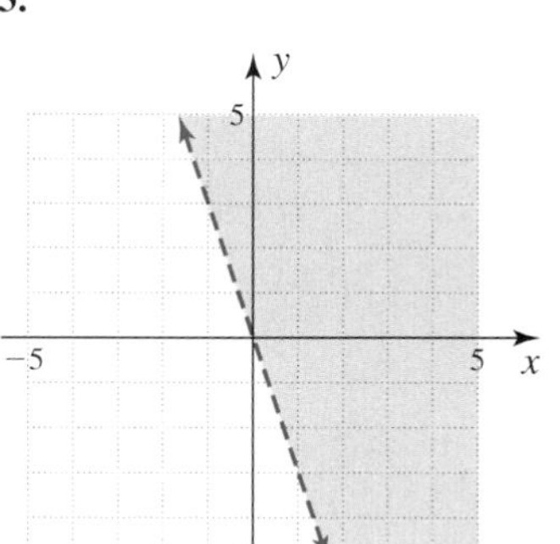

EXAMPLE 4 Graphing linear inequalities involving horizontal or vertical lines

Graph:

a. $x \geq -2$ **b.** $y - 2 < 0$

SOLUTION

a. The two-step procedure applies here also.

1. Graph the vertical line $x = -2$ as solid, since the inequality involved is $\geq$.
2. Use (0, 0) as a test point. When $x = 0$, $x \geq -2$ becomes $0 \geq -2$, a true statement. So we shade all the points to the *right* of the line $x = -2$, as shown in Figure 11.

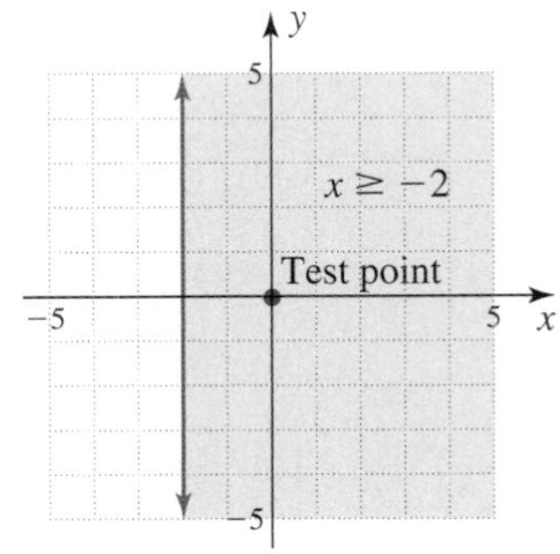

Figure 11

b. We again follow the steps.

1. First, we write the inequality as $y < 2$, and then we graph the horizontal line $y = 2$ as dashed, since the inequality involved is $<$.
2. Use (0, 0) as a test point. When $y = 0$, $y < 2$ becomes $0 < 2$, a true statement. So we shade all the points *below* the line $y = 2$, as shown in Figure 12.

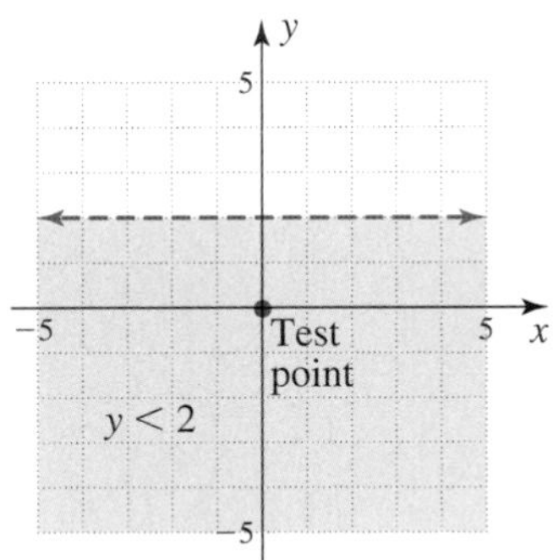

Figure 12

PROBLEM 4

Graph:

a. $x \geq 2$ **b.** $y + 2 > 0$

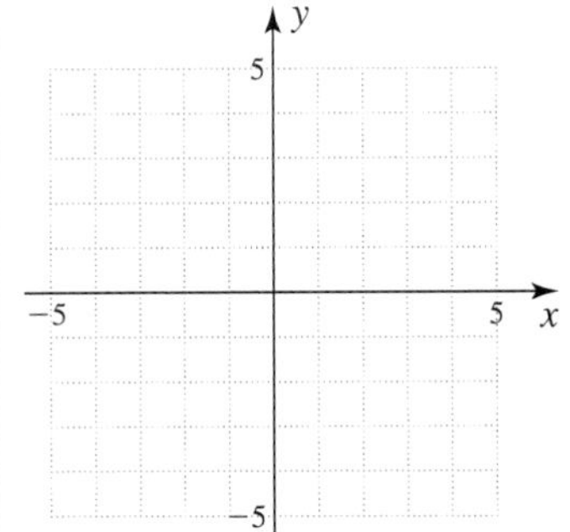

Answers

4. a.

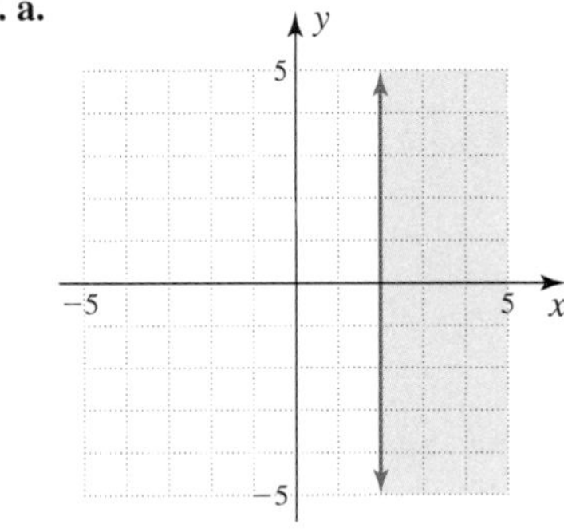

b.

Web It

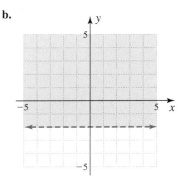

For a lesson on how to graph inequalities, go to link 7-3-1 on the Bello Website at mhhe.com/bello.

To practice with inequalities, try link 7-3-2.

Calculate It Solving Linear Inequalities

Your calculator can help solve linear inequalities, but you still have to know the algebra. Thus to do Example 1, you must first solve $2x - 4y = -8$ for y to obtain $y = \frac{1}{2}x + 2$. Now graph this equation using the Zinteger window, a window in which the x- and y-coordinates are integers. You can do this by pressing ZOOM 8. Now graph $y = \frac{1}{2}x + 2$.

```
Shade(.5X+2,50,
-50,2)
```

Window 1

To decide whether you need to shade above or below the line, move the cursor *above* the line. In the integer window, you can see the values for x and y. For $x = 2$ and $y = 5$, $2x - 4y < -8$ becomes

$$2(2) - 4(5) < -8 \quad \text{or} \quad 4 - 20 < -8$$

a true statement. Thus you need to shade *above* the line. You can do this with the DRAW feature. Press 2nd PRGM 7 and enter the line above which you want to shade—that is, $\frac{1}{2}x + 2$. Now press , and enter the x-range that you want to shade. Let the calculator shade below 50 and above -50 by entering 5 0 , (−) 5 0 ,. Finally, enter the resolution you want for the shading, 2, close the parentheses, and press ENTER.

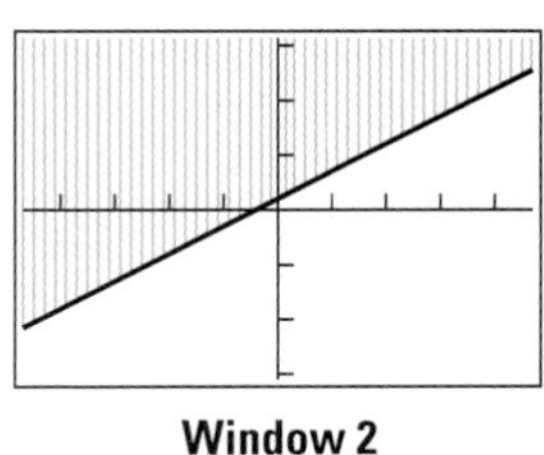

Window 2

The instructions we asked you to enter are shown in Window 1, and the resulting graph is shown in Window 2. What is missing? *You* should know the line itself is not part of the graph, since the inequality $<$ is involved.

Exercises 7.3

A In Problems 1–32, graph the inequalities.

1. $2x + y > 4$

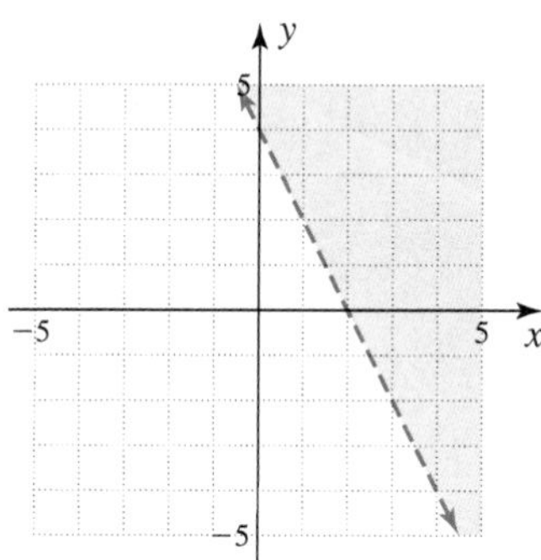

2. $y + 3x > 3$

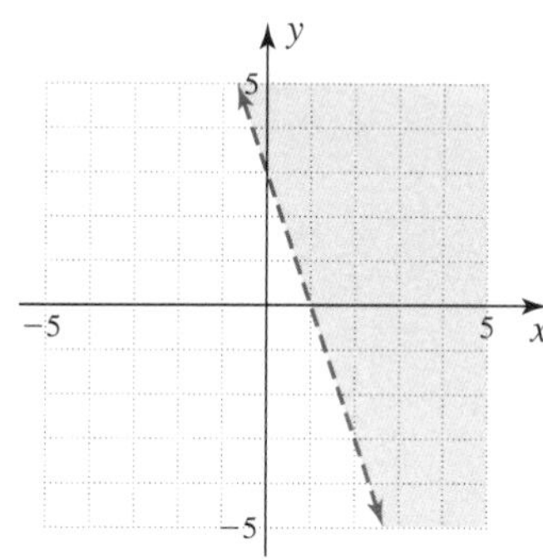

3. $-2x - 5y \leq 10$

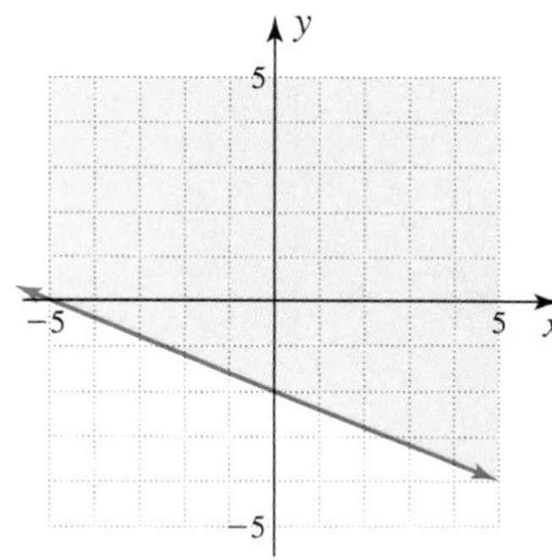

4. $-3x - 2y \leq -6$

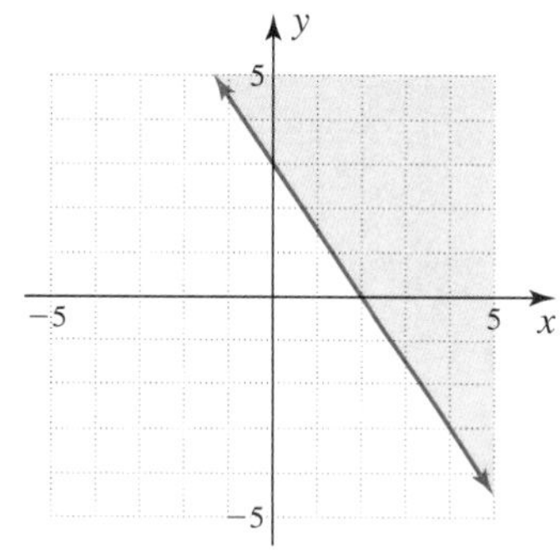

5. $y \geq 3x - 3$

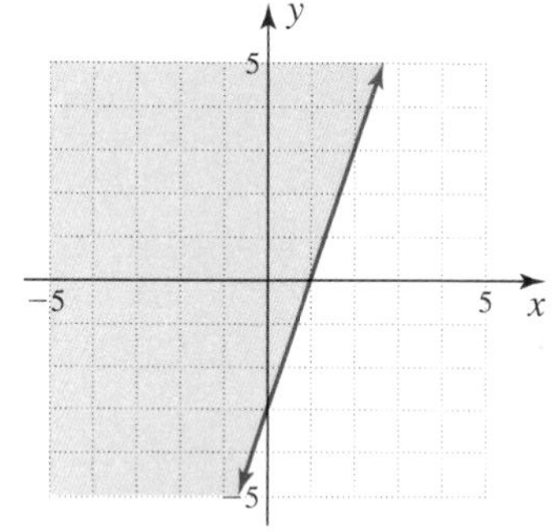

6. $y \geq -2x + 4$

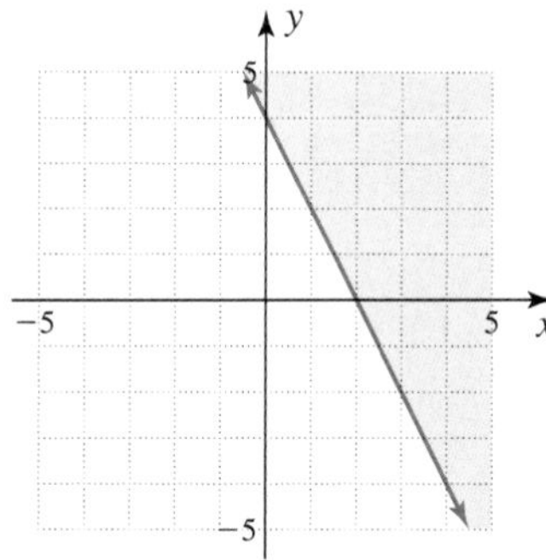

7. $6 < 3x - 6y$

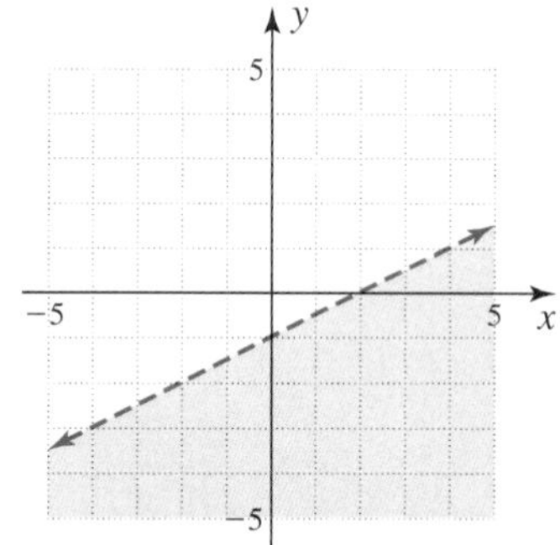

8. $6 < 2x - 3y$

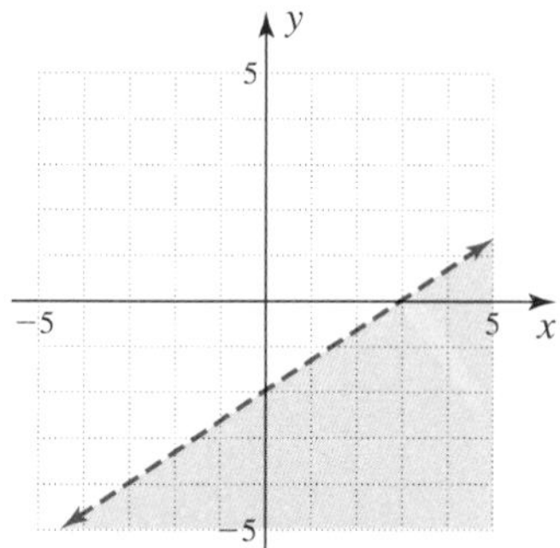

9. $3x + 4y \geq 12$

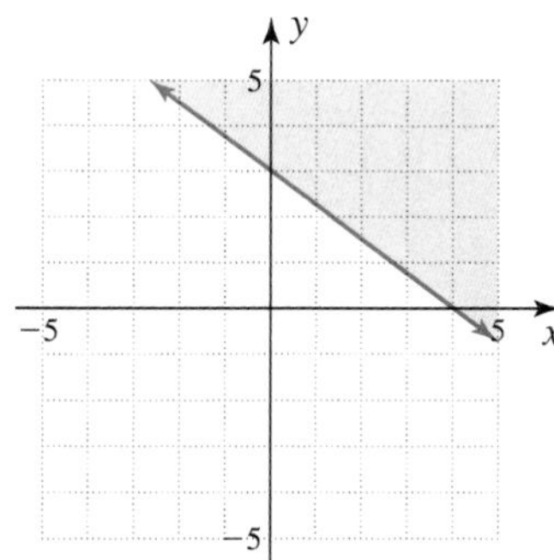

10. $-3y \geq 6x + 6$

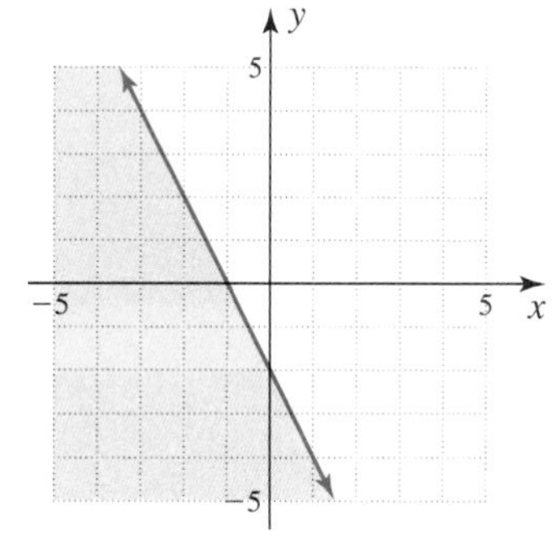

11. $10 < -2x + 5y$

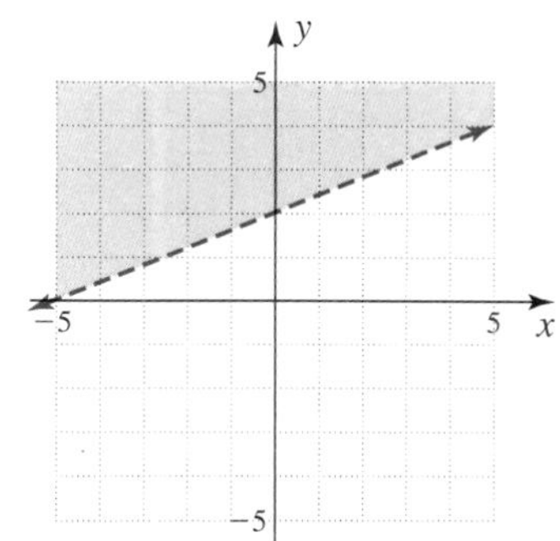

12. $4 < x - y$

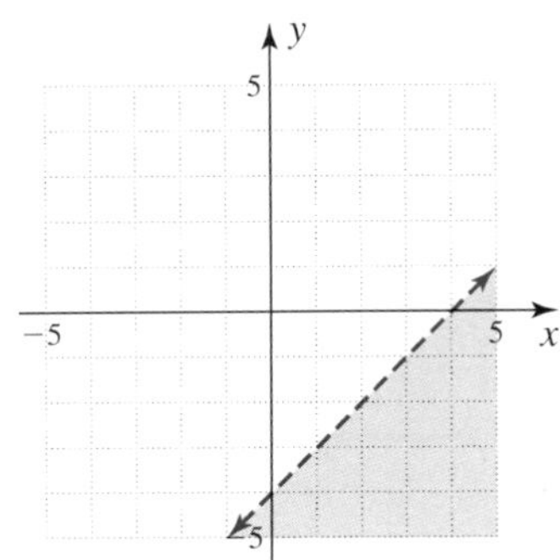

13. $x \geq 2y - 4$

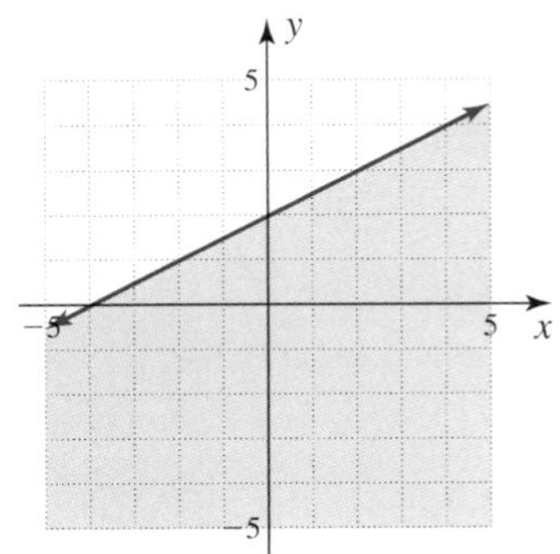

14. $2x \geq 4y + 2$

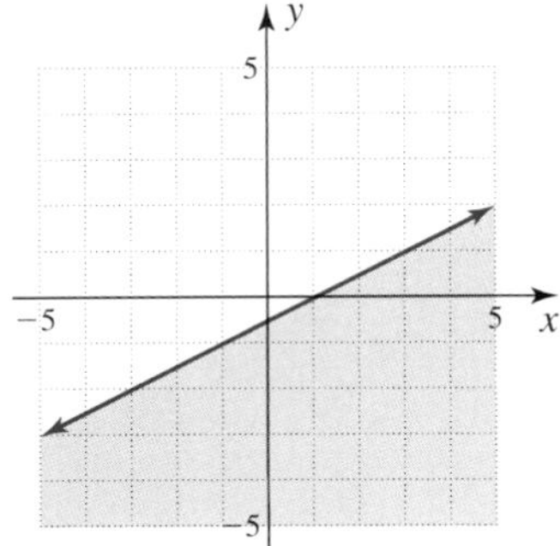

15. $y < -x + 5$

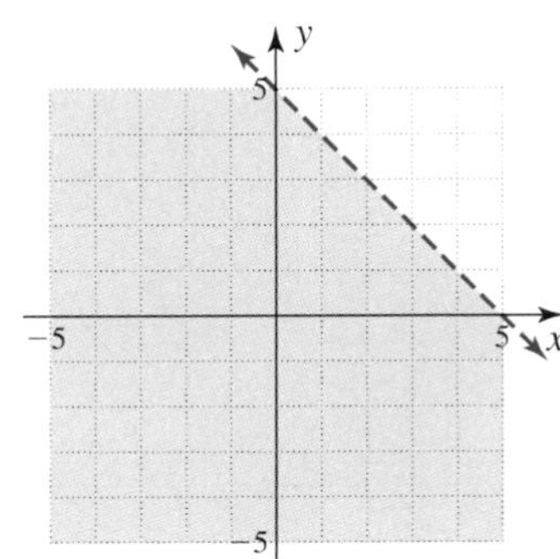

16. $2y < 4x - 8$

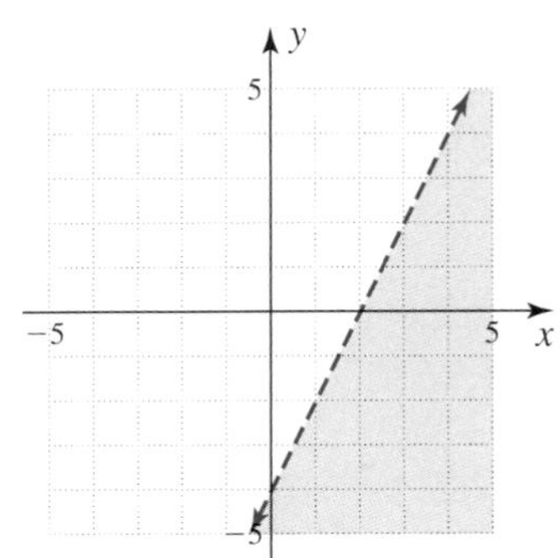

17. $2y < 4x + 5$

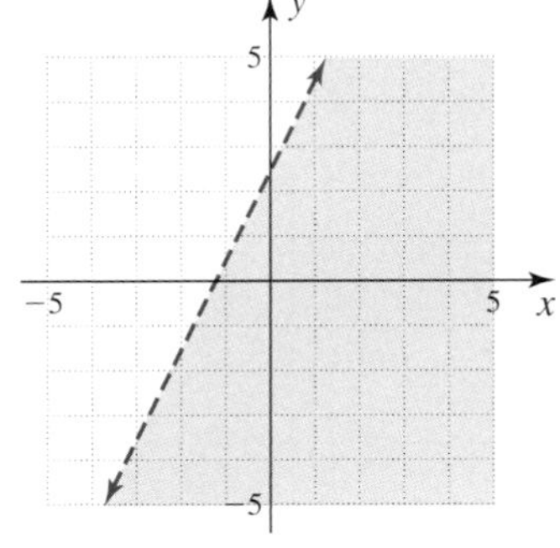

18. $2y \geq 3x + 5$

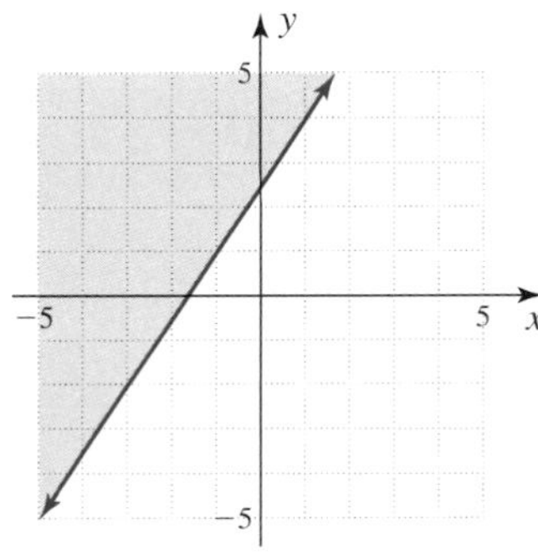

19. $x > 1$

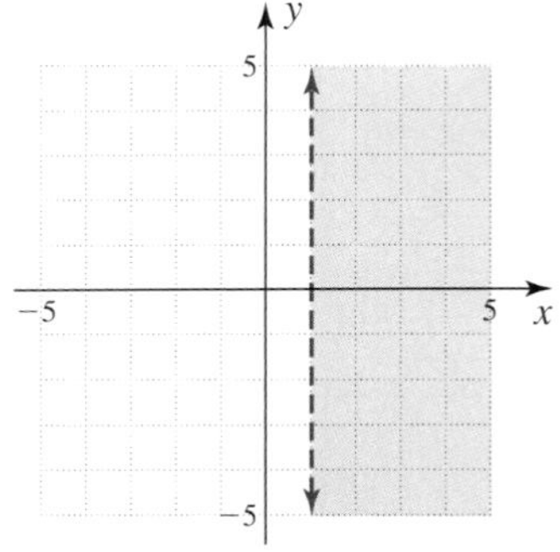

20. $y \leq \frac{3}{2}$

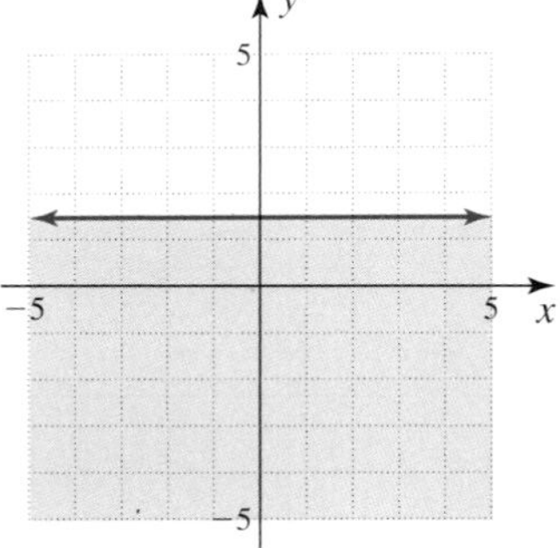

21. $x \leq \frac{5}{2}$

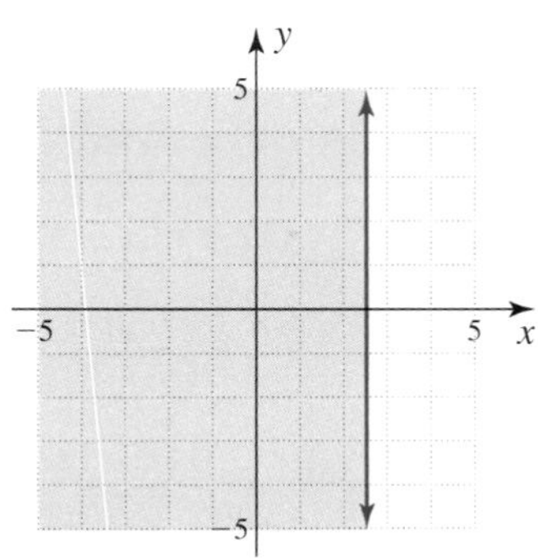

22. $x \geq -\frac{2}{3}$

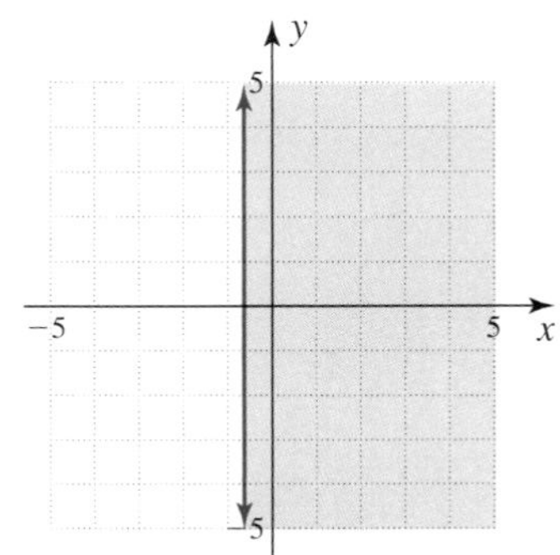

23. $y \leq -\frac{3}{2}$

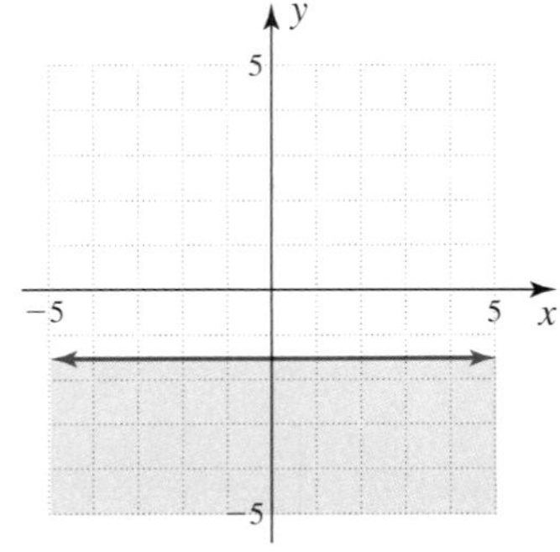

24. $x - \frac{1}{3} \geq 0$

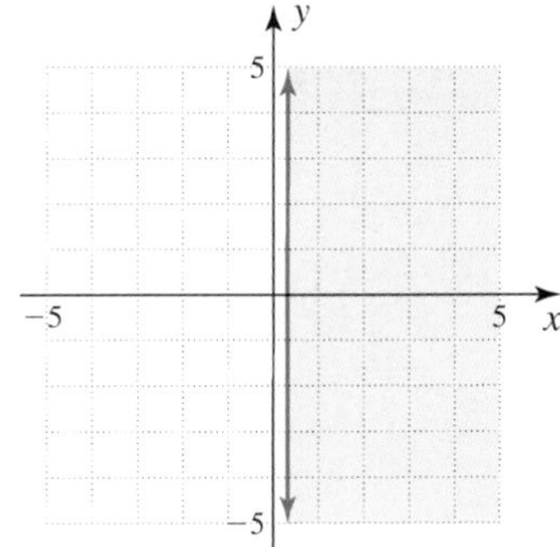

25. $x - \frac{2}{3} > 0$

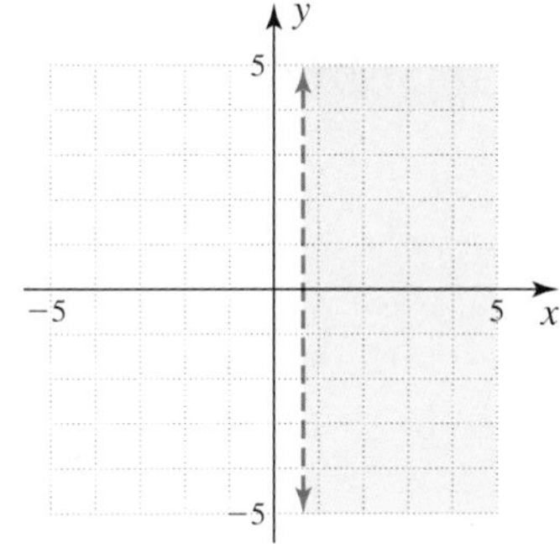

26. $y - \frac{5}{2} > \frac{1}{2}$

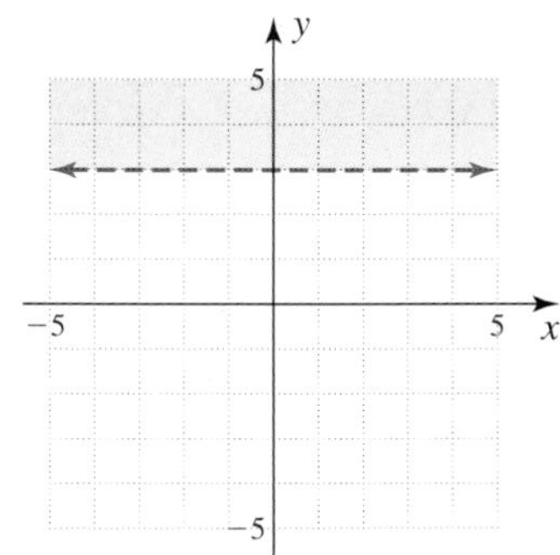

27. $y + \frac{1}{3} \geq \frac{2}{3}$

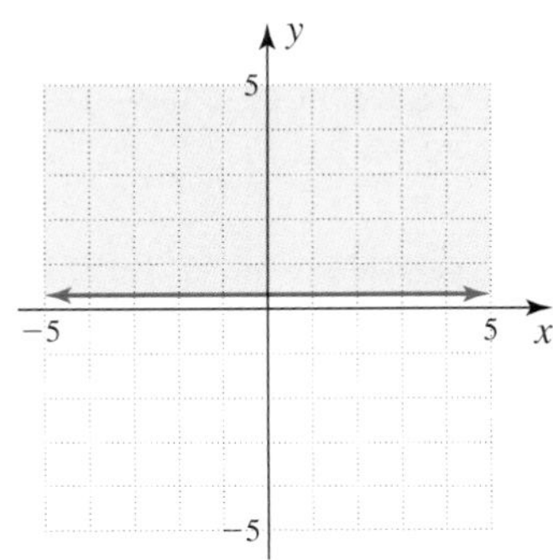

28. $x + \frac{1}{5} < \frac{6}{5}$

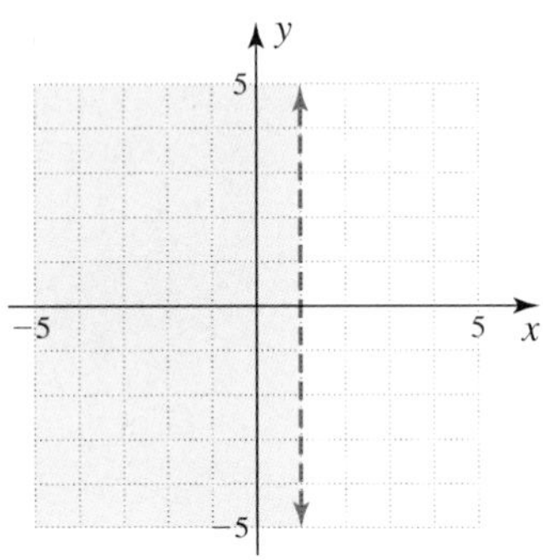

29. $2x + y < 0$

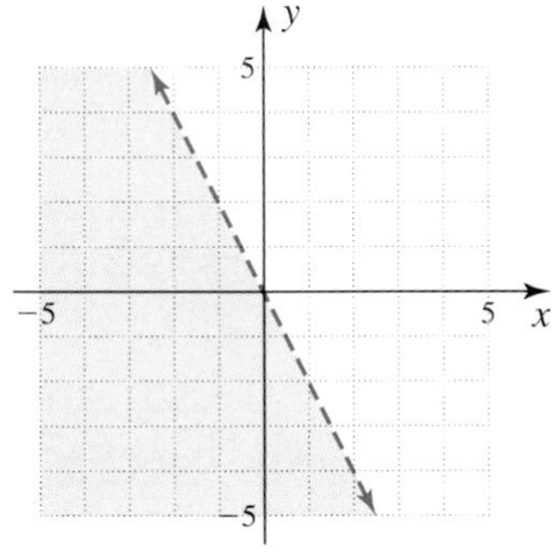

30. $2y + x \geq 0$

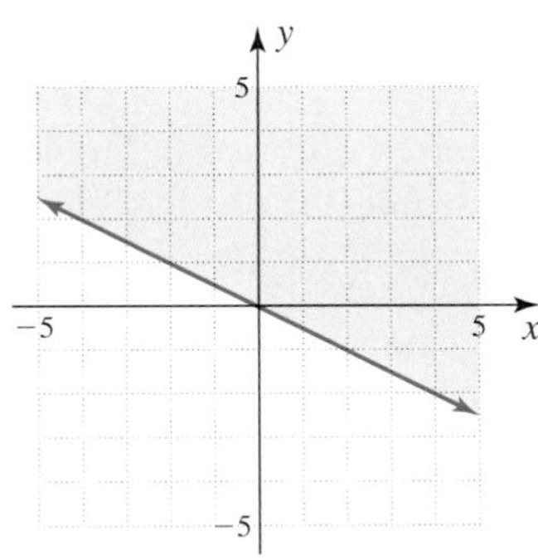

31. $y - 3x > 0$

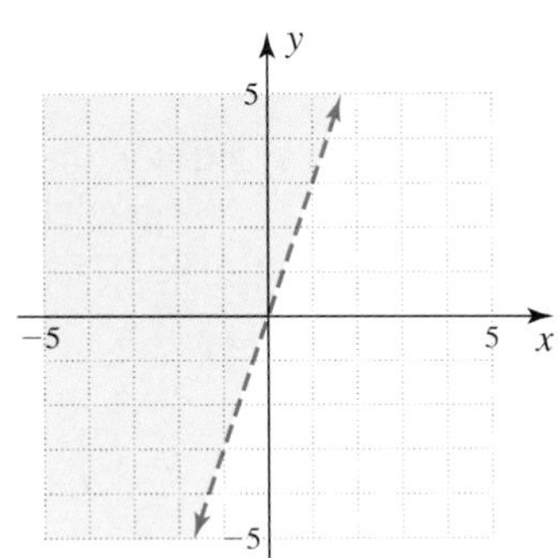

32. $2x - y \leq 0$

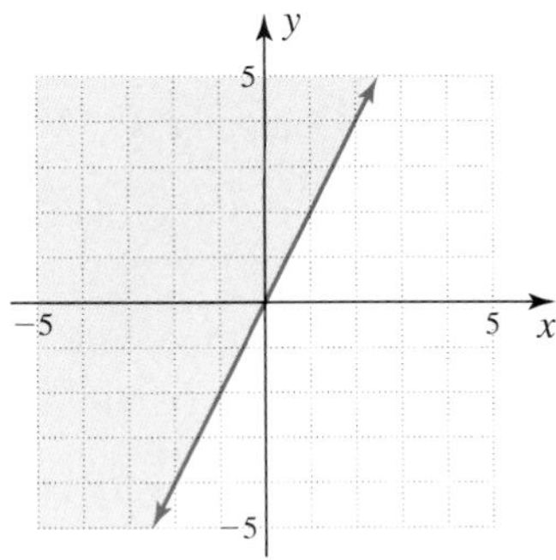

In Problems 33 and 34, the graph of the given system of inequalities defines a geometric figure. Graph the inequalities on the same set of axes and identify the figure.

33. $-x \leq -3, \quad -y \geq -4, \quad x \leq 4, \quad y \geq 2$

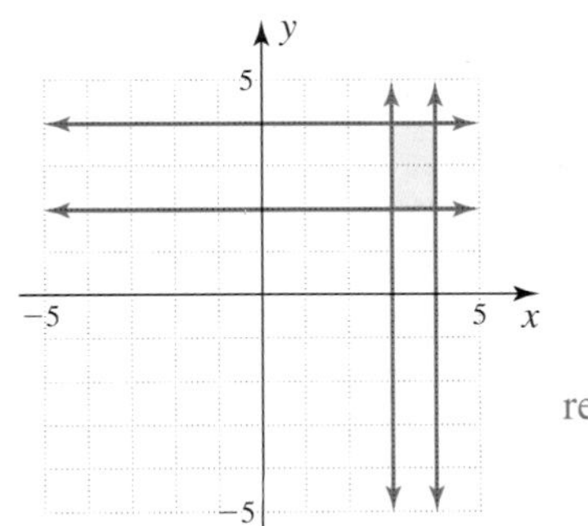

rectangle

34. $x \geq 2, \quad -x \geq -5, \quad y \leq 5, \quad -y \leq -2$

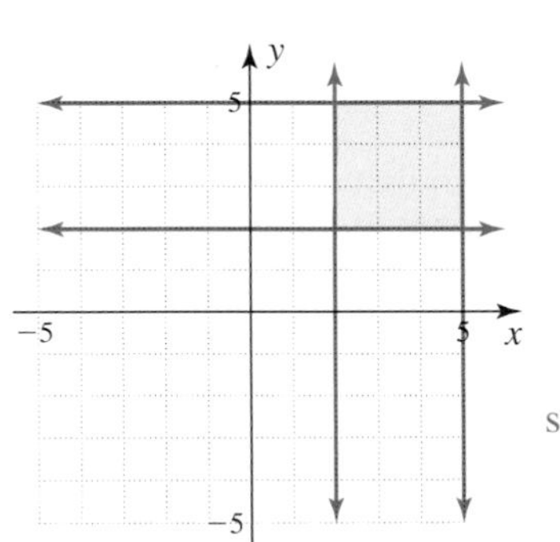

square

SKILL CHECKER

Try the Skill Checker Exercises so you'll be ready for the next section.

Solve for k.

35. $0.06 = \frac{k}{130}$ $k = 7.8$

36. $0.05 = \frac{k}{120}$ $k = 6$

37. $0.103 = 4k$ $k = 0.02575$

38. $0.052 = 2k$ $k = 0.026$

USING YOUR KNOWLEDGE

Savings on Rentals

The ideas we discussed in this section can save you money when you rent a car. Here's how.

Suppose you have the choice of renting a car from company A or from company B. The rates for these companies are as follows:

Company A: $20 a day plus $0.20 per mile

Company B: $15 a day plus $0.25 per mile

If x is the number of miles traveled in a day and y represents the cost for that day, the equations representing the cost for each company are

Company A: $y = 0.20x + 20$

Company B: $y = 0.25x + 15$

39. Graph the equation representing the cost for company A.

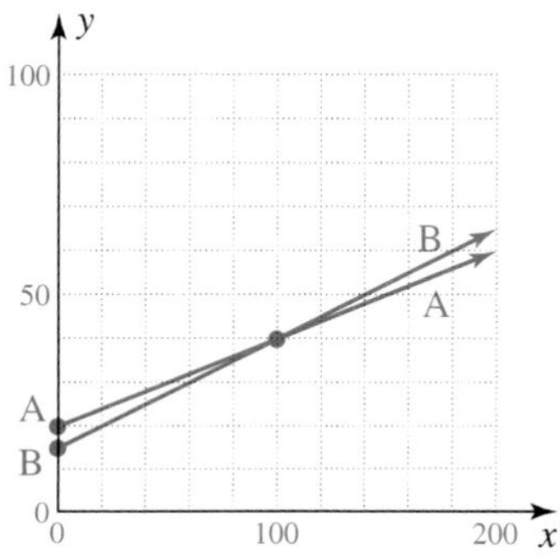

40. On the same coordinate axes you used in Problem 39, graph the equation representing the cost for company B.

41. When is the cost the same for both companies?
When they travel 100 miles

42. When is the cost less for company A?
When traveling more than 100 miles

43. When is the cost less for company B?
When traveling less than 100 miles

WRITE ON

44. Describe in your own words the graph of a linear inequality in two variables. How does it differ from the graph of a linear inequality in one variable?
Answers may vary.

45. Write the procedure you use to solve a linear inequality in two variables. How does the procedure differ from the one you use to solve a linear inequality in one variable? Answers may vary.

46. Explain how you decide whether the boundary line is solid or dashed when graphing a linear inequality in two variables. When the inequality is $\geq$ or $\leq$, use a solid line; when it is $>$ or $<$, use a dashed line.

47. Explain why a point on the boundary line cannot be used as a test point when graphing a linear inequality in two variables. Answers may vary.

MASTERY TEST

If you know how to do these problems, you have learned your lesson!

Graph:

48. $x - 4 \leq 0$

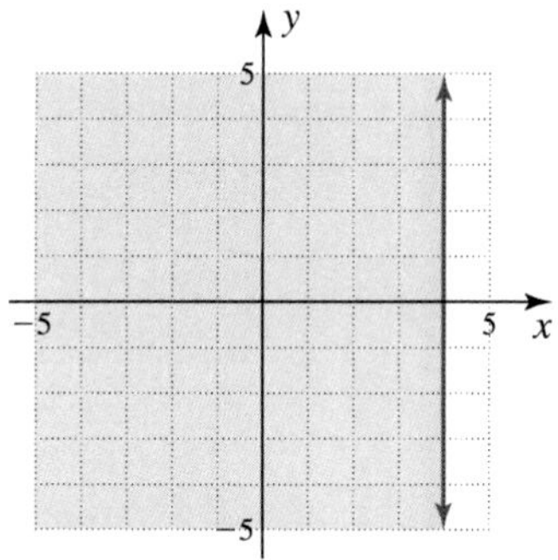

49. $y \geq -4$

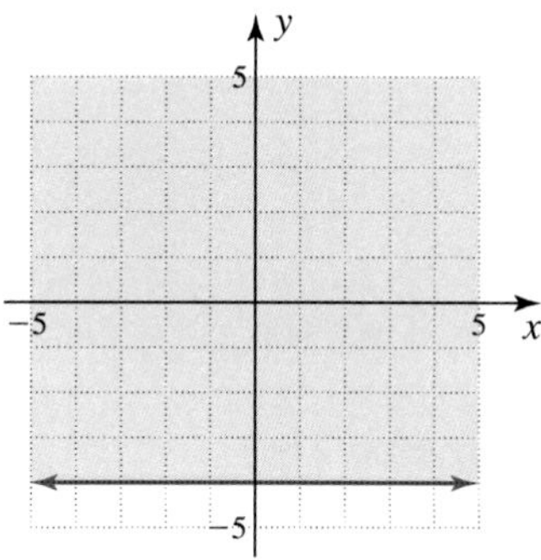

50. $x + 2 < 0$

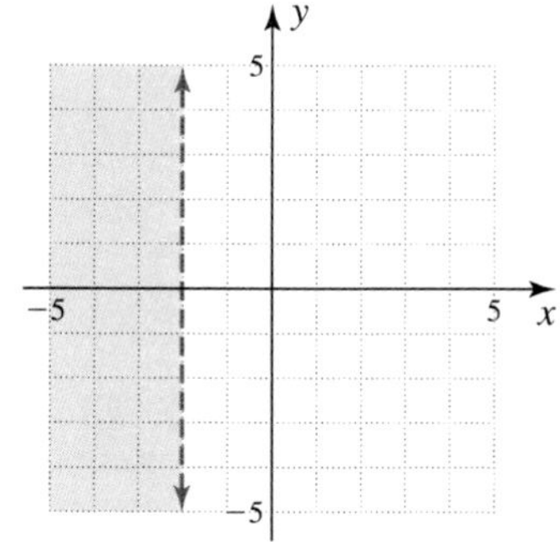

51. $y - 3 > 0$

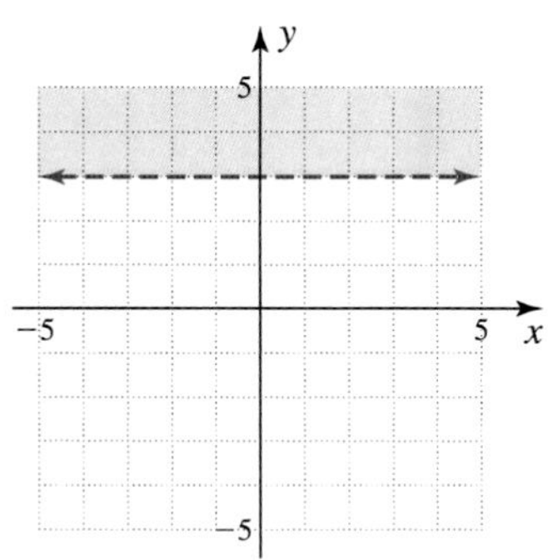

52. $3x - 2y < -6$

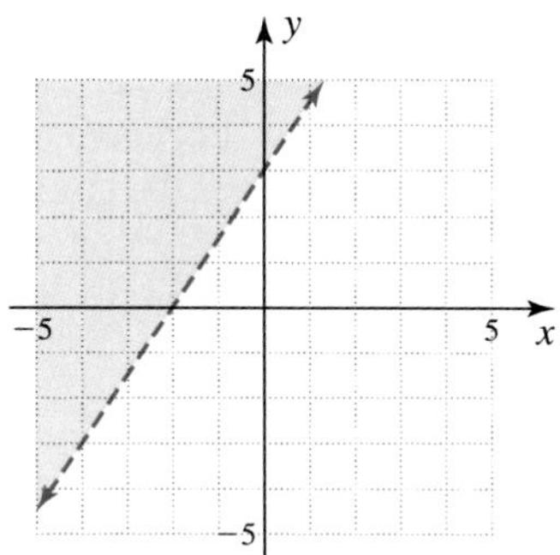

53. $y \leq -4x + 8$

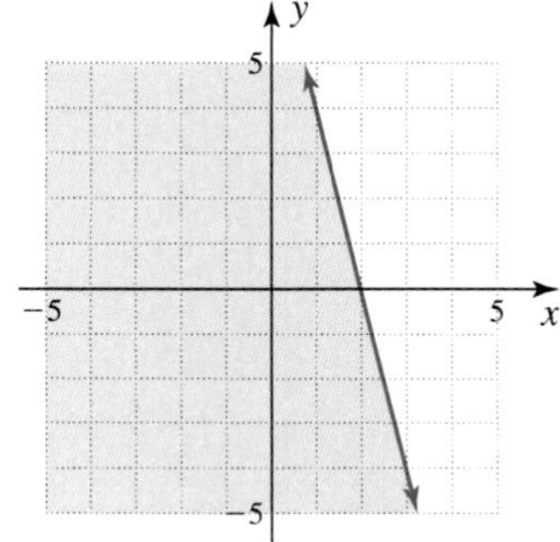

54. $y > 2x$

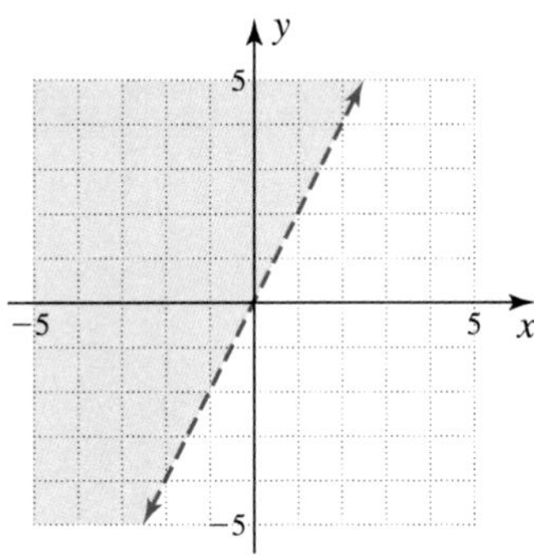

55. $y < 3x$

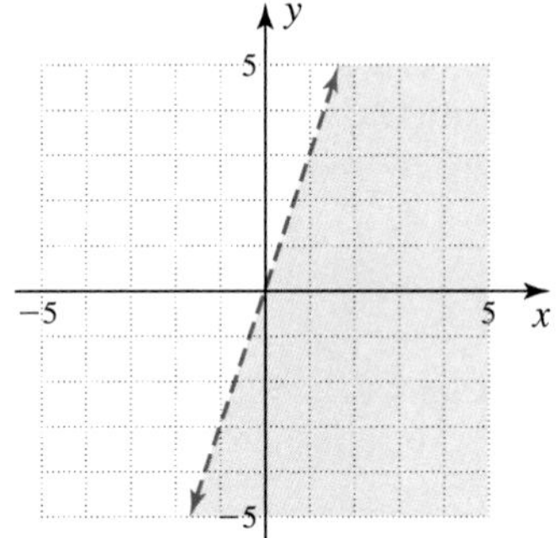

56. $x - 2y > 0$

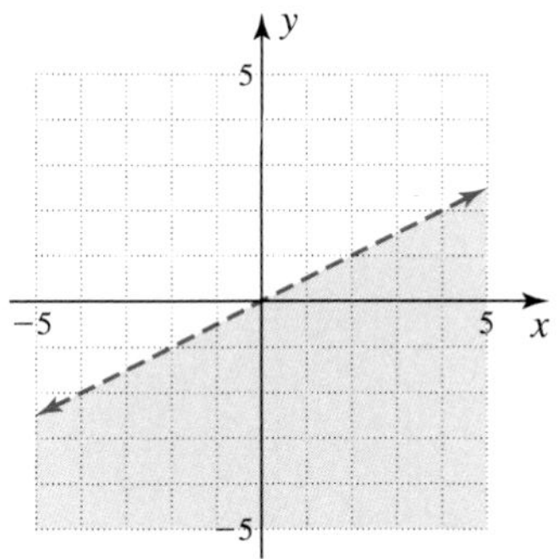

57. $3x + y \leq 0$

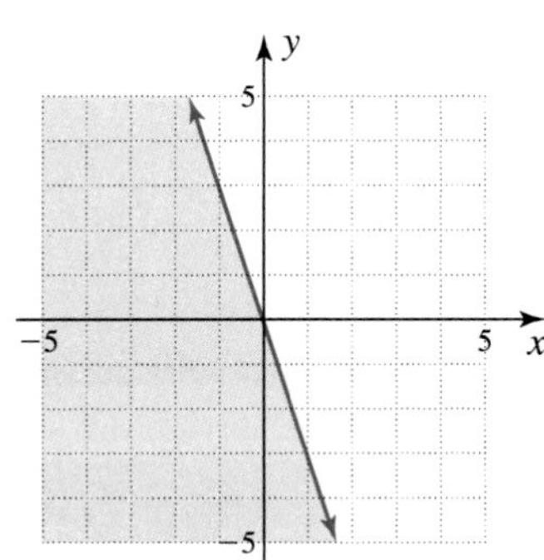

58. $y - 4x > 0$

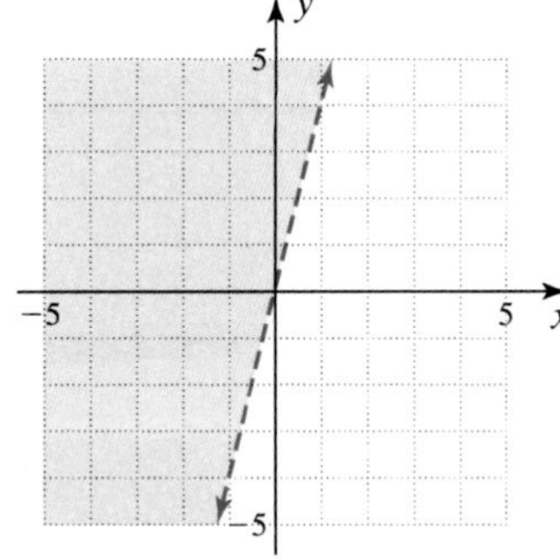

7.4 DIRECT AND INVERSE VARIATION

To Succeed, Review How To . . .

1. Solve linear equations (pp. 151–155).

Objectives

A Find and solve equations of direct variation given values of the variables.

B Find and solve equations of inverse variation given values of the variables.

C Solve an application involving variation.

GETTING STARTED

Don't Forget the Tip!

Jasmine is a server at CDB restaurant. Aside from her tips, she gets \$2.88/hour. In 1 hour, she earns \$2.88; in 2 hr, she earns \$5.76; in 3 hr, she earns \$8.64, and so on. We can form the set of ordered pairs (1, 2.88), (2, 5.76), (3, 8.64) using the number of hours she works as the first coordinate and the amount she earns as the second coordinate. Note that the ratio of second coordinates to first coordinates is the same number:

$$\frac{2.88}{1} = 2.88, \quad \frac{5.76}{2} = 2.88, \quad \frac{8.64}{3} = 2.88,$$

and so on.

When the ratio of ordered pairs of numbers is constant, we say that there is a **direct variation.** In this case, the earnings E **vary directly** (or are **directly proportional**) to the number of hours h, that is,

$$\frac{E}{h} = 2.88 \quad \text{(a constant)} \quad \text{or} \quad E = 2.88h$$

Note that in the graph of $E = 2.88h$ or the more familiar $y = 2.88x$, the constant $k = 2.88$ is the **slope** of the line.

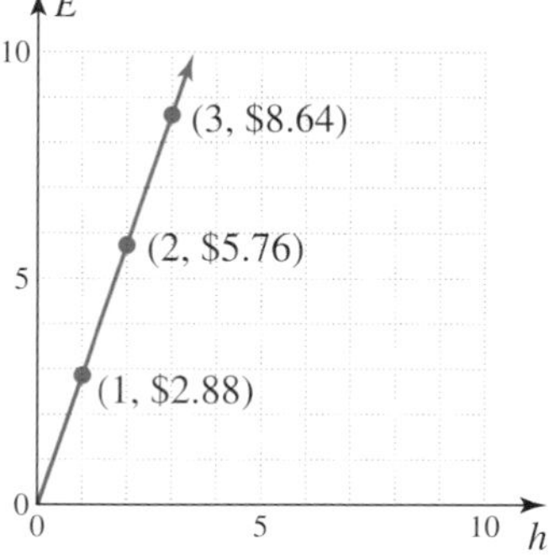

Direct Variation

In general, if we have a situation in which a variable **varies directly** or **is directly proportional** to another variable, we make the following definition.

DEFINITION

y* varies directly as *x if there is a constant k such that $y = kx$. The constant k is the constant of variation or proportionality.

In real life, we can find k by experimenting. Suppose you are dieting and want to eat some McDonald's® fries, but only 100 calories worth. The fries come in small, medium, or large sizes with 210, 450, or 540 calories, respectively. You could use **estimation** and eat approximately half of a small order ($\frac{210}{2} \approx 105$ calories), but you can estimate better! The numbers of fries in the small, medium, and large sizes are 45, 97, and 121; thus, the calories per fry are:

$$\frac{210}{45} \approx 4.7 \text{ (small)}, \quad \frac{450}{97} \approx 4.6 \text{ (medium)}, \quad \frac{540}{121} \approx 4.5 \text{ (large)}$$

Source: Author (counted and ate!).

EXAMPLE 1 **You want fries with that?**

The number C of calories in an order of McDonald's french fries varies directly as the number n of fries in the order. Suppose you buy a large order of fries (121 fries, 540 calories).

a. Write an equation of variation.

b. Find k.

c. If you want to eat 100 calories worth, how many fries would you eat?

SOLUTION

a. If C varies directly as n, then $C = kn$.

b. A large order of fries has $C = 540$ calories and $n = 121$ fries. Thus

$$C = kn \quad \text{becomes} \quad 540 = k(121) \quad \text{or} \quad k = \frac{540}{121} \approx 4.5$$

c. If you want to eat $C = 100$ calories worth,

$$C = kn = 4.5n \quad \text{becomes} \quad 100 = 4.5n$$

Solving for n by dividing both sides by 4.5, we get

$$n = \frac{100}{4.5} \approx 22$$

So, if you eat about 22 fries, you will have consumed about 100 calories. Note that this is indeed about half of a small order!

PROBLEM 1

The number C of calories in an order of Burger King® fries also varies directly as the number n of fries in the order. An order of king-size fries has 590 calories and 101 fries.

a. Write an equation of variation.

b. Find k.

c. If you want to eat 100 calories worth, how many fries would you eat?

Now that we have eaten some calories, let's see how we can spend (burn) some calories. Read on.

Answers

1. a. $C = kn$ **b.** $k = \frac{590}{101} \approx 5.8$

c. $\frac{100}{5.8} \approx 17$ (about 17 fries)

EXAMPLE 2 **Variation and bicycling**

If you weigh about 160 pounds and you jog (5 mi/hr) or ride a bicycle (12 mi/hr), the number C of calories used is proportional to the time t (in minutes).

a. Find an equation of variation.

b. If jogging for 15 minutes uses 150 calories, find k.

c. How many calories would you use if you jog for 20 minutes?

d. To lose a pound, you have to use about 3500 calories. How many minutes do you have to jog to lose 1 pound?

SOLUTION

a. Since the number C of calories used is proportional to the time t (in minutes), the equation of variation is $C = kt$.

b. If jogging for 15 minutes ($t = 15$) uses $C = 150$ calories, then

$C = kt$ becomes $\quad 150 = k(15)$

Dividing both sides by 15 $\quad 10 = k$

Note that now $C = kt$ becomes $\quad C = 10t$

c. We want to know how many calories C you would use if you jog for $t = 20$ minutes. Substitute 20 for t in $C = 10t$

obtaining $\quad C = 10(20) = 200$

Thus, you will use 200 calories if you jog for 20 minutes.

d. To lose 1 pound, you need to use $C = 3500$ calories.

$C = 10t$ becomes $\quad 3500 = 10t$

Dividing both sides by 10 $\quad 350 = t$

Thus, you need 350 minutes of jogging to lose 1 pound! By the way, if you jog for 350 minutes at 5 mi/hr, you will be 29 miles away!

PROBLEM 2

If you weigh 120 pounds and you jog (5 mi/hr) or ride a bicycle (12 mi/hr), the number C of calories used is proportional to the time t (in minutes).

a. Find an equation of variation.

b. If jogging for 15 minutes uses 105 calories, find k.

c. How many calories would you use if you jog for 20 minutes?

d. To lose a pound, you have to use about 3500 calories. How many minutes do you have to jog to lose 1 pound?

Web It

For more information on calories used with exercise, try links 7-4-1 and 7-4-2 on the Bello Website at mhhe.com/bello.

Web It

To learn about variation, go to links 7-4-3 and 7-4-4 on the Bello Website at mhhe.com/bello.

B Inverse Variation

Sometimes, as one quantity increases, a related quantity decreases proportionately. For example, the more time we spend practicing a task, the less time it will take us to do the task. In this case, we say that the quantities **vary inversely** as each other.

DEFINITION

y* varies inversely as *x if there is a constant k such that

$$y = \frac{k}{x}$$

Answers

2. a. $C = kt$ **b.** $C = 7t$
c. 140 calories **d.** 500 minutes

EXAMPLE 3 Inverse proportion problem

The rate of speed v at which a car travels is inversely proportional to the time t it takes to travel a given distance.

a. Write the equation of variation.

b. If a car travels at 60 mi/hr for 3 hr, what is k, and what does it represent?

SOLUTION

a. The equation is

$$v = \frac{k}{t}$$

b. We know that $v = 60$ when $t = 3$. Thus,

$$60 = \frac{k}{3}$$

$$k = 180$$

In this case, k represents the distance traveled, and the new equation of variation is

$$v = \frac{180}{t}$$

PROBLEM 3

Suppose a car travels at 55 miles per hour for 2 hours.

a. Find k and explain what it represents.

b. Write the new equation of variation.

EXAMPLE 4 Boom boxes and inverse proportion

Have you ever heard one of those loud "boom" boxes or a car sound system that makes your stomach tremble? The loudness L of sound is inversely proportional to the square of the distance d that you are from the source.

a. Write an equation of variation.

b. The loudness of rap music coming from a boom box 5 ft away is 100 dB (decibels). Find k.

c. If you move to 10 ft away from the boom box, how loud is the sound?

SOLUTION

a. The equation is

$$L = \frac{k}{d^2}$$

b. We know that $L = 100$ for $d = 5$, so that

$$100 = \frac{k}{5^2} = \frac{k}{25}$$

Multiplying both sides by 25, we find that $k = 2500$, and the new equation of variation is

$$L = \frac{2500}{d^2}$$

c. Since $k = 2500$,

$$L = \frac{2500}{d^2}$$

When $d = 10$,

$$L = \frac{2500}{10^2} = 25 \text{ dB}$$

PROBLEM 4

Suppose the loudness is 80 dB at 5 feet.

a. Find k.

b. Write the new equation of variation.

c. If you move 10 ft away, how loud is the sound now?

Web It

For applications dealing with inverse variation, go to link 7-4-5 on the Bello Website at mhhe.com/bello.

Answers

3. a. $k = 110$ and represents the distance traveled. **b.** $v = \frac{110}{t}$

4. a. $k = 2000$ **b.** $L = \frac{2000}{d^2}$ **c.** 20 dB

C Application

EXAMPLE 5 Water, water everywhere

Figure 13 shows the number of gallons of water, g (in millions), produced by an inch of snow in different cities. Note that the larger the area of the city, the more gallons of water are produced, so g is directly proportional to A, the area of the city (in square miles).

a. Write an equation of variation.

b. If the area of St. Louis is about 62 mi^2, what is k?

c. Find the amount of water produced by 1 in. of snow falling in Anchorage, Alaska, with an area of 1700 mi^2.

SOLUTION

a. Since g is directly proportional to A, $g = kA$.

b. From Figure 13 we can see that $g = 100$ (million) is the number of gallons of water produced by 1 in. of snow in St. Louis. Since it is given that $A = 62$, $g = kA$ becomes

$$100 = k \cdot 62 \quad \text{or} \quad k = \frac{100}{62} = \frac{50}{31}$$

c. For Anchorage, $A = 1700$, thus $g = \frac{50}{31} \cdot 1700 \approx 2742$ million gallons of water.

PROBLEM 5

If the area of the city of Boston is about 40 mi^2 and 60 (million) gallons of water are produced by 1 inch of snow, find an equation of variation for Boston.

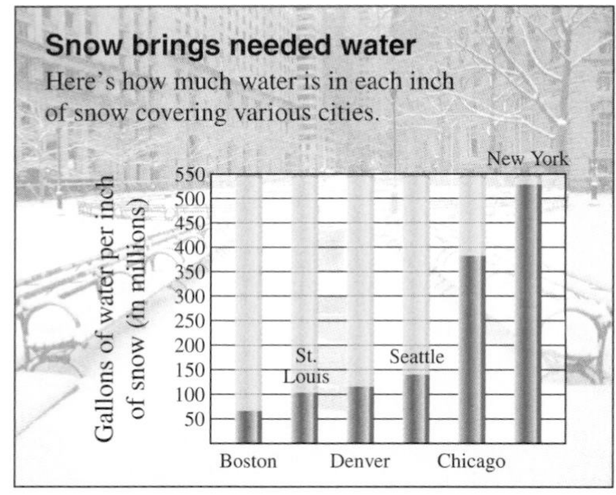

Figure 13

Source: Data from USA Today, 1994.

Answer

5. $g = \frac{3}{2}A$

Exercises 7.4

Boost *your* GRADE at mathzone.com!

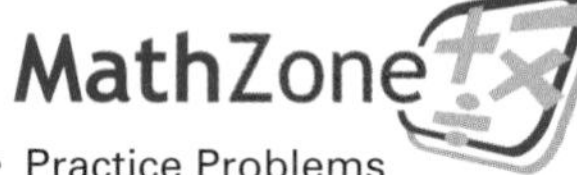

- Practice Problems
- Self-Tests
- Videos
- NetTutor
- e-Professors

A In Problems 1–7, solve the direct variation problem.

1. Do you know how much your skin weighs? Your weight W (in pounds) is proportional to the weight S of your skin, which can be obtained by dividing W by 16.

a. Write an equation of variation. $S = \frac{W}{16}$

b. What is k? $k = \frac{1}{16}$

c. If you weigh 160 pounds, what is the weight of your skin? 10 pounds

2. How much blood do you have? Your weight W (in kilograms) is directly proportional to your blood volume V (in liters) and can be obtained by dividing W by 12.

a. Write an equation of variation. $V = \frac{W}{12}$

b. What is k? $k = \frac{1}{12}$

c. If you weigh 64 kilograms (about 140 pounds), what is your blood volume? $5\frac{1}{3}$ liters

3. Your weight W (in pounds) is directly proportional to your basal metabolic rate R (number of calories you burn when at rest) and can be obtained by dividing W by 10.

a. Write an equation of variation. $R = \frac{W}{10}$

b. What is k? $k = \frac{1}{10}$

c. If you weigh 160 pounds, what is the value of your basal metabolic rate? 16

4. The amount I of annual interest received on a savings account is directly proportional to the amount m of money you have in the account.

a. Write an equation of variation. $I = km$

b. If \$480 produced \$26.40 in interest, what is k? $k = 0.055$

c. How much annual interest would you receive if the account had \$750? \$41.25

5. The number R of revolutions a record makes as it is being played varies directly with the time t that it is on the turntable.

a. Write an equation of variation. $R = kt$

b. A record that lasted $2\frac{1}{2}$ min made 112.5 revolutions. What is k? $k = 45$

c. If a record makes 108 revolutions, how long does it take to play it? 2.4 minutes

6. The distance d an automobile travels after the brakes have been applied varies directly as the square of its speed s.

a. Write an equation of variation. $d = ks^2$

b. If the stopping distance for a car going 30 mi/hr is 54 ft, what is k? $k = 0.06$

c. What is the stopping distance for a car going 60 mi/hr? 216 ft

7. The weight of a person varies directly as the cube of the person's height h (in inches). The **threshold weight** T (in pounds) for a person is defined as the "crucial weight, above which the mortality risk for the patient rises astronomically."

a. Write an equation of variation relating T and h. $T = kh^3$

b. If $T = 196$ when $h = 70$, find k to five decimal places. $k \approx 0.00057$

c. To the nearest pound, what is the threshold weight for a person 75 in. tall? 240 pounds

B In Problems 8–15, solve the inverse variation problem.

8. To remain popular, the number S of new songs a rock band needs to produce each year is inversely proportional to the number y of years the band has been in the business.

a. Write an equation of variation. $S = \frac{k}{y}$

b. If, after 3 years in the business, the band needs 50 new songs, how many songs will it need after 5 years? 30 songs

9. When a camera lens is focused at infinity, the f-stop on the lens varies inversely with the diameter d of the aperture (opening).

a. Write an equation of variation. $f = \frac{k}{d}$

b. If the f-stop on a camera is 8 when the aperture is $\frac{1}{2}$ in., what is k? $k = 4$

c. Find the f-stop when the aperture is $\frac{1}{4}$ in. 16

10. Boyle's law states that if the temperature is held constant, then the pressure P of an enclosed gas varies inversely as the volume V. If the pressure of the gas is 24 lb/in.^{2} when the volume is 18 in.^{3}, what is the pressure if the gas is compressed to 12 in.^{3}? 36 lb/in.^{2}

11. For the gas of Problem 10, if the pressure is 24 lb/in.^{2} when the volume is 18 in.^{3}, what is the volume if the pressure is increased to 40 lb/in.^{2}? 10.8 in.^{3}

12. The weight W of an object varies inversely as the square of its distance d from the center of the earth.

a. Write an equation of variation. $W = \frac{k}{d^2}$

b. An astronaut weighs 121 lb on the surface of the earth. If the radius of the earth is 3960 mi, find the value of k for this astronaut. (Do not multiply out your answer.) $k = 121 \times 3960^2$

c. What will this astronaut weigh when she is 880 mi above the surface of the earth? 81 lb

13. One of the manuscript pages of this book had about 600 words and was typed using a 12-point font. Suppose the average number w of words that can be printed on a manuscript page is inversely proportional to the font size s.

a. Write an equation of variation. $w = \frac{k}{s}$

b. What is k? $k = 7200$

c. How many words could be typed on the page if a 10-point font was used? 720 words

14. The price P of oil varies inversely with the supply S (in million barrels per day). In the year 2000, the price of one barrel of oil was \$26.00 and OPEC production was 24 million barrels per day.

a. Write an equation of variation. $P = \frac{k}{S}$

b. What is k? $k = 624$

c. If OPEC plans to increase production to 28 million barrels per day, what would the price of one barrel be? Answer to the nearest cent. \$22.29

15. According to the National Center for Health Statistics, the number b of births (per 1000 women) is inversely proportional to the age a of the woman. The number b of births (per 1000 women) for 27-year-olds is 110.

a. Write an equation of variation. $b = \frac{k}{a}$

b. What is k? $k = 2970$

c. What would you expect the number b of births (per 1000 women) to be for 33-year-old women? 90 (per 1000 women)

APPLICATIONS

16. *Miles per gallon* The number of miles m you can drive in your car is directly proportional to the amount of fuel g in your gas tank.

a. Write an equation of variation. $m = kg$

b. The greatest distance yet driven without refueling on a single fill in a standard vehicle is 1691.6 miles. If the twin tanks used to do this carried a total of 38.2 gal of fuel, what is k (round to two decimal places)? $k \approx 44.28$

c. How many miles per gallon is this? About 44.28 miles per gallon

17. *Distance and speed of car* The distance d (in miles) traveled by a car is directly proportional to the average speed s (in miles per hour) of the car, even when driving in reverse!

a. Write an equation of variation. $d = ks$

b. The highest average speed attained in any nonstop reverse drive of more than 500 mi is 28.41 mph. If the distance traveled was 501 mi, find k (round to two decimal places). $k \approx 17.63$

c. What does k represent? The time it took to drive d miles at s miles per hour

18. *Responses to radio call-in contest* Have you called in on a radio contest lately? According to Don Burley, a radio talk-show host in Kansas City, the listener response to a radio call-in contest is directly proportional to the size of the prize.

a. If 40 listeners call when the prize is \$100, write an equation of variation using N for the number of listeners and P for the prize in dollars. $N = 0.4P$

b. How many calls would you expect for a \$5000 prize? 2000 calls

19. *Cricket chirps and temperature* The number C of chirps a cricket makes each minute is directly proportional to 37 less than the temperature F in degrees Fahrenheit.

a. If a cricket chirps 80 times when the temperature is 57°F, what is the equation of variation? $C = 4(F - 37)$ or $C = 4F - 148$

b. How many chirps per minute would the cricket make when the temperature is 90°F? 212 chirps per minute

20. *Atmospheric carbon dioxide concentration* The concentration of carbon dioxide (CO_2) in the atmosphere has been increasing due to automobile emissions, electricity generation, and deforestation. In 1965, CO_2 concentration was 319.9 parts per million (ppm), and 23 years later, it had increased to 351.3 ppm. The *increase* I, of carbon dioxide concentration in the atmosphere is directly proportional to the number n of years elapsed since 1965.

a. Write an equation of variation for I. $I = kn$

b. Find k (round to two decimal places). $k \approx 1.37$

c. What would you predict the CO_2 concentration to be in the year 2000 (round to two decimal places)? 367.85 ppm

21. *Percent increase of college graduates* The *cumulative increase* I in the *percent* of college graduates in the United States after 1930 among persons 25 years and older is proportional to the square of the number of years after 1930. In 1940, the cumulative increase was about 5%.

a. Write an equation of variation for I if n is the number of years elapsed since 1930. $I = kn^2$

b. Find k. $k = 0.05$

c. What would you predict the cumulative percent increase to be in the year 2000? 245%

22. *Water depth and temperature* At depths of more than 1000 m (a kilometer), water temperature T (in degrees Celsius) in the Pacific Ocean varies inversely as the water depth d (in meters). If the water temperature at 4000 m is 1°C, what would it be at 8000 m? 0.5°C

23. *Blood alcohol concentration* You might think that the BAC is directly proportional to how many beers you drink in an hour. Strangely enough, for both males and females of a specific weight, the BAC is directly proportional to $(N - 1)$, which is one less than the number of beers consumed during the last hour.

a. Write an equation of variation. $\text{BAC} = k(N - 1)$

b. For a 150-pound man, the average BAC after 3 beers is 0.052. Find k. $k = 0.026$

c. What is the BAC after 5 beers? 0.104

d. How many beers can the man drink before going over the 0.08 limit? 4

24. *Blood alcohol concentration* For a 130-pound female, the average BAC is also directly proportional to one less than the number of beers consumed during the last hour $(N - 1)$.

a. Write an equation of variation. $\text{BAC} = k(N - 1)$

b. For a 130-pound female, the average BAC after 3 beers is 0.066. Find k. $k = 0.033$

c. What is the BAC after 5 beers? 0.132

d. How many beers can the woman drink before going over the 0.08 limit? 3

25. *BAC and weight* If you drank 3 beers in the last hour, your blood alcohol content (BAC) is inversely proportional to your weight W. For a 130-pound male, the BAC after drinking 3 beers is 0.06.

a. Write an equation of variation. $\text{BAC} = \frac{7.8}{W}$

b. What is the BAC of a 260-pound male after drinking 3 beers? 0.03

c. In most states, you are legally drunk if your BAC is 0.08 or higher. What is the weight of a male whose BAC is exactly 0.08 after drinking 3 beers in the last hour? 97.5 pounds

d. What would your BAC be if you weigh more than the male in part **c**? Less than 0.08

26. *BAC and weight* If you drank 3 beers in the last hour, your blood alcohol content (BAC) is inversely proportional to your weight W. For a 130-pound female, the BAC after drinking 3 beers is 0.066.

a. Write an equation of variation. $\text{BAC} = \frac{8.58}{W}$

b. What is the BAC of a 260-pound female after drinking those 3 beers? 0.033

c. In most states, you are legally drunk if your BAC is 0.08 or higher. What is the weight of a female whose BAC is exactly 0.08 after drinking 3 beers in the last hour? 107.25 pounds

d. What would your BAC be if you weigh more than the female in part **c**? Less than 0.08

SKILL CHECKER

Try the Skill Checker Exercises so you'll be ready for the next section.

In Problems 27–30, graph the equation.

27. $x + 2y = 4$

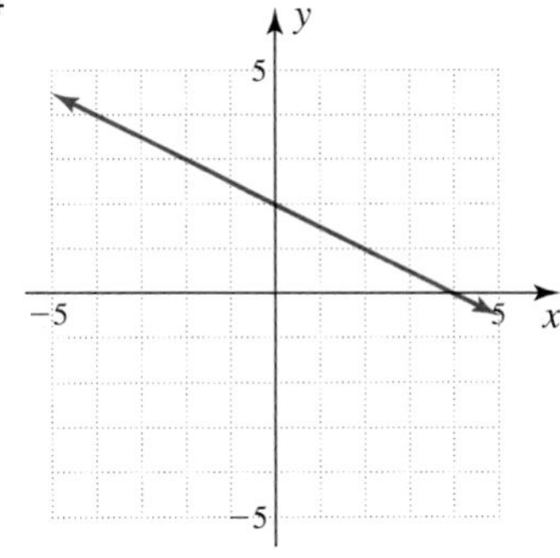

28. $2y - x = 0$

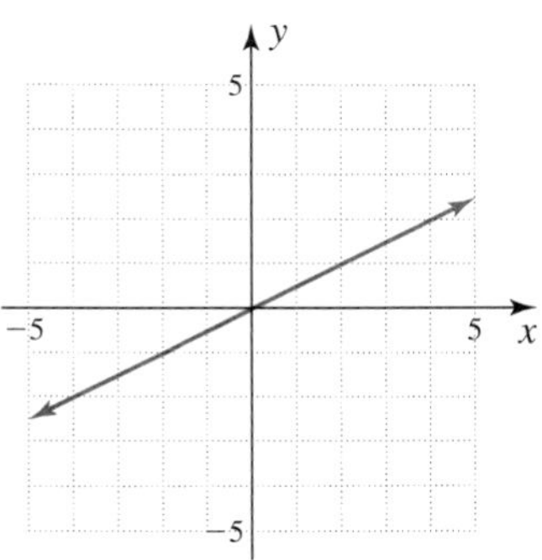

29. $y - 2x = 4$

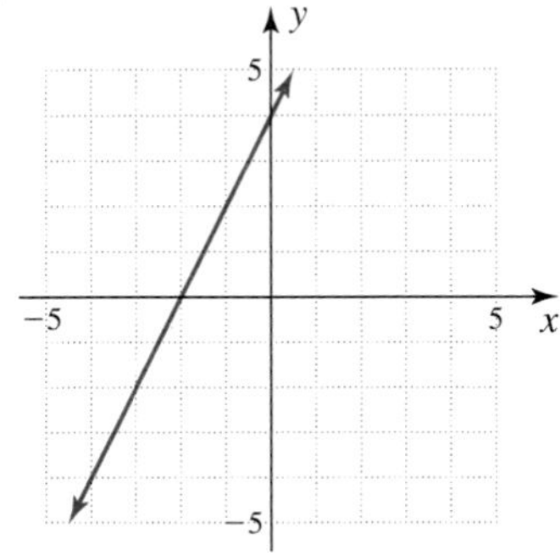

30. $2y - 4x = 4$

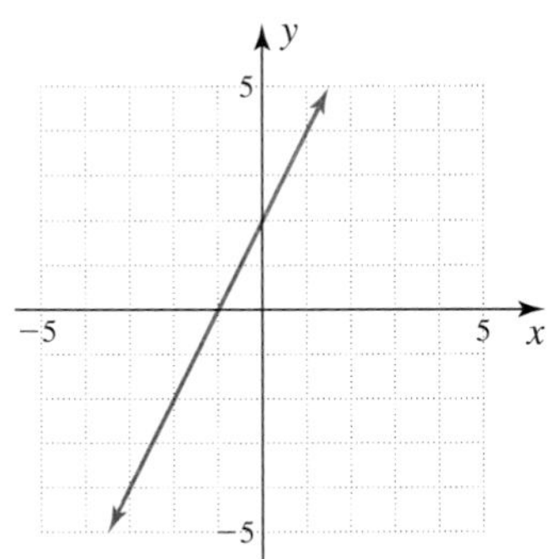

In Problems 31–32, determine whether the lines are parallel.

31. $x + y = 4$ Not parallel
$2x - y = -1$

32. $x + 2y = 4$ Parallel
$2x + 4y = 6$

USING YOUR KNOWLEDGE

33. Do you know what your Sun Protection Factor (SPF) is? The SPF of a sunscreen indicates the time period you can stay in the sun without burning, based on your complexion. The time T you are in the sun is directly proportional to your SPF (S). For example, suppose you can stay in the sun for 15 minutes without burning. Then, $T = 15S$ and we assume that your S is 1. If you want to stay in the sun for 30 minutes, $T = 30 = 15S$ and now you need a sunscreen with an SPF of 2, written as SPF2. Thus, you can stay out twice as much time in the sun ($2 \times 15 = 30$) without burning.

a. Write an equation of variation relating the time T and the SPF S. $T = 15S$

b. What SPF do you need if you want to stay out for an hour without burning? 4

WRITE ON

34. Write in your own words what it means for two variables to be directly proportional. Give examples of variables that are directly proportional. Answers may vary.

35. Write in your own words what it means for two variables to be inversely proportional. Give examples of variables that are inversely proportional. Answers may vary.

36. Explain in your own words the type of relationship (direct or inverse) that should exist between the blood alcohol concentration (BAC) and:

a. The weight of the person Answers may vary.

b. The number of beers the person has had in the last hour Answers may vary.

c. The gender of the person Answers may vary.

37. In Problems 23 and 24, we stated that for a specific weight the BAC is directly proportional to one less than the number of beers consumed in the last hour ($N - 1$). Why do you think it is not directly proportional to the number N of beers consumed in the last hour? (You can do a Web search to explore this further.) Answers may vary.

38. In the definition of direct variation, there is a constant of proportionality, k. In your own words, what is this k and what does it represent? Answers may vary.

COLLABORATIVE LEARNING

BAC Meters

In this section we have stated that the blood alcohol concentration (BAC) is directly proportional to one less than the number of beers consumed during the last hour ($N - 1$). No, there will be no beer drinking, but we want to corroborate this information.

1. Form several groups and find websites that will automatically tell you the BAC based on the number of beers consumed during the last hour. (Try an online search for "BAC meter.") How do the BAC values compare with those given for the 150-pound man and 130-pound woman mentioned in Problems 23 and 24?
2. Do the meters you found corroborate the information in Problems 23 and 24?
3. What are the factors considered by the BAC meter your group is using?
4. Compare the results you get with different online BAC meters. Are the results different? Why do you think that is?

Research Questions

1. Inequalities are used in solving "linear programming" problems in mathematics, economics, and many other fields. Find out what linear programming is and write a short paper about it. Include the techniques involved and the mathematicians and scientists who cooperated in the development of this field.
2. Linear programming problems are sometimes solved using the "simplex" method. Write a few paragraphs describing the simplex method, the people who developed it, and its uses.
3. For many years, scientists have tried to improve on the simplex method. As far back as 1979, the Soviet Academy of Sciences published a paper that did just that. Find the name of the author of the paper as well as the name of the other mathematicians who have supplied and then improved on the proof contained in the paper.

Summary

SECTION	ITEM	MEANING	EXAMPLE
7.1A	Point-slope form	The point-slope form of an equation for the line passing through (x_1, x_2) and with slope m is $y - y_1 = m(x - x_1)$.	The point-slope form of an equation for the line passing through $(2, -5)$ and with slope -3 is $y - (-5) = -3(x - 2)$.
7.1B	Slope-intercept form	The slope-intercept form of an equation for the line with slope m and y-intercept b is $y = mx + b$.	The slope-intercept form of an equation for the line with slope -5 and y-intercept 2 is $y = -5x + 2$.
7.1C	Two-point form	The two-point form of an equation for a line going through (x_1, y_1) and (x_2, y_2) is $y - y_1 = m(x - x_1)$, where $m = \dfrac{y_2 - y_1}{x_2 - x_1}$	The two-point form of a line going through (2, 3) and (7, 13) is $y - 3 = 2(x - 2)$.
7.3A	Linear inequality	An inequality that can be written in the form $Ax + By < C$ or $Ax + By > C$ (substituting $\leq$ for $<$ or $\geq$ for $>$ also yields a linear inequality).	$3x < 2y - 6$ is a linear inequality because it can be written as $3x - 2y < -6$.

(Continued)

SECTION	ITEM	MEANING	EXAMPLE
7.3A	Graph of a linear inequality	1. Find the boundary. Draw the line solid for $\le$ or $\ge$, dashed for $<$ or $>$. 2. Use (a, b) as a test point and shade the side of the line containing the test point if the resulting inequality is a true statement; otherwise, shade the region that does not contain the test point.	Graph of $3x - 2y < 6$. Using (0, 0) as the test point for graphing $3x - 2y < 6$, we get $3(0) - 2(0) < 6$, true. Shade the region containing the test point, which is the region above the dashed line.
7.4A	Direct variation	y varies directly as x if $y = kx$. We also say y is proportional to x. k is the constant of proportionality.	The cost C of bagels is proportional to the number n you buy. $C = kn$ and k is the price of one bagel.
7.4B	Inverse variation	y varies inversely as x if $y = \frac{k}{x}$; k is the constant of proportionality.	The acceleration a of an object is inversely proportional to its mass m: $a = \frac{k}{m}$

Review Exercises

(If you need help with these exercises, look in the section indicated in brackets.)

1. [7.1A] Find an equation of the line going through the point (3, − 5) and with the given slope.

a. $m = -2$ $2x + y = 1$

b. $m = -3$ $3x + y = 4$

c. $m = -4$ $4x + y = 7$

2. [7.1B] Find an equation of the line with the given slope and intercept.

a. Slope 5, y-intercept -2 $y = 5x - 2$

b. Slope 4, y-intercept 7 $y = 4x + 7$

c. Slope 6, y-intercept -4 $y = 6x - 4$

3. [7.1C] Find an equation for the line passing through the given points.

a. $(-1, 2)$ and $(4, 7)$ $x - y = -3$

b. $(-3, 1)$ and $(7, 6)$ $x - 2y = -5$

c. $(1, 2)$ and $(7, -2)$ $2x + 3y = 8$

4. [7.2A]

a. The cost C of a long-distance call that lasts for m minutes is \$3 plus \$0.20 for each minute. If a 10-minute call costs \$5, write an equation for the cost C and find the cost of a 15-minute call. $C = 0.2m + 3$; \$6

b. Long-distance rates for m minutes are \$5 plus \$0.20 for each minute. If a 10-minute call costs \$7, write an equation for the cost C and find the cost of a 15-minute call. $C = 0.2m + 5$; \$8

c. Long-distance rates for m minutes are \$5 plus \$0.30 for each minute. If a 10-minute call costs \$8, write an equation for the cost C and find the cost of a 15-minute call. $C = 0.3m + 5$; \$9.50

5. [7.2B]

a. A cell phone plan costs $30 per month with 500 free minutes and $0.40 for each additional minute. Find an equation for the total cost C of the plan when m minutes are used after the first 500. What is the cost when 800 minutes are used? $C = 0.4m + 30$ for $m > 500$; $150

b. Find an equation when the cost is $40 per month and $0.30 for each additional minute. What is the cost when 600 minutes are used? $C = 0.3m + 40$ for $m > 500$; $70

c. Find an equation when the cost is $50 per month and $0.20 for each additional minute. What is the cost when 900 minutes are used? $C = 0.2m + 50$ for $m > 500$; $130

6. [7.2C]

a. The bills for two long-distance calls are $3 for 5 minutes and $5 for 10 minutes. Find an equation for the total cost C of the calls when m minutes are used. $C = 0.40m + 1$

b. Repeat part **a** if the charges are $4 for 5 minutes and $6 for 10 minutes. $C = 0.40m + 2$

c. Repeat part **a** if the charges are $5 for 5 minutes and $7 for 10 minutes. $C = 0.40m + 3$

7. [7.3A] Graph.

a. $2x - 4y < -8$

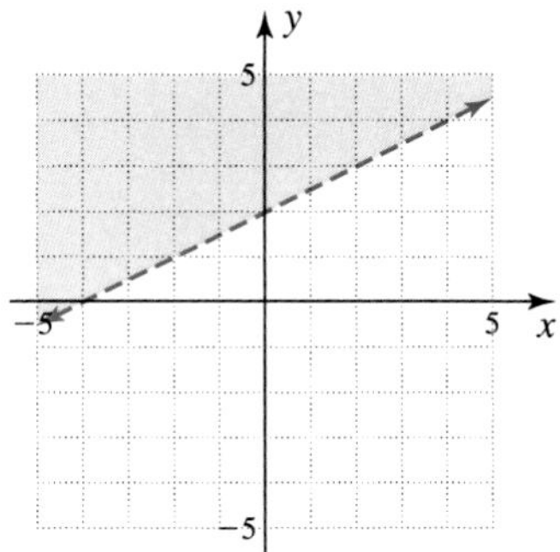

b. $3x - 6y < -12$

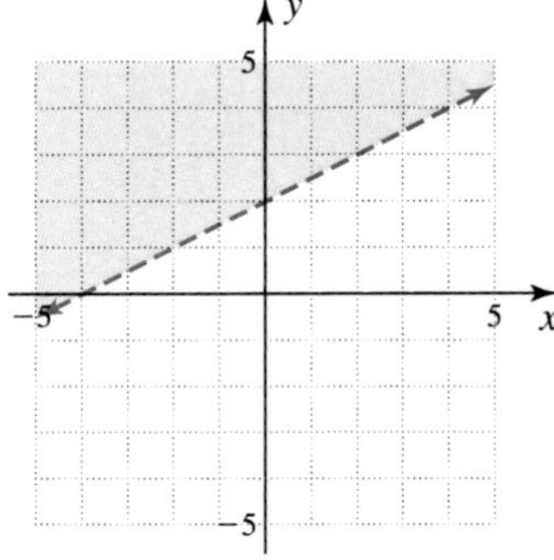

c. $4x - 2y < -8$

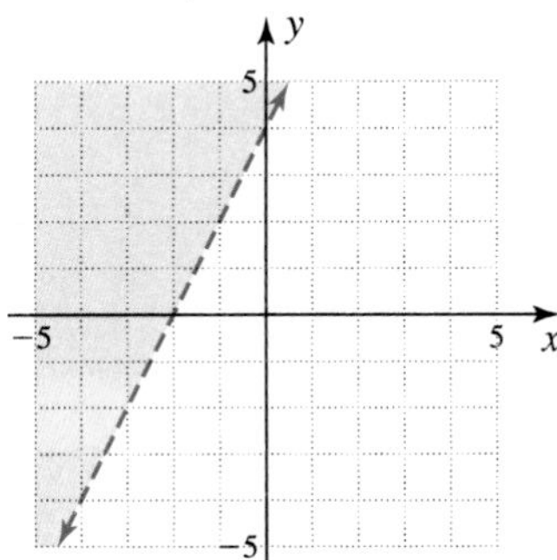

8. [7.3A] Graph.

a. $-y \leq -2x + 2$

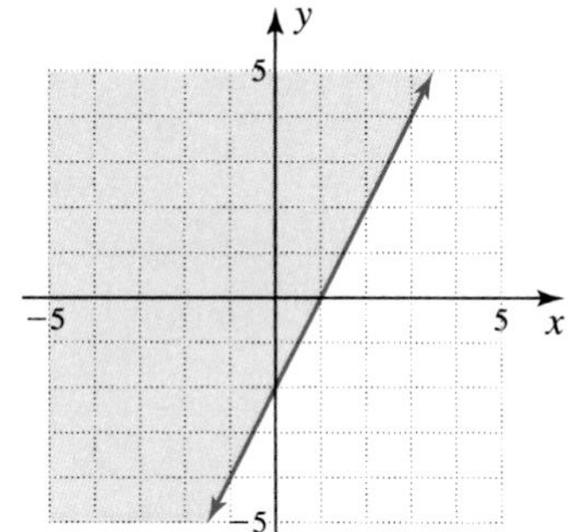

b. $-y \leq -2x + 4$

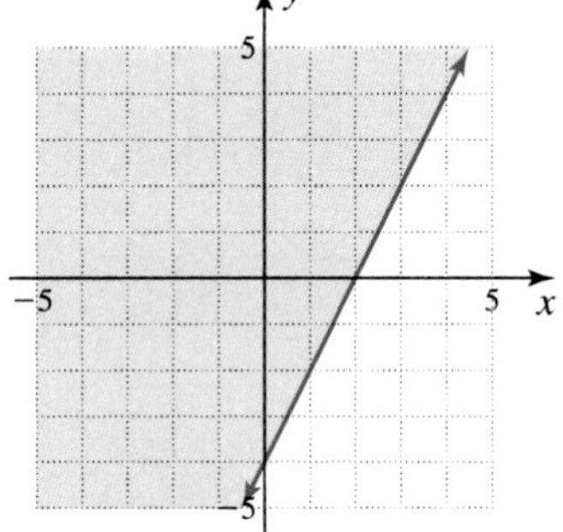

c. $-y \leq -x + 3$

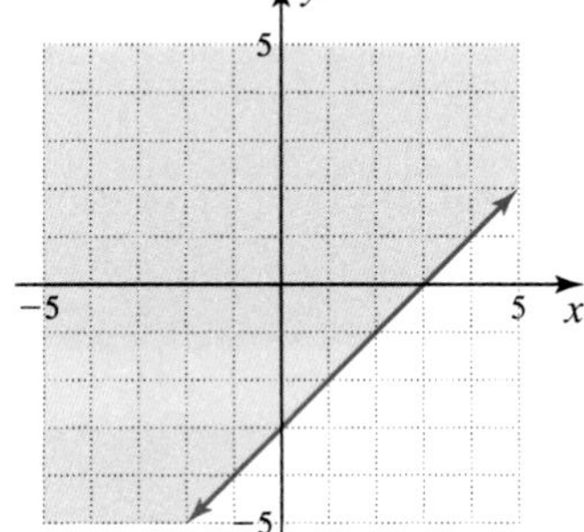

9. [7.3A] Graph.

a. $2x + y > 0$

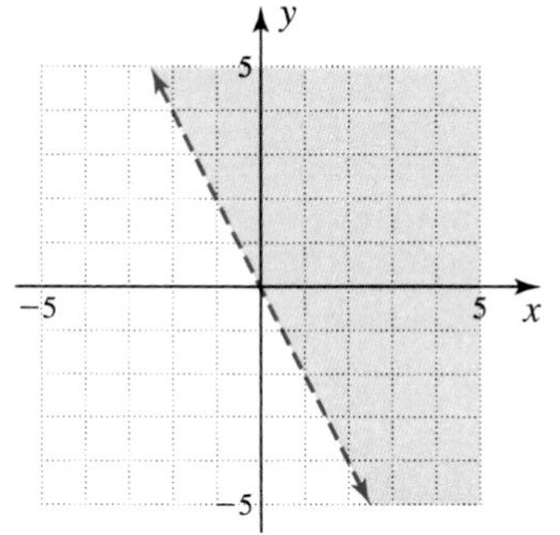

b. $3x + y > 0$

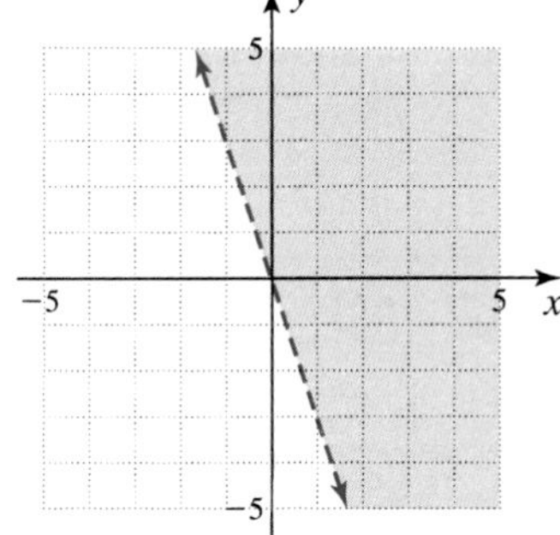

c. $3x - y < 0$

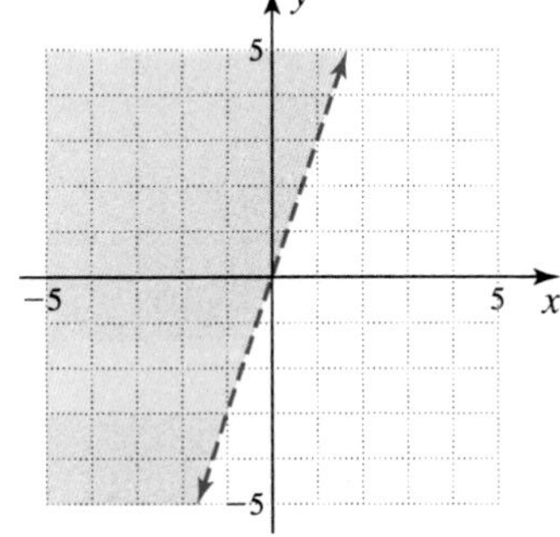

10. [7.3A] Graph.

a. $x \geq -4$

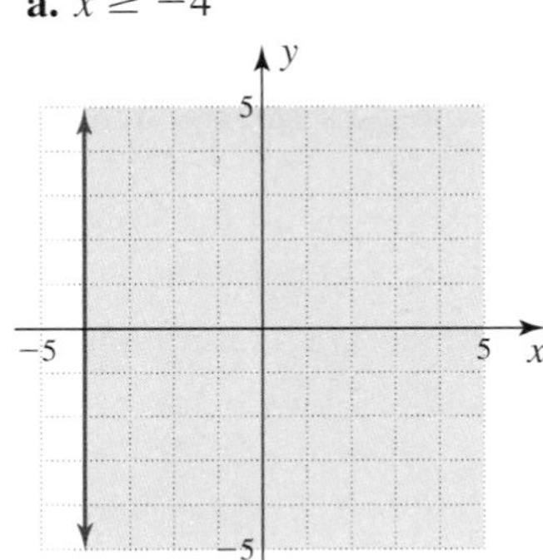

b. $y - 4 < 0$

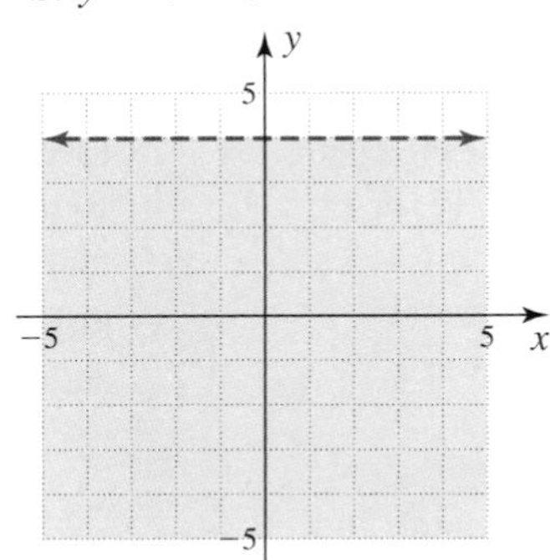

c. $2y - 4 \geq 0$

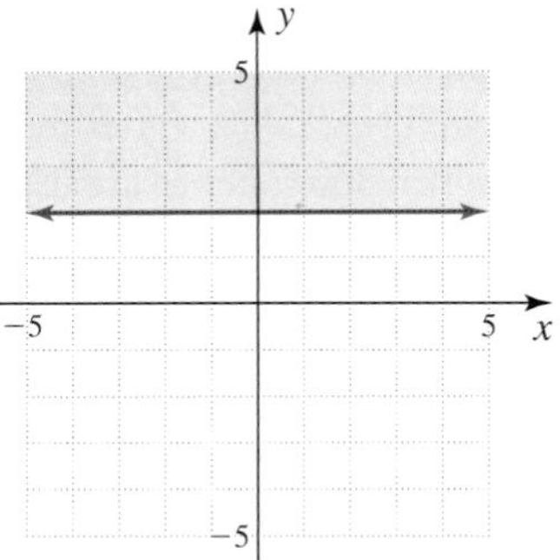

11. [7.4A]

a. If you walk for m minutes, the number C of calories used is proportional to the time you walk. If 30 calories are used when you walk for 12 minutes, write an equation of variation and find the number of calories used when you walk for an hour.
$C = km$; 150 calories

b. If you row for m minutes, the number C of calories used is proportional to the time you row. If 140 calories are used when you row for 20 minutes, write an equation of variation and find the number of calories used when you row for an hour.
$C = km$; 420 calories

c. If you play tennis for m minutes, the number C of calories used is proportional to the time you play. If 180 calories are used when you play for 180 minutes, write an equation of variation and find the number of calories used when you play for 45 minutes. $C = km$; 45 calories

12. [7.4B]

a. The amount F of force you exert on a wrench handle to loosen a rusty bolt varies inversely with the length L of the handle. If $k = 30$, write an equation of variation and find the force needed when the handle is 6 inches long. $F = \frac{30}{L}$; $F = 5$

b. What force is needed when the handle is 10 inches long? $F = 3$

c. What force is needed when the handle is 15 inches long? $F = 2$

Practice Test 7

(Answers on pages 573–575)

1. Find an equation of the line going through the point $(2, -6)$ with slope -5. Write the answer in point-slope form and then graph the line.

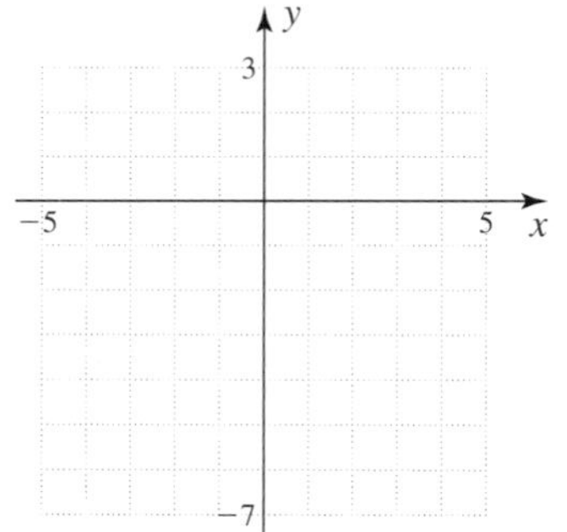

2. Find an equation of the line with slope 5 and y-intercept -4. Write the answer in slope-intercept form and then graph the line.

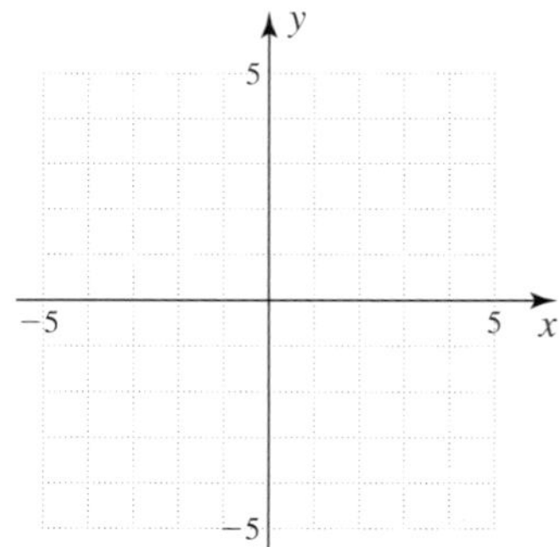

3. Find an equation of the line passing through the points $(2, -8)$ and $(-4, -2)$. Write the answer in standard form.

4. Long-distance rates for m minutes are \$10 plus \$0.20 for each minute. If a 10-minute call costs \$12, write an equation for the total cost C and find the cost of a 15-minute call.

5. A cell phone plan costs \$40 per month with 500 free minutes and \$0.50 for each additional minute. Find an equation for the total cost C of the plan when m minutes are used after the first 500. What is the cost when 800 total minutes are used?

6. The bills for two long-distance calls are \$7 for 5 minutes and \$10 for 10 minutes. Find an equation for the total cost C of calls when m minutes are used.

7. Graph $3x - 2y < -6$.

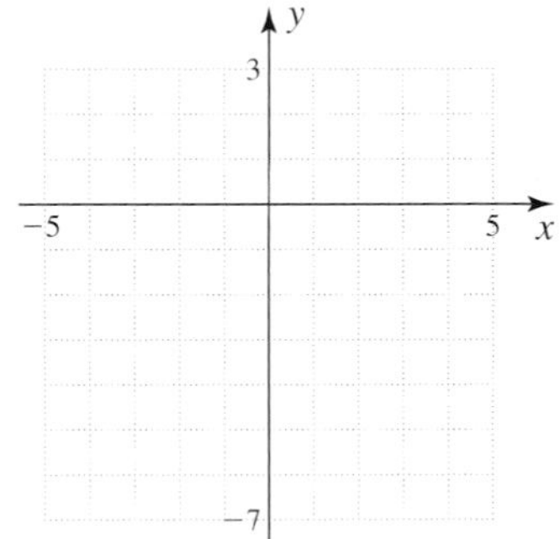

8. Graph $-y \le -3x + 3$.

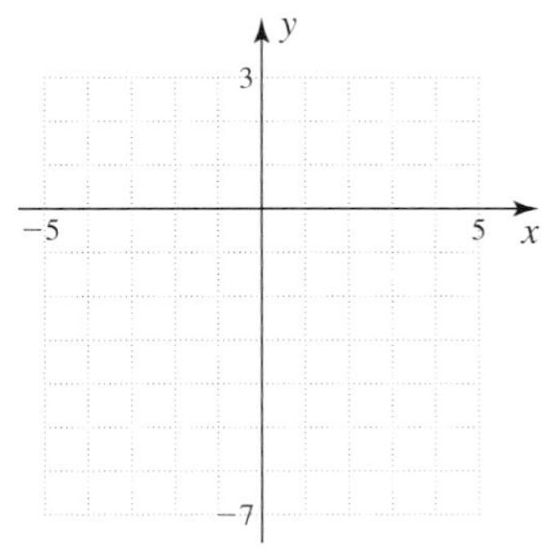

9. Graph $4x - y > 0$.

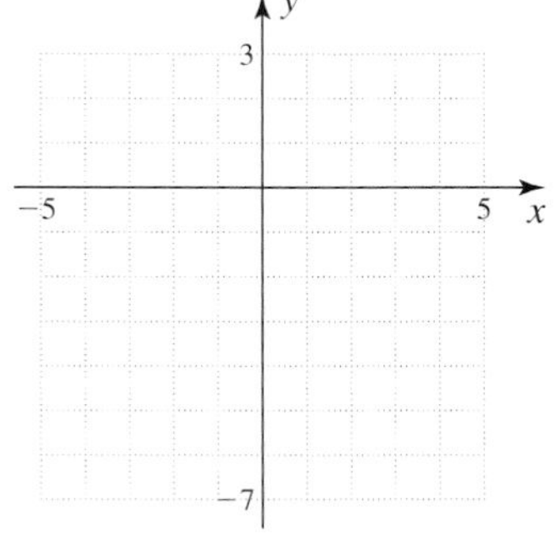

10. Graph $2y - 8 \ge 0$.

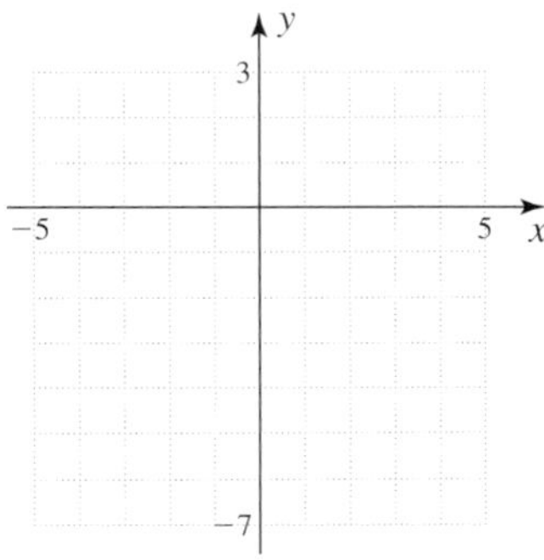

11. Have you raked leaves lately? If you rake leaves for m minutes, the number C of calories used is proportional to the time you rake. If 60 calories are used when you rake for 30 minutes, write an equation of variation and find the number of calories used when you rake for $2\frac{1}{2}$ hours.

12. The maximum weight W that can be supported by a 2-by-4-inch piece of pinewood varies inversely with its length L. If the maximum weight W that can be supported by a 10-foot-long 2-by-4 piece of pine is 500 pounds, find an equation of variation and the maximum weight W that can be supported by a 25-foot length of 2-by-4 pine.

Answers to Practice Test

ANSWER	IF YOU MISSED	REVIEW		
	QUESTION	SECTION	EXAMPLES	PAGE
1. $y + 6 = -5(x - 2)$ 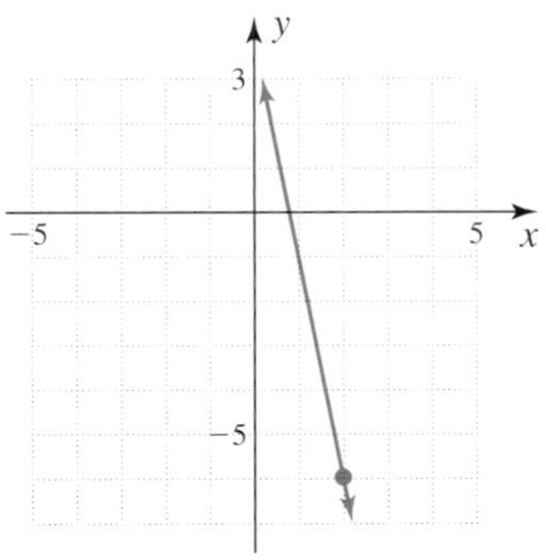	1	7.1	1	536

ANSWER	IF YOU MISSED	REVIEW		
	QUESTION	SECTION	EXAMPLES	PAGE
2. $y = 5x - 4$ 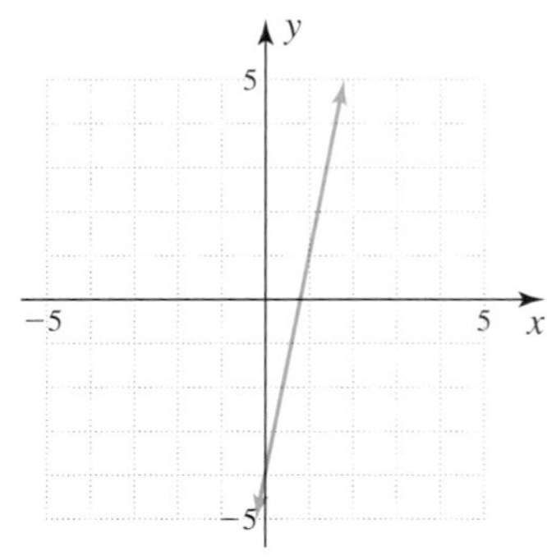	2	7.1	2	537
3. $x + y = -6$	3	7.1	3	538
4. $C = 0.20m + 10$; \$13	4	7.2	1	544
5. $C = 40 + 0.50m$; 190	5	7.2	2	544
6. $C = 0.60m + 4$	6	7.2	3	545
7. 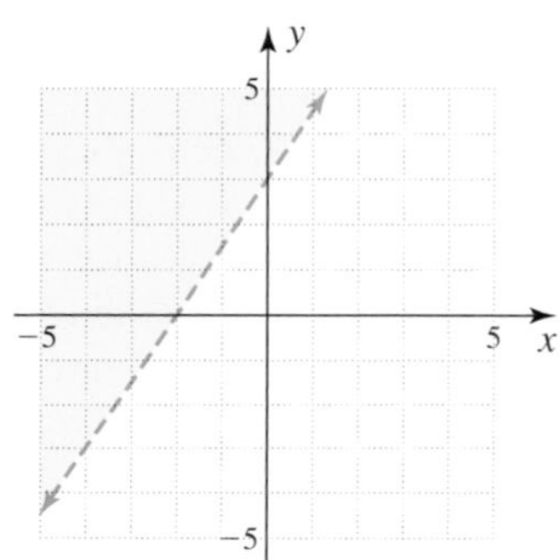	7	7.3	1	551
8. 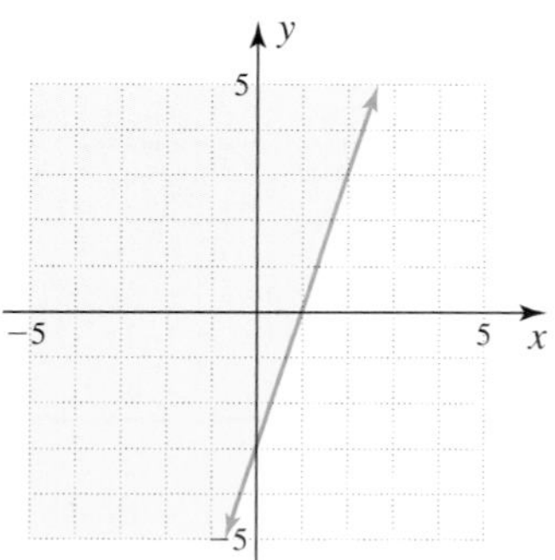	8	7.3	2	552
9. 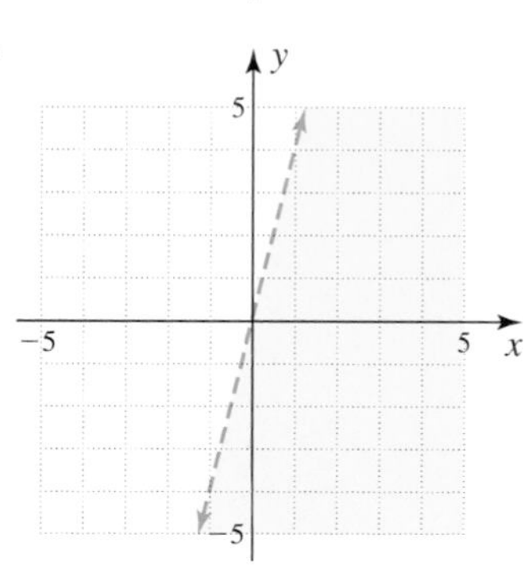	9	7.3	3	553

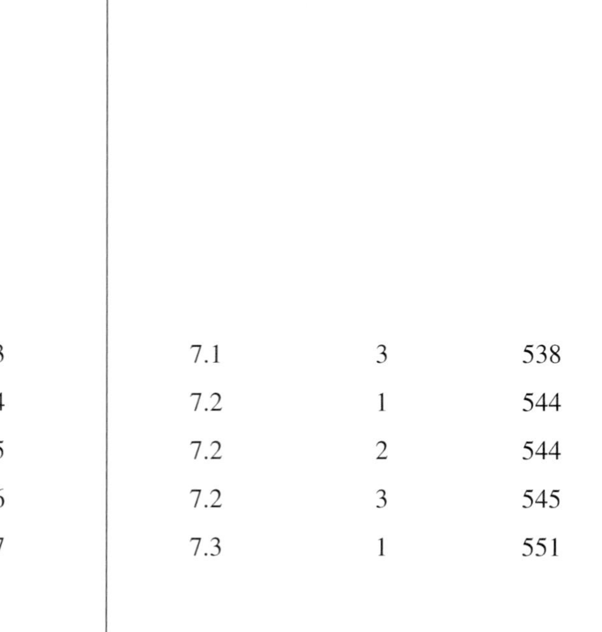

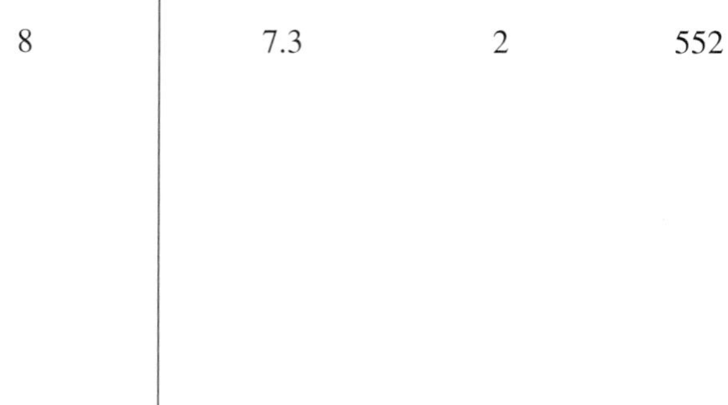

ANSWER	IF YOU MISSED	REVIEW		
	QUESTION	SECTION	EXAMPLES	PAGE
10. 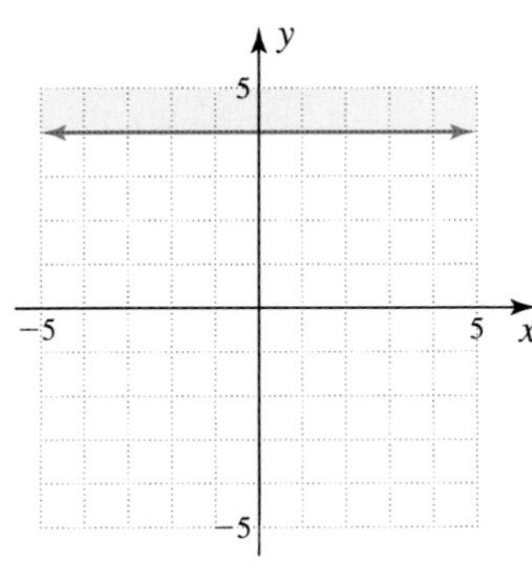	10	7.3	4	554
11. $C = 2m$; 300	11	7.4	1, 2	561, 562
12. $W = \frac{5000}{L}$; 200 pounds	12	7.4	3, 4	563

Cumulative Review Chapters 1–7

1. Find: $-\frac{2}{9} + \left(-\frac{1}{8}\right)$ $-\frac{25}{72}$

2. Find: $4.3 - (-3.9)$ 8.2

3. Find: $(-2)^4$ 16

4. Find: $-\frac{1}{5} \div \left(-\frac{1}{10}\right)$ 2

5. Evaluate $y \div 2 \cdot x - z$ for $x = 2$, $y = 8$, $z = 3$. 5

6. Simplify: $x + 4(x - 2) + (x - 3)$. $6x - 11$

7. Write in symbols: The quotient of $(d + 4e)$ and f. $\frac{d + 4e}{f}$

8. Solve for x: $5 = 3(x - 1) + 5 - 2x$ $x = 3$

9. Solve for x: $\frac{x}{7} - \frac{x}{9} = 2$ $x = 63$

10. The sum of two numbers is 110. If one of the numbers is 40 more than the other, what are the numbers? 35 and 75

11. Susan purchased some municipal bonds yielding 11% annually and some certificates of deposit yielding 14% annually. If Susan's total investment amounts to \$9000 and the annual income is \$1140, how much money is invested in bonds and how much is invested in certificates of deposit? \$4000 in bonds; \$5000 in certificates of deposit

12. Graph: $-\frac{x}{7} + \frac{x}{4} \ge \frac{x - 4}{4}$

0 7

13. Graph the point $C(-2, -3)$.

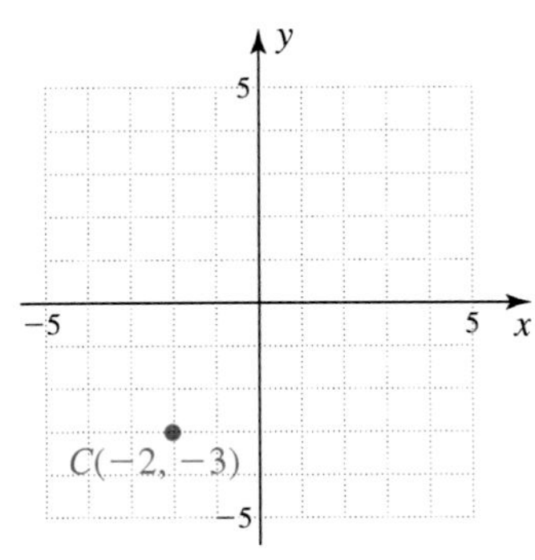

14. Determine whether the ordered pair $(-3, -4)$ is a solution of $5x - y = -19$. No

15. Find x in the ordered pair $(x, -1)$ so that the ordered pair satisfies the equation $4x + 2y = -10$. $x = -2$

16. Graph: $3x + y = 3$

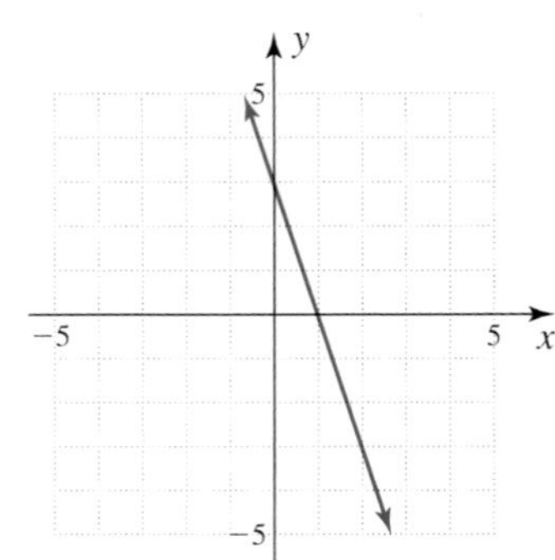

17. Graph: $3x + 9 = 0$

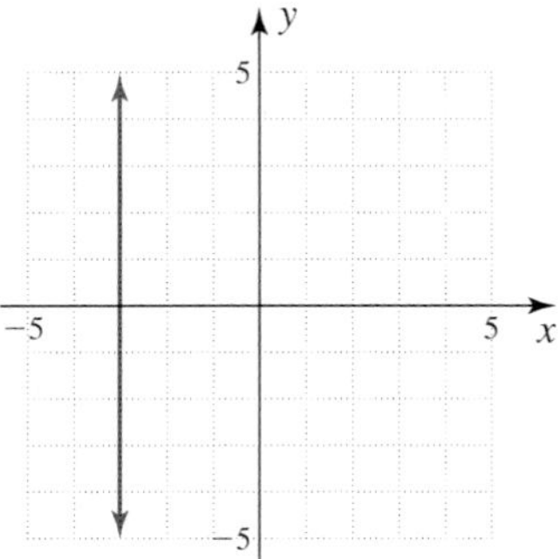

18. Find the slope of the line going through the points $(-4, -6)$ and $(-7, 1)$. $-\frac{7}{3}$

19. What is the slope of the line $12x - 4y = -14$? 3

20. Find the pair of parallel lines. (1) and (3)

(1) $20x - 5y = 2$ **(2)** $5y + 20x = 2$

(3) $-y = -4x + 2$

21. Find: $\frac{x^{-4}}{x^{-8}}$ x^4

22. Multiply and simplify: $x^{-7} \cdot x^{-2}$ $\frac{1}{x^9}$

23. Simplify: $(2x^3y^{-5})^{-4}$ $\frac{y^{20}}{16x^{12}}$

24. Write in scientific notation: 3,400,000 3.4×10^6

25. Divide and express the answer in scientific notation: $(8.33 \times 10^3) \div (1.7 \times 10^3)$ 4.9

26. Find the value of $3x^3 + 2x^2 - 5$ when $x = -3$. -68

27. Add $(-7x^2 - 7x^5 - 6)$ and $(3x^5 + 6 + 8x^2)$. $-4x^5 + x^2$

28. Find (expand): $\left(3x^2 - \frac{1}{2}\right)^2$ $9x^4 - 3x^2 + \frac{1}{4}$

29. Find: $(5x^2 + 2)(5x^2 - 2)$ $25x^4 - 4$

30. Divide $(3x^3 - 10x^2 + 3x + 12)$ by $(x - 2)$. $(3x^2 - 4x - 5)$ R 2

31. Factor completely: $15x^4 - 35x^7$ $5x^4(3 - 7x^3)$

32. Factor completely: $\frac{5}{7}x^8 - \frac{2}{7}x^7 + \frac{2}{7}x^6 - \frac{3}{7}x^4$ $\frac{1}{7}x^4(5x^4 - 2x^3 + 2x^2 - 3)$

33. Factor completely: $x^2 - 3x + 2$ $(x - 1)(x - 2)$

34. Factor completely: $9x^2 - 24xy + 16y^2$ $(3x - 4y)(3x - 4y)$

35. Factor completely: $4x^2 - 9y^2$ $(2x + 3y)(2x - 3y)$

36. Factor completely: $-6x^4 + 24x^2$ $-6x^2(x + 2)(x - 2)$

37. Factor completely: $2x^3 - 2x^2 - 4x$ $2x(x + 1)(x - 2)$

38. Factor completely: $4x^2 + 3x + 4x + 3$ $(4x + 3)(x + 1)$

39. Factor completely: $25kx^2 + 10kx + k$ $k(5x + 1)^2$

40. Solve for x: $4x^2 + 5x = 6$ $x = -2$; $x = \frac{3}{4}$

41. Write $\frac{3x}{2y}$ with a denominator of $8y^3$. $\frac{12xy^2}{8y^3}$

42. Reduce to lowest terms: $\frac{-6(x^2 - y^2)}{3(x - y)}$ $-2(x + y)$

43. Reduce to lowest terms: $\frac{x^2 + 5x - 14}{2 - x}$ $-(x + 7)$

44. Multiply: $(x - 9) \cdot \frac{x - 4}{x^2 - 81}$ $\frac{x - 4}{x + 9}$

45. Divide: $\frac{x + 9}{x - 9} \div \frac{x^2 - 81}{9 - x}$ $\frac{1}{9 - x}$

46. Add: $\frac{4}{3(x - 2)} + \frac{8}{3(x - 2)}$ $\frac{4}{x - 2}$

47. Subtract: $\frac{x + 2}{x^2 + x - 6} - \frac{x + 3}{x^2 - 4}$ $\frac{-2x - 5}{(x + 3)(x + 2)(x - 2)}$

48. Simplify: $\dfrac{\frac{4}{x} - \frac{1}{2x}}{\frac{1}{3x} - \frac{3}{4x}}$ $-\frac{42}{5}$

49. Solve for x: $\frac{5x}{x - 5} - 2 = \frac{x}{x - 5}$ $x = -5$

50. Solve for x: $\dfrac{x}{x^2 - 64} + \dfrac{8}{x - 8} = \dfrac{1}{x + 8}$ $x = -9$

51. Solve for x: $\dfrac{x}{x + 1} - \dfrac{1}{2} = \dfrac{-1}{x + 1}$ No solution

52. Solve for x: $1 + \dfrac{4}{x - 4} = \dfrac{32}{x^2 - 16}$ $x = -8$

53. A van travels 120 miles on 8 gallons of gas. How many gallons will it need to travel 210 miles? 14 gal

54. Solve for x: $\dfrac{x - 6}{3} = \dfrac{3}{10}$ $x = \dfrac{69}{10}$

55. Janet can paint a kitchen in 3 hours and James can paint the same kitchen in 5 hours. How long would it take for both working together to paint the kitchen? $1\frac{7}{8}$ hr

56. Find an equation of the line that goes through the point (2, 0) and has slope $m = -3$. $y = -3x + 6$

57. Find an equation of the line having slope 3 and y-intercept -1. $y = 3x - 1$

58. Graph: $6x - y < -6$

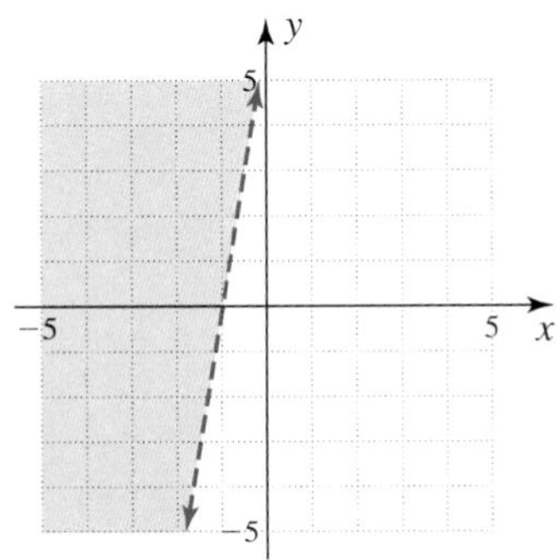

59. Graph: $-y \geq -6x - 6$

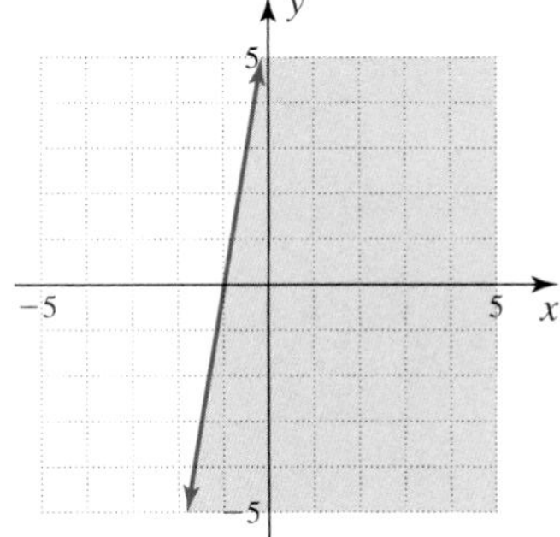

60. An enclosed gas exerts a pressure P on the walls of the container. This pressure is directly proportional to the temperature T of the gas. If the pressure is 5 lb/in.^{2} when the temperature is 250°F, find k.
$k = \frac{1}{50}$

61. If the temperature of a gas is held constant, the pressure P varies inversely as the volume V. A pressure of 1960 lb/in.^{2} is exerted by 7 ft^3 of air in a cylinder fitted with a piston. Find k. $k = 13{,}720$

Solving Systems of Linear Equations and Inequalities

8

The Human Side of Algebra

The first evidence of a systematic method of solving systems of linear equations is provided in the *Nine Chapters of the Mathematical Arts,* the oldest arithmetic textbook in existence. The method for solving a system of three equations with three unknowns occurs in the 18 problems of the eighth chapter, entitled "The Way of Calculating by Arrays." Unfortunately, the original copies of the *Nine Chapters* were destroyed in 213 B.C. However, the Chinese mathematician Liu Hui wrote a commentary on the *Nine Chapters* in A.D. 263, and information concerning the original work comes to us through this commentary.

In modern times, when a large number of equations or inequalities has to be solved, a method called the **simplex method** is used. This method, based on the **simplex algorithm,** was developed in the 1940s by George B. Dantzig. It was first used by the Allies of World War II to solve logistics problems dealing with obtaining, maintaining, and transporting military equipment and personnel.

Pretest for Chapter 8

(Answers on page 581)

1. Use the graphical method to solve the system

$$x + 2y = 6$$
$$4y - x = 0$$

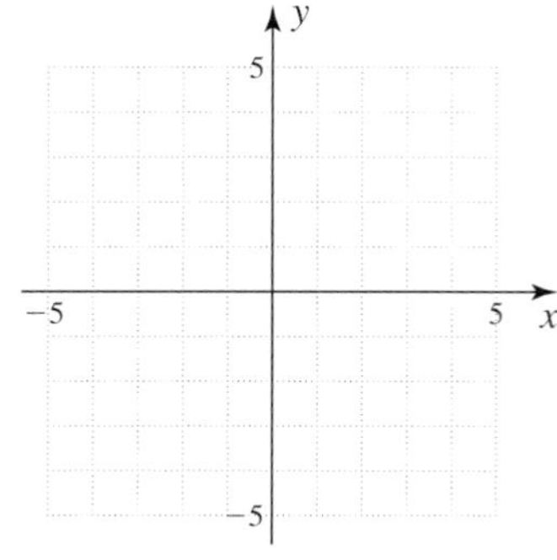

2. Use the graphical method to solve the system

$$y - 2x = 2$$
$$3y - 6x = 12$$

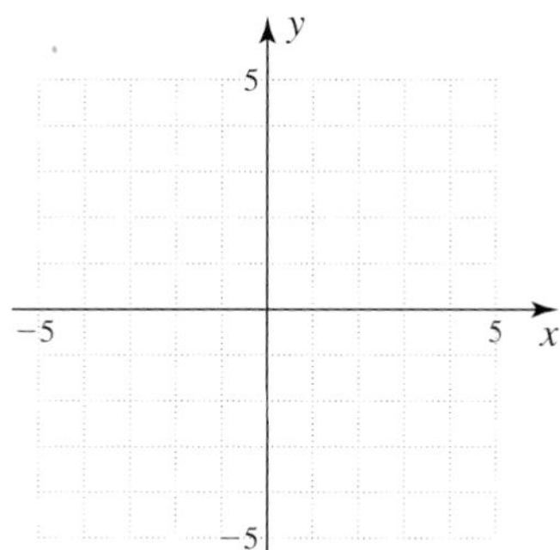

3. Use the substitution method to solve the system (if possible).

$$x + 3y = 6$$
$$2x + 6y = 10$$

4. Use the substitution method to solve the system (if possible).

$$x + 3y = 6$$
$$2x + 6y = 12$$

5. Use the elimination method to solve the system (if possible).

$$2x + 3y = -7$$
$$3x + y = -7$$

6. Use the elimination method to solve the system (if possible).

$$3x - 2y = 6$$
$$-9x + 6y = -8$$

7. Use the elimination method to solve the system (if possible).

$$2y + 3x = -13$$
$$6x + 4y = -26$$

8. Eva has \$3 in nickels and dimes. She has twice as many dimes as nickels. How many nickels and how many dimes does she have?

9. The sum of two numbers is 140. Their difference is 80. What are the numbers?

10. A plane flies 600 miles with a tailwind in 2 hours. It takes the same plane 3 hours to fly the 600 miles when flying against the wind. What is the plane's speed in still air?

11. Herbert invests \$10,000, part at 5% and part at 4%. How much money is invested at each rate if his annual interest is \$432?

12. Graph the solution set of the system.

$$x - 3y > 6$$
$$2x - y \le 6$$

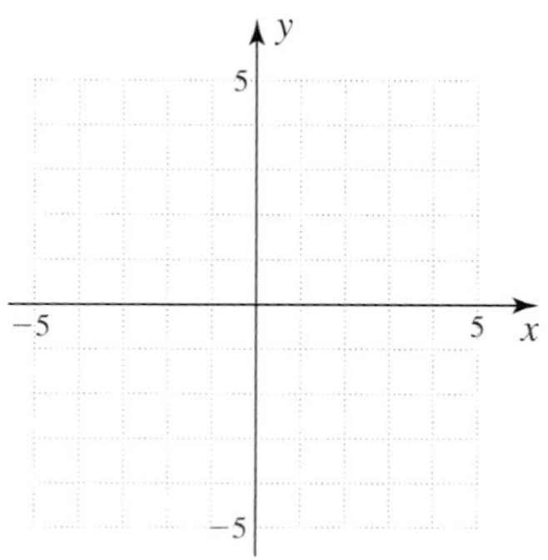

Answers to Pretest

ANSWER	IF YOU MISSED	REVIEW		
	QUESTION	SECTION	EXAMPLES	PAGE
1. The solution is (4, 1). 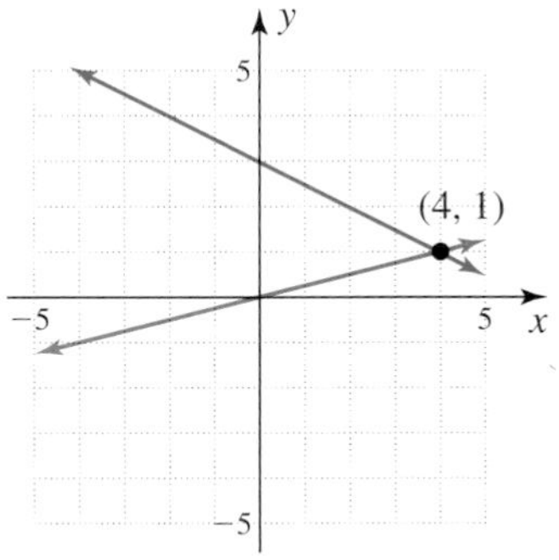	1	8.1	1	583–584
2. The lines are parallel; there is no solution 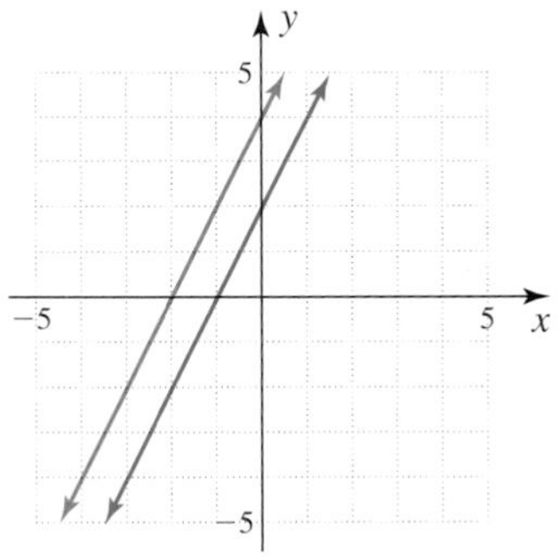	2	8.1	2, 3	584, 585
3. No solution (inconsistent)	3	8.2	1, 2	597–598
4. Dependent (infinitely many solutions)	4	8.2	3	598–599
5. $(-2, -1)$	5	8.3	1	606
6. No solution (inconsistent)	6	8.3	1, 2	606, 607
7. Dependent (infinitely many solutions)	7	8.3	3	607–608
8. 12 nickels; 24 dimes	8	8.4	1, 2	614–615
9. 110 and 30	9	8.4	3	616
10. 250 mi/hr	10	8.4	4	617
11. \$3200 at 5%; \$6800 at 4%	11	8.4	5	618
12. 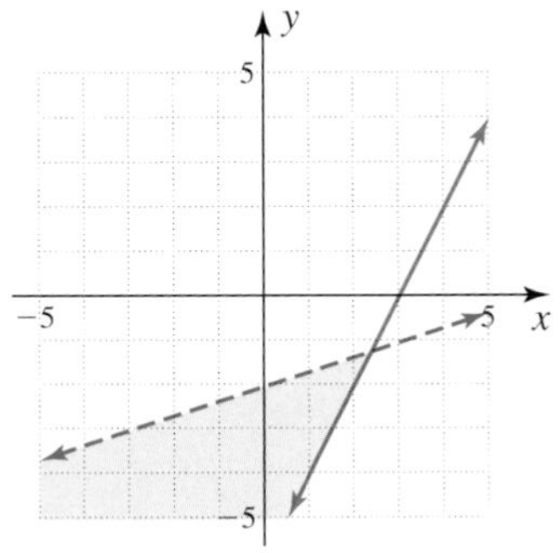	12	8.5	1, 2, 3	623–624

8.1 SOLVING SYSTEMS OF EQUATIONS BY GRAPHING

To Succeed, Review How To . . .

1. Graph the equation of a line (pp. 245–247).
2. Determine whether an ordered pair is a solution of an equation (pp. 242–243).

Objectives

A Solve a system of two equations in two variables by graphing.

B Determine whether a system of equations is consistent, inconsistent, or dependent.

C Solve an application.

GETTING STARTED

Supply, Demand, and Intersections

Can you tell from the graph when the energy supply and the demand were about the same? This happened where the line representing the supply and the line representing the demand intersect, or around 1980. When the demand and the supply are *equal,* prices reach *equilibrium.* If x is the year and y the number of millions of barrels of oil per day, the point (x, y) at which the demand is the same as the supply—that is, the point at which the graphs *intersect*—is (1980, 85). We can graph a pair of linear equations and find a point of intersection if it exists. This point of intersection is an ordered pair of numbers such as (1980, 85) and is a **solution** of both equations. This means that when you substitute 1980 for x and 85 for y in the original equations, the results are true statements. In this section we learn how to find the solution of a system of two equations in two variables by using the graphical method, which involves graphing the equations and finding their point of intersection, if it exists.

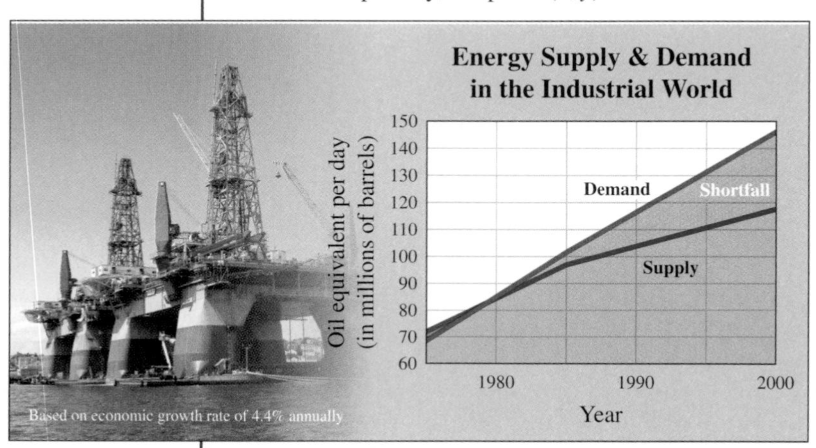

A Solving a System by Graphing

The solution of a linear equation in one variable, as studied in Chapter 2, was a single number. Thus if we solve *two* linear equations in two variables simultaneously, we expect to get two numbers. We write this solution as an ordered pair. For example, the solution of

$$x + 2y = 4$$
$$2y - x = 0$$

is (2, 1). This can be checked by letting $x = 2$ and $y = 1$ in both equations:

$$\begin{aligned} x + 2y &= 4 \qquad & 2y - x &= 0 \\ 2 + 2(1) &= 4 & 2(1) - 2 &= 0 \\ 2 + 2 &= 4 & 2 - 2 &= 0 \\ 4 &= 4 & 0 &= 0 \end{aligned}$$

Web It

To learn how to solve systems of equations by graphing, go to link 8-1-1 on the Bello Website at mhhe.com/bello.

To practice doing so, go to link 8-1-2.

Clearly, a true statement results in both cases. We call a system of two linear equations a **system of simultaneous equations.** To solve one of these systems, we need to find (if possible) all ordered pairs of numbers that satisfy both equations. Thus to *solve* the system

$$x + 2y = 4$$
$$2y - x = 0$$

This system is called a system of *simultaneous* equations because we have to find a solution that satisfies both equations.

we graph each of the equations in the usual way. To graph $x + 2y = 4$, we find the intercepts using the following table:

x	y
0	2
4	0

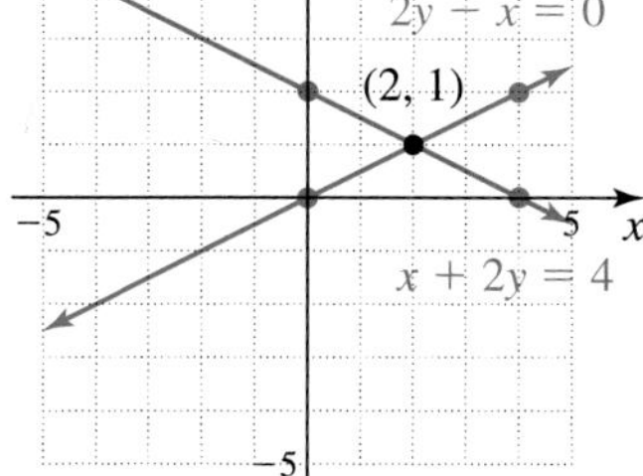

Figure 1

The intercepts (0, 2) and (4, 0), as well as the completed graph, are shown in red in Figure 1.

The equation $2y - x = 0$ is graphed similarly using the following table:

x	y
0	0

Teaching Tip

After Example 1, have students sketch examples of systems that don't intersect at one point. Have them discuss all cases of possible solution sets to a system of two equations.

Note that we need another point in the table. We can pick any x we want and then find y. If we pick $x = 4$, then $2y - 4 = 0$, or $y = 2$, giving us the point (4, 2). The graph of the equation $2y - x = 0$ is shown in blue. The lines intersect at (2, 1), which is the solution of the system of equations. (Recall that we checked that this point satisfies the equations by substituting 2 for x and 1 for y in each equation.)

EXAMPLE 1 Using the graphical method to solve a system

Use the graphical method to find the solution of the system:

$$2x + y = 4$$
$$y - 2x = 0$$

SOLUTION We first graph the equation $2x + y = 4$ using the following table:

x	y
0	4
2	0

The two points and the complete graph are shown in blue in Figure 2. We then graph $y - 2x = 0$ using the following table:

x	y
0	0
2	4

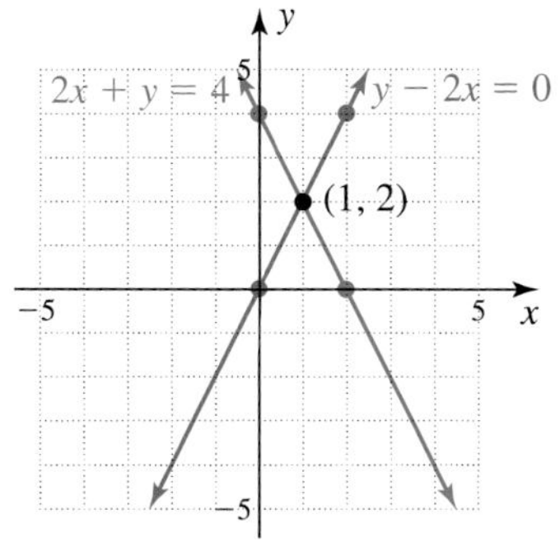

Figure 2

PROBLEM 1

Use the graphical method to solve the system:

$$x + 2y = 4$$
$$2x - 4y = 0$$

Answer

1.

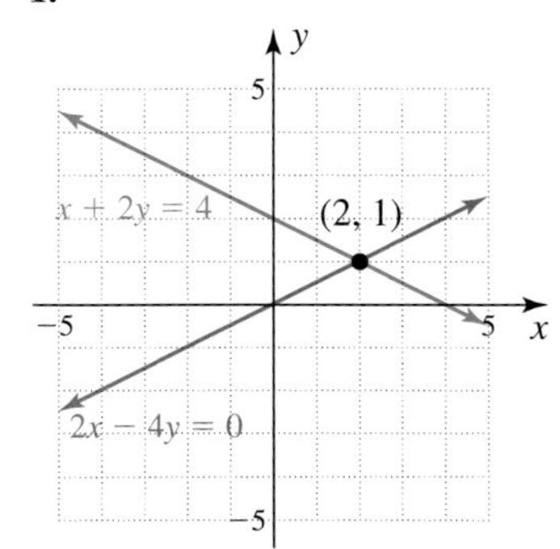

The graph of $y - 2x = 0$ is shown in red. The lines intersect at (1, 2), which is the solution of the system of equations. We can check this by letting $x = 1$ and $y = 2$ in

$$2x + y = 4$$
$$y - 2x = 0$$

thus obtaining the true statements

$$2(1) + 2 = 4$$
$$2 - 2(1) = 0$$

EXAMPLE 2 Solving an inconsistent system

Use the graphical method to find the solution of the system:

$$y - 2x = 4$$
$$2y - 4x = 12$$

SOLUTION We first graph the equation $y - 2x = 4$ using the following table:

x	y
0	4
−2	0

The two points, as well as the completed graph, are shown in blue in Figure 3. We then graph $2y - 4x = 12$ using the following table:

x	y
0	6
−3	0

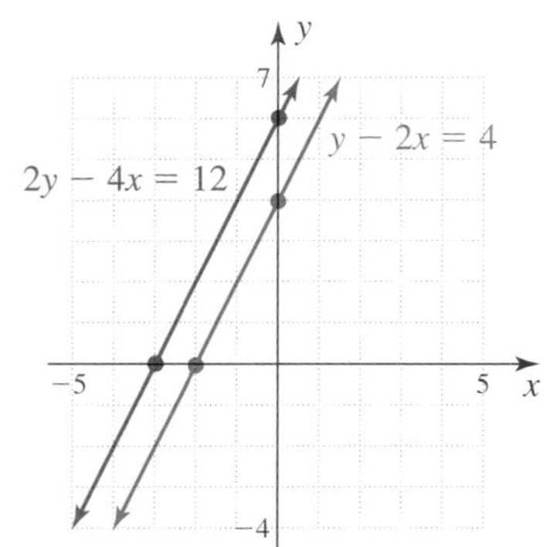

Figure 3

The graph of $2y - 4x = 12$ is shown in red. The two lines appear to be parallel; they do not intersect. If we examine the equations more carefully, we see that by dividing both sides of the second equation by 2, we get $y - 2x = 6$. Thus one equation says $y - 2x = 4$, and the other says that $y - 2x = 6$. Hence both equations cannot be true at the same time, and their graphs cannot intersect.

To confirm this, note that $y - 2x = 6$ is equivalent to $y = 2x + 6$ and $y - 2x = 4$ is equivalent to $y = 2x + 4$. Since $y = 2x + 6$ and $y = 2x + 4$ both have slope 2 but different y-intercepts, their graphs are **parallel lines.** Thus there is *no solution* for this system, since the two lines do not have any points in common; the system is said to be **inconsistent.**

PROBLEM 2

Use the graphical method to solve the system:

$$y - 3x = 3$$
$$2y - 6x = 12$$

Answer

2. No solution

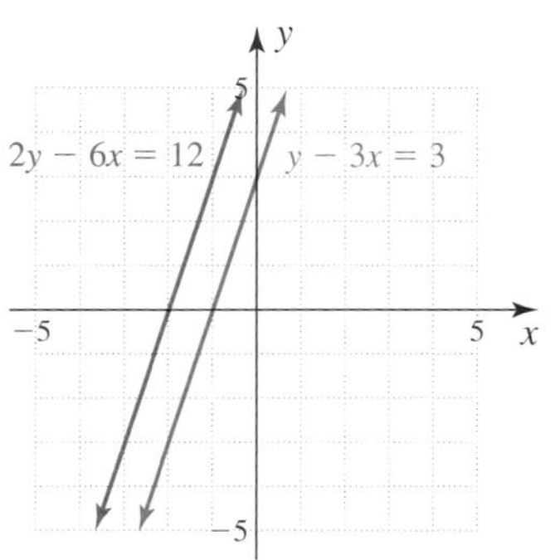

EXAMPLE 3 **Solving a dependent system**

Use the graphical method to sove the system:

$$2x + y = 4$$
$$2y + 4x = 8$$

SOLUTION We use the table

x	y
0	4
2	0

to graph $2x + y = 4$, which is shown in color in Figure 4. To graph $2y + 4x = 8$, we first let $x = 0$ to obtain $2y = 8$ or $y = 4$. For $y = 0$, $4x = 8$ or $x = 2$. Thus the two points in our second table will be

x	y
0	4
2	0

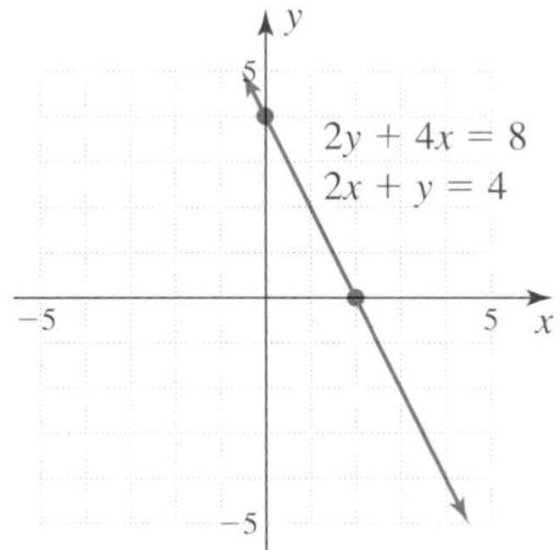

Figure 4

But these points are exactly the same as those obtained in the first table! What does this mean? It means that the graphs of the lines $2x + y = 4$ and $2y + 4x = 8$ **coincide** (are the same). Thus a solution of one equation is *automatically* a solution for the other. In fact, there are *infinitely many* solutions; every point on the graph is a solution of the system. Such a system is said to be **dependent.** In a dependent system, one of the equations is a **constant multiple** of the other. (If you multiply both sides of the first equation by 2, you get the second equation.)

PROBLEM 3

Use the graphical method to solve the system:

$$x + 2y = 4$$
$$4y + 2x = 8$$

Web It

For an excellent discussion of consistent, inconsistent, and dependent equations, go to link 8-1-3 on the Bello Website at mhhe.com/bello.

B Finding Consistent, Inconsistent, and Dependent Equations

As you can see from the examples we've given, a system of equations can have exactly *one* solution (when the lines *intersect,* as in Figure 5), *no* solution (when the lines are *parallel,* as in Figure 6), or *infinitely many* solutions (when the graphs of the two lines are *identical,* as in Figure 7). These examples illustrate the three possible solutions to a system of simultaneous equations.

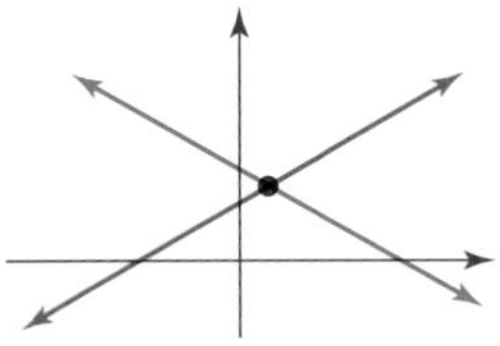

Consistent and independent (one solution)

Figure 5

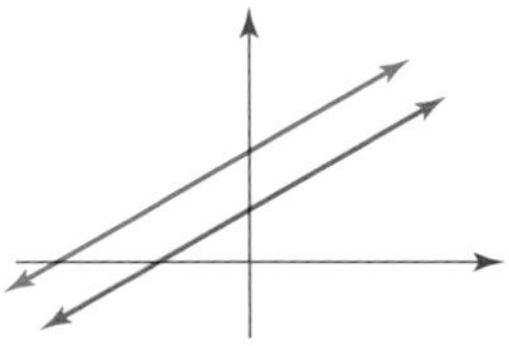

Inconsistent; parallel lines (no solution)

Figure 6

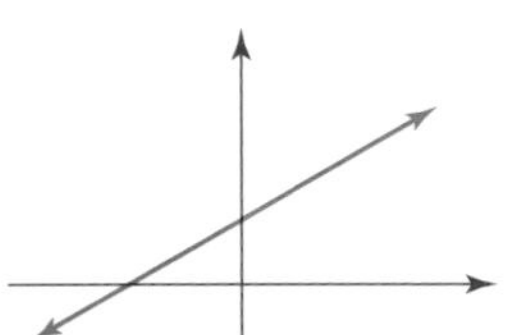

Dependent; lines coincide (infinitely many solutions)

Figure 7

Answer

3. Infinitely many solutions

$x + 2y = 4$
$4y + 2x = 8$

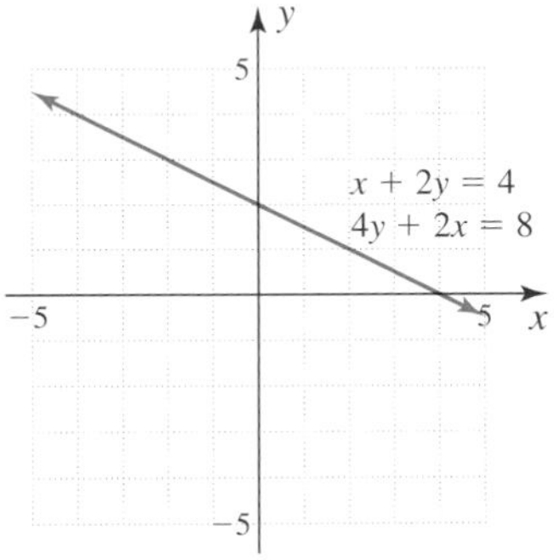

POSSIBLE SOLUTIONS TO A SYSTEM OF SIMULTANEOUS EQUATIONS

1. **Consistent and independent equations:** The graphs of the equations intersect at *one* point, whose coordinates give the solution of the system.
2. **Inconsistent equations:** The graphs of the equations are *parallel* lines; there is *no* solution for the system.
3. **Dependent equations:** The graphs of the equations *coincide* (are the same). There are *infinitely many* solutions for the system.

The following table will help you further.

Type of Lines	Slopes	y-Intercept	Number of Solutions	Type of System
Intersecting	Different	Same or different	One	Consistent
Parallel	Same	Different	None	Inconsistent
Coinciding	Same	Same	Infinite	Dependent

EXAMPLE 4 Classifying a system by graphing

Use the graphical method to solve the given system of equations. Classify each system as consistent (one solution), inconsistent (no solution), or dependent (infinitely many solutions).

a. $x + y = 4$
$2y - x = -1$

b. $x + 2y = 4$
$2x + 4y = 6$

c. $x + 2y = 4$
$4y + 2x = 8$

SOLUTION

a. The respective tables for $x + y = 4$ and $2y - x = -1$ are

x	y
0	4
4	0

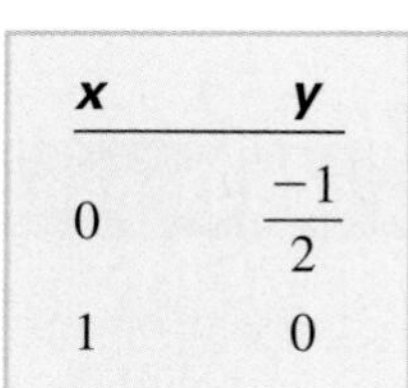

x	y
0	$-\frac{1}{2}$
1	0

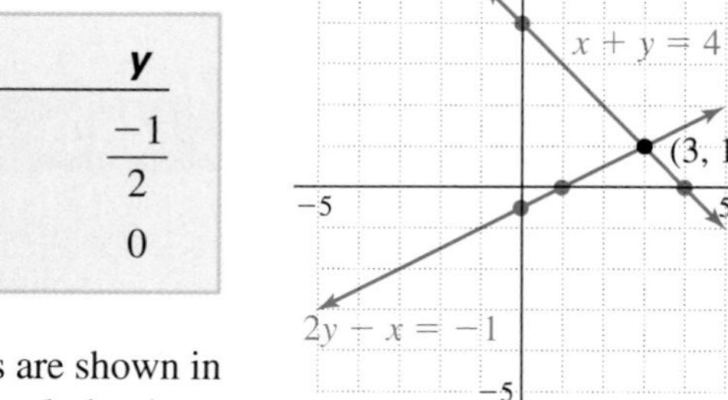

The graphs of these two lines are shown in Figure 8. As you can see, the solution is (3, 1). (Check this!) The system is *consistent*.

Figure 8

b. The respective tables for $x + 2y = 4$ and $2x + 4y = 6$ are

x	y
0	2
4	0

x	y
0	$\frac{3}{2}$
3	0

PROBLEM 4

Use the graphical method to solve the given system. Classify each system as consistent (one solution), inconsistent (no solution), or dependent (infinitely many solutions).

a. $x + y = 4$
$2y - x = 2$

b. $2x + y = 4$
$2y + 4x = 6$

c. $2x + y = 4$
$2y + 4x = 8$

Answers

4. a. Consistent; solution (2, 2)

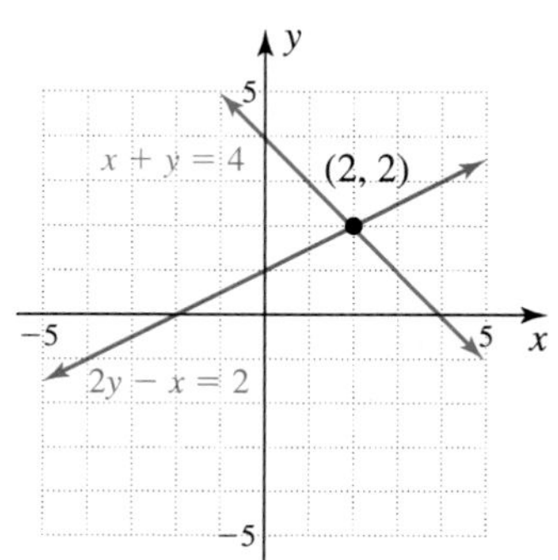

The graphs of the two lines are shown in Figure 9. There is no solution because the lines are parallel. To see this, we solve $2x + 4y = 6$ and $x + 2y = 4$ for y to obtain

$$y = -\frac{1}{2}x + \frac{3}{2} \qquad \text{and} \qquad y = -\frac{1}{2}x + 2$$

These equations represent two lines with the same slope and different y-intercepts. Thus the lines are parallel, there is no solution, and the system is *inconsistent.*

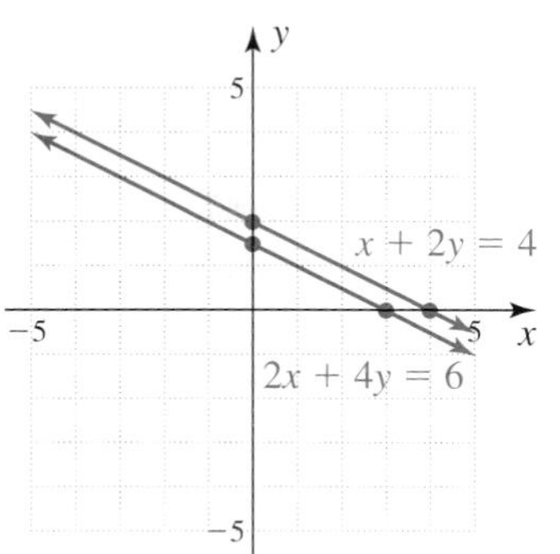

Figure 9

c. The respective tables for $x + 2y = 4$ and $4y + 2x = 8$ are

x	y
0	2
4	0

x	y
0	2
4	0

and they are identical. So, there are infinitely many solutions because the lines coincide (see Figure 10). The system is *dependent,* and the solutions are all the points on the graph. For example, (0, 2), (4, 0), and (2, 1) are solutions.

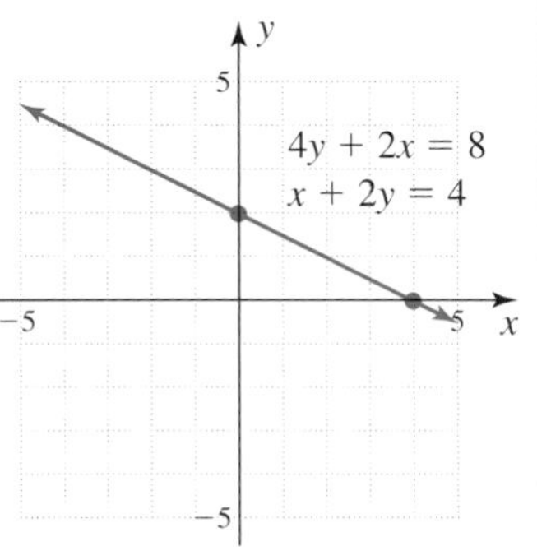

Figure 10

Answers (*contd.*)

4. b. Inconsistent; no solution

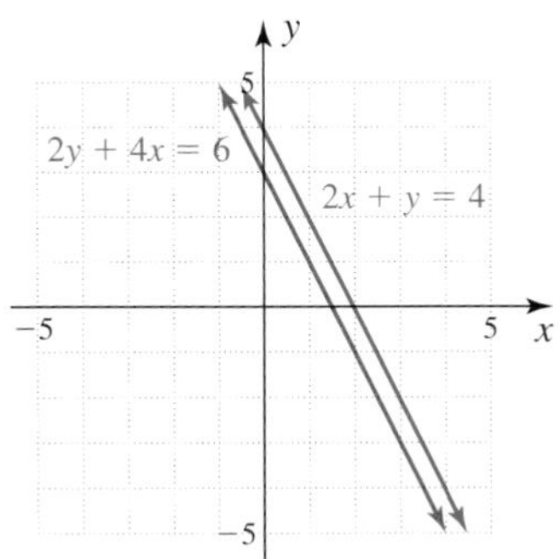

c. Dependent; infinitely many solutions

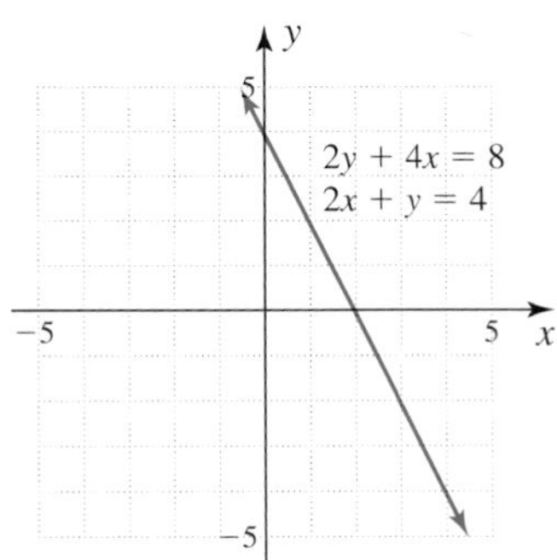

Note that in a *dependent* system, one equation is a constant *multiple* of the other. Thus

$$x + 2y = 4 \qquad \text{and} \qquad 4y + 2x = 8$$

are dependent because $4y + 2x = 8$ is a constant *multiple* of $x + 2y = 4$. Note that

$$2(x + 2y) = 2(4)$$

becomes

$$2x + 4y = 8 \qquad \text{or} \qquad 4y + 2x = 8$$

which is the second equation.

A HELPFUL HINT

If you want to know what *type* of solutions you are going to have, solve both equations for y to obtain the system

$$y = m_1x + b_1$$
$$y = m_2x + b_2$$

When $m_1 \neq m_2$, the lines intersect (there is **one** solution).
When $m_1 = m_2$ and $b_1 = b_2$, there is only one line (**infinitely many** solutions).
When $m_1 = m_2$ and $b_1 \neq b_2$, the lines are parallel (there is **no** solution).

C Solving an Application

Most of the problems that we've discussed use x- and y-values ranging from -10 to 10. This is not the case when working real-life applications! For example, the prices for connecting to the Internet using America Online® (AOL) and CompuServe® (CS) are identical: \$9.95 per month plus \$2.95 for every hour over 5 hours. If you are planning on using more than 5 hours, however, both companies have a special plan.

AOL: \$19.95 for 20 hours plus \$2.95 per hour after 20 hours*

CS: \$24.95 for 20 hours plus \$1.95 per hour after 20 hours

For convenience, let's round the numbers as follows:

AOL: \$20 for 20 hours plus \$3 per hour after 20 hours

CS: \$25 for 20 hours plus \$2 per hour after 20 hours

EXAMPLE 5 "Graphing" the Internet

Make a graph of both prices to see which is a better deal.

SOLUTION Let's start by making a table where h represents the number of hours used and p the monthly price. Keep in mind that when the number of hours is 20 or less the price is fixed: \$20 for AOL and \$25 for CS.

Price for AOL

h	p	
0	20	
5	20	
10	20	
15	20	
20	20	
25	35	$20 + (25 - 20)3 = 35$
30	50	$20 + (30 - 20)3 = 50$
35	65	$20 + (35 - 20)3 = 65$

Price for CS

h	p	
0	25	
5	25	
10	25	
15	25	
20	25	
25	35	$25 + (25 - 20)2 = 35$
30	45	$25 + (30 - 20)2 = 45$
35	55	$25 + (35 - 20)2 = 55$

To graph these two functions, we let h run from 0 to 50 and p run from 0 to 100 in increments of 5. After 20 hours, the price p for h hours with AOL is

$$p = 20 + (h - 20) \cdot 3 = 20 + 3h - 60 = 3h - 40$$

The price for CS is

$$p = 25 + (h - 20) \cdot 2 = 25 + 2h - 40 = 2h - 15$$

The graph is shown in Figure 11. As you can see, if you plan on using 20 hours or less, the price is fixed for both AOL and CS. From 20 to 25 hours, AOL has a lower price, but if you are using more than 25 hours, the price for CS is lower. At exactly 25 hours, the price is the same for both, \$35. Note that the complete graph is in quadrant I, since neither the number of hours h nor the price p is negative.

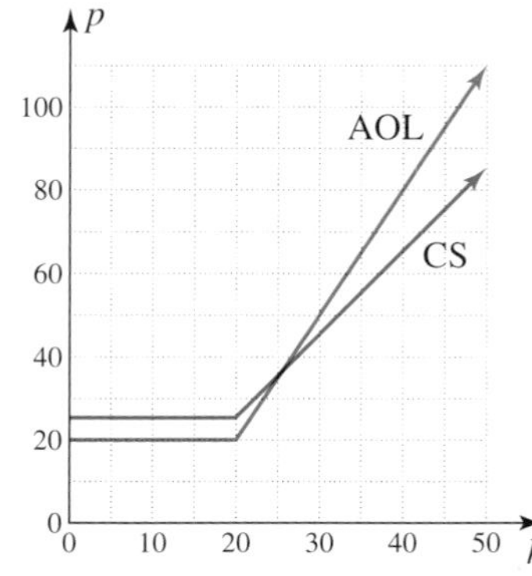

Figure 11

PROBLEM 5

One Internet provider charges \$20 for 20 hours plus \$1.50 per hour after 20 hours. Another Internet provider charges \$15 for 20 hours plus \$2.00 per hour after 20 hours. Make a graph of both prices to see which is a better deal.

Answer

5.

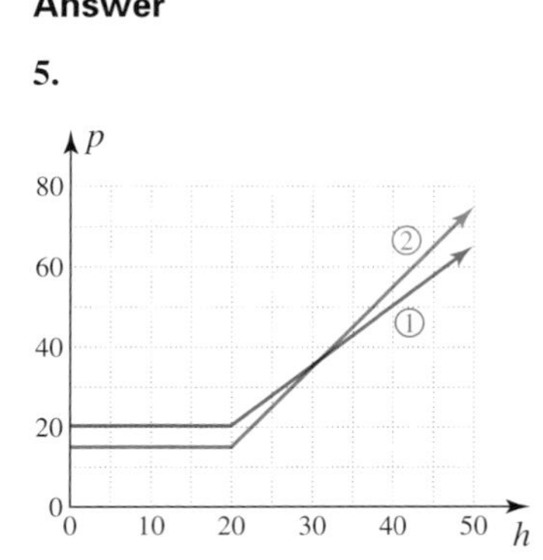

*As of this writing, AOL has a new plan costing \$23.90 per month with no limit on the number of hours used, and CompuServe has been acquired by AOL.

Calculate It Solving Systems of Equations

If you have a calculator, you can solve a linear system of equations very easily. However, you must know how to solve an equation for a specified variable. Thus in Example 1, you can solve $2x + y = 4$ for y to obtain $y = -2x + 4$, and graph $Y_1 = -2x + 4$. Next, solve $y - 2x = 0$ for y to obtain $y = 2x$ and graph $Y_2 = 2x$. To find the intersection, use a decimal window and the trace and zoom features of your calculator, or better yet, if you have an intersection feature, press 2nd TRACE 5 and follow the prompts. (The graph is shown in Window 1.)

Example 2 is done similarly. Solve $y - 2x = 4$ for y to obtain $y = 2x + 4$. Next, solve $2y - 4x = 12$ for y to get $y = 2x + 6$. You have the equations $y = 2x + 4$ and $y = 2x + 6$, but you don't need a graph to know that there's no solution! Since the lines have the same slope, algebra tells you that the lines are parallel! (Sometimes algebra is better than your calculator.) To verify this, graph $y = 2x + 4$ and $y = 2x + 6$. The results are shown in Window 2.

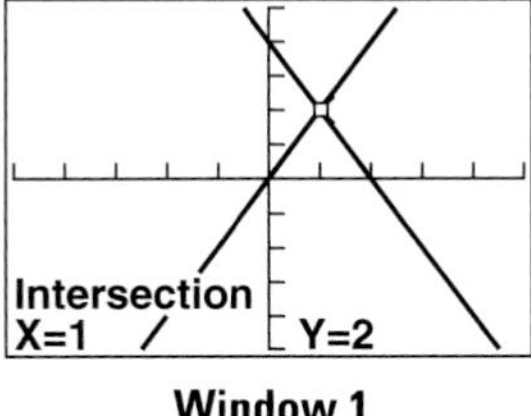

Window 1

To solve Example 3, solve $2x + y = 4$ for y to obtain $y = -2x + 4$. Next, solve $2y + 4x = 8$ for y to get $y = -2x + 4$. Again, you don't need a calculator to see that the lines are the same! Their graph appears in Window 3.

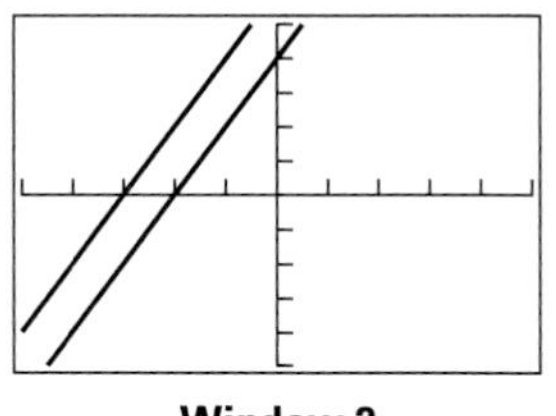
Window 2

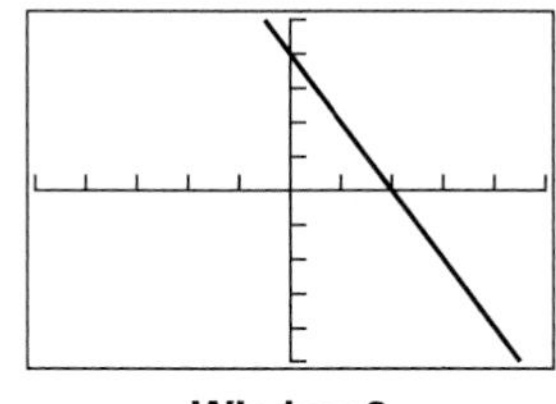
Window 3

Exercises 8.1

A B In Problems 1–30, solve by graphing. Label each system as consistent (write the solution), inconsistent (no solution), or dependent (infinitely many solutions).

1. $x + y = 4$
$x - y = -2$

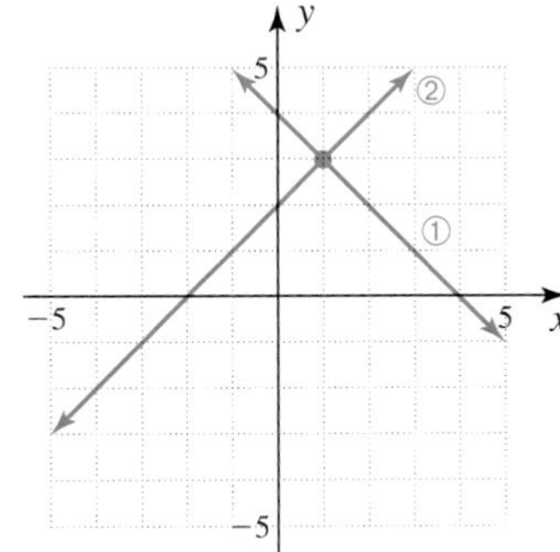

Consistent; (1, 3)

2. $x + y = 3$
$x - y = -5$

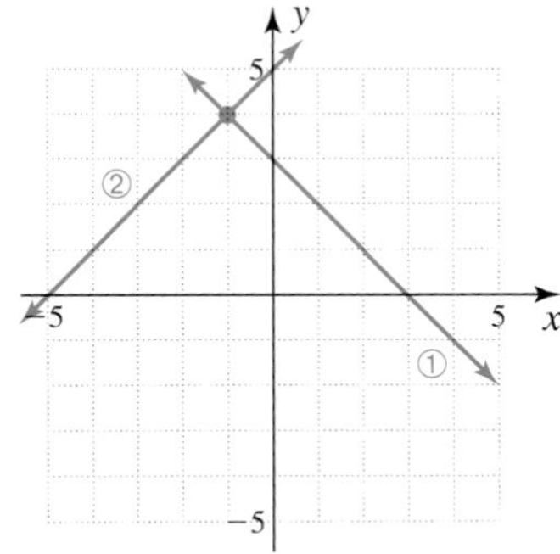

Consistent; (−1, 4)

3. $x + 2y = 0$
$x - y = -3$

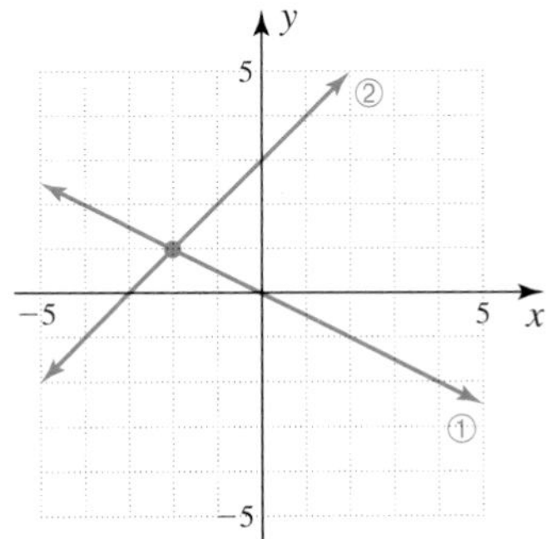

Consistent; (−2, 1)

4. $y + 2x = -3$
$y - x = 3$

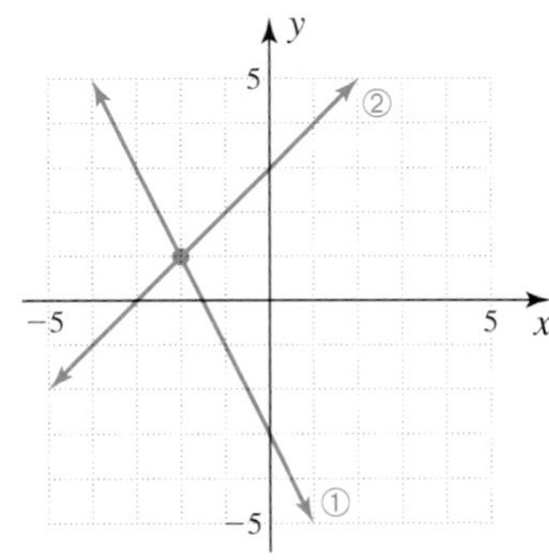

Consistent; (−2, 1)

5. $3x - 2y = 6$
$6x - 4y = 12$

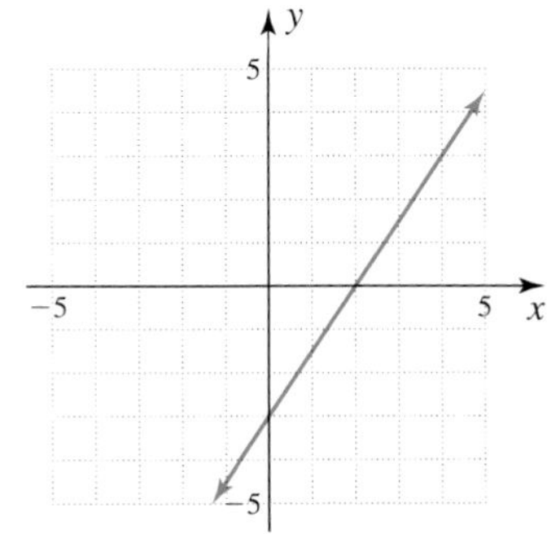

Dependent

6. $2x + y = -2$
$8x + 4y = 8$

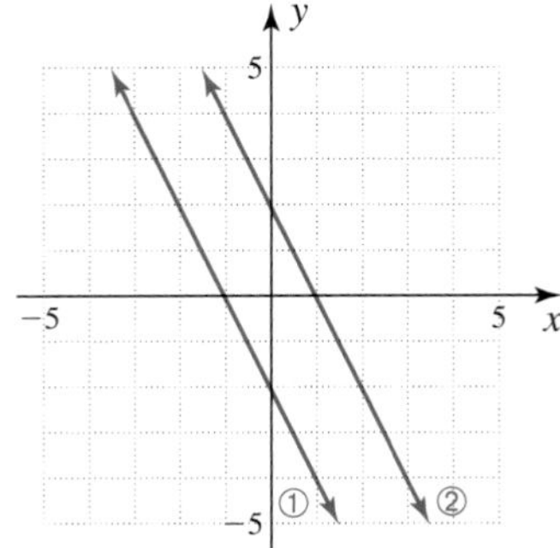

Inconsistent

7. $3x - y = -3$
$y - 3x = 3$

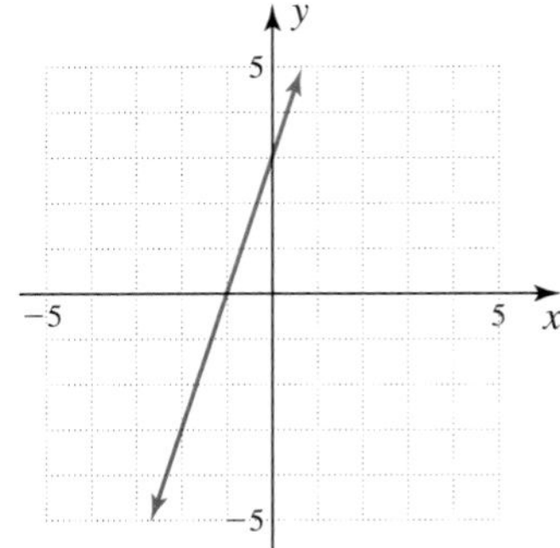

Dependent

8. $4x - 2y = 8$
$y - 2x = -4$

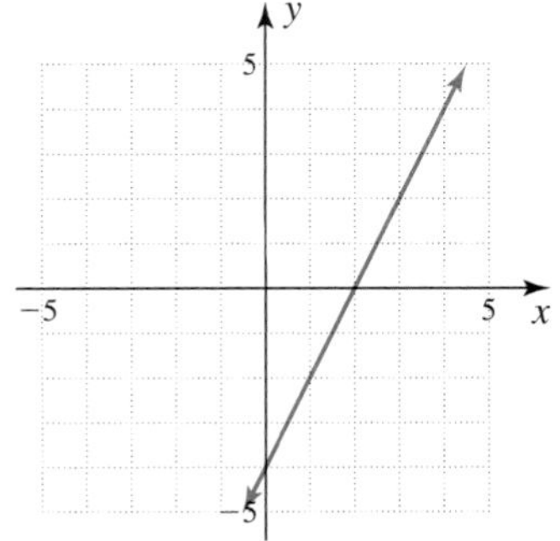

Dependent

9. $2x - y = -2$
$y = 2x + 4$

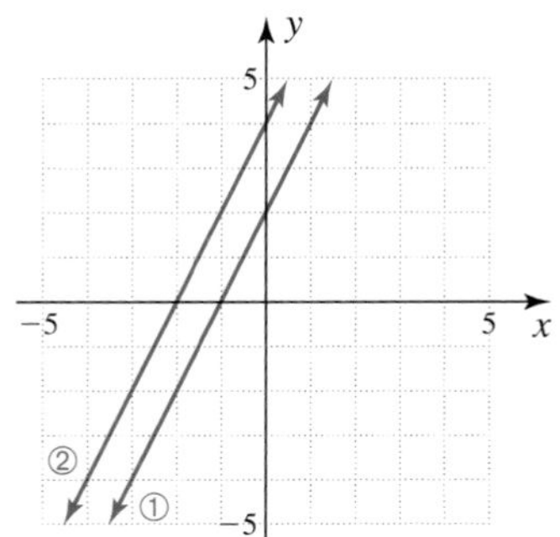

Inconsistent

10. $2x + y = -2$
$y = -2x + 4$

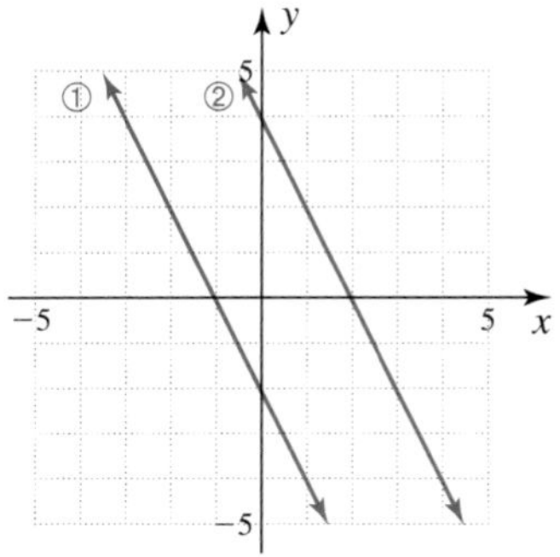

Inconsistent

11. $y = -2$
$2y = x - 2$

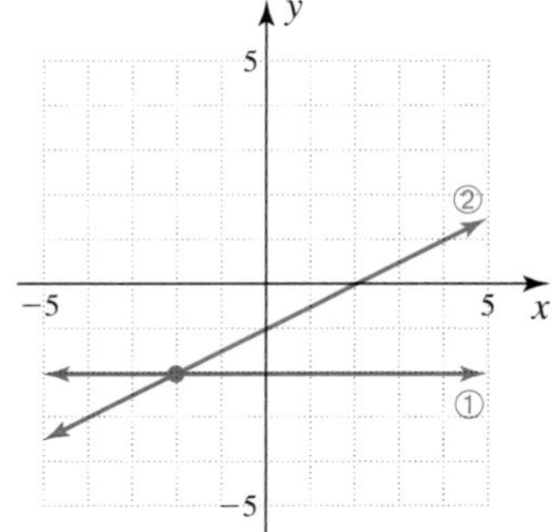

Consistent; $(-2, -2)$

12. $3y = 6 - x$
$y = 3$

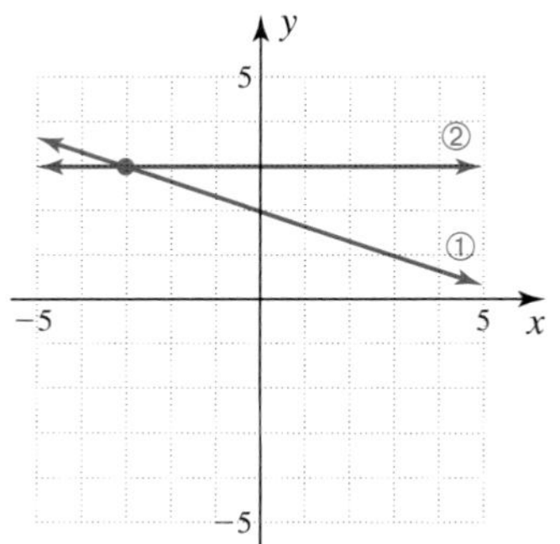

Consistent; $(-3, 3)$

13. $x = 3$
$y = 2x - 4$

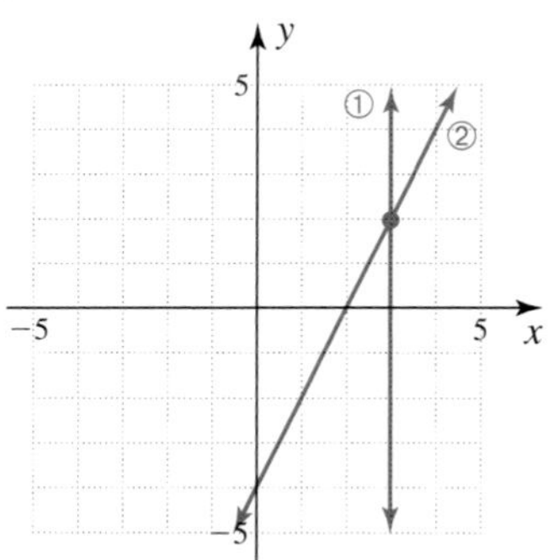

Consistent; $(3, 2)$

14. $y = -x + 2$
$x = -1$

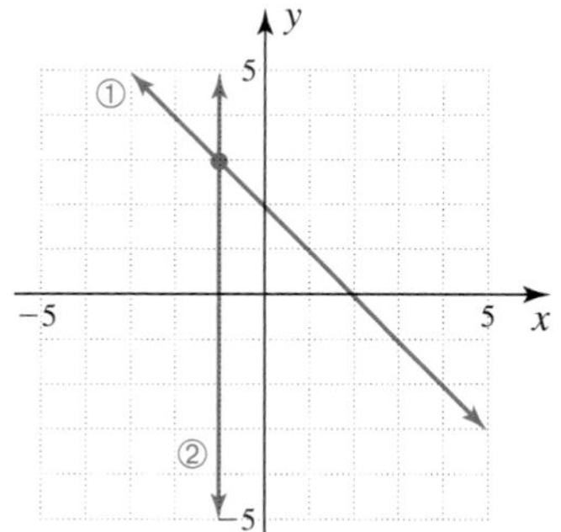

Consistent; $(-1, 3)$

15. $x + y = 3$
$2x - y = 0$

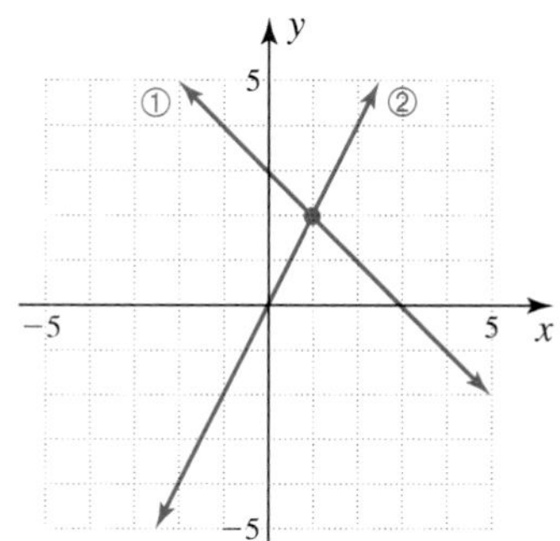

Consistent; $(1, 2)$

16. $x + y = 5$
$x - 4y = 0$

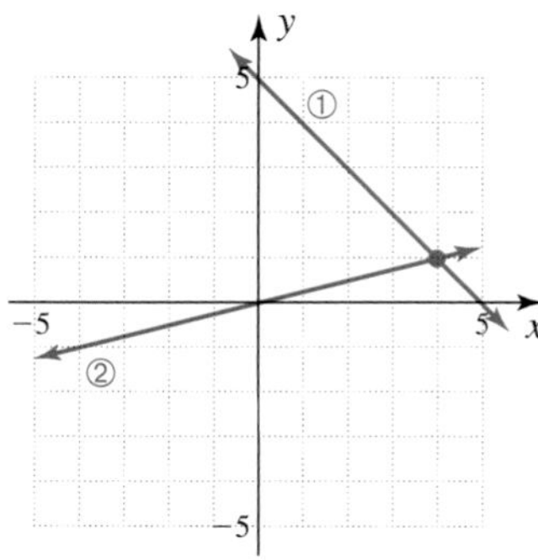

Consistent; $(4, 1)$

17. $5x + y = 5$
$5x = 15 - 3y$

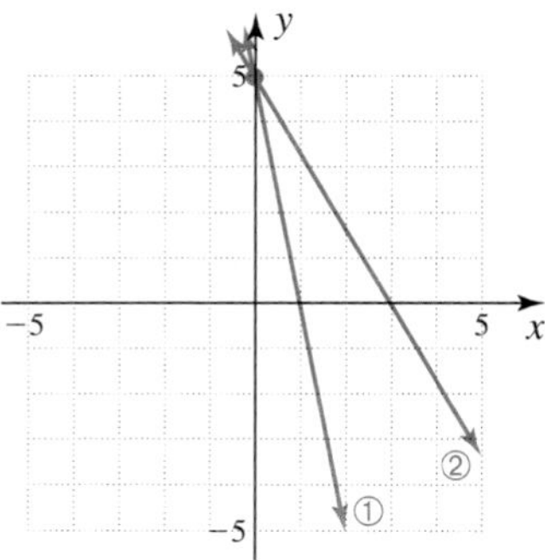

Consistent; $(0, 5)$

18. $2x - y = -4$
$4x = 4 + 2y$

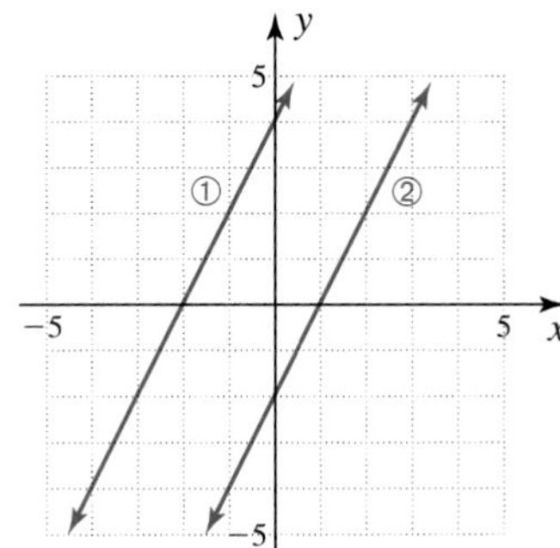

Inconsistent

19. $3x + 4y = 12$
$8y = 24 - 6x$

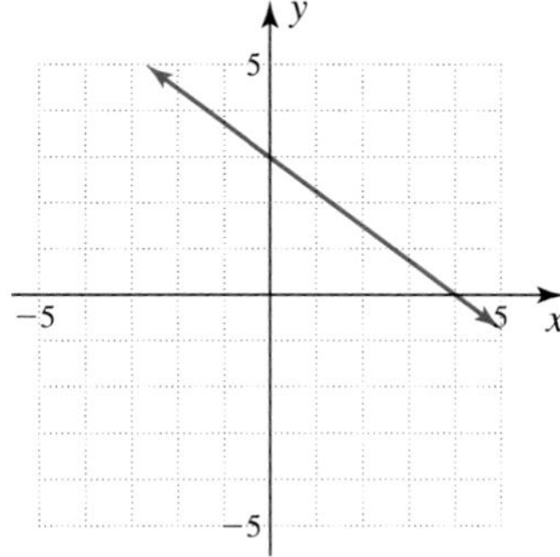

Dependent

20. $2x - 3y = 6$
$6x = 18 + 9y$

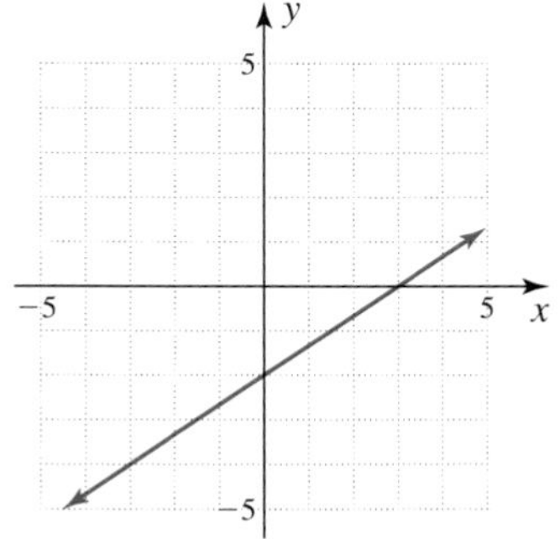

Dependent

21. $y = x + 3$
$y = -x + 3$

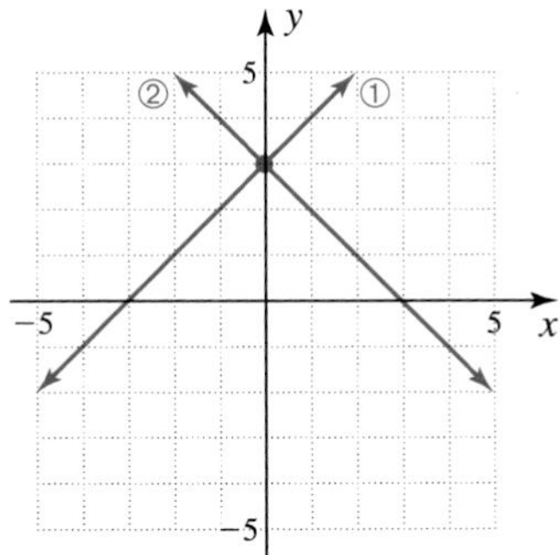

Consistent; (0, 3)

22. $y = 3x + 6$
$y = -2x - 4$

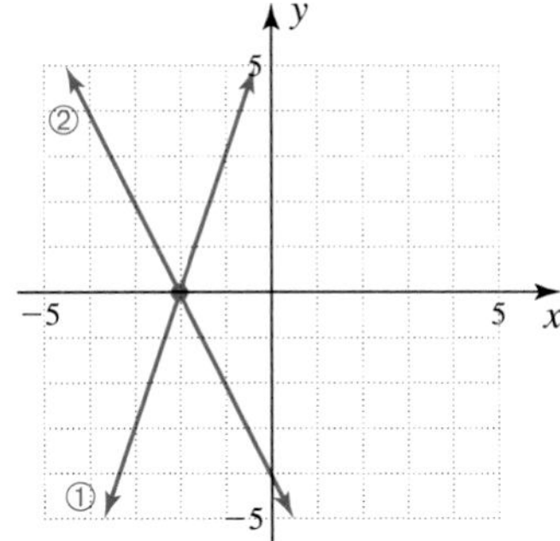

Consistent; (−2, 0)

23. $y = 2x - 2$
$y = -3x + 3$

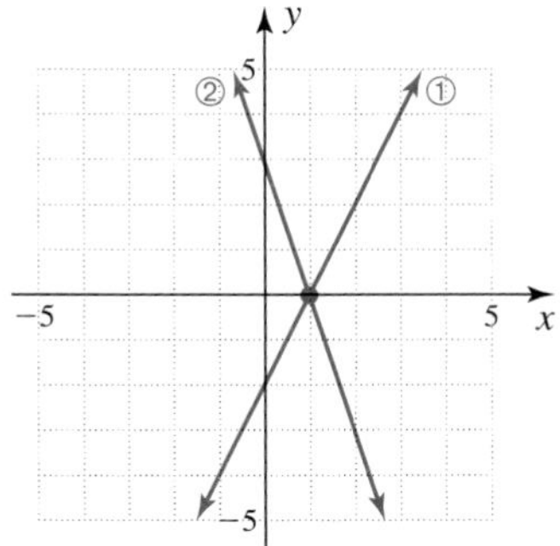

Consistent; (1, 0)

24. $3x = 6$
$y = -2$

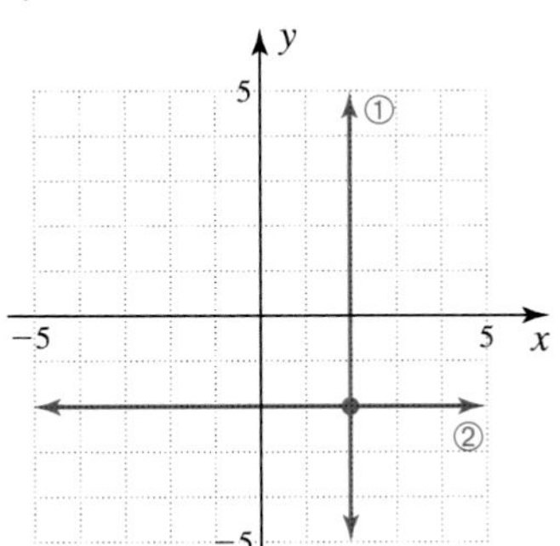

Consistent; (2, −2)

25. $-2x = 4$
$y = -3$

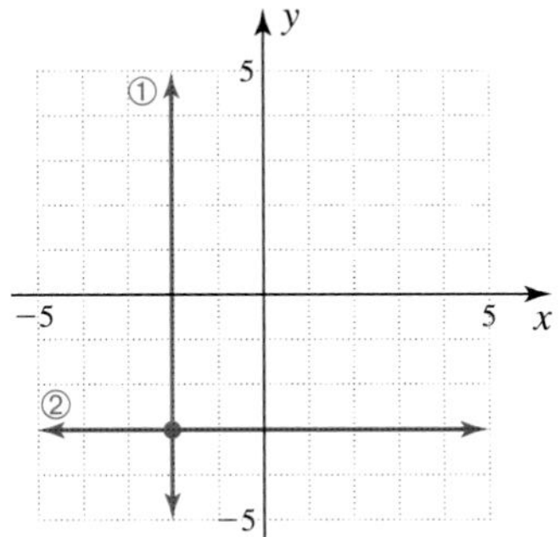

Consistent; (−2, −3)

26. $y = 2$
$y = 2x - 4$

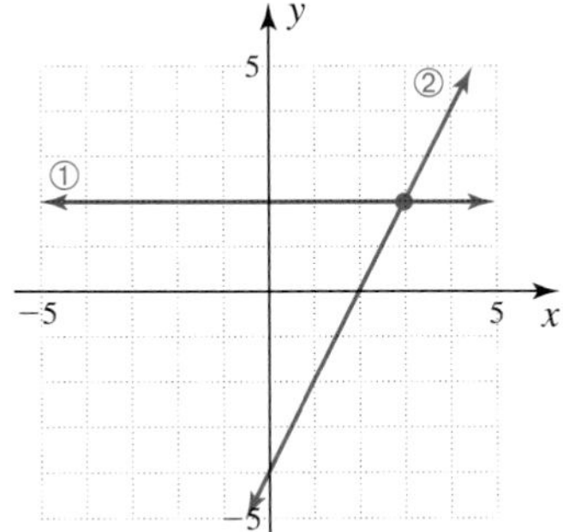

Consistent; (3, 2)

27. $y = -3$
$y = -3x + 6$

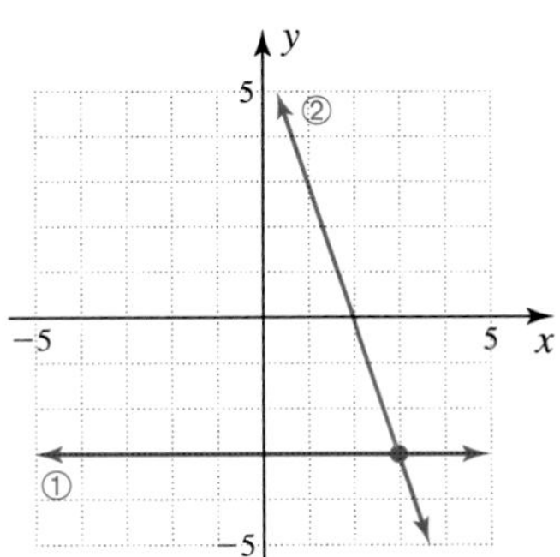

Consistent; (3, −3)

28. $y = -\frac{1}{3}x + 2$
$3y + x = 6$

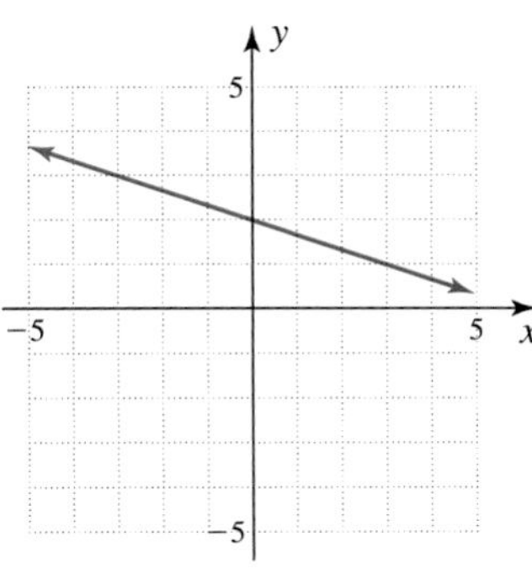

Dependent

29. $x + 4y = 4$

$y = -\frac{1}{4}x + 2$

Inconsistent

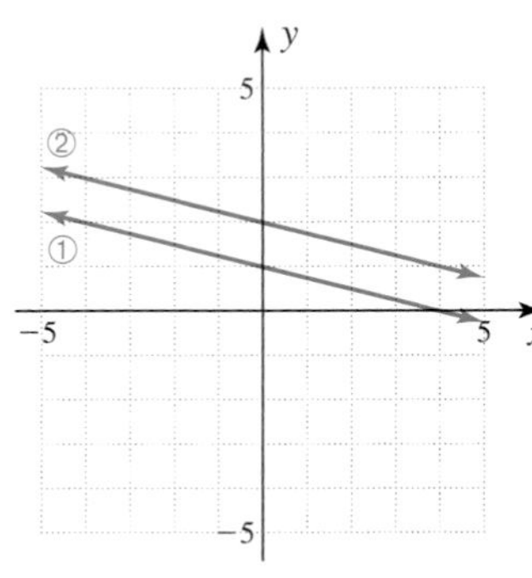

30. $2x - y = 2$

$y = \frac{1}{2}x + 1$

Consistent; (2, 2)

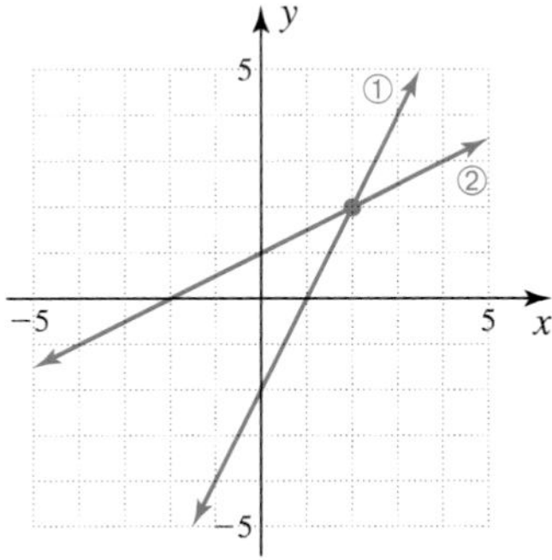

APPLICATIONS

In Problems 31–35, use the following information. You want to watch 10 movies at home each month. You have two options:

OPTION 1 Get cable service. The cost is \$20 for the installation fee and \$35 per month.

OPTION 2 Buy a VCR and rent movies. The cost is \$200 for a VCR and \$25 a month for movie rental fees.

31. *Cost of cable service*

a. If C is the cost of installing cable service plus the monthly fee for m months, write an equation for C in terms of m. $C = 20 + 35m$

b. Complete the following table where C is the cost of cable service for m months:

m	C
6	230
12	440
18	650

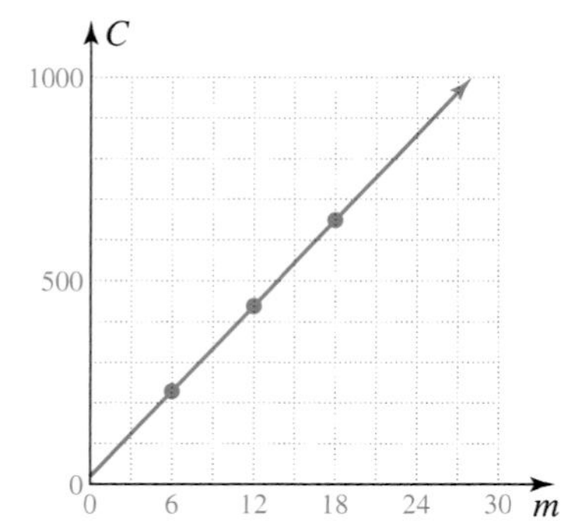

c. Graph the information obtained in parts **a** and **b**. (*Hint:* Let m run from 1 to 24 and C run from 0 to 1000 in increments of 100.)

32. *Cost of renting movies*

a. If C is the total cost of buying the VCR plus renting the movies for m months, write an equation for C in terms of m. $C = 200 + 25m$

b. Complete the following table where C is the cost of buying a VCR and renting movies for m months:

m	C
6	350
12	500
18	650

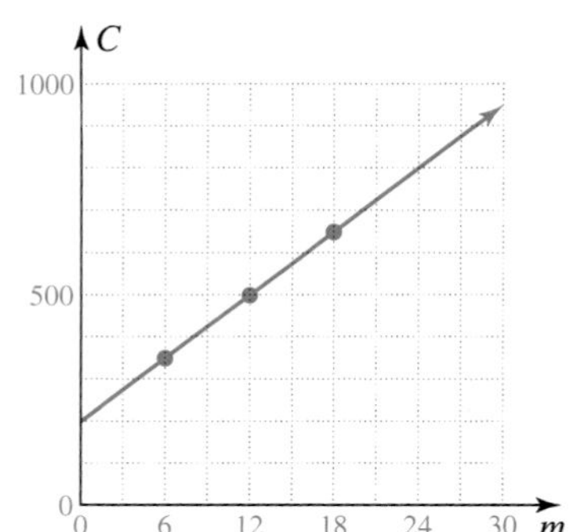

c. Graph the information obtained in parts **a** and **b**.

33. *Graphical comparison* Make a graph of the information obtained in Problems 31 and 32 on the same coordinate axes.

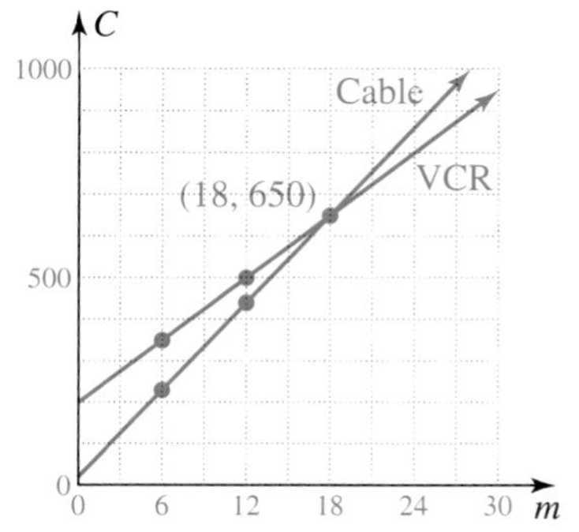

34. *When is cable cheaper?* Based on the graph for Problem 33, when is the cable service cheaper?
Cable service is cheaper if used less than 18 months.

35. *When is renting movies cheaper?* Based on the graph for Problem 33, when is the VCR and rental option cheaper? VCR and rental option is cheaper if used more than 18 months. (The options are equal if used for 18 months.)

36. *Lower wages, higher tips* At Grady's restaurant, servers earn \$80 a week plus tips, which amount to \$5 per table.

a. Write an equation for the weekly wages W based on serving t tables. $W = 80 + 5t$

b. Complete the following table where W is the wages and t is the number of tables served:

t	W
5	105
10	130
15	155
20	180

c. Graph the information obtained in parts **a** and **b**.

37. *Higher wages, lower tips* At El Centro restaurant, servers earn \$100 a week, but the average tip per table is only \$3.

a. Write an equation for the weekly wages W based on serving t tables. $W = 100 + 3t$

b. Complete the following table where W is the wages and t is the number of tables served:

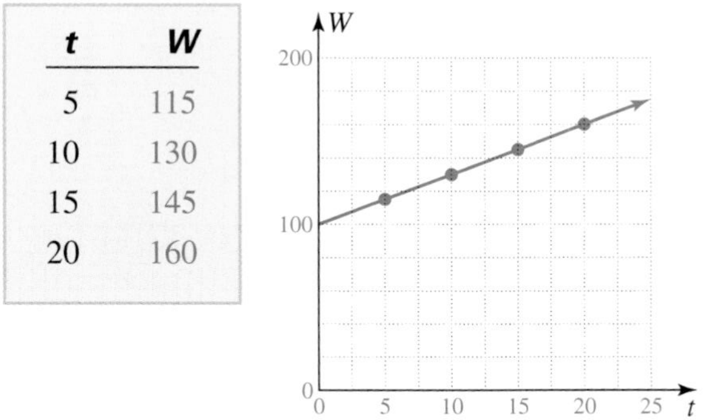

t	W
5	115
10	130
15	145
20	160

c. Graph the information obtained in parts **a** and **b**.

38. *Graphical comparison* Graph the information from Problems 36 and 37 on the same coordinate axes. Based on the graph, answer the following questions.

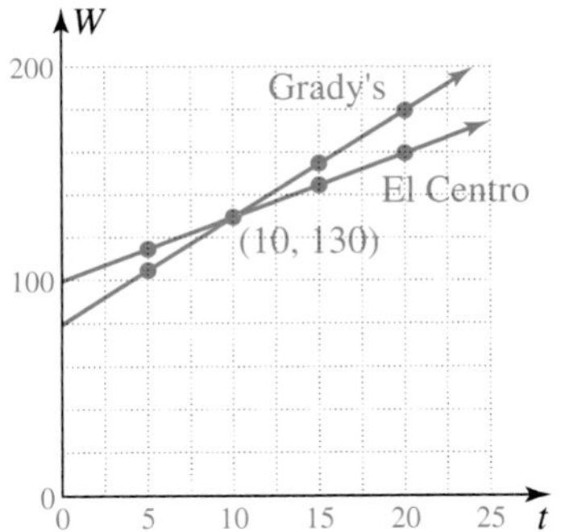

a. When does a server at Grady's make more money than a server at El Centro? A server at Grady's earns more when more than 10 tables are served.

b. When does a server at El Centro make more money than a server at Grady's? A server at El Centro earns more when fewer than 10 tables are served.

39. *Do you have a cell phone?* How much do you pay a month? At the present time, two companies have cell phone plans that cost \$19.95 per month. However, plan A costs \$0.60 per minute of airtime during peak hours, while plan B costs \$0.45 per minute of airtime during peak hours. Plan A offers a free phone with its plan, while plan B's phone costs \$45. For comparison purposes, since the monthly cost is the same for both plans, the cost C is based on the price of the phone plus the number m of minutes of airtime used.

a. Write an equation for the cost C of plan A. $C = 0.60m$

b. Write an equation for the cost C of plan B. $C = 0.45m + 45$

c. Using the same coordinate axes, make a graph for the costs of plans A and B. (*Hint:* Let m and C run from 0 to 500.)

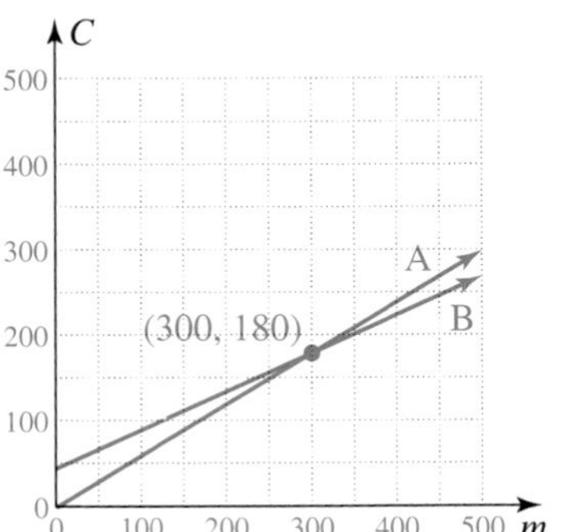

40. *Which is better?* Based on the graphs obtained in Problem 39, which plan would you buy, A or B? Explain. Purchase plan A if you plan to use fewer than 300 minutes of peak airtime. Purchase plan B if you plan to use more than 300 minutes of peak airtime.

SKILL CHECKER

Try the Skill Checker Exercises so you'll be ready for the next section.

Solve:

41. $3x + 72 = 2.5x + 74$ $x = 4$

42. $5x + 20 = 3.5x + 26$ $x = 4$

Determine whether the given point is a solution of the equation:

43. $(3, 5)$; $2x + y = 11$ Yes

44. $(-1, 4)$; $2x - y = -6$ Yes

45. $(-1, 2)$; $2x - y = 0$ No

46. $(-2, 6)$; $3x - y = 0$ No

USING YOUR KNOWLEDGE

Art Books for Sale!

According to the *Statistical Abstract of the United States,* the prices for art books and the number of books *supplied* in 1980, 1985, and 1990 are as shown in the following tables:

Demand Function		Supply Function	
Price ($)	**Quantity (hundreds)**	**Price ($)**	**Quantity (hundreds)**
$28	17	$26	12
$35	15	$35	15
$42	13	$44	18

Note that *fewer* books are *demanded* by consumers as the price of the books goes up. (Consumers are not willing to pay that much for the book.)

47. Graph the points corresponding to the demand function in the coordinate system shown.

48. Draw a line passing through the points. The line is an approximation to the demand function.

On the other hand, book sellers *supply fewer* books when prices go down and more books when prices go up.

49. Graph the points corresponding to the supply function on the same coordinate system.

50. Draw a line passing through the points. The line is an approximation to the supply function.

51. What is the point of intersection of the two lines? (35, 15)

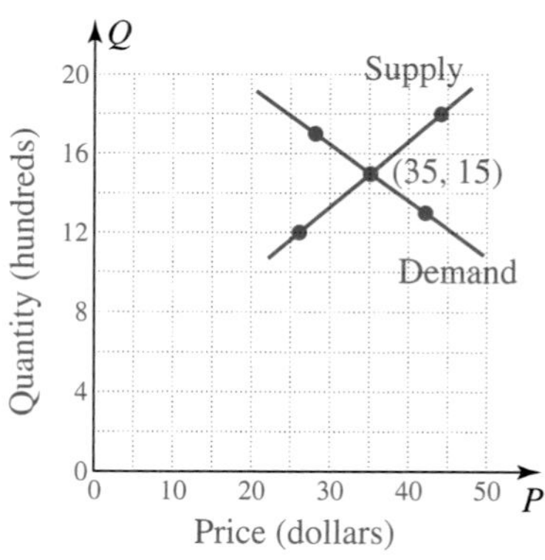

52. At the point of intersection, what is the price? $35

53. At the point of intersection, what is the quantity of books sold? 1500

WRITE ON

54. What does the solution of a system of linear equations represent? Answers may vary.

55. Define a consistent, an inconsistent, and a dependent system of equations. Answers may vary.

56. How can you tell graphically whether a system is consistent, inconsistent, or dependent? Answers may vary.

Suppose you have a system of equations and you solve both equations for y to obtain

$$y = m_1x + b_1$$
$$y = m_2x + b_2$$

What can you say about the graph of the system when

57. $m_1 = m_2$ and $b_1 \neq b_2$? How many solutions do you have? Explain. No solution; answers may vary.

58. $m_1 = m_2$ and $b_1 = b_2$? How many solutions do you have? Explain. Infinitely many solutions; answers may vary.

59. $m_1 \neq m_2$? How many solutions do you have? Explain. One solution; answers may vary.

MASTERY TEST

If you know how to do these problems, you have learned your lesson!

Use the graphical method to find the solution of the system of equations (if it exists):

60. $x + 2y = 4$
$2x - 4y = 0$

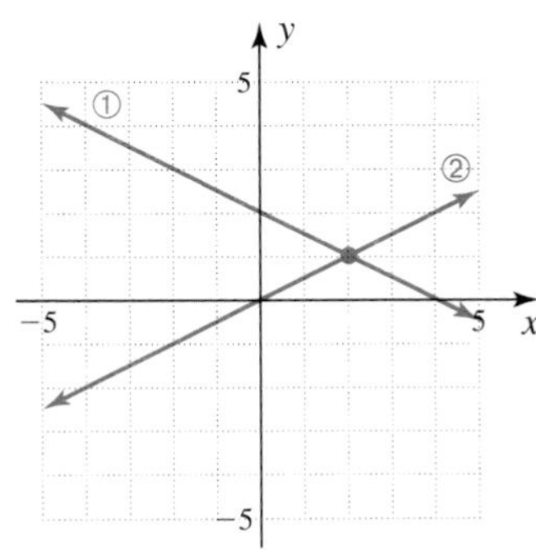

Solution (2, 1)

61.

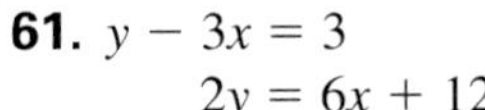

$y - 3x = 3$
$2y = 6x + 12$

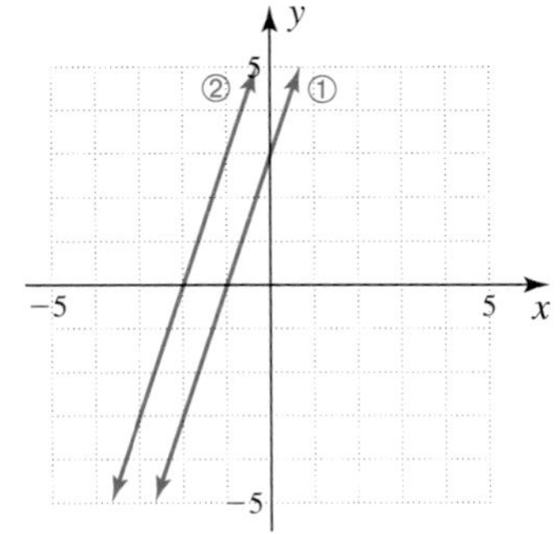

No solution

62. $x + 2y = 4$
$4y = -2x + 8$

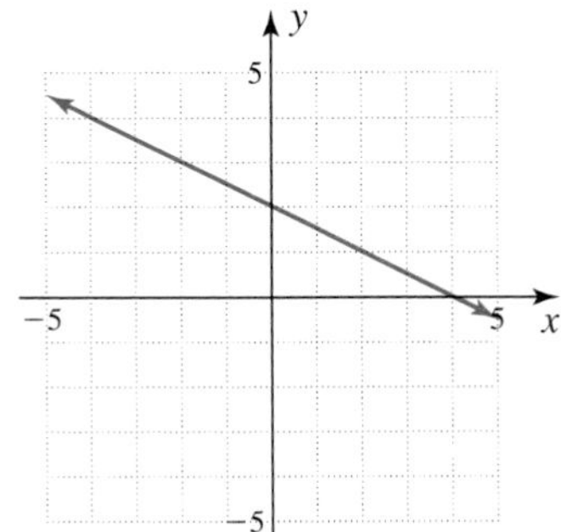

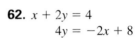

Infinitely many solutions

Classify each system as consistent, inconsistent, or dependent; if the system is consistent, find the solution:

63. $x + y = 4$
$2x - y = 2$
Consistent; (2, 2)

64. $2x + y = 4$
$2y + 4x = 6$
Inconsistent; no solution

65. $2x + y = 4$
$2y + 4x = 8$
Dependent; infinitely many solutions

66. The monthly cost of Internet provider A is \$10 for the first 5 hours and \$3 for each hour after 5. Provider B's cost is \$15 for the first 5 hours and \$2 for each hour after 5. Make a graph of the cost C for providers A and B when h hours of airtime are used. (*Hint:* Let h run from 0 to 20 and C run from 0 to 60.)

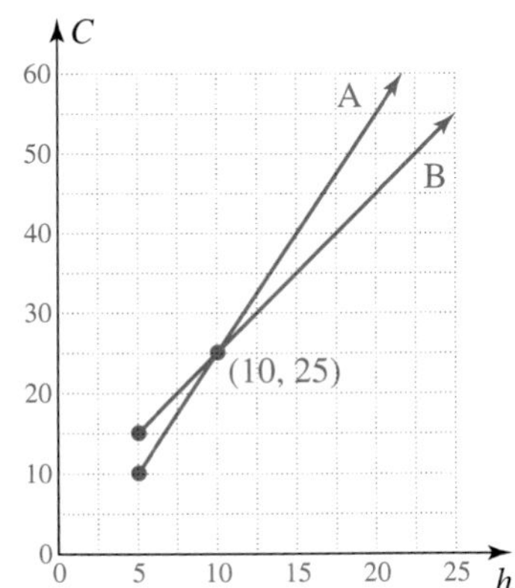

8.2 SOLVING SYSTEMS OF EQUATIONS BY SUBSTITUTION

To Succeed, Review How To . . .

1. Solve linear equations (pp. 151–155).
2. Determine whether an ordered pair satisfies an equation (pp. 242–243).

Objectives

Use the substitution method to:

A Solve a system of equations in two variables.

B Determine whether a system of equations is consistent, inconsistent, or dependent.

C Solve an application.

GETTING STARTED

Supply, Demand, and Substitution

Does this graph look familiar? As we noted in the preceding section's *Getting Started,* the supply and the demand were about the same in the year 1980. However, this solution is only an *approximate* one because the x-scale representing the years is numbered at 5-year intervals, and it's hard to pinpoint the exact point at which the lines intersect. Suppose we have the equations for the supply and the demand between 1975 and 1985. These equations are

$$\text{Supply: } y = 2.5x + 72 \qquad x = \text{the number of years elapsed after 1975}$$
$$\text{Demand: } y = 3x + 70$$

Can we now tell *exactly* where the lines meet? Not graphically! For one thing, if we let $x = 0$ in the first equation, we obtain $y = 2.5 \cdot 0 + 72$, or $y = 72$. Thus we either need a piece of graph paper with 72 units, or we have to make each division on the graph paper 10 units, thereby *losing* accuracy. But there's a way out. We can use an *algebraic* method rather than a *graphical* one. Since we are looking for the point at which the supply (y) is the same as the demand (y), we may *substitute* the expression for y in the demand equation—that is, $3x + 70$—into the supply equation. Thus we have

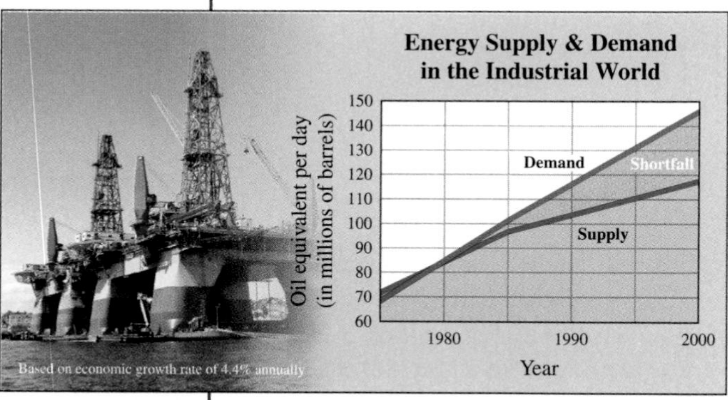

$$\underbrace{\text{Demand }(y)}\ = \underbrace{\text{Supply }(y)}$$
$$3x + 70 = 2.5x + 72$$
$$3x = 2.5x + 2 \qquad \text{Subtract 70.}$$
$$0.5x = 2 \qquad \text{Subtract } 2.5x.$$
$$\frac{0.5x}{0.5} = \frac{2}{0.5} \qquad \text{Divide by 0.5.}$$
$$x = 4 \qquad \text{Simplify.}$$

Thus 4 years after 1975 (or in 1979), the supply equaled the demand. At this time the demand was

$$y = 3(4) + 70 = 12 + 70 = 82 \text{ (million barrels)}$$

In this section we learn to solve equations by the *substitution method.* This method is recommended for solving systems in which *one* equation is solved, or can be easily solved, for one of the variables.

A Using the Substitution Method to Solve a System of Equations

Here is a summary of the substitution method, which we just used in the *Getting Started.*

Web It

To study how to solve systems of equations using the substitution method, go to link 8-2-1 on the Bello Website at mhhe.com/bello.

Teaching Tip

Have students discuss how they would choose which variable to solve for when using the substitution method.

PROCEDURE

Solving a System of Equations by the Substitution Method

1. Solve one of the equations for x or y.
2. Substitute the resulting expression into the other equation. (Now you have an equation in one variable.)
3. Solve the new equation for the variable.
4. Substitute the value of that variable into one of the original equations and solve this equation to get the value for the second variable.
5. Check the solution by substituting the numerical values of the variables in both equations.

EXAMPLE 1 Solving a system by substitution

Solve the system:

$$x + y = 8$$
$$2x - 3y = -9$$

SOLUTION We use the five-step procedure.

1. Solve one of the equations for x or y (we solve the first equation for y). $\quad y = 8 - x$
2. Substitute $8 - x$ for y in $2x - 3y = -9$. $\quad 2x - 3(8 - x) = -9$
3. Solve the new equation for the variable.

$$2x - 3(8 - x) = -9$$
$$2x - 24 + 3x = -9 \quad \text{Simplify.}$$
$$5x - 24 = -9 \quad \text{Combine like terms.}$$
$$5x = 15 \quad \text{Add 24 to both sides.}$$
$$x = 3 \quad \text{Divide by 5.}$$

4. Substitute the value of the variable $x = 3$ into one of the original equations. (We substitute in the equation $x + y = 8$.) Then solve for the second variable. Our solution is the ordered pair (3, 5).

$$3 + y = 8$$
$$y = 5$$

5. **CHECK** When $x = 3$ and $y = 5$,

$$x + y = 8$$

becomes

$$3 + 5 = 8$$
$$8 = 8$$

which is true. Then the second equation

$$2x - 3y = -9$$

PROBLEM 1

Use the five-step procedure to solve the system:

$$x + y = 5$$
$$2x - 4y = -8$$

Answer

1. The solution is (2, 3).

becomes

$$2(3) - 3(5) = -9$$
$$6 - 15 = -9$$
$$-9 = -9$$

which is also true. Thus our solution (3, 5) is correct.

EXAMPLE 2 **Solving an inconsistent system**

Solve the system:

$$x + 2y = 4$$
$$2x = -4y + 6$$

SOLUTION We use the five-step procedure.

1. Solve one of the equations for one of the variables (we solve the first equation for x). $x = 4 - 2y$
2. Substitute $x = 4 - 2y$ into $2x = -4y + 6$.

$$2(4 - 2y) = -4y + 6$$
$$8 - 4y = -4y + 6 \quad \text{Simplify.}$$
$$8 - 4y + 4y = -4y + 4y + 6 \quad \text{Add 4y.}$$
$$8 = 6$$

3. There is no equation to solve. The result, $8 = 6$, is never true. It is a contradiction. Since our procedure is correct, we conclude that the given system has *no solution;* it is *inconsistent.*
4. We do not need step 4.
5. **CHECK** Note that if you divide the second equation by 2, you get $x = -2y + 3$ or $x + 2y = 3$ which *contradicts* the first equation, $x + 2y = 4$.

PROBLEM 2

Solve the system:

$$x - 3y = 6$$
$$2x - 6y = 8$$

EXAMPLE 3 **Solving a dependent system**

Solve the system:

$$x + 2y = 4$$
$$4y + 2x = 8$$

SOLUTION As before, we use the five-step procedure.

1. Solve the first equation for x. $x = 4 - 2y$
2. Substitute $x = 4 - 2y$ into $4y + 2x = 8$. $4y + 2(4 - 2y) = 8$
3. There is no equation to solve. Note that in this case we have obtained the true statement $8 = 8$, regardless of the value we assign to either x or to y.

$$4y + 8 - 4y = 8 \quad \text{Simplify.}$$
$$8 = 8$$

4. We do not need step 4 because the equations are *dependent;* that is, there are infinitely many solutions.
5. **CHECK** If we let $x = 0$ in the equation $x + 2y = 4$, we obtain $2y = 4$, or $y = 2$. Similarly, if we let $x = 0$ in the equation $4y + 2x = 8$, we obtain $4y = 8$, or $y = 2$, so (0, 2) is a solution for both equations. It can also be

PROBLEM 3

Solve the system:

$$x - 3y = 6$$
$$6y - 2x = -12$$

Answers

2. No solution; the system is inconsistent. **3.** Infinitely many solutions. The equations are dependent. Some solutions are (0, −2), (6, 0), and (3, −1).

shown that $x = 2, y = 1$ satisfies both equations. Therefore, (2, 1) is another solution, and so on. Note that if you divide the second equation by 2 and rearrange, you get $x + 2y = 4$, which is identical to the first equation. Thus any solution of the first equation is also a solution of the second equation; that is, the solution consists of all points satisfying $x + 2y = 4$.

EXAMPLE 4 **Simplifying and solving a system by substitution**

Solve the system:

$$-2x = -y + 2$$
$$6 - 3x + y = -4x + 5$$

SOLUTION The second equation has x's and constants on both sides, so we first simplify it by adding $4x$ and subtracting 6 from both sides to obtain

$$6 - 3x + y + 4x - 6 = -4x + 5 + 4x - 6$$
$$x + y = -1$$

We now have the equivalent system

$$-2x = -y + 2$$
$$x + y = -1$$

Solving the second equation for x, we get $x = -y - 1$. Substituting $-y - 1$ for x in the first equation, we have

$$-2(-y - 1) = -y + 2$$
$$2y + 2 = -y + 2$$
$$3y = 0 \quad \text{Add } y \text{, subtract 2.}$$
$$y = 0 \quad \text{Divide by 3.}$$

Since $-2x = -y + 2$ and $y = 0$, we have

$$-2x = 0 + 2$$
$$x = -1$$

Thus the system is *consistent* and its solution is $(-1, 0)$. You can verify this by substituting -1 for x and 0 for y in the two original equations.

PROBLEM 4

Solve the system:

$$-3x = -y + 6$$
$$6 - 3x + y = -5x + 2$$

If a system has equations that contain fractions, we clear the fractions by multiplying each side by the LCD (Remember? LCD is the lowest common denominator), and then we solve the resulting system, as shown in Example 5.

EXAMPLE 5 **Solving a system involving fractions**

Solve the system:

$$2x + \frac{y}{4} = -1$$
$$\frac{x}{4} + \frac{3y}{8} = \frac{5}{4}$$

PROBLEM 5

Solve the system:

$$2x + \frac{y}{3} = -1$$
$$\frac{x}{4} + \frac{y}{6} = \frac{1}{4}$$

Answers

4. $(-2, 0)$; consistent system
5. $(-1, 3)$; consistent system

SOLUTION Multiply both sides of the first equation by 4, and both sides of the second equation by 8 (the LCM of 4 and 8) to obtain

$$4\left(2x + \frac{y}{4}\right) = (4)(-1) \quad \text{or equivalently} \quad 8x + y = -4$$

$$8\left(\frac{x}{4} + \frac{3y}{8}\right) = 8\left(\frac{5}{4}\right) \quad \text{or equivalently} \quad 2x + 3y = 10$$

Solving the first equation for y, we have $y = -8x - 4$. Now we substitute $-8x - 4$ for y in $2x + 3y = 10$:

$$2x + 3(-8x - 4) = 10$$

$$2x - 24x - 12 = 10 \quad \text{Simplify.}$$

$$-22x = 22 \quad \text{Simplify and add 12.}$$

$$x = -1 \quad \text{Divide by } -22.$$

Substituting -1 for x in $2x + \frac{y}{4} = -1$, we get $2(-1) + \frac{y}{4} = -1$ or $y = 4$. Thus the system is *consistent* and its solution is $(-1, 4)$. Verify this!

B Consistent, Inconsistent, and Dependent Systems

Web It

To study consistent, inconsistent, and dependent systems as they relate to solving systems of equations using substitution, go to link 8-2-3 on the Bello Website at mhhe.com/bello.

When we use the substitution method, one of three things can occur:

1. The equations are *consistent;* there is only *one* solution (x, y).
2. The equations are *inconsistent;* we get a contradictory (false) statement, and there will be *no* solution.
3. The equations are *dependent;* we get a statement that is true for all values of the remaining variable, and there will be *infinitely many* solutions.

Keep this in mind when you do the exercise set, and be very careful with your arithmetic!

C Solving an Application

Remember the prices for connecting to the Internet via America Online (AOL) and CompuServe (CS)? The cost for each of the plans is as follows:

AOL: \$20 for 20 hours plus \$3 for each hour after 20 hours

CS: \$25 for 20 hours plus \$2 for each hour after 20 hours

This means that the price p for h hours of service when $h > 20$ is

$$\text{AOL: } p = \underbrace{20}_{\$20} + \underbrace{3(h - 20)}_{\$3 \text{ for each hour after } 20}$$

$$\text{CS: } p = \underbrace{25}_{\$25} + \underbrace{2(h - 20)}_{\$2 \text{ for each hour after } 20}$$

EXAMPLE 6 **Substitution and Internet prices**

When is the price for both services the same?

SOLUTION To find when the price p is the same for both services, we substitute $p = 20 + 3(h - 20)$ into the second equation to obtain

$$20 + 3(h - 20) = 25 + 2(h - 20)$$

$$20 + 3h - 60 = 25 + 2h - 40 \quad \text{Use the distributive property.}$$

$$3h - 40 = 2h - 15 \quad \text{Simplify.}$$

$$3h = 2h + 25 \quad \text{Add 40 to both sides.}$$

$$h = 25 \quad \text{Subtract } h \text{ from both sides.}$$

Thus if you use 25 hours, the price is the same for AOL and CS.

CHECK The price for 25 hours of AOL service is

$$p = 20 + 3(25 - 20) = 20 + 15 = \$35$$

The price for 25 hours of CS service is

$$p = 25 + 2(25 - 20) = 25 + 10 = \$35$$

Thus when using 25 hours, the price is the same for both services: \$35.

PROBLEM 6

Quickie Internet charges \$25 for 10 hours plus \$2.50 for each hour after 10. When is this price the same as the AOL price, $20 + 3(h - 20)$?

Answer

6. When 80 hours are used

Exercises 8.2

A B In Problems 1–32, use the substitution method to find the solution. Label each system as consistent (one solution), inconsistent (no solution), or dependent (infinitely many solutions). If the system is consistent, give the solution.

1. $y = 2x - 4$
$-2x = y - 4$
Consistent; (2, 0)

2. $y = 2x + 2$
$-x = y + 1$
Consistent; (−1, 0)

3. $x + y = 5$
$3x + y = 9$
Consistent; (2, 3)

4. $x + y = 5$
$3x + y = 3$
Consistent; (−1, 6)

5. $y - 4 = 2x$
$y = 2x + 2$
Inconsistent; no solution

6. $y + 5 = 4x$
$y = 4x + 7$
Inconsistent; no solution

7. $x = 8 - 2y$
$x + 2y = 4$
Inconsistent; no solution

8. $x = 4 - 2y$
$x - 2y = 0$
Consistent; (2, 1)

9. $x + 2y = 4$
$x = -2y + 4$ Dependent; infinitely many solutions

10. $x + 3y = 6$
$x = -3y + 6$ Dependent; infinitely many solutions

11. $x = 2y + 1$
$y = 2x + 1$
Consistent; (−1, −1)

12. $y = 3x + 2$
$x = 3y + 2$
Consistent; (−1, −1)

13. $2x - y = -4$
$4x = 4 + 2y$
Inconsistent; no solution

14. $5x + y = 5$
$5x = 15 - 3y$
Consistent; (0, 5)

15. $x = 5 - y$
$0 = x - 4y$
Consistent; (4, 1)

16. $x = 3 - y$
$0 = 2x - y$
Consistent; (1, 2)

17. $x + 1 = y + 3$
$x - 3 = 3y - 7$
Consistent; (5, 3)

18. $x - 1 = 2y + 12$
$x + 6 = 3 - 6y$
Consistent; (9, −2)

19. $2y = -x + 4$
$8 + x - 4y = -2y + 4$
Consistent; (0, 2)

20. $y - 1 = 2x + 1$
$3x + y + 2 = 5x + 6$
Inconsistent; no solution

21. $3x + y - 5 = 7x + 2$
$y + 3 = 4x - 2$
Inconsistent; no solution

22. $4x + 2y + 1 = 4 + 3x + 5$
$x - 3 = 5 - 2y$
Dependent; infinitely many solutions

23. $4x - 2y - 1 = 3x - 1$
$x + 2 = 6 - 2y$
Consistent; (2, 1)

24. $8 + y - 4x = -2x + 4$
$2x + 3 = -y + 7$
Consistent; (2, 0)

25. $\frac{x}{6} + \frac{y}{2} = 1$
$5x - 2y = 13$
Consistent; (3, 1)

26. $\frac{x}{8} - \frac{5y}{8} = 1$
$-7x + 8y = 25$
Consistent; $(-7, -3)$

27. $3x - y = 12$
$-\frac{x}{2} + \frac{y}{6} = -2$
Dependent; infinitely many solutions

28. $x - 3y = -4$
$-\frac{x}{6} + \frac{y}{2} = \frac{2}{3}$
Dependent; infinitely many solutions

29. $\frac{y}{4} + x = \frac{3}{8}$
$y = 8 - 4x$
Inconsistent; no solution

30. $y = 1 - 5x$
$\frac{y}{5} + x = \frac{3}{10}$
Inconsistent; no solution

31. $3x + \frac{y}{3} = 5$
$\frac{x}{2} - \frac{2y}{3} = 3$ Consistent; $(2, -3)$

32. $3y + \frac{x}{3} = 5$
$\frac{y}{2} - \frac{2x}{3} = 3$ Consistent; $(-3, 2)$

APPLICATIONS

33. *Internet service costs* The Information Network charges a $20 fee for 15 hours of Internet service plus $3 for each additional hour while InterServe Communications charges $20 for 15 hours plus $2 for each additional hour.

a. Write an equation for the price p when you use h hours of Internet service with The Information Network. $p = 20 + 3(h - 15)$ when $h > 15$

b. Write an equation for the price p when you use h hours of Internet service with InterServe Communications. $p = 20 + 2(h - 15)$ when $h > 15$

c. When is the price p for both services the same? When $h \leq 15$ hours, $p = \$20$.

34. *Internet service costs* TST On Ramp charges $10 for 10 hours of Internet service plus $2 for each additional hour.

a. Write an equation for the price p when you use h hours of Internet service with TST On Ramp. $p = 10 + 2(h - 10)$ when $h > 10$

b. When is the price of TST the same as that for InterServe Communications (see Problem 33)? (*Hint:* The algebra won't tell you—try a graph!) When $h \geq 15$, both services are equal.

35. *Cell phone costs* Phone Company A has a plan costing $20 per month plus 60¢ for each minute m of airtime, while Company B charges $50 per month plus 40¢ for each minute m of airtime. When is the cost for both companies the same? When 150 minutes are used

36. *Cell phone charges* Sometimes phone companies charge an activation fee to "turn on" your cell phone. One company charges $50 for the activation fee, $40 for your cell phone, and 60¢ per minute m of airtime. Another company charges $100 for your cell phone and 40¢ a minute of airtime. When is the cost for both companies the same? When 50 minutes are used

37. *Wages and tips* Le Bon Ton restaurant pays its servers $50 a week plus tips, which average $10 per table. Le Magnifique pays $100 per week but tips average only $5 per table. How many tables t have to be served so that the weekly income of a server is the same at both restaurants? 10 tables

38. *Fitness center costs* The Premier Fitness Center has a $200 initiation fee plus $25 per month. Bodies by Jacques has an initial charge of $500 but charges only $20 per month. At the end of which month is the cost the same? 60th

39. *Cable company costs* One cable company charges $35 for the initial installation plus $20 per month. Another company charges $20 for the initial installation plus $35 per month. At the end of which month is the cost the same for both companies? 1st

40. *Plumber charges* A plumber charges $20 an hour plus $60 for the house call. Another plumber charges $25 an hour, but the house call is only $50. What is the least number of hours for which the costs for both plumbers are the same? 2 hours

41. *Temperature conversions* The formula for converting degrees Celsius C to degrees Fahrenheit F is

$$F = \frac{9}{5}C + 32$$

When is the temperature in degrees Fahrenheit the same as that in degrees Celsius? At $-40°$, $F = C$.

42. *Temperature conversions* The formula for converting degrees Fahrenheit F to degrees Celsius C is

$$C = \frac{5}{9}(F - 32)$$

When is the temperature in degrees Celsius the same as that in degrees Fahrenheit? At $-40°$, $C = F$.

43. *Supply and demand* The supply y of a certain item is given by the equation $y = 2x + 8$, where x is the number of days elapsed. If the demand is given by $y = 4x$, how many days will the supply equal the demand? 4 days

44. *Supply and demand* The supply of a certain item is $y = 3x + 8$, where x is the number of days elapsed. If the demand is given by $y = 4x$, in how many days will the supply equal the demand? 8 days

45. *Supply and demand* A company has 10 units of a certain item and can manufacture 5 items each day; thus the supply is $y = 5x + 10$. If the demand for the item is $y = 7x$, where x is the number of days elapsed, in how many days will the demand equal the supply? 5 days

46. *Supply and demand* Clonker Manufacturing has 12 clonkers in stock. The company manufactures 3 more clonkers each day. If the clonker demand is 7 each day, in how many days will the supply equal the demand? 3 days

SKILL CHECKER

Solve:

47. $-0.5x = -4$ $x = 8$ **48.** $-0.2y = -6$ $y = 30$ **49.** $9x = -9$ $x = -1$ **50.** $11y = -11$ $y = -1$

USING YOUR KNOWLEDGE

The Three R's: Regeneration, Resistors, and Revenue

The ideas presented in this section are important in many fields. Use your knowledge to solve the following problems.

51. The total inductance of the inductors L_1 and L_2 in an oscillator must be 400 microhenrys. Thus

$$L_1 = -L_2 + 400$$

To provide the correct regeneration for the oscillator circuit requires

$$\frac{L_2}{L_1} = 4, \qquad \text{that is,} \qquad L_2 = 4L_1$$

Solve the system

$$L_1 = -L_2 + 400$$
$$L_2 = 4L_1$$

by substitution. $L_1 = 80$; $L_2 = 320$

52. The equations for the resistors in a voltage divider must be such that

$$R_1 = 3R_2$$
$$R_1 + R_2 = 400$$

Solve for R_1 and R_2 using the substitution method. $R_1 = 300$; $R_2 = 100$

53. The total revenue R for a certain manufacturer is

$$R = 5x$$

the total cost C is

$$C = 4x + 500$$

where x represents the number of units produced and sold.

a. Use your knowledge of the substitution method to write the equation that will result when

$$R = C \quad 5x = 4x + 500$$

b. The point at which $R = C$ is called the *break-even point.* Find the number of units the manufacturer must produce and sell in order to break even. 500 units

WRITE ON

54. If you are solving the system

$$2x + y = -7$$
$$3x - 2y = 7$$

which variable would you solve for in the first step of the five-step procedure given in the text? Answers may vary.

55. When solving a system of equations using the substitution method, how can you tell whether the system is

a. consistent? **b.** inconsistent? **c.** dependent? Answers may vary.

56. In Example 5, we multiplied both sides of the first equation by 4 and both sides of the second equation by 8. Would you get the same answer if you multiplied both sides of the first equation by 8 and both sides of the second equation by 4? Why is it not a good idea to do that? Yes; answers may vary.

MASTERY TEST

If you know how to do these problems, you have learned your lesson!

Solve and label the system as consistent, inconsistent, or dependent. If the system is consistent, write the solution.

57. $x - 3y = 6$
$2x - 6y = 8$
Inconsistent

58. $x - 3y = 6$
$6y - 2x = -12$
Dependent

59. $x + y = 5$
$2x - 3y = -5$
Consistent; (2, 3)

60. $2y + x = 3$
$x - 3y = 0$
Consistent; $(\frac{9}{5}, \frac{3}{5})$

61. $3x - 3y + 1 = 5 + 2x$
$x - 3y = 4$
Dependent

62. $5x + y = 5$
$5x + y - 10 = 5 - 2y$
Consistent; (0, 5)

63. $\frac{x}{2} - \frac{y}{4} = -1$
$2x = 2 + y$
Inconsistent

64. $x + \frac{y}{5} = 1$
$\frac{x}{3} + \frac{y}{5} = 1$
Consistent; (0, 5)

65. A store is selling a Sony® DSS system for \$300. The basic monthly charge is \$50. An RCA® system is selling for \$500 with a \$30 monthly charge. What is the least number of months for which the prices of both systems are the same? 10 months

8.3 SOLVING SYSTEMS OF EQUATIONS BY ELIMINATION

To Succeed, Review How To . . .

1. Use the RSTUV method for solving word problems (p. 162).
2. Solve linear equations (pp. 151–155).

Objectives

Use the elimination method to:

A Solve a system of equations in two variables.

B Determine whether a system is consistent, inconsistent, or dependent.

C Solve an application.

GETTING STARTED

Using Elimination When Buying Coffee

We've studied two methods for solving systems of equations. The graphical method gives us a visual model of the system and allows us to find approximate solutions. The substitution method gives exact solutions but is best used when either of the given equations has at least one coefficient of 1 or -1. If graphing or substitution is not desired or feasible, there is another method we can use: the *elimination method,* sometimes called the **addition** or **subtraction method.**

The man in the photo is selling coffee, ground to order. A customer wants 10 pounds of a mixture of coffee A costing \$6 per pound and coffee B costing \$4.50 per pound. If the price for the purchase is \$54, how many pounds of each of the coffees will the customer get?

To solve this problem, we use an idea that we've already learned: solving a system of equations. In this problem we want a precise answer, so we use the elimination method, which we will learn next.

Solving Systems of Equations by Elimination

To solve the coffee problem in the *Getting Started,* we first need to organize the information. The information can be summarized in a table like this:

	Price	Pounds	Total Price
Coffee A	6.00	a	$6a$
Coffee B	4.50	b	$4.50b$
Totals		$a + b$	$6a + 4.50b$

Web It

For a colorful website dealing with solving systems of equations by elimination, try link 8-3-1 on the Bello Website at mhhe.com/bello.

Since the customer bought a total of 10 pounds of coffee, we know that

$$a + b = 10$$

Also, since the purchase came to \$54, we know that

$$6a + 4.50b = 54$$

So, the system of equations we need to solve is

$$\begin{aligned} a + b &= 10 \\ 6a + 4.50b &= 54 \end{aligned}$$

To solve this system, we shall use the **elimination method,** which consists of replacing the given system by equivalent systems until we get a system with an obvious solution. To do this, we first write the equations in the form $Ax + By = C$. Recall that an *equivalent* system is one that has the *same* solution as the given one. For example, the equation

$$A = B$$

is equivalent to the equation

$$kA = kB \quad (k \neq 0)$$

This means that you can multiply both sides of the equation $A = B$ by the same nonzero expression k and obtain the equivalent equation $kA = kB$.

Also, the system

$$\begin{aligned} A &= B \\ C &= D \end{aligned}$$

is equivalent to the system

$$\begin{aligned} A &= B \\ k_1A + k_2C &= k_1B + k_2D \quad (k_1 \text{ and } k_2 \text{ not both } 0) \end{aligned}$$

We can check this by multiplying both sides of $A = B$ by k_1 and both sides of $C = D$ by k_2 and adding.

Now let's return to the coffee problem. We multiply the first equation in the given system by -6; we then get the equivalent system

$$\begin{aligned} -6a - 6b &= -60 \\ (+)\ 6a + 4.50b &= 54 \end{aligned}$$

Multiply by -6 because we want the coefficients of a to be opposites (like -6 and 6).

Add the equations.

$$0 - 1.50b = -6$$

$$-1.50b = -6$$

$$b = 4 \qquad \text{Divide by } -1.50.$$

$$a + 4 = 10 \qquad \text{Substitute 4 for } b \text{ in } a + b = 10.$$

$$a = 6 \qquad \text{Solve for } a.$$

Thus the customer bought 6 pounds of coffee A and 4 pounds of coffee B. This answer can be verified. If the customer bought 6 pounds of coffee A and 4 of B, she did indeed buy 10 pounds. Her price for coffee A was $6 \cdot 6 = \$36$ and for coffee B, $4 \cdot 4.50 = \$18$. Thus the entire cost was $\$36 + \$18 = \$54$, as stated.

What have we done here? Well, this technique depends on the fact that one (or both) of the equations in a system can be multiplied by a nonzero number to obtain two equivalent equations with opposite coefficients of x (or y). Here is the idea.

ELIMINATION METHOD

One or both of the equations in a system of simultaneous equations can be multiplied (or divided) by any nonzero number to obtain an *equivalent* system in which the coefficients of the x's (or of the y's) are opposites, thus *eliminating* x or y when the equations are added.

EXAMPLE 1 Solving a consistent system by elimination

Solve the system:

$$2x + y = 1$$
$$3x - 2y = -9$$

SOLUTION Remember the idea: we multiply one or both of the equations by a number or numbers that will cause either the coefficients of x or the coefficients of y to be opposites. We can do this by multiplying the first equation by 2:

$$2x + y = 1 \xrightarrow{\text{Multiply by 2.}} 4x + 2y = 2$$
$$3x - 2y = -9 \xrightarrow{\text{Leave as is.}} \underline{3x - 2y = -9}$$

Add the equations. $7x + 0 = -7$

$7x = -7$

Divide by 7. $x = -1$

Substitute -1 for x in $2x + y = 1$. $2(-1) + y = 1$

$-2 + y = 1$

Add 2. $y = 3$

Thus the solution of the system is $(-1, 3)$.

CHECK When $x = -1$ and $y = 3$, $2x + y = 1$ becomes

$$2(-1) + 3 = 1$$
$$-2 + 3 = 1$$
$$1 = 1$$

a true statement, and $3x - 2y = -9$ becomes

$$3(-1) - 2(3) = -9$$
$$-3 - 6 = -9$$
$$-9 = -9$$

which is also true.

PROBLEM 1

Solve the system:

$$3x + y = 1$$
$$3x - 4y = 11$$

Teaching Tip

Have students graph Examples 1, 2, and 3 before using the elimination method. Then, have students discuss the solution sets and how to interpret the solution when both variables cancel out.

Ex 2: $0 \neq -12$ false, no solution, parallel lines

Ex 3: $0 = 0$ true, infinitely many solutions, same line

Answer

1. $(1, -2)$

Determining Whether a System Is Consistent, Inconsistent, or Dependent

As we have seen, not all systems have solutions. How do we find out if they don't? Let's look at the next example, which shows a contradiction for a system that has no solution.

EXAMPLE 2 Solving an inconsistent system by elimination

Solve the system:

$$2x + 3y = 3$$
$$4x + 6y = -6$$

SOLUTION In this case, we try to eliminate the variable x by multiplying the first equation by -2.

$$\begin{array}{lcr} 2x + 3y = 3 & \xrightarrow{\text{Multiply by } -2.} & -4x - 6y = -6 \\ 4x + 6y = -6 & \xrightarrow{\text{Leave as is.}} & \underline{4x + 6y = -6} \\ & & 0 + 0 = -12 \quad \text{Add.} \\ & & 0 = -12 \end{array}$$

Of course, this is a contradiction, so there is no solution; the system is *inconsistent.*

PROBLEM 2

Solve:

$$3x + 2y = 1$$
$$6x + 4y = 12$$

EXAMPLE 3 Solving a dependent system by elimination

Solve the system:

$$2x - 4y = 6$$
$$-x + 2y = -3$$

SOLUTION Here we try to eliminate the variable x by multiplying the second equation by 2. We obtain

$$\begin{array}{lcr} 2x - 4y = 6 & \xrightarrow{\text{Leave as is.}} & 2x - 4y = 6 \\ -x + 2y = -3 & \xrightarrow{\text{Multiply by 2.}} & \underline{-2x + 4y = -6} \\ & & 0 + 0 = 0 \quad \text{Add.} \\ & & 0 = 0 \end{array}$$

Lo and behold, we've eliminated both variables! However, notice that if we had multiplied the second equation in the original system by -2, we would have obtained

$$\begin{array}{lcr} 2x - 4y = 6 & \xrightarrow{\text{Leave as is.}} & 2x - 4y = 6 \\ -x + 2y = -3 & \xrightarrow{\text{Multiply by } -2.} & 2x - 4y = 6 \end{array}$$

This means that the first equation is a constant multiple of the second one; that is, they are equivalent equations. When a system of equations consists of

PROBLEM 3

Solve:

$$3x - 6y = 9$$
$$-x + 2y = -3$$

Answers

2. No solution; inconsistent
3. Infinitely many solutions; dependent

two equivalent equations, the system is said to be *dependent,* and any solution of one equation is a solution of the other. Because of this, the system has *infinitely many* solutions. For example, if we let x be 0 in the first equation, then $y = -\frac{3}{2}$, and $(0, -\frac{3}{2})$ is a solution of the system. Similarly, if we let y be 0 in the first equation, then $x = 3$, and we obtain the solution (3, 0). Many other solutions are possible; try to find some of them.

Finally, in some cases, we cannot multiply just one of the equations by an integer that will cause the coefficients of one of the variables to be opposites. For example, to solve the system

$$2x + 3y = 3$$
$$5x + 2y = 13$$

we must multiply both equations by integers chosen so that the coefficients of one of the variables will be opposites. We can do this using either of the following methods.

METHOD 1

Solving by elimination: x or y?

To eliminate x, multiply the first equation by 5 and the second one by -2 to obtain an equivalent system.

$2x + 3y = 3$	Multiply by 5. →	$10x + 15y = 15$
$5x + 2y = 13$	Multiply by -2. →	$-10x - 4y = -26$
Add.		$0 + 11y = -11$
		$11y = -11$
Divide by 11.		$y = -1$
Substitute -1 for y in $2x + 3y = 3$.		$2x + 3(-1) = 3$
Simplify.		$2x - 3 = 3$
Add 3.		$2x = 6$
Divide by 2.		$x = 3$

Thus the solution of the system is $(3, -1)$. This time we eliminated the x and solved for y. Alternatively, we can eliminate the y first, as shown next.

METHOD 2

This time, we eliminate the y:

$2x + 3y = 3$	Multiply by -2. →	$-4x - 6y = -6$
$5x + 2y = 13$	Multiply by 3. →	$15x + 6y = 39$
Add.		$11x + 0 = 33$
		$11x = 33$
Divide by 11.		$x = 3$
Substitute 3 for x in $2x + 3y = 3$.		$2(3) + 3y = 3$
Simplify.		$6 + 3y = 3$
Subtract 6.		$3y = -3$
Divide by 3.		$y = -1$

Thus the solution is $(3, -1)$, as before.

Before we continue, here are two important reminders.

NOTE

1. There are *three* possibilities when solving simultaneous linear equations.
   A. *Consistent* and *independent* equations have *one* solution.
   B. *Inconsistent* equations have *no* solution. You can recognize them when you get a contradiction (a false statement) in your work, as we did in Example 2. (In Example 2, we got $0 = -12$, a contradiction.)
   C. *Dependent* equations have *infinitely many* solutions. You can recognize them when you get a true statement such as $0 = 0$ in Example 3. Remember that any solution of one of these equations is a solution of the other.
2. Look at the positions of the variables in the equations. All the equations with which we have worked except those we solved by substitution were written in the form

$$\left.\begin{aligned} ax + by &= c \\ dx + ey &= f \end{aligned}\right\} \text{This is the standard form.}$$

x column, y column, Constant terms

If the equations are not in this form, and you are not using the substitution method, rewrite them using this form. It helps to keep things straight!

EXAMPLE 4 Writing in standard form and solving by elimination

Solve the system:

$$5y + 2x = 9$$
$$2y = 8 - 3x$$

SOLUTION We first write the system in standard form—that is, the x's first, then the y's, and then the constants. The result is the equivalent system

$$2x + 5y = 9$$
$$3x + 2y = 8$$

Now we multiply the first equation by 3 and the second one by -2 so that, upon addition, the x's will be eliminated.

$$2x + 5y = 9 \xrightarrow{\text{Multiply by 3.}} 6x + 15y = 27$$
$$3x + 2y = 8 \xrightarrow{\text{Multiply by } -2.} -6x - 4y = -16$$

Add.	$0 + 11y = 11$
	$11y = 11$
Divide by 11.	$y = 1$
Substitute 1 for y in $2x + 5y = 9$.	$2x + 5(1) = 9$
Simplify.	$2x + 5 = 9$
Subtract 5.	$2x = 4$
Divide by 2.	$x = 2$

Thus the solution is (2, 1). You should verify this result to make sure it satisfies both equations.

PROBLEM 4

Solve:

$$5x + 4y = 6$$
$$3y = 4x - 11$$

Teaching Tip

If either of the equations has a common factor, divide each side by that number.

Example:

$$2x + 6y = 12$$

common factor 2

$$\frac{2x}{2} + \frac{6y}{2} = \frac{12}{2}$$
$$x + 3y = 6$$

Answer

4. $(2, -1)$

Solving an Application

EXAMPLE 5 Peak, off-peak, and elimination

Do you have a cell phone? The time the phone is used, called airtime, is usually charged by the minute at two different rates: peak and off-peak. Suppose your plan charges $0.60 for each peak-time minute and $0.45 for each off-peak minute. If your airtime cost $54 and you've used 100 minutes of airtime, how many minutes p of peak time and how many minutes n of off-peak time did you use?

SOLUTION To solve this problem, we need two equations involving the two unknowns, p and n. We know that the airtime amounts to $54 and that 100 minutes of airtime were used. How can we accumulate $54 of airtime?

Since peak time costs $0.60 per minute, peak times cost $0.60p$.

Since off-peak time costs $0.45 per minute, off-peak times cost $0.45n$.

The total cost is $54, so we add peak and off-peak costs:

$$0.60p + 0.45n = 54$$

Also, the total number of minutes is 100. Thus $p + n = 100$. To try to *eliminate n*, we multiply both sides of the second equation by -0.45 and then add:

$$0.60p + 0.45n = 54 \quad \xrightarrow{\text{Leave as is.}} \quad 0.60p + 0.45n = 54$$

$$p + n = 100 \quad \xrightarrow{\text{Multiply by } -0.45.} \quad -0.45p - 0.45n = -45$$

$$\text{Add.} \quad 0.15p = 9$$

$$\text{Divide both sides by 0.15.} \quad p = \frac{9}{0.15} = 60$$

Thus $p = 60$ minutes of peak time were used and the rest of the 100 minutes used—that is, $100 - 60 = 40$—were off-peak minutes. This means that $n = 40$. You can check that $p = 60$ and $n = 40$ by substituting in the original equations.

PROBLEM 5

Big Cell Phone has a plan that charges $0.50 for each peak-time minute and $0.40 for each off-peak minute. If airtime cost $44 and 100 minutes were used, how many minutes p of peak time and how many minutes n of off-peak time were used?

Teaching Tip

Use the following box to help organize Example 5.

	How Many Minutes	Cost Per Minute	Cost
Peak	p	$0.60	$0.60p$
Off-peak	n	$0.45	$0.45n$
Totals	100		$54

Add down the first column to get one equation and add down the last column to get the other equation.

1st column: $p + n = 100$

last column: $0.60p + 0.45n = 54$

Answer

5. 40 minutes of peak and 60 minutes of off-peak

Exercises 8.3

Boost *your* GRADE at mathzone.com!

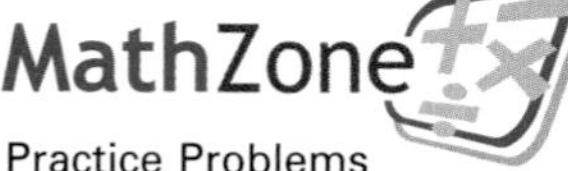

- Practice Problems
- Self-Tests
- Videos
- NetTutor
- e-Professors

A B In Problems 1–30, use the elimination method to solve each system. If the system is not consistent, state whether it is inconsistent or dependent.

1. $x + y = 3$
$x - y = -1$
(1, 2)

2. $x + y = 5$
$x - y = 1$
(3, 2)

3. $x + 3y = 6$
$x - 3y = -6$
(0, 2)

4. $x + 2y = 4$
$x - 2y = 8$
(6, −1)

5. $2x + y = 4$
$4x + 2y = 0$
Inconsistent

6. $3x + 5y = 2$
$6x + 10y = 5$
Inconsistent

7. $2x + 3y = 6$
$4x + 6y = 2$
Inconsistent

8. $3x - 5y = 4$
$-6x + 10y = 0$
Inconsistent

9. $x - 5y = 15$
$x + 5y = 5$
$(10, -1)$

10. $-3x + 2y = 1$
$2x + y = 4$
$(1, 2)$

11. $x + 2y = 2$
$2x + 3y = -10$
$(-26, 14)$

12. $3x - 2y = -1$
$x + 7y = -8$
$(-1, -1)$

13. $3x - 4y = 10$
$5x + 2y = 34$
$(6, 2)$

14. $5x - 4y = 6$
$3x + 2y = 8$
$(2, 1)$

15. $11x - 3y = 25$
$5x + 8y = 2$
$(2, -1)$

16. $12x + 8y = 8$
$7x - 5y = 24$
$(2, -2)$

17. $2x + 3y = 21$
$3x = y + 4$
$(3, 5)$

18. $2x - 3y = 16$
$x = y + 7$
$(5, -2)$

19. $x = 1 + 2y$
$-y = x + 5$
$(-3, -2)$

20. $3y = 1 - 2x$
$3x = -4y - 1$
$(-7, 5)$

21. $\frac{x}{4} + \frac{y}{3} = 4$
$\frac{x}{2} - \frac{y}{6} = 3$ $(8, 6)$
(*Hint:* Multiply by the LCD first.)

22. $\frac{x}{5} + \frac{y}{6} = 5$
$\frac{2x}{5} + \frac{y}{3} = -2$ Inconsistent
(*Hint:* Multiply by the LCD first.)

23. $\frac{1}{4}x - \frac{1}{3}y = -\frac{5}{12}$
$\frac{1}{5}x + \frac{2}{5}y = 1$ $(1, 2)$
(*Hint:* Multiply by the LCD first.)

24. $\frac{x}{2} + \frac{y}{2} = \frac{5}{2}$
$\frac{x}{2} - \frac{y}{3} = \frac{5}{2}$ $(5, 0)$
(*Hint:* Multiply by the LCD first.)

25. $\frac{x}{8} + \frac{y}{8} = 1$
$\frac{x}{2} - \frac{y}{2} = 1$ $(5, 3)$

26. $\frac{x}{5} + \frac{y}{5} = 1$
$\frac{x}{4} - \frac{y}{4} = \frac{1}{4}$ $(3, 2)$

27. $\frac{x}{2} - \frac{y}{3} = 1$
$\frac{x}{2} + \frac{y}{2} = \frac{7}{2}$
$(4, 3)$

28. $\frac{x}{3} + \frac{y}{2} = \frac{7}{3}$
$\frac{x}{3} - \frac{y}{2} = 0$
$\left(\frac{7}{2}, \frac{7}{3}\right)$

29. $\frac{2x}{9} - \frac{y}{2} = -1$
$x - \frac{9y}{4} = -\frac{9}{2}$
Dependent

30. $-\frac{8x}{49} + \frac{5y}{49} = -1$
$\frac{2x}{3} - \frac{5y}{12} = \frac{49}{12}$
Dependent

APPLICATIONS

31. *Coffee blends* The Holiday House blends Costa Rican coffee that sells for \$8 a pound and Indian Mysore coffee that sells for \$9 a pound to make 1-pound bags of its Gourmet Blend coffee, which sells for \$8.20 a pound. How much Costa Rican and how much Indian coffee should go into each pound of the Gourmet Blend?
0.8 lb Costa Rican; 0.2 lb Indian Mysore

32. *Coffee blends* The Holiday House also blends Colombian Swiss Decaffeinated coffee that sells for \$11 a pound and High Mountain coffee that sells for \$9 a pound to make its Lower Caffeine coffee, 1-pound bags of which sell for \$10 a pound. How much Colombian and how much High Mountain should go into each pound of the Lower Caffeine mixture? $\frac{1}{2}$ lb of each

33. *Tea blends* Oolong tea that sells for \$19 per pound is blended with regular tea that sells for \$4 per pound to produce 50 pounds of tea that sells for \$7 per pound. How much oolong and how much regular should go into the mixture? 10 lb oolong; 40 lb regular tea

34. *Metal alloys* If the price of copper is 65¢ per pound and the price of zinc is 30¢ per pound, how many pounds of copper and zinc should be mixed to make 70 pounds of brass, which sells for 45¢ per pound?
30 lb copper; 40 lb zinc

SKILL CHECKER

Try the Skill Checker Exercises so you'll be ready for the next section.

Write an expression corresponding to the given sentence:

35. The sum of the numbers of nickels (n) and dimes (d) equals 300. $n + d = 300$

36. The difference of h and w is 922. $h - w = 922$

37. The product of 4 and $(x - y)$ is 48. $4(x - y) = 48$

38. The quotient of x and y is 80. $\frac{x}{y} = 80$

39. The number m is 3 less than the number n. $m = n - 3$

40. The number m is 5 more than the number n. $m = n + 5$

USING YOUR KNOWLEDGE

Tweedledee and Tweedledum

Have you ever read *Alice in Wonderland?* Do you know who the author is? It's Lewis Carroll, of course. Although better known as the author of *Alice in Wonderland,* Lewis Carroll was also an accomplished mathematician and logician. Certain parts of his second book, *Through the Looking Glass,* reflect his interest in mathematics. In this book, one of the characters, Tweedledee, is talking to Tweedledum. Here is the conversation.

Tweedledee: The sum of your weight and twice mine is 361 pounds.

Tweedledum: Contrariwise, the sum of your weight and twice mine is 360 pounds.

41. If Tweedledee weighs x pounds and Tweedledum weighs y pounds, find their weights using the ideas of this section. Tweedledee: $120\frac{2}{3}$ lb; Tweedledum: $119\frac{2}{3}$ lb

WRITE ON

42. When solving a system of equations by elimination, how would you recognize if the pair of equations is

a. consistent? Answers may vary.
b. inconsistent? Answers may vary.
c. dependent? Answers may vary.

43. Explain why the system

$$2x + 5y = 9$$
$$3x + 2y = 8$$

is easier to solve by elimination rather than substitution. Answers may vary.

44. Write the procedure you use to solve a system of equations by the elimination method. Answers may vary.

MASTERY TEST

If you know how to do these problems, you have learned your lesson!

Solve the system; if the system is not consistent, state whether it is inconsistent or dependent.

45. $2x + 5y = 9$
$4x - 3y = 11$ $\left(\frac{41}{13}, \frac{7}{13}\right)$

46. $5x - 4y = 7$
$4x + 2y = 16$ (3, 2)

47. $2x - 5y = 5$
$2x - y = 4 + x$ (5, 1)

48. $3x - 2y = 6$
$-6x = -4y - 12$ Dependent

49. $\frac{x}{6} - \frac{y}{2} = 1$
$-\frac{x}{4} + \frac{3y}{4} = -\frac{3}{4}$ Inconsistent

50. A 10-pound bag of coffee sells for \$114 and contains a mixture of coffee A, which costs \$12 a pound, and coffee B, which costs \$10 a pound. How many pounds of each of the coffees does the bag contain? 7 lb A; 3 lb B

8.4 COIN, GENERAL, MOTION, AND INVESTMENT PROBLEMS

To Succeed, Review How To . . .

1. Use the RSTUV method to solve word problems (p. 162).
2. Solve a system of two equations with two unknowns (pp. 605–609).

Objectives

Solve word problems:

A Involving coins.

B Of a general nature.

C Using the distance formula $D = RT$.

D Involving the interest formula $I = PR$.

GETTING STARTED

Money Problems

Patty's upset; she needs *help!* Why? Because *she* hasn't learned about systems of equations, but we have, so we can help her! In the preceding sections we studied systems of equations. We now use that knowledge to solve word problems involving two variables.

PEANUTS reprinted by permission of United Feature Syndicate, Inc.

Before we tackle Patty's problem, let's get down to nickels, dimes, and quarters! Suppose you are down to your last nickel: You have 5¢.

Follow the pattern:

$5 \cdot 1$

If you have 2 nickels, you have $5 \cdot 2 = 10$ cents. $5 \cdot 2$

If you have 3 nickels, you have $5 \cdot 3 = 15$ cents. $5 \cdot 3$

If you have n nickels, you have $5 \cdot n = 5n$ cents. $5 \cdot n$

The same thing can be done with dimes.

Follow the pattern:

If you have 1 dime, you have $10 \cdot 1 = 10$ cents. $10 \cdot 1$

If you have 2 dimes, you have $10 \cdot 2 = 20$ cents. $10 \cdot 2$

If you have n dimes, you have $10 \cdot n = 10n$ cents. $10 \cdot n$

We can construct a table that will help us summarize the information:

	Value (cents)	×	How Many	=	Total Value
Nickels	5		n		$5n$
Dimes	10		d		$10d$
Quarters	25		q		$25q$
Half-dollars	50		h		$50h$

In this section we shall use information like this and systems of equations to solve word problems.

Solving Coin Problems

Now we are ready to help poor Patty! As usual, we use the RSTUV method. If you've forgotten how that goes, this is a good time to review it (see p. 162).

EXAMPLE 1 Patty's coin problem

Read the cartoon in the *Getting Started* again for the details of Patty's problem.

SOLUTION

Teaching Tip

Remind students that money can be expressed in dollars ($) or cents (¢).

Example:

$0.90 = 90¢

In coin problems, it is easier to solve the system when money is in cents (¢) so we don't have to work in decimals.

1. **Read the problem.**
   Patty is asked how many dimes and quarters the man has.
2. **Select the unknowns.**
   Let d be the number of dimes the man has and q the number of quarters.
3. **Think of a plan.**
   We translate each of the sentences in the cartoon:

   a. "A man has 20 coins consisting of dimes and quarters."

   $$20 = d + q$$

   b. The next sentence seems hard to translate. So, let's look at the easy part first—how much money he has now. Since he has d dimes, the table in the *Getting Started* tells us that he has $10d$ (cents). He also has q quarters, which are worth $25q$ (cents). Thus he has

   $$(10d + 25q) \text{ cents}$$

   What would happen if the dimes were quarters and the quarters were dimes? We simply would change the amount the coins are worth, and he would have

   $$(25d + 10q) \text{ cents}$$

   Now let's translate the sentence:

   $$\underbrace{25d + 10q}_{\text{If the dimes were quarters and the quarters were dimes,}} \quad \underbrace{=}_{\text{he would have}} \quad \underbrace{(10d + 25q) + 90}_{\text{90¢ more than he has now.}}$$

If we put the information from parts **a** and **b** together, we have the following system of equations:

$$d + q = 20$$
$$25d + 10q = 10d + 25q + 90$$

Now we need to write this system in standard form—that is, with all the variables and constants in the proper columns. We do this by subtracting $10d$ and $25q$ from both sides of the second equation to obtain

$$d + q = 20$$
$$15d - 15q = 90$$

We then divide each term in the second equation by 15 to get

$$d + q = 20$$
$$d - q = 6$$

4. Use the elimination method to solve the problem.
To eliminate q, we simply add the two equations:

$$\begin{aligned} d + q &= 20 \\ d - q &= 6 \\ \hline 2d \quad &= 26 \quad \text{Add.} \\ \text{Number of dimes:} \quad d &= 13 \quad \text{Divide by 2.} \\ 13 + q &= 20 \quad \text{Substitute 13 for } d \text{ in } d + q = 20. \\ \text{Number of quarters:} \quad q &= 7 \end{aligned}$$

Thus the man has 13 dimes ($1.30) and 7 quarters ($1.75), a total of $3.05.

5. Verify the solution.
If the dimes were quarters and the quarters were dimes, the man would have 13 quarters ($3.25) and 7 dimes ($0.70), a total of $3.95, which is indeed $0.90 more than the $3.05 he now has. Patty, you got your help!

Let's solve another coin problem.

Web It

For some more coin problems using two variables for their solution, go to link 8-4-1 on the Bello Website at mhhe.com/bello.

EXAMPLE 2 **Jack's coin problem**

Jack has $3 in nickels and dimes. He has twice as many nickels as he has dimes. How many nickels and how many dimes does he have?

SOLUTION As usual, we use the RSTUV method.

1. Read the problem.
We are asked to find the numbers of nickels and dimes.

2. Select the unknowns.
Let n be the number of nickels and d the number of dimes.

3. Think of a plan.
If we translate the problem and use the table in the *Getting Started,* Jack has $3 (300 cents) in nickels and dimes:

$$300 = 5n + 10d$$

He has twice as many nickels as he has dimes:

$$n = 2d$$

We then have the system

$$\begin{aligned} 5n + 10d &= 300 \\ n &= 2d \end{aligned}$$

4. Use the substitution method to solve the problem.
This time it's easy to use the substitution method.

$$5n + 10d = 300 \xrightarrow{\text{Letting } n = 2d.} 5(2d) + 10d = 300$$

$$\begin{aligned} \text{Simplify.} \qquad 10d + 10d &= 300 \\ \text{Combine like terms.} \qquad 20d &= 300 \\ \text{Divide by 20.} \qquad d &= 15 \\ \text{Substitute 15 for } d \text{ in } n = 2d. \qquad n = 2(15) &= 30 \end{aligned}$$

Thus Jack has 15 dimes ($1.50) and 30 nickels ($1.50).

5. Verify the solution.
Since Jack has $3 ($1.50 + $1.50) and he does have twice as many nickels as dimes, the answer is correct.

PROBLEM 2

Jill has $1.50 in nickels and dimes. She has twice as many dimes as she has nickels. How many nickels and how many dimes does she have?

Answer

2. 6 nickels, 12 dimes

Solving a General Problem

We can use systems of equations to solve many problems. Here is an interesting one.

EXAMPLE 3 **A heavy marriage**

The greatest weight difference recorded for a married couple is 922 pounds (Mills Darden of North Carolina and his wife Mary). Their combined weight is 1118 pounds. What is the weight of each of the Dardens? (He is the heavy one.)

SOLUTION

1. **Read the problem.**
   We are asked to find the weight of each of the Dardens.
2. **Select the unknowns.**
   Let h be the weight of Mills and w be the weight of Mary.
3. **Think of a plan.**
   We translate the problem. The weight difference is 922 pounds:
   $$h - w = 922$$
   Their combined weight is 1118 pounds:
   $$h + w = 1118$$
   We then have the system
   $$h - w = 922$$
   $$h + w = 1118$$
4. **Use the elimination method to solve the problem.**
   Using the elimination method, we have
   $$h - w = 922$$
   $$\underline{h + w = 1118}$$
   $$2h = 2040 \quad \text{Add.}$$
   $$h = 1020 \quad \text{Divide by 2.}$$
   $$1020 + w = 1118 \quad \text{Substitute 1020 for } h \text{ in } h + w = 1118.$$
   $$w = 98 \quad \text{Subtract 1020.}$$
   Thus Mary weighs 98 pounds and Mills weighs 1020 pounds.
5. **Verify the solution.**
   You can verify this in the *Guinness Book of Records.*

PROBLEM 3

When a couple of astronauts stood on a scale together before a mission, their combined weight was 320 pounds. The difference in their weights was 60 pounds, and the woman was lighter than the man. What was the weight of each?

C Solving a Motion Problem

Remember the motion problems we solved in Section 2.5? They can also be done using two variables. The procedure is about the same. We write the given information in a chart labeled $R \times T = D$ and then use our RSTUV method, as demonstrated in Example 4.

Answer

3. Man: 190 pounds; woman: 130 pounds

EXAMPLE 4 **Current boating**

The world's strongest current is the Saltstraumen in Norway. The current is so strong that a boat that travels 48 miles downstream (with the current) in 1 hour takes 4 hours to go the same 48 miles upstream (against the current). How fast is the current flowing?

SOLUTION

1. **Read the problem.**
   We are asked to find the speed of the current. Note that the speed of the boat downstream has two components: the boat speed and the current speed.

2. **Select the unknowns.**
   Let x be the speed of the boat in still water and y be the speed of the current. Then $(x + y)$ is the speed of the boat going downstream; $(x - y)$ is the speed of the boat going upstream.

3. **Think of a plan.**
   We enter this information in a chart:

	R	$\times$	T	$=$	D	
Downstream:	$x + y$		1		48	$\longrightarrow$ $x + y = 48$
Upstream:	$x - y$		4		48	$\longrightarrow$ $4(x - y) = 48$

4. **Use the elimination method to solve the problem.**
   Our system of equations can be simplified as follows:

$$x + y = 48 \xrightarrow{\text{Leave as is.}} x + y = 48$$

$$4(x - y) = 48 \xrightarrow{\text{Divide by 4.}} x - y = 12$$

Add.	$2x = 60$
Divide by 2.	$x = 30$
Substitute 30 for x in $x + y = 48$.	$30 + y = 48$
Subtract 30.	$y = 18$

   Thus the speed of the boat in still water is $x = 30$ miles per hour, and the speed of the current is 18 miles per hour.

5. **Verify the solution.**
   We leave the verification to you.

PROBLEM 4

A plane travels 800 miles against a storm in 4 hours. Giving up, the pilot turns around and flies back 800 miles to the airport in only 2 hours with the aid of the tailwind. Find the speed of the wind and the speed of the plane in still air.

Web It

To practice with motion problems using two variables, try link 8-4-2 on the Bello Website at mhhe.com/bello.

D Solving an Investment Problem

The investment problems we solved in Section 2.5 can also be worked using two variables. These problems use the formula $I = PR$ to find the *annual* interest I on a principal P at a rate R. The procedure is similar to that used to solve distance problems and uses the same strategy: use a table to enter the information, obtain a system of two equations and two unknowns, and solve the system. We show this strategy next.

Answer

4. Plane speed: 300 mi/hr; wind speed: 100 mi/hr

EXAMPLE 5 Clayton's credit card problem

Clayton owes a total of $10,900 on two credit cards with annual interest rates of 12% and 18%. If he pays a total of $1590 in interest for the year, how much does he owe on each card?

12%

THRIFT E BANK
4221 7032 7124 3004
CLAYTON D SIZTION
PASS PORT

18%

GLOBE BANK
4221 0732 2174 0377
CLAYTON D SIZTION
PASS PORT

SOLUTION

1. **Read the problem.**
   We are asked to find the amount owed on *each* card.
2. **Select the unknowns.**
   Let x be the amount Clayton owes on the first card and y be the amount he owes on the second card.
3. **Think of a plan.**
   We make a table similar to the one in Example 4 but using the heading $P \times R = I$.

	$P\ \times$	$R\ =$	I
Card A	x	0.12	$0.12x$
Card B	y	0.18	$0.18y$

Since the total amount owed is $10,900: $x + y = 10{,}900$

Since the total interest paid is $1590: $0.12x + 0.18y = 1590$

Thus we have to solve the system

$$x + y = 10{,}900$$
$$0.12x + 0.18y = 1590$$

4. **Use the substitution method to solve the problem.**
   We solve for x in the first equation to obtain $x = 10{,}900 - y$. Now we substitute $x = 10{,}900 - y$ in

$$0.12x + 0.18y = 1590$$
$$0.12(10{,}900 - y) + 0.18y = 1590$$
$$1308 - 0.12y + 0.18y = 1590 \quad \text{Simplify.}$$
$$0.06y = 282 \quad \text{Subtract 1308 and combine } y\text{'s.}$$
$$y = 4700 \quad \text{Divide by 0.06.}$$

Since

$$x = 10{,}900 - y$$
$$x = 10{,}900 - 4700$$
$$x = 6200$$

Thus Clayton owes $6200 on card A and $4700 on card B.

5. **Verify the solution.**
   Since $0.12 \cdot \$6200 + 0.18 \cdot \$4700 = \$744 + \$846 = \$1590$ (the amount Clayton paid in interest), the amounts of $6200 and $4700 are correct.

PROBLEM 5

Dorothy owes $11,000 on two credit cards with annual interest rates of 9% and 12%. If she pays $1140 in interest for the year, how much does she owe on each card?

Teaching Tip

Suggest to students that they can multiply the second equation by 100 and rewrite it without decimals.

Example:

$$0.12x + 0.18y = 1590$$

(Multiplying by 100 moves decimal two places right.)

$$0.12x + 0.18y = 1590.00$$
$$12x + 18y = 159{,}000$$

Web It

For investment problems using two variables, go to link 8-4-3 on the Bello Website at mhhe.com/bello.

Answer

5. $6000 and $5000, respectively

Exercises 8.4

A B In Problems 1–6, solve the money problems.

1. Mida has \$2.25 in nickels and dimes. She has four times as many dimes as nickels. How many dimes and how many nickels does she have? 5 nickels; 20 dimes

2. Dora has \$5.50 in nickels and quarters. She has twice as many quarters as she has nickels. How many of each coin does she have? 10 nickels; 20 quarters

3. Mongo has 20 coins consisting of nickels and dimes. If the nickels were dimes and the dimes were nickels, he would have 50¢ more than he now has. How many nickels and how many dimes does he have? 15 nickels; 5 dimes

4. Desi has 10 coins consisting of pennies and nickels. Strangely enough, if the nickels were pennies and the pennies were nickels, she would have the same amount of money as she now has. How many pennies and nickels does she have? 5 pennies; 5 nickels

5. Don had \$26 in his pocket. If he had only \$1 bills and \$5 bills, and he had a total of 10 bills, how many of each of the bills did he have? 4 fives; 6 ones

6. A person went to the bank to deposit \$300. The money was in \$10 and \$20 bills, 25 bills in all. How many of each did the person have? 5 twenties; 20 tens

In Problems 7–14, find the solution.

7. The sum of two numbers is 102. Their difference is 16. What are the numbers? 59; 43

8. The difference between two numbers is 28. Their sum is 82. What are the numbers? 55; 27

9. The sum of two integers is 126. If one of the integers is 5 times the other, what are the integers? 105; 21

10. The difference between two integers is 245. If one of the integers is 8 times the other, find the integers. 280; 35

11. The difference between two numbers is 16. One of the numbers exceeds the other by 4. What are the numbers? Impossible

12. The sum of two numbers is 116. One of the numbers is 50 less than the other. What are the numbers? 83 and 33

13. Longs Peak is 145 feet higher than Pikes Peak. If you were to put these two peaks on top of each other, you would still be 637 feet short of reaching the elevation of Mount Everest, 29,002 feet. Find the elevations of Longs Peak and Pikes Peak. Pikes Peak = 14,110 ft; Longs Peak = 14,255 ft

14. Two brothers had a total of \$7500 in separate bank accounts. One of the brothers complained, and the other brother took \$250 and put it in the complaining brother's account. They now had the same amount of money! How much did each of the brothers have in the bank before the transfer? \$4000; \$3500

C In Problems 15–20, solve the motion problems.

15. A plane flying from city A to city B at 300 miles per hour arrives $\frac{1}{2}$ hour later than scheduled. If the plane had flown at 350 miles per hour, it would have made the scheduled time. How far apart are cities A and B? 1050 miles

16. A plane flies 540 miles with a tailwind in $2\frac{1}{4}$ hours. The plane makes the return trip against the same wind and takes 3 hours. Find the speed of the plane in still air and the speed of the wind. Plane speed = 210 mi/hr; wind speed = 30 mi/hr

17. A motorboat runs 45 miles downstream in $2\frac{1}{2}$ hours and 39 miles upstream in $3\frac{1}{4}$ hours. Find the speed of the boat in still water and the speed of the current. Boat speed = 15 mi/hr; current speed = 3 mi/hr

18. A small plane travels 520 miles with the wind in 3 hours, 20 minutes ($3\frac{1}{3}$ hours), the same time that it takes to travel 460 miles against the wind. What is the plane's speed in still air? 147 mi/hr

19. If Bill drives from his home to his office at 40 miles per hour, he arrives 5 minutes early. If he drives at 30 miles per hour, he arrives 5 minutes late. How far is it from his home to his office? 20 miles

20. An unidentified plane approaching the U.S. coast is sighted on radar and determined to be 380 miles away and heading straight toward the coast at 600 miles per hour. Five minutes ($\frac{1}{12}$ hour) later, a U.S. jet, flying at 720 miles per hour, scrambles from the coastline to meet the plane. How far from the coast does the interceptor meet the plane? 180 miles

D In Problems 21–23, solve the investment problems.

21. Fred invested \$20,000, part at 6% and the rest at 8%. Find the amount invested at each rate if the annual income from the two investments is \$1500.
\$5000 at 6%; \$15,000 at 8%

22. Maria invested \$25,000, part at 7.5% and the rest at 6%. If the annual interest from the two investments amounted to \$1620, how much money was invested at each rate? \$8000 at 7.5%; \$17,000 at 6%

23. Dominic has a savings account that pays 5% annual interest and some certificates of deposit that pay 7% annually. His total interest from the two investments is \$1100 and the total amount invested is \$18,000. How much money does he have in the savings account?
\$8000

APPLICATIONS

24. *Hurricane damages* The two costliest hurricanes in U.S. history were Andrew (1992) and Hugo (1989), in that order; together they caused \$27 billion in damages. If the difference in damages caused by Andrew and Hugo was \$13 billion, how much damage did each of them cause? Andrew: \$20 billion; Hugo: \$7 billion

Source: University of Colorado Natural Hazards Center.

25. *Higher education enrollments* In a recent year, the total enrollment in public and private institutions of higher education was 15 million students. If there were 8.4 million more students enrolled in public institutions than in private, how many students were enrolled in public and how many were enrolled in private institutions? Public: 11.7 million; private: 3.3 million

Source: U.S. Department of Education.

26. *Education expenditures* In a recent year, the education expenditures for public and private institutions amounted to \$187 billion. If \$49 billion more was spent in public institutions than in private, what were the education expenditures for public and for private institutions? Public: \$118 billion; private: \$69 billion

Source: U.S. Department of Education.

27. *Premium cable services* The total number of subscribers for Home Box Office and Showtime in a recent year was 28,700,000. If Home Box Office had 7300 more subscribers than Showtime, how many subscribers did each of the services have?
HBO: 14,353,650; Showtime: 14,346,350

Source: National Cable Television Association.

28. *Automobile and home accidents* In a recent year, the cost of motor vehicle and home accidents reached \$241.7 billion. If motor vehicle accidents caused losses that were \$85.1 billion more than those caused by home accidents, what were the losses in each category?
Motor vehicles: \$163.4 billion; home: \$78.3 billion

Source: National Safety Council Accident Facts.

SKILL CHECKER

Try the Skill Checker Exercises so you'll be ready for the next section.

Graph:

29. $x + 2y < 4$

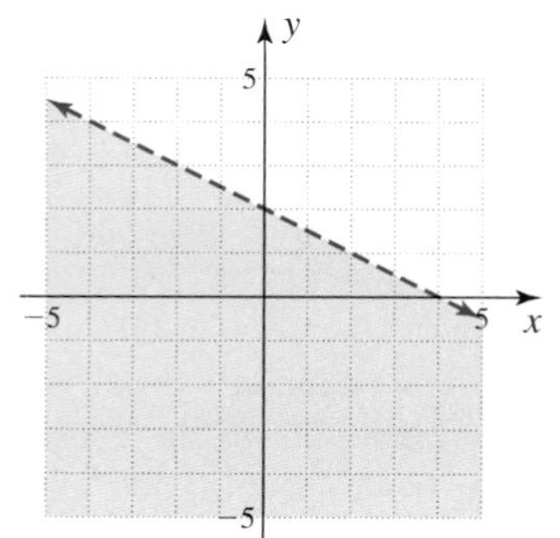

30. $x - 2y > 6$

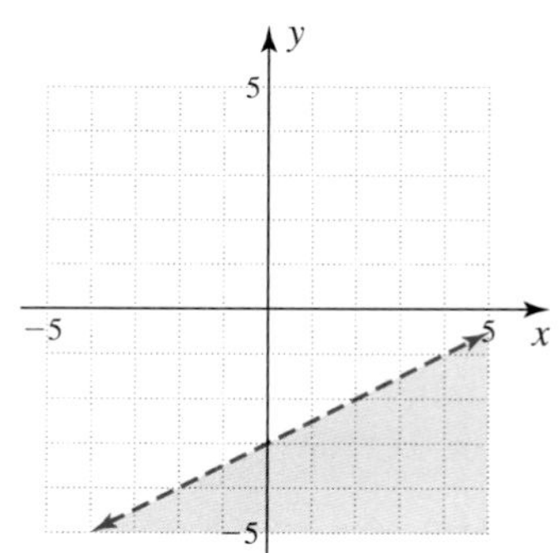

31. $x > y$

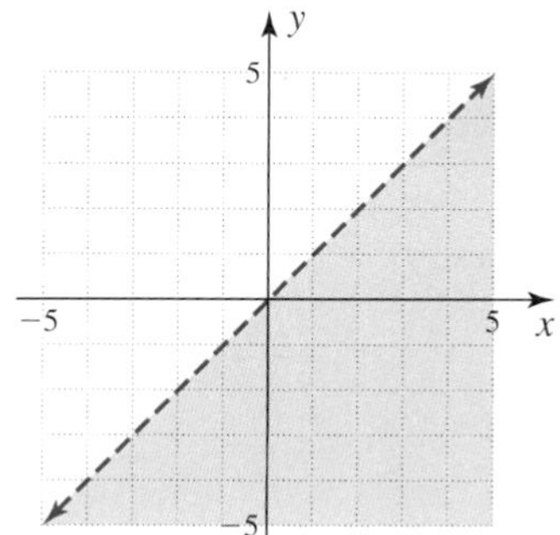

32. $x \leq 2y$

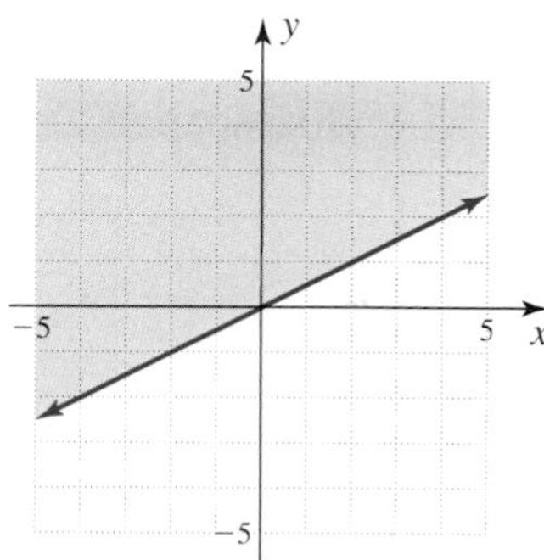

WRITE ON

33. Make up a problem involving coins, and write a solution for it using the RSTUV procedure.
Answers may vary.

34. Make up a problem whose solution involves the distance formula $D = RT$, and write a solution for it using the RSTUV procedure. Answers may vary.

35. Make up a problem whose solution involves the interest formula $I = PR$, and write a solution for it using the RSTUV procedure. Answers may vary.

36. The problems in Examples 1–4 have precisely the information you need to solve them. In real life, however, irrelevant information is often present. This type of information is called a **red herring.** Find some problems with red herrings and point them out.
Answers may vary.

MASTERY TEST

If you know how to do these problems, you have learned your lesson!

37. Jill has \$2 in nickels and dimes. She has twice as many nickels as she has dimes. How many nickels and how many dimes does she have? 20 nickels; 10 dimes

38. At birth, the Stimson twins weighed a total of 35 ounces. If their weight difference was 3 ounces, what was the weight of each of the twins? 16 oz; 19 oz

39. A plane travels 1200 miles with a tailwind in 3 hours. It takes 4 hours to travel the same distance against the wind. Find the speed of the wind and the speed of the plane in still air.
Wind speed = 50 mi/hr; plane speed = 350 mi/hr

40. Harper makes two investments totaling \$10,000. The first investment pays 8% annually, and the second investment pays 5%. If the annual return from both investments is \$600, how much has Harper invested at each rate? \$3333.33 at 8%; \$6666.67 at 5%

41. In the 1992 presidential election, the 84 million people who voted gave the Democratic ticket of Clinton and Gore 6 million (to the nearest million) more votes than the Republican ticket of Bush and Quayle. How many votes (in millions) did each ticket get?
Democrats: 45 million; Republicans: 39 million

8.5 SYSTEMS OF LINEAR INEQUALITIES

To Succeed, Review How To . . .

1. Graph a linear inequality (pp. 199–206).

Objective

A Solve a system of linear inequalities by graphing.

GETTING STARTED

Inequalities and Hospital Stays

According to the American Hospital Association, between 1970 and 1990 (inclusive), the average length of a hospital stay was $y = 8.03 - 0.08x$ (days), where x is the number of years after 1970, as shown in the first graph.

The graph of $y > 8.03 - 0.08x$ represents those longer-than-average stays. If we graph those longer-than-average stays *between* 1970 ($x = 0$) and 1990 ($x = 20$), we get the shaded area *above* the line $y = 8.03 - 0.08x$ and *between* $x = 0$ and $x = 20$. Since the line $y = 8.03 - 0.08x$ is *not* part of the graph, the line itself is shown dashed. The region satisfying the inequalities

$$y > 8.03 - 0.08x, \qquad x > 0, \qquad \text{and} \qquad x < 20$$

is shown shaded in the second graph.

In this section we learn how to solve linear inequalities graphically by finding the set of points that satisfy *all* the inequalities in the system.

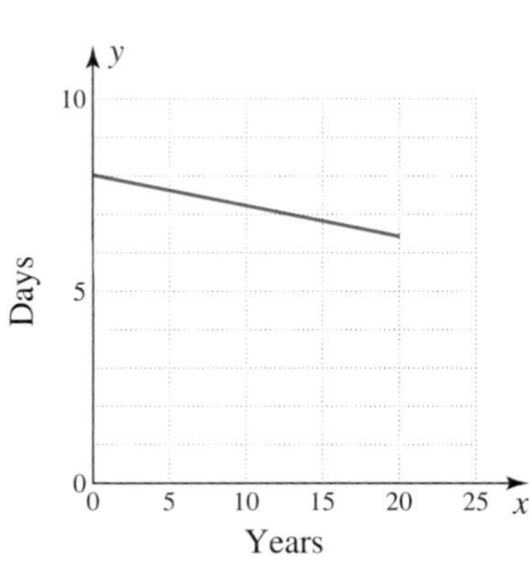

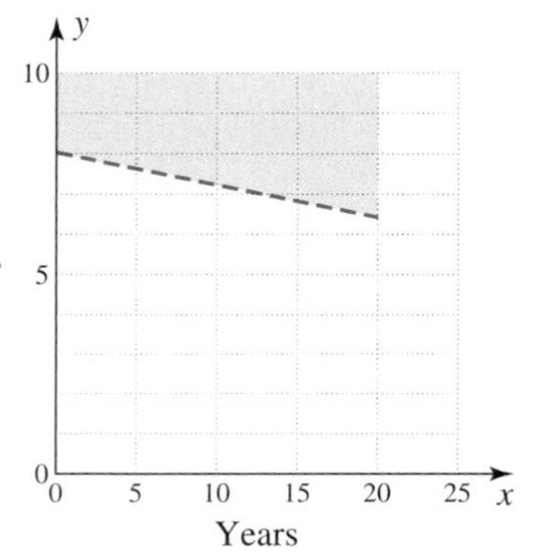

Solving a System of Linear Inequalities by Graphing

It turns out that we can use the procedure we studied in Section 7.3 to solve a system of linear inequalities.

Teaching Tip

The solution set is the area on the graph where the shadings overlap. This part is the intersection of the two inequalities and should be shaded heavier than the rest.

Web It

Try this procedure and see a lesson on how to graph a system of inequalities at link 8-5-1 on the Bello Website at mhhe.com/bello.

PROCEDURE

Solving a System of Inequalities

Graph each inequality on the *same* set of axes using the following steps:

1. Graph the line that is the boundary of the region. If the inequality involves $\leq$ or $\geq$, draw a **solid** line; if it involves $<$ or $>$, draw a **dashed** line.
2. Use any point (a, b) *not* on the line as a test point. Substitute the values of a and b for x and y in the inequality. If a *true* statement results, shade the side of the line containing the test point. If a *false* statement results, shade the other side.

The **solution set** is the set of points that satisfies *all* the inequalities in the system.

EXAMPLE 1 **Solving systems of inequalities involving horizontal and vertical lines by graphing**

Graph the solution of the system: $x \leq 0$ and $y \geq 2$

SOLUTION Since $x = 0$ is a vertical line corresponding to the y-axis, $x \leq 0$ consists of the graph of the line $x = 0$ and all points to the *left,* as shown in Figure 12. The condition $y \geq 2$ defines all points on the line $y = 2$ and *above,* as shown in Figure 13.

The solution set is the set satisfying *both* conditions, that is, where $x \leq 0$ *and* $y \geq 2$. This set is the darker area in Figure 14.

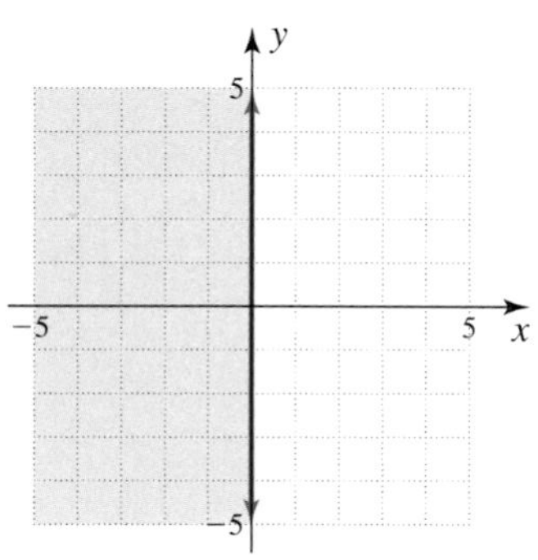

Figure 12

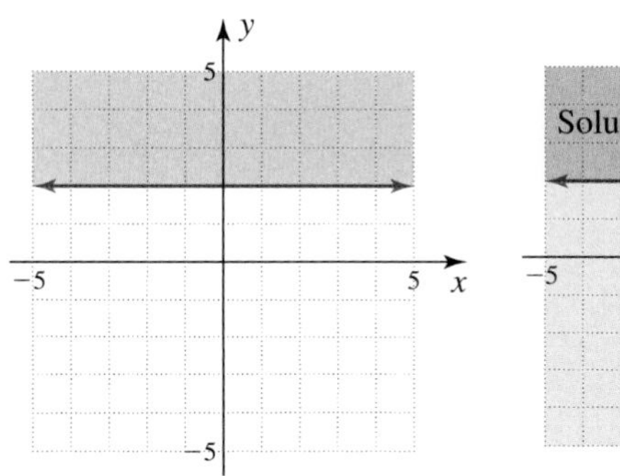

Figure 13

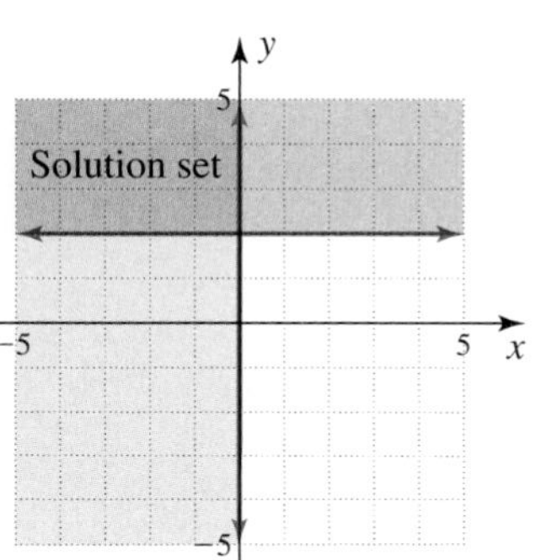

Figure 14

PROBLEM 1

Graph the solution of the system:

$x \leq 2$ and $y \geq 0$

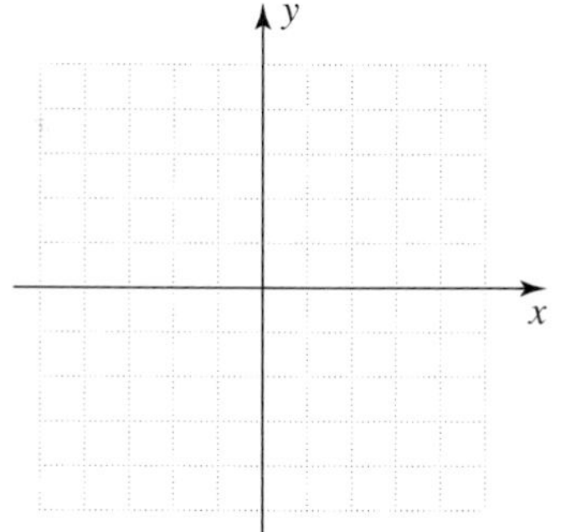

Answer

1.

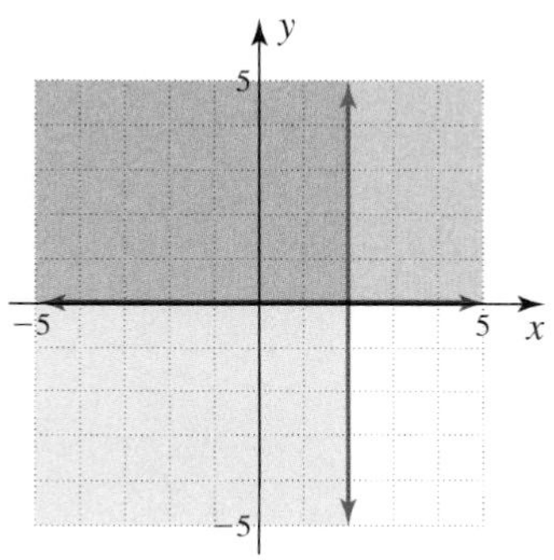

EXAMPLE 2 **Solving systems of inequalities using a test point**

Graph the solution of the system: $x + 2y \leq 5$ and $x - y < 2$

SOLUTION First we graph the lines $x + 2y = 5$ and $x - y = 2$. Using (0, 0) as a test point, $x + 2y \leq 5$ becomes $0 + 2 \cdot 0 \leq 5$, a true statement. So we shade the region containing (0, 0): the points *on or below* the line $x + 2y = 5$. (See Figure 15.) The inequality $x - y < 2$ is also satisfied by the test point (0, 0), so we shade the points *above* the line $x - y = 2$. This line is drawn dashed to indicate that the points on it do *not* satisfy the inequality $x - y < 2$. (See Figure 16.) The solution set of the system is

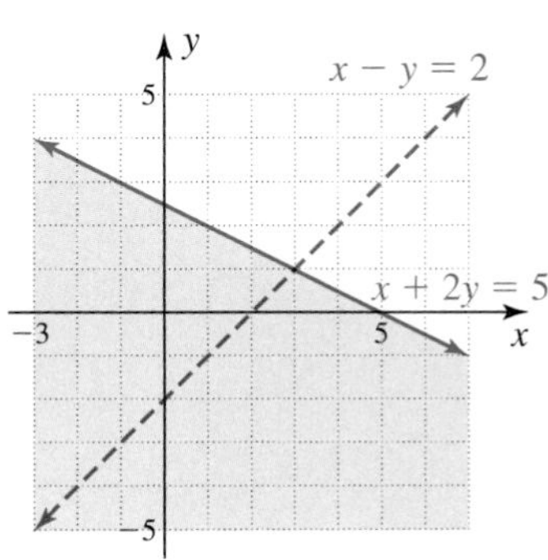

Figure 15

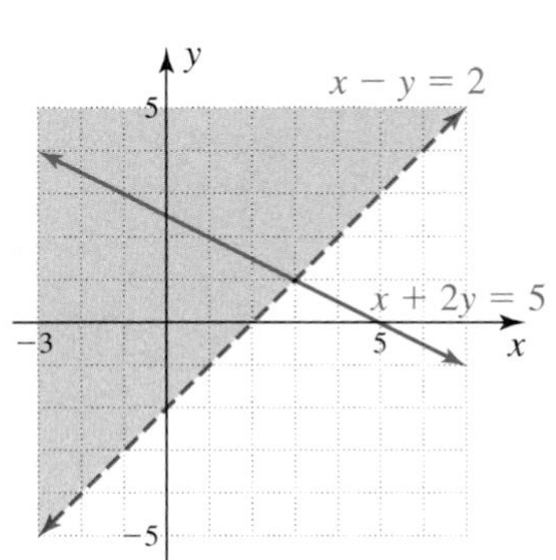

Figure 16

PROBLEM 2

Graph the solution of the system:

$2x + y \leq 5$ and $x - y < 3$

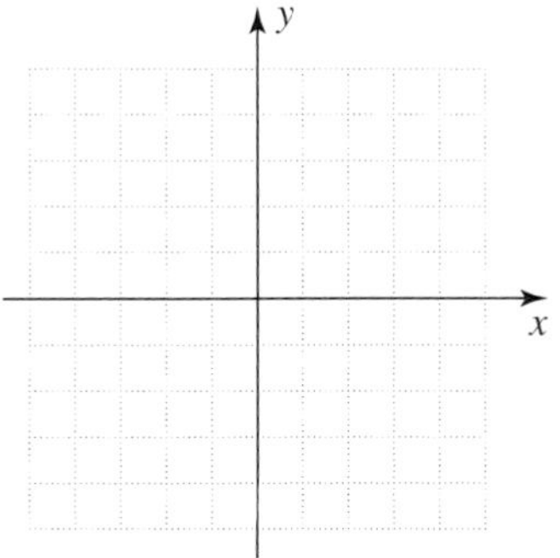

(Answer on next page)

shown in Figure 17 by the darker region and the portion of the solid line forming one boundary of the region.

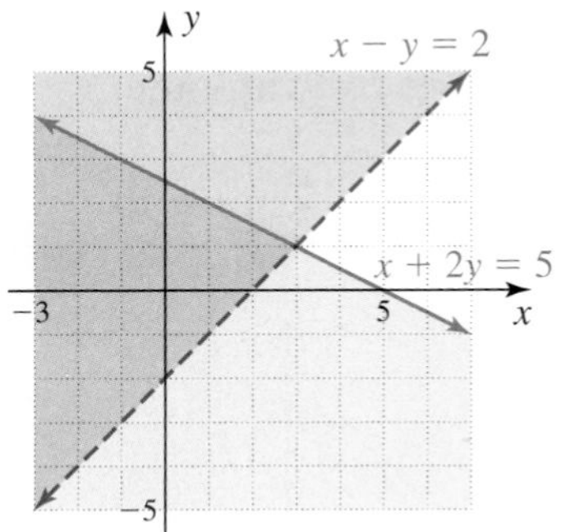

Figure 17

Answer

2.

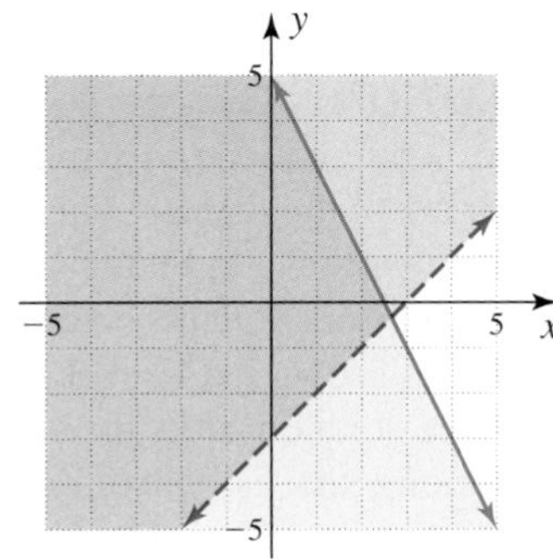

EXAMPLE 3 Solving systems of inequalities using a test point that does not satisfy both inequalities

Graph the solution of the system: $y + x \geq 2$ and $y - x \leq 2$

SOLUTION Graph the lines $y + x = 2$ and $y - x = 2$. Since the test point (0, 0) does *not* satisfy the inequality $y + x \geq 2$, we shade the region that does *not* contain (0, 0): the points *on or above* the line $y + x = 2$. (See Figure 18.) The test point (0, 0) *does* satisfy the inequality $y - x \leq 2$, so we shade the points *on or below* the line $y - x = 2$. (See Figure 19.) The solution set of the system is the darker region in Figure 20 and includes parts of both lines.

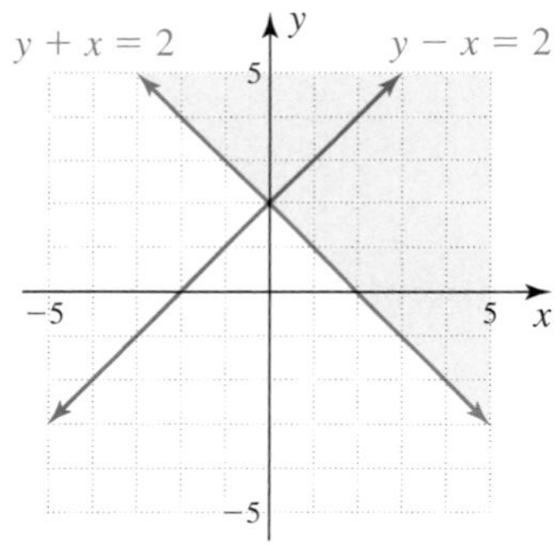

Figure 18

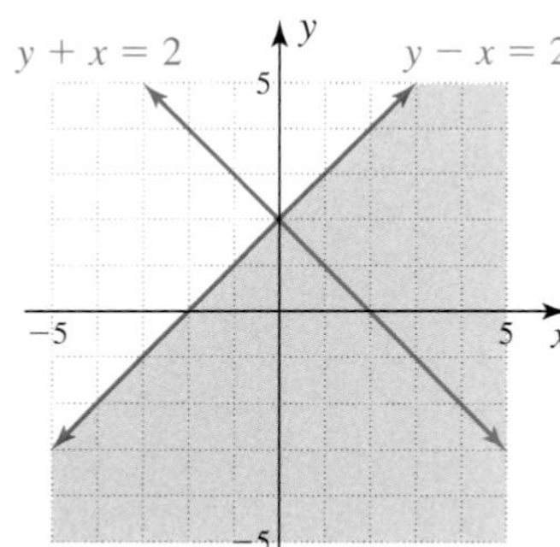

Figure 19

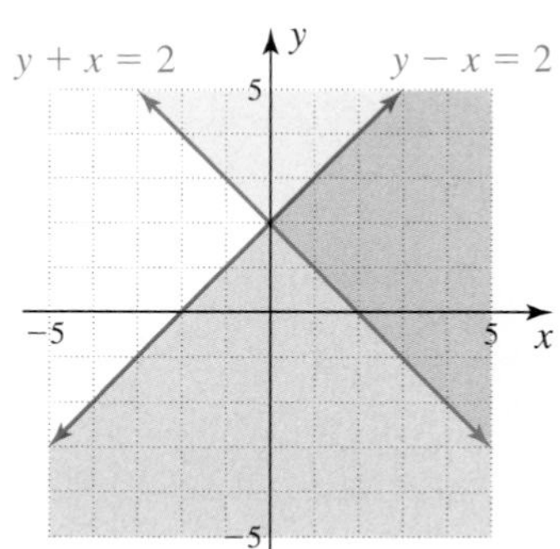

Figure 20

PROBLEM 3

Graph the solution of the system:

$$y + x \leq 3 \text{ and } y - x \geq 3$$

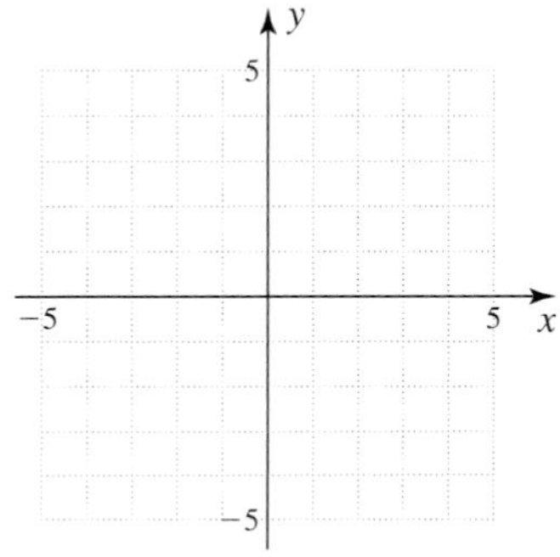

Answer

3.

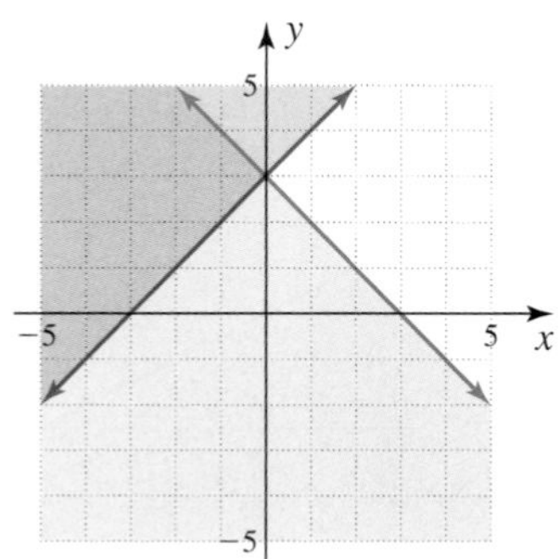

Exercises 8.5

A In Problems 1–15, graph the solution set of the system of inequalities.

1. $x \geq 0$ and $y \leq 2$

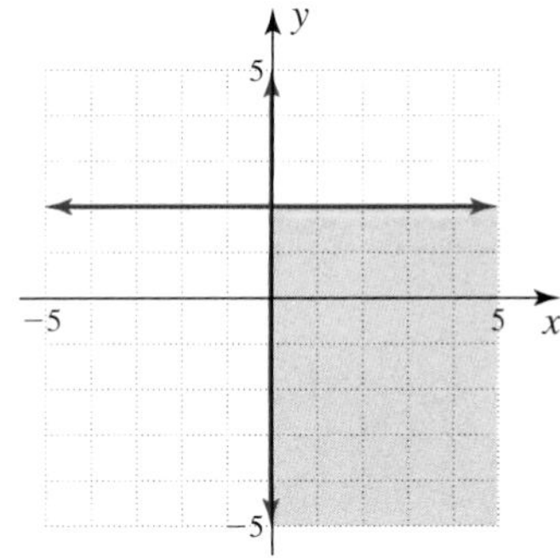

2. $x > 1$ and $y < 3$

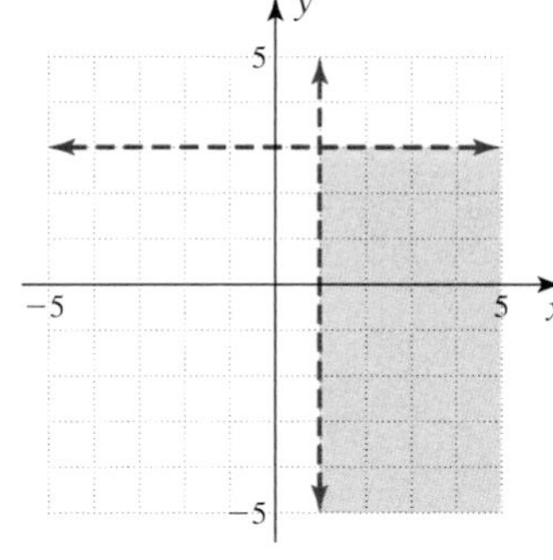

3. $x < -1$ and $y > -2$

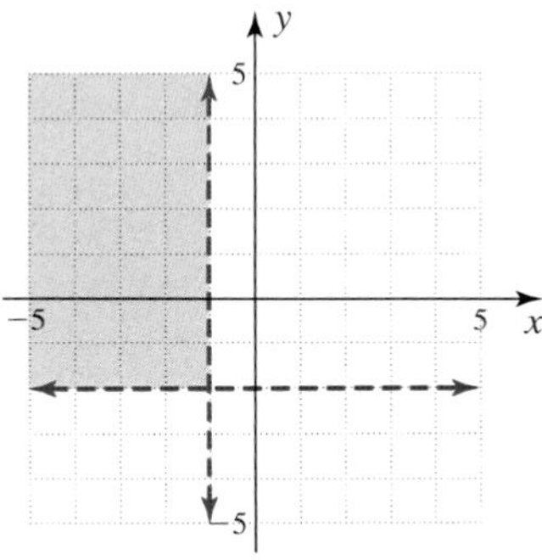

4. $x - y \geq 2$
$x + y \leq 6$

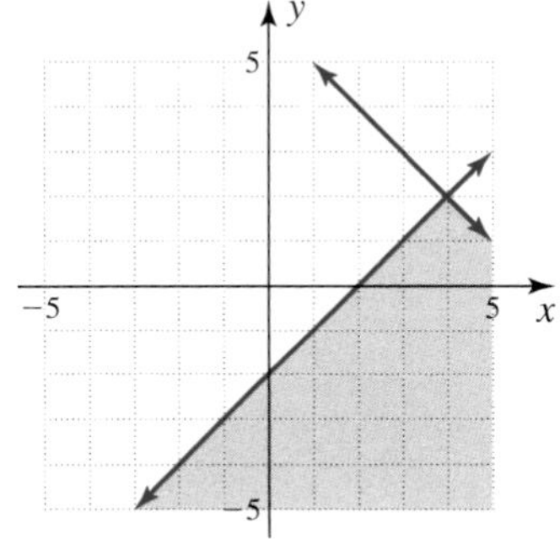

5. $x + 2y \leq 3$
$x < y$

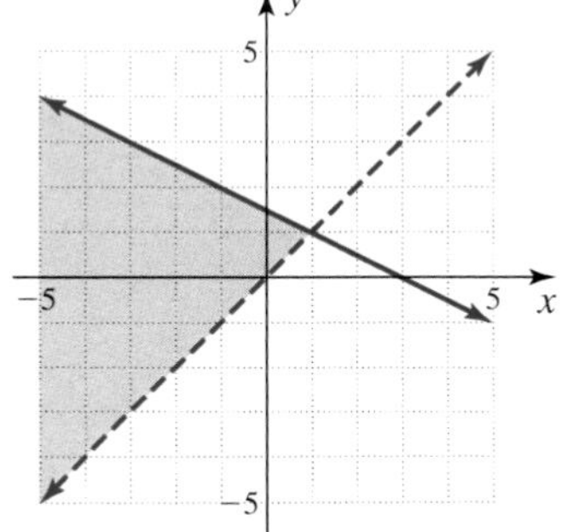

6. $5x - y > -1$
$-x + 2y \leq 6$

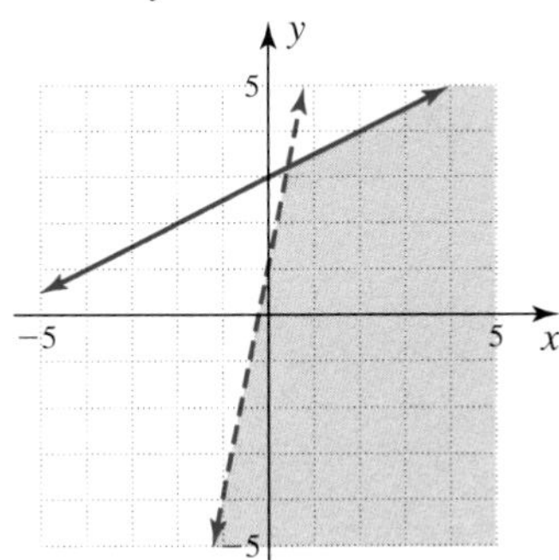

7. $4x - y > -1$
$-2x - y \leq -3$

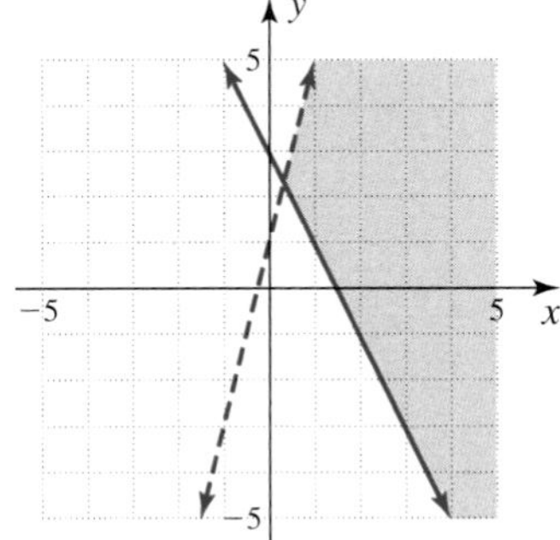

8. $2x - 3y < 6$
$4x - 3y > 12$

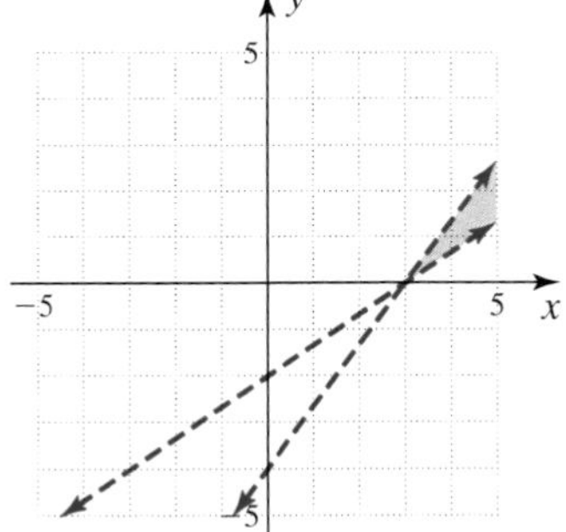

9. $-2x + y > 3$
$5x - y \leq -10$

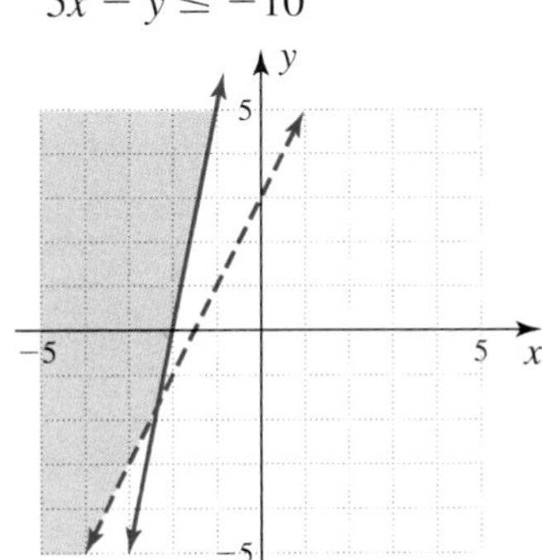

10. $2x - 5y \leq 10$
$3x + 2y < 6$

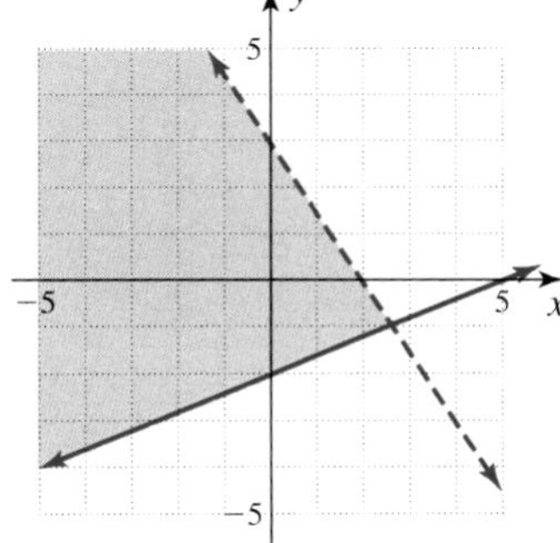

11. $2x - 3y < 5$
$x \geq y$

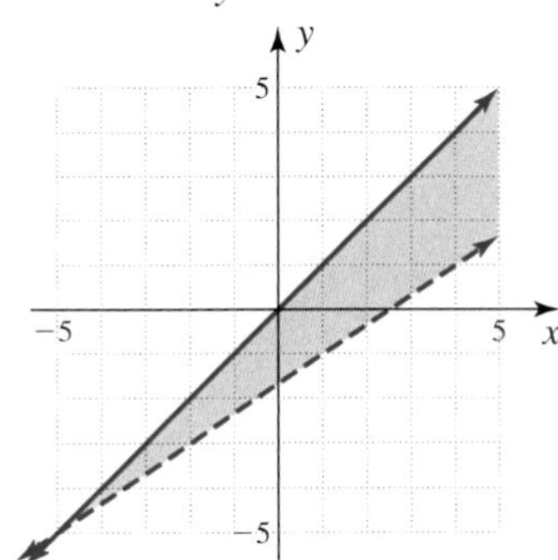

12. $x \le 2y$
$x + y < 4$

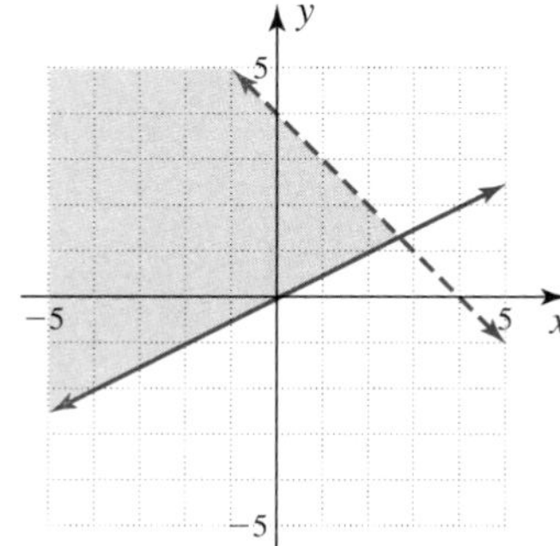

13. $x + 3y \le 6$
$x > y$

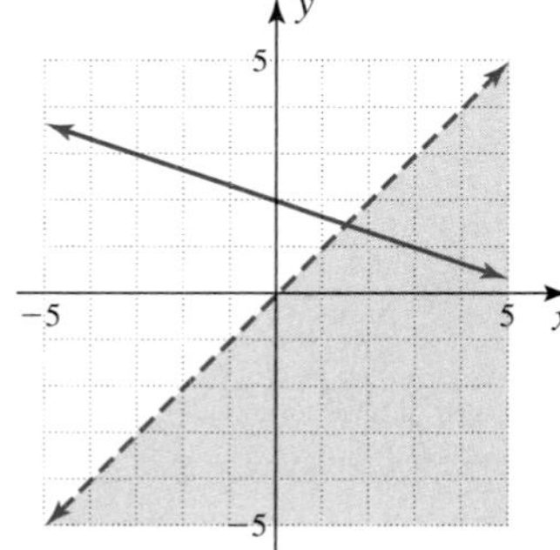

14. $2x - y < 2$
$x \le y$

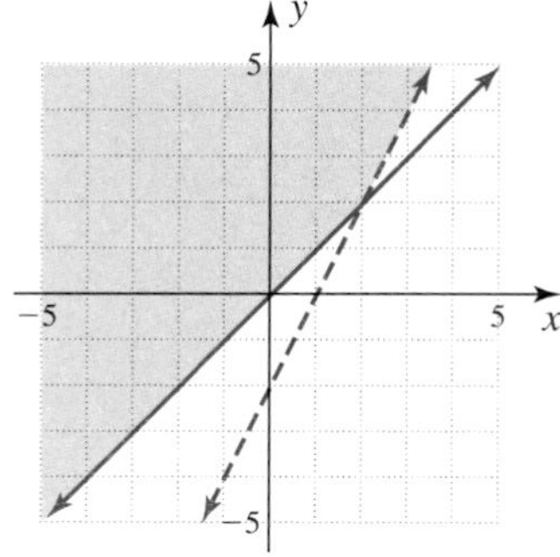

15. $3x + y > 6$
$x \le y$

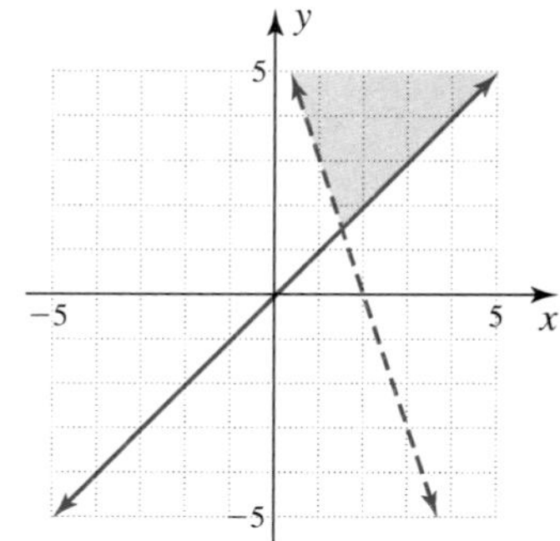

SKILL CHECKER

Try the Skill Checker Exercises so you'll be ready for the next section.

Find:

16. $\sqrt{9}$ 3

17. $\sqrt{64}$ 8

18. $\sqrt{144}$ 12

USING YOUR KNOWLEDGE

Inequalities in Exercise

The target zone used to gauge your effort when performing aerobic exercises is determined by your pulse rate p and your age a. Graph the given inequalities on the set of axes provided.

19. $p \ge -\frac{2a}{3} + 150$

20. $p \le -a + 190$

21. $10 \le a \le 70$

The target zone is the solution set of the inequalities in Problems 19–21. What is your target zone?

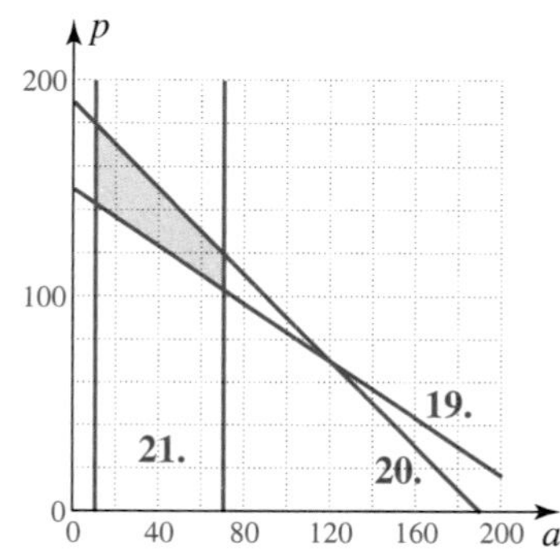

WRITE ON

22. Write the steps you would take to graph the inequality

$$ax + by > c \quad (a \neq 0)$$ Answers may vary.

23. Describe the solution set of the inequality $x \geq k$.
Answers may vary.

24. Describe the solution set of the inequality $y < k$.
Answers may vary.

MASTERY TEST

If you know how to do these problems, you have learned your lesson!

Graph the solution set:

25. $x > 2$ and $y < 3$

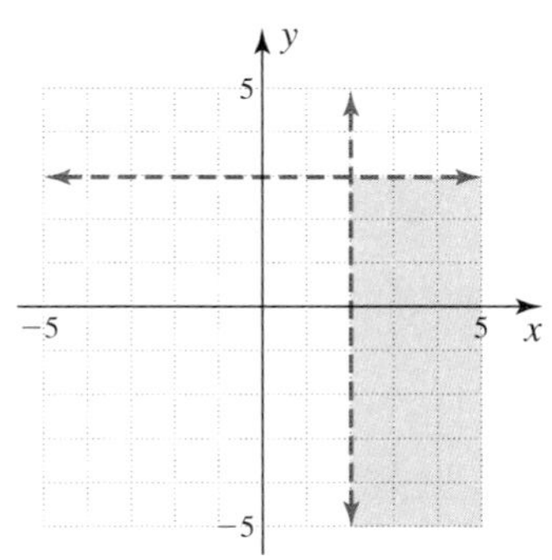

26. $x + y > 4$
$x - y \leq 2$

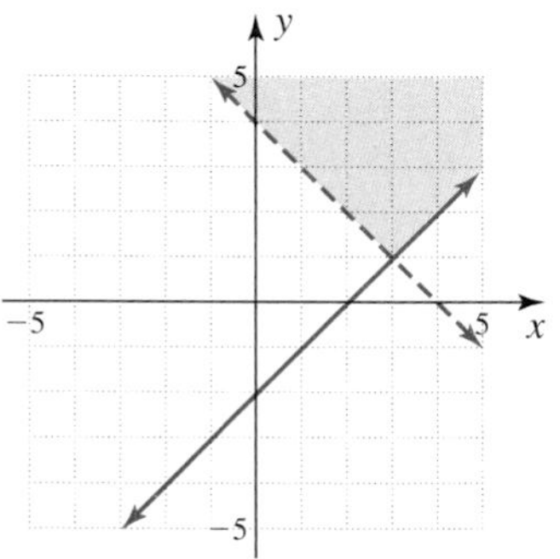

27. $3x - y < -1$
$x + 2y \leq 2$

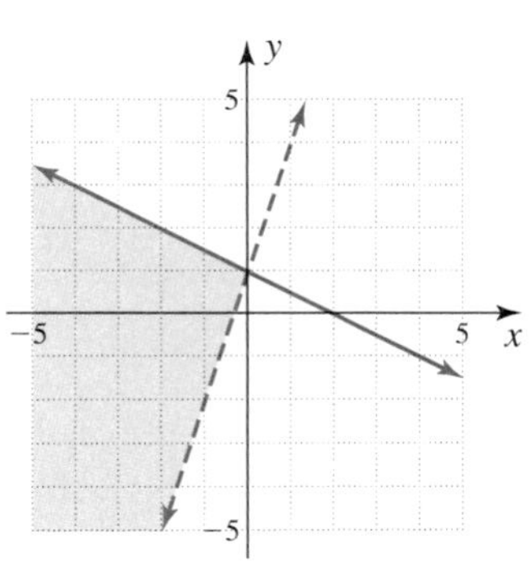

28. $2x - 3y \geq 6$
$-x + 2y < -4$

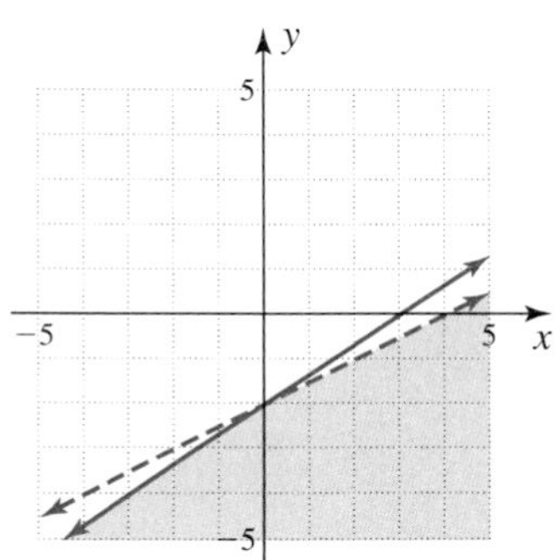

COLLABORATIVE LEARNING

Women and Men in the Workforce

The figure shows the percentages of women (W) and men (M) in the workforce since 1955. Can you tell from the graph in which year the percentages will be the same? It will happen after 2005! If t is the number of years after 1955, W the percentage of women in the workforce, and M the percentage of men, the graphs will *intersect* at a point (a, b). The coordinates of the point of intersection (if there is one) will be the common solution of both equations. If the percentages W and M are approximated by

$$W = 0.55t + 34.4 \text{ (percent)}$$

$$M = -0.20t + 83.9 \text{ (percent)}$$

you can find the year after 1955 when the same percentages of women and men are in the workforce by graphing M and W and locating the point of intersection.

Form three teams of students: **Team 1,** *The Estimators;* **Team 2,** *The Graphicals;* and **Team 3,** *The Substituters.*

Women and Men in the Workforce

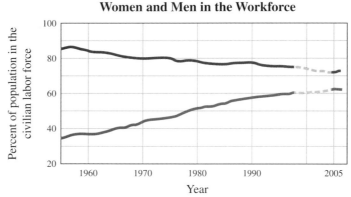

Source: U.S. Bureau of Labor Statistics

Here are the assignments:

Team 1 Extend the lines for the percentage of men (M), the percentage of women (W), and the horizontal axis.

1. In what year do the lines seem to intersect?
2. What would be the percentage of men and women in the labor force in that year?

Team 2 Get some graph paper with a 20-by-20 grid, each grid representing 1 unit. Graph the lines

$$M = -0.20t + 83.9 \text{ (percent)}$$
$$W = 0.55t + 34.4 \text{ (percent)}$$

where t is the number of years after 1955.

1. Where do the lines intersect?
2. What does the point of intersection represent?

Team 3 The point at which the lines intersect will satisfy both equations; consequently at that point $M = W$. Substitute $-0.20t + 83.9$ for M and $0.55t + 34.4$ for W and solve for t.

1. What value did you get?
2. What is the point of intersection for the two lines, and what does it represent?

Compare the results of teams 1, 2, and 3. Do they agree? Why or why not? Which is the most accurate method for obtaining an answer for this problem?

Research Questions

1. Write a report on the content of the *Nine Chapters of the Mathematical Arts.*
2. Write a paragraph detailing how the copies of the *Nine Chapters* were destroyed.
3. Write a short biography on the Chinese mathematician Liu Hui.
4. Write a report explaining the relationship between systems of linear equations and inequalities and the simplex method.
5. Write a report on the simplex algorithm and its developer.

Summary

SECTION	ITEM	MEANING	EXAMPLE
8.1A	System of simultaneous equations	A set of equations that may have a common solution	$x + y = 2$ $x - y = 4$ is a system of equations.
	Inconsistent system	A system with no solution	$x + y = 2$ $x + y = 3$ is an inconsistent system.
	Dependent system	A system in which both equations are equivalent	$x + y = 2$ $2x + 2y = 4$ is a dependent system.
8.2A	Substitution method	A method used to solve systems of equations by solving one equation for one variable and substituting this result in the other equation	To solve the system $x + y = 2$ $2x + 3y = 6$ solve the first equation for x: $x = 2 - y$ and substitute in the other equation: $2(2 - y) + 3y = 6$ $4 - 2y + 3y = 6$ $y = 2$ $x = 2 - 2 = 0$
8.3A	Elimination method	A method used to solve systems of equations by multiplying by numbers that will cause the coefficients of one of the variables to be opposites	To solve the system $x + 2y = 5$ $x - y = -1$ multiply the second equation by 2 and add to the first equation.
8.4	RSTUV method	A method for solving word problems consisting of **R**eading, **S**electing the variables, **T**hinking of a plan to solve the problem, **U**sing algebra to solve, and **V**erifying the answer	
8.5A	Solution set of a system of inequalities	The set of points that satisfy all inequalities in the system	The solution set of the system $x + y \geq 2$ $-x + y \leq -1$ is 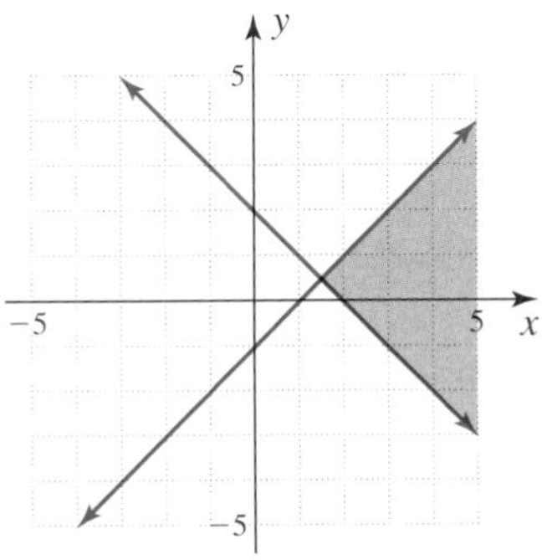

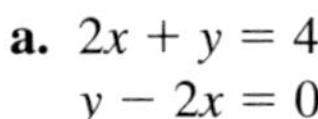

Review Exercises

(If you need help with these exercises, look in the section indicated in brackets.)

1. [8.1A, B] Use the graphical method to solve the system.

a. $2x + y = 4$
$y - 2x = 0$

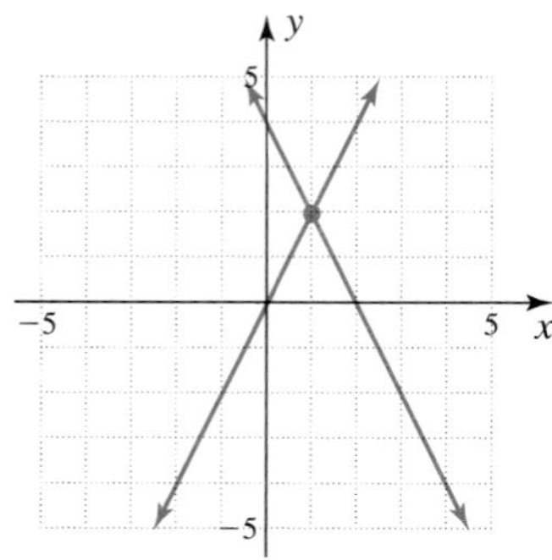

Solution: (1, 2)

b. $x + y = 4$
$y - x = 0$

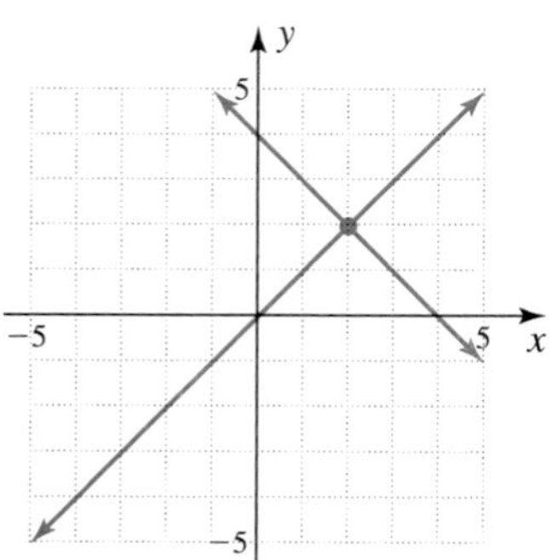

Solution: (2, 2)

c. $x + y = 4$
$y - 3x = 0$

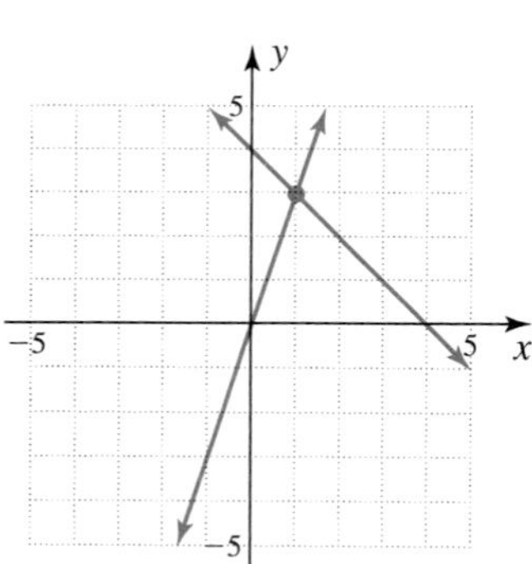

Solution: (1, 3)

2. [8.1A, B] Use the graphical method to solve the system (if possible).

a. $y - 3x = 3$
$2y - 6x = 12$

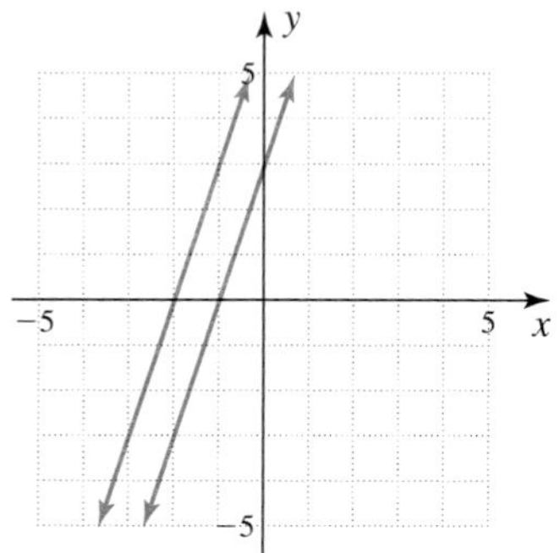

Inconsistent; no solution

b. $y - 2x = 2$
$2y - 4x = 8$

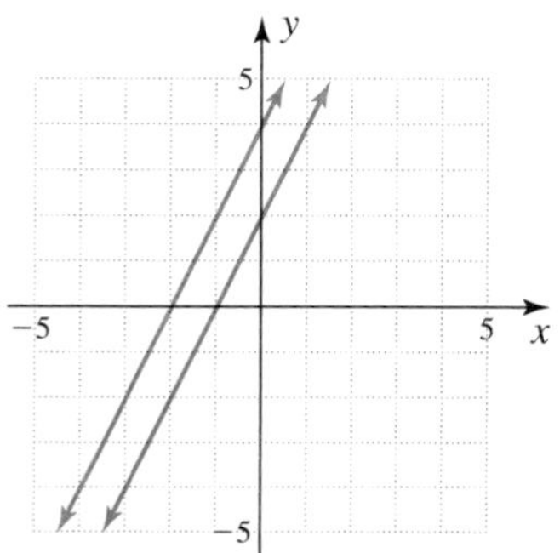

Inconsistent; no solution

c. $y - 3x = 6$
$2y - 6x = 6$

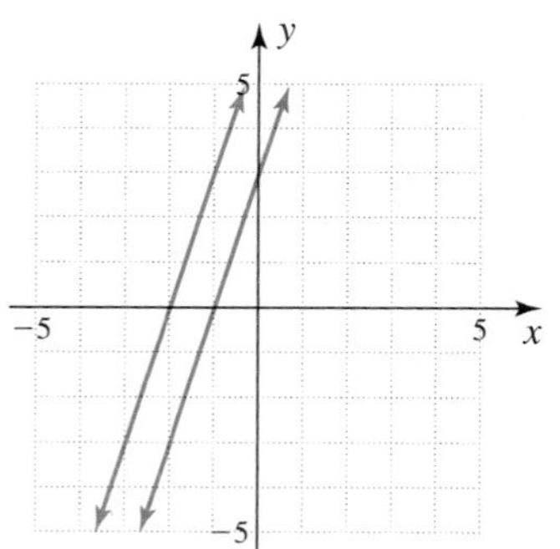

Inconsistent; no solution

3. [8.2A, B] Use the substitution method to solve the system (if possible).

a. $x + 4y = 5$
$2x + 8y = 15$ Inconsistent; no solution

b. $x + 3y = 6$
$3x + 9y = 12$ Inconsistent; no solution

c. $x + 4y = 5$
$2x + 13y = 15$ (1, 1)

4. [8.2A, B] Use the substitution method to solve the system (if possible).

a. $x + 4y = 5$
$2x + 8y = 10$ Dependent; infinitely many solutions

b. $x + 3y = 6$
$3x + 9y = 18$ Dependent; infinitely many solutions

c. $x + 4y = 5$
$-2x - 8y = -10$ Dependent; infinitely many solutions

5. [8.3A] Use the elimination method to solve the system (if possible).

a. $3x + 2y = 1$
$2x + y = 0$ $(-1, 2)$

b. $3x + 2y = 4$
$2x + y = 3$ $(2, -1)$

c. $3x + 2y = -7$
$2x + y = -4$ $(-1, -2)$

6. [8.3B] Use the elimination method to solve the system (if possible).

a. $2x - 3y = 6$
$-4x + 6y = -2$ Inconsistent; no solution

b. $3x - 2y = 8$
$-9x + 6y = -4$ Inconsistent; no solution

c. $3x - 5y = 6$
$-3x + 5y = -12$ Inconsistent; no solution

7. [8.3B] Use the elimination method to solve the system (if possible).

a. $3y + 2x = 1$
$6y + 4x = 2$ Dependent; infinitely many solutions

b. $2y + 3x = 1$
$6x + 4y = 2$ Dependent; infinitely many solutions

c. $3y + 4x = -11$
$8x + 6y = -22$ Dependent; infinitely many solutions

8. [8.4A] Desi has $3 in nickels and dimes. How many nickels and how many dimes does she have if

a. she has the same number of nickels and dimes?
20 nickels; 20 dimes

b. she has 4 times as many nickels as she has dimes?
40 nickels; 10 dimes

c. she has 10 times as many nickels as she has dimes?
50 nickels; 5 dimes

9. [8.4B] The sum of two numbers is 180. What are the numbers if

a. their difference is 40? 70; 110

b. their difference is 60? 60; 120

c. their difference is 80? 50; 130

10. [8.4C] A plane flew 2400 miles with a tailwind in 3 hours. What was the plane's speed in still air if the return trip took

a. 8 hours? 550 mi/hr

b. 10 hours? 520 mi/hr

c. 12 hours? 500 mi/hr

11. [8.4D] An investor bought some municipal bonds yielding 5% annually and some certificates of deposit yielding 10% annually. If the total investment amounts to $20,000, how much money is invested in bonds and how much in certificates of deposit if the annual interest is

a. $1750? Bonds: $5000; CDs: $15,000

b. $1150? Bonds: $17,000; CDs: $3000

c. $1500? Bonds: $10,000; CDs: $10,000

12. [8.5] Graph the solution set of the system.

a. $x > 4$ and $y < -1$

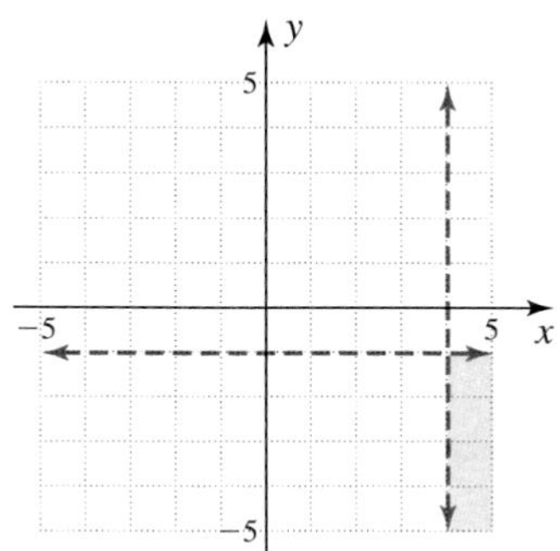

b. $x + y > 3$
$x - y < 4$

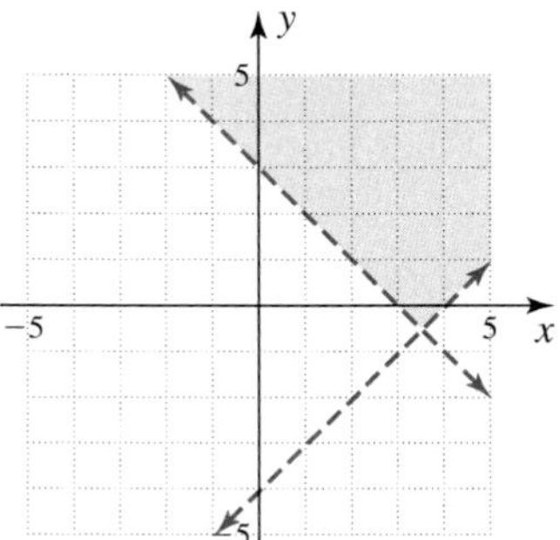

c. $2x + y \le 4$
$x - 2y > 2$

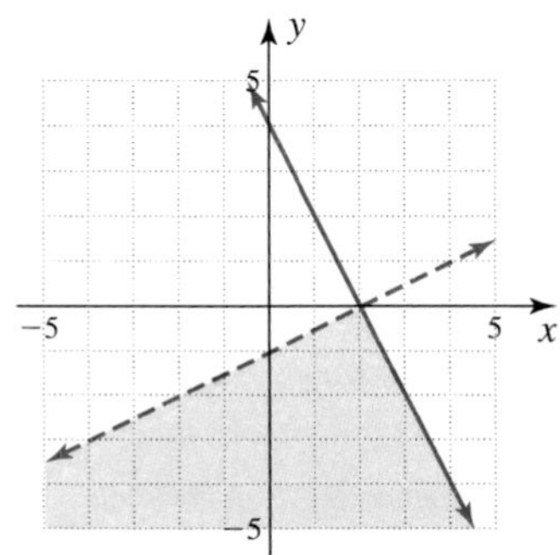

Practice Test 8

(Answers on page 633)

1. Use the graphical method to solve the system.

$$x + 2y = 4$$
$$2y - x = 0$$

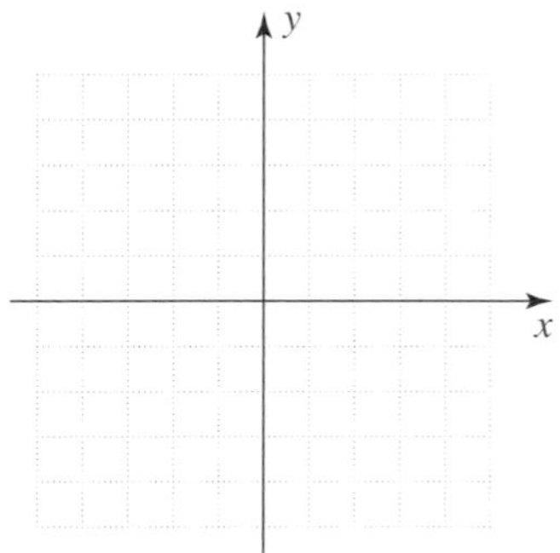

2. Use the graphical method to solve the system.

$$y - 2x = 2$$
$$2y - 4x = 8$$

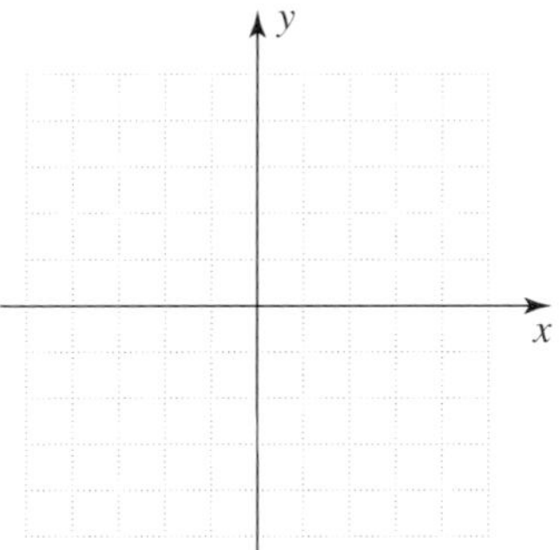

3. Use the substitution method to solve the system (if possible).

$$x + 3y = 6$$
$$2x + 6y = 8$$

4. Use the substitution method to solve the system (if possible).

$$x + 3y = 6$$
$$3x + 9y = 18$$

5. Use the elimination method to solve the system (if possible).

$$2x + 3y = -8$$
$$3x + y = -5$$

6. Use the elimination method to solve the system (if possible).

$$3x - 2y = 6$$
$$-6x + 4y = -2$$

7. Use the elimination method to solve the system (if possible).

$$2y + 3x = -12$$
$$6x + 4y = -24$$

8. Eva has \$2 in nickels and dimes. She has twice as many dimes as nickels. How many nickels and how many dimes does she have?

9. The sum of two numbers is 140. Their difference is 90. What are the numbers?

10. A plane flies 600 miles with a tailwind in 2 hours. It takes the same plane 3 hours to fly the 600 miles when flying against the wind. What is the plane's speed in still air?

11. Herbert invests \$10,000, part at 5% and part at 6%. How much money is invested at each rate if his annual interest is \$568?

12. Graph the solution set of the system.

$$x - 2y > 4$$
$$2x - y \le 6$$

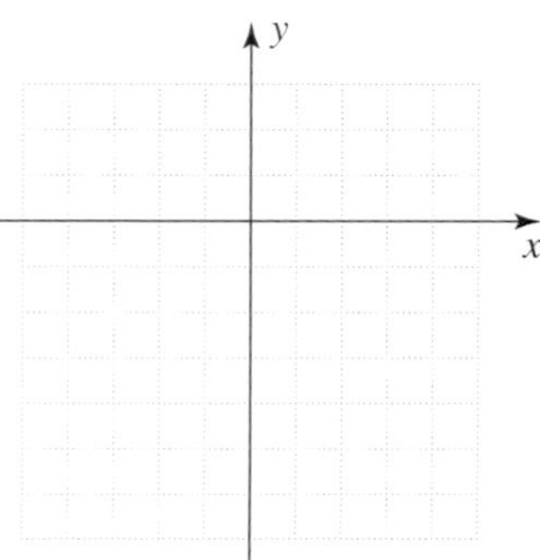

Answers to Practice Test

ANSWER	IF YOU MISSED	REVIEW		
	QUESTION	SECTION	EXAMPLES	PAGE
1. The solution is (2, 1).	1	8.1	1	583–584
2. The lines are parallel; there is no solution (inconsistent).	2	8.1	2, 3	584, 585
3. No solution (inconsistent)	3	8.2	1, 2	597–598
4. Dependent (infinitely many solutions)	4	8.2	3	598–599
5. $(-1, -2)$	5	8.3	1	606
6. No solution (inconsistent)	6	8.3	1, 2	606, 607
7. Dependent (infinitely many solutions)	7	8.3	3	607–608
8. 8 nickels; 16 dimes	8	8.4	1, 2	614–615
9. 115 and 25	9	8.4	3	616
10. 250 miles per hour	10	8.4	4	617
11. \$3200 at 5%; \$6800 at 6%	11	8.4	5	618
12.	12	8.5	1, 2, 3	623–624

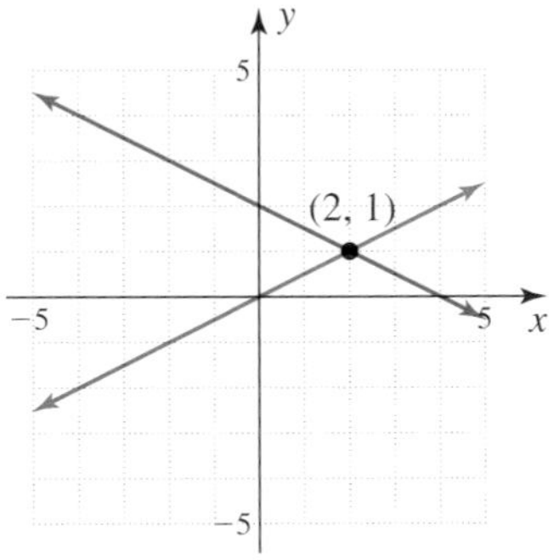

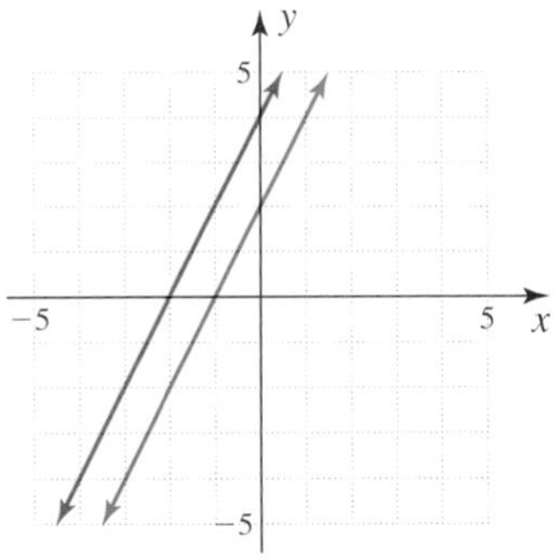

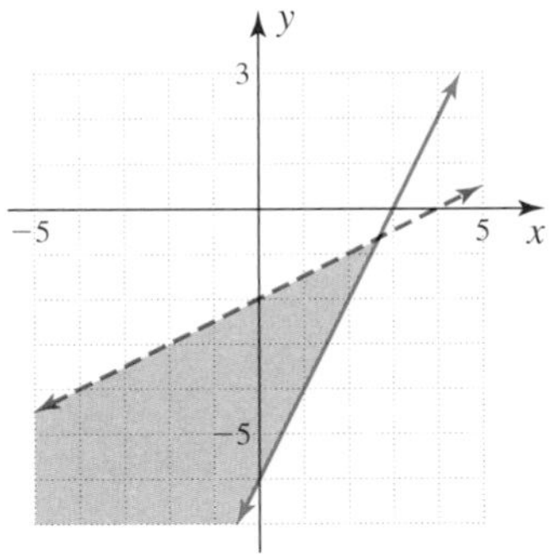

Cumulative Review Chapters 1–8

1. Find: $-\frac{3}{7} + \left(-\frac{1}{6}\right)$ $-\frac{25}{42}$

2. Find: $(-4)^4$ 256

3. Find: $-\frac{1}{4} \div \left(-\frac{1}{8}\right)$ 2

4. Evaluate $y \div 2 \cdot x - z$ for $x = 6$, $y = 24$, $z = 3$. 69

5. Simplify: $2x - (x + 4) - 2(x + 3)$ $-x - 10$

6. Write in symbols: The quotient of $(x - 5y)$ and z $\frac{x - 5y}{z}$

7. Solve for x: $3 = 3(x - 2) + 3 - 2x$ $x = 6$

8. Solve for x: $\frac{x}{4} - \frac{x}{9} = 5$ $x = 36$

9. Graph: $-\frac{x}{3} + \frac{x}{8} \geq \frac{x - 8}{8}$

10. Graph the point $C(1, -2)$.

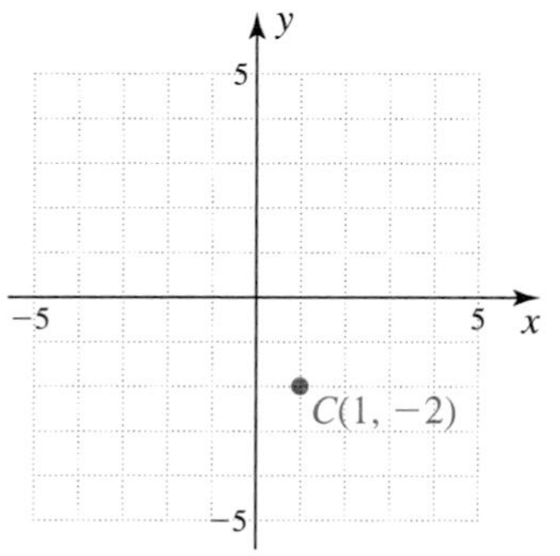

11. Determine whether the ordered pair $(-1, -1)$ is a solution of $4x + 3y = -1$. No

12. Find x in the ordered pair $(x, 2)$ so that the ordered pair satisfies the equation $3x - 2y = -13$. $x = -3$

13. Graph: $x + y = 1$

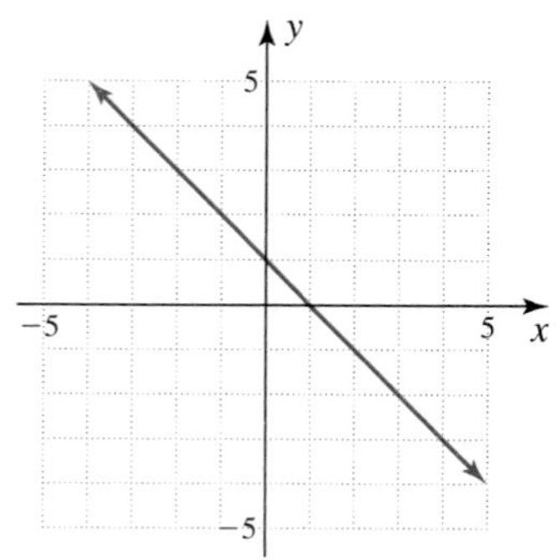

14. Graph: $2y + 8 = 0$

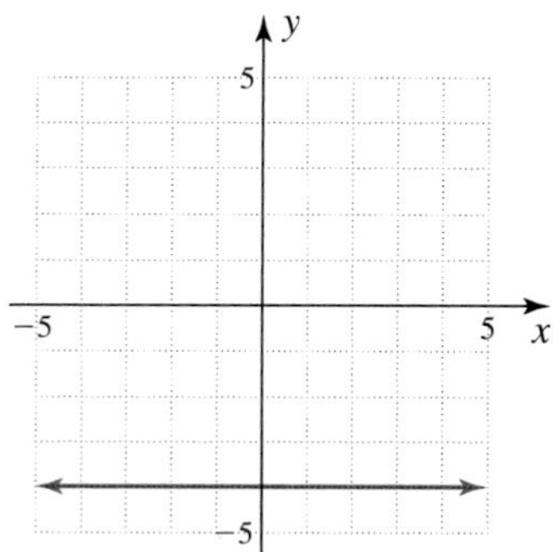

15. Find the slope of the line going through the points $(-2, -3)$ and $(5, -4)$. $-\frac{1}{7}$

16. What is the slope of the line $12x - 4y = 14$? 3

17. Find the pair of parallel lines.

(1) $2y = -x + 6$

(2) $-4x - 8y = 6$

(3) $8y - 4x = 6$ (1) and (2)

18. Simplify: $(2x^4y^{-3})^{-3}$ $\frac{y^9}{8x^{12}}$

19. Write in scientific notation: 0.000035 3.5×10^{-5}

20. Divide and express the answer in scientific notation: $(5.46 \times 10^{-3}) \div (2.6 \times 10^{4})$ 2.10×10^{-7}

21. Find (expand): $\left(6x^2 - \frac{1}{4}\right)^2$ $36x^4 - 3x^2 + \frac{1}{16}$

22. Divide $(2x^3 - x^2 - 2x + 4)$ by $(x - 1)$. $(2x^2 + x - 1)$ R 3

23. Factor completely: $x^2 - 11x + 30$ $(x - 6)(x - 5)$

24. Factor completely: $15x^2 - 29xy + 12y^2$ $(3x - 4y)(5x - 3y)$

25. Factor completely: $81x^2 - 64y^2$ $(9x + 8y)(9x - 8y)$

26. Factor completely: $-3x^4 + 12x^2$ $-3x^2(x + 2)(x - 2)$

27. Factor completely: $4x^3 - 4x^2 - 8x$ $4x(x + 1)(x - 2)$

28. Factor completely: $4x^2 + 3x + 4x + 3$ $(4x + 3)(x + 1)$

29. Factor completely: $25kx^2 - 30kx + 9k$ $k(5x - 3)^2$

30. Solve for x: $3x^2 + 4x = 15$ $x = -3; x = \frac{5}{3}$

31. Write $\frac{4x}{3y}$ with a denominator of $15y^2$. $\frac{20xy}{15y^2}$

32. Reduce to lowest terms: $\frac{-9(x^2 - y^2)}{3(x - y)}$ $-3(x + y)$

33. Reduce to lowest terms: $\frac{x^2 - 4x - 12}{6 - x}$ $-(x + 2)$

34. Multiply: $(x - 5) \cdot \frac{x - 2}{x^2 - 25}$ $\frac{x - 2}{x + 5}$

35. Divide: $\frac{x + 4}{x - 4} \div \frac{x^2 - 16}{4 - x}$ $\frac{1}{4 - x}$

36. Add: $\frac{9}{2(x - 9)} + \frac{3}{2(x - 9)}$ $\frac{6}{x - 9}$

37. Subtract: $\frac{x + 2}{x^2 + x - 6} - \frac{x + 3}{x^2 - 4}$ $\frac{-2x - 5}{(x + 3)(x + 2)(x - 2)}$

38. Simplify: $\dfrac{\frac{4}{x} - \frac{3}{4x}}{\frac{2}{3x} - \frac{1}{2x}}$ $\frac{39}{2}$

39. Solve for x: $\frac{5x}{x - 5} - 2 = \frac{x}{x - 5}$ $x = -5$

40. Solve for x: $\frac{x}{x^2 - 4} + \frac{2}{x - 2} = \frac{1}{x + 2}$ $x = -3$

41. Solve for x: $\frac{x}{x + 4} - \frac{1}{5} = \frac{-4}{x + 4}$ No solution

42. Solve for x: $2 + \frac{8}{x - 2} = \frac{32}{x^2 - 4}$ $x = -6$

43. A van travels 60 miles on 4 gallons of gas. How many gallons will it need to travel 195 miles? 13 gallons

44. Solve for x: $\frac{x + 1}{4} = \frac{9}{5}$ $x = \frac{31}{5}$

45. Sandra can paint a kitchen in 5 hours, and Roger can paint the same kitchen in 4 hours. How long would it take for both working together to paint the kitchen? $2\frac{2}{9}$ hr

46. Find an equation of the line that goes through the point $(-6, -2)$ and has slope $m = -4$. $y = -4x - 26$

47. Find an equation of the line having slope 5 and y-intercept 2. $y = 5x + 2$

48. Graph: $4x - 3y > -12$

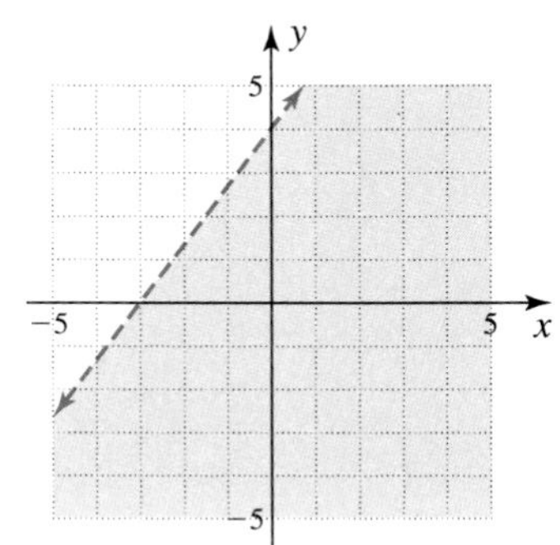

49. Graph: $-y \leq -3x + 6$

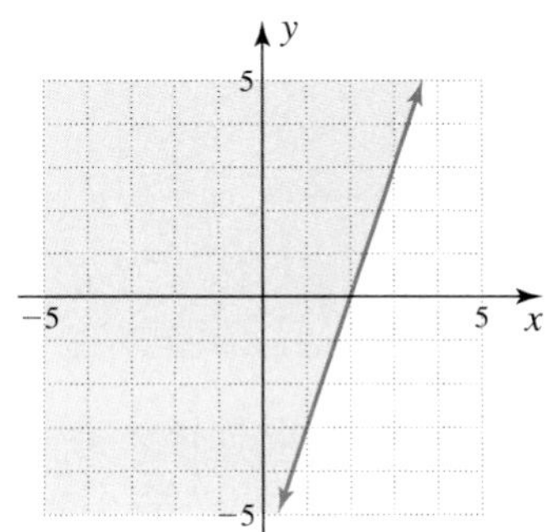

50. An enclosed gas exerts a pressure P on the walls of the container. This pressure is directly proportional to the temperature T of the gas. If the pressure is 7 lb/in.^{2} when the temperature is 350°F, find k. $\frac{1}{50}$

51. If the temperature of a gas is held constant, the pressure P varies inversely as the volume V. A pressure of 1560 lb/in.^{2} is exerted by 4 ft^3 of air in a cylinder fitted with a piston. Find k. 6240

52. Graph the system and find the solution (if possible):

$$x + 2y = 6$$
$$2y - x = -2$$

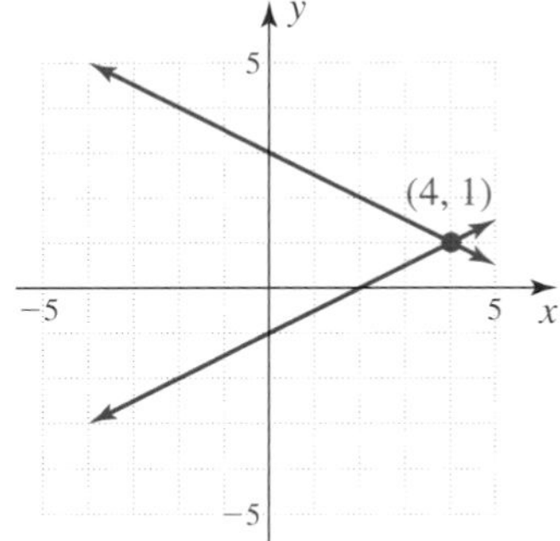

53. Graph the system and find the solution (if possible):

$$y - x = -1$$
$$2y - 2x = -4$$

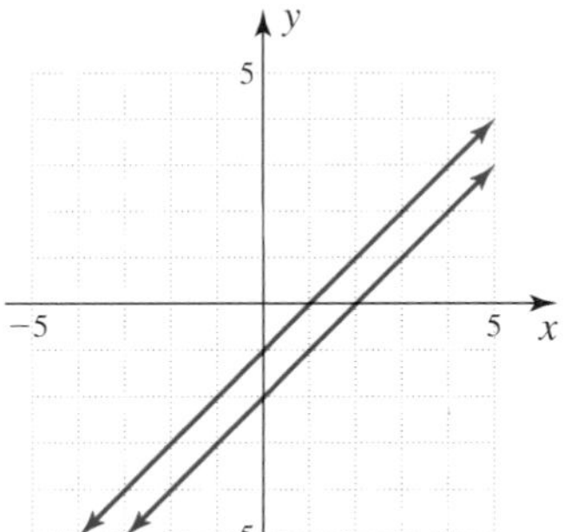

No solution (inconsistent)

54. Solve by substitution (if possible):

$$x - 4y = 5$$
$$-3x + 12y = -17$$

Inconsistent; no solution

55. Solve by substitution (if possible):

$$x - 4y = -13$$
$$3x - 12y = -39$$

Dependent; infinitely many solutions

56. Solve the system (if possible):

$$3x + 4y = 7$$
$$2x + 3y = 5$$

(1, 1)

57. Solve the system (if possible):

$$4x - 3y = 1$$
$$4x - 3y = -4$$

Inconsistent; no solution

58. Solve the system (if possible):

$$2y - x = 0$$
$$-2x + 4y = 0$$

Dependent; infinitely many solutions

59. Kaye has $4.20 in nickels and dimes. She has three times as many dimes as nickels. How many nickels and how many dimes does she have?
12 nickels and 36 dimes

60. The sum of two numbers is 165. Their difference is 95. What are the numbers? 35 and 130

Roots and Radicals

9

The Human Side of Algebra

Beginning in the tenth century and extending into the fourteenth century, Chinese mathematics shifted to arithmetical algebra. A Chinese mathematician discovered the relation between extracting roots and the array of binomial coefficients in Pascal's triangle. This discovery and iterated multiplications were used to extend root extraction and to solve equations of higher degree than cubic, an effort that reached its apex with the work of four prominent thirteenth-century Chinese algebraists.

The square root sign $\sqrt{}$ can be traced back to Christoff Rudolff (1500–1545), who wrote it as $\surd$ with only two strokes. Rudolff thought that $\surd$ resembled the small letter r, the first letter in the word *radix,* which means root. One way of computing square roots was developed by the English mathematician Isaac Newton (1642–1727), and the process is aptly named Newton's method. With the advent of calculators and computers, Newton's method, as well as the square root tables popular a few years ago, are seldom used anymore.

Pretest for Chapter 9

(Answers on page 639)

1. Find.

a. $\sqrt{144}$ b. $-\sqrt{\frac{36}{81}}$

2. Find the square of each radical expression.

a. $-\sqrt{81}$ b. $\sqrt{x^2 + 9}$

3. Classify each number as rational, irrational, or not a real number, and simplify if possible.

a. $\sqrt{19}$ b. $-\sqrt{81}$ c. $\sqrt{-100}$ d. $\sqrt{\frac{100}{36}}$

4. Find each root if possible.

a. $\sqrt[4]{16}$ b. $-\sqrt[4]{625}$ c. $\sqrt[3]{-27}$ d. $\sqrt[4]{-16}$

5. A diver jumps from a cliff 45 meters high. If the time t (in seconds) it takes an object dropped from a distance d (in meters) to reach the ground is given by

$$t = \sqrt{\frac{d}{5}}$$

how long does it take the diver to reach the water?

6. Simplify.

a. $\sqrt{180}$ b. $\sqrt{24}$

7. Multiply.

a. $\sqrt{2} \cdot \sqrt{11}$ b. $\sqrt{11} \cdot \sqrt{y}$, $y > 0$

8. Simplify.

a. $\sqrt{\frac{7}{25}}$ b. $\frac{21\sqrt{50}}{7\sqrt{10}}$

9. Simplify.

a. $\sqrt{169n^2}$, $n > 0$ b. $\sqrt{48y^7}$, $y > 0$

10. Simplify.

a. $\sqrt[4]{80}$ b. $\sqrt[3]{\frac{-125}{27}}$

11. Simplify.

a. $9\sqrt{13} + 8\sqrt{13}$ b. $24\sqrt{6} - 3\sqrt{6}$

12. Simplify.

a. $\sqrt{28} + \sqrt{112}$ b. $\sqrt{40} + \sqrt{90} - \sqrt{160}$

13. Simplify.

a. $\sqrt{3}(\sqrt{18} - \sqrt{2})$ b. $\sqrt{5}(\sqrt{5} - \sqrt{3})$

14. Write $\sqrt{\frac{7}{20}}$ with a rationalized denominator.

15. Write $\sqrt{\frac{y^2}{50}}$, $y > 0$, with a rationalized denominator.

16. Simplify.

a. $9\sqrt{14} - \sqrt{7} \cdot \sqrt{2}$ b. $\frac{\sqrt{20x^3}}{\sqrt{4x^2}}$, $x > 0$

17. Simplify.

a. $\frac{\sqrt[3]{500}}{\sqrt[3]{2}}$ b. $\frac{3}{\sqrt[3]{16}}$

18. Simplify.

a. $(\sqrt{3} + 5\sqrt{2})(\sqrt{3} - 2\sqrt{2})$

b. $(\sqrt{10} - 3\sqrt{20})(\sqrt{10} + 3\sqrt{20})$

19. Simplify.

a. $\frac{9}{\sqrt{3} + 1}$ b. $\frac{5}{\sqrt{5} - \sqrt{2}}$

20. Simplify.

a. $\frac{-9 + \sqrt{18}}{3}$ b. $\frac{-8 + \sqrt{8}}{4}$

21. Solve.

a. $\sqrt{x + 8} = 3$ b. $\sqrt{x + 5} = x + 5$

22. Solve $\sqrt{x + 9} - x = 3$

23. Solve $\sqrt{y + 4} = \sqrt{2y + 2}$

24. Solve $\sqrt{y + 7} - 3\sqrt{3y - 5} = 0$

25. The average length L of a long-distance call (in minutes) has been approximated by $L = \sqrt{t + 9}$, where t is the number of years after 2000. In how many years would you expect the average length of a call to be 4 minutes?

Answers to Pretest

ANSWER		IF YOU MISSED	REVIEW		
		QUESTION	SECTION	EXAMPLES	PAGE
1. a. 12	**b.** $-\frac{2}{3}$	1	9.1	1	641
2. a. 81	**b.** $x^2 + 9$	2	9.1	2	642
3. a. Irrational; **c.** Not real	**b.** -9; rational; **d.** $\frac{5}{3}$; rational	3	9.1	3	642
4. a. 2; **c.** -3	**b.** -5; **d.** Not real	4	9.1	4	643
5. 3 seconds		5	9.1	5	643–644
6. a. $6\sqrt{5}$	**b.** $2\sqrt{6}$	6	9.2	1	648
7. a. $\sqrt{22}$	**b.** $\sqrt{11y}$	7	9.2	2	648
8. a. $\frac{\sqrt{7}}{5}$	**b.** $3\sqrt{5}$	8	9.2	3	649
9. a. $13n$	**b.** $4y^3\sqrt{3y}$	9	9.2	4	649
10. a. $2\sqrt[4]{5}$	**b.** $-\frac{5}{3}$	10	9.2	5	650
11. a. $17\sqrt{13}$	**b.** $21\sqrt{6}$	11	9.3	1	654
12. a. $6\sqrt{7}$	**b.** $\sqrt{10}$	12	9.3	2	654
13. a. $2\sqrt{6}$	**b.** $5 - \sqrt{15}$	13	9.3	3	655
14. $\frac{\sqrt{35}}{10}$		14	9.3	4	656
15. $\frac{y\sqrt{2}}{10}$		15	9.3	5	657
16. a. $8\sqrt{14}$	**b.** $\sqrt{5x}$	16	9.4	1	660
17. a. $5\sqrt[3]{2}$	**b.** $\frac{3\sqrt[3]{4}}{4}$	17	9.4	2	660-661
18. a. $-17 + 3\sqrt{6}$	**b.** -170	18	9.4	3	661
19. a. $\frac{9\sqrt{3} - 9}{2}$	**b.** $\frac{5\sqrt{5} + 5\sqrt{2}}{3}$	19	9.4	4	662
20. a. $-3 + \sqrt{2}$	**b.** $\frac{-4 + \sqrt{2}}{2}$	20	9.4	5	663
21. a. $x = 1$	**b.** $x = -4$ or $x = -5$	21	9.5	1	668–669
22. $x = 0$		22	9.5	2	669
23. $y = 2$		23	9.5	3	670–671
24. $y = 2$		24	9.5	3	670–671
25. In $t = 7$ years		25	9.5	4	671

9.1 FINDING ROOTS

To Succeed, Review How To . . .

1. Find the square of a number (p. 73).
2. Raise a number to a power (p. 72).

Objectives

A Find the square root of a number.

B Square a radical expression.

C Classify the square root of a number and approximate it with a calculator.

D Find higher roots of numbers.

E Solve an application involving square roots.

GETTING STARTED

Square Roots and Round Wheels

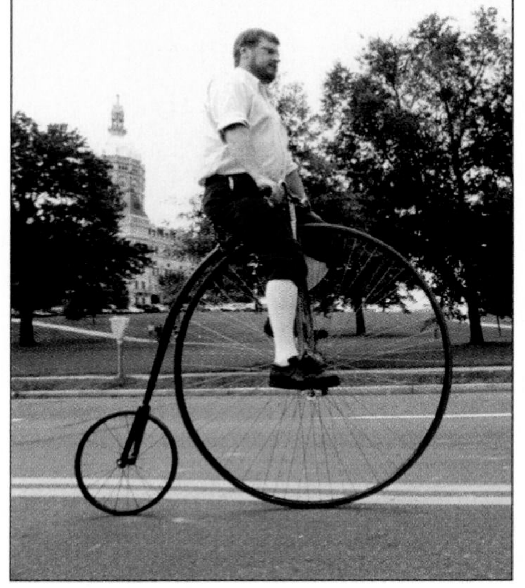

The first bicycles were built in 1839 by Kirkpatrick Macmillan of Scotland. These bicycles were heavy and unstable, so in 1870, James Starley of Great Britain set out to reduce their weight. The result, shown in the photo, was a lighter bicycle but one that was likely to tip over when going around corners. As a matter of fact, even now, the greatest speed s (in miles per hour) at which a cyclist can safely take a corner of radius r (in feet) is given by

$s = 4\sqrt{r}$ Read "s equals 4 times the square root of r."

The concept of a *square root* is related to the concept of squaring a number. The number 36, for example, is the square of 6 because $6^2 = 36$. Six, on the other hand, is the square root of 36—that is, $\sqrt{36} = 6$. Similarly,

$$\sqrt{16} = 4 \quad \text{because} \quad 4^2 = 16$$

$$\sqrt{\frac{4}{9}} = \frac{2}{3} \quad \text{because} \quad \left(\frac{2}{3}\right)^2 = \left(\frac{4}{9}\right)$$

Thus to find $\sqrt{a}$, we find the number b so that $b^2 = a$; that is,

$$\sqrt{a} = b \quad \text{is equivalent to} \quad b^2 = a$$

Note that since $b^2 = a$, a must be nonnegative.

For example,

$$\sqrt{4} = 2 \quad \text{because} \quad 2^2 = 4$$

$$\sqrt{9} = 3 \quad \text{because} \quad 3^2 = 9$$

$$\sqrt{\frac{1}{4}} = \frac{1}{2} \quad \text{because} \quad \left(\frac{1}{2}\right)^2 = \frac{1}{4}$$

In this section we learn how to find square roots.

Finding Square Roots

Web It

For a lesson on square roots (and even higher roots), go to link 9-1-1 on the Bello Website at mhhe.com/bello.

Teaching Tip

Let students know that when both square roots are required for 4, we symbolize it by ± 2, which is read "plus or minus 2."

The symbol $\sqrt{}$ is called a **radical sign,** and it indicates the **positive** square root of a number (except $\sqrt{0} = 0$). The expression under the radical sign is called the **radicand.** An algebraic expression containing a radical is called a **radical expression.** We can summarize our discussion as follows.

SQUARE ROOT

If a is a positive real number,

$\sqrt{a} = b$ is the **positive** square root of a so that $b^2 = a$.

$-\sqrt{a} = b$ is the **negative** square root of a so that $b^2 = a$.

If $a = 0$, $\sqrt{a} = 0$ and $\sqrt{0} = 0$ since $0^2 = 0$.

Note that the symbol $-\sqrt{}$ represents the *negative* square root. Thus the square root of 4 is $\sqrt{4} = 2$, but the *negative* square root of 4 is $-\sqrt{4} = -2$.

EXAMPLE 1 **Finding square roots**

Find:

a. $\sqrt{121}$ **b.** $-\sqrt{100}$ **c.** $\sqrt{\frac{25}{36}}$

SOLUTION

a. $\sqrt{121} = 11$ Since $11^2 = 121$

b. $-\sqrt{100} = -10$ Since $10^2 = 100$

c. $\sqrt{\frac{25}{36}} = \frac{5}{6}$ Since $\left(\frac{5}{6}\right)^2 = \frac{25}{36}$

PROBLEM 1

Find:

a. $\sqrt{169}$ **b.** $-\sqrt{64}$ **c.** $\sqrt{\frac{81}{25}}$

B

Squaring Radical Expressions

Since $b = \sqrt{a}$ means that $b^2 = a$,

$$\sqrt{a} \cdot \sqrt{a} = (\sqrt{a})^2 = a$$

Similarly, since $b = -\sqrt{a}$ means that $b^2 = a$,

$$(-\sqrt{a}) \cdot (-\sqrt{a}) = (-\sqrt{a})^2 = a$$

Thus we have the following rule.

RULE

Squaring a Square Root

When the square root of a nonnegative real number a is squared, the result is that positive real number; that is, $(\sqrt{a})^2 = a$ and $(-\sqrt{a})^2 = a$. Also, $(\sqrt{0})^2 = 0$.

Answers

1. a. 13 **b.** -8 **c.** $\frac{9}{5}$

EXAMPLE 2 **Squaring expressions involving radicals**

Find the square of each radical expression:

a. $\sqrt{7}$ **b.** $-\sqrt{49}$ **c.** $\sqrt{x^2 + 3}$

SOLUTION

a. $(\sqrt{7})^2 = 7$

b. $(-\sqrt{49})^2 = 49$

c. $(\sqrt{x^2 + 3})^2 = x^2 + 3$

PROBLEM 2

Find the square of each radical expression:

a. $\sqrt{11}$

b. $-\sqrt{36}$

c. $\sqrt{x^2 + 7}$

C Classifying and Approximating Square Roots

If a number has a rational number for its square root, it's called a **perfect square.** Thus 1, 4, 9, 16, $\frac{4}{25}$, $\frac{9}{4}$, and $\frac{81}{16}$ are perfect squares. If a number is *not* a perfect square, its square root is *irrational.* Thus $\sqrt{2}$, $\sqrt{3}$, $\sqrt{5}$, $\sqrt{\frac{3}{4}}$, and $\sqrt{\frac{5}{6}}$ are irrational. As you recall, an irrational number *cannot* be written as the ratio of two integers. Its decimal representation is nonrepeating and nonterminating. A decimal approximation for such a number can be obtained by using the 2nd x^2 on your calculator. Using the symbol $\approx$ (which means "is approximately equal to"), we have

$$\sqrt{2} \approx 1.4142136$$
$$\sqrt{3} \approx 1.7320508$$
$$\sqrt{5} \approx 2.236068$$

Of course, not all real numbers have a *real-number* square root. As you recall, $\sqrt{a} = b$ is *equivalent* to $b^2 = a$. Since b^2 is nonnegative, a is nonnegative.

Teaching Tip

In Example 2, remind students that they are squaring a square root, which is not to be confused with squaring a nonradical.

Example:

$(\sqrt{7})^2 = 7$ square the radical $\sqrt{7}$

$(7)^2 = 49$ square the number 7

SQUARE ROOT OF A NEGATIVE NUMBER

If a is **negative,** $\sqrt{a}$ is not a real number.

For example, $\sqrt{-7}$, $\sqrt{-4}$, and $\sqrt{-\frac{4}{5}}$ are *not* real numbers.

EXAMPLE 3 **Classifying and approximating real numbers**

Classify the given numbers as rational, irrational, or not a real number. Approximate the irrational numbers using a calculator.

a. $\sqrt{\frac{81}{25}}$ **b.** $-\sqrt{64}$ **c.** $\sqrt{13}$ **d.** $\sqrt{-64}$

SOLUTION

a. $\frac{81}{25} = \left(\frac{9}{5}\right)^2$ is a perfect square, so $\sqrt{\frac{81}{25}} = \frac{9}{5}$ is a rational number.

b. $64 = 8^2$ is a perfect square, so $-\sqrt{64} = -8$ is a rational number.

c. 13 is *not* a perfect square, so $\sqrt{13} \approx 3.6055513$ is irrational.

d. $\sqrt{-64}$ is *not* a real number, since there is no number that we can square and get -64 for an answer.

PROBLEM 3

Classify as rational, irrational, or not a real number.

a. $\sqrt{\frac{36}{25}}$ **b.** $-\sqrt{81}$

c. $\sqrt{19}$ **d.** $\sqrt{-9}$

Answers

2. a. 11 **b.** 36 **c.** $x^2 + 7$
3. a. $\frac{6}{5}$; Ra **b.** -9; Ra **c.** Ir
d. Not real

Finding Higher Roots of Numbers

We've already seen that

$$\sqrt{a} = b \quad \text{is equivalent to} \quad b^2 = a$$

Similarly,

$$\sqrt[3]{a} = b \quad \text{is equivalent to} \quad b^3 = a$$

$$\sqrt[4]{a} = b \quad \text{is equivalent to} \quad b^4 = a$$

Because of these relationships, finding the square root of a number can be regarded as the **inverse** of finding the square of the number. Similarly, the inverse of finding the **cube** or **fourth power** of a number a is to find $\sqrt[3]{a}$, the **cube root** of a, or $\sqrt[4]{a}$, the **fourth root** of a. In general, $\sqrt[n]{a}$ is called the ***n*th root of *a*** and the number n is the **index** or **order** of the radical. Note that we could write $\sqrt[2]{a}$, but the simpler notation $\sqrt{a}$ is customary because the square root is the most widely performed operation.

Teaching Tip

Have students find the following roots and compare the results in the two columns.

Even Index	Odd Index
$\sqrt{-16}$ = not real	$\sqrt[3]{-8} = -2$
$\sqrt[4]{-16}$ = not real	$\sqrt[5]{-32} = -2$
$\sqrt[6]{-64}$ = not real	$\sqrt[7]{-128} = -2$

Now, have them draw a conclusion about how the index affects the result when the radicand is negative.

EXAMPLE 4 Finding higher roots of real numbers

Find each root if possible:

a. $\sqrt[4]{16}$ **b.** $\sqrt[4]{-81}$ **c.** $\sqrt[3]{8}$

d. $\sqrt[3]{-27}$ **e.** $-\sqrt[4]{16}$

SOLUTION

a. We need to find a number whose fourth power is 16, that is, $(?)^4 = 16$. Since $2^4 = 16$, $\sqrt[4]{16} = 2$.

b. We need to find a number whose fourth power is -81. There is no such real number, so $\sqrt[4]{-81}$ is *not* a real number.

c. We need to find a number whose cube is 8. Since $2^3 = 8$, $\sqrt[3]{8} = 2$.

d. We need to find a number whose cube is -27. Since $(-3)^3 = -27$, $\sqrt[3]{-27} = -3$.

e. We need to find a number whose fourth power is 16. Since $2^4 = 16$, $-\sqrt[4]{16} = -2$.

PROBLEM 4

Find each root if possible:

a. $\sqrt[4]{81}$

b. $\sqrt[4]{-16}$

c. $\sqrt[3]{-64}$

d. $\sqrt[4]{-81}$

Solving an Application

EXAMPLE 5 Radical divers

The time t (in seconds) it takes an object dropped from a distance d (in feet) to reach the ground is given by

$$t = \sqrt{\frac{d}{16}}$$

The highest regularly performed dive used to be at La Quebrada in Acapulco. How long does it take the divers to reach the water 100 feet below?

PROBLEM 5

How long does it take the diver to travel 81 feet?

Answers

4. a. 3 **b.** Not real **c.** -4 **d.** Not real **5.** 2.25 sec

SOLUTION Using

$$t = \sqrt{\frac{d}{16}} \quad \text{and} \quad d = 100$$

we have

$$t = \sqrt{\frac{100}{16}} = \frac{10}{4} = 2.5 \text{ sec}$$

By the way, the water is only 12 feet deep, but the height of the dive has been given as high as 118 feet and as low as 87.5 feet!

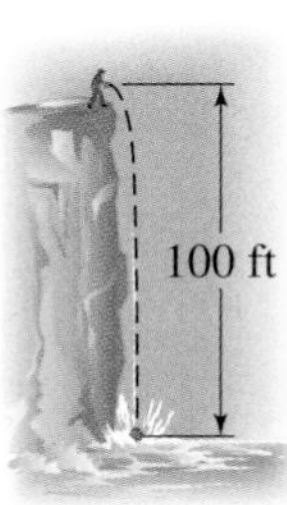

Exercises 9.1

A In Problems 1–8, find the square root.

1. $\sqrt{25}$ 5
2. $\sqrt{36}$ 6
3. $-\sqrt{9}$ -3
4. $-\sqrt{81}$ -9
5. $\sqrt{\frac{16}{9}}$ $\frac{4}{3}$
6. $\sqrt{\frac{25}{4}}$ $\frac{5}{2}$
7. $-\sqrt{\frac{4}{81}}$ $-\frac{2}{9}$
8. $-\sqrt{\frac{9}{100}}$ $-\frac{3}{10}$

In Problems 9–12, find all the square roots (positive and negative) of each number.

9. $\frac{25}{81}$ $\frac{5}{9}$ and $-\frac{5}{9}$
10. $\frac{36}{49}$ $\frac{6}{7}$ and $-\frac{6}{7}$
11. $\frac{49}{100}$ $\frac{7}{10}$ and $-\frac{7}{10}$
12. $\frac{1}{9}$ $\frac{1}{3}$ and $-\frac{1}{3}$

B In Problems 13–20, find the square of each radical expression.

13. $\sqrt{5}$ 5
14. $\sqrt{8}$ 8
15. $-\sqrt{11}$ 11
16. $-\sqrt{13}$ 13
17. $\sqrt{x^2 + 1}$ $x^2 + 1$
18. $-\sqrt{a^2 + 2}$ $a^2 + 2$
19. $-\sqrt{3y^2 + 7}$ $3y^2 + 7$
20. $\sqrt{8z^2 + 1}$ $8z^2 + 1$

C In Problems 21–32, find and classify the given numbers as rational, irrational, or not a real number. Use a calculator to approximate the irrational numbers. *Note:* The number of decimals in your calculator display may be different.

21. $\sqrt{36}$ 6; rational
22. $-\sqrt{4}$ -2; rational
23. $\sqrt{-100}$ Not a real number
24. $\sqrt{-9}$ Not a real number
25. $-\sqrt{64}$ -8; rational
26. $\sqrt{\frac{9}{49}}$ $\frac{3}{7}$; rational
27. $\sqrt{\frac{16}{9}}$ $\frac{4}{3}$; rational
28. $\sqrt{-\frac{4}{81}}$ Not a real number
29. $-\sqrt{6}$ -2.449489743; irrational
30. $\sqrt{7}$ 2.645751311; irrational
31. $-\sqrt{2}$ -1.414213562; irrational
32. $-\sqrt{12}$ -3.464101615; irrational

D In Problems 33–40, find each root if possible.

33. $\sqrt[4]{81}$ 3
34. $\sqrt[4]{-16}$ Not a real number
35. $-\sqrt[4]{81}$ -3
36. $\sqrt[3]{27}$ 3
37. $-\sqrt[3]{64}$ -4
38. $\sqrt[3]{-64}$ -4
39. $-\sqrt[3]{-125}$ 5
40. $-\sqrt[3]{-8}$ 2

APPLICATIONS

41. *Ribbon Falls* Ribbon Falls in California is the tallest continuous waterfall in the United States, with a drop of about 1600 feet. How long does it take the water to travel from the top of the waterfall to the bottom? (*Hint:* See Example 5.) (Actually, no time at all from late July to early April when the waterfall is dry!) 10 sec

42. *Jump from airship* Have you heard about the *Hindenburg* airship? On June 22, 1936, Colonel Harry A. Froboess of Switzerland jumped almost 400 feet from the *Hindenburg* into the Bodensee. How long did his jump take? (*Hint:* See Example 5.) 5 sec

43. *Highest dive into airbag* The time t (in seconds) it takes an object dropped from a distance d (in meters) to reach the ground is given by

$$t = \sqrt{\frac{d}{5}}$$

The highest reported dive into an airbag is 100 meters from the top of Vegas World Hotel and Casino by stuntman Dan Koko. How long did it take Koko to reach the airbag below? $\sqrt{20}$ sec ≈ 4.5 sec

44. *Time of highest dive* Colonel Froboess of Problem 42 jumped about 125 meters from the *Hindenburg* to the Bodensee. Using the formula in Problem 43, how long did his jump take? Is your answer consistent with that of Problem 42? 5 sec; yes

45. *Pythagorean Theorem* The Greek mathematician Pythagoras discovered a theorem that states that the lengths a, b, and c of the sides of a right triangle (a triangle with a 90° angle) are related by the formula $a^2 + b^2 = c^2$. (See the diagram.) This means that $c = \sqrt{a^2 + b^2}$. If the shorter sides of a right triangle are 4 inches and 3 inches, find the length of the longest side c (called the hypotenuse). 5 in.

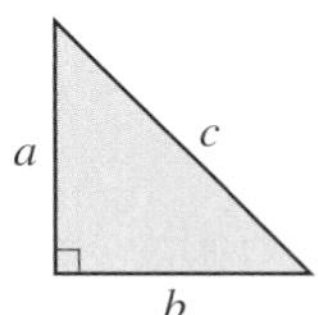

46. *Pythagorean Theorem at work* Use the formula of Problem 45 to find how high on the side of a building a 13-foot ladder will reach if the base of the ladder is 5 feet from the wall on which the ladder is resting. 12 ft

47. *View from in-flight airplane* How far can you see from an airplane in flight? It depends on how high the airplane is. As a matter of fact, your view V_m (in miles) is $V_m = 1.22\sqrt{a}$, where a is the altitude of the plane in feet. What is your view when your plane is cruising at an altitude of 40,000 feet? 244 mi

48. *Dimensions of a square* The area of a square is 144 square inches. How long is each side of the square? 12 in.

49. *Dimensions of square room* A square room has an area of 121 square feet. What are the dimensions of the room? 11×11 ft

50. *Dimensions of a quilt* A square quilt has an area of 169 square units. What are its dimensions? 13×13 units

SKILL CHECKER

Try the Skill Checker Exercises so you'll be ready for the next section.

Find $\sqrt{b^2 - 4ac}$ given:

51. $a = 1, b = -4, c = 3$ 2

52. $a = 1, b = -1, c = -2$ 3

53. $a = 2, b = 5, c = -3$ 7

54. $a = 6, b = -7, c = -3$ 11

USING YOUR KNOWLEDGE

Interpolation

Suppose you want to approximate $\sqrt{18}$ *without* using your calculator. Since $\sqrt{16} = 4$ and $\sqrt{25} = 5$, you have $\sqrt{16} = 4 < \sqrt{18} < \sqrt{25} = 5$. This means that $\sqrt{18}$ is between 4 and 5. To find a better approximation of $\sqrt{18}$, you can use a method that mathematicians call **interpolation.** Don't be scared by this name! The process is *really* simple. If you want to find an approximation for $\sqrt{18}$, follow the steps in the diagram; $\sqrt{20}$ and $\sqrt{22}$ are also approximated.

$18 - 16 = 2$	$20 - 16 = 4$	$22 - 16 = 6$
$\sqrt{16} = 4$	$\sqrt{16} = 4$	$\sqrt{16} = 4$
$\sqrt{18} \approx 4 + \frac{2}{9}$	$\sqrt{20} \approx 4 + \frac{4}{9}$	$\sqrt{22} \approx 4 + \frac{6}{9}$
$\sqrt{25} = 5$	$\sqrt{25} = 5$	$\sqrt{25} = 5$
$25 - 16 = 9$	$25 - 16 = 9$	$25 - 16 = 9$

As you can see,

$$\sqrt{18} \approx 4\frac{2}{9} \qquad \sqrt{20} \approx 4\frac{4}{9} \qquad \text{and} \qquad \sqrt{22} \approx 4\frac{6}{9} = 4\frac{2}{3}$$

Do you see a pattern? What would $\sqrt{24}$ be? Using a calculator, we obtain

$$\sqrt{18} \approx 4.2, \qquad \sqrt{20} \approx 4.5, \qquad \text{and} \qquad \sqrt{22} \approx 4.7$$

$$\sqrt{18} \approx 4\frac{2}{9} \approx 4.2, \qquad \sqrt{20} \approx 4\frac{4}{9} \approx 4.4, \qquad \sqrt{22} \approx 4\frac{6}{9} \approx 4.7$$

As you can see, we can get very close approximations!

Use this knowledge to approximate the following roots. Give the answer as a mixed number.

55. **a.** $\sqrt{26}$ $5\frac{1}{11}$ **b.** $\sqrt{28}$ $5\frac{3}{11}$ **c.** $\sqrt{30}$ $5\frac{5}{11}$

56. **a.** $\sqrt{67}$ $8\frac{3}{17}$ **b.** $\sqrt{70}$ $8\frac{6}{17}$ **c.** $\sqrt{73}$ $8\frac{9}{17}$

WRITE ON

57. If a is a *nonnegative* number, how many square roots does a have? Explain. One; answers may vary.

58. If a is *negative,* how many square roots does a have? Explain. None; answers may vary.

59. If $a = 0$, how many square roots does a have? One

60. How many real-number cube roots does a positive number have? One

61. How many real-number cube roots does a negative number have? One

62. How many cube roots does zero have? One

63. Suppose you want to find $\sqrt{a}$. What assumption must you make about the variable a? $a \geq 0$

64. If you are given a whole number less than 200, how would you determine whether the square root of the number is rational or irrational without a calculator? Explain. Answers may vary.

MASTERY TEST

If you know how to do these problems, you have learned your lesson!

Find:

65. $\sqrt{\frac{16}{49}}$ $\frac{4}{7}$

66. $-\sqrt{\frac{36}{25}}$ $-\frac{6}{5}$

67. $\sqrt{144}$ 12

Find the square of each radical expression:

68. $-\sqrt{28}$ 28

69. $\sqrt{17}$ 17

70. $-\sqrt{x^2 + 9}$ $x^2 + 9$

Find and classify as rational, irrational, or not a real number and approximate the irrational numbers using a calculator:

71. $-\sqrt{\frac{49}{121}}$ $-\frac{7}{11}$; rational

72. $\sqrt{-100}$ Not a real number

73. $\sqrt{15}$ 3.872983346; irrational

Find if possible:

74. $\sqrt[4]{81}$ 3

75. $\sqrt[3]{-125}$ -5

76. $\sqrt[4]{-1}$ Not a real number

77. $-\sqrt[3]{-27}$ 3

78. The time t in seconds it takes an object dropped from a distance d (in feet) to reach the ground is given by $t = \sqrt{\frac{d}{16}}$. How long does it take an object dropped from a distance of 81 feet to reach the ground? $\frac{9}{4}$ sec

9.2 MULTIPLICATION AND DIVISION OF RADICALS

To Succeed, Review How To . . .

1. Write a number as a product of primes (p. 8).
2. Write a number using specified powers as factors (pp. 17–18).

Objectives

A Multiply and simplify radicals using the product rule.

B Divide and simplify radicals using the quotient rule.

C Simplify radicals involving variables.

D Simplify higher roots.

GETTING STARTED

Skidding Is No Accident!

Have you seen your local police measuring skid marks at the scene of an accident? The speed s (in miles per hour) a car was traveling if it skidded d feet on a dry concrete road is given by $s = \sqrt{24d}$. If a car left a 50-foot skid mark at the scene of an accident and the speed limit was 30 miles per hour, was the driver speeding? To find the answer, we need to use the formula $s = \sqrt{24d}$. Letting $d = 50$, $s = \sqrt{24 \cdot 50} = \sqrt{1200}$. How can we simplify $\sqrt{1200}$? If we use factors that are perfect squares, we get

$$\begin{aligned} \sqrt{1200} &= \sqrt{100 \cdot 4 \cdot 3} \\ &= \sqrt{100} \cdot \sqrt{4} \cdot \sqrt{3} \\ &= 10 \cdot 2 \cdot \sqrt{3} \\ &= 20\sqrt{3} \end{aligned}$$

which is about 35 miles per hour. So the driver *was* exceeding the speed limit! In solving this problem, we have made the assumption that $\sqrt{a \cdot b} = \sqrt{a} \cdot \sqrt{b}$. Is this assumption always true? In this section you will learn that this is indeed the case!

Using the Product Rule for Radicals

Web It

For a lesson on operations with radicals, go to link 9-2-1 on the Bello Website at mhhe.com/bello.

Is it true that $\sqrt{a \cdot b} = \sqrt{a} \cdot \sqrt{b}$? Let's check some examples.

$$\sqrt{4 \cdot 9} = \sqrt{36} = 6, \quad \sqrt{4} \cdot \sqrt{9} = 2 \cdot 3 = 6, \quad \text{thus} \quad \sqrt{4 \cdot 9} = \sqrt{4} \cdot \sqrt{9}$$

Similarly,

$$\sqrt{9 \cdot 16} = \sqrt{144} = 12, \quad \sqrt{9} \cdot \sqrt{16} = 3 \cdot 4 = 12, \quad \text{thus} \quad \sqrt{9 \cdot 16} = \sqrt{9} \cdot \sqrt{16}$$

In general, we have the following rule

PRODUCT RULE FOR RADICALS

If a and b are nonnegative numbers,

$$\sqrt{a \cdot b} = \sqrt{a} \cdot \sqrt{b}$$

The product rule can be used to *simplify* radicals—that is, to make sure that no perfect squares remain under the radical sign. Thus $\sqrt{15}$ is simplified, but $\sqrt{18}$ is not because $\sqrt{18} = \sqrt{9 \cdot 2}$, and $\sqrt{9 \cdot 2} = \sqrt{9} \cdot \sqrt{2} = 3\sqrt{2}$ using the product rule.

EXAMPLE 1 **Simplifying expressions using the product rule**

Simplify:

a. $\sqrt{75}$ **b.** $\sqrt{48}$

SOLUTION To use the product rule, you have to remember the square root of a few perfect square numbers such as 4, 9, 16, 25, 36, 49, 64, and 81 and try to write the number under the radical using one of these numbers as a factor.

a. Since $\sqrt{75} = \sqrt{25 \cdot 3}$, we write $\sqrt{75} = \sqrt{25 \cdot 3} = \sqrt{25} \cdot \sqrt{3} = 5\sqrt{3}$.

b. Similarly, $\sqrt{48} = \sqrt{16 \cdot 3}$. Thus $\sqrt{48} = 4\sqrt{3}$.

PROBLEM 1

Simplify:

a. $\sqrt{72}$ **b.** $\sqrt{98}$

We can also use the product rule to actually multiply radicals. Thus

$$\sqrt{3} \cdot \sqrt{5} = \sqrt{3 \cdot 5} = \sqrt{15} \quad \text{and} \quad \sqrt{8} \cdot \sqrt{2} = \sqrt{16} = 4$$

EXAMPLE 2 **Multiplying expressions using the product rule**

Multiply:

a. $\sqrt{6} \cdot \sqrt{5}$ **b.** $\sqrt{20} \cdot \sqrt{5}$ **c.** $\sqrt{5} \cdot \sqrt{x}, x > 0$

SOLUTION

a. $\sqrt{6} \cdot \sqrt{5} = \sqrt{30}$ **b.** $\sqrt{20} \cdot \sqrt{5} = \sqrt{100} = 10$

c. $\sqrt{5} \cdot \sqrt{x} = \sqrt{5x}$

PROBLEM 2

Multiply:

a. $\sqrt{5} \cdot \sqrt{7}$

b. $\sqrt{40} \cdot \sqrt{10}$

c. $\sqrt{6} \cdot \sqrt{x}, x > 0$

B Using the Quotient Rule for Radicals

We already know that

$$\sqrt{\frac{81}{16}} = \frac{9}{4}$$

It's also true that

$$\sqrt{\frac{81}{16}} = \frac{\sqrt{81}}{\sqrt{16}} = \frac{9}{4}$$

This rule can be stated as follows.

QUOTIENT RULE FOR RADICALS

If a and b are positive numbers,

$$\sqrt{\frac{a}{b}} = \frac{\sqrt{a}}{\sqrt{b}}$$

Teaching Tip

An alternate way to simplify $\sqrt{75}$: Use a vertical scheme rather than a horizontal one.

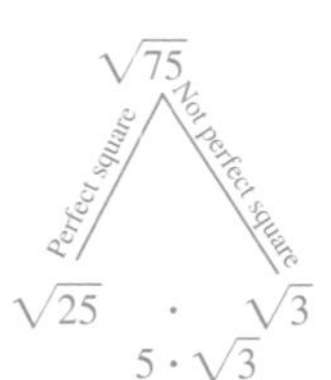

Answers

1. a. $6\sqrt{2}$ **b.** $7\sqrt{2}$
2. a. $\sqrt{35}$ **b.** 20 **c.** $\sqrt{6x}$

EXAMPLE 3 **Simplifying expressions using the quotient rule**

Simplify:

a. $\sqrt{\frac{5}{9}}$ **b.** $\frac{\sqrt{18}}{\sqrt{6}}$ **c.** $\frac{14\sqrt{20}}{7\sqrt{10}}$

SOLUTION

a. $\sqrt{\frac{5}{9}} = \frac{\sqrt{5}}{\sqrt{9}} = \frac{\sqrt{5}}{3}$ **b.** $\frac{\sqrt{18}}{\sqrt{6}} = \sqrt{\frac{18}{6}} = \sqrt{3}$

c. $\frac{14\sqrt{20}}{7\sqrt{10}} = \frac{2\sqrt{20}}{\sqrt{10}} = 2\sqrt{\frac{20}{10}} = 2\sqrt{2}$

PROBLEM 3

Simplify:

a. $\sqrt{\frac{7}{16}}$

b. $\frac{\sqrt{30}}{\sqrt{6}}$

c. $\frac{16\sqrt{30}}{8\sqrt{10}}$

C Simplifying Radicals Involving Variables

We can simplify radicals involving variables as long as the expressions under the radical sign are defined. This means that these expressions must *not* be negative. What about the answers? Note that $\sqrt{2^2} = 2$ and $\sqrt{(-2)^2} = \sqrt{4} = 2$. Thus the square root of a squared nonzero number is always positive. We use absolute values to indicate this.

ABSOLUTE VALUE OF A RADICAL

For any real number a,

$$\sqrt{a^2} = |a|$$

Of course, in examples and problems where the variables are assumed to be positive, the absolute-value bars are not necessary.

Teaching Tip

In Example 3c, have students think of the $\frac{14}{7}$ as the coefficient part and $\frac{\sqrt{20}}{\sqrt{10}}$ as the radical part. Simplify each and write the result as a product. Since

$$\frac{14}{7} = 2 \text{ and } \frac{\sqrt{20}}{\sqrt{10}} = \sqrt{2},$$

$$\text{then } \frac{14\sqrt{20}}{7\sqrt{10}} = 2\sqrt{2}.$$

EXAMPLE 4 **Simplifying radical expressions involving variables**

Simplify (assume all variables represent positive real numbers):

a. $\sqrt{49x^2}$ **b.** $\sqrt{81n^4}$ **c.** $\sqrt{50y^8}$ **d.** $\sqrt{x^{11}}$

SOLUTION

a. $\sqrt{49x^2} = \sqrt{49} \cdot \sqrt{x^2}$ Use the product rule.

$= 7x$

b. $\sqrt{81n^4} = \sqrt{81} \cdot \sqrt{n^4}$ Use the product rule.

$= 9n^2$ Since $(n^2)^2 = n^4$, $\sqrt{n^4} = n^2$.

c. $\sqrt{50y^8} = \sqrt{50} \cdot \sqrt{y^8}$ Use the product rule.

$= \sqrt{25 \cdot 2} \cdot \sqrt{y^8}$ Factor 50 using a perfect square.

$= 5\sqrt{2} \cdot y^4$ Since $\sqrt{25} = 5$ and $\sqrt{y^8} = y^4$

$= 5y^4\sqrt{2}$

d. $\sqrt{x^{11}} = \sqrt{x^{10} \cdot x}$ Since $x^{10} \cdot x = x^{11}$

$= x^5\sqrt{x}$ Since $\sqrt{x^{10}} = x^5$

PROBLEM 4

Simplify (all variables are positive):

a. $\sqrt{81y^2}$ **b.** $\sqrt{9n^4}$

c. $\sqrt{72x^{10}}$ **d.** $\sqrt{y^{13}}$

Teaching Tip

In Example 4c, help students notice that when a product involves a radical, the numerical coefficient and variables are written first and the radical last. See if they can name the property that allows reordering for multiplication.

Answers

3. a. $\frac{\sqrt{7}}{4}$ **b.** $\sqrt{5}$ **c.** $2\sqrt{3}$
4. a. $9y$ **b.** $3n^2$ **c.** $6x^5\sqrt{2}$
d. $y^6\sqrt{y}$

Simplifying Higher Roots

The product and quotient rules can be generalized for higher roots as shown in the box.

PROPERTIES OF RADICALS

For all real numbers where the indicated roots exist,

$$\sqrt[n]{a \cdot b} = \sqrt[n]{a} \cdot \sqrt[n]{b} \qquad \text{and} \qquad \sqrt[n]{\frac{a}{b}} = \frac{\sqrt[n]{a}}{\sqrt[n]{b}}$$

The idea when simplifying *cube* roots is to find factors that are *perfect cubes*. Similarly, to simplify *fourth roots,* we look for factors that are *perfect fourth powers*. This means that to simplify $\sqrt[4]{48}$, we need to find a perfect fourth-power factor of 48. Since $48 = 16 \cdot 3 = 2^4 \cdot 3$,

$$\sqrt[4]{48} = \sqrt[4]{2^4 \cdot 3} = 2 \cdot \sqrt[4]{3}$$

EXAMPLE 5 **Simplifying radical expressions involving higher roots**

Simplify:

a. $\sqrt[3]{54}$ **b.** $\sqrt[4]{162}$ **c.** $\sqrt[3]{\frac{27}{8}}$

SOLUTION

a. We are looking for a *cube* root, so we need to find a *perfect cube* factor of 54. Since $2^3 = 8$ is *not* a factor of 54, we try $3^3 = 27$:

$$\sqrt[3]{54} = \sqrt[3]{27 \cdot 2} = \sqrt[3]{3^3 \cdot 2} = 3\sqrt[3]{2}$$

b. This time we need a *perfect fourth* factor of 162. Since $2^4 = 16$ is *not* a factor of 162, we try $3^4 = 81$:

$$\sqrt[4]{162} = \sqrt[4]{81 \cdot 2} = \sqrt[4]{3^4 \cdot 2} = 3\sqrt[4]{2}$$

c. This time we are looking for *perfect cube* factors of 27 and 8. Now $3^3 = 27$ and $2^3 = 8$, so

$$\sqrt[3]{\frac{27}{8}} = \frac{\sqrt[3]{3^3}}{\sqrt[3]{2^3}} = \frac{3}{2}$$

PROBLEM 5

Simplify:

a. $\sqrt[3]{16}$ **b.** $\sqrt[4]{112}$ **c.** $\sqrt[3]{\frac{64}{27}}$

Teaching Tip

Remind students that when they are simplifying radicals involving higher roots they must carry the index along each time they rewrite the radical symbol.

Example:

$\sqrt{16} \neq \sqrt[4]{16}$ since $4 \neq 2$

NOTE

You can find $\sqrt[3]{\frac{27}{8}}$ *without* using the quotient rule. Since $3^3 = 27$ and $2^3 = 8$, you can write

$$\sqrt[3]{\frac{27}{8}} = \sqrt[3]{\frac{3^3}{2^3}} = \sqrt[3]{\left(\frac{3}{2}\right)^3} = \frac{3}{2}$$

Answers

5. a. $2\sqrt[3]{2}$ **b.** $2\sqrt[4]{7}$ **c.** $\frac{4}{3}$

Calculate It Finding Your Roots

Some years ago, it was quite tedious to find the root of a number, and extensive tables were consequently provided for that purpose. With the advent of scientific calculators, the task of finding roots was greatly reduced. But there's an even better way to find roots: using your calculator! Since we are mainly interested in finding the roots of integers, use the integer or decimal window in your calculator. When you use an integer window, the cursor moves in increments of 1. In the case of a decimal window, the cursor moves in increments of 0.1.

Now for the fun part: start by graphing $y = \sqrt{x}$. To find $\sqrt{25}$, use TRACE to move the cursor slowly through the x-values—$x = 1$, $x = 2$, $x = 3$, and so on until you get to $x = 25$. The y-value for $x = 5$ is $y = \sqrt{25} = 5$. (See Window 1.) Did you notice as you traced to the right on the curve $y = \sqrt{x}$ that the y-values for $\sqrt{1}$, $\sqrt{2}$, $\sqrt{3}$, and so on were displayed? In fact, what you have here is a built-in square root table! How would you obtain $\sqrt{8}$ as displayed in Window 2?

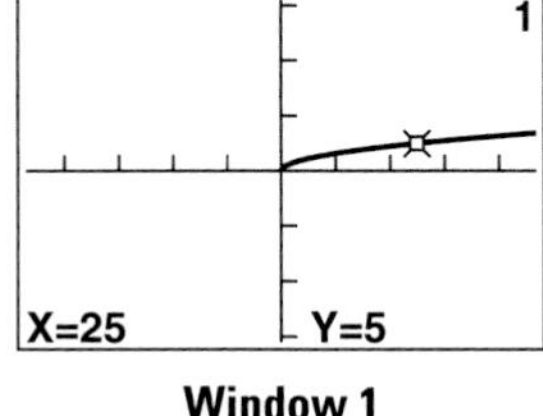

Window 1

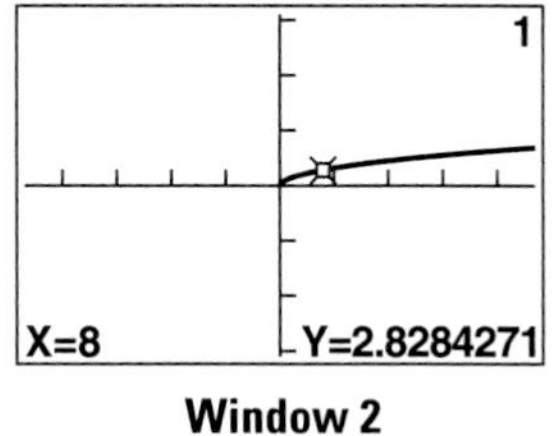

Window 2

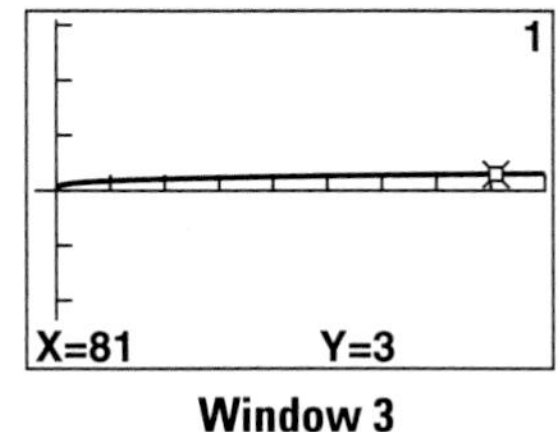

Window 3

Finally, there are some other things to learn from the graph in Window 1. Can you obtain $\sqrt{-2}$? Why not? Can you find fourth roots by graphing $y = \sqrt[4]{x}$? Note that the calculator doesn't have a fourth-root button. If you know that $\sqrt[4]{x} = x^{1/4}$, then you can graph $y = x^{1/4}$ and use a similar procedure to find the fourth root of a number.

Now look at Window 3, where $\sqrt[4]{81}$ is shown. What is the value for $\sqrt[4]{81}$? Can you get $\sqrt[4]{-1}$? Why not? Can you now discover how to obtain $\sqrt[3]{8}$ and $\sqrt[3]{-8}$ with your calculator? Do you get the same answer as you would doing it with paper and pencil?

Exercises 9.2

A In Problems 1–20, simplify.

1. $\sqrt{45}$ $3\sqrt{5}$ **2.** $\sqrt{72}$ $6\sqrt{2}$ **3.** $\sqrt{125}$ $5\sqrt{5}$

4. $\sqrt{175}$ $5\sqrt{7}$ **5.** $\sqrt{180}$ $6\sqrt{5}$ **6.** $\sqrt{162}$ $9\sqrt{2}$

7. $\sqrt{200}$ $10\sqrt{2}$ **8.** $\sqrt{245}$ $7\sqrt{5}$ **9.** $\sqrt{384}$ $8\sqrt{6}$

10. $\sqrt{486}$ $9\sqrt{6}$ **11.** $\sqrt{75}$ $5\sqrt{3}$ **12.** $\sqrt{80}$ $4\sqrt{5}$

13. $\sqrt{600}$ $10\sqrt{6}$ **14.** $\sqrt{324}$ 18 **15.** $\sqrt{361}$ 19 **16.** $\sqrt{648}$ $18\sqrt{2}$

17. $\sqrt{700}$ $10\sqrt{7}$ **18.** $\sqrt{726}$ $11\sqrt{6}$ **19.** $\sqrt{432}$ $12\sqrt{3}$ **20.** $\sqrt{507}$ $13\sqrt{3}$

In Problems 21–30, find the product. Assume all variables represent positive numbers.

21. $\sqrt{3} \cdot \sqrt{5}$ $\sqrt{15}$ **22.** $\sqrt{5} \cdot \sqrt{7}$ $\sqrt{35}$ **23.** $\sqrt{27} \cdot \sqrt{3}$ 9 **24.** $\sqrt{2} \cdot \sqrt{32}$ 8

25. $\sqrt{7} \cdot \sqrt{7}$ 7 **26.** $\sqrt{15} \cdot \sqrt{15}$ 15 **27.** $\sqrt{3} \cdot \sqrt{x}$ $\sqrt{3x}$ **28.** $\sqrt{5} \cdot \sqrt{y}$ $\sqrt{5y}$

29. $\sqrt{2a} \cdot \sqrt{18a}$ $6a$ **30.** $\sqrt{3b} \cdot \sqrt{12b}$ $6b$

B In Problems 31–40, find the quotient. Assume all variables represent positive numbers.

31. $\sqrt{\frac{2}{25}}$ $\frac{\sqrt{2}}{5}$ **32.** $\sqrt{\frac{5}{81}}$ $\frac{\sqrt{5}}{9}$ **33.** $\frac{\sqrt{20}}{\sqrt{5}}$ 2 **34.** $\frac{\sqrt{27}}{\sqrt{9}}$ $\sqrt{3}$

35. $\frac{\sqrt{27}}{\sqrt{3}}$ 3 **36.** $\frac{\sqrt{72}}{\sqrt{2}}$ 6 **37.** $\frac{15\sqrt{30}}{3\sqrt{10}}$ $5\sqrt{3}$ **38.** $\frac{52\sqrt{28}}{26\sqrt{4}}$ $2\sqrt{7}$

39. $\frac{18\sqrt{40}}{3\sqrt{10}}$ 12 **40.** $\frac{22\sqrt{40}}{11\sqrt{10}}$ 4

C In Problems 41–50, simplify. Assume all variables represent positive numbers.

41. $\sqrt{100a^2}$ $10a$ **42.** $\sqrt{16b^2}$ $4b$ **43.** $\sqrt{49a^4}$ $7a^2$ **44.** $\sqrt{64x^8}$ $8x^4$

45. $-\sqrt{32a^6}$ $-4a^3\sqrt{2}$ **46.** $-\sqrt{18b^{10}}$ $-3b^5\sqrt{2}$ **47.** $\sqrt{m^{13}}$ $m^6\sqrt{m}$ **48.** $\sqrt{n^{17}}$ $n^8\sqrt{n}$

49. $-\sqrt{27m^{11}}$ $-3m^5\sqrt{3m}$ **50.** $-\sqrt{50n^7}$ $-5n^3\sqrt{2n}$

D In Problems 51–60, simplify. Assume all variables represent positive numbers.

51. $\sqrt[3]{40}$ $2\sqrt[3]{5}$ **52.** $\sqrt[3]{108}$ $3\sqrt[3]{4}$ **53.** $\sqrt[3]{-16}$ $-2\sqrt[3]{2}$ **54.** $\sqrt[3]{-48}$ $-2\sqrt[3]{6}$

55. $\sqrt[3]{\frac{8}{27}}$ $\frac{2}{3}$ **56.** $\sqrt[3]{\frac{27}{64}}$ $\frac{3}{4}$ **57.** $\sqrt[4]{48}$ $2\sqrt[4]{3}$ **58.** $\sqrt[4]{243}$ $3\sqrt[4]{3}$

59. $\sqrt[3]{\frac{64}{27}}$ $\frac{4}{3}$ **60.** $\frac{\sqrt[4]{16}}{\sqrt[4]{81}}$ $\frac{2}{3}$

SKILL CHECKER

Try the Skill Checker Exercises so you'll be ready for the next section.

Combine like terms:

61. $5x + 7x$ $12x$ **62.** $8x^2 - 3x^2$ $5x^2$ **63.** $9x^3 + 7x^3 - 2x^3$ $14x^3$

WRITE ON

64. Explain why $\sqrt{(-2)^2}$ is not -2. Answers may vary.

65. Explain why $\sqrt[3]{(-2)^3} = -2$. Answers may vary.

66. Which of the following is in simplified form? Explain.

a. $-\sqrt{18}$ **b.** $\sqrt{41}$ **c.** $\sqrt{144}$

b; answers may vary.

67. Assume that p is a prime number.

a. Is $\sqrt{p}$ rational or irrational? irrational

b. Is $\sqrt{p}$ in simplified form? Explain.
Yes; answers may vary.

MASTERY TEST

If you know how to do these problems, you have learned your lesson!

Simplify (assume all variables represent positive numbers):

68. $\sqrt[3]{80}$ $2\sqrt[3]{5}$ **69.** $\sqrt[3]{135}$ $3\sqrt[3]{5}$ **70.** $\sqrt[4]{16x^4}$ $2x$ **71.** $\sqrt[3]{\frac{64}{27}}$ $\frac{4}{3}$

72. $\frac{\sqrt[4]{81}}{\sqrt[4]{16}}$ $\frac{3}{2}$ **73.** $\sqrt{100x^6}$ $10x^3$ **74.** $\sqrt{32x^7}$ $4x^3\sqrt{2x}$ **75.** $\frac{\sqrt{28}}{\sqrt{14}}$ $\sqrt{2}$

76. $\frac{\sqrt{72}}{\sqrt{2}}$ 6 **77.** $\sqrt{\frac{4}{9}}$ $\frac{2}{3}$ **78.** $\frac{\sqrt{17}}{\sqrt{36}}$ $\frac{\sqrt{17}}{6}$ **79.** $\sqrt{72} \cdot \sqrt{2}$ 12

80. $\sqrt{18} \cdot \sqrt{3x^2}$ $3x\sqrt{6}$

9.3 ADDITION AND SUBTRACTION OF RADICALS

To Succeed, Review How To . . .

1. Combine like terms (pp. 98–101).
2. Simplify radicals (pp. 649–650).
3. Use the distributive property (p. 101).

Objectives

A Add and subtract like radicals.

B Use the distributive property to simplify radicals.

C Rationalize the denominator in an expression.

GETTING STARTED

A Broken Pattern

In the preceding section we learned the product rule for radicals: $\sqrt{a \cdot b} = \sqrt{a} \cdot \sqrt{b}$, which we know means that the square root of a product is the product of the square roots. Similarly, the square root of a quotient is the quotient of the square roots; that is,

$$\sqrt{\frac{a}{b}} = \frac{\sqrt{a}}{\sqrt{b}}$$

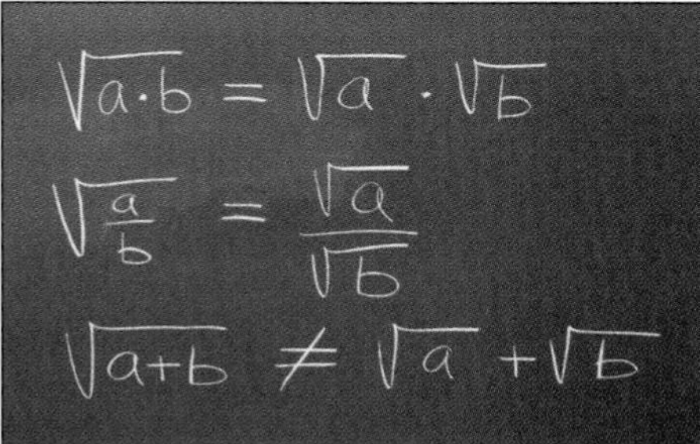

Is the square root of a sum or difference the sum or difference of the square roots? Let's look at an example. Is $\sqrt{9 + 16}$ the same as $\sqrt{9} + \sqrt{16}$? First, $\sqrt{9 + 16} = \sqrt{25} = 5$. But $\sqrt{9} + \sqrt{16} = 3 + 4 = 7$. Thus

$$\sqrt{9 + 16} \neq \sqrt{9} + \sqrt{16} \qquad \text{because} \qquad 5 \neq 3 + 4$$

So, what can be done with sums and differences involving radicals? We will learn that next.

A Adding and Subtracting Radicals

Expressions involving radicals can be handled using simple arithmetic rules. For example, since like terms can be combined,

$$3x + 7x = 10x$$

and

$$3\sqrt{6} + 7\sqrt{6} = 10\sqrt{6}$$

Similarly,

$$9x - 2x = 7x$$

and

$$9\sqrt{3} - 2\sqrt{3} = 7\sqrt{3}$$

Test your understanding of this idea in the next example.

Web It

To practice operations with radicals, go to the Bello Website at mhhe.com/bello and try link 9-3-1.

EXAMPLE 1 Adding and subtracting expressions involving radicals

Simplify:

a. $6\sqrt{7} + 9\sqrt{7}$ **b.** $8\sqrt{5} - 2\sqrt{5}$

SOLUTION

a. $6\sqrt{7} + 9\sqrt{7} = 15\sqrt{7}$ (just like $6x + 9x = 15x$)

b. $8\sqrt{5} - 2\sqrt{5} = 6\sqrt{5}$ (just like $8x - 2x = 6x$)

PROBLEM 1

Simplify:

a. $8\sqrt{5} + 2\sqrt{5}$ **b.** $7\sqrt{6} - 2\sqrt{6}$

Of course, you may have to simplify before combining **like radical terms—**that is, terms in which the *radical factors* are *exactly* the same. (For example, $4\sqrt{3}$ and $5\sqrt{3}$ are like radical terms.) Here is a problem that may seem difficult:

$$\sqrt{48} + \sqrt{75}$$

In this case, $\sqrt{48}$ and $\sqrt{75}$ are *not* like terms, and only like terms can be combined. But wait!

$$\sqrt{48} = \sqrt{16 \cdot 3} = \sqrt{16} \cdot \sqrt{3} = 4\sqrt{3}$$
$$\sqrt{75} = \sqrt{25 \cdot 3} = \sqrt{25} \cdot \sqrt{3} = 5\sqrt{3}$$

Now,

$$\begin{aligned}\sqrt{48} + \sqrt{75} &= 4\sqrt{3} + 5\sqrt{3} \\ &= 9\sqrt{3}\end{aligned}$$

Here are more complicated problems.

EXAMPLE 2 Adding and subtracting expressions involving radicals

Simplify:

a. $\sqrt{80} + \sqrt{20}$ **b.** $\sqrt{75} + \sqrt{12} - \sqrt{147}$

SOLUTION

a.
$$\sqrt{80} = \sqrt{16 \cdot 5} = \sqrt{16} \cdot \sqrt{5} = 4\sqrt{5}$$
$$\sqrt{20} = \sqrt{4 \cdot 5} = \sqrt{4} \cdot \sqrt{5} = 2\sqrt{5}$$
$$\sqrt{80} + \sqrt{20} = 4\sqrt{5} + 2\sqrt{5} = 6\sqrt{5}$$

b.
$$\sqrt{75} = \sqrt{25 \cdot 3} = \sqrt{25} \cdot \sqrt{3} = 5\sqrt{3}$$
$$\sqrt{12} = \sqrt{4 \cdot 3} = \sqrt{4} \cdot \sqrt{3} = 2\sqrt{3}$$
$$\sqrt{147} = \sqrt{49 \cdot 3} = \sqrt{49} \cdot \sqrt{3} = 7\sqrt{3}$$
$$\sqrt{75} + \sqrt{12} - \sqrt{147} = 5\sqrt{3} + 2\sqrt{3} - 7\sqrt{3} = 0$$

PROBLEM 2

Simplify:

a. $\sqrt{150} + \sqrt{24}$

b. $\sqrt{20} + \sqrt{80} - \sqrt{45}$

Answers

1. a. $10\sqrt{5}$ **b.** $5\sqrt{6}$
2. a. $7\sqrt{6}$ **b.** $3\sqrt{5}$

Using the Distributive Property to Simplify Expressions

Now let's see how we can use the distributive property to simplify an expression.

EXAMPLE 3 **Multiplying expressions involving radicals**

Simplify:

a. $\sqrt{5}(\sqrt{40} - \sqrt{2})$

b. $\sqrt{2}(\sqrt{2} - \sqrt{3})$

SOLUTION

a. Using the distributive property,

$$\sqrt{5}(\sqrt{40} - \sqrt{2}) = \sqrt{5}\sqrt{40} - \sqrt{5}\sqrt{2}$$

$$= \sqrt{200} - \sqrt{10}$$ Since $\sqrt{5}\sqrt{40} = \sqrt{200}$ and $\sqrt{5}\sqrt{2} = \sqrt{10}$

$$= \sqrt{100 \cdot 2} - \sqrt{10}$$ Since $\sqrt{100 \cdot 2} = \sqrt{100} \cdot \sqrt{2} = 10\sqrt{2}$

$$= 10\sqrt{2} - \sqrt{10}$$

b. $\sqrt{2}(\sqrt{2} - \sqrt{3}) = \sqrt{2}\sqrt{2} - \sqrt{2}\sqrt{3}$ Use the distributive property.

$$= 2 - \sqrt{6}$$ Since $\sqrt{2}\sqrt{2} = 2$

PROBLEM 3

Simplify:

a. $\sqrt{3}(\sqrt{45} - \sqrt{2})$

b. $\sqrt{5}(\sqrt{5} - \sqrt{3})$

Teaching Tip

Before Example 3, do an example of distributing without radicals.

Example:

$$5(x - 2) = 5 \cdot x - 5 \cdot 2$$
$$= 5x - 10$$

C Rationalizing Denominators

In the next chapter the solution of some quadratic equations will be of the form

$$\sqrt{\frac{9}{5}}$$

If we use the quotient rule for radicals, we obtain

$$\sqrt{\frac{9}{5}} = \frac{\sqrt{9}}{\sqrt{5}} = \frac{3}{\sqrt{5}}$$

The expression $\frac{3}{\sqrt{5}}$ contains the square root of a nonperfect square, which is an irrational number. To simplify $\frac{3}{\sqrt{5}}$, we **rationalize the denominator.** This means we remove all radicals from the denominator.

PROCEDURE

Rationalizing Denominators

Method 1. Multiply *both* the *numerator* and *denominator* of the fraction by the **square root** in the denominator; or

Method 2. Multiply numerator and denominator by the square root of a number that makes the denominator the square root of a perfect square.

Web It

To read about rationalizing denominators, go to link 9-3-2 on the Bello Website at mhhe.com/bello.

Then try practice problems at link 9-3-3.

Teaching Tip

Have students note that in Example 3a and 3b the final two terms did not contain "like" radicals. Thus, the terms cannot be subtracted and remain $10\sqrt{2} - \sqrt{10}$ and $2 - \sqrt{6}$.

Answers

3. a. $3\sqrt{15} - \sqrt{6}$ **b.** $5 - \sqrt{15}$

Thus to rationalize the denominator in $\frac{3}{\sqrt{5}}$, we multiply the numerator and denominator by $\sqrt{5}$ to obtain

$$\frac{3}{\sqrt{5}} = \frac{3 \cdot \sqrt{5}}{5 \cdot \sqrt{5}}$$

$$= \frac{3\sqrt{5}}{5} \quad \text{Since } \sqrt{5} \cdot \sqrt{5} = 5$$

Note that the idea in rationalizing

$$\sqrt{\frac{a}{b}} = \frac{\sqrt{a}}{\sqrt{b}}$$

is to make the denominator $\sqrt{b}$ a square root of a perfect square. Multiplying numerator and denominator by $\sqrt{b}$ *always* works, but you can save time if you find a factor smaller than $\sqrt{b}$ that will make the denominator a square root of a perfect square. Thus when rationalizing $\frac{\sqrt{3}}{\sqrt{8}}$, you could *first* multiply numerator and denominator by $\sqrt{8}$. However, it's *better* to multiply numerator and denominator by $\sqrt{2}$, as shown in the next example.

EXAMPLE 4 **Rationalizing denominators: Two methods**

Write with a rationalized denominator: $\sqrt{\frac{3}{8}}$

SOLUTION

Method 1. Use the quotient rule and then multiply numerator and denominator by $\sqrt{8}$.

$$\sqrt{\frac{3}{8}} = \frac{\sqrt{3}}{\sqrt{8}} \quad \text{Use the quotient rule.}$$

$$= \frac{\sqrt{3} \cdot \sqrt{8}}{\sqrt{8} \cdot \sqrt{8}} \quad \text{Multiply numerator and denominator by } \sqrt{8}.$$

$$= \frac{\sqrt{24}}{8} \quad \text{Since } \sqrt{3} \cdot \sqrt{8} = \sqrt{24} \text{ and } \sqrt{8} \cdot \sqrt{8} = 8$$

$$= \frac{\sqrt{4 \cdot 6}}{8} \quad \text{Since } 24 = 4 \cdot 6 \text{ and 4 is a perfect square}$$

$$= \frac{2 \cdot \sqrt{6}}{8} \quad \text{Since } \sqrt{4} = 2$$

$$= \frac{\sqrt{6}}{4} \quad \text{Divide numerator and denominator by 2.}$$

Method 2. If you noticed that multiplying numerator and denominator by $\sqrt{2}$ yields $\sqrt{16} = 4$ in the denominator, you could have obtained

$$\sqrt{\frac{3}{8}} = \frac{\sqrt{3}}{\sqrt{8}} \quad \text{Use the quotient rule.}$$

$$= \frac{\sqrt{3} \cdot \sqrt{2}}{\sqrt{8} \cdot \sqrt{2}} \quad \text{Multiply by } \sqrt{2} \text{ so the denominator is } \sqrt{16} = 4.$$

$$= \frac{\sqrt{6}}{\sqrt{16}} \quad \text{Since } \sqrt{3} \cdot \sqrt{2} = \sqrt{6}$$

$$= \frac{\sqrt{6}}{4} \quad \text{Since } \sqrt{16} = 4$$

You get the same answer, worked with smaller numbers, and did it in fewer steps! *You can save time by looking for factors that make the denominator the square root of a perfect square.*

PROBLEM 4

Write with a rationalized denominator: $\sqrt{\frac{5}{12}}$

Teaching Tip

Sometimes students have difficulty knowing when they have finished simplifying a fraction involving radicals. Here is a check list.

1. There are no radicals in the denominator.
2. If there is a radical in the numerator, it has been simplified.
3. The nonradical numbers in the numerator and denominator are reduced to lowest terms.

Answer

4. $\frac{\sqrt{15}}{6}$

EXAMPLE 5 **Rationalizing denominators by making them perfect squares**

Write with a rationalized denominator: $\sqrt{\frac{x^2}{32}}$ $(x > 0)$

SOLUTION Since

$$\sqrt{\frac{x^2}{32}} = \frac{\sqrt{x^2}}{\sqrt{32}}$$

we could multiply numerator and denominator by $\sqrt{32}$. However, multiplying by $\sqrt{2}$ gives a denominator of $\sqrt{64} = 8$. Thus

$$\sqrt{\frac{x^2}{32}} = \frac{\sqrt{x^2} \cdot \sqrt{2}}{\sqrt{32} \cdot \sqrt{2}} = \frac{x\sqrt{2}}{\sqrt{64}} = \frac{x\sqrt{2}}{8}$$

PROBLEM 5

Write with a rationalized denominator:

$$\sqrt{\frac{x^2}{50}} \quad x > 0$$

Answer

5. $\frac{x\sqrt{2}}{10}$

Exercises 9.3

Boost your GRADE at mathzone.com!

- Practice Problems
- Self-Tests
- Videos
- NetTutor
- e-Professors

A In Problems 1–16, perform the indicated operations (and simplify).

1. $6\sqrt{7} + 4\sqrt{7}$ $10\sqrt{7}$
2. $4\sqrt{11} + 9\sqrt{11}$ $13\sqrt{11}$
3. $9\sqrt{13} - 4\sqrt{13}$ $5\sqrt{13}$
4. $6\sqrt{10} - 2\sqrt{10}$ $4\sqrt{10}$
5. $\sqrt{32} + \sqrt{50} - \sqrt{72}$ $3\sqrt{2}$
6. $\sqrt{12} + \sqrt{27} - \sqrt{75}$ 0
7. $\sqrt{162} + \sqrt{50} - \sqrt{200}$ $4\sqrt{2}$
8. $\sqrt{48} + \sqrt{75} - \sqrt{363}$ $-2\sqrt{3}$
9. $9\sqrt{48} - 5\sqrt{27} + 3\sqrt{12}$ $27\sqrt{3}$
10. $3\sqrt{32} - 5\sqrt{8} + 4\sqrt{50}$ $22\sqrt{2}$
11. $5\sqrt{7} - 3\sqrt{28} - 2\sqrt{63}$ $-7\sqrt{7}$
12. $3\sqrt{28} - 6\sqrt{7} - 2\sqrt{175}$ $-10\sqrt{7}$
13. $-5\sqrt{3} + 8\sqrt{75} - 2\sqrt{27}$ $29\sqrt{3}$
14. $-6\sqrt{99} + 6\sqrt{44} - \sqrt{176}$ $-10\sqrt{11}$
15. $-3\sqrt{45} + \sqrt{20} - \sqrt{5}$ $-8\sqrt{5}$
16. $-5\sqrt{27} + \sqrt{12} - 5\sqrt{48}$ $-33\sqrt{3}$

B In Problems 17–30, simplify.

17. $\sqrt{10}(\sqrt{20} - \sqrt{3})$ $10\sqrt{2} - \sqrt{30}$
18. $\sqrt{10}(\sqrt{30} - \sqrt{2})$ $10\sqrt{3} - 2\sqrt{5}$
19. $\sqrt{6}(\sqrt{14} + \sqrt{5})$ $2\sqrt{21} + \sqrt{30}$
20. $\sqrt{14}(\sqrt{18} + \sqrt{3})$ $6\sqrt{7} + \sqrt{42}$
21. $\sqrt{3}(\sqrt{3} - \sqrt{2})$ $3 - \sqrt{6}$
22. $\sqrt{6}(\sqrt{6} - \sqrt{5})$ $6 - \sqrt{30}$
23. $\sqrt{5}(\sqrt{2} + \sqrt{5})$ $\sqrt{10} + 5$
24. $\sqrt{3}(\sqrt{2} + \sqrt{3})$ $\sqrt{6} + 3$
25. $\sqrt{6}(\sqrt{2} - \sqrt{3})$ $2\sqrt{3} - 3\sqrt{2}$
26. $\sqrt{5}(\sqrt{15} - \sqrt{27})$ $5\sqrt{3} - 3\sqrt{15}$
27. $2(\sqrt{2} - 5)$ $2\sqrt{2} - 10$
28. $\sqrt{5}(3 - \sqrt{5})$ $3\sqrt{5} - 5$
29. $\sqrt{2}(\sqrt{6} - 3)$ $2\sqrt{3} - 3\sqrt{2}$
30. $\sqrt{3}(\sqrt{6} - 4)$ $3\sqrt{2} - 4\sqrt{3}$

C In Problems 31–50, rationalize the denominator. Assume all variables are positive real numbers.

31. $\frac{3}{\sqrt{6}}$ $\frac{\sqrt{6}}{2}$
32. $\frac{6}{\sqrt{7}}$ $\frac{6\sqrt{7}}{7}$
33. $\frac{-10}{\sqrt{5}}$ $-2\sqrt{5}$
34. $\frac{-9}{\sqrt{3}}$ $-3\sqrt{3}$
35. $\frac{\sqrt{8}}{\sqrt{2}}$ 2
36. $\frac{\sqrt{48}}{\sqrt{3}}$ 4
37. $\frac{-\sqrt{2}}{\sqrt{5}}$ $\frac{-\sqrt{10}}{5}$
38. $\frac{-\sqrt{3}}{\sqrt{7}}$ $\frac{-\sqrt{21}}{7}$
39. $\frac{\sqrt{2}}{\sqrt{8}}$ $\frac{1}{2}$
40. $\frac{\sqrt{3}}{\sqrt{12}}$ $\frac{1}{2}$
41. $\frac{\sqrt{x^2}}{\sqrt{18}}$ $\frac{x\sqrt{2}}{6}$
42. $\frac{\sqrt{a^4}}{\sqrt{32}}$ $\frac{a^2\sqrt{2}}{8}$

43. $\frac{\sqrt{a^2}}{\sqrt{b}}$ $\frac{a\sqrt{b}}{b}$

44. $\frac{\sqrt{x^4}}{\sqrt{y}}$ $\frac{x^2\sqrt{y}}{y}$

45. $\sqrt{\frac{3}{10}}$ $\frac{\sqrt{30}}{10}$

46. $\sqrt{\frac{2}{27}}$ $\frac{\sqrt{6}}{9}$

47. $\sqrt{\frac{x^2}{32}}$ $\frac{x\sqrt{2}}{8}$

48. $\sqrt{\frac{x}{18}}$ $\frac{\sqrt{2x}}{6}$

49. $\sqrt{\frac{x^4}{20}}$ $\frac{x^2\sqrt{5}}{10}$

50. $\sqrt{\frac{x^6}{72}}$ $\frac{x^3\sqrt{2}}{12}$

SKILL CHECKER

Try the Skill Checker Exercises so you'll be ready for the next section.

Multiply:

51. $(x + 3)(x - 3)$ $x^2 - 9$

52. $(a + b)(a - b)$ $a^2 - b^2$

Simplify:

53. $\frac{6x + 12}{3}$ $2x + 4$

54. $\frac{4x - 8}{2}$ $2x - 4$

USING YOUR KNOWLEDGE

"Radical" Shortcuts

Suppose you want to rationalize the denominator in the expression $\sqrt{\frac{3}{32}}$. Using the quotient rule, we can write

$$\sqrt{\frac{3}{32}} = \frac{\sqrt{3}}{\sqrt{32}} = \frac{\sqrt{3} \cdot \sqrt{32}}{\sqrt{32} \cdot \sqrt{32}} = \frac{\sqrt{96}}{32}$$

$$= \frac{\sqrt{16 \cdot 6}}{32} = \frac{4 \cdot \sqrt{6}}{32} = \frac{\sqrt{6}}{8}$$

A shorter way is as follows:

$$\sqrt{\frac{3}{32}} = \sqrt{\frac{3 \cdot 2}{32 \cdot 2}} = \sqrt{\frac{6}{64}} = \frac{\sqrt{6}}{8}$$

55. Use this shorter procedure to do Problems 45–50.

WRITE ON

56. In Problems 31–50, we specified that *all* variables should be positive real numbers. Specify which variables *have to be* positive real numbers and in which problems this must occur. Answers may vary.

57. In Example 4, we rationalized the denominator in the expression

$$\sqrt{\frac{3}{8}} = \frac{\sqrt{3}}{\sqrt{8}}$$

by *first* using the quotient rule and writing

$$\sqrt{\frac{3}{8}} = \frac{\sqrt{3}}{\sqrt{8}}$$

and then multiplying numerator and denominator by $\sqrt{2}$. How can you do the example without first using the quotient rule? Answers may vary.

58. Write an explanation of what is meant by like radicals. Answers may vary.

MASTERY TEST

If you know how to do these problems, you have learned your lesson!

Simplify:

59. $8\sqrt{3} + 5\sqrt{3}$ $13\sqrt{3}$

60. $7\sqrt{5} - 2\sqrt{5}$ $5\sqrt{5}$

61. $\sqrt{18} + 3\sqrt{2}$ $6\sqrt{2}$

62. $\sqrt{32} - 3\sqrt{2}$ $\sqrt{2}$

63. $7\sqrt{18} + 5\sqrt{2} - 7\sqrt{8}$ $12\sqrt{2}$

64. $3(\sqrt{5} - 2)$ $3\sqrt{5} - 6$

65. $\sqrt{3}(5 - \sqrt{3})$ $5\sqrt{3} - 3$

66. $\sqrt{18}(\sqrt{2} - 2)$ $6 - 6\sqrt{2}$

Rationalize the denominator, assuming all variables are positive real numbers:

67. $\frac{3}{\sqrt{7}}$ $\frac{3\sqrt{7}}{7}$

68. $\sqrt{\frac{x^6}{10}}$ $\frac{x^3\sqrt{10}}{10}$

69. $\frac{\sqrt{x}}{\sqrt{2}}$ $\frac{\sqrt{2x}}{2}$

70. $\frac{\sqrt{x^2}}{\sqrt{20}}$ $\frac{x\sqrt{5}}{10}$

9.4 SIMPLIFYING RADICALS

To Succeed, Review How To . . .

1. Find the root of an expression (pp. 641, 643).
2. Add, subtract, multiply, and divide radicals (pp. 647–649, 653–655).
3. Rationalize the denominator in an expression (pp. 655–657).
4. Expand $(x \pm y)^2$ (pp. 357–358).

Objectives

A Simplify a radical expression involving products, quotients, sums, or differences.

B Use the conjugate of a number to rationalize the denominator of an expression.

C Reduce a fraction involving a radical by factoring.

GETTING STARTED

Breaking the Sound Barrier

How fast can this plane travel? The answer is classified information, but it exceeds twice the speed of sound (747 miles per hour). It is said that the plane's speed is more than Mach 2. The formula for calculating the Mach number is

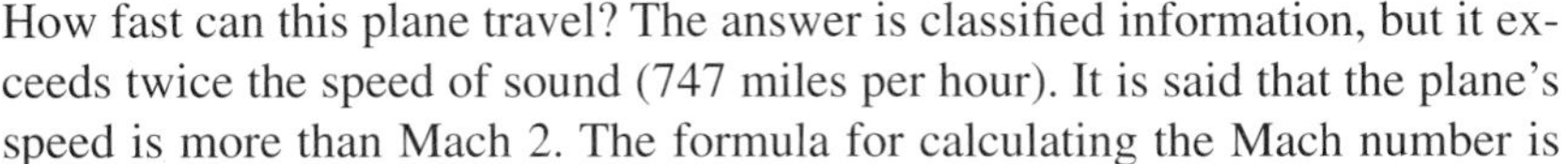

$$M = \sqrt{\frac{2}{\gamma}}\sqrt{\frac{P_2 - P_1}{P_1}}$$

where P_1 and P_2 are air pressures. This expression can be simplified by multiplying the radical expressions and then rationalizing the denominator (we discuss how to do this in the *Using Your Knowledge*). In this section you will learn how to simplify more complicated radical expressions such as the one above by using the techniques we've discussed in the last three sections. What did we do in those sections? We found the roots of algebraic expressions, multiplied and divided radicals, added and subtracted radicals, and rationalized denominators. You will see that the procedure used to *completely* simplify a radical involves steps that perform these tasks in precisely the order in which the topics were studied.

A Simplifying Radical Expressions

In the preceding sections you were asked to "simplify" expressions involving radicals. To make this idea more precise and to help you simplify radical expressions, we use the following rules.

RULES

Simplifying Radical Expressions

1. Whenever possible, write the *rational-number* representation of a radical expression. For example, write

$$\sqrt{81} \text{ as } 9, \qquad \sqrt{\frac{4}{9}} \text{ as } \frac{2}{3} \qquad \text{and} \qquad \sqrt[3]{\frac{1}{8}} \text{ as } \frac{1}{2}$$

2. Use the product rule $\sqrt{x} \cdot \sqrt{y} = \sqrt{xy}$ to write indicated products as a single radical. For example, write

$$\sqrt{6} \text{ instead of } \sqrt{2} \cdot \sqrt{3} \qquad \text{and} \qquad \sqrt{2ab} \text{ instead of } \sqrt{2a} \cdot \sqrt{b}$$

Web It

For a review and practice simplifying many types of expressions containing radicals, go to link 9-4-1 on the Bello Website at mhhe.com/bello.

3. Use the quotient rule

$$\frac{\sqrt{x}}{\sqrt{y}} = \sqrt{\frac{x}{y}}$$

to write indicated quotients as a single radical. For example, write

$$\frac{\sqrt{6}}{\sqrt{2}} \text{ as } \sqrt{3} \quad \text{and} \quad \frac{\sqrt[3]{10}}{\sqrt[3]{5}} \text{ as } \sqrt[3]{2}$$

4. If a radicand has a perfect square as a factor, write the radical expression as the product of the square root of the perfect square and the radical of the other factor. A similar statement applies to cubes and higher roots. For example, write

$$\sqrt{18} = \sqrt{9 \cdot 2} \text{ as } 3\sqrt{2} \quad \text{and} \quad \sqrt[3]{54} = \sqrt[3]{27 \cdot 2} \text{ as } 3\sqrt[3]{2}$$

5. Combine like radicals whenever possible. For example,

$$2\sqrt{5} + 8\sqrt{5} = (2 + 8)\sqrt{5} = 10\sqrt{5}$$

$$9\sqrt{11} - 2\sqrt{11} = (9 - 2)\sqrt{11} = 7\sqrt{11}$$

6. Rationalize the denominator of algebraic expressions. For example,

$$\frac{3}{\sqrt{2}} = \frac{3 \cdot \sqrt{2}}{\sqrt{2} \cdot \sqrt{2}} = \frac{3 \cdot \sqrt{2}}{2}$$

Now let's use these rules.

EXAMPLE 1 **Simplifying radicals: Sums and differences**

Simplify:

a. $\sqrt{9 + 16} - \sqrt{4}$ b. $9\sqrt{6} - \sqrt{2} \cdot \sqrt{3}$ c. $\frac{\sqrt{6x^3}}{\sqrt{2x^2}}$, $x > 0$

SOLUTION In each case, a rule applies.

a. Since $\sqrt{9 + 16} = \sqrt{25} = 5$ and $\sqrt{4} = 2$,

$\sqrt{9 + 16} - \sqrt{4} = 5 - 2 = 3$ Use Rules 1 and 5.

b. $9\sqrt{6} - \sqrt{2} \cdot \sqrt{3} = 9\sqrt{6} - \sqrt{6} = (9 - 1)\sqrt{6} = 8\sqrt{6}$ Use Rules 2 and 5.

c. $\frac{\sqrt{6x^3}}{\sqrt{2x^2}} = \sqrt{\frac{6x^3}{2x^2}} = \sqrt{3x}$ Use Rule 3.

PROBLEM 1

Simplify:

a. $\sqrt{60 + 4} - \sqrt{9}$

b. $8\sqrt{10} - \sqrt{2}\sqrt{5}$

c. $\frac{\sqrt{10x^3}}{\sqrt{5x^2}}$, $x > 0$

EXAMPLE 2 **Simplifying quotients**

Simplify:

a. $\frac{\sqrt[3]{256}}{\sqrt[3]{2}}$ b. $\frac{9}{\sqrt[3]{4}}$

PROBLEM 2

Simplify:

a. $\frac{\sqrt[3]{108}}{\sqrt[3]{2}}$ b. $\frac{7}{\sqrt[3]{9}}$

Answers

1. a. 5 b. $7\sqrt{10}$ c. $\sqrt{2x}$

2. a. $3\sqrt[3]{2}$ b. $\frac{7\sqrt[3]{3}}{3}$

SOLUTION A rule applies in each case.

a. $\dfrac{\sqrt[3]{256}}{\sqrt[3]{2}} = \sqrt[3]{128}$ Use Rule 3.

$= \sqrt[3]{64 \cdot 2}$ Since $64 \cdot 2 = 128$ and 64 is a perfect cube

$= 4\sqrt[3]{2}$ Use Rule 4.

b. We have to make the denominator the cube root of a perfect cube. To do this, we multiply numerator and denominator by $\sqrt[3]{2}$. Note that the denominator will be $\sqrt[3]{4} \cdot \sqrt[3]{2} = \sqrt[3]{8} = 2$. Thus we have

$$\frac{9}{\sqrt[3]{4}} = \frac{9 \cdot \sqrt[3]{2}}{\sqrt[3]{4} \cdot \sqrt[3]{2}} = \frac{9 \cdot \sqrt[3]{2}}{\sqrt[3]{8}} = \frac{9\sqrt[3]{2}}{2}$$

Teaching Tip

In Example 2b, rewriting $\sqrt[3]{4}$ as $\sqrt[3]{2^2}$ might make it more obvious to students that multiplying by $\sqrt[3]{2}$ will make the denominator into the cube root of a perfect cube.

Teaching Tip

Before Example 3, do the following examples without radicals to refresh the technique for multiplying binomials.

Examples:

$(x + 5)(x - 4) = x^2 - 4x + 5x - 20$
$= x^2 + x - 20$

$(x + 2)(x - 2) = x^2 - 4$

Do you recall the FOIL method? It and the special products can also be used to simplify expressions involving radicals. We do this next.

EXAMPLE 3 Using FOIL to simplify products

Simplify:

a. $(\sqrt{2} + 5\sqrt{3})(\sqrt{2} - 4\sqrt{3})$ **b.** $(\sqrt{3} + 2\sqrt{5})(\sqrt{3} - 2\sqrt{5})$

SOLUTION Using the FOIL method, we have

a. $(\sqrt{2} + 5\sqrt{3})(\sqrt{2} - 4\sqrt{3})$

$$= \underbrace{\sqrt{2} \cdot \sqrt{2}}_{F} + \underbrace{\sqrt{2}(-4\sqrt{3})}_{O} + \underbrace{5\sqrt{3}(\sqrt{2})}_{I} + \underbrace{(5\sqrt{3})(-4\sqrt{3})}_{L}$$

$= 2 - 4\sqrt{6} + 5\sqrt{6} - 20 \cdot 3$ Use the product rule.

$= 2 + \sqrt{6} - 60$ Combine radicals.

$= -58 + \sqrt{6}$ Since $2 - 60 = -58$

b. Using SP4 (p. 359), we have

$$(X + A)(X - A) = X^2 - A^2$$

$$\begin{aligned}(\sqrt{3} + 2\sqrt{5})(\sqrt{3} - 2\sqrt{5}) &= (\sqrt{3})^2 - (2\sqrt{5})^2 \\ &= 3 - [(2)^2(\sqrt{5})^2] \\ &= 3 - (4)(5) \\ &= 3 - 20 \\ &= -17\end{aligned}$$

PROBLEM 3

Simplify:

a. $(\sqrt{3} + 5\sqrt{2})(\sqrt{3} - 4\sqrt{2})$

b. $(2 + 3\sqrt{5})(2 - 3\sqrt{5})$

Answers

3. a. $-37 + \sqrt{6}$ **b.** -41

B Using Conjugates to Rationalize Denominators

Web It

To get some more practice with radicals and rationalizing the denominator using conjugates, go to link 9-4-2 on the Bello Website at mhhe.com/bello.

In Example 3(b), the product of the sum $\sqrt{3} + 2\sqrt{5}$ and the difference $\sqrt{3} - 2\sqrt{5}$ is the rational number -17. This is no coincidence! The expressions $\sqrt{3} + 2\sqrt{5}$ and $\sqrt{3} - 2\sqrt{5}$ are **conjugates** of each other. In general, the expressions $a\sqrt{b} + c\sqrt{d}$ and $a\sqrt{b} - c\sqrt{d}$ are conjugates of each other. Their product is obtained by using SP4 and always results in a rational number. Here is one way of using conjugates to simplify radical expressions.

PROCEDURE

Using Conjugates to Simplify Radical Expressions

To simplify an algebraic expression with two terms in the denominator, at least one of which is a square root, multiply both numerator and denominator by the conjugate of the denominator.

EXAMPLE 4 **Using conjugates to rationalize denominators**

Simplify:

a. $\dfrac{7}{\sqrt{5} + 1}$ **b.** $\dfrac{3}{\sqrt{5} - \sqrt{3}}$

SOLUTION

a. The denominator $\sqrt{5} + 1$ has two terms, one of which is a radical. To simplify

$$\frac{7}{\sqrt{5} + 1}$$

multiply numerator and denominator by the conjugate of the denominator, which is $\sqrt{5} - 1$. [*Note:* $(\sqrt{5} + 1)(\sqrt{5} - 1) = (\sqrt{5})^2 - (1)^2$.] Thus,

$$\frac{7}{\sqrt{5} + 1} = \frac{7 \cdot (\sqrt{5} - 1)}{(\sqrt{5} + 1)(\sqrt{5} - 1)}$$ Multiply the numerator and denominator by $\sqrt{5} - 1$.

$$= \frac{7\sqrt{5} - 7}{(\sqrt{5})^2 - (1)^2}$$ Use the distributive property and SP4.

$$= \frac{7\sqrt{5} - 7}{5 - 1}$$ Since $(\sqrt{5})^2 = 5$

$$= \frac{7\sqrt{5} - 7}{4}$$

b. This time we multiply the numerator and denominator of the fraction by the conjugate of $\sqrt{5} - \sqrt{3}$, which is $\sqrt{5} + \sqrt{3}$.

$$\frac{3}{\sqrt{5} - \sqrt{3}} = \frac{3(\sqrt{5} + \sqrt{3})}{(\sqrt{5} - \sqrt{3})(\sqrt{5} + \sqrt{3})}$$ Multiply the numerator and denominator by $\sqrt{5} + \sqrt{3}$.

$$= \frac{3\sqrt{5} + 3\sqrt{3}}{(\sqrt{5})^2 - (\sqrt{3})^2}$$ Use the distributive property and SP4.

$$= \frac{3\sqrt{5} + 3\sqrt{3}}{5 - 3}$$ Since $(\sqrt{5})^2 = 5$ and $(\sqrt{3})^2 = 3$

$$= \frac{3\sqrt{5} + 3\sqrt{3}}{2}$$

PROBLEM 4

Simplify:

a. $\dfrac{5}{\sqrt{7} + 1}$ **b.** $\dfrac{5}{\sqrt{6} - \sqrt{2}}$

Answers

4. a. $\frac{5\sqrt{7} - 5}{6}$ **b.** $\frac{5\sqrt{6} + 5\sqrt{2}}{4}$

C Reducing Fractions Involving Radicals by Factoring

In the next chapter we shall encounter solutions of quadratic equations written as

$$\frac{8 + \sqrt{20}}{4}$$

To simplify these expressions, first note that $\sqrt{20} = \sqrt{4 \cdot 5} = 2\sqrt{5}$. Thus

$$\frac{8 + \sqrt{20}}{4} = \frac{8 + 2\sqrt{5}}{4}$$ Since $\sqrt{20} = 2\sqrt{5}$

$$= \frac{2 \cdot (4 + \sqrt{5})}{2 \cdot 2}$$ Factor the numerator and denominator.

$$= \frac{4 + \sqrt{5}}{2}$$ Divide by 2.

Teaching Tip

For Example 5a, we can also reduce

$$\frac{-4 + \sqrt{8}}{2} = \frac{-4 + 2\sqrt{2}}{2}$$

by writing

$$\frac{-4 + 2\sqrt{2}}{2} = \frac{-4}{2} + \frac{2\sqrt{2}}{2} = -2 + \sqrt{2}$$

EXAMPLE 5 Reducing fractions by factoring first

Simplify:

a. $\frac{-4 + \sqrt{8}}{2}$ **b.** $\frac{-8 + \sqrt{28}}{4}$

SOLUTION

a. Since $\sqrt{8} = \sqrt{4 \cdot 2} = 2\sqrt{2}$, we have

$$\frac{-4 + \sqrt{8}}{2} = \frac{-4 + 2\sqrt{2}}{2}$$

$$= \frac{2 \cdot (-2 + \sqrt{2})}{2}$$ Factor the numerator and denominator.

$$= -2 + \sqrt{2}$$ Divide numerator and denominator by 2.

b. Since $\sqrt{28} = \sqrt{4 \cdot 7} = 2\sqrt{7}$, we have

$$\frac{-8 + \sqrt{28}}{4} = \frac{-8 + 2\sqrt{7}}{4}$$

$$= \frac{2(-4 + \sqrt{7})}{2 \cdot 2}$$ Factor the numerator and denominator.

$$= \frac{-4 + \sqrt{7}}{2}$$ Divide numerator and denominator by 2.

PROBLEM 5

Simplify:

a. $\frac{-9 + \sqrt{18}}{3}$

b. $\frac{-12 + \sqrt{24}}{6}$

Answers

5. a. $-3 + \sqrt{2}$ **b.** $\frac{-6 + \sqrt{6}}{3}$

Exercises 9.4

A In Problems 1–36, simplify. Assume all variables represent positive real numbers.

1. $\sqrt{36} + \sqrt{100}$ 16

2. $\sqrt{9} + \sqrt{25}$ 8

3. $\sqrt{144 + 25}$ 13

4. $\sqrt{24^2 + 10^2}$ 26

5. $\sqrt{4} - \sqrt{36}$ -4

6. $\sqrt{64} - \sqrt{121}$ -3

7. $\sqrt{13^2 - 12^2}$ 5

8. $\sqrt{17^2 - 8^2}$ 15

9. $15\sqrt{10} + \sqrt{90}$ $18\sqrt{10}$

10. $8\sqrt{7} + \sqrt{28}$ $10\sqrt{7}$

11. $14\sqrt{11} - \sqrt{44}$ $12\sqrt{11}$

12. $3\sqrt{13} - \sqrt{52}$ $\sqrt{13}$

13. $\sqrt[3]{54} - \sqrt[3]{8}$ $3\sqrt[3]{2} - 2$

14. $\sqrt[3]{81} - \sqrt[3]{16}$ $3\sqrt[3]{3} - 2\sqrt[3]{2}$

15. $5\sqrt[3]{16} - 3\sqrt[3]{54}$ $\sqrt[3]{2}$

16. $\sqrt[3]{250} - \sqrt[3]{128}$ $\sqrt[3]{2}$

17. $\sqrt{\dfrac{9x^2}{x}}$ $3\sqrt{x}$

18. $\sqrt{\dfrac{16x^5}{x^2}}$ $4x\sqrt{x}$

19. $\sqrt{\dfrac{81y^7}{16y^5}}$ $\dfrac{9y}{4}$

20. $\sqrt{\dfrac{4x^3y^4}{3z}}$ $\dfrac{2xy^2\sqrt{3xz}}{3z}$

21. $\sqrt{\dfrac{64a^4b^6}{3ab^4}}$ $\dfrac{8ab\sqrt{3a}}{3}$

22. $\sqrt{\dfrac{25a^5b^6}{7a^2b^4c^2}}$ $\dfrac{5ab\sqrt{7a}}{7c}$

23. $\sqrt[3]{\dfrac{8b^6c^{10}}{27bc}}$ $\dfrac{2bc^3\sqrt[3]{b^2}}{3}$

24. $\sqrt[3]{\dfrac{64ab^4}{125a^4b}}$ $\dfrac{4b}{5a}$

25. $\dfrac{\sqrt[3]{500}}{\sqrt[3]{2}}$ $5\sqrt[3]{2}$

26. $\dfrac{\sqrt[3]{243}}{\sqrt[3]{3}}$ $3\sqrt[3]{3}$

27. $\dfrac{6}{\sqrt[3]{9}}$ $2\sqrt[3]{3}$

28. $\dfrac{7}{\sqrt[3]{2}}$ $\dfrac{7\sqrt[3]{4}}{2}$

29. $(\sqrt{3} + 6\sqrt{5})(\sqrt{3} - 4\sqrt{5})$ $2\sqrt{15} - 117$

30. $(\sqrt{5} + 6\sqrt{2})(\sqrt{5} - 3\sqrt{2})$ $3\sqrt{10} - 31$

31. $(\sqrt{2} + 3\sqrt{3})(\sqrt{2} + 3\sqrt{3})$ $6\sqrt{6} + 29$

32. $(3\sqrt{2} + \sqrt{3})(3\sqrt{2} + \sqrt{3})$ $21 + 6\sqrt{6}$

33. $(5\sqrt{2} - 3\sqrt{3})(5\sqrt{2} - \sqrt{3})$ $59 - 20\sqrt{6}$

34. $(7\sqrt{5} - 4\sqrt{2})(7\sqrt{5} - 4\sqrt{2})$ $277 - 56\sqrt{10}$

35. $(\sqrt{13} + 2\sqrt{2})(\sqrt{13} - 2\sqrt{2})$ 5

36. $(\sqrt{17} + 3\sqrt{5})(\sqrt{17} - 3\sqrt{5})$ -28

B In Problems 37–50, rationalize the denominator.

37. $\dfrac{3}{\sqrt{2} + 1}$ $3\sqrt{2} - 3$

38. $\dfrac{5}{\sqrt{5} + 1}$ $\dfrac{5\sqrt{5} - 5}{4}$

39. $\dfrac{4}{\sqrt{7} - 1}$ $\dfrac{2\sqrt{7} + 2}{3}$

40. $\dfrac{6}{\sqrt{7} - 2}$ $2\sqrt{7} + 4$

41. $\dfrac{\sqrt{2}}{2 + \sqrt{3}}$ $2\sqrt{2} - \sqrt{6}$

42. $\dfrac{\sqrt{3}}{3 + \sqrt{2}}$ $\dfrac{3\sqrt{3} - \sqrt{6}}{7}$

43. $\dfrac{\sqrt{5}}{2-\sqrt{3}}$ $2\sqrt{5}+\sqrt{15}$

44. $\dfrac{\sqrt{6}}{3-\sqrt{5}}$ $\dfrac{3\sqrt{6}+\sqrt{30}}{4}$

45. $\dfrac{\sqrt{5}}{\sqrt{2}+\sqrt{3}}$ $-\sqrt{10}+\sqrt{15}$

46. $\dfrac{\sqrt{2}}{\sqrt{5}+\sqrt{3}}$ $\dfrac{\sqrt{10}-\sqrt{6}}{2}$

47. $\dfrac{\sqrt{3}}{\sqrt{5}-\sqrt{2}}$ $\dfrac{\sqrt{15}+\sqrt{6}}{3}$

48. $\dfrac{6}{\sqrt{6}-\sqrt{2}}$ $\dfrac{3\sqrt{6}+3\sqrt{2}}{2}$

49. $\dfrac{\sqrt{3}+\sqrt{2}}{\sqrt{3}-\sqrt{2}}$ $5+2\sqrt{6}$

50. $\dfrac{\sqrt{5}-\sqrt{2}}{\sqrt{5}+\sqrt{2}}$ $\dfrac{7-2\sqrt{10}}{3}$

C In Problems 51–64, reduce the fraction.

51. $\dfrac{-8+\sqrt{16}}{2}$ -2

52. $\dfrac{-4+\sqrt{36}}{4}$ $\dfrac{1}{2}$

53. $\dfrac{-6-\sqrt{4}}{6}$ $\dfrac{-4}{3}$

54. $\dfrac{-8-\sqrt{16}}{8}$ $\dfrac{-3}{2}$

55. $\dfrac{2+2\sqrt{3}}{6}$ $\dfrac{1+\sqrt{3}}{3}$

56. $\dfrac{6+2\sqrt{7}}{8}$ $\dfrac{3+\sqrt{7}}{4}$

57. $\dfrac{-2+2\sqrt{23}}{4}$ $\dfrac{-1+\sqrt{23}}{2}$

58. $\dfrac{-6-2\sqrt{6}}{4}$ $\dfrac{-3-\sqrt{6}}{2}$

59. $\dfrac{-6+3\sqrt{10}}{9}$ $\dfrac{-2+\sqrt{10}}{3}$

60. $\dfrac{-20+5\sqrt{10}}{15}$ $\dfrac{-4+\sqrt{10}}{3}$

61. $\dfrac{-8+\sqrt{28}}{6}$ $\dfrac{-4+\sqrt{7}}{3}$

62. $\dfrac{15-\sqrt{189}}{6}$ $\dfrac{5-\sqrt{21}}{2}$

63. $\dfrac{-9+\sqrt{243}}{6}$ $\dfrac{-3+3\sqrt{3}}{2}$

64. $\dfrac{-6+\sqrt{180}}{9}$ $\dfrac{-2+2\sqrt{5}}{3}$

SKILL CHECKER

Try the Skill Checker Exercises so you'll be ready for the next section.

Find the square of each radical expression:

65. $\sqrt{x-1}$ $x-1$

66. $\sqrt{x+7}$ $x+7$

67. $2\sqrt{x}$ $4x$

68. $-3\sqrt{y}$ $9y$

69. $\sqrt{x^2+2x+1}$ x^2+2x+1

70. $\sqrt{x^2-2x+7}$ x^2-2x+7

Factor completely:

71. x^2-3x $x(x-3)$

72. x^2+4x $x(x+4)$

73. x^2-3x+2 $(x-2)(x-1)$

74. x^2+4x+3 $(x+3)(x+1)$

USING YOUR KNOWLEDGE

Simplifying Mach Numbers

The Mach number M mentioned in the *Getting Started* is given by

$$\sqrt{\frac{2}{\gamma}}\sqrt{\frac{P_2-P_1}{P_1}}$$

75. Write this expression as a single radical. $\sqrt{\dfrac{2(P_2-P_1)}{\gamma P_1}}$

76. Rationalize the denominator of the expression obtained in Problem 75. $\dfrac{\sqrt{2\gamma P_1(P_2-P_1)}}{\gamma P_1}$

WRITE ON

77. Write in your own words the procedure you use to simplify any expression containing radicals.
Answers may vary.

78. Explain the difference between rationalizing the denominator in an algebraic expression whose denominator has only one term involving a radical and one whose denominator has two terms, at least one of which involves a radical. Answers may vary.

79. Suppose you wish to rationalize the denominator in the expression

$$\frac{1}{\sqrt{2}+1}$$

and you decide to multiply numerator and denominator by $\sqrt{2}+1$. Would you obtain a rational denominator? What should you multiply by? No; $\sqrt{2}-1$

MASTERY TEST

If you know how to do these problems, you have learned your lesson!

Simplify:

80. $\frac{4+\sqrt{36}}{8}$ $\frac{5}{4}$

81. $\frac{-4+\sqrt{28}}{2}$ $-2+\sqrt{7}$

82. $\frac{-6-\sqrt{72}}{2}$ $-3-3\sqrt{2}$

83. $\frac{3}{\sqrt{2}+2}$ $\frac{6-3\sqrt{2}}{2}$

84. $\frac{\sqrt{2}+\sqrt{3}}{\sqrt{5}-\sqrt{2}}$ $\frac{2+\sqrt{10}+\sqrt{15}+\sqrt{6}}{3}$

85. $\frac{3}{\sqrt[3]{2}}$ $\frac{3\sqrt[3]{4}}{2}$

86. $\frac{5}{\sqrt[3]{9}}$ $\frac{5\sqrt[3]{3}}{3}$

87. $\frac{\sqrt[3]{500}}{\sqrt[3]{2}}$ $5\sqrt[3]{2}$

88. $\frac{\sqrt{12x^4}}{\sqrt{3x^2}}$ $2x$

89. $\frac{\sqrt[3]{16a^5}}{\sqrt[3]{2a^3}}$ $2\sqrt[3]{a^2}$

90. $(\sqrt{7}+\sqrt{3})(\sqrt{7}-\sqrt{3})$ 4

91. $(\sqrt{5}+2\sqrt{3})(\sqrt{5}-3\sqrt{3})$
$-13-\sqrt{15}$

92. $(\sqrt{2}+\sqrt{3})(3\sqrt{2}-5\sqrt{3})$
$-9-2\sqrt{6}$

9.5 APPLICATIONS

To Succeed, Review How To . . .

1. Square a radical expression (pp. 641–642).
2. Square a binomial (pp. 357–358).
3. Solve quadratic equations by factoring (pp. 431–438).

Objectives

A Solve equations with one square root term containing the variable.

B Solve equations with two square root terms containing the variable.

C Solve an application.

GETTING STARTED

Your Weight and Your Life

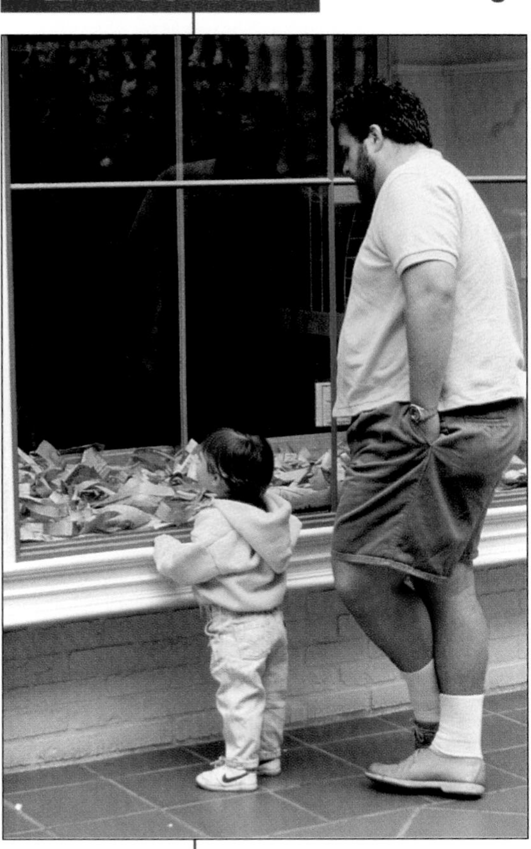

Has your doctor said that you are a little bit overweight? What does that mean? Can it be quantified? The "threshold weight" T (in pounds) for a man between 40 and 49 years of age is defined as "the crucial weight above which the mortality risk rises astronomically." In plain language, this means that if you get too fat, you are almost certainly going to die as a result! The formula linking T and the height h in inches is

$$12.3\sqrt[3]{T} = h$$

Can you solve for T in this equation? To start, we divide both sides of the equation by 12.3 so that the radical term $\sqrt[3]{T}$ is *isolated* (by itself) on one side of the equation. We obtain

$$\sqrt[3]{T} = \frac{h}{12.3}$$

Now we can cube each side of the equation to get

$$(\sqrt[3]{T})^3 = \left(\frac{h}{12.3}\right)^3$$

or

$$T = \left(\frac{h}{12.3}\right)^3$$

Thus, if a 40-year-old man is 73.8 inches tall, his threshold weight should be

$$T = \left(\frac{73.8}{12.3}\right)^3 = 6^3 = 216 \text{ lb}$$

In this section we shall use a new technique to solve equations involving radicals—raising both sides of an equation to a power. We did just that when we *cubed* both sides of the equation

$$\sqrt[3]{T} = \frac{h}{12.3}$$

so we could solve for T. We shall use this new method many times in this section.

Solving Equations with One Square Root Term Containing the Variable

Web It

For a lesson and procedure for solving equations involving radicals, visit link 9-5-1 on the Bello Website at mhhe.com/bello.

Teaching Tip

Have students discuss why $\sqrt{x+3} = -4$ can never have a real-number solution.

Hint:

$$\sqrt{9} = 3$$
$$-\sqrt{9} = -3$$

$\sqrt{-9}$ is not a real number.

The properties of equality studied in Chapter 2 are not enough to solve equations such as $\sqrt{x+1} = 2$. A new property we need can be stated as follows.

RAISING BOTH SIDES OF AN EQUATION TO A POWER

If both sides of the equation $A = B$ are *squared,* all solutions of $A = B$ are *among* the solutions of the new equation $A^2 = B^2$.

Note that this property can yield a new equation that has *more* solutions than the original equation. For example, the equation $x = 3$ has one solution, 3. If we square both sides of the equation $x = 3$, we have

$$x^2 = 3^2$$
$$x^2 = 9$$

which has *two* solutions, $x = 3$ and $x = -3$. The -3 *does not* satisfy the original equation $x = 3$, and it is called an **extraneous** solution. Because of this, we must check our answers carefully when we solve equations with radicals by substituting the answers in the *original* equation and discarding any extraneous solutions.

EXAMPLE 1 **Solving equations in which the radical expression is isolated**

Solve:

a. $\sqrt{x+3} = 4$ **b.** $\sqrt{x+3} = x + 3$

SOLUTION

a. We shall proceed in steps.

1. $\sqrt{x+3} = 4$ The square root term is isolated.
2. $(\sqrt{x+3})^2 = 4^2$ Square each side.
3. $x + 3 = 16$ Simplify.
4. $x = 13$ Subtract 3.
5. Now we check this answer in the original equation:

$\sqrt{13+3} \overset{?}{=} 4$ Substitute $x = 13$ in the original equation.

$\sqrt{16} = 4$ A true statement

Thus $x = 13$ is the only solution of $\sqrt{x+3} = 4$.

b. We again proceed in steps.

1. $\sqrt{x+3} = x + 3$ The square root term is isolated.
2. $(\sqrt{x+3})^2 = (x+3)^2$ Square each side.
3. $x + 3 = x^2 + 6x + 9$ Simplify.

PROBLEM 1

Solve:

a. $\sqrt{x+1} = 5$

b. $\sqrt{x+1} = x + 1$

Answers

1. a. 24 **b.** -1; 0

4. $0 = x^2 + 5x + 6$ Subtract $x + 3$.

$0 = (x + 3)(x + 2)$ Factor.

$x + 3 = 0$ or $x + 2 = 0$ Set each factor equal to zero.

$x = -3$ $x = -2$ Solve each equation.

Thus the proposed solutions are -3 and -2.

5. We check these proposed solutions in the original equation.

If $x = -3$,

$$\sqrt{-3 + 3} \stackrel{?}{=} -3 + 3$$

$$\sqrt{0} = 0 \quad \text{True.}$$

If $x = -2$,

$$\sqrt{-2 + 3} \stackrel{?}{=} -2 + 3$$

$$\sqrt{1} = 1 \quad \text{True.}$$

Thus -3 and -2 are the solutions of $\sqrt{x + 3} = x + 3$.

EXAMPLE 2 Solving equations by first isolating the radical expression

Solve: $\sqrt{x + 1} - x = -1$

SOLUTION First we must isolate the radical term $\sqrt{x + 1}$ by adding x to both sides of the equation.

$\sqrt{x + 1} - x = -1$ Given.

1. $\sqrt{x + 1} = x - 1$ Add x.

2. $(\sqrt{x + 1})^2 = (x - 1)^2$ Square each side.

3. $x + 1 = x^2 - 2x + 1$ Simplify.

4. $0 = x^2 - 3x$ Subtract $x + 1$.

$0 = x(x - 3)$ Factor.

$x = 0$ or $x - 3 = 0$ Set each factor equal to zero.

$x = 0$ $x = 3$ Solve each equation.

Thus the proposed solutions are 0 and 3.

5. The check is as follows:

If $x = 0$,

$$\sqrt{0 + 1} - 0 \stackrel{?}{=} -1$$

$$\sqrt{1} - 0 \stackrel{?}{=} -1 \quad \text{False.}$$

If $x = 3$,

$$\sqrt{3 + 1} - 3 \stackrel{?}{=} -1$$

$$\sqrt{4} - 3 = -1 \quad \text{True.}$$

Thus the equation $\sqrt{x + 1} - x = -1$ has *one* solution, 3.

PROBLEM 2

Solve: $\sqrt{x + 3} - x = -3$

Answer

2. 6

We can now generalize the steps we've been using to solve equations that have one square root term containing the variable.

PROCEDURE

Solving Radical Equations

1. Isolate the square root terms containing the variable.
2. Square both sides of the equation.
3. Simplify and repeat steps 1 and 2 if there is a square root term containing the variable.
4. Solve the resulting linear or quadratic equation.
5. Check all proposed solutions in the original equation.

B Solving Equations with Two Square Root Terms Containing the Variable

The equations $\sqrt{y+4} = \sqrt{2y+3}$ and $\sqrt{y+7} - 3\sqrt{2y-3} = 0$ are different from the equations we have just solved because they have *two* square root terms containing the variable. However, we can still solve them using the five-step procedure.

EXAMPLE 3 Solving equations using the five-step procedure

Solve:

a. $\sqrt{y+4} = \sqrt{2y+3}$

b. $\sqrt{y+7} - 3\sqrt{2y-3} = 0$

SOLUTION

a. Since the square root terms containing the variable are isolated, we first square each side of the equation and then solve for y.

1. $\sqrt{y+4} = \sqrt{2y+3}$ The radicals are isolated.
2. $(\sqrt{y+4})^2 = (\sqrt{2y+3})^2$ Square both sides.
3. $y + 4 = 2y + 3$ Simplify.
4. $4 = y + 3$ Subtract y.
   $1 = y$ Subtract 3.

Thus the proposed solution is 1.

5. Let's check this: If $y = 1$,

$$\sqrt{1+4} \stackrel{?}{=} \sqrt{2 \cdot 1 + 3}$$

$$\sqrt{5} = \sqrt{5} \qquad \text{True.}$$

Thus the solution of $\sqrt{y+4} = \sqrt{2y+3}$ is 1.

PROBLEM 3

Solve:

a. $\sqrt{x+4} = \sqrt{2x+1}$

b. $\sqrt{x+3} - 3\sqrt{2x-11} = 0$

Answers

3. a. 3 **b.** 6

b. We start by isolating the square root terms containing the variable by adding $3\sqrt{2y - 3}$ to both sides.

	$\sqrt{y + 7} - 3\sqrt{2y - 3} = 0$	Given.
1.	$\sqrt{y + 7} = 3\sqrt{2y - 3}$	Add $3\sqrt{2y - 3}$.
2.	$(\sqrt{y + 7})^2 = (3\sqrt{2y - 3})^2$	Square both sides.
3.	$y + 7 = 3^2(2y - 3)$	Simplify.
4.	$y + 7 = 9(2y - 3)$	Since $3^2 = 9$
	$y + 7 = 18y - 27$	Simplify.
	$7 = 17y - 27$	Subtract y.
	$34 = 17y$	Add 27.
	$2 = y$	Divide by 17.

5. And our check: If $y = 2$,

$$\sqrt{2 + 7} - 3\sqrt{2 \cdot 2 - 3} \stackrel{?}{=} 0$$

$$\sqrt{9} - 3\sqrt{1} \stackrel{?}{=} 0$$

$$3 - 3 = 0 \quad \text{True.}$$

Thus the solution of $\sqrt{y + 7} - 3\sqrt{2y - 3} = 0$ is 2.

Solving an Application

Let's see how square roots can be used in a real-world application.

EXAMPLE 4 Cell phones and radicals

If you have a cell phone, how long are your calls? According to the Cellular Telecommunications Industry Association, the average length L of a local call in 2000 was 2.56 minutes, and can be approximated by $L = \sqrt{t + 5}$, where L is the length of the call in minutes t years after 2000. In what year would you expect the average length of a call to be 3 minutes?

SOLUTION To predict when the length of a call will be 3 minutes, we have to find t when $L = 3$. Thus we have to solve the equation $\sqrt{t + 5} = 3$. We will use our five-step procedure.

1.	$\sqrt{t + 5} = 3$	The radical is isolated.
2.	$(\sqrt{t + 5})^2 = 3^2$	Square both sides.
3.	$t + 5 = 9$	Simplify.
4.	$t = 4$	Subtract 5 from both sides.

5. We check this: If $t = 4$,

$$\sqrt{4 + 5} = 3 \quad \text{True.}$$

Thus 4 years after 2000—that is, in 2004—the average length of a cellular call is 3 minutes.

PROBLEM 4

When would you expect the average length of a call to be 4 min?

Answer

4. 11 years after 2000, that is, in the year 2011.

Exercises 9.5

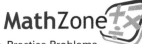

A In Problems 1–20, solve the given equations.

1. $\sqrt{x} = 4$ $x = 16$

2. $\sqrt{x} = -3$ No solution

3. $\sqrt{x-1} = -2$ No solution

4. $\sqrt{x+1} = 2$ $x = 3$

5. $\sqrt{y} - 2 = 0$ $y = 4$

6. $\sqrt{y} + 3 = 0$ No solution

7. $\sqrt{y+1} - 3 = 0$ $y = 8$

8. $\sqrt{y-1} + 2 = 0$ No solution

9. $\sqrt{x+1} = x - 5$ $x = 8$

10. $\sqrt{x+14} = x + 2$ $x = 2$

11. $\sqrt{x+4} = x + 2$ $x = 0$

12. $\sqrt{x+9} = x - 3$ $x = 7$

13. $\sqrt{x-1} - x = -3$ $x = 5$

14. $\sqrt{x-2} - x = -4$ $x = 6$

15. $y - 10 - \sqrt{5y} = 0$ $y = 20$

16. $y - 3 - \sqrt{4y} = 0$ $y = 9$

17. $\sqrt{y} + 20 = y$ $y = 25$

18. $\sqrt{y} + 12 = y$ $y = 16$

19. $4\sqrt{y} = y + 3$ $y = 9$ or $y = 1$

20. $6\sqrt{y} = y + 5$ $y = 25$ or $y = 1$

B In Problems 21–30, solve the given equation.

21. $\sqrt{y+3} = \sqrt{2y-3}$ $y = 6$

22. $\sqrt{y+7} = \sqrt{3y+3}$ $y = 2$

23. $\sqrt{3x+1} = \sqrt{2x+6}$ $x = 5$

24. $\sqrt{4x-3} = \sqrt{3x-2}$ $x = 1$

25. $2\sqrt{x+5} = \sqrt{8x+4}$ $x = 4$

26. $3\sqrt{x+2} = 2\sqrt{x+7}$ $x = 2$

27. $\sqrt{4x-1} - \sqrt{x+10} = 0$ $x = \frac{11}{3}$

28. $\sqrt{3x+6} - \sqrt{5x+4} = 0$ $x = 1$

29. $\sqrt{3y-2} - \sqrt{2y+3} = 0$ $y = 5$

30. $\sqrt{5y+7} - \sqrt{3y+11} = 0$ $y = 2$

APPLICATIONS

31. *Radius of a sphere* The radius r of a sphere is given by

$$r = \sqrt{\frac{S}{4\pi}}$$

where S is the surface area. If the radius of a sphere is 2 feet, what is its surface area? Use $\pi \approx 3.14$.
$S = 50.24$ sq ft

32. *Radius of a cone* The radius r of a cone is given by

$$r = \sqrt{\frac{3V}{\pi h}}$$

where V is the volume of the cone and h is its height. If a 10-centimeter-high ice cream cone has a radius of 2 centimeters, what is the volume of the ice cream in the cone? Use $\pi \approx 3.14$ and round to one decimal place.
$V = 41.9$ cm^3

33. *Time to fall* The time t (in seconds) it takes a body to fall d feet is given by

$$t = \sqrt{\frac{d}{16}}$$

How far would a body fall in 3 seconds? 144 ft

34. *Velocity of a falling body* After traveling d feet, the velocity v (in feet per second) of a falling body starting from rest is given by $v = \sqrt{64d}$. If a body that started from rest is traveling at 44 feet per second, how far has it fallen? 30.25 ft

35. *Length of pendulum cycle* A pendulum of length L feet takes

$$t = 2\pi\sqrt{\frac{L}{32}} \text{ (seconds)}$$

to go through a complete cycle. If a pendulum takes 2 seconds to go through a complete cycle, how long is the pendulum? Use $\pi \approx \frac{22}{7}$ and round to two decimal places.
3.24 ft

36. *Vehicle emissions* According to Environmental Protection Agency (EPA) figures, the estimated amount A of particulate matters (in short tons) emitted by transportation sources (cars, buses, and so on) is $A = \sqrt{2 + 0.2y}$, where y is the number of years after 1984. In what year would you expect the amount of particulate matters emitted by transportation sources to reach 2 short tons?
1994

37. *Sulfur oxide emissions* According to EPA figures, the amount A of sulfur oxides (in short tons) emitted by transportation sources is $A = \sqrt{1 + 0.04y}$, where y is the number of years after 1989. In what year would you expect the amount of sulfur oxides emitted by transportation sources to reach 1.2 short tons? 2000

38. *Distance seen from tall buildings* The maximum distance d in kilometers you can see from a tall building is $d = 110\sqrt{h}$, where h is the height of the building in kilometers. If the maximum distance you can see from the tallest building in the world is 77 kilometers, how high is the building? Write your answer in decimal form. 0.49 km

SKILL CHECKER

Try the Skill Checker Exercises so you'll be ready for the next section.

Find:

39. $\sqrt{49}$ 7

40. $\sqrt{\frac{81}{16}}$ $\frac{9}{4}$

41. $\sqrt{\frac{5}{16}}$ $\frac{\sqrt{5}}{4}$

42. $\sqrt{\frac{7}{4}}$ $\frac{\sqrt{7}}{2}$

USING YOUR KNOWLEDGE

Working with Higher Roots

Step 2 in the five-step procedure for solving radical equations directs us to "square each side of the equation." Thus to solve $\sqrt{x} = 2$, we square each side of the equation to obtain 4 as our solution. If we have an equation of the form $\sqrt[3]{x} = 2$, we can **cube** each side of the equation to obtain $2^3 = 8$. You can check that 8 is the correct solution by substituting 8 for x in $\sqrt[3]{x} = 2$ to obtain $\sqrt[3]{8} = 2$, a true statement.

Use this knowledge to solve the following equations.

43. $\sqrt[3]{x} = 3$ $x = 27$

44. $\sqrt[3]{x} = -4$ $x = -64$

45. $\sqrt[3]{x+1} = 2$ $x = 7$

46. $\sqrt[3]{x-1} = -2$ $x = -7$

Generalize the idea used in Problems 43–46 to solve the following problems.

47. $\sqrt[4]{x} = 2$ $x = 16$

48. $\sqrt[4]{x+1} = 1$ $x = 0$

49. $\sqrt[4]{x-1} = -2$ No solution

50. $\sqrt[4]{x-1} = 2$ $x = 17$

WRITE ON

51. Consider the equation $\sqrt{x} + 3 = 0$. What should the first step be in solving this equation? If you follow the rest of the steps in the procedure to solve radical equations, you should conclude that this equation has no real-number solution. Can you write an explanation of why this is so *after the first step* in the procedure? Answers may vary.

52. Consider the equation $\sqrt{x+1} = -\sqrt{x+2}$. Write an explanation of why this equation has no real-number solutions; then follow the procedure given in the text to prove that this is the case. Answers may vary.

53. Write your explanation of a "proposed" solution for solving equations involving radicals. Why do you think they are called "proposed" solutions? Answers may vary.

54. Why is it necessary to check proposed solutions in the original equation when solving equations involving radicals? Answers may vary.

MASTERY TEST

If you know how to do these problems, you have learned your lesson!

55. The total monthly cost C (in millions of dollars) of running daily flights between two cities is $C = \sqrt{0.3p + 1}$, where p is the number of passengers in thousands. If the monthly cost C for a certain month was \$2 million, what was the number of passengers for the month? 10 thousand

Solve if possible:

56. $\sqrt{y+6} = \sqrt{2y+3}$ $y = 3$

57. $\sqrt{3y+10} = \sqrt{y+14}$ $y = 2$

58. $\sqrt{4x-3} - \sqrt{x+3} = 0$ $x = 2$

59. $\sqrt{5x+1} - \sqrt{x+9} = 0$ $x = 2$

60. $\sqrt{x+1} - x = -5$ $x = 8$

61. $\sqrt{x+2} - x = -4$ $x = 7$

62. $\sqrt{x+2} = 3$ $x = 7$

63. $\sqrt{x-1} = -2$ No solution

64. $\sqrt{x-3} = x - 3$ $x = 3$ or $x = 4$

65. $\sqrt{x-2} = x - 2$ $x = 2$ or $x = 3$

COLLABORATIVE LEARNING

The Golden Ratio

Form two groups of students. Each group should find a different picture of the Mona Lisa and compute the ratio of length to width of the frame. Are the answers about the same for all groups?

Each group should compare their answer to the Golden Ratio, $\frac{\sqrt{5}+1}{2} \approx 1.618$. Are the answers close?

Let us now construct a Golden Rectangle:

Group 1 Start with a 1-by-1 square. What is the ratio of length to width?

Group 2 Add another 1-by-1 square to the right of the original square. What is the ratio of length to width?

Group 1 Add a 2-by-2 square under the previous rectangle. What is the ratio of length to width?

Group 2 Add a 3-by-3 square to the right of the previous rectangle. What is the ratio of length to width?

Group 1 Add a 5-by-5 square under the previous rectangle. What is the ratio of length to width?

Are the ratios approximating the Golden Ratio?

Now, let us take some measurements. Group 1 will take the smaller measurements (x) and Group 2 the larger measurements (y). Record all measurements to the nearest tenth of a centimeter.

Group 1 (x) Smaller Measurement	Group 2 (y) Larger Measurement
Height of belly button from the floor	Total height
Belly button to top of head	Belly button height from floor
Chin to top of head	Belly button to chin

Look at the ratio of each individual measurement y to x in each row. Are the ratios close to the Golden Ratio? Compute the average of the individual measurements in each row (for example, the average of the heights of belly buttons in row 1 and the average of total heights in row 1). Look at the ratios of the averages of the y (larger) measurements to the averages of the x (smaller) measurements (for example, average of total heights to average of heights of belly buttons). Are they now closer to the Golden Ratio?

Research Questions

1. Write a paragraph about Christoff Rudolff, the inventor of the square root sign, and indicate where the square root sign was first used.
2. A Chinese mathematician discovered the relationship between extracting roots and the array of binomial coefficients in Pascal's triangle. Write a paragraph about this mathematician.
3. Name the four Chinese algebraists who discovered the relationship between root extraction and the coefficients of Pascal's triangle and who then extended the idea to solve higher-than-cubic equations.
4. Write a paragraph about Newton's method for finding square roots and illustrate its use by finding the square root of 11, for example.

Summary

SECTION	ITEM	MEANING	EXAMPLE
9.1A	$\sqrt{a}$ $-\sqrt{a}$	$\sqrt{a} = b$ is equivalent to $b^2 = a$. $-\sqrt{a} = b$ is equivalent to $b^2 = a$.	$\sqrt{4} = 2$ because $2^2 = 4$. $-\sqrt{4} = -2$ because $(-2)^2 = 4$.
9.1C	$\sqrt{a}$ if a is negative	If a is negative, $\sqrt{a}$ is not a real number.	$\sqrt{-16}$ and $\sqrt{-7}$ are not real numbers.
9.1D	$\sqrt[n]{a}$	The nth root of a	$\sqrt[3]{8} = 2, \sqrt[4]{81} = 3$
9.2A	Product rule for radicals	$\sqrt{a \cdot b} = \sqrt{a} \cdot \sqrt{b}$	$\sqrt{16 \cdot 9} = \sqrt{16} \cdot \sqrt{9} = 4 \cdot 3 = 12$ $\sqrt{18} = \sqrt{9 \cdot 2} = \sqrt{9} \cdot \sqrt{2} = 3\sqrt{2}$
9.2B	Quotient rule for radicals	$\sqrt{\frac{a}{b}} = \frac{\sqrt{a}}{\sqrt{b}}$	$\sqrt{\frac{9}{4}} = \frac{\sqrt{9}}{\sqrt{4}} = \frac{3}{2}$
9.2C	$\sqrt{a^2} = \lvert a \rvert$	The square root of a real number a is the absolute value of a.	$\sqrt{5^2} = \lvert 5 \rvert, \sqrt{(-3)^2} = \lvert -3 \rvert = 3$
9.2D	Properties of radicals	For real numbers where the roots exist, $\sqrt[n]{a \cdot b} = \sqrt[n]{a}\sqrt[n]{b}$ $\sqrt[n]{\frac{a}{b}} = \frac{\sqrt[n]{a}}{\sqrt[n]{b}}$	$\sqrt[4]{80} = \sqrt[4]{2^4 \cdot 5} = 2\sqrt[4]{5}$ $\sqrt[3]{\frac{8}{27}} = \frac{\sqrt[3]{8}}{\sqrt[3]{27}} = \frac{\sqrt[3]{2^3}}{\sqrt[3]{3^3}} = \frac{2}{3}$
9.3C	Rationalizing the denominator	Multiply the numerator and denominator of the fraction by the square root in the denominator.	$\frac{1}{\sqrt{3}} = \frac{1 \cdot \sqrt{3}}{\sqrt{3} \cdot \sqrt{3}} = \frac{\sqrt{3}}{3}$
9.4B, C	Conjugate	$a + b$ and $a - b$ are conjugates.	To rationalize the denominator of $\frac{1}{\sqrt{5} - \sqrt{3}}$, multiply the numerator and denominator of $\frac{1}{\sqrt{5} - \sqrt{3}}$ by the conjugate of $\sqrt{5} - \sqrt{3}$, which is $\sqrt{5} + \sqrt{3}$.
9.5A, B	Raising both sides of an equation to a power	If both sides of the equation $A = B$ are squared, all solutions of $A = B$ are among the solutions of the new equation $A^2 = B^2$.	If both sides of the equation $\sqrt{x} = 3$ are squared, all solutions of $\sqrt{x} = 3$ are among the solutions of $(\sqrt{x})^2 = 3^2$, that is, $x = 9$.

Review Exercises

(If you need help with these exercises, look in the section indicated in brackets.)

1. [9.1A, C] Find the root if possible.

a. $\sqrt{81}$ 9 **b.** $\sqrt{-64}$ Not a real number

c. $\sqrt{\frac{36}{25}}$ $\frac{6}{5}$

2. [9.1A, C] Find the root if possible.

a. $-\sqrt{36}$ -6 **b.** $-\sqrt{\frac{64}{25}}$ $-\frac{8}{5}$

c. $\sqrt{-\frac{9}{4}}$ Not a real number

3. [9.1B] Find the square of each radical expression.

a. $\sqrt{8}$ 8 **b.** $\sqrt{25}$ 25 **c.** $\sqrt{17}$ 17

4. [9.1B] Find the square of each radical expression.

a. $-\sqrt{36}$ 36 **b.** $-\sqrt{17}$ 17 **c.** $-\sqrt{64}$ 64

5. [9.1B] Find the square of each radical expression.

a. $\sqrt{x^2+1}$ x^2+1 **b.** $\sqrt{x^2+4}$ x^2+4

c. $-\sqrt{x^2+5}$ x^2+5

6. [9.1C] Find and classify each number as rational, irrational, or not a real number. Approximate the irrational numbers using a calculator.

a. $\sqrt{11}$ Irrational; 3.3166 **b.** $-\sqrt{25}$ Rational; -5

c. $\sqrt{-9}$ Not a real number

7. [9.1C] Classify each number as rational, irrational, or not a real number.

a. $\sqrt{\frac{9}{4}}$ Rational **b.** $-\sqrt{\frac{9}{4}}$ Rational

c. $\sqrt{-\frac{9}{4}}$ Not a real number

8. [9.1D] Find each root if possible.

a. $\sqrt[3]{64}$ 4 **b.** $\sqrt[3]{-8}$ -2 **c.** $-\sqrt[4]{81}$ -3

9. [9.1D] Find each root if possible.

a. $\sqrt[4]{16}$ 2 **b.** $-\sqrt[4]{16}$ -2 **c.** $\sqrt[4]{-16}$ Not a real number

10. [9.1E] If an object is dropped from a distance d (in feet), it takes

$$t = \sqrt{\frac{d}{16}}$$

seconds to reach the ground. How long does it take an object to reach the ground if it is dropped from

a. 121 feet? $2\frac{3}{4}$ sec **b.** 144 feet? 3 sec **c.** 169 feet? $3\frac{1}{4}$ sec

11. [9.2A] Simplify.

a. $\sqrt{32}$ $4\sqrt{2}$ **b.** $\sqrt{48}$ $4\sqrt{3}$ **c.** $\sqrt{196}$ 14

12. [9.2A] Multiply.

a. $\sqrt{3}\cdot\sqrt{7}$ $\sqrt{21}$ **b.** $\sqrt{12}\cdot\sqrt{3}$ 6 **c.** $\sqrt{5}\sqrt{y}$, $y > 0$ $\sqrt{5y}$

13. [9.2B] Simplify.

a. $\sqrt{\frac{3}{16}}$ $\frac{\sqrt{3}}{4}$ **b.** $\sqrt{\frac{5}{36}}$ $\frac{\sqrt{5}}{6}$ **c.** $\sqrt{\frac{9}{4}}$ $\frac{3}{2}$

14. [9.2B] Simplify.

a. $\frac{\sqrt{8}}{\sqrt{2}}$ 2 **b.** $\frac{\sqrt{21}}{\sqrt{3}}$ $\sqrt{7}$ **c.** $\frac{6\sqrt{50}}{2\sqrt{10}}$ $3\sqrt{5}$

15. [9.2C] Simplify. Assume all variables represent positive real numbers.

a. $\sqrt{36x^2}$ $6x$ **b.** $\sqrt{100y^4}$ $10y^2$ **c.** $\sqrt{81n^8}$ $9n^4$

16. [9.2C] Simplify. Assume all variables represent positive real numbers.

a. $\sqrt{72y^{10}}$ $6y^5\sqrt{2}$ **b.** $\sqrt{147z^8}$ $7z^4\sqrt{3}$ **c.** $\sqrt{48x^{12}}$ $4x^6\sqrt{3}$

17. [9.2C] Simplify. Assume all variables represent positive real numbers.

a. $\sqrt{y^{15}}$ $y^7\sqrt{y}$ **b.** $\sqrt{y^{13}}$ $y^6\sqrt{y}$ **c.** $\sqrt{50n^7}$ $5n^3\sqrt{2n}$

18. [9.2D] Simplify.

a. $\sqrt[3]{24}$ $2\sqrt[3]{3}$ **b.** $\sqrt[3]{\frac{8}{27}}$ $\frac{2}{3}$ **c.** $\sqrt[3]{-\frac{125}{64}}$ $-\frac{5}{4}$

19. [9.2D] Simplify.

a. $\sqrt[4]{81}$ 3 **b.** $\sqrt[4]{48}$ $2\sqrt[4]{3}$ **c.** $\sqrt[4]{80}$ $2\sqrt[4]{5}$

20. [9.3A] Simplify.

a. $7\sqrt{3} + 8\sqrt{3}$ $15\sqrt{3}$ **b.** $\sqrt{32} + 5\sqrt{2}$ $9\sqrt{2}$ **c.** $\sqrt{12} + \sqrt{48}$ $6\sqrt{3}$

21. [9.3A] Simplify.

a. $9\sqrt{11} - 6\sqrt{11}$ $3\sqrt{11}$ **b.** $\sqrt{50} - 4\sqrt{2}$ $\sqrt{2}$

c. $\sqrt{108} - \sqrt{75}$ $\sqrt{3}$

22. [9.3B] Simplify.

a. $\sqrt{3}(\sqrt{20} - \sqrt{2})$ $2\sqrt{15} - \sqrt{6}$ **b.** $\sqrt{5}(\sqrt{5} - \sqrt{3})$ $5 - \sqrt{15}$

c. $\sqrt{7}(\sqrt{7} - \sqrt{98})$ $7 - 7\sqrt{14}$

23. [9.3C] Write with a rationalized denominator.

a. $\sqrt{\dfrac{5}{8}}$ $\dfrac{\sqrt{10}}{4}$ **b.** $\sqrt{\dfrac{x^2}{50}}$, $x > 0$ $\dfrac{x\sqrt{2}}{10}$ **c.** $\sqrt{\dfrac{y^2}{27}}$, $y > 0$ $\dfrac{y\sqrt{3}}{9}$

24. [9.4A] Simplify.

a. $\sqrt{32 + 4} - \sqrt{9}$ 3 **b.** $\sqrt{18 + 7} - \sqrt{4}$ 3

c. $\sqrt{60 + 4} - \sqrt{16}$ 4

25. [9.4A] Simplify.

a. $8\sqrt{15} - \sqrt{3} \cdot \sqrt{5}$ $7\sqrt{15}$ **b.** $7\sqrt{6} - \sqrt{2} \cdot \sqrt{3}$ $6\sqrt{6}$

c. $9\sqrt{14} - 2\sqrt{7} \cdot \sqrt{2}$ $7\sqrt{14}$

26. [9.4A] Simplify.

a. $\dfrac{\sqrt[3]{162}}{\sqrt[3]{2}}$ $3\sqrt[3]{3}$ **b.** $\dfrac{\sqrt[3]{135}}{\sqrt[3]{5}}$ 3 **c.** $\dfrac{\sqrt[3]{192}}{\sqrt[3]{24}}$ 2

27. [9.4A] Simplify.

a. $\dfrac{7}{\sqrt[3]{4}}$ $\dfrac{7\sqrt[3]{2}}{2}$ **b.** $\dfrac{5}{\sqrt[3]{9}}$ $\dfrac{5\sqrt[3]{3}}{3}$ **c.** $\dfrac{9}{\sqrt[3]{25}}$ $\dfrac{9\sqrt[3]{5}}{5}$

28. [9.4A] Simplify.

a. $(\sqrt{3} + 3\sqrt{2})(\sqrt{3} - 5\sqrt{2})$ $-27 - 2\sqrt{6}$

b. $(\sqrt{7} + 3\sqrt{5})(\sqrt{7} - 2\sqrt{5})$ $-23 + \sqrt{35}$

29. [9.4A] Simplify.

a. $(\sqrt{7} + 2\sqrt{3})(\sqrt{7} - 2\sqrt{3})$ -5

b. $(\sqrt{11} + 3\sqrt{5})(\sqrt{11} - 3\sqrt{5})$ -34

30. [9.4B] Simplify.

a. $\dfrac{3}{\sqrt{3} + 1}$ $\dfrac{3\sqrt{3} - 3}{2}$ **b.** $\dfrac{5}{\sqrt{2} - 1}$ $5\sqrt{2} + 5$

31. [9.4B] Simplify.

a. $\dfrac{7}{\sqrt{3} - \sqrt{2}}$ $7\sqrt{3} + 7\sqrt{2}$ **b.** $\dfrac{2}{\sqrt{5} - \sqrt{2}}$ $\dfrac{2\sqrt{5} + 2\sqrt{2}}{3}$

32. [9.4C] Simplify.

a. $\dfrac{-8 + \sqrt{8}}{2}$ $-4 + \sqrt{2}$ **b.** $\dfrac{-16 + \sqrt{12}}{4}$ $\dfrac{-8 + \sqrt{3}}{2}$

33. [9.5A] Solve.

a. $\sqrt{x + 2} = 3$ $x = 7$ **b.** $\sqrt{x - 2} = -2$ No solution

34. [9.5A] Solve.

a. $\sqrt{x + 5} = x - 1$ $x = 4$ **b.** $\sqrt{x + 10} = x - 2$ $x = 6$

35. [9.5A] Solve.

a. $\sqrt{x + 4} - x = -2$ $x = 5$ **b.** $\sqrt{x + 2} - x = -4$ $x = 7$

36. [9.5B] Solve.

a. $\sqrt{y + 5} = \sqrt{3y - 3}$ $y = 4$

b. $\sqrt{y + 5} = \sqrt{2y + 5}$ $y = 0$

37. [9.5B] Solve.

a. $\sqrt{y + 8} - 3\sqrt{2y - 1} = 0$ $y = 1$

b. $\sqrt{y + 9} - 3\sqrt{2y + 1} = 0$ $y = 0$

38. [9.5C] The total daily cost C (thousand dollars) of producing a certain product is given by $C = \sqrt{0.2x + 1}$, where x is the number of items in hundreds. How many items were produced on a day in which the cost was

a. \$3(thousand)? 40 thousand or 40,000 **b.** \$7(thousand)? 240 thousand or 240,000

Practice Test 9

(*Answers on page 679*)

1. Find.

a. $\sqrt{169}$ b. $-\sqrt{\frac{49}{81}}$

2. Find the square of each radical expression.

a. $-\sqrt{121}$ b. $\sqrt{x^2 + 7}$

3. Classify each number as rational, irrational, or not a real number, and simplify if possible.

a. $\sqrt{17}$ b. $-\sqrt{36}$

c. $\sqrt{-100}$ d. $\sqrt{\frac{100}{49}}$

4. Find each root if possible.

a. $\sqrt[4]{81}$ b. $-\sqrt[4]{625}$

c. $\sqrt[3]{-8}$ d. $\sqrt[4]{-16}$

5. A diver jumps from a cliff 20 meters high. If the time t (in seconds) it takes an object dropped from a distance d (in meters) to reach the ground is given by

$$t = \sqrt{\frac{d}{5}}$$

how long does it take the diver to reach the water?

6. Simplify.

a. $\sqrt{125}$ b. $\sqrt{54}$

7. Multiply.

a. $\sqrt{3} \cdot \sqrt{11}$ b. $\sqrt{11} \cdot \sqrt{y}$, $y > 0$

8. Simplify.

a. $\sqrt{\frac{7}{16}}$ b. $\frac{21\sqrt{50}}{7\sqrt{5}}$

9. Simplify.

a. $\sqrt{144n^2}$, $n > 0$ b. $\sqrt{32y^7}$, $y > 0$

10. Simplify.

a. $\sqrt[4]{96}$ b. $\sqrt[3]{\frac{-125}{8}}$

11. Simplify.

a. $9\sqrt{13} + 7\sqrt{13}$ b. $14\sqrt{6} - 3\sqrt{6}$

12. Simplify.

a. $\sqrt{28} + \sqrt{63}$ b. $\sqrt{40} + \sqrt{90} - \sqrt{160}$

13. Simplify.

a. $\sqrt{3}(\sqrt{18} - \sqrt{5})$ b. $\sqrt{5}(\sqrt{5} - \sqrt{7})$

14. Write $\sqrt{\frac{3}{20}}$ with a rationalized denominator.

15. Write $\sqrt{\frac{y^2}{50}}$, $y > 0$ with a rationalized denominator.

16. Simplify.

a. $8\sqrt{14} - \sqrt{7} \cdot \sqrt{2}$ b. $\frac{\sqrt{12x^3}}{\sqrt{4x^2}}$, $x > 0$

17. Simplify.

a. $\frac{\sqrt[3]{500}}{\sqrt[3]{2}}$ b. $\frac{3}{\sqrt[3]{25}}$

18. Simplify.

a. $(\sqrt{3} + 6\sqrt{2})(\sqrt{3} - 2\sqrt{2})$

b. $(\sqrt{10} - 2\sqrt{20})(\sqrt{10} + 2\sqrt{20})$

19. Simplify.

a. $\frac{11}{\sqrt{3} + 1}$ b. $\frac{2}{\sqrt{5} - \sqrt{2}}$

20. Simplify.

a. $\frac{-6 + \sqrt{18}}{3}$ b. $\frac{-8 + \sqrt{8}}{4}$

21. Solve.

a. $\sqrt{x + 1} = 2$ b. $\sqrt{x + 6} = x + 6$

22. Solve $\sqrt{x + 4} - x = 2$.

23. Solve $\sqrt{y + 3} = \sqrt{2y + 1}$.

24. Solve $\sqrt{y + 6} - 3\sqrt{2y - 5} = 0$.

25. The average length L of a long-distance call (in minutes) has been approximated by $L = \sqrt{t + 4}$, where t is the number of years after 1995. In how many years would you expect the average length of a call to be 3 minutes?

Answers to Practice Test

ANSWER		IF YOU MISSED	REVIEW		
		QUESTION	SECTION	EXAMPLES	PAGE
1. a. 13	**b.** $-\frac{7}{9}$	1	9.1	1	641
2. a. 121	**b.** $x^2 + 7$	2	9.1	2	642
3. a. Irrational	**b.** Rational; -6	3	9.1	3	642
c. Not real	**d.** Rational; $\frac{10}{7}$				
4. a. 3	**b.** -5	4	9.1	4	643
c. -2	**d.** Not real				
5. 2 seconds		5	9.1	5	643–644
6. a. $5\sqrt{5}$	**b.** $3\sqrt{6}$	6	9.2	1	648
7. a. $\sqrt{33}$	**b.** $\sqrt{11y}$	7	9.2	2	648
8. a. $\frac{\sqrt{7}}{4}$	**b.** $3\sqrt{10}$	8	9.2	3	649
9. a. $12n$	**b.** $4y^3\sqrt{2y}$	9	9.2	4	649
10. a. $2\sqrt[4]{6}$	**b.** $-\frac{5}{2}$	10	9.2	5	650
11. a. $16\sqrt{13}$	**b.** $11\sqrt{6}$	11	9.3	1	654
12. a. $5\sqrt{7}$	**b.** $\sqrt{10}$	12	9.3	2	654
13. a. $3\sqrt{6} - \sqrt{15}$	**b.** $5 - \sqrt{35}$	13	9.3	3	655
14. $\frac{\sqrt{15}}{10}$		14	9.3	4	656
15. $\frac{y\sqrt{2}}{10}$		15	9.3	5	657
16. a. $7\sqrt{14}$	**b.** $\sqrt{3x}$	16	9.4	1	660
17. a. $5\sqrt[3]{2}$	**b.** $\frac{3\sqrt[3]{5}}{5}$	17	9.4	2	660–661
18. a. $-21 + 4\sqrt{6}$	**b.** -70	18	9.4	3	661
19. a. $\frac{11\sqrt{3} - 11}{2}$	**b.** $\frac{2\sqrt{5} + 2\sqrt{2}}{3}$	19	9.4	4	662
20. a. $-2 + \sqrt{2}$	**b.** $\frac{-4 + \sqrt{2}}{2}$	20	9.4	5	663
21. a. $x = 3$	**b.** $x = -5$ or $x = -6$	21	9.5	1	668–669
22. $x = 0$		22	9.5	2	669
23. $y = 2$		23	9.5	3	670–671
24. $y = 3$		24	9.5	3	670–671
25. In $t = 5$ years		25	9.5	4	671

Cumulative Review Chapters 1–9

1. Find: $-\frac{2}{9} + \left(-\frac{1}{8}\right)$ $-\frac{25}{72}$

2. Find: $(-3)^4$ 81

3. Find: $-\frac{1}{6} \div \left(-\frac{7}{12}\right)$ $\frac{2}{7}$

4. Evaluate $y \div 5 \cdot x - z$ for $x = 5$, $y = 50$, $z = 3$. 47

5. Simplify: $2x - (x + 4) - 2(x + 3)$ $-x - 10$

6. Write in symbols: The quotient of $(m + 3n)$ and p $\frac{m + 3n}{p}$

7. Solve for x: $2 = 5(x - 3) + 1 - 4x$ $x = 16$

8. Solve for x: $\frac{x}{2} - \frac{x}{3} = 1$ $x = 6$

9. Graph: $-\frac{x}{6} + \frac{x}{2} \leq \frac{x - 2}{2}$

0 6

10. Graph the point $C(-1, -3)$.

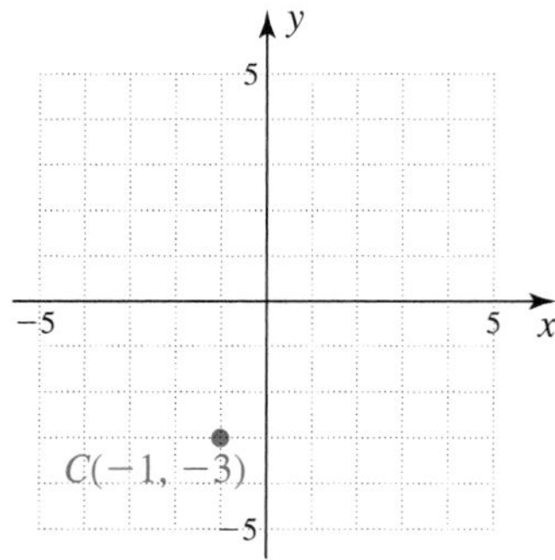

11. Determine whether the ordered pair $(-1, 3)$ is a solution of $4x - y = -1$. No

12. Find x in the ordered pair $(x, 3)$ so that the ordered pair satisfies the equation $2x - 4y = -10$. $x = 1$

13. Graph: $5x + y = 5$

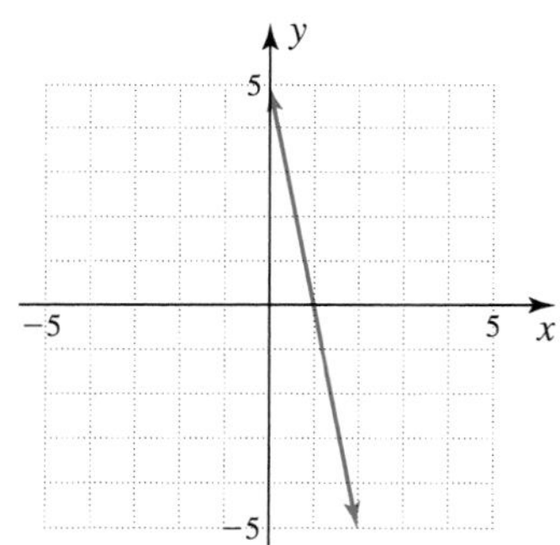

14. Graph: $4y - 8 = 0$

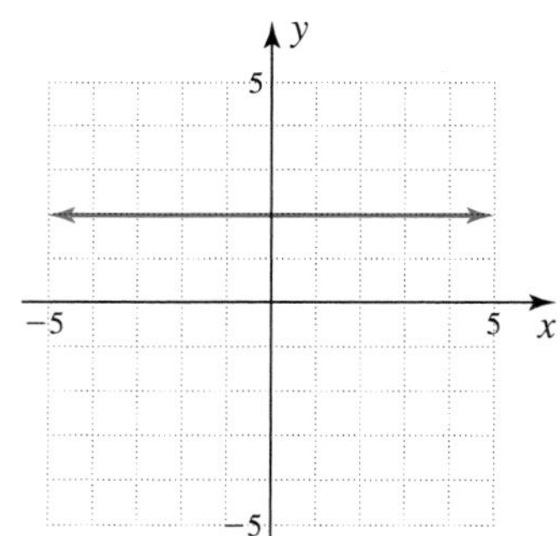

15. Find the slope of the line going through the points $(0, 4)$ and $(6, 1)$. $-\frac{1}{2}$

16. What is the slope of the line $3x - 3y = -8$? 1

17. Find the pair of parallel lines. (1) and (3)

(1) $15x - 12y = 4$

(2) $12y + 15x = 4$

(3) $-4y = -5x + 4$

18. Simplify: $(2x^4y^{-3})^{-4}$ $\frac{y^{12}}{16x^{16}}$

19. Write in scientific notation: 0.00000025 2.5×10^{-7}

20. Divide and express the answer in scientific notation: $(2.72 \times 10^{-4}) \div (1.6 \times 10^4)$ 1.7×10^{-8}

21. Find (expand): $\left(4x^2 - \frac{1}{2}\right)^2$ $16x^4 - 4x^2 + \frac{1}{4}$

22. Divide $(2x^3 - 7x^2 + x + 9)$ by $(x - 3)$. $(2x^2 - x - 2)$ R 3

23. Factor completely: $x^2 - 4x + 3$ $(x - 1)(x - 3)$

24. Factor completely: $9x^2 - 27xy + 20y^2$ $(3x - 5y)(3x - 4y)$

25. Factor completely: $4x^2 - 25y^2$ $(2x + 5y)(2x - 5y)$

26. Factor completely: $-5x^4 + 5x^2$ $-5x^2(x + 1)(x - 1)$

27. Factor completely: $4x^3 - 8x^2 - 12x$
$4x(x + 1)(x - 3)$

28. Factor completely: $3x^2 + 4x + 9x + 12$
$(3x + 4)(x + 3)$

29. Factor completely: $16kx^2 + 8kx + k$ $k(4x + 1)^2$

30. Solve for x: $4x^2 + 17x = 15$ $x = -5; x = \frac{3}{4}$

31. Write $\frac{2x}{5y}$ with a denominator of $15y^2$. $\frac{6xy}{15y^2}$

32. Reduce to lowest terms: $\frac{-16(x^2 - y^2)}{4(x - y)}$ $-4(x + y)$

33. Reduce to lowest terms: $\frac{x^2 + 4x - 21}{3 - x}$ $-(x + 7)$

34. Multiply: $(x - 8) \cdot \frac{x + 4}{x^2 - 64}$ $\frac{x + 4}{x + 8}$

35. Divide: $\frac{x + 5}{x - 5} \div \frac{x^2 - 25}{5 - x}$ $\frac{1}{5 - x}$

36. Add: $\frac{7}{3(x + 5)} + \frac{5}{3(x + 5)}$ $\frac{4}{x + 5}$

37. Subtract: $\frac{x + 4}{x^2 + x - 20} - \frac{x + 5}{x^2 - 16}$
$\frac{-2x - 9}{(x + 5)(x + 4)(x - 4)}$

38. Simplify: $\dfrac{\frac{2}{3x} + \frac{1}{2x}}{\frac{1}{x} + \frac{1}{4x}}$ $\frac{14}{15}$

39. Solve for x: $\frac{4x}{x - 4} + 1 = \frac{3x}{x - 4}$ $x = 2$

40. Solve for x: $\frac{x}{x^2 - 4} + \frac{2}{x - 2} = \frac{1}{x + 2}$ $x = -3$

41. Solve for x: $\frac{x}{x + 6} - \frac{1}{7} = \frac{-6}{x + 6}$ No solution

42. Solve for x: $1 + \frac{2}{x - 5} = \frac{20}{x^2 - 25}$ $x = -7$

43. A van travels 100 miles on 4 gallons of gas. How many gallons will it need to travel 625 miles? 25 gal

44. Solve for x: $\frac{x - 5}{5} = \frac{5}{4}$ $x = \frac{45}{4}$

45. Janet can paint a kitchen in 3 hours and James can paint the same kitchen in 4 hours. How long would it take for both working together to paint the kitchen? $1\frac{5}{7}$ hr

46. Find an equation of the line that goes through the points (6, 4) and has slope $m = 5$. $y = 5x - 26$

47. Find an equation of the line having slope 4 and y-intercept 3. $y = 4x + 3$

48. Graph: $x - 5y < -5$

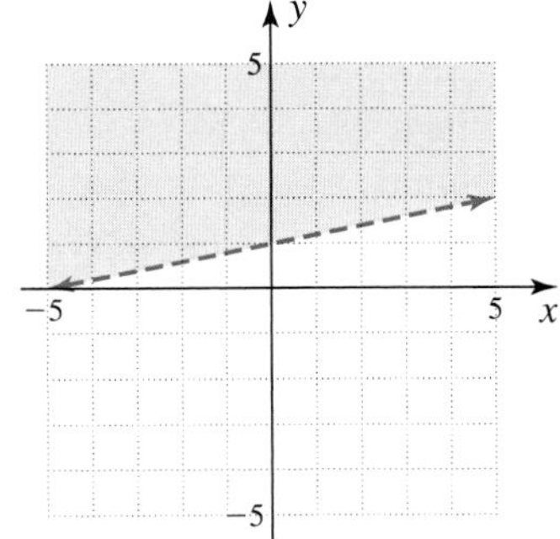

49. Graph:
$-y \geq -5x - 5$

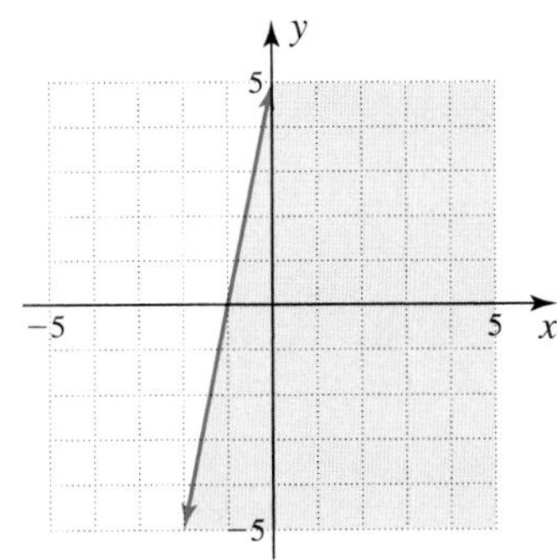

50. An enclosed gas exerts a pressure P on the walls of the container. This pressure is directly proportional to the temperature T of the gas. If the pressure is 3 lb/in.^{2} when the temperature is 240°F, find k. $\frac{1}{80}$

51. If the temperature of a gas is held constant, the pressure P varies inversely as the volume V. A pressure of 1800 lb/in.^{2} is exerted by 6 ft^3 of air in a cylinder fitted with a piston. Find k. 10,800

52. Graph the system and find the solution (if possible):

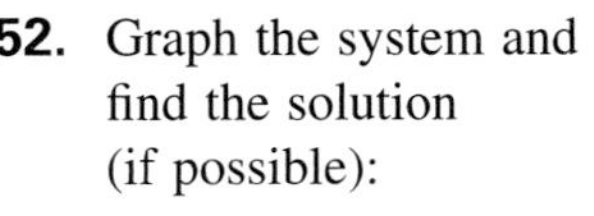

$$x + 4y = 16$$

$$4y - x = 12$$

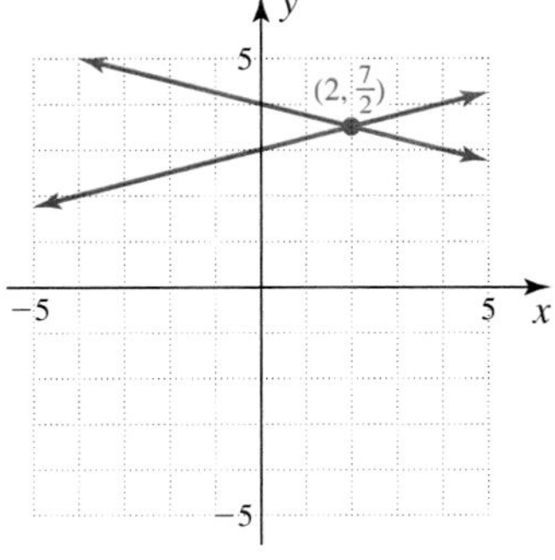

53. Graph the system and find the solution if possible:

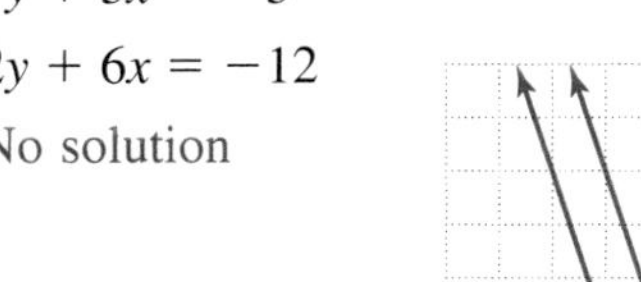

$$y + 3x = -3$$

$$2y + 6x = -12$$

No solution

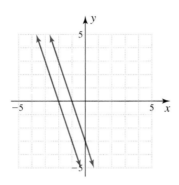

54. Solve by substitution (if possible):

$$x + 4y = -18$$

$$-2x - 8y = 32$$

Inconsistent; no solution

55. Solve by substitution (if possible):

$$x + 3y = 10$$

$$-2x - 6y = -20$$

Dependent; infinitely many solutions

56. Solve the system (if possible):

$$x - 2y = 3$$

$$2x - y = 0 \quad (-1, -2)$$

57. Solve the system (if possible):

$$5x + 4y = -18$$

$$-10x - 8y = 3$$ Inconsistent; no solution

58. Solve the system (if possible):

$$4y + 3x = 11$$

$$6x + 8y = 22$$

Dependent; infinitely many solutions

59. Sara has \$3.50 in nickels and dimes. She has three times as many dimes as nickels. How many nickels and how many dimes does she have?
10 nickels and 30 dimes

60. The sum of two numbers is 215. Their difference is 85. What are the numbers? 65 and 150

61. Evaluate: $\sqrt[3]{-64}$ -4

62. Simplify: $\sqrt[7]{(-3)^7}$ -3

63. Simplify: $\sqrt{\frac{7}{243}}$ $\frac{\sqrt{21}}{27}$

64. Simplify: $\sqrt[3]{128a^8b^9}$ $4a^2b^3\sqrt[3]{2a^2}$

65. Simplify: $\sqrt{48} + \sqrt{12}$ $6\sqrt{3}$

66. Perform the indicated operations:
$\sqrt[3]{2x}\,(\sqrt[3]{4x^2} - \sqrt[3]{81x})$ $2x - 3\sqrt[3]{6x^2}$

67. Rationalize the denominator: $\frac{\sqrt{3}}{\sqrt{2p}}$ $\frac{\sqrt{6p}}{2p}$

68. Find the product: $(\sqrt{125} + \sqrt{343})(\sqrt{245} + \sqrt{175})$
$420 + 74\sqrt{35}$

69. Rationalize the denominator: $\frac{\sqrt{x}}{\sqrt{x} - \sqrt{5}}$ $\frac{x + \sqrt{5x}}{x - 5}$

70. Reduce: $\frac{6 + \sqrt{18}}{3}$ $2 + \sqrt{2}$

71. Solve: $\sqrt{x + 3} = -2$ No real-number solution

72. Solve: $\sqrt{x - 6} - x = -6$ $x = 6$; $x = 7$

Quadratic Equations

10

The Human Side of Algebra

The study of quadratic equations dates back to antiquity. Scores of clay tablets indicate that the Babylonians of 2000 B.C. were already familiar with the quadratic formula that you will study in this chapter. Their solutions using the formula are actually verbal instructions that amounted to using the formula

$$x = \sqrt{\left(\frac{a}{2}\right)^2 + b} - \frac{a}{2}$$

to solve the equation $x^2 + ax = b$.

By Euclid's time (circa 300 B.C.), Greek geometry had reached a stage of development where geometric algebra could be used to solve quadratic equations. This was done by reducing them to the geometric equivalent of one of the following forms:

$$x(x + a) = b^2$$

$$x(x - a) = b^2$$

$$x(a - x) = b^2$$

These equations were then solved by applying different theorems dealing with specific areas.

Later, the Arabian mathematician Muhammed ibn Musa al-Khowarizmi (circa A.D. 820) divided quadratic equations into three types:

$$x^2 + ax = b$$

$$x^2 + b = ax$$

$$x^2 = ax + b$$

with only positive coefficients admitted. All of these developments form the basis for our study of the quadratic equation.

Pretest for Chapter 10

(Answers on pages 685–686)

1. Solve $x^2 = 81$.

2. Solve $49x^2 - 36 = 0$.

3. Solve $8x^2 + 81 = 0$.

4. Solve $36(x + 1)^2 - 5 = 0$.

5. Solve $9(x - 3)^2 + 6 = 0$.

6. Find the missing term in the expression.
$(x + 3)^2 = x^2 + 6x + \square$.

7. The missing terms in the expression $x^2 - 6x + \square = (\ \ )^2$ are ___ and ___, respectively.

8. To solve the equation $7x^2 - 10x = -3$ by completing the square, the first step will be to divide each term by ___ .

9. To solve the equation $x^2 - 5x = -6$ by completing the square, one has to add ___ to both sides of the equation.

10. The solution of $ax^2 + bx + c = 0$ is ___ .

11. Solve $3x^2 - 8x - 3 = 0$.

12. Solve $x^2 = 4x - 3$.

13. Solve $9x = x^2$.

14. Solve $\dfrac{x^2}{2} + \dfrac{5}{4}x = -\dfrac{1}{2}$.

15. Graph $y = 2x^2$.

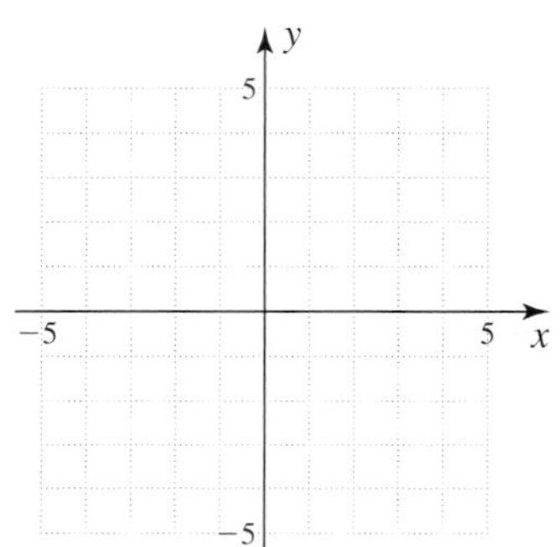

16. Graph $y = (x - 1)^2 + 1$.

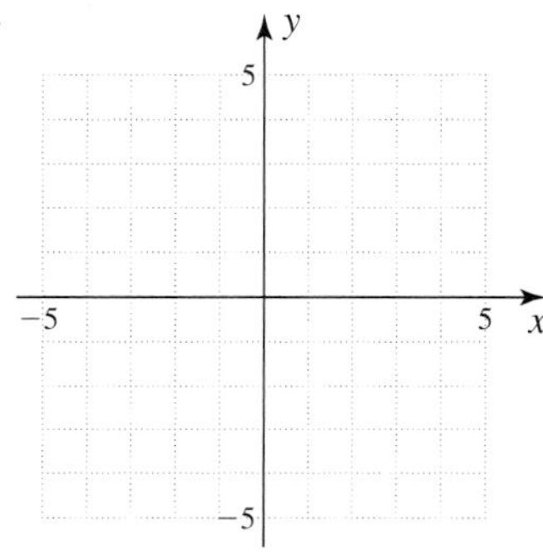

17. Graph the equation
$y = -(x - 1)^2 + 1$.

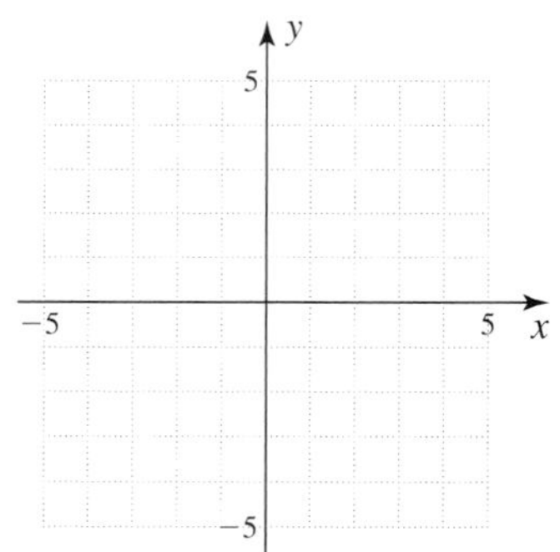

18. Graph the equation
$y = -x^2 - 2x + 8$.
Label the vertex and intercepts.

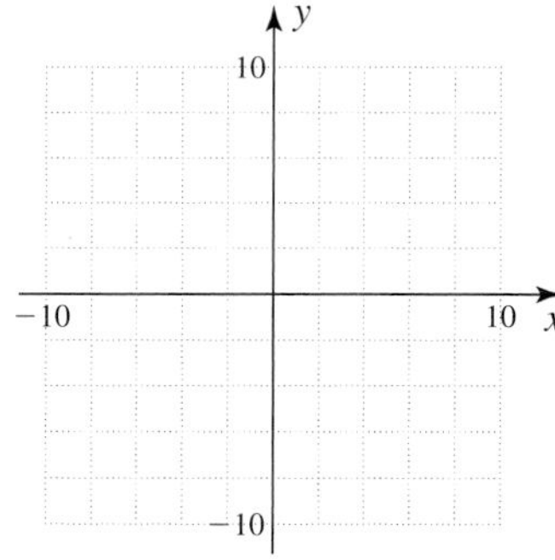

19. Find the length of the hypotenuse of a right triangle if the lengths of the two sides are 3 inches and 5 inches.

20. The formula $d = 5t^2 + v_0t$ gives the distance d (in meters) an object thrown downward with an initial velocity v_0 will have gone after t seconds. How long would it take an object dropped ($v_0 = 0$) from a distance of 80 meters to hit the ground?

21. Find the domain and range of

$$\{(-1, 1), (-2, 1), (-3, 2)\}$$

22. Find the domain and range of

$$\left\{(x, y) \,\middle|\, y = \frac{1}{x - 4}\right\}$$

23. State whether each of the following is a function.

a. $\{(2, -4), (-1, 3), (-1, 4)\}$

b. $\{(2, -4), (-1, 3), (1, -3)\}$

24. If $f(x) = x^3 + 2x^2 - x - 3$, find $f(-2)$.

25. The average price $P(n)$ of books depends on the number n of millions of books sold and is given by the function

$$P(n) = 25 - 0.4n \text{ (dollars)}$$

Find the average price of a book when 30 million copies are sold.

Answers to Pretest

ANSWER	IF YOU MISSED	REVIEW		
	QUESTION	SECTION	EXAMPLES	PAGE
1. ± 9	1	10.1	1	688–689
2. $\pm\frac{6}{7}$	2	10.1	1, 2	688–689, 690
3. No real-number solution	3	10.1	3	691
4. $-1 \pm \frac{\sqrt{5}}{6} = \frac{-6 \pm \sqrt{5}}{6}$	4	10.1	4	692–693
5. No real-number solution	5	10.1	5	693
6. 9	6	10.2	1	699
7. $9; x - 3$	7	10.2	2	700
8. 7	8	10.2	3, 4	702
9. $\frac{25}{4}$	9	10.2	3, 4	702
10. $x = \frac{-b \pm \sqrt{b^2 - 4ac}}{2a}$	10	10.3	1	707
11. $3; -\frac{1}{3}$	11	10.3	1	707
12. 1; 3	12	10.3	2	708
13. 0; 9	13	10.3	3	708–709
14. $-\frac{1}{2}; -2$	14	10.3	4	709–710
15. 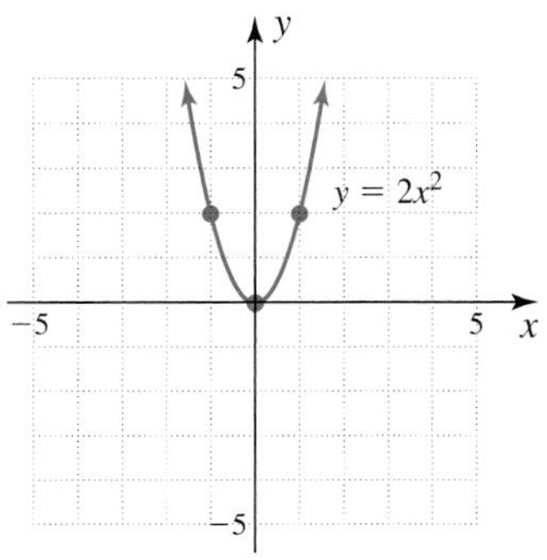	15	10.4	1	714
16. 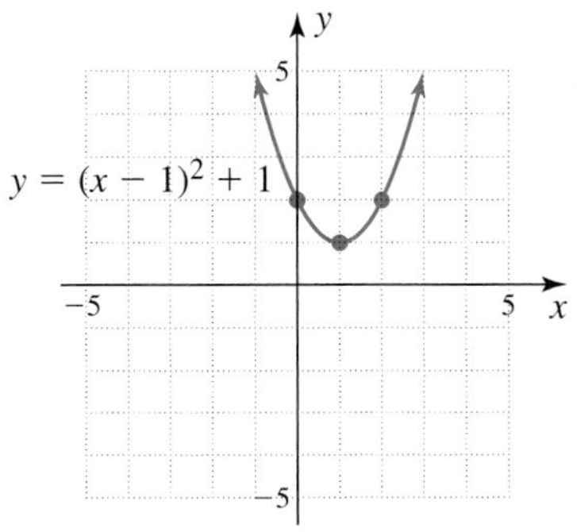	16	10.4	3	716

ANSWER	IF YOU MISSED	REVIEW		
	QUESTION	SECTION	EXAMPLES	PAGE
17. 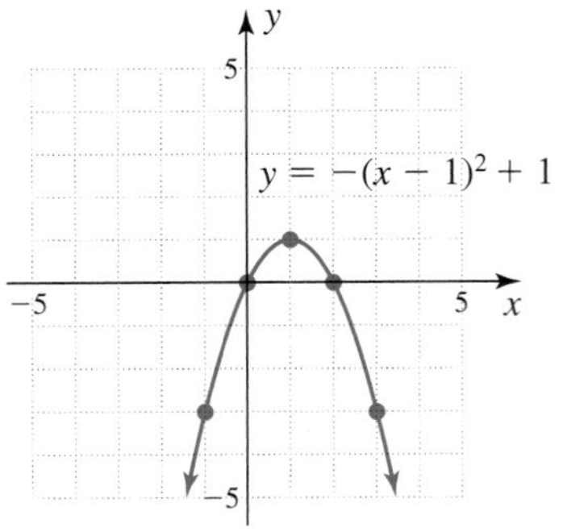 $y = -(x - 1)^2 + 1$	17	10.4	3	716
18. 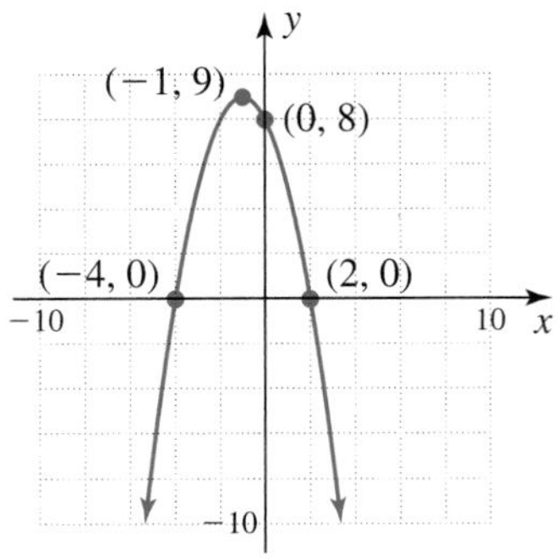 (−1, 9); (0, 8); (−4, 0); (2, 0)	18	10.4	4	718
19. $\sqrt{34}$ inches	19	10.5	1	724–725
20. 4 seconds	20	10.5	2, 3	725–726
21. $D = \{-1, -2, -3\}$; $R = \{1, 2\}$	21	10.6	1	730
22. Domain: All real numbers except 4. Range: All real numbers except 0.	22	10.6	2	731
23. **a.** No **b.** Yes	23	10.6	3	731
24. $f(-2) = -1$	24	10.6	4, 5	732
25. \$13	25	10.6	7	733

10.1 SOLVING QUADRATIC EQUATIONS BY THE SQUARE ROOT PROPERTY

To Succeed, Review How To . . .

1. Find the square roots of a number (pp. 641, 643).
2. Simplify expressions involving radicals (pp. 649–650).

Objectives

Solve quadratic equations of the form

A $X^2 = A$

B $(AX \pm B)^2 = C$

GETTING STARTED

Square Chips

The Pentium chip is used in many computers. Because of space limitations, the chip is very small and covers an area of only 324 square millimeters. If the chip is square, how long is its side? As you recall, the area of a square is obtained by multiplying the length X of its side by itself. If we assume that the length of the side of the chip is X millimeters, then its area is X^2. The area is also 324 square millimeters; we thus have the equation $X^2 = 324$.

Suppose the side of the square is 5 units. Then the area is 5^2 or 25 square units. Now, suppose the side of the square is X units. What is the area? X^2. Now, let's go backwards. If we know that the area of a square is 100 square units, what is the length of the side? The corresponding equation is $X^2 = 100$ and the solution is $X = 10$.

In this section we shall learn how to solve quadratic equations that can be written in the form $X^2 = A$ or $(AX \pm B)^2 = C$ by introducing a new property called the *square root property of equations.*

5 units	X units
Area 25 square units	Area X^2 square units

A Solving Quadratic Equations of the Form $X^2 = A$

The equation $X^2 = 16$ is a *quadratic equation.* In general, a **quadratic equation** is an equation that can be written in the form $ax^2 + bx + c = 0$. How can we solve $X^2 = 16$? First, the equation tells us that a certain number X multiplied by itself gives 16 as a result. Obviously, one possible answer is $X = 4$ because $4^2 = 16$. But wait, what about $X = -4$? It's also true that $(-4)^2 = (-4)(-4) = 16$. Thus the solutions of the equation $X^2 = 16$ are 4 and -4.

In mathematics, the number 4 is called the **positive square root** of 16, and -4 is called the **negative square root** of 16. These roots are usually denoted by

$$\sqrt{16} = 4$$ Read "the positive square root of 16 is 4."

and

$$-\sqrt{16} = -4$$ Read "the negative square root of 16 is -4."

Web It

For a short lesson on solving quadratic equations using the square root property, go to link 10-1-1 on the Bello Website at mhhe.com/bello.

Many of the equations we are about to study have irrational numbers for their solutions. For example, the equation $x^2 = 3$ has two *irrational* solutions. How do we obtain these solutions? By taking the square root of both sides of the equation:

$$x^2 = 3$$

Then

$$x = \pm\sqrt{3} \quad \text{Note that } (\sqrt{3})^2 = 3 \text{ and } (-\sqrt{3})^2 = 3.$$

The notation $\pm\sqrt{3}$ is a shortcut to indicate that x can be $\sqrt{3}$ or $-\sqrt{3}$.

NOTE

With a calculator, the answers to the equation $x^2 = 3$ can be approximated as $x \approx \pm 1.7320508$.

On the other hand, the equation $x^2 = 100$ has *rational* roots. To solve this equation, we proceed as follows:

$$x^2 = 100$$

Then

$$x = \pm\sqrt{100}$$

$$x = \pm 10$$

Thus the solutions are 10 and -10. Here is the property we just used.

SQUARE ROOT PROPERTY OF EQUATIONS

If A is a positive number and $X^2 = A$, then $X = \pm\sqrt{A}$; that is,

$$X = \sqrt{A} \quad \text{or} \quad X = -\sqrt{A}$$

EXAMPLE 1 **Solving quadratic equations using the square root property**

Solve:

a. $x^2 = 36$ **b.** $x^2 - 49 = 0$ **c.** $x^2 = 10$

SOLUTION

a. Given: $x^2 = 36$. Then

$$x = \pm\sqrt{36} \quad \text{Use the square root property.}$$

$$x = \pm 6$$

Thus the solutions of the equation $x^2 = 36$ are 6 and -6, since $6^2 = 36$ and $(-6)^2 = 36$. (Both solutions are rational numbers.)

b. Given: $x^2 - 49 = 0$. Unfortunately, this equation is not of the form $X^2 = A$. However, by adding 49 to both sides of the equation, we can remedy this situation.

$$x^2 - 49 = 0 \quad \text{Given.}$$

$$x^2 = 49 \quad \text{Add 49.}$$

$$x = \pm\sqrt{49} \quad \text{Use the square root property.}$$

$$x = \pm 7$$

Thus the solutions of $x^2 - 49 = 0$ are 7 and -7. (Both solutions are rational numbers.)

PROBLEM 1

Solve:

a. $x^2 = 81$

b. $x^2 - 1 = 0$

c. $x^2 = 13$

Answers

1. a. 9 and -9 **b.** 1 and -1 **c.** $\sqrt{13}$ and $-\sqrt{13}$

Note that the equation $x^2 - 49 = 0$ can also be solved by factoring to write $x^2 - 49 = 0$ as $(x + 7)(x - 7) = 0$. Thus $x + 7 = 0$ or $x - 7 = 0$; that is, $x = -7$ or $x = 7$, as before.

c. Given: $x^2 = 10$. Then, using the square root property,

$$x = \pm\sqrt{10}$$

Since 10 does not have a rational square root, the solutions of the equation $x^2 = 10$ are written as $\sqrt{10}$ and $-\sqrt{10}$. (Here, both solutions are irrational.)

In solving Example 1(b), we added 49 to both sides of the equation to obtain an equivalent equation of the form $X^2 = A$. Does this method work for more complicated examples? The answer is yes. In fact, when solving any quadratic equation in which the *only* exponent of the variable is 2, we can always transform the equation into an equivalent one of the form $X^2 = A$. Here is the idea.

PROCEDURE

Solving Quadratic Equations of the Form $AX^2 - B = 0$

To solve any equation of the form

$$AX^2 - B = 0$$

write

$$AX^2 = B \quad \text{Add } B.$$

and

$$X^2 = \frac{B}{A} \quad \text{Divide by } A.$$

so that, using the square root property,

$$X = \pm\sqrt{\frac{B}{A}}, \quad \frac{B}{A} \geq 0$$

Thus to solve the equation

$$16x^2 - 81 = 0$$

we write it in the form $AX^2 = B$ or $X^2 = \frac{B}{A}$:

$$16x^2 - 81 = 0 \quad \text{Given.}$$

$$16x^2 = 81 \quad \text{Add 81.}$$

$$x^2 = \frac{81}{16} \quad \text{Divide by 16.}$$

$$x = \pm\sqrt{\frac{81}{16}} \quad \text{Use the square root property.}$$

$$x = \pm\frac{9}{4}$$

The solutions are $\frac{9}{4}$ and $-\frac{9}{4}$. Since $16 \cdot (\frac{9}{4})^2 - 81 = 16 \cdot \frac{81}{16} - 81 = 0$ and $16(\frac{-9}{4})^2 - 81 = 16 \cdot \frac{81}{16} - 81 = 0$, our result is correct. Just remember to first rewrite the equation in the form $X^2 = \frac{B}{A}$.

EXAMPLE 2 **Solving a quadratic equation of the form $AX^2 - B = 0$**

Solve: $36x^2 - 25 = 0$

SOLUTION

$$36x^2 - 25 = 0 \quad \text{Given.}$$

$$36x^2 = 25 \quad \text{Add 25.}$$

$$x^2 = \frac{25}{36} \quad \text{Divide by 36.}$$

$$x = \pm\sqrt{\frac{25}{36}} \quad \text{Use the square root property.}$$

$$x = \pm\frac{5}{6}$$

The solutions are $\frac{5}{6}$ and $-\frac{5}{6}$. (Here, both solutions are rational numbers.)

PROBLEM 2

Solve: $9x^2 - 16 = 0$

Of course, not all equations of the form $X^2 = A$ have solutions that are real numbers. For example, to solve the equation $x^2 + 64 = 0$, we write

$$x^2 + 64 = 0 \quad \text{Given.}$$

$$x^2 = -64 \quad \text{Subtract 64.}$$

But there is no *real* number whose square is -64. If you square a nonzero real number, the answer is always positive. Thus x^2 is positive and can never equal -64. The equation $x^2 + 64 = 0$ has *no real-number* solution.

As we have seen, not all equations have *rational-number* solutions—that is, solutions of the form $\frac{a}{b}$, where a and b are integers, $b \neq 0$. For example, the equation

$$16x^2 - 5 = 0$$

is solved as follows:

$$16x^2 - 5 = 0 \quad \text{Given.}$$

$$16x^2 = 5 \quad \text{Add 5.}$$

$$x^2 = \frac{5}{16} \quad \text{Divide by 16.}$$

$$x = \pm\sqrt{\frac{5}{16}} \quad \text{Use the square root property.}$$

However, $\sqrt{\frac{5}{16}}$ is *not* a rational number even though the denominator, 16, is the square of 4. As you recall, using the quotient rule for radicals, we may write

$$\pm\sqrt{\frac{5}{16}} = \pm\frac{\sqrt{5}}{\sqrt{16}} = \pm\frac{\sqrt{5}}{4}$$

Thus the solutions of the equation $16x^2 - 5 = 0$ are

$$\frac{\sqrt{5}}{4} \quad \text{and} \quad -\frac{\sqrt{5}}{4}$$

Both solutions are irrational numbers.

Web It

For a lesson on how to solve quadratic equations by "taking roots," go to link 10-1-2 on the Bello Website at mhhe.com/bello and click on "Taking Roots."

Answer

2. $\frac{4}{3}$ and $-\frac{4}{3}$

EXAMPLE 3 **Solving a quadratic equation using the quotient rule**

Solve:

a. $4x^2 - 7 = 0$

b. $8x^2 + 49 = 0$

SOLUTION

a. $4x^2 - 7 = 0$ Given.

$4x^2 = 7$ Add 7.

$x^2 = \frac{7}{4}$ Divide by 4.

$x = \pm\sqrt{\frac{7}{4}}$ Use the square root property.

$x = \pm\frac{\sqrt{7}}{\sqrt{4}}$ Use the quotient rule for radicals.

$x = \pm\frac{\sqrt{7}}{2}$

Thus the solutions of $4x^2 - 7 = 0$ are $\frac{\sqrt{7}}{2}$ and $-\frac{\sqrt{7}}{2}$. They are both irrational numbers.

b. $8x^2 + 49 = 0$ Given.

$8x^2 = -49$ Subtract 49.

$x^2 = -\frac{49}{8}$ Divide by 8.

But since the square of a real number x cannot be negative and $-\frac{49}{8}$ is negative, this equation has *no* real-number solution.

PROBLEM 3

Solve:

a. $49x^2 - 3 = 0$

b. $10x^2 + 9 = 0$

B Solving Quadratic Equations of the Form $(AX \pm B)^2 = C$

We've already mentioned that to solve an equation in which the only exponent of the variable is 2, we must transform the equation into an equivalent one of the form $X^2 = A$. Now consider the equation

$$(x - 2)^2 = 9$$

If we think of $(x - 2)$ as X, we have an equation of the form $X^2 = 9$, which we just learned how to solve! Thus we have the following.

$(x - 2)^2 = 9$ Given.

$X^2 = 9$ Write $x - 2$ as X.

$X = \pm\sqrt{9}$ Use the square root property.

$X = \pm 3$

$x - 2 = \pm 3$ Write X as $x - 2$.

$x = 2 \pm 3$ Add 2.

Hence

$$x = 2 + 3 \quad \text{or} \quad x = 2 - 3$$

The solutions are $x = 5$ and $x = -1$.

Answers

3. a. $\frac{\sqrt{3}}{7}$ and $-\frac{\sqrt{3}}{7}$

b. No real-number solution

Clearly, by thinking of $(x - 2)$ as X, we can solve a more complicated equation. In the same manner, we can solve $9(x - 2)^2 - 5 = 0$:

$9(x - 2)^2 - 5 = 0$ Given.

$9(x - 2)^2 = 5$ Add 5.

$(x - 2)^2 = \frac{5}{9}$ Divide by 9.

$x - 2 = \pm\sqrt{\frac{5}{9}}$ *Think* of $x - 2$ as X and use the square root property.

$x - 2 = \pm\frac{\sqrt{5}}{\sqrt{9}}$ Use the quotient rule for radicals.

$x - 2 = \pm\frac{\sqrt{5}}{3}$

$x = 2 \pm \frac{\sqrt{5}}{3}$ Add 2.

The solutions are

$$2 + \frac{\sqrt{5}}{3} = \frac{6 + \sqrt{5}}{3} \quad \text{and} \quad 2 - \frac{\sqrt{5}}{3} = \frac{6 - \sqrt{5}}{3}$$

Web It

For a lesson on solving $(AX + B)^2 = C$, try link 10-1-3 on the Bello Website at mhhe.com/bello.

EXAMPLE 4 **Solving quadratic equations of the form $(AX + B)^2 = C$**

Solve:

a. $(x + 3)^2 = 9$ **b.** $(x + 1)^2 - 4 = 0$ **c.** $25(x + 2)^2 - 3 = 0$

SOLUTION

a. $(x + 3)^2 = 9$ Given.

$x + 3 = \pm\sqrt{9}$ *Think* of $(x + 3)$ as X.

$x + 3 = \pm 3$

$x = -3 \pm 3$ Subtract 3.

$x = -3 + 3$ or $x = -3 - 3$

$x = 0$ or $x = -6$

The solutions are 0 and -6.

b. $(x + 1)^2 - 4 = 0$ Given.

$(x + 1)^2 = 4$ Add 4 (to have an equation of the form $X^2 = A$).

$x + 1 = \pm\sqrt{4}$ *Think* of $(x + 1)$ as X.

$x + 1 = \pm 2$

$x = -1 \pm 2$ Subtract 1.

$x = -1 + 2$ or $x = -1 - 2$

$x = 1$ or $x = -3$

The solutions are 1 and -3.

PROBLEM 4

Solve:

a. $(x + 6)^2 = 36$

b. $(x + 2)^2 - 9 = 0$

c. $16(x + 1)^2 - 5 = 0$

Answers

4. a. The solutions are 0 and -12. **b.** The solutions are 1 and -5. **c.** The solutions are

$$-1 + \frac{\sqrt{5}}{4} = \frac{-4 + \sqrt{5}}{4}$$

and

$$-1 - \frac{\sqrt{5}}{4} = \frac{-4 - \sqrt{5}}{4}$$

c. $25(x + 2)^2 - 3 = 0$ Given.

$25(x + 2)^2 = 3$ Add 3.

$(x + 2)^2 = \frac{3}{25}$ Divide by 25.

$x + 2 = \pm\sqrt{\frac{3}{25}}$ *Think* of $(x + 2)$ as X.

$x + 2 = \pm\frac{\sqrt{3}}{\sqrt{25}} = \pm\frac{\sqrt{3}}{5}$ Since $\sqrt{\frac{3}{25}} = \frac{\sqrt{3}}{\sqrt{25}}$

$x = -2 \pm \frac{\sqrt{3}}{5}$ Subtract 2.

The solutions are

$$-2 + \frac{\sqrt{3}}{5} = \frac{-10 + \sqrt{3}}{5} \quad \text{and} \quad -2 - \frac{\sqrt{3}}{5} = \frac{-10 - \sqrt{3}}{5}$$

NOTE

It is easy to spot quadratic equations that have no real-number solution. Here is how: if the equation can be written in the form Expression^2 = Negative number [that is, $(\quad)^2$ = Negative number], the equation has no solution.

As we mentioned before, if we square any real number, the result is not negative. For this reason, an equation such as

$(x - 4)^2 = -5$ If $(x - 4)$ represents a real number, $(x - 4)^2$ *cannot* be negative. But -5 is negative, so $(x - 4)^2$ and -5 can *never* be equal.

has *no* real-number solution. Similarly,

$$(x - 3)^2 + 8 = 0$$

has no real-number solution, since

$$(x - 3)^2 + 8 = 0 \quad \text{is equivalent to} \quad (x - 3)^2 = -8$$

by subtracting 8. We use this idea in the next example.

EXAMPLE 5 **Solving a quadratic equation when there is no real-number solution**

Solve: $9(x - 5)^2 + 1 = 0$

SOLUTION

$9(x - 5)^2 + 1 = 0$ Given.

$9(x - 5)^2 = -1$ Subtract 1.

$(x - 5)^2 = -\frac{1}{9}$ Divide by 9.

Since $(x - 5)$ is to be a real number, $(x - 5)^2$ *can never* be negative. But $-\frac{1}{9}$ is negative. Thus the equation $(x - 5)^2 = -\frac{1}{9}$ [which is equivalent to $9(x - 5)^2 + 1 = 0$] has *no* real-number solution.

PROBLEM 5

Solve: $16(x - 3)^2 + 7 = 0$

Answer

5. No real-number solution

EXAMPLE 6 Solving a quadratic equation of the form $(AX - B)^2 = C$

Solve: $3(2x - 3)^2 = 54$

SOLUTION We want to write the equation in the form $X^2 = A$.

$$3(2x - 3)^2 = 54 \quad \text{Given.}$$

$$(2x - 3)^2 = \frac{54}{3} = 18 \quad \text{Divide by 3.}$$

$$X^2 = 18 \quad \text{Write } 2x - 3 \text{ as } X.$$

$$X = \pm\sqrt{18} \quad \text{Use the square root property.}$$

$$X = \pm 3\sqrt{2} \quad \text{Simplify the radical } (\sqrt{18} = \sqrt{9 \cdot 2} = 3\sqrt{2}).$$

$$2x - 3 = \pm 3\sqrt{2} \quad \text{Write } X \text{ as } 2x - 3.$$

$$2x = 3 \pm 3\sqrt{2} \quad \text{Add 3.}$$

$$x = \frac{3 \pm 3\sqrt{2}}{2} \quad \text{Divide by 2.}$$

The solutions are

$$\frac{3 + 3\sqrt{2}}{2} \quad \text{and} \quad \frac{3 - 3\sqrt{2}}{2}$$

PROBLEM 6

Solve: $2(3x - 2)^2 = 36$

EXAMPLE 7 Application: BMI and weight

Your Body Mass Index (BMI) is a measurement of your healthy weight relative to your height. According to the National Heart and Lung Institute, the formula for your BMI is given by $\text{BMI} = \frac{705W}{H^2}$, where W is your weight in pounds and H is your height in inches. If a person has a BMI of 20, which is in the "normal" range, and weighs 140 pounds, how tall is the person?

SOLUTION Substituting 20 for BMI and 140 for W, we have

$$20 = \frac{705 \cdot 140}{H^2}$$

Multiplying both sides by H^2 $\quad 20H^2 = 705 \cdot 140$

Dividing both sides by 20 $\quad H^2 = 705 \cdot 7$

Or $\quad H^2 = 4935$

Thus, $\quad H = \sqrt{4935} \approx 70$ inches

PROBLEM 7

What is the height of a person weighing 180 pounds with a BMI of 30, which is in the "overweight" range?

Answers

6. The solutions are $\frac{2 \pm 3\sqrt{2}}{3}$.
7. $\sqrt{4230} \approx 65$ inches

Exercises 10.1

A In Problems 1–20, solve the given equation.

1. $x^2 = 100$ $x = \pm 10$

2. $x^2 = 1$ $x = \pm 1$

3. $x^2 = 0$ $x = 0$

4. $x^2 = 121$ $x = \pm 11$

5. $y^2 = -4$ No real-number solution

6. $y^2 = -16$ No real-number solution

7. $x^2 = 7$ $x = \pm\sqrt{7}$

8. $x^2 = 3$ $x = \pm\sqrt{3}$

9. $x^2 - 9 = 0$ $x = \pm 3$

10. $x^2 - 64 = 0$ $x = \pm 8$

11. $x^2 - 3 = 0$ $x = \pm\sqrt{3}$

12. $x^2 - 5 = 0$ $x = \pm\sqrt{5}$

13. $25x^2 - 1 = 0$ $x = \pm\frac{1}{5}$

14. $36x^2 - 49 = 0$ $x = \pm\frac{7}{6}$

15. $100x^2 - 49 = 0$ $x = \pm\frac{7}{10}$

16. $81x^2 - 36 = 0$ $x = \pm\frac{2}{3}$

17. $25y^2 - 17 = 0$ $y = \pm\frac{\sqrt{17}}{5}$

18. $9y^2 - 11 = 0$ $y = \pm\frac{\sqrt{11}}{3}$

19. $25x^2 + 3 = 0$
No real-number solution

20. $49x^2 + 1 = 0$
No real-number solution

B In Problems 21–60, solve the given equation.

21. $(x + 1)^2 = 81$
$x = 8$ or $x = -10$

22. $(x + 3)^2 = 25$
$x = 2$ or $x = -8$

23. $(x - 2)^2 = 36$
$x = 8$ or $x = -4$

24. $(x - 3)^2 = 16$
$x = 7$ or $x = -1$

25. $(z - 4)^2 = -25$
No real-number solution

26. $(z + 2)^2 = -16$
No real-number solution

27. $(x - 9)^2 = 81$
$x = 18$ or $x = 0$

28. $(x - 6)^2 = 36$
$x = 12$ or $x = 0$

29. $(x + 4)^2 = 16$
$x = 0$ or $x = -8$

30. $(x + 7)^2 = 49$
$x = 0$ or $x = -14$

31. $25(x + 1)^2 - 1 = 0$
$x = -\frac{4}{5}$ or $x = -\frac{6}{5}$

32. $16(x + 2)^2 - 1 = 0$
$x = -\frac{7}{4}$ or $x = -\frac{9}{4}$

33. $36(x - 3)^2 - 49 = 0$
$x = \frac{25}{6}$ or $x = \frac{11}{6}$

34. $9(x - 1)^2 - 25 = 0$
$x = \frac{8}{3}$ or $x = -\frac{2}{3}$

35. $4(x + 1)^2 - 25 = 0$
$x = \frac{3}{2}$ or $x = -\frac{7}{2}$

36. $49(x + 2)^2 - 16 = 0$
$x = -\frac{10}{7}$ or $x = -\frac{18}{7}$

37. $9(x - 1)^2 - 5 = 0$
$x = 1 \pm \frac{\sqrt{5}}{3} = \frac{3 \pm \sqrt{5}}{3}$

38. $4(x - 2)^2 - 3 = 0$
$x = 2 \pm \frac{\sqrt{3}}{2} = \frac{4 \pm \sqrt{3}}{2}$

39. $16(x + 1)^2 + 1 = 0$
No real-number solution

40. $25(x + 2)^2 + 16 = 0$
No real-number solution

41. $x^2 = \frac{1}{81}$ $x = \pm\frac{1}{9}$

42. $x^2 = \frac{1}{9}$ $x = \pm\frac{1}{3}$

43. $x^2 - \frac{1}{16} = 0$ $x = \pm\frac{1}{4}$

44. $x^2 - \frac{1}{36} = 0$ $x = \pm\frac{1}{6}$

45. $6x^2 - 24 = 0$
$x = \pm 2$

46. $3x^2 - 75 = 0$
$x = \pm 5$

47. $2(v + 1)^2 - 18 = 0$
$v = 2$ or $v = -4$

48. $3(v - 2)^2 - 48 = 0$
$v = 6$ or $v = -2$

49. $8(x - 1)^2 - 18 = 0$
$x = \frac{5}{2}$ or $x = -\frac{1}{2}$

50. $50(x + 3)^2 - 72 = 0$
$x = -\frac{9}{5}$ or $x = -\frac{21}{5}$

51. $4(2y - 3)^2 = 32$
$y = \frac{3}{2} \pm \sqrt{2} = \frac{3 \pm 2\sqrt{2}}{2}$

52. $2(3y - 1)^2 = 24$
$y = \frac{1}{3} \pm \frac{2\sqrt{3}}{3} = \frac{1 \pm 2\sqrt{3}}{3}$

53. $8(2x - 3)^2 - 64 = 0$
$x = \frac{3}{2} \pm \sqrt{2} = \frac{3 \pm 2\sqrt{2}}{2}$

54. $5(3x - 1)^2 - 60 = 0$
$x = \frac{1}{3} \pm \frac{2\sqrt{3}}{3} = \frac{1 \pm 2\sqrt{3}}{3}$

55. $3\left(\frac{1}{2}x + 1\right)^2 = 54$
$x = -2 \pm 6\sqrt{2}$

56. $4\left(\frac{1}{3}x - 1\right)^2 = 80$
$x = 3 \pm 6\sqrt{5}$

57. $2\left(\frac{1}{3}x - 1\right)^2 - 40 = 0$
$x = 3 \pm 6\sqrt{5}$

58. $3\left(\frac{1}{3}x - 2\right)^2 - 54 = 0$
$x = 6 \pm 9\sqrt{2}$

59. $5\left(\frac{1}{2}y - 1\right)^2 + 60 = 0$
No real-number solution

60. $6\left(\frac{1}{3}y - 2\right)^2 + 72 = 0$
No real-number solution

SKILL CHECKER

Try the Skill Checker Exercises so you'll be ready for the next section.

Expand:

61. $(x + 7)^2$
$x^2 + 14x + 49$

62. $(x + 5)^2$
$x^2 + 10x + 25$

63. $(x - 3)^2$
$x^2 - 6x + 9$

64. $(x - 5)^2$
$x^2 - 10x + 25$

USING YOUR KNOWLEDGE

Going in Circles

The area A of a circle of radius r is given by $A = \pi r^2$. Find the radius of a circle whose area measures

65. 25π square inches 5 in.

66. 12π square feet $2\sqrt{3}$ ft

The surface area A of a sphere of radius r is given by $A = 4\pi r^2$. Find the radius of a sphere whose surface area measures

67. 49π square feet $3\frac{1}{2}$ ft

68. 81π square inches $4\frac{1}{2}$ in.

WRITE ON

69. Explain why the equation $x^2 + 6 = 0$ has no real-number solution. Answers may vary.

70. Explain why the equation $(x + 1)^2 + 3 = 0$ has no real-number solution. Answers may vary.

Consider the equation $X^2 = A$.

71. What can you say about A if the equation has no real-number solution? $A < 0$

72. What can you say about A if the equation has exactly one solution? $A = 0$

73. What can you say about A if the equation has two solutions? $A > 0$

74. What types of solution does the equation have if A is a prime number? Irrational

75. What types of solution does the equation have if A is a positive perfect square? Rational

MASTERY TEST

If you know how to do these problems, you have learned your lesson!

Solve if possible:

76. $16(x - 3)^2 - 7 = 0$
$x = 3 \pm \frac{\sqrt{7}}{4} = \frac{12 \pm \sqrt{7}}{4}$

77. $3(x - 1)^2 - 24 = 0$
$x = 1 \pm 2\sqrt{2}$

78. $(x + 5)^2 = 25$
$x = 0$ or $x = -10$

79. $(x + 2)^2 - 9 = 0$
$x = 1$ or $x = -5$

80. $16(x + 1)^2 - 5 = 0$
$x = -1 \pm \frac{\sqrt{5}}{4} = \frac{-4 \pm \sqrt{5}}{4}$

81. $49x^2 - 3 = 0$
$x = \pm\frac{\sqrt{3}}{7}$

82. $10x^2 + 9 = 0$
No real-number solution

83. $9x^2 - 16 = 0$
$x = \pm\frac{4}{3}$

84. $x^2 = 121$ $x = \pm 11$

85. $x^2 - 1 = 0$ $x = \pm 1$

86. $x^2 = 13$ $x = \pm\sqrt{13}$

10.2 SOLVING QUADRATIC EQUATIONS BY COMPLETING THE SQUARE

To Succeed, Review How To . . .

1. Recognize a quadratic equation (p. 434).
2. Expand $(x \pm a)^2$ (pp. 357–358).

Objective

A Solve a quadratic equation by completing the square.

GETTING STARTED

Completing the Square for Round Baseballs

The man has just batted the ball straight up at 96 feet per second. At the end of t seconds, the height h of the ball will be

$$h = -16t^2 + 96t$$

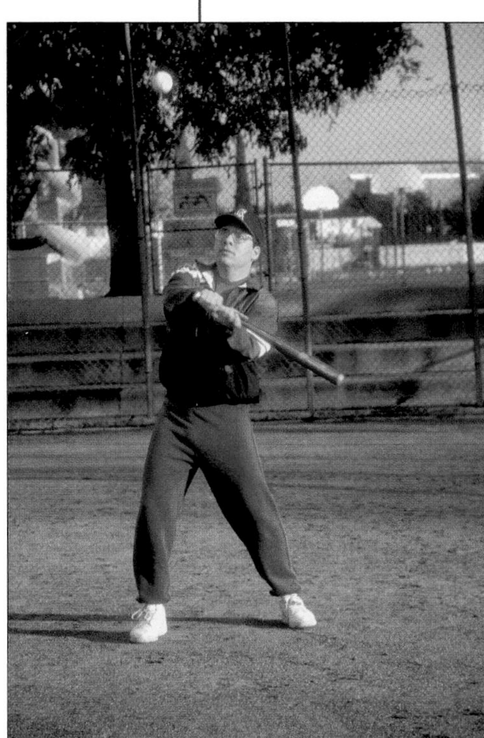

How long will it be before the ball reaches 44 feet? To solve this problem, we let $h = 44$ to obtain

$$-16t^2 + 96t = 44$$

$$t^2 - 6t = -\frac{44}{16} = -\frac{11}{4} \qquad \text{Divide by } -16 \text{ and simplify.}$$

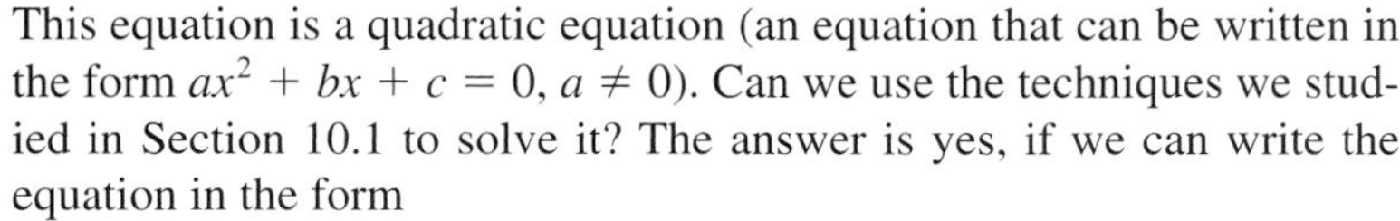

This equation is a quadratic equation (an equation that can be written in the form $ax^2 + bx + c = 0$, $a \neq 0$). Can we use the techniques we studied in Section 10.1 to solve it? The answer is yes, if we can write the equation in the form

$$(t - N)^2 = A \qquad N \text{ and } A \text{ are the numbers we need to find to solve the problem.}$$

To do this, however, we should know a little bit more about a technique used in algebra called **completing the square.** We will present several examples and then generalize the results. Why? Because when we generalize the process and learn how to solve the quadratic equation $ax^2 + bx + c = 0$ by completing the square, we can use the formula obtained, called the *quadratic formula,* to solve *any* equation of the form $ax^2 + bx + c = 0$, by simply substituting the values of a, b, and c in the quadratic formula. We will do this in the next section, but right now we need to review how to expand binomials, because completing the square involves that process. So let's start this section by doing just that!

A Solving Quadratic Equations by Completing the Square

Recall that

$$(\underbrace{X}_{\text{First term}} + \underbrace{A}_{\text{Second term}})^2 = \underbrace{X^2}_{\text{First term squared}} + \underbrace{2A}_{\text{Coefficient of } X}X + \underbrace{A^2}_{\text{Second term squared}}$$

Thus

$$(x + 7)^2 = x^2 + 14x + 7^2$$

$$(x + 2)^2 = x^2 + 4x + 2^2$$

$$(x + 5)^2 = x^2 + 10x + 5^2$$

Do you see any relationship between the coefficients of x (14, 4, and 10, respectively) and the last terms? Perhaps you will see it better if we write it in a table.

Coefficient of X	Last Term Squared
14	7^2
4	2^2
10	5^2

It seems that half the coefficient of x gives the number to be squared for the last term. Thus

$$\frac{14}{2} = 7$$
$$\frac{4}{2} = 2$$
$$\frac{10}{2} = 5$$

Now, what numbers would you add to complete the given squares?

$$(x + 3)^2 = x^2 + 6x + \square$$
$$(x + 4)^2 = x^2 + 8x + \square$$
$$(x + 6)^2 = x^2 + 12x + \square$$

The correct answers are $\left(\frac{6}{2}\right)^2 = 3^2 = 9$, $\left(\frac{8}{2}\right)^2 = 4^2 = 16$, and $\left(\frac{12}{2}\right)^2 = 6^2 = 36$. We then have

$$(x + 3)^2 = x^2 + 6x + 3^2$$
$6 \div 2 = 3$

The last term is always the square of half of the coefficient of the middle term.

$$(x + 4)^2 = x^2 + 8x + 4^2$$
$8 \div 2 = 4$

$$(x + 6)^2 = x^2 + 12x + 6^2$$
$12 \div 2 = 6$

You can also find the missing number to complete the square by using a diagram. For example, to complete the square in $(x + 3)^2$ complete the diagram:

	x	$+\ 3$
x	x^2	$3x$
$+\ 3$	$3x$	?

You need 3^2 or 9 in the lower right corner, so $(x + 3)^2 = x^2 + 6x + 9$, as before.

Here is the procedure we have just used to complete the square.

PROCEDURE

Completing the Square $x^2 + bx + \square$

1. Find the coefficient of the x term. (b)
2. Divide the coefficient by 2. $\left(\frac{b}{2}\right)$
3. Square this number to obtain the last term. $\left(\frac{b}{2}\right)^2 = \frac{b^2}{4}$

Thus to complete the square in

$$x^2 + 16x + \square$$

We proceed as follows:

1. Find the coefficient of the x term $\longrightarrow$ 16.
2. Divide the coefficient by 2 $\longrightarrow$ 8.
3. Square this number to obtain the last term $\longrightarrow$ 8^2.

Hence,

$$x^2 + 16x + \boxed{8^2} = (x + 8)^2$$

Now consider

$$x^2 - 18x + \square$$

Our steps to fill in the blank are as before:

1. Find the coefficient of the x term $\longrightarrow$ -18.
2. Divide the coefficient by 2 $\longrightarrow$ -9.
3. Square this number to obtain the last term $\longrightarrow$ $(-9)^2$.

Hence

$$\begin{aligned} &x^2 - 18x + (-9)^2 \\ &= x^2 - 18x + 9^2 \\ &= x^2 - 18x + 81 = (x - 9)^2 \end{aligned}$$

Recall that $(-9)^2 = (9)^2$; they are both 81.

EXAMPLE 1 **Completing the square**

Find the missing term to complete the square:

a. $x^2 + 20x + \square$ **b.** $x^2 - x + \square$

SOLUTION

a. We use the three-step procedure:

1. The coefficient of x is 20.
2. $\frac{20}{2} = 10$
3. The missing term is $\boxed{10^2} = \boxed{100}$. Hence, $x^2 + 20x + 100 = (x + 10)^2$.

b. Again we use the three-step procedure:

1. The coefficient of x is -1.
2. $\frac{-1}{2} = -\frac{1}{2}$
3. The missing term is $\boxed{(-\frac{1}{2})^2} = \boxed{\frac{1}{4}}$. Hence, $x^2 - x + \frac{1}{4} = (x - \frac{1}{2})^2$.

PROBLEM 1

Find the missing terms:

a. $(x + 11)^2 = x^2 + 22x + \square$

b. $\left(x - \frac{1}{4}\right)^2 = x^2 - \frac{1}{2}x + \square$

Can we use the patterns we've just studied to look for further patterns? Of course! For example, how would you fill in the blanks in

$$x^2 + 16x + \square = (\quad)^2$$

Here the coefficient of x is 16, so $(\frac{16}{2})^2 = 8^2$ goes in the box. Since

Same

$$X^2 + 2AX + A^2 = (X + A)^2$$

Same

$$x^2 + 16x + 8^2 = (x + 8)^2$$

Answers

1. a. $(\frac{22}{2})^2 = 11^2 = 121$

b. $\left(\frac{-\frac{1}{2}}{2}\right)^2 = (-\frac{1}{4})^2 = \frac{1}{16}$

Similarly,

$$x^2 - 6x + \square = (\quad)^2$$

is completed by reasoning that the coefficient of x is -6, so

$$\left(\frac{-6}{2}\right)^2 = (-3)^2 = (3)^2$$

goes in the box. Now

$$X^2 - 2AX + A^2 = (X - A)^2$$

(Same: A^2 and A; Same: X^2 and X)

Since the middle term on the left has a negative sign, the sign inside the parentheses must be negative.

Hence

$$x^2 - 6x + \square = (\quad)^2$$

becomes

$$x^2 - 6x + 3^2 = (x - 3)^2$$

EXAMPLE 2 More practice completing the square

Find the missing terms:

a. $x^2 - 10x + \square = (\quad)^2$ **b.** $x^2 + 3x + \square = (\quad)^2$

SOLUTION

a. The coefficient of x is -10; thus the number in the box should be $(\frac{-10}{2})^2 = (-5)^2 = 5^2$, and we should have

Same

$$x^2 - 10x + 5^2 = (x - 5)^2$$

b. The coefficient of x is 3; thus we have to add $(\frac{3}{2})^2$. Then

$$x^2 + 3x + \left(\frac{3}{2}\right)^2 = \left(x + \frac{3}{2}\right)^2$$

PROBLEM 2

Find the missing terms:

a. $x^2 - 12x + \square = (\quad)^2$

b. $x^2 + 5x + \square = (\quad)^2$

We are finally ready to solve the equation from the *Getting Started* that involves the time it takes the ball to reach 44 feet, that is,

$$t^2 - 6t = -\frac{11}{4}$$

As you recall, t represents the time it takes the ball to reach 44 feet (p. 697).

Since the coefficient of t is -6, we must add $(\frac{-6}{2})^2 = (-3)^2 = 3^2$ to both sides of the equation. We then have

$$t^2 - 6t + 3^2 = -\frac{11}{4} + 3^2$$

$$(t - 3)^2 = -\frac{11}{4} + \frac{36}{4}$$

Note that $3^2 = 9 = \frac{36}{4}$.

$$(t - 3)^2 = \frac{25}{4}$$

Answers

2. a. $x^2 - 12x + 6^2 = (x - 6)^2$
b. $x^2 + 5x + (\frac{5}{2})^2 = (x + \frac{5}{2})^2$

Web It

You can see a lesson on how to complete the square at link 10-2-1 on the Bello Website at mhhe.com/bello. Then go to link 10-2-2 and practice with an interactive system for completing the square.

Then

$$(t - 3) = \pm\sqrt{\frac{25}{4}} = \pm\frac{5}{2}$$

$$t = 3 \pm \frac{5}{2}$$

$$t = 3 + \frac{5}{2} = \frac{11}{2} \quad \text{or} \quad t = 3 - \frac{5}{2} = \frac{1}{2}$$

This means that the ball reaches 44 feet after $\frac{1}{2}$ second (on the way up) and after $\frac{11}{2} = 5\frac{1}{2}$ seconds (on the way down).

Here is a summary of the steps needed to solve a quadratic equation by completing the square. The solution of our original equation follows so you can see how the steps are carried out.

PROCEDURE

Solving a Quadratic Equation by Completing the Square

1. Write the equation with the variables in descending order on the left and the constants on the right.
2. If the coefficient of the squared term is not 1, divide each term by this coefficient.
3. Add the square of one-half of the coefficient of the first-degree term to both sides.
4. Rewrite the left-hand side as a perfect square binomial.
5. Use the square root property to solve the resulting equation.

For the equation from the *Getting Started,* we have:

1. $$-16t^2 + 96t = 44$$

2. $$\frac{-16t^2}{-16} + \frac{96t}{-16} = \frac{44}{-16}$$

$$t^2 - 6t = -\frac{11}{4}$$

3. $$t^2 - 6t + \left(\frac{-6}{2}\right)^2 = -\frac{11}{4} + \left(\frac{-6}{2}\right)^2$$

$$t^2 - 6t + 3^2 = -\frac{11}{4} + 3^2$$

4. $$(t - 3)^2 = \frac{25}{4}$$

5. $$(t - 3) = \pm\sqrt{\frac{25}{4}}$$

$$t - 3 = \pm\frac{5}{2}$$

$$t = 3 \pm \frac{5}{2}$$

Thus $t = 3 + \frac{5}{2} = 5\frac{1}{2}$ or $t = 3 - \frac{5}{2} = \frac{1}{2}$.

We now use this procedure in another example.

EXAMPLE 3 **Solving a quadratic equation by completing the square**

Solve: $4x^2 - 16x + 7 = 0$

SOLUTION We use the five-step procedure just given.

$$4x^2 - 16x + 7 = 0 \quad \text{Given.}$$

1. Subtract 7. $\quad 4x^2 - 16x = -7$
2. Divide by 4 so the coefficient of x^2 is 1. $\quad x^2 - 4x = -\frac{7}{4}$
3. Add $(\frac{-4}{2})^2 = (-2)^2 = 2^2$. $\quad x^2 - 4x + 2^2 = -\frac{7}{4} + 2^2$
4. Rewrite. $\quad (x-2)^2 = -\frac{7}{4} + \frac{16}{4} = \frac{9}{4}$

   $(x-2)^2 = \frac{9}{4}$
5. Use the square root property to solve the resulting equation. $\quad (x-2) = \pm\sqrt{\frac{9}{4}}$

   $x - 2 = \pm\frac{3}{2}$

   $x = 2 \pm \frac{3}{2}$

That is,

$$x = \frac{4}{2} + \frac{3}{2} = \frac{7}{2} \quad \text{or} \quad x = \frac{4}{2} - \frac{3}{2} = \frac{1}{2}$$

Thus, the solutions of $4x^2 - 16x + 7 = 0$ are $\frac{7}{2}$ and $\frac{1}{2}$.

PROBLEM 3

Solve by completing the square:

$$4x^2 - 24x + 27 = 0$$

Finally, we must point out that in many cases the answers you will obtain when you solve quadratic equations by completing the square are *not* rational numbers. (Remember? A rational number can be written as $\frac{a}{b}$, a and b integers, $b \neq 0$.) As a matter of fact, quadratic equations with rational solutions can be solved by factoring, often a simpler method.

EXAMPLE 4 **Solving a quadratic equation by completing the square**

Solve: $36x + 9x^2 + 31 = 0$

SOLUTION We proceed using the five-step procedure.

$$36x + 9x^2 + 31 = 0 \quad \text{Given.}$$

1. Subtract 31 and write in descending order. $\quad 9x^2 + 36x = -31$
2. Divide by 9. $\quad x^2 + 4x = \frac{-31}{9}$
3. Add $(\frac{4}{2})^2 = 2^2$. $\quad x^2 + 4x + 2^2 = \frac{-31}{9} + 2^2 = -\frac{31}{9} + \frac{36}{9}$
4. Rewrite as a perfect square. $\quad (x+2)^2 = \frac{5}{9}$
5. Use the square root property. $\quad (x+2) = \pm\sqrt{\frac{5}{9}}$

   $x + 2 = \pm\frac{\sqrt{5}}{3}$

   $x = -2 \pm \frac{\sqrt{5}}{3} = \frac{-6 \pm \sqrt{5}}{3}$

Thus, the solutions of $36x + 9x^2 + 31 = 0$ are $\frac{-6+\sqrt{5}}{3}$ and $\frac{-6-\sqrt{5}}{3}$.

PROBLEM 4

Solve:

$$4x^2 + 24x + 31 = 0$$

Answers

3. The solutions are $3 \pm \frac{3}{2}$, that is, $\frac{9}{2}$ and $\frac{3}{2}$. **4.** The solutions are $-3 \pm \frac{\sqrt{5}}{2} = \frac{-6 \pm \sqrt{5}}{2}$.

Exercises 10.2

A In Problems 1–20, find the missing term(s) to make the expression a perfect square.

1. $x^2 + 18x + \square$ 81
2. $x^2 + 2x + \square$ 1
3. $x^2 - 16x + \square$ 64
4. $x^2 - 4x + \square$ 4
5. $x^2 + 7x + \square$ $\frac{49}{4}$
6. $x^2 + 9x + \square$ $\frac{81}{4}$
7. $x^2 - 3x + \square$ $\frac{9}{4}$
8. $x^2 - 7x + \square$ $\frac{49}{4}$
9. $x^2 + x + \square$ $\frac{1}{4}$
10. $x^2 - x + \square$ $\frac{1}{4}$
11. $x^2 + 4x + \square = (\quad)^2$ $4; x + 2$
12. $x^2 + 6x + \square = (\quad)^2$ $9; x + 3$
13. $x^2 + 3x + \square = (\quad)^2$ $\frac{9}{4}; x + \frac{3}{2}$
14. $x^2 + 9x + \square = (\quad)^2$ $\frac{81}{4}; x + \frac{9}{2}$
15. $x^2 - 6x + \square = (\quad)^2$ $9; x - 3$
16. $x^2 - 24x + \square = (\quad)^2$ $144; x - 12$
17. $x^2 - 5x + \square = (\quad)^2$ $\frac{25}{4}; x - \frac{5}{2}$
18. $x^2 - 11x + \square = (\quad)^2$ $\frac{121}{4}; x - \frac{11}{2}$
19. $x^2 - \frac{3}{2}x + \square = (\quad)^2$ $\frac{9}{16}; x - \frac{3}{4}$
20. $x^2 - \frac{5}{2}x + \square = (\quad)^2$ $\frac{25}{16}; x - \frac{5}{4}$

In Problems 21–40, solve the given equation.

21. $x^2 + 2x + 7 = 0$ No real-number solution
22. $x^2 + 4x + 1 = 0$ $x = -2 \pm \sqrt{3}$
23. $x^2 + x - 1 = 0$ $x = -\frac{1}{2} \pm \frac{\sqrt{5}}{2} = \frac{-1 \pm \sqrt{5}}{2}$
24. $x^2 + 2x - 1 = 0$ $x = -1 \pm \sqrt{2}$
25. $x^2 + 3x - 1 = 0$ $x = -\frac{3}{2} \pm \frac{\sqrt{13}}{2} = \frac{-3 \pm \sqrt{13}}{2}$
26. $x^2 - 3x - 4 = 0$ $x = 4$ or $x = -1$
27. $x^2 - 3x - 3 = 0$ $x = \frac{3}{2} \pm \frac{\sqrt{21}}{2} = \frac{3 \pm \sqrt{21}}{2}$
28. $x^2 - 3x - 1 = 0$ $x = \frac{3}{2} \pm \frac{\sqrt{13}}{2} = \frac{3 \pm \sqrt{13}}{2}$
29. $4x^2 + 4x - 3 = 0$ $x = \frac{1}{2}$ or $x = -\frac{3}{2}$
30. $2x^2 + 10x - 1 = 0$ $x = -\frac{5}{2} \pm \frac{3\sqrt{3}}{2} = \frac{-5 \pm 3\sqrt{3}}{2}$
31. $4x^2 - 16x = 15$ $x = 2 \pm \frac{\sqrt{31}}{2} = \frac{4 \pm \sqrt{31}}{2}$
32. $25x^2 - 25x = -6$ $x = \frac{3}{5}$ or $x = \frac{2}{5}$
33. $4x^2 - 7 = 4x$ $x = \frac{1}{2} \pm \sqrt{2} = \frac{1 \pm 2\sqrt{2}}{2}$
34. $2x^2 - 18 = -9x$ $x = \frac{3}{2}$ or $x = -6$
35. $2x^2 + 1 = 4x$ $x = 1 \pm \frac{\sqrt{2}}{2} = \frac{2 \pm \sqrt{2}}{2}$
36. $2x^2 + 3 = 6x$ $x = \frac{3}{2} \pm \frac{\sqrt{3}}{2} = \frac{3 \pm \sqrt{3}}{2}$
37. $(x + 3)(x - 2) = -4$ $x = 1$ or $x = -2$
38. $(x + 4)(x - 1) = -6$ $x = -1$ or $x = -2$
39. $2x(x + 5) - 1 = 0$ $x = -\frac{5}{2} \pm \frac{3\sqrt{3}}{2} = \frac{-5 \pm 3\sqrt{3}}{2}$
40. $2x(x - 4) = 2(9 - 8x) - x$ $x = \frac{3}{2}$ or $x = -6$

SKILL CHECKER

Try the Skill Checker Exercises so you'll be ready for the next section.

Write each equation in the standard form $ax^2 + bx + c = 0$, where a, b, and c are integers:

41. $10x^2 + 5x = 12$ $10x^2 + 5x - 12 = 0$
42. $x^2 = 2x + 2$ $x^2 - 2x - 2 = 0$
43. $9x = x^2$ $x^2 - 9x = 0$
44. $\frac{x^2}{2} + \frac{3}{5} = \frac{x}{4}$ $10x^2 - 5x + 12 = 0$
45. $\frac{x}{4} = \frac{3}{5} - \frac{x^2}{2}$ $10x^2 + 5x - 12 = 0$
46. $\frac{x}{3} = \frac{3}{4}x^2 - \frac{1}{6}$ $9x^2 - 4x - 2 = 0$

(*Hint:* First multiply each term by the LCD.)

USING YOUR KNOWLEDGE

Finding a Maximum or a Minimum

Many applications of mathematics require finding the maximum or the minimum of certain algebraic expressions. Thus a certain business may wish to find the price at which a product will bring *maximum* profits, whereas engineers may be interested in *minimizing* the amount of carbon monoxide produced by automobiles.

Now suppose you are the manufacturer of a certain product whose average manufacturing cost $\overline{C}$ (in dollars), based on producing x (thousand) units, is given by the expression

$$\overline{C} = x^2 - 8x + 18$$

How many units should be produced to *minimize* the cost per unit? If we consider the right-hand side of the equation, we can complete the square and leave the equation unchanged by adding and subtracting the appropriate number. Thus

$$\overline{C} = x^2 - 8x + 18$$
$$\overline{C} = (x^2 - 8x + \quad) + 18$$
$$\overline{C} = (x^2 - 8x + 4^2) + 18 - 4^2$$

Note that we have added and subtracted the square of one-half of the coefficient of x, $\left(\frac{-8}{2}\right)^2 = 4^2$.

Then

$$\overline{C} = (x - 4)^2 + 2$$

Since the smallest possible value that $(x - 4)^2$ can have is 0 for $x = 4$, let $x = 4$ to give the minimum cost, $\overline{C} = 2$.

Use your knowledge about completing the square to solve the following problems.

47. A manufacturer's average cost $\overline{C}$ (in dollars), based on manufacturing x (thousand) items, is given by

$$\overline{C} = x^2 - 4x + 6$$

a. How many units should be produced to minimize the cost per unit? 2 (thousand) = 2000

b. What is the minimum average cost per unit? \$2

48. The demand D for a certain product depends on the number x (in thousands) of units produced and is given by

$$D = x^2 - 2x + 3$$

What is the number of units that have to be produced so that the demand is at its lowest? 1 (thousand) = 1000

49. Have you seen people adding chlorine to their pools? This is done to reduce the number of bacteria present in the water. Suppose that after t days, the number of bacteria per cubic centimeter is given by the expression

$$B = 20t^2 - 120t + 200$$

In how many days will the number of bacteria be at its lowest? 3 days

WRITE ON

50. What is the first step you would perform to solve the equation $6x^2 + 9x = 3$ by completing the square? Divide by 6

51. Can you solve the equation $x^3 + 2x^2 = 5$ by completing the square? Explain. No; x is cubed in the equation.

52. Describe the procedure you would use to find the number that has to be added to $x^2 + 5x$ to make the expression a perfect square trinomial. Divide the coefficient of x (5) by 2, then square the result.

53. Make a perfect square trinomial that has a term of $-6x$ and explain how you constructed your trinomial. $x^2 - 6x + 9$; answers may vary.

MASTERY TEST

If you know how to do these problems, you have learned your lesson!

Find the missing term:

54. $(x + 11)^2 = x^2 + 22x + \square$ 121

55. $\left(x - \frac{1}{4}\right)^2 = x^2 - \frac{1}{2}x + \square$ $\frac{1}{16}$

56. $x^2 - 12x + \square = (\quad)^2$ 36; $x - 6$

57. $x^2 + 5x + \square = (\quad)^2$ $\frac{25}{4}$; $x + \frac{5}{2}$

Solve:

58. $4x^2 - 24x + 27 = 0$ $x = \frac{9}{2}$ or $x = \frac{3}{2}$

59. $4x^2 + 24x + 31 = 0$ $x = -3 \pm \frac{\sqrt{5}}{2} = \frac{-6 \pm \sqrt{5}}{2}$

10.3 SOLVING QUADRATIC EQUATIONS BY THE QUADRATIC FORMULA

To Succeed, Review How To . . .

1. Solve a quadratic equation by completing the square (pp. 697–702).
2. Write a quadratic equation in standard form (p. 434).

Objectives

A Write a quadratic equation in the form $ax^2 + bx + c = 0$ and identify a, b, and c.

B Solve a quadratic equation using the quadratic formula.

GETTING STARTED

Name That Formula!

"My final recommendation is that we reject all these proposals and continue to call it the quadratic formula."
Courtesy of Ralph Castellano

As you were going through the preceding sections of this chapter, you probably wondered whether there is a surefire method for solving any quadratic equation. Fortunately, there is! As a matter of fact, this new technique incorporates the method of completing the square. As you recall, in each of the problems we followed the same procedure. Why not use the method of completing the square and solve the equation $ax^2 + bx + c = 0$ once and for all? We shall do that now. (You can refer to the procedure we gave to solve equations by completing the square on page 701.) Here is how the quadratic formula is derived:

$$ax^2 + bx + c = 0, \quad a \neq 0 \qquad \text{Given.}$$

Rewrite with the constant c on the right. $$ax^2 + bx = -c$$

Divide each term by a. $$x^2 + \frac{b}{a}x = -\frac{c}{a}$$

Add the square of one-half the coefficient of x. $$x^2 + \frac{b}{a}x + \left(\frac{b}{2a}\right)^2 = \left(\frac{b}{2a}\right)^2 - \frac{c}{a}$$

Rewrite the left side as a perfect square. $$\left(x + \frac{b}{2a}\right)^2 = \frac{b^2}{4a^2} - \frac{c}{a}$$

$$\left(x + \frac{b}{2a}\right)^2 = \frac{b^2}{4a^2} - \frac{4ac}{4a^2}$$

Write the right side as a single fraction. $$\left(x + \frac{b}{2a}\right)^2 = \frac{b^2 - 4ac}{4a^2}$$

Now, take the square root of both sides. $$x + \frac{b}{2a} = \frac{\pm\sqrt{b^2 - 4ac}}{2a}$$

Subtract $\frac{b}{2a}$ from both sides. $$x = -\frac{b}{2a} \pm \frac{\sqrt{b^2 - 4ac}}{2a}$$

Combine fractions. $$x = \frac{-b \pm \sqrt{b^2 - 4ac}}{2a}$$

There you have it, *the* surefire method for solving quadratic equations! We call this the *quadratic formula,* and we shall use it in this section to solve quadratic equations.

Do you realize what we've just done? We've given ourselves a powerful tool! *Any* time we have a quadratic equation in the standard form $ax^2 + bx + c = 0$, we can solve the equation by simply substituting a, b, and c in the formula.

Web It

For a lesson explaining the definitions involved and a proof of the quadratic formula, go to link 10-3-1 on the Bello Website at mhhe.com/bello.

THE QUADRATIC FORMULA

The solutions of $ax^2 + bx + c = 0 \quad (a \neq 0)$ are

$$\frac{-b \pm \sqrt{b^2 - 4ac}}{2a}$$

that is,

$$\frac{-b + \sqrt{b^2 - 4ac}}{2a} \quad \text{and} \quad \frac{-b - \sqrt{b^2 - 4ac}}{2a}$$

This formula is so important that you should memorize it right now. Before using it, however, you must remember to do two things:

1. First, write the given equation in the standard form $ax^2 + bx + c = 0$; then
2. Determine the values of a, b, and c.

Writing Quadratic Equations in Standard Form

If we are given the equation $x^2 = 5x - 2$, then follow the steps we just listed:

1. We must *first* write it in the standard form $ax^2 + bx + c = 0$ by subtracting $5x$ and adding 2; we then have

$$x^2 - 5x + 2 = 0 \qquad \text{In standard form}$$

2. When the equation is in standard form, it's very easy to find the values of a, b, and c:

$$\underbrace{x^2}_{a = 1} \; - \; \underbrace{5x}_{b = -5} \; + \; \underbrace{2}_{c = 2} = 0$$

Recall that $x^2 = 1x^2$; thus the coefficient of x^2 is 1.

If the equation contains fractions, multiply each term by the least common denominator (LCD) to clear the fractions. For example, consider the equation

$$\frac{x}{4} = \frac{3}{5} - \frac{x^2}{2}$$

Since the LCD of 4, 5, and 2 is 20, we multiply each term by 20:

$$20 \cdot \frac{x}{4} = \frac{3}{5} \cdot 20 - \frac{x^2}{2} \cdot 20$$

$$5x = 12 - 10x^2$$

$$10x^2 + 5x = 12 \qquad \text{Add } 10x^2.$$

$$10x^2 + 5x - 12 = 0 \qquad \text{Subtract 12.}$$

Now the equation is in standard form:

$$\underbrace{10x^2}_{a = 10} \; + \; \underbrace{5x}_{b = 5} \; - \; \underbrace{12}_{c = -12} = 0$$

You can get a lot of practice by completing this table:

Given	Standard Form	*a*	*b*	*c*
1. $2x^2 + 7x - 4 = 0$				
2. $x^2 = 2x + 2$				
3. $9x = x^2$				
4. $\frac{x^2}{4} + \frac{2}{3}x = -\frac{1}{3}$				

STOP! Do not proceed until you complete the table!

We will now use these four equations as examples.

B Solving Quadratic Equations with the Quadratic Formula

EXAMPLE 1 Solving a quadratic equation using the quadratic formula

Solve using the quadratic formula: $2x^2 + 7x - 4 = 0$

SOLUTION

1. The equation is already written in standard form:

$$\underbrace{2x^2}_{a=2} + \underbrace{7x}_{b=7} - \underbrace{4}_{c=-4} = 0$$

2. As step 1 shows, it is clear that $a = 2$, $b = 7$, and $c = -4$.

3. Substituting the values of a, b, and c in the quadratic formula, we obtain

$$x = \frac{-7 \pm \sqrt{(7)^2 - 4(2)(-4)}}{2(2)}$$

$$x = \frac{-7 \pm \sqrt{49 + 32}}{4}$$

$$x = \frac{-7 \pm \sqrt{81}}{4}$$

$$x = \frac{-7 \pm 9}{4}$$

Thus

$$x = \frac{-7 + 9}{4} = \frac{2}{4} = \frac{1}{2} \quad \text{or} \quad x = \frac{-7 - 9}{4} = \frac{-16}{4} = -4$$

The solutions are $\frac{1}{2}$ and -4.

PROBLEM 1

Solve using the quadratic formula:
$3x^2 + 2x - 5 = 0$

In Example 1, you could also solve $2x^2 + 7x - 4 = 0$ by factoring $2x^2 + 7x - 4$ and writing $(2x - 1)(x + 4) = 0$. You will get the same answers!

NOTE

Try factoring first unless otherwise specified.

Answer

1. The solutions are 1 and $-\frac{5}{3}$.

EXAMPLE 2 Writing in standard form before using the formula

Solve using the quadratic formula: $x^2 = 2x + 2$

SOLUTION We proceed by steps as in Example 1.

1. To write the equation in standard form, subtract $2x$ and then subtract 2 to obtain

$$\underbrace{x^2}_{a=1} - \underbrace{2x}_{b=-2} - \underbrace{2}_{c=-2} = 0$$

2. As step 1 shows, $a = 1$, $b = -2$, and $c = -2$.
3. Substituting these values in the quadratic formula, we have

$$x = \frac{-(-2) \pm \sqrt{(-2)^2 - 4(1)(-2)}}{2(1)}$$

$$x = \frac{2 \pm \sqrt{4 + 8}}{2}$$

$$x = \frac{2 \pm \sqrt{12}}{2}$$

$$x = \frac{2 \pm \sqrt{4 \cdot 3}}{2}$$

$$x = \frac{2 \pm 2\sqrt{3}}{2}$$

Thus

$$x = \frac{2 + 2\sqrt{3}}{2} = \frac{2}{2} + \frac{2\sqrt{3}}{2} = 1 + \sqrt{3}$$

or

$$x = \frac{2 - 2\sqrt{3}}{2} = \frac{2}{2} - \frac{2}{2}\sqrt{3} = 1 - \sqrt{3}$$

The solutions are $1 + \sqrt{3}$ and $1 - \sqrt{3}$.

Note that $x^2 - 2x - 2$ is *not* factorable. You *must* know the quadratic formula or you must complete the square to solve this equation!

PROBLEM 2

Solve: $x^2 = 4x + 4$

EXAMPLE 3 Rewriting a quadratic equation before using the quadratic formula

Solve using the quadratic formula: $9x = x^2$

SOLUTION We proceed in steps.

1. Subtracting $9x$, we have

$$0 = x^2 - 9x$$

or

$$\underbrace{x^2}_{a=1} - \underbrace{9x}_{b=-9} + \underbrace{0}_{c=0} = 0$$

2. As step 1 shows, $a = 1$, $b = -9$, and $c = 0$ (because the c term is missing).

PROBLEM 3

Solve: $6x = x^2$

Answers

2. The solutions are $2 \pm 2\sqrt{2}$.
3. The solutions are 0 and 6.

3. Substituting these values in the quadratic formula, we obtain

$$x = \frac{-(-9) \pm \sqrt{(-9)^2 - 4(1)(0)}}{2(1)}$$

$$x = \frac{9 \pm \sqrt{81 - 0}}{2}$$

$$x = \frac{9 \pm \sqrt{81}}{2}$$

$$x = \frac{9 \pm 9}{2}$$

Thus

$$x = \frac{9 + 9}{2} = \frac{18}{2} = 9 \quad \text{or} \quad x = \frac{9 - 9}{2} = \frac{0}{2} = 0$$

The solutions are 9 and 0.

NOTE

$x^2 - 9x = x(x - 9) = 0$ could have been solved by factoring. Always try factoring first unless otherwise specified!

How do we know that the original equation could have been solved by factoring? We know because the answers are rational numbers (fractions or whole numbers). If your answer is a rational number, the equation could have been solved by factoring. Can we determine this in advance? Yes, and we will explain how to do so before the end of the section.

EXAMPLE 4 **Solving a quadratic equation involving fractions**

Solve using the quadratic formula:

$$\frac{x^2}{4} + \frac{2}{3}x = -\frac{1}{3}$$

SOLUTION

1. We have to write the equation in standard form, but first we clear fractions by multiplying by the LCM of 4 and 3, that is, by 12:

$$12 \cdot \frac{x^2}{4} + 12 \cdot \frac{2}{3}x = -\frac{1}{3} \cdot 12$$

$$3x^2 + 8x = -4$$

We then add 4 to obtain

$$\underbrace{3x^2}_{a=3} + \underbrace{8x}_{b=8} + \underbrace{4}_{c=4} = 0$$

2. As step 1 shows, $a = 3$, $b = 8$, and $c = 4$.

3. Substituting in the quadratic formula gives

$$x = \frac{-8 \pm \sqrt{(8)^2 - 4(3)(4)}}{2(3)}$$

$$x = \frac{-8 \pm \sqrt{64 - 48}}{6}$$

$$x = \frac{-8 \pm \sqrt{16}}{6}$$

$$x = \frac{-8 \pm 4}{6}$$

PROBLEM 4

Solve:

$$\frac{x^2}{4} - \frac{3}{8}x = \frac{1}{4}$$

Answer

4. The solutions are 2 and $-\frac{1}{2}$.

Thus

$$x = \frac{-8+4}{6} = \frac{-4}{6} = -\frac{2}{3} \quad \text{or} \quad x = \frac{-8-4}{6} = \frac{-12}{6} = -2$$

The solutions are $-\frac{2}{3}$ and -2. (Can you tell whether the original equation was factorable by looking at the answers?)

Now, a final word of warning. As you recall, some quadratic equations do not have real-number solutions. This is still true, even when you use the quadratic formula. The next example shows how this happens.

EXAMPLE 5 **Recognizing a quadratic equation with no real-number solution**

Solve using the quadratic formula: $2x^2 + 3x = -3$

SOLUTION

1. We add 3 to write the equation in standard form. We then have

$$\underbrace{2x^2}_{a=2} + \underbrace{3x}_{b=3} + \underbrace{3}_{c=3} = 0$$

2. As step 1 shows, $a = 2$, $b = 3$, and $c = 3$. Now,

$$x = \frac{-3 \pm \sqrt{(3)^2 - 4(2)(3)}}{2(2)}$$

$$x = \frac{-3 \pm \sqrt{9 - 24}}{4}$$

$$x = \frac{-3 \pm \sqrt{-15}}{4}$$

Thus

$$x = \frac{-3 + \sqrt{-15}}{4} \quad \text{or} \quad x = \frac{-3 - \sqrt{-15}}{4}$$

But wait! If we check the definition of square root, we can see that $\sqrt{a}$ is defined only for $a \geq 0$. Hence $\sqrt{-15}$ is not a real number, and the equation $2x^2 + 3x = -3$ has no real-number solution.

PROBLEM 5

Solve: $3x^2 + 2x = -1$

Web It

If you want a calculator to do the work for you (after you find a, b, and c), go to link 10-3-2 on the Bello Website at mhhe.com/bello.

As you can see from Example 5, the number -15 under the radical determines the type of roots the equation has. The expression $b^2 - 4ac$ under the radical in the quadratic formula is called the discriminant.

If $b^2 - 4ac \geq 0$, the equation has real-number solutions.

If $b^2 - 4ac < 0$, the equation has no real-number solutions.

Note that in Example 5, $a = 2$, $b = 3$, and $c = 3$. Thus,

$$b^2 - 4ac = 3^2 - 4(2)(3) = 9 - 24 < 0$$

so the equation has no real-number solutions.

Is the discriminant good for anything else? Of course. (They do not call it the *discriminant* for nothing!) The discriminant can also suggest the method you use to solve quadratic equations. Here are the suggestions:

Answer

5. No real-number solution

Start by writing the equation in standard form. Then, if $b^2 - 4ac \geq 0$ is a perfect square, try factoring. (*Note:* Some perfect squares are 0, 1, 4, 9, 16, and so on.)

In Example 1, $2x^2 + 7x - 4 = 0$ and $b^2 - 4ac = 81$, so we could try to solve the equation by factoring.

In Example 2, $x^2 - 2x - 2 = 0$ and $b^2 - 4ac = 12$. The equation cannot be factored!

In Example 3, $x^2 - 9x = 0$ and $b^2 - 4ac = 81$. Try factoring!

In Example 4, $3x^2 + 8x + 4 = 0$ and $b^2 - 4ac = 16$. Try to solve by factoring.

In Example 5, $2x^2 + 3x + 3 = 0$ and $b^2 - 4ac = -15$, so you know there are no real-number solutions.

But, you may ask, which is the easiest method to use? Look at the earlier sections!

Type of Equation	**Method to Solve**
$X^2 = A$	Square root property
$(AX \pm B)^2 = C$	Square root property
$ax^2 + bx + c = 0$	Factor if $b^2 - 4ac$ is a perfect square.
$ax^2 + bx + c = 0$	Quadratic formula if $b^2 - 4ac$ is not a perfect square

Note: If $b^2 - 4ac < 0$, there is no real-number solution. You do not even have to try a method!

Exercises 10.3

Boost *your* GRADE at mathzone.com!

- Practice Problems
- Self-Tests
- Videos
- NetTutor
- e-Professors

A B In Problems 1–30, write the given equation in standard form; then solve the equation using the quadratic formula.

1. $x^2 + 3x + 2 = 0$
$x = -2$ or $x = -1$

2. $x^2 + 4x + 3 = 0$
$x = -3$ or $x = -1$

3. $x^2 + x - 2 = 0$
$x = -2$ or $x = 1$

4. $x^2 + x - 6 = 0$
$x = -3$ or $x = 2$

5. $2x^2 + x - 2 = 0$
$x = \frac{-1 \pm \sqrt{17}}{4}$

6. $2x^2 + 7x + 3 = 0$
$x = -\frac{1}{2}$ or $x = -3$

7. $3x^2 + x = 2$
$x = \frac{2}{3}$ or $x = -1$

8. $3x^2 + 2x = 5$
$x = -\frac{5}{3}$ or $x = 1$

9. $2x^2 + 7x = -6$
$x = -\frac{3}{2}$ or $x = -2$

10. $2x^2 - 7x = -6$
$x = \frac{3}{2}$ or $x = 2$

11. $7x^2 = 12x - 5$
$x = \frac{5}{7}$ or $x = 1$

12. $-5x^2 = 16x + 8$
$x = \frac{-8 \pm 2\sqrt{6}}{5}$

13. $5x^2 = 11x - 4$
$x = \frac{11 \pm \sqrt{41}}{10}$

14. $7x^2 = 12x - 3$
$x = \frac{6 \pm \sqrt{15}}{7}$

15. $\frac{x^2}{5} - \frac{x}{2} = -\frac{3}{10}$
$x = \frac{3}{2}$ or $x = 1$

16. $\frac{x^2}{7} + \frac{x}{2} = -\frac{3}{4}$
No real-number solution

17. $\frac{x^2}{2} = \frac{3x}{4} - \frac{1}{8}$
$x = \frac{3 \pm \sqrt{5}}{4}$

18. $\frac{x^2}{10} = \frac{x}{5} + \frac{3}{2}$
$x = 5$ or $x = -3$

19. $\frac{x^2}{8} = -\frac{x}{4} - \frac{1}{8}$
$x = -1$

20. $\frac{x^2}{3} = -\frac{x}{3} - \frac{1}{12}$
$x = -\frac{1}{2}$

21. $6x = 4x^2 + 1$
$x = \frac{3 \pm \sqrt{5}}{4}$

22. $6x = 9x^2 - 4$
$x = \frac{1 \pm \sqrt{5}}{3}$

23. $3x = 1 - 3x^2$
$x = \frac{-3 \pm \sqrt{21}}{6}$

24. $3x = 2x^2 - 5$
$x = \frac{5}{2}$ or $x = -1$

25. $x(x + 2) = 2x(x + 1) - 4$
$x = 2$ or $x = -2$

26. $x(4x - 7) - 10 = 6x^2 - 7x$
No real-number solution

27. $6x(x + 5) = (x + 15)^2$
$x = \pm 3\sqrt{5}$

28. $6x(x + 1) = (x + 3)^2$
$x = \pm \frac{3\sqrt{5}}{5}$

29. $(x - 2)^2 = 4x(x - 1)$
$x = \pm \frac{2\sqrt{3}}{3}$

30. $(x - 4)^2 = 4x(x - 2)$
$x = \pm \frac{4\sqrt{3}}{3}$

SKILL CHECKER

Try the Skill Checker Exercises so you'll be ready for the next section.

Find:

31. $(-2)^2$ 4

32. -2^2 -4

33. $(-1)^2$ 1

34. -1^2 -1

USING YOUR KNOWLEDGE

Deriving the Quadratic Formula and Deciding When to Use It

In this section we derived the quadratic formula by completing the square. The procedure depends on making the x coefficient a equal to 1. But there's another way to derive the quadratic formula. See whether you can identify the change being made in each step.

$$ax^2 + bx + c = 0 \quad \text{Given.}$$

35. $4a^2x^2 + 4abx + 4ac = 0$ Multiply each term by $4a$.

36. $4a^2x^2 + 4abx = -4ac$ Subtract $4ac$ on both sides.

37. $4a^2x^2 + 4abx + b^2 = b^2 - 4ac$ Add b^2 on both sides.

38. $(2ax + b)^2 = b^2 - 4ac$ Factor the left side.

39. $2ax + b = \pm\sqrt{b^2 - 4ac}$
Take the square root of each side.

40. $2ax = -b \pm \sqrt{b^2 - 4ac}$ Subtract b on both sides.

41. $x = \dfrac{-b \pm \sqrt{b^2 - 4ac}}{2a}$ Divide each side by $2a$.

The quadratic formula is not the only way to solve quadratic equations. We've actually studied four methods for solving them. The following table lists these methods and suggests the best use for each.

Method	When Used
1. The square root property	When you can write the equation in the form $X^2 = A$
2. Factoring	When the equation is factorable. Use the *ac* test (from Section 5.3) to find out!
3. Completing the square	You can always use this method, but it requires more steps than the other methods.
4. The quadratic formula	This formula can always be used, but you should try factoring first.

In Problems 42–56, solve the equation by the method of your choice.

42. $x^2 = 144$ $x = \pm 12$

43. $x^2 - 17 = 0$ $x = \pm\sqrt{17}$

44. $x^2 - 2x = -1$ $x = 1$

45. $x^2 + 4x = -4$ $x = -2$

46. $x^2 - x - 1 = 0$ $x = \frac{1 \pm \sqrt{5}}{2}$

47. $y^2 - y = 0$ $y = 0$ or $y = 1$

48. $y^2 + 5y - 6 = 0$
$y = -6$ or $y = 1$

49. $x^2 + 5x + 6 = 0$
$x = -3$ or $x = -2$

50. $5y^2 = 6y - 1$
$y = \frac{1}{5}$ or $y = 1$

51. $(z + 2)(z + 4) = 8$
$z = 0$ or $z = -6$

52. $(y - 3)(y + 4) = 18$
$y = -6$ or $y = 5$

53. $y^2 = 1 - \frac{8}{3}y$ $y = \frac{1}{3}$ or $y = -3$

54. $x^2 = \frac{3}{2}x + 1$ $x = -\frac{1}{2}$ or $x = 2$

55. $3z^2 + 1 = z$
No real-number solution

56. $\frac{1}{2}x^2 - \frac{1}{2}x = \frac{1}{4}$ $x = \frac{1 \pm \sqrt{3}}{2}$

WRITE ON

In the quadratic formula, the expression $D = b^2 - 4ac$ appears under the radical, and it is called the **discriminant** of the equation $ax^2 + bx + c = 0$. How many solutions does the equation have if

57. $D = 0$? Is (are) the solution(s) rational or irrational? One; rational

58. $D > 0$? Two; rational if D is a perfect square, irrational otherwise

59. $D < 0$? None

60. Suppose that one solution of $ax^2 + bx + c = 0$ is

$$x = \frac{-b + \sqrt{D}}{2a}$$

What is the other solution? $x = \frac{-b - \sqrt{D}}{2a}$

MASTERY TEST

If you know how to do these problems, you have learned your lesson!

Solve if possible:

61. $3x^2 + 2x = -1$ No real-number solution

62. $6x = x^2$ $x = 0$ or $x = 6$

63. $x^2 = 4x + 4$ $x = 2 \pm 2\sqrt{2}$

64. $3x^2 + 2x - 5 = 0$ $x = 1$ or $x = -\frac{5}{3}$

65. $\frac{x^2}{4} - \frac{3}{8}x = \frac{1}{4}$ $x = -\frac{1}{2}$ or $x = 2$

66. $\frac{2}{3}x^2 + x = -1$ No real-number solution

10.4 GRAPHING QUADRATIC EQUATIONS

To Succeed, Review How To . . .

1. Evaluate an expression (p. 81).
2. Graph points in the plane (p. 228).
3. Factor a quadratic (pp. 399, 437).

Objectives

A Graph quadratic equations.

B Find the intercepts and the vertex, and graph parabolas involving factorable quadratic expressions.

GETTING STARTED

Parabolic Water Streams

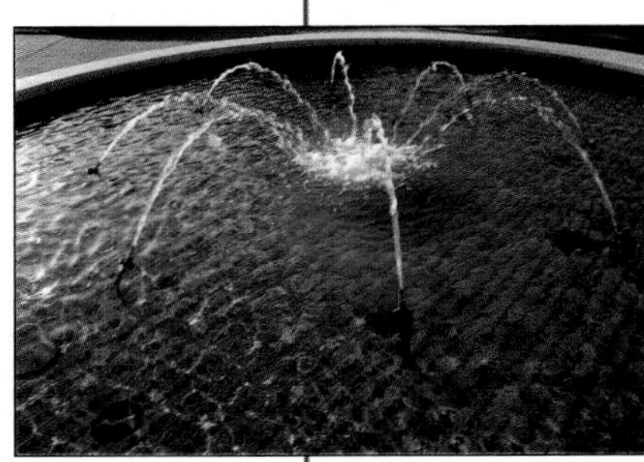

Look at the streams of water from the fountain. What shape do they have? This shape is called a **parabola.** The *graph* of a quadratic equation of the form $y = ax^2 + bx + c$ is a parabola. The simplest of these equations is $y = x^2$. This equation can be graphed in the same way as lines were graphed—that is, by selecting values for x and then finding the corresponding y-values as shown in the table on the left. The usual shortened version is shown in the table on the right.

x-value	y-value
$x = -2$	$y = x^2 = (-2)^2 = 4$
$x = -1$	$y = x^2 = (-1)^2 = 1$
$x = 0$	$y = x^2 = (0)^2 = 0$
$x = 1$	$y = x^2 = (1)^2 = 1$
$x = 2$	$y = x^2 = (2)^2 = 4$

x	y
-2	4
-1	1
0	0
1	1
2	4

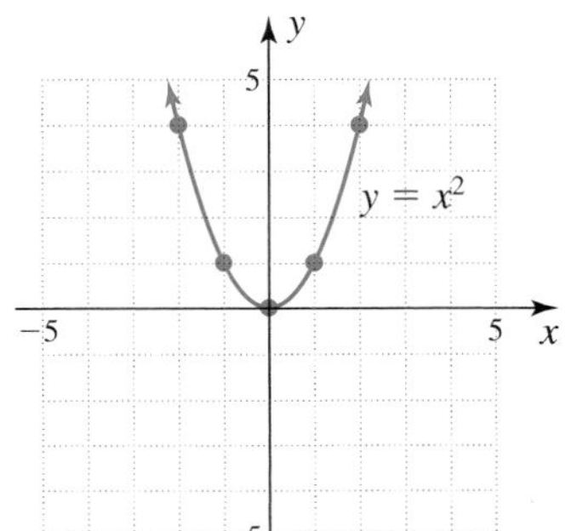

We graph these points on a coordinate system and draw a smooth curve through them. The result is the graph of the parabola $y = x^2$. Note that the arrows at the end indicate that the curve goes on indefinitely.

In this section we shall learn how to graph parabolas.

Graphing Quadratic Equations

Now that we know how to graph $y = x^2$, what would happen if we graph $y = -x^2$? To start, we could alter the table in the *Getting Started* by using the negative of the y-value on $y = x^2$, as shown next.

EXAMPLE 1 Graphing $y = ax^2$ when a is negative

Graph: $y = -x^2$

SOLUTION We could always make a table of x- and y-values as before. However, note that for any x-value, the y-value will be the *negative* of the y-value on the parabola $y = x^2$. (If you don't believe this, go ahead and make the table and check it.) Thus the parabola $y = -x^2$ has the same shape as $y = x^2$, but it's turned in the *opposite* direction (opens *downward*). The graph of $y = -x^2$ is shown in Figure 1.

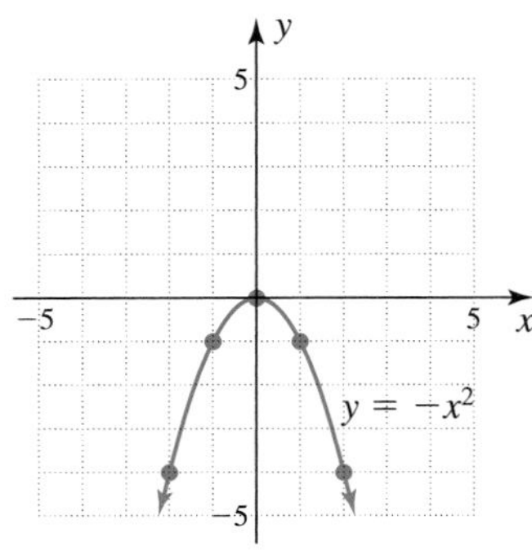

Figure 1

PROBLEM 1

Graph: $y = -2x^2$

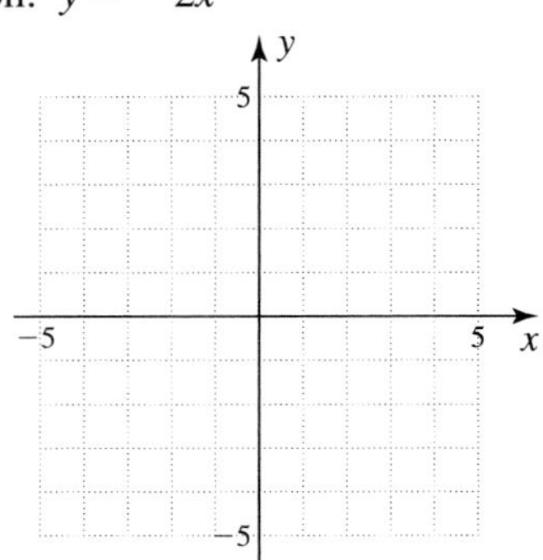

As you can see from the two preceding examples, when the coefficient of x^2 is positive (as in $y = x^2 = 1x^2$), the parabola opens *upward,* but when the coefficient of x^2 is negative (as in $y = -x^2 = -1x^2$), the parabola opens *downward.* In general, we have the following definition.

GRAPH OF A QUADRATIC EQUATION

The graph of a quadratic equation of the form $y = ax^2 + bx + c$ is a parabola that

1. Opens upward if $a > 0$
2. Opens downward if $a < 0$

y
7
$y = x^2 + 1$
−5
5
x
−3

Figure 2

Now, what do you think will happen if we graph the parabola $y = x^2 + 1$? Two things: First, the parabola opens upward, since the coefficient of x^2 is understood to be 1. Second, all of the points will be 1 unit higher than those for the same value of x on the parabola $y = x^2$. Thus we can make the graph of $y = x^2 + 1$ by following the pattern of $y = x^2$. The graphs of $y = x^2 + 1$ and $y = x^2 + 2$ are shown in Figures 2 and 3, respectively.

You can verify that these graphs are correct by graphing several of the points for $y = x^2 + 1$ and $y = x^2 + 2$ that are shown in the following tables:

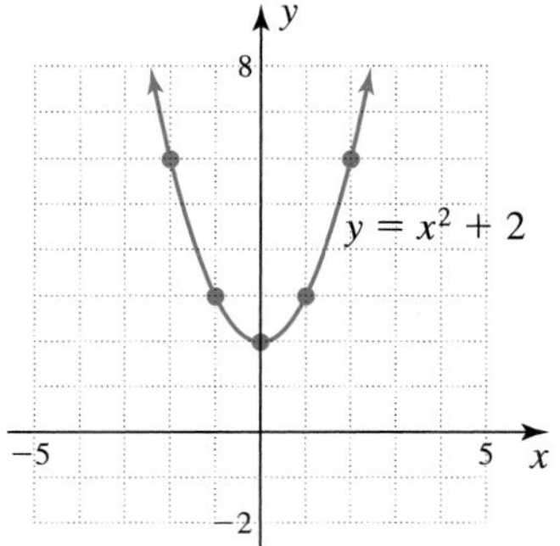

Figure 3

$y = x^2 + 1$

x	y
0	1
1	2
−1	2

$y = x^2 + 2$

x	y
0	2
1	3
−1	3

Answer

1.

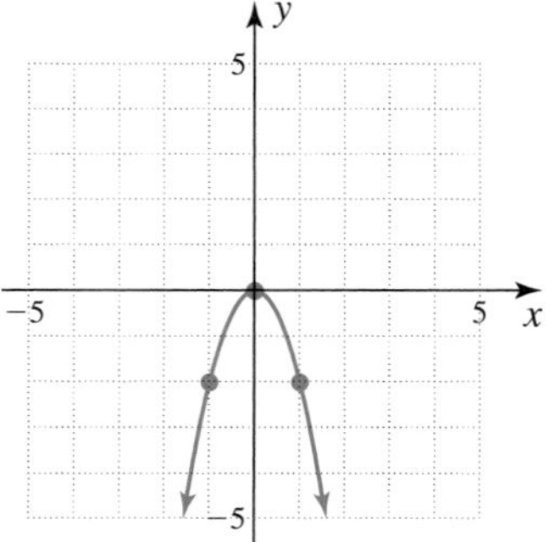

EXAMPLE 2 Graphing a parabola that opens downward

Graph: $y = -x^2 - 2$

SOLUTION Since the coefficient of x^2 (which is understood to be -1) is *negative,* the parabola opens downward. It is also 2 units *lower* than the graph of $y = -x^2$. Thus the graph of $y = -x^2 - 2$ is as shown in Figure 4.

You can verify the graph by checking the points in the table for $y = -x^2 - 2$:

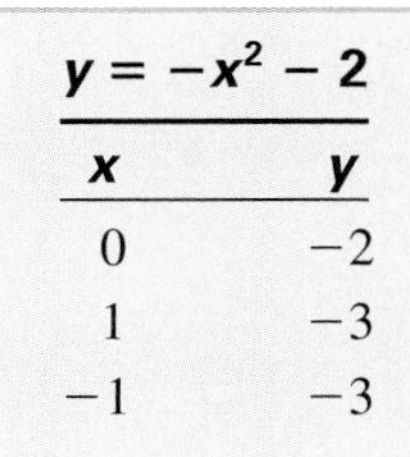

$y = -x^2 - 2$	
x	**y**
0	−2
1	−3
−1	−3

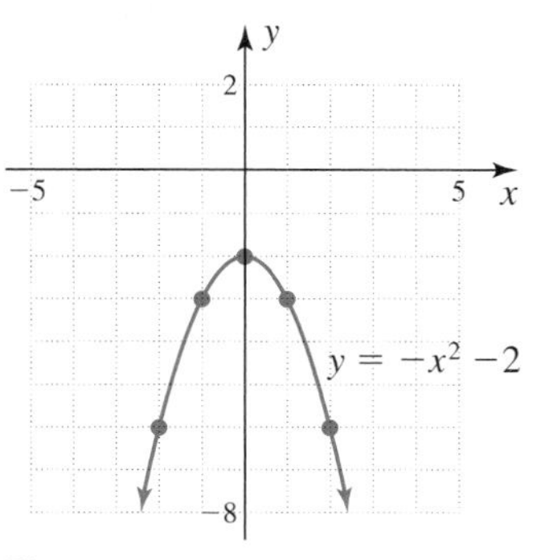

Figure 4

PROBLEM 2

Graph: $y = -x^2 - 1$

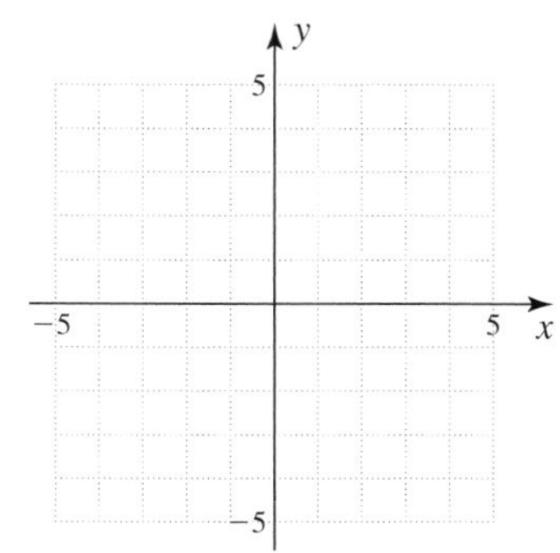

So far, we've graphed parabolas only of the form $y = ax^2 + b$. How do you think the graph of $y = (x - 1)^2$ looks? As before, we make a table of values. For example,

For $x = -1$, $y = (-1 - 1)^2 = (-2)^2 = 4$

For $x = 0$, $y = (0 - 1)^2 = (-1)^2 = 1$

For $x = 1$, $y = (1 - 1)^2 = (0)^2 = 0$

For $x = 2$, $y = (2 - 1)^2 = 1^2 = 1$

$y = (x - 1)^2$	
x	**y**
−1	4
0	1
1	0
2	1

Web It

For a complete lesson on how to graph quadratic equations, go to the Bello Website at mhhe.com/bello and check link 10-4-1.

For a different version with a procedure you can use, try link 10-4-2.

The completed graph is shown in Figure 5. Note that the shape of the graph is identical to that of $y = x^2$, but it is shifted 1 unit to the *right.*

Similarly, the graph of $y = -(x + 1)^2$ is identical to that of $y = -x^2$ but shifted 1 unit to the *left,* as shown in Figure 6.

You can verify this using the table for $y = -(x + 1)^2$:

$y = -(x + 1)^2$	
x	**y**
0	−1
−1	0
−2	−1

Answer

2.

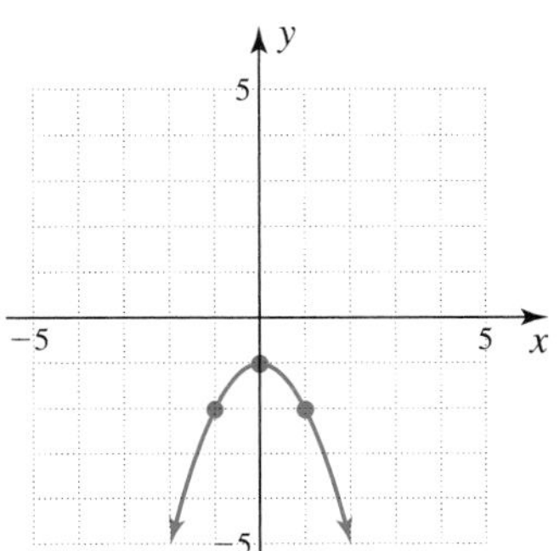

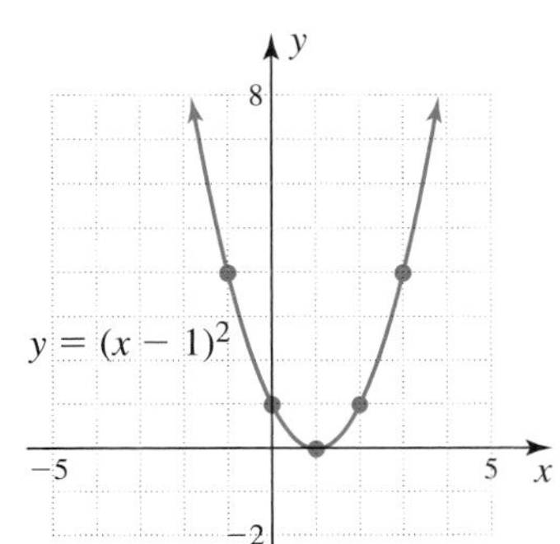

Figure 5

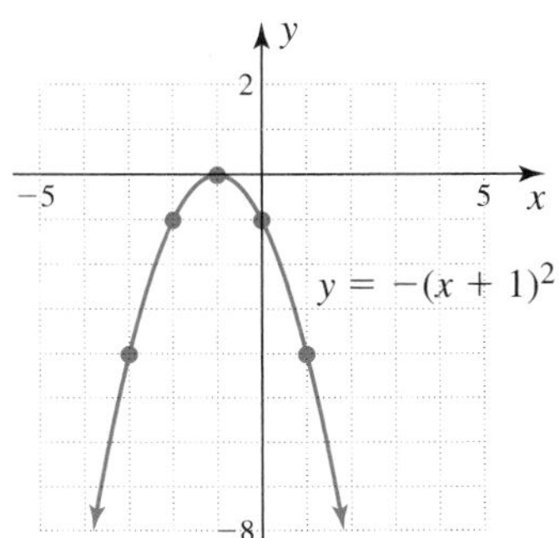

Figure 6

EXAMPLE 3 Graphing by shifting and moving up

Graph: $y = -(x - 1)^2 + 2$

SOLUTION The graph of this equation is identical to the graph of $y = -x^2$ except for its position. The parabola $y = -(x - 1)^2 + 2$ opens downward (because of the first negative sign), and it's shifted 1 unit to the right (because of the -1) and 2 units up (because of the $+2$):

$$y = -(x - 1)^2 + 2$$

- $-$: Opens downward (negative)
- $(x-1)$: Shifted 1 unit right
- $+2$: Shifted 2 units up

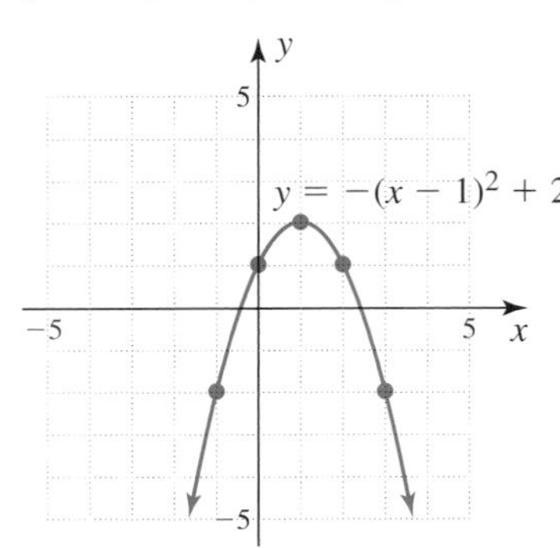

Figure 7

Figure 7 shows the finished graph of $y = -(x - 1)^2 + 2$.

As usual, you can verify this using the table for $y = -(x - 1)^2 + 2$:

$y = -(x - 1)^2 + 2$

x	y
0	1
1	2
2	1

PROBLEM 3

Graph: $y = -(x - 2)^2 - 1$

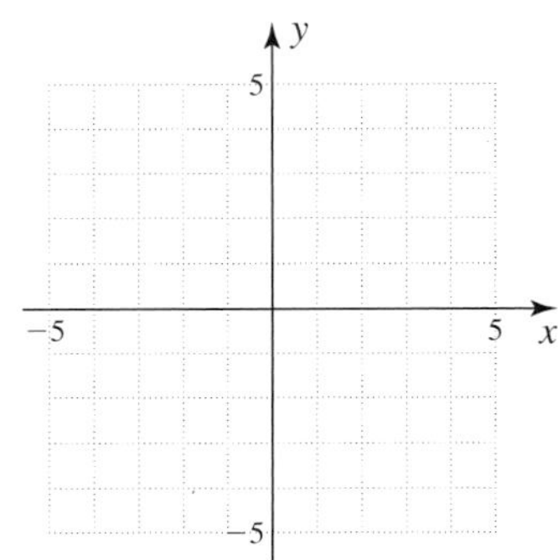

In conclusion, we follow these directions for changing the graph of $y = ax^2$:

$$y = a(x \pm b)^2 \pm c$$

b and *c* positive.

- a: Opens upward for $a > 0$, downward for $a < 0$
- $\pm b$: Shifts the graph right $(-b)$ or left $(+b)$
- $\pm c$: Moves the graph up $(+c)$ or down $(-c)$

Note that as a increases, the parabola "stretches." (We shall examine this fact in the *Calculate It.*)

B Graphing Parabolas Using Intercepts and the Vertex

You've probably noticed that the graph of a parabola is **symmetric;** that is, if you draw a vertical line through the **vertex** (the high or low point on the parabola) and fold the graph along this line, the two halves of the parabola coincide. If a parabola crosses the x-axis, we can use the x-intercepts to find the vertex. For example, to graph $y = x^2 + 2x - 8$, we start by finding the x- and y-intercepts:

For $x = 0$, $\quad y = 0^2 + 2 \cdot 0 - 8 = -8$ The y-intercept

For $y = 0$, $\quad 0 = x^2 + 2x - 8$

$$0 = (x + 4)(x - 2)$$

Thus

$x = -4$ or $x = 2$ The x-intercepts

Answer

3.

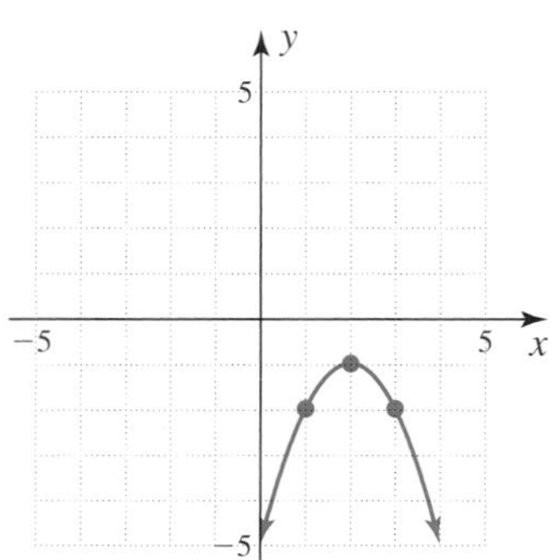

Web It

To graph parabolas using intercepts and vertices, go to link 10-4-3 on the Bello Website at mhhe.com/bello.

We enter the points $(0, -8)$, $(-4, 0)$, and $(2, 0)$ in a table like this:

x	y	
0	−8	y-intercept
−4	0	x-intercepts
2	0	
?	?	Vertex

How can we find the vertex? Since the parabola is symmetric, the x-coordinate of the vertex is *exactly* halfway between -4 and 2, so

$$x = \frac{-4 + 2}{2} = \frac{-2}{2} = -1$$

We then find the y-coordinate by letting $x = -1$ in $y = x^2 + 2x - 8$ to obtain

$$y = (-1)^2 + 2 \cdot (-1) - 8 = -9$$

Thus the vertex is at $(-1, -9)$. The table now looks like this:

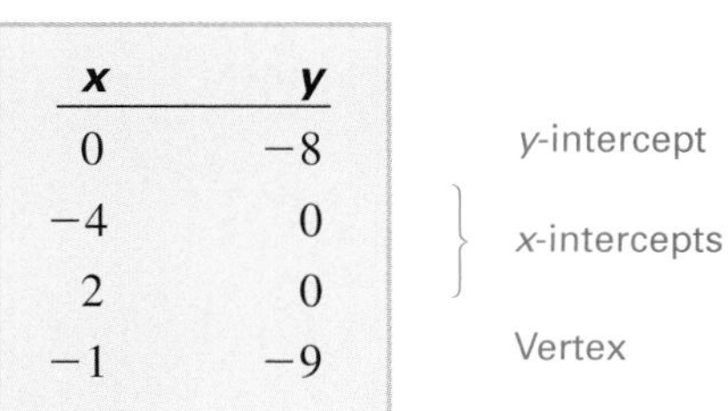

x	y	
0	−8	y-intercept
−4	0	x-intercepts
2	0	
−1	−9	Vertex

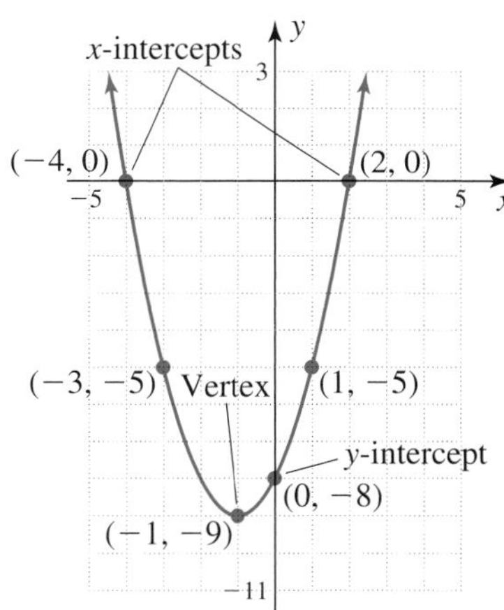

Figure 8

We can draw the graph using these four points or plot one or two more points. We plotted $(1, -5)$ and $(-3, -5)$. The graph is shown in Figure 8.

Here is the complete procedure.

PROCEDURE

Graphing a Factorable Quadratic Equation

1. Find the y-intercept by letting $x = 0$ and then finding y.
2. Find the x-intercepts by letting $y = 0$, factoring the equation, and solving for x.
3. Find the vertex by averaging the solutions of the equation found in step 2 (the average is the x-coordinate of the vertex) and substituting in the equation to find the y-coordinate of the vertex.
4. Plot the points found in steps 1–3 and one or two more points, if desired. The curve drawn through the points found in steps 1–4 is the graph.

Now let's use this procedure to graph a parabola.

EXAMPLE 4 Graphing parabolas using the intercepts and vertex

Graph: $y = -x^2 + 2x + 8$

SOLUTION We use the four steps discussed.

1. Since $-x^2 = -1 \cdot x^2$ and -1 is negative, the parabola opens downward. We then let $x = 0$ to obtain $y = -(0)^2 + 2 \cdot 0 + 8 = 8$. Thus $(0, 8)$ is the y-intercept.

2. We find the x-intercepts by letting $y = 0$ and solving for x. We have

$$0 = -x^2 + 2x + 8$$

$$0 = x^2 - 2x - 8 \quad \text{Multiply both sides by } -1 \text{ to make the factorization easier.}$$

$$0 = (x - 4)(x + 2)$$

$$x = 4 \quad \text{or} \quad x = -2$$

We now have the x-intercepts: $(4, 0)$ and $(-2, 0)$.

3. The x-coordinate of the vertex is found by averaging 4 and -2 to obtain

$$x = \frac{4 + (-2)}{2} = \frac{2}{2} = 1$$

Letting $x = 1$ in $y = -x^2 + 2x + 8$, we find $y = -(1)^2 + 2 \cdot (1) + 8 = 9$, so the vertex is at $(1, 9)$.

4. We plot these points and the two additional solutions $(2, 8)$ and $(3, 5)$. The completed graph is shown in Figure 9.

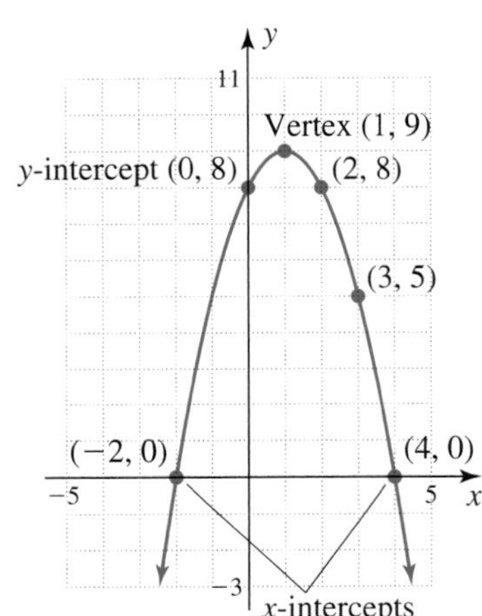

Figure 9

PROBLEM 4

Graph: $y = -x^2 - 2x + 8$

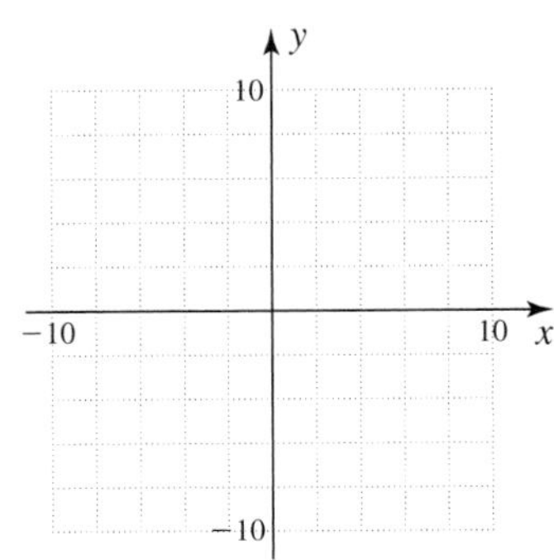

Answer

4.

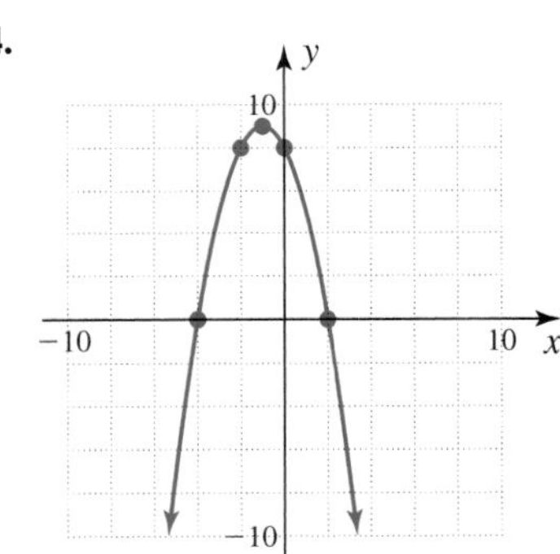

Calculate It Graphing Parabolas

If you have a calculator, you can graph parabolas very easily. Let's explore the effect of different values of a, b, and c in the equation

$$y = a(x \pm b) \pm c$$

Using a standard window, graph $y = x^2$, $y = x^2 + 3$, and $y = x^2 - 3$. Clearly, adding the positive number k "moves" the graph of the parabola k units *up*. Similarly, subtracting the positive number k "moves" the graph of the parabola k units *down*, as shown in Window 1. Graph $y = -x^2$, $y = -x^2 + 3$, and $y = -x^2 - 3$ to check the effect of having a negative sign in front of x^2.

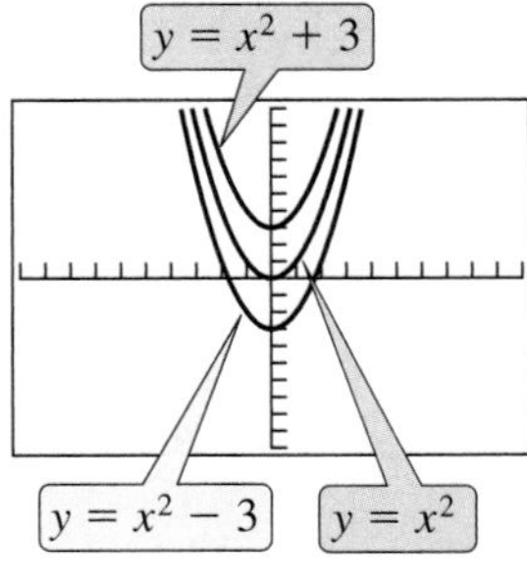

Window 1

Now, let's try some stretching exercises! Look at the graphs of $y = x^2$, $y = 2x^2$, and $y = 5x^2$ using a decimal window. As you can see in Window 2, as the coefficient of x^2 increases, the parabola "stretches"!

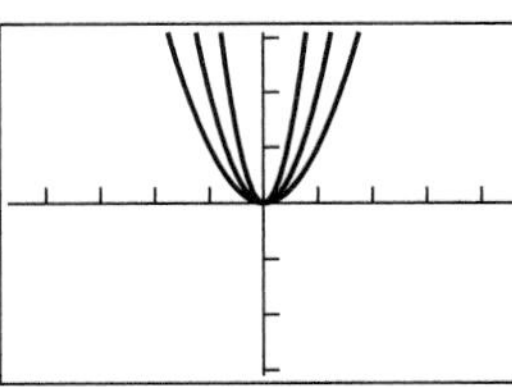

Window 2

Can you find the vertex and intercepts with your calculator? Yes, they're quite easy to find using a calculator. For example, let's use a decimal window to graph

$$y = x^2 - x - 2$$

which is shown in Window 3. You can use the TRACE key to find the

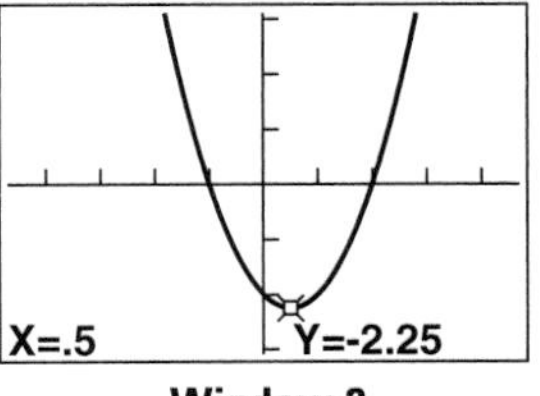

Window 3

vertex (0.5, −2.25). Better yet, some calculators find the minimum of a function by pressing 2nd TRACE 3 and following the prompts. The **minimum,** of course, occurs at the vertex, as shown in Window 4. On the other hand, the **maximum** of $y = -x^2 + x + 2$ occurs at the vertex. Can you find this maximum?

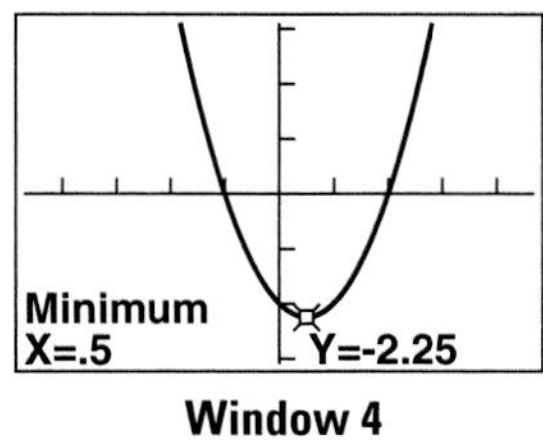

Window 4

Let's return to the parabola $y = x^2 - x - 2$. This parabola has two x-intercepts, as shown in Window 5. Using your TRACE key and a decimal window, you can find both of them! Window 5 shows that one of the intercepts occurs at $x = -1$. The other x-intercept (not shown) occurs when $x = 2$. Now solve the equation $0 = x^2 - x - 2$. What are the solutions of this equation? Do you see a relationship between the graph of $y = ax^2 + bx + c$ and the solutions of $0 = ax^2 + bx + c$? Can you devise a procedure so that you can solve quadratic equations using a calculator?

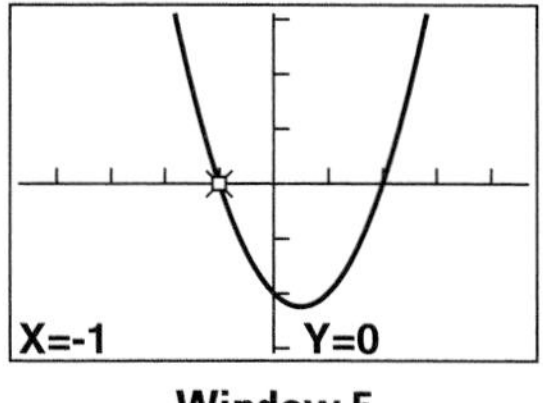

Window 5

If you're lucky, your calculator also has a "root" or "zero" feature. With this feature, you can graph $y = x^2 - x - 2$ and ask the calculator to find the root. The prompts ask "Left bound?" Use the ◀ key to move the cursor to a point left of the intercept (where this graph is above the x-axis). Find a point below the x-axis for the "Right bound" and press ENTER when asked for a "Guess." The root −1 is shown in Window 6.

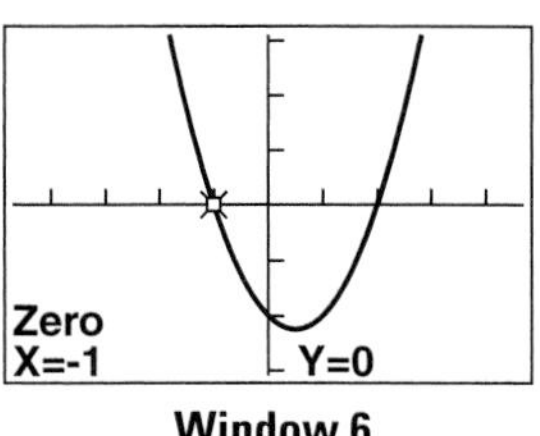

Window 6

Exercises 10.4

Boost *your* GRADE at mathzone.com!

- Practice Problems
- Self-Tests
- Videos
- NetTutor
- e-Professors

A In Problems 1–16, graph the given equation.

1. $y = 2x^2$

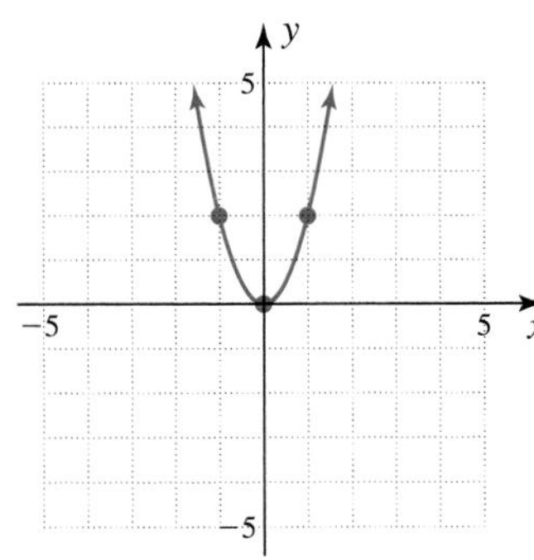

2. $y = 2x^2 + 1$

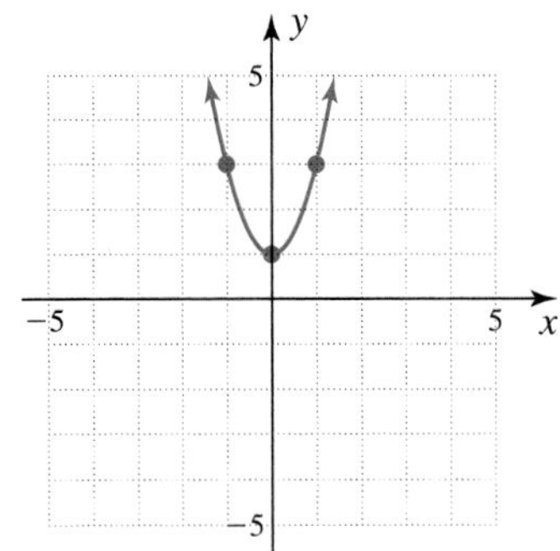

3. $y = 2x^2 - 1$

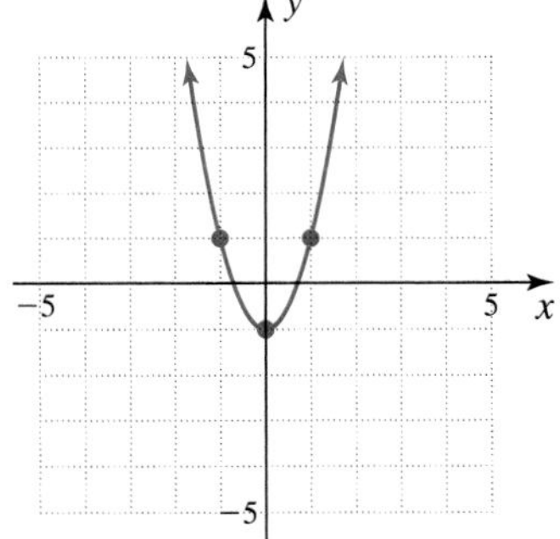

4. $y = 2x^2 - 2$

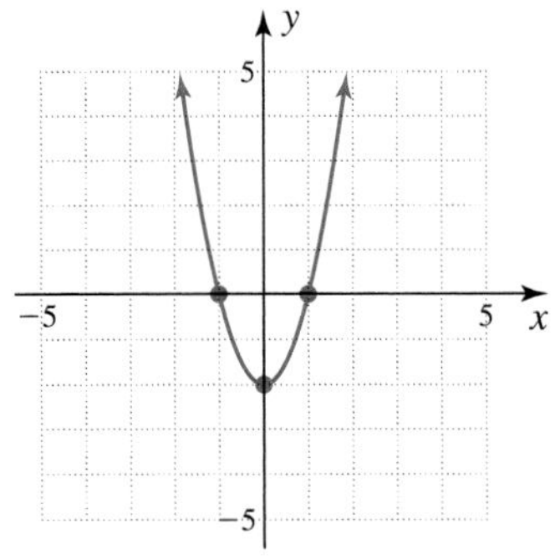

5. $y = -2x^2 + 2$

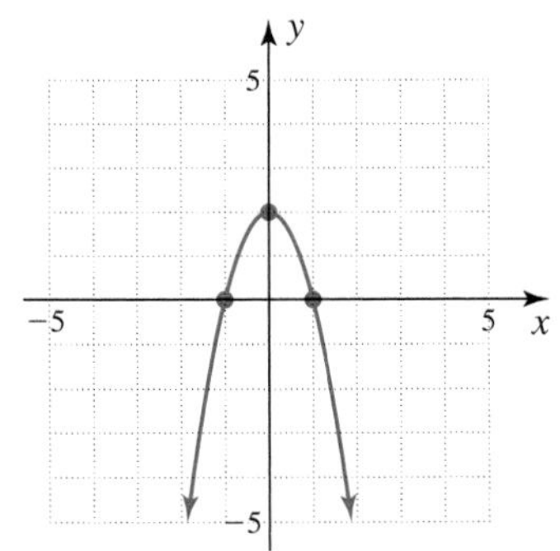

6. $y = -2x^2 - 1$

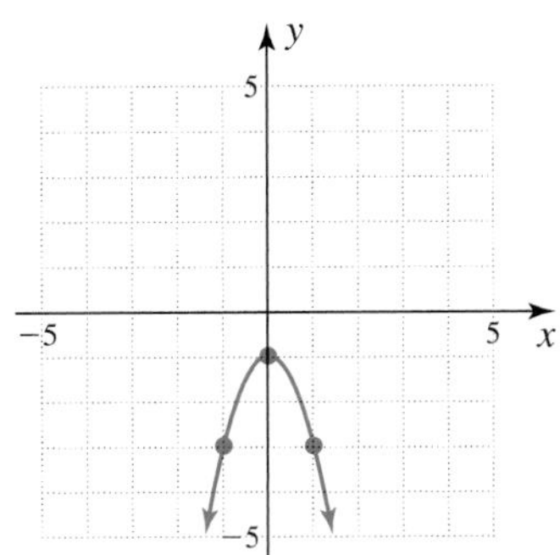

7. $y = -2x^2 - 2$

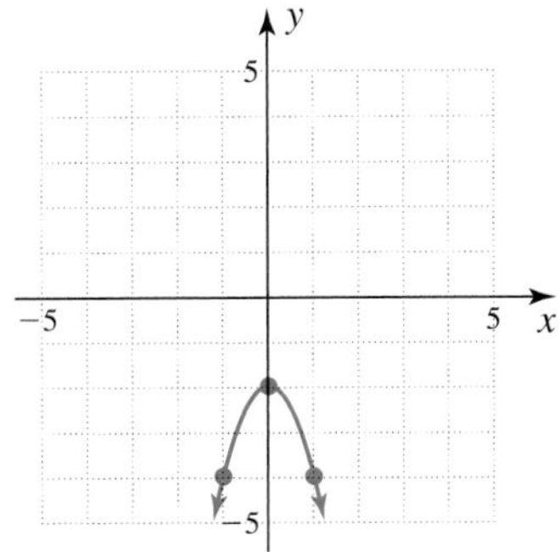

8. $y = -2x^2 + 1$

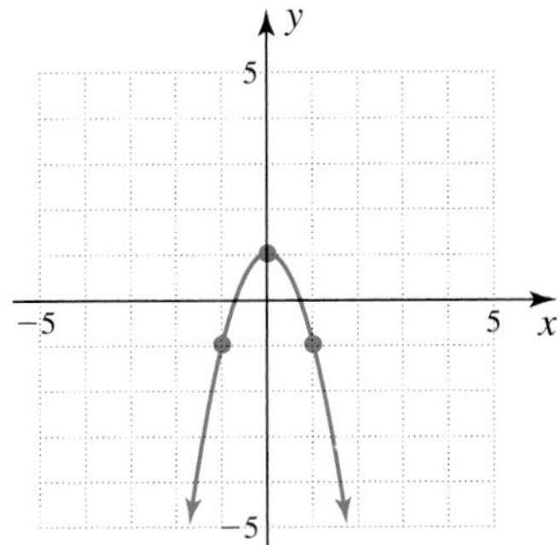

9. $y = (x - 2)^2$

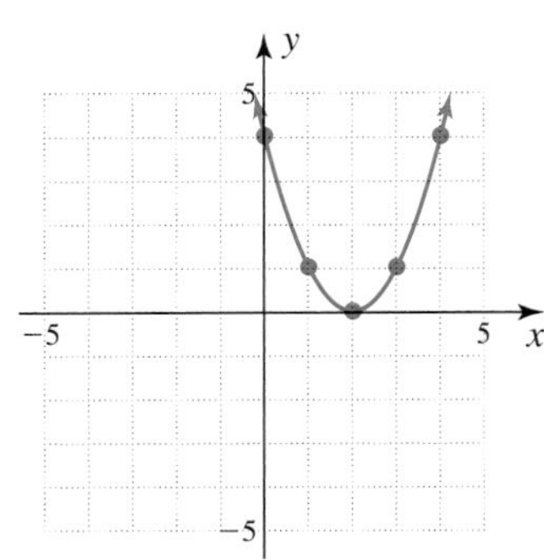

10. $y = (x - 2)^2 + 2$

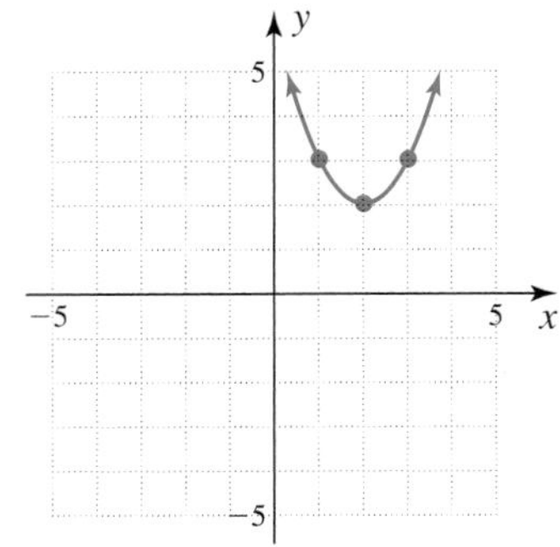

11. $y = (x - 2)^2 - 2$

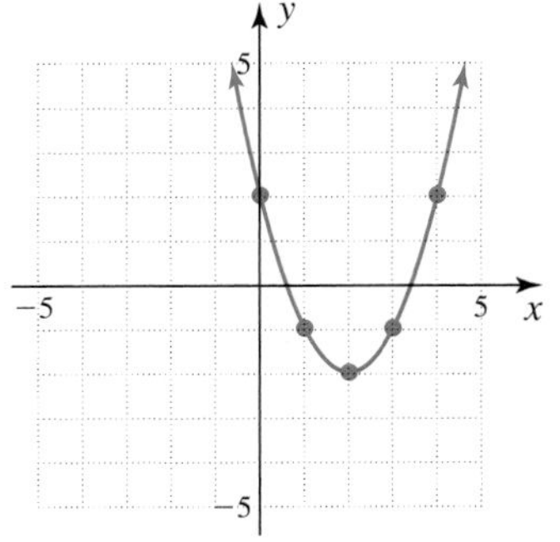

12. $y = (x - 2)^2 - 1$

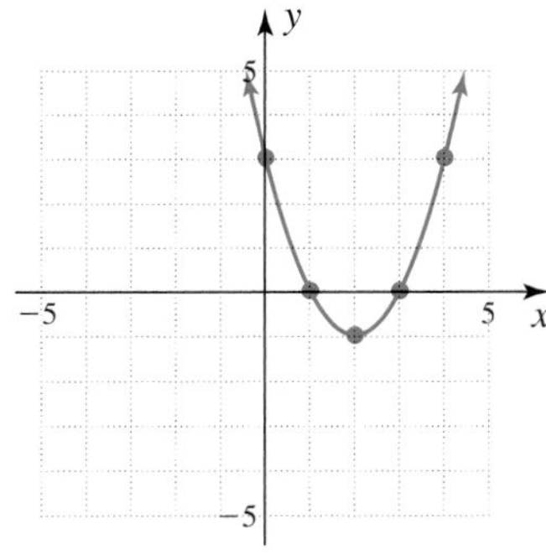

13. $y = -(x - 2)^2$

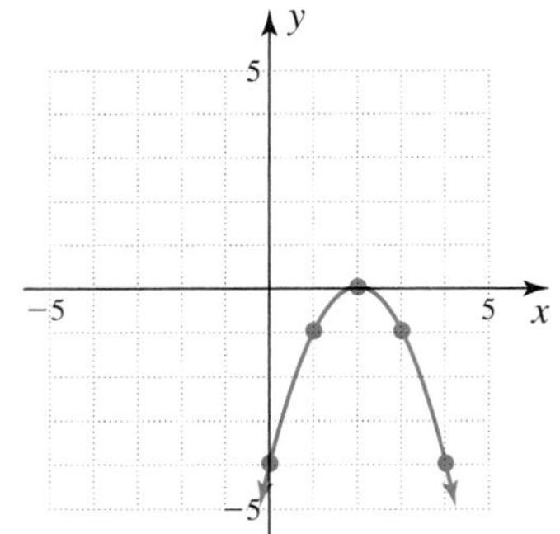

14. $y = -(x - 2)^2 + 2$

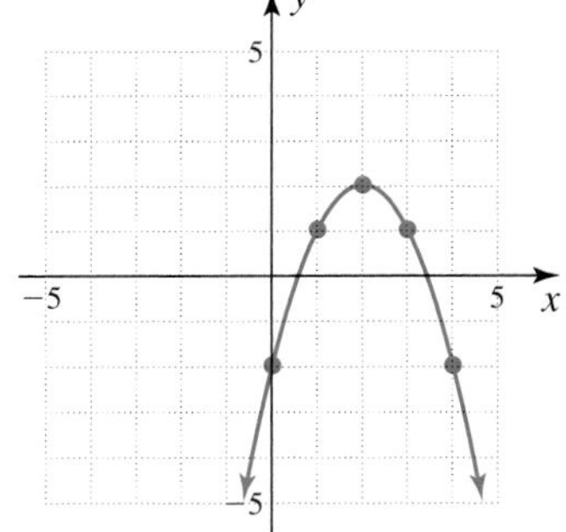

15. $y = -(x - 2)^2 - 2$

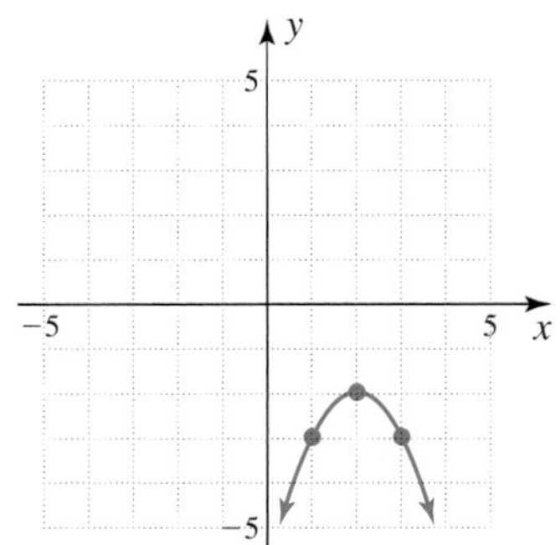

16. $y = -(x - 2)^2 - 1$

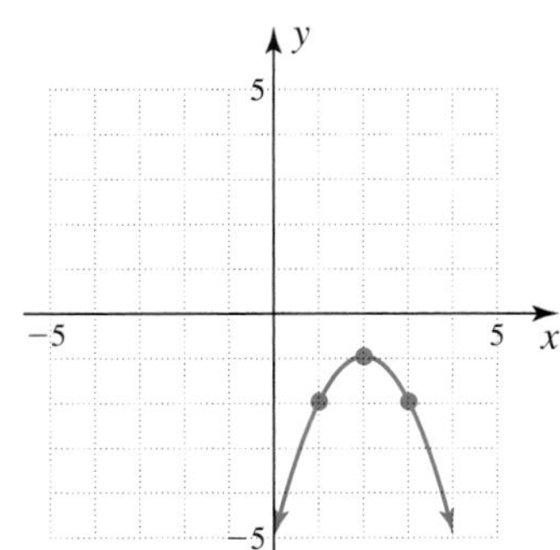

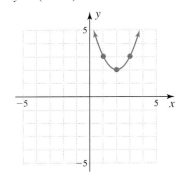

B In Problems 17–22, find the x-intercepts, the y-intercept, and the vertex; then draw the graph.

17. $y = x^2 + 4x + 3$

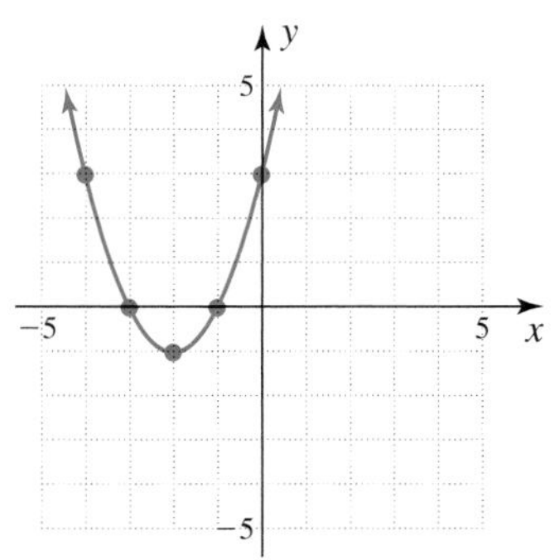

x-int: $(-3, 0)$, $(-1, 0)$; y-int: $(0, 3)$; $V(-2, -1)$

18. $y = x^2 - 4x + 4$

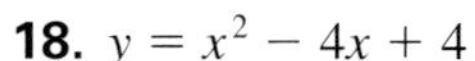

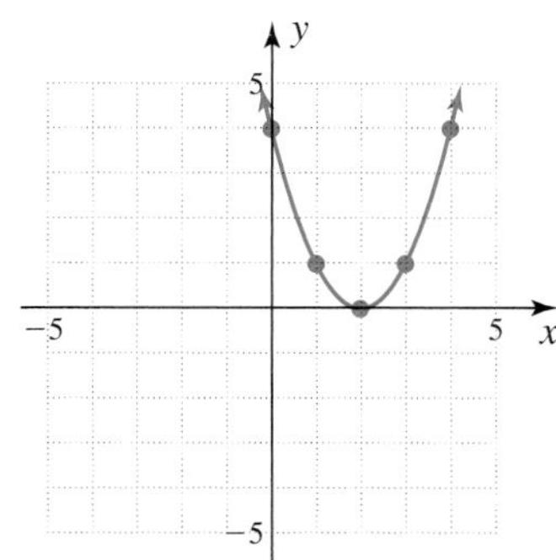

x-int: $(2, 0)$; y-int: $(0, 4)$; $V(2, 0)$

19. $y = x^2 + 2x - 3$

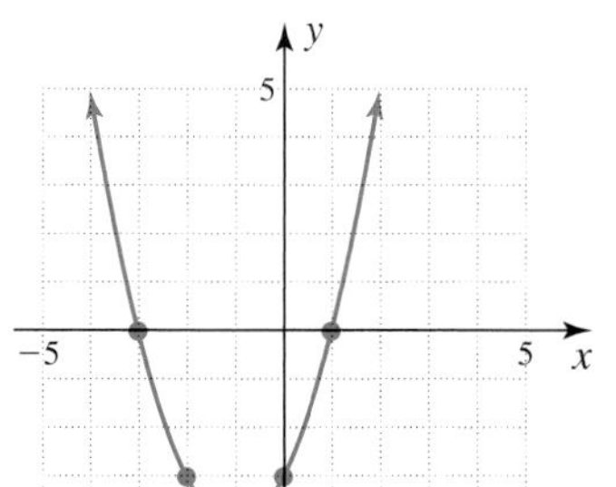

x-int: $(-3, 0)$, $(1, 0)$; y-int: $(0, -3)$; $V(-1, -4)$

20. $y = -x^2 - 4x - 3$

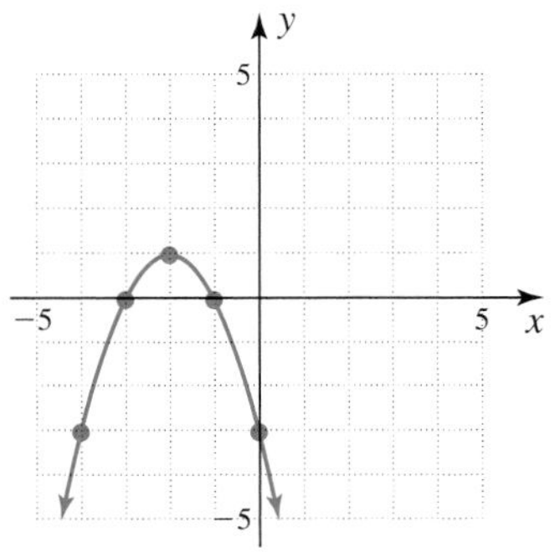

x-int: $(-3, 0)$, $(-1, 0)$; y-int: $(0, -3)$; $V(-2, 1)$

21. $y = -x^2 + 4x - 3$

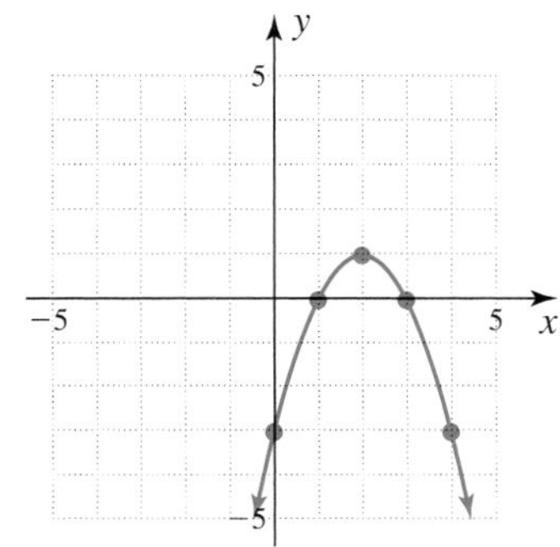

x-int: $(3, 0)$, $(1, 0)$; y-int: $(0, -3)$; $V(2, 1)$

22. $y = -x^2 - 2x + 3$

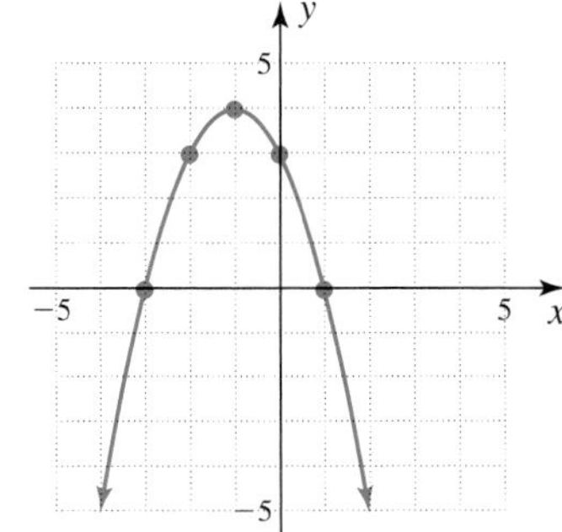

x-int: $(-3, 0)$, $(1, 0)$; y-int: $(0, 3)$; $V(-1, 4)$

APPLICATIONS

Travel from Earth to Moon

The following graph looks somewhat like one-half of a parabola. It indicates the time it takes to travel from the earth to the moon at different speeds. Use it to solve Problems 23–26.

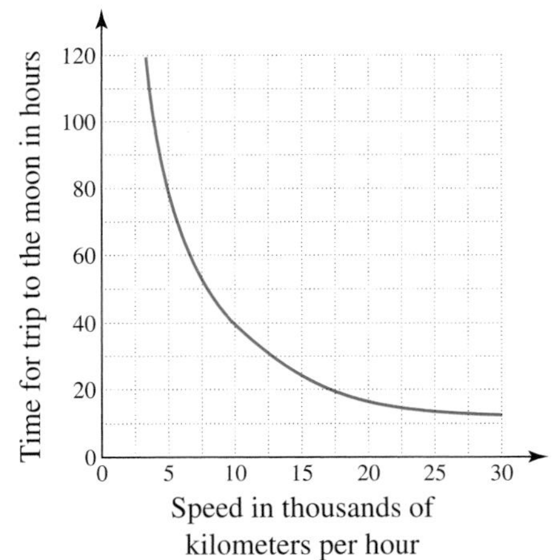

23. How long does the trip take if you travel at 10,000 kilometers per hour? About 40 hours

24. How long does the trip take if you travel at 25,000 kilometers per hour? About 15 hours

25. *Apollo 11* averaged about 6400 kilometers per hour when returning from the moon. How long was the trip? About 60 hours

26. If you wish to make the trip in 50 hours, what should the speed be? About 7500 km/hr

SKILL CHECKER

Try the Skill Checker Exercises so you'll be ready for the next section.

Find:

27. $3^2 + 4^2$ 25

28. $5^2 + 12^2$ 169

29. $\frac{120}{0.003}$ 40,000

30. $\frac{150}{0.005}$ 30,000

USING YOUR KNOWLEDGE

Maximizing Profits!

The ideas studied in this section are used in business to find ways to maximize profits. For example, if a manufacturer can produce a certain item for \$10 each, and then sell the item for x dollars, the profit per item will be $x - 10$ dollars. If it is then estimated that consumers will buy $60 - x$ items per month, the total profit will be

$$\text{Total profit} = \begin{pmatrix}\text{Number of} \\ \text{items sold}\end{pmatrix}\begin{pmatrix}\text{Profit per} \\ \text{item}\end{pmatrix}$$

$$= (60 - x)(x - 10)$$

The graph for the total profit is this parabola:

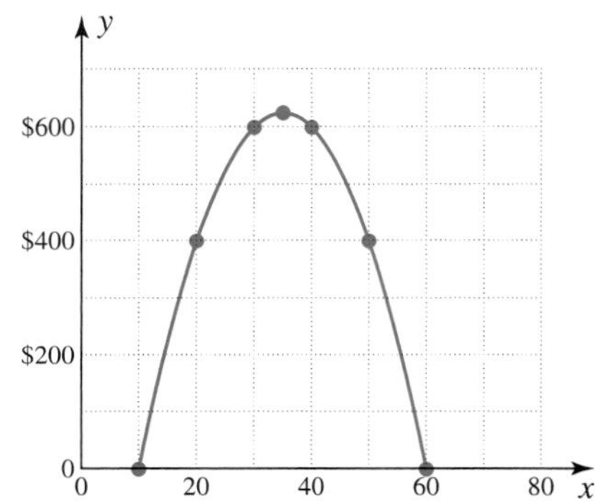

When will the profits be at a maximum? When the manufacturer produces 35 items. Note that 35 is exactly halfway between

$$10 \quad \text{and} \quad 60 \qquad \left(\frac{10 + 60}{2} = 35\right)$$

At this price, the total profits will be

$$P_T = (60 - 35)(35 - 10)$$
$$= (25)(25)$$
$$= \$625$$

Use your knowledge to answer the following questions.

31. What price would maximize the profits of a certain item costing \$20 each if $60 - x$ items are sold each month (where x is the selling price of each item)? \$40

32. Sketch the graph of the resulting parabola.

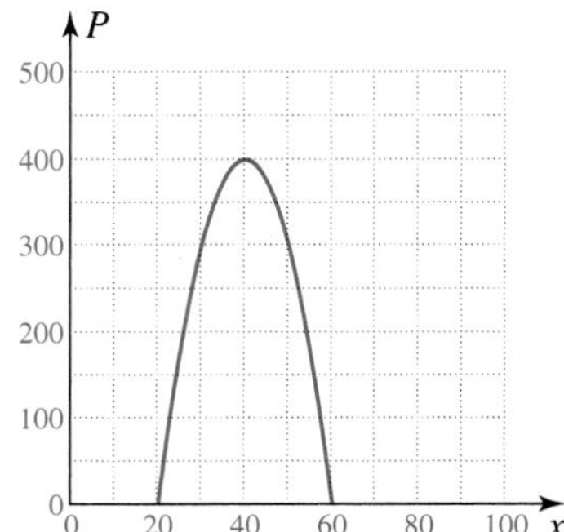

33. What will the maximum profits be? \$400

WRITE ON

34. What is the vertex of a parabola? Answers may vary.

Consider the graph of the parabola $y = ax^2$.

35. What can you say if $a > 0$? The parabola opens upward.

36. What can you say if $a < 0$? The parabola opens downward.

37. If a is a positive integer, what happens to the graph of $y = ax^2$ as a increases? The graph "stretches" upward.

38. What causes the graph of $y = ax^2$ to be wider or narrower than that of $y = x^2$? Answers may vary.

39. What happens to the graph of $y = ax^2$ if you add the positive constant k? What happens if k is negative? The graph is shifted upward; the graph is shifted downward.

40. If a parabola has two x-intercepts and the vertex is at (1, 1), does the parabola open upward or downward? Explain. Downward; answers may vary.

MASTERY TEST

If you know how to do these problems, you have learned your lesson!

Graph:

41. $y = -(x - 2)^2 - 1$

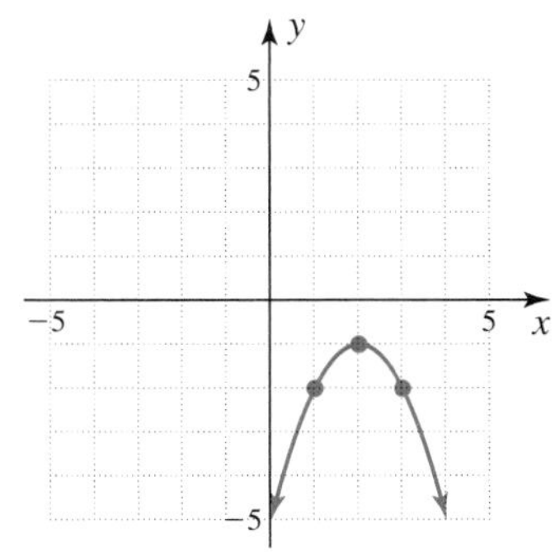

42. $y = -x^2 - 1$

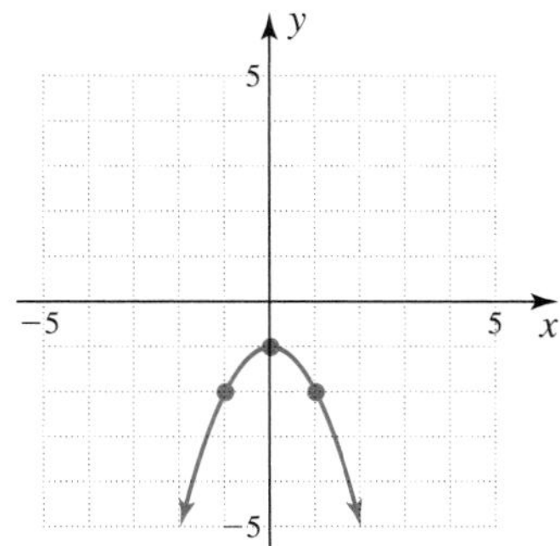

43. $y = -2x^2$

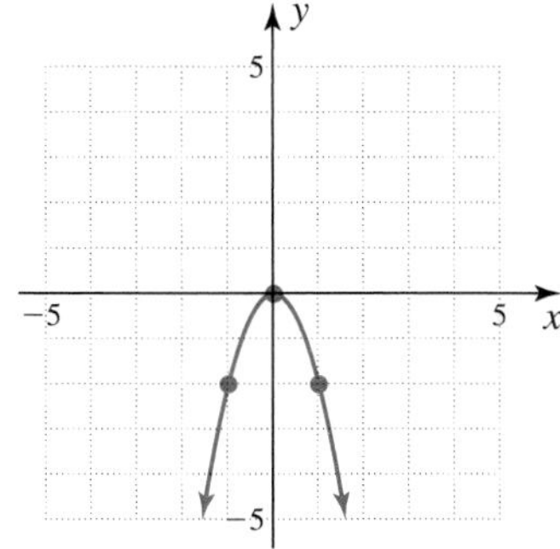

44. $y = -2x^2 + 3$

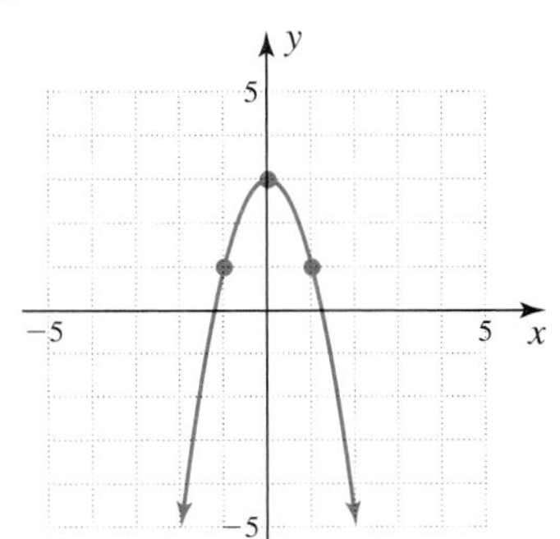

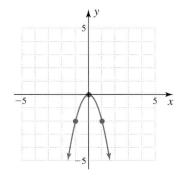

Graph and label the vertex and intercepts:

45. $y = -x^2 - 2x + 3$

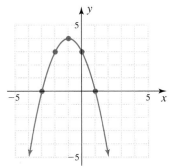

x-int: $(-3, 0)$, $(1, 0)$; y-int: $(0, 3)$; $V(-1, 4)$

46. $y = x^2 + 3x + 2$

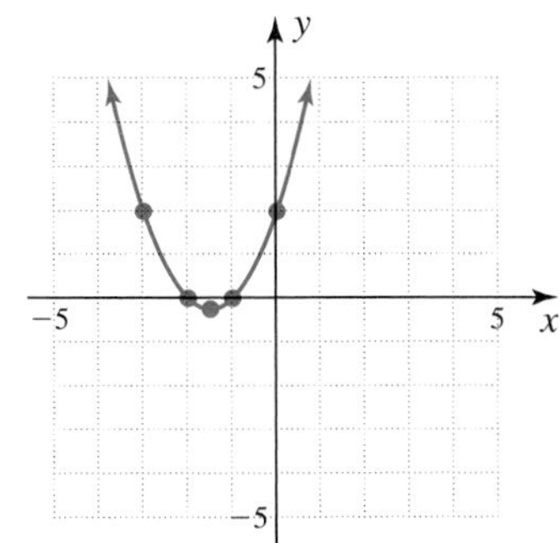

x-int: $(-2, 0)$, $(-1, 0)$; y-int: $(0, 2)$; $V\left(-\frac{3}{2}, -\frac{1}{4}\right)$

10.5 THE PYTHAGOREAN THEOREM AND OTHER APPLICATIONS

To Succeed, Review How To . . .

1. Find the square of a number (p. 73).
2. Divide by a decimal (pp. 70–75).
3. Solve a quadratic by factoring (pp. 432–437).

Objectives

A Use the Pythagorean Theorem to solve right triangles.

B Solve word problems involving quadratic equations.

GETTING STARTED A Theorem Not Dumb Enough!

B.C.
by johnny hart
CAN I ASK YOU A REAL DUMB QUESTION, DAD?
CERTAINLY, THAT'S HOW YOU LEARN ABOUT THINGS.
WHY IS THE SQUARE OF THE HYPOTENUSE EQUAL TO THE SUM OF THE SQUARES OF THE OTHER TWO SIDES?
THAT'S NOT DUMB ENOUGH.

By Permission of John L. Hart FLP, and Creators Syndicate, Inc.

In this section we discuss several applications involving quadratic equations. One of them pertains to the theorem mentioned in the cartoon. This theorem, first proved by Pythagoras, will be stated next.

Using the Pythagorean Theorem

Web It

For applications of the Pythagorean Theorem, go to link 10-5-1 on the Bello Website at mhhe.com/bello.

Before using the Pythagorean Theorem, you have to know what a right triangle is! A right triangle is a triangle containing one right angle (90°), and the hypotenuse is the side opposite the 90° angle. If that is the case, here is what the theorem says:

> **Pythagorean Theorem**
>
> The square of the hypotenuse of a right triangle equals the sum of the squares of the other two sides; that is,
>
> $$h^2 = a^2 + b^2$$
>
> where h is the hypotenuse of the triangle, and a and b are the two remaining sides (called the legs).

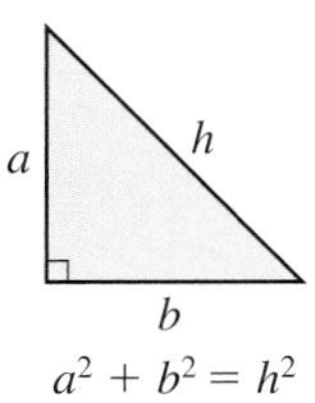

Figure 10

Figure 10 shows a right triangle with sides a, b, and h. So if $a = 3$ units and $b = 4$ units, then the hypotenuse h is such that

$$h^2 = 3^2 + 4^2$$
$$h^2 = 9 + 16$$
$$h^2 = 25$$
$$h = \pm\sqrt{25} = \pm 5$$

Since h represents length, h cannot be negative, so the length of the hypotenuse is 5 units.

EXAMPLE 1 **Climbing the wall**

The distance from the bottom of the wall to the base of the ladder in Figure 11 is 5 feet, and the distance from the floor to the top of the ladder is 12 feet. Can you find the length of the ladder?

Figure 11

SOLUTION We use the RSTUV method given earlier.

1. Read the problem.
We are asked to find the length of the ladder.

2. Select the unknown.
Let L represent the length of the ladder.

3. Think of a plan.
First we draw a diagram (see Figure 12) that gives the appropriate dimensions. Then we translate the problem using the Pythagorean Theorem:

$$\underbrace{\begin{pmatrix}\text{Distance from}\\ \text{wall to base of}\\ \text{ladder}\end{pmatrix}^2}_{5^2} + \underbrace{\begin{pmatrix}\text{Distance from}\\ \text{floor to top}\\ \text{of ladder}\end{pmatrix}^2}_{12^2} = \underbrace{\begin{pmatrix}\text{Length of}\\ \text{ladder}\end{pmatrix}^2}_{L^2}$$

PROBLEM 1

Find the length of the ladder if the distance from the wall to the base of the ladder is 5 ft but the height of the top of the ladder is 14 ft.

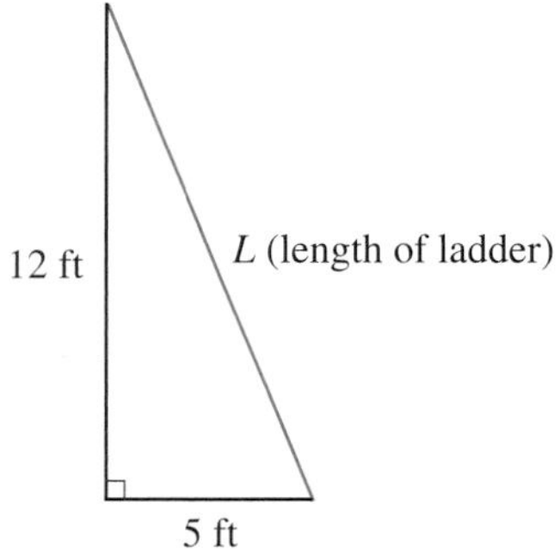

Figure 12

Answer

1. $\sqrt{221}$ ft

4. Use algebra to solve the problem.
We use algebra to solve this equation:

$$5^2 + 12^2 = L^2$$
$$25 + 144 = L^2$$
$$L^2 = 169$$
$$L = \pm 13$$

Since L is the length of the ladder, L is positive, so $L = 13$.

5. Verify the solution.
This solution is correct because $5^2 + 12^2 = 13^2$ since $25 + 144 = 169$.

B Solving Word Problems

Many applications of quadratic equations come from the field of engineering. For example, it is known that the pressure p (in pounds per square foot) exerted by a wind blowing at v miles per hour is given by the equation

$$p = 0.003v^2$$ Since v is raised to the second power, the equation is a quadratic.

If the pressure p is known, the equation $p = 0.003v^2$ is an example of a quadratic equation in v. We show how this information is used in the next example.

EXAMPLE 2 Gauging the gale

A wind pressure gauge at Commonwealth Bay registered 120 pounds per square foot during a gale. What was the wind speed at that time?

SOLUTION We will again use the RSTUV method.

1. Read the problem.
We are asked to find the wind speed.

2. Select the unknown.
Since we know that $p = 120$ and that $p = 0.003v^2$, we let v be the wind speed.

3. Think of a plan.
We substitute for p to obtain

$$0.003v^2 = 120$$

4. Use algebra to solve the problem.
The easiest way to solve this equation is to divide by 0.003 first. We then have

$$v^2 = \frac{120}{0.003}$$

$$v^2 = \frac{120 \cdot 1000}{0.003 \cdot 1000} = \frac{120{,}000}{3} = 40{,}000$$

$$v = \pm\sqrt{40{,}000} = \pm 200$$

You can also divide 120 by 0.003 to obtain

$$0.003\overline{)120.000} = 40{,}000$$
12

Thus the wind speed v was 200 miles per hour. (We discard the negative answer as not suitable.)

5. Verify the solution.
We leave the verification to you.

PROBLEM 2

Find the wind speed at the Golden Gate Bridge when a pressure gauge is registering 7.5 lb/ft^2.

Answer

2. 50 mi/hr

It can be shown in physics that when an object is dropped or thrown downward with an initial velocity v_0, the distance d (in meters) traveled by the object in t seconds is given by the formula

$$d = 5t^2 + v_0t$$

This is an approximate formula. A more exact formula is $d = 4.9t^2 + v_0t$.

Suppose an object is dropped from a height of 125 meters. How long would it be before the object hits the ground? The solution to this problem is given in the next example.

EXAMPLE 3 Falling the distance

Use the formula $d = 5t^2 + v_0t$ to find the time it takes an object dropped from a height of 125 meters to hit the ground.

SOLUTION Since the object is dropped, the initial velocity is $v_0 = 0$, and we are given $d = 125$; we then have

$$125 = 5t^2$$

$$5t^2 = 125$$

$$t^2 = 25 \quad \text{Divide by 5.}$$

$$t = \pm\sqrt{25} = \pm 5$$

Since the time must be positive, the correct answer is $t = 5$; that is, it takes 5 seconds for the object to hit the ground.

PROBLEM 3

Find the time it takes an object dropped from a height of 180 m to hit the ground.

Finally, quadratic equations are also used in business. Perhaps you already know what it means to "break even." In business, the break-even point is the point at which the revenue R equals the cost C of the manufactured goods. In symbols,

$$R = C$$

Now suppose a company produces x (thousand) items. If each item sells for \$3, the revenue R (in thousands of dollars) can be expressed by

$$R = 3x$$

If the manufacturing cost C (in thousands of dollars) is

$$C = x^2 - 3x + 5$$

and it is known that the company always produces more than 1000 items, the break-even point occurs when

$$R = C$$

$$3x = x^2 - 3x + 5$$

$$x^2 - 6x + 5 = 0$$

$$(x - 5)(x - 1) = 0 \quad \text{Factor.}$$

$$x - 5 = 0 \quad \text{or} \quad x - 1 = 0 \quad \text{Use the principle of zero product.}$$

$$x = 5 \qquad x = 1$$

But since it's known that the company produces more than 1 (thousand) item(s), the break-even point occurs when the company manufactures 5 (thousand) items.

Answer

3. 6 sec

EXAMPLE 4 Breaking even quadratically

Find the number x of items that have to be produced to break even when the revenue R (in thousands of dollars) of a company is given by $R = 3x$ and the cost (also in thousands of dollars) is $C = 2x^2 - 3x + 4$.

SOLUTION In order to break even,

$$R = C$$
$$3x = 2x^2 - 3x + 4$$
$$2x^2 - 6x + 4 = 0$$
$$x^2 - 3x + 2 = 0 \quad \text{Divide by 2.}$$
$$(x - 1)(x - 2) = 0 \quad \text{Factor.}$$
$$x - 1 = 0 \quad \text{or} \quad x - 2 = 0 \quad \text{Use the principle of zero product.}$$
$$x = 1 \qquad x = 2$$

This means that the company breaks even when it produces either 1 (thousand) or 2 (thousand) items.

PROBLEM 4

Find the number of items that have to be produced to break even when the revenue R of a company (in thousands of dollars) is given by $R = 4x$ and the cost C (also in thousands of dollars) is $C = x^2 - 3x + 6$.

Answer

4. 1 (thousand) or 6 (thousand) items

Exercises 10.5

Boost *your* GRADE at mathzone.com!

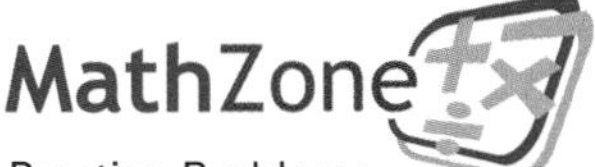

- Practice Problems
- Self-Tests
- Videos
- NetTutor
- e-Professors

A In Problems 1–10, let a and b represent the lengths of the sides of a right triangle and h the length of the hypotenuse. Find the missing side.

1. $a = 12, h = 13$ $b = 5$

2. $a = 4, h = 5$ $b = 3$

3. $a = 5, h = 15$ $b = 10\sqrt{2}$

4. $a = b, h = 4$ $a = b = 2\sqrt{2}$

5. $b = 3, a = \sqrt{6}$ $h = \sqrt{15}$

6. $b = 7, a = 9$ $h = \sqrt{130}$

7. $a = \sqrt{5}, b = 2$ $h = 3$

8. $a = 3, b = \sqrt{7}$ $h = 4$

9. $h = \sqrt{13}, a = 3$ $b = 2$

10. $h = \sqrt{52}, a = 4$ $b = 6$

B In Problems 11–24, solve the resulting quadratic equation.

11. How long is a wire extending from the top of a 40-foot telephone pole to a point on the ground 30 feet from the base of the pole? 50 ft

12. Repeat Problem 11 where the point on the ground is 16 feet away from the pole. $8\sqrt{29}$ ft

13. The pressure p (in pounds per square foot) exerted on a surface by a wind blowing at v miles per hour is given by

$$p = 0.003v^2$$ 100 mi/hr

Find the wind speed when a wind pressure gauge recorded a pressure of 30 pounds per square foot.

14. Repeat Problem 13 where the pressure is 2.7 pounds per square foot. 30 mi/hr

15. An object is dropped from a height of 320 meters. How many seconds does it take for this object to hit the ground? (See Example 3.) 8 sec

16. Repeat Problem 15 where the object is dropped from a height of 45 meters. 3 sec

17. An object is thrown downward with an initial velocity of 3 meters per second. How long does it take for the object to travel 8 meters? (See Example 3.) 1 sec

18. The revenue of a company is given by $R = 2x$, where x is the number of units produced (in thousands). If the cost is $C = 4x^2 - 2x + 1$, how many units have to be produced before the company breaks even? 500 units

19. Repeat Problem 18 where the revenue is $R = 5x$ and the cost is $C = x^2 - x + 9$. 3000 units

20. If P dollars are invested at r percent compounded annually, then at the end of 2 years the amount will have grown to $A = P(1 + r)^2$. At what rate of interest r will \$1000 grow to \$1210 in 2 years? (*Hint:* $A = 1210$ and $P = 1000$.) 10%

21. A rectangle is 2 feet wide and 3 feet long. Each side is increased by the same amount to give a rectangle with twice the area of the original one. Find the dimensions of the new rectangle. (*Hint:* Let x feet be the amount by which each side is increased.) 3 ft by 4 ft

22. The hypotenuse of a right triangle is 4 centimeters longer than the shortest side and 2 centimeters longer than the remaining side. Find the dimensions of the triangle. 6 cm; 8 cm; 10 cm

23. Repeat Problem 22 where the hypotenuse is 16 centimeters longer than the shortest side and 2 centimeters longer than the remaining side. 10 cm; 24 cm; 26 cm

24. The square of a certain positive number is 5 more than 4 times the number itself. Find this number. 5

SKILL CHECKER

Try the Skill Checker Exercises so you'll be ready for the next section.

Solve:

25. $A + 20 = 32$ $A = 12$

26. $A + 9 = 17$ $A = 8$

27. $18 + B = 30$ $B = 12$

28. $30 + B = 45$ $B = 15$

USING YOUR KNOWLEDGE

Solving for Variables

Many formulas require the use of some of the techniques we've studied when solving for certain unknowns in these formulas. For example, consider the formula

$$v^2 = 2gh$$

How can we solve for v in this equation? In the usual way, of course! Thus we have

$$v^2 = 2gh$$
$$v = \pm\sqrt{2gh}$$

Since the speed v is positive, we simply write

$$v = \sqrt{2gh}$$

Use this idea to solve for the indicated variables in the given formulas.

29. Solve for c in the formula $E = mc^2$. $c = \sqrt{\frac{E}{m}} = \frac{\sqrt{Em}}{m}$

30. Solve for r in the formula $A = \pi r^2$. $r = \sqrt{\frac{A}{\pi}} = \frac{\sqrt{\pi A}}{\pi}$

31. Solve for r in the formula $F = \frac{GMm}{r^2}$. $r = \sqrt{\frac{GMm}{F}} = \frac{\sqrt{FGMm}}{F}$

32. Solve for v in the formula $\text{KE} = \frac{1}{2}mv^2$. $v = \sqrt{\frac{2\text{KE}}{m}} = \frac{\sqrt{2\text{KE}m}}{m}$

33. Solve for P in the formula $I = kP^2$. $P = \sqrt{\frac{I}{k}} = \frac{\sqrt{Ik}}{k}$

WRITE ON

34. When solving for the hypotenuse h in a right triangle whose sides were 3 and 4 units, respectively, a student solved the equation $3^2 + 4^2 = h^2$ and obtained the answer 5. Later, the same student was given the equation $3^2 + 4^2 = x^2$ and gave the same answer. The instructor said that the answer was not complete. Explain. Answers may vary.

35. In Example 2, one of the answers was discarded. Explain why. Answers may vary.

36. Referring to Example 4, explain why you "break even" when $R = C$. Answers may vary.

MASTERY TEST

If you know how to do these problems, you have learned your lesson!

37. Find the number x of hundreds of items that have to be produced to break even when the revenue R of a company (in thousands of dollars) is given by $R = 4x$ and the cost C (also in thousands of dollars) is $C = x^2 - 3x + 6$. 1 or 6 (hundred)

38. Find the time t (in seconds) it takes an object dropped from a height of 245 meters to hit the ground by using the formula $d = 5t^2 + v_0t$ (meters). 7 sec

39. The pressure p (in pounds per square foot) exerted by a wind blowing at v miles per hour is given by

$$p = 0.003v^2$$

Find the wind speed at the Golden Gate Bridge when a pressure gauge is registering 7.5 pounds per square foot. 50 mi/hr

40. Find the length L of the ladder in the diagram if the distance from the wall to the base of the ladder is 5 feet and the height to the top of the ladder is 13 feet. $\sqrt{194}$ ft

10.6 FUNCTIONS

To Succeed, Review How To . . .

1. Evaluate an expression (pp. 72, 81, 98–101).
2. Use set notation (p. 55).

Objectives

A Find the domain and range of a relation.

B Determine whether a given relation is a function.

C Use function notation.

D Solve an application.

GETTING STARTED

Functions for Fashions

Did you know that women's clothing sizes are getting smaller? According to a recent J. C. Penney catalog, "Simply put, you will wear one size smaller than before." Is there a relationship between the new sizes and waist size? The table gives a $1\frac{1}{2}$-inch leeway for waist sizes. Let's use this table to consider the first number in the waist sizes corresponding to different dress sizes—that is, 30, 32, 34, 36. For Petite sizes 14–22, the waist size is 16 inches more than the dress size. If we wish to formalize this relationship, we can write

$$w(s) = s + 16 \qquad \text{Read "}w\text{ of }s\text{ equals }s\text{ plus 16."}$$

What would be the waist size of a woman who wears size 14? It would be

$$w(14) = 14 + 16 = 30$$

For size 16,

$$w(16) = 16 + 16 = 32$$

and so on. It works! Can you do the same for hip sizes? If you get

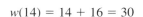

$$h(s) = s + 26.5$$

you are on the right track. As you can see, there is a *relationship* between hip size and dress size.

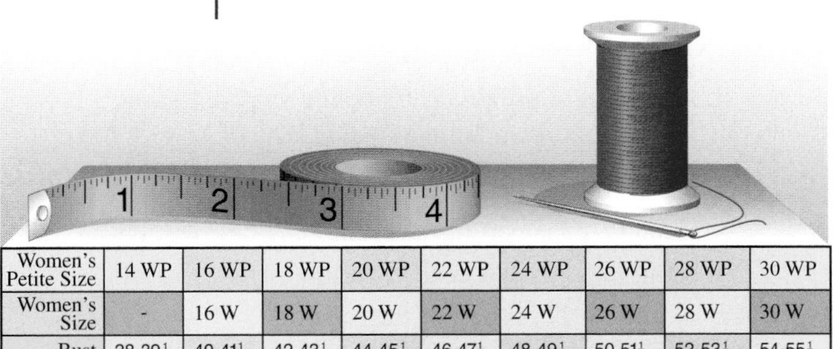

Women's Petite Size	14 WP	16 WP	18 WP	20 WP	22 WP	24 WP	26 WP	28 WP	30 WP
Women's Size	-	16 W	18 W	20 W	22 W	24 W	26 W	28 W	30 W
Bust	38-$39\frac{1}{2}$	40-$41\frac{1}{2}$	42-$43\frac{1}{2}$	44-$45\frac{1}{2}$	46-$47\frac{1}{2}$	48-$49\frac{1}{2}$	50-$51\frac{1}{2}$	52-$53\frac{1}{2}$	54-$55\frac{1}{2}$
Waist	30-$31\frac{1}{2}$	32-$33\frac{1}{2}$	34-$35\frac{1}{2}$	36-$37\frac{1}{2}$	38-40	$40\frac{1}{2}$-$42\frac{1}{2}$	43-45	$45\frac{1}{2}$-$47\frac{1}{2}$	48-50
Hips	$40\frac{1}{2}$-42	$42\frac{1}{2}$-44	$44\frac{1}{2}$-46	$46\frac{1}{2}$-48	$48\frac{1}{2}$-50	$50\frac{1}{2}$-52	$52\frac{1}{2}$-54	$54\frac{1}{2}$-56	$56\frac{1}{2}$-58

The word *relation* might remind you of members of your family—parents, brothers, sisters, and so on—but in mathematics, relations are expressed using ordered pairs. Thus the relationship between Petite sizes 14–22 in the table and the waist size can be expressed by the set of ordered pairs:

$$w = \{(14, 30), (16, 32), (18, 34), (20, 36), (22, 38)\}$$

Note that we've specified that the waist size corresponding to the dress size is the first number in the row labeled "Waist." Thus for every dress size, there is only *one* waist size. Relations that have this property are called *functions.* In this section we shall learn about *relations* and *functions,* and how to evaluate them using notation such as $w(s) = s + 16$ and $h(s) = s + 26.5$.

Finding Domains and Ranges

Web It

For an example of finding domain and range, go to link 10-6-1 on the Bello Website at mhhe.com/bello.

In previous chapters, we studied ordered pairs that satisfy linear equations, graphed these ordered pairs, and then found the graph of the corresponding line. Thus to find the graph of $y = 3x - 1$, we assigned the value 0 to x to obtain the corresponding y-value -1. Thus a y-value of -1 is paired to an x-value of 0, which results in the ordered pair $(0, -1)$ as assigned by the rule $y = 3x - 1$.

RELATION, DOMAIN, AND RANGE

In mathematics, a **relation** is a set of ordered pairs. The **domain** of the relation is the set of all possible x-values, and the **range** of the relation is the set of all possible y-values.

EXAMPLE 1 Finding the domain and range of a relation given as a set of ordered pairs

Find the domain and range:

$$S = \{(4, -3), (2, -5), (-3, 4)\}$$

SOLUTION The domain D is the set of all possible x-values. Thus $D = \{4, 2, -3\}$.

The range R is the set of all possible y-values. Thus $R = \{-3, -5, 4\}$.

PROBLEM 1

Find the domain and range:

$T = \{(6, -4), (4, -6), (-1, 3)\}$

In most cases, relations are defined by a **rule** for finding the y-value for a given x-value. Thus the relation

$$\underbrace{S = \{}_{\text{"The set}} \quad \underbrace{(x, y)}_{\text{of all } (x, y)} \quad \underbrace{|}_{\text{such that}} \quad \underbrace{y = 3x - 1, x \text{ a natural number less than } 5\}}_{y = 3x - 1 \text{ and } x \text{ is a natural number less than 5."}}$$

defines—that is, gives a *rule* about—a set of ordered pairs obtained by assigning a natural number less than 5 to x and obtaining the corresponding y-value. Thus

For $x = 1$, $y = 3(1) - 1 = 2$ and $(1, 2)$ is part of the relation.
For $x = 2$, $y = 3(2) - 1 = 5$ and $(2, 5)$ is part of the relation.
For $x = 3$, $y = 3(3) - 1 = 8$ and $(3, 8)$ is part of the relation.
For $x = 4$, $y = 3(4) - 1 = 11$ and $(4, 11)$ is part of the relation.

Can you see the pattern? For $S = \{(1, 2), (2, 5), (3, 8), (4, 11)\}$, the domain is $\{1, 2, 3, 4\}$, and the range is $\{2, 5, 8, 11\}$.

Teaching Tip

In Example 1, the relation can be illustrated using a mapping diagram that would clearly separate the domain and range.

Example:

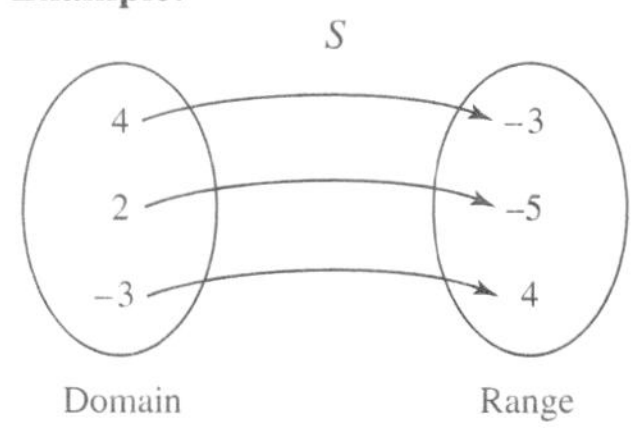

NOTE

Unless otherwise specified, the *domain* of a relation is the largest set of real numbers that can be substituted for x that result in a real number for y. The *range* is then determined by the rule of the relation.

For example, for the relation

$$S = \left\{(x, y) \,\middle|\, y = \frac{1}{x}\right\}$$

the domain consists of all real numbers *except zero* since you can substitute any real number for x in $y = \frac{1}{x}$ *except zero* because division by zero is not defined. The range is also the set of real numbers *except zero* because the rule $y = \frac{1}{x}$ will never yield a value of zero for any given value of x.

Answer

1. $D = \{6, 4, -1\}$; $R = \{-4, -6, 3\}$

EXAMPLE 2 **Finding the domain and range of a relation given as an equation**

Find the domain and range of the relation:

$$\left\{(x, y) \,\middle|\, y = \frac{1}{x - 1}\right\}$$

SOLUTION The domain is the set of real numbers *except* 1, since

$$\frac{1}{1 - 1} = \frac{1}{0} \quad \text{which is undefined}$$

The range is the set of real numbers *except* 0, since

$$y = \frac{1}{x - 1}$$

is never 0.

PROBLEM 2

Find the domain and range:

$$\left\{(x, y) \,\middle|\, y = \frac{1}{x + 1}\right\}$$

B Determining Whether a Relation Is a Function

In mathematics, an important type of relation is one where, for each element in the domain, there corresponds one and only one element in the range. Such relations are called *functions*.

FUNCTION

A **function** is a set of ordered pairs in which each domain value has exactly one range value; that is, no two different ordered pairs have the same first coordinate.

Teaching Tip

Remind students that the domain always starts with the reals, unless there is an operation that has exclusions, such as never dividing by zero.

Example:

Find the domain for $y = 3x - 1$.
Answer: Reals

EXAMPLE 3 **Determining whether a relation is a function**

Determine whether the given relations are functions:

a. $\{(3, 4), (4, 3), (4, 4), (5, 4)\}$ **b.** $\{(x, y) \mid y = 5x + 1\}$

SOLUTION

a. The relation is not a function because the two ordered pairs (4, 3) and (4, 4) have the same first coordinate.

b. The relation $\{(x, y) \mid y = 5x + 1\}$ is a function because if x is any real number, the expression $y = 5x + 1$ yields one and only one value, so there will never be two ordered pairs with the same first coordinate.

PROBLEM 3

Determine whether the relations are functions:

a. $\{(5, 6), (6, 5), (6, 6), (7, 6)\}$

b. $\{(x, y) \mid y = 3x + 2\}$

Answers

2. The domain is the set of all real numbers except -1. The range is the set of all real numbers except 0.
3. a. Not a function because (6, 5) and (6, 6) have the same first coordinate. **b.** A function. If x is a real number, $y = 3x + 2$ yields one and only one value.

C Using Function Notation

We often use letters such as f, F, g, G, h, and H to designate functions. Thus for the relation in Example 3(b), we use set notation to write

$$f = \{(x, y) \mid y = 5x + 1\}$$

because we know this relation to be a function. Another very commonly used notation for denoting the range value that corresponds to a given domain value x is $f(x)$. (This is usually read, "f of x.")

The $f(x)$ notation, called **function notation,** is quite convenient because it denotes the value of the function for the given value of x. For example, if

$$f(x) = 2x + 3$$

then

$$f(1) = 2(1) + 3 = 5$$
$$f(0) = 2(0) + 3 = 3$$
$$f(-6) = 2(-6) + 3 = -9$$
$$f(4) = 2(4) + 3 = 11$$
$$f(a) = 2(a) + 3 = 2a + 3$$
$$f(w + 2) = 2(w + 2) + 3 = 2w + 7$$

and so on. Whatever appears between the parentheses in $f(\ \ )$ is to be substituted for x in the rule that defines $f(x)$.

Instead of describing a function in set notation, we frequently say, "the function defined by $f(x) = \ldots$" where the three dots are replaced by the expression for the value of the function. For instance, "the function defined by $f(x) = 5x + 1$" has the same meaning as "the function $f = \{(x, y) \mid y = 5x + 1\}$."

Web It

For a lesson on function notation, go to link 10-6-2 on the Bello Website at mhhe.com/bello.

Teaching Tip

The notation "$f(x)$" can also be read "f *at* x" or "f *of* x," which is short for finding the value of the f rule *at* a particular value of x.

Example:

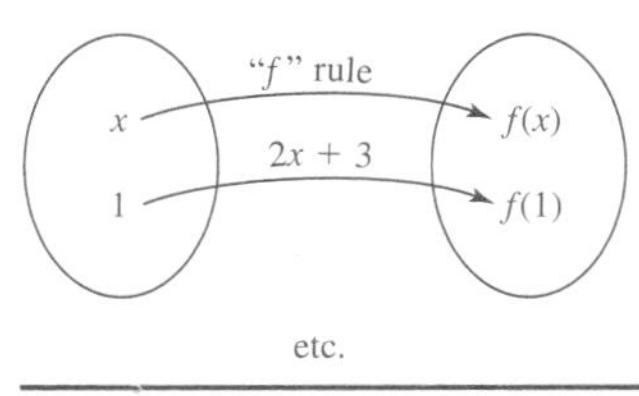

EXAMPLE 4 Evaluating functions

Let $f(x) = 3x + 5$. Find:

a. $f(4)$ **b.** $f(2)$ **c.** $f(2) + f(4)$ **d.** $f(x + 1)$

SOLUTION Since $f(x) = 3x + 5$,

a. $f(4) = 3 \cdot 4 + 5 = 12 + 5 = 17$

b. $f(2) = 3 \cdot 2 + 5 = 6 + 5 = 11$

c. Since $f(2) = 11$ and $f(4) = 17$,

$$f(2) + f(4) = 11 + 17 = 28$$

d. $f(x + 1) = 3(x + 1) + 5 = 3x + 8$

PROBLEM 4

Let $f(x) = 4x - 5$. Find:

a. $f(3)$ **b.** $f(4)$

c. $f(3) + f(4)$ **d.** $f(x - 1)$

EXAMPLE 5 More practice evaluating functions

A function g is defined by $g(x) = x^3 - 2x^2 + 3x - 4$. Find:

a. $g(2)$ **b.** $g(-3)$ **c.** $g(2) - g(-3)$

SOLUTION Since $g(x) = x^3 - 2x^2 + 3x - 4$,

a. $g(2) = 2^3 - 2(2)^2 + 3(2) - 4$
$= 8 - 8 + 6 - 4 = 2$

b. $g(-3) = (-3)^3 - 2(-3)^2 + 3(-3) - 4$
$= -27 - 18 - 9 - 4 = -58$

c. $g(2) - g(-3) = 2 - (-58) = 60$

PROBLEM 5

If $g(x) = x^3 - 3x^2 + 2x - 1$, find:

a. $g(3)$ **b.** $g(-2)$

c. $g(3) - g(-2)$

Answers

4. a. 7 **b.** 11 **c.** 18 **d.** $4x - 9$
5. a. 5 **b.** -25 **c.** 30

In the preceding examples, we evaluated a specified function. Sometimes, as was the case in *Getting Started,* we must find the function, as shown in the next example.

EXAMPLE 6 **Finding the relationship between ordered pairs**

Consider the ordered pairs (2, 6), (3, 9), (1.2, 3.6), and $(\frac{2}{5}, \frac{6}{5})$. There is a functional relationship $y = f(x)$ between the numbers in each pair. Find $f(x)$ and use it to fill in the missing numbers in the pairs (__, 12), (__, 3.3), and (5, __).

SOLUTION The given pairs are of the form (x, y). A close examination reveals that each of the y's in the pairs is 3 times the corresponding x; that is, $y = 3x$ or $f(x) = 3x$. Now in each of the ordered pairs (__, 12), (__, 3.3), and (5, __), the y-value must be three times the x-value. Thus

$$(\_\_, 12) = (4, 12)$$
$$(\_\_, 3.3) = (1.1, 3.3)$$
$$(5, \_\_) = (5, 15)$$

PROBLEM 6

Consider the ordered pairs (3, 6), (4, 8), (1.1, 2.2), and $(\frac{3}{5}, \frac{6}{5})$. There is a functional relationship $y = f(x)$ between the numbers in each pair. Find $f(x)$ and use it to fill in the missing numbers in the pairs (__, 10), (__, 4.4), and (6, __).

D Solving an Application

Web It

For a target zone calculator, go to link 10-6-3 on the Bello Website at mhhe.com/bello.

In recent years, aerobic exercises such as jogging, swimming, bicycling, and roller blading have been taken up by millions of Americans. To see whether you are exercising too hard (or not hard enough), you should stop from time to time and take your pulse to determine your heart rate. The idea is to keep your rate within a range known as the **target zone,** which is determined by your age. The next example explains how to find the **lower limit** of your target zone.

EXAMPLE 7 **Exercises and functions**

The lower limit L (heartbeats per minute) of your target zone is a function of your age a (in years) and is given by

$$L(a) = -\frac{2}{3}a + 150$$

Find the value of L for a person who is

a. 30 years old **b.** 45 years old

SOLUTION

a. We need to find $L(30)$, and because

$$L(a) = -\frac{2}{3}a + 150,$$
$$L(30) = -\frac{2}{3}(30) + 150$$
$$= -20 + 150 = 130$$

This result means that a 30-year-old person should try to attain at least 130 heartbeats per minute while exercising.

b. Here, we want to find $L(45)$. Proceeding as before, we obtain

$$L(45) = -\frac{2}{3}(45) + 150$$
$$= -30 + 150 = 120$$

(Find the value of L for your own age.)

PROBLEM 7

a. Find L for a 21-year-old person.

b. Find L for a 33-year-old person.

Answers

6. $f(x) = 2x$; (5, 10), (2.2, 4.4), (6, 12)
7. a. $L = 136$ **b.** $L = 128$

Calculate It Entering Functions

The idea of a function is so important in mathematics that even your calculator will accept only functions to be entered for Y=. Thus if you are given an equation with variables x and y and you can solve for y, you have a function. What does this mean?

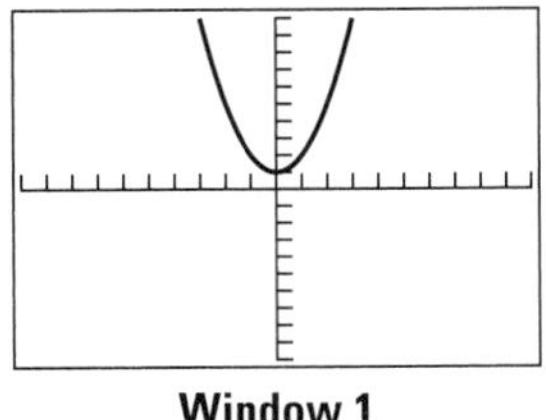

Window 1

Let's take $2x + 4y = 4$. Solving for y, we obtain $y = -\frac{1}{2}x + 1$, which can be entered and graphed. Similarly, $y = x^2 + 1$ can be entered and graphed (see Window 1). Can we solve for y in $x^2 + y^2 = 4$? Not with the techniques we've learned so far! The equation $x^2 + y^2 = 4$, whose graph happens to be a circle of radius 2, does not describe a function.

Your calculator can also evaluate functions. For example, to find the value $L(30)$ in Example 7, let's start by using x's instead of a's. Store the 30 in the calculator's memory by pressing 30 STO▸ X,T,θ,n ENTER, then enter the function $-\frac{2}{3}x + 150$ ENTER; the answer, 130, is then displayed as shown in Window 2.

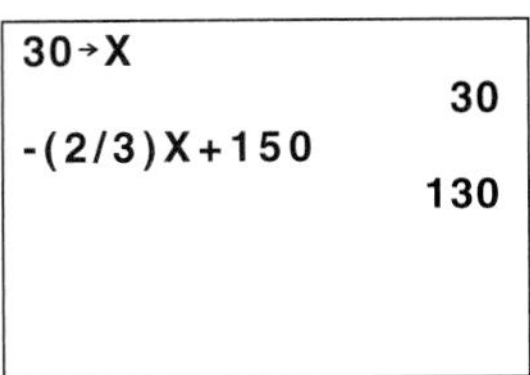

Window 2

Exercises 10.6

A In Problems 1–14, find the domain and the range of the given relation. [*Hint:* Remember that you cannot divide by zero.]

1. $\{(1, 2), (2, 3), (3, 4)\}$
$D = \{1, 2, 3\}; R = \{2, 3, 4\}$

2. $\{(3, 1), (2, 1), (1, 1)\}$
$D = \{1, 2, 3\}; R = \{1\}$

3. $\{(1, 1), (2, 2), (3, 3)\}$
$D = \{1, 2, 3\}; R = \{1, 2, 3\}$

4. $\{(4, 1), (5, 2), (6, 1)\}$
$D = \{4, 5, 6\}; R = \{1, 2\}$

5. $\{(x, y) \mid y = 3x\}$
$D =$ All real numbers;
$R =$ All real numbers

6. $\{(x, y) \mid y = 2x + 1\}$
$D =$ All real numbers;
$R =$ All real numbers

7. $\{(x, y) \mid y = x + 1\}$
$D =$ All real numbers;
$R =$ All real numbers

8. $\{(x, y) \mid y = 1 - 2x\}$
$D =$ All real numbers;
$R =$ All real numbers

9. $\{(x, y) \mid y = x^2\}$
$D =$ All real numbers;
$R =$ All nonnegative real numbers

10. $\{(x, y) \mid y = 2 + x^2\}$
$D =$ All real numbers;
$R =$ All real numbers ≥ 2

11. $\{(x, y) \mid y^2 = x\}$
(*Hint:* y^2 is never negative.)
$D =$ All nonnegative real numbers;
$R =$ All real numbers

12. $\{(x, y) \mid x = 1 + y^2\}$
$D =$ All real numbers ≥ 1;
$R =$ All real numbers

13. $\left\{(x, y) \,\middle|\, y = \frac{1}{x - 3}\right\}$
$D =$ All real numbers except 3;
$R =$ All real numbers except 0

14. $\left\{(x, y) \,\middle|\, y = \frac{1}{x - 2}\right\}$ $D =$ All real numbers except 2;
$R =$ All real numbers except 0

Find the domain and the range of the relation given in each of Problems 15–22. List the ordered pairs in each relation.

15. $\{(x, y) \mid y = 2x, x$ an integer between -1 and 2, inclusive$\}$
$D = \{-1, 0, 1, 2\}; R = \{-2, 0, 2, 4\}$; $(-1, -2), (0, 0), (1, 2), (2, 4)$

16. $\{(x, y) \mid y = 2x - 1, x$ a counting number not greater than 5$\}$ $D = \{1, 2, 3, 4, 5\}; R = \{1, 3, 5, 7, 9\}$; $(1, 1), (2, 3), (3, 5), (4, 7), (5, 9)$

17. $\{(x, y) \mid y = 2x - 3, x$ an integer between 0 and 4, inclusive$\}$ $D = \{0, 1, 2, 3, 4\}; R = \{-3, -1, 1, 3, 5\}$; $(0, -3), (1, -1), (2, 1), (3, 3), (4, 5)$

18. $\{(x, y) \mid y = \frac{1}{x}, x$ an integer between 1 and 5, inclusive$\}$
$D = \{1, 2, 3, 4, 5\}; R = \left\{1, \frac{1}{2}, \frac{1}{3}, \frac{1}{4}, \frac{1}{5}\right\}$; $(1, 1), \left(2, \frac{1}{2}\right), \left(3, \frac{1}{3}\right), \left(4, \frac{1}{4}\right), \left(5, \frac{1}{5}\right)$

19. $\{(x, y) \mid y = \sqrt{x}, x = 0, 1, 4, 9, 16, \text{ or } 25\}$
$D = \{0, 1, 4, 9, 16, 25\}$; $R = \{0, 1, 2, 3, 4, 5\}$; (0, 0), (1, 1), (4, 2), (9, 3), (16, 4), (25, 5)

20. $\{(x, y) \mid y \le x + 1, x \text{ and } y \text{ positive integers less than } 4\}$
$D = \{1, 2, 3\}$; $R = \{1, 2, 3\}$; (1, 1), (1, 2), (2, 1), (2, 2), (2, 3), (3, 1), (3, 2), (3, 3)

21. $\{(x, y) \mid y > x, x \text{ and } y \text{ positive integers less than } 5\}$
$D = \{1, 2, 3\}$; $R = \{2, 3, 4\}$; (1, 2), (1, 3), (1, 4), (2, 3), (2, 4), (3, 4)

22. $\{(x, y) \mid 0 < x + y < 5, x \text{ and } y \text{ positive integers less than } 4\}$ $D = \{1, 2, 3\}$; $R = \{1, 2, 3\}$; (1, 1), (1, 2), (1, 3), (2, 1), (2, 2), (3, 1)

B In Problems 23–30, decide whether the given relation is a function. State the reason for your answer in each case.

23. {(0, 1), (1, 2), (2, 3)}
A function; one y-value for each x-value

24. {(1, 2), (2, 1), (1, 3), (3, 1)}
Not a function; two y-values for $x = 1$

25. {(−1, 1), (−2, 2), (−3, 3)}
A function; one y-value for each x-value

26. {(−1, 1), (−1, 2), (−1, 3)}
Not a function; three y-values for $x = -1$

27. $\{(x, y) \mid y = 5x + 6\}$
A function; one y-value for each x-value

28. $\{(x, y) \mid y = 3 - 2x\}$
A function; one y-value for each x-value

29. $\{(x, y) \mid x = y^2\}$
Not a function; two y-values for each positive x-value

30. $\{(x, y) \mid y = x^2\}$
A function; one y-value for each x-value

C In Problems 31–35, find the given value.

31. A function f is defined by $f(x) = 3x + 1$. Find:
a. $f(0)$ $f(0) = 1$ **b.** $f(2)$ $f(2) = 7$ **c.** $f(-2)$ $f(-2) = -5$

32. A function g is defined by $g(x) = -2x + 1$. Find:
a. $g(0)$ $g(0) = 1$ **b.** $g(1)$ $g(1) = -1$ **c.** $g(-1)$ $g(-1) = 3$

33. A function F is defined by $F(x) = \sqrt{x - 1}$. Find:
a. $F(1)$ $F(1) = 0$ **b.** $F(5)$ $F(5) = 2$ **c.** $F(26)$ $F(26) = 5$

34. A function G is defined by $G(x) = x^2 + 2x - 1$. Find:
a. $G(0)$ $G(0) = -1$ **b.** $G(2)$ $G(2) = 7$ **c.** $G(-2)$ $G(-2) = -1$

35. A function f is defined by $f(x) = 3x + 1$. Find:
a. $f(x + h)$ $f(x + h) = 3x + 3h + 1$
b. $f(x + h) - f(x)$ $f(x + h) - f(x) = 3h$
c. $\dfrac{f(x + h) - f(x)}{h}$, $h \ne 0$ $\dfrac{f(x + h) - f(x)}{h} = 3, h \ne 0$

36. Given are the ordered pairs: (2, 1), (6, 3), (9, 4.5), and (1.6, 0.8). There is a simple functional relationship, $y = f(x)$, between the numbers in each pair. What is $f(x)$? Use this to fill in the missing number in the pairs (___, 7.5), (___, 2.4), and (___, $\frac{1}{7}$). $y = f(x) = \frac{1}{2}x$; (15, 7.5), (4.8, 2.4), $\left(\frac{2}{7}, \frac{1}{7}\right)$

37. Given are the ordered pairs: $\left(\frac{1}{2}, \frac{1}{4}\right)$, (1.2, 1.44), (5, 25), and (7, 49). There is a simple functional relationship, $y = g(x)$, between the numbers in each pair. What is $g(x)$? Use this to fill in the missing number in the pairs ($\frac{1}{4}$, ___), (2.1, ___), and (___, 64). $y = g(x) = x^2$; $\left(\frac{1}{4}, \frac{1}{16}\right)$, (2.1, 4.41), (±8, 64)

38. Given that $f(x) = x^3 - x^2 + 2x$, find:
a. $f(-1)$ $f(-1) = -4$ **b.** $f(-3)$ $f(-3) = -42$ **c.** $f(2)$ $f(2) = 8$

39. If $g(x) = 2x^3 + x^2 - 3x + 1$, find:
a. $g(0)$ $g(0) = 1$ **b.** $g(-2)$ $g(-2) = -5$ **c.** $g(2)$ $g(2) = 15$

APPLICATIONS

40. *Fahrenheit and Celsius temperatures* The Fahrenheit temperature reading F is a function of the Celsius temperature reading C. This function is given by

$$F(C) = \frac{9}{5}C + 32$$

a. If the temperature is 15°C, what is the Fahrenheit temperature? 59°F

b. Water boils at 100°C. What is the corresponding Fahrenheit temperature? 212°F

c. The freezing point of water is 0°C or 32°F. How many Fahrenheit degrees below freezing is a temperature of −10°C? 18° below freezing

d. The lowest temperature attainable is −273°C; this is the zero point on the absolute temperature scale. What is the corresponding Fahrenheit temperature? −459.4°F

41. *Upper limit of target zone* Refer to Example 7. The **upper limit** U of your target zone when exercising is also a function of your age a (in years), and is given by

$$U(a) = -a + 190$$

Find the highest safe heart rate for a person who is

a. 50 years old 140 **b.** 60 years old 130

42. *Target zone* Refer to Example 7 and Problem 41. The target zone for a person a years old consists of all the heart rates between $L(a)$ and $U(a)$, inclusive. Thus if a person's heart rate is R, that person's target zone is described by $L(a) \le R \le U(a)$. Find the target zone for a person who is

a. 30 years old $130 \le R \le 160$ **b.** 45 years old $120 \le R \le 145$

43. *Ideal weight for men* The ideal weight w (in pounds) of a man is a function of his height h (in inches). This function is defined by

$$w(h) = 5h - 190$$

a. If a man is 70 inches tall, what should his weight be? 160 lb

b. If a man weighs 200 pounds, what should his height be? 78 in.

44. *Car rental costs* The cost C in dollars of renting a car for 1 day is a function of the number m of miles traveled. For a car renting for \$20 per day and 20¢ per mile, this function is given by

$$C(m) = 0.20m + 20$$

a. Find the cost of renting a car for 1 day and driving 290 miles. \$78

b. If an executive paid \$60.60 after renting a car for 1 day, how many miles did she drive? 203 mi

45. *Pressure below ocean surface* The pressure P (in pounds per square foot) at a depth of d ft below the surface of the ocean is a function of the depth. This function is given by

$$P(d) = 63.9d$$

Find the pressure on a submarine at a depth of:

a. 10 feet 639 lb/ft^2 **b.** 100 feet 6390 lb/ft^2

46. *Distance traveled by dropped ball* If a ball is dropped from a point above the surface of the earth, the distance s (in meters) that the ball falls in t seconds is a function of t. This function is given by

$$s(t) = 4.9t^2$$

Find the distance that the ball falls in:

a. 2 seconds 19.6 m **b.** 5 seconds 122.5 m

47. *Distance traveled by falling object* The function $S(t) = \frac{1}{2}gt^2$ gives the distance that an object falls from rest in t seconds. If S is measured in feet, then the gravitational constant, g, is approximately 32 feet/second per second. Find the distance that the object will fall in:

a. 3 seconds 144 ft **b.** 5 seconds 400 ft

48. *Estimation of gravitational constant* An experiment, carefully carried out, showed that a ball dropped from rest fell 64.4 feet in 2 seconds. What is a more accurate value of g than that given in Problem 47? 32.2 feet/second per second

49. *Kinetic energy of moving object* The kinetic energy (E) of a moving object is measured in joules and is given by the function $E(v) = \frac{1}{2}mv^2$, where m is the mass in kilograms and v is the velocity in meters per second. An automobile of mass 1200 kilograms is going at a speed of 36 kilometers per hour. What is the kinetic energy of the automobile? 60,000 joules

50. *Kinetic energy of automobile* If the kinetic energy of the automobile in Problem 49 is 93,750 joules, how fast (in kilometers per hour) is the automobile traveling? 45 km/hr

USING YOUR KNOWLEDGE

The Vertical Line Test

The graph of an equation can be used to determine whether a relation is a function by applying the **vertical line test.** Since a function is a set of ordered pairs in which no two ordered pairs have the same first coordinate, the graph of a function will never have two points "stacked" vertically; that is, the graph of a function *never* contains two points that can be connected with a vertical line.

Thus the first two graphs shown here represent functions (no two points on the graph can be connected with a vertical line), but the last two do not.

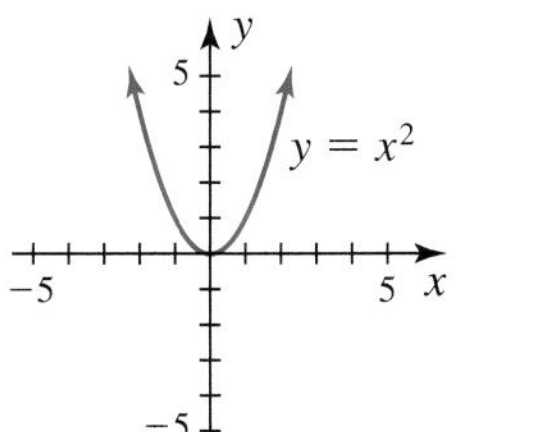

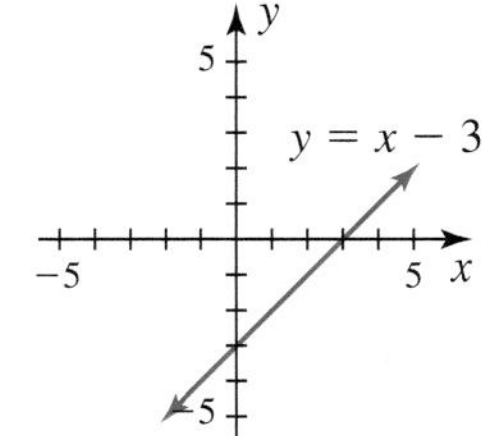

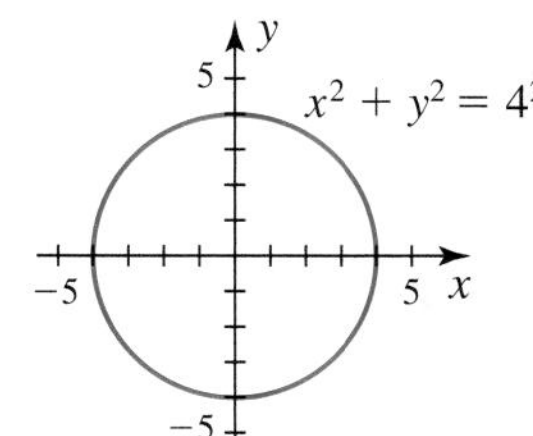

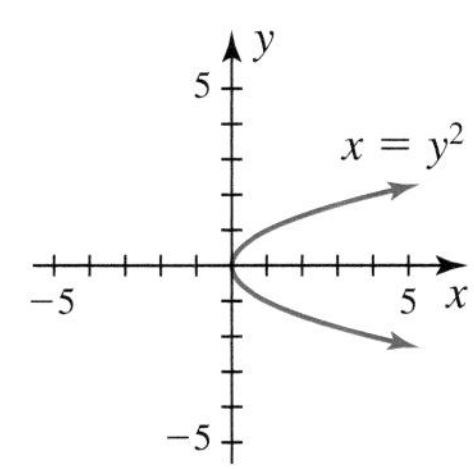

In Problems 51–54, use the vertical line test to determine whether the graph represents a function.

51. A function

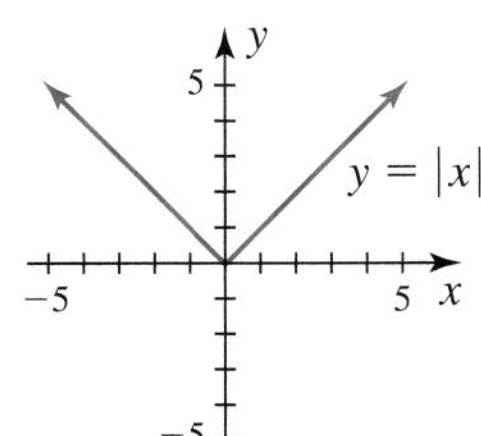

52. Not a function

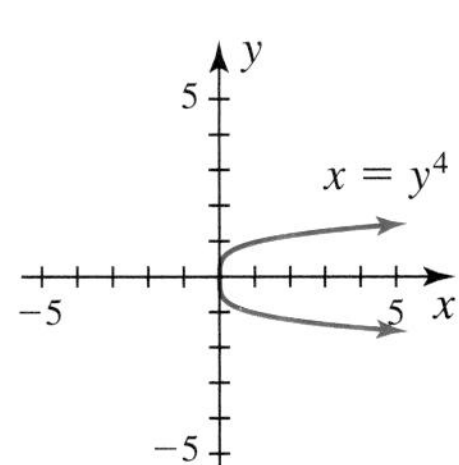

53. A function

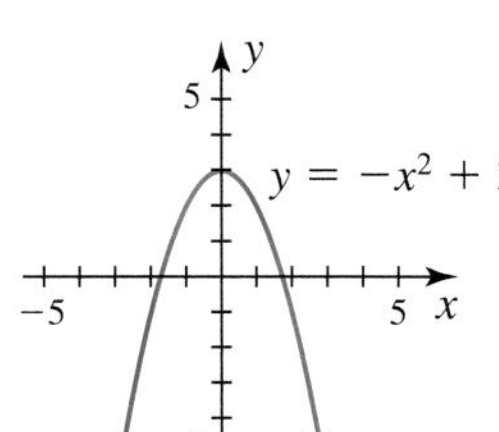

54. Not a function

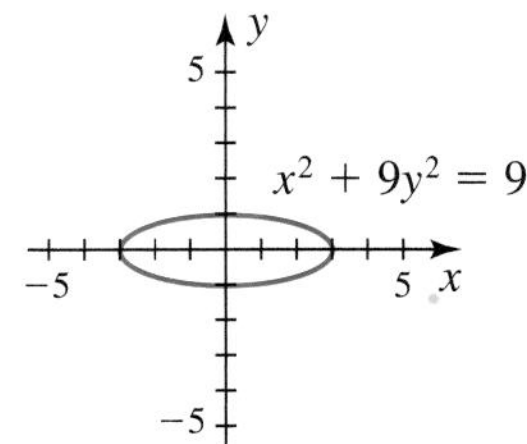

WRITE ON

55. Is every relation a function? Explain.
No; answers may vary.

56. Is every function a relation? Explain.
Yes; answers may vary.

57. Explain how you determine the domain of a function defined by a set of ordered pairs. Answers may vary.

58. Explain how you determine the domain of a function defined by a formula. Answers may vary.

MASTERY TEST

If you know how to do these problems, you have learned your lesson!

In Problems 59–62, find the domain and range of the relation:

59. $\{(-5, 5), (-6, 6), (-7, 7)\}$
$D = \{-5, -6, -7\}$; $R = \{5, 6, 7\}$

60. $\{(5, -5), (-5, 5), (3, -3), (-3, 3)\}$
$D = \{5, -5, 3, -3\}$; $R = \{-5, 5, -3, 3\}$

61. $\left\{(x, y) \middle| y = \dfrac{1}{x - 3}\right\}$ D = All real numbers except 3; R = All real numbers except 0

62. $\{(x, y) \mid y = -x + 2\}$
D = All real numbers; R = All real numbers

In Problems 63–67, determine whether the relation is a function, and explain why or why not.

63. $\{(-4, 5), (-5, 5), (-6, 5)\}$
A function; one y-value for each x-value

64. $\{(5, -3), (5, -4), (5, -5)\}$
Not a function; three y-values for $x = 5$

65. $\{(x, y) \mid y = 2x + 3\}$
A function; one y-value for each x-value

66. $\{(x, y) \mid y = x^2\}$
A function; one y-value for each x-value

67. $\{(x, y) \mid x = y^2\}$
Not a function; two y-values for each positive x-value

68. If $h(x) = x^2 - 3$, find $h(-2)$. $h(-2) = 1$

69. If $g(x) = x^3 - 1$, find $g(-1)$. $g(-1) = -2$

70. The monthly cost for searches (questions) in an online service is \$9.95 with 100 free searches. After that, the cost is \$0.10 per search. The monthly cost for the service (in dollars) can be represented by the following function, where q is the number of searches:

For $q \le 100$: $f(q) = 9.95$

For $q > 100$: $f(q) = 9.95 + 0.10(q - 100)$

a. If you made 99 searches during the month, what would your monthly cost be? \$9.95

b. If you made 130 searches during the month, what would your monthly cost be? \$12.95

COLLABORATIVE LEARNING

This exercise concerns an emergency ejection from a fighter jet and how it could relate to the present discussion of completing the square.

The height H (in feet relative to the ground) of a pilot after t seconds is given by

$$H = -16t^2 + 608t + 4482$$

At about 10,000 feet a malfunction occurs. Eject! Eject!

We want to answer three questions:

1. What is the maximum height (relative to the ground) reached by the pilot after being ejected?
2. How many feet above the jet was the pilot ejected?
3. When did the pilot reach the ground with the aid of his or her parachute?

```
WINDOW
 Xmin=0
 Xmax=50
 Xscl=10
 Ymin=-2000
 Ymax=12000
 Yscl=2000
 Xres=1
```

Form three teams to investigate the incident.

Team 1 is in charge of graphing the path of the pilot and the point of ejection using a graphing calculator. A possible viewing window to graph the path is shown. What expression Y_1 does the team have to graph? What is the equation of the line Y_2 shown in the diagram and representing the point at which the pilot ejected?

Team 2 is in charge of answering the first two questions.

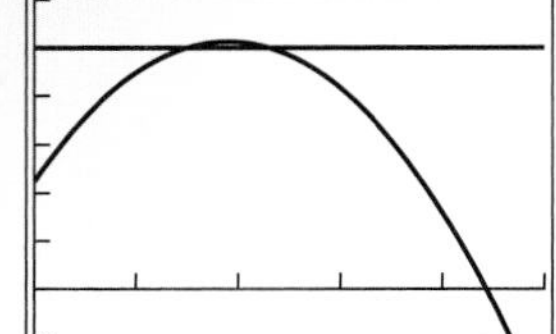

Start with: $H = -16(\quad)^2 + C$

Here is a hint: $H = -16(t^2 - 38t) + 4482$

Now, complete the square inside the parentheses to put the equation in the standard form for a parabola. What is the maximum for this parabola? This answers question 1. Since we know that the pilot ejected at 10,000 feet, how many feet above the jet was the pilot ejected? This answers question 2.

Team 3 has to answer the third question. When the pilot reaches the ground, what is H? The point (t, H) is a solution (zero) of the equation whose graph is shown and it occurs when $H = 0$. To find this point press 2nd TRACE 2. The value of t is the time it took the pilot to land back on earth!

Research Questions

1. Write a short paper detailing the work and techniques used by the Babylonians to solve quadratic equations.
2. Show that the solution of the equation $x^2 + ax = b$ obtained by the Babylonians and the solution obtained by using the quadratic formula developed in this chapter are identical.
3. Write a short paper detailing the methods used by Euler to solve quadratic equations.
4. Write a short paper detailing the methods used by the Arabian mathematician al-Khowarizmi to solve quadratic equations.

Summary

SECTION	ITEM	MEANING	EXAMPLE
10.1A	Quadratic equation	An equation that can be written in the form $ax^2 + bx + c = 0$	$x^2 = 16$ is a quadratic equation since it can be written as $x^2 - 16 = 0$.
	$\sqrt{a}$, the principal square root of a, $a \geq 0$	A nonnegative number b such that $b^2 = a$	$\sqrt{16} = 4$ since $4^2 = 16$
	Irrational number	A number that cannot be written in the form $\frac{a}{b}$, where a and b are integers and b is not 0	$\sqrt{2}$, $\sqrt{5}$, and $3\sqrt{7}$ are irrational.
	Real numbers	The rational and the irrational numbers	$-7, -\frac{3}{2}, 0, \sqrt{5}$, and 17 are real numbers.
	Square root property of equations	If A is a positive number and $X^2 = A$, then $X = \pm\sqrt{A}$.	If $X^2 = 7$, then $X = \pm\sqrt{7}$.
10.2A	Completing the square	A method used to solve quadratic equations	To solve $2x^2 + 4x - 1 = 0$ **1.** Add 1. $2x^2 + 4x = 1$ **2.** Divide by 2. $x^2 + 2x = \frac{1}{2}$ **3.** Add 1^2. $x^2 + 2x + 1 = \frac{3}{2}$ **4.** Rewrite. $(x + 1)^2 = \frac{3}{2}$ **5.** Solve. $x + 1 = \pm\sqrt{\frac{3}{2}}$ $x + 1 = \pm\frac{\sqrt{6}}{2}$ $x = -1 \pm \frac{\sqrt{6}}{2} = \frac{-2 \pm \sqrt{6}}{2}$
10.3	Quadratic formula	The solutions of $ax^2 + bx + c = 0$ are $x = \frac{-b \pm \sqrt{b^2 - 4ac}}{2a}$	The solutions of $2x^2 + 3x + 1 = 0$ are $\frac{-3 \pm \sqrt{3^2 - 4 \cdot 2 \cdot 1}}{2 \cdot 2}$, that is, -1 and $-\frac{1}{2}$.
10.3A	Standard form of a quadratic equation	$ax^2 + bx + c = 0$ is the standard form of a quadratic equation.	The standard form of $x^2 + 2x = -11$ is $x^2 + 2x + 11 = 0$.
10.4	Graph of a quadratic equation	The graph of $y = ax^2 + bx + c$ $(a \neq 0)$ is a parabola.	

(*Continued*)

SECTION	ITEM	MEANING	EXAMPLE
10.5	Pythagorean Theorem	The square of the hypotenuse of a right triangle equals the sum of the squares of the other two sides.	If a and b are the lengths of the sides and h is the length of the hypotenuse, $h^2 = a^2 + b^2$.
10.6A	Relation	A set of ordered pairs	$\{(2, 5), (2, 6), (3, 7)\}$ is a relation.
	Domain	The domain of a relation is the set of all possible x-values.	The domain of the relation $\{(2, 5), (2, 6), (3, 7)\}$ is $\{2, 3\}$.
	Range	The range of a relation is the set of all possible y-values.	The range of the relation $\{(2, 5), (2, 6), (3, 7)\}$ is $\{5, 6, 7\}$.
10.6B	Function	A set of ordered pairs in which no two ordered pairs have the same x-value.	The relation $\{(1, 2), (3, 4)\}$ is a function.

Review Exercises

(If you need help with these exercises, look at the section indicated in brackets.)

1. [10.1A] Solve.

a. $x^2 = 1$ $x = \pm 1$
b. $x^2 = 100$ $x = \pm 10$
c. $x^2 = 81$ $x = \pm 9$

2. [10.1A] Solve.

a. $16x^2 - 25 = 0$ $x = \pm\frac{5}{4}$
b. $25x^2 - 9 = 0$ $x = \pm\frac{3}{5}$
c. $64x^2 - 25 = 0$ $x = \pm\frac{5}{8}$

3. [10.1A] Solve.

a. $7x^2 + 36 = 0$ No real-number solution
b. $8x^2 + 49 = 0$ No real-number solution
c. $3x^2 + 81 = 0$ No real-number solution

4. [10.1B] Solve.

a. $49(x + 1)^2 - 3 = 0$ $x = -1 \pm \frac{\sqrt{3}}{7} = \frac{-7 \pm \sqrt{3}}{7}$
b. $25(x + 2)^2 - 2 = 0$ $x = -2 \pm \frac{\sqrt{2}}{5} = \frac{-10 \pm \sqrt{2}}{5}$
c. $16(x + 1)^2 - 5 = 0$ $x = -1 \pm \frac{\sqrt{5}}{4} = \frac{-4 \pm \sqrt{5}}{4}$

5. [10.2A] Find the missing term in the given expression.

a. $(x + 3)^2 = x^2 + 6x + \square$ 9
b. $(x + 7)^2 = x^2 + 14x + \square$ 49
c. $(x + 6)^2 = x^2 + 12x + \square$ 36

6. [10.2A] Find the missing terms in the given expression.

a. $x^2 - 6x + \square = (\quad)^2$ $9; x - 3$
b. $x^2 - 10x + \square = (\quad)^2$ $25; x - 5$
c. $x^2 - 12x + \square = (\quad)^2$ $36; x - 6$

7. [10.2A] Find the number that must divide each term in the given equation so that the equation can be solved by the method of completing the square.

a. $7x^2 - 14x = -4$ 7
b. $6x^2 - 18x = -2$ 6
c. $5x^2 - 15x = -3$ 5

8. [10.2A] Find the term that must be added to both sides of the given equation so that the equation can be solved by the method of completing the square.

a. $x^2 - 4x = -4$ 4
b. $x^2 - 6x = -9$ 9
c. $x^2 - 12x = -3$ 36

9. [10.3B] Solve.

a. $dx^2 + ex + f = 0, d \neq 0$ $x = \frac{-e \pm \sqrt{e^2 - 4df}}{2d}$
b. $gx^2 + hx + i = 0, g \neq 0$ $x = \frac{-h \pm \sqrt{h^2 - 4gi}}{2g}$
c. $jx^2 + kx + m = 0, j \neq 0$ $x = \frac{-k \pm \sqrt{k^2 - 4jm}}{2j}$

10. [10.3B] Solve.

a. $2x^2 - x - 1 = 0$ $x = 1$ or $x = -\frac{1}{2}$
b. $2x^2 - 2x - 5 = 0$ $x = \frac{1 \pm \sqrt{11}}{2}$
c. $2x^2 - 3x - 3 = 0$ $x = \frac{3 \pm \sqrt{33}}{4}$

11. [10.3B] Solve.

a. $3x^2 - x = 1$ $x = \frac{1 \pm \sqrt{13}}{6}$

b. $3x^2 - 2x = 2$ $x = \frac{1 \pm \sqrt{7}}{3}$

c. $3x^2 - 3x = 2$ $x = \frac{3 \pm \sqrt{33}}{6}$

12. [10.3B] Solve.

a. $9x = x^2$ $x = 0$ or $x = 9$

b. $4x = x^2$ $x = 0$ or $x = 4$

c. $25x = x^2$ $x = 0$ or $x = 25$

13. [10.3B] Solve.

a. $\frac{x^2}{9} - x = -\frac{4}{9}$ $x = \frac{9 \pm \sqrt{65}}{2}$

b. $\frac{x^2}{5} - \frac{x}{2} = \frac{3}{10}$ $x = -\frac{1}{2}$ or $x = 3$

c. $\frac{x^2}{3} + \frac{x}{6} = -\frac{1}{2}$ No real-number solution

14. [10.4A] Graph.

a. $y = x^2 + 1$

b. $y = x^2 + 2$

c. $y = x^2 + 3$

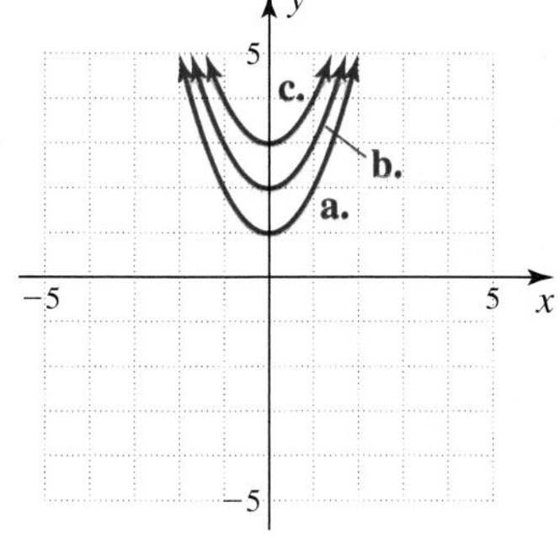

15. [10.4A] Graph.

a. $y = -x^2 - 1$

b. $y = -x^2 - 2$

c. $y = -x^2 - 3$

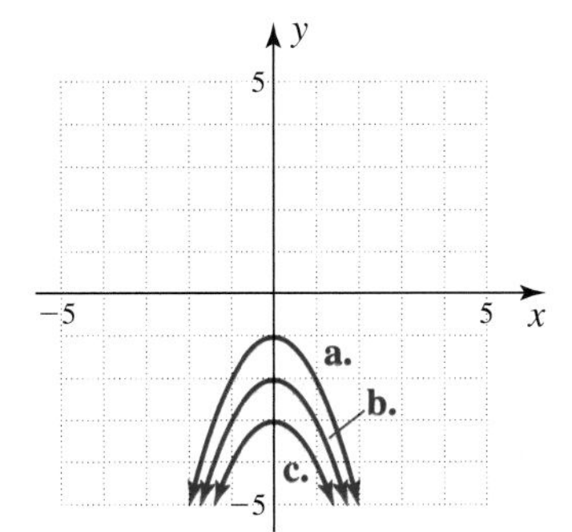

16. [10.4A] Graph.

a. $y = -(x - 2)^2$

b. $y = -(x - 3)^2$

c. $y = -(x - 4)^2$

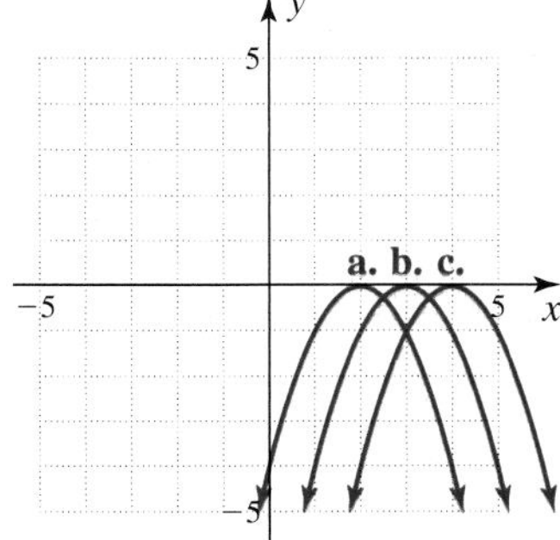

17. [10.4A] Graph.

a. $y = (x - 2)^2 + 1$

b. $y = (x - 2)^2 + 2$

c. $y = (x - 2)^2 + 3$

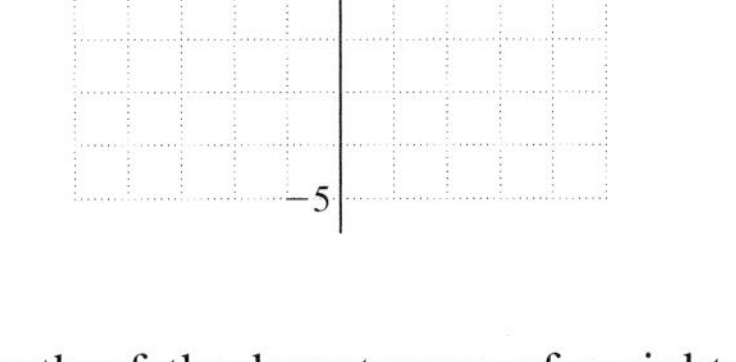

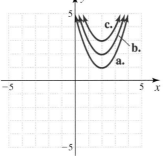

18. [10.4A] Graph.

a. $y = -(x - 2)^2 - 1$

b. $y = -(x - 2)^2 - 2$

c. $y = -(x - 2)^2 - 3$

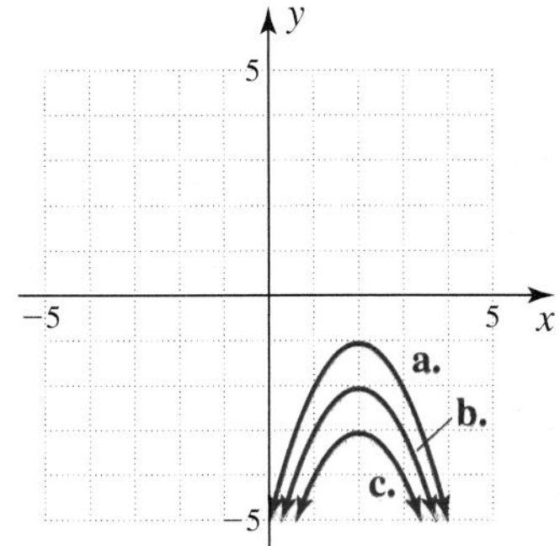

19. [10.5A] Find the length of the hypotenuse of a right triangle if the lengths of the two sides are

a. 5 inches and 12 inches 13 in.

b. 2 inches and 3 inches $\sqrt{13}$ in.

c. 4 inches and 5 inches $\sqrt{41}$ in.

20. [10.5B] After t seconds, the distance d (in meters) traveled by an object thrown downward with an initial velocity v_0 is given by

$$d = 5t^2 + v_0t$$

Find the number of seconds it takes an object to hit the ground if the object is dropped ($v_0 = 0$) from a height of

a. 125 meters 5 sec

b. 245 meters 7 sec

c. 320 meters 8 sec

21. [10.6A] Find the domain and range.

a. $\{(-3, 1), (-4, 1), (-5, 2)\}$
$D = \{-3, -4, -5\}; R = \{1, 2\}$

b. $\{(2, -4), (-1, 3), (-1, 4)\}$
$D = \{2, -1\}; R = \{-4, 3, 4\}$

c. $\{(-1, 2), (-1, 3), (-1, 4)\}$
$D = \{-1\}; R = \{2, 3, 4\}$

22. [10.6A] Find the domain and range.

a. $\{(x, y) \mid y = 2x - 3\}$
D = All real numbers; R = All real numbers

b. $\{(x, y) \mid y = -3x - 2\}$
D = All real numbers; R = All real numbers

c. $\{(x, y) \mid y = x^2\}$
D = All real numbers; R = All nonnegative real numbers

23. [10.6B] Which of the following are functions?

a. $\{(-3, 1), (-4, 1), (-5, 2)\}$ A function

b. $\{(2, -4), (-1, 3), (-1, 4)\}$ Not a function

c. $\{(-1, 2), (-1, 3), (-1, 4)\}$ Not a function

24. [10.6C] If $f(x) = x^3 - 2x^2 + x - 1$, find

a. $f(2)$ $f(2) = 1$

b. $f(-2)$ $f(-2) = -19$

c. $f(1)$ $f(1) = -1$

25. [10.6D] The average price $P(n)$ of books depends on the number n of millions of books sold and is given by the function

$$P(n) = 25 - 0.3n \text{ (dollars)}$$

Find the average price of a book

a. when 10 million copies are sold. \$22

b. when 20 million copies are sold. \$19

c. when 30 million copies are sold. \$16

Practice Test 10

(Answers on pages 744–745)

1. Solve $x^2 = 64$.

2. Solve $49x^2 - 25 = 0$.

3. Solve $6x^2 + 49 = 0$.

4. Solve $36(x + 1)^2 - 7 = 0$.

5. Solve $9(x - 3)^2 + 1 = 0$.

6. Find the missing term in the expression $(x + 4)^2 = x^2 + 8x + \square$.

7. The missing terms in the expressions $x^2 - 8x + \square = (\quad)^2$ are ______ and ______, respectively.

8. To solve the equation $8x^2 - 32x = -5$ by completing the square, the first step will be to divide each term by ______.

9. To solve the equation $x^2 - 8x = -15$ by completing the square, one has to add ______ to both sides of the equation.

10. The solution of $ax^2 + bx + c = 0$ is ______.

11. Solve $2x^2 - 3x - 2 = 0$.

12. Solve $x^2 = 3x - 2$.

13. Solve $16x = x^2$.

14. Solve $\frac{x^2}{2} + \frac{5}{4}x = -\frac{1}{2}$.

15. Graph $y = 3x^2$.

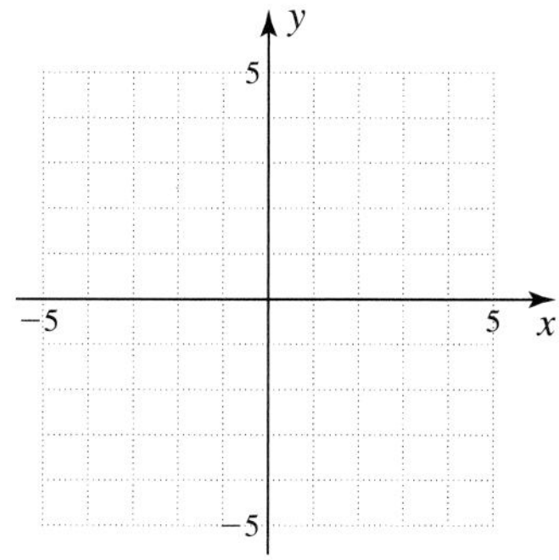

16. Graph $y = (x - 1)^2 + 1$.

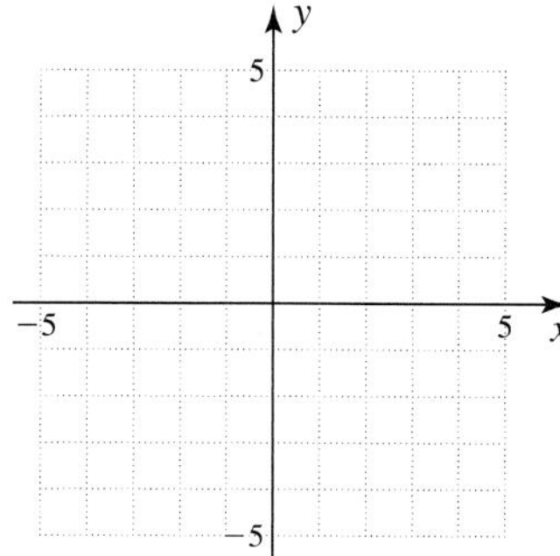

17. Graph the equation $y = -(x - 1)^2 + 1$.

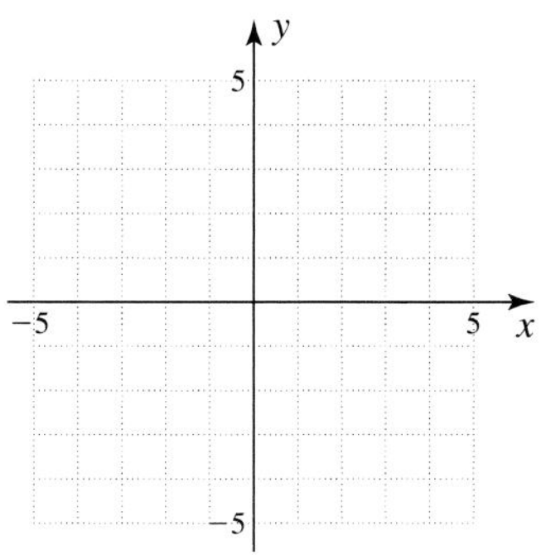

18. Graph the equation $y = -x^2 - 2x + 8$. Label the vertex and intercepts.

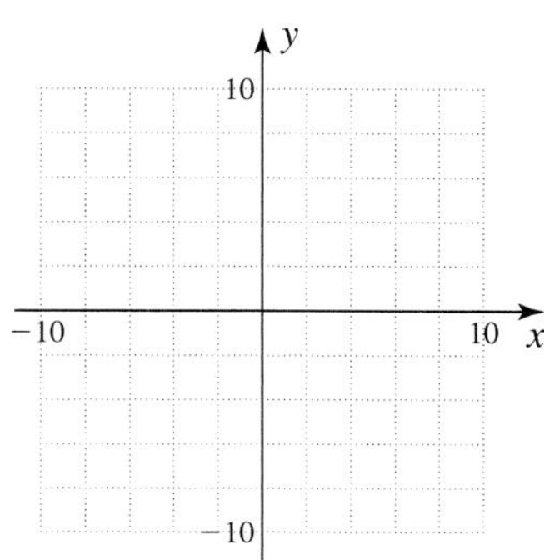

19. Find the length of the hypotenuse of a right triangle if the lengths of the two sides are 2 inches and 5 inches.

20. The formula $d = 5t^2 + v_0t$ gives the distance d (in meters) an object thrown downward with an initial velocity v_0 will have gone after t seconds. How long would it take an object dropped ($v_0 = 0$) from a distance of 180 meters to hit the ground?

21. Find the domain and range of

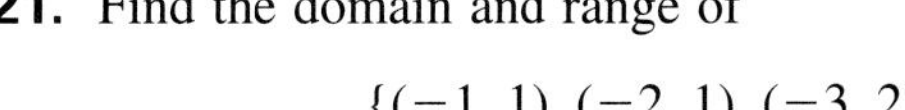

$$\{(-1, 1), (-2, 1), (-3, 2)\}$$

22. Find the domain and range of

$$\left\{(x, y) \,\middle|\, y = \frac{1}{x - 8}\right\}$$

23. State whether each of the following is a function.

a. $\{(2, -4), (-1, 3), (-1, 4)\}$

b. $\{(2, -4), (-1, 3), (1, -3)\}$

24. If $f(x) = x^3 + 2x^2 - x - 1$, find $f(-2)$.

25. The average price $P(n)$ of books depends on the number n of millions of books sold and is given by the function

$$P(n) = 25 - 0.4n \text{ (dollars)}$$

Find the average price of a book when 20 million copies are sold.

Answers to Practice Test

ANSWER	IF YOU MISSED: QUESTION	REVIEW: SECTION	REVIEW: EXAMPLES	REVIEW: PAGE
1. ± 8	1	10.1	1	688–689
2. $\pm\frac{5}{7}$	2	10.1	1, 2	688–689, 690
3. No real-number solution	3	10.1	3	691
4. $-1 \pm \frac{\sqrt{7}}{6} = \frac{-6 \pm \sqrt{7}}{6}$	4	10.1	4	692–693
5. No real-number solution	5	10.1	5	693
6. 16	6	10.2	1	699
7. 16; $x - 4$	7	10.2	2	700
8. 8	8	10.2	3, 4	702
9. 16	9	10.2	3, 4	702
10. $x = \frac{-b \pm \sqrt{b^2 - 4ac}}{2a}$	10	10.3	1	707
11. 2; $-\frac{1}{2}$	11	10.3	1	707
12. 1; 2	12	10.3	2	708
13. 0; 16	13	10.3	3	708–709
14. $-\frac{1}{2}$; -2	14	10.3	4	709–710
15. 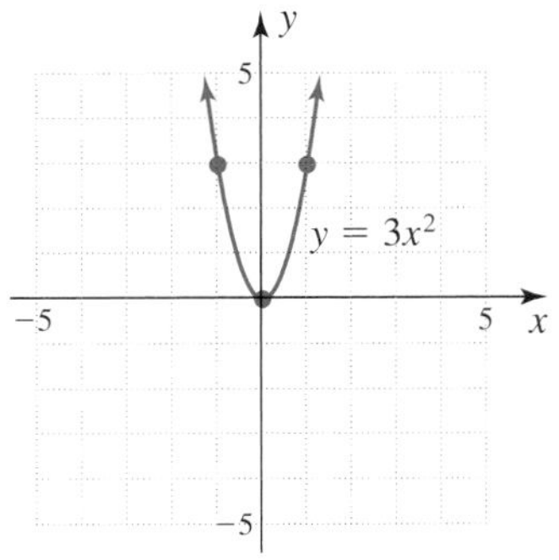 $y = 3x^2$	15	10.4	1	714
16. 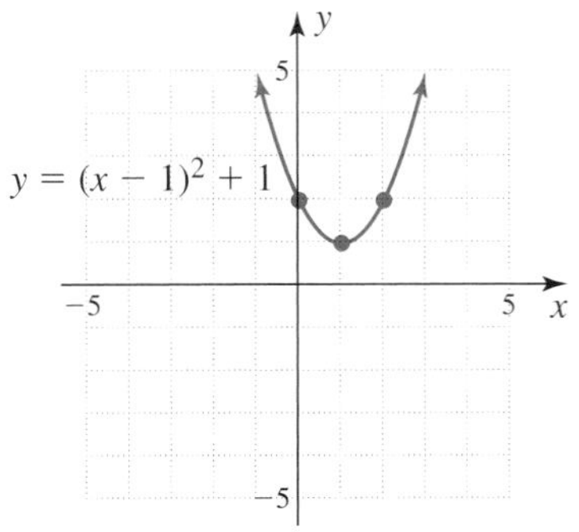 $y = (x - 1)^2 + 1$	16	10.4	3	716

ANSWER	IF YOU MISSED	REVIEW		
	QUESTION	SECTION	EXAMPLES	PAGE
17. $y = -(x-1)^2 + 1$	17	10.4	3	716
18. $(-1, 9)$; $(0, 8)$; $(-4, 0)$; $(2, 0)$	18	10.4	4	718
19. $\sqrt{29}$ inches	19	10.5	1	724–725
20. 6 seconds	20	10.5	2, 3	725–726
21. $D = \{-1, -2, -3\}$; $R = \{1, 2\}$	21	10.6	1	730
22. Domain: All real numbers except 8; Range: All real numbers except zero.	22	10.6	2	731
23. **a.** Not a function **b.** A function	23	10.6	3	731
24. $f(-2) = 1$	24	10.6	4, 5	732
25. \$17	25	10.6	7	733

Cumulative Review Chapters 1–10

1. Find: $-\frac{3}{8} + \left(-\frac{1}{7}\right)$ $-\frac{29}{56}$

2. Find: $(-4)^4$ 256

3. Find: $-\frac{3}{8} \div \left(-\frac{5}{24}\right)$ $\frac{9}{5}$

4. Evaluate $y \div 4 \cdot x - z$ for $x = 4$, $y = 16$, $z = 2$. 14

5. Simplify: $2x - (x + 4) - 2(x + 3)$
$-x - 10$

6. Write in symbols: The quotient of $(x - 5y)$ and z
$\frac{x - 5y}{z}$

7. Solve for x: $5 = 3(x - 1) + 5 - 2x$ $x = 3$

8. Solve for x: $\frac{x}{8} - \frac{x}{9} = 1$ $x = 72$

9. Graph: $-\frac{x}{4} + \frac{x}{8} \geq \frac{x-8}{8}$

0 4

10. Graph the point $C(-1, -2)$.

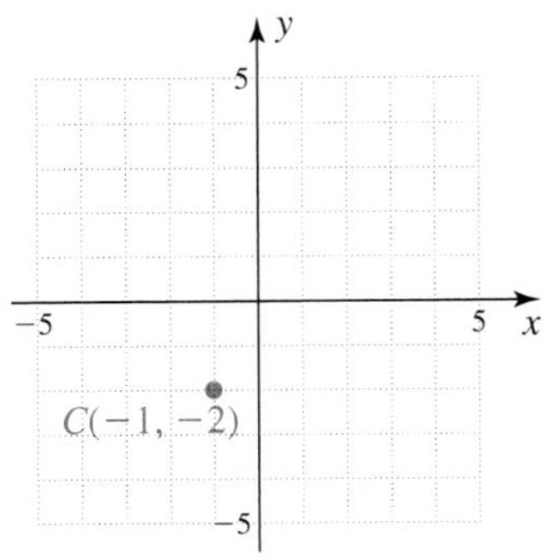

11. Determine whether the ordered pair $(-2, -3)$ is a solution of $2x - y = -1$. Yes

12. Find x in the ordered pair $(x, 2)$ so that the ordered pair satisfies the equation $2x - 3y = -10$. $x = -2$

13. Graph: $2x + y = 4$

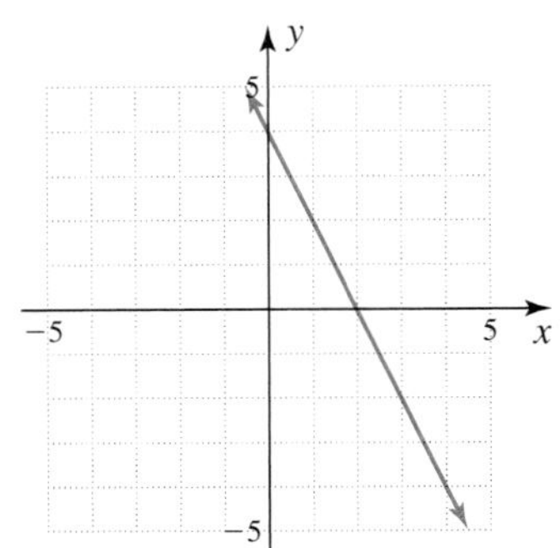

14. Graph: $2y - 8 = 0$

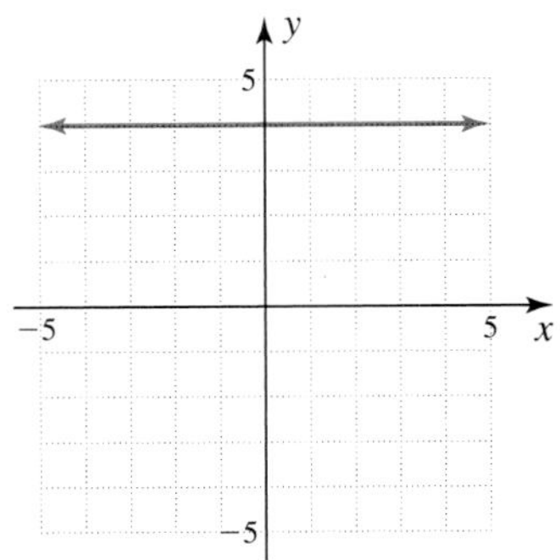

15. Find the slope of the line going through the points $(-5, -8)$ and $(6, 6)$. $\frac{14}{11}$

16. What is the slope of the line $8x - 4y = -13$? 2

17. Find the pair of parallel lines. (2) and (3)

(1) $8y + 12x = 7$

(2) $12x - 8y = 7$

(3) $-2y = -3x + 7$

18. Simplify: $(3x^3y^{-5})^4$ $\frac{81x^{12}}{y^{20}}$

19. Write in scientific notation: 0.0050 5.0×10^{-3}

20. Divide and express the answer in scientific notation: $(1.65 \times 10^{-4}) \div (1.1 \times 10^3)$ 1.5×10^{-7}

21. Find (expand): $\left(2x^2 - \frac{1}{5}\right)^2$ $4x^4 - \frac{4}{5}x^2 + \frac{1}{25}$

22. Divide $(2x^3 + 5x^2 - 4x - 2)$ by $(x + 3)$. $(2x^2 - x - 1)$ R 1

23. Factor completely: $x^2 - 8x + 15$ $(x - 5)(x - 3)$

24. Factor completely: $12x^2 - 25xy + 12y^2$ $(4x - 3y)(3x - 4y)$

25. Factor completely: $25x^2 - 36y^2$ $(5x + 6y)(5x - 6y)$

26. Factor completely: $-5x^4 + 80x^2$ $-5x^2(x + 4)(x - 4)$

27. Factor completely: $3x^3 - 3x^2 - 18x$ $3x(x + 2)(x - 3)$

28. Factor completely: $4x^2 + 12x + 5x + 15$ $(x + 3)(4x + 5)$

29. Factor completely: $25kx^2 + 20kx + 4k$ $k(5x + 2)^2$

30. Solve for x: $3x^2 + 13x = 10$ $x = -5$; $x = \frac{2}{3}$

31. Write $\frac{6x}{5y}$ with a denominator of $10y^3$. $\frac{12xy^2}{10y^3}$

32. Reduce to lowest terms: $\frac{-4(x^2 - y^2)}{4(x - y)}$ $-(x + y)$

33. Reduce to lowest terms: $\frac{x^2 - 4x - 21}{7 - x}$ $-(x + 3)$

34. Multiply: $(x - 4) \cdot \frac{x - 3}{x^2 - 16}$ $\frac{x-3}{x+4}$

35. Divide: $\frac{x + 2}{x - 2} \div \frac{x^2 - 4}{2 - x}$ $\frac{1}{2-x}$

36. Add: $\frac{5}{4(x + 1)} + \frac{7}{4(x + 1)}$ $\frac{3}{x+1}$

37. Subtract: $\frac{x + 7}{x^2 + x - 56} - \frac{x + 8}{x^2 - 49}$ $\frac{-2x-15}{(x+8)(x+7)(x-7)}$

38. Simplify: $\dfrac{\frac{1}{x} + \frac{3}{2x}}{\frac{2}{3x} - \frac{1}{4x}}$ 6

39. Solve for x: $\frac{x}{x - 3} + 6 = \frac{5x}{x - 3}$ $x = 9$

40. Solve for x: $\frac{x}{x^2 - 16} + \frac{4}{x - 4} = \frac{1}{x + 4}$ $x = -5$

41. Solve for x: $\frac{x}{x + 4} - \frac{1}{5} = \frac{-4}{x + 4}$ No solution

42. Solve for x: $2 + \frac{8}{x - 2} = \frac{32}{x^2 - 4}$ $x = -6$

43. A bus travels 250 miles on 10 gallons of gas. How many gallons will it need to travel 725 miles? 29 gal

44. Solve for x: $\frac{x - 8}{4} = \frac{6}{5}$ $x = \frac{64}{5}$

45. Janet can paint a kitchen in 4 hours and James can paint the same kitchen in 5 hours. How long would it take for both working together to paint the kitchen? $2\frac{2}{9}$ hr

46. Find an equation of the line that goes through the point $(-2, -4)$ and has slope $m = 6$. $y = 6x + 8$

47. Find an equation of the line having slope -2 and y-intercept -5. $y = -2x - 5$

48. Graph: $3x - 2y < -6$

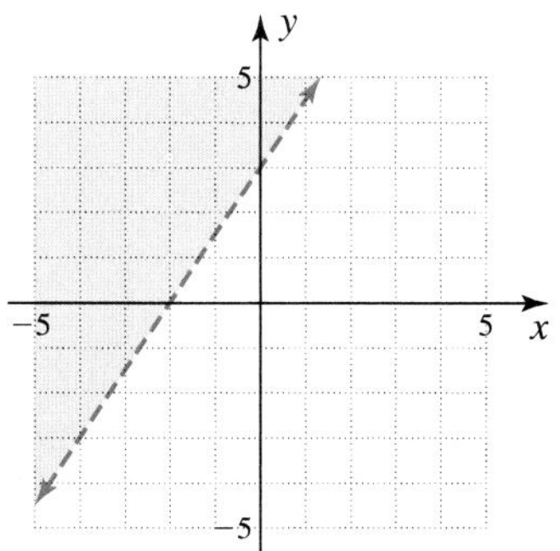

49. Graph: $-y \geq -5x + 5$

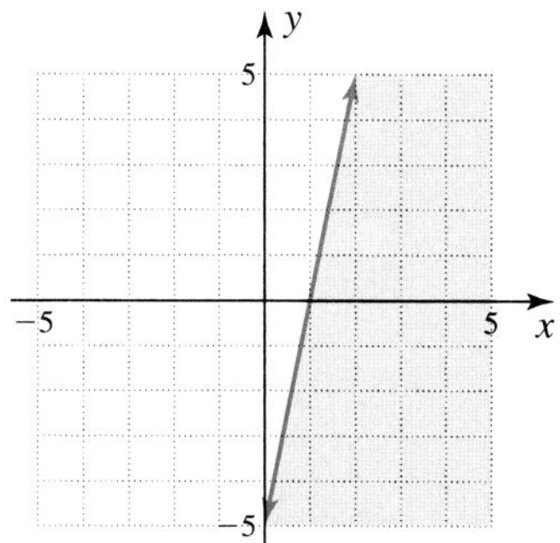

50. An enclosed gas exerts a pressure P on the walls of the container. This pressure is directly proportional to the temperature T of the gas. If the pressure is 7 lb/in.^{2} when the temperature is 490°F, find k. $\frac{1}{70}$

51. If the temperature of a gas is held constant, the pressure P varies inversely as the volume V. A pressure of 1960 lb/in.^{2} is exerted by 7 ft^3 of air in a cylinder fitted with a piston. Find k. 13,720

52. Graph the system:

$x + 3y = -12$
$2y - x = -2$

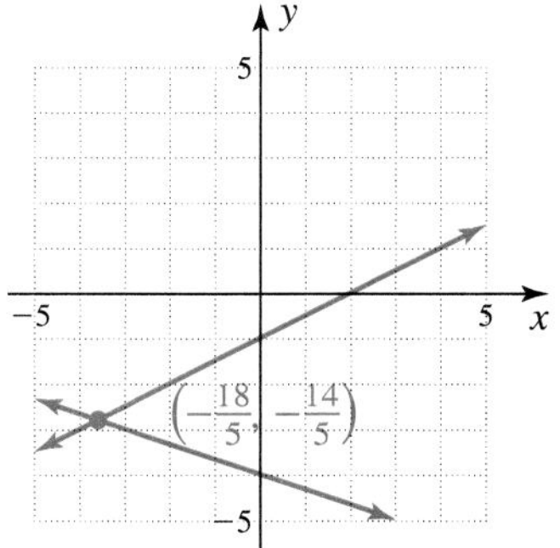

53. Graph the system and find the solution if possible:

$y + 3x = 3$ No solution
$3y + 9x = 18$

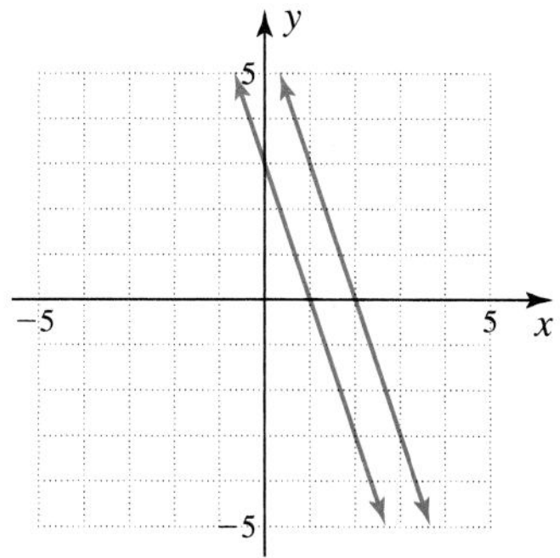

54. Solve (if possible) by substitution:

$x + 3y = 10$
$4x + 12y = 42$ Inconsistent; no solution

55. Solve (if possible) by substitution:

$x - 3y = 3$
$-3x + 9y = -9$ Dependent; infinitely many solutions

56. Solve the system (if possible):

$3x - 4y = 3$
$2x - 3y = 3$ $(-3, -3)$

57. Solve the system (if possible):

$4x + 3y = 18$
$8x + 6y = -3$ Inconsistent; no solution

58. Solve the system (if possible):

$3y + x = -13$
$-4x - 12y = 52$
Dependent; infinitely many solutions

59. Kaye has \$2.50 in nickels and dimes. She has twice as many dimes as nickels. How many nickels and how many dimes does she have? 10 nickels and 20 dimes

60. The sum of two numbers is 170. Their difference is 110. What are the numbers? 30 and 140

61. Evaluate: $\sqrt[5]{243}$ 3

62. Simplify: $\sqrt[4]{(-5)^4}$ 5

63. Simplify: $\sqrt{\dfrac{5}{243}}$ $\dfrac{\sqrt{15}}{27}$

64. Rationalize the denominator: $\dfrac{\sqrt{2}}{\sqrt{7j}}$ $\dfrac{\sqrt{14j}}{7j}$

65. Simplify: $\sqrt{32} + \sqrt{18}$ $7\sqrt{2}$

66. Perform the indicated operations: $\sqrt[3]{3x}\,(\sqrt[3]{9x^2} - \sqrt[3]{16x})$
$3x - 2\sqrt[3]{6x^2}$

67. Find the product: $(\sqrt{125} + \sqrt{63})(\sqrt{245} + \sqrt{175})$
$280 + 46\sqrt{35}$

68. Reduce: $\dfrac{20 - \sqrt{32}}{4}$ $5 - \sqrt{2}$

69. Solve: $\sqrt{x + 4} = -3$ No real-number solution

70. Solve: $\sqrt{x - 4} - x = -4$ $x = 4; x = 5$

71. Solve for x: $9x^2 - 4 = 0$ $x = \dfrac{2}{3}; x = -\dfrac{2}{3}$

72. Solve for x: $64x^2 + 9 = 0$ No real-number solution

73. Solve for x: $36(x - 4)^2 - 7 = 0$
$x = 4 + \dfrac{\sqrt{7}}{6}; x = 4 - \dfrac{\sqrt{7}}{6}$

74. Find the missing term: $(x + 2)^2 = x^2 + 4x +$ ___4___

75. Find the number that must divide each term in the equation so that the equation can be solved by completing the square: $4x^2 + 8x = 14$ 4

76. Find the number that must be added to both sides of the equation so that the equation can be solved by the method of completing the square: $x^2 + 4x = 5$ 4

77. Solve for x: $jx^2 + kx + m = 0$
$x = \dfrac{-k + \sqrt{k^2 - 4jm}}{2j}; x = \dfrac{-k - \sqrt{k^2 - 4jm}}{2j}$

78. Solve for x: $2x^2 + 3x - 9 = 0$ $x = \dfrac{3}{2}; x = -3$

79. Solve for x: $16x = x^2$ $x = 0; x = 16$

80. Graph: $y = -(x + 3)^2 + 2$

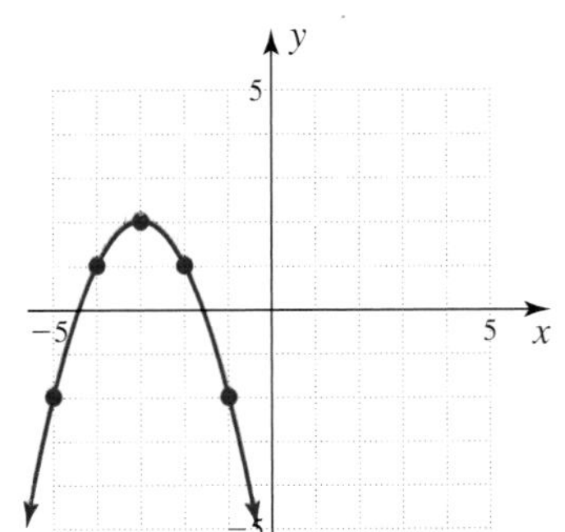

81. The distance d (in meters, m) traveled by an object thrown downward with an initial velocity of v_0 after t seconds is given by the formula $d = 5t^2 + v_0t$. Find the number of seconds it takes an object to hit the ground if the object is dropped ($v_0 = 0$) from a height of 20 m. 2 sec

Selected Answers

The brackets preceding answers for the Chapter Review Exercises indicate the Chapter, Section, and Objective for you to review for further study. For example, [3.4C] appearing before answers means those exercises correspond to Chapter 3, Section 4, Objective C.

Chapter R

Exercises R.1

1. $\frac{28}{1}$ **3.** $\frac{-42}{1}$ **5.** $\frac{0}{1}$ **7.** $\frac{-1}{1}$ **9.** 3 **11.** 42 **13.** 25 **15.** 21 **17.** 32 **19.** 6 **21.** 14 **23.** 30 **25.** 3 **27.** 1 **29.** 2 **31.** $\frac{5}{4}$ **33.** $\frac{1}{4}$ **35.** $\frac{7}{3}$ **37.** $\frac{2}{3}$ **39.** $\frac{4}{1} = 4$ **41. (a)** $\frac{1000}{45}$; **(b)** $\frac{200}{9}$; **(c)** $22\frac{2}{9}$; **(d)** $\frac{1000}{24} = \frac{125}{3} = 41\frac{2}{3}$ **43.** $\frac{41}{100}$ **45. (a)** $\frac{2}{6}$; **(b)** 2 pieces; **(c)** $\frac{2}{6} = \frac{1}{3}$; **(d)** They ate the same amount. **47.** $\frac{1}{8}$ **49.** $\frac{5}{8}$ **51. (a)** 2880 seconds; **(b)** 120 shots; **(c)** 24 **53.** $\frac{2}{15}$ **55.** Nat'l & Int'l News, $\frac{1}{2}$ **57.** $\frac{7}{30}$ **59.** Most: News & Beyond the Bay, $\frac{13}{30}$; Least: Traffic, $\frac{1}{10}$

Exercises R.2

1. $\frac{14}{9}$ **3.** $\frac{7}{5}$ **5.** $\frac{2}{1}$ or 2 **7.** $\frac{8}{1}$ or 8 **9.** $\frac{39}{7}$ **11.** $\frac{35}{3}$ **13.** $\frac{2}{3}$ **15.** $\frac{3}{2}$ **17.** $\frac{16}{5}$ **19.** $\frac{2}{5}$ **21.** $\frac{3}{5}$ **23.** $\frac{1}{1}$ or 1 **25.** $\frac{13}{8}$ **27.** $\frac{17}{15}$ **29.** $\frac{23}{6}$ **31.** $\frac{31}{10}$ **33.** $\frac{193}{28}$ **35.** $\frac{3}{8}$ **37.** $\frac{1}{6}$ **39.** $\frac{11}{20}$ **41.** $\frac{34}{75}$ **43.** $\frac{9}{20}$ **45.** $\frac{5}{4}$ **47.** $\frac{11}{18}$ **49.** $\frac{37}{15}$ **51.** 75 lb **53.** 5600 **55.** $16\frac{5}{8}$ **57.** $\frac{3}{10}$ **59.** $7\frac{7}{10}$ **61.** 4 pkg of hot dogs, 5 pkg of buns **67.** 8 **69.** $\frac{1}{5}$

Exercises R.3

1. $4 + \frac{7}{10}$ **3.** $5 + \frac{6}{10} + \frac{2}{100}$ **5.** $10 + 6 + \frac{1}{10} + \frac{2}{100} + \frac{3}{1000}$ **7.** $40 + 9 + \frac{1}{100} + \frac{2}{1000}$ **9.** $50 + 7 + \frac{1}{10} + \frac{4}{1000}$ **11.** $\frac{9}{10}$ **13.** $\frac{3}{50}$ **15.** $\frac{3}{25}$ **17.** $\frac{27}{500}$ **19.** $\frac{213}{100}$ **21.** $\$989.07$ **23.** 919.154 **25.** 182.103 **27.** 4.077 **29.** 26.85 **31.** $\$2.38$ **33.** 3.024 **35.** 6.844 **37.** 9.0946 **39.** 12.0735 **41.** 5.6396 **43.** 95.7 **45.** 0.024605 **47.** 12.90516 **49.** 0.002542 **51.** 0.6 **53.** 6.4 **55.** 1700 **57.** 80 **59.** 0.046 **61.** 0.2 **63.** 0.875 **65.** 0.1875 **67.** $0.\overline{2}$ **69.** $0.\overline{54}$ **71.** $0.\overline{27}$ **73.** $0.1\overline{6}$ **75.** $1.\overline{1}$ **77.** 0.33 **79.** 0.05 **81.** 3 **83.** 0.118 **85.** 0.005 **87.** 5% **89.** 39% **91.** 41.6% **93.** 0.3% **95.** 100% **97.** $\frac{3}{10}$ **99.** $\frac{3}{50}$ **101.** $\frac{7}{100}$ **103.** $\frac{9}{200}$ **105.** $\frac{1}{75}$ **107.** 60% **109.** 50% **111.** $83\frac{1}{3}\%$ **113.** 50% **115.** $133\frac{1}{3}\%$ **117.** 27.6 **119.** 26.7467 **121.** 35.250 **123.** 52.38 **125.** 74.84601 **127.** $\frac{9}{25}$ **129.** $\frac{12}{25}$ **131.** 70% **133. (a)** 40%; **(b)** $\frac{2}{5}$ **135. (a)** $\frac{49}{100}$; **(b)** 0.49 **137.** 5% **139.** 1%

Chapter 1

Exercises 1.1

1. $a + c$ **3.** $3x + y$ **5.** $9x + 17y$ **7.** $3a - 2b$ **9.** $-2x - 5$ **11.** $7a$ **13.** $\frac{1}{7}a$ **15.** bd **17.** xyz **19.** $-b(c + d)$ **21.** $(a - b)x$ **23.** $(x - 3y)(x + 7y)$ **25.** $(c - 4d)(x + y)$ **27.** $\frac{y}{3x}$ **29.** $\frac{2b}{a}$ **31.** $\frac{a}{x + y}$ **33.** $\frac{a - b}{c}$ **35.** $\frac{y}{x}$ **37.** $\frac{p - q}{p + q}$ **39.** $\frac{x + 2y}{x - 2y}$ **41.** 16 **43.** 61 **45.** 9 **47.** 3 **49.** 56 **51.** 6 **53.** 50 **55.** 1 **57.** 4 **59.** $\frac{1}{4}$ **61.** $V = IR$ **63.** $P = P_A + P_B + P_C$ **65.** $D = RT$ **67.** $E = mc^2$ **69.** $c^2 = a^2 + b^2$ **75.** $3xy$ **77.** $\frac{3x}{2y}$ **79.** $\frac{b - c}{b + c}$ **81.** 2

Exercises 1.2

1. -4 **3.** 49 **5.** $-\frac{7}{3}$ **7.** 6.4 **9.** $-3\frac{1}{7}$ **11.** -0.34 **13.** $0.\overline{5}$ **15.** $-\sqrt{7}$ **17.** $-\pi$ **19.** 2 **21.** 48 **23.** 3 **25.** $\frac{4}{5}$ **27.** 3.4 **29.** $1\frac{1}{2}$ **31.** $-\frac{3}{4}$ **33.** $-0.\overline{5}$ **35.** $-\sqrt{3}$ **37.** $-\pi$ **39.** Natural, whole, integer, rational, real **41.** Rational, real **43.** Whole, integer, rational, real **45.** Rational, real **47.** Rational, real **49.** Irrational, real **51.** Rational, real **53.** Rational, real **55.** 8 **57.** 8 **59.** 0, 8 **61.** $-5, \frac{1}{5}, 0, 8, 0.\overline{1}, 3.666\ldots$ **63.** True **65.** False; $|0| = 0$ **67.** True **69.** False; $\frac{3}{5}$ is not an integer. **71.** False; 0.12345. . . is not rational. **73.** True **75.** +20 yards **77.** -1312 feet from sea level **79.** -4°F to $+45$°F **81.** 5.2 **83.** $\frac{9}{40}$ **85.** 13.02 **87.** $\frac{15}{8}$ **89.** -43°F **91.** -40°F **99.** $\sqrt{19}$ **101.** $8\frac{1}{4}$ **103.** $0.\overline{4}$ **105.** $-\frac{1}{2}$ **107.** Integer, rational, real **109.** Rational, real **111.** Whole, integer, rational, real

Exercises 1.3

1. 6 **3.** -4 **5.** 1 **7.** -7 **9.** 0 **11.** 3 **13.** 13 **15.** 2 **17.** -9 **19.** -12 **21.** 3.1 **23.** -4.7 **25.** -5.4 **27.** -8.6 **29.** $\frac{3}{7}$ **31.** $-\frac{1}{2}$ **33.** $-\frac{1}{12}$ **35.** $\frac{7}{12}$ **37.** $-\frac{13}{21}$ **39.** $-\frac{31}{18}$ **41.** -16 **43.** -20 **45.** -6 **47.** 16 **49.** -4 **51.** -2.6 **53.** 5.2 **55.** -3.7 **57.** $\frac{4}{7}$ **59.** $-\frac{29}{12}$ **61.** -7 **63.** 9 **65.** -4 **67.** 3500°C **69.** $\$46$ **71.** 14°C **73.** $\frac{703}{20}$ or $35\frac{3}{20}$ **75.** $\frac{36}{121}$ **77.** 799 years **79.** 284 years **81.** 2262 years **87.** -3.1 **89.** -7.7 **91.** -20 **93.** 1.2 **95.** -0.3 **97.** -5.7 **99.** $\frac{14}{5}$

Exercises 1.4

1. 36 **3.** −40 **5.** −81 **7.** −36 **9.** −54 **11.** −7.26 **13.** 2.86 **15.** $-\frac{25}{42}$ **17.** $\frac{1}{4}$ **19.** $-\frac{15}{4}$ **21.** −16 **23.** 25 **25.** −125 **27.** 1296 **29.** $-\frac{1}{32}$ **31.** 7 **33.** −5 **35.** −3 **37.** 0 **39.** Undefined **41.** 0 **43.** 5 **45.** 5 **47.** −2 **49.** −6 **51.** $-\frac{21}{20}$ **53.** $\frac{4}{7}$ **55.** $-\frac{5}{7}$ **57.** −0.5 or $-\frac{1}{2}$ **59.** $0.1\overline{6}$ or $\frac{1}{6}$ **61.** 15 calories; gain **63.** 5 calories; loss **65.** 6 minutes **67.** 15.4 mi/hr each second **69.** \$4.45 **71.** 1 **73.** 0 **75.** −1.35 **77.** −3.045 **79.** 3.875 **85.** 0.25 or $\frac{1}{4}$ **87.** −33 **89.** 55 **91.** 64 **93.** −7.04 **95.** $\frac{12}{35}$ **97.** $-\frac{21}{20}$ **99.** $-\frac{1}{2}$

Exercises 1.5

1. 26 **3.** 13 **5.** 53 **7.** 5 **9.** 3 **11.** 10 **13.** 7 **15.** 3 **17.** 8 **19.** 10 **21.** 10 **23.** 20 **25.** 27 **27.** −31 **29.** 36 **31.** −5 **33.** −5 **35.** 25 **37.** −15 **39.** −24 **41.** 87 **43.** (a) 144; (b) 126 **45.** 1 **47.** 0 **49.** 1 **51.** 5 milligrams **53.** $1\frac{1}{3}$ tablets every 12 hours **57.** 0 **59.** 18 **61.** 38 **63.** 10 **65.** 119

Exercises 1.6

1. Commutative property of addition
3. Commutative property of multiplication **5.** Distributive property
7. Commutative property of multiplication
9. Associative property of addition
11. Associative property of addition; 5
13. Commutative property of multiplication; 7
15. Commutative property of addition; 6.5
17. Associative property of multiplication; 2 **19.** $13 + 2x$ **21.** 0 **23.** 1 **25.** Multiplicative inverse property **27.** Identity property for addition **29.** 3 **31.** 5 **33.** a **35.** −2 **37.** −2 **39.** $24 + 6x$ **41.** $8x + 8y + 8z$ **43.** $6x + 42$ **45.** $ab + 5b$ **47.** $30 + 6b$ **49.** $-4x - 4y$ **51.** $-9a - 9b$ **53.** $-12x - 6$ **55.** $-\frac{3a}{2} + \frac{6}{7}$ **57.** $-2x + 6y$ **59.** $-2.1 - 3y$ **61.** $-4a - 20$ **63.** $-6x - xy$ **65.** $-8x + 8y$ **67.** $-6a + 21b$ **69.** $0.5x + 0.5y - 1.0$ **71.** $\frac{6}{5}a - \frac{6}{5}b + 6$ **73.** $-2x + 2y - 8$ **75.** $-0.3x - 0.3y + 1.8$ **77.** $-\frac{5}{2}a + 5b - \frac{5}{2}c + \frac{5}{2}$ **79.** 81 **81.** 81 **83.** −27 **85.** 64 **87.** 266 **89.** 276 **97.** $-4a + 20$ **99.** $4a + 24$ **101.** 7 **103.** $16 + 4x$ **105.** 0 **107.** 1 **109.** 2
111. Identity property for multiplication
113. Identity property for addition
115. Associative property for multiplication
117. Associative property for addition
119. Associative property for multiplication; 2
121. Commutative property for multiplication; 1
123. Associative property for addition; 3

Exercises 1.7

1. $11a$ **3.** $-5c$ **5.** $12n^2$ **7.** $-7ab^2$ **9.** $3abc$ **11.** $1.3ab$ **13.** $0.2x^2y - 0.9xy^2$ **15.** $3ab^2c$ **17.** $-6ab + 7xy$ **19.** $\frac{3}{5}a + \frac{4}{7}a^2b$ **21.** $11x$ **23.** $8ab$ **25.** $-7a^2b$ **27.** 0 **29.** 0 **31.** $10xy - 4$ **33.** $-2R + 6$ **35.** $-L - 2W$ **37.** $-3x - 1$ **39.** $\frac{x}{9} + 2$ **41.** $6a + 2b$ **43.** $3x - 4y$ **45.** $x - 5y + 36$ **47.** $-2x^3 + 9x^2 - 3x + 12$ **49.** $x^2 - \frac{2}{5}x + \frac{1}{2}$ **51.** $10a - 18$ **53.** $-23a + 18b + 9$ **55.** $-4.8x + 3.4y + 5$ **57.** $\frac{1}{f} = \frac{1}{u} + \frac{1}{v}$; 2 terms **59.** $F = \frac{n}{4} + 40$; 2 terms **61.** $I = Prt$; 1 term **63.** $P = a + b + c$; 3 terms **65.** $A = \frac{bh}{2}$; 1 term **67.** $K = Cmv^2$; 1 term **69.** $d = 16t^2$; 1 term **71.** $\frac{5}{8}$ **73.** $\frac{11}{20}$ **75.** $4S$; $P = 4S$ **77.** 18 ft, 6 in. **79.** $2h + 2w + L$ **81.** $(7x + 2)$ in. **87.** $A = LW$; 1 term **89.** $14y - 11$ **91.** $-3ab^3$ **93.** $-3x^2 + 12x + 6$

Review Exercises

1. [1.1A] **(a)** $a + b$; **(b)** $a - b$; **(c)** $7a + 2b - 8$ **2.** [1.1A] **(a)** $3m$; **(b)** $-mnr$; **(c)** $\frac{1}{7}m$; **(d)** $8m$ **3.** [1.1A] **(a)** $\frac{m}{9}$; **(b)** $\frac{9}{n}$ **4.** [1.1A] **(a)** $\frac{m+n}{r}$; **(b)** $\frac{m+n}{m-n}$ **5.** [1.1B] **(a)** 12; **(b)** 6; **(c)** 36 **6.** [1.1B] **(a)** 3; **(b)** 9; **(c)** 7 **7.** [1.2A] **(a)** 5; **(b)** $-\frac{2}{3}$; **(c)** −0.37 **8.** [1.2B] **(a)** 8; **(b)** $3\frac{1}{2}$; **(c)** −0.76 **9.** [1.2C] **(a)** Integer, rational, real; **(b)** Whole, integer, rational, real; **(c)** Rational, real; **(d)** Irrational, real; **(e)** Irrational, real; **(f)** Rational, real **10.** [1.3A] **(a)** 2; **(b)** −0.8; **(c)** $-\frac{11}{20}$; **(d)** −2.2; **(e)** $\frac{1}{4}$ **11.** [1.3B] **(a)** −20; **(b)** −2.4; **(c)** $-\frac{17}{12}$ **12.** [1.3C] **(a)** 30; **(b)** −15 **13.** [1.4A] **(a)** −35; **(b)** −18.4; **(c)** −19.2; **(d)** $\frac{12}{35}$ **14.** [1.4B] **(a)** 16; **(b)** −9; **(c)** $-\frac{1}{27}$; **(d)** $\frac{1}{27}$ **15.** [1.4C] **(a)** −4; **(b)** 2; **(c)** $-\frac{2}{3}$; **(d)** $\frac{3}{2}$ **16.** [1.5A] **(a)** 10; **(b)** 0 **17.** [1.5B] **(a)** 8; **(b)** −41 **18.** [1.5C] 152 **19.** [1.6A] **(a)** Commutative property of addition; **(b)** Associative property of multiplication; **(c)** Associative property of addition **20.** [1.6B] **(a)** $4x - 4$; **(b)** $-8x + 2$ **21.** [1.6C, E] **(a)** 1; **(b)** $-\frac{3}{4}$; **(c)** −3.7; **(d)** $\frac{2}{3}$; **(e)** 0; **(f)** −5 **22.** [1.6E] **(a)** $-3a - 24$; **(b)** $-4x + 20$; **(c)** $-x + 4$ **23.** [1.7A] **(a)** $-5x$; **(b)** $-10x$; **(c)** $-8x$; **(d)** $3a^2b$ **24.** [1.7B] **(a)** $12a - 5$; **(b)** $5x - 9$ **25.** [1.7C] **(a)** $m = \frac{p}{n}(2.75)$; one term; **(b)** $W = \frac{11h}{2} - 220$; two terms

Chapter 2

Exercises 2.1

1. Yes **3.** Yes **5.** Yes **7.** No **9.** No **11.** $x = 14$ **13.** $m = 19$ **15.** $y = \frac{10}{3}$ **17.** $k = 21$ **19.** $z = \frac{11}{12}$ **21.** $x = \frac{7}{2}$ **23.** $c = 0$ **25.** $x = -3$ **27.** $y = 0$ **29.** $c = -2.5$ **31.** $p = -9$ **33.** $x = -3$ **35.** $m = 6$ **37.** $y = 18$ **39.** $a = -7$ **41.** $c = -8$ **43.** $x = -2$ **45.** $g = -3$ **47.** No solution **49.** All real numbers **51.** $b = 6$ **53.** $p = \frac{20}{3}$ **55.** $r = \frac{9}{8}$ **57.** \$9.41 **59.** 184.7 **61.** 87.8 million tons; 142.1 million tons **63.** 21% **65.** 27% **67.** −20 **69.** $-\frac{1}{2}$ **71.** $\frac{2}{3}$ **73.** 48 **75.** 40 **77.** No **79.** No **85.** No solution **87.** $x = 9$ **89.** $x = 5.7$ **91.** $x = 4$ **93.** $x = \frac{7}{9}$ **95.** $z = 7$ **97.** No solution **99.** No

Exercises 2.2

1. $x = 35$ **3.** $x = -8$ **5.** $b = -15$ **7.** $f = 6$ **9.** $v = \frac{4}{3}$ **11.** $x = -\frac{15}{4}$ **13.** $z = 11$ **15.** $x = -7$ **17.** $c = -7$ **19.** $x = 7$ **21.** $y = -\frac{11}{3}$ **23.** $a = -0.6$ **25.** $t = \frac{3}{2}$ **27.** $x = -2.25$

29. $C = -8$ **31.** $a = 12$ **33.** $y = -0.5$ **35.** $p = 0$ **37.** $t = -30$ **39.** $x = -0.02$ **41.** $y = 12$ **43.** $x = 21$ **45.** $x = 20$ **47.** $t = 24$ **49.** $x = 1$ **51.** $c = 15$ **53.** $W = 10$ **55.** $x = 4$ **57.** $x = 10$ **59.** $x = 3$ **61.** 12 **63.** 28 **65.** 50% **67.** 150 **69.** 20 **71.** 7.6% **73.** 9.5% **75.** 3.8% **77.** 5.3% **79.** 3.8% **81.** \$24 **83.** \$12 **85.** 150 calories **87.** $40 - 5y$ **89.** $54 - 27y$ **91.** $-15x + 20$ **93.** 4 **95.** 3 **97.** \$16,000 **103.** 20% **105.** $x = 6$ **107.** $y = 30$ **109.** $y = 10$ **111.** $y = 14$ **113.** $x = -14$

Exercises 2.3

1. $x = 4$ **3.** $y = 1$ **5.** $z = 2$ **7.** $y = \frac{14}{5}$ **9.** $x = 3$ **11.** $x = -4$ **13.** $v = -8$ **15.** $m = -4$ **17.** $z = -\frac{2}{3}$ **19.** $x = 0$ **21.** $a = \frac{1}{13}$ **23.** $c = 35.6$ **25.** $x = -\frac{11}{80}$ **27.** $x = \frac{9}{2}$ **29.** $x = -20$ **31.** $x = -5$ **33.** $h = 0$ **35.** $w = -1$ **37.** $x = -3$ **39.** $x = \frac{5}{2}$ **41.** All real numbers **43.** $x = 2$ **45.** $x = 10$ **47.** $x = \frac{27}{20}$ **49.** $x = \frac{69}{7}$ **51.** $r = \frac{C}{2\pi}$ **53.** $y = \frac{6 - 3x}{2}$ **55.** $s = \frac{A - \pi r^2}{\pi r}$ **57.** $V_2 = \frac{P_1V_1}{P_2}$ **59.** $H = \frac{f + Sh}{S}$ **61. (a)** $L = \frac{S + 22}{3}$; **(b)** $L = \frac{S + 21}{3}$ **63.** $H = \frac{W + 200}{5}$ **65.** 11 inches **67.** Size 8 **69.** 15 hr **71.** $\frac{a + b}{c}$ **73.** $a(b + c)$ **75.** $a - bc$ **77.** $66\frac{2}{3}$ miles **79. (a)** $S = 1.2C$; **(b)** \$9.60 **85.** $m = 150$ **87.** $x = -1$ **89.** $x = 1$ **91.** $x = 1$

Exercises 2.4

1. 44, 46, 48 **3.** −10, −8, −6 **5.** −13, −12 **7.** 7, 8, 9 **9.** Any consecutive integers **11.** \$20, \$44 **13.** 39 to 94 **15.** \$104, \$43 **17.** 37, 111, 106 **19.** 21, 126 **21.** 38% from home, 13% from work **23.** \$392 per week **25.** 222 feet **27.** 3 seconds **29.** 304 and 676 **31.** 10 minutes **33. (a)** 22 miles; **(b)** The limo is cheaper. **35.** 20° **37.** 45° **39.** 42° **41.** $T = \frac{20}{11}$ **43.** $T = 8$ **45.** $x = 8$ **47.** $P = 3500$ **49.** $x = 1000$ **51.** 84 **55.** 81, 83, 85 **57.** Pizza: 230 calories, Shake: 300 calories

Exercises 2.5

1. 50 mi/hr **3.** 2 meters/min **5.** $\frac{1}{2}$ hr **7.** 4 hr **9.** 10 mi **11.** 6 hr **13.** 1920 mi **15.** 24 in./sec first 180 sec, 1.2 in./sec last 60 sec **17.** 6 liters **19.** 30 lb copper, 40 lb zinc **21.** 20 pounds **23.** 32 ounces of each **25.** 12 quarts **27.** Impossible; it would need 120% of concentrate. **29.** \$5000 at 6%, \$15,000 at 8% **31.** \$8000 **33.** \$6000 at 5%, \$4000 at 6% **35.** \$25,000 at 6%, \$15,000 at 10% **37.** 54 **39.** 4 **45.** 10 gallons **47.** 5 hr

Exercises 2.6

1. (a) 120; **(b)** 275 mi; **(c)** $R = \frac{D}{T}$; **(d)** 60 mi/hr **3. (a)** 110 pounds; **(b)** $H = \frac{W + 190}{5}$; **(c)** 78 inches tall (6 ft, 6 in.) **5. (a)** 59°F; **(b)** $C = \frac{5}{9}(F - 32)$; **(c)** 10°C **7. (a)** EER $= \frac{\text{Btu}}{w}$; **(b)** 9; **(c)** Btu = (EER)(w); **(d)** 20,000 **9. (a)** $S = C + M$; **(b)** \$67; **(c)** $M = S - C$; **(d)** \$8.25 **11. (a)** 60 cm; **(b)** 90 cm **13. (a)** 62.8 in.; **(b)** $r = \frac{C}{2\pi}$; **(c)** 10 in. **15. (a)** 13.02 m^2; **(b)** $W = \frac{A}{L}$; **(c)** 6 m **17.** 35° each **19.** 140° each **21.** 175° each **23.** 70°, 20° **25.** 23°, 67° **27.** 142°, 38° **29.** 69°, 111° **31.** 75 m **33.** 4.5 in. **35. (a)** $A = C - 10$; **(b)** 40 **37. (a)** 13.74 million; **(b)** $t = \frac{N - 9.74}{0.40}$ or $t = \frac{5(N - 9.74)}{2}$; **(c)** 1995 **39.** $x = 4$ **41.** $x = 3$ **43.** $x = -6$ **45.** 90 ft **53.** 44°, 46° **55.** 128°, 52° **57.** 46°, 134° **59. (a)** 126.48 cm; **(b)** $h = \frac{H - 71.48}{2.75}$; **(c)** 25 cm **61. (a)** 21 thousand; **(b)** $t = \frac{N - 15}{0.60}$ or $t = \frac{5(N - 15)}{3}$; **(c)** 1995 **63. (a)** 5°C; **(b)** $F = \frac{9}{5}C + 32$; **(c)** 68°F

Exercises 2.7

1. < **3.** > **5.** < **7.** > **9.** <

11. $x \le 1$

13. $y \le 2$

15. $x > 3$

17. $a \le 3$

19. $z \le -4$

21. $x \le -1$

23. $x < 0$

25. $x \le -\frac{11}{80}$

27. $x \ge -20$

29. $x \ge -2$

31. $2 < x < 3$

33. $1 < x < 3$

35. $-2 < x < 5$

37. $-5 < x < 1$

39. $3 < x < 5$

41. 20°F < t < 40°F **43.** \$12,000 < s < \$13,000 **45.** $2 \le e \le 7$ **47.** \$3.50 < c < \$4.00 **49.** a < 41 ft **51.** After 1996 **53.** 1991 **55.** 1997 **57.** −40 **59.** 40 **61.** −2 **63.** 7 **65.** J = 5 ft = 60 in. **67.** $F = S - 3$ **69.** S = 6 ft 5 in. = 77 in. **71.** B > 74 in. or B > 6 ft 2 in.

77. $-3 \le x \le -1$

79. $x > 3$

81. $x > 3$

83. $x \ge 2$

85. $x < 1$

87. < **89.** >

Review Exercises

1. [2.1A] **(a)** No; **(b)** Yes; **(c)** Yes **2.** [2.1B] **(a)** $x = \frac{2}{3}$; **(b)** $x = 1$; **(c)** $x = \frac{2}{3}$ **3.** [2.1B] **(a)** $x = \frac{2}{9}$; **(b)** $x = \frac{4}{7}$; **(c)** $x = \frac{1}{6}$ **4.** [2.1C] **(a)** $x = 5$; **(b)** $x = 0$; **(c)** $x = 3$ **5.** [2.1C] **(a)** No solution;

(b) $x = -\frac{1}{8}$; **(c)** No solution **6.** [2.1C] **(a)** All real numbers; **(b)** All real numbers; **(c)** All real numbers **7.** [2.2A] **(a)** $x = -15$; **(b)** $x = -14$; **(c)** $x = -2$ **8.** [2.2B] **(a)** $x = 12$; **(b)** $x = 15$; **(c)** $x = 9$ **9.** [2.2C] **(a)** $x = 6$; **(b)** $x = \frac{24}{7}$; **(c)** $x = 20$ **10.** [2.2C] **(a)** $x = 12$; **(b)** $x = 28$; **(c)** $x = 40$ **11.** [2.2C] **(a)** $x = 17$; **(b)** $x = 7$; **(c)** $x = 9$ **12.** [2.2D] **(a)** 20%; **(b)** 10%; **(c)** 20% **13.** [2.2D] **(a)** 50; **(b)** $33\frac{1}{3}$; **(c)** $33\frac{1}{3}$ **14.** [2.3A] **(a)** $x = -3$; **(b)** $x = -4$; **(c)** $x = -5$ **15.** [2.3B] **(a)** $h = \frac{2A}{b}$; **(b)** $r = \frac{C}{2\pi}$; **(c)** $b = \frac{3V}{h}$ **16.** [2.4A] **(a)** 32, 52; **(b)** 14, 33; **(c)** 29, 52 **17.** [2.4B] **(a)** Chicken: 278, pie: 300; **(b)** Chicken: 291, pie: 329; **(c)** Chicken: 304, pie: 346 **18.** [2.4C] **(a)** 35°; **(b)** 30°; **(c)** 25° **19.** [2.5A] **(a)** 4 hours; **(b)** $1\frac{1}{2}$ hours; **(c)** 2 hours **20.** [2.5B] **(a)** 10 pounds; **(b)** 15 pounds; **(c)** 25 pounds **21.** [2.5C] **(a)** \$10,000 at 6%, \$20,000 at 5%; **(b)** \$20,000 at 7%, \$10,000 at 9%; **(c)** \$25,000 at 6%, \$5000 at 10% **22.** [2.6A] **(a)** $m = \frac{C-3}{3.05}$; 8 minutes; **(b)** $m = \frac{C-3}{3.15}$; 10 minutes; **(c)** $m = \frac{C-2}{3.25}$; 6 minutes **23.** [2.6C] **(a)** 100°, 80°; **(b)** Both 60°; **(c)** 65°, 25° **24.** [2.7A] **(a)** $<$; **(b)** $>$; **(c)** $<$

25. [2.7B] **(a)** $x < 3$ 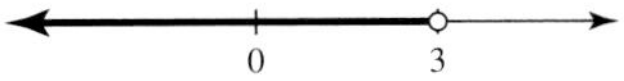

;

(b) $x < 2$ 

;

(c) $x < 1$ 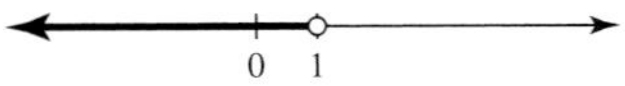

26. [2.7B] **(a)** $x \geq 4$ 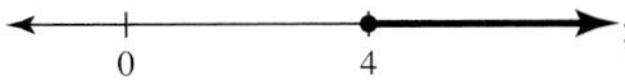

;

(b) $x \geq 2$ 

;

(c) $x \geq 5$ 

27. [2.7B] **(a)** $x \geq \frac{1}{2}$ 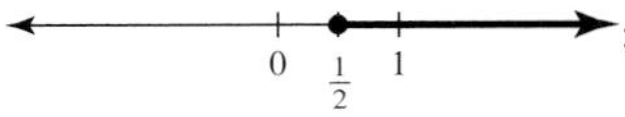

;

(b) $x \geq \frac{4}{7}$ 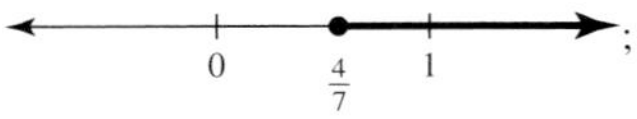

;

(c) $x \geq \frac{5}{3}$ 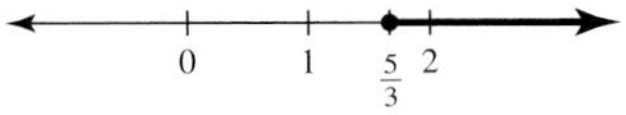

28. [2.7C] **(a)** $-3 < x \leq 2$ 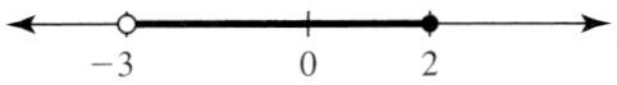

;

(b) $-3 < x \leq 2$ 

;

(c) $-2 < x \leq 1$ 

Cumulative Review Chapters 1–2

1. 7 **2.** $9\frac{9}{10}$ **3.** $-\frac{32}{63}$ **4.** 8.2 **5.** -8.64 **6.** -16 **7.** $\frac{21}{5}$ **8.** 69 **9.** Commutative law of multiplication **10.** $30x + 42$ **11.** cd **12.** $-3x - 11$ **13.** $\frac{a - 4b}{c}$ **14.** Yes **15.** $x = 13$ **16.** $x = 9$ **17.** $x = 15$ **18.** $x = 8$ **19.** $b = \frac{S}{6a^2}$ **20.** 60 and 95 **21.** \$350 from stocks and \$245 from bonds **22.** 24 hr **23.** \$2000 in bonds and \$3000 in certificates of deposit **24.** $x \geq 6$ 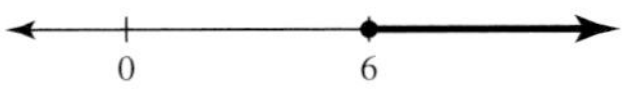

Chapter 3

Exercises 3.1

1.

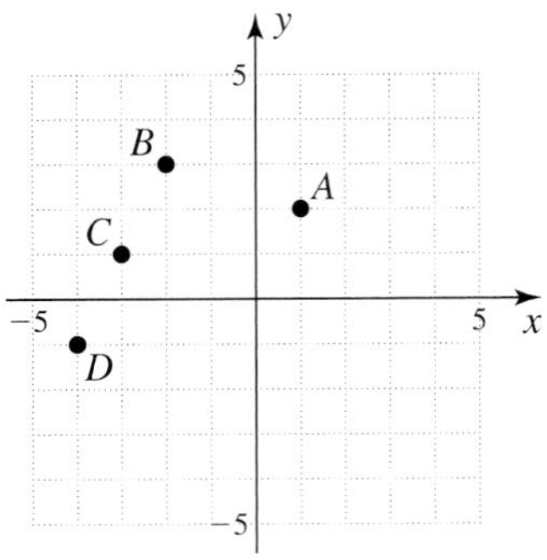

3.

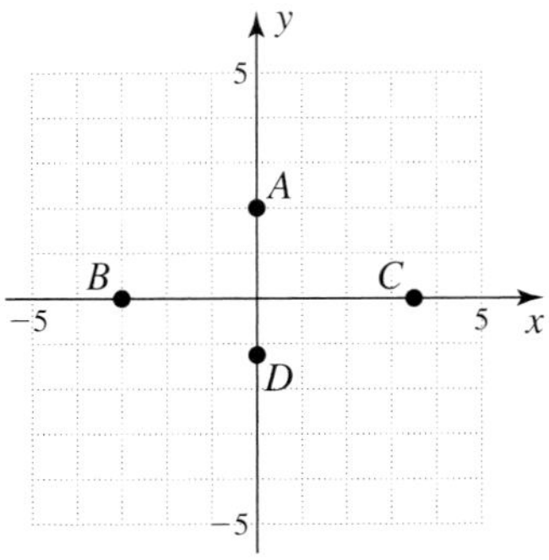

5.

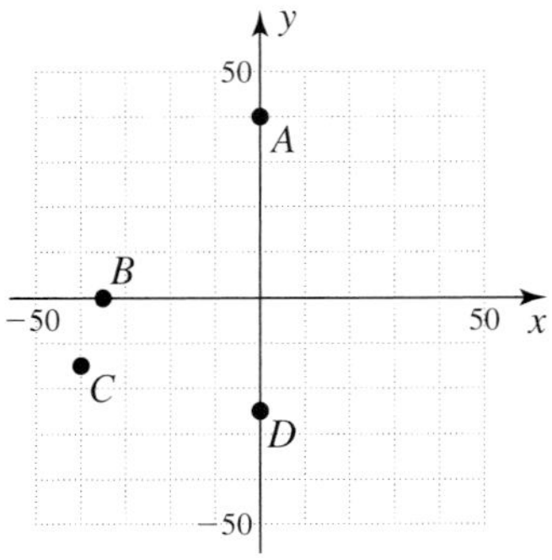

7. $A(0, 2\frac{1}{2})$; $B(-3, 1)$; $C(-2, -2)$; $D(0, -3\frac{1}{2})$; $E(1\frac{1}{2}, -2)$; Quadrant II: B; Quadrant III: C; Quadrant IV: E; y-axis: A, D
9. $A(10, 10)$; $B(-30, 20)$; $C(-25, -20)$; $D(0, -40)$; $E(15, -15)$; Quadrant I: A; Quadrant II: B; Quadrant III: C; Quadrant IV: E; y-axis: D
11. (20, 140) **13.** (45, 150) **15.** 2.7 **17.** 1.9 **19.** 4.7 **21.** 229.9 **23.** 24.7 **25.** \$950 **27.** \$800 **29.** 70 **31.** 3 A.M. **33.** 55 **35.** Before 3 A.M.: 195; after 3 A.M.: 288 **37.** Before 3 A.M.: 162; after 3 A.M.: 13 **39.** **(a)** Auto batteries; **(b)** Tires **41.** 26.5; 1.43% **43.** \$500 **45.** \$25

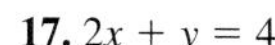

47 and **49.**

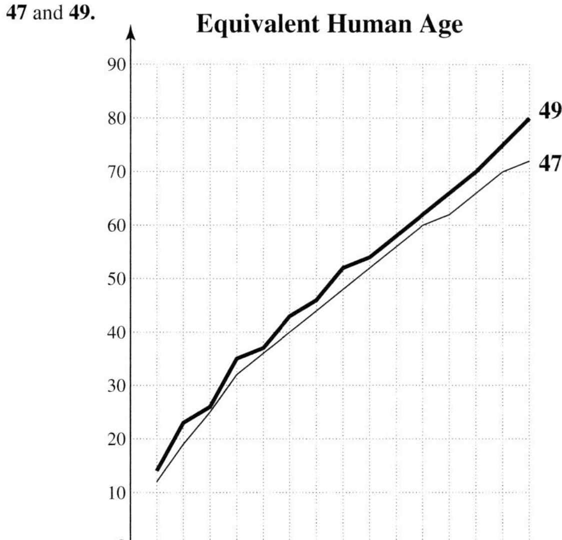

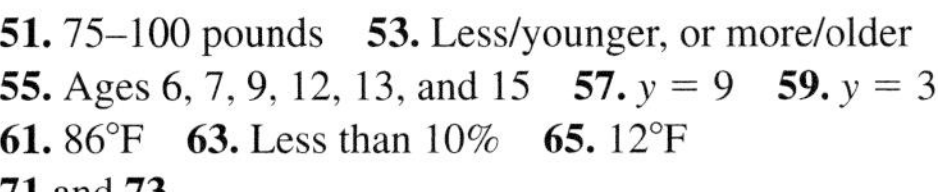

51. 75–100 pounds **53.** Less/younger, or more/older
55. Ages 6, 7, 9, 12, 13, and 15 **57.** $y = 9$ **59.** $y = 3$
61. 86°F **63.** Less than 10% **65.** 12°F
71 and **73.**

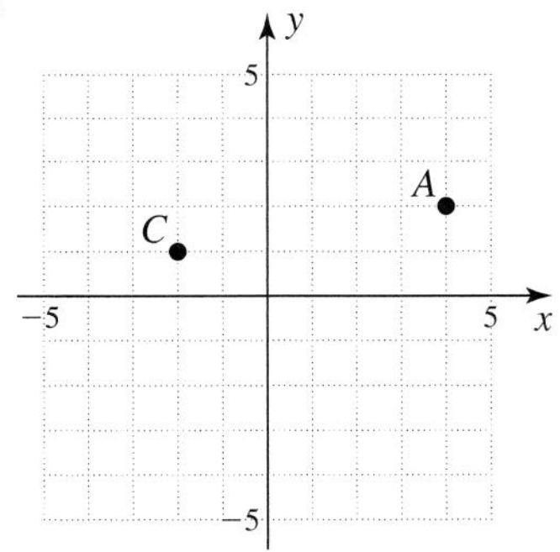

75.

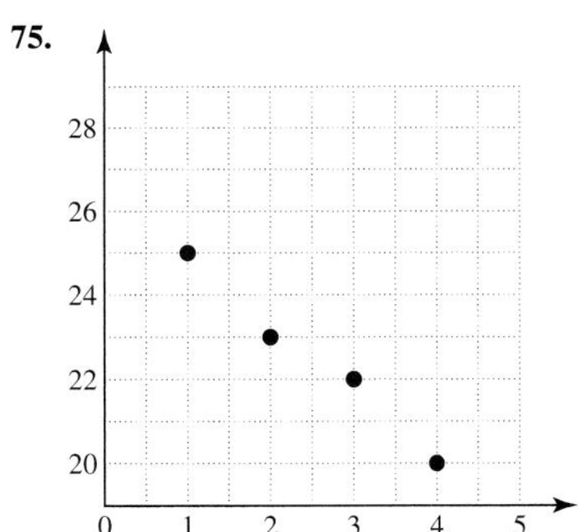

77. 11 human years **79.** **(a)** Quadrant II; **(b)** Quadrant IV; **(c)** Quadrant I; **(d)** Quadrant III

Exercises 3.2

1. Yes **3.** Yes **5.** No **7.** 0 **9.** −2 **11.** −3 **13.** 3 **15.** 2

17. $2x + y = 4$

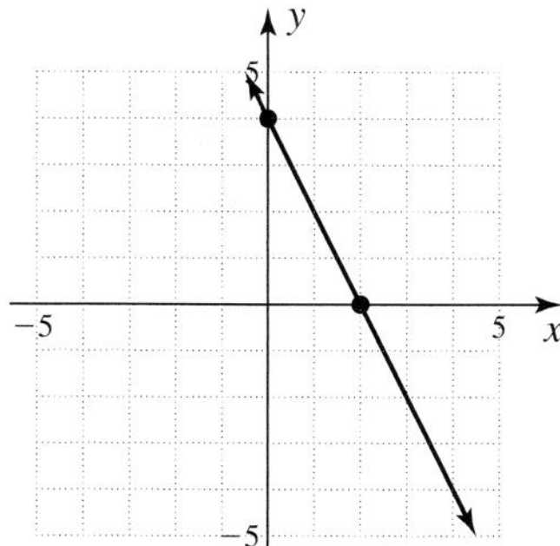

19. $-2x - 5y = -10$

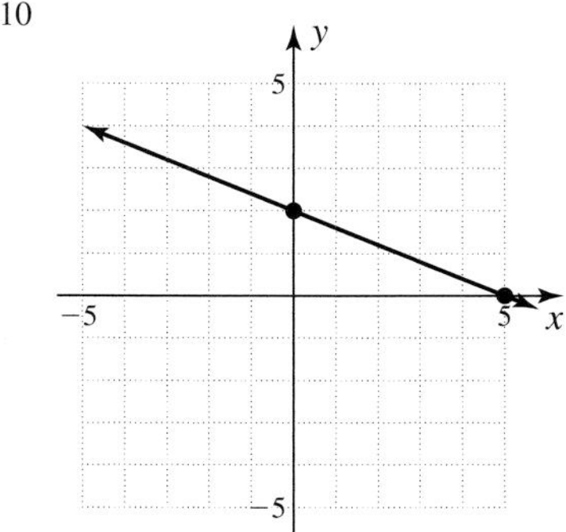

21. $y + 3 = 3x$

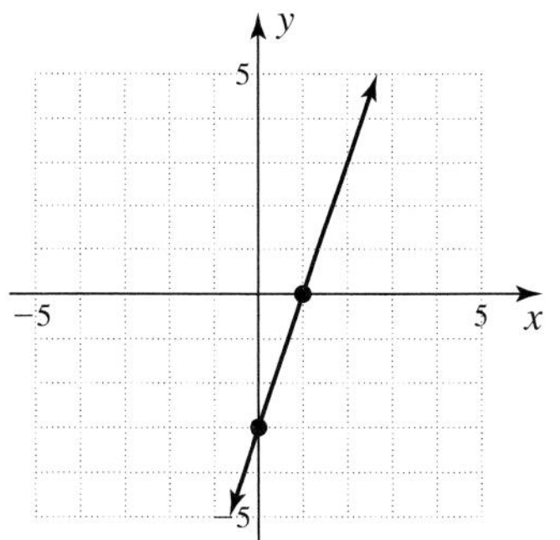

23. $6 = 3x - 6y$

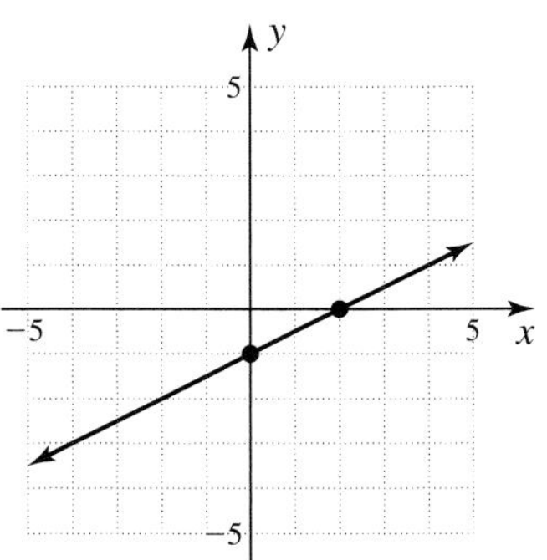

25. $-3y = 4x + 12$

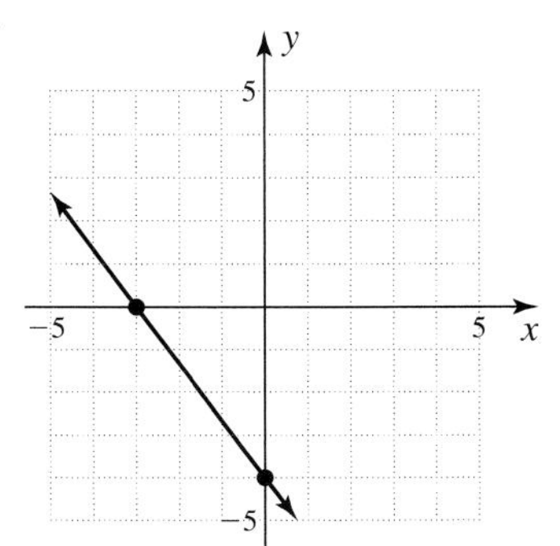

27. $-2y = -x + 4$

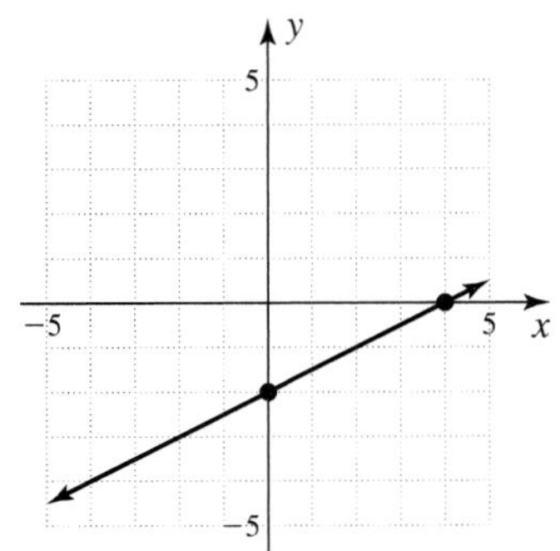

29. $-3y = -6x + 3$

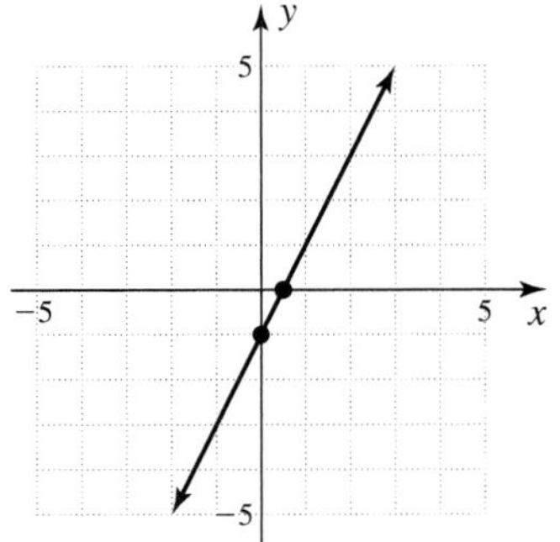

31. $y = 2x + 4$

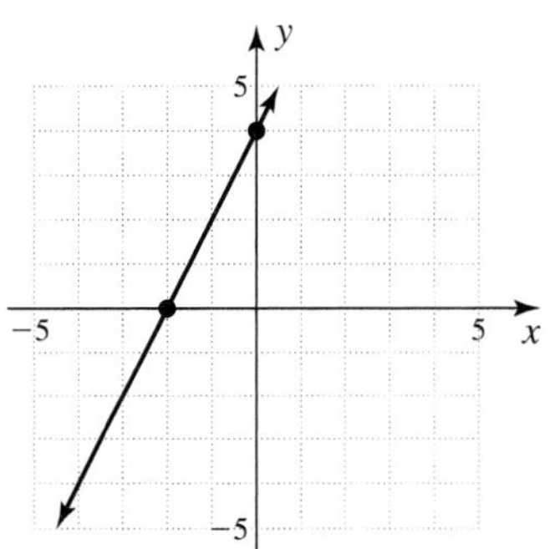

33. $y = -2x + 4$

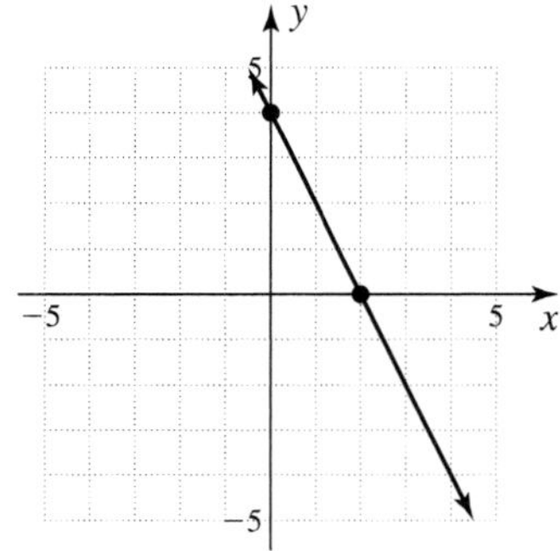

35. $y = -3x - 6$

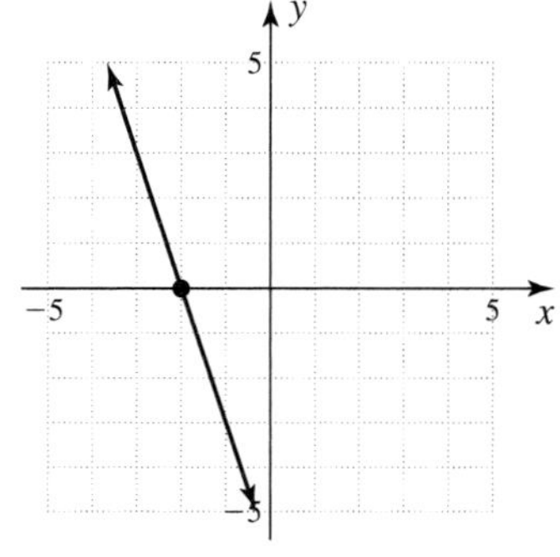

37. $y = \frac{1}{2}x - 2$

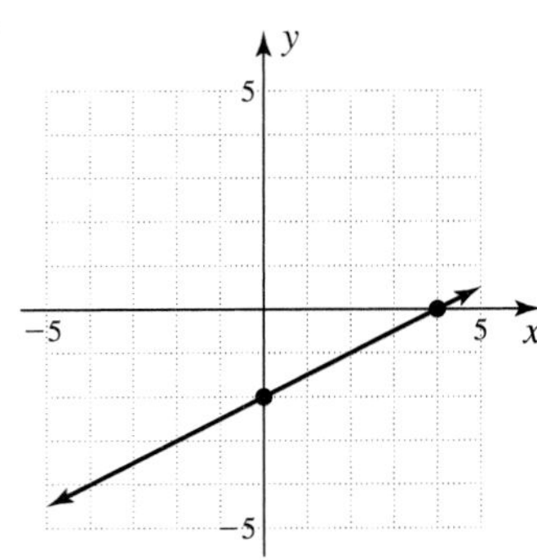

39. $y = -\frac{1}{2}x - 2$

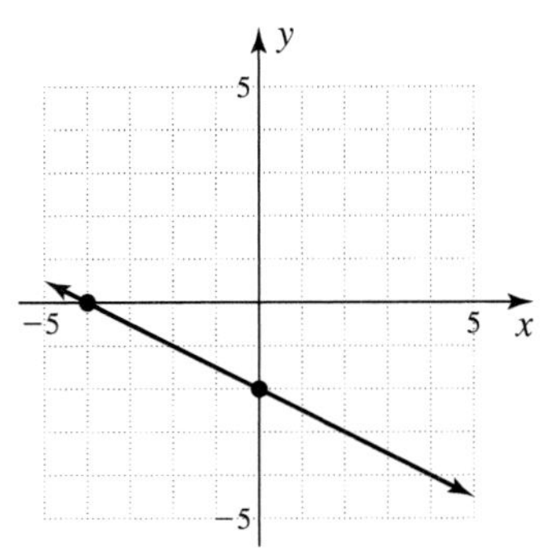

41. **(a)** −9; **(b)** 2;
(c)

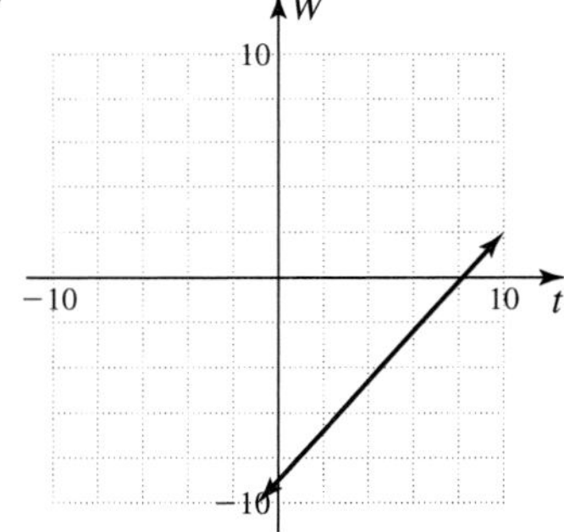

43. **(a)** \$7000; **(b)** \$8000; **(c)** \$9000;
(d)

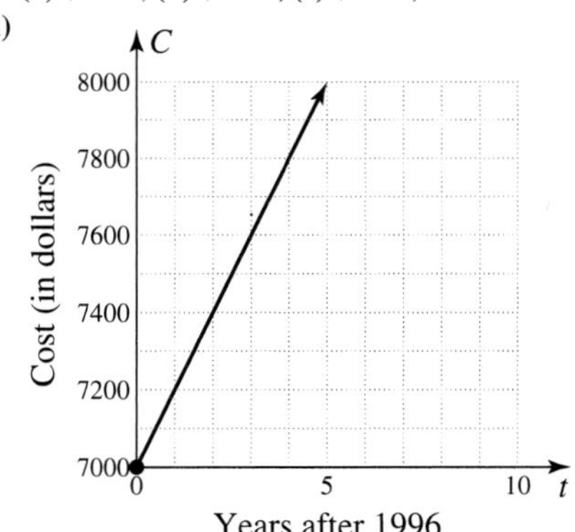

45. $x = -2$ **47.** $t = 0$ **49.** Yes **51.** No; 16 in. **53.** Yes
59. **(a)** No; **(b)** No

61. $y = -3x + 6$

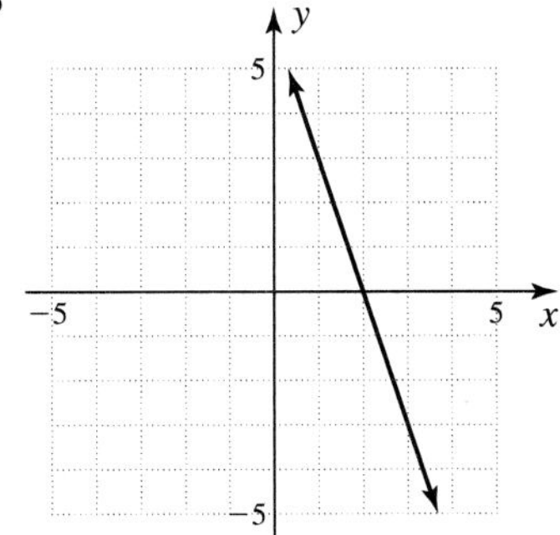

63. **(a)** -18; **(b)** -5;
(c)

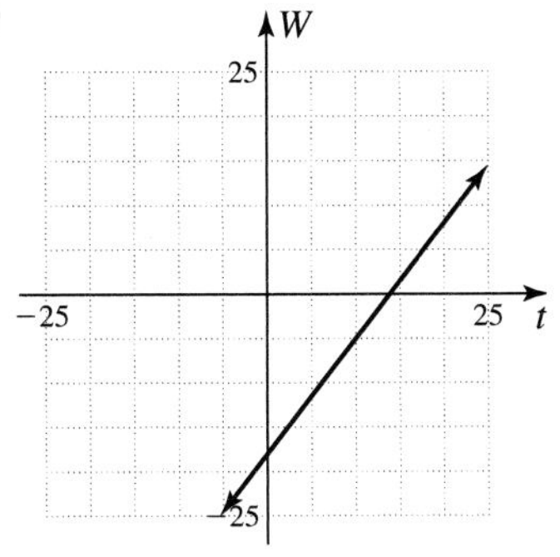

Exercises 3.3

1. $x + 2y = 4$

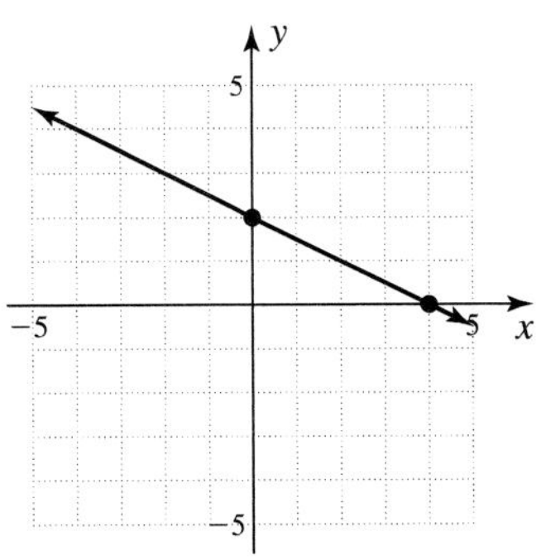

3. $-5x - 2y = -10$

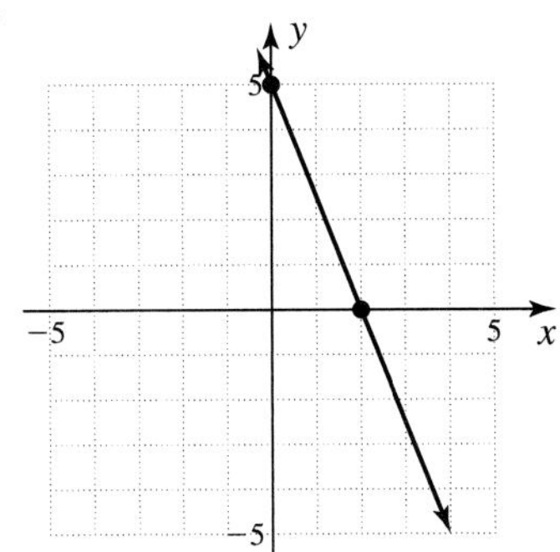

5. $y - 3x - 3 = 0$

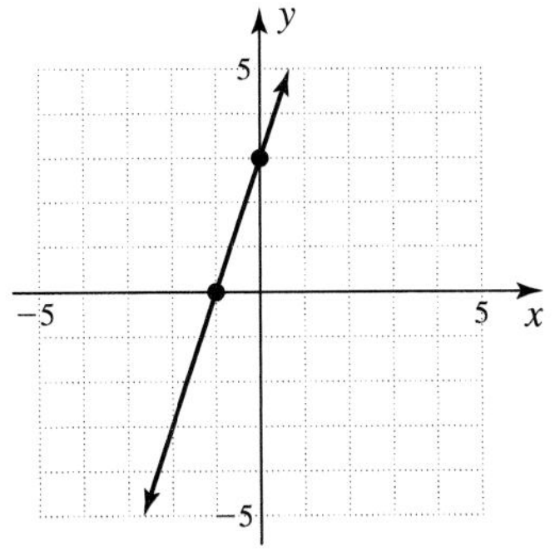

7. $6 = 6x - 3y$

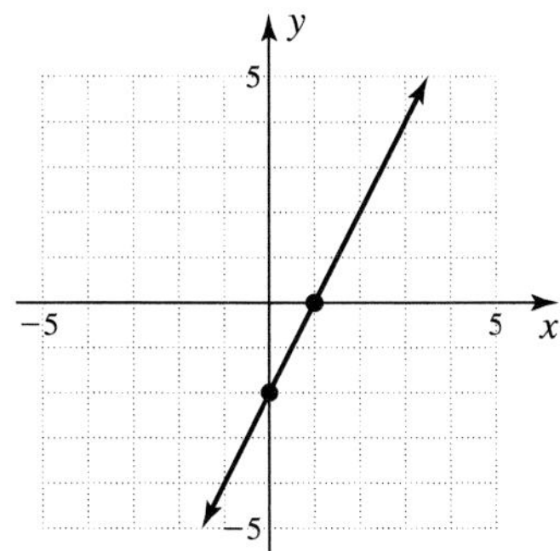

9. $3x + 4y + 12 = 0$

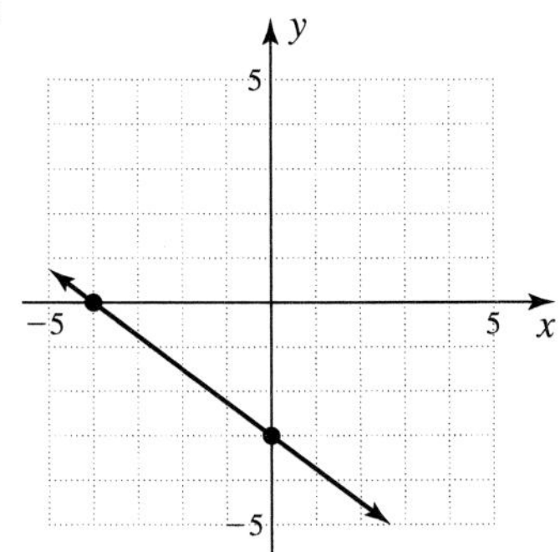

11. $3x + y = 0$

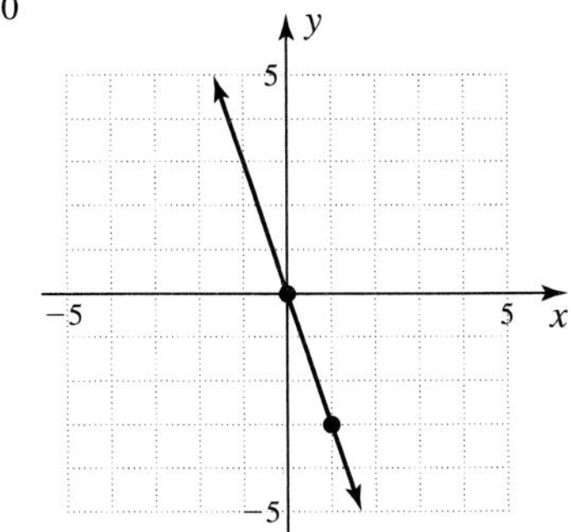

13. $2x + 3y = 0$

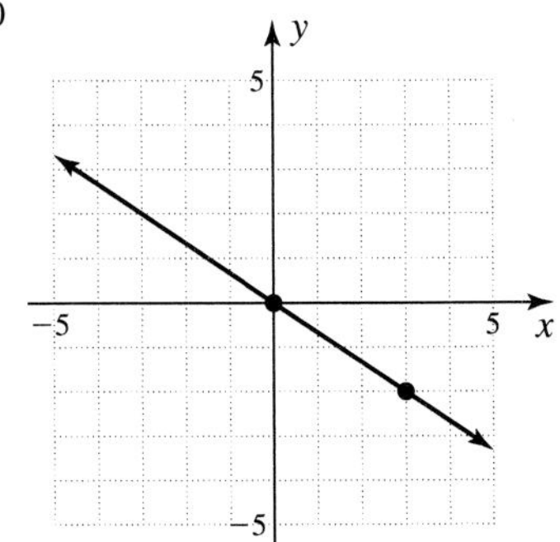

15. $-2x + y = 0$

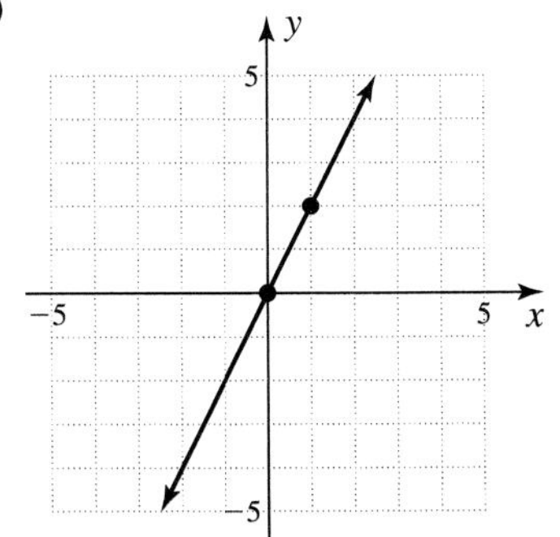

17. $2x - 3y = 0$

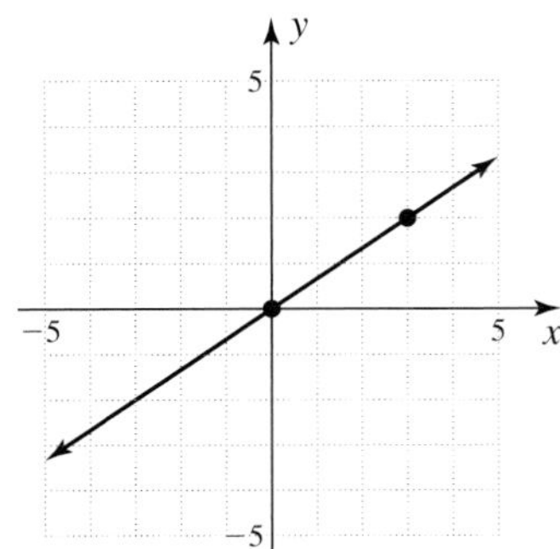

19. $-3x = -2y$

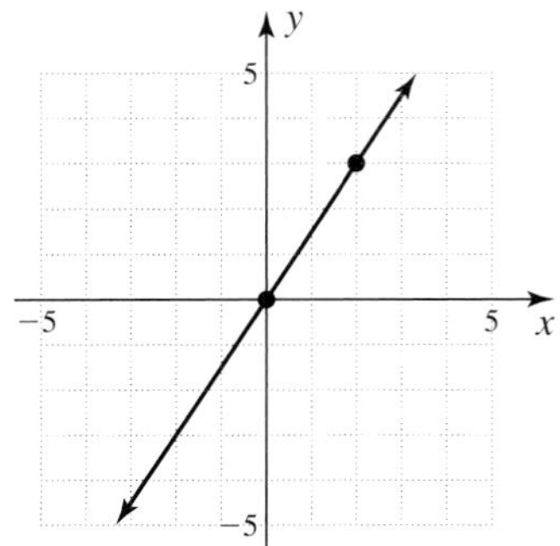

21. $y = -4$

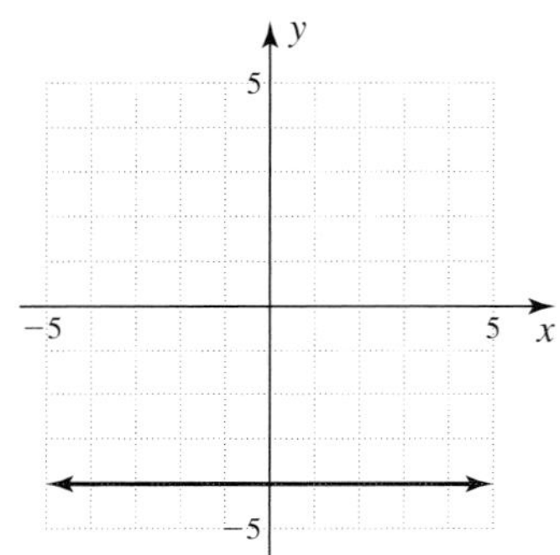

23. $2y + 6 = 0$

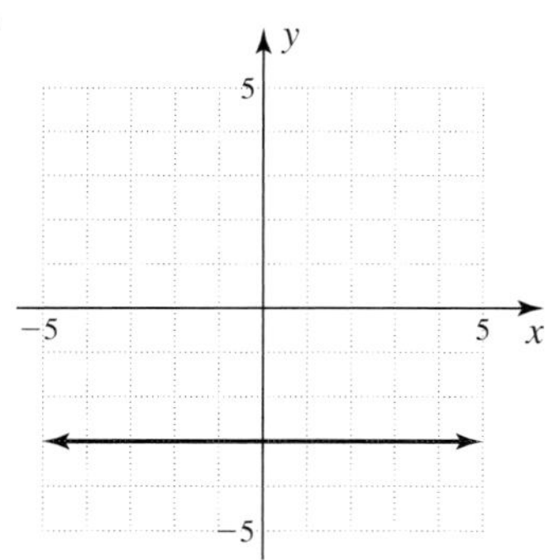

25. $x = -\frac{5}{2}$

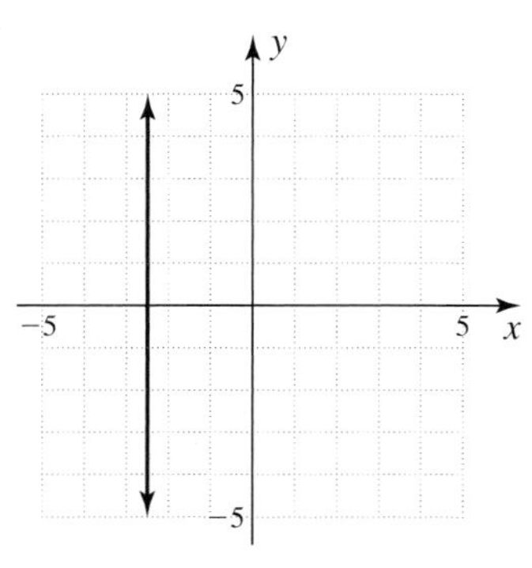

27. $2x + 4 = 0$

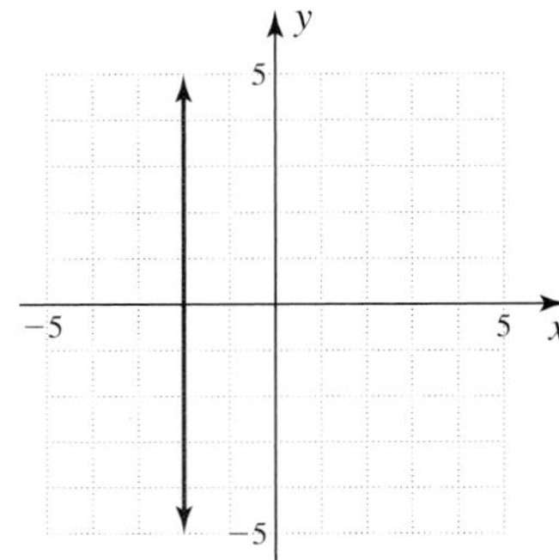

29. $2x - 9 = 0$

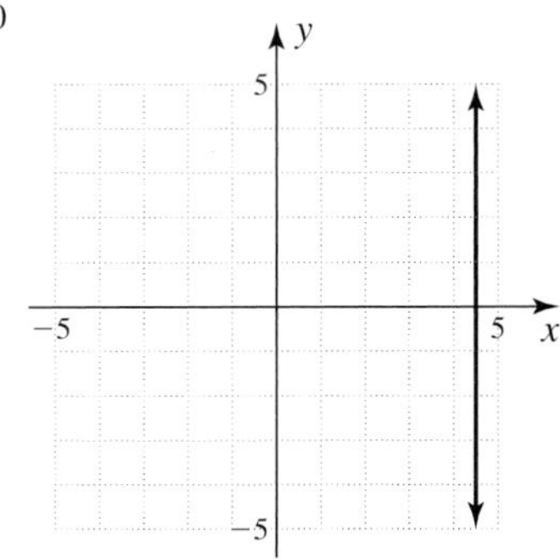

31. **(a)** 5.6; **(b)** 70; **(c)**

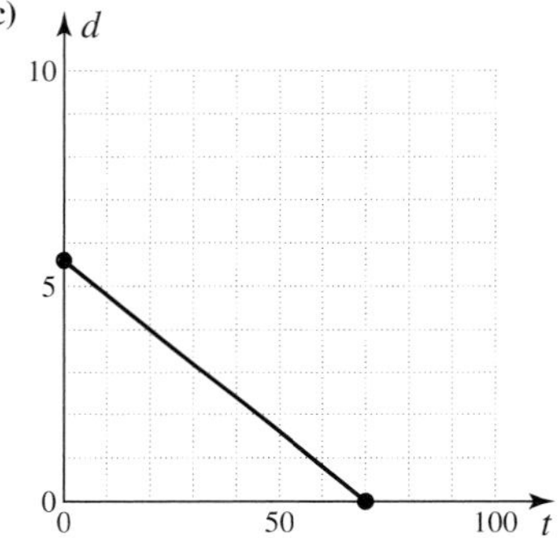

33. **(a)** 140 g; **(b)** 180 g; **(c)** 190 g; **(d)**

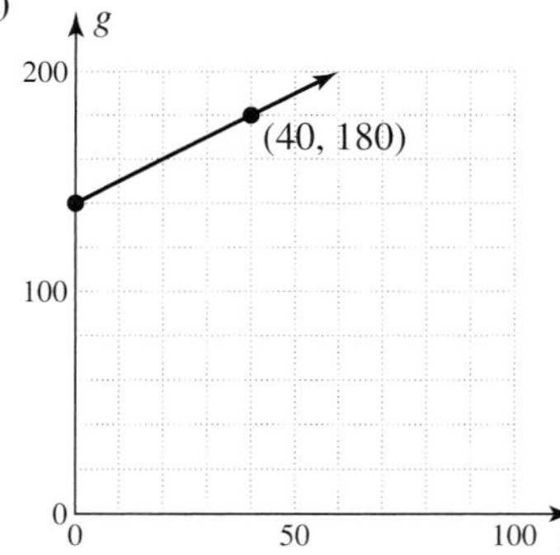

35. 9 **37.** −9 **39.** 215 **41.**

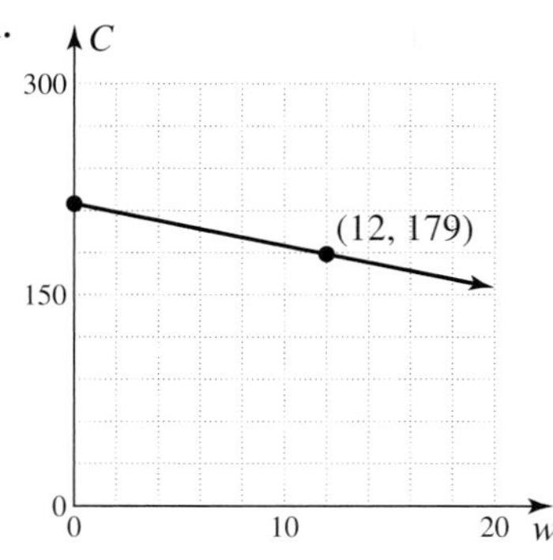

49. $-2x + y = 4$

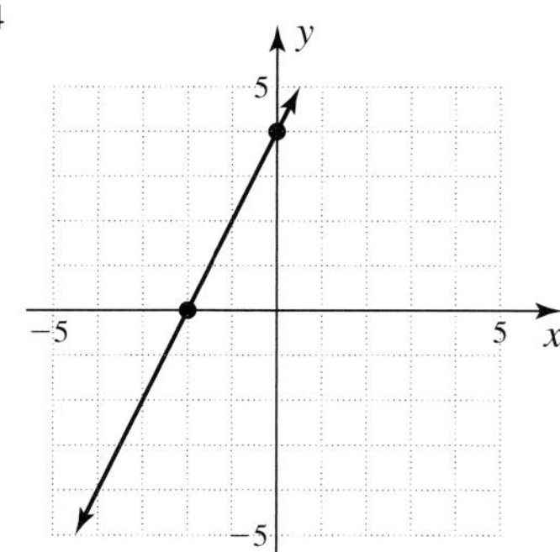

51. $4 + 2y = 0$

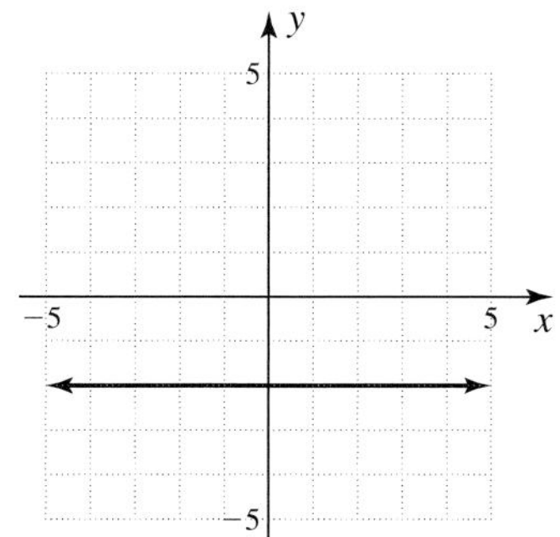

53. $-x + 4y = 0$

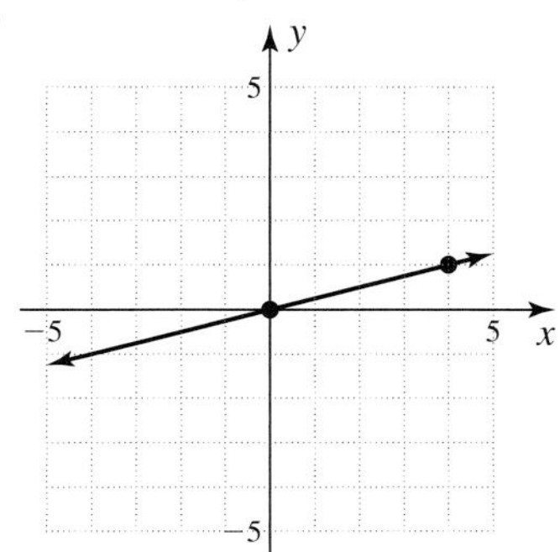

55. **(a)** C: 30, A: −30,

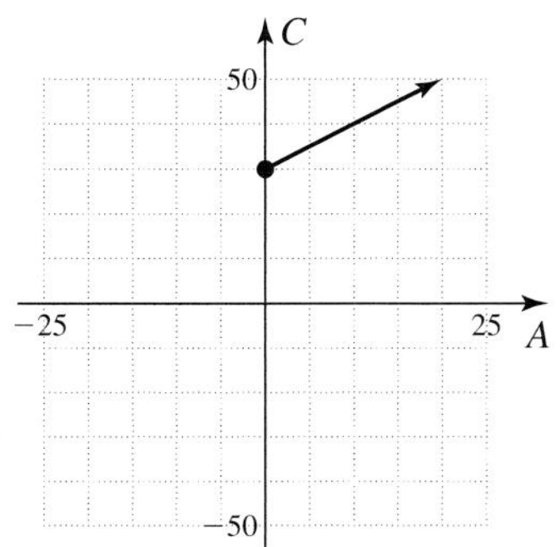

Each vertical unit = 10; each horizontal unit = 5
(b) Quadrant I

Exercises 3.4

1. $m = 1$ **3.** $m = -1$ **5.** $m = -\frac{1}{8}$ **7.** $m = \frac{1}{4}$ **9.** $m = 0$

11. $m = 0$ **13.** $m = 0$ **15.** $m = 3$ **17.** $m = -\frac{2}{3}$

19. $m = -\frac{1}{3}$ **21.** $m = \frac{2}{5}$ **23.** $m = 0$ **25.** Undefined

27. Parallel **29.** Perpendicular **31.** Parallel
33. Parallel **35.** Neither; the lines coincide **37.** Perpendicular
39. **(a)** $m = -0.1$; **(b)** Decreasing; **(c)** The rate at which the average hospital stay is changing.

41. **(a)** 0.15; **(b)** Increasing; **(c)** The slope 0.15 represents the annual increase in the life span of American women.
43. **(a)** −32; **(b)** Decreasing; **(c)** The slope −32 represents the decrease in the velocity of the ball.
45. **(a)** 0.4; **(b)** Increasing; **(c)** The slope 0.4 represents the annual increase in the consumption of fat.
47. $2x + 8$ **49.** $-2x - 2$

51.

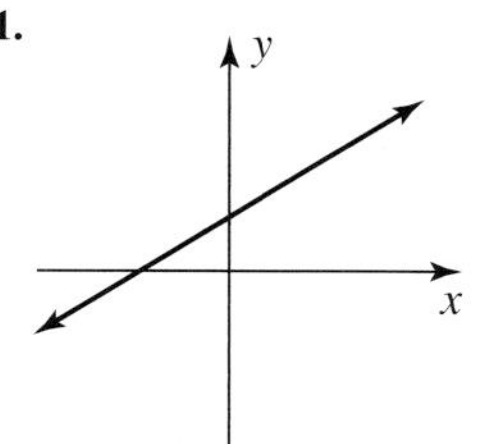

53.

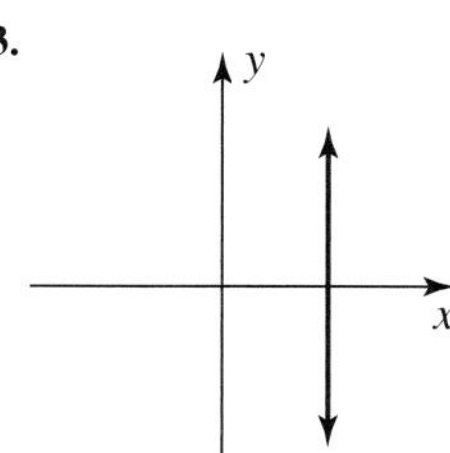

57. $m = -1$ **59.** Undefined **61.** $m = -\frac{3}{2}$ **63.** Parallel
65. Perpendicular

Review Exercises

1. [3.1A]

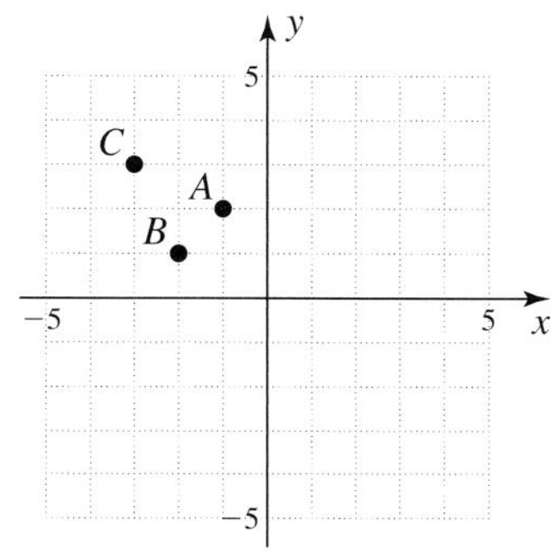

2. [3.1B] A: (1, 1); B: (−1, −2); C: (3, −3)
3. [3.1C] **(a)** For a wind speed of 10 mi/hr, the wind chill temperature is −10°F; **(b)** −22°F; **(c)** −45°F
4. [3.1D] **(a)** \$10; **(b)** \$12; **(c)** \$20
5. [3.1E] **(a)** Quadrant II; **(b)** Quadrant III; **(c)** Quadrant IV
6. [3.1F] **(a)**

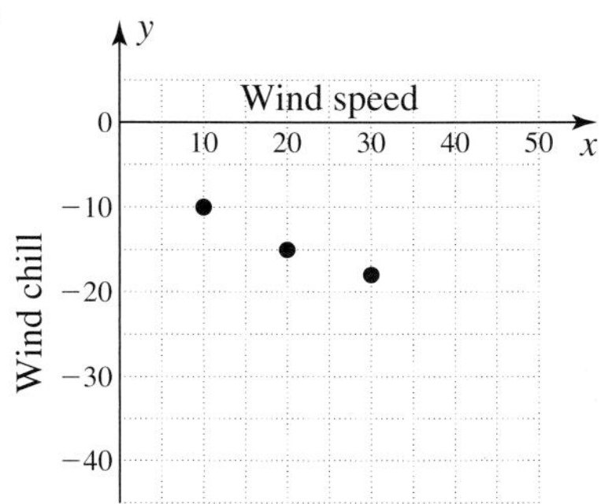

(b) For a wind speed of 10 mi/hr, the wind chill temperature is −10°F;
(c) $s = 20$ **7.** [3.2A] **(a)** No; **(b)** No; **(c)** Yes
8. [3.2B] **(a)** $x = 3$; **(b)** $x = 4$; **(c)** $x = 2$
9. [3.2C]

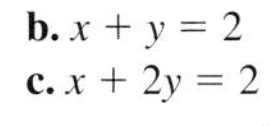

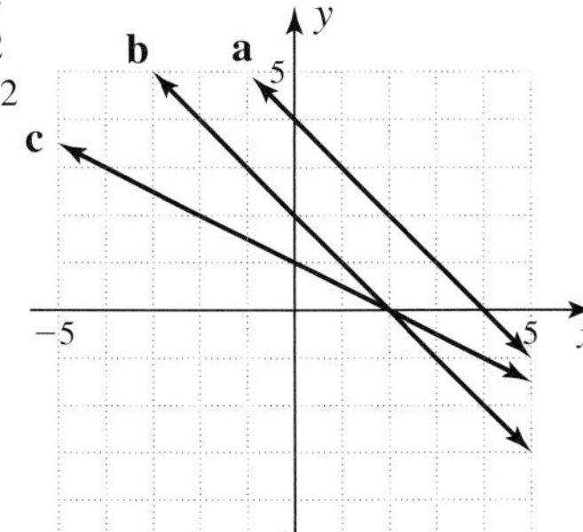

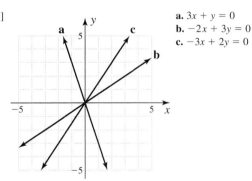

10. [3.2C]

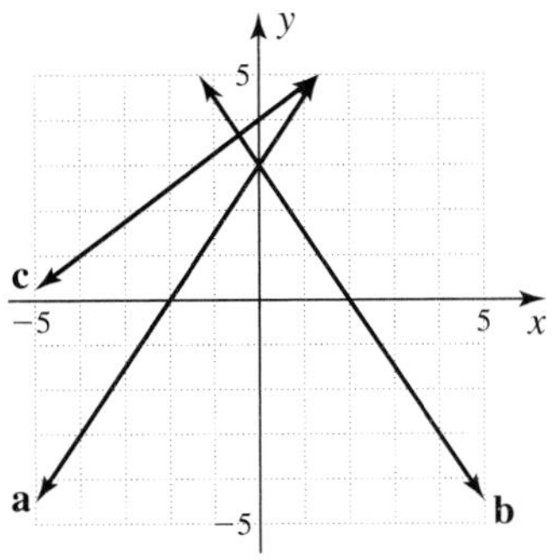

a. $y = \frac{3}{2}x + 3$ **b.** $y = -\frac{3}{2}x + 3$ **c.** $y = \frac{3}{4}x + 4$

11. [3.2D]

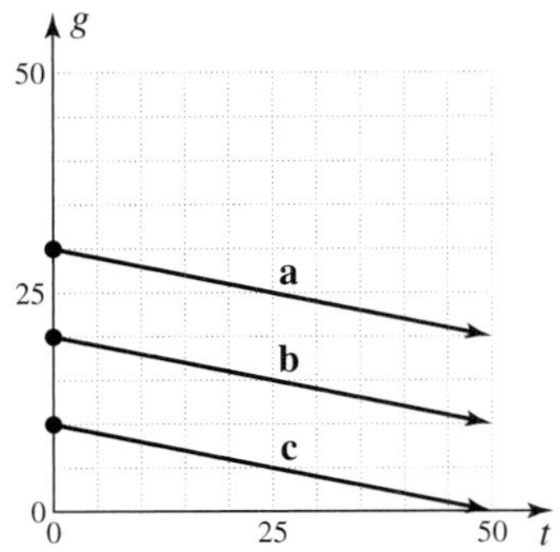

a. $g = 30 - 0.2t$ **b.** $g = 20 - 0.2t$ **c.** $g = 10 - 0.2t$

12. [3.3A]

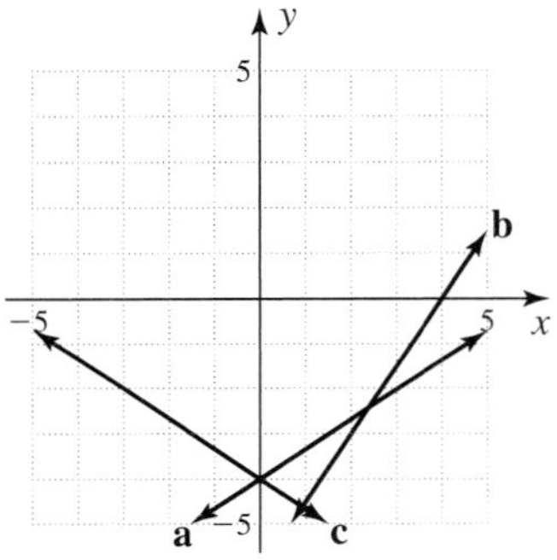

a. $2x - 3y - 12 = 0$ **b.** $3x - 2y - 12 = 0$ **c.** $2x + 3y + 12 = 0$

13. [3.3B]

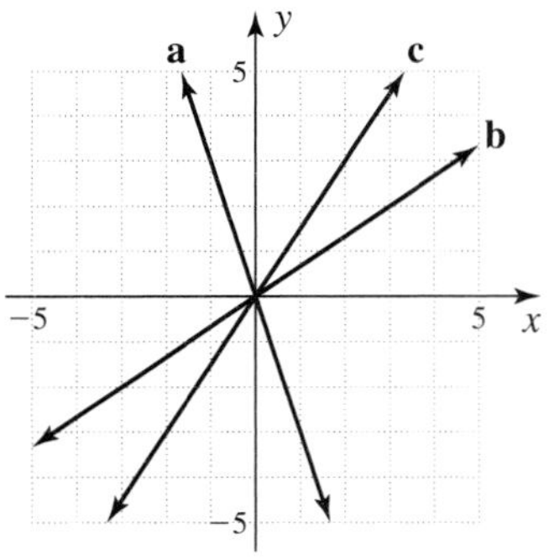

a. $3x + y = 0$ **b.** $-2x + 3y = 0$ **c.** $-3x + 2y = 0$

14. [3.3C]

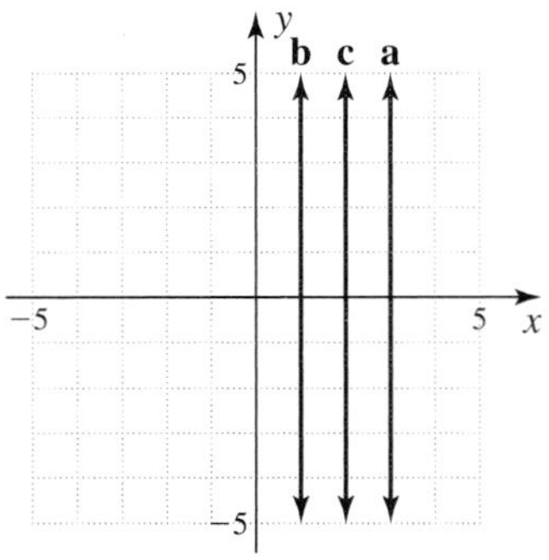

a. $2x - 6 = 0$ **b.** $2x - 2 = 0$ **c.** $2x - 4 = 0$

15. [3.3C]

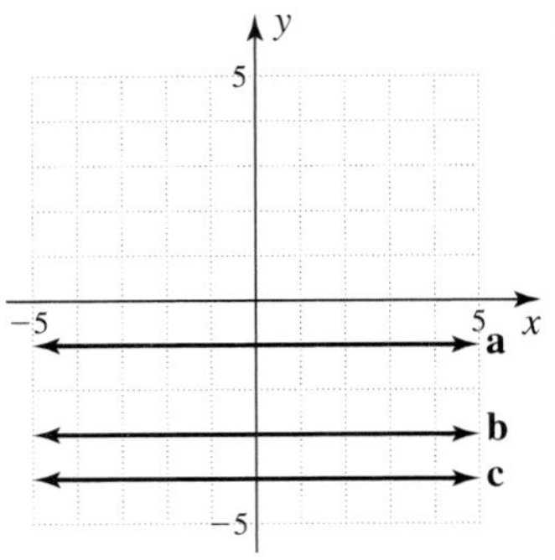

a. $y = -1$ **b.** $y = -3$ **c.** $y = -4$

16. [3.4A] **(a)** $m = 1$; **(b)** $m = \frac{7}{3}$; **(c)** $m = -4$

17. [3.4A] **(a)** Undefined; **(b)** 0; **(c)** Undefined

18. [3.4B] **(a)** $m = -\frac{3}{2}$; **(b)** $m = -\frac{1}{4}$; **(c)** $m = \frac{2}{3}$

19. [3.4C] **(a)** Neither; **(b)** Perpendicular; **(c)** Parallel

20. [3.4D] **(a)** $m = 0.6$; **(b)** The change (increase) in the number of theaters per year; **(c)** 600

Cumulative Review Chapters 1–3

1. 1 **2.** $3\frac{1}{7}$ **3.** $-\frac{13}{24}$ **4.** 13.0 **5.** −19.76 **6.** 1296 **7.** 12

8. 5 **9.** Commutative law of addition **10.** $18x - 21$ **11.** cd^2

12. $-x - 14$ **13.** $\frac{m + n}{p}$ **14.** No **15.** $x = 6$ **16.** $x = 9$

17. $x = 54$ **18.** $x = 10$ **19.** $d = \frac{S}{7c^2}$ **20.** 65 and 105

21. \$435 from stocks and \$190 from bonds **22.** 2 hr

23. \$12,000 in bonds, \$13,000 in certificates of deposit

24.

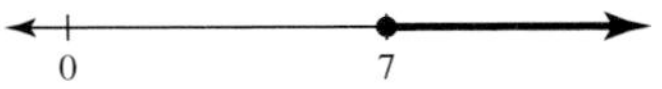

25.

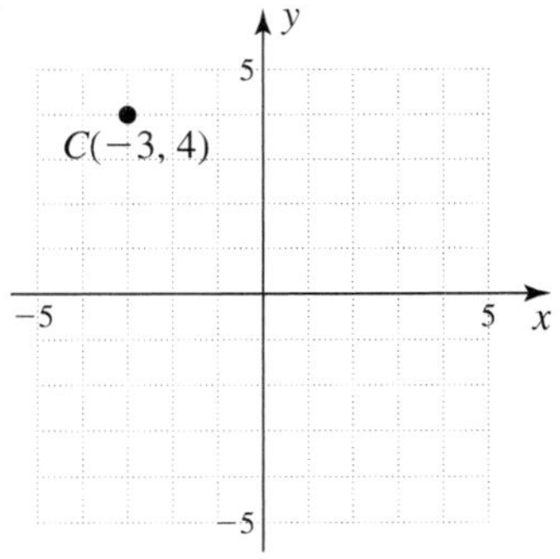

26. (2, −4) **27.** No **28.** $x = -2$

29. $2x + y = 8$

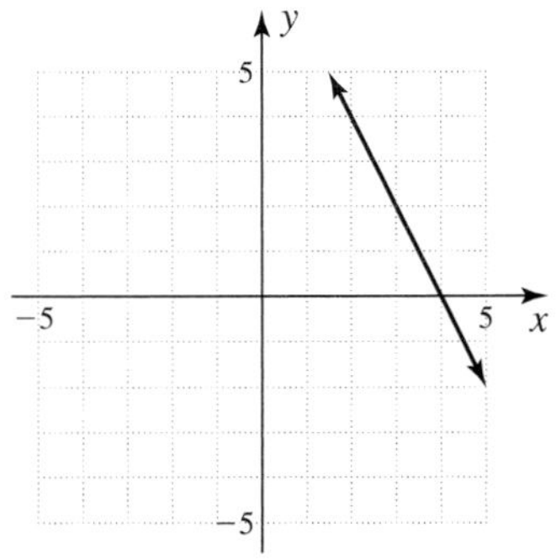

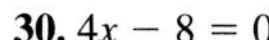

30. $4x - 8 = 0$

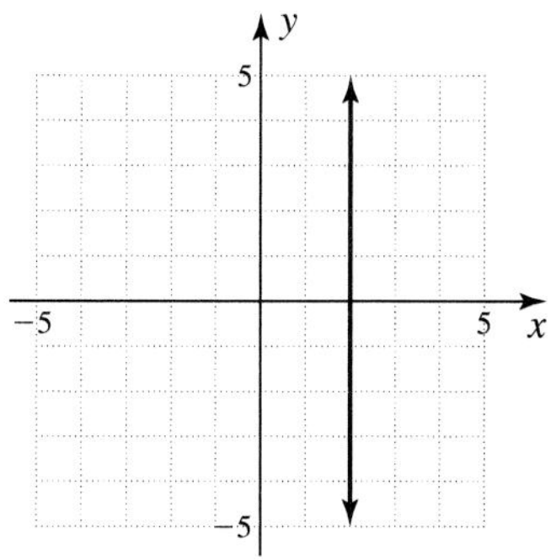

31. 9 **32.** 2 **33.** (1) and (3)

Chapter 4

Exercises 4.1

1. $24x^3$ **3.** $30a^4b^3$ **5.** $3x^3y^3$ **7.** $-\frac{b^5c}{5}$ **9.** $\frac{3x^3y^3z^6}{2}$ **11.** $-30x^5y^3z^8$ **13.** $15a^4b^3c^4$ **15.** $24a^3b^5c^5$ **17.** x^4 **19.** $-\frac{a^2}{2}$ **21.** $2x^3y^2$ **23.** $-\frac{x^3y^2}{2}$ **25.** $\frac{2a^3y^4}{3}$ **27.** $\frac{3ab^3c}{4}$ **29.** $\frac{3a^4}{2}$ **31.** $-x^6y$ **33.** $-x^2y^2$ **35.** $2^6 = 64$ **37.** $3^2 = 9$ **39.** x^9 **41.** y^6 **43.** $-a^6$ **45.** $8x^9y^6$ **47.** $4x^4y^6$ **49.** $-27x^9y^6$ **51.** $9x^{12}y^6$ **53.** $-8x^{12}y^{12}$ **55.** $\frac{16}{81}$ **57.** $\frac{27x^6}{8y^9}$ **59.** $\frac{16x^8}{81y^{12}}$ **61.** $12x^6y^3$ **63.** $-576a^2b^3$ **65.** $-144a^4b^6$ **67.** 96 **69.** x^5y^2 **71.** \$1259.71 **73.** \$1464.10 **75.** $V = \frac{2}{3}x^3$ **77.** 144 sandwiches **79.** \$0.0075 or 0.75¢ **81.** $V = x^3$ **83.** 16 in.^{3} **85.** ≈ \$23.63 **87.** $V = \frac{9}{16}x^3$ in.^{3} **89.** 37.5 servings **91.** 81 **93.** 9142 **95. (a)** 123,454,321; **(b)** 12,345,654,321 **97. (a)** $5^2 = 25$; **(b)** $7^2 = 49$ **103.** $30a^{11}$ **105.** $-12x^4yz^5$ **107.** $-5x^2y^6$ **109.** $\frac{ab^7c}{4}$ **111.** y^{20} **113.** $\frac{27y^6}{x^{15}}$ **115.** 4

Exercises 4.2

1. $\frac{1}{4^2} = \frac{1}{16}$ **3.** $\frac{1}{5^3} = \frac{1}{125}$ **5.** $8^2 = 64$ **7.** x^7 **9.** $\frac{3^3}{4^2} = \frac{27}{16}$ **11.** $\frac{3^4}{5^2} = \frac{81}{25}$ **13.** $\frac{b^6}{a^5}$ **15.** $\frac{y^9}{x^9}$ **17.** 2^{-3} **19.** y^{-5} **21.** q^{-5} **23.** 3 **25.** 4 **27.** $\frac{1}{16}$ **29.** $\frac{1}{216}$ **31.** $\frac{1}{64}$ **33.** x^2 **35.** y^2 **37.** $\frac{1}{a^5}$ **39.** $\frac{1}{x^2}$ **41.** $\frac{1}{x^2}$ **43.** $\frac{1}{a^5}$ **45.** 1 **47.** $3^5 = 243$ **49.** $\frac{1}{4^3} = \frac{1}{64}$ **51.** $\frac{1}{y^2}$ **53.** x^3 **55.** $\frac{1}{x^2}$ **57.** $\frac{1}{x^7}$ **59.** x^3 **61.** $\frac{a^2}{b^6}$ **63.** $-\frac{8b^6}{27a^3}$ **65.** a^8b^4 **67.** $\frac{1}{x^{15}y^6}$ **69.** $x^{27}y^6$ **71.** 839 **73.** 0.0816 **75.** 6.242 billion **77.** 4.922 billion **79. (a)** \$412; **(b)** \$463.71 **81.** \$61.68 **83.** \$180,405.36 **85.** \$106.14 **87. (a)** 50% per day; **(b)** $A = 40(1 + 0.50)^n$; **(c)** 203; **(d)** 11,677; **(e)** Answers may vary. **93.** $\frac{6^2}{7^2} = \frac{36}{49}$ **95.** $9^2 = 81$ **97.** 7^{-6} **99.** $5^2 = 25$ **101.** z^2 **103.** $\frac{81x^8}{16y^{24}}$ **105. (a)** \$3645.00; **(b)** \$6172.84

Exercises 4.3

1. 5.4×10^7 **3.** 2.8719052×10^8 **5.** 1.9×10^9 **7.** 2.4×10^{-4} **9.** 2×10^{-9} **11.** 153 **13.** 8,000,000 **15.** 6,850,000,000 **17.** 0.23 **19.** 0.00025 **21.** 1.5×10^{10} **23.** 3.06×10^4 **25.** 1.24×10^{-4} **27.** 2×10^3 **29.** 2.5×10^{-3} **31. (a)** 2.32×10^{10}; **(b)** 23,200,000,000 **33.** 2.844827586 **35.** 2×10^7 **37.** Now: 2,000,000; later: 12,000,000 **39.** 45.625 **41.** 8 **43.** 4 **45.** 2.99792458×10^8 **47.** 3.09×10^{13} **49.** 3.27 **53.** Distance: 2.89×10^5; mass: 1.2456×10^{-1} **55.** $6 \times 10^0 = 6$ **57.** $1.32 \times 10^8 = 132{,}000{,}000$

Exercises 4.4

1. B; 1 **3.** M; 1 **5.** T; 2 **7.** M; 0 **9.** B; 3 **11.** $8x^3 - 3x$; 3 **13.** $8x^2 + 4x - 7$; 2 **15.** $x^2 + 5x$; 2 **17.** $x^3 - x^2 + 3$; 3 **19.** $4x^5 - 3x^3 + 2x^2$; 5 **21. (a)** 4; **(b)** −8 **23. (a)** 7; **(b)** 7 **25. (a)** 9; **(b)** 13 **27. (a)** 9; **(b)** −3 **29. (a)** 3; **(b)** −1 **31. (a)** $(-16t^2 + 150)$ ft; **(b)** 134 ft; **(c)** 86 ft **33. (a)** $(-4.9t^2 + 200)$ m; **(b)** 195.1 m; **(c)** 180.4 m **35. (a)** 251; **(b)** About 610; about 1318 **37. (a)** 30%; **(b)** 25%; **(c)** Decreasing; **(d)** \$2.71 billion; **(e)** \$3.25 billion **39. (a)** 28°; **(b)** −122°; 122° below 0 is unreasonable **41. (a)** 34 million; **(b)** $N(5) = 34.6$; the result is close **43.** $-7ab$ **45.** $3x^2y$ **47.** $8xy^2$ **49.** $(-32t - 10)$ ft/sec **51. (a)** −11.8 m/sec; **(b)** −21.6 m/sec **59.** 12 **61.** 18 **63.** 8 **65.** No degree **67.** $5x^2 - 3x - 8$ **69.** Monomial **71.** Polynomial **73. (a)** 0.115; **(b)** 0.105; **(c)** 0.1149; very close; **(d)** 0.1057; very close

Exercises 4.5

1. $12x^2 + 5x + 6$ **3.** $-2x^2 - x - 8$ **5.** $-3x^2 + 7x - 5$ **7.** $-x^2 - 5$ **9.** $x^3 - 2x^2 - x - 2$ **11.** $2x^4 - 8x^3 + 2x^2 + x + 5$ **13.** $-\frac{3}{5}x^2 + x + 1$ **15.** $0.4x^2 + 0.3x - 0.4$ **17.** $-2x^2 - 2x + 3$ **19.** $3x^4 + x^3 - 5x - 1$ **21.** $-5x^4 + 5x^3 - 2x^2 - 2$ **23.** $5x^3 - 4x^2 + 5x - 1$ **25.** $-\frac{1}{7}x^3 - \frac{2}{3}x^2 + 5x$ **27.** $-\frac{3}{7}x^3 + \frac{1}{9}x^2 + x - 6$ **29.** $-4x^4 - 4x^3 - 3x^2 - x + 2$ **31.** $4x^2 + 7$ **33.** $-x^2 - 4x - 6$ **35.** $4x^2 - 7x + 6$ **37.** $-3x^3 + 6x^2 - 2x$ **39.** $7x^3 - x^2 + 2x - 6$ **41.** $3x^2 - 7x + 7$ **43.** 0 **45.** $4x^3 - 3x^2 - 7x + 6$ **47.** $3x^3 - 2x^2 + x - 8$ **49.** $-5x^3 - 5x^2 + 4x - 9$ **51.** $2x^2 + 6x$ **53.** $3x^2 + 4x$ **55.** $27x^2 + 10x$ **57. (a)** $P(t) = 0.05t^3 - 0.85t^2 + 0.6t + 45$; **(b)** $P(10) = 16$ lb; $P(20) = 117$ lb **59. (a)** $T(t) + B(t) + R(t) = 77t^2 + 337.5t + 8766$; **(b)** \$8766; **(c)** \$12,378.50 **61.** $T(t) + B(t) + R(t) - S(t) = -485.5t^2 + 25t + 6141$ **63.** $-6x^9$ **65.** $6y - 24$ **67.** $R = x - 50$ **69.** $R = 2x$ **71.** 30 **77.** $5x^2 + 3x + 1$ **79.** $4x^3 + 10x^2 + 1$ **81.** $4x^2 + 8x$

Exercises 4.6

1. $45x^5$ **3.** $-10x^3$ **5.** $6y^3$ **7.** $3x + 3y$ **9.** $10x - 5y$ **11.** $-8x^2 + 12x$ **13.** $x^5 + 4x^4$ **15.** $4x^2 - 4x^3$ **17.** $3x^2 + 3xy$ **19.** $-8xy^2 + 12y^3$ **21.** $x^2 + 3x + 2$ **23.** $y^2 - 5y - 36$ **25.** $x^2 - 5x - 14$ **27.** $x^2 - 12x + 27$ **29.** $y^2 - 6y + 9$ **31.** $6x^2 + 7x + 2$ **33.** $6y^2 + y - 15$ **35.** $10z^2 + 43z - 9$ **37.** $6x^2 - 34x + 44$ **39.** $16z^2 + 8z + 1$ **41.** $6x^2 + 11xy + 3y^2$ **43.** $2x^2 + xy - 3y^2$ **45.** $10z^2 + 13yz - 3y^2$ **47.** $12x^2 - 11xz + 2z^2$ **49.** $4x^2 - 12xy + 9y^2$ **51.** $6 + 17x + 12x^2$ **53.** $6 - 7x - 3x^2$ **55.** $8 - 16x - 10x^2$ **57.** $(x^2 + 7x + 10)$ square units **59.** $(96t - 16t^2)$ ft **61.** $V_2CP + V_2PR - V_1CP - V_1PR$ **63.** 1200 ft^2 **65.** $800 + 160 + 40 + 200 = 1200$ ft^2 **67. (a)** $S_1 = 20x$ ft^2; **(b)** $S_2 = xy$ ft^2; **(c)** $S_3 = 40y$ ft^2 **69.** $800 + 20x + xy + 40y$; yes

71. $9x^2$ **73.** $9A^2$ **75.** $P = -3p^2 + 65p - 100$ **77.** $18 **83.** $-35x^6$ **85.** $x^2 + 4x - 21$ **87.** $9x^2 + 9x - 4$ **89.** $10x^2 - 19xy + 6y^2$ **91.** $6x - 18y$

Exercises 4.7

1. $x^2 + 2x + 1$ **3.** $4x^2 + 4x + 1$ **5.** $9x^2 + 12xy + 4y^2$ **7.** $x^2 - 2x + 1$ **9.** $4x^2 - 4x + 1$ **11.** $9x^2 - 6xy + y^2$ **13.** $36x^2 - 60xy + 25y^2$ **15.** $4x^2 - 28xy + 49y^2$ **17.** $x^2 - 4$ **19.** $x^2 - 16$ **21.** $9x^2 - 4y^2$ **23.** $x^2 - 36$ **25.** $x^2 - 144$ **27.** $9x^2 - y^2$ **29.** $4x^2 - 49y^2$ **31.** $x^4 + 7x^2 + 10$ **33.** $x^4 + 2x^2y + y^2$ **35.** $9x^4 - 12x^2y^2 + 4y^4$ **37.** $x^4 - 4y^4$ **39.** $4x^2 - 16y^4$ **41.** $x^3 + 4x^2 + 8x + 15$ **43.** $x^3 + 3x^2 - x + 12$ **45.** $x^3 + 2x^2 - 5x - 6$ **47.** $x^3 - 8$ **49.** $-x^3 + 2x^2 - 3x + 2$ **51.** $-x^3 + 8x^2 - 15x - 4$ **53.** $2x^3 + 6x^2 + 4x$ **55.** $3x^3 + 3x^2 - 6x$ **57.** $4x^3 - 12x^2 + 8x$ **59.** $5x^3 - 20x^2 - 25x$ **61.** $x^3 + 15x^2 + 75x + 125$ **63.** $8x^3 + 36x^2 + 54x + 27$ **65.** $8x^3 + 36x^2y + 54xy^2 + 27y^3$ **67.** $16t^4 + 24t^2 + 9$ **69.** $16t^4 + 24t^2u + 9u^2$ **71.** $9t^4 - 2t^2 + \frac{1}{9}$ **73.** $9t^4 - 2t^2u + \frac{1}{9}u^2$ **75.** $9x^4 - 25$ **77.** $9x^4 - 25y^4$ **79.** $16x^6 - 25y^6$

81. $\frac{(x^3 + 8x^2 + 20x + 16)\pi}{3} = \frac{1}{3}\pi x^3 + \frac{8}{3}\pi x^2 + \frac{20}{3}\pi x + \frac{16}{3}\pi$

83. $\frac{4(x^3 + 3x^2 + 3x + 1)\pi}{3} = \frac{4}{3}\pi x^3 + 4\pi x^2 + 4\pi x + \frac{4}{3}\pi$

85. $(x^3 + 4x^2 + 5x + 2)\pi = \pi x^3 + 4\pi x^2 + 5\pi x + 2\pi$ **87.** $I_D = 0.004 + 0.004V + 0.001V^2$ **89.** $T_1^4 - T_2^4$ **91.** $Kt_n^2 - 2Kt_nt_a + Kt_a^2$ **93.** $-\frac{1}{2}$ **95.** (a) 9; (b) 5; (c) No **97.** (a) x^2; (b) xy; (c) y^2; (d) xy **99.** They are equal. **105.** $x^2 + 14x + 49$ **107.** $4x^2 + 20xy + 25y^2$ **109.** $x^4 + 4x^2y + 4y^2$ **111.** $9x^2 - 4y^2$ **113.** $15x^2 + 2x - 24$ **115.** $21x^2 - 34xy + 8y^2$ **117.** $x^3 + 3x^2 + 3x + 2$ **119.** $x^4 + x^3 - x^2 - 2x - 2$ **121.** $x^3 + 6x^2y + 12xy^2 + 8y^3$ **123.** $3x^5 - 3x$

Exercises 4.8

1. $x + 3y$ **3.** $2x - y$ **5.** $-2y + 8 - \frac{4}{y}$ **7.** $10x + 8$ **9.** $3x - 2$ **11.** $x + 2$ **13.** $(y - 2)$ R -1 **15.** $x + 6$ **17.** $3x - 4$ **19.** $(2y - 5)$ R -1 **21.** $x^2 - x - 1$ **23.** $y^2 - y - 1$ **25.** $(4x^2 + 3x + 7)$ R 12 **27.** $x^2 + 2x + 4$ **29.** $4y^2 + 8y + 16$ **31.** $(x^2 + x + 1)$ R 3 **33.** $x^3 + x^2 - x + 1$ **35.** $(m^2 - 8)$ R 10 **37.** $x^2 + xy + y^2$ **39.** $(x^2 + 2x + 4)$ R 16 **41.** 180 **43.** 120 **45.** 180 **47.** $\overline{C(x)} = 3x + 5$ **49.** $\overline{P(x)} = 50 + x - \frac{7000}{x}$ **55.** $2x^2 + 2x + 3$ **57.** $4x^2 - 3x$ **59.** $2y - 4 - \frac{1}{y^2}$

Review Exercises

1. [4.1A] (a) (i) $-15a^3b^4$; (ii) $-24a^3b^5$; (iii) $-35a^3b^4$; (b) (i) $6x^3y^3z^5$; (ii) $12x^3y^4z^3$; (iii) $20x^2y^3z^4$ **2.** [4.1B] (a) (i) $-2x^5y^4$; (ii) $-6x^6y^3$; (iii) $-2x^7y^3$; (b) (i) $\frac{x^5y^6}{2}$; (ii) $\frac{xy^7}{2}$; (iii) $\frac{xy^6}{3}$ **3.** [4.1C] (a) $2^6 = 64$; (b) $2^4 = 16$; (c) $3^4 = 81$ **4.** [4.1C] (a) y^6; (b) x^6; (c) a^{20} **5.** [4.1C] (a) $16x^2y^6$; (b) $8x^6y^3$; (c) $27x^6y^6$ **6.** [4.1C] (a) $-8x^3y^9$; (b) $9x^4y^6$; (c) $-8x^6y^6$ **7.** [4.1C] (a) $\frac{4y^4}{x^8}$; (b) $\frac{27x^3}{y^9}$; (c) $\frac{16x^8}{y^{16}}$ **8.** [4.1C] (a) $32x^{12}y^4$; (b) $-72x^4y^9$; (c) $256x^6y^{16}$ **9.** [4.2A] (a) $\frac{1}{2^3} = \frac{1}{8}$; (b) $\frac{1}{3^4} = \frac{1}{81}$; (c) $\frac{1}{5^2} = \frac{1}{25}$ **10.** [4.2A] (a) x^4; (b) y^3; (c) z^5 **11.** [4.2B] (a) x^{-5}; (b) y^{-7}; (c) z^{-8} **12.** [4.2C] (a) (i) 8; (ii) 4; (iii) 27; (b) (i) $\frac{1}{y^8}$; (ii) $\frac{1}{y^5}$; (iii) $\frac{1}{y^6}$ **13.** [4.2C] (a) (i) $\frac{1}{x^4}$; (ii) $\frac{1}{x^6}$; (iii) $\frac{1}{x^8}$; (b) (i–iii) 1; (c) (i) x; (ii) x^3; (iii) x^2 **14.** [4.2C] (a) $\frac{9x^2}{4y^{14}}$; (b) $\frac{8x^6y^{21}}{27}$; (c) $\frac{4}{9x^2y^8}$ **15.** [4.3A] (a) (i) 4.4×10^7; (ii) 4.5×10^6; (iii) 4.6×10^5; (b) (i) 1.4×10^{-3}; (ii) 1.5×10^{-4}; (iii) 1.6×10^{-5} **16.** [4.3B] (a) (i) 2.2×10^5; (ii) 9.3×10^6; (iii) 1.24×10^8; (b) (i) 5; (ii) 6; (iii) 7 **17.** [4.4A] (a) T; (b) M; (c) B **18.** [4.4B] (a) 4; (b) 2; (c) 2 **19.** [4.4C] (a) $9x^4 + 4x^2 - 8x$; (b) $4x^2 - 3x - 3$; (c) $-4x^2 + 3x + 8$ **20.** [4.4D] (a) 284; (b) 156; (c) -100 **21.** [4.5A] (a) $5x^2 - x - 10$; (b) $-5x^2 + 15x + 2$; (c) $9x^2 - 7x + 1$ **22.** [4.5B] (a) $-x^2 - 7x + 4$; (b) $7x^2 - 7x + 3$; (c) $-5x^2 + 4x - 11$ **23.** [4.6A] (a) $-18x^7$; (b) $-40x^9$; (c) $-27x^{11}$ **24.** [4.6B] (a) $-2x^3 - 4x^2y$; (b) $-6x^4 - 9x^3y$; (c) $-20x^4 - 28x^3y$ **25.** [4.6C] (a) $x^2 + 15x + 54$; (b) $x^2 + 5x + 6$; (c) $x^2 + 16x + 63$ **26.** [4.6C] (a) $x^2 + 4x - 21$; (b) $x^2 + 4x - 12$; (c) $x^2 + 4x - 5$ **27.** [4.6C] (a) $x^2 - 4x - 21$; (b) $x^2 - 4x - 12$; (c) $x^2 - 4x - 5$ **28.** [4.6C] (a) $6x^2 - 13xy + 6y^2$; (b) $20x^2 - 27xy + 9y^2$; (c) $8x^2 - 26xy + 15y^2$ **29.** [4.7A] (a) $4x^2 + 12xy + 9y^2$; (b) $9x^2 + 24xy + 16y^2$; (c) $16x^2 + 40xy + 25y^2$ **30.** [4.7B] (a) $4x^2 - 12xy + 9y^2$; (b) $9x^2 - 12xy + 4y^2$; (c) $25x^2 - 20xy + 4y^2$ **31.** [4.7C] (a) $9x^2 - 25y^2$; (b) $9x^2 - 4y^2$; (c) $9x^2 - 16y^2$ **32.** [4.7D] (a) $x^3 + 4x^2 + 5x + 2$; (b) $x^3 + 5x^2 + 8x + 4$; (c) $x^3 + 6x^2 + 11x + 6$ **33.** [4.7E] (a) $3x^3 + 9x^2 + 6x$; (b) $4x^3 + 12x^2 + 8x$; (c) $5x^3 + 15x^2 + 10x$ **34.** [4.7E] (a) $x^3 + 6x^2 + 12x + 8$; (b) $x^3 + 9x^2 + 27x + 27$; (c) $x^3 + 12x^2 + 48x + 64$ **35.** [4.7E] (a) $25x^4 - 5x^2 + \frac{1}{4}$; (b) $49x^4 - 7x^2 + \frac{1}{4}$; (c) $81x^4 - 9x^2 + \frac{1}{4}$ **36.** [4.7E] (a) $9x^4 - 4$; (b) $9x^4 - 16$; (c) $4x^4 - 25$ **37.** [4.8A] (a) $2x^2 - x$; (b) $4x^2 - 2x$; (c) $4x^2 - 2x$ **38.** [4.8B] (a) $x + 6$; (b) $x + 7$; (c) $x + 8$ **39.** [4.8B] (a) $4x^2 - 4x - 4$; (b) $6x^2 - 6x - 6$; (c) $2x^2 - 2x - 2$ **40.** [4.8B] (a) $(2x^2 + 6x - 2)$ R 2; (b) $(2x^2 + 6x - 3)$ R 3; (c) $(3x^2 + 3x - 1)$ R 4 **41.** [4.8B] (a) $(x^2 + x)$ R $(4x + 1)$; (b) $(x^2 + x)$ R $(5x + 1)$; (c) $(x^2 + x)$ R $(6x + 1)$

Cumulative Review Chapters 1–4

1. 7 **2.** $9\frac{9}{10}$ **3.** $-\frac{5}{18}$ **4.** 13.6 **5.** -6.24 **6.** -16 **7.** $\frac{6}{7}$ **8.** 47 **9.** Associative law of multiplication **10.** $6x - 24$ **11.** $-7xy^3$ **12.** $4x + 10$ **13.** $\frac{m + 3n}{p}$ **14.** No **15.** $x = 15$ **16.** $x = 49$ **17.** $x = 36$ **18.** $x = 6$ **19.** $b = \frac{S}{6a^2}$ **20.** 40 and 60 **21.** $430 from stocks and $185 from bonds **22.** 18 hr **23.** $8000 in bonds; $9000 in certificates of deposit

24.

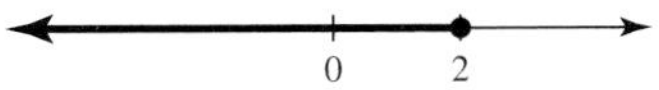

25.

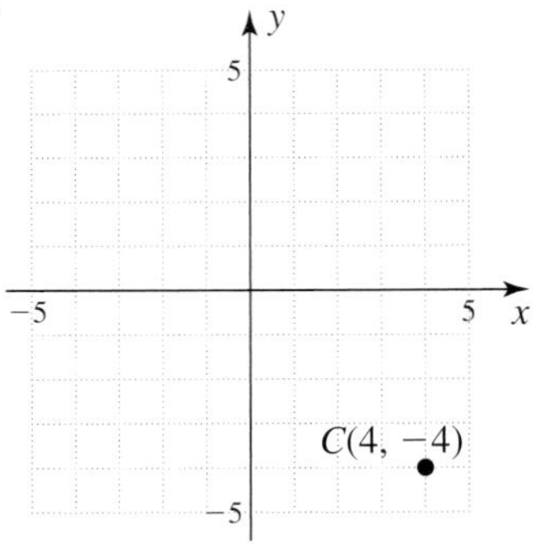

26. $(3, -2)$ **27.** Yes **28.** $x = 1$

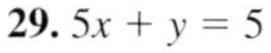

29. $5x + y = 5$

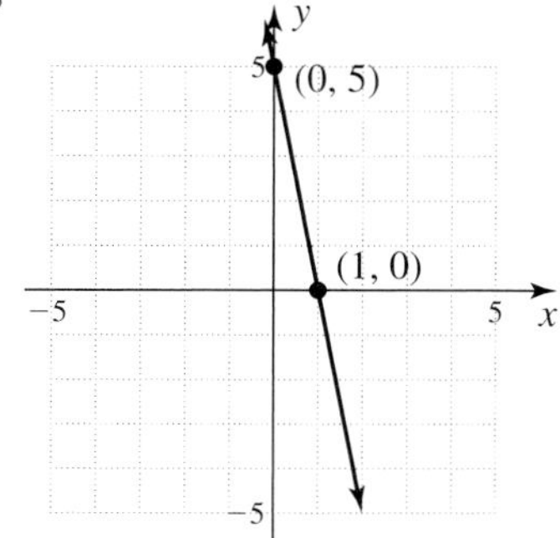

30. $3y - 15 = 0$

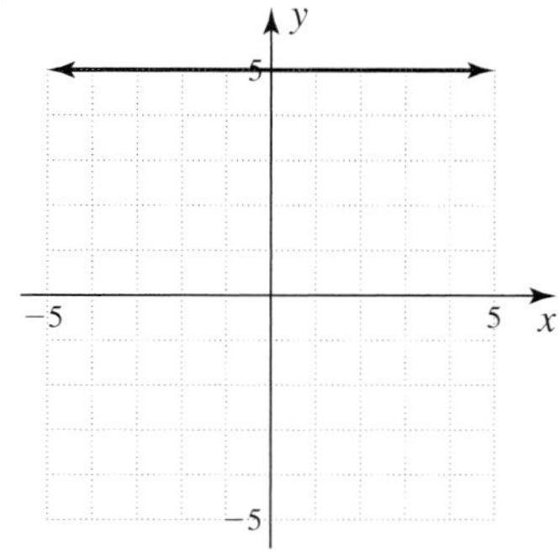

31. $\frac{7}{5}$ **32.** 2 **33.** (2) and (3) **34.** $18x^7y^3$ **35.** $-\frac{5x}{y^2}$ **36.** x^3
37. x^5 **38.** $\frac{64x^{12}}{y^{12}}$ **39.** 8.0×10^6 **40.** 4.2×10^{-8} **41.** Monomial
42. 2 **43.** $-3x^3 - 3x^2 - x + 4$ **44.** 32 **45.** $4x^3 + 9x - 3$
46. $-9x^3 - 12xy$ **47.** $9x^2 + 12xy + 4y^2$ **48.** $16x^2 - 9y^2$
49. $25x^4 - 5x^2 + \frac{1}{4}$ **50.** $16x^4 - 81$ **51.** $(3x^2 - 5x + 4)$ R 3

Chapter 5

Exercises 5.1

1. 4 **3.** 8 **5.** 1 **7.** a^3 **9.** x^3 **11.** $5y^6$ **13.** $2x^3$ **15.** $3bc$
17. 3 **19.** $9ab^3z$ **21.** $3(x + 5)$ **23.** $9(y - 2)$ **25.** $-5(y - 4)$
27. $-3(x + 9)$ **29.** $4x(x + 8)$ **31.** $6x(1 - 7x)$
33. $-5x^2(1 + 5x^2)$ **35.** $3x(x^2 + 2x + 3)$ **37.** $9y(y^2 - 2y + 3)$
39. $6x^2(x^4 + 2x^3 - 3x^2 + 5)$ **41.** $8y^3(y^5 + 2y^2 - 3y + 1)$
43. $\frac{1}{7}(4x^3 + 3x^2 - 9x + 3)$ **45.** $\frac{1}{8}y^2(7y^7 + 3y^4 - 5y^2 + 5)$
47. $(x + 4)(3 - y)$ **49.** $(y - 2)(x - 1)$ **51.** $(t + s)(c - 1)$
53. $4x^3(1 + x - 3x^2)$ **55.** $6y^7(1 - 2y^2 - y^4)$ **57.** $(x + 2)(x^2 + 1)$
59. $(y - 3)(y^2 + 1)$ **61.** $(2x + 3)(2x^2 + 1)$ **63.** $(3x - 1)(2x^2 + 1)$
65. $(y + 2)(4y^2 + 1)$ **67.** $(2a + 3)(a^2 + 1)$ **69.** $(x^2 + 4)(3x^2 + 1)$
71. $(2y^2 + 3)(3y^2 + 1)$ **73.** $(y^2 + 3)(4y^2 + 1)$ **75.** $(a^2 - 2)(3a^2 - 2)$
77. $(6a - 5b)(1 + 2d)$ **79.** $(x^2 - y)(1 - 3z)$ **81.** $\alpha L(t_2 - t_1)$
83. $(R - 1)(R - 1)$ **85.** $2\pi r(h + r)$ **87.** $x^2 + 8x + 15$
89. $x^2 - 5x + 4$ **91.** $2x^2 - x - 3$ **93.** $-w(\ell - z)$ **95.** $a(a + 2s)$
97. $-16(t^2 - 5t - 15)$ **101.** $(3x^2 + 1)(2x^2 - 3)$
103. $(2x - 3)(3x^2 - 1)$ **105.** $3x^2(x^4 - 2x^3 + 4x^2 + 9)$
107. $(2x + y)(6x - 5y)$ **109.** $-3(y - 7)$ **111.** $(x + b)(5x + 6y)$
113. 2

Exercises 5.2

1. $(y + 2)(y + 4)$ **3.** $(x + 2)(x + 5)$ **5.** $(y + 5)(y - 2)$
7. $(x + 7)(x - 2)$ **9.** $(x + 1)(x - 7)$ **11.** $(y + 2)(y - 7)$
13. $(y - 2)(y - 1)$ **15.** $(x - 4)(x - 1)$ **17.** Not factorable
19. Not factorable **21.** $(x + a)(x + 2a)$ **23.** $(z + 3b)(z + 3b)$
25. $(r + 4a)(r - 3a)$ **27.** $(x + 10a)(x - a)$ **29.** $-(b - y)(b + 3y)$
31. $(m - 2a)(m + a)$ **33.** $2t(t + 1)(t + 4)$ **35.** $ax^2(ax + 3a + 2)$
37. $b^3x^5(x - 3)(x + 4)$ **39.** $2c^5z^4(z - 3)(z + 5)$
41. **(a)** $-5t^2 + 5t + 10$; **(b)** $-5(t - 2)(t + 1)$; **(c)** $t = 2$ and $t = -1$;
(d) 2 seconds **43.** **(a)** $-16(t - 5)(t + 3)$; **(b)** $t = 5$ and $t = -3$;
(c) 5 seconds **45.** $x^2 + 8x + 16$ **47.** $x^2 - 6x + 9$
49. $9x^2 + 12xy + 4y^2$ **51.** $25x^2 - 20xy + 4y^2$ **53.** $9x^2 - 16y^2$
63. $(y - 2)(3y^2 - 1)$ **65.** $(y - 3)(2y^2 - 1)$ **67.** $(x + 4y)(x + 3y)$
69. $(x - 2y)(x - 5y)$ **71.** not factorable **73.** $2y^3(1 + 3y^2 + 5y^4)$

Exercises 5.3

1. $(2x + 3)(x + 1)$ **3.** $(2x + 3)(3x + 1)$ **5.** $(2x + 1)(3x + 4)$
7. $(x + 2)(2x - 1)$ **9.** $(x + 6)(3x - 2)$ **11.** $(4y - 3)(y - 2)$
13. $2(2y^2 - 4y + 3)$ **15.** $2(3y + 1)(y - 2)$ **17.** $(3y + 2)(4y - 3)$
19. $3(3y + 1)(2y - 3)$ **21.** $(3x + 1)(x + 2)$ **23.** $(5x + 1)(x + 2)$
25. Not factorable **27.** $(3x + 1)(x - 2)$ **29.** $(5x + 2)(3x - 1)$
31. $4(2x + y)(x + 2y)$ **33.** $(2x + 3y)(3x - y)$ **35.** $(7x - 3y)(x - y)$
37. $(3x + y)(5x - 2y)$ **39.** Not factorable **41.** $(3r + 5)(4r - 1)$
43. $(11t + 2)(2t - 3)$ **45.** $3(3x - 2)(2x - 1)$ **47.** $a(3b + 1)(2b + 1)$
49. $x^3y(6x + y)(x + 4y)$ **51.** $-(3x + 2)(2x + 1)$
53. $-(3x + 2)(3x - 1)$ **55.** $-(4m + n)(2m - 3n)$
57. $-(8x - y)(x - y)$ **59.** $-x(x + 3)(x + 2)$ **61.** $(2g + 9)(g - 4)$
63. $(2R - 1)(R - 1)$ **65.** **(a)** $(0.02t + 1)(0.2t + 2.72)$; **(b)** \$2.72;
(c) \$3.72 **67.** $x^2 + 16x + 64$ **69.** $9x^2 - 12x + 4$
71. $4x^2 + 12xy + 9y^2$ **73.** $9x^2 - 25y^2$ **75.** $x^4 - 16$
77. $(2L - 3)(L - 3)$ **79.** $(5t - 7)(t - 1)$ **83.** $(3x + 2)(x - 2)$
85. $(2x + 3y)(x - 2y)$ **87.** $2(4x - 1)(2x + 1)$ **89.** $x^2(3x^2 + 5x - 3)$

Exercises 5.4

1. Yes **3.** No **5.** Yes **7.** No **9.** Yes **11.** $(x + 1)^2$ **13.** $3(x + 5)^2$
15. $(3x + 1)^2$ **17.** $(3x + 2)^2$ **19.** $(4x + 5y)^2$ **21.** $(5x + 2y)^2$
23. $(y - 1)^2$ **25.** $3(y - 4)^2$ **27.** $(3x - 1)^2$ **29.** $(4x - 7)^2$
31. $(3x - 2y)^2$ **33.** $(5x - y)^2$ **35.** $(x + 7)(x - 7)$
37. $(3x + 7)(3x - 7)$ **39.** $(5x + 9y)(5x - 9y)$
41. $(x^2 + 1)(x + 1)(x - 1)$ **43.** $(4x^2 + 1)(2x + 1)(2x - 1)$
45. $\left(\frac{1}{3}x + \frac{1}{4}\right)\left(\frac{1}{3}x - \frac{1}{4}\right)$ **47.** $\left(\frac{1}{2}z + 1\right)\left(\frac{1}{2}z - 1\right)$ **49.** $\left(1 + \frac{1}{2}s\right)\left(1 - \frac{1}{2}s\right)$
51. $\left(\frac{1}{2} + \frac{1}{3}y\right)\left(\frac{1}{2} - \frac{1}{3}y\right)$ **53.** Not factorable **55.** $3x(x + 2)(x - 2)$
57. $5t(t + 2)(t - 2)$ **59.** $5t(1 + 2t)(1 - 2t)$ **61.** $(7x + 2)^2$
63. $(x + 10)(x - 10)$ **65.** $(x + 10)^2$ **67.** $(3 + 4m)(3 - 4m)$
69. $(3x - 5y)^2$ **71.** $(x^2 + 4)(x + 2)(x - 2)$ **73.** $3x(x + 5)(x - 5)$
75. $R^2 - r^2$ **77.** $6(x + 1)(x - 4)$ **79.** $2(x + 3)(x - 3)$
81. $D(x) = (x - 7)^2$ **83.** $C(x) = (x + 6)^2$ **89.** $(x + 1)(x - 1)$
91. $(3x + 5y)(3x - 5y)$ **93.** $(3x - 4y)^2$ **95.** $(4x + 3y)^2$ **97.** $(3x + 5)^2$
99. Not factorable **101.** $\left(\frac{1}{9} + \frac{1}{2}x\right)\left(\frac{1}{9} - \frac{1}{2}x\right)$
103. $2x(3x + 5y)(3x - 5y)$ **105.** Yes; $(x + 3)^2$ **107.** No
109. Yes; $(2x - 5y)^2$

Exercises 5.5

1. $(x+2)(x^2-2x+4)$ **3.** $(2m-3)(4m^2+6m+9)$ **5.** $(3m-2n)(9m^2+6mn+4n^2)$ **7.** $s^3(4-s)(16+4s+s^2)$ **9.** $x^4(3+2x)(9-6x+4x^2)$ **11.** $3(x-3)(x+2)$ **13.** $(5x+1)(x+2)$ **15.** $3x(x^2+2x+7)$ **17.** $2x^2(x^2-2x-5)$ **19.** $2x^2(2x^2+6x+9)$ **21.** $(x+2)(3x^2+1)$ **23.** $(x+1)(3x^2+2)$ **25.** $(x+1)(2x^2-1)$ **27.** $3(x+4)^2$ **29.** $k(x+2)^2$ **31.** $4(x-3)^2$ **33.** $k(x-6)^2$ **35.** $3x(x+2)^2$ **37.** $2x(3x+1)^2$ **39.** $3x^2(2x-3)^2$ **41.** $(x^2+1)(x+1)(x-1)$ **43.** $(x^2+y^2)(x+y)(x-y)$ **45.** $(x^2+4y^2)(x+2y)(x-2y)$ **47.** $-(x+3)^2$ **49.** $-(x+2)^2$ **51.** $-(2x+y)^2$ **53.** $-(3x+2y)^2$ **55.** $-(2x-3y)^2$ **57.** $-2x(3x+2y)^2$ **59.** $-2x(3x+5y)^2$ **61.** $-x(x+1)(x-1)$ **63.** $-x^2(x+2)(x-2)$ **65.** $-x^2(2x+3)(2x-3)$ **67.** $-2x(x-2)(x^2+2x+4)$ **69.** $-2x^2(2x+1)(4x^2-2x+1)$ **71.** $(2x+3)(5x-1)$ **73.** $(2x+1)(x-3)$ **75.** $\frac{2\pi A}{360}(R_1+Kt)$ **77.** $\frac{3S}{2bd^3}(d+2z)(d-2z)$ **83.** $5x^2(x^2-2x+4)$ **85.** $(3x+7)(2x-5)$ **87.** $(3t-4)(9t^2+12t+16)$ **89.** $(x^2+9)(x+3)(x-3)$ **91.** $-(3x+5y)^2$ **93.** $(4y+3x)(16y^2-12xy+9x^2)$ **95.** $-x^2(x+y)(x^2-xy+y^2)$

Exercises 5.6

1. $x=-3$ or $x=-\frac{1}{2}$ **3.** $x=1$ or $x=-\frac{3}{2}$ **5.** $y=3$ or $y=\frac{2}{3}$ **7.** $y=1$ or $y=-\frac{1}{3}$ **9.** $x=-\frac{4}{3}$ or $x=-\frac{1}{2}$ **11.** $x=2$ or $x=-\frac{1}{3}$ **13.** $x=\frac{4}{5}$ or $x=-2$ **15.** $x=1$ or $x=\frac{8}{5}$ **17.** $y=5$ or $y=\frac{2}{3}$ **19.** $y=-2$ or $y=-\frac{1}{2}$ **21.** $x=-\frac{1}{3}$ **23.** $y=4$ **25.** $x=-\frac{2}{3}$ **27.** $y=\frac{5}{2}$ **29.** $x=-5$ **31.** $x=4$ or $x=1$ **33.** $x=-2$ or $x=-\frac{5}{2}$ **35.** $x=1$ **37.** $x=\frac{1}{2}$ or $x=-\frac{1}{2}$ **39.** $y=\frac{5}{2}$ or $y=-\frac{5}{2}$ **41.** $z=3$ or $z=-3$ **43.** $x=\frac{7}{5}$ or $x=-\frac{7}{5}$ **45.** $m=0$ or $m=5$ **47.** $n=0$ or $n=5$ **49.** $y=-3$ or $y=-8$ **51.** $y=9$ or $y=7$ **53.** $v=2$ or $v=-1$ or $v=-2$ **55.** $m=2$ or $m=1$ or $m=4$ **57.** $n=4$ or $n=-1$ or $n=-2$ **59.** $x=1$ or $x=-3$ **61.** $2H^2+6H+9$ **63.** $H^2-6H-27$ **65.** $x=7$ **71.** $x=3$ or $x=-\frac{2}{3}$ **73.** $x=-2$ or $x=\frac{1}{5}$ **75.** $x=0$ or $x=1$ **77.** $x=4$ or $x=-5$ or $x=1$ **79.** $m=-\frac{2}{3}$ or $m=-1$

Exercises 5.7

1. 8, 9 or 1, 2 **3.** 3, 5 or −3, −1 **5.** 4, 6 or 2, 4 **7.** $b=10$ in., $h=8$ in. **9.** $L=15$ in., $h=10$ in. **11.** $x=4$; $r=7$ units **13.** $L=50$ ft, $W=5$ ft **15.** About 247 ft **17.** 6 in., 8 in., 10 in. **19.** 9 in., 12 in., 15 in. **21.** 30 mi/hr **23.** **(a)** 35 mi/hr; **(b)** Yes **25.** 40 mi/hr **27.** $\frac{15}{18}$ **29.** $18(x-2y)$ **31.** $(x+6)(x-1)$ **33.** 1 second **35.** 1 second **41.** 3, 5 or −7, −5 **43.** 5 in., 12 in., 13 in. **45.** 50 mi/hr

Review Exercises

1. [5.1A] **(a)** 30; **(b)** 6; **(c)** 1 **2.** [5.1B] **(a)** $6x^5$; **(b)** $2x^8$; **(c)** x^3 **3.** [5.1C] **(a)** $5x^3(4-11x^2)$; **(b)** $7x^4(2-5x^2)$; **(c)** $8x^7(2-5x^2)$ **4.** [5.1C] **(a)** $\frac{1}{7}x^2(3x^4-5x^3+2x^2-1)$; **(b)** $\frac{1}{9}x^3(4x^4-2x^3+2x^2-1)$; **(c)** $\frac{1}{8}x^5(3x^4-7x^3+3x^2-1)$ **5.** [5.1D] **(a)** $(x-7)(3x^2-1)$; **(b)** $(x+6)(3x^2+1)$; **(c)** $(x-2y)(4x^2+1)$ **6.** [5.2A] **(a)** $(x+7)(x+1)$; **(b)** $(x-9)(x+1)$; **(c)** $(x+5)(x+1)$ **7.** [5.2A] **(a)** $(x-5)(x-2)$; **(b)** $(x-7)(x-2)$; **(c)** $(x+4)(x-2)$ **8.** [5.3B] **(a)** $(2x+3)(3x-2)$; **(b)** $(2x+1)(3x-1)$; **(c)** $(2x+5)(3x-1)$ **9.** [5.3B] **(a)** $(3x-y)(2x-5y)$; **(b)** $(3x-2y)(2x-y)$; **(c)** $(3x-4y)(2x-y)$ **10.** [5.4A] **(a)** $(x+2)^2$; **(b)** $(x+5)^2$; **(c)** $(x+4)^2$ **11.** [5.4B] **(a)** $(3x+2y)^2$; **(b)** $(3x+5y)^2$; **(c)** $(3x+4y)^2$ **12.** [5.4B] **(a)** $(x-2)^2$; **(b)** $(x-3)^2$; **(c)** $(x-6)^2$ **13.** [5.4B] **(a)** $(2x-3y)^2$; **(b)** $(2x-5y)^2$; **(c)** $(2x-7y)^2$ **14.** [5.4C] **(a)** $(x+6)(x-6)$; **(b)** $(x+7)(x-7)$; **(c)** $(x+9)(x-9)$ **15.** [5.4C] **(a)** $(4x+9y)(4x-9y)$; **(b)** $(5x+8y)(5x-8y)$; **(c)** $(3x+10y)(3x-10y)$ **16.** [5.5A] **(a)** $(m+5)(m^2-5m+25)$; **(b)** $(n+4)(n^2-4n+16)$; **(c)** $(y+2)(y^2-2y+4)$ **17.** [5.5A] **(a)** $(2y-3x)(4y^2+6xy+9x^2)$; **(b)** $(4y-5x)(16y^2+20xy+25x^2)$; **(c)** $(2m-5n)(4m^2+10mn+25n^2)$ **18.** [5.5B] **(a)** $3x(x^2-2x+9)$; **(b)** $3x(x^2-2x+10)$; **(c)** $4x(x^2-2x+8)$ **19.** [5.5B] **(a)** $2x(x-2)(x+1)$; **(b)** $3x(x-3)(x+1)$; **(c)** $4x(x-4)(x+1)$ **20.** [5.5B] **(a)** $(x+4)(2x^2+1)$; **(b)** $(x+5)(2x^2+1)$; **(c)** $(x+6)(2x^2+1)$ **21.** [5.5B] **(a)** $k(3x+2)^2$; **(b)** $k(3x+5)^2$; **(c)** $k(2x+5)^2$ **22.** [5.5B] **(a)** $-3x^2(x+3)(x-3)$; **(b)** $-4x^2(x+4)(x-4)$; **(c)** $-5x^2(x+2)(x-2)$ **23.** [5.5C] **(a)** $-(x+y)(x^2-xy+y^2)$; **(b)** $-(2m+3n)(4m^2-6mn+9n^2)$; **(c)** $-(4n+m)(16n^2-4mn+m^2)$ **24.** [5.5C] **(a)** $-(y-x)(y^2+xy+x^2)$; **(b)** $-(2m-3n)(4m^2+6mn+9n^2)$; **(c)** $-(4t-5s)(16t^2+20st+25s^2)$ **25.** [5.5C] **(a)** $-(4x^2+12xy-9y^2)$; **(b)** $-(25x^2+30xy-9y^2)$; **(c)** $-(16x^2+24xy-9y^2)$ **26.** [5.6A] **(a)** $x=5$ or $x=-1$; **(b)** $x=6$ or $x=-1$; **(c)** $x=7$ or $x=-1$ **27.** [5.6A] **(a)** $x=-\frac{5}{2}$ or $x=2$; **(b)** $x=-\frac{5}{2}$ or $x=1$; **(c)** $x=-\frac{3}{2}$ or $x=1$ **28.** [5.6A] **(a)** $x=1$ or $x=\frac{4}{3}$; **(b)** $x=3$ or $x=\frac{7}{2}$; **(c)** $x=3$ or $x=-1$ **29.** [5.7A] **(a)** 4, 6; **(b)** 2, 4 or −2, 0; **(c)** 10, 12 **30.** [5.7C] 18 in., 24 in., 30 in.

Cumulative Review Chapters 1–5

1. $-\frac{13}{24}$ **2.** 16.4 **3.** −31.35 **4.** −25 **5.** 3 **6.** 69 **7.** Associative law of multiplication **8.** xy^4 **9.** $-x-11$ **10.** $\frac{d+5e}{f}$ **11.** $x=3$ **12.** $x=63$ **13.** $x=6$ **14.** 40 and 65 **15.** 10 hr **16.** \$4000 in bonds; \$2000 in certificates of deposit

17.

18.

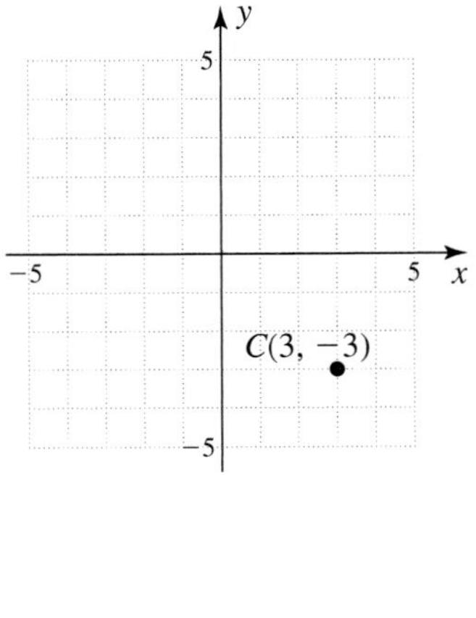

19. No **20.** $x = 2$

21. $x + y = 4$

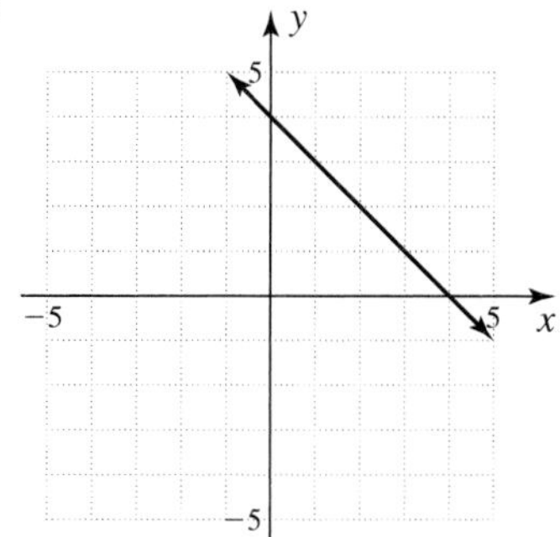

22. $4y - 20 = 0$

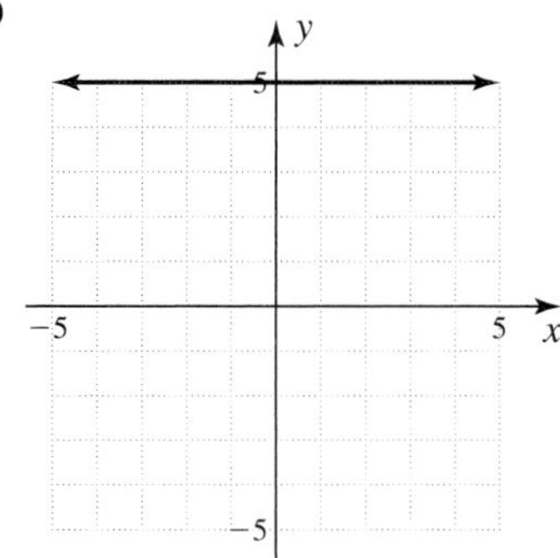

23. $\frac{2}{11}$ **24.** 3 **25.** (1) and (2) **26.** $-\frac{4x^2}{y^4}$ **27.** x^2 **28.** $\frac{1}{x^4}$ **29.** $\frac{y^9}{27x^9}$
30. 4.8×10^{-5} **31.** 4.60×10^{-5} **32.** 2 **33.** 1 **34.** $3x^3 - 7x^2 - 13$
35. $-32x^6 - 16x^4y$ **36.** $9x^2 + 12xy + 4y^2$ **37.** $25x^2 - 49y^2$
38. $4x^4 - \frac{4}{5}x^2 + \frac{1}{25}$ **39.** $25x^4 - 81$ **40.** $2x^2 + 5x + 5$ R 1
41. $2x^6(6 - 7x^3)$ **42.** $\frac{1}{5}x^3(4x^4 - 3x^3 + 4x^2 - 2)$ **43.** $(x - 3)(x - 9)$
44. $(5x - 2y)(4x - 3y)$ **45.** $(5x + 7y)(5x - 7y)$
46. $-5x^2(x + 4)(x - 4)$ **47.** $3x(x + 1)(x - 3)$ **48.** $(2x + 5)(x + 3)$
49. $k(3x + 1)^2$ **50.** $x = -5$ or $x = \frac{3}{4}$

Chapter 6

Exercises 6.1

1. $x = 7$ **3.** $y = -4$ **5.** $x = 3$ or $x = -3$ **7.** None **9.** $y = 2$ or $y = 4$
11. $x = -2$ **13.** $x = 3$ **15.** $x = -3$ **17.** $y = -2$ or $y = 8$
19. $x = 0$ or $x = -1$ **21.** $\frac{9}{21}$ **23.** $\frac{-16}{22}$ **25.** $\frac{20xy}{24y^3}$ **27.** $\frac{-9xy^3}{21y^4}$
29. $\frac{4x(x-2)}{x^2 - x - 2}$ or $\frac{4x^2 - 8x}{x^2 - x - 2}$ **31.** $\frac{-5x(x-2)}{x^2 + x - 6}$ or $\frac{-5x^2 + 10x}{x^2 + x - 6}$
33. $\frac{x^2}{2y^3}$ **35.** $\frac{-3y^4}{x}$ **37.** $\frac{1}{2xy^3}$ **39.** $\frac{5x}{y^2}$ **41.** $\frac{x - y}{3}$
43. $-3(x - y)$ or $3y - 3x$ **45.** $\frac{-1}{4(x - y)}$ or $\frac{1}{4y - 4x}$
47. $\frac{1}{2(x + 2)}$ or $\frac{1}{2x + 4}$ **49.** $\frac{1}{x + y}$ **51.** $\frac{1}{2}$ **53.** $\frac{1}{3}$ **55.** $\frac{-1}{1 + 2y}$
57. $\frac{-1}{1 + 2y}$ **59.** $\frac{-1}{x + 2}$ **61.** -1 **63.** $-(x + 5)$ or $-x - 5$
65. $-(x - 2)$ or $2 - x$ **67.** $\frac{-1}{x + 6}$ **69.** $\frac{1}{x - 2}$
71. **(a)** \$50 million; \$130 million; **(b)** \$30 million; \$78 million;
(c) \$20 million; \$52 million;
(d) The fraction of the total advertising that is spent on national advertising

73. **(a)** 10%; **(b)** 21.25%;
(c) The fraction of the total advertising that is spent on TV advertising
75. $\frac{2}{3}$ **77.** $(x + 3)(x - 1)$ **79.** $(x - 5)(x - 2)$ **81.** $\frac{2}{3}$
83. **(a)** 4 to 1; **(b)** 2500 rpm **89.** $x - 3$ **91.** $\frac{5}{2}$ **93.** $\frac{-1}{x - 4}$ or $\frac{1}{4 - x}$
95. $2(x + y)$ or $2x + 2y$ **97.** $\frac{9xy}{24y^3}$ **99.** $x = 2$ or $x = -2$

Exercises 6.2

1. $\frac{8x}{3y}$ **3.** $\frac{-4xy}{3}$ **5.** $\frac{3y}{2}$ **7.** $-8xy$ **9.** $\frac{x + 1}{x + 7}$ **11.** $\frac{2 - 2x}{x - 2}$
13. $\frac{3x + 15}{x + 1}$ **15.** $-(x + 2)$ or $-x - 2$ **17.** $\frac{x + 3}{x - 3}$ **19.** $\frac{1}{4}$
21. $\frac{2x^2 - 2x}{x + 4}$ **23.** $\frac{y^2 + 5y + 6}{y + 6}$ **25.** $\frac{y^2}{2 - 3y}$ **27.** $\frac{5x - 2}{x - 2}$ **29.** $2y - 5$
31. $\frac{x - 1}{x + 2}$ **33.** $\frac{x - 5}{5x - 15}$ **35.** $\frac{x + 4}{x - 3}$ **37.** $\frac{-5x - 20}{2x + 6}$ **39.** $\frac{x - 2}{x - 1}$
41. $\frac{x + 3}{1 - x^2}$ **43.** $\frac{2x - 4}{x - 3}$ **45.** $\frac{x^2 + 11x + 30}{4}$ **47.** $\frac{x^2 - 4x + 4}{x^2 + 2x - 3}$
49. $\frac{x + 6}{5}$ **51.** $x - 1$ **53.** $\frac{x^2 - 6x + 5}{x^2 - 6x + 8}$ **55.** 1 **57.** $\frac{x + y}{x}$
59. $\frac{xy + 2x + 6y + 3y^2}{xy - 2x - 12y + 6y^2}$ **61.** $\frac{x^2 - x - 12}{2x^2 - x - 1}$ **63.** $\frac{a}{a + 3}$ **65.** $\frac{51}{40}$
67. $\frac{19}{40}$ **69.** $\frac{7}{3}$ **71.** $\frac{RR_T}{R - R_T}$ **73.** $C_R = \frac{60{,}000 + 9000x}{x}$
79. $\frac{x^2 + 4x + 3}{x^2 - 4}$ **81.** $\frac{1}{6 - x}$ **83.** $x^2 - 6x + 9$ **85.** $2 - x$
87. $12x$ **89.** $\frac{x^2 - x - 12}{2x^2 - x - 1}$

Exercises 6.3

1. **(a)** $\frac{5}{7}$; **(b)** $\frac{11}{x}$ **3.** **(a)** $\frac{2}{3}$; **(b)** $\frac{4}{x}$ **5.** **(a)** 2; **(b)** $\frac{5}{x}$
7. **(a)** 2; **(b)** $\frac{2}{3(x + 1)}$ **9.** **(a)** $\frac{4}{3}$; **(b)** $\frac{5x}{2(x + 1)}$
11. **(a)** $\frac{5}{12}$; **(b)** $\frac{56 - 3x}{8x}$ **13.** **(a)** $\frac{12}{35}$; **(b)** $\frac{x^2 + 36}{9x}$
15. **(a)** $\frac{2}{15}$; **(b)** $\frac{5}{14(x - 1)}$ **17.** **(a)** $\frac{53}{56}$; **(b)** $\frac{8x - 1}{(x + 1)(x - 2)}$
19. **(a)** $\frac{13}{24}$; **(b)** $\frac{3x + 12}{(x - 2)(x + 1)}$ **21.** $\frac{2x^2 - 2x - 6}{(x + 4)(x - 4)(x - 1)}$
23. $\frac{5x^2 + 19x}{(x + 5)(x - 2)(x + 3)}$ **25.** $\frac{6x - 4y}{(x + y)^2(x - y)}$
27. $\frac{10 - x}{(x + 5)(x - 5)}$ **29.** $\frac{-6x - 10}{(x + 2)(x + 1)(x + 3)}$ **31.** $\frac{y^2}{(y + 1)(y - 1)}$
33. $\frac{5 - 4y - 2y^2}{(y + 4)(y - 4)}$ **35.** $\frac{2x^2 - x + 3}{(x - 2)(x + 1)^2}$ **37.** 0
39. $\frac{a^2 + 3a + 3}{(a + 2)(a^2 - 2a + 4)}$
41. **(a)** $\frac{P(t)}{R(t)} = \frac{(0.04t^3 - 0.28t^2 + 3.2t + 23)\text{ billion}}{(0.01t^3 - 0.12t^2 + 0.28t + 22)\text{ million}}$;
(b) \$1045.45; **(c)** \$4949.66; \$4922.04
43. \$616.88 **45.** $\frac{9}{20}$ **47.** 42 **49.** $x^2 - 1$
51. 1.5 **53.** $1.41\overline{6}$ **55.** 0.0025
59. $\frac{R(t) - P(t)}{G(t)} = \frac{0.02t^2 - 0.34t + 1.42}{0.04t^2 + 2.34t + 90}$ **61.** $\frac{2x + 18}{(x - 1)(x + 3)}$
63. $\frac{2}{x - 2}$ **65.** $\frac{7x + 9}{(x + 2)(x - 2)(x + 1)}$

Exercises 6.4

1. $\frac{1}{5}$ **3.** $\frac{1}{3}$ **5.** $\frac{ab-a}{b+a}$ **7.** $\frac{b+a}{b-a}$ **9.** $\frac{6b+4a}{48b-9a}$ **11.** $\frac{26}{5}$

13. $\frac{x}{2x-1}$ **15.** 2 **17.** $\frac{1}{x-y}$ **19.** $\frac{x+2}{x+1}$ **21.** $\frac{x-5}{4}$

23. $\frac{1}{2(x-4)}$ **25.** $w = 124$ **27.** $x = -3$ **29.** $x = 3$

31. $\frac{6}{25}$ yr **33.** $11\frac{43}{50}$ yr **41.** $\frac{11}{6}$ **43.** $\frac{w+4}{w+1}$ **45.** $\frac{m+5}{m-4}$

Exercises 6.5

1. $x = 6$ **3.** $x = 4$ **5.** $x = -6$ **7.** $x = 12$ **9.** $x = 4$

11. $x = 5$ **13.** $x = 2$ **15.** $x = 2$ **17.** $x = -11$ **19.** $x = -2$

21. No solution **23.** $x = \frac{-3}{2}$ **25.** $x = -6$ **27.** $x = -8$

29. No solution **31.** No solution **33.** $x = -3$ **35.** $x = \frac{-5}{2}$

37. $x = \frac{1}{7}$ **39.** $x = \frac{-3}{5}$ **41.** $z = \frac{45}{2}$ **43.** $y = \frac{5}{7}$ **45.** $v = \frac{2}{5}$

47. $z = 3$ **49.** $x = 4$ **51.** $x = \frac{33}{4}$ **53.** $h = \frac{12}{5}$ **55.** $R = 15$

57. $h = \frac{2A}{b_1 + b_2}$ **59.** $Q_1 = \frac{PQ_2}{P+1}$ **61.** $f = \frac{ab}{a+b}$ **65.** $x = -3$

67. $x = -4$ **69.** $x = -2$ **71.** $F = \frac{9}{5}C + 32$

Exercises 6.6

1. 9250 **3.** 100 **5.** 6 **7.** 3.6 **9.** **(a)** 2250; **(b)** 36

11. **(a)** 3 to 1; **(b)** 35 to 1; **(c)** Minneapolis; climate **13.** 70,000

15. $5\frac{1}{7}$ hr **17.** $1\frac{7}{8}$ min **19.** 6 hr **21.** 10 mi/hr **23.** 100 mi/hr

25. **(a)** 58; **(b)** No **27.** 189; no **29.** $x = 5\frac{1}{3}$; $y = 6\frac{2}{3}$

31. $DE = 13\frac{5}{7}$ in. **33.** $x = 16$; $y = 8$ **35.** $x = 12$; $y = 12$

37. $a = 24$; $b = 30$ **39.** $x = 8$; $y = 8\frac{4}{5}$ **41.** $x = 10$; $y = 25$

43. -8 **45.** $x = 2$ **47.** $18\frac{2}{3}$ lb **49.** 6 ft **51.** 2650

57. $3\frac{1}{13}$ hr **59.** 200

Review Exercises

1. [6.1B] **(a)** $\frac{10xy}{16y^2}$; **(b)** $\frac{12xy}{16y^3}$; **(c)** $\frac{10x^2y}{15x^5}$

2. [6.1C] **(a)** $-3(x-y)$ or $3y - 3x$; **(b)** $-2(x-y)$ or $2y - 2x$; **(c)** $4(x-y)$ or $4x - 4y$

3. [6.1A, C] **(a)** $\frac{-1}{x+1}$; 0, -1; **(b)** $\frac{-1}{x-1}$ or $\frac{1}{1-x}$; 0, 1; **(c)** $\frac{-1}{1-x}$ or $\frac{1}{x-1}$; 0, 1 **4.** [6.1C] **(a)** $-(x+3)$ or $-x-3$; **(b)** $-(x+2)$ or $-x-2$; **(c)** $-(x+3)$ or $-x-3$

5. [6.2A] **(a)** $\frac{2xy}{3}$; **(b)** $\frac{3xy}{2}$; **(c)** $8xy^2$

6. [6.2A] **(a)** $\frac{x+2}{x+3}$; **(b)** $\frac{x+1}{x+5}$; **(c)** $\frac{x+5}{x+4}$

7. [6.2B] **(a)** $\frac{x-3}{x+2}$; **(b)** $\frac{x-4}{x+1}$; **(c)** $\frac{x-5}{x+4}$

8. [6.2B] **(a)** $\frac{-1}{x-5}$ or $\frac{1}{5-x}$; **(b)** $\frac{-1}{x-1}$ or $\frac{1}{1-x}$; **(c)** $\frac{-1}{x-2}$ or $\frac{1}{2-x}$

9. [6.3A] **(a)** $\frac{2}{x-1}$; **(b)** $\frac{2}{x-2}$; **(c)** $\frac{1}{x+1}$

10. [6.3A] **(a)** $\frac{2}{x+1}$; **(b)** $\frac{2}{x+2}$; **(c)** $\frac{2}{x+3}$

11. [6.3B] **(a)** $\frac{3x-2}{(x+2)(x-2)}$; **(b)** $\frac{4x-2}{(x+1)(x-1)}$; **(c)** $\frac{5x-9}{(x+3)(x-3)}$

12. [6.3B] **(a)** $\frac{-6x-10}{(x+1)(x+2)(x+3)}$; **(b)** $\frac{7x+1}{(x-2)(x+1)^2}$; **(c)** $\frac{-4x}{(x+2)(x+1)(x-1)}$ **13.** [6.4A] **(a)** $\frac{6}{17}$; **(b)** $\frac{14}{27}$; **(c)** $\frac{6}{25}$

14. [6.5A] **(a)** $x = 3$; **(b)** $x = 7$; **(c)** $x = 6$

15. [6.5A] **(a)** $x = -3$; **(b)** $x = -5$; **(c)** $x = -6$

16. [6.5A] **(a)** No solution; **(b)** No solution; **(c)** No solution

17. [6.5A] **(a)** $x = -6$ or $x = \frac{13}{3}$; **(b)** $x = -7$ or $x = \frac{11}{2}$; **(c)** $x = -8$ or $x = \frac{33}{5}$

18. [6.5B] **(a)** $a_1 = \frac{A(1-b^n)}{1-b}$; **(b)** $b_1 = \frac{B(1-c)}{1-c^n}$; **(c)** $c_1 = \frac{C(d+1)}{(1-d)^n}$

19. [6.6A] **(a)** $10\frac{1}{2}$ gal; **(b)** $13\frac{1}{2}$ gal; **(c)** 18 gal

20. [6.6A] **(a)** $x = 18$; **(b)** $x = 10\frac{2}{5}$; **(c)** $x = \frac{-13}{5}$

21. [6.6B] **(a)** $3\frac{3}{7}$ hr; **(b)** $4\frac{4}{9}$ hr; **(c)** $3\frac{3}{5}$ hr

22. [6.6B] **(a)** 4 mi/hr; **(b)** 8 mi/hr; **(c)** 12 mi/hr

23. [6.6B] **(a)** 32; **(b)** 35; **(c)** 40

24. [6.6B] **(a)** 1125; **(b)** 1250; **(c)** 1375

25. [6.6B] **(a)** $10\frac{1}{2}$; **(b)** $10\frac{2}{3}$; **(c)** $2\frac{2}{3}$

Cumulative Review Chapters 1–6

1. $-\frac{17}{30}$ **2.** 2.7 **3.** 25 **4.** 2 **5.** 69 **6.** $6x - 14$

7. $\frac{a-b}{c}$ **8.** $x = 4$ **9.** $x = 63$ **10.** 30 and 65

11. \$9000 in bonds; \$3000 in certificates of deposit

12. 0 2

13.

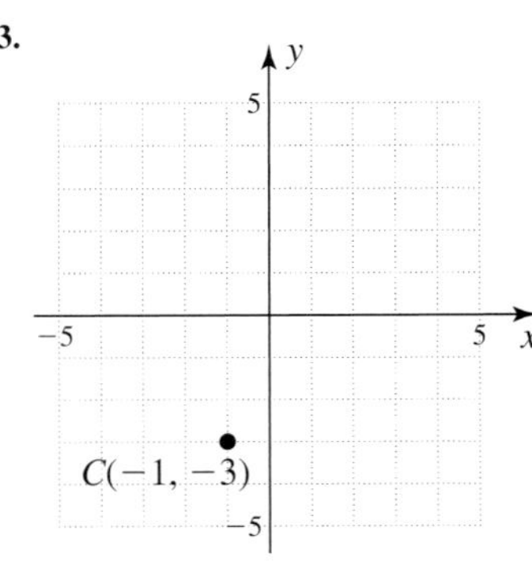

14. Yes **15.** $x = 3$

16.

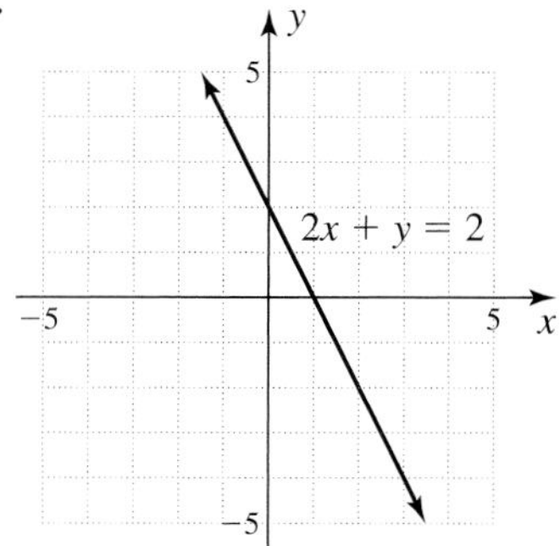

17.

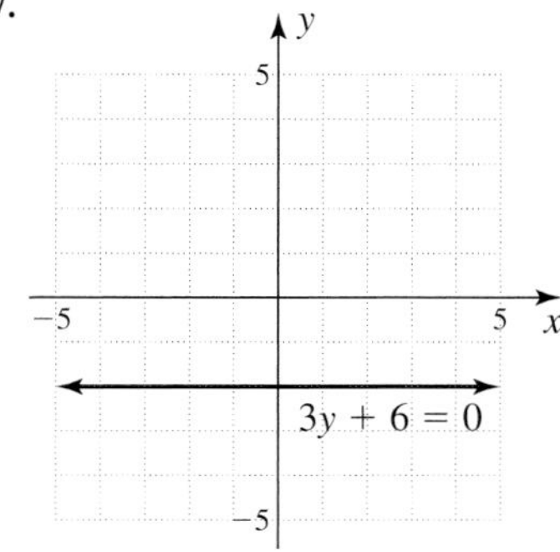

18. $\frac{7}{16}$ **19.** 2 **20.** (2) and (3) **21.** x **22.** $\frac{1}{x^2}$

23. $\frac{y^8}{81x^{12}}$ **24.** 3.6×10^{-4} **25.** 1.80×10^{-5}

26. -4 **27.** $-12x^6 + 9x^4 - 4$ **28.** $25x^4 - 2x^2 + \frac{1}{25}$

29. $81x^4 - 4$ **30.** $(3x^2 - x + 4)$ R 2 **31.** $3x^2(2 - 3x^2)$

32. $\frac{1}{5}x^3(2x^4 - 4x^3 + 4x^2 - 1)$ **33.** $(x - 7)(x - 8)$

34. $(5x - 4y)(3x - 5y)$ **35.** $(4x + 3y)(4x - 3y)$

36. $-4x^2(x + 1)(x - 1)$ **37.** $3x(x + 1)(x - 3)$ **38.** $(x + 1)(4x + 3)$

39. $k(4x - 1)^2$ **40.** $x = -3, x = \frac{5}{2}$ **41.** $\frac{14xy^2}{12y^3}$

42. $-2(x + y)$ or $-2x - 2y$ **43.** $-(x + 5)$ or $-x - 5$ **44.** $\frac{x + 3}{x + 7}$

45. $\frac{-1}{x - 3}$ or $\frac{1}{3 - x}$ **46.** $\frac{6}{x + 8}$ **47.** $\frac{-2x - 11}{(x + 6)(x + 5)(x - 5)}$ **48.** 42

49. 3 **50.** $x = -8$ **51.** No solution **52.** $x = -5$ **53.** 23 gal

54. $-\frac{70}{11}$ **55.** $3\frac{3}{13}$ hr

Chapter 7

Exercises 7.1

1. $y = \frac{1}{2}x + \frac{3}{2}$

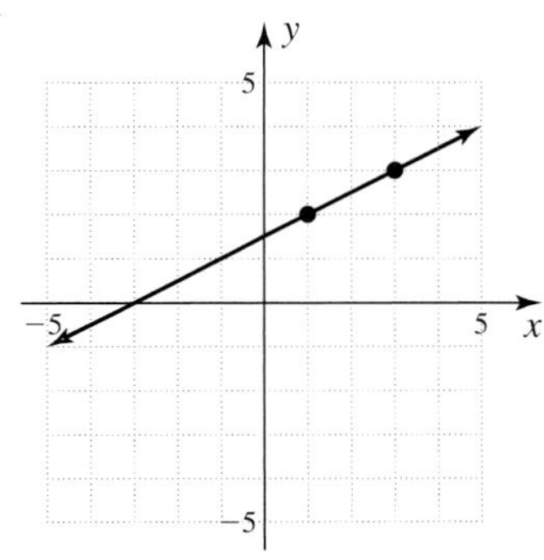

3. $y = -x + 6$

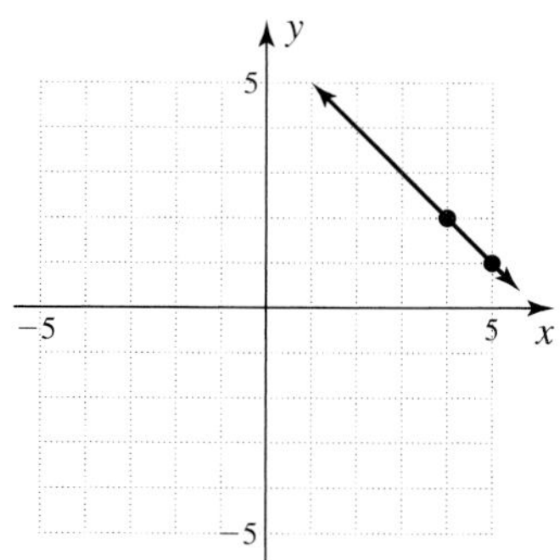

5. $y = 5$

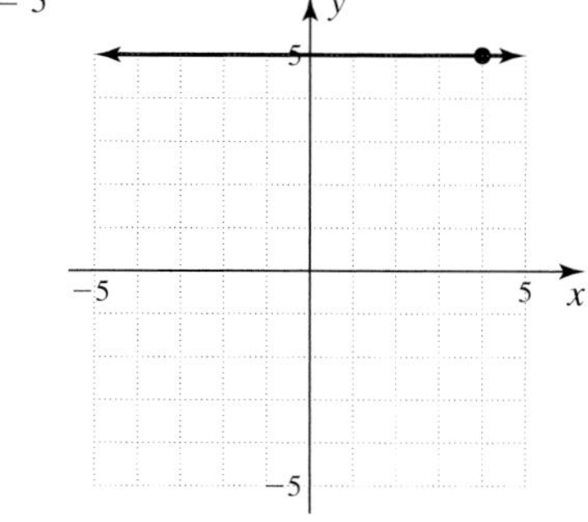

7. $y = 2x - 3$ **9.** $y = -4x + 6$ **11.** $y = \frac{3}{4}x + \frac{7}{8}$ **13.** $y = 2.5x - 4.7$
15. $y = -3.5x + 5.9$

17. $y = \frac{1}{4}x + 3$

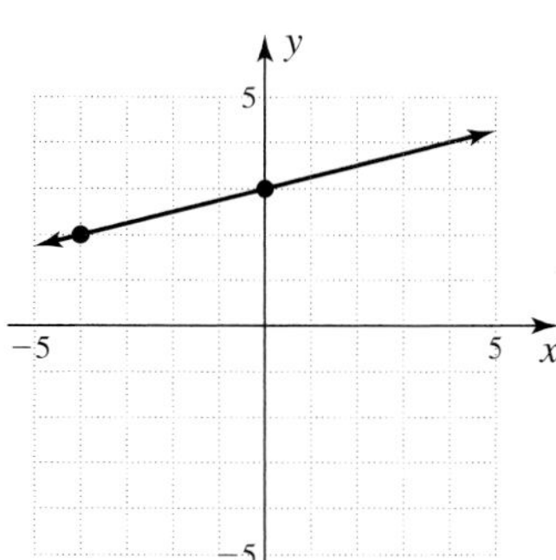

19. $y = -\frac{3}{4}x - 2$

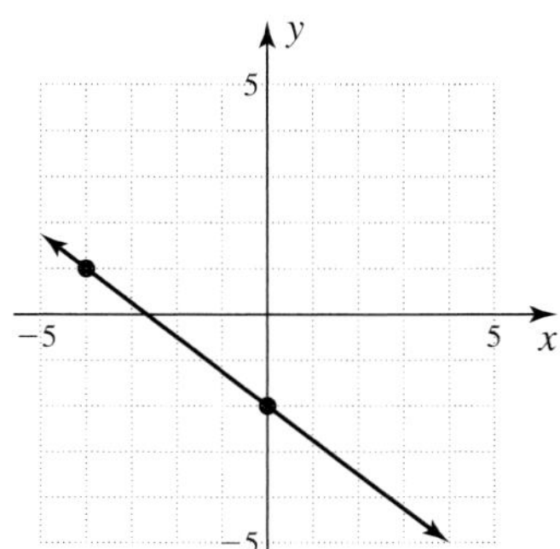

21. $x - y = -1$

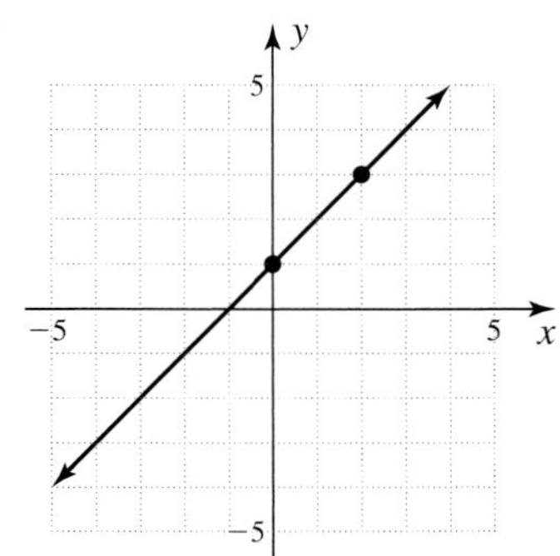

23. $3x - y = 4$

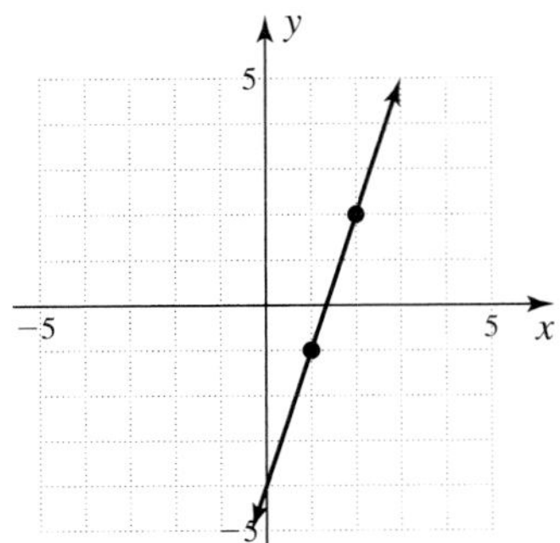

25. $4x + 3y = 12$

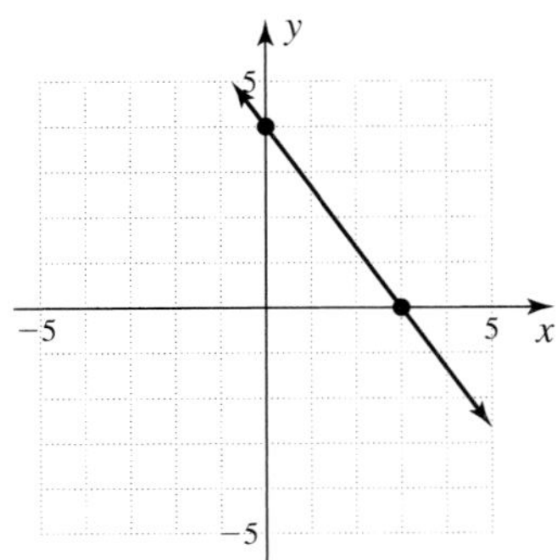

27. $x = 3$

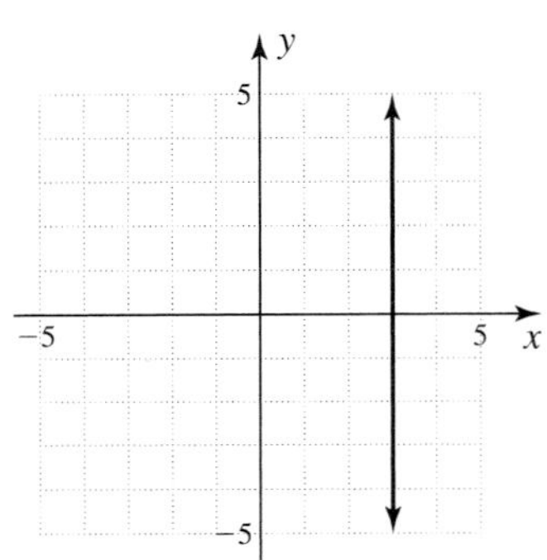

29. $y = -3$

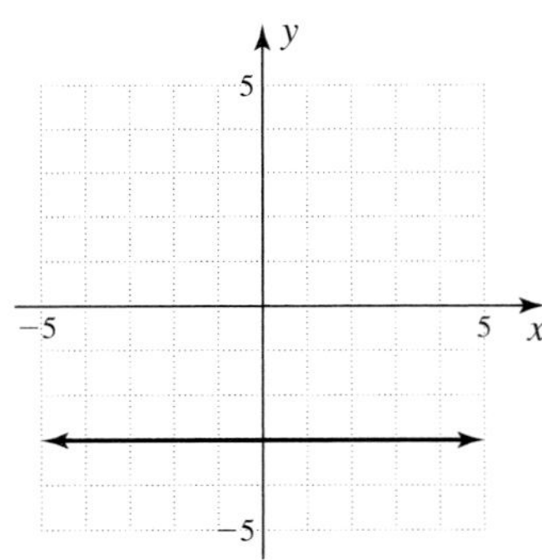

31. $y = x + 3$ **33.** $y = 2x$ **35.** $y = -2x - 3$ **37.** $y = 2x + 50$

39. (a) \$2; **(b)** \$75

47. $y = -\frac{3}{4}x + 2$

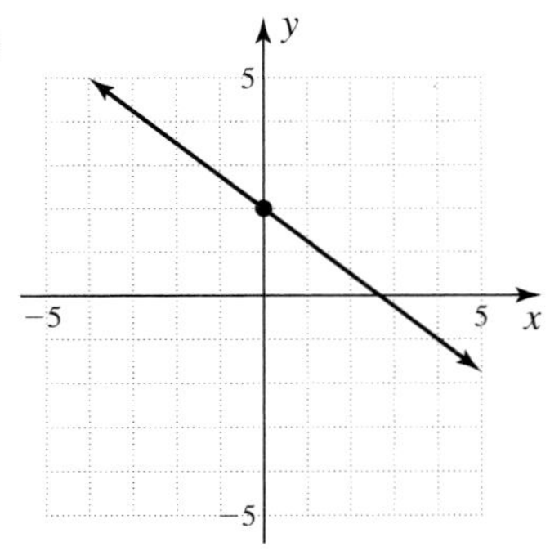

49. $2x - 3y = 9$

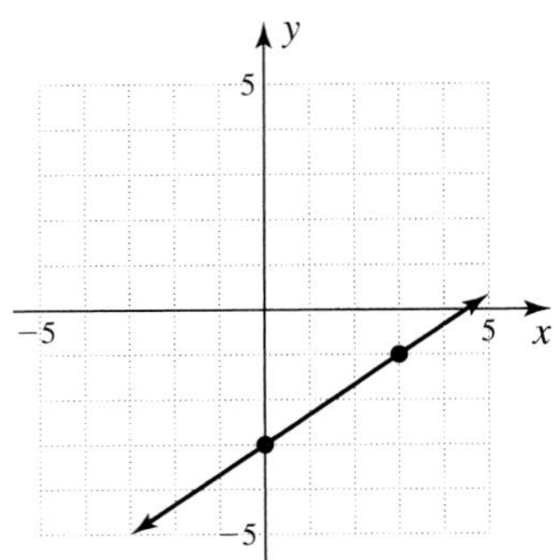

51. $5x + y = -11$

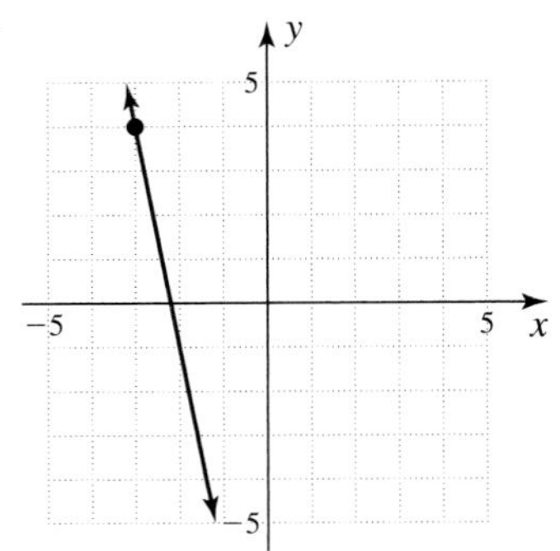

Exercises 7.2

1. $C = 1.7m + 0.3$; \$51.30 **3.** 6 miles **5. (a)** \$2; **(b)** $m - 1$; **(c)** \$1.70; **(d)** $1.70(m - 1)$; **(e)** $C = 2 + 1.70(m - 1)$ or $C = 1.7m + 0.3$; yes **7. (a)** $C = 175$, if $m \le 60$; **(b)** $C = 175$, if $m \le 60$; **(c)** $C = \frac{175}{m}$, if $m \le 60$ **9. (a)** $m = \frac{6}{5}$; **(b)** $m = \frac{6}{5}$; **(c)** Yes; **(d)** $y = \frac{6}{5}x - 11$; **(e)** $y = \frac{6}{5}x - 11$; yes; **(f)** -5

11. (a) $C = 37.5h + 100$; **(b)** 3 hours **13. (a)** $C = 2.3m + 40$; **(b)** 70 minutes; **(c)** $C = 7.2m$; **(d)** About 8 minutes

15. (a) $C = 0.35m + 50$; **(b)** 110 minutes **17.** 100 minutes; after 100 minutes **19. (a)** $y = 1.95x + 29$; **(b)** $m = 1.95$; **(c)** $b = 29$

21. $y = -0.1x + 8$; 5 days; 4 days

23. (a) 3 human years corresponds to 30 dog years; 9 human years corresponds to 60 dog years; **(b)** $m = 5$; **(c)** $d = 5h + 15$; **(d)** 35 dog years; **(e)** 10 years; **(f)** 1.2 years **25. (a)** $m = 0.0255$; **(b)** $c = 0.0255b - 0.0245$; **(c)** 0.0775; 0.1285; **(d)** 4 **27.** $>$ **29.** $>$ **31.** $<$

37.

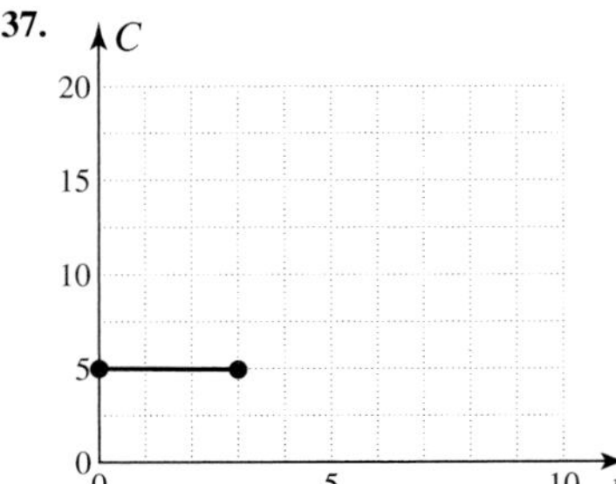

39.

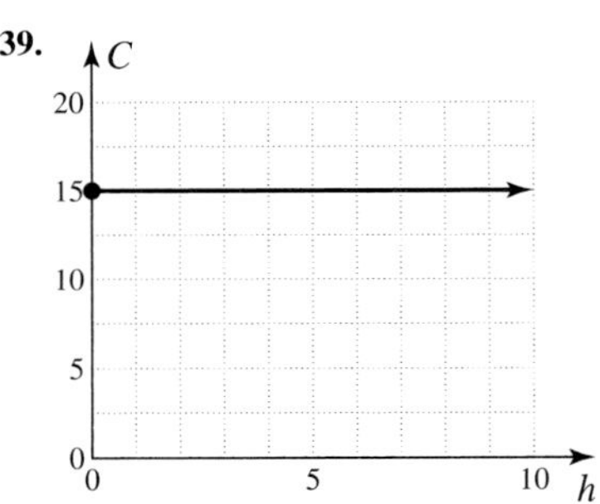

41. $C = 0.15m + 40$; \$107.50 **43.** **(a)** 30 hours cost \$24; **(b)** 33 hours cost \$26.70; **(c)** $C = 0.9x - 3$

Exercises 7.3

1. $2x + y > 4$

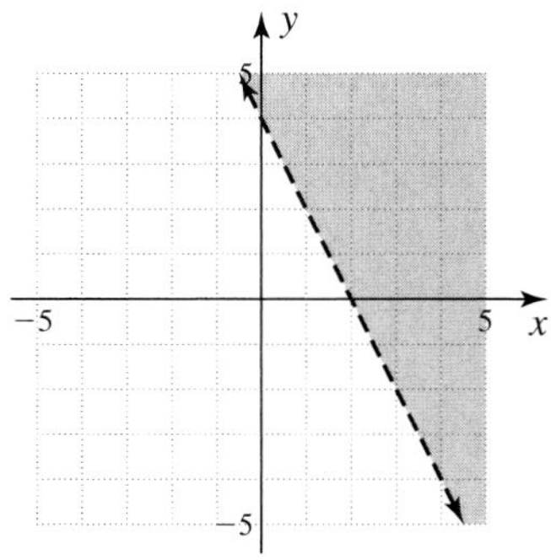

3. $-2x - 5y \le 10$

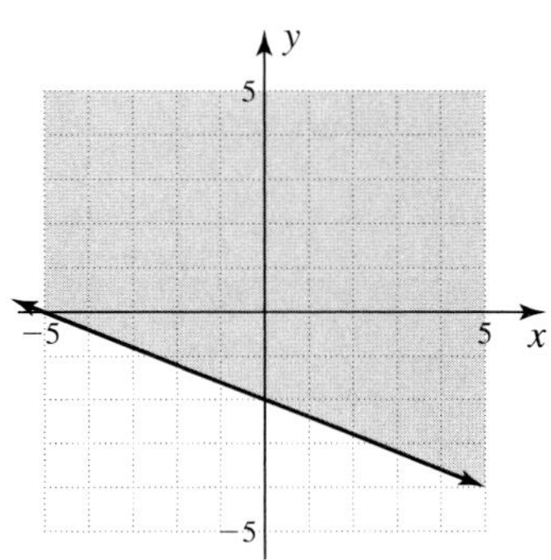

5. $y \ge 3x - 3$

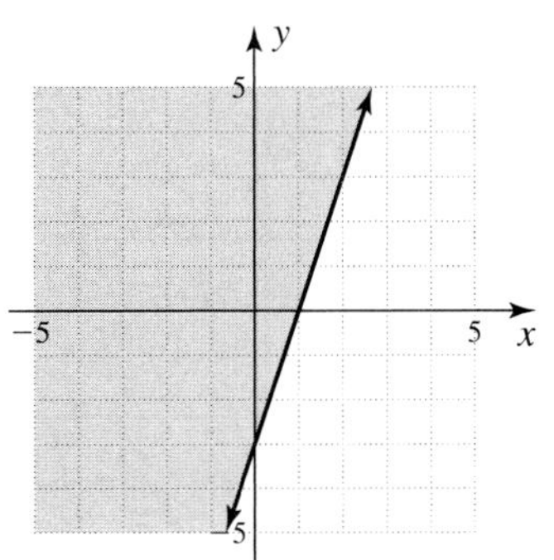

7. $6 < 3x - 6y$

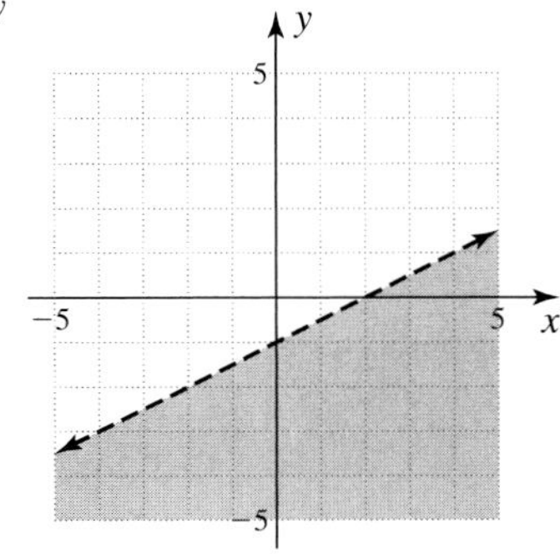

9. $3x + 4y \ge 12$

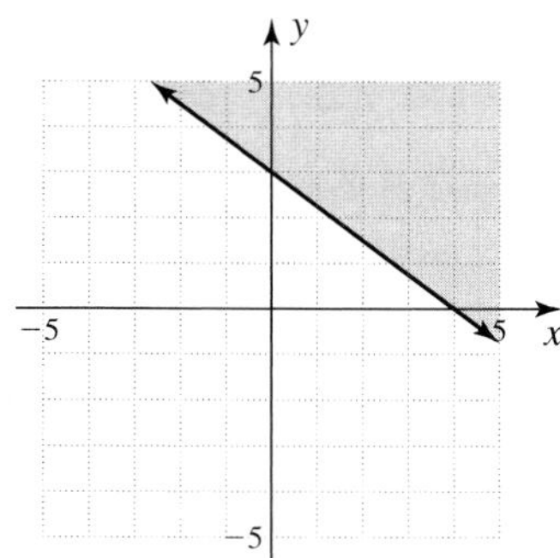

11. $10 < -2x + 5y$

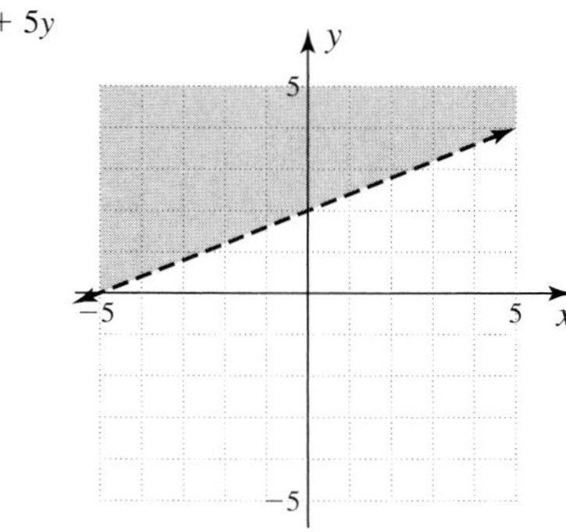

13. $x \ge 2y - 4$

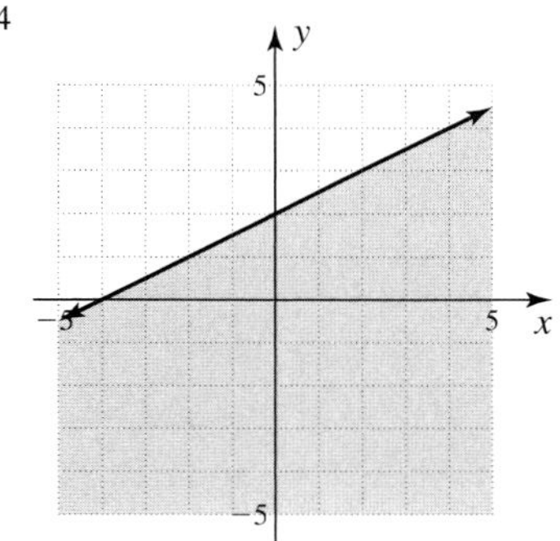

15. $y < -x + 5$

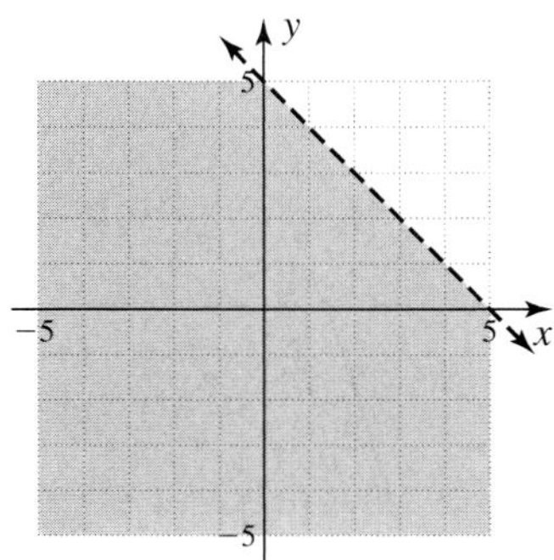

17. $2y < 4x + 5$

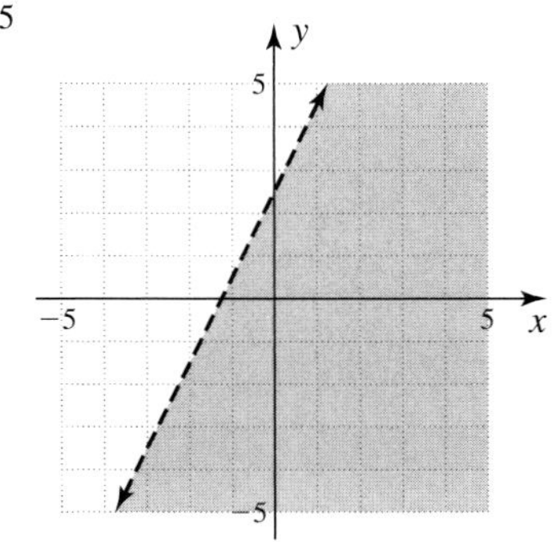

19. $x > 1$

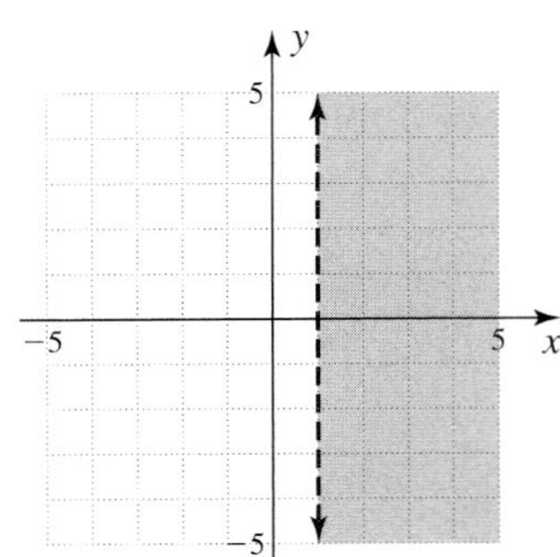

21. $x \le \frac{5}{2}$

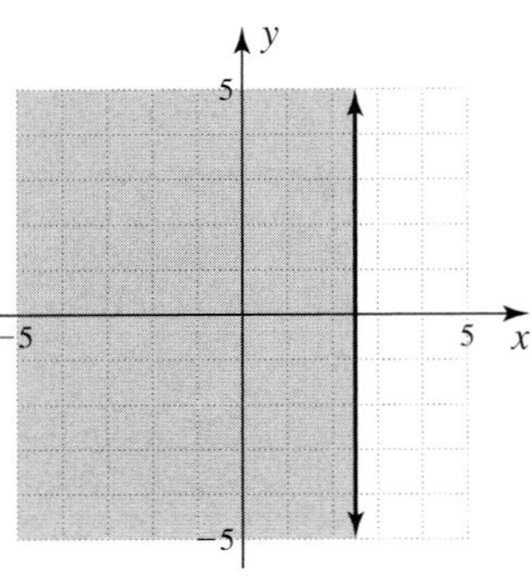

23. $y \le -\frac{3}{2}$

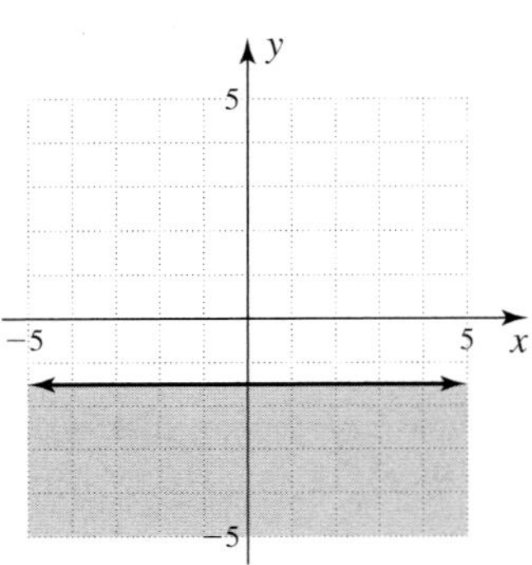

25. $x - \frac{2}{3} > 0$

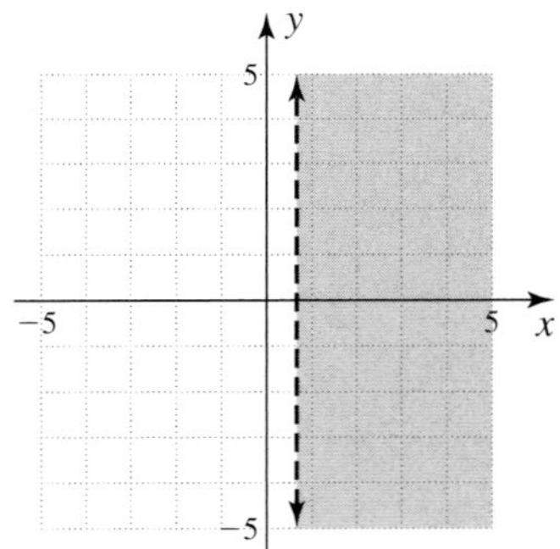

27. $y + \frac{1}{3} \ge \frac{2}{3}$

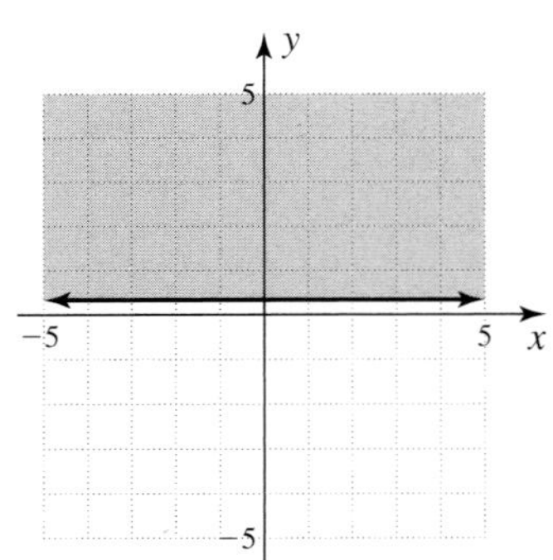

29. $2x + y < 0$

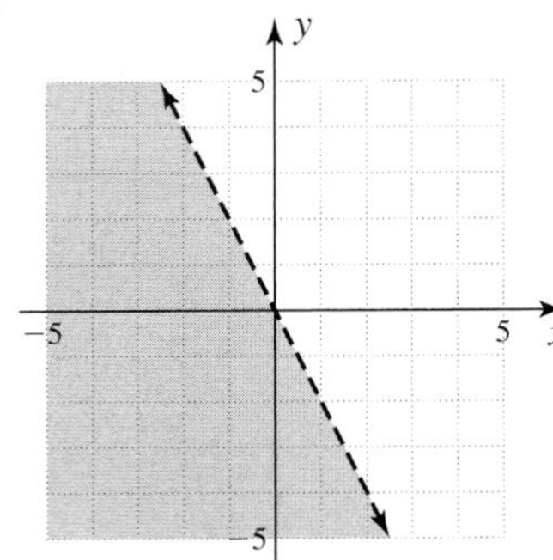

31. $y - 3x > 0$

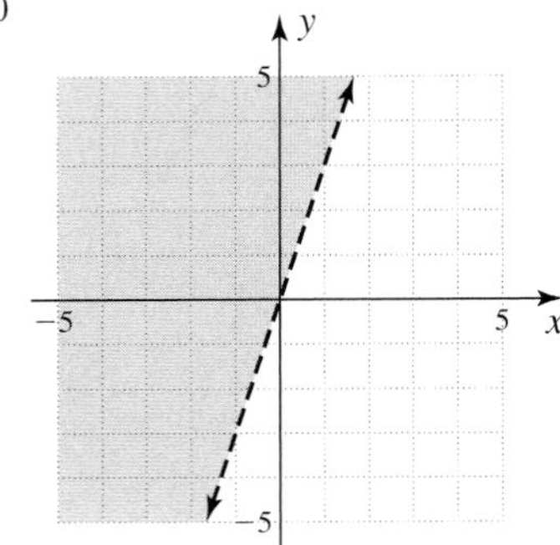

33. rectangle

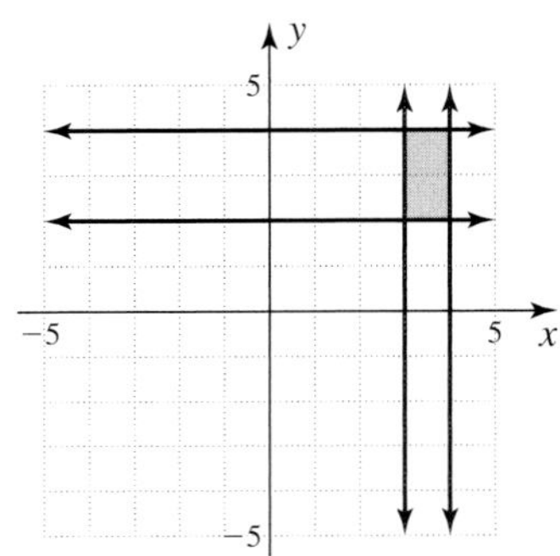

35. $k = 7.8$
37. $k = 0.02575$

39.

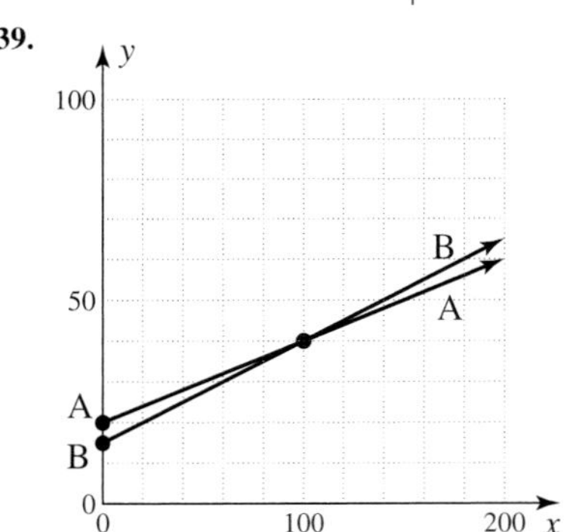

41. When they travel 100 miles **43.** When traveling less than 100 miles

49. $y \ge -4$

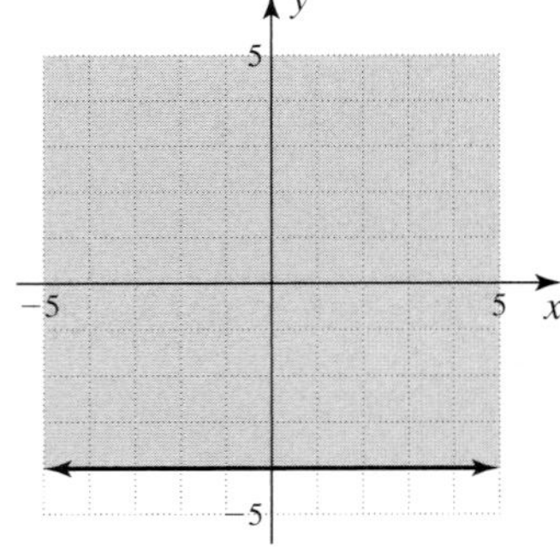

51. $y - 3 > 0$

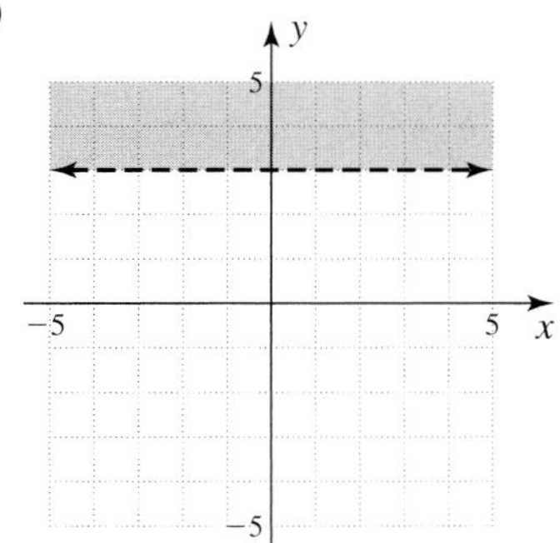

53. $y \leq -4x + 8$

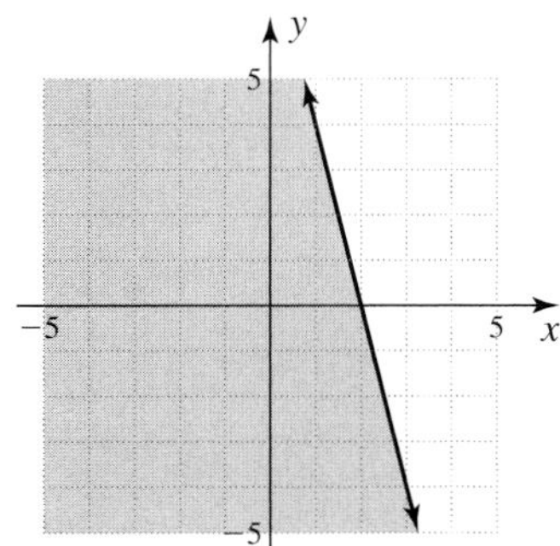

55. $y < 3x$

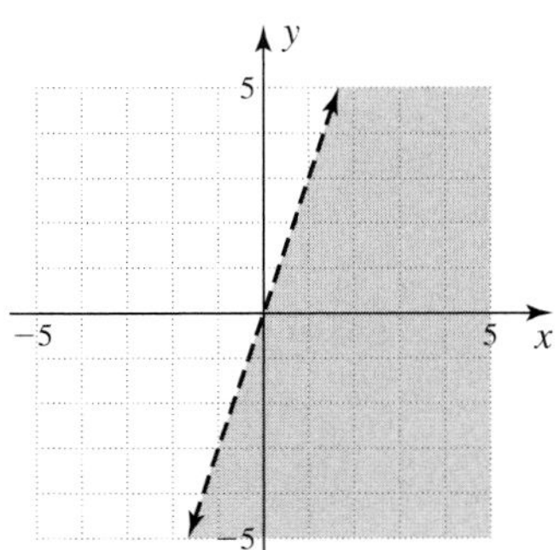

57. $3x + y \leq 0$

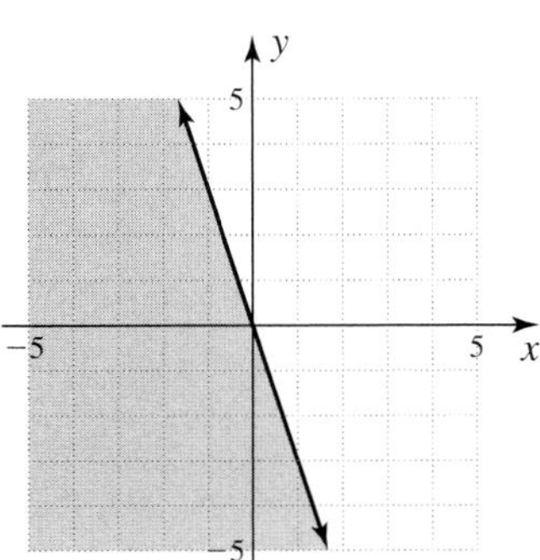

Exercises 7.4

1. **(a)** $S = \frac{W}{16}$; **(b)** $k = \frac{1}{16}$; **(c)** 10 pounds **3.** **(a)** $R = \frac{W}{10}$; **(b)** $k = \frac{1}{10}$; **(c)** 16 **5.** **(a)** $R = kt$; **(b)** $k = 45$; **(c)** 2.4 minutes **7.** **(a)** $T = kh^3$; **(b)** $k \approx 0.00057$; **(c)** 240 pounds **9.** **(a)** $f = \frac{k}{d}$; **(b)** $k = 4$; **(c)** 16 **11.** 10.8 in.^{3} **13.** **(a)** $w = \frac{k}{s}$; **(b)** $k = 7200$; **(c)** 720 words **15.** **(a)** $b = \frac{k}{a}$; **(b)** $k = 2970$; **(c)** 90 (per 1000 women) **17.** **(a)** $d = ks$; **(b)** $k \approx 17.63$; **(c)** The time it took to drive d miles at s miles per hour **19.** **(a)** $C = 4(F - 37)$ or $C = 4F - 148$; **(b)** 212 chirps per minute **21.** **(a)** $I = kn^2$; **(b)** $k = 0.05$; **(c)** 245% **23.** **(a)** $\text{BAC} = k(N - 1)$; **(b)** $k = 0.026$; **(c)** 0.104; **(d)** 4 **25.** **(a)** $\text{BAC} = \frac{7.8}{W}$; **(b)** 0.03 **(c)** 97.5 pounds; **(d)** Less than 0.08

27. $x + 2y = 4$

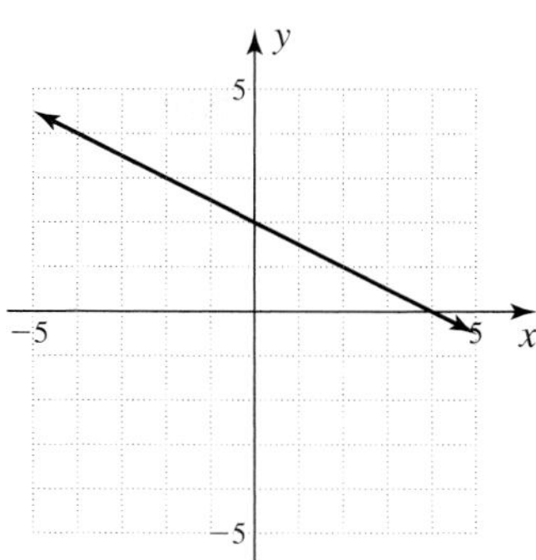

29. $y - 2x = 4$

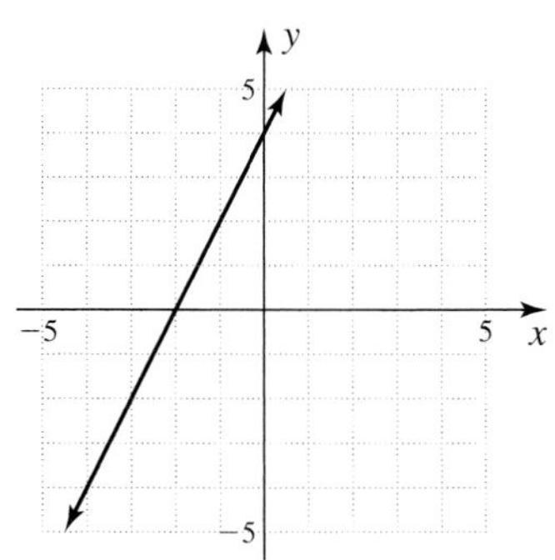

31. Not parallel **33.** **(a)** $T = 15S$; **(b)** 4

Review Exercises

1. [7.1A] **(a)** $2x + y = 1$; **(b)** $3x + y = 4$; **(c)** $4x + y = 7$
2. [7.1B] **(a)** $y = 5x - 2$; **(b)** $y = 4x + 7$; **(c)** $y = 6x - 4$
3. [7.1C] **(a)** $x - y = -3$; **(b)** $x - 2y = -5$; **(c)** $2x + 3y = 8$
4. [7.2A] **(a)** $C = 0.2m + 3$; \$6; **(b)** $C = 0.2m + 5$; \$8; **(c)** $C = 0.3m + 5$; \$9.50 **5.** [7.2B] **(a)** $C = 0.4m + 30$ for $m > 500$; \$150; **(b)** $C = 0.3m + 40$ for $m > 500$; \$70; **(c)** $C = 0.2m + 50$ for $m > 500$; \$130
6. [7.2C] **(a)** $C = 0.40m + 1$; **(b)** $C = 0.40m + 2$; **(c)** $C = 0.40m + 3$
7. [7.3A] **(a)** $2x - 4y < -8$

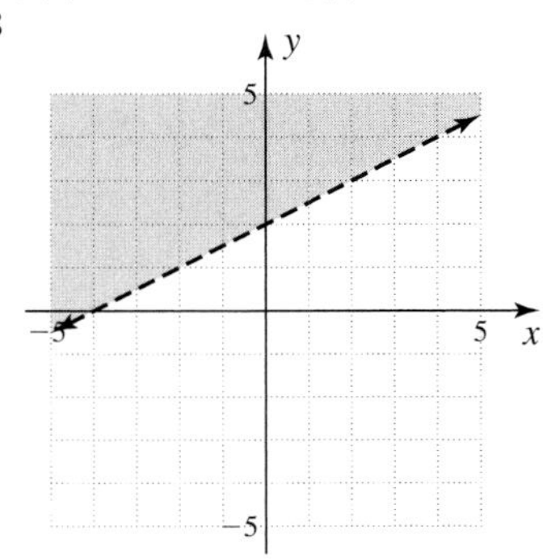

(b) $3x - 6y < -12$

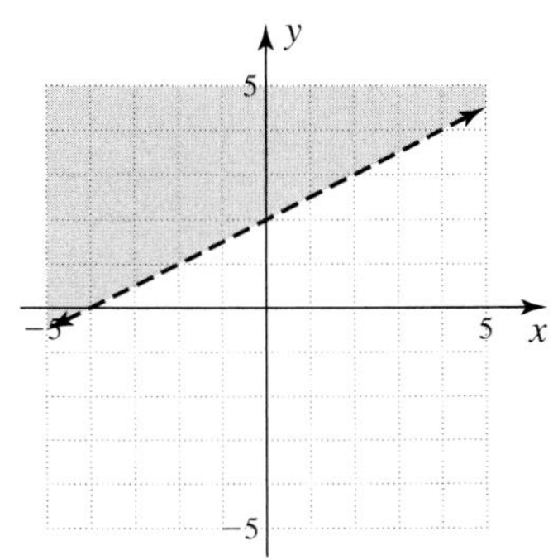

(c) $4x - 2y < -8$

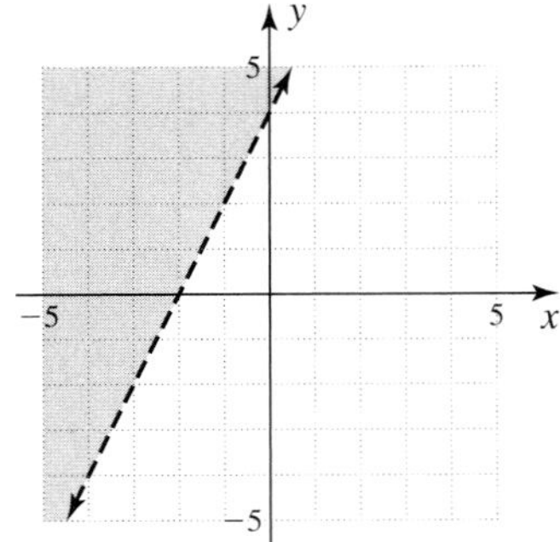

8. [7.3A] **(a)** $-y \leq -2x + 2$

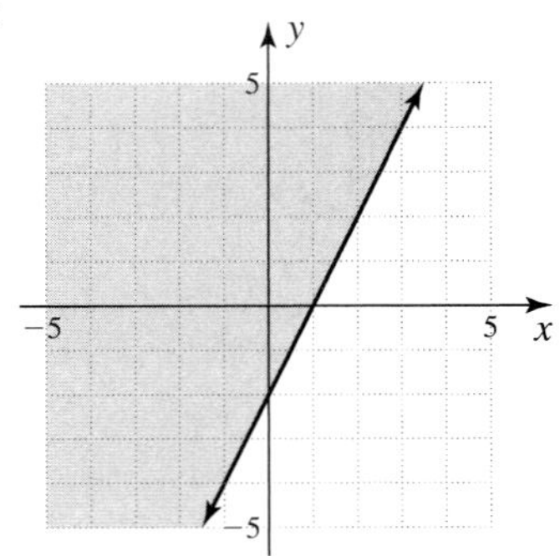

(b) $-y \leq -2x + 4$

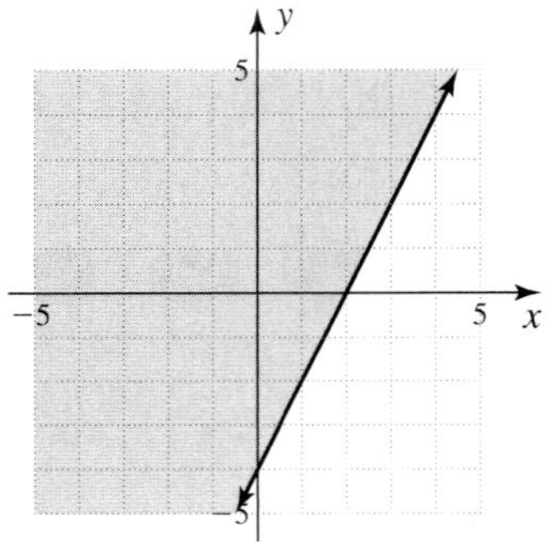

(c) $-y \leq -x + 3$

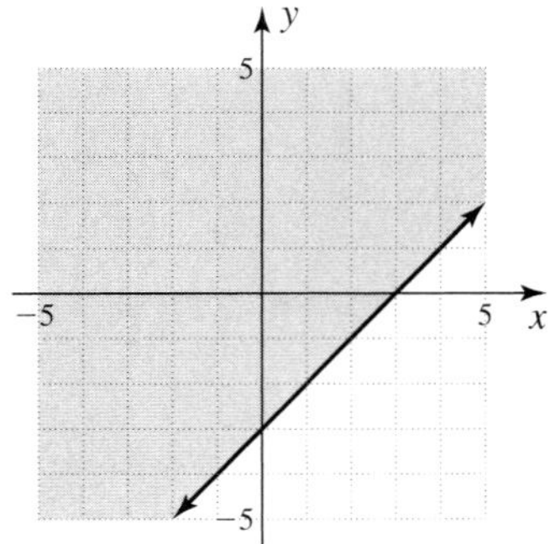

9. [7.3A] **(a)** $2x + y > 0$

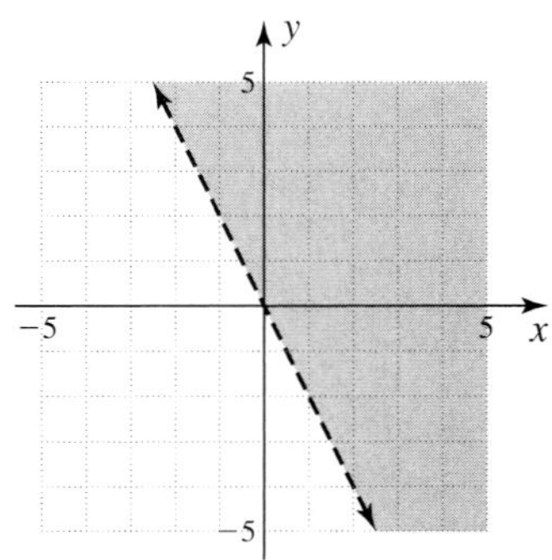

(b) $3x + y > 0$

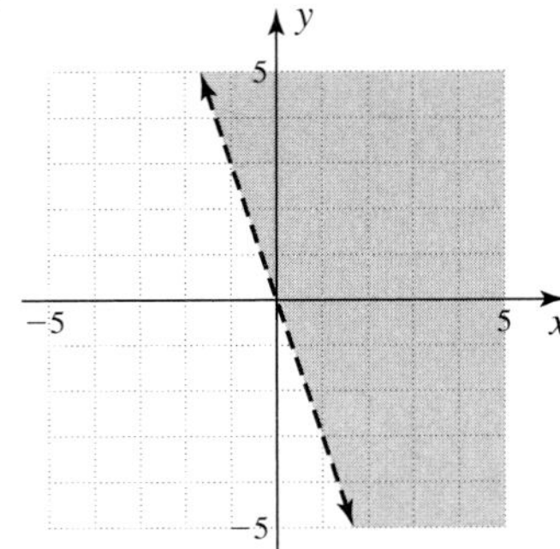

(c) $3x - y < 0$

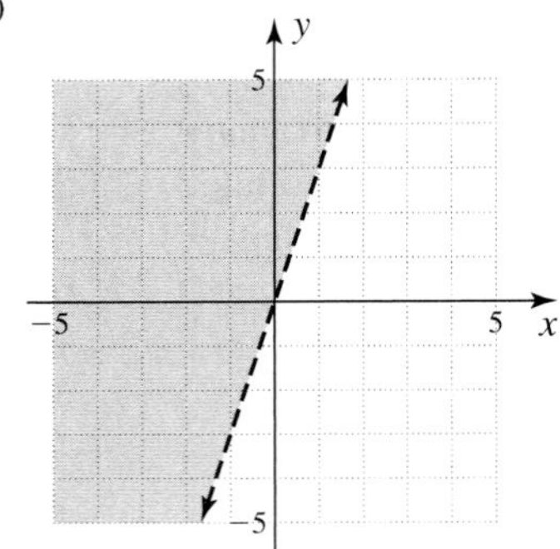

10. [7.3A] **(a)** $x \geq -4$

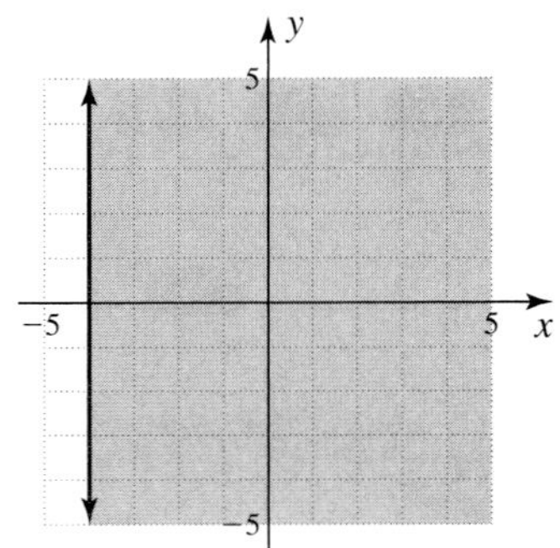

(b) $y - 4 < 0$

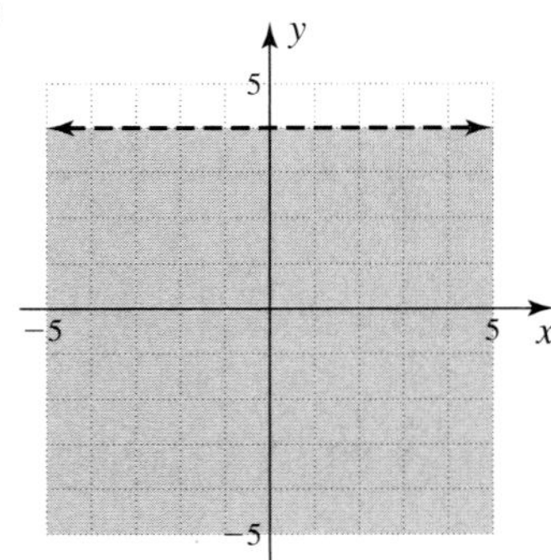

(c) $2y - 4 \geq 0$

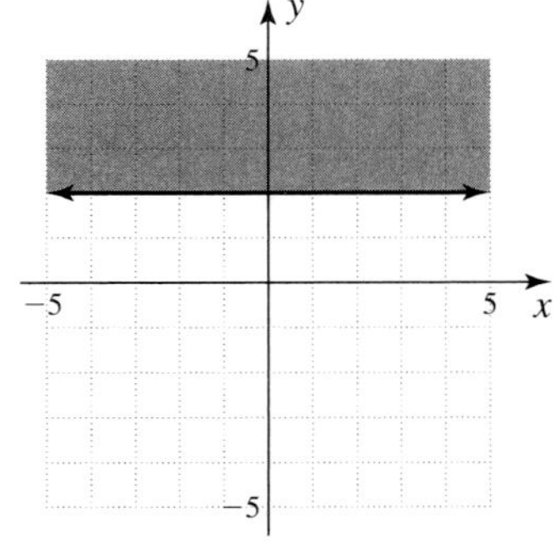

11. [7.4A] **(a)** $C = km$; 150 calories; **(b)** $C = km$; 420 calories; **(c)** $C = km$; 45 calories

12. [7.4B] **(a)** $F = \frac{30}{L}$; $F = 5$; **(b)** $F = 3$; **(c)** $F = 2$

Cumulative Review Chapters 1–7

1. $-\frac{25}{72}$ **2.** 8.2 **3.** 16 **4.** 2 **5.** 5 **6.** $6x - 11$ **7.** $\frac{d + 4e}{f}$

8. $x = 3$ **9.** $x = 63$ **10.** 35 and 75 **11.** \$4000 in bonds; \$5000 in certificates of deposit

12. 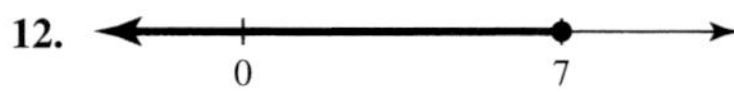

13.

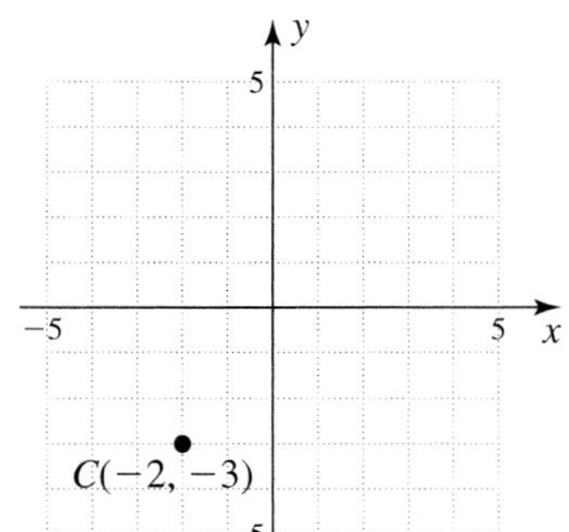

14. No **15.** $x = -2$

16. $3x + y = 3$

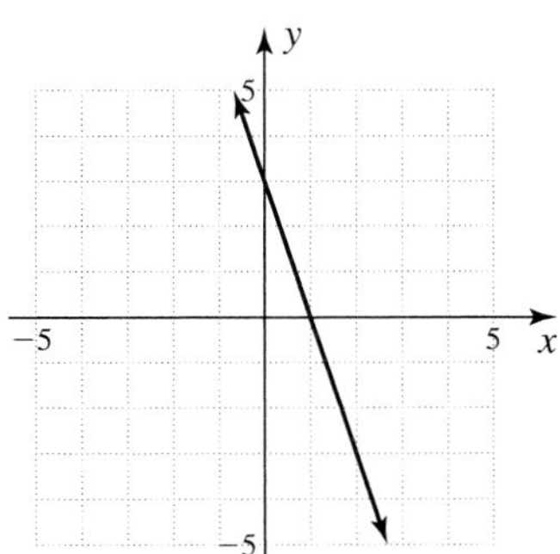

17. $3x + 9 = 0$

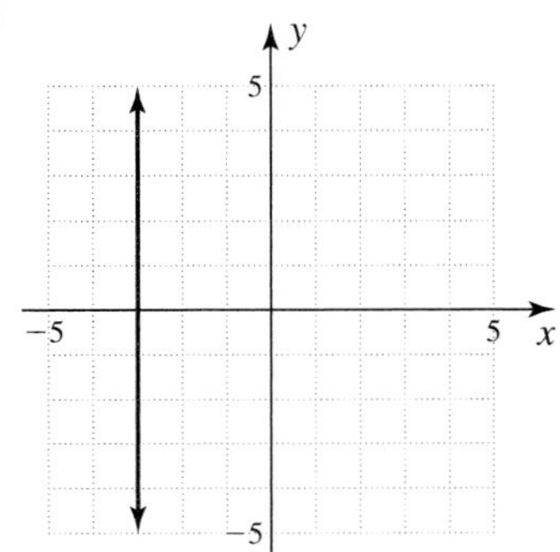

18. $-\frac{7}{3}$ **19.** 3 **20.** (1) and (3) **21.** x^4 **22.** $\frac{1}{x^9}$ **23.** $\frac{y^{20}}{16x^{12}}$

24. 3.4×10^6 **25.** 4.9 **26.** -68 **27.** $-4x^5 + x^2$

28. $9x^4 - 3x^2 + \frac{1}{4}$ **29.** $25x^4 - 4$ **30.** $(3x^2 - 4x - 5)$ R 2

31. $5x^4(3 - 7x^3)$ **32.** $\frac{1}{7}x^4(5x^4 - 2x^3 + 2x^2 - 3)$ **33.** $(x - 1)(x - 2)$

34. $(3x - 4y)(3x - 4y)$ **35.** $(2x + 3y)(2x - 3y)$

36. $-6x^2(x + 2)(x - 2)$ **37.** $2x(x + 1)(x - 2)$ **38.** $(4x + 3)(x + 1)$

39. $k(5x + 1)^2$ **40.** $x = -2$; $x = \frac{3}{4}$ **41.** $\frac{12xy^2}{8y^3}$ **42.** $-2(x + y)$

43. $-(x + 7)$ **44.** $\frac{x - 4}{x + 9}$ **45.** $\frac{1}{9 - x}$ **46.** $\frac{4}{x - 2}$

47. $\frac{-2x - 5}{(x + 3)(x + 2)(x - 2)}$ **48.** $-\frac{42}{5}$ **49.** $x = -5$ **50.** $x = -9$

51. No solution **52.** $x = -8$ **53.** 14 gal **54.** $x = \frac{69}{10}$ **55.** $1\frac{7}{8}$ hr

56. $y = -3x + 6$ **57.** $y = 3x - 1$

58. $6x - y < -6$

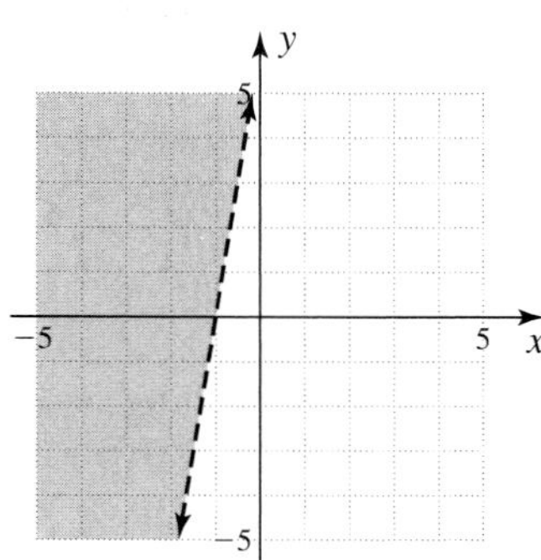

59. $-y \geq -6x - 6$

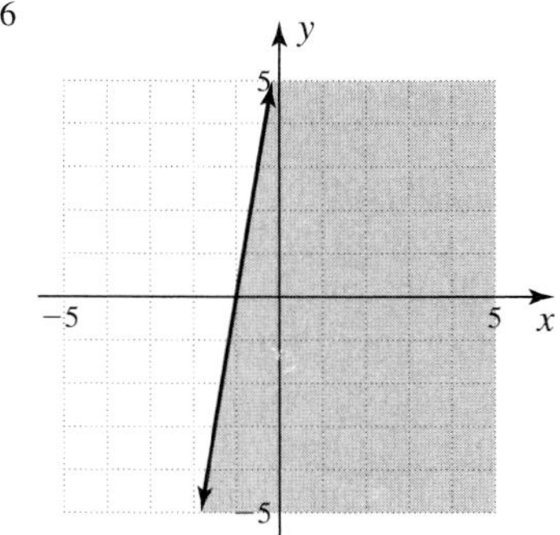

60. $k = \frac{1}{50}$ **61.** $k = 13{,}720$

Chapter 8

Exercises 8.1

1. Consistent; (1, 3)
① $x + y = 4$; ② $x - y = -2$

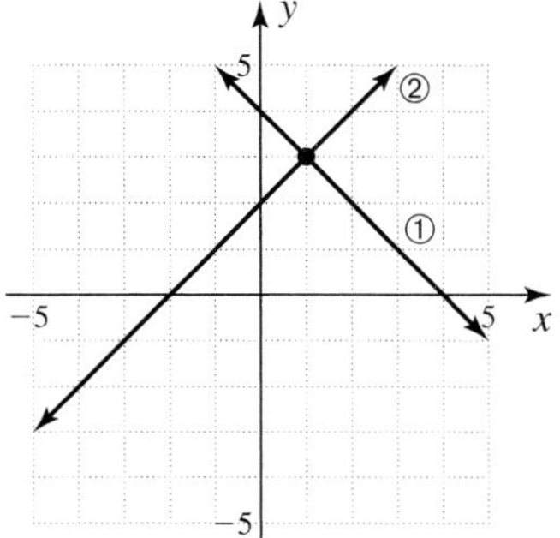

3. Consistent; (−2, 1)
① $x + 2y = 0$; ② $x - y = -3$

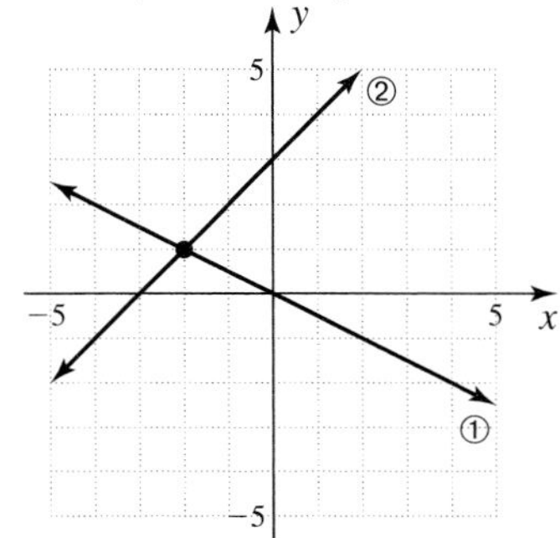

5. Dependent
$3x - 2y = 6$; $6x - 4y = 12$

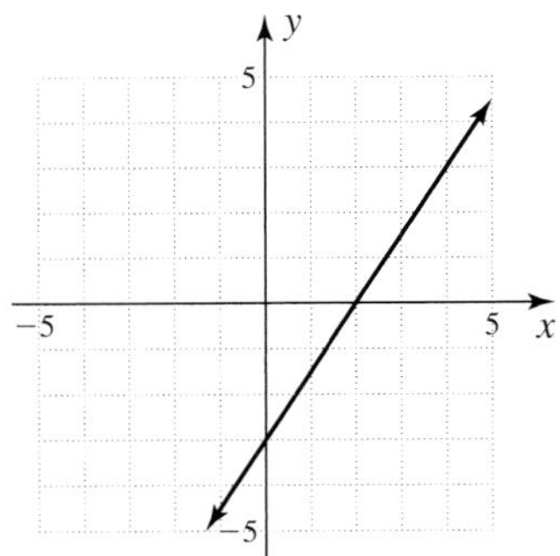

7. Dependent
$3x - y = -3$; $y - 3x = 3$

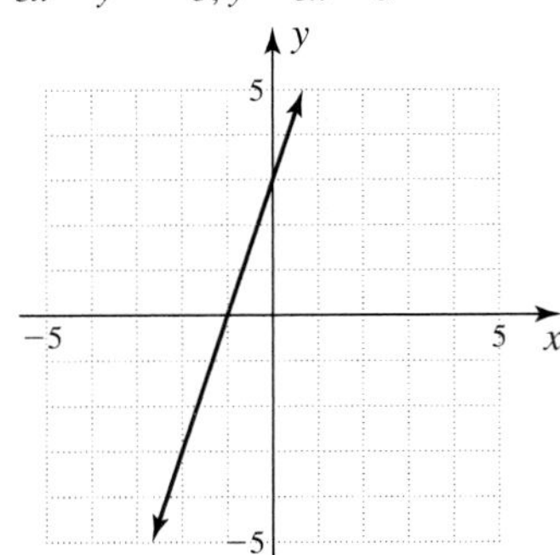

9. Inconsistent
① $2x - y = -2$; ② $y = 2x + 4$

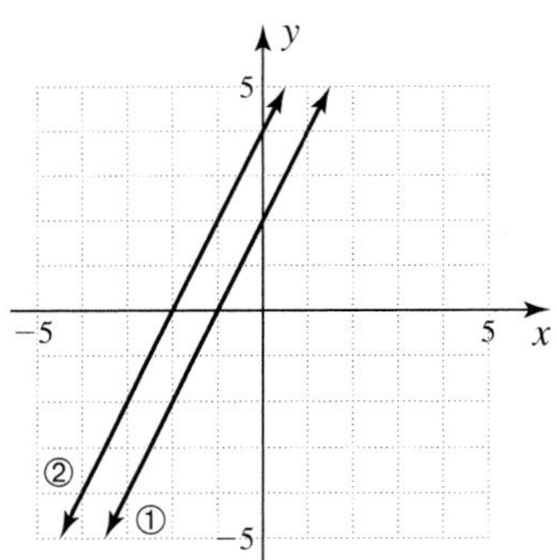

11. Consistent; $(-2, -2)$
① $y = -2$; ② $2y = x - 2$

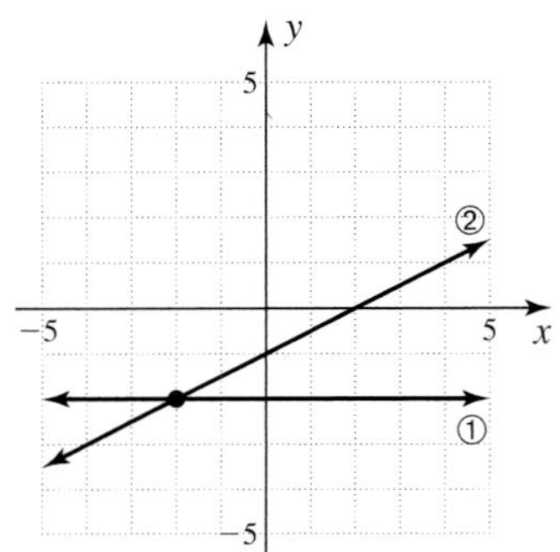

13. Consistent; $(3, 2)$
① $x = 3$; ② $y = 2x - 4$

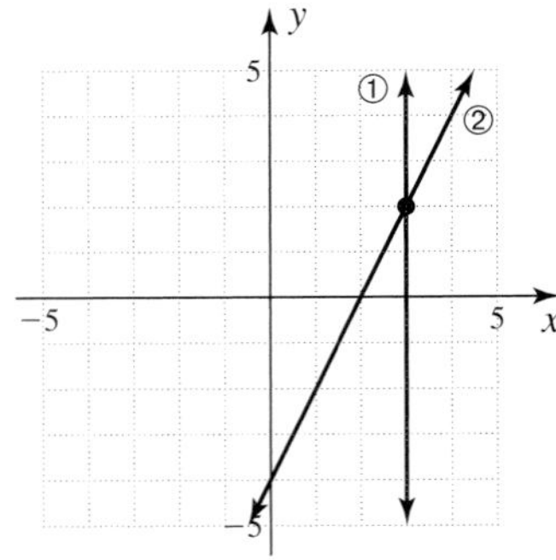

15. Consistent; $(1, 2)$
① $x + y = 3$; ② $2x - y = 0$

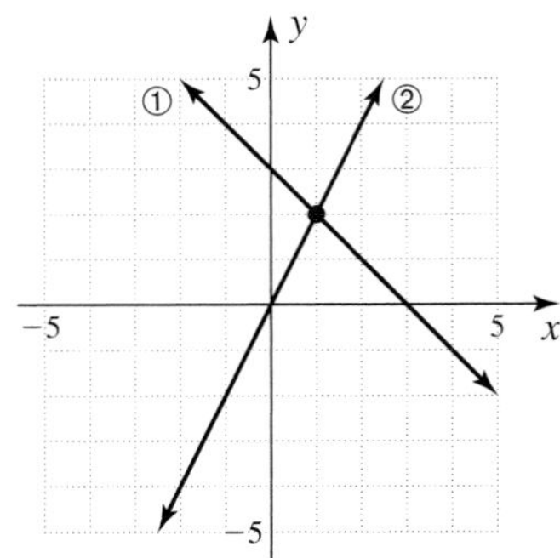

17. Consistent; $(0, 5)$
① $5x + y = 5$; ② $5x = 15 - 3y$

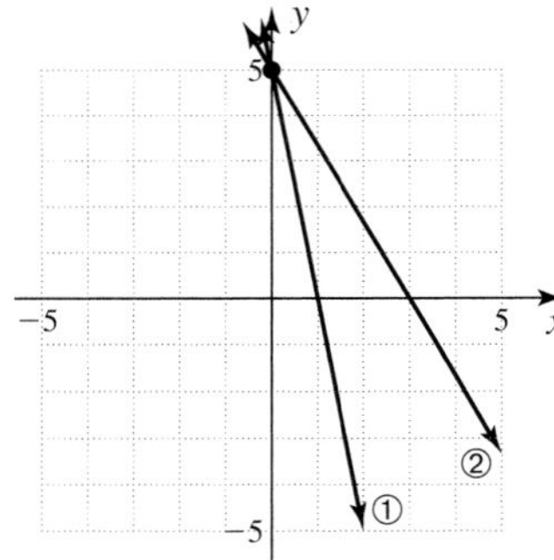

19. Dependent
$3x + 4y = 12$; $8y = 24 - 6x$

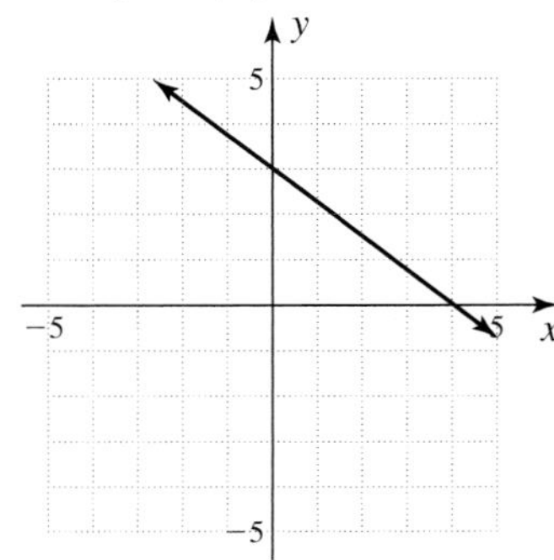

21. Consistent; (0, 3)
① $y = x + 3$; ② $y = -x + 3$

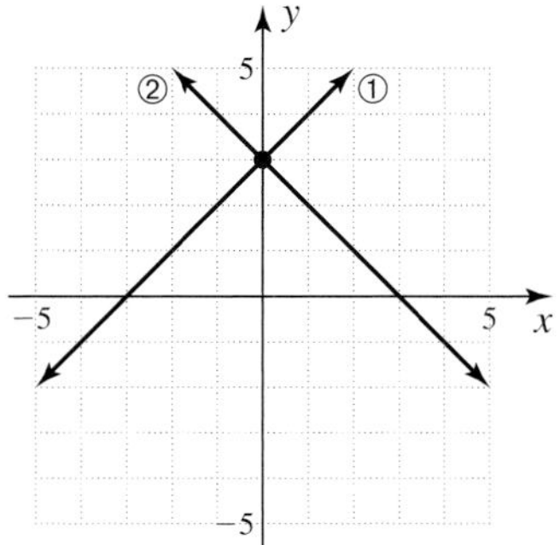

23. Consistent; (1, 0)
① $y = 2x - 2$; ② $y = -3x + 3$

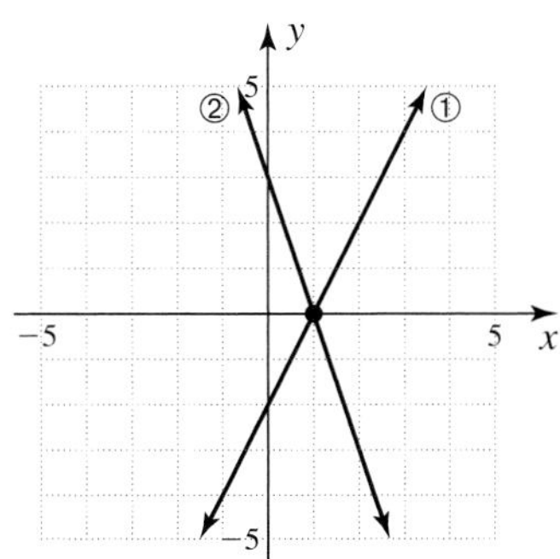

25. Consistent; (−2, −3)
① $-2x = 4$; ② $y = -3$

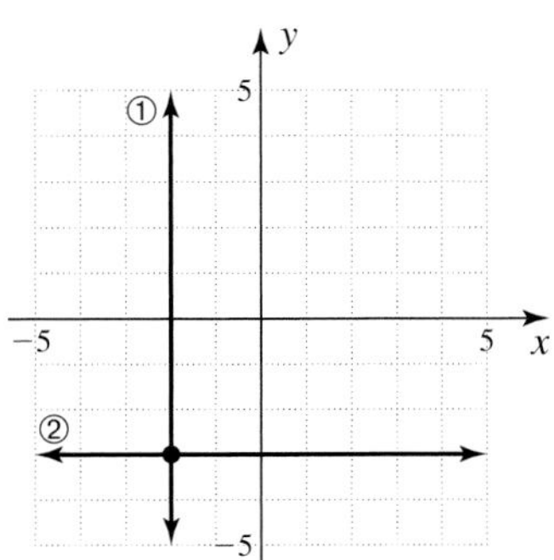

27. Consistent; (3, −3)
① $y = -3$; ② $y = -3x + 6$

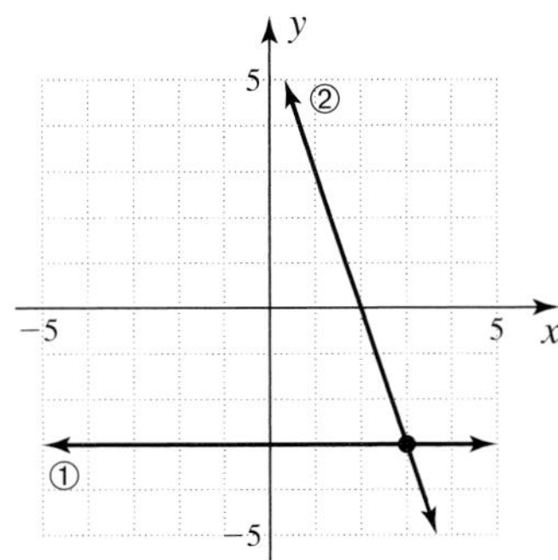

29. Inconsistent
① $x + 4y = 4$; ② $y = -\frac{1}{4}x + 2$

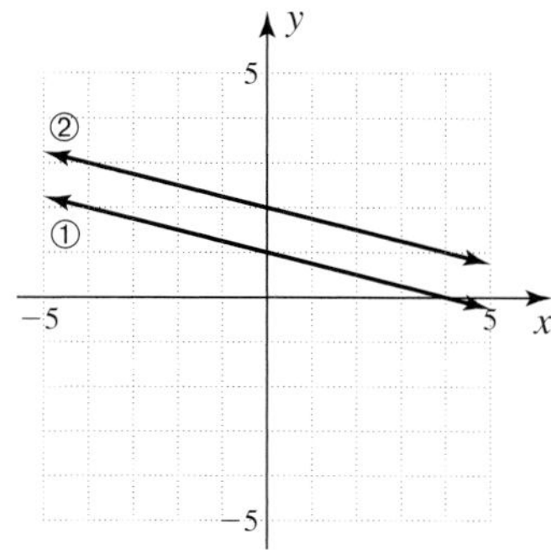

31. (a) $C = 20 + 35m$; **(b)**

m	*C*
6	230
12	440
18	650

(c)

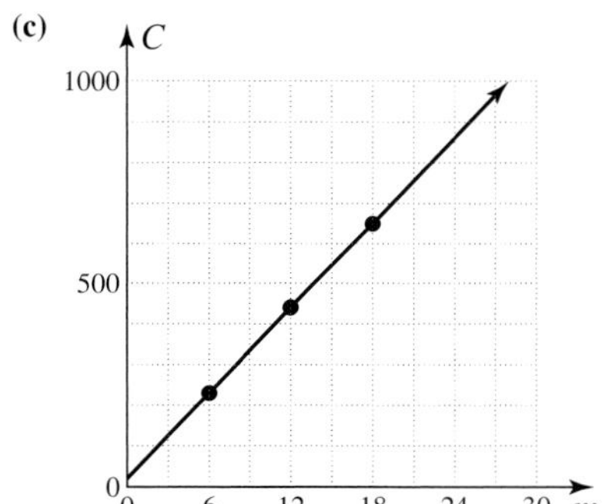

33.

Cable
VCR
(18, 650)

35. VCR and rental option is cheaper if used more than 18 months. (The options are equal if used for 18 months.)

37. (a) $W = 100 + 3t$; **(b)**

t	*W*
5	115
10	130
15	145
20	160

;

(c)

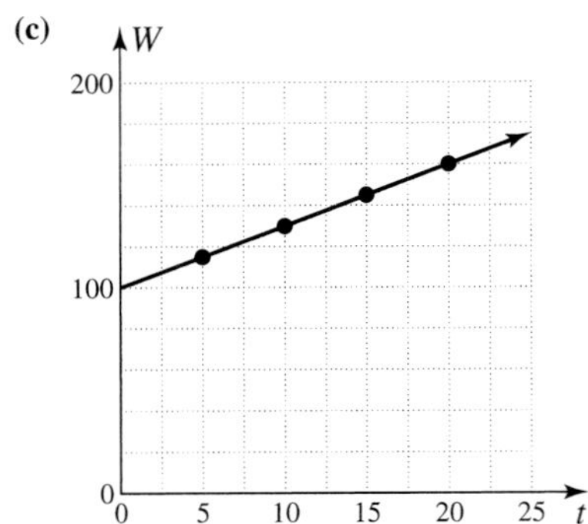

39. **(a)** $C = 0.60m$; **(b)** $C = 0.45m + 45$;
(c)

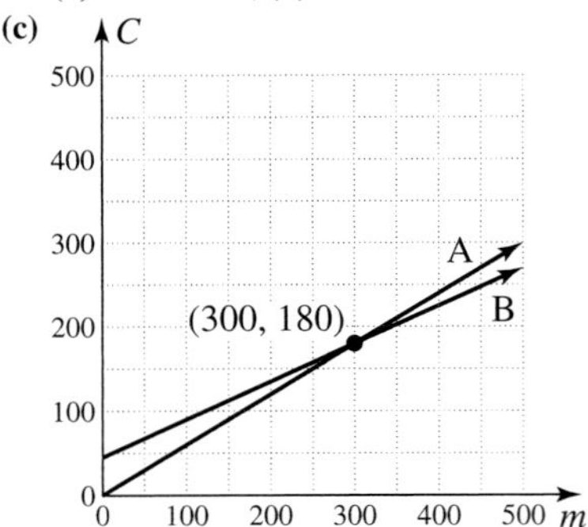

41. $x = 4$ **43.** Yes **45.** No
47. & 49.

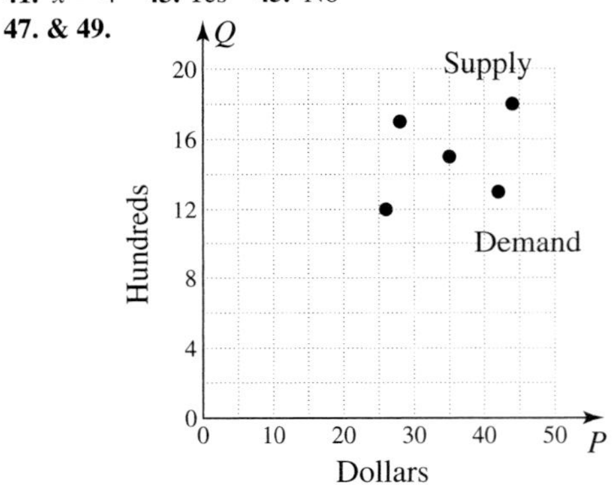

51. (35, 15) **53.** 1500
61. no solution; ① $y - 3x = 3$; ② $2y = 6x + 12$

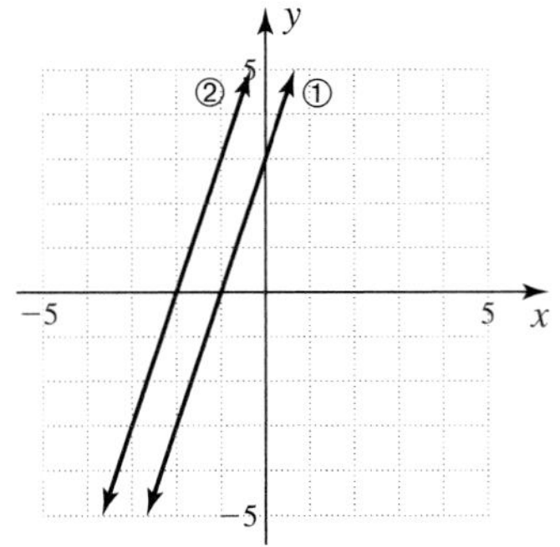

63. Consistent; (2, 2)
65. Dependent; infinitely many solutions

Exercises 8.2

1. Consistent; (2, 0) **3.** Consistent; (2, 3) **5.** Inconsistent; no solution
7. Inconsistent; no solution **9.** Dependent; infinitely many solutions
11. Consistent; (−1, −1) **13.** Inconsistent; no solution
15. Consistent; (4, 1) **17.** Consistent; (5, 3) **19.** Consistent; (0, 2)
21. Inconsistent; no solution **23.** Consistent; (2, 1)
25. Consistent; (3, 1) **27.** Dependent; infinitely many solutions
29. Inconsistent; no solution **31.** Consistent; (2, −3)
33. **(a)** $p = 20 + 3(h - 15)$ when $h > 15$; **(b)** $p = 20 + 2(h - 15)$ when $h > 15$; **(c)** When $h \le 15$ hours, $p = \$20$. **35.** When 150 minutes are used
37. 10 tables **39.** 1st **41.** At −40°, $F = C$. **43.** 4 days **45.** 5 days
47. $x = 8$ **49.** $x = -1$ **51.** $L_1 = 80$; $L_2 = 320$
53. **(a)** $5x = 4x + 500$; **(b)** 500 units
57. Inconsistent **59.** Consistent; (2, 3) **61.** Dependent
63. Inconsistent **65.** 10 months

Exercises 8.3

1. (1, 2) **3.** (0, 2) **5.** Inconsistent **7.** Inconsistent **9.** (10, −1)
11. (−26, 14) **13.** (6, 2) **15.** (2, −1) **17.** (3, 5) **19.** (−3, −2)
21. (8, 6) **23.** (1, 2) **25.** (5, 3) **27.** (4, 3) **29.** Dependent
31. 0.8 lb Costa Rican; 0.2 lb Indian Mysore
33. 10 lb oolong; 40 lb regular tea **35.** $n + d = 300$
37. $4(x - y) = 48$ **39.** $m = n - 3$ **41.** Tweedledee $120\frac{2}{3}$ lb; Tweedledum $119\frac{2}{3}$ lb
45. $\left(\frac{41}{13}, \frac{7}{13}\right)$ **47.** (5, 1) **49.** Inconsistent

Exercises 8.4

1. 5 nickels; 20 dimes **3.** 15 nickels; 5 dimes **5.** 4 fives; 6 ones
7. 59; 43 **9.** 105; 21 **11.** Impossible **13.** Pikes Peak = 14,110 ft; Longs Peak = 14,255 ft **15.** 1050 miles
17. Boat speed = 15 mi/hr; current speed = 3 mi/hr **19.** 20 miles
21. \$5000 at 6%; \$15,000 at 8% **23.** \$8000
25. Public: 11.7 million; private: 3.3 million
27. HBO: 14,353,650; Showtime: 14,346,350
29. $x + 2y < 4$

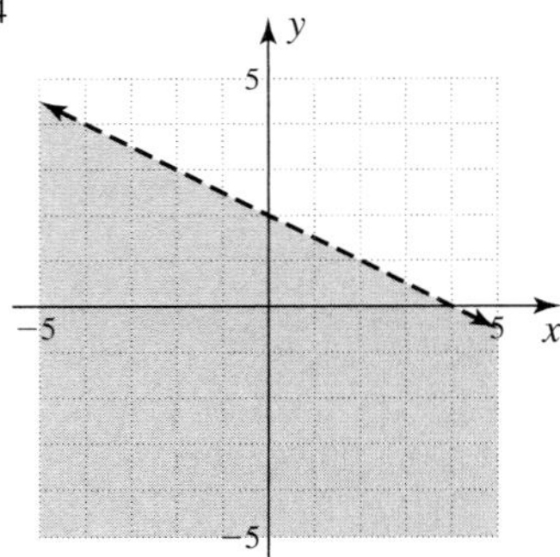

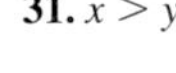
31. $x > y$

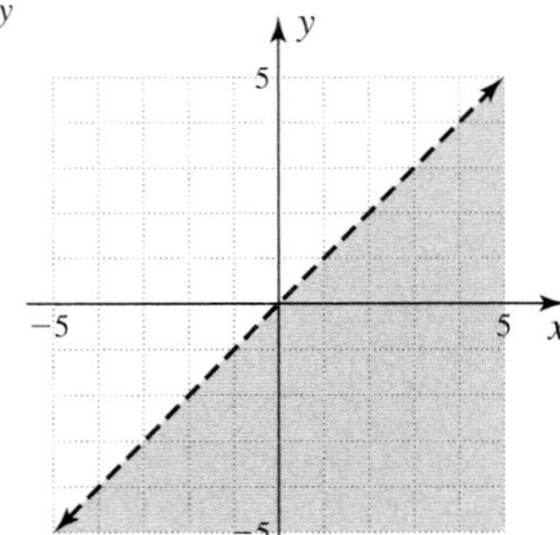

37. 20 nickels; 10 dimes
39. Wind speed = 50 mi/hr; plane speed = 350 mi/hr
41. Democrats: 45 million; Republicans: 39 million

Exercises 8.5

1. $x \geq 0$ and $y \leq 2$

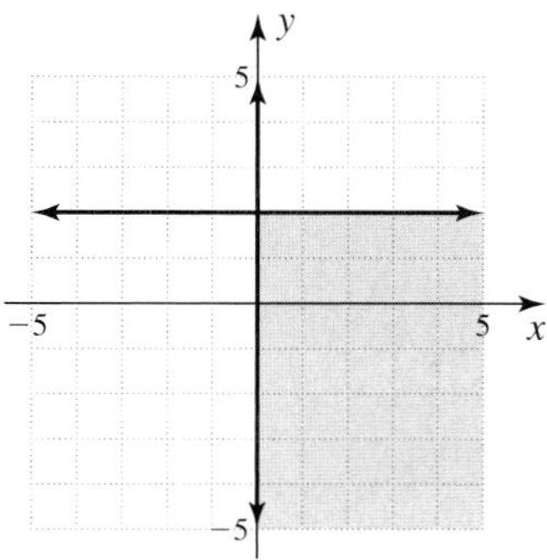

3. $x < -1$ and $y > -2$

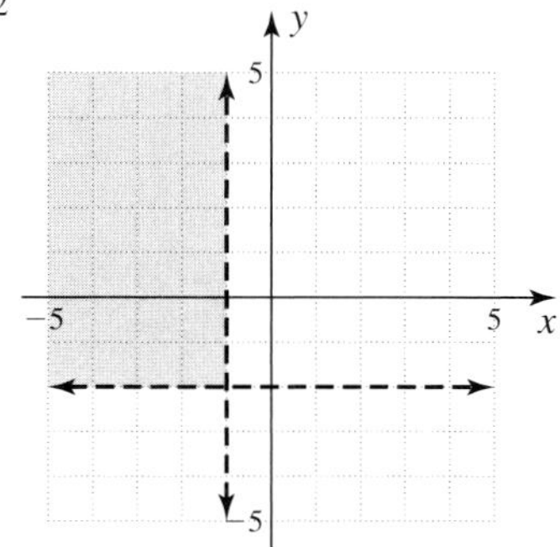

5. $x + 2y \leq 3$; $x < y$

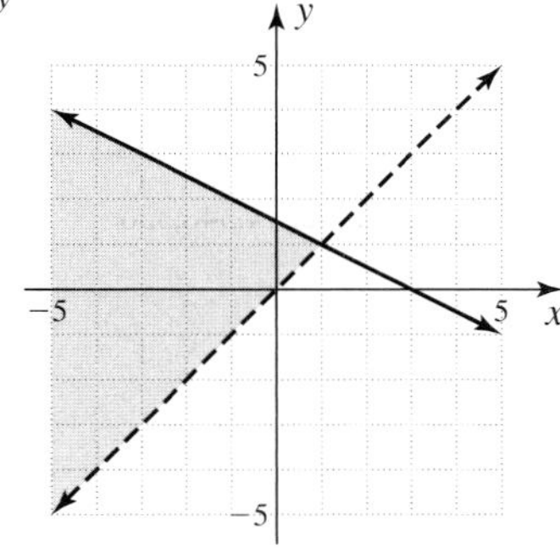

7. $4x - y > -1$; $-2x - y \leq -3$

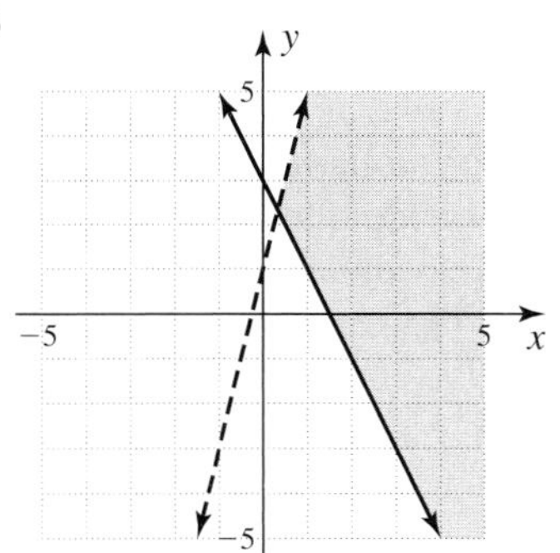

9. $-2x + y > 3$; $5x - y \leq -10$

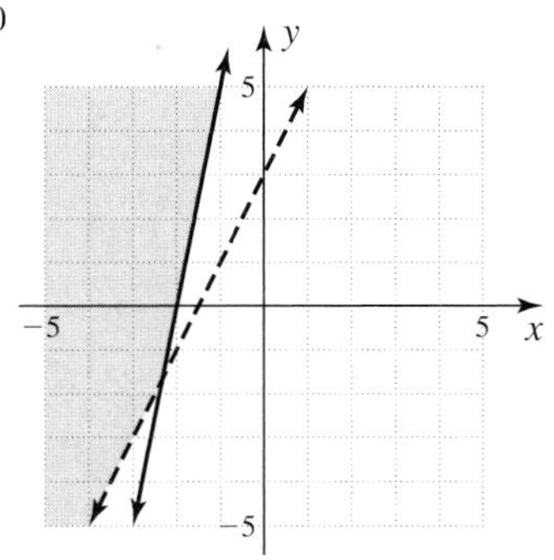

11. $2x - 3y < 5$; $x \geq y$

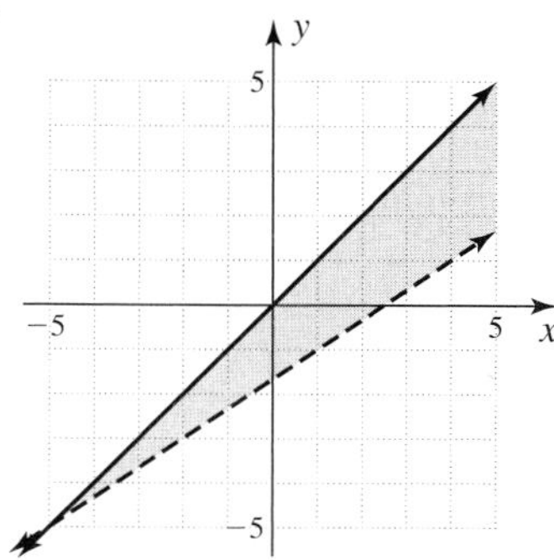

13. $x + 3y \leq 6$; $x > y$

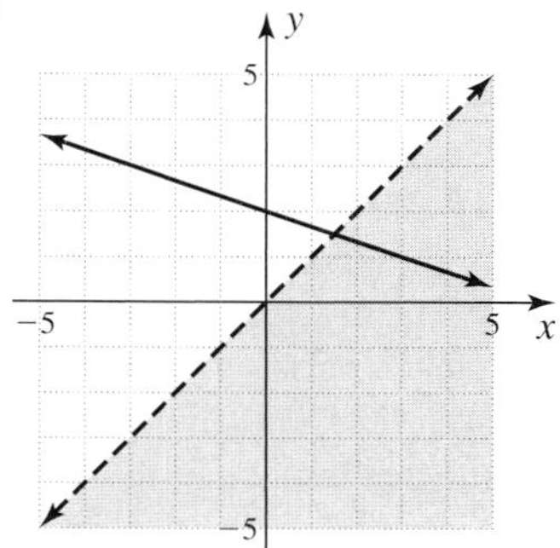

15. $3x + y > 6$; $x \leq y$

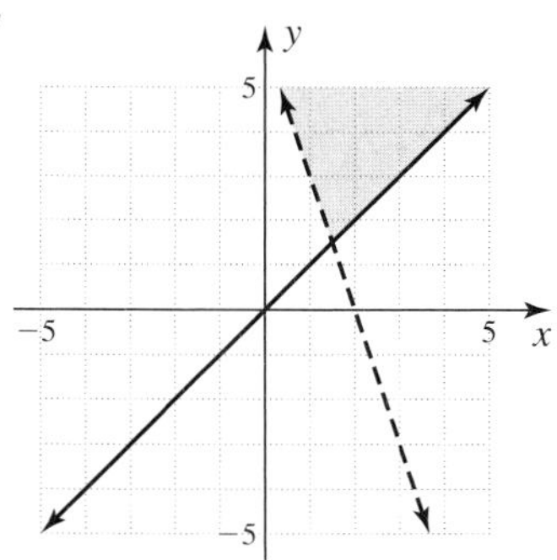

17. 8

19. & 21.

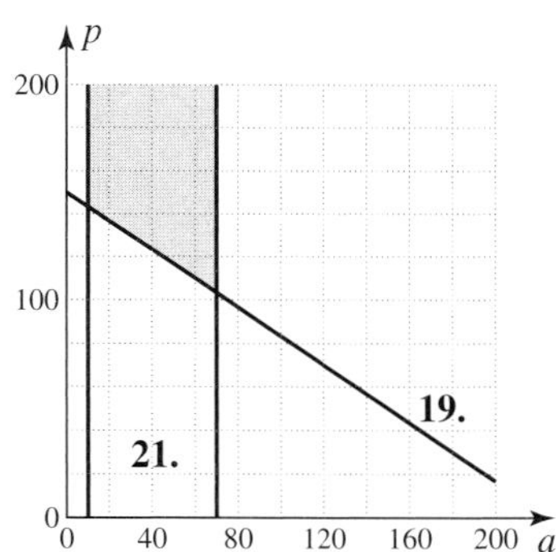

25. $x > 2$ and $y < 3$

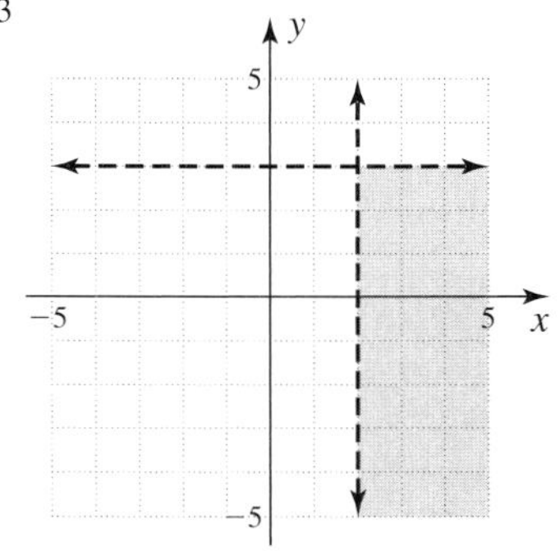

27. $3x - y < -1; x + 2y \leq 2$

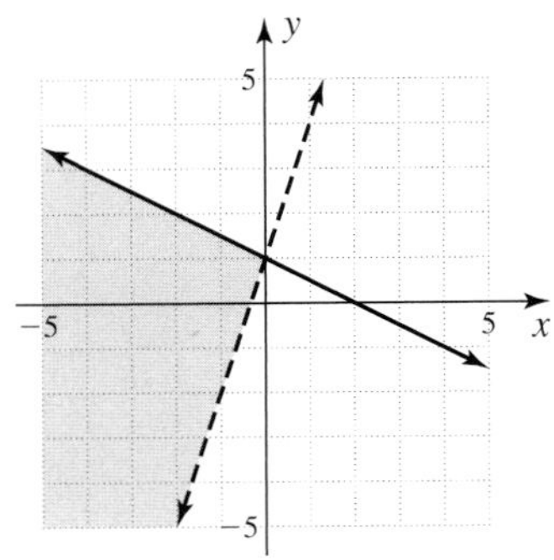

Review Exercises

1. [8.1A, B]

(a) Solution: (1, 2)
$2x + y = 4; y - 2x = 0$

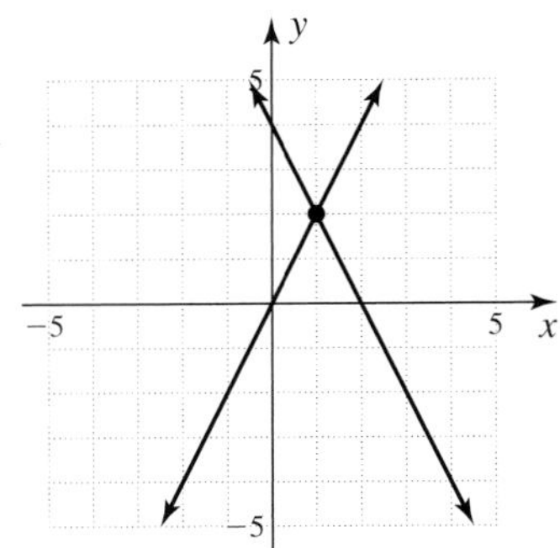

b. Solution: (2, 2)
$x + y = 4; y - x = 0$

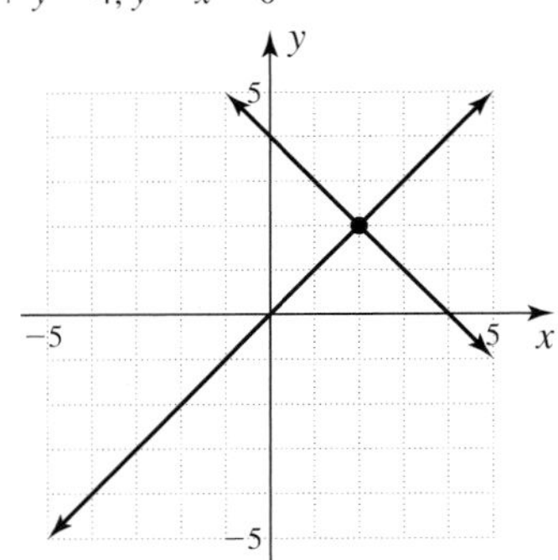

c. Solution: (1, 3)
$x + y = 4; y - 3x = 0$

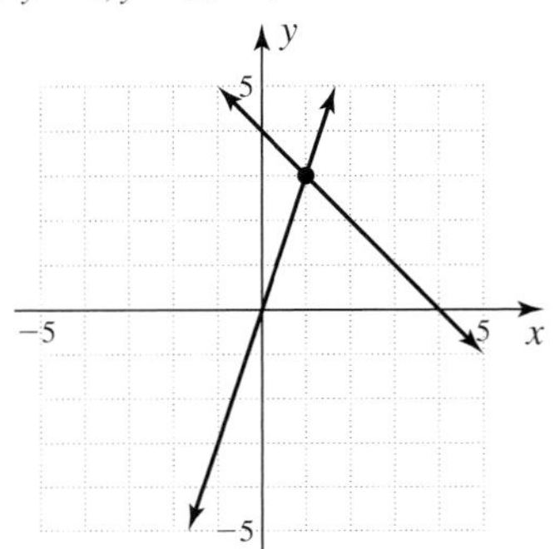

2. [8.1A, B]

(a) Inconsistent; no solution
$y - 3x = 3; 2y - 6x = 12$

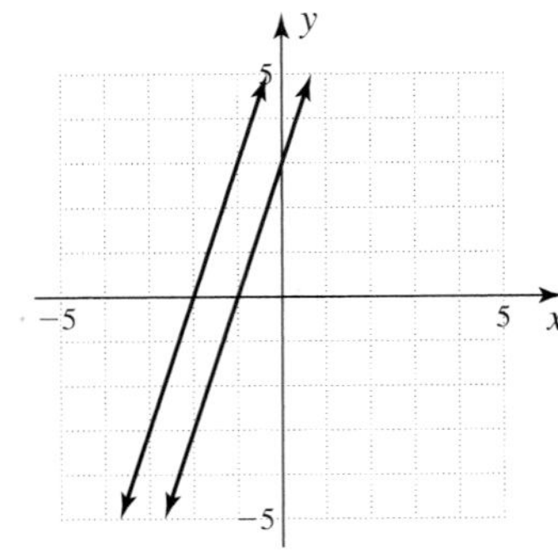

(b) Inconsistent; no solution
$y - 2x = 2; 2y - 4x = 8$

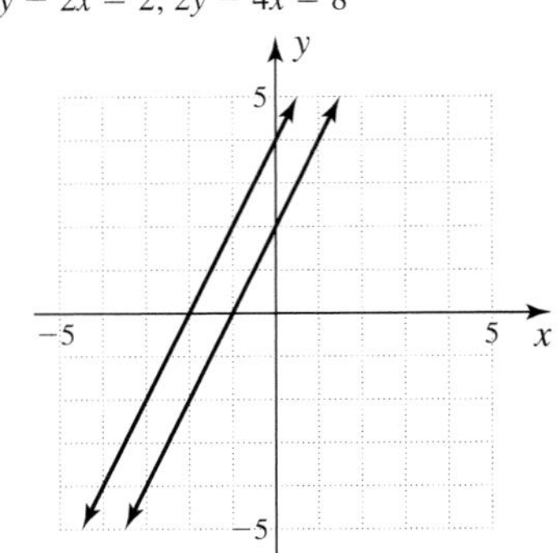

(c) Inconsistent; no solution
$y - 3x = 6; 2y - 6x = 6$

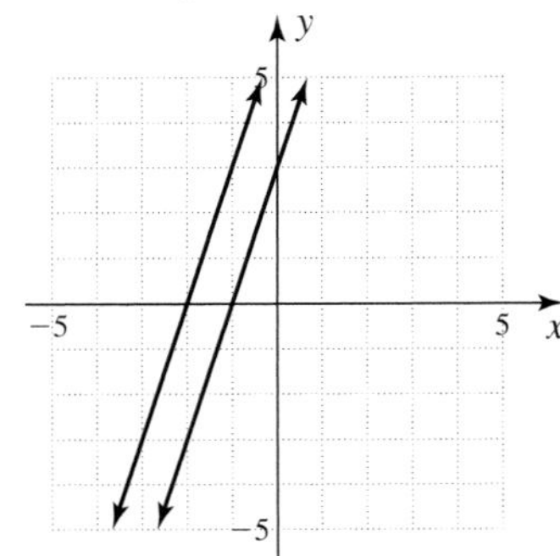

3. [8.2A, B] **(a)** Inconsistent; no solution; **(b)** Inconsistent; no solution; **(c)** (1, 1)
4. [8.2A, B] **(a)** Dependent; infinitely many solutions; **(b)** Dependent; infinitely many solutions; **(c)** Dependent; infinitely many solutions
5. [8.3A] **(a)** $(-1, 2)$; **(b)** $(2, -1)$; **(c)** $(-1, -2)$
6. [8.3A] **(a)** Inconsistent; no solution; **(b)** Inconsistent; no solution; **(c)** Inconsistent; no solution
7. [8.3B] **(a)** Dependent; infinitely many solutions; **(b)** Dependent; infinitely many solutions; **(c)** Dependent; infinitely many solutions
8. [8.4A] **(a)** 20 nickels; 20 dimes; **(b)** 40 nickels; 10 dimes; **(c)** 50 nickels; 5 dimes
9. [8.4B] **(a)** 70; 110; **(b)** 60; 120; **(c)** 50; 130

10. [8.4C] **(a)** 550 mi/hr; **(b)** 520 mi/hr; **(c)** 500 mi/hr

11. [8.4D] **(a)** Bonds: \$5000; CDs: \$15,000; **(b)** Bonds: \$17,000; CDs: \$3000; **(c)** Bonds: \$10,000; CDs: \$10,000

12. [8.5]

(a) $x > 4$ and $y < -1$;

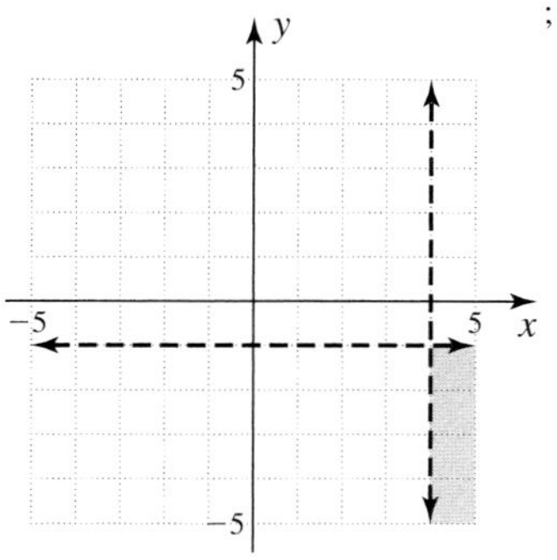

(b) $x + y > 3$; $x - y < 4$

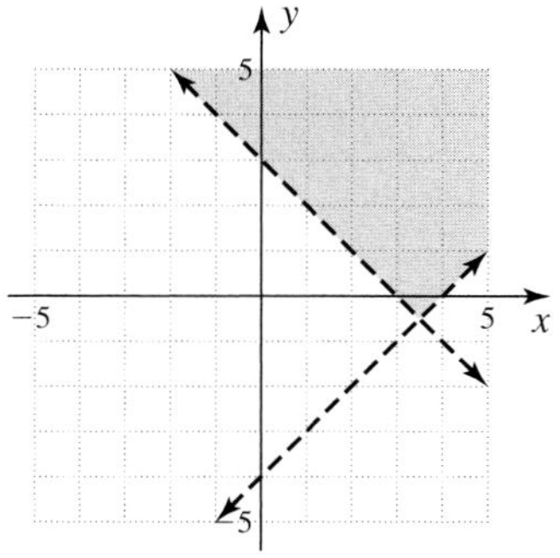

(c) $2x + y \le 4$; $x - 2y > 2$

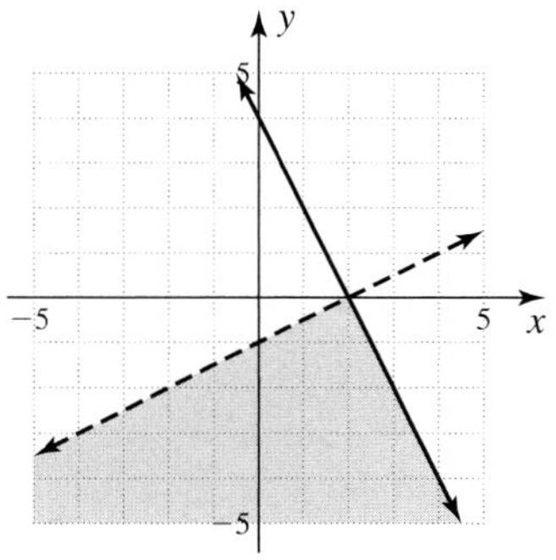

Cumulative Review Chapters 1–8

1. $-\frac{25}{42}$ **2.** 256 **3.** 2 **4.** 69 **5.** $-x - 10$ **6.** $\frac{x - 5y}{z}$ **7.** $x = 6$

8. $x = 36$ **9.**

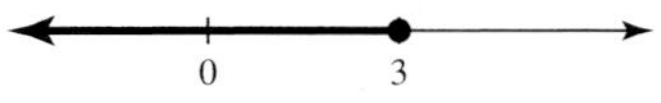

10.

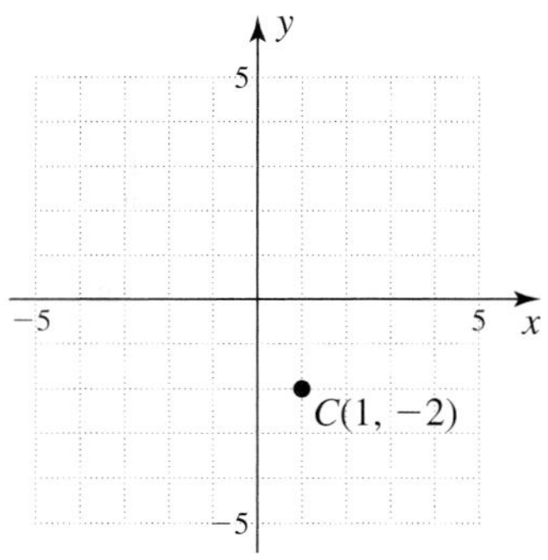

11. No **12.** $x = -3$

13. $x + y = 1$

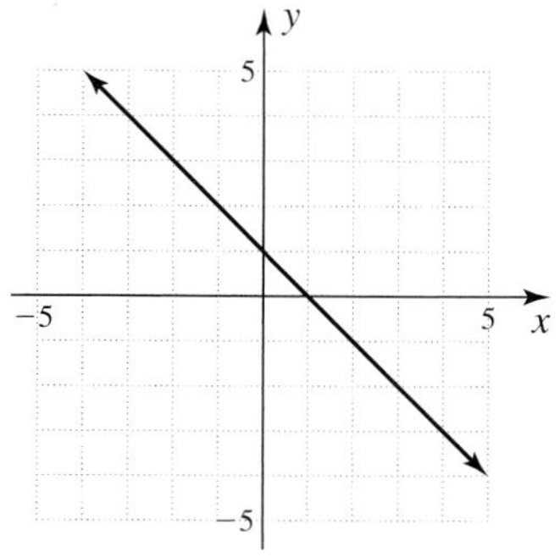

14. $2y + 8 = 0$

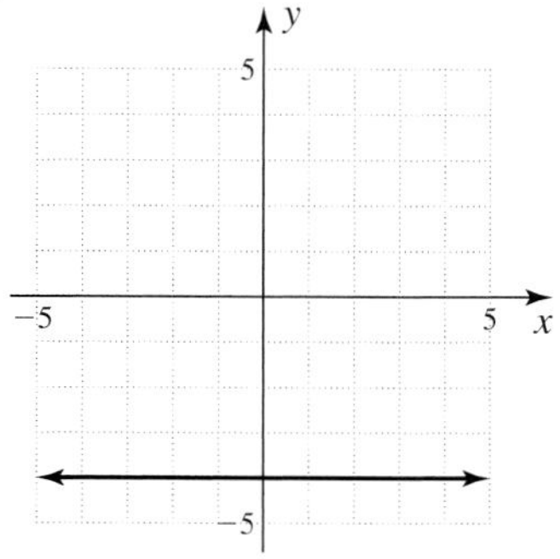

15. $-\frac{1}{7}$ **16.** 3 **17.** (1) and (2) **18.** $\frac{y^9}{8x^{12}}$ **19.** 3.5×10^{-5}

20. 2.10×10^{-7} **21.** $36x^4 - 3x^2 + \frac{1}{16}$ **22.** $(2x^2 + x - 1)$ R 3

23. $(x - 6)(x - 5)$ **24.** $(3x - 4y)(5x - 3y)$

25. $(9x + 8y)(9x - 8y)$ **26.** $-3x^2(x + 2)(x - 2)$

27. $4x(x + 1)(x - 2)$ **28.** $(4x + 3)(x + 1)$ **29.** $k(5x - 3)^2$

30. $x = -3$; $x = \frac{5}{3}$ **31.** $\frac{20xy}{15y^2}$ **32.** $-3(x + y)$ **33.** $-(x + 2)$

34. $\frac{x - 2}{x + 5}$ **35.** $\frac{1}{4 - x}$ **36.** $\frac{6}{x - 9}$ **37.** $\frac{-2x - 5}{(x + 3)(x + 2)(x - 2)}$

38. $\frac{39}{2}$ **39.** $x = -5$ **40.** $x = -3$ **41.** No solution **42.** $x = -6$

43. 13 gallons **44.** $x = \frac{31}{5}$ **45.** $2\frac{2}{9}$ hr **46.** $y = -4x - 26$

47. $y = 5x + 2$

48. $4x - 3y > -12$

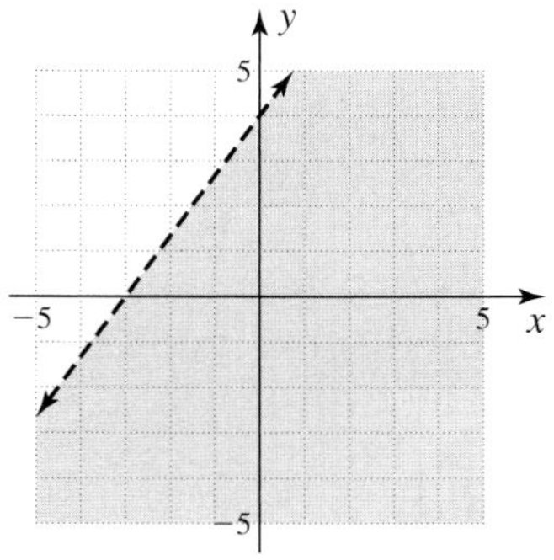

49. $-y \le -3x + 6$

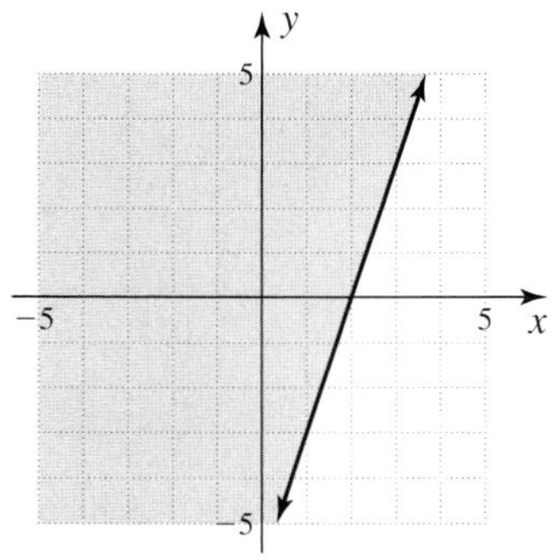

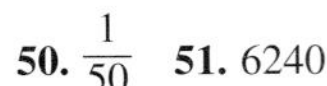
50. $\frac{1}{50}$ **51.** 6240

52. $x + 2y = 6$; $2y - x = -2$

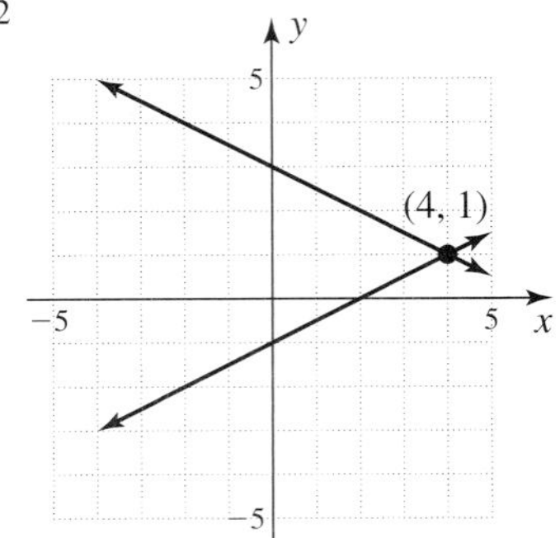

53. $y - x = -1$; $2y - 2x = -4$; No solution

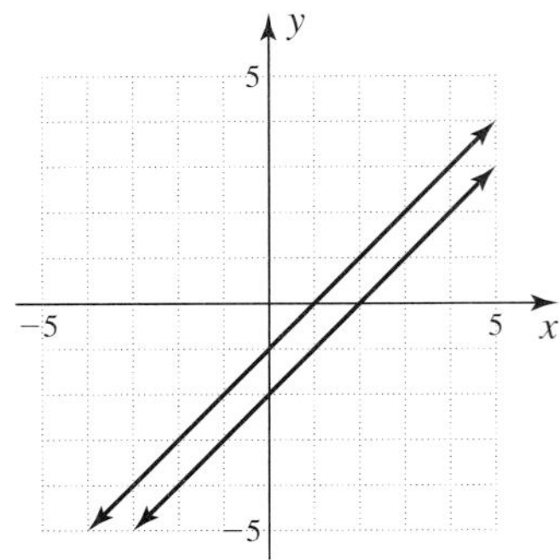

54. Inconsistent; no solution
55. Dependent; infinitely many solutions **56.** (1, 1)
57. Inconsistent; no solution **58.** Dependent; infinitely many solutions
59. 12 nickels and 36 dimes **60.** 35 and 130

Chapter 9

Exercises 9.1

1. 5 **3.** −3 **5.** $\frac{4}{3}$ **7.** $-\frac{2}{9}$ **9.** $\frac{5}{9}$ and $-\frac{5}{9}$ **11.** $\frac{7}{10}$ and $-\frac{7}{10}$ **13.** 5 **15.** 11 **17.** $x^2 + 1$ **19.** $3y^2 + 7$ **21.** 6; rational **23.** Not a real number **25.** −8; rational **27.** $\frac{4}{3}$; rational **29.** −2.449489743; irrational **31.** −1.414213562; irrational **33.** 3 **35.** −3 **37.** −4 **39.** 5 **41.** 10 sec **43.** $\sqrt{20}$ sec ≈ 4.5 sec **45.** 5 in. **47.** 244 mi **49.** 11 × 11 ft **51.** 2 **53.** 7 **55.** (a) $5\frac{1}{11}$; (b) $5\frac{3}{11}$; (c) $5\frac{5}{11}$ **65.** $\frac{4}{7}$ **67.** 12 **69.** 17 **71.** $-\frac{7}{11}$; rational **73.** 3.872983346; irrational **75.** −5 **77.** 3

Exercises 9.2

1. $3\sqrt{5}$ **3.** $5\sqrt{5}$ **5.** $6\sqrt{5}$ **7.** $10\sqrt{2}$ **9.** $8\sqrt{6}$ **11.** $5\sqrt{3}$ **13.** $10\sqrt{6}$ **15.** 19 **17.** $10\sqrt{7}$ **19.** $12\sqrt{3}$ **21.** $\sqrt{15}$ **23.** 9 **25.** 7 **27.** $\sqrt{3x}$ **29.** $6a$ **31.** $\frac{\sqrt{2}}{5}$ **33.** 2 **35.** 3 **37.** $5\sqrt{3}$ **39.** 12 **41.** $10a$ **43.** $7a^2$ **45.** $-4a^3\sqrt{2}$ **47.** $m^6\sqrt{m}$ **49.** $-3m^5\sqrt{3m}$ **51.** $2\sqrt[3]{5}$ **53.** $-2\sqrt[3]{2}$ **55.** $\frac{2}{3}$ **57.** $2\sqrt[4]{3}$ **59.** $\frac{4}{3}$ **61.** $12x$ **63.** $14x^3$ **69.** $3\sqrt[3]{5}$ **71.** $\frac{4}{3}$ **73.** $10x^3$ **75.** $\sqrt{2}$ **77.** $\frac{2}{3}$ **79.** 12

Exercises 9.3

1. $10\sqrt{7}$ **3.** $5\sqrt{13}$ **5.** $3\sqrt{2}$ **7.** $4\sqrt{2}$ **9.** $27\sqrt{3}$ **11.** $-7\sqrt{7}$ **13.** $29\sqrt{3}$ **15.** $-8\sqrt{5}$ **17.** $10\sqrt{2} - \sqrt{30}$ **19.** $2\sqrt{21} + \sqrt{30}$ **21.** $3 - \sqrt{6}$ **23.** $\sqrt{10} + 5$ **25.** $2\sqrt{3} - 3\sqrt{2}$ **27.** $2\sqrt{2} - 10$ **29.** $2\sqrt{3} - 3\sqrt{2}$ **31.** $\frac{\sqrt{6}}{2}$ **33.** $-2\sqrt{5}$ **35.** 2 **37.** $\frac{-\sqrt{10}}{5}$ **39.** $\frac{1}{2}$ **41.** $\frac{x\sqrt{2}}{6}$ **43.** $\frac{a\sqrt{b}}{b}$ **45.** $\frac{\sqrt{30}}{10}$ **47.** $\frac{x\sqrt{2}}{8}$ **49.** $\frac{x^2\sqrt{5}}{10}$ **51.** $x^2 - 9$ **53.** $2x + 4$ **59.** $13\sqrt{3}$ **61.** $6\sqrt{2}$ **63.** $12\sqrt{2}$ **65.** $5\sqrt{3} - 3$ **67.** $\frac{3\sqrt{7}}{7}$ **69.** $\frac{\sqrt{2x}}{2}$

Exercises 9.4

1. 16 **3.** 13 **5.** −4 **7.** 5 **9.** $18\sqrt{10}$ **11.** $12\sqrt{11}$ **13.** $3\sqrt[3]{2} - 2$ **15.** $\sqrt[3]{2}$ **17.** $3\sqrt{x}$ **19.** $\frac{9y}{4}$ **21.** $\frac{8ab\sqrt{3a}}{3}$ **23.** $\frac{2bc^3\sqrt[3]{b^2}}{3}$ **25.** $5\sqrt[3]{2}$ **27.** $2\sqrt[3]{3}$ **29.** $2\sqrt{15} - 117$ **31.** $6\sqrt{6} + 29$ **33.** $59 - 20\sqrt{6}$ **35.** 5 **37.** $3\sqrt{2} - 3$ **39.** $\frac{2\sqrt{7} + 2}{3}$ **41.** $2\sqrt{2} - \sqrt{6}$ **43.** $2\sqrt{5} + \sqrt{15}$ **45.** $-\sqrt{10} + \sqrt{15}$ **47.** $\frac{\sqrt{15} + \sqrt{6}}{3}$ **49.** $5 + 2\sqrt{6}$ **51.** −2 **53.** $\frac{-4}{3}$ **55.** $\frac{1 + \sqrt{3}}{3}$ **57.** $\frac{-1 + \sqrt{23}}{2}$ **59.** $\frac{-2 + \sqrt{10}}{3}$ **61.** $\frac{-4 + \sqrt{7}}{3}$ **63.** $\frac{-3 + 3\sqrt{3}}{2}$ **65.** $x - 1$ **67.** $4x$ **69.** $x^2 + 2x + 1$ **71.** $x(x - 3)$ **73.** $(x - 2)(x - 1)$ **75.** $\frac{\sqrt{2\gamma P_1(P_2 - P_1)}}{\gamma P_1}$ **81.** $-2 + \sqrt{7}$ **83.** $\frac{6 - 3\sqrt{2}}{2}$ **85.** $\frac{3\sqrt[3]{4}}{2}$ **87.** $5\sqrt[3]{2}$ **89.** $2\sqrt[3]{a^2}$ **91.** $-13 - \sqrt{15}$

Exercises 9.5

1. $x = 16$ **3.** No solution **5.** $y = 4$ **7.** $y = 8$ **9.** $x = 8$ **11.** $x = 0$ **13.** $x = 5$ **15.** $y = 20$ **17.** $y = 25$ **19.** $y = 9$ or $y = 1$ **21.** $y = 6$ **23.** $x = 5$ **25.** $x = 4$ **27.** $x = \frac{11}{3}$ **29.** $y = 5$ **31.** $S = 50.24$ sq ft **33.** 144 ft **35.** 3.24 ft **37.** 2000 **39.** 7 **41.** $\frac{\sqrt{5}}{4}$ **43.** $x = 27$ **45.** $x = 7$ **47.** $x = 16$ **49.** No solution **55.** 10 thousand **57.** $y = 2$ **59.** $x = 2$ **61.** $x = 7$ **63.** No solution **65.** $x = 2$ or $x = 3$

Review Exercises

1. [9.1A, C] **(a)** 9; **(b)** Not a real number; **(c)** $\frac{6}{5}$ **2.** [9.1A, C] **(a)** −6; **(b)** $-\frac{8}{5}$; **(c)** Not a real number **3.** [9.1B] **(a)** 8; **(b)** 25; **(c)** 17 **4.** [9.1B] **(a)** 36; **(b)** 17; **(c)** 64 **5.** [9.1B] **(a)** $x^2 + 1$; **(b)** $x^2 + 4$; **(c)** $x^2 + 5$ **6.** [9.1C] **(a)** Irrational; 3.3166; **(b)** Rational; −5; **(c)** Not a real number **7.** [9.1C] **(a)** Rational; **(b)** Rational; **(c)** Not a real number **8.** [9.1D] **(a)** 4; **(b)** −2; **(c)** −3 **9.** [9.1D] **(a)** 2; **(b)** −2; **(c)** Not a real number **10.** [9.1E] **(a)** $2\frac{3}{4}$ sec; **(b)** 3 sec; **(c)** $3\frac{1}{4}$ sec **11.** [9.2A] **(a)** $4\sqrt{2}$; **(b)** $4\sqrt{3}$; **(c)** 14 **12.** [9.2A] **(a)** $\sqrt{21}$; **(b)** 6; **(c)** $\sqrt{5y}$ **13.** [9.2B] **(a)** $\frac{\sqrt{3}}{4}$; **(b)** $\frac{\sqrt{5}}{6}$; **(c)** $\frac{3}{2}$ **14.** [9.2B] **(a)** 2; **(b)** $\sqrt{7}$; **(c)** $3\sqrt{5}$ **15.** [9.2C] **(a)** $6x$; **(b)** $10y^2$; **(c)** $9n^4$ **16.** [9.2C] **(a)** $6y^5\sqrt{2}$; **(b)** $7z^4\sqrt{3}$; **(c)** $4x^6\sqrt{3}$ **17.** [9.2C] **(a)** $y^7\sqrt{y}$; **(b)** $y^6\sqrt{y}$; **(c)** $5n^3\sqrt{2n}$ **18.** [9.2D] **(a)** $2\sqrt[3]{3}$; **(b)** $\frac{2}{3}$; **(c)** $-\frac{5}{4}$ **19.** [9.2D] **(a)** 3; **(b)** $2\sqrt[4]{3}$; **(c)** $2\sqrt[4]{5}$ **20.** [9.3A] **(a)** $15\sqrt{3}$; **(b)** $9\sqrt{2}$; **(c)** $6\sqrt{3}$ **21.** [9.3A] **(a)** $3\sqrt{11}$; **(b)** $\sqrt{2}$; **(c)** $\sqrt{3}$ **22.** [9.3B] **(a)** $2\sqrt{15} - \sqrt{6}$; **(b)** $5 - \sqrt{15}$; **(c)** $7 - 7\sqrt{14}$ **23.** [9.3C] **(a)** $\frac{\sqrt{10}}{4}$; **(b)** $\frac{x\sqrt{2}}{10}$; **(c)** $\frac{y\sqrt{3}}{9}$ **24.** [9.4A] **(a)** 3; **(b)** 3; **(c)** 4 **25.** [9.4A] **(a)** $7\sqrt{15}$; **(b)** $6\sqrt{6}$; **(c)** $7\sqrt{14}$ **26.** [9.4A] **(a)** $3\sqrt[3]{3}$; **(b)** 3; **(c)** 2 **27.** [9.4A] **(a)** $\frac{7\sqrt[3]{2}}{2}$; **(b)** $\frac{5\sqrt[3]{3}}{3}$; **(c)** $\frac{9\sqrt[3]{5}}{5}$ **28.** [9.4A] **(a)** $-27 - 2\sqrt{6}$; **(b)** $-23 + \sqrt{35}$ **29.** [9.4A] **(a)** −5; **(b)** −34 **30.** [9.4B] **(a)** $\frac{3\sqrt{3} - 3}{2}$; **(b)** $5\sqrt{2} + 5$

31. [9.4B] **(a)** $7\sqrt{3} + 7\sqrt{2}$; **(b)** $\dfrac{2\sqrt{5} + 2\sqrt{2}}{3}$
32. [9.4C] **(a)** $-4 + \sqrt{2}$; **(b)** $\dfrac{-8 + \sqrt{3}}{2}$ **33.** [9.5A] **(a)** $x = 7$; **(b)** No solution **34.** [9.5A] **(a)** $x = 4$; **(b)** $x = 6$ **35.** [9.5A] **(a)** $x = 5$; **(b)** $x = 7$ **36.** [9.5B] **(a)** $y = 4$; **(b)** $y = 0$ **37.** [9.5B] **(a)** $y = 1$; **(b)** $y = 0$ **38.** [9.5C] **(a)** 40 thousand or 40,000; **(b)** 240 thousand or 240,000

Cumulative Review Chapters 1–9

1. $-\dfrac{25}{72}$ **2.** 81 **3.** $\dfrac{2}{7}$ **4.** 47 **5.** $-x - 10$ **6.** $\dfrac{m + 3n}{p}$ **7.** $x = 16$
8. $x = 6$ **9.**

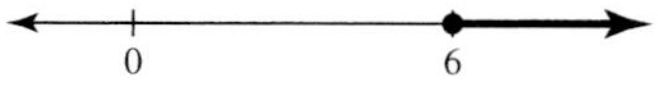

10.

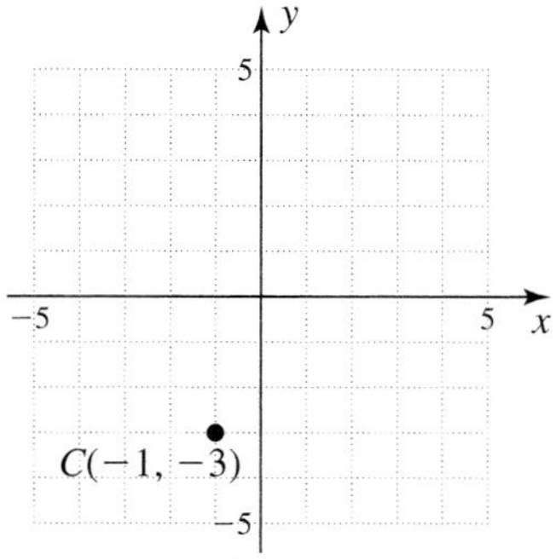

11. No **12.** $x = 1$

13. $5x + y = 5$

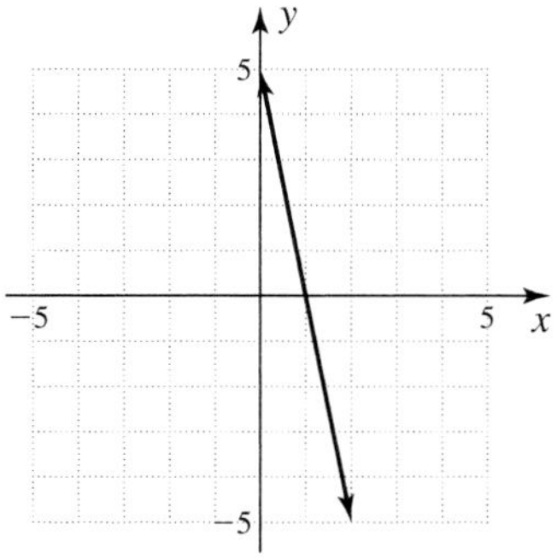

14. $4y - 8 = 0$

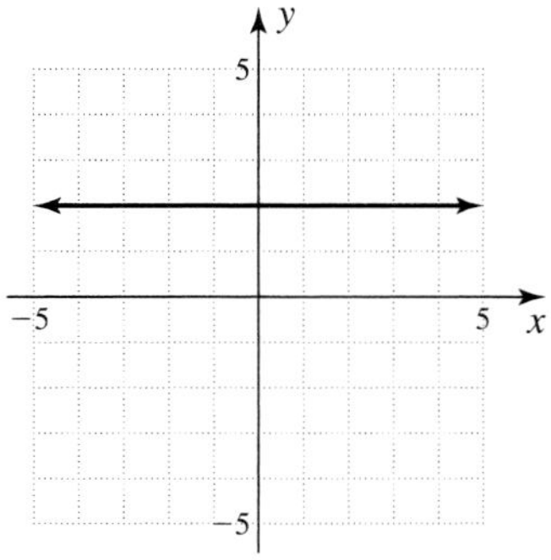

15. $-\dfrac{1}{2}$ **16.** 1

17. (1) and (3) **18.** $\dfrac{y^{12}}{16x^{16}}$ **19.** 2.5×10^{-7} **20.** 1.7×10^{-8}
21. $16x^4 - 4x^2 + \dfrac{1}{4}$ **22.** $(2x^2 - x - 2)$ R 3 **23.** $(x - 1)(x - 3)$
24. $(3x - 5y)(3x - 4y)$ **25.** $(2x + 5y)(2x - 5y)$
26. $-5x^2(x + 1)(x - 1)$ **27.** $4x(x + 1)(x - 3)$ **28.** $(3x + 4)(x + 3)$
29. $k(4x + 1)^2$ **30.** $x = -5; x = \dfrac{3}{4}$ **31.** $\dfrac{6xy}{15y^2}$ **32.** $-4(x + y)$
33. $-(x + 7)$ **34.** $\dfrac{x + 4}{x + 8}$ **35.** $\dfrac{1}{5 - x}$ **36.** $\dfrac{4}{x + 5}$
37. $\dfrac{-2x - 9}{(x + 5)(x + 4)(x - 4)}$ **38.** $\dfrac{14}{15}$ **39.** $x = 2$ **40.** $x = -3$

41. No solution **42.** $x = -7$ **43.** 25 gal **44.** $x = \dfrac{45}{4}$ **45.** $1\frac{5}{7}$ hr
46. $y = 5x - 26$ **47.** $y = 4x + 3$
48. $x - 5y < -5$

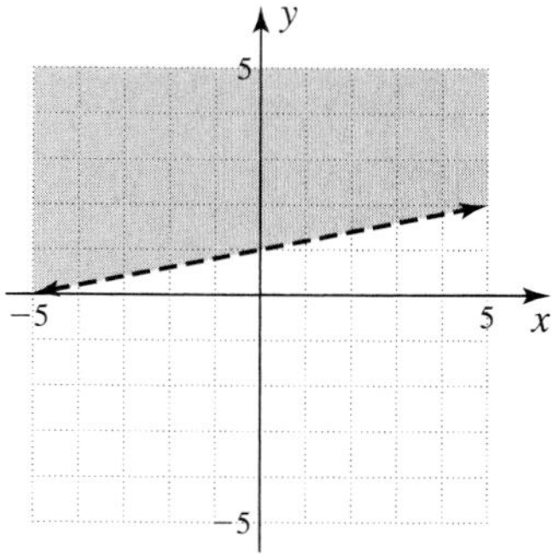

49. $-y \geq -5x - 5$

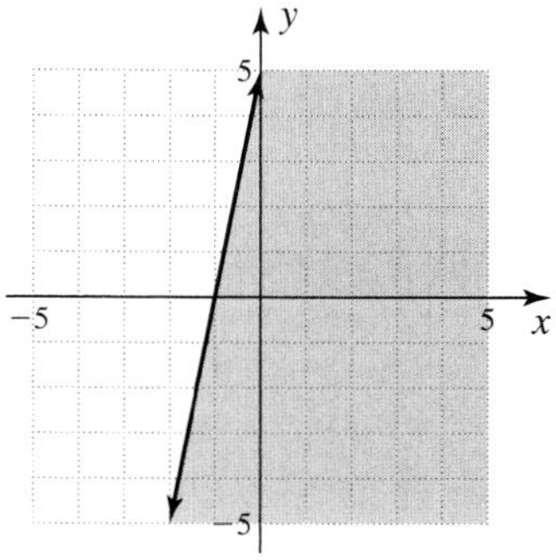

50. $\dfrac{1}{80}$ **51.** 10,800

52. $x + 4y = 16$; $4y - x = 12$

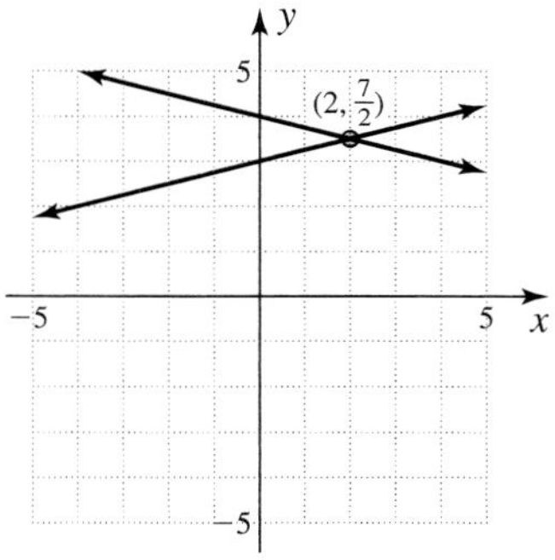

53. $y + 3x = -3$; $2y + 6x = -12$;
No solution

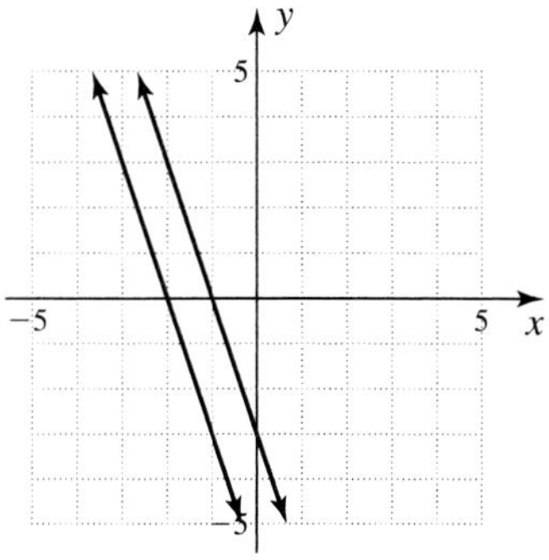

54. Inconsistent; no solution **55.** Dependent; infinitely many solutions **56.** $(-1, -2)$ **57.** Inconsistent; no solution **58.** Dependent; infinitely many solutions **59.** 10 nickels and 30 dimes **60.** 65 and 150
61. -4 **62.** -3 **63.** $\dfrac{\sqrt{21}}{27}$ **64.** $4a^2b^3\sqrt[3]{2a^2}$ **65.** $6\sqrt{3}$
66. $2x - 3\sqrt[3]{6x^2}$ **67.** $\dfrac{\sqrt{6p}}{2p}$ **68.** $420 + 74\sqrt{35}$ **69.** $\dfrac{x + \sqrt{5x}}{x - 5}$
70. $2 + \sqrt{2}$ **71.** No real-number solution **72.** $x = 6; x = 7$

Chapter 10

Exercises 10.1

1. $x = \pm 10$ **3.** $x = 0$ **5.** No real-number solution **7.** $x = \pm\sqrt{7}$ **9.** $x = \pm 3$ **11.** $x = \pm\sqrt{3}$ **13.** $x = \pm\frac{1}{5}$ **15.** $x = \pm\frac{7}{10}$ **17.** $y = \pm\frac{\sqrt{17}}{5}$ **19.** No real-number solution **21.** $x = 8$ or $x = -10$ **23.** $x = 8$ or $x = -4$ **25.** No real-number solution **27.** $x = 18$ or $x = 0$ **29.** $x = 0$ or $x = -8$ **31.** $x = -\frac{4}{5}$ or $x = -\frac{6}{5}$ **33.** $x = \frac{25}{6}$ or $x = \frac{11}{6}$ **35.** $x = \frac{3}{2}$ or $x = -\frac{7}{2}$ **37.** $x = 1 \pm \frac{\sqrt{5}}{3} = \frac{3 \pm \sqrt{5}}{3}$ **39.** No real-number solution **41.** $x = \pm\frac{1}{9}$ **43.** $x = \pm\frac{1}{4}$ **45.** $x = \pm 2$ **47.** $v = 2$ or $v = -4$ **49.** $x = \frac{5}{2}$ or $x = -\frac{1}{2}$ **51.** $y = \frac{3}{2} \pm \sqrt{2} = \frac{3 \pm 2\sqrt{2}}{2}$ **53.** $x = \frac{3}{2} \pm \sqrt{2} = \frac{3 \pm 2\sqrt{2}}{2}$ **55.** $x = -2 \pm 6\sqrt{2}$ **57.** $x = 3 \pm 6\sqrt{5}$ **59.** No real-number solution **61.** $x^2 + 14x + 49$ **63.** $x^2 - 6x + 9$ **65.** 5 in. **67.** $3\frac{1}{2}$ ft **77.** $x = 1 \pm 2\sqrt{2}$ **79.** $x = 1$ or $x = -5$ **81.** $x = \pm\frac{\sqrt{3}}{7}$ **83.** $x = \pm\frac{4}{3}$ **85.** $x = \pm 1$

Exercises 10.2

1. 81 **3.** 64 **5.** $\frac{49}{4}$ **7.** $\frac{9}{4}$ **9.** $\frac{1}{4}$ **11.** 4; $x + 2$ **13.** $\frac{9}{4}$; $x + \frac{3}{2}$ **15.** 9; $x - 3$ **17.** $\frac{25}{4}$; $x - \frac{5}{2}$ **19.** $\frac{9}{16}$; $x - \frac{3}{4}$ **21.** No real-number solution **23.** $x = -\frac{1}{2} \pm \frac{\sqrt{5}}{2} = \frac{-1 \pm \sqrt{5}}{2}$ **25.** $x = -\frac{3}{2} \pm \frac{\sqrt{13}}{2} = \frac{-3 \pm \sqrt{13}}{2}$ **27.** $x = \frac{3}{2} \pm \frac{\sqrt{21}}{2} = \frac{3 \pm \sqrt{21}}{2}$ **29.** $x = \frac{1}{2}$ or $x = -\frac{3}{2}$ **31.** $x = 2 \pm \frac{\sqrt{31}}{2} = \frac{4 \pm \sqrt{31}}{2}$ **33.** $x = \frac{1}{2} \pm \sqrt{2} = \frac{1 \pm 2\sqrt{2}}{2}$ **35.** $x = 1 \pm \frac{\sqrt{2}}{2} = \frac{2 \pm \sqrt{2}}{2}$ **37.** $x = 1$ or $x = -2$ **39.** $x = -\frac{5}{2} \pm \frac{3\sqrt{3}}{2} = \frac{-5 \pm 3\sqrt{3}}{2}$ **41.** $10x^2 + 5x - 12 = 0$ **43.** $x^2 - 9x = 0$ **45.** $10x^2 + 5x - 12 = 0$ **47.** **(a)** 2 (thousand) = 2000; **(b)** \$2 **49.** 3 days **55.** $\frac{1}{16}$ **57.** $\frac{25}{4}$; $x + \frac{5}{2}$ **59.** $x = -3 \pm \frac{\sqrt{5}}{2} = \frac{-6 \pm \sqrt{5}}{2}$

Exercises 10.3

1. $x = -2$ or $x = -1$ **3.** $x = -2$ or $x = 1$ **5.** $x = \frac{-1 \pm \sqrt{17}}{4}$ **7.** $x = \frac{2}{3}$ or $x = -1$ **9.** $x = -\frac{3}{2}$ or $x = -2$ **11.** $x = \frac{5}{7}$ or $x = 1$ **13.** $x = \frac{11 \pm \sqrt{41}}{10}$ **15.** $x = \frac{3}{2}$ or $x = 1$ **17.** $x = \frac{3 \pm \sqrt{5}}{4}$ **19.** $x = -1$ **21.** $x = \frac{3 \pm \sqrt{5}}{4}$ **23.** $x = \frac{-3 \pm \sqrt{21}}{6}$ **25.** $x = 2$ or $x = -2$ **27.** $x = \pm 3\sqrt{5}$ **29.** $x = \pm\frac{2\sqrt{3}}{3}$ **31.** 4 **33.** 1 **35.** Multiply each term by $4a$. **37.** Add b^2 on both sides. **39.** Take the square root of each side. **41.** Divide each side by $2a$. **43.** $x = \pm\sqrt{17}$ **45.** $x = -2$ **47.** $y = 0$ or $y = 1$ **49.** $x = -3$ or $x = -2$ **51.** $z = 0$ or $z = -6$ **53.** $y = \frac{1}{3}$ or $y = -3$ **55.** No real-number solution **61.** No real-number solution **63.** $x = 2 \pm 2\sqrt{2}$ **65.** $x = -\frac{1}{2}$ or $x = 2$

Exercises 10.4

1. $y = 2x^2$

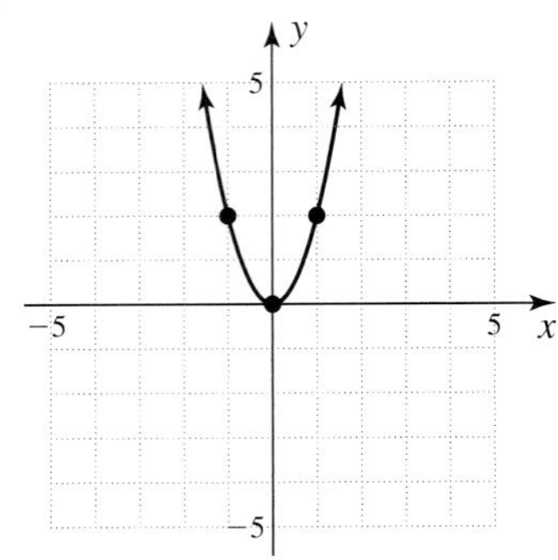

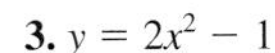

3. $y = 2x^2 - 1$

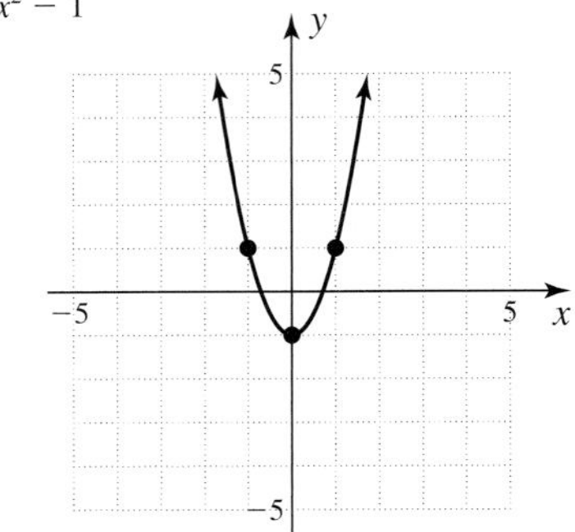

5. $y = -2x^2 + 2$

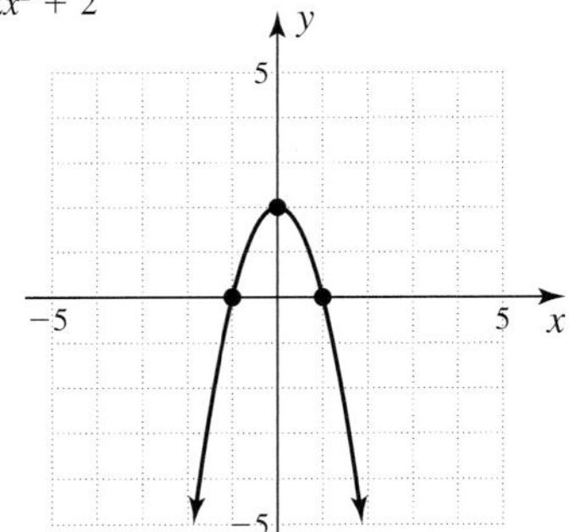

7. $y = -2x^2 - 2$

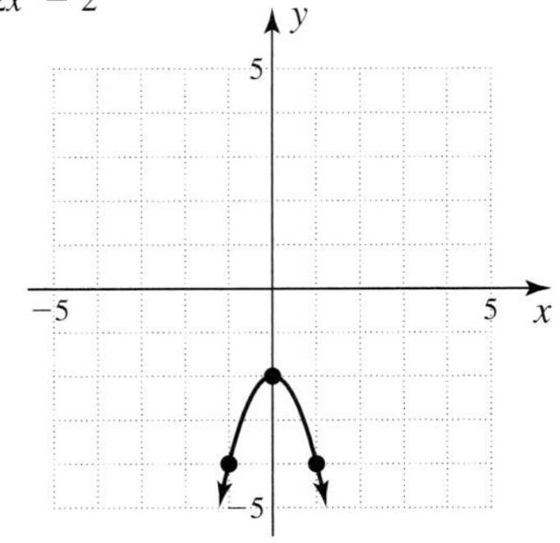

9. $y = (x + 2)^2$

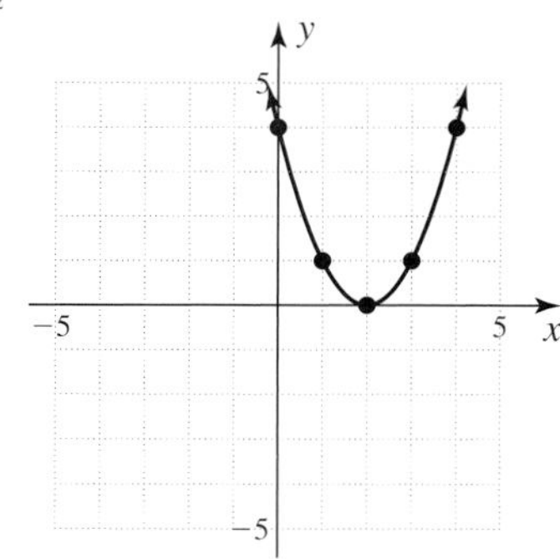

11. $y = (x - 2)^2 - 2$

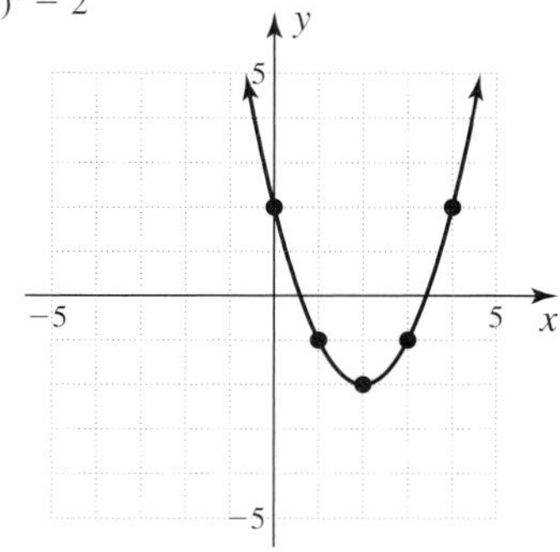

13. $y = -(x - 2)^2$

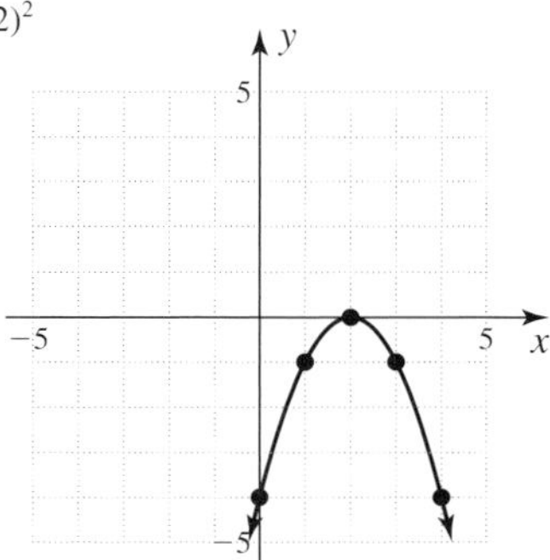

15. $y = -(x - 2)^2 - 2$

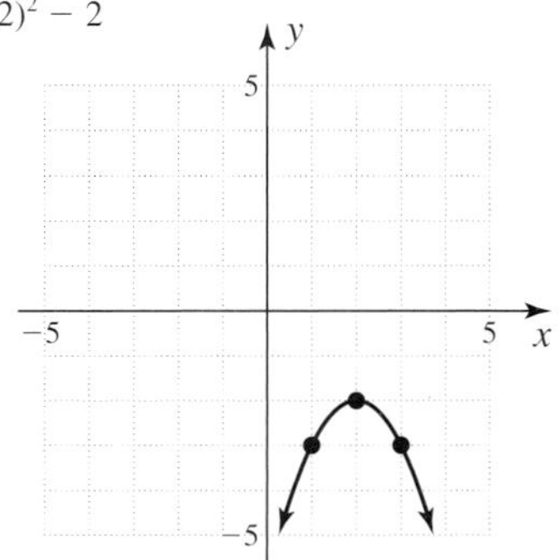

17. $y = x^2 + 4x + 3$

x-int: $(-3, 0)$, $(-1, 0)$; y-int: $(0, 3)$; $V(-2, -1)$

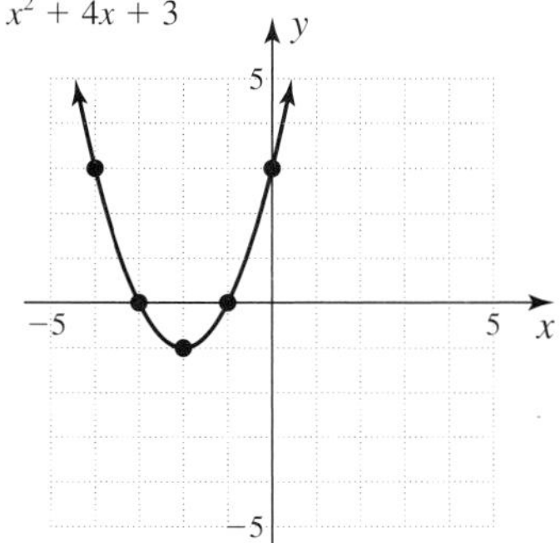

19. $y = x^2 + 2x - 3$

x-int: $(-3, 0)$, $(1, 0)$; y-int: $(0, -3)$; $V(-1, -4)$

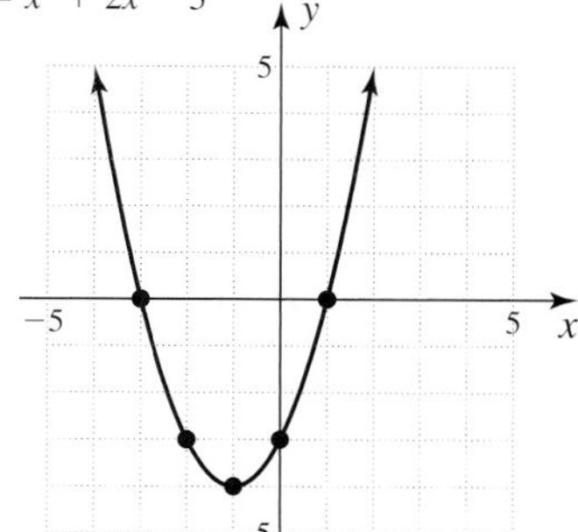
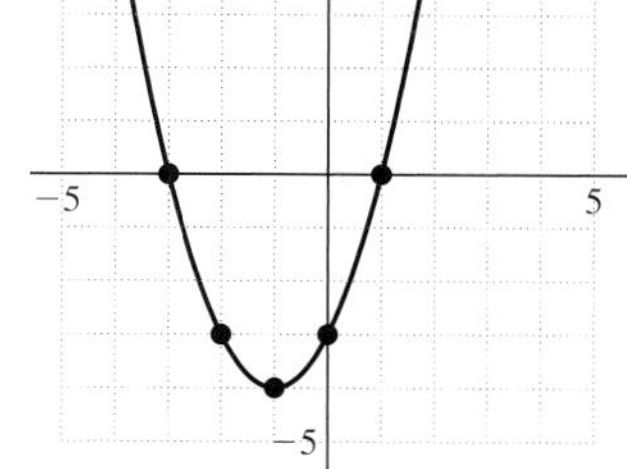

21. $y = -x^2 + 4x - 3$

x-int: $(3, 0)$, $(1, 0)$; y-int: $(0, -3)$; $V(2, 1)$

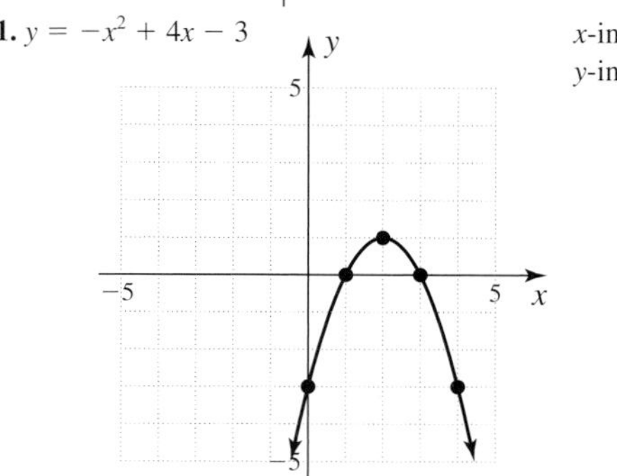

23. About 40 hours **25.** About 60 hours **27.** 25 **29.** 40,000 **31.** \$40 **33.** \$400

41. $y = -(x - 2)^2 - 1$

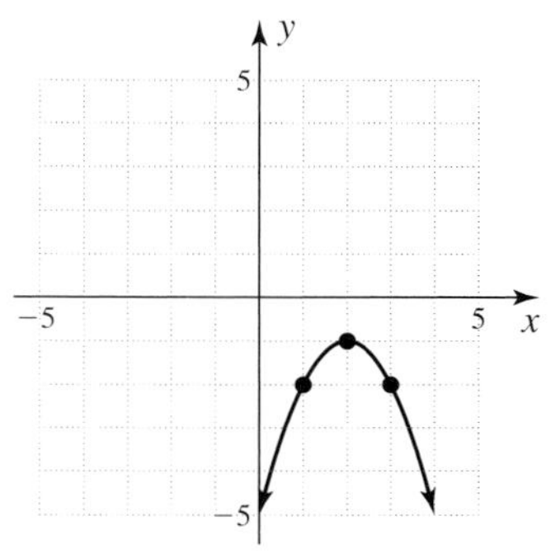

43. $y = -2x^2$

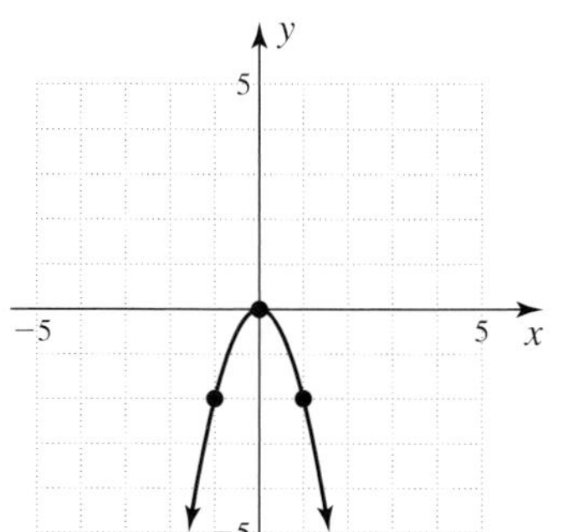

45. $y = -x^2 - 2x + 3$

x-int: $(-3, 0)$, $(1, 0)$; y-int: $(0, 3)$; $V(-1, 4)$

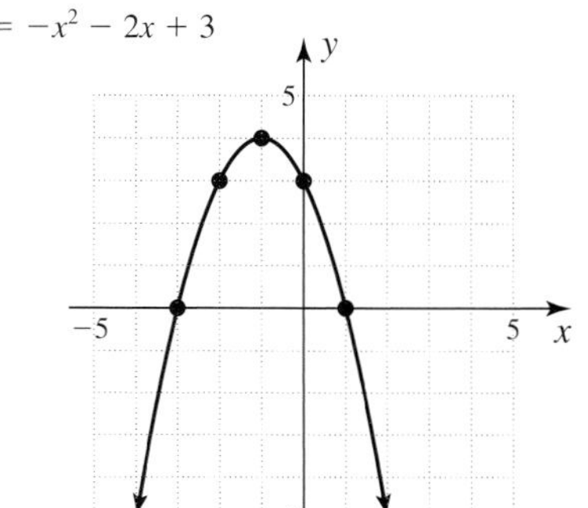

Exercises 10.5

1. $b = 5$ **3.** $b = 10\sqrt{2}$ **5.** $h = \sqrt{15}$ **7.** $h = 3$ **9.** $b = 2$ **11.** 50 ft **13.** 100 mi/hr **15.** 8 sec **17.** 1 sec **19.** 3000 units **21.** 3 ft by 4 ft **23.** 10 cm; 24 cm; 26 cm **25.** $A = 12$ **27.** $B = 12$ **29.** $c = \sqrt{\frac{E}{m}} = \frac{\sqrt{Em}}{m}$ **31.** $r = \sqrt{\frac{GMm}{F}} = \frac{\sqrt{FGMm}}{F}$ **33.** $P = \sqrt{\frac{I}{k}} = \frac{\sqrt{Ik}}{k}$ **37.** 1 or 6 (hundred) **39.** 50 mi/hr

Exercises 10.6

1. $D = \{1, 2, 3\}$; $R = \{2, 3, 4\}$ **3.** $D = \{1, 2, 3\}$; $R = \{1, 2, 3\}$ **5.** $D =$ All real numbers; $R =$ All real numbers **7.** $D =$ All real numbers; $R =$ All real numbers **9.** $D =$ All real numbers; $R =$ All nonnegative real numbers **11.** $D =$ All nonnegative real numbers; $R =$ All real numbers **13.** $D =$ All real numbers except 3; $R =$ All real numbers except 0 **15.** $D = \{-1, 0, 1, 2\}$; $R = \{-2, 0, 2, 4\}$; $(-1, -2), (0, 0), (1, 2), (2, 4)$ **17.** $D = \{0, 1, 2, 3, 4\}$; $R = \{-3, -1, 1, 3, 5\}$; $(0, -3), (1, -1), (2, 1), (3, 3), (4, 5)$ **19.** $D = \{0, 1, 4, 9, 16, 25\}$; $R = \{0, 1, 2, 3, 4, 5\}$; $(0, 0), (1, 1), (4, 2), (9, 3), (16, 4), (25, 5)$ **21.** $D = \{1, 2, 3\}$; $R = \{2, 3, 4\}$; $(1, 2), (1, 3), (1, 4), (2, 3), (2, 4), (3, 4)$ **23.** A function; one y-value for each x-value **25.** A function; one y-value for each x-value **27.** A function; one y-value for each x-value **29.** Not a function; two y-values for each positive x-value **31.** **(a)** $f(0) = 1$; **(b)** $f(2) = 7$; **(c)** $f(-2) = -5$ **33.** **(a)** $F(1) = 0$; **(b)** $F(5) = 2$; **(c)** $F(26) = 5$ **35.** **(a)** $f(x + h) = 3x + 3h + 1$; **(b)** $f(x + h) - f(x) = 3h$; **(c)** $\frac{f(x + h) - f(x)}{h} = 3, h \neq 0$ **37.** $y = g(x) = x^2$; $\left(\frac{1}{4}, \frac{1}{16}\right)$, $(2.1, 4.41)$, $(\pm 8, 64)$ **39.** **(a)** $g(0) = 1$; **(b)** $g(-2) = -5$; **(c)** $g(2) = 15$ **41.** **(a)** 140; **(b)** 130 **43.** **(a)** 160 lb; **(b)** 78 in. **45.** **(a)** 639 lb/ft^2; **(b)** 6390 lb/ft^2 **47.** **(a)** 144 ft; **(b)** 400 ft **49.** 60,000 joules **51.** A function **53.** A function **59.** $D = \{-5, -6, -7\}$; $R = \{5, 6, 7\}$ **61.** $D =$ All real numbers except 3; $R =$ All real numbers except 0 **63.** A function; one y-value for each x-value **65.** A function; one y-value for each x-value **67.** Not a function; two y-values for each positive x-value **69.** $g(-1) = -2$

Review Exercises

1. [10.1A] **(a)** $x = \pm 1$; **(b)** $x = \pm 10$; **(c)** $x = \pm 9$

2. [10.1A] **(a)** $x = \pm\frac{5}{4}$; **(b)** $x = \pm\frac{3}{5}$; **(c)** $x = \pm\frac{5}{8}$

3. [10.1A] **(a)** No real-number solution; **(b)** No real-number solution; **(c)** No real-number solution

4. [10.1B] **(a)** $x = -1 \pm \frac{\sqrt{3}}{7} = \frac{-7 \pm \sqrt{3}}{7}$; **(b)** $x = -2 \pm \frac{\sqrt{2}}{5} = \frac{-10 \pm \sqrt{2}}{5}$; **(c)** $x = -1 \pm \frac{\sqrt{5}}{4} = \frac{-4 \pm \sqrt{5}}{4}$

5. [10.2A] **(a)** 9; **(b)** 49; **(c)** 36 **6.** [10.2A] **(a)** 9; $x - 3$; **(b)** 25; $x - 5$; **(c)** 36; $x - 6$ **7.** [10.2A] **(a)** 7; **(b)** 6; **(c)** 5 **8.** [10.2A] **(a)** 4; **(b)** 9; **(c)** 36 **9.** [10.3B] **(a)** $x = \frac{-e \pm \sqrt{e^2 - 4df}}{2d}$; **(b)** $x = \frac{-h \pm \sqrt{h^2 - 4gi}}{2g}$; **(c)** $x = \frac{-k \pm \sqrt{k^2 - 4jm}}{2j}$

10. [10.3B] **(a)** $x = 1$ or $x = -\frac{1}{2}$; **(b)** $x = \frac{1 \pm \sqrt{11}}{2}$; **(c)** $x = \frac{3 \pm \sqrt{33}}{4}$

11. [10.3B] **(a)** $x = \frac{1 \pm \sqrt{13}}{6}$; **(b)** $x = \frac{1 \pm \sqrt{7}}{3}$; **(c)** $x = \frac{3 \pm \sqrt{33}}{6}$

12. [10.3B] **(a)** $x = 0$ or $x = 9$; **(b)** $x = 0$ or $x = 4$; **(c)** $x = 0$ or $x = 25$

13. [10.3B] **(a)** $x = \frac{9 \pm \sqrt{65}}{2}$; **(b)** $x = -\frac{1}{2}$ or $x = 3$; **(c)** No real-number solution

14. [10.4A]

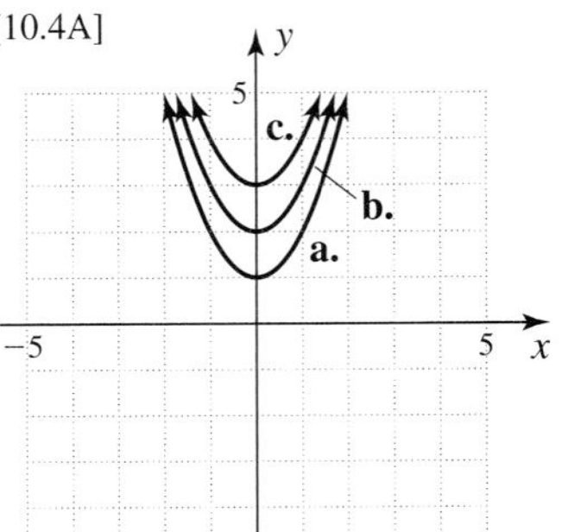

(a) $y = x^2 + 1$; **(b)** $y = x^2 + 2$; **(c)** $y = x^2 + 3$

15. [10.4A]

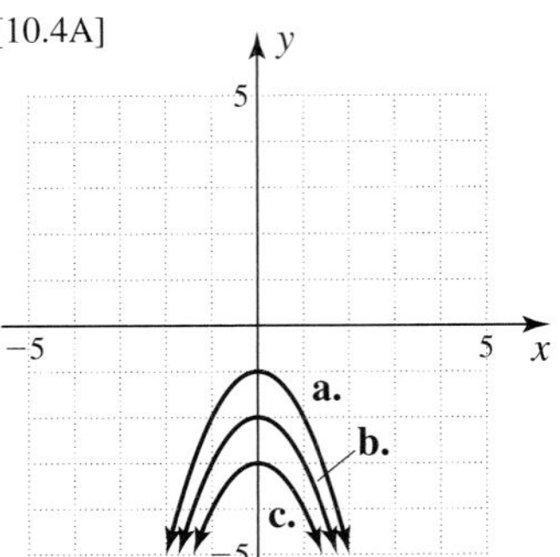

(a) $y = -x^2 - 1$; **(b)** $y = -x^2 - 2$; **(c)** $y = -x^2 - 3$

16. [10.4A]

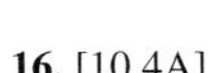

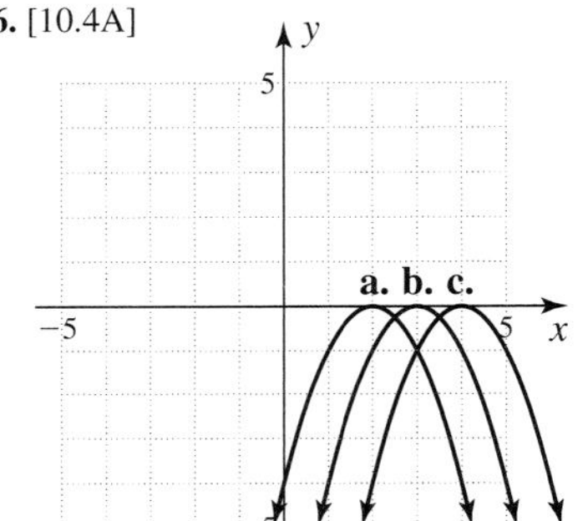

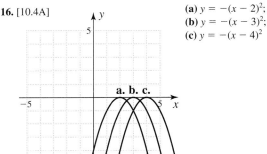

(a) $y = -(x - 2)^2$; **(b)** $y = -(x - 3)^2$; **(c)** $y = -(x - 4)^2$

17. [10.4A]

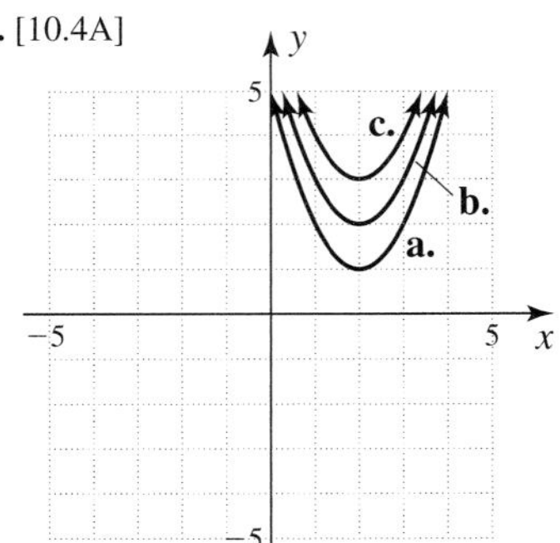

(a) $y = (x - 2)^2 + 1$; **(b)** $y = (x - 2)^2 + 2$; **(c)** $y = (x - 2)^2 + 3$

18. [10.4A]

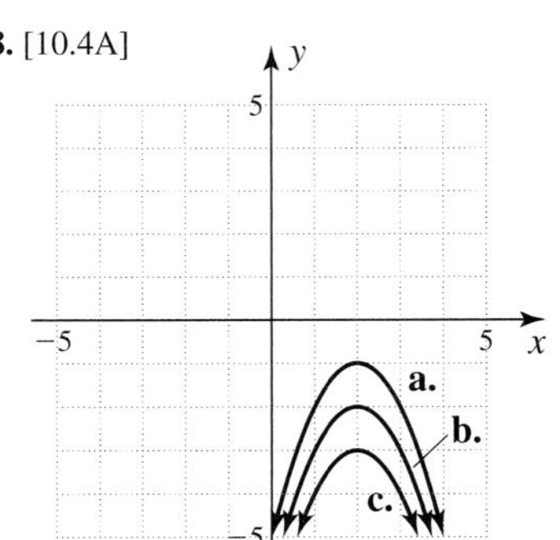

(a) $y = -(x - 2)^2 - 1$; **(b)** $y = -(x - 2)^2 - 2$; **(c)** $y = -(x - 2)^2 - 3$

19. [10.5A] **(a)** 13 in.; **(b)** $\sqrt{13}$ in.; **(c)** $\sqrt{41}$ in. **20.** [10.5B] **(a)** 5 sec; **(b)** 7 sec; **(c)** 8 sec **21.** [10.6A] **(a)** $D = \{-3, -4, -5\}$; $R = \{1, 2\}$; **(b)** $D = \{2, -1\}$; $R = \{-4, 3, 4\}$; **(c)** $D = \{-1\}$; $R = \{2, 3, 4\}$ **22.** [10.6A] **(a)** D = All real numbers; R = All real numbers; **(b)** D = All real numbers; R = All real numbers; **(c)** D = All real numbers; R = All nonnegative real numbers **23.** [10.6B] **(a)** A function; **(b)** Not a function; **(c)** Not a function **24.** [10.6C] **(a)** $f(2) = 1$; **(b)** $f(-2) = -19$; **(c)** $f(1) = -1$ **25.** [10.6D] **(a)** \$22; **(b)** \$19; **(c)** \$16

Cumulative Review Chapters 1–10

1. $-\frac{29}{56}$ **2.** 256 **3.** $\frac{9}{5}$ **4.** 14 **5.** $-x - 10$ **6.** $\frac{x - 5y}{z}$ **7.** $x = 3$ **8.** $x = 72$ **9.** 

0 4

10.

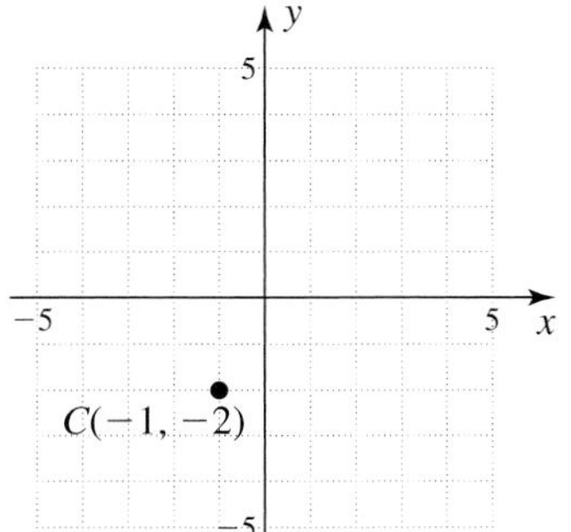

11. Yes **12.** $x = -2$

13. $2x + y = 4$

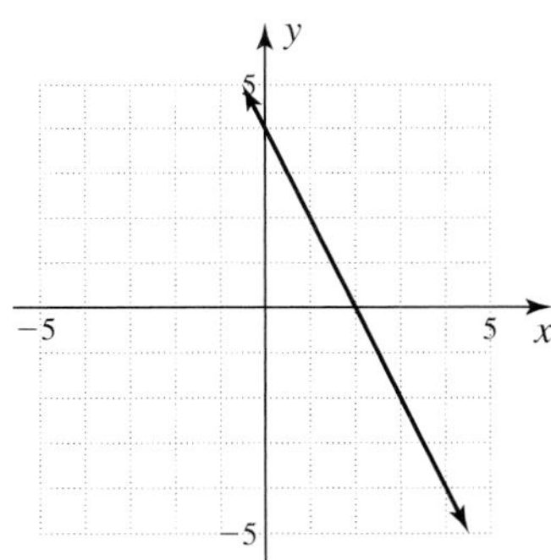

14. $2y - 8 = 0$

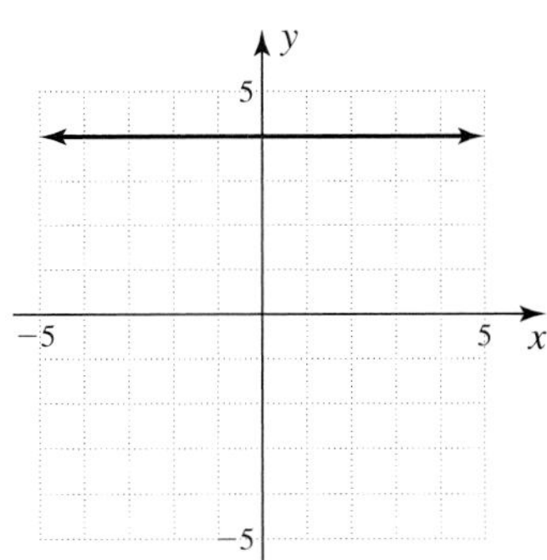

15. $\frac{14}{11}$ **16.** 2 **17.** (2) and (3) **18.** $\frac{81x^{12}}{y^{20}}$ **19.** 5.0×10^{-3} **20.** 1.5×10^{-7} **21.** $4x^4 - \frac{4}{5}x^2 + \frac{1}{25}$ **22.** $(2x^2 - x - 1)$ R 1 **23.** $(x - 5)(x - 3)$ **24.** $(4x - 3y)(3x - 4y)$ **25.** $(5x + 6y)(5x - 6y)$ **26.** $-5x^2(x + 4)(x - 4)$ **27.** $3x(x + 2)(x - 3)$ **28.** $(x + 3)(4x + 5)$ **29.** $k(5x + 2)^2$ **30.** $x = -5$; $x = \frac{2}{3}$ **31.** $\frac{12xy^2}{10y^3}$ **32.** $-(x + y)$ **33.** $-(x + 3)$ **34.** $\frac{x - 3}{x + 4}$ **35.** $\frac{1}{2 - x}$ **36.** $\frac{3}{x + 1}$ **37.** $\frac{-2x - 15}{(x + 8)(x + 7)(x - 7)}$ **38.** 6 **39.** $x = 9$ **40.** $x = -5$

41. No solution **42.** $x = -6$ **43.** 29 gal **44.** $x = \frac{64}{5}$ **45.** $2\frac{2}{9}$ hr **46.** $y = 6x + 8$ **47.** $y = -2x - 5$ **48.** $3x - 2y < -6$

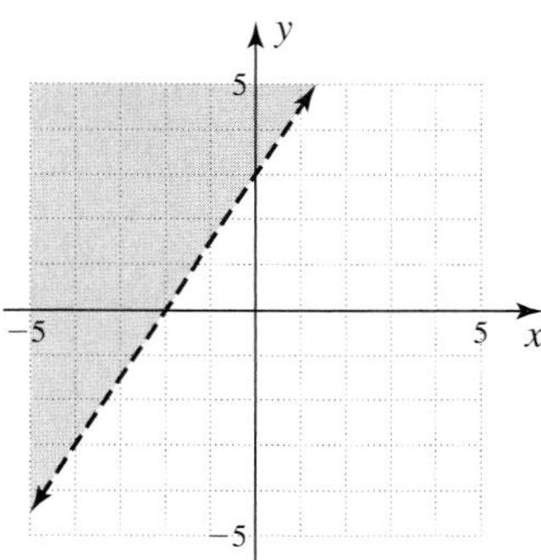

49. $-y \geq -5x + 5$

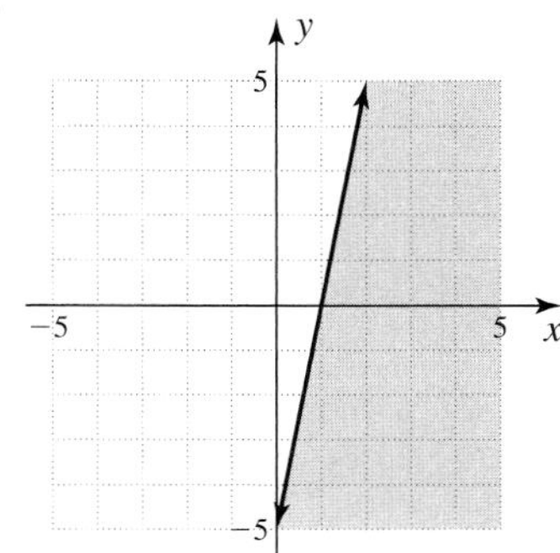

50. $\frac{1}{70}$ **51.** 13,720

52. $x + 3y = -12$; $2y - x = -2$

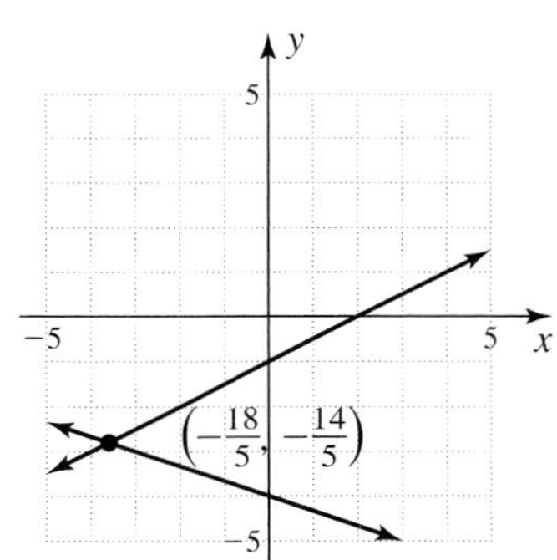

53. $y + 3x = 3$; $3y + 9x = 18$ No solution

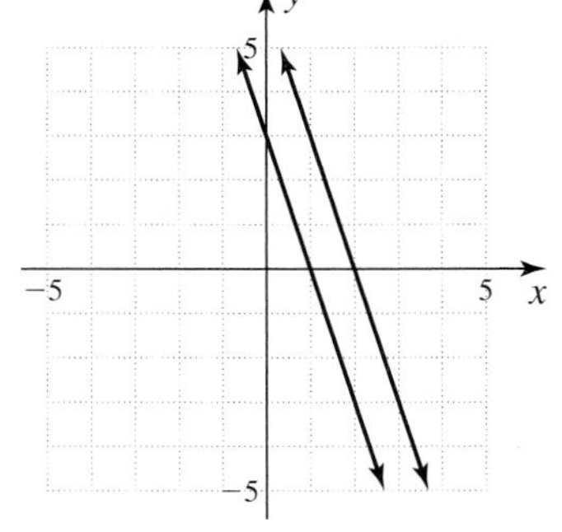

54. Inconsistent; no solution **55.** Dependent; infinitely many solutions **56.** $(-3, -3)$ **57.** Inconsistent; no solution **58.** Dependent; infinitely many solutions **59.** 10 nickels and 20 dimes **60.** 30 and 140 **61.** 3 **62.** 5 **63.** $\frac{\sqrt{15}}{27}$ **64.** $\frac{\sqrt{14j}}{7j}$ **65.** $7\sqrt{2}$ **66.** $3x - 2\sqrt[3]{6x^2}$ **67.** $280 + 46\sqrt{35}$ **68.** $5 - \sqrt{2}$ **69.** No real-number solution

70. $x = 4; x = 5$ **71.** $x = \frac{2}{3}; x = -\frac{2}{3}$ **72.** No real-number solution

73. $x = 4 + \frac{\sqrt{7}}{6}; x = 4 - \frac{\sqrt{7}}{6}$ **74.** 4 **75.** 4 **76.** 4

77. $x = \frac{-k + \sqrt{k^2 - 4jm}}{2j}; x = \frac{-k - \sqrt{k^2 - 4jm}}{2j}$

78. $x = \frac{3}{2}; x = -3$ **79.** $x = 0; x = 16$

80. $y = -(x + 3)^2 + 2$

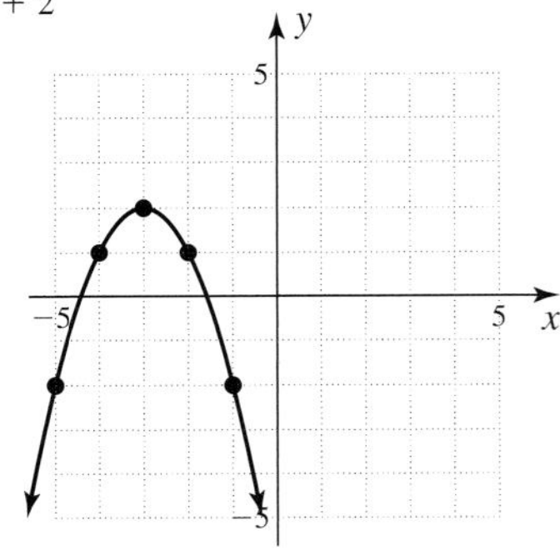

81. 2 sec

Photo Credits

Review Chapter:

Page 4(left): © Royalty-Free/CORBIS; **p. 4(right):** © Thinkstock/Getty Images; **p. 10(left):** © Brand X Pictures/Getty Images; **p. 10(right):** © Chris Speedie/Stone/Getty Images; **p. 12, 15:** © Ignacio Bello; **p. 24:** © McGraw-Hill Higher Education, photographer Dr. Thomas D. Porter

Chapter 1

Page 70: © Royalty-Free/CORBIS

Chapter 2

Opener: © The British Museum; **p. 122:** © Royalty-Free/CORBIS; **p. 150:** © Jeff Greenberg/Photo Edit; **p. 161:** © Mary Kate Denny/Photo Edit; **p. 191:** © Fred Bellet/Tampa Tribune

Chapter 3

Opener: © Stapleton Collection/CORBIS

Chapter 4

Opener: © Bibiliotheque nationale de France; **p. 302:** © PhotoDisc Vol. OS23/Getty CD; **p. 304, 305 (top & bottom):** © Ignacio Bello; **307:** © Doug Struthers/Stone/Getty Images; **p. 318:** © Ignacio Bello; **p. 326:** © Dallas and John Heaton/Stock Connection/PictureQuest; **p. 347:** © Ignacio Bello; **p. 369:** © Royalty-Free/CORBIS

Chapter 5

Opener: © Archivo Iconografico, S.A./CORBIS; **p. 406:** © Ignacio Bello; **p. 417:** © Morton Beebe/CORBIS; **p. 424:** © Martin M. Rotker/Photo Researchers; **p. 431:** © Ignacio Bello

Chapter 6

Opener: © The British Museum/Topham-HIP/The Image Works; **p. 464, 476:** © Ignacio Bello; **p. 486:** © Tony Freeman/Photo Edit; **p. 497:** Courtesy Adler Planetarium and Astronomy Museum, Chicago, Illinois; **p. 502:** © Ignacio Bello

Chapter 7

Opener: © Bettmann/CORBIS; **p. 543:** © Eunice Harris/Index Stock; **p. 560:** © Ignacio Bello

Chapter 8

Page 604: © Ignacio Bello

Chapter 9

Opener: © Stock Montage/Hulton Archive/Getty Images; **p. 640:** © AP/Wide World Photos; **p. 645:** © Hal Whipple; **p. 647, 659:** © Ignacio Bello; **p. 667:** © Tony Freeman/Photo Edit

Chapter 10

Opener: © Bettmann/CORBIS; **p. 687, 697:** © Tony Freeman/Photo Edit; **p. 713:** © Ignacio Bello; **p. 738:** Courtesy NASA Dryden Flight Research Center, Tony Landis photographer

Index

Index of Calculate It Topics

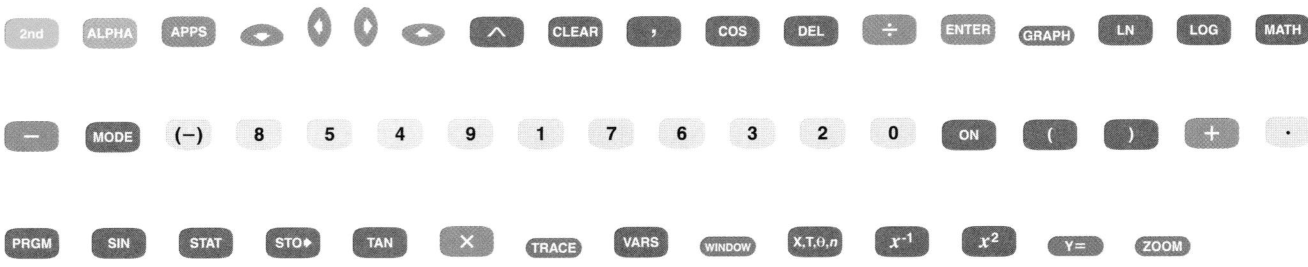

Introductory Algebra

A

B

C

D

E

F

G

H

I

J

K

L

M

N

O

P

Q

R

S

T

U

V

W

X

Y

Z